Glia

SECOND EDITION

A subject collection from *Cold Spring Harbor Perspectives in Biology*

Glia

SECOND EDITION

A subject collection from *Cold Spring Harbor Perspectives in Biology*

EDITED BY

Beth Stevens
Boston Children's Hospital
Broad Institute of MIT and Harvard

Kelly R. Monk
Oregon Health and Science University
The Vollum Institute

Marc R. Freeman
Oregon Health and Science University
The Vollum Institute

COLD SPRING HARBOR LABORATORY PRESS
Cold Spring Harbor, New York • www.cshlpress.org

Glia, Second Edition

A subject collection from *Cold Spring Harbor Perspectives in Biology*
Articles online at www.cshperspectives.org

Executive Editor	Richard Sever
Project Supervisor	Barbara Acosta
Editorial Assistant	Danett Gil
Permissions Administrator	Carol Brown
Production Editor	Diane Schubach
Production Manager/Cover Designer	Denise Weiss
Publisher	John Inglis

Front cover artwork: Individual glial cells from the adult fly brain (astrocyte, bottom left) and larval zebrafish (oligodendrocyte, *top*; oligodendrocyte precursor cell, *bottom right*) labeled to highlight their cytoskeletal architecture and plasma membranes imaged by confocal microscopy. (Images created by Jaikun Chen and Cody Call, Oregon Health & Science University.)

Library of Congress Cataloging-in-Publication Data

Names: Stevens, Beth, 1970– editor. | Monk, Kelly R., editor. | Freeman, Marc R., 1970– editor.
Title: Glia / edited by Beth Stevens, Boston Children's Hospital, Harvard Medical School, Kelly R. Monk, Oregon Health and Science University, The Vollum Institute and Marc R. Freeman, Oregon Health and Science University, The Vollum Institute.
Other titles: Cold Spring Harbor perspectives in biology.
Description: Second edition. | Cold Spring Harbor, New York : Cold Spring Harbor Laboratory Press, [2025] | "A subject collection from Cold Spring Harbor perspectives in biology." | Includes bibliographical references and index. | Summary: "Glial cells are non-neuronal cells that support and protect neurons in the nervous systems. Once thought to be passive bystanders, they are increasingly appreciated for their active roles in nervous system development, function, and disease. This second edition will explore the major progress made in the field over the past few years"-- Provided by publisher.
Identifiers: LCCN 2023048447 (print) | LCCN 2023048448 (ebook) | ISBN 9781621824640 (hardcover) | ISBN 9781621824657 (epub)
Subjects: LCSH: Neuroglia. | MESH: Neuroglia--Collected Works.
Classification: LCC QP363.2 .G562 2024 (print) | LCC QP363.2 (ebook) | DDC 612.8/1046--dc23/eng/20231107
LC record available at https://lccn.loc.gov/2023048447
LC ebook record available at https://lccn.loc.gov/2023048448

All World Wide Web addresses are accurate to the best of our knowledge at the time of printing.

For a complete catalog of all Cold Spring Harbor Laboratory Press publications, visit our website at www.cshlpress.org.

Contents

Contents

Preface

SINCE THE DISCOVERY OF GLIAL CELLS AS a major class of cells in the nervous system more than 100 years ago, their functions have been the subject of great mystery and debate. Glial cells are generally considered to consist of all neuroectoderm-derived cell types that are not electrically excitable neurons. Glia are present in both invertebrates and vertebrates, and according to most estimates they constitute the majority of cells in the mammalian nervous system.

But what do glial cells do? Progress had been limited by a lack of tools to identify and genetically manipulate them, to purify and culture them, and to study their physiology. Although there is much more work to be done, profound technical advances over the past 30 years have finally made it possible to begin to really understand glial development and function.

Our goal in this monograph is to review recent progress in our understanding of the major classes of glial cells: astrocytes, oligodendrocytes, Schwann cells, microglia, as well as invertebrate glia. How are they generated, how do they develop, and what are their functions both normally and in disease?

We review this progress in seven main book sections. In Section 1, we consider the types of glial cells and their functions in worms, flies, and fish, which have all emerged as powerful new genetic model systems for understanding glial cell function. Diverse aspects of glial biology, from astrocyte specification, growth and function, to axon–glia interactions and glial phagocytic functions are well conserved in these animals, and can be subjected to powerful forward- and molecular-genetic approaches to characterize conserved glial signaling pathways.

Next, in Section 2, we cover astrocytes. Chapters in this section review their development and regional specialization. They also review their surprisingly active roles in synapse formation, function, plasticity, and elimination, as well as in controlling blood flow. Intracellular pathways involved in calcium signaling and metabolism differ strikingly in astrocytes compared to neurons; several chapters review these differences and their possible functional implications. Last, the roles of reactive astrocytes and their roles in central nervous system (CNS) disease are considered.

In Sections 3 and 4, we review progress in understanding the biology of myelinating glial cells and their precursors, beginning with oligodendrocyte precursor cells (OPCs), and also covering oligodendrocytes and Schwann cells. Our knowledge of functional roles for OPCs has been expanding rapidly beyond simply their ability to make new oligodendrocytes, and now includes their shaping of neural circuits. Oligodendrocyte generation continues into adulthood, in which these cells may play important roles in adaptive myelination and even certain kinds of learning. An enormous amount has been learned about how myelinating cells are specified and how they myelinate axons, but also how they then support axonal function. Several chapters review how oligodendrocytes and Schwann cells help to organize nodal, paranodal, and internodal axon domains, which is critical for rapid and faithful electrical conduction of action potentials. As for astrocytes in the CNS, Schwann cells are now known to have multiple active roles in the control of synapse formation and function in the peripheral nervous system (PNS). Roles for perisynaptic Schwann at the neuromuscular junction are discussed, along with how Schwan cells help orchestrate nerve assembly and repair.

In Sections 5 and 6, we review the exciting recent progress on the origin and functions of microglia and other neuroimmune interactions mediated by glia at the boundaries of neuronal and non-neuronal tissues. Section 5 begins with an introduction to microglial roles in coordinating CNS neuroimmune function and in the spatial patterning and synaptic wiring throughout the healthy, developing, and adult CNS; additional roles for microglia are discussed in chapters throughout the book. Section 6 goes on to explore how glia play roles in forming barriers or regulating peripheral function, ranging from the blood–brain barrier to the long-overlooked satellite glial cells found in peripheral ganglia and enteric glia that support neuronal function in internal organs. The morphology and function of glia at CNS–PNS transition zones are also considered, along with the cellular and functional basis of the glymphatic and lymphatic waste-clearance system in the brain and its potential roles in disease.

Finally, in Section 7, we consider recent studies that implicate glial cells as critical players in diverse aspects of disease, repair, and regeneration. Chapters focus on the critical roles of glial cells in promoting and hindering axon regeneration after injury in the PNS and CNS. Included is a discussion of the recent explosion of exciting work revealing the likely origin of and neuronal regulatory roles in driving glial malignancies such as glioma. This is followed by a discussion of the biology of how astrocytes and microglia might help or hinder (or both) in neurodegenerative disease. The book closes with chapters exploring demyelinating diseases in the CNS and PNS, white matter disorders, and diseases of the peripheral nerve.

What emerges from this work is new insight into the importance of glial cells, especially an appreciation that the development, function, and malfunction of our brains can only be understood as a signaling interplay between neurons and glial cells. Rather than being passive support cells as long thought, glial cells are highly active participants in the vast majority, if not all, neurobiological processes. In complex animals, there is very little that neurons do without talking to glia.

Equally important, this monograph highlights the reality that many mysteries remain. Trainees considering entering the field should be excited by the vast open space to be explored, key challenges to be overcome on our journey, and complex functional roles to be demystified. To name just a few: how do glia keep neurons alive, what is the overall role of glia in circuit function, what neuroactive substances do glia secrete and how do they regulate neuronal function, have human astrocytes evolved in their abilities to control synapse formation or function, and does that contribute to the enhanced cognitive capacities of humans? What is the functional significance of regional astrocyte specialization, do astrocytes function individually or as a syncytium (or both), what other unexpected functions do oligodendrocyte lineage cells have beyond myelination, what exactly are the roles of microglia in health and disease, and could glial cells be important targets for new drugs? Clearly there is much work left to be done! Our hope is that readers of this book will be stimulated to join the chase.

We wish to thank the staff of Cold Spring Harbor Laboratory Press, namely (the ever-so-patient), Barbara Acosta for inviting us to edit this volume and managing the project, and Editorial Assistant Danett Gil and Production Editor Diane Schubach for all their hard work in producing this volume. For the cover photographs, we are indebted to Jiakun Chen (UNC) and Cody Call (OHSU) for their beautiful images. Last, we thank all of the authors for their superb contributions to this volume, and for their patience with the many Covid-related and other delays we experienced along the way.

Finally, we want to close by acknowledging two colleagues we recently lost far too soon. Laura Feltri (MD) was a widely respected neuroscientist and mentor, whose research focused on peripheral nerves, Charcot-Marie-Tooth disease, and multiple sclerosis. Laura died December 25, 2023. She made a number of seminal discoveries in the area of myelination, many of which were in collaboration with her husband, Lawrence Wrabetz (MD). Laura was a fixture at meetings, and was a pioneer in pushing creative and difficult molecular-genetic approaches to explore the molecules that regulate myelin biology. Her approach to hard questions was relentless and rigorous. Laura was a member and then-chair of the National Institutes of Health (NIH) study section on the Cellular and Molecular Biology of Glia, President of the Peripheral Nerve Society, an AAAS fellow, and a guiding voice for many scientific organizations, including the Muscular Dystrophy Association and the National Multiple Sclerosis Society. It is hard to think of a kinder, friendlier face one might bump into at a meeting, and then have a remarkably stimulating discussion with. She is sorely missed.

The first edition of this book, like many things in our field (e.g., the CSHL Glia in Health and Disease Meeting), was the product of Ben Barres' (MD/PhD) passion for the field of glial biology, dedication to academic science, and his desire to bring trainees and faculty together for meaningful scientific discussion. Ben studied nearly all corners of glial cell biology during his career—astrocyte development and function, OPCs, myelin biology, and microglia. He died on December 27, 2017 at the age of 63. It is hard to imagine any contemporary researcher that had a greater impact on the field of glial biology, from his own scientific breakthroughs, to his relentless mentoring of his trainees (and anyone else within earshot), to his remarkably tuned moral compass and outspoken nature on social issues. Ben made a lasting impact on the quality of our science, and how we treat each other in the scientific community through his efforts. This continues through his own trainees and friends still working in our field. He was a bright light that is sorely missed. Amazingly, he had all of his massive impact having only had a laboratory for ~24 years.

BETH STEVENS
KELLY R. MONK
MARC R. FREEMAN

Glia Development and Function in the Nematode *Caenorhabditis elegans*

Aakanksha Singhvi,[1,2,7] Shai Shaham,[3,7] and Georgia Rapti[4,5,6,7]

[1]Division of Basic Sciences, Fred Hutchinson Cancer Center, Seattle, Washington 98109, USA

[2]Department of Biological Structure, University of Washington School of Medicine, Seattle, Washington 98195, USA

[3]Laboratory of Developmental Genetics, The Rockefeller University, New York, New York 10065, USA

[4]Developmental Biology Unit, European Molecular Biology Laboratory, Heidelberg 69117, Germany

[5]Epigenetics and Neurobiology Unit, European Molecular Biology Laboratory, Monterotondo, Rome 00015, Italy

[6]Interdisciplinary Center of Neurosciences, Heidelberg University, Heidelberg, Germany

Correspondence: asinghvi@fredhutch.org; shaham@rockefeller.edu; grapti@embl.de

The nematode *Caenorhabditis elegans* is a powerful experimental setting for uncovering fundamental tenets of nervous system organization and function. Its nearly invariant and simple anatomy, coupled with a plethora of methodologies for interrogating single-gene functions at single-cell resolution in vivo, have led to exciting discoveries in glial cell biology and mechanisms of glia–neuron interactions. Findings over the last two decades reinforce the idea that insights from *C. elegans* can inform our understanding of glial operating principles in other species. Here, we summarize the current state-of-the-art, and describe mechanistic insights that have emerged from a concerted effort to understand *C. elegans* glia. The remarkable acceleration in the pace of discovery in recent years paints a portrait of striking molecular complexity, exquisite specificity, and functional heterogeneity among glia. Glial cells affect nearly every aspect of nervous system development and function, from generating neurons, to promoting neurite formation, to animal behavior, and to whole-animal traits, including longevity. We discuss emerging questions where *C. elegans* is poised to fill critical knowledge gaps in our understanding of glia biology.

The nematode *Caenorhabditis elegans* is a powerful experimental setting in which to uncover biological principles in molecular detail (Brenner 1974; Goldstein 2016). An extensive genetic toolkit coupled with optical transparency, enabling facile in vivo microscopy and optogenetics, has allowed the nervous system of this animal to be probed at unprecedented single-gene and single-synapse resolution. *C. elegans* uses its structurally invariant neural network to perform complex and flexible behaviors, including sensory preference choice, locomotion, sleep, mating, and decision-making, and to store information of different qualities and on different

---

[7]All authors contributed equally to this work.

timescales (White et al. 1986; Bargmann 1993; Sengupta and Samuel 2009; Bargmann and Marder 2013; Emmons 2018; Schafer 2018; Cook et al. 2019; Goodman and Sengupta 2019; Witvliet et al. 2021).

C. elegans glia are molecularly and anatomically diverse (Cao et al. 2017; Singhvi and Shaham 2019; Purice et al. 2023), arising primarily from ectodermal precursors, and associating with sense organs and with the brain neuropil, the nerve ring. One glial class is mesodermally derived and also abuts the nerve ring (White et al. 1986). Gene expression and functional studies of *C. elegans* glia have revealed extensive similarities to vertebrate glia (Shaham 2005; Heiman and Shaham 2007; Katz et al. 2019; Singhvi and Shaham 2019; Purice et al. 2023).

C. elegans glia, however, differ from vertebrate glia in one important aspect: while vertebrate glia provide neurons with trophic support, *C. elegans* glia do not (Shaham 2005; Wagner et al. 2006; Barres 2008; Singhvi and Shaham 2019; Chiareli et al. 2021). Trophic support in vertebrates may serve to ensure precision in neural cell number and circuit formation during development (Barres 2008; Clarke and Barres 2013; Farhy-Tselnicker and Allen 2018). In *C. elegans*, however, development and circuit formation are stereotyped, predetermined by cell lineage (Sulston and Horvitz 1977; Sulston et al. 1980, 1983; White et al. 1986; Varshney et al. 2011; Jarrell et al. 2012; Doroquez et al. 2014). Further, neuronal cell bodies are not known to be metabolically privileged by a restrictive blood–brain barrier. Thus, glial support of neuron survival may be unnecessary (Shaham 2015; Singhvi et al. 2016; Singhvi and Shaham 2019; Rapti 2021). *C. elegans*, therefore, provides a unique in vivo arena for perturbing glial cell functions without the concern that associated neuron survival is affected (Singhvi and Shaham 2019; Rapti 2021).

GENERAL PROPERTIES OF *C. elegans* GLIA

Glial Cell Types

Like *C. elegans* neurons, whose numbers are invariant between individuals (302/387 in hermaphrodites/males), *C. elegans* glia numbers and developmental origins are constant, with 50 sex-shared glia, 36 male-specific neuroectoderm-derived glia, and six sex-shared, mesoderm-derived glial cells (Fig. 1; Table 1; Sulston and Horvitz 1977; Sulston et al. 1980, 1983; White et al. 1986). Whole-animal, single-cell transcriptome profiling (RNA-seq) initially suggested that even the handful of glia interrogated in those studies were not molecularly identical (Cao et al. 2017; Packer et al. 2019; Fung et al. 2020; Taylor et al. 2021). This has now been conclusively demonstrated in a complete molecular atlas of *C. elegans* glia across the male and hermaphrodite nervous systems (Purice et al. 2023). Single-nuclear RNA-seq transcriptome profiling, coupled with custom computational/machine-learning analytics and in vivo validation studies, show that glia are molecularly heterogeneous, with even anatomically similar glia having distinct molecular signatures.

Glia associate with every level of information transfer in *C. elegans* circuits (Ward et al. 1975; White et al. 1986). Ten bilateral sheath-glia pairs (ADEsh, AMsh, ILsh, ILshD/V, OLLsh, OLQshV/D, PDEsh, PHsh) and 13 bilateral socket glia pairs (ADEso, AMso, CEPsoD/V, ILso, ILsoD/V, OLLso, OLQsoD/V, PDEso, PHso1/2) fasciculate with neuron dendrites, and enwrap their sensory tips, forming environment-accessible compartments (Fig. 1A). Like epithelia, socket glia also secrete cuticles (Chisholm and Hsiao 2012). Four CEPsh glia and six mesoderm-derived GLR glia associate with both dendrites and axons. Anterior CEPsh glia processes fasciculate with dendrites and form compartments around sensory tips of CEP and male-specific CEM neurons. Distal GLR glia processes fasciculate with IL1 neuron dendritic bundles without ensheathing their endings. CEPsh glia posterior processes surround the outer aspect of the nerve ring (brain neuropil) and penetrate it to contact synapses. GLR glia proximal processes expand sheet-like structures that surround the inner aspect of the nerve ring. Here, GLR glia form gap junctions with GABAergic RME neurons and muscles. GLR glia sheets may create a seal between the nerve ring and the pseudocoelom, perhaps resembling a blood–brain barrier.

Thirty six male-specific neuroectodermal glia—seven sheath, 11 socket, and 18 sheath–

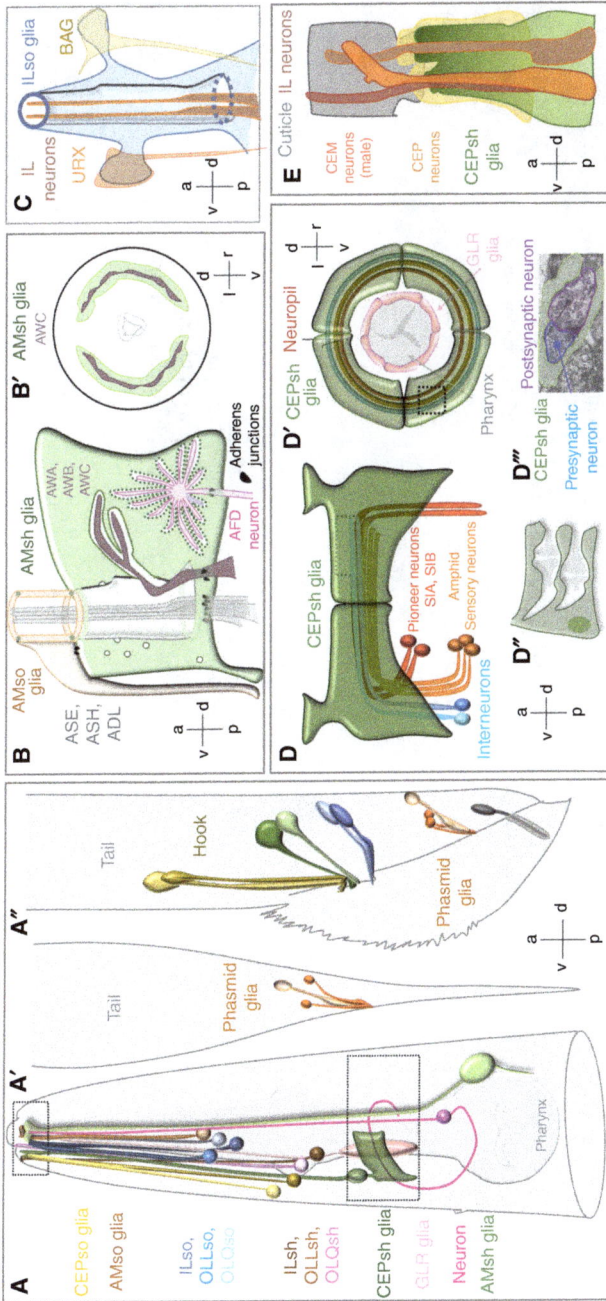

Figure 1. Subtypes and anatomy of *Caenorhabditis elegans* glia. (*A*) A schematic representation of each glia type in the head, (*A'*) hermaphrodite tail, (*A''*) and male tail. Anterior and posterior deirid glia (ADEsh/so, PDEsh/so) are not depicted. Glia–neuron associations are magnified in *B–E* as follows: amphid glia (*B*), IL socket glia (*C*), cephalic sensilla (*E*), and cephalic posterior membrane sheaths enveloping the brain neuropil (*D*). (*B–B'*): (*B*) Amphid sensilla schematic showing AMso–AMsh sense organ glia forming a channel lumen associating with dendrite tips, neuron-receptive endings (NREs) that traverse the glial channel (ASE, ASH, ADL), and embedded NREs (AFD and AWA/B/C neurons). (*B'*) Cross section of bilateral AMsh glia–AWC pairs, (*C*) ILso glia interact with NREs of different neurons (URX, IL, BAG) at distinct contact sites, with only IL neurons traversing a channel made by the glia. (*D*) Schematic of CEPsh glial processes (green) ensheathing different axon commissures. (*D'*) A schematic cross-section view of the brain neuropil shows the relative location of axon commissures with CEPsh and GLR glia. (*D''*) Glial processes also infiltrate between neuron processes in the neuropil. (*D'''*) Electron micrograph showing the CEPsh glia–ALA neuron–AVE neuron tripartite synapse. (*E*) Schematic of the cephalic sensilla of a male animal, noting relative localization and glia–neuron contacts of the sex-shared CEP neuron and male-specific CEM neuron. Sensory NREs in *B*, *C*, and *E* are depicted without dendrites and cell bodies for simplicity. Neurons in *D* are depicted without dendrites for simplicity. (Schematics in *A*, *B*, and *D* are reprinted with permission from Singhvi and Shaham 2019 with permission from the author. Electron microscope [EM] image in *D'''* is adapted with permission from White et al. 1986. Schematic in *C* is based on data in Wang et al. 2015 and Sulston et al. 1980.)

Table 1. *Caenorhabditis elegans* glia, associated neurons, and behaviors affected by respective glia–neuron interactions

Organ	Glia	Associated neuron(s)	Number of sensilla	Documented functions of glia–neuron interactions	Animal behavior regulated by glia–neuron interactions
Sensilla glia					
Amphid	AMso, AMsh	Contact AMsh only: AFD, AWA, AWB, AWC; AMsh–AMso Channel neurons: ASE, ASG, ASH, ASI, ASK, ASJ, ADF, ADL	2 (L/R)	AMsh: NRE shape and plasticity, glial compartment, dendrite outgrowth, ion regulation, engulfment, interaction with epithelia; AMso: adult neurogenesis	Olfaction, gustation, tactile sensation, thermosensation
Phasmid	PHsh, PHso1, PHso2	PHA, PHB, PHC; Contact PHso only: PQR	2 (L/R)	PHso: adult neurogenesis	n.d.
Anterior deirid	ADEsh, ADEso	ADE	2 (L/R)	n.d.	n.d.
Posterior deirid	PDEsh, PDEso	PDE	2 (L/R)	n.d.	n.d.
Inner labial	ILsh, ILso	IL1, IL2; Contact ILso only: BAG, URX, FLP	6 (L/R, D/V/L)	IL/OL nerve outgrowth, neuropil placement	Mechanosensation
Outer labial	OL(Q/L)sh, OL(Q/L)so	OLQ or OLL	6 (L/R, D/V/L)	n.d.	Mechanosensation
Cephalic	CEPsh, CEPso	CEP (±CEM in males)	4 (D/V, L/R)	CEPso: aECM pore formation	n.d.
Hook	HOsh, HOso	HOA, HOB	1	n.d.	Male mating
Ray	R(N)st	R(N)A, R(N)B	18 (L/R), (N) = 1–9	n.d.	Male mating
Postcloacal	PCsh, PCso	PCA; PCsh only: PCB, PCC	2 (L/R)	n.d.	Male mating
Spicules	SPsh syncytium (two cells), SPso syncytium (four cells)	SPD, SPV	2 (L/R)	NRE development. SPso: extrasynaptic source of dopamine	Male mating
Neuropil glia					
Neuropil	CEPsh	Many neuropil axons	4 (D/V, L/R)	Neuropil assembly, AIY synapse positioning, neurotransmitter clearance	Locomotion during sleep, repetitive behavior, swimming-induced paralysis
Mesodermal-lineage glia					
Mesodermal	GLR	RME, others (?)	6 (D/V/L, L/R)	RME axon specification	Locomotion

Reproduced from Singhvi and Shaham 2019 with permission from the author.

Cite this article as *Cold Spring Harb Perspect Biol* doi: 10.1101/cshperspect.a041346

socket hybrid structural cells (HOsh/HOso, bilateral pairs of PCsh, PCso, R1-9, SPso1-4, SPshD/V)—associate with the cloaca and with sensory rays and hook sensillum copulation structures (Sulston et al. 1980; Emmons 2005; Cook et al. 2019).

Sexual Dimorphism

Several *C. elegans* glial cells exhibit structural sexual dimorphism. The four CEPsh and CEPso glia form a compartment around dendritic endings of CEP and CEM neurons in males, but only around CEP neurons in hermaphrodites, as CEMs die during development (Sulston et al. 1983; White et al. 1986). CEPso glial cells control the formation of this pore through a male-specific transcriptional switch visualized by secretion of a Hedgehog-related protein, to differentially pattern the apical extracellular matrix (ECM) (Fung et al. 2023). In the tail phasmid sense organ, the phasmid channel is built by PHso1 glia in hermaphrodites, while PHso2 serves this function in males (Hall 1977; Sulston et al. 1980). Finally, AMso glia and PHso1 glia are neurogenic only in males, as described in detail below.

Molecular profiling of glia from sexually mature day 1 adults using shRNA-seq also reveals sex dimorphism (wormglia.org). Specifically, profiles of functionally analogous hermaphrodite PHso1 and male PHso2 socket glia are different (Purice et al. 2023; wormglia.org). Furthermore, some anatomically sex-shared glia like PHsh, OLsh, and OLso glia also exhibit divergent identities. In contrast, there is no discernable molecular sex dimorphism in either CEPsh, CEPso, or AMso glia that persists into the adult stage after dimorphic development is complete.

Other Heterogeneity

Despite physical association and similar contributions to forming sensory organ channels, sheath and socket glia, exhibit both morphological diversity (White et al. 1986) and distinct molecular signatures (Purice et al. 2023). Sheath glia of different sensory organs resemble each other more than lineally or anatomically related socket glia, implying functional convergence during development. Within each class (sheath/socket), glia are also different across sense organs (Bacaj et al. 2008; Katz et al. 2019; Fung et al. 2020; Purice et al. 2023). These differences likely reflect the identities of neurons with which they associate. For example, ILso and AMsh glia promote dendrite elongation of associated neurons through retrograde extension, but using distinct physical structures and molecular mechanisms (Cebul et al. 2020). Molecular heterogeneity between and within glial cell types is also evident (Oikonomou and Shaham 2011; Mizeracka and Heiman 2015; Singhvi and Shaham 2019). For example, ventral and dorsal CEPsh glia develop and specify their gene expression through molecularly distinguishable pathways (see below) (Yoshimura et al. 2008).

Glial Membrane Subdomains

Each *C. elegans* glial cell associates with a defined number of neurons, whose identities and contact sites are invariant, allowing for studies of glia–neuron interactions at the resolution of single contact sites. Often, the glial membranes apposing a given neuron are enriched for specific proteins. For example, the apical membranes of a single AMsh glia can be divided into at least three molecularly distinct domains contacting different sensory neurons (Ray et al. 2024). Membranes associated with AFD neurons accumulate the KCC-3 K/Cl transporter (Singhvi et al. 2016; Ray et al. 2024), domains associated with amphid channel neurons accumulate DAF-6/Patched-related, VAP-1/secreted protein, and LIT-1/Nemo-like kinase (Perens and Shaham 2005; Oikonomou et al. 2011), and those around the AWC neuron are devoid of these proteins (Fig. 1B). Cilia of non-AFD neurons dictate KCC-3 localization to a microdomain around AFD, revealing cross talk across these molecularly distinct glia–neuron contacts (Ray et al. 2024). Likewise, URX and BAG neuron dendrite endings contact ILso glia at distinct sites, suggesting molecularly distinct membrane contact regions (Fig. 1C; Cebul et al. 2020). Such separation of function is revealed in the extreme for CEPsh glia; CEPsh anterior processes wrap around dendritic endings of CEP sensory neurons (also of CEM in males), while posterior ramifications envelop the ~180 neurite processes of the nerve-ring neuropil and penetrate the neuropil to contact synaptic sites and engage in tripartite synapses (Fig. 1D,E; White et al. 1986; Doroquez et al. 2014; Katz et al. 2019).

Similarities to Vertebrate Glia

CEPsh glia development suggests homology with vertebrate radial glia and astrocytes. Embryonically, CEPsh glia guide midline-crossing of nerve-ring axons using Netrin, just as radial glia pial branches direct axon guidance in the vertebrate spinal cord (Dominici et al. 2017; Rapti et al. 2017; Varadarajan et al. 2017). Vertebrate radial glia eventually transform into astrocytes (Schmechel and Rakic 1979; Noctor et al. 2008), and CEPsh glia undergo similar remodeling (Rapti et al. 2017). In both settings, glia abut synapses (White et al. 1986) and direct synaptogenesis (Christopherson et al. 2005; Colón-Ramos et al. 2007; Eroglu et al. 2009; Allen et al. 2012; Shao et al. 2013). Furthermore, astrocytes and CEPsh glia cover nonoverlapping neural domains, respecting unknown tiling rules (White et al. 1986; Bushong et al. 2002). Gene expression profiles reveal that CEPsh glia are more similar to mouse astrocytes than to any other murine brain cell (Katz et al. 2019). Finally, astrocytes exhibit Ca^{2+} transients, and gap junctions allow Ca^{2+} flow between astrocytes (Shigetomi et al. 2016; Khakh 2019; Nagai et al. 2021). CEPsh glia exhibit similar responses (M Katz and S Shaham, unpubl. data). Finally, multiple glia express gap junction proteins like mammalian glia (Cuadras et al. 1985; White et al. 1986; Nedergaard 1994), whose functions are not yet explored (Altun et al. 2015).

Sensory organ socket and sheath glia are related in function and molecules to vertebrate sensory organ glia and to astrocytes. For example, similar to glia-like retinal pigment epithelium cells in the eye and astrocytes, AMsh glia regulate contacting neurons by pruning their endings, regulating their ionic milieu, and deploying thrombospondin-domain proteins (Bacaj et al. 2008; Singhvi et al. 2016; Allen and Eroglu 2017; Raiders et al. 2021a,b; Ray and Singhvi 2021). Like olfactory epithelium sustentacular

cells, they also express xenobiotic metabolism gene batteries (Wallace et al. 2021). Lastly, sensory cues that activate neurons ensheathed by AMsh glia promote changes in AMsh glia intracellular Ca^{2+} concentration, similar to glia in other animals, suggesting that this property may be a conserved glial feature (Rousse and Robitaille 2006; Han et al. 2013; Shigetomi et al. 2016; Duan et al. 2020; Fernandez-Abascal et al. 2022).

GLR glia, like vertebrate microglia, arise mesodermally. Like microglia, GLRs express GABAergic signaling effectors and regulators and may also engulf dying cell debris (Nass et al. 2002; Gendrel et al. 2016; Wilton et al. 2019; Favuzzi et al. 2021).

SPECIFICATION OF *C. elegans* GLIA

Pan-glia Cell-Fate Specification

Most *C. elegans* glia are born embryonically and diversify using transcription factors expressed either after or before progenitor cell division. The Zn-finger transcription factor LIN-26 is expressed in neuroectoderm-derived glia using enhancer elements embedded within the *lin-26* genomic locus (Landmann et al. 2004). LIN-26 loss promotes glial cell degeneration and/or adoption of nonglial fates (Ferguson et al. 1987; Labouesse et al. 1996). Although there is no obvious sequence similarity, LIN-26 functions similarly to *Drosophila* glial-cells-missing (Gcm), a Zn-finger protein required for glia specification (Hosoya et al. 1995; Jones et al. 1995; Vincent et al. 1996). In both *lin-26* and *gcm* mutants, cells slated to become glia can become transformed into neurons. More direct molecular conservation is evident with the *C. elegans* transcription factor PROS-1, a homolog of Prospero/Prox1 that specifies *Drosophila* and vertebrate glia, respectively (Bunk et al. 2016; Peco et al. 2016). PROS-1 is expressed in a large subset of glia to regulate the expression of many glial genes, and defects in *pros-1* animals are rescued by the expression of human Prox1 (Kage-Nakadai et al. 2016; Wallace et al. 2016).

While they are expressed in most, if not all glia, both embryonic LIN-26 and postembryonic MIR-228, a microRNA expressed in most glia, are also present in nonglial cells. Indeed, RNA-seq studies suggest that there are no panglial genes expressed exclusively in glia (Purice et al. 2023).

Specification of Glial Subtypes

AMsh Glia

AMsh glial fate specification requires a cascade of transcription factors. Early on, AMsh glia express and require the gliogenic transcription factor LIN-26/Zn finger and the UNC-130/FKHD repressor (Labouesse et al. 1996; Mizeracka et al. 2021). UNC-130 specifies the identities of AMsh and other glia and epidermal cells (but not neurons) that arise from a discrete developmental lineage. The roles of UNC-130 in AMsh glia can also be executed by human FOXD3, a neural crest glial lineage regulator, suggesting possibly conserved programs of peripheral glia fate specification (Kastriti and Adameyko 2017; Dawes and Kelsh 2021; Mizeracka et al. 2021).

To ensure that only two AMsh glia are generated, this fate is suppressed in other lineages. Conserved Atoh/NeuroD family proneural bHLH genes restrict AMsh glial fate expression. Loss of *lin-32*/Atonal/Atoh results in misexpression of AMsh markers in CEPsh glia, and possibly other cells, leading to supernumerary AMsh glia. LIN-32 acts in parallel to CND-1/NeuroG1 and NGN-1/NeuroD1 (Zhang et al. 2020c) to prevent ectopic expression of AMsh glia fate. LIN-32 also regulates AMso glia specification, although its roles here are unknown (Zhang et al. 2020c). Finally, once AMsh glial fate is specified, PROS-1/Prox1/Prospero acts to maintain expression of the AMsh glia secretome until the adult stage (Kage-Nakadai et al. 2016; Wallace et al. 2016).

AMso/PHso

AMso and PHso glia express ALR-1, the *C. elegans* ortholog of Aristaless, a Paired homeodomain transcription factor that drives neural fate and sense organ specification in flies and mammals (Meijlink et al. 1999). ALR-1 regulates glial cell shape and adhesion to overlying epithe-

lia (Tucker et al. 2005). AMso, PHso, and other glial cells express hedgehog-related genes like GRL-12, although their roles in glial development await inquiry (Melkman and Sengupta 2005; Hao et al. 2006).

CEPsh Glia

CEPsh glia express the transcription factor HLH-17 (McMiller and Johnson 2005; Yoshimura et al. 2008), the *C. elegans* protein most similar to vertebrate Olig2, which is expressed in precursors of oligodendrocytes and motor neurons, and in subsets of astrocytes (Masahira et al. 2006; Tatsumi et al. 2018). VAB-3/Pax6/7 cell-autonomously controls HLH-17 expression in CEPsh glia. With the Nkx/Hmx-related protein MLS-2, VAB-3 activates HLH-17/Olig2 in ventral CEPsh glia; but does so independently of MLS-2 in dorsal CEPsh glia (Yoshimura et al. 2008). This gene-expression pathway is reminiscent of specification events in the mouse spinal cord, where Olig2 expression depends on Nkx6 in ventral regions and Pax6/Pax7 in ventral/dorsal regions (Rowitch 2004; Miller 2005). HLH-17 expression in dorsal CEPsh glia is also affected by the loss of LIN-32/Atoh1 (see above) (Zhang et al. 2020c).

ILso Glia

Like CEPsh glia, the six ILso glia arise from distinct lineages. UNC-130/FoxD3, acts as a repressor, promoting generation of the two dorsal, but not ventral or lateral, ILso glia (Mizeracka et al. 2021). UNC-130, acting through its DNA-binding domain, functions in progenitor and newly born terminal cells to also specify other cells related by lineage to ILsoD glia. These studies reveal that different transcriptional programs operate in different ILso developmental lineages to specify similar cell fates.

Ray Structural Glia

Ray neuroblast progenitors give rise to RnB neurons and to their sister glia—the ray structural cells. This asymmetric division requires asymmetric Wnt and POP-1/LEF-β-catenin signaling to specify anterior–posterior neuron–glia fates. Loss of the Wnt receptor LIN-17/Frizzled transforms posterior glial cell daughters into anterior neurons through the aberrant expression of LIN-32/Atoh in these cells (Sulston et al. 1980; Portman and Emmons 2000; Emmons 2005; Miller and Portman 2011). Thus, LIN-32/Atoh1/Atonal can have context-specific pro- or antigliogenic fate specification roles (see above). Finally, VAB-3/Pax6/7 regulates fate specification here through interactions between a subset of developing Rnst and their contacting neurons and epithelia (Zhang and Emmons 1995), highlighting context-specific gene functions.

GLR Glia

The six GLR glia derive from the MS mesodermal lineage (Sulston et al. 1983; White et al. 1986). GLRs express the *C. elegans* myoD homolog HLH-1; type IV collagen, like muscles; and mesoderm-lineage-enriched genes like DIG-1 and EMB-9 (Krause et al. 1994; Graham et al. 1997), and contain GABA and the GABA transporter SNF-10 (Gendrel et al. 2016). Recent transcriptome studies suggest that GLR glia merge astrocytic and endothelial characteristics relegated to separate cell types in vertebrates (Stefanakis et al. 2024). Combined fate acquisition is orchestrated by LET-381/FoxF, a fate specification/maintenance transcription factor expressed in glia and endothelia of other animals. Among LET-381/FoxF targets, UNC-30/Pitx2 transcription factor controls GLR glia morphology and represses an alternative mesodermal fate of the HMC cell, a GABA-containing, muscle-related cell. LET-381 and UNC-30 coexpression in naive cells is sufficient for GLR glia gene expression.

MORPHOGENESIS AND CELL BIOLOGY OF *C. elegans* GLIA

Glial Cell Polarity

C. elegans neuroectodermal glia are lineally related to epithelia and neurons, which are both polarized. Sheath glia make tight junctions with socket glia and with associated neurons (Ward et al. 1975; Perkins et al. 1986). For AMsh glia, whose apical domain faces neuron contact sites,

apical–basolateral domain segregation can be visualized by expression of subcellular domain markers, including ERM-1/ERM and PIP2 (Low et al. 2019; Martin et al. 2024; Ray et al. 2024). Like vertebrate ZP proteins, DYF-7/ZP is also an apical ECM component of these glia (Fig. 2A; Low et al. 2019). Similar organization likely characterizes all sense organ–ensheathing processes of *C. elegans* (Lee et al. 2021b).

Glial Cell Size Control

The size of AMsh glia cells is regulated by the conserved *cis*-Golgi membrane protein EAS-1/GOLT1B, the loss of which causes enlarged AMsh glial cell bodies (Fig. 2B; Zhang et al. 2020b). EAS-1/GOLT1B, through the E3 ubiquitin-ligase gene *rnf-145*/RNF145, promotes nuclear activation of *sbp-1*/SREBP, a sterol and fatty-acid synthesis regulator. Long-chain polyunsaturated fatty acids may be relevant products of this pathway, although how they affect cell size is not understood.

Glial Cell Morphogenesis

AMsh Glia Process Extension

Adult amphid sensory dendrites and glial processes extend ~100 μm toward the nose tip (Ward et al. 1975; White et al. 1986). Elongation of these processes occurs in embryogenesis through a mechanism termed retrograde extension (Sulston et al. 1983; Heiman and Shaham 2009; Lamkin and Heiman 2017). Newly born amphid sensory neurons and AMsh glia extend short projections, anchored at the nose tip with an ECM composed of the zona-pellucida (ZP) domain protein DYF-7, secreted by the neurons, and DEX-1, a zonadhesin domain-containing protein secreted by nonneuronal neighboring cells (Fig. 2B′; Heiman and Shaham 2009; Oikonomou and Shaham 2011). DEX-1 and DYF-7 resemble α- and β-tectorins, proteins comprising the tectorial membrane that anchors hair cell cilia in the inner ear (Legan et al. 1997). Subsequent migration of neuron and glia cell bodies extends dendrites and glial processes, respectively. *dyf-7* or *dex-1* mutations result in short dendrites and AMsh glia processes. AMso glia, which connect anteriorly to AMsh glia, are generally unaffected;

Cite this article as *Cold Spring Harb Perspect Biol* doi: 10.1101/cshperspect.a041346

however, abnormal AMso glia posterior processes attached to AMsh glia are occasionally observed, suggesting that AMsh–AMso adhesion is independent of DYF-7 and DEX-1 (Heiman and Shaham 2009).

Genes controlling glial cell body migration, and presumably relevant for retrograde extension, are known. Lesions in the VAB-3/Pax-6 transcription factor block posterior migration of several anterior glia and neuron cell bodies (Yoshimura et al. 2008). A cleaved, secreted form of SAX-3/Robo can interact with SYG-1/Neph, expressed in AMsh glia, to promote glia migration (Qu et al. 2020). Likewise, dietary vitamin B12, acting through PTP-3/LAR PRTP and NID-1/Nidogen on glia, controls glial migration (Zhang et al. 2020a).

AMsh Glia Process Tip Morphogenesis

Eight amphid neurons extend ciliated dendrites passing through a matrix-filled channel open to the environment (Ward et al. 1975; Perkins et al. 1986). AMso and AMsh glial membranes comprise the anterior and posterior channel sections, respectively, and are joined by tight junctions to form a continuous tube (Fig. 1B; Martin et al. 2024). This compartment resembles synaptic compartments surrounded by astrocyte end-feet (Shaham 2010). The anterior region of AMsh glia in this compartment is decorated by a web of apical β_H-spectrin (Martin et al. 2024). Glial compartment formation occurs in the embryo before neuronal cilia enter it (Oikonomou and Shaham 2011). Subsequent morphogenesis requires DAF-6/Patched-related and the secreted DAF-6-binding protein DYF-4, both of which restrict amphid channel expansion (Fig. 2C–C'; Perens and Shaham 2005; Oikonomou et al. 2011; Hong et al. 2021). Bloated channels in *daf-6* or *dyf-4* mutants can be rescued by mutations in the *lit-1*, *snx-1*, or *igdb-2* genes (Oikonomou and Shaham 2011; Oikonomou et al. 2012; Wang et al. 2017; Hong et al. 2021), suggesting that these genes antagonize DAF-6 and DYF-4 functions and promote channel growth. LIT-1/NEMO-like kinase, SNX-1, a retromer component, and IGBD-2, an Ig/FNIII protein, act in parallel. IGDB-2 functions in AMso glia and its loss can be partially compensated for by the

loss of LGC-34, a predicted ligand-gated ion channel (Wang et al. 2017), indicating that extracellular ion levels and/or glia:glia coordination influence amphid channel dimensions. *daf-6*, *lit-1*, and *snx-1* also interact genetically with *che-14*/Dispatched-related, required for apical secretion from AMsh glia (Michaux et al. 2000; Perens and Shaham 2005; Oikonomou and Shaham 2011; Oikonomou et al. 2012). All identified channel size regulators function in AMsh/so glia; nonetheless, sensory-neuron cilia defects affect DAF-6 localization (Perens and Shaham 2005), suggesting cross talk between these cell types (Martin et al. 2024).

CEPsh Glia Morphogenesis

CEPsh glia are born in the embryo (Sulston et al. 1983) where, in addition to extending a process associated with CEP dendrites, they also extend nonbranching processes that mark the location of the presumptive nerve ring and that promote nerve-ring assembly. Later, CEPsh glia processes become highly ramified, extending membrane sheets that envelop the nerve ring, and finer processes that penetrate it and abut synapses (Fig. 2D–D'; White et al. 1986; Rapti et al. 2017; Katz et al. 2019). The CEPsh glial cell morphology does not appear disrupted after ablation of nerve-ring pioneer neurons SIA and SIB, suggesting that these neurons may be dispensable for the development of CEPsh glia membrane sheaths.

Glia–Epithelia Interactions in Glial Shape Maintenance

C. elegans glia are mechanically coupled to overlying epithelia, Briefly, epithelial UNC-23/BAG2, an Hsp cochaperone, maintains epithelial cell shape against mechanical stress (Martin et al. 2024). Loss of *unc-23*, through misregulated HSP-1-DNJ-13 chaperone cycling, induces progressive age-dependent deformation in the epithelia cell shape of adult animals. This, through FGFR signaling, disrupts the glial apical cytoskeleton protein SMA-1/β-spectrin and F-actin, resulting in the loss of shape of AMsh glial apical domains that contact neuron endings. This leads to the consequent loss of neuron-ending shapes

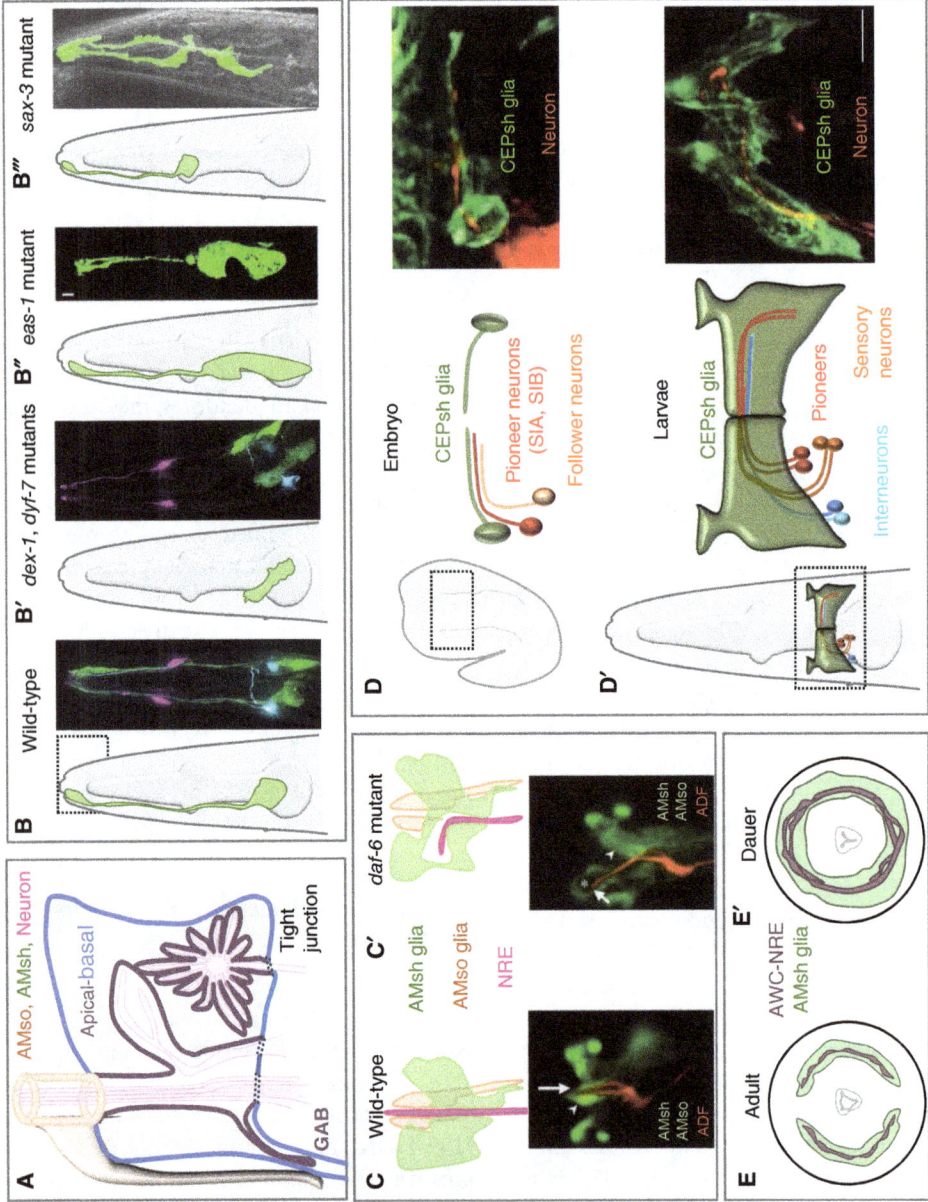

Figure 2. (*See following page for legend.*)

and function (Martin et al. 2024). This epithelia–AMsh glia coupling only occurs at the L4-adult critical developmental window (Martin et al. 2024). Further, this coupling only affects glia in the anterior head region, like AMsh and CEPsh glia (Martin et al. 2024), but tail PHsh glia or phasmid neurons are unaffected (Martin et al. 2024). Thus, epithelia–glia mechanical coupling is regulated with exquisite spatial and temporal specificity (Martin et al. 2024). UNC-23 also affects tissue viscoelasticity and integrity of the Perlecan ECM, in relation to temperature and mechanical stress, to affect the architecture of CEPsh glia sheaths in an age-progressive manner (Rahmani et al. 2015; Coraggio et al. 2023; Martin et al. 2024).

Similarly, loss of the ECM protein DIG-1, secreted by muscle and epithelia to regulate basement membrane architecture, induces fragmentation of AMsh glia anterior endings and abnormal cell position, shape, and fasciculation of neurons associated with ILso glia, and other neuronal processes (Bénard et al. 2006; Chong et al. 2021). Loss of either UNC-23 or DIG-1 results in progressive defects in adult animals, indicating lifelong roles for epithelia in glia-shape maintenance and neural aging (Chong et al. 2021; Martin et al. 2024).

Morphology of postembryonic CEPsh glia posterior processes is also maintained by CIMA-1/SLC17A5 transporter, which controls EGL-15/FGFR levels in epithelia, and MIG-17/ADAMTS protease, which influences basement membrane composition (Fig. 3A; Shao et al. 2013; Fan et al. 2020).

Finally, RAM-5/ZP domain protein and DPY-18/prolyl-4-hydroxylase subunit are expressed in the male tail-specific Rnst glia and epithelia, respectively, and may act in a glycosylation-dependent pathway to mediate epithelia–glial interactions and Rnst cell shape (Yu et al. 2000).

Recurrent epithelia–glia juxtaposition across species raises the possibility that analogous signals may broadly maintain glial cell shape and polarity (Salzer 2003; Derouiche et al. 2012).

Glia Remodeling in Dauer Animals

Upon stress, *C. elegans* larvae enter a developmental state called dauer through steroid hormone signaling. In dauers, the bilateral AMsh glia expand, fuse, and exchange cytoplasm (Fig. 2E–E'; Cassada and Russell 1975; Albert and Riddle 1983; Procko et al. 2011). AMsh glia expansion occurs concomitantly with the expansion of AWC neuron dendritic endings ensheathed by these glia. Behavior studies reveal that remodeling facilitates dauer exit upon exposure to favorable conditions (Lee et al. 2021a). Dauer-induced glial fusion is driven by AMsh glia-expressed REMO-1/G-protein-coupled receptor; transcription factors TTX-1/Otx and ZTF-16/Ikaros; fusogen AFF-1; and the stress-responsive factor VER-1/RTK (Procko et al. 2011, 2012; Lee et al. 2021a). REMO-1 localizes to AMsh glia anterior tips and, with TTX-1 and ZTF-16, is required for

---

Figure 2. Glial cell development and morphogenesis. (*A*) Schematic depicting apical–basal polarity of AMsh and AMso glia at the amphid channel lumen. (*B–B'''*) Diagram and images of AMsh glia in wild-type (*B*) and different mutant backgrounds (*B'–B'''*). AMsh glia anterior processes are collapsed in *dyf-7* and *dex-1* mutant (*B'*), glial cell size is enlarged in *eas-1* mutant (*B''*) and glial cell body migration is disrupted in *sax-3* mutant (*B'''*) animals. (*C–C'*) Diagram and images of AMsh glia in wild-type (*C*) and *daf-6* (*C'*) mutant animals showing aberrant sensory compartment lumen in mutants, which impedes ADF-NRE from accessing the outside environment. (*D–D'*) Diagram and image of CEPsh and pioneer/follower neurons growing into the neuropil for brain assembly in the embryo (*D*) and larvae (*D'*). Scale bar, 10 μm. Micrograph shows CEPsh glia process guiding pioneer axon processes. (*E, E'*) Schematic of AMsh glia and AWC neuron remodeling in non-dauer (*E*) and post-dauer animals (*E'*). (AMsh) Amphid sheath glia, (AMso) amphid socket glia, (CEPsh) cephalic sheath glia, (NRE) neuronal receptive ending. (Fluorescence images as follows: *B,B'* reprinted from Heiman and Shaham 2009 with permission; *B''* from Zhang et al. 2020b, reprinted under the terms of the Creative Commons CC BY 4.0 License; *C,C''* from Qu et al. 2020, reprinted under the terms of the Creative Commons Attribution License; *D,D'* from Rapti et al. 2017, reprinted with permission from the author. Schematics from Singhvi and Shaham 2019, reprinted with permission from the author.)

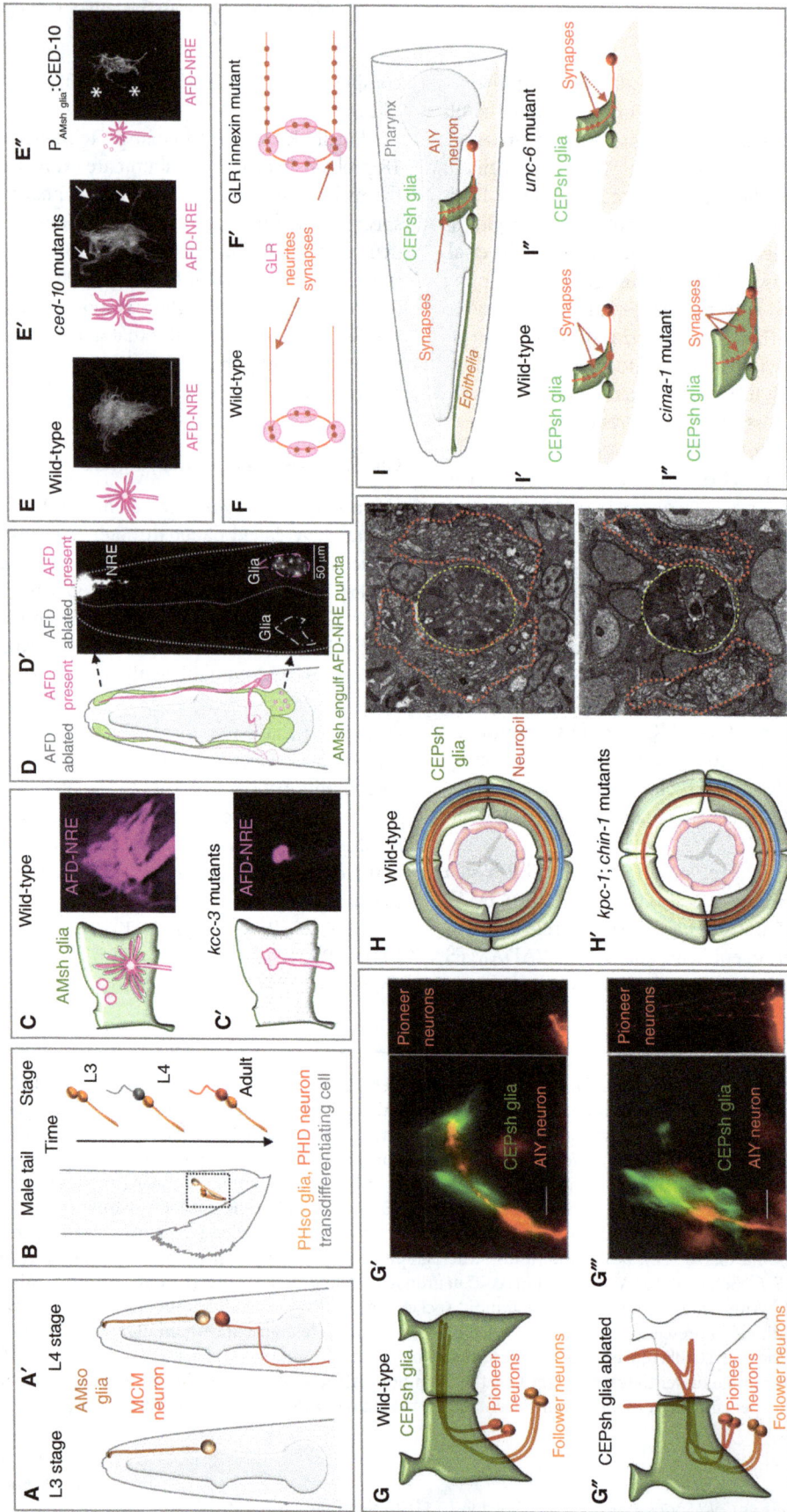

Figure 3. (*See following page for legend.*)

the expression of VER-1/RTK (Procko et al. 2011, 2012; Lee et al. 2021a). While glial REMO-1 impacts the remodeling of AWC neurons, loss of AWC neurons does not alter glia remodeling, suggesting that glial cues dictate downstream neuron remodeling events (Procko et al. 2011, 2012; Lee et al. 2021a). Nonetheless, some dauer-neuron remodeling events are independent of the REMO-1 glial pathway (Lee et al. 2021a).

GLIAL REGULATION OF NEURON GENERATION AND DEGENERATION

Glia as Neuronal Progenitors

The embryonic and postembryonic lineages of *C. elegans* elucidated four decades ago, report that glia are terminally differentiated cells (Sulston and Horvitz 1977; Sulston et al. 1983). Recent studies, however, reveal that some glial cell divisions were missed. Indeed, male AMso glia are a postembryonic source of neurons (Fig. 3A). During male sexual maturation, each AMso divides asymmetrically in a budding-like division to generate an AMso cell and a male-specific MCM neuron that expresses neuronal gene batteries, forms brain synapses, and regulates male-specific behaviors (Sammut et al. 2015). Remarkably, the AMso daughter

retains glial markers, cilia-ensheathing projections, and polarized morphology of the precursor cell through the division, reminiscent of vertebrate radial glia divisions. Concomitantly, during sexual maturation, two other (*trans*)differentiation events occur in the male. The male glial cell PHso2 transforms to make the pore for tail sensory sensilla, formed by socket glia PHso1 in hermaphrodites, while male PHso1 socket glia *trans*-differentiate to generate sensory PHD neurons, which contribute to male copulation circuits (Fig. 3B; Sulston et al. 1980; Molina-García et al. 2020). This glia-to-neuron fate change generally happens without cell division and is molecularly distinct from Y-to-PDA differentiation, a different *trans*-differentiation event (Jarriault et al. 2008; Rashid et al. 2022).

Glial Roles in Neurodegeneration

Mutations in the gene *swip-10* result in swimming-induced paralysis, mediated by hyperexcitability of dopaminergic CEP neurons (Hardaway et al. 2015; Gibson et al. 2018). SWIP-10, a metallo-β-lactamase-domain protein, is expressed in CEP-ensheathing CEPsh glia, and other glia. Its action may be indirect and mediated through its effects on glutamate signaling.

Figure 3. *Caenorhabditis elegans* glial cells in neuronal generation and morphogenesis. (*A–A'*) Schematic of AMso glia in males in L3 (*A*) and L4 (*A'*) larval stages showing its cell division to generate the MCM neuron. (*B*) Schematic of the male PHso glia transdifferentiating into a PHD neuron during developmental L3–L4-adult transition stages. (*C–C'*) Diagram and image of AFD-NRE in wild-type animals with intact AMsh glial ensheathment (*C*), and in *kcc-3* mutant animals (*C'*). (*D–D'*) Schematic (*D*) and image (*D'*) of bilateral AMsh glia–AFD. AFD-NRE staining (*top* arrow) is also seen as punctate fragments in the AMsh glia cell body (*bottom* arrow) on the side with AFD neurons present and lost in the AMsh glial cell body on the side with AFD neuron ablated. (*E–E''*) Pruning by AMsh glia regulates AFD-NRE shape. Reduced pruning (*ced-10* mutants) causes elongated AFD-NRE, and excess pruning (overexpress CED-10 in AMsh) causes shorter AFD-NRE. (*F–F'*) RME synapses localized to a specific region of the neurite in wild-type (*F*) are misrouted along the neurite processes in GLR innexin mutant animals (*F'*). (*G–G'''*) Diagram and image of CEPsh glial posterior membrane sheaths directing pioneer and follower neuronal axons (*G, G'*), and their misdirection in glia-ablated-animals (*G'–G'''*). (*H–H'*) Diagram and image of glial cell and axon processes in the brain neuropil of wild-type animals (*H*), which become truncated or misguided in *kpc-1; chin-1* double mutant animals with abnormal trafficking in CEPsh glial cells, leading to reduced brain neuropil size. (*I–I''*) Schematic of epithelia and AIY neuron synapses within CEPsh glia posterior process zone. Densities of AIY synapses apposing specific CEPsh glia posterior membrane sheath process regions in wild-type animals (*I, I'*) are decreased in animals defective for glia-secreted UNC-6/Netrin (*I''*) and ectopically positioned in *cima-1* animals with aberrant CEPsh posterior sheath processes. (Panel *C* is reprinted, with permission, from Singhvi et al. 2016; *D, E* reprinted from Raiders et al. 2021a under the terms of the Creative Commons Attribution License; *G,H* reprinted from Rapti et al. 2017 with permission from the author. Schematics adapted from Singhvi and Shaham 2019 with permission from the author.)

Of note, *swip-10* loss promotes age-dependent dopaminergic-neuron degeneration, suggesting that glia play important roles in maintaining cell numbers by preventing neurodegeneration.

GLIAL CONTROL OF NEURON MORPHOLOGY

Dendrite Outgrowth

Studies of retrograde extension in amphid development (see above) suggest important roles for glia in sensory-neuron dendrite extension (Fig. 2B–B'). AMsh glia ablation during embryogenesis results in short dendrites, resembling those of *dyf-7* or *dex-1* mutants (Singhal and Shaham 2017). Similarly, CEPsh glia ablation results in CEP neuron dendrite extension defects (Yoshimura et al. 2008). The URX and BAG sensory dendrites also form through retrograde extension (Chong et al. 2021). Unlike the amphid, dendritic tips here contact a protrusion of ILso glia to form a dendritic anchor independent of DYF-7/DEX-1. Instead, SAX-7/L1CAM, acting in neurons and ILso glia, and GRDN-1/Girdin/ CCDC88C, acting in ILso glia, anchor BAG and URX neurite tips as the neuron cell bodies migrate (Lamkin and Heiman 2017).

Neuron-Receptive-Ending (NRE) Shape

Sensory NREs are specialized subcellular domains housing sensory-transduction machinery and are proposed as analogs of postsynaptic dendritic spines (Shaham 2010). *C. elegans* sense-organ glia regulate NRE shape, sensory-neuron function, or both; resulting in behavioral deficits (Singhvi and Shaham 2019; Ray and Singhvi 2021; Martin et al. 2024; Ray et al. 2024). Glial regulation of NRE shape has been predominantly investigated for AMsh glia, which associate with sensory NREs of 12 neurons and regulate the shape of many (Bacaj et al. 2008). Molecular studies reveal that a single AMsh glial cell uses distinct mechanisms to regulate each neuron.

The glial-secretome regulator PROS-1/Prox regulates the shapes of multiple amphid NREs (Wallace et al. 2016). In contrast, the glial K/Cl cotransporter KCC-3 regulates only AFD neuron-receptive-endings shape (Fig. 3C–C'; Singhvi et al.

2016; Yoshida et al. 2016). KCC-3 controls chloride levels in the glia–neuron intercellular milieu. Chloride inhibits the AFD neuron receptor-guanylyl-cyclase GCY-8. GCY-8-dependent control of neuronal cGMP signaling, in turn, regulates NRE shape through the actin-polymerization factor WASP-1/nWSP (Singhvi et al. 2016). *kcc-3* mutants exhibit impaired AFD-dependent thermosensory behavior.

AMsh glia also regulate NRE shape by phagocytosis of AFD neuron NRE fragments (Fig. 3D–E''; Raiders et al. 2021a). Briefly, AMsh glia dynamically tune pruning rates based on neuron activity. This pruning requires molecular cues regulating exposure of phosphatidylserine on neurons similar to apoptotic cells and recognized by glia using apoptotic cell engulfment components (Raiders et al. 2021a). Similar pruning is documented at *Drosophila* and vertebrate synapses, supporting the notion that synapses and sensory endings are functionally related (Shaham 2010; Wilton et al. 2019; Hilu-Dadia and Kurant 2020; Raiders et al. 2021b).

AMsh glia also uptake NRE-derived extracellular vesicles (Ohkura and Bürglin 2011; Razzauti and Laurent 2021). This may impact NRE shape and function for some neurons, but the physiological significance remains to be determined.

Finally, maintenance of NRE shape also requires appropriate glial cell cytoskeleton maintenance downstream from epithelia–glia signaling and epithelial UNC-23/BAG2 Hsp cochaperone activity (see glia-shape section above) (Martin et al. 2024).

Neurite Specification

The mesodermal-lineage GLR glia regulate the specification of RME motoneuron neurites. Through gap junctions with RME neurons, GLR glia regulate calcium concentration, a CDK-5 pathway, and microtubule polarity to control the placement of synaptic proteins (Fig. 3F–F'; Meng et al. 2016). Glia–neuron gap junctions are also reported in other species including humans, suggesting that these mechanistic insights may be broadly relevant (Cuadras et al. 1985; White et al. 1986; Nedergaard 1994).

Cite this article as *Cold Spring Harb Perspect Biol* doi: 10.1101/cshperspect.a041346

Axon Guidance and Brain Assembly

In the embryo, CEPsh glia extend processes along the dorsoventral axis, demarcating the presumptive nerve ring. These processes guide sublateral commissure pioneer neurons (primarily SIA and SIB) into the nerve ring, and a combination of CEPsh glia and pioneer neuron signals directs follower neuron entry (Fig. 3G–H′; Rapti et al. 2017). CEPsh glia ablation, mutations inactivating axon-guidance factors released from these cells or blocking trafficking of these cues in Chimaerin/Furin double mutants disrupts pioneer axon pathfinding and brain assembly (Rapti et al. 2017). CEPsh glia communicate with pioneers via UNC-6/Netrin and with follower axons using MAB-20/Semaphorin and FMI-1/CELSR (Rapti et al. 2017). Since the loss of MLS-2/Nkx/Hmx and VAB-3/Pax6/7 affects CEPsh fate specification and nerve-ring axons guidance defects (Yoshimura et al. 2008), it is possible that these transcription factors regulate the expression of guidance genes in CEPsh glia.

GLR glia, which line the inner surface of the nerve ring, are required for maintaining the nerve-ring position between the anterior and posterior pharyngeal bulbs. Embryonic ablation of GLR-glia precursor cells causes postembryonic anterior displacement and defasciculated of the nerve ring (Shah et al. 2017).

Synapse Formation and Maintenance

Like astrocytes, which mediate synapses formation and maturation (Ullian et al. 2001; Christopherson et al. 2005; Allen et al. 2012; Chung et al. 2013, 2016), CEPsh glia may also control synaptic placement. UNC-6/Netrin expressed in these glia appears to promote the enrichment of its receptor, UNC-40/DCC, in presynaptic regions of AIY neurons, as well as process guidance of its postsynaptic partner RIA (Fig. 3I–I″; Colón-Ramos et al. 2007). Whether CEPsh glia affect synaptogenesis directly or by fine-tuning AIY/RIA process guidance remains unclear, since their processes in the neuropil contact axons of these neurons but not their synapses directly (White et al. 1986; Witvliet et al. 2021).

CEPsh glia may also affect synapse maintenance (Fig. 3I″). Aberrant elongation of CEPsh

glia upon disruption of epithelial CIMA-1/SLC17A5, EGL-15/FGFR, and MIG-17/ADAMTS, promotes ectopic presynaptic sites in AIY axons (Shao et al. 2013). These sites are adjacent to mispositioned postsynaptic RIA axons, correlate with ectopic axon–glia contacts, and are independent of Netrin (Shao et al. 2013; Fan et al. 2020). Whether these synapses result from new synapse assembly or abnormal density due to altered axon morphology remains unclear. Postembryonic maintenance of axon shape and AIY synaptic sites also requires maintenance of CEPsh glia sheath integrity by UNC-23/BAG2 (Shao et al. 2013; Coraggio et al. 2023). By contrast, CDC-42 GTPase and its effector PAS-7/IQGAP act downstream from glia, within the neuron, to promote ectopic sites, without regulating glia morphology (Dong et al. 2020).

Ray Structural Glia in Ray Sensilla Morphogenesis

The male Rnst structural glial cells express RAM-5, a transmembrane protein that acts with epidermal MAB-7/SNED during tissue remodeling that leads to ray neuron morphogenesis (Yu et al. 2000).

GLIA IN NEURAL CIRCUIT FUNCTION AND ANIMAL BEHAVIOR

Glia as Sensory Cells

AMsh glia can detect sensory cues independently of associated neurons. They exhibit Ca^{2+} transients upon isoamyl alcohol or octanol exposure using glia-specific G-protein-coupled receptors, which triggers olfactory adaptation by glia-driven GABA signaling to sensory neurons (Duan et al. 2020). AMsh glia Ca^{2+} transients upon tactile nose-touch stimulation also modulates behavioral adaptation through chloride channels CLH-1 and GABA signaling (Fig. 4A; Ding et al. 2015; Fernandez-Abascal et al. 2022). Glial Ca^{2+} transients require EGL-19/L-type Ca^{2+} channel α1 subunit activity, and loss of *egl-19* causes defects in olfactory adaptation (Chen et al. 2022). They also respond to environmental stress by inducing expression of VER-1/VEGFR to drive glia-neuron structural remodeling (reviewed above) (Procko et al. 2011; Duan et al. 2020; Fernandez-Abascal et al. 2022).

Glial Regulation of Sensory-Neuron Activity

AMsh Glia

Besides modulating NRE shape to effect neuron functional changes, AMsh glia also use signaling pathways to influence neuron activity. AMsh glia FIG-1/thrombospondin domain protein is required for octanol detection and amphid neuron dye filling (Bacaj et al. 2008; Wallace et al. 2016). Astrocyte thrombospondins regulate vertebrate synapse assembly and function (Christopherson et al. 2005), perhaps suggesting similar activities. The Na^+-selective AMsh glia DEG/EnaC subunit ACD-1 regulates AWC chemosensory neuron responses to specific odors (Wang et al. 2008, 2012), relying on glial acidification by the CLH-1/ClC chloride channel (Grant et al. 2015). CLH-

1 also promotes glia-dependent GABA activation and cAMP signaling in ASH neurons (Fig. 4B; Grant et al. 2015; Park et al. 2021; Fernandez-Abascal et al. 2022). Human Clc2, expressed in glia, also regulates ion homeostasis and GABA signaling (Sík et al. 2000; Depienne et al. 2013).

OLQso/ILso Glia

These glia express DEG/EnaC channel subunits DELM-1/2 and NaK^+-ATPase α-subunits EAT-6 and CATP-1, which regulate nose-touch sensitivity and foraging behaviors, likely through modulating neuron excitability (Fig. 4B; Han et al. 2013; Johnson et al. 2020; Ray and Singhvi 2021).

Glial Regulation of Synaptic Activity

Sleep Regulation

Postembryonic ablation of CEPsh glia does not alter nerve-ring morphology. Nonetheless, ani-

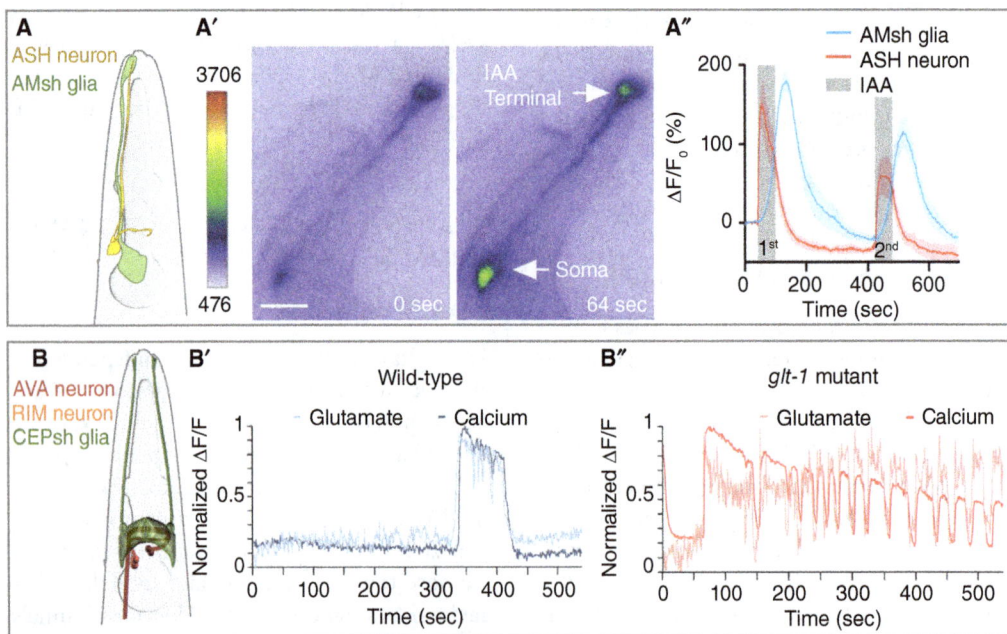

Figure 4. *Caenorhabditis elegans* glial functions in animal behavior. (*A–A″*) Schematic (*A*), representative micrograph (*A′*) (scale bar, 20 μm), and Ca^{2+} transient quantification (*A″*) in AMsh glia and ASH neuron upon two pulses of isoamylalcohol (IAA) stimulation. (*B–B″*) Schematic (*B*) depicting the region where CEPsh posterior processes contact AVA and RIM neurons. Representative traces of spontaneous glutamate (light, measured by iGluSnFR) and calcium (dark line, measured by GCaMP) dynamics near the AVA neuron in wild-type (*B′*) and *glt-1* mutant (*B″*) animals. (Images in *A* are reprinted with permission from Duan et al. 2020. Images in *B* are reprinted from Katz et al. 2019 under a Creative Commons Attribution 4.0 International License.)

mals move at half speed, along circular trajectories, and lapse into ectopic sleep bouts (Katz et al. 2018, 2019). In vivo, Ca^{2+} imaging during sleep reveals that while most neurons are silent, ALA neurons exhibit calcium transients (Nichols et al. 2017). These neurons form inhibitory synapses onto AVE locomotion interneurons, and these synapses are inactivated by CEPsh glia that wrap around them (Katz et al. 2018). AVE activity precedes backward movement; however, CEPsh glia ablation results in prolonged AVE Ca^{2+} signals uncoupled from movement. Astrocyte regulation of sleep is conserved in *Drosophila* and in mice (Frank 2013; Poskanzer and Yuste 2016; Artiushin and Sehgal 2020; Blum et al. 2021).

Repetitive Behavior

Animals lacking CEPsh glia or GLT-1, a conserved glutamate transporter expressed in CEPsh glia and vertebrate astrocytes, exhibit repeated backward movement initiations (Mano et al. 2007; Katz et al. 2019). Dual-color imaging of extracellular glutamate and intracellular Ca^{2+} signals in AVA (a backward-locomotion interneuron) in *glt-1* mutants, reveals oscillations of glutamate release near AVA and of AVA firing (Fig. 4B). These studies suggest that in the absence of glial GLT-1, glutamate diffuses away from AVA postsynaptic sites and engages an extrasynaptic glutamate receptor, MGL-2/mGluR5, on presynaptic neurons. This leads to unevoked release of glutamate, mediated by presynaptic EGL-30/Gαq, driving an autocrine feedforward loop that causes AVA to fire repeatedly (Katz et al. 2019). Conditional knockout of GLT1 in mouse astrocytes results in repetitive grooming behavior (Aida et al. 2015), and murine mGluR5 inhibition prevents repetitive grooming and head tics in mouse models of autism spectrum and other repetitive behavior disorders (Silverman et al. 2010). Thus, it is possible that mammalian repetitive behavior (Ting and Feng 2008) also originates from synaptic glutamate control defects.

Locomotion and Salt Resistance

let-381 mutants, which block postembryonic maintenance of GLR fate, as well as animals in which GLR glia are genetically ablated after ner-

vous system development is largely complete, exhibit specific defects in locomotory behavior resembling those seen in CEPsh ablated animals (Katz et al. 2018, 2019). Among other defects, GLR glia-defective animals have reduced locomotion speed and increased reversal probability. In addition, animals with disrupted GLR glia are hypersensitive to salt, arresting locomotion for longer than wild-type animals, and recovering more slowly once normo-osmotic conditions are restored, suggesting important roles for these glia in coordinating neuronal activity.

GLIAL ROLES IN STRESS AND AGING

Immunity

C. elegans encounters microorganisms in its environment, adapting physiology and behavior accordingly. Transcriptome studies reveal that co-culture with *Penicillium brevicompactum*, an ecologically relevant mold, up-regulates stress-response genes, including xenobiotic metabolizing enzymes (XMEs), in the intestine and AMsh glia. The nuclear-hormone receptors NHR-45 and NHR-156 are induction regulators, and mutants that cannot induce XMEs in the intestine when exposed to *P. brevicompactum* experience mitochondrial stress and exhibit developmental defects. Wild isolates of *C. elegans* harbor sequence polymorphisms in *nhr-156*, resulting in phenotypic diversity in AMsh glia responses to microbes. Thus, as in flies and mammals, *C. elegans* glia may also mediate immunity (Wallace et al. 2021).

Longevity

C. elegans glia may also regulate aging. Expression in CEPsh glia of constitutively active XBP-1, a transcription factor mediating responses to endoplasmic reticulum stress, extends life span and has ameliorating effects on distal tissues (Apfeld and Kenyon 1999; Arey and Murphy 2017; Frakes et al. 2020). This response may be mediated in part through changes in neurotransmitters (Wang and Bianchi 2021). Loss of RGBA-1, a neuropeptide-like protein expressed in glia, or of the neuropeptide receptor NPR-28 influences age-related decline in worm mating behavior

(Fig. 4D; Yin et al. 2017). However, since glia lack the canonical dense core vesicle release factor UNC-31/CAPS and EGL-3/convertase (Purice et al. 2023), how they secrete neuropeptides like RGBA-1 remains to be determined.

GLIAL FUNCTIONS OF EPITHELIAL CELLS

A *C. elegans* Model for Radial Glia-to-Motoneuron Differentiation

Canonical radial glia stem cells are absent in *C. elegans*; nonetheless, the cell division-independent transformation of the epithelial tube lining Y cell into the PDA motoneuron has been informative in understanding motoneuron generation (Jarriault et al. 2008; Zuryn et al. 2014; Rashid et al. 2022). This *trans*-differentiation event requires LIN-12/Notch acting through NGN-1/Ngn and its regulator HLH-16/Olig. *lin-12* loss blocks transformation, while *lin-12* (gf) promotes precocious PDA formation. Early basal expression of *ngn-1*/Ngn and *hlh-16*/Olig depends on *sem-4*/Sall and *egl-5*/Hox. Later, coincident with Y-cell morphological changes, *ngn-1*/Ngn expression is up-regulated in a *sem-4*/Sall and *egl-5*/Hox-dependent but *hlh-16*/Olig-independent manner. Control of histone methylation by JMJD-3.1, an H3K27me3/me2 demethylase, and the SET-2/Set1 H3K4 methylation complex ensures robustness of this *trans*-differentiation. Homologous proteins regulate motoneuron generation from radial glia in the vertebrate spinal cord (Jessell 2000; Dasen and Jessell 2009), suggesting that *C. elegans* genetics can help identify additional genes and interactions mediating these events.

Epithelia-Mediated Neurite Morphogenesis and Maintenance

In addition to maintaining glia–neuron architecture with age (Chong et al. 2021; Coraggio et al. 2023; Martin et al. 2024), epithelia assume glial-like roles and provide substrates for peripheral neurites and regulate their form and integrity. Epithelial EGL-15/FGFR guides the outgrowth of specific axons (PVP, PVQ, PVT, DA/DB) along the anteroposterior and dorsoventral axes in a kinase-independent manner via LET-60/Ras GTPase and adaptors SOC-1/2 (Bülow et al. 2004). Epithelia-expressed DRAG-1 regulates axon branching via UNC-40/DCC in hermaphrodite-specific neurons (HSNs) (Tsutsui et al. 2021). COL-99/ColA1, and DPY-18/P4HA2 affect longitudinal axons and male tail ray morphology. They are also expressed by epithelia but their tissue requirement is unclear (Baird and Emmons 1990; Hill et al. 2000; Soete et al. 2007; Taylor et al. 2018). Epithelia also guide dendrite morphogenesis. Epithelial adhesion molecules MNR-1/Fam151 and SAX-7/L1CAM form a coligand complex, bind the LRR transmembrane receptor DMA-1 on PVD neurons and instruct PVD dendritic branching (Liu and Shen 2012; Dong et al. 2013; Salzberg et al. 2013).

Epithelia Protect Axons

Epithelial cells can also guide synaptogenesis and synaptic maintenance in the PNS. Hemidesmosome attachments couple PLM neuron axons to the epidermis (Emtage et al. 2004). Hemidesmosome components (LET-805/Myotactin, VAB-10/Plakin) localize periodically for this attachment, are required to protect axons from damage, and are disrupted with age (Coakley et al. 2020; Bonacossa-Pereira et al. 2022). Epidermal UNC-70/β-Spectrin, TBC-10/GAP, and RAB-35/GTPase synergize to preserve hemidesmosomes, axon-epidermal attachments, and axon integrity against breakage (Coakley et al. 2020). Thus, the adhesion of peripheral axons to epithelia ensures their mechanical resilience.

Epithelia Regulate Synapse Assembly

Two immunoglobulin-fibronectin-domain adhesion proteins, SYG-1 and SYG-2, expressed in HSNs and vulval epithelia, respectively, interact to specify synapses (Shen and Bargmann 2003; Shen et al. 2004). SYG-2 instructs SYG-1 localization at presynaptic sites. The Ig-domain transmembrane protein ZIG-10 controls synapse maintenance. ZIG-10, localized by MAGU-2 near neuromuscular junctions, is required in the epidermis and motor neurons for synapse maintenance (Cherra and Jin 2016; Cherra et al. 2020).

Epithelial ZIG-10 regulates CED-1-mediated phagocytosis to constrain the cholinergic synaptic apparatus. This is reminiscent of MGEF10/CED-1-mediated synaptic pruning by astrocytes, microglia, and *Drosophila* glia (Stevens et al. 2007; Fuentes-Medel et al. 2009; Awasaki and Lee 2011; Chung et al. 2013; Raiders et al. 2021b).

LOOKING AHEAD

Studies of *C. elegans* already reveal extensive conservation across species in glia development and function, even at the molecular level. The invariant lineage and contacts each glial cell makes with neurons and nonneural tissue enable an understanding of the roles of these conserved molecules at single cell and single contact resolution. With such a powerful molecular-genetic toolkit, *C. elegans* is poised to shed light on many unresolved and exciting aspects of glia biology. For example, the molecular basis of specificity in glia–neuron interactions is generally unexplored. Findings on AMsh, CEPsh, and ILso glia interactions with their neuron partners provide excellent foundations to interrogate in vivo mechanisms and logic behind interaction specificity. The roles of Ca^{2+} transients in vertebrate glia remain highly debated. With a completely mapped connectome and molecular tools for selectively labeling glia, *C. elegans* may finally reveal whether glia Ca^{2+} serves roles in information processing.

The invariance in *C. elegans* glial cell numbers and anatomy provides a powerful opportunity to understand glia heterogeneity, across sex, age, position, circuit activity, stress, and other variables. Further, the complete molecular atlas of glia across both sexes in *C. elegans* has been compiled by snRNA-seq, the first such map for glia of a multicellular nervous system (Purice et al. 2023). Coupled with the animal's invariant glia–neuron development and connectome, this now provides unparalleled resolution to dissect glia biology molecularly. Further work extending such analyses to how different glia are tuned to different variables should emerge over the next few years.

Caenorhabditis elegans also provides an appropriate setting to understand how glia acquire their fates, maintain their elaborate morphologies, and organize in tiled configuration, questions that still remain unanswered for most vertebrate glia.

The discovery of neuron-receptive-ending pruning by AMsh glia also places *C. elegans* at the forefront of understanding this disease-relevant but molecularly enigmatic ability. Likewise, the observation that CEPsh astrocytes contribute to sleep, and control repetitive behavior, using machinery conserved with vertebrates, suggests that insights into glial control of behavior are likely to emerge from future studies of the worm.

The last two decades of research on *C. elegans* glia have borne fruit to early promise by rapidly delivering surprising and novel insights into glia biology. This, however, is just the beginning. Building on this exciting momentum, the coming years are likely to reveal fundamental insights into how glia govern nearly every aspect of the nervous system.

ACKNOWLEDGMENTS

The authors apologize to those whose work was not cited due to unintentional oversight or space concerns. The authors thank members of the Singhvi, Shaham, and Rapti groups for discussions. G.R. thanks colleagues at the FKNE-Kavli and Interdisciplinary Center for Neuroscience for scientific discussions. A.S. sincerely thanks all the generous philanthropic support to her laboratory including from Stephanus, Brown, and Van Sloun Foundations. This work was funded by Simons Foundation/SFARI grant (488574), Esther A. and Joseph Klingenstein Fund and the Simons Foundation Award in Neuroscience (227823), Brain Research Foundation Seed grant (BRFSG-2023-10) and NIH/NINDS funding (NS114222) to A.S., and NIH grant R35NS105094 to S.S. A.S. thanks the Glenn Foundation for Medical Research and AFAR Junior Faculty Grant for support. G.R. was supported by the European Molecular Biology Laboratory (EMBL).

REFERENCES

Aida T, Yoshida J, Nomura M, Tanimura A, Iino Y, Soma M, Bai N, Ito Y, Cui W, Aizawa H, et al. 2015. Astroglial glutamate transporter deficiency increases synaptic excitability and leads to pathological repetitive behaviors in mice. *Neuropsychopharmacology* **40:** 1569–1579. doi:10.1038/npp.2015.26

Albert PS, Riddle DL. 1983. Developmental alterations in sensory neuroanatomy of the *Caenorhabditis elegans* dauer larva. *J Comp Neurol* **219:** 461–481. doi:10.1002/cne.902190407

Allen NJ, Eroglu C. 2017. Cell biology of astrocyte-synapse interactions. *Neuron* **96:** 697–708. doi:10.1016/j.neuron.2017.09.056

Allen NJ, Bennett ML, Foo LC, Wang GX, Chakraborty C, Smith SJ, Barres BA. 2012. Astrocyte glypicans 4 and 6 promote formation of excitatory synapses via GluA1 AMPA receptors. *Nature* **486:** 410–414. doi:10.1038/nature11059

Altun ZF, Chen B, Wang ZW, Hall DH. 2009. High resolution map of Caenorhabditis elegans gap junction proteins. *Dev Dyn* **238:** 1936–1950. doi:10.1002/dvdy.22025

Apfeld J, Kenyon C. 1999. Regulation of lifespan by sensory perception in *Caenorhabditis elegans*. *Nature* **402:** 804–809. doi:10.1038/45544

Arey RN, Murphy CT. 2017. Conserved regulators of cognitive aging: from worms to humans. *Behav Brain Res* **322:** 299–310. doi:10.1016/j.bbr.2016.06.035

Artiushin G, Sehgal A. 2020. The glial perspective on sleep and circadian rhythms. *Annu Rev Neurosci* **43:** 119–140. doi:10.1146/annurev-neuro-091819-094557

Awasaki T, Lee T. 2011. New tools for the analysis of glial cell biology in *Drosophila*. *Glia* **59:** 1377–1386. doi:10.1002/glia.21133

Bacaj T, Tevlin M, Lu Y, Shaham S. 2008. Glia are essential for sensory organ function in *C. elegans*. *Science* **322:** 744–747. doi:10.1126/science.1163074

Baird SE, Emmons SW. 1990. Properties of a class of genes required for ray morphogenesis in *Caenorhabditis elegans*. *Genetics* **126:** 335–344. doi:10.1093/genetics/126.2.335

Bargmann CI. 1993. Genetic and cellular analysis of behavior in *C. elegans*. *Annu Rev Neurosci* **16:** 47–71. doi:10.1146/annurev.ne.16.030193.000403

Bargmann CI, Marder E. 2013. From the connectome to brain function. *Nat Methods* **10:** 483–490. doi:10.1038/nmeth.2451

Barres BA. 2008. The mystery and magic of glia: a perspective on their roles in health and disease. *Neuron* **60:** 430–440. doi:10.1016/j.neuron.2008.10.013

Bénard CY, Boyanov A, Hall DH, Hobert O. 2006. DIG-1, a novel giant protein, non-autonomously mediates maintenance of nervous system architecture. *Development* **133:** 3329–3340. doi:10.1242/dev.02507

Blum ID, Keleş MF, Baz ES, Han E, Park K, Luu S, Issa H, Brown M, Ho MCW, Tabuchi M, et al. 2021. Astroglial calcium signaling encodes sleep need in *Drosophila*. *Curr Biol* **31:** 150–162.e7. doi:10.1016/j.cub.2020.10.012

Bonacossa-Pereira I, Coakley S, Hilliard MA. 2022. Neuron-epidermal attachment protects hyper-fragile axons from mechanical strain. *Cell Rep* **38:** 110501. doi:10.1016/j.celrep.2022.110501

Brenner S. 1974. The genetics of *Caenorhabditis elegans*. *Genetics* **77:** 71–94. doi:10.1093/genetics/77.1.71

Bülow HE, Boulin T, Hobert O. 2004. Differential functions of the *C. elegans* FGF receptor in axon outgrowth and main-

tenance of axon position. *Neuron* **42:** 367–374. doi:10.1016/s0896-6273(04)00246-6

Bunk EC, Ertaylan G, Ortega F, Pavlou MA, Gonzalez Cano L, Stergiopoulos A, Safaiyan S, Völs S, van Cann M, Politis PK, et al. 2016. Prox1 is required for oligodendrocyte cell identity in adult neural stem cells of the subventricular zone. *Stem Cells* **34:** 2115–2129. doi:10.1002/stem.2374

Bushong EA, Martone ME, Jones YZ, Ellisman MH. 2002. Protoplasmic astrocytes in CA1 stratum radiatum occupy separate anatomical domains. *J Neurosci* **22:** 183–192. doi:10.1523/JNEUROSCI.22-01-00183.2002

Cao J, Packer JS, Ramani V, Cusanovich DA, Huynh C, Daza R, Qiu X, Lee C, Furlan SN, Steemers FJ, et al. 2017. Comprehensive single-cell transcriptional profiling of a multicellular organism. *Science* **357:** 661–667. doi:10.1126/science.aam8940

Cassada RC, Russell RL. 1975. The dauerlarva, a post-embryonic developmental variant of the nematode *Caenorhabditis elegans*. *Dev Biol* **46:** 326–342. doi:10.1016/0012-1606(75)90109-8

Cebul ER, McLachlan IG, Heiman MG. 2020. Dendrites with specialized glial attachments develop by retrograde extension using SAX-7 and GRDN-1. *Development* **147:** dev180448. doi:10.1242/dev.180448

Chen D, Cheng H, Liu S, Al-Sheikh U, Fan Y, Duan D, Zou W, Zhu L, Kang L. 2022. The voltage-gated calcium channel EGL-19 acts on glia to drive olfactory adaptation. *Front Mol Neurosci* **15:** 907064. doi:10.3389/fnmol.2022.907064

Cherra SJ III, Jin Y. 2016. A two-immunoglobulin-domain transmembrane protein mediates an epidermal-neuronal interaction to maintain synapse density. *Neuron* **89:** 325–336. doi:10.1016/j.neuron.2015.12.024

Cherra SJ III, Goncharov A, Boassa D, Ellisman M, Jin Y. 2020. *C. elegans* MAGU-2/Mpp5 homolog regulates epidermal phagocytosis and synapse density. *J Neurogenet* **34:** 298–306. doi:10.1080/01677063.2020.1726915

Chiareli RA, Carvalho GA, Marques BL, Mota LS, Oliveira-Lima OC, Gomes RM, Birbair A, Gomez RS, Simão F, Klempin F, et al. 2021. The role of astrocytes in the neuro-repair process. *Front Cell Dev Biol* **9:** 665795. doi:10.3389/fcell.2021.665795

Chisholm AD, Hsiao TI. 2012. The *Caenorhabditis elegans* epidermis as a model skin. I: Development, patterning, and growth. *Wiley Interdiscip Rev Dev Biol* **1:** 861–878. doi:10.1002/wdev.79

Chong MK, Cebul ER, Mizeracka K, Heiman MG. 2021. Loss of the extracellular matrix protein DIG-1 causes glial fragmentation, dendrite breakage, and dendrite extension defects. *J Dev Biol* **9:** 42. doi:10.3390/jdb9040042

Christopherson KS, Ullian EM, Stokes CC, Mullowney CE, Hell JW, Agah A, Lawler J, Mosher DF, Bornstein P, Barres BA. 2005. Thrombospondins are astrocyte-secreted proteins that promote CNS synaptogenesis. *Cell* **120:** 421–433. doi:10.1016/j.cell.2004.12.020

Chung WS, Clarke LE, Wang GX, Stafford BK, Sher A, Chakraborty C, Joung J, Foo LC, Thompson A, Chen C, et al. 2013. Astrocytes mediate synapse elimination through MEGF10 and MERTK pathways. *Nature* **504:** 394–400. doi:10.1038/nature12776

Chung WS, Verghese PB, Chakraborty C, Joung J, Hyman BT, Ulrich JD, Holtzman DM, Barres BA. 2016. Novel allele-dependent role for APOE in controlling the rate of synapse pruning by astrocytes. *Proc Natl Acad Sci* **113:** 10186–10191. doi:10.1073/pnas.1609896113

Clarke LE, Barres BA. 2013. Emerging roles of astrocytes in neural circuit development. *Nat Rev Neurosci* **14:** 311–321. doi:10.1038/nrn3484

Coakley S, Ritchie FK, Galbraith KM, Hilliard MA. 2020. Epidermal control of axonal attachment via β-spectrin and the GTPase-activating protein TBC-10 prevents axonal degeneration. *Nat Commun* **11:** 133. doi:10.1038/s41467-019-13795-x

Colón-Ramos DA, Margeta MA, Shen K. 2007. Glia promote local synaptogenesis through UNC-6 (netrin) signaling in *C. elegans*. *Science* **318:** 103–106. doi:10.1126/science.1143762

Cook SJ, Jarrell TA, Brittin CA, Wang Y, Bloniarz AE, Yakovlev MA, Nguyen KCQ, Tang LT, Bayer EA, Duerr JS, et al. 2019. Whole-animal connectomes of both *Caenorhabditis elegans* sexes. *Nature* **571:** 63–71. doi:10.1038/s41586-019-1352-7

Coraggio F, Bhushan M, Roumeliotis S, Caroti F, Bevilacqua C, Prevedel R, Rapti G. 2023. An interplay of HSP-proteostasis, biomechanics and ECM-cell junctions ensures *C. elegans* astroglial architecture. bioRxiv doi:10.1101/2023.10.28.564505

Cuadras J, Martin G, Czternasty G, Bruner J. 1985. Gap-like junctions between neuron cell bodies and glial cells of crayfish. *Brain Res* **326:** 149–151. doi:10.1016/0006-8993(85)91394-0

Dasen JS, Jessell TM. 2009. Chapter six hox networks and the origins of motor neuron diversity. *Curr Top Dev Biol* **88:** 169–200. doi:10.1016/S0070-2153(09)88006-X

Dawes JHP, Kelsh RN. 2021. Cell fate decisions in the neural crest, from pigment cell to neural development. *Int J Mol Sci* **22:** 13531. doi:10.3390/ijms222413531

Depienne C, Bugiani M, Dupuits C, Galanaud D, Touitou V, Postma N, van Berkel C, Polder E, Tollard E, Darios F, et al. 2013. Brain white matter oedema due to ClC-2 chloride channel deficiency: an observational analytical study. *Lancet Neurol* **12:** 659–668. doi:10.1016/S1474-4422(13)70053-X

Derouiche A, Pannicke T, Haseleu J, Blaess S, Grosche J, Reichenbach A. 2012. Beyond polarity: functional membrane domains in astrocytes and Müller cells. *Neurochem Res* **37:** 2513–2523. doi:10.1007/s11064-012-0824-z

Ding G, Zou W, Zhang H, Xue Y, Cai Y, Huang G, Chen L, Duan S, Kang L. 2015. In vivo tactile stimulation-evoked responses in *Caenorhabditis elegans* amphid sheath glia. *PLoS ONE* **10:** e0117114. doi:10.1371/journal.pone.0117114

Dominici C, Moreno-Bravo JA, Puiggros SR, Rappeneau Q, Rama N, Vieugue P, Bernet A, Mehlen P, Chédotal A. 2017. Floor-plate-derived netrin-1 is dispensable for commissural axon guidance. *Nature* **545:** 350–354. doi:10.1038/nature22331

Dong X, Liu OW, Howell AS, Shen K. 2013. An extracellular adhesion molecule complex patterns dendritic branching and morphogenesis. *Cell* **155:** 296–307. doi:10.1016/j.cell.2013.08.059

Dong X, Jin S, Shao Z. 2020. Glia promote synaptogenesis through an IQGAP PES-7 in *C. elegans*. *Cell Rep* **30:** 2614–2626.e2. doi:10.1016/j.celrep.2020.01.102

Doroquez DB, Berciu C, Anderson JR, Sengupta P, Nicastro D. 2014. A high-resolution morphological and ultrastructural map of anterior sensory cilia and glia in *Caenorhabditis elegans*. *eLife* **3:** e01948. doi:10.7554/eLife.01948

Duan D, Zhang H, Yue X, Fan Y, Xue Y, Shao J, Ding G, Chen D, Li S, Cheng H, et al. 2020. Sensory glia detect repulsive odorants and drive olfactory adaptation. *Neuron* **108:** 707–721.e8. doi:10.1016/j.neuron.2020.08.026

Emmons SW. 2005. Male development. WormBook 1–22. doi:10.1895/wormbook.1.33.1

Emmons SW. 2018. Neural circuits of sexual behavior in *Caenorhabditis elegans*. *Annu Rev Neurosci* **41:** 349–369. doi:10.1146/annurev-neuro-070815-014056

Emtage L, Gu G, Hartwieg E, Chalfie M. 2004. Extracellular proteins organize the mechanosensory channel complex in *C. elegans* touch receptor neurons. *Neuron* **44:** 795–807. doi:10.1016/j.neuron.2004.11.010

Eroglu C, Allen NJ, Susman MW, O'Rourke NA, Park CY, Özkan E, Chakraborty C, Mulinyawe SB, Annis DS, Huberman AD, et al. 2009. Gabapentin receptor α2δ-1 is a neuronal thrombospondin receptor responsible for excitatory CNS synaptogenesis. *Cell* **139:** 380–392. doi:10.1016/j.cell.2009.09.025

Fan J, Ji T, Wang K, Huang J, Wang M, Manning L, Dong X, Shi Y, Zhang X, Shao Z, et al. 2020. A muscle-epidermis-glia signaling axis sustains synaptic specificity during allometric growth in *Caenorhabditis elegans*. *eLife* **9:** e55890. doi:10.7554/eLife.55890

Farhy-Tselnicker I, Allen NJ. 2018. Astrocytes, neurons, synapses: a tripartite view on cortical circuit development. *Neural Dev* **13:** 7. doi:10.1186/s13064-018-0104-y

Favuzzi E, Huang S, Saldi GA, Binan L, Ibrahim LA, Fernández-Otero M, Cao Y, Zeine A, Sefah A, Zheng K, et al. 2021. GABA-receptive microglia selectively sculpt developing inhibitory circuits. *Cell* **184:** 4048–4063.e32. doi:10.1016/j.cell.2021.06.018

Ferguson EL, Sternberg PW, Horvitz HR. 1987. A genetic pathway for the specification of the vulval cell lineages of *Caenorhabditis elegans*. *Nature* **326:** 259–267. doi:10.1038/326259a0

Fernandez-Abascal J, Johnson CK, Graziano B, Wang L, Encalada N, Bianchi L. 2022. A glial ClC Cl⁻ channel mediates nose touch responses in *C. elegans*. *Neuron* **110:** 470–485.e7. doi:10.1016/j.neuron.2021.11.010

Frakes AE, Metcalf MG, Tronnes SU, Bar-Ziv R, Durieux J, Gildea HK, Kandahari N, Monshietehadi S, Dillin A. 2020. Four glial cells regulate ER stress resistance and longevity via neuropeptide signaling in *C. elegans*. *Science* **367:** 436–440. doi:10.1126/science.aaz6896

Frank MG. 2013. Astroglial regulation of sleep homeostasis. *Curr Opin Neurobiol* **23:** 812–818. doi:10.1016/j.conb.2013.02.009

Fuentes-Medel Y, Logan MA, Ashley J, Ataman B, Budnik V, Freeman MR. 2009. Glia and muscle sculpt neuromuscular arbors by engulfing destabilized synaptic boutons and shed presynaptic debris. *PLoS Biol* **7:** e1000184. doi:10.1371/journal.pbio.1000184

Fung W, Wexler L, Heiman MG. 2020. Cell-type-specific promoters for *C. elegans* glia. *J Neurogenet* **34:** 335–346. doi:10.1080/01677063.2020.1781851

Fung W, Tan TM, Kolotuev I, Heiman MG. 2023. A sex-specific switch in a single glial cell patterns the apical extracellular matrix. *Curr Biol* **33:** 4174–4186 e7. doi:10.1016/j.cub.2023.08.046

Gendrel M, Atlas EG, Hobert O. 2016. A cellular and regulatory map of the GABAergic nervous system of *C. elegans*. *eLife* **5:** e17686. doi:10.7554/eLife.17686

Gibson CL, Balbona JT, Niedzwiecki A, Rodriguez P, Nguyen KCQ, Hall DH, Blakely RD. 2018. Glial loss of the metallo β-lactamase domain containing protein, SWIP-10, induces age- and glutamate-signaling dependent, dopamine neuron degeneration. *PLoS Genet* **14:** e1007269. doi:10.1371/journal.pgen.1007269

Goldstein B. 2016. Sydney Brenner on the genetics of *Caenorhabditis elegans*. *Genetics* **204:** 1–2. doi:10.1534/genetics.116.194084

Goodman MB, Sengupta P. 2019. How *Caenorhabditis elegans* senses mechanical stress, temperature, and other physical stimuli. *Genetics* **212:** 25–51. doi:10.1534/genetics.118.300241

Graham PL, Johnson JJ, Wang S, Sibley MH, Gupta MC, Kramer JM. 1997. Type IV collagen is detectable in most, but not all, basement membranes of *Caenorhabditis elegans* and assembles on tissues that do not express it. *J Cell Biol* **137:** 1171–1183. doi:10.1083/jcb.137.5.1171

Grant J, Matthewman C, Bianchi L. 2015. A novel mechanism of pH buffering in *C. elegans* glia: bicarbonate transport via the voltage-gated ClC Cl⁻ channel CLH-1. *J Neurosci* **35:** 16377–16397. doi:10.1523/JNEUROSCI.3237-15.2015

Hall DH. 1977. *The posterior nervous system of the nematode* Caenorhabditis elegans. California Institute of Technology, Pasadena, CA.

Han L, Wang Y, Sangaletti R, D'Urso G, Lu Y, Shaham S, Bianchi L. 2013. Two novel DEG/ENaC channel subunits expressed in glia are needed for nose-touch sensitivity in *Caenorhabditis elegans*. *J Neurosci* **33:** 936–949. doi:10.1523/JNEUROSCI.2749-12.2013

Hao L, Johnsen R, Lauter G, Baillie D, Bürglin TR. 2006. Comprehensive analysis of gene expression patterns of hedgehog-related genes. *BMC Genomics* **7:** 280. doi:10.1186/1471-2164-7-280

Hardaway JA, Sturgeon SM, Snarrenberg CL, Li Z, Xu XZ, Bermingham DP, Odiase P, Spencer WC, Miller DM, Carvelli L, et al. 2015. Glial expression of the *Caenorhabditis elegans* gene swip-10 supports glutamate dependent control of extrasynaptic dopamine signaling. *J Neurosci* **35:** 9409–9423. doi:10.1523/JNEUROSCI.0800-15.2015

Heiman MG, Shaham S. 2007. Ancestral roles of glia suggested by the nervous system of *Caenorhabditis elegans*. *Neuron Glia Biol* **3:** 55–61. doi:10.1017/S1740925X07000609

Heiman MG, Shaham S. 2009. DEX-1 and DYF-7 establish sensory dendrite length by anchoring dendritic tips during cell migration. *Cell* **137:** 344–355. doi:10.1016/j.cell.2009.01.057

Hill KL, Harfe BD, Dobbins CA, L'Hernault SW. 2000. *dpy-18* encodes an α-subunit of prolyl-4-hydroxylase in *Caenorhabditis elegans*. *Genetics* **155:** 1139–1148. doi:10.1093/genetics/155.3.1139

Hilu-Dadia R, Kurant E. 2020. Glial phagocytosis in developing and mature *Drosophila* CNS: tight regulation for a healthy brain. *Curr Opin Immunol* **62:** 62–68. doi:10.1016/j.coi.2019.11.010

Hong H, Chen H, Zhang Y, Wu Z, Zhang Y, Zhang Y, Hu Z, Zhang JV, Ling K, Hu J, et al. 2021. DYF-4 regulates patched-related/DAF-6-mediated sensory compartment formation in *C. elegans*. *PLoS Genet* **17:** e1009618. doi:10.1371/journal.pgen.1009618

Hosoya T, Takizawa K, Nitta K, Hotta Y. 1995. Glial cells missing: a binary switch between neuronal and glial determination in *Drosophila*. *Cell* **82:** 1025–1036. doi:10.1016/0092-8674(95)90281-3

Jarrell TA, Wang Y, Bloniarz AE, Brittin CA, Xu M, Thomson JN, Albertson DG, Hall DH, Emmons SW. 2012. The connectome of a decision-making neural network. *Science* **337:** 437–444. doi:10.1126/science.1221762

Jarriault S, Schwab Y, Greenwald I. 2008. A *Caenorhabditis elegans* model for epithelial–neuronal transdifferentiation. *Proc Natl Acad Sci* **105:** 3790–3795. doi:10.1073/pnas.0712159105

Jessell TM. 2000. Neuronal specification in the spinal cord: inductive signals and transcriptional codes. *Nat Rev Genet* **1:** 20–29. doi:10.1038/35049541

Johnson CK, Fernandez-Abascal J, Wang Y, Wang L, Bianchi L. 2020. The Na⁺-K⁺-ATPase is needed in glia of touch receptors for responses to touch in *C. elegans*. *J Neurophysiol* **123:** 2064–2074. doi:10.1152/jn.00636.2019

Jones BW, Fetter RD, Tear G, Goodman CS. 1995. Glial cells missing: a genetic switch that controls glial versus neuronal fate. *Cell* **82:** 1013–1023. doi:10.1016/0092-8674(95)90280-5

Kage-Nakadai E, Ohta A, Ujisawa T, Sun S, Nishikawa Y, Kuhara A, Mitani S. 2016. *Caenorhabditis elegans* homologue of Prox1/Prospero is expressed in the glia and is required for sensory behavior and cold tolerance. *Genes Cells* **21:** 936–948. doi:10.1111/gtc.12394

Kastriti ME, Adameyko I. 2017. Specification, plasticity and evolutionary origin of peripheral glial cells. *Curr Opin Neurobiol* **47:** 196–202. doi:10.1016/j.conb.2017.11.004

Katz M, Corson F, Iwanir S, Biron D, Shaham S. 2018. Glia modulate a neuronal circuit for locomotion suppression during sleep in *C. elegans*. *Cell Rep* **22:** 2575–2583. doi:10.1016/j.celrep.2018.02.036

Katz M, Corson F, Keil W, Singhal A, Bae A, Lu Y, Liang Y, Shaham S. 2019. Glutamate spillover in *C. elegans* triggers repetitive behavior through presynaptic activation of MGL-2/mGluR5. *Nat Commun* **10:** 1882. doi:10.1038/s41467-019-09581-4

Khakh BS. 2019. Astrocyte–neuron interactions in the striatum: insights on identity, form, and function. *Trends Neurosci* **42:** 617–630. doi:10.1016/j.tins.2019.06.003

Krause M, Harrison SW, Xu SQ, Chen L, Fire A. 1994. Elements regulating cell- and stage-specific expression of the *C. elegans* MyoD family homolog hlh-1. *Dev Biol* **166:** 133–148. doi:10.1006/dbio.1994.1302

Labouesse M, Hartwieg E, Horvitz HR. 1996. The *Caenorhabditis elegans* LIN-26 protein is required to specify and/or maintain all non-neuronal ectodermal cell fates. *Development* **122:** 2579–2588. doi:10.1242/dev.122.9.2579

Lamkin ER, Heiman MG. 2017. Coordinated morphogenesis of neurons and glia. *Curr Opin Neurobiol* **47:** 58–64. doi:10.1016/j.conb.2017.09.011

Landmann F, Quintin S, Labouesse M. 2004. Multiple regulatory elements with spatially and temporally distinct activities control the expression of the epithelial differentiation gene lin-26 in *C. elegans*. *Dev Biol* **265**: 478–490. doi:10.1016/j.ydbio.2003.09.009

Lee IH, Procko C, Lu Y, Shaham S. 2021a. Stress-induced neural plasticity mediated by glial GPCR REMO-1 promotes *C. elegans* adaptive behavior. *Cell Rep* **34**: 108607. doi:10.1016/j.celrep.2020.108607

Lee J, Magescas J, Fetter RD, Feldman JL, Shen K. 2021b. Inherited apicobasal polarity defines the key features of axon-dendrite polarity in a sensory neuron. *Curr Biol* **31**: 3768–3783.e3. doi:10.1016/j.cub.2021.06.039

Liu OW, Shen K. 2012. The transmembrane LRR protein DMA-1 promotes dendrite branching and growth in *C. elegans*. *Nat Neurosci* **15**: 57–63. doi:10.1038/nn.2978

Low IIC, Williams CR, Chong MK, McLachlan IG, Wierbowski BM, Kolotuev I, Heiman MG. 2019. Morphogenesis of neurons and glia within an epithelium. *Development* **146**: dev171124. doi:10.1242/dev.171124

Mano I, Straud S, Driscoll M. 2007. *Caenorhabditis elegans* glutamate transporters influence synaptic function and behavior at sites distant from the synapse. *J Biol Chem* **282**: 34412–34419. doi:10.1074/jbc.M704134200

Martin CG, Bent JS, Hill T, Topalidou I, Singhvi A. 2024. Epithelia delimits glial apical polarity against mechanical shear to maintain glia-neuron—architecture. *Devel Cell* (in press).

Masahira N, Takebayashi H, Ono K, Watanabe K, Ding L, Furusho M, Ogawa Y, Nabeshima Y-I, Alvarez-Buylla A, Shimizu K, et al. 2006. Olig2-positive progenitors in the embryonic spinal cord give rise not only to motoneurons and oligodendrocytes, but also to a subset of astrocytes and ependymal cells. *Dev Biol* **293**: 358–369. doi:10.1016/j.ydbio.2006.02.029

McMiller TL, Johnson CM. 2005. Molecular characterization of HLH-17, a *C. elegans* bHLH protein required for normal larval development. *Gene* **356**: 1–10. doi:10.1016/j.gene.2005.05.003

Meijlink F, Beverdam A, Bruower A, Oosterveen TC, Berge DT. 1999. Vertebrate aristaless-related genes. *Int J Dev Biol* **43**: 651–663.

Melkman T, Sengupta P. 2005. Regulation of chemosensory and GABAergic motor neuron development by the *C. elegans* Aristaless/Arx *homolog alr-1*. *Development* **132**: 1935–1949. doi:10.1242/dev.01788

Meng L, Zhang A, Jin Y, Yan D. 2016. Regulation of neuronal axon specification by glia-neuron gap junctions in *C. elegans*. *eLife* **5**: e19510. doi:10.7554/eLife.19510

Michaux G, Gansmuller A, Hindelang C, Labouesse M. 2000. CHE-14, a protein with a sterol-sensing domain, is required for apical sorting in *C. elegans* ectodermal epithelial cells. *Curr Biol* **10**: 1098–1107. doi:10.1016/s0960-9822(00)00695-3

Miller RH. 2005. Dorsally derived oligodendrocytes come of age. *Neuron* **45**: 1–3. doi:10.1016/j.neuron.2004.12.032

Miller RM, Portman DS. 2011. The Wnt/β-catenin asymmetry pathway patterns the *atonal ortholog lin-32* to diversify cell fate in a *Caenorhabditis elegans* sensory lineage. *J Neurosci* **31**: 13281–13291. doi:10.1523/JNEUROSCI.6504-10.2011

Mizeracka K, Heiman MG. 2015. The many glia of a tiny nematode: studying glial diversity using *Caenorhabditis*

elegans. *Wiley Interdiscip Rev Dev Biol* **4**: 151–160. doi:10.1002/wdev.171

Mizeracka K, Rogers JM, Rumley JD, Shaham S, Bulyk ML, Murray JI, Heiman MG. 2021. Lineage-specific control of convergent differentiation by a Forkhead repressor. *Development* **148**: dev199493. doi:10.1242/dev.199493

Molina-García L, Lloret-Fernández C, Cook SJ, Kim B, Bonnington RC, Sammut M, O'Shea JM, Gilbert SP, Elliott DJ, Hall DH, et al. 2020. Direct glia-to-neuron transdifferentiation gives rise to a pair of male-specific neurons that ensure nimble male mating. *eLife* **9**: e48361. doi:10.7554/eLife.48361

Nagai J, Yu X, Papouin T, Cheong E, Freeman MR, Monk KR, Hastings MH, Haydon PG, Rowitch D, Shaham S, et al. 2021. Behaviorally consequential astrocytic regulation of neural circuits. *Neuron* **109**: 576–596. doi:10.1016/j.neuron.2020.12.008

Nass R, Hall DH, Miller DM III, Blakely RD. 2002. Neurotoxin-induced degeneration of dopamine neurons in *Caenorhabditis elegans*. *Proc Natl Acad Sci* **99**: 3264–3269. doi:10.1073/pnas.042497999

Nedergaard M. 1994. Direct signaling from astrocytes to neurons in cultures of mammalian brain cells. *Science* **263**: 1768–1771. doi:10.1126/science.8134839

Nichols ALA, Eichler T, Latham R, Zimmer M. 2017. A global brain state underlies *C. elegans* sleep behavior. *Science* **356**: eaam6851. doi:10.1126/science.aam6851

Noctor SC, Martínez-Cerdeño V, Kriegstein AR. 2008. Distinct behaviors of neural stem and progenitor cells underlie cortical neurogenesis. *J Comp Neurol* **508**: 28–44. doi:10.1002/cne.21669

Ohkura K, Bürglin TR. 2011. Dye-filling of the amphid sheath glia: implications for the functional relationship between sensory neurons and glia in *Caenorhabditis elegans*. *Biochem Biophys Res Commun* **406**: 188–193. doi:10.1016/j.bbrc.2011.02.003

Oikonomou G, Shaham S. 2011. The glia of *Caenorhabditis elegans*. *Glia* **59**: 1253–1263. doi:10.1002/glia.21084

Oikonomou G, Perens EA, Lu Y, Watanabe S, Jorgensen EM, Shaham S. 2011. Opposing activities of LIT-1/NLK and DAF-6/patched-related direct sensory compartment morphogenesis in *C. elegans*. *PLoS Biol* **9**: e1001121. doi:10.1371/journal.pbio.1001121

Oikonomou G, Perens EA, Lu Y, Shaham S. 2012. Some, but not all, retromer components promote morphogenesis of *C. elegans* sensory compartments. *Dev Biol* **362**: 42–49. doi:10.1016/j.ydbio.2011.11.009

Packer JS, Zhu Q, Huynh C, Sivaramakrishnan P, Preston E, Dueck H, Stefanik D, Tan K, Trapnell C, Kim J, et al. 2019. A lineage-resolved molecular atlas of *C. elegans* embryogenesis at single-cell resolution. *Science* **365**: eaax1971. doi:10.1126/science.aax1971

Park C, Sakurai Y, Sato H, Kanda S, Iino Y, Kunitomo H. 2021. Roles of the ClC chloride channel CLH-1 in food-associated salt chemotaxis behavior of *C. elegans*. *eLife* **10**: e55701. doi:10.7554/eLife.55701

Peco E, Davla S, Camp D, Stacey SM, Landgraf M, van Meyel DJ. 2016. *Drosophila* astrocytes cover specific territories of the CNS neuropil and are instructed to differentiate by Prospero, a key effector of Notch. *Development* **143**: 1170–1181. doi:10.1242/dev.133165

Perens EA, Shaham S. 2005. *C. elegans* daf-6 encodes a patched-related protein required for lumen formation. *Dev Cell* **8**: 893–906. doi:10.1016/j.devcel.2005.03.009

Perkins LA, Hedgecock EM, Thomson JN, Culotti JG. 1986. Mutant sensory cilia in the nematode *Caenorhabditis elegans*. *Dev Biol* **117**: 456–487. doi:10.1016/0012-1606(86)90314-3

Portman DS, Emmons SW. 2000. The basic helix-loop-helix transcription factors LIN-32 and HLH-2 function together in multiple steps of a *C. elegans* neuronal sublineage. *Development* **127**: 5415–5426. doi:10.1242/dev.127.24.5415

Poskanzer KE, Yuste R. 2016. Astrocytes regulate cortical state switching in vivo. *Proc Natl Acad Sci* **113**: E2675–E2684. doi:10.1073/pnas.1520759113

Procko C, Lu Y, Shaham S. 2011. Glia delimit shape changes of sensory neuron receptive endings in *C. elegans*. *Development* **138**: 1371–1381. doi:10.1242/dev.058305

Procko C, Lu Y, Shaham S. 2012. Sensory organ remodeling in *Caenorhabditis elegans* requires the zinc-finger protein ZTF-16. *Genetics* **190**: 1405–1415. doi:10.1534/genetics.111.137786

Purice M, Quitevis E, Manning R, Severs L, Tran N, Sorrentino V, Setty M, Singhvi A. 2023. Molecular atlas of *C. elegans* glia across sexes reveals heterogeneity, variable sex-dimorphism, and glial properties. bioRxiv doi:10.1101/2023.03.21.533668

Qu Z, Zhang A, Yan D. 2020. Robo functions as an attractive cue for glial migration through SYG-1/Neph. *eLife* **9**: e57921. doi:10.7554/eLife.57921

Rahmani P, Rogalski T, Moerman DG. 2015. The *C. elegans* UNC-23 protein, a member of the BCL-2-associated athanogene (BAG) family of chaperone regulators, interacts with HSP-1 to regulate cell attachment and maintain hypodermal integrity. *Worm* **4**: e1023496. doi:10.1080/21624054.2015.1023496

Raiders S, Black EC, Bae A, MacFarlane S, Klein M, Shaham S, Singhvi A. 2021a. Glia actively sculpt sensory neurons by controlled phagocytosis to tune animal behavior. *eLife* **10**: e63532. doi:10.7554/eLife.63532

Raiders S, Han T, Scott-Hewitt N, Kucenas S, Lew D, Logan MA, Singhvi A. 2021b. Engulfed by glia: glial pruning in development, function, and injury across species. *J Neurosci* **41**: 823–833. doi:10.1523/JNEUROSCI.1660-20.2020

Rapti G. 2021. Open frontiers in neural cell type investigations; lessons from *Caenorhabditis elegans* and beyond, toward a multimodal integration. *Front Neurosci* **15**: 787753. doi:10.3389/fnins.2021.787753

Rapti G, Li C, Shan A, Lu Y, Shaham S. 2017. Glia initiate brain assembly through noncanonical Chimaerin-Furin axon guidance in *C. elegans*. *Nat Neurosci* **20**: 1350–1360. doi:10.1038/nn.4630

Rashid A, Tevlin M, Lu Y, Shaham S. 2022. A developmental pathway for epithelial-to-motoneuron transformation in *C. elegans*. *Cell Rep* **40**: 111414. doi:10.1016/j.celrep.2022.111414

Ray S, Singhvi A. 2021. Charging up the periphery: glial ionic regulation in sensory perception. *Front Cell Dev Biol* **9**: 687732. doi:10.3389/fcell.2021.687732

Ray S, Gurung P, Manning RS, Kravchuk AA, Singhvi A. 2024. Neuron cilia restrain glial KCC-3 to a microdomain to regulate multi-sensory processing. *Cell Rep* **43**: 113844. doi:10.1016/j.celrep.2024.113844

Razzauti A, Laurent P. 2021. Ectocytosis prevents accumulation of ciliary cargo in *C. elegans* sensory neurons. *eLife* **10**: e67670. doi:10.7554/eLife.67670

Rousse I, Robitaille R. 2006. Calcium signaling in Schwann cells at synaptic and extra-synaptic sites: active glial modulation of neuronal activity. *Glia* **54**: 691–699. doi:10.1002/glia.20388

Rowitch DH. 2004. Glial specification in the vertebrate neural tube. *Nat Rev Neurosci* **5**: 409–419. doi:10.1038/nrn1389

Salzberg Y, Díaz-Balzac CA, Ramirez-Suarez NJ, Attreed M, Tecle E, Desbois M, Kaprielian Z, Bülow HE. 2013. Skin-derived cues control arborization of sensory dendrites in *Caenorhabditis elegans*. *Cell* **155**: 308–320. doi:10.1016/j.cell.2013.08.058

Salzer JL. 2003. Polarized domains of myelinated axons. *Neuron* **40**: 297–318. doi:10.1016/s0896-6273(03)00628-7

Sammut M, Cook SJ, Nguyen KCQ, Felton T, Hall DH, Emmons SW, Poole RJ, Barrios A. 2015. Glia-derived neurons are required for sex-specific learning in *C. elegans*. *Nature* **526**: 385–390. doi:10.1038/nature15700

Schafer WR. 2018. The worm connectome: back to the future. *Trends Neurosci* **41**: 763–765. doi:10.1016/j.tins.2018.09.002

Schmechel DE, Rakic P. 1979. A Golgi study of radial glial cells in developing monkey telencephalon: morphogenesis and transformation into astrocytes. *Anat Embryol (Berl)* **156**: 115–152. doi:10.1007/BF00300010

Sengupta P, Samuel AD. 2009. *Caenorhabditis elegans*: a model system for systems neuroscience. *Curr Opin Neurobiol* **19**: 637–643. doi:10.1016/j.conb.2009.09.009

Shah PK, Santella A, Jacobo A, Siletti K, Hudspeth AJ, Bao Z. 2017. An in toto approach to dissecting cellular interactions in complex tissues. *Dev Cell* **43**: 530–540.e4. doi:10.1016/j.devcel.2017.10.021

Shaham S. 2005. Glia–neuron interactions in nervous system function and development. *Curr Top Dev Biol* **69**: 39–66. doi:10.1016/S0070-2153(05)69003-5

Shaham S. 2010. Chemosensory organs as models of neuronal synapses. *Nat Rev Neurosci* **11**: 212–217. doi:10.1038/nrn2740

Shaham S. 2015. Glial development and function in the nervous system of *Caenorhabditis elegans*. *Cold Spring Harb Perspect Biol* **7**: a020578. doi:10.1101/cshperspect.a020578

Shao Z, Watanabe S, Christensen R, Jorgensen EM, Colón-Ramos DA. 2013. Synapse location during growth depends on glia location. *Cell* **154**: 337–350. doi:10.1016/j.cell.2013.06.028

Shen K, Bargmann CI. 2003. The immunoglobulin superfamily protein SYG-1 determines the location of specific synapses in *C. elegans*. *Cell* **112**: 619–630. doi:10.1016/s0092-8674(03)00113-2

Shen K, Fetter RD, Bargmann CI. 2004. Synaptic specificity is generated by the synaptic guidepost protein SYG-2 and its receptor, SYG-1. *Cell* **116**: 869–881. doi:10.1016/S0092-8674(04)00251-X

Shigetomi E, Patel S, Khakh BS. 2016. Probing the complexities of astrocyte calcium signaling. *Trends Cell Biol* **26**: 300–312. doi:10.1016/j.tcb.2016.01.003

Sík A, Smith RL, Freund TF. 2000. Distribution of chloride channel-2-immunoreactive neuronal and astrocytic processes in the hippocampus. *Neuroscience* **101:** 51–65. doi:10.1016/S0306-4522(00)00360-2

Silverman JL, Tolu SS, Barkan CL, Crawley JN. 2010. Repetitive self-grooming behavior in the BTBR mouse model of autism is blocked by the mGluR5 antagonist MPEP. *Neuropsychopharmacology* **35:** 976–989. doi:10.1038/npp.2009.201

Singhal A, Shaham S. 2017. Infrared laser-induced gene expression for tracking development and function of single *C. elegans* embryonic neurons. *Nat Commun* **8:** 14100. doi:10.1038/ncomms14100

Singhvi A, Shaham S. 2019. Glia-neuron interactions in *Caenorhabditis elegans. Annu Rev Neurosci* **42:** 149–168. doi:10.1146/annurev-neuro-070918-050314

Singhvi A, Liu B, Friedman CJ, Fong J, Lu Y, Huang XY, Shaham S. 2016. A glial K/Cl transporter controls neuronal receptive ending shape by chloride inhibition of an rGC. *Cell* **165:** 936–948. doi:10.1016/j.cell.2016.03.026

Soete G, Betist MC, Korswagen HC. 2007. Regulation of *Caenorhabditis elegans* body size and male tail development by the novel gene lon-8. *BMC Dev Biol* **7:** 20. doi:10.1186/1471-213X-7-20

Stefanakis N, Jiang J, Liang Y, Shaham S. 2024. LET-381/FoxF and UNC-30/Pitx2 control the development of *C. elegans* mesodermal glia that regulate motor behavior. *EMBO J* (in press).

Stevens B, Allen NJ, Vazquez LE, Howell GR, Christopherson KS, Nouri N, Micheva KD, Mehalow AK, Huberman AD, Stafford B, et al. 2007. The classical complement cascade mediates CNS synapse elimination. *Cell* **131:** 1164–1178. doi:10.1016/j.cell.2007.10.036

Sulston JE, Horvitz HR. 1977. Post-embryonic cell lineages of the nematode, *Caenorhabditis elegans. Dev Biol* **56:** 110–156. doi:10.1016/0012-1606(77)90158-0

Sulston JE, Albertson DG, Thomson JN. 1980. The *Caenorhabditis elegans* male: postembryonic development of nongonadal structures. *Dev Biol* **78:** 542–576. doi:10.1016/0012-1606(80)90352-8

Sulston JE, Schierenberg E, White JG, Thomson JN. 1983. The embryonic cell lineage of the nematode *Caenorhabditis elegans. Dev Biol* **100:** 64–119. doi:10.1016/0012-1606(83)90201-4

Tatsumi K, Isonishi A, Yamasaki M, Kawabe Y, Morita-Takemura S, Nakahara K, Terada Y, Shinjo T, Okuda H, Tanaka T, et al. 2018. Olig2-Lineage astrocytes: a distinct subtype of astrocytes that differs from GFAP astrocytes. *Front Neuroanat* **12:** 8. doi:10.3389/fnana.2018.00008

Taylor J, Unsoeld T, Hutter H. 2018. The transmembrane collagen COL-99 guides longitudinally extending axons in *C. elegans. Mol Cell Neurosci* **89:** 9–19. doi:10.1016/j.mcn.2018.03.003

Taylor SR, Santpere G, Weinreb A, Barrett A, Reilly MB, Xu C, Varol E, Oikonomou P, Glenwinkel L, McWhirter R, et al. 2021. Molecular topography of an entire nervous system. *Cell* **184:** 4329–4347.e23. doi:10.1016/j.cell.2021.06.023

Ting JT, Feng G. 2008. Glutamatergic synaptic dysfunction and obsessive-compulsive disorder. *Curr Chem Genomics* **2:** 62–75. doi:10.2174/1875397300802010062

Tsutsui K, Kim HS, Yoshikata C, Kimura K, Kubota Y, Shibata Y, Tian C, Liu J, Nishiwaki K. 2021. Repulsive guidance molecule acts in axon branching in *Caenorhabditis elegans. Sci Rep* **11:** 22370. doi:10.1038/s41598-021-01853-8

Tucker M, Sieber M, Morphew M, Han M. 2005. The *Caenorhabditis elegans* aristaless orthologue, *alr-1*, is required for maintaining the functional and structural integrity of the amphid sensory organs. *Mol Biol Cell* **16:** 4695–4704. doi:10.1091/mbc.e05-03-0205

Ullian EM, Sapperstein SK, Christopherson KS, Barres BA. 2001. Control of synapse number by glia. *Science* **291:** 657–661. doi:10.1126/science.291.5504.657

Varadarajan SG, Kong JH, Phan KD, Kao TJ, Panaitof SC, Cardin J, Eltzschig H, Kania A, Novitch BG, Butler SJ. 2017. Netrin1 produced by neural progenitors, not floor plate cells, is required for axon guidance in the spinal cord. *Neuron* **94:** 790–799.e3. doi:10.1016/j.neuron.2017.03.007

Varshney LR, Chen BL, Paniagua E, Hall DH, Chklovskii DB. 2011. Structural properties of the *Caenorhabditis elegans* neuronal network. *PLoS Comput Biol* **7:** e1001066. doi:10.1371/journal.pcbi.1001066

Vincent S, Vonesch JL, Giangrande A. 1996. *Glide* directs glial fate commitment and cell fate switch between neurones and glia. *Development* **122:** 131–139. doi:10.1242/dev.122.1.131

Wagner B, Natarajan A, Grünaug S, Kroismayr R, Wagner EF, Sibilia M. 2006. Neuronal survival depends on EGFR signaling in cortical but not midbrain astrocytes. *EMBO J* **25:** 752–762. doi:10.1038/sj.emboj.7600988

Wallace SW, Singhvi A, Liang Y, Lu Y, Shaham S. 2016. PROS-1/Prospero is a major regulator of the glia-specific secretome controlling sensory-neuron shape and function in *C. elegans. Cell Rep* **15:** 550–562. doi:10.1016/j.celrep.2016.03.051

Wallace SW, Lizzappi MC, Magemizoğlu E, Hur H, Liang Y, Shaham S. 2021. Nuclear hormone receptors promote gut and glia detoxifying enzyme induction and protect *C. elegans* from the mold *P. brevicompactum. Cell Rep* **37:** 110166. doi:10.1016/j.celrep.2021.110166

Wang L, Bianchi L. 2021. Maintenance of protein homeostasis in glia extends lifespan in *C. elegans. Exp Neurol* **339:** 113648. doi:10.1016/j.expneurol.2021.113648

Wang Y, Apicella A, Lee SK, Ezcurra M, Slone RD, Goldmit M, et al. 2008. A glial DEG/ENaC channel functions with neuronal channel DEG-1 to mediate specific sensory functions in *C. elegans. EMBO J* **27:** 2388–2399. doi:10.1038/emboj.2008.161

Wang Y, D'Urso G, Bianch L. 2012. iKnockout of glial channel ACD-1 exacerbates sensory deficits in a *C. elegans* mutant by regulating calcium levels of sensory neurons. *J Neurophysiol* **107:** 148–158. doi:10.1152/jn.00299.2011

Wang J, Kaletsky R, Silva M, Williams A, Haas LA, Androwski RJ, Landis JN, Patrick C, Rashid A, Santiago-Martinez D, et al. 2015. Cell-specific transcriptional profiling of ciliated sensory neurons reveals regulators of behavior and extracellular vesicle biogenesis. *Curr Biol* **25:** 3232–3238. doi:10.1016/j.cub.2015.10.057

Wang W, Perens EA, Oikonomou G, Wallace SW, Lu Y, Shaham S. 2017. IGDB-2, an Ig/FNIII protein, binds the ion channel LGC-34 and controls sensory compartment morphogenesis in *C. elegans. Dev Biol* **430:** 105–112. doi:10 .1016/j.ydbio.2017.08.009

Ward S, Thomson N, White JG, Brenner S. 1975. Electron microscopical reconstruction of the anterior sensory anatomy of the nematode *Caenorhabditis elegans. J Comp Neurol* **160:** 313–337. doi:10.1002/cne.901600305

White JG, Southgate E, Thomson JN, Brenner S. 1986. The structure of the nervous system of the nematode *Caenorhabditis elegans. Philos Trans R Soc Lond B Biol Sci* **314:** 1– 340. doi:10.1098/rstb.1986.0056

Wilton DK, Dissing-Olesen L, Stevens B. 2019. Neuron-glia signaling in synapse elimination. *Annu Rev Neurosci* **42:** 107–127. doi:10.1146/annurev-neuro-070918-050306

Witvliet D, Mulcahy B, Mitchell JK, Meirovitch Y, Berger DR, Wu Y, Liu Y, Koh WX, Parvathala R, Holmyard D, et al. 2021. Connectomes across development reveal principles of brain maturation. *Nature* **596:** 257–261. doi:10.1038/ s41586-021-03778-8

Yin JA, Gao G, Liu XJ, Hao ZQ, Li K, Kang XL, Li H, Shan YH, Hu WL, Li HP, et al. 2017. Genetic variation in glia–neuron signalling modulates ageing rate. *Nature* **551:** 198–203. doi:10.1038/nature24463

Yoshida A, Nakano S, Suzuki T, Ihara K, Higashiyama T, Mori I. 2016. A glial K(+)/Cl(−) cotransporter modifies temperature-evoked dynamics in Caenorhabditis elegans sensory neurons. *Genes Brain Behav* **15:** 429–440. doi:10.1111/gbb .12260

Yoshimura S, Murray JI, Lu Y, Waterston RH, Shaham S. 2008. *mls-2* and *vab-3* control glia development, *hlh-17/*

Olig expression and glia-dependent neurite extension in *C. elegans. Development* **135:** 2263–2275. doi:10.1242/dev .019547

Yu RY, Nguyen CQ, Hall DH, Chow KL. 2000. Expression of *ram-5* in the structural cell is required for sensory ray morphogenesis in *Caenorhabditis elegans* male tail. *EMBO J* **19:** 3542–3555. doi:10.1093/emboj/19.14.3542

Zhang Y, Emmons SW. 1995. Specification of sense-organ identity by a *Caenorhabditis elegans* Pax-6 homologue. *Nature* **377:** 55–59. doi:10.1038/377055a0

Zhang A, Ackley BD, Yan D. 2020a. Vitamin B12 regulates glial migration and synapse formation through isoform-specific control of PTP-3/LAR PRTP expression. *Cell Rep* **30:** 3981–3988.e3. doi:10.1016/j.celrep.2020.02 .113

Zhang A, Guan Z, Ockerman K, Dong P, Guo J, Wang Z, Yan D. 2020b. Regulation of glial size by eicosapentaenoic acid through a novel Golgi apparatus mechanism. *PLoS Biol* **18:** e3001051. doi:10.1371/journal.pbio.3001 051

Zhang A, Noma K, Yan D. 2020c. Regulation of gliogenesis by *lin-32/Atoh1* in *Caenorhabditis elegans. G3* **10:** 3271– 3278. doi:10.1534/g3.120.401547

Zuryn S, Ahier A, Portoso M, White ER, Morin MC, Margueron R, Jarriault S. 2014. Transdifferentiation. Sequential histone-modifying activities determine the robustness of transdifferentiation. *Science* **345:** 826–829. doi:10.1126/ science.1255885

Cite this article as *Cold Spring Harb Perspect Biol* doi: 10.1101/cshperspect.a041346

Glial Regulation of Circuit Wiring, Firing, and Expiring in the *Drosophila* Central Nervous System

Jaeda Coutinho-Budd,[1] Marc R. Freeman,[2] and Sarah Ackerman[3]

[1]Department of Neuroscience, Center for Brain Immunology and Glia, University of Virginia, Charlottesville, Virginia 22903, USA

[2]Vollum Institute, Oregon Health and Science University, Portland, Oregon 97239, USA

[3]Department of Pathology and Immunology, Brain Immunology and Glia Center, and Department of Developmental Biology, Washington University School of Medicine, Saint Louis, Missouri 63110, USA

Correspondence: sarah.ackerman@wustl.edu; jaeda.coutinho-budd@virginia.edu

Molecular genetic approaches in small model organisms like *Drosophila* have helped to elucidate fundamental principles of neuronal cell biology. Much less is understood about glial cells, although interest in using invertebrate preparations to define their in vivo functions has increased significantly in recent years. This review focuses on our current understanding of the three major neuron-associated glial cell types found in the *Drosophila* central nervous system (CNS)—astrocytes, cortex glia, and ensheathing glia. Together, these cells act like mammalian astrocytes and microglia; they associate closely with neurons including surrounding neuronal cell bodies and proximal neurites, regulate synapses, and engulf neuronal debris. Exciting recent work has shown critical roles for these CNS glial cells in neural circuit formation, function, plasticity, and pathology. As we gain a more firm molecular and cellular understanding of how *Drosophila* CNS glial cells interact with neurons, it is clear that they share significant molecular and functional attributes with mammalian glia and will serve as an excellent platform for mechanistic studies of glial function.

Invertebrate models have contributed enormously to our understanding of fundamental principles of nervous system biology, including the genetic, chemical, and electrophysiological basis of the action potential, synaptic vesicle release, neural cell fate specification, and axon pathfinding.[4] This is largely thanks to the high experimental accessibility, ease of maintenance, rapid growth and panoply of molecular genetic tools for in vivo and cell-specific genetic manipulation in organisms like *Drosophila* and *Caenorhabditis elegans*. The focus of many neuroscientists has shifted in recent years toward careful exploration of how glial cells participate in nervous system development, neural circuit function and plasticity, refinement and debris clearance, and neurological disease. Based on the remarkable success with which invertebrates

---

[4]This is an update to a previous article published in *Cold Spring Harbor Perspectives in Biology* [Freeman (2015). *Cold Spring Harb Perspect Biol* 8: a020552. doi:10.1101/cshperspect.a020552].

Cite this article as *Cold Spring Harb Perspect Biol* doi: 10.1101/cshperspect.a041347

were used to dissect fundamental aspects of neuronal cell biology, interest in the potential of small genetic model organisms to contribute to unraveling the mysteries of glial cells has grown significantly. This article provides a brief overview of *Drosophila* glial cell biology, then focuses on three glial cell subtypes that are directly associated with neurons in the central nervous system (CNS)—astrocytes, ensheathing glia, and cortex glia. A growing body of work argues strongly that these glia share a range of morphological and functional features with mammalian astrocytes and microglia, and recent molecular studies indicate that conservation of basic glial cell biology extends, perhaps not surprisingly, to the molecular level. Discussion of additional *Drosophila* glial subtypes (perineurial, subperineurial, and wrapping glia) can be found in Fernandes et al. (2024).

OVERVIEW OF *DROSOPHILA* NERVOUS SYSTEM HISTOLOGY

In total, the adult fly CNS including the brain and thoracic ganglion (the fly equivalent of the mammalian spinal cord) houses ~100,000 neurons (Lacin et al. 2019; Allen et al. 2020; Scheffer et al. 2020). *Drosophila* neurons are quite similar in terms of electrophysiological properties to mammalian neurons. They fire Na^+/K^+-based action potentials, and they use highly conserved mechanisms for synaptic vesicle release of conserved neurotransmitters such as GABA, glutamate, and acetylcholine, as well as neuromodulators such as biogenic amines and neuropeptides. Just as with their vertebrate counterparts, *Drosophila* neurons drive a diverse behavioral repertoire that can be studied in the intact organism that shows both electrophysiological and behavioral plasticity. The histology of the larval and adult *Drosophila* nervous system is relatively complex, but still simple enough to be highly accessible experimentally. The brain houses anatomically distinct regions, which are connected to one another by fasciculated nerves. The CNS can be subdivided into two histological regions: the neuronal cell cortex, where all CNS neuronal cell bodies reside, and the neuropil, to which axons and dendrites project and form neural circuits (Fig. 1A).

As in mammals, glial cells in *Drosophila* are characterized in large part by their morphology and relationship with neurons (Fig. 1B; Lago-Baldaia et al. 2023). The precise number of glia in the fly nervous system remains unclear, but likely represents ~10% of the total population of cells within the CNS (Kremer et al. 2017). The outermost layer of cells associated with the surface of the CNS is composed of a subset of glia termed perineural glia (PG), which together with hemocytes are thought to secrete a dense carbohydrate-rich lamella that covers the CNS and peripheral nerves and acts as a chemical and physical barrier for the CNS (Carlson et al. 2000; Leiserson et al. 2000). The PG layer is discontinuous, with small gaps through which some molecules can still pass. Below the PG cells is a layer of subperineural glial cells (SPGs), which show a flattened morphology, cover the entire CNS surface, and establish a blood–brain barrier (BBB) by forming pleated septate junctions with one another (Auld et al. 1995; Baumgartner et al. 1996; Schwabe et al. 2005). SPGs make contact with only the most superficial neuronal cell bodies in the cortex; whether they make any contact with neurites has not been carefully studied. Deeper in the CNS, a number of specialized glial subtypes—cortex glia, ensheathing glia, and astrocytes—associate closely with neurons. These are the focus of this review and discussed in detail below, along with a comparison of these cells to their mammalian counterparts.

Drosophila also have a number of glial subtypes outside of the CNS that ensheath, support, and modulate the development and function of peripheral sensory neurons and motor neuron axons and terminals (Fig. 1A–D; Stork et al. 2012; Freeman 2015). Peripheral nerves are covered by the PG- and SPG-based blood–nerve barrier (BNB) similar to the CNS, but additionally house a population of glia termed wrapping glia that ensheath motor and sensory axons and whose histology is very similar to that of mammalian nonmyelinating Schwann cells that make up Remak bundles (Leiserson et al. 2000; Beckervordersandforth et al. 2008; Stork et al. 2008). At the neuromuscular junction (NMJ), SPGs can (at some, but not all NMJs)

Cite this article as *Cold Spring Harb Perspect Biol* doi: 10.1101/cshperspect.a041347

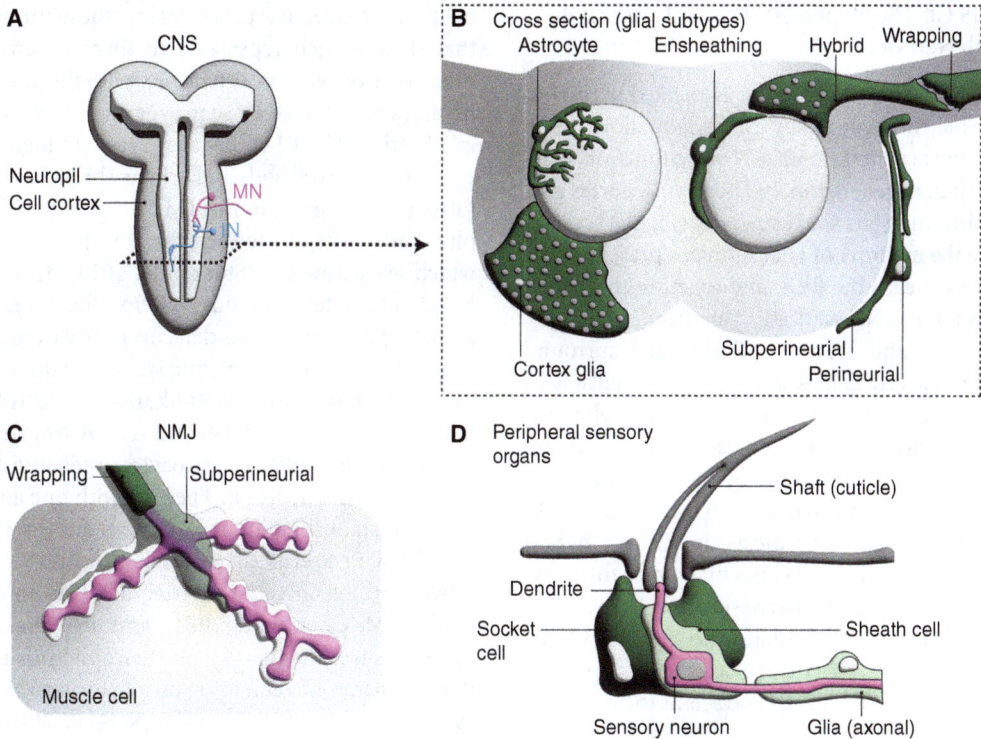

Figure 1. Organization of the *Drosophila* central nervous system (CNS). (*A*) Overview of the CNS. Neuronal cell bodies are positioned in the cell cortex (gray), which surrounds the synaptic neuropil (white). All CNS synapses are found within the neuropil. (*B*) Cross-sectional view of CNS glial subtypes. A single example for each subtype is shown. Hybrid ensheathing/wrapping glial cells at nerve roots ensheath axons and neuronal cell bodies. For other cell types, see text for details. (*C*) Neuromuscular junction (NMJ) organization. Wrapping glia ensheath axons up to the point of the muscle insertion site of the motor neuron ending, and synaptic boutons form inside the muscle. Subperineurial glia either stop at the muscle entry point, or in some cases invade the NMJ and associate with synaptic boutons. (*D*) Peripheral sensory organ glia. Socket cells help form the sensory organ shaft. Sheath cells surround the neuron and proximal dendrite/axon. Additional glia surround the axon as it projects toward the CNS. (MN) Motoneuron, (IN) interneuron.

extend processes that interact with motor neuron synaptic contacts on muscles (Fig. 1C) where they perform many key functions, including recycling neurotransmitters (Rival et al. 2004; Danjo et al. 2011), sculpting growing presynaptic morphology by engulfing shed axonal/synaptic debris during development (Fuentes-Medel et al. 2009), secreting transforming growth factor β (TGF-β) molecules that modulate retrograde muscle to presynapse signaling and thereby NMJ growth (Fuentes-Medel et al. 2012), and regulating synaptic physiology by secreting Wnts that modulate postsynaptic glutamate receptor clustering (Kerr et al. 2014).

Finally, sensory organ neurons on the external portions of the animal that are responsible for receiving mechanical, chemical, or other stimuli from the environment are closely associated with socket glial cells, which help form the sensory hairs, sheath glial cells that wrap the neuronal dendrite and cell body, and an axon-associated glial cell (Fig. 1D). The biology of these sensory organ glia will likely be very similar to sensory organ glia/support cells of *C. elegans*, zebrafish, and mammals (Grant et al. 2005; Shaham 2006; Chitnis et al. 2012; Jang et al. 2022), but their functions have not been studied extensively.

CNS Glial Subtypes that Primarily Associate with Neurons

Glial cells that are directly associated with neurons likely mediate key events that allow glia to modulate neural circuit assembly, function, plasticity, or degeneration. In *Drosophila*, cortex glia, ensheathing glia, and astrocytes (Fig. 1A) constitute the majority of glial subtypes present in the CNS beneath the BBB, and together these fully cover the CNS scaffold of neuron cell bodies, neurites, and synapses. Cortex glia surround neuronal cell bodies and proximal neurites as they enter the neuropil, ensheathing glia surround and compartmentalize the neuropil and nerves as they project out of the CNS, and astrocytes densely infiltrate the synaptic neuropil. As the peripheral nerve meets the CNS, a hybrid ensheathing glial cell exists that wraps the proximal portions of the nerve similarly to wrapping glia, but also encapsulates the dorsalmost cell bodies similarly to cortex glia, as well as a portion of the neuropil (Fig. 1B) and seems to have properties of each of these glial subtypes. We will not cover this cell type in this review as it is currently not well studied. We note that an additional mesectodermally derived subset of glia, termed midline glia, are also present in the *Drosophila* CNS. Midline glia play a central role in early nervous system development, including in axon pathfinding, separation of the major commissures of the CNS axon scaffold, and, ultimately, they ensheath axons at the CNS midline. These have been the subject of excellent reviews (Jacobs 2000; Crews 2010) and will only be covered briefly here.

Astrocytes

Drosophila astrocytes exhibit remarkable morphological, molecular, and functional similarities to mammalian protoplasmic astrocytes (Fig. 2), suggesting this is an ancient cell type in complex metazoans (Awasaki et al. 2008; Doherty et al. 2009; Muthukumar et al. 2014; Stork et al. 2014; Tasdemir-Yilmaz and Freeman 2014). Of the three glial subtypes discussed in this work, astrocytes are the most heavily studied. Astrocyte cell bodies reside at the cortex/neuropil interface,

but they extend major processes into the neuropil that then branch repeatedly to form a dense meshwork of very fine extensions, with the finest membranes situated close to synapses (Figs. 1B, 2A; Stork et al. 2014). Fly astrocytes are highly polarized. The cell body and primary branches of astrocytes are microtubule (MT)-rich with MT plus ends oriented toward the fine processes, which are actin-rich (Stork et al. 2014). In the larval and adult nervous system, these cells seem to extend processes that cover the vast majority of the neuropil synaptic space, but do not overlap with synapses (Muthukumar et al. 2014; Stork et al. 2014; Kremer et al. 2017). Astrocytes appear to talk to adjacent astrocytes to ensure full coverage of the neuropil. They tile with one another to establish unique spatial domains (Stork et al. 2014; Peco et al. 2016), similar to astrocyte–astrocyte tiling observed in vertebrates (Bushong et al. 2004; Chen et al. 2020), and ablation of astrocytes from regions of the neuropil leads to the expansion of remaining astrocytes into the astrocyte-depleted regions (Stork et al. 2014). The molecular basis of how astrocytes tile with each other remains unclear in any organism. In fact, all *Drosophila* glia subsets form similar tiled domains (Kremer et al. 2017; Salazar et al. 2022), similar to most if not all vertebrate glia (including astrocytes, microglia, oligodendrocyte precursor cells, Müller glia, satellite glia, etc.). That *Drosophila* glia show tiling behavior like their vertebrate counterparts opens the door to a forward genetic analysis of the mechanisms of glial tiling and its importance in neural circuit function.

Most of what astrocytes are thought to do in the CNS depends on their close physical relationship with synapses. Understanding how astrocytes acquire their remarkable morphology and closely associate with synapses remains a major challenge for the field. Stork and colleagues found that early astrocyte morphogenesis critically depends on a neuron to astrocyte fibroblast growth factor (FGF) signaling cascade (Stork et al. 2014). The *Drosophila* FGF receptor (FGFR) *heartless* is expressed early in astrocyte development (Shishido et al. 1997). Interestingly, *heartless* mutants showed defects in the migration of astrocyte cell bodies to their appropriate positions around the neuropil and a failure of

Cite this article as *Cold Spring Harb Perspect Biol* doi: 10.1101/cshperspect.a041347

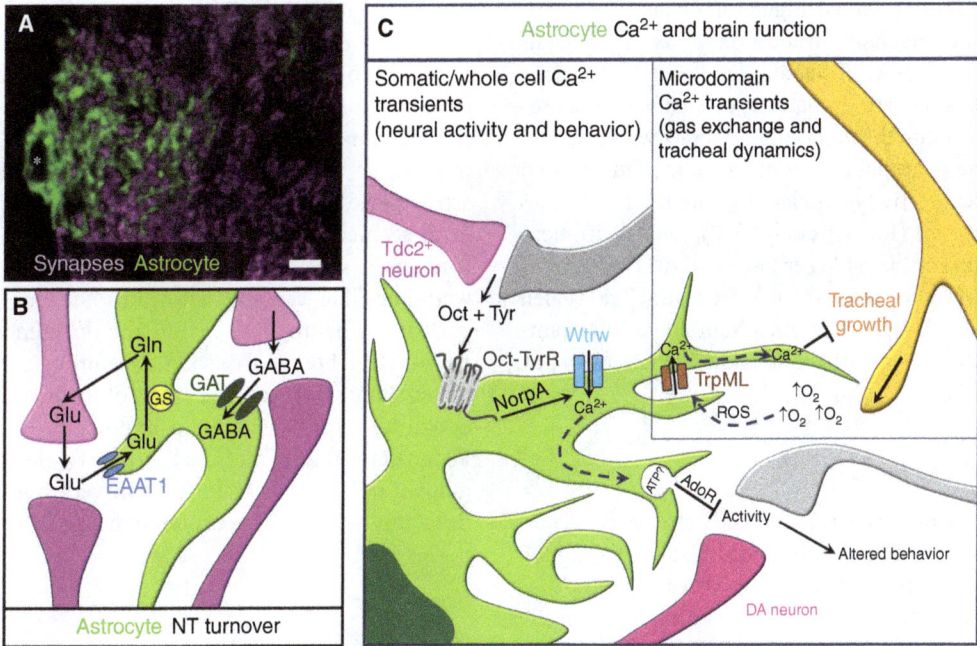

Figure 2. *Drosophila* astrocytes regulate brain function. (*A*) A single *Drosophila* astrocyte clone. Astrocyte processes (green) infiltrate the synaptic neuropil (magenta), while their cell bodies (white asterisk) reside at the neuropil/cell cortex interface. Scale bar, 5 μm. (*B*) Neurotransmitter recycling in fly astrocytes. *Drosophila* astrocytes express transporters for GABA (GAT) and glutamate (EAAT) along with metabolic enzymes for their breakdown (e.g., glutamine synthetase [GS]). (*C*) Astrocytic Ca^{2+} and brain function. (*Left*) Astrocyte somatic calcium signaling influences *Drosophila* behavior. Mechanosensory stimuli trigger $Tdc2^+$ neurons to release octopamine (Oct) and tyramine (Tyr) to activate Oct-Tyr receptors (Oct-TyrR) on astrocytes. Activation of Oct-TyrR stimulates astrocyte Ca^{2+} influx through the TRP channel Wtrw via NorpA/phosphoinositide phospholipase C, which is sufficient to silence dopaminergic (DA) neuron activity and suppress normal chemotaxis and startle-induced behaviors. (*Right*) Astrocyte microdomain Ca^{2+} signaling influences tracheal growth and gas exchange. Microdomain localization of TrpML results in local increases in Ca^{2+} in response to ROS, which in turn suppresses tracheal growth to decrease local O_2 concentrations.

astrocyte membrane extension into the neuropil (Stork et al. 2014). Thus, without Heartless/FGFR signaling, astrocytes are born but fail to elaborate their tufted morphology. The level of Heartless/FGFR signaling appears to have a strong regulatory effect on astrocyte growth rates, as expression of an activated version of this receptor in a single astrocyte led to an increase in its size relative to its wild-type neighbors, and partial blockade of Heartless signaling in a single astrocyte by RNAi-mediated knockdown had the opposite effect, making the Heartless-deficient astrocyte smaller than its neighbors. Elimination of the ligands for Heartless, Pyramus (Pyr) and Thisbe (Ths), led to similar defects in astrocyte morphogenesis, and based on RNA in

situ hybridizations and rescue experiments, it is believed that Pyr and Ths are released by neurons (Stork et al. 2014). It is interesting to note that mouse astrocytes also express high levels of FGFR3 (Pringle et al. 2003; Cahoy et al. 2008), the ortholog of *Drosophila* Heartless. Initial studies of an FGFR3 mutant mouse suggested that it negatively regulated GFAP expression (Cahoy et al. 2008), but it would be interesting to revisit its role in astrocyte morphological elaboration based on the role for Heartless in *Drosophila* astrocytes. Indeed, a recent report demonstrated that CRISPR/Cas9-mediated knockdown of FGF receptors in zebrafish astrocytes resulted in a significant reduction in astrocyte size (Chen et al. 2020), suggesting that FGFR signaling is a critical, conserved regulator of astrocyte–synapse interactions. In addition to Heartless signaling, a number of signaling molecules have since been shown to regulate the extension of fine processes into the neuropil to interact with and regulate neuronal functions including leucine-rich repeat activity–regulated protein at synapses (Lapsyn) (Richier et al. 2017), neuroligin–neurexin signaling (Ackerman et al. 2021), the lipid-binding G-protein-coupled receptor, Tre1 (Chen et al. 2024), and Orion, a homolog of the mammalian chemokine, fractalkine (Boulanger et al. 2021).

Ensheathing Glia

Ensheathing glial cells extend flattened processes along the edges of the neuropil, subdividing the neuropil from the cortex, as well as dividing major regions and commissures into anatomically discrete compartments (Hartenstein 2011). They are polarized glial cells, with PIP_3-enriched basolateral membranes facing outward toward the cortex and a PIP_2-enriched apical membrane toward the neuropil (Pogodalla et al. 2021). Like astrocytes, their cell bodies lie at the interface of the neuropil and cortex, but unlike astrocytes, the majority of their processes remain on the outside of the neuropil under normal conditions (Fig. 1). Furthermore, like astrocytes, ensheathing glia also form tight tiling boundaries with themselves (Pogodalla et al. 2021; Salazar et al. 2022), and their development has been shown to be regulated by FGF signaling (Wu et al. 2017). Loss of either the neuronally derived Pyr or Ths ligands, or reduction of the Heartless receptor in ensheathing glia, alters ensheathing glial extension around the developing neuropil of the antennal lobe, subsequently affecting the shape, size, and synaptic targeting of the glomeruli within the lobe (Wu et al. 2017). While ensheathing glia and astrocytes comprise the two types of neuropil glia, are derived from the same precursor cells, and are situated in close proximity, the two can be differentiated from one another based on a number of molecular markers. During development, astrocyte fates are specified by a Notch signaling protein, Prospero, which promotes astrocyte morphogenesis and suppresses expression of en-

sheathing glial genes. Ensheathing glia can be marked by expression of the excitatory amino acid transporter 2 (EAAT2), whereas astrocytes instead express EAAT1 (Peco et al. 2016). Ensheathing glia do not express the GABA transporter GAT, which is highly enriched in astrocytes (Stork et al. 2014). Additionally, each hemi-segment of the ventral nerve cord (VNC) contains a separate hybrid ensheathing glial/wrapping glial cell that ensheaths the dorsal part of the neuropil as well as the dorsalmost neuronal cell bodies of the cortex, before extending out along the proximal axons of the peripheral nerve (Fig. 1B; Coutinho-Budd et al. 2017; Pogodalla et al. 2021). This cell is often labeled by many if not all ensheathing glial drivers, but its functions have not been explored in detail.

Cortex Glia

Cortex glial cells densely infiltrate the neuronal cell cortex, and by early larval stages appear to individually wrap each neuronal cell body (Fig. 1A,B; Coutinho-Budd et al. 2017). The intricate association with neuronal cell bodies is similar to satellite glia of the vertebrate peripheral sensory ganglia (Hanani 2005) and the cell body contacts made by protoplasmic astrocytes (Salm 2000; Cui et al. 2018), satellite microglia (Cunningham et al. 2013; Baalman et al. 2015; Stowell et al. 2018; Cserép et al. 2020), perineuronal oligodendrocytes (Takasaki et al. 2010), and other vertebrate glial subtypes in the CNS (Cheng et al. 2018; Charlton-Perkins et al. 2019). Impressively, a single cortex glial cell can encase around 50 neuronal cell bodies in the larval CNS (Coutinho-Budd et al. 2017) and up to 100 in the adult (Awasaki et al. 2008). During mid-larval stages, cortex glia begin to expand their nuclei (Coutinho-Budd et al. 2017) through a combination of endocycling and endomitosis along with fusion of their processes as they wrap newly born neurons (Rujano et al. 2022). Together, mature cortex glia form the "trophospongium"—the honeycomb-like structure of glial membranes that surround and presumably support neuronal cell bodies and the proximal regions of neurites as they extend toward the neuropil. Early ablation of cortex glial cells using a genetically driven ap-

Cite this article as *Cold Spring Harb Perspect Biol* doi: 10.1101/cshperspect.a041347

optotic gene, *hid*, results in premature lethality (Coutinho-Budd et al. 2017), demonstrating that cortex glia are critical for CNS function and health. Even without total ablation, morphological changes that result in the loss of cortex glia interactions with neuronal soma later in life, such as during the transition from the second to third larval instar stage, lead to increased neuronal cell death, impaired locomotion (Coutinho-Budd et al. 2017), and seizures (Kunduri et al. 2018). Cortex glial cells form tiled domains like other glia, but also associate closely with other glial subtypes including astrocytes, ensheathing glia, and the SPGs that form the BBB. This close cellular juxtaposition and intercellular tiling are believed to help form cell–cell conduits for the efficient transfer of nutrients from the blood-like hemolymph fluid surrounding the brain to neuronal cell bodies (Volkenhoff et al. 2015). Gas exchange is also likely occurring through cortex glia and astrocytes as these glial cell types make significant contact with the *Drosophila* tracheal system (the fly's blood vessel-like extensions of air tubes) as it penetrates the CNS (Fig. 2C; Pereanu et al. 2007).

Cortex glia play a crucial role in establishing stem cell niches. Following their birth from neuroblasts, immature neurons are wrapped by newly extending cortex glial cellular processes. At the same time, cortex glia associate closely with neuroblasts, forming a critical glial niche for continued stem cell divisions. Neuroblasts and cortex glia signal reciprocally to regulate the formation and maintenance of this niche. For instance, expression of a dominant-negative DE-cadherin in cortex glia led to misplacement of neuroblasts and neuronal cell bodies, which in turn altered fiber tract morphology (Dumstrei et al. 2003). Cortex glia also receive nutrition-dependent cues in the form of *Drosophila* insulin-like peptide 2 (dILP 2) that increases intracellular PI3K and Akt, resulting in both cortex glial membrane expansion and reactivation of neuroblast proliferation (Yuan et al. 2020). Neuroblasts in turn release PDGF- and VEGF-related factors 1–3 (Pvf1–3) to signal back to their receptor, PVR, on cortex glial cells to enhance cortex glial membrane extension, which in turn supports neuroblast reactivation and maintenance (Read 2018).

While we are only beginning to discover the ways in which cortex glia communicate with neuroblasts and neuronal cell bodies, cortex glial studies have increased dramatically in recent years, thanks to the development of new markers, opening up new avenues for exciting future findings on the regulation of their development, function, and plasticity.

GLIAL REGULATION OF NEURON AND NEURAL CIRCUIT STRUCTURE

Although many aspects of neuronal identity (e.g., neuropil location, neurotransmitter identity) are pre-determined by neuroblast lineage (Doe 2017), neural circuit architecture is heavily influenced by neighboring glial cells (Lago-Baldaia et al. 2020; Perez-Catalan et al. 2021). *Drosophila* glia heavily regulate each step of neural circuit development, from neuroblast proliferation and the birth of new neurons (for more see Fernandes et al. 2024), to axon guidance, synaptogenesis, and synapse remodeling. Recent data indicates a prominent role for glia–neuron signaling in establishing dendrite/axon structure through stabilization/destabilization of growing neurites, and glia can directly remodel neurons through developmental pruning. In this section, we summarize recent studies that uncovered important roles for glia in determining both neuronal and neural circuit structure.

Axon and Dendrite Patterning Is Governed by Glial Signaling

Given the rapid developmental progression of invertebrates like *Drosophila*, the neurons that make up the larval (~10,000 neurons at the first instar larval stage [Brunet Avalos et al. 2019]) and adult (~100,000 neurons [Zheng et al. 2018]) nervous systems must quickly extend their axons to the appropriate targets and form functional neural circuits. This is highly dependent on glia–neuron signaling via classic axon guidance cues including Slits and Netrins, many of which were first discovered and functionally characterized in *Drosophila* (Lemke 2001; Learte and Hidalgo 2007). Several distinct glial cell populations in *Drosophila*, including midline glia, astrocytes, ensheathing glia, and cortex glia, are essential for proper axon guidance and, in turn, organization of the CNS. Developing astrocytes and ensheathing glia in the VNC (Peco et al. 2016) are critical for axon pathfinding and fasciculation (Griffiths and Hidalgo 2004; Learte and Hidalgo 2007). Ablation of these glia causes complete loss of longitudinal axon tracks in the developing VNC (Hidalgo et al. 1995). In addition to longitudinal organization of the VNC, midline crossing and lateral positioning of axons/dendrites are also regulated by a specialized population of "midline" glia. Midline glia are akin to the vertebrate floor plate (for review, see Lemke 2001). The role of midline glia in the organization of the VNC has been particularly well-studied in the motor circuit. During development, motoneuron dendrites tile the CNS in a myotopic map, similar to their organization in the periphery. Changes in Slit/Robo and Netrin/Frazzled signaling between midline glia/motoneurons are sufficient to change dendritic targeting and perturb the myotopic map (Mauss et al. 2009). Together, astrocyte and ensheathing glial precursors, along with midline glia, orchestrate proper organization of axons throughout the embryonic VNC.

Drosophila neurons require additional glial-derived positional cues or input following embryogenesis. The axons of many embryonic-born neurons are remodeled during metamorphosis to accommodate the new body plan and functional needs of the adult fly (Watts et al. 2003; Williams and Truman 2005; Chen et al. 2017). Additionally, new populations of neurons and glia are born during larval life, which ultimately innervate their targets during metamorphosis as the adult body plan is elaborated (Doe 2017). Here, distinct glial populations similarly regulate axon guidance to help organize the adult *Drosophila* brain. As described above, astrocytes tile the synaptic neuropil, ensheathing glia establish neuropil borders (discussed further in Fernandes et al. 2024), and cortex glia functionally segregate neuronal cell bodies. Spindler et al. (2009) found that ablation of cortex glia, astrocytes, and ensheathing glia caused brain-wide defects in secondary axon tract morphology, ranging from misguided axons to missing fascicles. This was particularly pronounced in the mushroom body (MB), a learning and memory

center within the fly brain (Spindler et al. 2009). A recent report also identified cortex glia as critical regulators of axon extension in the developing visual system (Takechi et al. 2021). Interestingly, this latter study identified glial-derived insulin signaling, through direct modulation of guidance receptor signaling, as a key regulator of axon pathfinding (Takechi et al. 2021). Thus, axon pathfinding throughout the *Drosophila* life cycle is controlled by many types of glia that use circuit-specific signaling mechanisms. Similar roles have been defined for vertebrate astrocyte-like cells in the developing CNS (Barresi et al. 2005; Lago-Baldaia et al. 2020) and during regeneration (Mokalled et al. 2016).

Glial Regulation of Synapse Number

After neurons have projected to the proper position in the nervous system, they must establish appropriate synaptic connections. Astrocytes are the primary synapse-associated glial cell type in the invertebrate and vertebrate CNS, where they are key modulators of neural circuit function (see below). The last two decades have revealed that metazoan astrocytes not only support existing synapses, but directly regulate synapse number through synaptogenic and antisynaptogenic proteins (Allen and Eroglu 2017). Astrocyte morphogenesis is conspicuously timed with initial waves of synaptogenesis in the vertebrate and invertebrate brains (Muthukumar et al. 2014; Stork et al. 2014; Stogsdill et al. 2017; Ackerman et al. 2021). In the fly, for example, Muthukumar et al. (2014) examined the timing of synapse formation compared with astrocyte infiltration of the adult *Drosophila* CNS. At the transition from larval to pupal life, defined as 0 h after pupal formation (APF), larvae build a pupal case and initiate metamorphosis. The major wave of pupal synaptogenesis (scored by classical electron microscopy-based identification of synapses) began ~72 h APF, coincident with the initiation of astrocyte infiltration into the developing adult CNS. Synapse numbers continued to increase over developmental time but largely plateaued by late metamorphosis, about a day before eclosion (i.e., adult emergence from the pupal case). Immature astrocyte membranes were also initially found in

Cite this article as *Cold Spring Harb Perspect Biol* doi: 10.1101/cshperspect.a041347

the neuropil at 72 h APF, and they continued to infiltrate the neuropil more densely over time such that, by eclosion as adults (~96 h APF), astrocytes were found throughout the neuropil and had taken on their mature morphology. Interestingly, ablation of astrocytes just prior to the major wave of synaptogenesis led to a 40%–50% decrease in the number of synapses formed in the late pupal brain, although gross brain histology and neuronal numbers remained largely unchanged (Muthukumar et al. 2014). These data argue that *Drosophila* astrocytes, like their mammalian counterparts, are important for CNS synapse formation. Whether similar presynaptic molecules, such as thrombospondins (Christopherson et al. 2005), Hevin (Kucukdereli et al. 2011), glypicans (Allen et al. 2012), or HepaCAMs (Baldwin et al. 2021)—known astrocyte-derived synaptogenic molecules in mammals—are secreted by *Drosophila* astrocytes to promote synapse formation remains to be explored.

Astrocyte Regulation of Developing Dendrites

Dendrites are the major sites of circuit integration in the brain. Dendrite growth is a dynamic process of extension, branching, retraction, and, ultimately, stabilization, resulting in relatively stereotyped arborization patterns within a single neuron type (Lefebvre 2021). Dendrite formation and remodeling must be tightly regulated because dendrite structure is directly correlated to neuronal function, and even subtle changes in dendrite geometry can affect overall neural circuit function (Hayashi and Majewska 2005). Proper dendrite development is dependent on a panoply of factors, both intrinsic (transcriptionally regulated) and extrinsic (patterning from the environment and neighboring cells). At the cellular level, there are two primary processes that ultimately shape dendrite architecture: membrane availability and cytoskeletal stability (Lefebvre 2021). Recent data from *Drosophila* and other model systems have identified glia as key modulators of dendritogenesis, and, unsurprisingly, many of these studies link glia to altered lipid and/or cytoskeletal dynamics in developing dendrites.

Neurite development and subsequent synaptic firing is energy intensive, yet, neurons cannot meet this energetic demand on their own. Instead, neurons rely on metabolic support by glia (e.g., shuttling of lipids and lactate) for their development and maintenance. In *Drosophila*, the protein ApoD (encoded by *Glaz* and *Nlaz*) is functionally homologous with Apolipoprotein E (APOE), a protein that shuttles cholesterol, amyloid-β, and other lipids to neurons. Altered function of GLaz in *Drosophila* astrocytes or of APOE in vertebrate astrocytes is strongly linked to changes in synapse stability and neurodegeneration (Liu et al. 2013, 2017; Lane-Donovan and Herz 2017; Wang et al. 2021; Mahan et al. 2022), but roles in development are underexplored. Recent reports found that in addition to its role in neuronal homeostasis (Liu et al. 2017; Yin et al. 2021), astrocyte-derived GLaz, as well as two other previously uncharacterized lipoproteins (CG12926 and CG30392), are essential for proper growth of larval ventral lateral neuron dendrites (LNvs), visual projection neurons (PNs) that develop in larval life. The authors found that GLaz physically interacts with a brain-specific short isoform of lipophorin receptor 1 (LpR1-short) to facilitate astrocyte-neuron lipid shuttling. Moreover, they found that LpR1-short was specifically localized to the dendrites and soma of LNvs, and LNv-specific LpR1-short CRISPR/Cas9-mediated knockout impaired dendrite development. In cultured rat retinal ganglion cells, the visual PNs of vertebrates, astrocyte-derived APOE shuttles cholesterol to neurons to regulate the rate of dendritogenesis, which in turn facilitates synapse formation (Mauch et al. 2001; Goritz et al. 2005). Furthermore, astrocyte-specific knockout of SREBP2, a transcription factor that governs most cholesterol synthesis, resulted in global reductions in postsynaptic PSD95 in vivo and reduced dendrite growth in vitro (Ferris et al. 2017). Thus, astrocyte-derived lipids are crucial conserved mediators of dendrite development.

Glial Regulation of Neurite Structural Plasticity

In addition to membrane lipid availability, another primary determinant of axon and dendrite architecture is the underlying organization of the actin and MT cytoskeletons (Lefebvre 2021). Here, we summarize how glial signaling may shape the neuronal cytoskeleton and, in turn, neurite structural plasticity.

LNv neurons (described above) first develop in the larva, persist into adulthood, and function in the adult brain as pacemakers to control rest–wake cycles (Lear et al. 2009; Yin et al. 2021). LNv neurons are particularly well-studied in structural plasticity, as they undergo a dramatic circadian phenomenon: each night, they scale back their axonal projections, which are regrown during the following day in an activity-dependent manner (Fernández et al. 2008). This remodeling is dependent on the cycling activity of Rho1, which drives retraction of LNv axons at dusk via contraction of the actin cytoskeleton and myosin phosphorylation (Petsakou et al. 2015). For many years, the cycling pattern of LNv growth and retraction was thought to be entirely neuron-intrinsic. Recent data showed that expression of a dominant-negative isoform of the clock gene *cyc* in all glia was sufficient to impair daily remodeling of LNv terminals (Herrero et al. 2017); which glial subtype(s) are most important for this process were not identified. Interestingly, altering the glial clock impairs circadian remodeling without disrupting the expression of pacemaker proteins within LNv neurons themselves (Herrero et al. 2017); thus, glia do not set the neuronal clock, but specifically regulate circadian remodeling of the LNv dendrites. How glia alter neuronal cytoskeletal dynamics involved in this remodeling, and whether they directly alter neuronal Rho1 signaling to affect cycling remodeling, remain to be tested.

Astrocytes Regulate Developmental Dendrite Plasticity

Following initial circuit establishment, neuronal structure can be extensively remodeled in response to environmental experience. Although this remodeling, or plasticity, can occur in some areas of the adult brain (Fernández et al. 2008), structural plasticity is often limited to developmental windows called critical periods (Hensch 2004). In a recent study, Ackerman et al. (2021) demonstrated that there is a critical period of dendrite plasticity within the developing *Drosophila* motor circuit. Within this window (embryonic stage 17 through 8 h after larval hatching), *Drosophila* aCC/RP2 motor neurons undergo activity-dependent remodeling of their dendrites and accompanying synapses; motor neuron activation resulted in a reduction in dendrite volume and of excitatory inputs, whereas silencing resulted in expansion of the dendritic domain, increased excitatory inputs, and reduced inhibitory inputs. Closure of this critical period coincided with a period of accelerated astrocyte–motor synapse contact, suggesting astrocyte ingrowth restricts plasticity, and, indeed, ablation of astrocytes was sufficient to extend the critical period. Through RNAi screening, several receptor–ligand pairs expressed in motor neurons and astrocytes were found to be required to drive timely critical period closure. These pathways included astrocyte-derived neuroligins, which bind neurexin on motor neurons, and overexpression of these pathways was sufficient to stabilize MTs within motor dendrites and drive precocious critical period closure (Ackerman et al. 2021). Importantly, in mammals, neuronal neurexins bind astrocytic neuroligins to regulate both neuronal and astrocyte morphology (Stogsdill et al. 2017; Tan and Eroglu 2021). This supports the notion that *Drosophila* will be useful for uncovering functionally conserved neuron–glia signaling mechanisms.

Although the link between membrane-bound neurexin and cytoskeletal remodeling remains unclear, a recent study found that signaling from reactive oxygen species (ROS) is critical for activity-dependent remodeling for motor dendrites (Oswald et al. 2018). Interestingly, changes in astrocytic EAAT1 (a glutamate transporter) lead to hyperexcitability, accumulation of ROS in motor neurons, and remodeling of presynaptic boutons at the NMJ (Peng et al. 2019). ROS is known to affect MT stability (Wilson and González-Billault 2015); thus, it is possible that neurexin signaling influences local ROS concentrations to regulate MT stability and, in turn, dendritic remodeling.

ROLES FOR GLIAL CELLS IN NEURONAL CIRCUIT REMODELING AND PHAGOCYTOSIS

Among the most essential roles for glial cells is sculpting and refining the nervous system, often

Cite this article as *Cold Spring Harb Perspect Biol* doi: 10.1101/cshperspect.a041347

by clearing unwanted cells, projections, and synaptic connections during development. Whether glial cells actively select targets for pruning or merely respond to neuronal cues and assist in cell suicide remains an open question. The full array of signals that specify how neurons or their compartments (i.e., axons, dendrites, or synapses) are selected for elimination and clearance by glia in the healthy brain remains a fundamental, unanswered question, and defining these mechanisms is a crucial goal for the field. Furthermore, aberrant activation of glial engulfment has been proposed to contribute to neurodegenerative diseases such as Alzheimer's disease and Parkinson's disease, perhaps through inappropriate elimination of synapses (Hong et al. 2016).

Microglia, the resident immune cells of the vertebrate brain, are often the focus of these studies given their well-known roles in phagocytosis of debris during development, injury, and disease (Schafer and Stevens 2013). However, most glial subtypes in the vertebrate nervous system can act as "nonprofessional" phagocytes and engulf cellular debris, including astrocytes (Lööv et al. 2012; Damisah et al. 2020; Konishi et al. 2020; Pradhan et al. 2022; Zhou et al. 2022), oligodendrocytes (Nguyen and Pender 1998), Schwann cells (Brosius Lutz et al. 2017; Cunningham et al. 2020), and oligodendrocyte precursor cells (Auguste et al. 2022; Buchanan et al. 2022; Xiao et al. 2022), among others (Wu et al. 2009; Sakami et al. 2019). The same is true in *Drosophila*, where most subtypes of *Drosophila* glia seem to have the capacity for phagocytosing cellular debris and remodeling neuronal circuitry.

A central regulator of phagocytic activity in any *Drosophila* glial subtype is the phagocytic receptor, Draper, the homolog of mammalian MEGF10 and JEDI-1. Work on Draper represents a good example of how basic cellular mechanisms can be elucidated in *Drosophila* glia and teach us about mammalian glial cell biology. This receptor was first identified in *C. elegans* (CED-1), and later shown in *Drosophila* to mediate glial engulfment of cell corpses (Etchegaray et al. 2016; McLaughlin et al. 2019), degenerating axons after injury (MacDonald et al. 2006), pruned neurites (Tasdemir-Yilmaz and Freeman 2014),

and to encode an ancient immunoreceptor that activated downstream Src/Syk family kinase signaling (Ziegenfuss et al. 2008). Later, dorsal root ganglia (DRG) satellite cell precursors were shown to also use MEGF10/JEDI signaling to engulf cell corpses during mammalian nervous system development (Wu et al. 2009), and mammalian astrocytes have been shown to undergo MEGF10-mediated synaptic phagocytosis in both the developing and adult brains (Chung et al. 2013). Like Draper, MEGF10 also activates a downstream Src family signaling cascade similar to Src42a and Shark (Scheib et al. 2012), and both MEGF10 and Draper appear to be activated by complement proteins, C1q and Mcr, respectively (Iram et al. 2016; Lin et al. 2017). This signaling pathway therefore represents an ancient and widespread mediator of glial phagocytic functions.

Glial phagocytosis has recently emerged as an important factor in neurodegeneration. First, removing neuronal debris is believed to be imperative for reducing inflammation and minimizing inflammation-driven secondary degeneration. Whether aberrant activation of phagocytic function can drive neurodegeneration is an interesting possibility and is consistent with the observation that overexpression of *draper* in adult brains has been reported to lead to the phagocytosis and removal of nonapoptotic, presumably live neurons by hyperphagocytic glia (Hakim-Mishnaevski et al. 2019). Additionally, glial engulfment plays a role in removing protein aggregates in the context of neurodegenerative diseases. For instance, the loss of Draper has been shown to exacerbate life span reduction and further impair motor activity in an amyloid fly model of Alzheimer's disease (Ray et al. 2017), similar to glial roles in removing amyloid plaques in the mammalian brain (Ries and Sastre 2016). Furthermore, the ability of glia to engulf debris through Draper activity is a driving force in the spread of mutant Huntingtin protein from presynaptic to postsynaptic neurons in a fly model of Huntington's disease (Pearce et al. 2015). How glia participate in the transfer of mutant proteins from pre- to postsynaptic cells remains an interesting and open question.

Astrocytes Transform into Phagocytes to Drive Developmental Neuronal Remodeling

During metamorphosis, the *Drosophila* CNS undergoes a dramatic transformation from simple larval brain lobes and VNC to the much more architecturally complex adult brain and thoracic ganglion, respectively. Neural circuit reorganization entails the elimination of a significant number of neurons by apoptosis and the pruning of many larval-specific neurites, followed by the wiring of new adult and retained larval neurons into adult-specific neural circuits (Truman 1996). Astrocytes, with their cellular processes spread across the synaptic neuropil (Figs. 1B, 2A,C), are primed for the removal of synaptic debris and neurite pruning. Prior to metamorphosis, astrocytes in the larva express very low levels of *draper*. However, astrocytes transform at the initiation of metamorphosis from a cell that nourishes neurons and synapses to a highly phagocytic cell type that engulfs the vast majority of pruned debris within the neuropil (Hakim et al. 2014; Tasdemir-Yilmaz and Freeman 2014). The first 2 d of *Drosophila* metamorphosis is the time frame during which larval neural circuits are deconstructed. Within 2 h APF, steroid-dependent signaling events in astrocytes result in their dramatic increase in Draper expression and activity, transformation into phagocytes, and initiation of engulfment of pruned axons, dendrites, and synapses (Tasdemir-Yilmaz and Freeman 2014). Astrocytes actively prune neurites and synapses during this developmental window, with their processes first developing phagocytic cups as they remove debris, and their membranes subsequently become progressively less prominent in the neuropil as debris is cleared. By 48 h APF, astrocytic processes are absent from the neuropil, and almost no synapses are present in the CNS (Muthukumar et al. 2014). As described above, the reappearance of synapses corresponds to the emergence of new astrocytic processes in late pupal stages (Muthukumar et al. 2014). Interestingly, astrocytes in the late pupa display high levels of Draper and contribute to synaptic refinement simultaneously during this period of reconnection, as loss of astrocytic *draper* results in altered synaptic size and shape of olfactory glomeruli in the adult antennal lobe (Jindal

et al. 2023). As synaptic development continues into the early adult, *draper* is down-regulated in astrocytes and expressed mostly in ensheathing glia surrounding these glomeruli. Correspondingly, loss of *draper* in ensheathing glia also leads to altered glomeruli synaptic refinement (Jindal et al. 2023).

Drosophila MB γ neurons have served as a useful model for local neurite pruning—where only selected neurites and their synapses are eliminated but the parent neuron is retained and reorganized (Lee et al. 1999). In the larva, MB γ neurons extend both medial and dorsal axonal projections into the MB. At metamorphosis, MB axons and their synapses fragment and are cleared from the CNS, and subsequently adult-specific MB axonal projections are elaborated. The secreted CX$_3$C motif-containing chemokine, Orion, originates from these same MB γ neurons, and acts as a required signal for astrocyte process infiltration into the MB γ neuron axon bundle and subsequent pruning and removal of neurite debris (Boulanger et al. 2021). Intriguingly, the mammalian CX$_3$C-containing chemokine, fractalkine, has similarly been shown to regulate synaptic pruning in the visual cortex (Gunner et al. 2019). As glial cells invade the MB lobes at the initiation of axon pruning (Awasaki and Ito 2004; Watts et al. 2004), they prime the MB γ neurons for pruning through secretion of the TGF-β molecule Myoglianin (Myo). Elimination of Myo from glial cells leads to a blockade of MB γ neuron pruning, providing direct evidence that *Drosophila* glial cells actively signal to neurons to make them competent to prune (Awasaki et al. 2011).

As *Drosophila* MB γ axons fragment, glial cells engulf and degrade degenerating axonal debris. Genetic blockade of glial engulfing activity potently prevents clearance of degenerating axons (Awasaki and Ito 2004), and genetically labeled axon fragments can be found within phagocytic glial cells (Watts et al. 2004). Interestingly, Chung et al. (2013) discovered a conserved role for MEGF10, the mammalian ortholog of Draper, in astrocyte engulfment of synapses during neural circuit refinement in the mammalian visual system, where it acts in parallel with MERTK.

Blockade of astrocyte signaling through Draper or the ecdysone receptor (EcR), a nuclear hormone receptor, suppresses axon clearance of the MB γ neurons (Hakim et al. 2014; Tasdemir-Yilmaz and Freeman 2014; Wang et al. 2019b). While glial EcR and Myo signaling regulates both MB γ neuron and ventral corazonin-expressing (vCrz) neuronal neurites (Wang et al. 2019b), surprisingly, careful genetic studies reveal that there is not a single engulfment pathway responsible for clearance of all neurite debris during pruning. It appears that astrocytes can also engage unique context-dependent molecular programs to engulf different subsets of neurites. For instance, clearance of MB γ neuron axons requires Draper signaling (Awasaki et al. 2006; Hoopfer et al. 2006; Hakim et al. 2014; Tasdemir-Yilmaz and Freeman 2014) and the guanine nucleotide exchange factor (GEF) Crk/Mbc/dCed12 in a partially redundant fashion (Tasdemir-Yilmaz and Freeman 2014). In contrast, the clearance of neurites from vCrz neurons requires Crk/Mbc/dCed12, while Draper is dispensable (Tasdemir-Yilmaz and Freeman 2014). The latter result was unexpected and was the first example of a glial-mediated engulfment event that occurs in a Draper-independent fashion in *Drosophila*.

Ensheathing Glia Respond to Axon Injury and Phagocytose Neuronal Debris

While astrocytes are the primary phagocyte in the synaptic neuropil during development, in the adult brain, ensheathing glia that normally lay at the interface between the neuropil and cortex potently respond to injury, invading more deeply into the neuropil, and clearing axonal and synaptic debris. MacDonald et al. (2006) examined glial responses to axonal injury using a simple nerve injury assay in which olfactory receptor neurons (ORNs) were severed by removal of the antennae. Within hours after axon injury, ensheathing glial cells up-regulated Draper, extended membranes directly to degenerating axons and phagocytosed axonal debris. Elimination of Draper signaling blocked all glial responses to axonal injury, indicating that Draper is a central regulator of glial clearance of ax-

onal debris after axotomy. Astrocytes in the adult brain do not appear to respond morphologically to axon injury, nor do they up-regulate *draper* expression postinjury (Doherty et al. 2009). While this difference in cellular response between the two neuropil glia has not been investigated deeply, it could arise from different basal levels of Draper in the adult, distinct functional roles for each cell type at this time, or simply diverging compositions of other molecular machinery and signaling pathways that regulate engulfment between the two cell types.

Upstream of Draper, insulin receptor (InR) signaling has been shown to regulate glial engulfment in both injury (through Akt1 and STAT92E; Musashe et al. 2016) and in developmental debris clearance in a disease model of Fragile X syndrome (Vita et al. 2021; Song and Broadie 2023). Signaling mechanisms downstream of Draper have been delineated in a series of mechanistic studies. Draper-dependent signaling of glial responses to axonal injury occurs through activation of a Src-family signaling cascade composed of Src42a and Shark, which, together with the PTB-domain protein dCed-6, promote engulfment of axon debris (Awasaki et al. 2006; Ziegenfuss et al. 2008; Doherty et al. 2009). Additional signaling molecules required for glial clearance of degenerating axonal debris include Rac1 and the Rac1 GEF Crk/Mbc/dCed-12, which is required for glial internalization of axonal debris (Ziegenfuss et al. 2012). The Rac1 GEF Drk/Dos/Sos appears to be partially redundant with Crk/Mbc/dCed-12, as it is able to activate Rac1 downstream from Draper (Lu et al. 2014) and protein phosphatase 4 (PP4) (Winfree et al. 2017). PP4 then drives further activation of Drk/Dos/Sos and links Draper activation to cytoskeleton remodeling. Interestingly, Draper activation results in a positive feedback loop to further promote glial engulfment. Draper-dependent, transcriptional activation of genes required for engulfment involves signaling through TNF receptor–associated factor 4 (TRAF4) (Lu et al. 2017), the c-Jun kinase (JNK) cascade, and the transcriptional activators dAP-1 (Macdonald et al. 2013) and STAT92E (Doherty et al. 2014). This leads to the release of matrix metalloprotease 1 (MMP-1), recruitment of ensheathing glia to the injured axons, and ultimately debris clearance (Purice et al.

2016). That Draper signaling can regulate transcriptional activation of the *draper* gene itself suggests a simple model for how glia might gauge their activation state in accordance with the severity of axonal injury—a more severe injury should lead to the production of more axonal debris, which in turn will activate Draper signaling more strongly and ultimately lead to enhanced activation of engulfment gene expression (Doherty et al. 2014).

Axonal injury can lead to remarkable synaptic plasticity in adult *Drosophila*, which is mediated by glial cells. ORNs project into the antennal lobe of the brain from either the antennae or maxillary palps, where they synapse onto highly stereotyped target PNs within defined glomerular structures. Local interneurons also interconnect different glomeruli within the antennal lobe, although excitatory connections between glomeruli are normally weak. Axotomy of ORN sensory afferents from the antenna strongly potentiated interglomerular excitatory connections, indicating that injury somehow induced plasticity of interglomerular PNs (Kazama et al. 2011). Silencing ORN activity was not sufficient to induce this remodeling, rather the degeneration of the severed ORN axon terminals turned out to be essential. Blocking ORN axon degeneration by ectopic expression of neuroprotective *Wld^s* suppressed the refinement of interglomerular PNs after axotomy. What is the cellular mechanism for this plasticity? Blockade of endocytic function (by expressing a dominant-negative *dynamin*) in local ensheathing glia suppressed induction of PN plasticity after ORN axotomy, indicating that ensheathing glia somehow mediate injury-induced remodeling of excitatory PN connections in the olfactory circuit (Kazama et al. 2011). Interestingly, the capacity of ensheathing glia to engulf is dependent on brain state. Sleep–wake cycles in *Drosophila* have been shown to regulate glial morphological plasticity, Draper levels, and engulfment after injury. Sleep deprivation also led to slower Wallerian degeneration and persistence of axonal debris, while promoting sleep increased glial infiltration and phagocytic clearance of axonal debris (Stanhope et al. 2020). Defining the mechanistic basis of this circadian regulation of glial phagocytic function is an important future goal to better understand glial biology and how it regulates neural circuits in this context.

Cortex Glia Detect Cell Corpses and Up-Regulate Phagocytic Function

In the absence of Draper signaling, neuronal corpses appear throughout the CNS due to phagocytic clearance defects starting in development and accumulating into metamorphosis and adulthood (Etchegaray et al. 2016; Hilu-Dadia et al. 2018). Given their intimate association with neuronal cell bodies, cortex glia are perfectly positioned in the CNS to remove neuronal cell bodies throughout development, disease, or injury (Fig. 1). Indeed, when Draper signaling is reduced within these cells, death caspase-1-positive (DCP-1$^+$) dead and dying neurons accumulate (Etchegaray et al. 2016; Hilu-Dadia et al. 2018; McLaughlin et al. 2019; Nakano et al. 2019), resulting in neuronal degeneration in adults (Etchegaray et al. 2016; McLaughlin et al. 2019). In the optic lobe, cortex glia use Shark and Ced-6 signaling for the engulfment of young neurons, with the Crk/Mbc/dCed-12 pathway playing a minor role (Nakano et al. 2019). Loss of the candidate Draper ligands calcium-binding protein 1 (CaBP1) or Pretaporter, or manipulation of the proposed eat-me signal phosphatidylserine (PS) in this region also reduced glial engulfment and increased the accumulation of dead neurons in the optic lobes (Nakano et al. 2019; Song and Broadie 2023). Furthermore, DCP-1$^+$ neurons were found to accumulate throughout the brain with pan-glial *draper* knockdown specifically during pupal stages using a temporal-specific control, suggesting that glial engulfment continues throughout metamorphosis (Hilu-Dadia et al. 2018). Interestingly, dying neurons throughout the CNS secrete a Spätzle family member, Spz5, to prime cortex glial phagocytic signaling through activation of the Toll-6 receptor and downstream signaling via Sarm1 and FoxO, ultimately leading to the upregulation of Draper expression (McLaughlin et al. 2019). Sarm1 has previously been a focus of neuronal degeneration signaling, as it is necessary for axon death signaling in Wallerian degeneration (Osterloh et al. 2012); however, we now know that Sarm1 is also necessary within glia for glial engulfment of neuronal corpses (McLaughlin et al. 2019; Herrmann et al. 2022). It is clear that modulating Draper in cortex glia is detrimental due to the accumulation of dead neuronal cell bodies; however, altered cortex glial development or even a failure to maintain cortex glial morphology can lead to similar issues.

Like astrocytes, cortex glia require FGFR/Heartless signaling for proper development, particularly with respect to proliferation, and this seems to require neuronally derived FGFs, Pyr, and Ths (Avet-Rochex et al. 2012). Not surprisingly, loss of Heartless in embryonic cortex glia leads to the impaired neuronal corpse clearance during the first wave of neuronal birth and the death of excess neurons in the embryo (Ayoub et al. 2023). Even when cortex glia are initially set up correctly, cortex glial loss of vesicular fusion machinery such as the NSF attachment protein α (αSNAP) or even just the loss of another Spätzle family member, the secreted neurotrophin Spätzle 3 (Spz3), impairs cell body wrapping starting in mid- to late-larval stages. Accordingly, this later disruption in cortex glial–somal interactions still results in increased levels of activated DCP-1 in neuronal cell bodies throughout the CNS (Coutinho-Budd et al. 2017). One interpretation of these data is that the loss of trophic signaling leads to poor neuronal support and subsequent neuron death, but equally possible is that the loss of glia at neuronal cell bodies results in reduced phagocytic clearance of neuronal corpses. It is hard to separate these two roles, as loss of somal wrapping likely results in alterations in both. Future studies that can parse out functional roles of support, wrapping, and engulfment will be key to more fully understand glial interactions with neuronal cell bodies.

ROLES FOR GLIA IN NEURAL CIRCUIT FUNCTION AND BEHAVIOR

The generation of highly specific drivers for different glial subtypes over the last decade has enabled a deep analysis of glial cell development, morphology, and function, particularly in the context of behavioral output. Excellent progress is now being made into how fly glia modulate neural activity and regulate complex behaviors, and we discuss a curated subset of these below.

Drosophila Astrocytes Regulate Neurotransmitter Tone and Generate Calcium Transients that Meaningfully Regulate Circuit Activity and Behavior

Drosophila astrocytes are important for clearance of neurotransmitters from the synaptic space, and loss of this activity can have profound effects on animal behavior and survival. Fly astrocytes have plasma membrane–localized neurotransmitter transporters for glutamate (EAATs; Rival et al. 2004; Stacey et al. 2010) and GABA (GAT; Stork et al. 2014) as well as enzymes such as glutamine synthetase (Freeman et al. 2003) and GABA transaminase (BDGP) (T Stork and M Freeman, unpubl.) for their metabolic breakdown (Fig. 2B). Depletion of EAAT1 from glial cells in adult *Drosophila* led to age-dependent behavioral defects and neuron loss that was rescued by drugs used to suppress excitotoxicity in humans (Rival et al. 2004). Interestingly, the loss of focal adhesion kinase (FAK) in astrocytes was able to rescue picrotoxin- and mechanosensitive-induced seizures through an upregulation of astrocytic EAAT and, in turn, glutamate modulation (Cho et al. 2018). Similarly, astrocyte-specific RNAi-depletion of *gat* led to severe motor defects in larvae and adults (Stork et al. 2014). Astrocytic expression of GAT is, in fact, essential for animal survival: null mutations in *gat* led to late embryonic or early larval lethality around the time these animals would emerge as larvae from the egg case, and these animals could be rescued to adulthood by resupplying GAT only in astrocytes (Stork et al. 2014).

Due to the fact that glia are not electrically excitable and do not fire action potentials like neurons, great interest has developed in the potential role of astrocyte Ca^{2+}-signaling events in the regulation of neural circuit function (Bazargani and Attwell 2016), as this appears to correlate with neuronal activity in some cases and allow for visualization of glial activity. Astrocytes in the *Drosophila* adult brain were found to show spontaneous Ca^{2+} activity in many brain regions, including the antennal lobe (Liu et al. 2014; Bajar et al. 2022). In the olfactory circuit, stimulation of astrocytes using channel rhodopsin 2 (ChR2), a cation channel that allows for the influx of Ca^{2+}, Na^+, and other cations into the cell, inhibited odor-evoked responses of second-order olfactory PNs. Activation of ChR2 in astrocytes decreased the amplitude and slope of excitatory postsynaptic potentials after antennal nerve stimulation (Liu et al. 2014; Bajar et al. 2022). These data begin to make a case for astrocytes directly regulating neuronal physiology, but caution should be taken in interpreting these results, as initiating astrocyte Ca^{2+} signaling via ChR2 activation is potentially nonphysiological.

What generates endogenous astrocyte Ca^{2+} signals, and do they meaningfully affect neural circuit function in vivo? An exciting, evolutionarily conserved role for norepinephrine (NE) (Paukert et al. 2014; Ma et al. 2016; Mu et al. 2019; Reitman et al. 2023) and octopamine (Oct)/tyramine (Tyr) (the fly versions of NE) has emerged in recent years (Fig. 2C). *Drosophila* Tdc2$^+$ neurons (the fly equivalent of locus coeruleus neurons) release Oct/Tyr onto astrocytes. This activates the astrocytic GPCR Oct-TyrR, which binds both Oct and Tyr and signals through PLCβ NorpA to activate the transient receptor potential (TRP)-like channel water witch (Wtrw). Wtrw activation leads to a dramatic rise in Ca^{2+} throughout the astrocyte (i.e., whole-cell Ca^{2+} signaling, rather than microdomain transients). This signaling ultimately leads to a suppression of dopaminergic (DA) neuron activity through the adenosine receptor AdoR. Mutant alleles of *Oct-TyrR* or *wtrw* blocked whole-cell Ca^{2+} signals in astrocytes, suppressed the ability of astrocyte Ca^{2+} signals to block DA neuron activity, and changed animal behavior; in a startle-induced escape response assay where larvae-initiated Oct/Tyr-dependent reversal behavior, selective elimination of *Oct-TyrR* or *wtrw* from astrocytes suppressed the initiation of the startle response and reversal behavior. This work provides direct evidence that astrocytes are integral components of neural circuits, specifically that Oct/Tyr silencing of DA neurons had to flow through astrocytes. It identified key endogenous molecules that are required for Oct/Tyr-dependent Ca^{2+} signaling in astrocytes (Oct-

TyrR and Wtrw), which are now excellent tools to probe the in vivo roles for whole-cell astrocyte Ca^{2+} signals and demonstrated a direct requirement for astrocyte-expressed endogenous Ca^{2+} signaling in animal behavior. Finally, it showed that fly astrocytes were sensitive to Oct/Tyr, exhibiting population responses to stimulation by these NTs, and that these responses were dependent on the fly orthologs of α1-adrenergic receptors (Fig. 2C; Ma et al. 2016). These whole-cell Ca^{2+} responses are remarkably similar to how mammalian astrocytes respond to NE in awake behaving animals (Ding et al. 2013; Paukert et al. 2014; Mu et al. 2019), supporting the notion that Oct/Tyr and NE signaling in astrocytes is an ancient feature of astrocytes.

The above study focuses on whole-cell Ca^{2+} signaling in astrocytes, but, like vertebrate astrocytes (Grosche et al. 1999; Wang et al. 2019a; Chen et al. 2020), *Drosophila* astrocytes also exhibit near-membrane microdomain transients (Fig. 2C; Ma and Freeman 2020). While whole-cell astrocyte Ca^{2+} signals in a semi-dissected preparation of *Drosophila* larvae were activity-dependent (i.e., they were blocked by TTX), microdomain transients were not. Furthermore, microdomain Ca^{2+} transients did not require signaling through Oct-TyrR or Wtrw. Rather, they were regulated by ROS and the TRP channel TrpML. Intriguingly, a subset of microdomain Ca^{2+} transients in astrocytes were in very close proximity to tracheal elements (the breathing apparatus of the larva); they were highly correlated with the initiation of retraction of tracheal filapodia, and loss of TrpML signaling led to increased ROS in the brain and tracheal overgrowth (Ma and Freeman 2020). The current model is that this subset of astrocyte microdomain Ca^{2+} transients occur when trachea deliver too much oxygen (i.e., initiating hyperoxia), thereby generating ROS, which can directly gate TrpML to drive increase microdomain transients and retraction of fine tracheal processes (Fig. 2C).

Additional roles for glial cells in regulating Ca^{2+} signaling have been explored in the fly. Cortex glial cells also exhibit somatic Ca^{2+} waves and smaller transients in their fine processes similar to astrocytes. In a screen for mutations that would

Cite this article as *Cold Spring Harb Perspect Biol* doi: 10.1101/cshperspect.a041347

induce a seizure at high temperatures, mutations in the Na^+/Ca^{2+}, K^+ exchanger *zydeco* were identified (Melom and Littleton 2013). Loss of *zydeco* led to alterations in glial microdomain Ca^{2+} oscillations, and enhanced seizure susceptibility, arguing that NCKX-mediated regulation of Ca^{2+} at the membrane contributes to glial Ca^{2+} signaling and directly modulates neuronal function, which was later linked to reduced levels of a K_{2P} channel and alterations in glial K^+ buffering (Weiss et al. 2019). More recently, an unexpected role for perineurial glia, which are found on the surface of the brain, has been described in regulating normal neuronal excitability in the brain; endoplasmic reticulum (ER) store-operated entry in perineurial glia led to diverse Ca^{2+} waves that varied by brain region, are propagated by gap junctions, and whose disruption led to stimulus-induced seizure-like bouts (Weiss et al. 2022). The visualization of cell signaling molecules is expanding rapidly with new biosensors for other intracellular signals such as cyclic adenosine monophosphate (cAMP), chloride, ROS, and many others, in addition to neurotransmitter biosensors for glutamate, ATP, serotonin, and dopamine (Greenwald et al. 2018), which will pave the way for a more thorough understanding of glial biology and neuron-glia signaling beyond merely calcium imaging.

Glia in the Adult Brain Regulate Complex Behaviors through Diverse Signaling Mechanisms—Circadian Rhythms and Sleep as a Model

State-dependent changes are emerging as a key area in which glial cells can broadly regulate neural activity. This phenomenon has been particularly well studied in the realm of circadian rhythms and sleep (Ingiosi and Frank 2023). The regulation of sleep by different glial subtypes seems complex even in *Drosophila*, both in the subtypes of glial cells that participate in sleep modulation, and the mechanisms they employ to exert their effects. Suh and Jackson (2007) made the landmark discovery that the β-alanyl transferase Ebony was expressed in CNS glia and was required for normal circadian rhythmicity. This was the first demonstration of a clear role for

glial-derived molecules in regulating complex behavior in the adult fly. Subsequent work showed that astrocyte-specific manipulation of vesicle dynamics or of Ca^{2+} signaling in the intact adult brain could reversibly alter circadian motor output (Ng et al. 2011).

More recently, several types of *Drosophila* glia have been implicated in sleep through modulation of neurotransmitter metabolism. Ensheathing glial cells promote wakefulness by limiting the length of sleep bouts through EAAT2, which, in addition to glutamate, can transport taurine. The latter promotes sleep in the fly, and the effect of EAAT2 on sleep may result from transport of taurine into cells, thereby increasing overall metabolic rate (Stahl et al. 2018). GABA is the main inhibitory neurotransmitter that promotes sleep in flies and mammals, and its levels are regulated by astrocyte-specific function of the sole GABA transporter GAT. Hypomorphic mutations in *gat* increase sleep by controlling sleep amount and consolidation; in wild-type conditions, GAT regulates GABA tone on wake-promoting I-LNv neurons (Chaturvedi et al. 2022). Finally, the arylalkylamine *N*-acetyltransferase 1 (AANAT1) enzyme, which acetylates and inactivates monoamines, is used by astroctyes and a small subset of neurons in *Drosophila*. Loss of AANAT1 from neurons had no effect on sleep, but selective elimination from astrocytes led to increased daytime sleep recovery after deprivation (Davla et al. 2020). Thus, an emerging mechanism used by astrocytes and astrocyte-like cells to regulate sleep is through modulation of neurotransmitters and, in turn, neuronal signaling.

In addition to regulation of neurotransmitter metabolism, recent data indicate that glia can secrete several cues to directly impact the function of sleep-regulating neurons. The secreted Ig-domain protein NKT is required in both astrocytes and neurons, where it selectively modulates night sleep (but not day sleep) and seems to be involved in communication between cells that regulate sleep (Sengupta et al. 2019). Astrocytes have also been shown to release DmMANF (*Drosophila* mesencephalic astrocyte-derived neurotrophic factor) that regulates sleep, along with clock neuron structural plasticity (Walkowicz et al. 2017, 2021). Likewise, the *Drosophila* TNF-α homolog Eiger is expressed in astrocytes and neurons, but loss in astrocytes alone was sufficient to reduce sleep duration. Whole animal mutants for *eiger* also had a reduced homeostatic response to sleep deprivation, which was mediated by the TNF-α receptor Wengen in astrocytes. *Eiger* is strongly expressed in cortex glia (Shklover et al. 2015) and therefore cortex glia could also play a potential role in Eiger-induced sleep regulation.

Apart from regulation of sleep need, glia can also regulate circadian patterns of sleep. Multiple types of glial cells in the central brain and visual system express clock components, and their disruption alters sleep in *Drosophila* and mice (Damulewicz et al. 2022). The steroid hormone nuclear receptors EcR and Eip75B are expressed in cortex glia and regulate daily timing and amount of sleep, potentially through systemic steroid hormone signaling from the body following lipid metabolism (Li et al. 2023b). The structural plasticity of circadian clock neuron neurite endings, which expand and contract over the course of the day (read more above), can be modulated by rhythmic sphingolipid catabolism by glial-expressed glucocerebrosidase (Vaughen et al. 2022). Part of this regulation could be through sleep-dependent endocytosis of lipids through the BBB, providing an additional layer of regulation of brain lipid usage (Li et al. 2023a).

One major question is how glia track sleep-need over the course of a day. Using genetically encoded sensors for Ca^{2+} signaling over time (CaMPARI2), Ca^{2+} signaling in astrocytes correlated with sleep need, which was mediated by the L-type Ca^{2+} channel subunit (Ca-α1D) in astrocytes. Knockdown of Ca-α1D led to decreases in astrocyte Ca^{2+} signaling over time, and disrupted sleep homeostasis. Activation of astrocyte Ca^{2+} signaling also changed sleep homeostasis, and the authors proposed a model where by astrocyte-derived Spätzle (similar to mammalian Il-1) conveys sleep need to key sleep regulatory neurons via Toll receptor signaling (Blum et al. 2021). Recent mammalian work also suggests that astrocytic Ca^{2+} encodes sleep need (Ingiosi et al. 2020); thus, *Drosophila* represents an excellent screening platform for identifying further signaling cascades required for sleep health in vertebrate systems.

In summary, an ever-expanding body of work points to a central role for glia in regulating circadian behavior and sleep, much of which appears to be conserved in mammals (Tso et al. 2017; Jackson et al. 2020). Deeper exploration of how glia regulate sleep in the fly, a system that pioneered our understanding of the mechanistic basis of circadian rhythms and sleep, should prove a fruitful approach to defining how glial cells regulate complex physiological events like sleep-related behaviors. An interesting line of future investigation lies in exploring how disease-related genes like the amyloid precursor protein App (Appl in *Drosophila*), which is localized to fly astrocytes and cortex glia, regulates sleep, as depletion or overexpression of Appl results in reciprocal increases or decrease in sleep, respectively (Farca Luna et al. 2017).

Beyond sleep, a recent report elegantly demonstrated that astrocytes modulate another important brain state change in adult *Drosophila*: thirst-driven behaviors (Park et al. 2022). Specifically, dehydration resulted in significant transcriptional changes in astrocytes above and beyond other cell types in the brain. The most significantly altered glial genes related to synthesis of D-serine, a well-known and evolutionarily conserved gliotransmitter (Araque et al. 2014). The authors found that astrocyte-specific depletion of *astray* (*aay*), which encodes a protein that synthesizes D-serine, suppressed water consumption. By contrast, increasing astrocyte Ca^{2+} via astrocyte-specific expression and activation of TrpA1 enhanced water consumption, regardless of need. D-serine is a co-agonist for NMDA-type glutamate receptors; thus, osmotic stress has been suggested to result in astrocyte release of D-serine to activate NMDA receptors on neurons that enhance drinking behavior. These data suggest that further exploration may unveil novel roles for astrocytes in a wide variety of brain state changes.

Glial Regulation of Memory

As discussed above, a primary function of glia in the nervous system is the metabolic support of their neighboring neurons, and this is necessary to build and maintain neuronal connections. While structural plasticity allows for broad remodeling of neural circuits, it is mostly limited to circuit development. In the mature brain, learning and memory is believed to be dependent on proper functional remodeling of synapses, perhaps as the mechanism underlying synaptic plasticity (Magee and Grienberger 2020). Recent reports found that cortex glia metabolically couple to neurons to facilitate long-term memory formation in the adult *Drosophila* brain (de Tredern et al. 2021; Silva et al. 2022). Under homeostatic conditions, glucose fuels the brain. A study by de Tredern et al. (2021) showed that during olfactory learning, cholinergic Kenyon cells of the *Drosophila* MB directly stimulated neighboring cortex glia. Following activation, cortex glia initiated an autocrine insulin signaling pathway to boost internal stores of glucose. The authors propose this glucose was then directly transferred from cortex glia to neighboring Kenyon cells via Glut1 transporters, which in turn facilitated long-term memory formation.

Under nutritional stress, brains can adapt their metabolism from glucose to other energy sources. During this stress, the metabolic basis of memory formation must therefore adapt as well. Using the olfactory learning center as a model, a recent report demonstrated that during starvation, Kenyon cells switched from glucose-based memory formation to the utilization of ketone bodies. Surprisingly, just as cortex glia are local sources of glucose during homeostasis, stressed cortex glia synthesized ketone bodies from internal stores of lipid droplets. The authors posit that starvation-induced ketone bodies were then transferred to Kenyon cells from cortex glia via the monocarboxylate transporters Sln and Chk, respectively, as cell-specific knockdown of these transporters suppressed associative memory formation. Furthermore, cortex glia-specific knockdown of *chk* resulted in an accumulation of internal lipid droplets following starvation, presumably due to a failure to oxidize fatty acids into ketone bodies for intercellular transport (Silva et al. 2022). Thus, metabolic support by cortex glia is required for

memory formation in homeostatic and conditions of environmental stress. We note that in both cases, memory formation was measured via behavioral performance. Whether cortex glia shape neural circuit structure and/or function to affect behavioral plasticity remains to be tested. Future efforts to link structural and functional plasticity in single circuits will be important to understand the full contribution of glia to animal behavior.

CONCLUDING REMARKS

In the last decade, it has become increasingly clear that *Drosophila* CNS glia are functionally quite similar to mammalian glia, from neurotransmitter clearance, regulation of synapse formation and plasticity, to the molecular pathways they use to engulf neuronal debris. A continued rigorous analysis of glial biology in *Drosophila*, along with direct comparisons to vertebrate glia, should highlight the similarities and the differences in their cellular and molecular biology and allow us to prioritize the study of ancient and conserved neuron–glia signaling pathways and cell–cell interactions. In many ways, the collection of *Drosophila* CNS glial cells discussed here—astrocytes, ensheathing glia, and cortex glia—might be thought of as having subdivided the functional roles of mammalian astrocytes (protoplasmic and fibrous), and they also share similar functions with mammalian microglia. Protoplasmic astrocytes associate closely with neuronal cell bodies and neural circuits, and fibrous astrocytes primarily make contact with axons at nodes of Ranvier in white matter axon tracts. Both astrocytes and microglia function to remove neuronal debris and refine synapses throughout development, after injury, and in aging and neurodegeneration. *Drosophila* cortex glia appear to provide support for neuronal cell bodies and proximal neurites in the cortex, removing their neuronal cell bodies similarly to microglia in the mammalian system (Damisah et al. 2020). Alternatively, fly astrocytes are confined to the synaptic neuropil and therefore associate closely with and regulate the vast majority of axons, dendrites, and synapses. How we should think of the ensheathing glial subtype in the context of mammalian glial subtypes is not entirely clear, although their phagocytic capabilities are reminiscent of both astrocytes and microglia. Ensheathing glia have a more flattened morphology and separate the cell cortex from the neuropil forming a diffusion barrier between the two (Pogodalla et al. 2021). Ensheathing glia may be astrocyte-like based on their roles in synaptic plasticity, their use of EAATs (Rival et al. 2004), and the fact that they engulf neurite debris through Draper/MEGF10 (MacDonald et al. 2006), similarly to mammalian astrocytes (Damisah et al. 2020). However, further studies will be needed to clarify these functional relationships. Nevertheless, we now have many examples of crucial glial signaling pathways that are well conserved in *Drosophila* and mammals (e.g., NE/Ca^{2+} signaling, Draper/MEGF10), and *Drosophila* and vertebrate glia share conserved roles in essential behaviors, including sleep. Will *Drosophila* glia prove to be exactly the same as mammalian glia? Probably not, but they do not have to be exactly the same for studies of fly glia to continue to be extremely useful to the glial field. *Drosophila* are well-positioned to catapult our mechanistic understanding of many glial functions rapidly forward, if we embrace the powerful molecular-genetic approaches available in flies to answer fundamental questions about conserved neuron–glia signaling events in the nervous system in health and disease.

ACKNOWLEDGMENTS

Our sincere apologies to colleagues in the field whose work we were not able to mention because of space limitations. Our thanks to the entire Ackerman, Coutinho-Budd, and Freeman laboratories for excellent discussions and Theresa Provitola for help with Figure 1. Work in our laboratories is supported by NIH grant NS121137 (to S.A.), NIH grants NS121101 and NS120689 (to J.C.-B.), and NIH grants NS053538, NS112215, and MH133004 (to M.R.F.). S.A. is also supported by Klingenstein Philanthropies in collaboration with the Simons Foundation as a 2023 Klingenstein-Simons Fellow.

REFERENCES

*Reference is also in this subject collection.

Ackerman SD, Perez-Catalan NA, Freeman MR, Doe CQ. 2021. Astrocytes close a motor circuit critical period. *Nature* **592:** 414–420. doi:10.1038/s41586-021-03441-2

Allen NJ, Eroglu C. 2017. Cell biology of astrocyte-synapse interactions. *Neuron* **96:** 697–708. doi:10.1016/j.neuron.2017.09.056

Allen NJ, Bennett ML, Foo LC, Wang GX, Chakraborty C, Smith SJ, Barres BA. 2012. Astrocyte glypicans 4 and 6 promote formation of excitatory synapses via GluA1 AMPA receptors. *Nature* **486:** 410–414. doi:10.1038/nature11059

Allen AM, Neville MC, Birtles S, Croset V, Treiber CD, Waddell S, Goodwin SF. 2020. A single-cell transcriptomic atlas of the adult *Drosophila* ventral nerve cord. *eLife* **9:** e54074. doi:10.7554/eLife.54074

Araque A, Carmignoto G, Haydon PG, Oliet SHR, Robitaille R, Volterra A. 2014. Gliotransmitters travel in time and space. *Neuron* **81:** 728–739. doi:10.1016/j.neuron.2014.02.007

Auguste YSS, Ferro A, Kahng JA, Xavier AM, Dixon JR, Vrudhula U, Nichitiu AS, Rosado D, Wee TL, Pedmale UV, et al. 2022. Oligodendrocyte precursor cells engulf synapses during circuit remodeling in mice. *Nat Neurosci* **25:** 1273–1278. doi:10.1038/s41593-022-01170-x

Auld VJ, Fetter RD, Broadie K, Goodman CS. 1995. Gliotactin, a novel transmembrane protein on peripheral glia, is required to form the blood–nerve barrier in *Drosophila*. *Cell* **81:** 757–767. doi:10.1016/0092-8674(95)90537-5

Avet-Rochex A, Kaul AK, Gatt AP, McNeill H, Bateman JM. 2012. Concerted control of gliogenesis by InR/TOR and FGF signalling in the *Drosophila* post-embryonic brain. *Development* **139:** 2763–2772. doi:10.1242/dev.074179

Awasaki T, Ito K. 2004. Engulfing action of glial cells is required for programmed axon pruning during *Drosophila* metamorphosis. *Curr Biol* **14:** 668–677. doi:10.1016/j.cub.2004.04.001

Awasaki T, Tatsumi R, Takahashi K, Arai K, Nakanishi Y, Ueda R, Ito K. 2006. Essential role of the apoptotic cell engulfment genes Draper and ced-6 in programmed axon pruning during *Drosophila* metamorphosis. *Neuron* **50:** 855–867. doi:10.1016/j.neuron.2006.04.027

Awasaki T, Lai SL, Ito K, Lee T. 2008. Organization and post-embryonic development of glial cells in the adult central brain of *Drosophila*. *J Neurosci* **28:** 13742–13753. doi:10.1523/JNEUROSCI.4844-08.2008

Awasaki T, Huang Y, O'Connor MB, Lee T. 2011. Glia instruct developmental neuronal remodeling through TGF-β signaling. *Nat Neurosci* **14:** 821–823. doi:10.1038/nn.2833

Ayoub M, David LM, Shklyar B, Hakim-Mishnaevski K, Kurant E. 2023. *Drosophila* FGFR/Htl signaling shapes embryonic glia to phagocytose apoptotic neurons. *Cell Death Discov* **9:** 90. doi:10.1038/s41420-023-01382-5

Baalman K, Marin MA, Ho TSY, Godoy M, Cherian L, Robertson C, Rasband MN. 2015. Axon initial segment-associated microglia. *J Neurosci* **35:** 2283–2292. doi:10.1523/JNEUROSCI.3751-14.2015

Bajar BT, Phi NT, Randhawa H, Akin O. 2022. Developmental neural activity requires neuron–astrocyte interactions. *Dev Neurobiol* **82:** 235–244. doi:10.1002/dneu.22870

Baldwin KT, Tan CX, Strader ST, Jiang C, Savage JT, Elorza-Vidal X, Contreras X, Rülicke T, Hippenmeyer S, Estévez R, et al. 2021. HepaCAM controls astrocyte self-organization and coupling. *Neuron* **109:** 2427–2442.e10. doi:10.1016/j .neuron.2021.05.025

Barresi MJF, Hutson LD, Chien CB, Karlstrom RO. 2005. Hedgehog regulated Slit expression determines commissure and glial cell position in the zebrafish forebrain. *Development* **132:** 3643–3656. doi:10.1242/dev.01929

Baumgartner S, Littleton JT, Broadie K, Bhat MA, Harbecke R, Lengyel JA, Chiquet-Ehrismann R, Prokop A, Bellen HJ. 1996. A *Drosophila* neurexin is required for septate junction and blood–nerve barrier formation and function. *Cell* **87:** 1059–1068. doi:10.1016/s0092-8674(00)81800-0

Bazargani N, Attwell D. 2016. Astrocyte calcium signaling: the third wave. *Nat Neurosci* **19:** 182–189. doi:10.1038/nn .4201

Beckervordersandforth RM, Rickert C, Altenhein B, Technau GM. 2008. Subtypes of glial cells in the *Drosophila* embryonic ventral nerve cord as related to lineage and gene expression. *Mech Dev* **125:** 542–557. doi:10.1016/j.mod .2007.12.004

Blum ID, Keleş MF, Baz ES, Han E, Park K, Luu S, Issa H, Brown M, Ho MCW, Tabuchi M, et al. 2021. Astroglial calcium signaling encodes sleep need in *Drosophila*. *Curr Biol* **31:** 150–162.e7. doi:10.1016/j.cub.2020.10.012

Boulanger A, Thinat C, Züchner S, Fradkin LG, Lortat-Jacob H, Dura JM. 2021. Axonal chemokine-like Orion induces astrocyte infiltration and engulfment during mushroom body neuronal remodeling. *Nat Commun* **12:** 1849. doi:10.1038/s41467-021-22054-x

Brosius Lutz A, Chung WS, Sloan SA, Carson GA, Zhou L, Lovelett E, Posada S, Zuchero JB, Barres BA. 2017. Schwann cells use TAM receptor-mediated phagocytosis in addition to autophagy to clear myelin in a mouse model of nerve injury. *Proc Natl Acad Sci* **114:** E8072–E8080. doi:10.1073/pnas.1710566114

Brunet Avalos C, Maier GL, Bruggmann R, Sprecher SG. 2019. Single cell transcriptome atlas of the *Drosophila* larval brain. *eLife* **8:** e50354. doi:10.7554/eLife.50354

Buchanan J, Elabbady L, Collman F, Jorstad NL, Bakken TE, Ott C, Glatzer J, Bleckert AA, Bodor AL, Brittain D, et al. 2022. Oligodendrocyte precursor cells ingest axons in the mouse neocortex. *Proc Natl Acad Sci* **119:** e2202580119. doi:10.1073/pnas.2202580119

Bushong EA, Martone ME, Ellisman MH. 2004. Maturation of astrocyte morphology and the establishment of astrocyte domains during postnatal hippocampal development. *Int J Dev Neurosci* **22:** 73–86. doi:10.1016/j.ijdevneu.2003 .12.008

Cahoy JD, Emery B, Kaushal A, Foo LC, Zamanian JL, Christopherson KS, Xing Y, Lubischer JL, Krieg PA, Krupenko SA, et al. 2008. A transcriptome database for astrocytes, neurons, and oligodendrocytes: a new resource for understanding brain development and function. *J Neurosci* **28:** 264–278. doi:10.1523/JNEUROSCI.4178-07.2008

Carlson SD, Juang JL, Hilgers SL, Garment MB. 2000. Blood barriers of the insect. *Annu Rev Entomol* **45:** 151–174. doi:10.1146/annurev.ento.45.1.151

Charlton-Perkins M, Almeida AD, MacDonald RB, Harris WA. 2019. Genetic control of cellular morphogenesis in Müller glia. *Glia* **67:** 1401–1411. doi:10.1002/glia.23615

Chaturvedi R, Stork T, Yuan C, Freeman MR, Emery P. 2022. Astrocytic GABA transporter controls sleep by modulating GABAergic signaling in *Drosophila* circadian neurons. *Curr Biol* **32:** 1895–1908.e5. doi:10.1016/j.cub.2022.02 .066

Chen D, Gu T, Pham TN, Zachary MJ, Hewes RS. 2017. Regulatory mechanisms of metamorphic neuronal remodeling revealed through a genome-wide modifier screen in *Drosophila melanogaster*. *Genetics* **206:** 1429–1443. doi:10.1534/genetics.117.200378

Chen J, Poskanzer KE, Freeman MR, Monk KR. 2020. Live-imaging of astrocyte morphogenesis and function in zebrafish neural circuits. *Nat Neurosci* **23:** 1297–1306. doi:10 .1038/s41593-020-0703-x

Chen J, Stork T, Kang Y, Nardone KAM, Auer F, Farrell RJ, Jay TR, Heo D, Sheehan A, Paton C, et al. 2024. Astrocyte growth is driven by the Tre1/S1pr1 phospholipid-binding G protein-coupled receptor. *Neuron* **112:** 93–112.e10. doi:10.1016/j.neuron.2023.11.008

Cheng FY, Fleming JT, Chiang C. 2018. Bergmann glial sonic hedgehog signaling activity is required for proper cerebellar cortical expansion and architecture. *Dev Biol* **440:** 152–166. doi:10.1016/j.ydbio.2018.05.015

Chitnis AB, Nogare DD, Matsuda M. 2012. Building the posterior lateral line system in zebrafish. *Dev Neurobiol* **72:** 234–255. doi:10.1002/dneu.20962

Cho S, Muthukumar AK, Stork T, Coutinho-Budd JC, Freeman MR. 2018. Focal adhesion molecules regulate astrocyte morphology and glutamate transporters to suppress seizure-like behavior. *Proc Natl Acad Sci* **115:** 11316–11321. doi:10.1073/pnas.1800830115

Christopherson KS, Ullian EM, Stokes CCA, Mullowney CE, Hell JW, Agah A, Lawler J, Mosher DF, Bornstein P, Barres BA. 2005. Thrombospondins are astrocyte-secreted proteins that promote CNS synaptogenesis. *Cell* **120:** 421–433. doi:10.1016/j.cell.2004.12.020

Chung WS, Clarke LE, Wang GX, Stafford BK, Sher A, Chakraborty C, Joung J, Foo LC, Thompson A, Chen C, et al. 2013. Astrocytes mediate synapse elimination through MEGF10 and MERTK pathways. *Nature* **504:** 394–400. doi:10.1038/nature12776

Coutinho-Budd JC, Sheehan AE, Freeman MR. 2017. The secreted neurotrophin Spätzle 3 promotes glial morphogenesis and supports neuronal survival and function. *Genes Dev* **31:** 2023–2038. doi:10.1101/gad.305888.117

Crews ST. 2010. Axon–glial interactions at the *Drosophila* CNS midline. *Cell Adh Migr* **4:** 67–71. doi:10.4161/cam.4 .1.10208

Cserép C, Pósfai B, Lénárt N, Fekete R, László ZI, Lele Z, Orsolits B, Molnár G, Heindl S, Schwarcz AD, et al. 2020. Microglia monitor and protect neuronal function through specialized somatic purinergic junctions. *Science* **367:** 528–537. doi:10.1126/science.aax6752

Cui Y, Yang Y, Ni Z, Dong Y, Cai G, Foncelle A, Ma S, Sang K, Tang S, Li Y, et al. 2018. Astroglial Kir4.1 in the lateral habenula drives neuronal bursts in depression. *Nature* **554:** 323–327. doi:10.1038/nature25752

Cunningham CL, Martínez-Cerdeño V, Noctor SC. 2013. Microglia regulate the number of neural precursor cells in the developing cerebral cortex. *J Neurosci* **33**: 4216–4233. doi:10.1523/JNEUROSCI.3441-12.2013

Cunningham ME, Meehan GR, Robinson S, Yao D, McGonigal R, Willison HJ. 2020. Perisynaptic Schwann cells phagocytose nerve terminal debris in a mouse model of Guillain–Barré syndrome. *J Peripher Nerv Syst* **25**: 143–151. doi:10.1111/jns.12373

Damisah EC, Hill RA, Rai A, Chen F, Rothlin CV, Ghosh S, Grutzendler J. 2020. Astrocytes and microglia play orchestrated roles and respect phagocytic territories during neuronal corpse removal in vivo. *Sci Adv* **6**: eaba3239. doi:10.1126/sciadv.aba3239

Damulewicz M, Doktór B, Baster Z, Pyza E. 2022. The role of glia clocks in the regulation of sleep in *Drosophila melanogaster*. *J Neurosci* **42**: 6848–6860. doi:10.1523/JNEUROSCI.2340-21.2022

Danjo R, Kawasaki F, Ordway RW. 2011. A tripartite synapse model in *Drosophila*. *PLoS ONE* **6**: e17131. doi:10.1371/journal.pone.0017131

Davla S, Artiushin G, Li Y, Chitsaz D, Li S, Sehgal A, van Meyel DJ. 2020. AANAT1 functions in astrocytes to regulate sleep homeostasis. *eLife* **9**: e53994. doi:10.7554/eLife.53994

de Tredern E, Rabah Y, Pasquer L, Minatchy J, Plaçais PY, Preat T. 2021. Glial glucose fuels the neuronal pentose phosphate pathway for long-term memory. *Cell Rep* **36**: 109620. doi:10.1016/j.celrep.2021.109620

Ding F, O'Donnell J, Thrane AS, Zeppenfeld D, Kang H, Xie L, Wang F, Nedergaard M. 2013. α1-Adrenergic receptors mediate coordinated Ca^{2+} signaling of cortical astrocytes in awake, behaving mice. *Cell Calcium* **54**: 387–394. doi:10.1016/j.ceca.2013.09.001

Doe CQ. 2017. Temporal patterning in the *Drosophila* CNS. *Annu Rev Cell Dev Biol* **33**: 219–240. doi:10.1146/annurev-cellbio-111315-125210

Doherty J, Logan MA, Taşdemir OE, Freeman MR. 2009. Ensheathing glia function as phagocytes in the adult *Drosophila* brain. *J Neurosci* **29**: 4768–4781. doi:10.1523/JNEUROSCI.5951-08.2009

Doherty J, Sheehan AE, Bradshaw R, Fox AN, Lu TY, Freeman MR. 2014. PI3K signaling and Stat92E converge to modulate glial responsiveness to axonal injury. *PLoS Biol* **12**: e1001985. doi:10.1371/journal.pbio.1001985

Dumstrei K, Wang F, Hartenstein V. 2003. Role of DE-cadherin in neuroblast proliferation, neural morphogenesis, and axon tract formation in *Drosophila* larval brain development. *J Neurosci* **23**: 3325–3335. doi:10.1523/JNEUROSCI.23-08-03325.2003

Etchegaray JI, Elguero EJ, Tran JA, Sinatra V, Feany MB, McCall K. 2016. Defective phagocytic corpse processing results in neurodegeneration and can be rescued by TORC1 activation. *J Neurosci* **36**: 3170–3183. doi:10.1523/JNEUROSCI.1912-15.2016

Farca Luna AJ, Perier M, Seugnet L. 2017. Amyloid precursor protein in *Drosophila* glia regulates sleep and genes involved in glutamate recycling. *J Neurosci* **37**: 4289–4300. doi:10.1523/JNEUROSCI.2826-16.2017

* Fernandes VM, Auld V, Klämbt C. 2024. Glia as functional barriers and signaling intermediaries. *Cold Spring Harb Perspect Biol* **16**: a041423. doi:10.1101/cshperspect.a041423

Fernández MP, Berni J, Ceriani MF. 2008. Circadian remodeling of neuronal circuits involved in rhythmic behavior. *PLoS Biol* **6**: e69. doi:10.1371/journal.pbio.0060069

Ferris HA, Perry RJ, Moreira GV, Shulman GI, Horton JD, Kahn CR. 2017. Loss of astrocyte cholesterol synthesis disrupts neuronal function and alters whole-body metabolism. *Proc Natl Acad Sci* **114**: 1189–1194. doi:10.1073/pnas.1620506114

Freeman MR. 2015. *Drosophila* central nervous system glia. *Cold Spring Harb Perspect Biol* **7**: a020552. doi:10.1101/cshperspect.a020552

Freeman MR, Delrow J, Kim J, Johnson E, Doe CQ. 2003. Unwrapping glial biology: Gcm target genes regulating glial development, diversification, and function. *Neuron* **38**: 567–580. doi:10.1016/s0896-6273(03)00289-7

Fuentes-Medel Y, Logan MA, Ashley J, Ataman B, Budnik V, Freeman MR. 2009. Glia and muscle sculpt neuromuscular arbors by engulfing destabilized synaptic boutons and shed presynaptic debris. *PLoS Biol* **7**: e1000184. doi:10.1371/journal.pbio.1000184

Fuentes-Medel Y, Ashley J, Barria R, Maloney R, Freeman M, Budnik V. 2012. Integration of a retrograde signal during synapse formation by glia-secreted TGF-β ligand. *Curr Biol* **22**: 1831–1838. doi:10.1016/j.cub.2012.07.063

Goritz C, Mauch DH, Pfrieger FW. 2005. Multiple mechanisms mediate cholesterol-induced synaptogenesis in a CNS neuron. *Mol Cell Neurosci* **29**: 190–201. doi:10.1016/j.mcn.2005.02.006

Grant KA, Raible DW, Piotrowski T. 2005. Regulation of latent sensory hair cell precursors by glia in the zebrafish lateral line. *Neuron* **45**: 69–80. doi:10.1016/j.neuron.2004.12.020

Greenwald EC, Mehta S, Zhang J. 2018. Genetically encoded fluorescent biosensors illuminate the spatiotemporal regulation of signaling networks. *Chem Rev* **118**: 11707–11794. doi:10.1021/acs.chemrev.8b00333

Griffiths RL, Hidalgo A. 2004. Prospero maintains the mitotic potential of glial precursors enabling them to respond to neurons. *EMBO J* **23**: 2440–2450. doi:10.1038/sj.emboj.7600258

Grosche J, Matyash V, Möller T, Verkhratsky A, Reichenbach A, Kettenmann H. 1999. Microdomains for neuron–glia interaction: parallel fiber signaling to Bergmann glial cells. *Nat Neurosci* **2**: 139–143. doi:10.1038/5692

Gunner G, Cheadle L, Johnson KM, Ayata P, Badimon A, Mondo E, Nagy MA, Liu L, Bemiller SM, Kim KW, et al. 2019. Sensory lesioning induces microglial synapse elimination via ADAM10 and fractalkine signaling. *Nat Neurosci* **22**: 1075–1088. doi:10.1038/s41593-019-0419-y

Hakim Y, Yaniv SP, Schuldiner O. 2014. Astrocytes play a key role in *Drosophila* mushroom body axon pruning. *PLoS ONE* **9**: e86178. doi:10.1371/journal.pone.0086178

Hakim-Mishnaevski K, Flint-Brodsly N, Shklyar B, Levy-Adam F, Kurant E. 2019. Glial phagocytic receptors promote neuronal loss in adult *Drosophila* brain. *Cell Rep* **29**: 1438–1448.e3. doi:10.1016/j.celrep.2019.09.086

Hanani M. 2005. Satellite glial cells in sensory ganglia: from form to function. *Brain Res Rev* **48**: 457–476. doi:10.1016/j.brainresrev.2004.09.001

Hartenstein V. 2011. Morphological diversity and development of glia in *Drosophila. Glia* 59: 1237–1252. doi:10.1002/glia.21162

Hayashi Y, Majewska AK. 2005. Dendritic spine geometry: functional implication and regulation. *Neuron* 46: 529–532. doi:10.1016/j.neuron.2005.05.006

Hensch TK. 2004. Critical period regulation. *Annu Rev Neurosci* 27: 549–579. doi:10.1146/annurev.neuro.27.070203.144327

Herrero A, Duhart JM, Ceriani MF. 2017. Neuronal and glial clocks underlying structural remodeling of pacemaker neurons in *Drosophila. Front Physiol* 8: 918. doi:10.3389/fphys.2017.00918

Herrmann KA, Liu Y, Llobet-Rosell A, McLaughlin CN, Neukomm LJ, Coutinho-Budd JC, Broihier HT. 2022. Divergent signaling requirements of dSARM in injury-induced degeneration and developmental glial phagocytosis. *PLoS Genet* 18: e1010257. doi:10.1371/journal.pgen.1010257

Hidalgo A, Urban J, Brand AH. 1995. Targeted ablation of glia disrupts axon tract formation in the *Drosophila* CNS. *Development* 121: 3703–3712. doi:10.1242/dev.121.11.3703

Hilu-Dadia R, Hakim-Mishnaevski K, Levy-Adam F, Kurant E. 2018. Draper-mediated JNK signaling is required for glial phagocytosis of apoptotic neurons during *Drosophila* metamorphosis. *Glia* 66: 1520–1532. doi:10.1002/glia.23322

Hong S, Beja-Glasser VF, Nfonoyim BM, Frouin A, Li S, Ramakrishnan S, Merry KM, Shi Q, Rosenthal A, Barres BA, et al. 2016. Complement and microglia mediate early synapse loss in Alzheimer mouse models. *Science* 352: 712–716. doi:10.1126/science.aad8373

Hoopfer ED, McLaughlin T, Watts RJ, Schuldiner O, O'Leary DDM, Luo L. 2006. Wlds protection distinguishes axon degeneration following injury from naturally occurring developmental pruning. *Neuron* 50: 883–895. doi:10.1016/j.neuron.2006.05.013

Ingiosi AM, Frank MG. 2023. Goodnight, astrocyte: waking up to astroglial mechanisms in sleep. *FEBS J* 290: 2553–2564. doi:10.1111/febs.16424

Ingiosi AM, Hayworth CR, Harvey DO, Singletary KG, Rempe MJ, Wisor JP, Frank MG. 2020. A role for astroglial calcium in mammalian sleep and sleep regulation. *Curr Biol* 30: 4373–4383.e7. doi:10.1016/j.cub.2020.08.052

Iram T, Ramirez-Ortiz Z, Byrne MH, Coleman UA, Kingery ND, Means TK, Frenkel D, El Khoury J. 2016. Megf10 is a receptor for C1Q that mediates clearance of apoptotic cells by astrocytes. *J Neurosci* 36: 5185–5192. doi:10.1523/JNEUROSCI.3850-15.2016

Jackson FR, You S, Crowe LB. 2020. Regulation of rhythmic behaviors by astrocytes. *Wiley Interdiscip Rev Dev Biol* 9: e372. doi:10.1002/wdev.372

Jacobs JR. 2000. The midline glia of *Drosophila*: a molecular genetic model for the developmental functions of glia. *Prog Neurobiol* 62: 475–508. doi:10.1016/s0301-0082(00)00016-2

Jang MW, Lim J, Park MG, Lee JH, Lee CJ. 2022. Active role of glia-like supporting cells in the organ of Corti: membrane proteins and their roles in hearing. *Glia* 70: 1799–1825. doi:10.1002/glia.24229

Jindal DA, Leier HC, Salazar G, Foden AJ, Seitz EA, Wilkov AJ, Coutinho-Budd JC, Broihier HT. 2023. Early Draper-mediated glial refinement of neuropil architecture and synapse number in the *Drosophila* antennal lobe. *Front Cell Neurosci* 17: 1166199. doi:10.3389/fncel.2023.1166199

Kazama H, Yaksi E, Wilson RI. 2011. Cell death triggers olfactory circuit plasticity via glial signaling in *Drosophila. J Neurosci* 31: 7619–7630. doi:10.1523/JNEUROSCI.5984-10.2011

Kerr KS, Fuentes-Medel Y, Brewer C, Barria R, Ashley J, Abruzzi KC, Sheehan A, Tasdemir-Yilmaz OE, Freeman MR, Budnik V. 2014. Glial wingless/Wnt regulates glutamate receptor clustering and synaptic physiology at the *Drosophila* neuromuscular junction. *J Neurosci* 34: 2910–2920. doi:10.1523/JNEUROSCI.3714-13.2014

Konishi H, Okamoto T, Hara Y, Komine O, Tamada H, Maeda M, Osako F, Kobayashi M, Nishiyama A, Kataoka Y, et al. 2020. Astrocytic phagocytosis is a compensatory mechanism for microglial dysfunction. *EMBO J* 39: e104464. doi:10.15252/embj.2020104464

Kremer MC, Jung C, Batelli S, Rubin GM, Gaul U. 2017. The glia of the adult *Drosophila* nervous system. *Glia* 65: 606–638. doi:10.1002/glia.23115

Kucukdereli H, Allen NJ, Lee AT, Feng A, Ozlu MI, Conatser LM, Chakraborty C, Workman G, Weaver M, Sage EH, et al. 2011. Control of excitatory CNS synaptogenesis by astrocyte-secreted proteins Hevin and SPARC. *Proc Natl Acad Sci* 108: E440–E449. doi:10.1073/pnas.1104977108

Kunduri G, Turner-Evans D, Konya Y, Izumi Y, Nagashima K, Lockett S, Holthuis J, Bamba T, Acharya U, Acharya JK. 2018. Defective cortex glia plasma membrane structure underlies light-induced epilepsy in *cpes* mutants. *Proc Natl Acad Sci* 115: E8919–E8928. doi:10.1073/pnas.1808463115

Lacin H, Chen HM, Long X, Singer RH, Lee T, Truman JW. 2019. Neurotransmitter identity is acquired in a lineage-restricted manner in the *Drosophila* CNS. *eLife* 8: e43701. doi:10.7554/eLife.43701

Lago-Baldaia I, Fernandes VM, Ackerman SD. 2020. More than mortar: glia as architects of nervous system development and disease. *Front Cell Dev Biol* 8: 611269. doi:10.3389/fcell.2020.611269

Lago-Baldaia I, Cooper M, Seroka A, Trivedi C, Powell GT, Wilson SW, Ackerman SD, Fernandes VM. 2023. A *Drosophila* glial cell atlas reveals a mismatch between transcriptional and morphological diversity. *PLoS Biol* 21: e3002328. doi:10.1371/journal.pbio.3002328

Lane-Donovan C, Herz J. 2017. The ApoE receptors Vldlr and Apoer2 in central nervous system function and disease. *J Lipid Res* 58: 1036–1043. doi:10.1194/jlr.R075507

Lear BC, Zhang L, Allada R. 2009. The neuropeptide PDF acts directly on evening pacemaker neurons to regulate multiple features of circadian behavior. *PLoS Biol* 7: e1000154. doi:10.1371/journal.pbio.1000154

Learte AR, Hidalgo A. 2007. The role of glial cells in axon guidance, fasciculation and targeting. *Adv Exp Med Biol* 621: 156–166. doi:10.1007/978-0-387-76715-4_12

Lee T, Lee A, Luo L. 1999. Development of the *Drosophila* mushroom bodies: sequential generation of three distinct types of neurons from a neuroblast. *Development* 126: 4065–4076. doi:10.1242/dev.126.18.4065

Lefebvre JL. 2021. Molecular mechanisms that mediate dendrite morphogenesis. *Curr Top Dev Biol* **142**: 233–282. doi:10.1016/bs.ctdb.2020.12.008

Leiserson WM, Harkins EW, Keshishian H. 2000. Fray, a *Drosophila* serine/threonine kinase homologous to mammalian PASK, is required for axonal ensheathment. *Neuron* **28**: 793–806. doi:10.1016/s0896-6273(00)00154-9

Lemke G. 2001. Glial control of neuronal development. *Annu Rev Neurosci* **24**: 87–105. doi:10.1146/annurev.neuro.24.1.87

Li F, Artiushin G, Sehgal A. 2023a. Modulation of sleep by trafficking of lipids through the *Drosophila* blood–brain barrier. *eLife* **12**: e86336. doi:10.7554/eLife.86336

Li Y, Haynes P, Zhang SL, Yue Z, Sehgal A. 2023b. Ecdysone acts through cortex glia to regulate sleep in *Drosophila*. *eLife* **12**: e81723. doi:10.7554/eLife.81723

Lin L, Rodrigues FSLM, Kary C, Contet A, Logan M, Baxter RHG, Wood W, Baehrecke EH. 2017. Complement-related regulates autophagy in neighboring cells. *Cell* **170**: 158–171.e8. doi:10.1016/j.cell.2017.06.018

Liu CC, Liu CC, Kanekiyo T, Xu H, Bu G. 2013. Apolipoprotein E and Alzheimer disease: risk, mechanisms and therapy. *Nat Rev Neurol* **9**: 106–118. doi:10.1038/nrneurol.2012.263

Liu H, Zhou B, Yan W, Lei Z, Zhao X, Zhang K, Guo A. 2014. Astrocyte-like glial cells physiologically regulate olfactory processing through the modification of ORN-PN synaptic strength in *Drosophila*. *Eur J Neurosci* **40**: 2744–2754. doi:10.1111/ejn.12646

Liu L, MacKenzie KR, Putluri N, Maletić-Savatić M, Bellen HJ. 2017. The glia–neuron lactate shuttle and elevated ROS promote lipid synthesis in neurons and lipid droplet accumulation in glia via APOE/D. *Cell Metab* **26**: 719–737.e6. doi:10.1016/j.cmet.2017.08.024

Lööv C, Hillered L, Ebendal T, Erlandsson A. 2012. Engulfing astrocytes protect neurons from contact-induced apoptosis following injury. *PLoS ONE* **7**: e33090. doi:10.1371/journal.pone.0033090

Lu TY, Doherty J, Freeman MR. 2014. DRK/DOS/SOS converge with Crk/Mbc/dCed-12 to activate Rac1 during glial engulfment of axonal debris. *Proc Natl Acad Sci* **111**: 12544–12549. doi:10.1073/pnas.1403450111

Lu TY, MacDonald JM, Neukomm LJ, Sheehan AE, Bradshaw R, Logan MA, Freeman MR. 2017. Axon degeneration induces glial responses through Draper-TRAF4-JNK signalling. *Nat Commun* **8**: 14355. doi:10.1038/ncomms14355

Ma Z, Freeman MR. 2020. TrpML-mediated astrocyte microdomain Ca^{2+} transients regulate astrocyte–tracheal interactions. *eLife* **9**: e58952. doi:10.7554/eLife.58952

Ma Z, Stork T, Bergles DE, Freeman MR. 2016. Neuromodulators signal through astrocytes to alter neural circuit activity and behaviour. *Nature* **539**: 428–432. doi:10.1038/nature20145

MacDonald JM, Beach MG, Porpiglia E, Sheehan AE, Watts RJ, Freeman MR. 2006. The *Drosophila* cell corpse engulfment receptor Draper mediates glial clearance of severed axons. *Neuron* **50**: 869–881. doi:10.1016/j.neuron.2006.04.028

Macdonald JM, Doherty J, Hackett R, Freeman MR. 2013. The c-Jun kinase signaling cascade promotes glial engulf-ment activity through activation of Draper and phagocytic function. *Cell Death Differ* **20**: 1140–1148. doi:10.1038/cdd.2013.30

Magee JC, Grienberger C. 2020. Synaptic plasticity forms and functions. *Annu Rev Neurosci* **43**: 95–117. doi:10.1146/annurev-neuro-090919-022842

Mahan TE, Wang C, Bao X, Choudhury A, Ulrich JD, Holtzman DM. 2022. Selective reduction of astrocyte apoE3 and apoE4 strongly reduces Aβ accumulation and plaque-related pathology in a mouse model of amyloidosis. *Mol Neurodegener* **17**: 13. doi:10.1186/s13024-022-00516-0

Mauch DH, Nägler K, Schumacher S, Göritz C, Müller EC, Otto A, Pfrieger FW. 2001. CNS synaptogenesis promoted by glia-derived cholesterol. *Science* **294**: 1354–1357. doi:10.1126/science.294.5545.1354

Mauss A, Tripodi M, Evers JF, Landgraf M. 2009. Midline signalling systems direct the formation of a neural map by dendritic targeting in the *Drosophila* motor system. *PLoS Biol* **7**: e1000200. doi:10.1371/journal.pbio.1000200

McLaughlin CN, Perry-Richardson JJ, Coutinho-Budd JC, Broihier HT. 2019. Dying neurons utilize innate immune signaling to prime glia for phagocytosis during development. *Dev Cell* **48**: 506–522.e6. doi:10.1016/j.devcel.2018.12.019

Melom JE, Littleton JT. 2013. Mutation of a NCKX eliminates glial microdomain calcium oscillations and enhances seizure susceptibility. *J Neurosci* **33**: 1169–1178. doi:10.1523/JNEUROSCI.3920-12.2013

Mokalled MH, Patra C, Dickson AL, Endo T, Stainier DYR, Poss KD. 2016. Injury-induced *ctgfa* directs glial bridging and spinal cord regeneration in zebrafish. *Science* **354**: 630–634. doi:10.1126/science.aaf2679

Mu Y, Bennett DV, Rubinov M, Narayan S, Yang CT, Tanimoto M, Mensh BD, Looger LL, Ahrens MB. 2019. Glia accumulate evidence that actions are futile and suppress unsuccessful behavior. *Cell* **178**: 27–43.e19. doi:10.1016/j.cell.2019.05.050

Musashe DT, Purice MD, Speese SD, Doherty J, Logan MA. 2016. Insulin-like signaling promotes glial phagocytic clearance of degenerating axons through regulation of Draper. *Cell Rep* **16**: 1838–1850. doi:10.1016/j.celrep.2016.07.022

Muthukumar AK, Stork T, Freeman MR. 2014. Activity-dependent regulation of astrocyte GAT levels during synaptogenesis. *Nat Neurosci* **17**: 1340–1350. doi:10.1038/nn.3791

Nakano R, Iwamura M, Obikawa A, Togane Y, Hara Y, Fukuhara T, Tomaru M, Takano-Shimizu T, Tsujimura H. 2019. Cortex glia clear dead young neurons via Drpr/dCed-6/Shark and Crk/Mbc/dCed-12 signaling pathways in the developing *Drosophila* optic lobe. *Dev Biol* **453**: 68–85. doi:10.1016/j.ydbio.2019.05.003

Ng FS, Tangredi MM, Jackson FR. 2011. Glial cells physiologically modulate clock neurons and circadian behavior in a calcium-dependent manner. *Curr Biol* **21**: 625–634. doi:10.1016/j.cub.2011.03.027

Nguyen KB, Pender MP. 1998. Phagocytosis of apoptotic lymphocytes by oligodendrocytes in experimental autoimmune encephalomyelitis. *Acta Neuropathol* **95**: 40–46. doi:10.1007/s004010050763

Osterloh JM, Yang J, Rooney TM, Fox AN, Adalbert R, Powell EH, Sheehan AE, Avery MA, Hackett R, Logan MA, et al.

2012. DSarm/Sarm1 is required for activation of an injury-induced axon death pathway. *Science* **337:** 481–484. doi:10.1126/science.1223899

Oswald MC, Brooks PS, Zwart MF, Mukherjee A, West RJ, Giachello CN, Morarach K, Baines RA, Sweeney ST, Landgraf M. 2018. Reactive oxygen species regulate activity-dependent neuronal plasticity in *Drosophila. eLife* **7:** e39393. doi:10.7554/eLife.39393

Park A, Croset V, Otto N, Agarwal D, Treiber CD, Meschi E, Sims D, Waddell S. 2022. Gliotransmission of D-serine promotes thirst-directed behaviors in *Drosophila. Curr Biol* **32:** 3952–3970.e8. doi:10.1016/j.cub.2022.07.038

Paukert M, Agarwal A, Cha J, Doze VA, Kang JU, Bergles DE. 2014. Norepinephrine controls astroglial responsiveness to local circuit activity. *Neuron* **82:** 1263–1270. doi:10.1016/j.neuron.2014.04.038

Pearce MMP, Spartz EJ, Hong W, Luo L, Kopito RR. 2015. Prion-like transmission of neuronal huntingtin aggregates to phagocytic glia in the *Drosophila* brain. *Nat Commun* **6:** 6768. doi:10.1038/ncomms7768

Peco E, Davla S, Camp D, Stacey SM, Landgraf M, van Meyel DJ. 2016. *Drosophila* astrocytes cover specific territories of the CNS neuropil and are instructed to differentiate by Prospero, a key effector of Notch. *Development* **143:** 1170–1181. doi:10.1242/dev.133165

Peng JJ, Lin SH, Liu YT, Lin HC, Li TN, Yao CK. 2019. A circuit-dependent ROS feedback loop mediates glutamate excitotoxicity to sculpt the *Drosophila* motor system. *eLife* **8:** e47372. doi:10.7554/eLife.47372

Pereanu W, Spindler S, Cruz L, Hartenstein V. 2007. Tracheal development in the *Drosophila* brain is constrained by glial cells. *Dev Biol* **302:** 169–180. doi:10.1016/j.ydbio.2006.09.022

Perez-Catalan NA, Doe CQ, Ackerman SD. 2021. The role of astrocyte-mediated plasticity in neural circuit development and function. *Neural Dev* **16:** 1. doi:10.1186/s13064-020-00151-9

Petsakou A, Sapsis TP, Blau J. 2015. Circadian rhythms in Rho1 activity regulate neuronal plasticity and network hierarchy. *Cell* **162:** 823–835. doi:10.1016/j.cell.2015.07.010

Pogodalla N, Kranenburg H, Rey S, Rodrigues S, Cardona A, Klämbt C. 2021. *Drosophila* ß_{Heavy}-Spectrin is required in polarized ensheathing glia that form a diffusion-barrier around the neuropil. *Nat Commun* **12:** 6357. doi:10.1038/s41467-021-26462-x

Pradhan AK, Shi Q, Tartler KJ, Rammes G. 2022. Quantification of astrocytic synaptic pruning in mouse hippocampal slices in response to ex vivo Aβ treatment via colocalization analysis with C1q. *STAR Protoc* **3:** 101687. doi:10.1016/j.xpro.2022.101687

Pringle NP, Yu WP, Howell M, Colvin JS, Ornitz DM, Richardson WD. 2003. Fgfr3 expression by astrocytes and their precursors: evidence that astrocytes and oligodendrocytes originate in distinct neuroepithelial domains. *Development* **130:** 93–102. doi:10.1242/dev.00184

Purice MD, Speese SD, Logan MA. 2016. Delayed glial clearance of degenerating axons in aged *Drosophila* is due to reduced PI3K/Draper activity. *Nat Commun* **7:** 12871. doi:10.1038/ncomms12871

Ray A, Speese SD, Logan MA. 2017. Glial Draper rescues aβ toxicity in a *Drosophila* model of Alzheimer's disease. *J Neurosci* **37:** 11881–11893. doi:10.1523/JNEUROSCI.0862-17.2017

Read RD. 2018. Pvr receptor tyrosine kinase signaling promotes post-embryonic morphogenesis, and survival of glia and neural progenitor cells in *Drosophila. Development* **145:** dev164285. doi:10.1242/dev.164285

Reitman ME, Tse V, Mi X, Willoughby DD, Peinado A, Aivazidis A, Myagmar BE, Simpson PC, Bayraktar OA, Yu G, et al. 2023. Norepinephrine links astrocytic activity to regulation of cortical state. *Nat Neurosci* **26:** 579–593. doi:10.1038/s41593-023-01284-w

Richier B, Vijandi CM, Mackensen S, Salecker I. 2017. Lapsyn controls branch extension and positioning of astrocyte-like glia in the *Drosophila* optic lobe. *Nat Commun* **8:** 317. doi:10.1038/s41467-017-00384-z

Ries M, Sastre M. 2016. Mechanisms of Aβ clearance and degradation by glial cells. *Front Aging Neurosci* **8:** 160. doi:10.3389/fnagi.2016.00160

Rival T, Soustelle L, Strambi C, Besson MT, Iché M, Birman S. 2004. Decreasing glutamate buffering capacity triggers oxidative stress and neuropil degeneration in the *Drosophila* brain. *Curr Biol* **14:** 599–605. doi:10.1016/j.cub.2004.03.039

Rujano MA, Briand D, Ðelić B, Marc J, Spéder P. 2022. An interplay between cellular growth and atypical fusion defines morphogenesis of a modular glial niche in *Drosophila. Nat Commun* **13:** 4999. doi:10.1038/s41467-022-32685-3

Sakami S, Imanishi Y, Palczewski K. 2019. Müller glia phagocytose dead photoreceptor cells in a mouse model of retinal degenerative disease. *FASEB J* **33:** 3680–3692. doi:10.1096/fj.201801662R

Salazar G, Ross G, Maserejian AE, Coutinho-Budd J. 2022. Quantifying glial–glial tiling using automated image analysis in *Drosophila. Front Cell Neurosci* **16:** 826483. doi:10.3389/fncel.2022.826483

Salm AK. 2000. Mechanisms of glial retraction in the hypothalamo–neurohypophysial system of the rat. *Exp Physiol* **85:** 197s–202s. doi:10.1111/j.1469-445x.2000.tb00024.x

Schafer DP, Stevens B. 2013. Phagocytic glial cells: sculpting synaptic circuits in the developing nervous system. *Curr Opin Neurobiol* **23:** 1034–1040. doi:10.1016/j.conb.2013.09.012

Scheffer LK, Xu CS, Januszewski M, Lu Z, Takemura SY, Hayworth KJ, Huang GB, Shinomiya K, Maitlin-Shepard J, Berg S, et al. 2020. A connectome and analysis of the adult *Drosophila* central brain. *eLife* **9:** e57443. doi:10.7554/eLife.57443

Scheib JL, Sullivan CS, Carter BD. 2012. Jedi-1 and MEGF10 signal engulfment of apoptotic neurons through the tyrosine kinase Syk. *J Neurosci* **32:** 13022–13031. doi:10.1523/JNEUROSCI.6350-11.2012

Schwabe T, Bainton RJ, Fetter RD, Heberlein U, Gaul U. 2005. GPCR signaling is required for blood-brain barrier formation in *Drosophila. Cell* **123:** 133–144. doi:10.1016/j.cell.2005.08.037

Sengupta S, Crowe LB, You S, Roberts MA, Jackson FR. 2019. A secreted Ig-domain protein required in both astrocytes and neurons for regulation of *Drosophila* night sleep. *Curr Biol* **29:** 2547–2554.e2. doi:10.1016/j.cub.2019.06.055

Shaham S. 2006. Glia–neuron interactions in the nervous system of *Caenorhabditis elegans*. *Curr Opin Neurobiol* **16:** 522–528. doi:10.1016/j.conb.2006.08.001

Shishido E, Ono N, Kojima T, Saigo K. 1997. Requirements of DFR1/heartless, a mesoderm-specific *Drosophila* FGF-receptor, for the formation of heart, visceral and somatic muscles, and ensheathing of longitudinal axon tracts in CNS. *Development* **124:** 2119–2128. doi:10.1242/dev.124.11.2119

Shklover J, Levy-Adam F, Kurant E. 2015. The role of *Drosophila* TNF Eiger in developmental and damage-induced neuronal apoptosis. *FEBS Lett* **589:** 871–879. doi:10.1016/j.febslet.2015.02.032

Silva B, Mantha OL, Schor J, Pascual A, Plaçais PY, Pavlowsky A, Preat T. 2022. Glia fuel neurons with locally synthesized ketone bodies to sustain memory under starvation. *Nat Metab* **4:** 213–224. doi:10.1038/s42255-022-00528-6

Song C, Broadie K. 2023. Fragile X mental retardation protein coordinates neuron-to-glia communication for clearance of developmentally transient brain neurons. *Proc Natl Acad Sci* **120:** e2216887120. doi:10.1073/pnas.2216887120

Spindler SR, Ortiz I, Fung S, Takashima S, Hartenstein V. 2009. *Drosophila* cortex and neuropile glia influence secondary axon tract growth, pathfinding, and fasciculation in the developing larval brain. *Dev Biol* **334:** 355–368. doi:10.1016/j.ydbio.2009.07.035

Stacey SM, Muraro NI, Peco E, Labbé A, Thomas GB, Baines RA, van Meyel DJ. 2010. *Drosophila* glial glutamate transporter Eaat1 is regulated by fringe-mediated notch signaling and is essential for larval locomotion. *J Neurosci* **30:** 14446–14457. doi:10.1523/JNEUROSCI.1021-10.2010

Stahl BA, Peco E, Davla S, Murakami K, Caicedo Moreno NA, van Meyel DJ, Keene AC. 2018. The taurine transporter Eaat2 functions in ensheathing glia to modulate sleep and metabolic rate. *Curr Biol* **28:** 3700–3708.e4. doi:10.1016/j.cub.2018.10.039

Stanhope BA, Jaggard JB, Gratton M, Brown EB, Keene AC. 2020. Sleep regulates glial plasticity and expression of the engulfment receptor Draper following neural injury. *Curr Biol* **30:** 1092–1101.e3. doi:10.1016/j.cub.2020.02.057

Stogsdill JA, Ramirez J, Liu D, Kim YH, Baldwin KT, Enustun E, Ejikeme T, Ji RR, Eroglu C. 2017. Astrocytic neuroligins control astrocyte morphogenesis and synaptogenesis. *Nature* **551:** 192–197. doi:10.1038/nature24638

Stork T, Engelen D, Krudewig A, Silies M, Bainton RJ, Klämbt C. 2008. Organization and function of the blood-brain barrier in *Drosophila*. *J Neurosci* **28:** 587–597. doi:10.1523/JNEUROSCI.4367-07.2008

Stork T, Bernardos R, Freeman MR. 2012. Analysis of glial cell development and function in *Drosophila*. *Cold Spring Harb Protoc* **2012:** pdb.top067587. doi:10.1101/pdb.top067587

Stork T, Sheehan A, Tasdemir-Yilmaz OE, Freeman MR. 2014. Neuron–glia interactions through the heartless FGF receptor signaling pathway mediate morphogenesis of *Drosophila* astrocytes. *Neuron* **83:** 388–403. doi:10.1016/j.neuron.2014.06.026

Stowell RD, Wong EL, Batchelor HN, Mendes MS, Lamantia CE, Whitelaw BS, Majewska AK. 2018. Cerebellar microglia are dynamically unique and survey Purkinje neurons in vivo. *Dev Neurobiol* **78:** 627–644. doi:10.1002/dneu.22572

Suh J, Jackson FR. 2007. *Drosophila* ebony activity is required in glia for the circadian regulation of locomotor activity. *Neuron* **55:** 435–447. doi:10.1016/j.neuron.2007.06.038

Takasaki C, Yamasaki M, Uchigashima M, Konno K, Yanagawa Y, Watanabe M. 2010. Cytochemical and cytological properties of perineuronal oligodendrocytes in the mouse cortex. *Eur J Neurosci* **32:** 1326–1336. doi:10.1111/j.1460-9568.2010.07377.x

Takechi H, Hakeda-Suzuki S, Nitta Y, Ishiwata Y, Iwanaga R, Sato M, Sugie A, Suzuki T. 2021. Glial insulin regulates cooperative or antagonistic Golden goal/Flamingo interactions during photoreceptor axon guidance. *eLife* **10:** e66718. doi:10.7554/eLife.66718

Tan CX, Eroglu C. 2021. Cell adhesion molecules regulating astrocyte-neuron interactions. *Curr Opin Neurobiol* **69:** 170–177. doi:10.1016/j.conb.2021.03.015

Tasdemir-Yilmaz OE, Freeman MR. 2014. Astrocytes engage unique molecular programs to engulf pruned neuronal debris from distinct subsets of neurons. *Genes Dev* **28:** 20–33. doi:10.1101/gad.229518.113

Truman JW. 1996. Steroid receptors and nervous system metamorphosis in insects. *Dev Neurosci* **18:** 87–101. doi:10.1159/000111398

Tso CF, Simon T, Greenlaw AC, Puri T, Mieda M, Herzog ED. 2017. Astrocytes regulate daily rhythms in the suprachiasmatic nucleus and behavior. *Curr Biol* **27:** 1055–1061. doi:10.1016/j.cub.2017.02.037

Vaughen JP, Theisen E, Rivas-Serna IM, Berger AB, Kalakuntla P, Anreiter I, Mazurak VC, Rodriguez TP, Mast JD, Hartl T, et al. 2022. Glial control of sphingolipid levels sculpts diurnal remodeling in a circadian circuit. *Neuron* **110:** 3186–3205.e7. doi:10.1016/j.neuron.2022.07.016

Vita DJ, Meier CJ, Broadie K. 2021. Neuronal fragile X mental retardation protein activates glial insulin receptor mediated PDF-Tri neuron developmental clearance. *Nat Commun* **12:** 1160. doi:10.1038/s41467-021-21429-4

Volkenhoff A, Weiler A, Letzel M, Stehling M, Klämbt C, Schirmeier S. 2015. Glial glycolysis is essential for neuronal survival in *Drosophila*. *Cell Metab* **22:** 437–447. doi:10.1016/j.cmet.2015.07.006

Walkowicz L, Kijak E, Krzeptowski W, Górska-Andrzejak J, Stratoulias V, Woznicka O, Chwastek E, Heino TI, Pyza EM. 2017. Downregulation of DmMANF in glial cells results in neurodegeneration and affects sleep and lifespan in *Drosophila melanogaster*. *Front Neurosci* **11:** 610. doi:10.3389/fnins.2017.00610

Walkowicz L, Krzeptowski W, Krzeptowska E, Warzecha K, Sałek J, Górska-Andrzejak J, Pyza E. 2021. Glial expression of DmMANF is required for the regulation of activity, sleep and circadian rhythms in the visual system of *Drosophila melanogaster*. *Eur J Neurosci* **54:** 5785–5797. doi:10.1111/ejn.15171

Wang Y, DelRosso NV, Vaidyanathan TV, Cahill MK, Reitman ME, Pittolo S, Mi X, Yu G, Ponzanxer KE. 2019a. Accurate quantification of astrocyte and neurotransmitter fluorescence dynamics for single-cell and population-level physiology. *Nat Neurosci* **22:** 1936–1944. doi:10.1038/s41593-019-0492-2

Wang Z, Lee G, Vuong R, Park JH. 2019b. Two-factor specification of apoptosis: TGF-β signaling acts cooperatively with ecdysone signaling to induce cell- and stage-specific apoptosis of larval neurons during metamorphosis in *Drosophila melanogaster*. *Apoptosis* **24**: 972–989. doi:10.1007/s10495-019-01574-4

Wang C, Xiong M, Gratuze M, Bao X, Shi Y, Andhey PS, Manis M, Schroeder C, Yin Z, Madore C, et al. 2021. Selective removal of astrocytic APOE4 strongly protects against tau-mediated neurodegeneration and decreases synaptic phagocytosis by microglia. *Neuron* **109**: 1657–1674.e7. doi:10.1016/j.neuron.2021.03.024

Watts RJ, Hoopfer ED, Luo L. 2003. Axon pruning during *Drosophila* metamorphosis: evidence for local degeneration and requirement of the ubiquitin-proteasome system. *Neuron* **38**: 871–885. doi:10.1016/s0896-6273(03)00295-2

Watts RJ, Schuldiner O, Perrino J, Larsen C, Luo L. 2004. Glia engulf degenerating axons during developmental axon pruning. *Curr Biol* **14**: 678–684. doi:10.1016/j.cub.2004.03.035

Weiss S, Melom JE, Ormerod KG, Zhang YV, Littleton JT. 2019. Glial Ca^{2+} signaling links endocytosis to K^+ buffering around neuronal somas to regulate excitability. *eLife* **8**: e44186. doi:10.7554/eLife.44186

Weiss S, Clamon LC, Manoim JE, Ormerod KG, Parnas M, Littleton JT. 2022. Glial ER and GAP junction mediated Ca^{2+} waves are crucial to maintain normal brain excitability. *Glia* **70**: 123–144. doi:10.1002/glia.24092

Williams DW, Truman JW. 2005. Remodeling dendrites during insect metamorphosis. *J Neurobiol* **64**: 24–33. doi:10.1002/neu.20151

Wilson C, González-Billault C. 2015. Regulation of cytoskeletal dynamics by redox signaling and oxidative stress: implications for neuronal development and trafficking. *Front Cell Neurosci* **9**: 381. doi:10.3389/fncel.2015.00381

Winfree LM, Speese SD, Logan MA. 2017. Protein phosphatase 4 coordinates glial membrane recruitment and phagocytic clearance of degenerating axons in *Drosophila*. *Cell Death Dis* **8**: e2623. doi:10.1038/cddis.2017.40

Wu HH, Bellmunt E, Scheib JL, Venegas V, Burkert C, Reichardt LF, Zhou Z, Fariñas I, Carter BD. 2009. Glial precursors clear sensory neuron corpses during development via Jedi-1, an engulfment receptor. *Nat Neurosci* **12**: 1534–1541. doi:10.1038/nn.2446

Wu B, Li J, Chou YH, Luginbuhl D, Luo L. 2017. Fibroblast growth factor signaling instructs ensheathing glia wrapping of *Drosophila* olfactory glomeruli. *Proc Natl Acad Sci* **114**: 7505–7512. doi:10.1073/pnas.1706533114

Xiao Y, Petrucco L, Hoodless LJ, Portugues R, Czopka T. 2022. Oligodendrocyte precursor cells sculpt the visual system by regulating axonal remodeling. *Nat Neurosci* **25**: 280–284. doi:10.1038/s41593-022-01023-7

Yin J, Spillman E, Cheng ES, Short J, Chen Y, Lei J, Gibbs M, Rosenthal JS, Sheng C, Chen YX, et al. 2021. Brain-specific lipoprotein receptors interact with astrocyte derived apolipoprotein and mediate neuron–glia lipid shuttling. *Nat Commun* **12**: 2408. doi:10.1038/s41467-021-22751-7

Yuan X, Sipe CW, Suzawa M, Bland ML, Siegrist SE. 2020. Dilp-2-mediated PI3-kinase activation coordinates reactivation of quiescent neuroblasts with growth of their glial stem cell niche. *PLoS Biol* **18**: e3000721. doi:10.1371/journal.pbio.3000721

Zheng Z, Lauritzen JS, Perlman E, Robinson CG, Nichols M, Milkie D, Torrens O, Price J, Fisher CB, Sharifi N, et al. 2018. A complete electron microscopy volume of the brain of adult *Drosophila melanogaster*. *Cell* **174**: 730–743.e22. doi:10.1016/j.cell.2018.06.019

Zhou T, Li Y, Li X, Zeng F, Rao Y, He Y, Wang Y, Liu M, Li D, Xu Z, et al. 2022. Microglial debris is cleared by astrocytes via C4b-facilitated phagocytosis and degraded via RUBICON-dependent noncanonical autophagy in mice. *Nat Commun* **13**: 6233. doi:10.1038/s41467-022-33932-3

Ziegenfuss JS, Biswas R, Avery MA, Hong K, Sheehan AE, Yeung YG, Stanley ER, Freeman MR. 2008. Draper-dependent glial phagocytic activity is mediated by Src and Syk family kinase signalling. *Nat* **453**: 935–939. doi:10.1038/nature06901

Ziegenfuss JS, Doherty J, Freeman MR. 2012. Distinct molecular pathways mediate glial activation and engulfment of axonal debris after axotomy. *Nat Neurosci* **15**: 979–987. doi:10.1038/nn.3135

Glia as Functional Barriers and Signaling Intermediaries

Vilaiwan M. Fernandes,[1] Vanessa Auld,[2] and Christian Klämbt[3]

[1]Department of Cell and Developmental Biology, University College London, London UC1E 6DE, United Kingdom

[2]Department of Zoology, University of British Columbia, Vancouver, British Columbia V6T 1Z4, Canada

[3]Institute for Neuro- and Behavioral Biology, University of Münster, Münster 48149, Germany

Correspondence: klaembt@uni-muenster.de

Glia play a crucial role in providing metabolic support to neurons across different species. To do so, glial cells isolate distinct neuronal compartments from systemic signals and selectively transport specific metabolites and ions to support neuronal development and facilitate neuronal function. Because of their function as barriers, glial cells occupy privileged positions within the nervous system and have also evolved to serve as signaling intermediaries in various contexts. The fruit fly, *Drosophila melanogaster*, has significantly contributed to our understanding of glial barrier development and function. In this review, we will explore the formation of the glial sheath, blood–brain barrier, and nerve barrier, as well as the significance of glia–extracellular matrix interactions in barrier formation. Additionally, we will delve into the role of glia as signaling intermediaries in regulating nervous system development, function, and response to injury.

A key evolutionarily conserved role of glia is to support neurons metabolically (Pellerin and Magistretti 2012; Volkenhoff et al. 2015). Glia control the flow of metabolites in the nervous system, ensuring that specific metabolites and ions reach neurons at the right time and place. Therefore, to support neurons as effectively as possible, glia first establish barrier properties to strictly separate neuronal compartments from external influences and, in addition, establish selective transport mechanisms for effective glia–neuron metabolic coupling.

Indeed, glia across different model systems and circuit types perform barrier functions, which require them to occupy key positions at biological interfaces across scales, from tissue to cellular to subcellular levels. In *Drosophila*, each glial cell type plays distinct barrier functions. Astrocytes shield synapses, ensheathing glia encase the entire neuropil, cortex glia wrap around neuronal cell bodies, wrapping glia shield peripheral axons and the surface glia or blood–brain barrier forming glia seal the whole nervous system from circulating hemolymph (Figs. 1 and 2; see also Coutinho-Budd et al. 2023). It is therefore unsurprising that in these specialized positions glia function not only as barriers, but also as signaling intermediaries to both regulate and integrate signals. Here we discuss the formation and function of the glial sheath, blood–brain barrier, and nerve

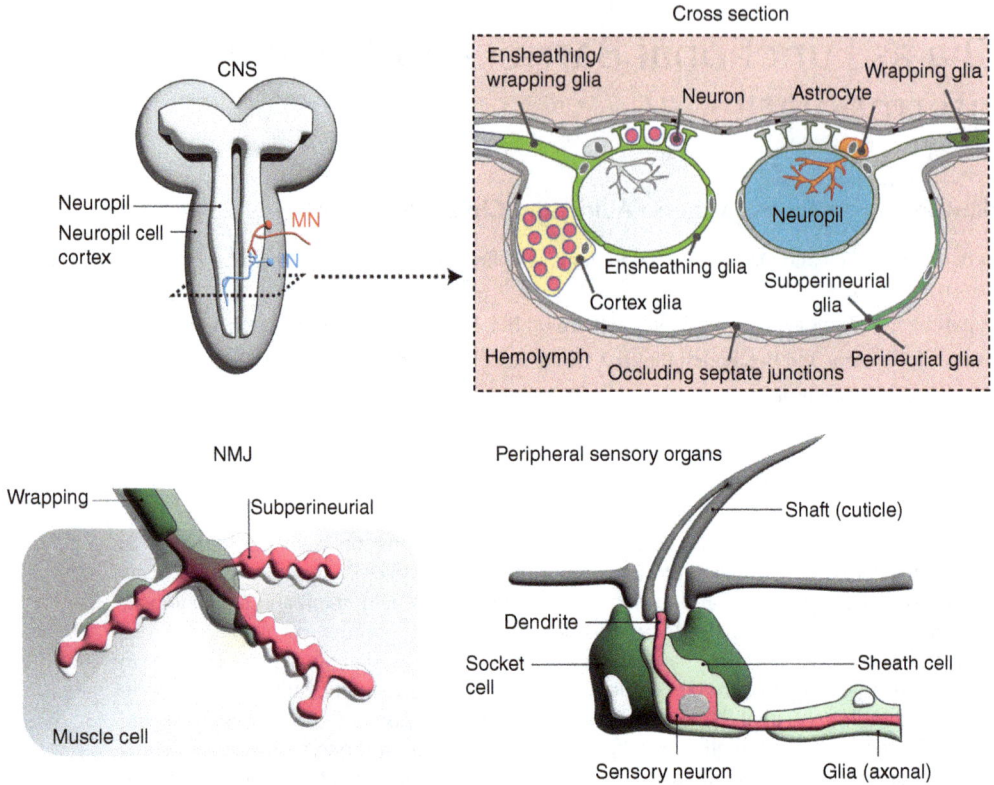

Figure 1. *Drosophila* glial classes.

barrier, as well as the role of glia–extracellular matrix (ECM) interactions in barrier formation. Finally, we discuss how their positions allow glia to act as signaling intermediaries to modulate nervous system development, function, and response to injury.

GLIA AS FUNCTIONAL BARRIERS

Formation of the Glial Sheath, Blood–Brain Barrier, and Nerve Barrier

The blood–brain barrier comprises the most established and best-studied barrier in the nervous system, as it represents the gatekeeper that regulates all entry into and also exit from the nervous system. In *Drosophila*, the blood–brain barrier is made up of two distinct cell types, perineurial glial cells and subperineurial glial cells; together called the surface glia (Figs. 1 and 2; Beckervor-

dersandforth et al. 2008; Stork et al. 2008). Here, it appears that the transport and barrier functions are, in part, divided across the two cell types. Although the subperineurial glial cells are very large and flat polarized cells that establish occluding junctions (Figs. 1 and 2A; Stork et al. 2008), perineurial glia appear responsible for taking up nutrients from the hemolymph and do not form occluding junctions (Volkenhoff et al. 2015).

Both blood–brain barrier–forming glial cells are born during embryonic stages. *Drosophila* neurogenesis is distinct from vertebrate neurogenesis, as individual progenitor cells delaminate into the interior of the animal in five waves beginning about 30 min after gastrulation. In total, 31 such neuroblasts are formed in each thoracic hemineuromer, one of which is a pure glioblast and six are neuroglioblasts (Doe and Technau 1993; Beckervordersandforth et al. 2008). The perineurial glial cells are formed by neuroglio-

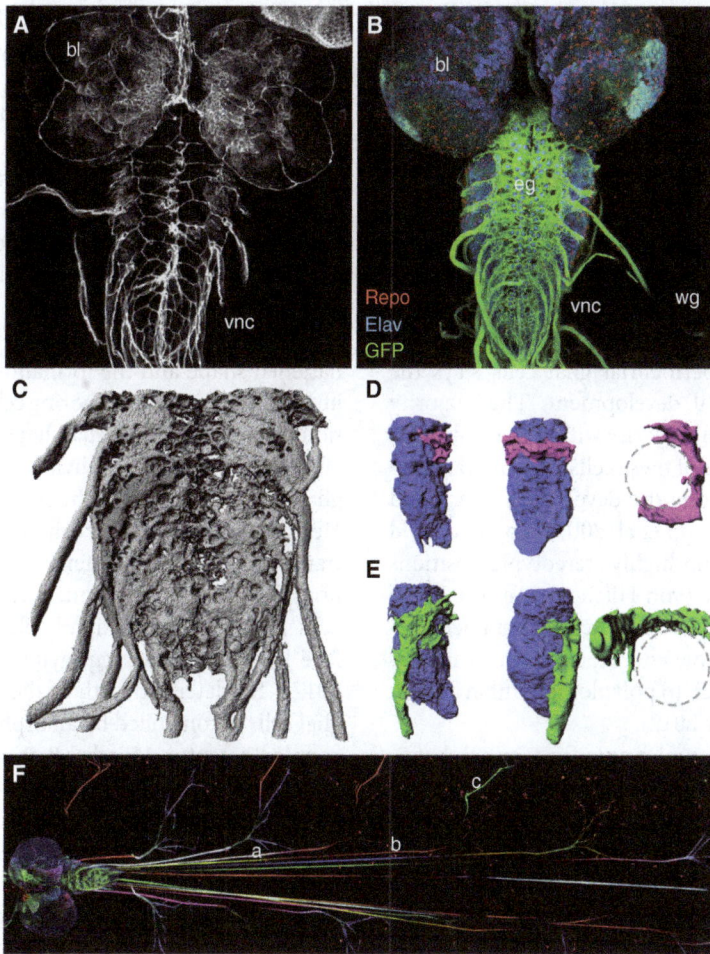

Figure 2. Glial cells as functional barriers. (*A*) The glial isoform of the cell adhesion protein NeurexinIV is specifically expressed by the subperineurial glial cells where it localizes to the pleated septate junctions. The tiling of subperineurial glial cells, which form the blood–brain barrier, becomes visible. (bl) Brain lobe, (vnc) ventral nerve cord. (*B*) Ensheathing glial cells are visualized by membrane-bound green fluorescent protein (GFP) expression directed by *83E12-Gal4*. (wg) Wrapping glia. (*C*) Surface reconstruction of the ensheathing glia (gray) using Imaris. Note the dense coverage of the neuropil. (*D*) Reconstruction of a single ensheathing glial cell (magenta) that covers part of the neuropil (blue). Note, that no projections into the neuropil are formed. (*E*) Reconstruction of a single ensheathing/wrapping glial cell (green) that covers part of the neuropil (blue) and also follows the nerve toward the periphery. No projections into the neuropil are formed. (*F*) Multicolor flipout labeling of the peripheral wrapping glia. In most segments three wrapping glial cells can be identified by a distinct color each (a,b,c). Wrapping glial cells are very large; polyploid cells can reach 2–3 mm in length.

blasts and cover the outermost surface of the nervous system, the peripheral nervous system (PNS), as well as the central nervous system (CNS). All subperineurial glial cells of the CNS also stem from distinct neuroglioblasts, whereas in the PNS some subperineurial glia are also generated by sensory organ precursor cells (Becker-

vordersandforth et al. 2008; von Hilchen et al. 2008).

The organization of the different surface glial cell types is best analyzed for the peripheral nerves in larvae (Figs. 1 and 2F). Along each of the abdominal nerves (except A8), three perineurial and three to four subperineurial glial cells are formed. Once specified, perineurial and subperineurial glial cells follow distinct modes of development. The perineurial glial cells divide and expand their cell numbers enormously. Depending on the length of the peripheral nerve, 20 to 100 perineurial glial cells are formed (Matzat et al. 2015). Therefore, it appears possible that perineurial glial cell division may be regulated in a cell–cell contact–dependent manner to generate the appropriate number of cells covering the nervous system. In contrast, the number of subperineurial glial cells stays the same until pupal development. The majority of subperineurial glia arise within the embryonic CNS and some of these cells then migrate into the periphery along the developing motor and sensory axons (Sepp et al. 2000; Sepp and Auld 2003) and take up highly stereotypic positions along each nerve (von Hilchen et al. 2008). To cover the entire nervous system, the subperineurial glia must grow enormously, which is supported by a switch to polyploidy (Unhavaithaya and Orr-Weaver 2012).

In a typical embryonic abdominal hemineuromere, eight subperineurial glial cells, and three to four perineurial glial cells can be identified (Beckervordersandforth et al. 2008; von Hilchen et al. 2008; Schwabe et al. 2017). The number of perineurial glial cells has not yet been determined for the brain lobes. The relatively few embryonic perineurial glial cells (~150 cells) divide to generate more than 2000 cells, which evenly cover the surface of the entire adult CNS (Awasaki et al. 2008; Kremer et al. 2017). The mitotic potential of the perineurial glial cells can be increased by the activation of receptor tyrosine kinase signaling pathways (Franzdóttir et al. 2009; Avet-Rochex et al. 2012).

The subperineurial glia covering the embryonic brain have not been counted but it is assumed to be 16 subperineurial glial cells per neuromere (Beckervordersandforth et al. 2008); 288 cells are expected by the end of embryogenesis (nine abdominal, three thoracic, and six head segments). This number matches the approximately 300 subperineurial glial cells that were reported to cover the adult nervous system (Kremer et al. 2017) suggesting that subperineurial glial cells do not divide, but instead grow in size and stay intact throughout life. In fact, lineage analysis failed to obtain any indication of cell division (Awasaki et al. 2008), and the use of photoconvertible dendra demonstrates that subperineurial glia stay intact until midpupal stages (Winkler et al. 2021). Almost 10 times as many perineurial cells as subperineurial glial cells cover the nervous system; thus, subperineurial cells are very large to match the surface area of the perineurial layer.

A key feature of subperineurial cells is their flattened shape and the formation of occluding junctions between neighboring cells. The formation of the characteristic flat shape starts at ~10 h after egg laying, after individual subperineurial glial cells have moved to the surface of the brain. Here, following a mesenchymal to epithelial transition, they grow extensively in a synchronous and isometric manner with neighboring cells, allowing them to tightly tile the brains surface by ~13 h after egg laying (Kremer et al. 2017). The lateral growth of the subperineurial glial cells is controlled by an orphan G-protein-coupled receptor Moody (Bainton et al. 2005; Schwabe et al. 2005, 2017). Moody acts via $G\alpha_i$, the RGS-protein Loco, and the cAMP effector PKA (Granderath et al. 1999; Schwabe et al. 2005; Li et al. 2021). Moody localizes to the domain of the subperineurial glia facing the cortex glia, suggesting that the signal-activating Moody might originate from the cortex glia or the CNS neurons. In absence of *moody* signaling, lateral expansion of the subperineurial glial cells is reduced during embryonic stages; however, a functional blood–brain barrier is established that allows survival of *moody* mutants until adulthood.

When subperineurial glial cells establish contacts with their neighbors, different junctional contacts are established. On the one hand, subperineurial glia form gap junctions with each other, which leads to metabolic coupling in the population and on the other hand, they also form occluding junctions, which separate the hemolymph from all neural cells. Gap junctional coupling is not needed for barrier formation, but, interestingly, is required for formation of neurons during larval stages and thus the growth of the brain lobes (discussed more below).

Cite this article as *Cold Spring Harb Perspect Biol* doi: 10.1101/cshperspect.a041423

Subperineurial glial cells form occluding pleated septate junctions (pSJs). These are generated by a bewilderingly large number of membrane and membrane-associated proteins that require endocytic recycling to properly form (Tepass and Hartenstein 1994; Baumgartner et al. 1996; Tiklová et al. 2010; Petri et al. 2019; Böhme et al. 2021). pSJs form belts surrounding the entire cell, which fence paracellular diffusion. To generate a fully tight barrier, these belts need to be formed in continuous bands that connect different cell vertices (Babatz et al. 2018). The process of septate strand generation requires vesicular traffic as septate junction components are preassembled before being integrated into the pSJ strands and the actin cytoskeleton (Hatan et al. 2011; Babatz et al. 2018; Li et al. 2021). It is possible that septate junction strand formation is initiated at a tricellular junction and spreads to link each cell vertex. In addition to its role in lateral subperineurial growth, Moody signaling is needed to ensure the formation of intact pSJ strands connecting the different cell vertices. Surprisingly, growth of pSJ strands is not needed as subperineurial glial cells increase their diameter during development, because strand formation during embryonic stages already matches the size requirements of later larval and adult stages. In *moody* mutants, septate junction formation initially proceeds normally; however, breaks can be seen in individual pSJ strands, which renders the blood–brain barrier slightly leaky (Babatz et al. 2018). Interestingly, subperineurial glial cells are able to sense the level of occluding junctions that are established and initiate compensatory growth and the formation of specific cellular interdigitations that increase the length of the diffusion path and thus counteract the reduced tightness of the barrier caused by misarranged septate junctions in *moody* mutants (Babatz et al. 2018). In conclusion, the blood–brain barrier adapts well to the growing brain tissue by regulation of cell division in the outer perineurial glial layer and by expansion of cell size by the subperineurial glia.

As aforementioned, subperineurial glia coordinate their own cell size with the perineurial glia cell number through polyploidization (Unhavaithaya and Orr-Weaver 2012). Polyploidization is common in excessively large cells (e.g., muscle cells). The level of polyploidy is controlled by the amino-terminal asparagine amidohydrolase homolog Öbek that controls the N-end rule protein degradation pathway (Zulbahar et al. 2018). In blood–brain barrier glia, Öbek counteracts fibroblast growth factor (FGF) and Hippo signaling to differentially affect cell growth and cell number. The Hippo pathway is a central regulator of cell growth and proliferation. In subperineurial glial cells, a double-negative feedback loop comprising the microRNA miR-285, the Hippo signaling effector Yorkie (vertebrate Yap), and the multiple ankyrin repeats single KH domain (Mask) regulates not only subperineurial glial growth but also pleated septate junction strand integrity (Li et al. 2017).

Ploidy can result from endocycling and from endomitosis—the former increases ploidy in mononucleated cells and the latter increases ploidy via multinucleation. The switch between endocycling to endomitosis is, in part, Notch-dependent, with Notch inhibiting this transition. Interestingly, subperineurial glial cells that cover the brain lobes undergo endomitosis (multinucleation) during larval development to accommodate brain growth (Von Stetina et al. 2018); whereas subperineurial glia in the ventral nerve cord (VNC) are polyploid through endocycling reduced Notch activity can also trigger endomitosis in subperineurial glia covering the VNC (Von Stetina et al. 2018). However, how Notch is differentially regulated in the VNC versus the brain lobe subperineurial glia and how this signaling contributes to later barrier functions is currently unknown.

To summarize, the blood–brain and nerve barriers are generated by the close apposition of two distinct glial cell types—perineurial and subperineurial. Although perineurial glia loosely tile with one another and have a direct interface with the nutrient-rich hemolymph, subperineurial glia play more classic "barrier" functions by tightly tiling the surface of the brain and sealing it from the external environment. Together, these cells perform the function of perineurial glia in mammals. In both cases, loss of barrier glia can have profound consequences for brain health and function, which will be discussed in more detail below.

Glial–ECM Interactions Drive External and Internal Glial Barrier Formation

A key glial interaction that mediates nervous system integrity is the interaction between glia and the ECM. The ECM surrounding the nervous system, or neural lamella, is composed of a range of ECM components including laminin, perlecan, collagen, nidogen, and SPARC (Hynes and Zhao 2000). These ECM components are secreted by the fat body, hemocytes, and the glia themselves (Broadie et al. 2011; Petley-Ragan et al. 2016). The surface glia interact with the ECM around the VNC during embryogenesis (subperineurial glia) and larval stages (perineurial glia), and these interactions are essential for nervous system morphology, function, and integrity. During embryogenesis, the CNS and, in particular, the VNC, drastically condenses in size. Although both glia and neurons contribute to the active contraction of the VNC, glia have a major contribution to this critical process (Karkali et al. 2022), which relies on the ECM surrounding the VNC and depends on the glia–ECM interface. The ECM at this stage is deposited by circulating hemocytes, a key function of hemocytes. Thus, hemocyte function is required for proper CNS morphogenesis, including condensation of the VNC (Sears et al. 2003; Olofsson and Page 2005; Martinek et al. 2008; Defaye et al. 2009; Evans et al. 2010). A key element to the organization and integrity of the ECM around the VNC is laminin. Loss of integrin, which is a major membrane-bound receptor of laminins, results in abnormal nervous system condensation (Brown 1994) and loss of integrin in glia causes the elongation of the VNC and deformation of the brain lobes (Xie and Auld 2011; Meyer et al. 2014). In the peripheral nerves, loss of integrin in the perineurial glia results in incomplete ensheathment of the nerve and individual perineurial glia appear to become detached from one another. Similarly, loss of laminins themselves disrupts collagen and perlecan accumulation around the VNC and blocks VNC condensation (Urbano et al. 2009) as does loss of collagen (Martinek et al. 2008). Furthermore, loss of expression of GlcAT-P, glucuronosyltransferase P, which is required for proper ECM formation,

causes an elongated larval VNC phenotype (Pandey et al. 2011). Beyond embryonic development, continued glia–ECM interactions are necessary to maintain the structure of the VNC. For example, directed degradation of the ECM in late larvae stages through expression of matrix metalloproteinases (MMPs) in the perineurial glia also leads to nerve cord extension and lethality (Xie and Auld 2011; Meyer et al. 2014). Modulation of the ECM is also important for nervous system integrity and function. MMP or protease remodeling of the ECM is key to the proper development of the nervous system including regulation of expression of MMPs, proteases such as kuzbanian, and ADAMTS-like family proteases (Meyer et al. 2014; Skeath et al. 2017; Calderon et al. 2022).

Although MMP activity is required for proper brain integrity, its levels must be tightly regulated. When Mmp1 activity is too high, septate junction strands are disrupted around peripheral nerves, compromising neurotransmitter release downstream at the neuromuscular junction (NMJ). Interestingly, matrix metalloproteinase 1 (*Mmp1*) expression is also regulated by Delta-dependent Notch activation, which inhibits c-Jun amino-terminal kinase (JNK) to reduce *Mmp1* levels. These results provide mechanistic insight into the regulation of neuronal health and function via glial-initiated signaling and open a framework for understanding the complex relationship between ECM regulation and the maintenance of barrier function (Calderon et al. 2022).

Glia–ECM interactions are also important for glial migration. During larval stages in the optic stalk, perineurial glia migrate between the basal ECM and cortex glia along the optic stalk into the eye disc (Silies et al. 2007). Pan-glial knockdown of integrin and talin, components of the focal adhesion complex, impaired perineurial glia migration along the optic stalk (Xie et al. 2014). ECM stiffness can be a modulator of glial migration. In a *Drosophila* glioma model, in which glial migration is increased by overexpression of PDGF receptor (PVR), glial overmigration is suppressed upon knockdown of *Drosophila* Lysyl oxidase (Lox), indicating that glia migrate less when ECM stiffness is reduced

(Kim et al. 2014). Lox oxidizes peptidyl lysine residues on ECM proteins such as collagen, a process critical for covalent cross-linking, and loss of Lox function leads to immature or less stiff ECM (Lucero and Kagan 2006; Kim et al. 2014). Of note, changes to ECM stiffness does not appear to alter normal migration of the perineurial glia during development but rather PVR activation changes integrin–ECM dynamics in this glioma model.

Beyond creating the external barriers that protect, insulate, and support nervous system integrity, glia play critical roles in creating internal barriers. For instance, the cortex–neuropil barrier is formed by ensheathing glia and contributes to glutamate homeostasis in the neuropil (Otto et al. 2018). For macrophage-like functions of the ensheathing glia, see Coutinho-Budd et al. (2023). As with the subperineurial glia, ensheathing glia have a polarized cell morphology but do not form special cell–cell junctions (Fig. 2B–E; Pogodalla et al. 2021). Within the CNS, ensheathing glia are necessary to separate the neuropil from the neuronal cell bodies, generating an internal diffusion barrier, a process that is complete by the larval third instar (Pogodalla et al. 2021). Integrin subunits and ECM components including the heparan sulfate proteoglycan Dally are enriched around the neuropil in larval stages, and in adults, both perlecan and collagens flank ensheathing glia (Pogodalla et al. 2021), suggesting the ensheathment of the neuropil requires glia–ECM interactions. Disruption of this internal glial barrier results in changes to larval locomotion pointing to an important role for internal glial barriers in both the CNS and peripheral nerves in maintaining nervous system function (Pogodalla et al. 2021).

The ensheathing glia comprises two related cell types. Two ensheathing glial cells are found in each abdominal hemineuromer that only encase the neuropil (Fig. 2B–D). In addition, two ensheathing/wrapping glial cells are found in each abdominal hemineuromer (Peco et al. 2016), which, in addition, encase dorsal neurons as well as they ensheath axons of the nerves connecting the neuropil with the periphery (Figs. 1 and 2F).

Ensheathment of axons in the PNS is done by the innermost layer of glia in the peripheral nerve generated by the wrapping glia. Similar to subperineurial glia, wrapping glia are polyploid and do not undergo mitosis; instead, three to four wrapping glia are present along the entire nerve length and send long extensive processes to cover the nerve interior (von Hilchen et al. 2013). In an MCFO-type single-cell labeling experiment, individual cells can be identified by their specific color (Nern et al. 2015; Kottmeier et al. 2020). A single wrapping glia can reach a length of 2 mm, demonstrating the extreme hypertrophic growth of this cell type resembling a nonmyelinating Schwann cell of a Remak fiber (von Hilchen et al. 2013; Matzat et al. 2015). Wrapping glia development occurs relatively late in comparison to the other glial types where differentiation of the wrapping glia begins in the first-instar larval stage and continues throughout the rest of the larval stages such that by the late third-instar stage, most individual axons and groups of axons are in contact with some wrapping glia processes (Matzat et al. 2015). Although loss of wrapping glial ensheathment is not lethal during larval stages, ablation does cause a reduction in axon diameter and nerve conduction velocity, along with larval locomotion defects (Kottmeier et al. 2020). Similarly, loss of nonmyelinating Schwann cells, the equivalent cell type in vertebrate peripheral nerves, impairs sensorimotor behaviors but is similarly viable (Harty and Monk 2017). During the pupal stage, axonal wrapping is reorganized (Subramanian et al. 2017). In adult leg nerves, a large number of wrapping glial cells wrap axons according to their size (Rey et al. 2023b). At the axon initial segment of large motor axons, glial cells form lacunar structures that possibly serve as an ion reservoir and participate in blocking ephaptic coupling between axons (Kottmeier et al. 2020; Rey et al. 2023a,b).

Several factors have been identified as important for wrapping glia differentiation and/or ensheathment. Inhibition of the epidermal growth factor (EGF) and FGF signaling reduces wrapping glia ensheathment (Matzat et al. 2015; Kottmeier et al. 2020). Proper ionic homeostasis, sphingolipid biosynthesis, and intracellular transport are all important for wrapping glia ensheathment, as demonstrated by the loss of the serine/threonine kinase Fray, the serine palmitoyltransferase subunit Lace, and kinesin heavy chain (Khc), respectively (Leiserson et al. 2000, 2011; Schmidt et al. 2012; Ghosh et al. 2013). Wrapping of axons by glia also requires the presence of ECM components and their receptors including laminins and the integrin receptor (Xie and Auld 2011; Petley-Ragan et al. 2016), the Discoidin domain receptor, and the XV/XVIII collagen Multiplexin (Corty et al. 2022).

GLIA AS SIGNALING INTERMEDIARIES

By virtue of their roles as barriers and partitions, glia occupy privileged positions in the nervous system and have been co-opted to also function as signaling intermediaries that gate and integrate signals during nervous system development and function and in response to injury.

Glia as Signaling Intermediaries during Nervous System Development

In *Drosophila*, neural stem cells or neuroblasts in the CNS proliferate during embryogenesis, but most enter a state of reversible cell-cycle arrest termed quiescence toward the end of embryogenesis. Following hatching and feeding during the first larval instar, neuroblasts exit quiescence and reenter the cell cycle in a process termed reactivation (Hartenstein et al. 1987; Truman and Bate 1988; Prokop and Technau 1991; Ito and Hotta 1992). These neuroblasts are isolated from the circulating hemolymph by several glial barriers including the perineurial glia and subperineurial glia, as well as the cortex glia, which envelop neuroblast and neuronal cell bodies within the CNS (Freeman 2015). Necessarily, systemic signals that promote neuroblast reactivation must act through these glia that shield neuroblasts. In *Drosophila*, the fat body, which performs both adipose and liver functions, senses nutrient status and dietary amino acids to stimulate larval growth (including neuroblast reactivation) (Britton and Edgar 1998). Tissue-specific genetic manipulations showed that an amino acid transporter called Slimfast (SLIF)

and target of rapamycin (TOR) signaling function autonomously within the fat body to produce yet-to-be-identified signal(s) required to stimulate larval growth and neuroblast reactivation (Sousa-Nunes et al. 2011). Blocking vesicular trafficking specifically in the fat body was sufficient to inhibit neuroblast reactivation, suggesting that the fat body–derived signal is likely secreted into the hemolymph where it acts systemically. Neuroblast reactivation was also shown to depend on phosphatidylinositol 3-kinase (PI3K) and TOR signaling pathway activity cell-autonomously in neuroblasts downstream of the insulin-like receptor (InR) (Chell and Brand 2010; Sousa-Nunes et al. 2011). Indeed, larvae deficient for various combinations of insulin-like peptides (ILPs) displayed defects in the timing of neuroblast reactivation (Chell and Brand 2010; Sousa-Nunes et al. 2011). Interestingly, ILPs are produced by multiple sources in *Drosophila* including the ILP-producing median neurosecretory cells in the brain as well as subsets of glia (Brogiolo et al. 2001; Ikeya et al. 2002; Rulifson et al. 2002). Although the most prominent source of ILPs is the median neurosecretory cells, which respond to signals from the fat body to secrete ILPs into the hemolymph, overexpression of ILPs in these cells under conditions of nutrient restriction rescued body growth but not neuroblast reactivation (Sousa-Nunes et al. 2011). In contrast, surface and cortex glia–specific overexpression of ILPs under nutrient restriction rescued neuroblast reactivation but not body growth (Chell and Brand 2010; Sousa-Nunes et al. 2011). Together these data support a model in which surface and cortex glia respond to a systemic fat body–derived signal(s) to produce and secrete ILPs, which stimulate neuroblast reactivation (Fig. 3A). Moreover, surface glial cells form an extensive network with each other through gap junctions, and this gap-junction coupling enables coordinated and synchronized calcium oscillations across the surface glial network, which in turn promotes synchronized secretion of ILPs for appropriately timing neuroblast reactivation (Spéder and Brand 2014). A recent characterization of cell-type-specific responses and growth over short intervals found that, initially, cortex glia do not fully envelop qui-

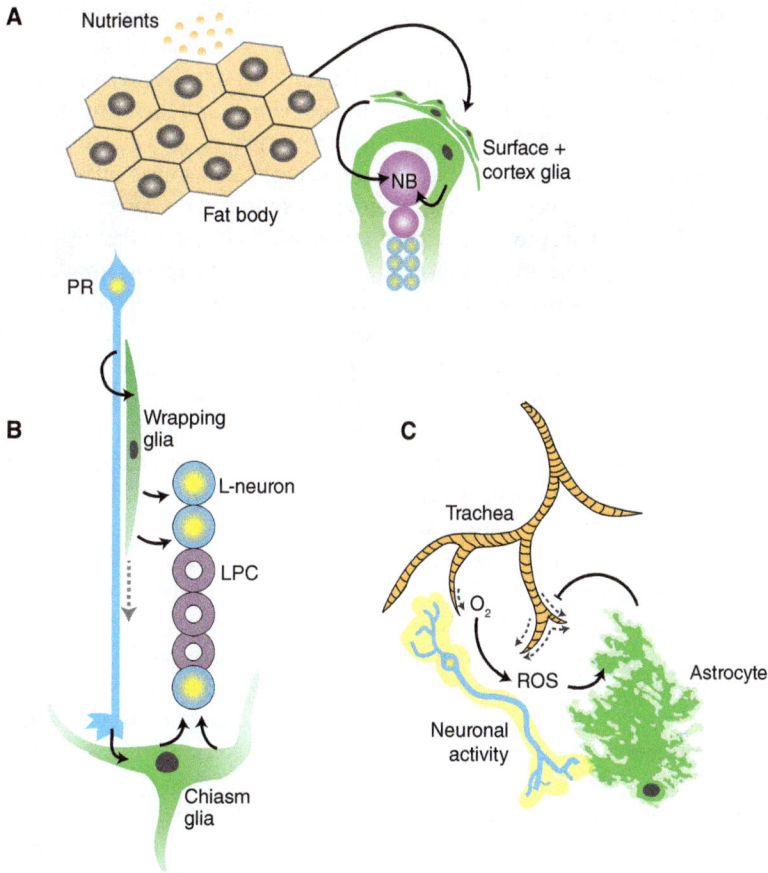

Figure 3. Examples of glia as signaling intermediaries. (*A*) Surface and cortex glia relay feeding-dependent signals from the fat body to reactivate quiescent neuroblasts (NBs) in the first-instar central nervous system (CNS). (*B*) Wrapping glia and outer chiasm giant glia respond to signals from photoreceptors (PRs) and relay these to lamina precursor cells (LPCs) to induce their differentiation into neurons. (*C*) Trachea are gas-filled vasculatures responsible for delivering oxygen to the brain. Astrocytes monitor their local environments and under local hyperoxic conditions signal to trachea to retract their branches and filopodia, which reduces local oxygen delivery.

escent neuroblasts, but instead grow their membranes in a nutrient-dependent manner to ensheath quiescent neuroblasts and promote their reactivation (Yuan et al. 2020). Thus, the full extent of interorgan and intercellular signals in this process is still ambiguous and may involve signaling relays between multiple glial cell types.

Another prime example of glia functioning as signaling intermediaries occurs in the developing visual system where glia maturing to establish barriers have been co-opted to act as signaling intermediaries that coordinate development across neuropils. During visual system develop-

ment, photoreceptor and wrapping glia morphogenesis are intricately coordinated. Photoreceptors from the developing eye disc are born sequentially as a wave of differentiation sweeps across the disc along the anteroposterior axis (Roignant and Treisman 2009). Photoreceptors grow their axons into the optic stalk, which connects the eye disc with the optic lobe, and into the developing lamina and medulla neuropils (Roignant and Treisman 2009). A population of wrapping glial cells ensheath photoreceptor axon bundles from individual ommatidial clusters (unit eyes) progressively from the eye disc through the optic stalk and into the lamina (Rangarajan et al. 1999; Franzdóttir et al. 2009). Wrapping glial morphogenesis (i.e., axonal ensheathment) is driven by photoreceptor-derived Thisbe, an FGF that activates the FGF receptor, Heartless, in wrapping glia (Franzdóttir et al. 2009). Thus, as photoreceptor axons arrive sequentially in the optic lobe, they are progressively ensheathed by wrapping glia from the eye disc with ensheathment delayed relative to axonal arrival.

In addition to promoting wrapping glial morphogenesis, photoreceptors also induce their neuronal target field, the lamina, such that every ommatidium has a corresponding lamina unit (cartridge or column) composed of five neuronal types (Huang and Kunes 1996, 1998; Huang et al. 1998). Photoreceptor-derived Hedgehog directly induces lamina precursors, their terminal divisions, and their assembly into columns (i.e., stacked ensembles of postmitotic precursors) (Huang and Kunes 1996, 1998). In addition, photoreceptors induce neuronal differentiation of postmitotic lamina precursors in columns indirectly through glial signaling intermediaries—the wrapping glia and a second ensheathing-like glial population positioned below the lamina called the outer chiasm giant glia (xg^O) (Fernandes et al. 2017; Rossi and Fernandes 2018; Prasad et al. 2022). The xg^O ensheaths neuronal projections between the lamina and medulla neuropils, including a subset of photoreceptor axons (Edwards and Meinertzhagen 2010). Both glial populations respond to Spitz, an EGF produced by photoreceptors, and relay this signal by producing either ILPs, in the case of the wrapping glia or Spitz and a type IV collagen, in the case of the xg^O (Fernandes et al. 2017; Prasad et al. 2022). These glial-derived signals nonautonomously activate mitogen-activated protein kinase (MAPK) signaling in lamina precursors, which drives their differentiation into neurons. The result is a striking spatiotemporal pattern of neuronal differentiation, which reflects the morphogenesis and positioning of both wrapping glia and xg^O (Fig. 3B). Interestingly, although each lamina column eventually contains five neurons, extra precursors, which do not differentiate, incorporate during column assembly (Huang and Kunes 1996). To ensure that only the correct number of neurons differentiate, differentiation signals from xg^O set up an additional relay between the neurons induced to differentiate and their neighboring undifferentiated precursors, such that the newly differentiating neurons antagonize differentiation signaling to prevent the "extra" precursors from differentiating, resulting in their death (Prasad et al. 2022). Thus, by relaying signals from photoreceptors to lamina precursors, wrapping glia and xg^O, coordinate neuronal differentiation between the developing eye disc and the developing lamina and set neuronal number and stoichiometry. Surprisingly, xg^O are born in the central brain from DL1 type II neuroblasts and migrate from the central brain and through the optic lobe before settling in their final position below the lamina (Viktorin et al. 2013; Ren et al. 2018). This raises the intriguing possibility that xg^O coordinates aspects of optic lobe development with that of central brain development, and that other glia, which migrate between brain regions, may also coordinate aspects of development. Although macroglia are not thought to be present during vertebrate embryonic neurogenesis, yolk sac–derived microglia, which migrate into the CNS during early embryonic development, have been shown to secrete mitogenic factors that induce neural stem cell proliferation in a PI3K and Notch-dependent manner in mammals (Morgan et al. 2004) and to regulate the timing of neural differentiation in the developing zebrafish retina (Huang et al. 2012). Whether microglia coordinate developmental processes by acting as signaling intermediaries in these contexts remains to be explored.

Glia as Signaling Intermediaries during Nervous System Function and during Injury

Beyond coordinating development, glia also maintain normoxic conditions in the brain by acting as intermediaries that regulate local gas exchange. In *Drosophila*, a dynamic branched tubular network called the tracheal system supplies oxygen to various organs including the brain. Astrocytes in the larval CNS exhibit TRP channel (TrpML)-mediated Ca^{2+} transients in microdomains that are spontaneous, activity independent, and regulated by reactive oxygen species (ROS) (Ma and Freeman 2020; see Coutinho-Budd et al. 2023). Interestingly, disrupting microdomain Ca^{2+} transients in astrocytes through loss of TrpML leads to tracheal overgrowth and increased ROS in the CNS (Ma and Freeman 2020). Thus, astrocytes monitor the brain environment, respond to local hyperoxic conditions, and signal to tracheal branches and filopodia for local branch retraction and reduction of local oxygen delivery (Fig. 3C). Importantly, vertebrate astrocytes are similarly coupled to vascular networks where they regulate metabolite and gas exchange, in addition to classic barrier functions (Petzold and Murthy 2011).

In addition to modulating local microenvironments, the compartmentalization of the brain by glia can also regulate circuit function and animal behavior. In addition to their role in neurovascular coupling, astrocytes tile with one another to establish a dense meshwork of gap junction–coupled processes, which individually envelop and compartmentalize synapses from their neighbors. In the larval CNS, tyrosine decarboxylase 2 (Tdc2)-expressing neurons are activated by olfactory neurons (Ma et al. 2016). Active Tdc2 neurons signal to astrocytes through the invertebrate analogs of norepinephrine, octopamine, and tyramine (Ma et al. 2016). In response, astrocytes, which are gap junction–coupled, increase whole-cell Ca^{2+} signaling synchronously (Ma et al. 2016). This, in turn, is required to inhibit downstream dopaminergic neuron activity with a substantial delay, because the astrocytic Ca^{2+} response occurs on a much slower timescale relative to Tdc2 neurons (see Singhvi et al. 2023 for a more extensive review). Parallel work in zebrafish showed that radial astrocytes also integrate information from neuromodulatory neurons over slower timescales to stop motor output (Mu et al. 2019).

Another context in which glia sense, integrate, and transmit information to neuronal circuitry occurs during sleep regulation. Sleep is defined by periods of reversible behavioral inactivity, which normally occur at set times of the day in a species-dependent manner, and which are subject to the homeostatic influence of sleep pressure, where sleep pressure builds up as a function of time spent awake (Artiushin and Seh-

gal 2020). Although the genetic, molecular, and cellular basis of sleep and sleep pressure are still being elucidated, it is increasingly clear that glia play a central role in sleep regulation (Artiushin and Sehgal 2020). Although numerous glial cell types, including astrocytes, ensheathing glia, cortex glia, and surface glia, have been implicated in homeostatic or baseline sleep regulation in *Drosophila* (reviewed extensively by Artiushin and Sehgal 2020), the role has been best studied in astrocytes (see Coutinho-Budd et al. 2023). Blum et al. (2021) showed that calcium dynamics in astrocytes encode sleep pressure and depend on Ca-α1D, an L-type voltage-gated calcium channel, which enables astrocytes to monitor neuronal activity. Increases in neuronal activity during wakefulness drives, in a calcium-dependent manner, astrocytic increases in expression of a monoamine-activated G-protein-coupled receptor called tyramine receptor II (TyrRII) (Blum et al. 2021). Together with Ca-α1D, TyrRII acts in a positive feedback loop to increase astrocytic calcium (and therefore its own expression), thus sensitizing astrocytes to extracellular monoamines that accumulate during wakefulness (Blum et al. 2021). In turn, astrocytes transmit sleep pressure information to a homeostatic sleep circuit by activating sleep-promoting neurons (R5 neurons) and inhibiting arousal-promoting neurons, in part, through Spätzle–Toll signaling. *spätzle (spz)*, which encodes a cytokine and signals through the Toll receptor, is transcriptionally up-regulated in astrocytes in response to elevated calcium signaling (Blum et al. 2021). Thus, astrocytes monitor their environments, integrate sleep pressure information, and regulate neuronal circuitry that controls state transitions during sleep.

Finally, the positioning of glia at specialized brain interfaces allows them to not only respond to neuronal damage rapidly for debris clearance (Doherty et al. 2009; Purice et al. 2017), but also to modulate the physiology of neighboring uninjured neurons called "bystander" neurons (Hsu et al. 2021). Axon transport is suppressed rapidly in both injured and bystander neurons following an injury. Moreover, bystander neurons also display reduced mechano- and chemosensory signal transduction following injury (Ma et al. 2003;

Meyer and Ringkamp 2008; Hsu et al. 2021). Although initially suppressed, bystander neurons eventually recover while injured axons degenerate (Hsu et al. 2021). Interestingly, glia (wrapping and subperineurial) appear to spread injury signals and suppress bystander neuron function transiently, in a process requiring Draper/MEGF10 signaling in glia (Hsu et al. 2021).

In summary, although glia play prominent roles as barriers, their unique positions in the nervous system are often repurposed to facilitate other diverse processes. In some instances, glia may function as signaling intermediaries out of necessity because glia insulate the CNS from systemic signals (Chell and Brand 2010; Sousa-Nunes et al. 2011; Spéder and Brand 2014). In addition to necessity, gap junction–coupled glial networks, such as the subperineurial glia involved in neuroblast reactivation or astrocytes in modulating circuit activity, may be uniquely poised to survey and coordinate a synchronous response to global metabolic/nutritional status or circuit activity more effectively. Thus, in both developing and functioning nervous systems, glia intercept, integrate, and relay signals both locally and globally, in a manner that is intimately tethered to their function as barriers and insulators.

REFERENCES

*Reference is also in this subject collection.

Artiushin G, Sehgal A. 2020. The glial perspective on sleep and circadian rhythms. *Annu Rev Neurosci* **43:** 119–140. doi:10.1146/annurev-neuro-091819-094557

Avet-Rochex A, Kaul AK, Gatt AP, McNeill H, Bateman JM. 2012. Concerted control of gliogenesis by InR/TOR and FGF signalling in the *Drosophila* post-embryonic brain. *Development* **139:** 2763–2772. doi:10.1242/dev.074179

Awasaki T, Lai SL, Ito K, Lee T. 2008. Organization and postembryonic development of glial cells in the adult central brain of *Drosophila*. *J Neurosci* **28:** 13742–13753. doi:10.1523/JNEUROSCI.4844-08.2008

Babatz F, Naffin E, Klämbt C. 2018. The *Drosophila* blood-brain barrier adapts to cell growth by unfolding of preexisting septate junctions. *Dev Cell* **47:** 697–710.e3. doi:10.1016/j.devcel.2018.10.002

Bainton RJ, Tsai LT, Schwabe T, DeSalvo M, Gaul U, Heberlein U. 2005. *moody* encodes two GPCRs that regulate cocaine behaviors and blood–brain barrier permeability in *Drosophila*. *Cell* **123:** 145–156. doi:10.1016/j.cell.2005.07.029

Baumgartner S, Littleton JT, Broadie K, Bhat MA, Harbecke R, Lengyel JA, Chiquet-Ehrismann R, Prokop A, Bellen HJ. 1996. A *Drosophila* neurexin is required for septate junction and blood–nerve barrier formation and function. *Cell* **87:** 1059–1068. doi:10.1016/S0092-8674(00)81800-0

Beckervordersandforth RM, Rickert C, Altenhein B, Technau GM. 2008. Subtypes of glial cells in the *Drosophila* embryonic ventral nerve cord as related to lineage and gene expression. *Mech Dev* **125:** 542–557. doi:10.1016/j.mod.2007.12.004

Blum ID, Keleş MF, Baz ES, Han E, Park K, Luu S, Issa H, Brown M, Ho MCW, Tabuchi M, et al. 2021. Astroglial calcium signaling encodes sleep need in *Drosophila*. *Curr Biol* **31:** 150–162.e7. doi:10.1016/j.cub.2020.10.012

Böhme MA, McCarthy AW, Blaum N, Berezeckaja M, Ponimaskine K, Schwefel D, Walter AM. 2021. Glial Synaptobrevin mediates peripheral nerve insulation, neural metabolic supply, and is required for motor function. *Glia* **69:** 1897–1915. doi:10.1002/glia.24000

Britton JS, Edgar BA. 1998. Environmental control of the cell cycle in *Drosophila*: nutrition activates mitotic and endoreplicative cells by distinct mechanisms. *Development* **125:** 2149–2158. doi:10.1242/dev.125.11.2149

Broadie K, Baumgartner S, Prokop A. 2011. Extracellular matrix and its receptors in *Drosophila* neural development. *Dev Neurobiol* **71:** 1102–1130. doi:10.1002/dneu.20935

Brogiolo W, Stocker H, Ikeya T, Rintelen F, Fernandez R, Hafen E. 2001. An evolutionarily conserved function of the *Drosophila* insulin receptor and insulin-like peptides in growth control. *Curr Biol* **11:** 213–221. doi:10.1016/S0960-9822(01)00068-9

Brown NH. 1994. Null mutations in the αPS2 and βPS integrin subunit genes have distinct phenotypes. *Development* **120:** 1221–1231. doi:10.1242/dev.120.5.1221

Calderon MR, Mori M, Kauwe G, Farnsworth J, Ulian-Benitez S, Maksoud E, Shore J, Haghighi AP. 2022. Delta/notch signaling in glia maintains motor nerve barrier function and synaptic transmission by controlling matrix metalloproteinase expression. *Proc Natl Acad Sci* **119:** e2110097119. doi:10.1073/pnas.2110097119

Chell JM, Brand AH. 2010. Nutrition-responsive glia control exit of neural stem cells from quiescence. *Cell* **143:** 1161–1173. doi:10.1016/j.cell.2010.12.007

Corty MM, Hulegaard AL, Hill JQ, Sheehan AE, Aicher SA, Freeman MR. 2022. Discoidin domain receptor regulates ensheathment, survival and caliber of peripheral axons. *Development* **149:** dev200636. doi:10.1242/dev.200636

* Coutinho-Budd J, Freeman MR, Ackerman S. 2023. *Drosophila* central nervous system glia. *Cold Spring Harb Perspect Biol* doi:10.1101/cshperspect.a041347

Defaye A, Evans I, Crozatier M, Wood W, Lemaitre B, Leulier F. 2009. Genetic ablation of *Drosophila* phagocytes reveals their contribution to both development and resistance to bacterial infection. *J Innate Immun* **1:** 322–334. doi:10.1159/000210264

Doe CQ, Technau GM. 1993. Identification and cell lineage of individual neural precursors in the *Drosophila* CNS. *Trends Neurosci* **16:** 510–514. doi:10.1016/0166-2236(93)90195-R

Doherty J, Logan MA, Taşdemir OE, Freeman MR. 2009. Ensheathing glia function as phagocytes in the adult *Drosophila* brain. *J Neurosci* **29:** 4768–4781. doi:10.1523/JNEUROSCI.5951-08.2009

Edwards TN, Meinertzhagen IA. 2010. The functional organisation of glia in the adult brain of *Drosophila* and other insects. *Prog Neurobiol* **90:** 471–497. doi:10.1016/j.pneurobio.2010.01.001

Evans IR, Hu N, Skaer H, Wood W. 2010. Interdependence of macrophage migration and ventral nerve cord development in *Drosophila* embryos. *Development* **137:** 1625–1633. doi:10.1242/dev.046797

Fernandes VM, Chen Z, Rossi AM, Zipfel J, Desplan C. 2017. Glia relay differentiation cues to coordinate neuronal development in *Drosophila*. *Science* **357:** 886–891. doi:10.1126/science.aan3174

Franzdóttir SR, Engelen D, Yuva-Aydemir Y, Schmidt I, Aho A, Klämbt C. 2009. Switch in FGF signalling initiates glial differentiation in the *Drosophila* eye. *Nature* **460:** 758–761. doi:10.1038/nature08167

Freeman MR. 2015. *Drosophila* central nervous system glia. *Cold Spring Harb Perspect Biol* **7:** a020552. doi:10.1101/cshperspect.a020552

Ghosh A, Kling T, Snaidero N, Sampaio JL, Shevchenko A, Gras H, Geurten B, Göpfert MC, Schulz JB, Voigt A, et al. 2013. A global in vivo *Drosophila* RNAi screen identifies a key role of ceramide phosphoethanolamine for glial ensheathment of axons. *PLoS Genet* **9:** e1003980. doi:10.1371/journal.pgen.1003980

Granderath S, Stollewerk A, Greig S, Goodman CS, OKane CJ, Klämbt C. 1999. loco encodes an RGS protein required for *Drosophila* glial differentiation. *Development* **126:** 1781–1791. doi:10.1242/dev.126.8.1781

Hartenstein V, Rudloff E, Campos-Ortega JA. 1987. The pattern of proliferation of the neuroblasts in the wildtype embryo of *Drosophila melanogaster*. *Rouxs Arch Dev Biol* **196:** 473–485. doi:10.1007/BF00399871

Harty BL, Monk KR. 2017. Unwrapping the unappreciated: recent progress in Remak Schwann cell biology. *Curr Opin Neurobiol* **47:** 131–137. doi:10.1016/j.conb.2017.10.003

Hatan M, Shinder V, Israeli D, Schnorrer F, Volk T. 2011. The *Drosophila* blood brain barrier is maintained by GPCR-dependent dynamic actin structures. *J Cell Biol* **192:** 307–319. doi:10.1083/jcb.201007095

Hsu JM, Kang Y, Corty MM, Mathieson D, Peters OM, Freeman MR. 2021. Injury-induced inhibition of bystander neurons requires dSarm and signaling from glia. *Neuron* **109:** 473–487.e5. doi:10.1016/j.neuron.2020.11.012

Huang Z, Kunes S. 1996. Hedgehog, transmitted along retinal axons, triggers neurogenesis in the developing visual centers of the *Drosophila* brain. *Cell* **86:** 411–422. doi:10.1016/S0092-8674(00)80114-2

Huang Z, Kunes S. 1998. Signals transmitted along retinal axons in *Drosophila*: hedgehog signal reception and the cell circuitry of lamina cartridge assembly. *Development* **125:** 3753–3764. doi:10.1242/dev.125.19.3753

Huang Z, Shilo BZ, Kunes S. 1998. A retinal axon fascicle uses spitz, an EGF receptor ligand, to construct a synaptic cartridge in the brain of *Drosophila*. *Cell* **95:** 693–703. doi:10.1016/S0092-8674(00)81639-6

Huang T, Cui J, Li L, Hitchcock PF, Li Y. 2012. The role of microglia in the neurogenesis of zebrafish retina. *Biochem Biophys Res Commun* **421:** 214–220. doi:10.1016/j.bbrc.2012.03.139

Hynes RO, Zhao Q. 2000. The evolution of cell adhesion. *J Cell Biol* **150:** F89–F96. doi:10.1083/jcb.150.2.F89

Ikeya T, Galic M, Belawat P, Nairz K, Hafen E. 2002. Nutrient-dependent expression of insulin-like peptides from neuroendocrine cells in the CNS contributes to growth regulation in *Drosophila*. *Curr Biol* **12:** 1293–1300. doi:10.1016/S0960-9822(02)01043-6

Ito K, Hotta Y. 1992. Proliferation pattern of postembryonic neuroblasts in the brain of *Drosophila melanogaster*. *Dev Biol* **149:** 134–148. doi:10.1016/0012-1606(92)90270-Q

Karkali K, Tiwari P, Singh A, Tlili S, Jorba I, Navajas D, Muñoz JJ, Saunders TE, Martin-Blanco E. 2022. Condensation of the *Drosophila* nerve cord is oscillatory and depends on coordinated mechanical interactions. *Dev Cell* **57:** 867–882.e5 doi:10.1016/j.devcel.2022.03.007

Kim SN, Jeibmann A, Halama K, Witte HT, Wälte M, Matzat T, Schillers H, Faber C, Senner V, Paulus W, et al. 2014. ECM stiffness regulates glial migration in *Drosophila* and mammalian glioma models. *Development* **141:** 3233–3242. doi:10.1242/dev.106039

Kottmeier R, Bittern J, Schoofs A, Scheiwe F, Matzat T, Pankratz M, Klämbt C. 2020. Wrapping glia regulates neuronal signaling speed and precision in the peripheral nervous system of *Drosophila*. *Nat Commun* **11:** 4491. doi:10.1038/s41467-020-18291-1

Kremer MC, Jung C, Batelli S, Rubin GM, Gaul U. 2017. The glia of the adult *Drosophila* nervous system. *Glia* **65:** 606–638. doi:10.1002/glia.23115

Leiserson WM, Harkins EW, Keshishian H. 2000. Fray, a *Drosophila* serine/threonine kinase homologous to mammalian PASK, is required for axonal ensheathment. *Neuron* **28:** 793–806. doi:10.1016/S0896-6273(00)00154-9

Leiserson WM, Forbush B, Keshishian H. 2011. *Drosophila* glia use a conserved cotransporter mechanism to regulate extracellular volume. *Glia* **59:** 320–332. doi:10.1002/glia.21103

Li D, Liu Y, Pei C, Zhang P, Pan L, Xiao J, Meng S, Yuan Z, Bi X. 2017. miR-285-Yki/Mask double-negative feedback loop mediates blood-brain barrier integrity in *Drosophila*. *Proc Natl Acad Sci* **114:** E2365–E2374.

Li X, Fetter R, Schwabe T, Jung C, Liu L, Steller H, Gaul U. 2021. The cAMP effector PKA mediates moody GPCR signaling in *Drosophila* blood-brain barrier formation and maturation. *eLife* **10:** e68275. doi:10.7554/eLife.68275

Lucero HA, Kagan HM. 2006. Lysyl oxidase: an oxidative enzyme and effector of cell function. *Cell Mol Life Sci* **63:** 2304–2316. doi:10.1007/s00018-006-6149-9

Ma Z, Freeman MR. 2020. TrpML-mediated astrocyte microdomain Ca^{2+} transients regulate astrocyte-tracheal interactions. *eLife* **9:** e58952. doi:10.7554/eLife.58952

Ma C, Shu Y, Zheng Z, Chen Y, Yao H, Greenquist KW, White FA, LaMotte RH. 2003. Similar electrophysiological changes in axotomized and neighboring intact dorsal root ganglion neurons. *J Neurophysiol* **89:** 1588–1602. doi:10.1152/jn.00855.2002

Ma Z, Stork T, Bergles DE, Freeman MR. 2016. Neuro-modulators signal through astrocytes to alter neural circuit activity and behaviour. *Nature* **539:** 428–432. doi:10.1038/nature20145

Martinek N, Shahab J, Saathoff M, Ringuette M. 2008. Haemocyte-derived SPARC is required for collagen-IV-dependent stability of basal laminae in *Drosophila* embryos. *J Cell Sci* **121:** 1671–1680. doi:10.1242/jcs.021931

Matzat T, Sieglitz F, Kottmeier R, Babatz F, Engelen D, Klambt C. 2015. Axonal wrapping in the *Drosophila* PNS is controlled by glia-derived neuregulin homolog vein. *Development* **142:** 1336–1345.

Meyer RA, Ringkamp M. 2008. A role for uninjured afferents in neuropathic pain. *Sheng Li Xue Bao* **60:** 605–609.

Meyer S, Schmidt I, Klämbt C. 2014. Glia ECM interactions are required to shape the *Drosophila* nervous system. *Mech Dev* **133:** 105–116. doi:10.1016/j.mod.2014.05.003

Morgan SC, Taylor DL, Pocock JM. 2004. Microglia release activators of neuronal proliferation mediated by activation of mitogen-activated protein kinase, phosphatidylinositol-3-kinase/Akt and delta-Notch signalling cascades. *J Neurochem* **90:** 89–101. doi:10.1111/j.1471-4159.2004.02461.x

Mu Y, Bennett DV, Rubinov M, Narayan S, Yang CT, Tanimoto M, Mensh BD, Looger LL, Ahrens MB. 2019. Glia accumulate evidence that actions are futile and suppress unsuccessful behavior. *Cell* **178:** 27–43.e19. doi:10.1016/j.cell.2019.05.050

Nern A, Pfeiffer BD, Rubin GM. 2015. Optimized tools for multicolor stochastic labeling reveal diverse stereotyped cell arrangements in the fly visual system. *Proc Natl Acad Sci* **112:** E2967–E2976. doi:10.1073/pnas.1506763112

Olofsson B, Page DT. 2005. Condensation of the central nervous system in embryonic *Drosophila* is inhibited by blocking hemocyte migration or neural activity. *Dev Biol* **279:** 233–243. doi:10.1016/j.ydbio.2004.12.020

Otto N, Marelja Z, Schoofs A, Kranenburg H, Bittern J, Yildirim K, Berh D, Bethke M, Thomas S, Rode S, et al. 2018. The sulfite oxidase Shopper controls neuronal activity by regulating glutamate homeostasis in *Drosophila* ensheathing glia. *Nature Commun* **9:** 3514. doi:10.1038/s41467-018-05645-z

Pandey R, Blanco J, Udolph G. 2011. The glucuronyltransferase GlcAT-P is required for stretch growth of peripheral nerves in *Drosophila*. *PLoS ONE* **6:** e28106. doi:10.1371/journal.pone.0028106

Peco E, Davla S, Camp D, Stacey SM, Landgraf M, van Meyel DJ. 2016. *Drosophila* astrocytes cover specific territories of the CNS neuropil and are instructed to differentiate by Prospero, a key effector of Notch. *Development* **143:** 1170–1181.

Pellerin L, Magistretti PJ. 2012. Sweet sixteen for ANLS. *J Cereb Blood Flow Metab* **32:** 1152–1166. doi:10.1038/jcbfm.2011.149

Petley-Ragan LM, Ardiel EL, Rankin CH, Auld VJ. 2016. Accumulation of laminin monomers in *Drosophila* glia leads to glial endoplasmic reticulum stress and disrupted larval locomotion. *J Neurosci* **36:** 1151–1164. doi:10.1523/JNEUROSCI.1797-15.2016

Petri J, Syed MH, Rey S, Klämbt C. 2019. Non-cell-autonomous function of the GPI-anchored protein Undicht during septate junction assembly. *Cell Rep* **26:** 1641–1653.e4. doi:10.1016/j.celrep.2019.01.046

Petzold GC, Murthy VN. 2011. Role of astrocytes in neurovascular coupling. *Neuron* **71:** 782–797. doi:10.1016/j.neuron.2011.08.009

Pogodalla N, Kranenburg H, Rey S, Rodrigues S, Cardona A, Klämbt C. 2021. *Drosophila* β (Heavy)-Spectrin is required in polarized ensheathing glia that form a diffusion-barrier around the neuropil. *Nat Commun* **12:** 6357. doi:10.1038/s41467-021-26462-x

Prasad AR, Lago-Baldaia I, Bostock MP, Housseini Z, Fernandes VM. 2022. Differentiation signals from glia are fine-tuned to set neuronal numbers during development. *eLife* **11:** e78092. doi:10.7554/eLife.78092

Prokop A, Technau GM. 1991. The origin of postembryonic neuroblasts in the ventral nerve cord of *Drosophila melanogaster*. *Development* **111:** 79–88. doi:10.1242/dev.111.1.79

Purice MD, Ray A, Münzel EJ, Pope BJ, Park DJ, Speese SD, Logan MA. 2017. A novel *Drosophila* injury model reveals severed axons are cleared through a draper/MMP-1 signaling cascade. *eLife* **6:** e23611. doi:10.7554/eLife.23611

Rangarajan R, Gong Q, Gaul U. 1999. Migration and function of glia in the developing *Drosophila* eye. *Development* **126:** 3285–3292. doi:10.1242/dev.126.15.3285

Ren Q, Awasaki T, Wang YC, Huang YF, Lee T. 2018. Lineage-guided notch-dependent gliogenesis by *Drosophila* multi-potent progenitors. *Development* **145:** dev160127. doi:10.1242/dev.160127

Rey S, Ohm H, Klämbt C. 2023a. Axonal ion homeostasis and glial differentiation. *FEBS J* **290:** 3737–3744. doi:10.1111/febs.16594

Rey S, Ohm H, Moschref F, Zeuschner D, Praetz M, Klämbt C. 2023b. Glial-dependent clustering of voltage-gated ion channels in *Drosophila* precedes myelin formation. *eLife* **12:** e85752. doi:10.7554/eLife.85752

Roignant JY, Treisman JE. 2009. Pattern formation in the *Drosophila* eye disc. *Int J Dev Biol* **53:** 795–804. doi:10.1387/ijdb.072483jr

Rossi AM, Fernandes VM. 2018. Wrapping glial morphogenesis and signaling control the timing and pattern of neuronal differentiation in the *Drosophila* lamina. *J Exp Neurosci* **12:** 117906951875929. doi:10.1177/1179069518759294

Rulifson EJ, Kim SK, Nusse R. 2002. Ablation of insulin-producing neurons in flies: growth and diabetic phenotypes. *Science* **296:** 1118–1120. doi:10.1126/science.1070058

Schmidt I, Thomas S, Kain P, Risse B, Naffin E, Klämbt C. 2012. Kinesin heavy chain function in *Drosophila* glial cells controls neuronal activity. *J Neurosci* **32:** 7466–7476. doi:10.1523/JNEUROSCI.0349-12.2012

Schwabe T, Bainton RJ, Fetter RD, Heberlein U, Gaul U. 2005. GPCR signaling is required for blood–brain barrier formation in *Drosophila*. *Cell* **123:** 133–144. doi:10.1016/j.cell.2005.08.037

Schwabe T, Li X, Gaul U. 2017. Dynamic analysis of the mesenchymal–epithelial transition of blood–brain barrier forming glia in *Drosophila*. *Biol Open* **6:** 232–243.

Sears HC, Kennedy CJ, Garrity PA. 2003. Macrophage-mediated corpse engulfment is required for normal *Drosoph*-

ila CNS morphogenesis. *Development* **130:** 3557–3565. doi:10.1242/dev.00586

Sepp KJ, Auld VJ. 2003. Reciprocal interactions between neurons and glia are required for *Drosophila* peripheral nervous system development. *J Neurosci* **23:** 8221–8230. doi:10.1523/JNEUROSCI.23-23-08221.2003

Sepp KJ, Schulte J, Auld VJ. 2000. Developmental dynamics of peripheral glia in *Drosophila melanogaster*. *Glia* **30:** 122–133. doi:10.1002/(SICI)1098-1136(200004)30:2<12 2::AID-GLIA2>3.0.CO;2-B

Silies M, Yuva Y, Engelen D, Aho A, Stork T, Klämbt C. 2007. Glial cell migration in the eye disc. *J Neurosci* **27:** 13130–13139. doi:10.1523/JNEUROSCI.3583-07.2007

* Singhvi A, Shaham S, Rapti G. 2023. Glia development and function in the nervous system of *Caenorhabditis elegans*. *Cold Spring Harb Perspect Biol* doi:10.1101/cshperspect .a041316

Skeath JB, Wilson BA, Romero SE, Snee MJ, Zhu Y, Lacin H. 2017. The extracellular metalloprotease AdamTS-A anchors neural lineages in place within and preserves the architecture of the central nervous system. *Development* **144:** 3102–3113.

Sousa-Nunes R, Yee LL, Gould AP. 2011. Fat cells reactivate quiescent neuroblasts via TOR and glial insulin relays in *Drosophila*. *Nature* **471:** 508–512. doi:10.1038/nature 09867

Spéder P, Brand AH. 2014. Gap junction proteins in the blood-brain barrier control nutrient-dependent reactivation of *Drosophila* neural stem cells. *Dev Cell* **30:** 309–321. doi:10.1016/j.devcel.2014.05.021

Stork T, Engelen D, Krudewig A, Silies M, Bainton RJ, Klämbt C. 2008. Organization and function of the blood–brain barrier in *Drosophila*. *J Neurosci* **28:** 587–597. doi:10.1523/JNEUROSCI.4367-07.2008

Subramanian A, Siefert M, Banerjee S, Vishal K, Bergmann KA, Clay C, Curts CM, Dorr M, Molina C, Fernandes J. 2017. Remodeling of peripheral nerve ensheathment during the larval-to-adult transition in *Drosophila*. *Dev Neurobiol* **77:** 1144–1160. doi:10.1002/dneu.22502

Tepass U, Hartenstein V. 1994. The development of cellular junctions in the *Drosophila* embryo. *Dev Biol* **161:** 563–596. doi:10.1006/dbio.1994.1054

Tiklová K, Senti KA, Wang S, Gräslund A, Samakovlis C. 2010. Epithelial septate junction assembly relies on melanotransferrin iron binding and endocytosis in *Drosophila*. *Nat Cell Biol* **12:** 1071–1077. doi:10.1038/ncb2111

Truman JW, Bate M. 1988. Spatial and temporal patterns of neurogenesis in the central nervous system of *Drosophila melanogaster*. *Dev Biol* **125:** 145–157. doi:10.1016/0012-1606(88)90067-X

Unhavaithaya Y, Orr-Weaver TL. 2012. Polyploidization of glia in neural development links tissue growth to blood–

brain barrier integrity. *Genes Dev* **26:** 31–36. doi:10.1101/ gad.177436.111

Urbano JM, Torgler CN, Molnar C, Tepass U, López-Varea A, Brown NH, de Celis JF, Martín-Bermudo MD. 2009. *Drosophila* laminins act as key regulators of basement membrane assembly and morphogenesis. *Development* **136:** 4165–4176. doi:10.1242/dev.044263

Viktorin G, Riebli N, Reichert H. 2013. A multipotent transit-amplifying neuroblast lineage in the central brain gives rise to optic lobe glial cells in *Drosophila*. *Dev Biol* **379:** 182–194. doi:10.1016/j.ydbio.2013.04.020

Volkenhoff A, Weiler A, Letzel M, Stehling M, Klämbt C, Schirmeier S. 2015. Glial glycolysis is essential for neuronal survival in *Drosophila*. *Cell Metab* **22:** 437–447. doi:10 .1016/j.cmet.2015.07.006

von Hilchen CM, Beckervordersandforth RM, Rickert C, Technau GM, Altenhein B. 2008. Identity, origin, and migration of peripheral glial cells in the *Drosophila* embryo. *Mech Dev* **125:** 337–352. doi:10.1016/j.mod.2007.10 .010

von Hilchen CM, Bustos AE, Giangrande A, Technau GM, Altenhein B. 2013. Predetermined embryonic glial cells form the distinct glial sheaths of the *Drosophila* peripheral nervous system. *Development* **140:** 3657–3668. doi:10 .1242/dev.093245

Von Stetina JR, Frawley LE, Unhavaithaya Y, Orr-Weaver TL. 2018. Variant cell cycles regulated by Notch signaling control cell size and ensure a functional blood–brain barrier. *Development* **145:** dev157115. doi:10.1242/dev .157115

Winkler B, Funke D, Benmimoun B, Spéder P, Rey S, Logan MA, Klämbt C. 2021. Brain inflammation triggers macrophage invasion across the blood-brain barrier in *Drosophila* during pupal stages. *Sci Adv* **7:** eabh0050. doi:10 .1126/sciadv.abh0050

Xie X, Auld VJ. 2011. Integrins are necessary for the development and maintenance of the glial layers in the *Drosophila* peripheral nerve. *Development* **138:** 3813–3822. doi:10.1242/dev.064816

Xie X, Gilbert M, Petley-Ragan L, Auld VJ. 2014. Loss of focal adhesions in glia disrupts both glial and photoreceptor axon migration in the *Drosophila* visual system. *Development* **141:** 3072–3083. doi:10.1242/dev.101972

Yuan X, Sipe CW, Suzawa M, Bland ML, Siegrist SE. 2020. Dilp-2-mediated PI3-kinase activation coordinates reactivation of quiescent neuroblasts with growth of their glial stem cell niche. *PLoS Biol* **18:** e3000721. doi:10.1371/jour nal.pbio.3000721

Zulbahar S, Sieglitz F, Kottmeier R, Altenhein B, Rumpf S, Klambt C. 2018. Differential expression of Obek controls ploidy in the *Drosophila* blood–brain barrier. *Development* **145:** dev164111. doi:10.1242/dev.164111

Glial Cell Development and Function in the Zebrafish Central Nervous System

Tim Czopka,[1] Kelly Monk,[2] and Francesca Peri[3]

[1]Centre for Clinical Brain Sciences, The University of Edinburgh, Edinburgh EH16 4SB, United Kingdom

[2]Vollum Institute, Oregon Health and Science University, Portland, Oregon 97239, USA

[3]Department of Molecular Life Sciences, University of Zürich, 8057 Zürich, Switzerland

Correspondence: Tim.Czopka@ed.ac.uk; monk@ohsu.edu; Francesca.peri@uzh.ch

Over the past decades the zebrafish has emerged as an excellent model organism with which to study the biology of all glial cell types in nervous system development, plasticity, and regeneration. In this review, which builds on the earlier work by Lyons and Talbot in 2015, we will summarize how the relative ease to manipulate the zebrafish genome and its suitability for intravital imaging have helped understand principles of glial cell biology with a focus on oligodendrocytes, microglia, and astrocytes. We will highlight recent findings on the diverse properties and functions of these glial cell types in the central nervous system and discuss open questions and future directions of the field.

All animals with a central nervous system (CNS) have glia, but only the vertebrate CNS contains three glial cell types: oligodendrocytes, microglia, and astrocytes. The zebrafish is among the simplest vertebrate model organisms used in biosciences and its popularity has increased steadily since its introduction in the 1980s by George Streisinger et al. (1981). Several properties make zebrafish a superb model for experimental research. They are highly fecund with a single pair giving rise to hundreds of offspring in each mating. Embryos develop externally making them easily accessible to the experimenter. As they develop from a fertilized egg to a freely swimming animal in <5 days, they are an ideal model for developmental studies. From 5 days postfertilization (dpf) onward, young zebrafish start hunting for prey, meaning that they have formed functional neural circuits and are able to carry out complex sensorimotor transformations. During all these early stages, zebrafish remain relatively small (under 1 cm in length) and optically transparent, which allows one to study glial development and function at unprecedented detail and without the need for surgical intervention. These combined features have made the zebrafish an exquisite model for genetic, pharmacological, cellular, physiological, and behavioral analyses in the intact living animal.

Although zebrafish represent evolutionary distant relatives to mammals with a CNS of lower complexity (about 100,000 neurons in a larval fish brain) and a different neuroanatomy, it is important to emphasize that principles of nervous system formation and function are highly conserved across species. A comparative study has revealed that at least 70% of human genes have at least one ortholog in zebrafish (Howe et al. 2013). Likewise, the past two decades have shown that development and function of zebrafish glia are highly conserved compared to mammals, from key transcription factors that regulate development to molecular signals and cellular dynamics that regulate the interaction of glia with other CNS cell types that surround them.

In this review, we aim to provide an update of the excellent contribution to the literature by David Lyons and William Talbot (Lyons and Talbot 2015). Since then, major progress has been made to understand properties and functions of glia due to the possibility to live image all glial cell types at single-cell resolution in the entire animal (Figs. 1 and 2). Here, we will summarize our current understanding of zebrafish glial biology, and discuss open questions and future directions.

OLIGODENDROCYTES

As in mammals, zebrafish oligodendrocyte lineage cells form an abundant population throughout the CNS where they coexist in different states from undifferentiated precursors to myelinating oligodendrocytes throughout development, adulthood, and aging. How individual oligoden-

drocytes progress through their lineage, how they communicate with surrounding neurons (and glia), when to differentiate, and which axons to select for myelination are fundamental questions that had remained unanswered for a long time. In addition to what we have learned from zebrafish about the genetic control of oligodendrocyte development (comprehensively, for reviews, see Lyons and Talbot 2015; Preston and Macklin 2015; Ackerman and Monk 2016; Czopka 2016), the suitability of young zebrafish for noninvasive live cell microscopy, along with the development of reagents and technologies to visualize and manipulate oligodendrocytes has allowed the study of oligodendrocyte biology in real time in the intact living animal. Indeed, zebrafish remains the only system in which one can live image oligodendrocyte–neuron interactions from the moment cells are specified to the point where they have formed mature myelin sheaths in vivo, and perform sophisticated genetic manipulations to understand mechanisms. The past 10 years have provided a substantial collection of studies in which in vivo imaging and cellular genetic manipulations have revealed fundamental properties of oligodendrocyte precursors (OPCs), oligodendrocytes, and myelin. In the following sections we will focus on summarizing these studies.

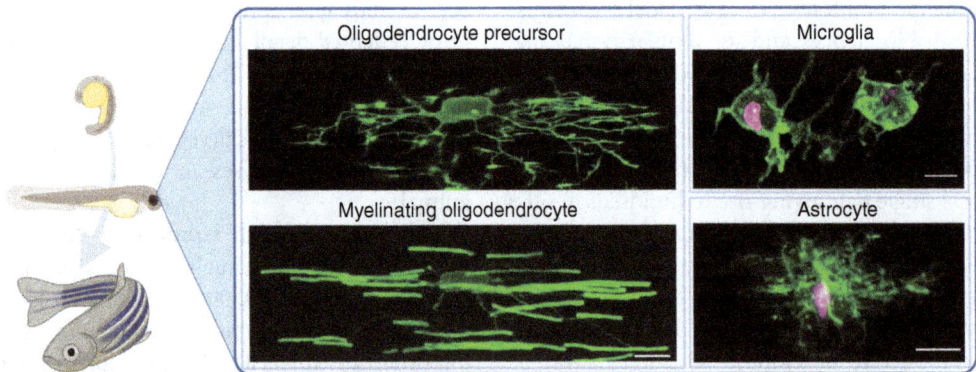

Figure 1. Visualizing different glial cell types in living zebrafish. (*Left*) Schematic of different stages of zebrafish development from embryo through adulthood. (*Right*) In vivo microscopy of different glial cell types in young zebrafish. The same oligodendrocyte at different stages of development labeled with an olig1:memEYFP transgenic reporter. (Image reproduced from Auer et al. 2018, with permission © 2018 the author(s). Published by Elsevier Ltd.) Microglia are labeled in a double transgenic line with membrane-targeted tagRFP (green) and nuclear nls-Crimson (magenta) (Tg[mpeg1:Gal4; UAS:lyn- tagRFPT]; Tg[spi1b:Gal4-UAS:NLS-Crimson]). A single astrocyte is labeled with membrane myrGFP (green) and nuclear H2A-mCherry (magenta) driven by the *glast* promoter. Cell was imaged from the larval spinal cord at 6 dpf (image credit: Jiakun Chen). Scale bars, 10 μm.

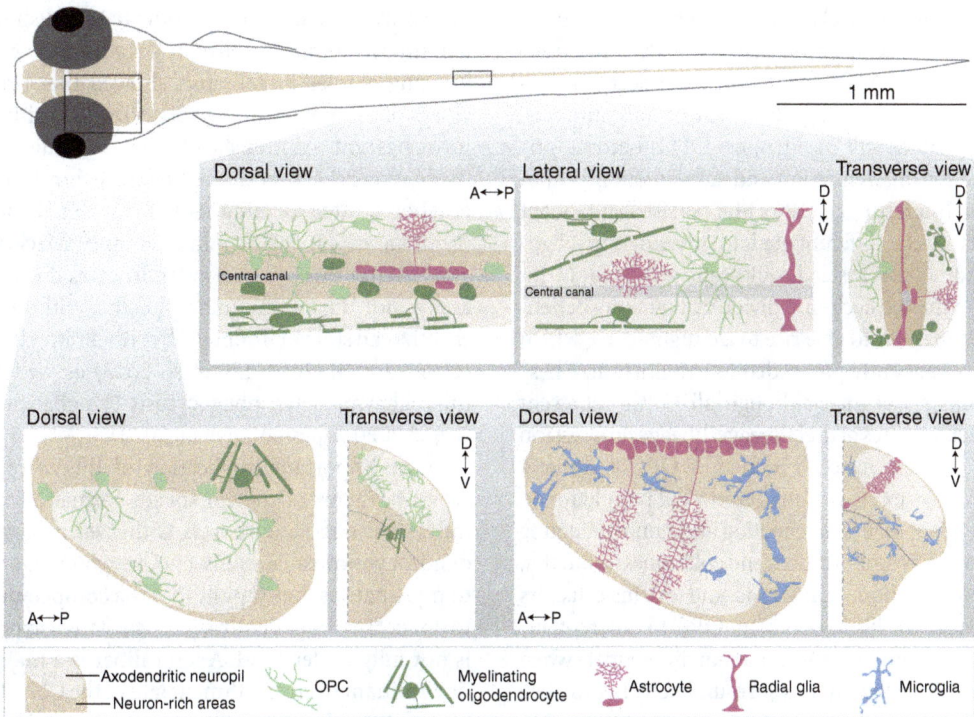

Figure 2. Distribution of morphologies of different glial cell types in zebrafish. (*Top*) Schematic dorsal view of a larval zebrafish and its central nervous system. Boxes indicate the positions of detailed zoom-ins underneath to outline positioning and morphologies of different glial cell types at the level of the optic tectum (*left* box with zoom-in at *bottom*) and the spinal cord (*right* box with zoom-in in the *middle*). (A) Anterior, (P) posterior, (D) dorsal, (V) ventral.

Formation of Oligodendrocyte Precursor Cells and Regulation of Their Lineage Progression

OPCs are specified in defined CNS regions from where they migrate and disperse throughout the CNS. In the spinal cord of zebrafish, like in all vertebrates, the first OPCs arise from the pMN domain defined by the *olig2* transcription factor, which initially gives rise to motor neurons followed by the generation of OPCs through the recruitment of a new wave of neural progenitors to the pMN domain (Park et al. 2002; Rowitch 2004; Ravanelli and Appel 2015). Throughout the CNS, OPCs display diverse properties with regard to gene expression profile, physiological properties, and ability to differentiate (Viganò et al. 2013; Marques et al. 2016; Spitzer et al. 2019). Consequently, it has been a long-standing question in the field whether the observed diversity of OPCs reflects intrinsically different types of OPCs, or

rather different states of the same cell (for reviews, see Dimou and Simons 2017; Foerster et al. 2019; Kamen et al. 2022). Marisca and colleagues (2020) addressed this question using an integrated approach in zebrafish to identify molecular, anatomical, and physiological differences between OPCs while monitoring their lineage formation and probing their function over time. They found that the zebrafish spinal cord contains a network of OPCs with different morphological complexities and process remodeling dynamics, depending on their local microenvironment. Although all these OPCs contact the same cohorts of myelination-competent axons, meaning that they have targets available that they can myelinate, only some OPCs differentiate readily while others do not. To test how these different groups of OPCs relate to another (i.e., if different OPCs seen at any point in time simply represent different states of the same cell), Marisca et al. generated clonal trees of OPC fates and interrelationships, which revealed a functional segregation between OPCs; some remain undifferentiated in either quiescent or proliferative states to regulate their overall numbers, while other still proliferative OPCs become primed for differentiation and subsequent myelination. Interestingly, OPCs that persist as quiescent cells rarely differentiated to myelinating oligodendrocytes. Instead, quiescent OPCs could reenter the cell cycle and divide in a calcium-dependent manner to give rise to a daughter cell, which then frequently proceeded to myelination. These results show that, although all OPCs represent different states of their lineage, lineage progression is not linear for each individual OPC and that a hierarchy exists within their overall population.

What makes myelinating and nonmyelinating OPCs different? Marisca and colleagues found that a combination of intrinsic and extrinsic factors regulate the likelihood of an OPC to differentiate. Extrinsic, because OPCs did not differentiate when the OPC cell body was surrounded by neuron cell bodies (regardless of the OPC processes contacting myelination competent axons). Intrinsic, because all myelinating oligodendrocytes that Marisca et al. identified in their clonal analyses were formed from a cell that had undergone a recent cell division. The finding that recently divided OPCs differentiate with a higher frequency is consistent with reports on oligodendrocyte generation from OPCs in the developing mouse cortex (Hill et al. 2014), as well as remyelination (Foerster et al. 2020), but it differs from the adult mouse cortex where direct differentiation of OPCs has been reported that have been persisting for long periods of time (Bacmeister et al. 2020). Future work will be needed to dissect whether readily differentiating OPCs in the adult animal are already somewhat primed and just need to have a break released to proceed to myelination, or whether fundamentally different mechanisms exist between developmental and adult oligodendrogenesis.

Choosing Axons for Myelination and Making the Right Number of Sheaths with the Right Length

Once an OPC has entered its terminal differentiation program, each individual cell appears to have only a narrow time window to establish its maximum number of myelin sheaths. Live cell imaging studies in zebrafish showed that processes of differentiating oligodendrocytes either form nascent axon ensheathments, or alternatively retract back to the cell body within just a few hours after forming its first myelin sheath (Czopka et al. 2013; Almeida and Macklin 2023). After this time, oligodendrocytes do generally not form any new sheaths, although sheaths can still be eliminated by either retraction (Czopka et al. 2013; Liu et al. 2013) as well as microglia-mediated phagocytosis (Hughes and Appel 2020; Djannatian et al. 2023).

One important implication of this stereotyped behavior of myelin sheath formation by individual oligodendrocytes is that each oligodendrocyte must carefully choose which axons to myelinate as they rapidly lose the competency to do so. How oligodendrocytes select their axons is not fully understood. Axon caliber is a major determinant of ensheathment fate in the CNS. In zebrafish, the largest caliber axon (the Mauthner axon) is also the first one myelinated, followed by other axons of smaller yet still relatively large caliber (Almeida et al. 2011; Koudelka et al. 2016). However, CNS axons of a very large range of calibers are ultimately myelinated, meaning that additional regulatory factors must be present.

Currently, the prevailing view is that there is no single determinant of myelination fate in the CNS, but that this process is under the influence of several factors that may be employed in a context-dependent manner. Over the past 10 years, the concept of "adaptive" myelination has emerged and is understood as the regulation of myelination in response to changes in nervous system activity, based on observations that white matter content, oligodendrogenesis, and myelination increase in response to changes to experience and learning (Scholz et al. 2009; Makinodan et al. 2012; McKenzie et al. 2014). Oligodendrocytes in mammals and zebrafish are perfectly equipped to sense neural activity using a wide range of neurotransmitter receptors and voltage-gated ion channels (Maldonado and Angulo 2015; Marisca et al. 2020). Several studies in zebrafish revealed that neuronal activity directly tunes myelin sheath formation at the level of individual cells down to single sheaths. Systemic blockade of axonal vesicle release reduces the overall number of myelin sheaths formed per oligodendrocyte, and vice versa, an increase of neural activity increases the number of sheaths per cell (Mensch et al. 2015). Furthermore, blocking vesicle release in single axons biases toward axon ensheathment of nonsilenced axons in choice situations (Hines et al. 2015). How are these effects mediated at the molecular level? They may involve direct neurotransmitter- and/or depolarization-induced signaling in oligodendrocytes as it was recently described that axon-OPC synaptic contacts can predict regions of sheath formation (Li et al. 2024), but also involve more indirect cascades. For example, one study using zebrafish and mice showed that myelin sheath numbers formed by individual oligodendrocytes involved signaling via endothelins released from the vasculature, possibly linking neurovascular communication to the regulation of oligodendrocyte behavior (Swire et al. 2019).

After differentiating oligodendrocytes have selected target axons for myelination, sheaths need to grow to the right length. Again, long-term in vivo imaging in zebrafish has revealed principles of myelin sheath dynamics by showing that nascent sheaths grow dynamically and at highly variable rates for the first 3 days after their respective initiation (Auer et al. 2018). After this phase, sheaths continued to extend at slow rates that are similar to the overall body growth of the larval zebrafish. Therefore, differences in the length between individual myelin sheaths are established during the first days after their respective formation (Auer et al. 2018). A series of related studies revealed that this early phase of variable sheath growth is regulated by dynamic neuron to oligodendrocyte communication. Newly formed sheaths exhibit intracellular calcium transients that can be raised by neuronal activity, and that can regulate their stabilization, extension, or shrinking (Baraban et al. 2017; Krasnow et al. 2018). Indeed, postsynaptic proteins have been detected in paranodal regions, which are sites of axonal vesicle fusion (Hughes and Appel 2019; Almeida et al. 2021), and the disruption of both axonal vesicle release as well as

synaptic and nonsynaptic adhesion molecules in paranodal regions impair myelin sheath extension (Djannatian et al. 2019; Hughes and Appel 2019; Klingseisen et al. 2019; Almeida et al. 2021). Together, these studies suggest that neuronal activity during the early phases of sheath growth may ultimately determine whether a long or a short myelin sheath will be formed.

Although axonal activity can directly regulate ensheathment fate and sheath growth, it should be noted that axonal activity is not an absolute requirement for myelination. Oligodendrocytes still myelinate when all action potentials are blocked by tetrodotoxin in zebrafish (Mensch et al. 2015). Furthermore, an activity-dependent control of sheath growth alone may not be sufficient to regulate how entire axons get myelinated along their length. Axon myelination patterns can be highly specific and form over long periods of time, which frequently leads to the formation of intermittent "patchy" myelination with sheaths that have no direct neighbors in zebrafish and mice (Tomassy et al. 2014; Auer et al. 2018). Therefore, growing myelin sheaths need to know (or be told) when and where to stop extending to form a heminode (and ultimately a node) in a desired place. Contact-mediated repulsion by neighboring sheaths is one mechanism to stop them growing (Auer et al. 2018). This process requires internodal and paranodal adhesions as their disruption results in myelin sheaths that overgrow each other (Djannatian et al. 2019). However, how do sheaths stop growing when they do not meet another sheath? One simple explanation would be that the axon itself provides stop signals. Evidence for such cues comes from two zebrafish imaging studies that showed that growing myelin sheaths frequently extend asymmetrically from the feeding cytoplasmic process (Auer et al. 2018). In some cases, this was due to the presence of axon collateral branches, which provide a physical barrier that stops sheaths extending further (Auer et al. 2018). In other cases, however, sheaths stopped growing in one but not the other direction even though no obvious physical barrier was present. Here, a later study revealed that the presence of prenodal clusters along unmyelinated axon stretches can serve as a stop signal for growing myelin and therefore a prefigure node of Ranvier position (Vagionitis et al. 2022). Another possibility to form myelin sheaths of a desired length with nodes in a specific position comes from very recent observations using zebrafish, mice, and human organoids where cytoplasmic bridges connecting adjacent myelin sheaths across a node of Ranvier have been observed (Call et al. 2022). Although the meaning of these paranodal bridges is presently unclear, it is tempting to speculate whether they represent a secondary constriction to split an existing myelin sheath into two, and thus an entirely new mechanism to regulate sheath length and node position. Together, these collective studies using zebrafish have revealed different ways of ongoing axon–oligodendrocyte cross talk to dynamically regulate whether and how axons get myelinated over time. Many of these processes are modulated by neuronal activity and thus adaptive, which opens new avenues to investigate how such adaptive myelination in turn changes axon and consequently circuit function.

Repairing a Demyelinated Axon

Damage to myelin and disease-mediated loss of myelin are hallmarks of CNS injury and demyelinating diseases like multiple sclerosis (MS), which have lasting and irreversible consequences for axonal health and function (Franklin and ffrench-Constant 2017). Although zebrafish do not get MS, just like any other nonhuman species, they are a valuable model to understand principles of regenerative oligodendrogenesis. Various models to demyelinate axons have been established and range from focal single-cell demyelination using photosensitizers (Auer et al. 2018), toxin-induced demyelination using cuprizone (Jaronen et al. 2022) and lysolecithin (Münzel et al. 2014; Cunha et al. 2020; Morris and Kucenas 2021), as well as chemogenetic models to induce oligodendrocyte death using targeted expression of nitroreductase (Karttunen et al. 2017) and TRPV1 channels (Neely et al. 2022). In vivo imaging of oligodendrocyte dynamics in these models has, for example, revealed that myelinating oligodendrocytes that survive experimental demyelination can sometimes form new myelin sheaths, but that

these sheaths are frequently mistargeted to non-axonal compartments (Neely et al. 2022). Inspired from these observations in zebrafish, the same study confirmed that such mistargeting can also be found in human MS lesions and may in fact impair neuronal function and hinder efficient myelin repair. This work showcases how discoveries from zebrafish can help us understand aspects of human disease without attempting to directly model the disease.

What Do OPCs Do in the CNS Beyond Making Myelin?

Owing to the fact that OPCs always form a constant number of resident CNS cells, the role of OPCs besides being the cellular source of myelinating oligodendrocytes has been a long-standing question. However, answers have remained largely elusive, primarily due to the circumstance that it is technically difficult to manipulate OPC function without indirectly affecting myelination. Several regions of the mammalian CNS contain OPCs but remain largely devoid of myelin, and would thus be suitable to specifically test OPC-specific functions without indirectly interfering with myelin formation (e.g., superficial layers of the cerebellar cortex and olfactory bulb glomeruli). However, reagents and assays to specifically target OPCs in these regions have remained sparse. Recently, Xiao et al. (2022) identified the optic tectum of larval zebrafish as a CNS region that allows the precise study of OPC functions without indirectly interfering with myelination. The zebrafish optic tectum is the region where retinal ganglion cell axons synapse to tectal interneurons. This region is easily accessible to the experimenter, densely interspersed with OPCs, but it contains hardly any myelin (Fig. 2). Importantly, during these stages, larval zebrafish have a functional visual system, thus allowing one to directly probe the roles of OPCs in a functional neural circuit. Using this model, Xiao et al. found through different perturbation methods that the absence of OPCs from the tectum impaired the precise formation and remodeling of retinal ganglion cell axon arbors, which consequently degraded the acuity of visual processing, thus providing a direct role for

OPCs in sculpting neural circuits (Xiao et al. 2022).

The finding that tissue-resident OPCs have mature roles over and above their canonical roles in myelin formation raises a vast range of open questions. First, how do OPCs exert their effects to fine-tuning circuit connectivity? They could either guide axons, as has been shown in the context of glial scar formation and CNS damage where OPCs inhibit axon growth (Tan et al. 2005). Alternatively, they could prune axons by phagocytosis. Indeed, it was recently shown in the mammalian visual system that OPCs can ingest axonal presynaptic compartments (Auguste et al. 2022; Buchanan et al. 2022). Regardless of the mechanism of action, by being an active player in neural circuit development, dysfunctional OPCs may likely contribute to a vast range of neurodevelopmental and neuropsychiatric disorders where the fine-tuning of circuit connectivity are dysregulated. For example, in a recent sequencing study of patients who suffered from major depressive disorders, about 50% of dysregulated genes were in fact encoded by OPCs (Nagy et al. 2020). In the light of the findings from zebrafish where OPCs directly regulate circuit connectivity (Xiao et al. 2022), it may thus be that OPCs themselves directly contribute to mental illness, which will be interesting research directions to address in the future.

MICROGLIA

Although microglia are immune cells that originate outside the brain parenchyma, many studies have demonstrated that they play essential roles in the development and homeostasis of the brain. Indeed, today we know that microglia have many functions besides fighting pathogens, including synaptic patterning, neurogenesis, neuronal removal, survival, and axon guidance. Moreover, the notion that microglia participate in many, if not all, neurodegenerative disorders affecting the CNS has generated a great deal of interest in these cells, pushing scientists to investigate how microglia respond to neuronal changes, with the zebrafish serving as an ideal model.

Intrinsic and Extrinsic Processes Contribute to Microglial Brain Colonization

Microglia come from yolk sac primitive macrophages that colonize the embryonic brain as highly migrating cells (for review, see Prinz et al. 2017). In mice and fish, this process relies on the tyrosine kinase colony-stimulating factor 1 receptor (Csf1r). In mammals, this receptor is responsible for both brain colonization and microglial survival (Erblich et al. 2011; Pridans et al. 2018; Rojo et al. 2019), and pharmacological inhibition of CSF1R can be used to deplete the microglial population (Elmore et al. 2014; Squarzoni et al. 2014). In mouse, this receptor has two ligands, Csf1 and interleukin 34 (Il34) (Lin et al. 2008), with distinct expression patterns and nonredundant functions (Cahoy et al. 2008; Greter et al. 2012; Wang et al. 2012; Zeisel et al. 2015; Easley-Neal et al. 2019; Kana et al. 2019). In contrast, zebrafish have two Csf receptor paralogs, Csf1ra and Csf1rb, resulting from genome duplication in teleosts (Braasch et al. 2006). There are no microglia in the absence of both paralogs (Oosterhof et al. 2018); however, less severe phenotypes are observed when only one of the two genes is mutated (Ferrero et al. 2020). Fish microglia colonize the brain in two waves; the first wave occurs during embryogenesis to establish primitive microglia, and the second occurs later to set up the adult population (Xu et al. 2015; Ferrero et al. 2018). Within this framework, Csf1ra and Csf1rb play distinct functions; Csf1ra is responsible for establishing primitive microglia, while Csf1rb is a regulator of microglial development in adults (Ferrero et al. 2020). Interestingly, in these mutants, when one population is absent, the other one is smaller, pointing to the fact that the primitive and adult microglial populations might be interdependent (Ferrero et al. 2020). There is also evidence that microglial progenitors infiltrate the mammalian cortex in multiple waves and via different routes (Swinnen et al. 2013; Smolders et al. 2019), and in humans microglia appear to colonize the brain in a stepwise manner during gestation (Menassa and Gomez-Nicola 2018). It is intriguing to speculate that, like in fish, these microglial colonization waves might be interdependent and account for the regional heterogeneity observed in mammalian microglia (see below).

Another interesting aspect of brain colonization is understanding how microglial precursors find their way to the brain. Studies in fish have shown that these cells are attracted by neuronal cell death, a key feature of brain development. Indeed, long-range signals from dying neurons attract microglial precursors into the CNS, highlighting the importance of neuronal cell death in shaping the brain's immune system (Casano et al. 2016; Xu et al. 2016). Reducing the rate of neuronal cell death leads to fewer microglia while an increase in apoptosis results in more microglia colonizing this organ (Casano et al. 2016). Research using zebrafish has also shown that lysosomes and their regulation in microglia influence brain colonization. Indeed, zebrafish microglia lacking components of the Rag regulatory complex—GTPases that function as heterodimers on lysosomes—have enlarged lysosomes and undigested apoptotic material (Shen et al. 2016). Moreover, these mutants have fewer microglia in the brain, suggesting that defects in lysosomes and cargo processing can affect essential microglial functions like migration and differentiation. In *raga* mutants, lysosomal genes are up-regulated, and microglial brain colonization defects can be rescued by ablating *tfeb* and *tfe3*, transcription factors required for activating lysosomal pathways (Iyer et al. 2022). Recent research on microglia that colonize the developing retina has shown that blood vessels provide ways for these cells to enter neurogenic eye regions, suggesting that guidance factors may be present on the surface of these blood vessels to facilitate microglial migration (Ranawat and Masai 2021).

The Interaction between Microglia and Other Glia in the Central Nervous System

Interactions between microglia and the local environment are of significant interest not only for understanding how microglia influence brain development and functionality but also for uncovering how changes in brain physiology affect key microglial behaviors. A recent study investigated how microglia engulf developing myelin sheaths, a function that could impact higher brain functions such as memory and learning (Hughes and Appel 2020; Djannatian et al. 2023). Fluorescent labeling of microglia and oligodendrocytes allowed visualization of cellular interactions and revealed the presence of engulfed myelin in microglia. Importantly, this study uncovered a functional link between the level of neuronal activity and the removal of myelin by microglia as optogenetic manipulations that make neurons less active led to more myelin in microglia (Hughes and Appel 2020). Depleting microglia did not affect the number or distribution of oligodendrocytes; however, it altered the morphology of myelin sheaths, which appeared shorter and often misshaped, suggesting a link between these two cell types and a role for microglia in myelination by oligodendrocytes (Hughes and Appel 2020). In line with this, a recent study using electron microscopy in mouse and in vivo confocal light microscopy in zebrafish has shown that during early development microglia engulf myelin fragments (Djannatian et al. 2023). This process depends on the presence of phosphatidylserine lipids on myelin, a signal that also mediates the engulfment of apoptotic cells and synaptic pruning (Mazaheri et al. 2014; Scott-Hewitt et al. 2020). Another study in this direction has shown that in humans, some *CSF1R* variants cause ALSP (adult-onset leukoencephalopathy with axonal spheroids and pigmented glia), a leukodystrophy characterized by fewer microglia and a cognitive decline (Ranawat and Masai 2021). Introducing these human ALSP-causing *CSF1R* variants in the fish genome recapitulates the microglial reduction seen in patients (Ranawat and Masai 2021). Interestingly, transcriptomic and proteomic approaches revealed up-regulation of genes in astrocytes associated with enhanced endocytosis, indicating that astrocytes might try to compensate for the loss of microglia in these mutants. This points to the existence of critical feedback compensatory mechanisms within the glial populations of the CNS.

Microglial Transcriptional and Functional Heterogeneity

Single-cell transcriptomics have demonstrated that microglia display a high degree of transcriptional heterogeneity (for review, see Masuda et al.

Cite this article as *Cold Spring Harb Perspect Biol* doi: 10.1101/cshperspect.a041350

2020). A big question in the field is how differences in gene expression translate into functional diversity. Understanding this will provide important insights into how the microglial population differs in its responses to challenges and changes in brain physiology. In the zebrafish, we can distinguish two adjacent brain regions, the synaptic-rich hindbrain (HB) and the neurogenic optic tectum (OT). Interestingly, in these two areas, microglia display different morphologies; hindbrain microglia are ramified, while optic tectum microglia are more ameboid (Wu et al. 2020; Silva et al. 2021). In addition, these cells exhibit regionally specific gene signatures; HB microglia are enriched for complement cascade components, whereas OT microglia are enriched for cathepsins and lysosomal enzymes. Interestingly, these cells also appear to perform different functions, and cathepsin-enriched microglia in the OT engulf apoptotic neurons, while complement-expressing microglia in the HB are likely to interact more with synapses (Wu et al. 2020; Silva et al. 2021). The direct comparison of these regional microglial populations represents a first step toward linking gene expression to function, an important goal in the field. While microglia have been seen to populate and adapt to specific areas within the CNS, time-lapse imaging in zebrafish has shown that these cells can also leave the CNS, for example, after spinal root injury (Green et al. 2019). Indeed, in response to damage in the periphery, microglia migrate out of the spinal root in a glutamatergic signaling-dependent manner to phagocytose debris. Green and colleagues discovered that once these microglia return to the CNS, they respond faster to a second injury and are more phagocytic than cells that remain in the spinal cord. Thus, live imaging of the fish shows the remarkable plasticity of these cells and the importance of investigating the spatiotemporal dynamics of microglial state transitions and adaptations.

Role for Microglia in Removing Neurons and Modulating Their Activity

Several studies have examined one of the main functions of microglia, which is the engulfment of neurons during brain development (Peri and Nüsslein-Volhard 2008). Engulfing an entire neu-

ron can be challenging, as microglia must also sort and recycle the products that derive from the degradation of this cell. These late steps in phagocytosis remain poorly understood, mainly due to the difficulty of studying these processes in vivo. However, understanding the mechanism by which the microglia process engulfed neurons is a fundamental goal, as many well-known Alzheimer's disease risk factors are genes that are required in microglia to degrade and transport lipids that derive from neuronal degradation (Thorlakur et al. 2013; Keren-Shaul et al. 2017; Nugent et al. 2020). Moreover, diseased microglia are often characterized by the presence of lipid aggregates (Marschallinger et al. 2020). A study in zebrafish tracked phagosomes inside microglia to follow the fate of the neuronal cargo in these cells (Villani et al. 2019). Live imaging showed that phagosomes containing dead neurons shrink progressively and fuse with the gastrosome, a previously undescribed cellular compartment that allows efficient processing of the apoptotic cargo (Villani et al. 2019). The gastrosome, also found in mammalian macrophages, contains membrane fragments and expands dramatically when phagocytosis increases, indicating that cells such as microglia must also limit neuronal uptake to allow digestion and maintain their shape. Indeed, a hallmark of microglia is their highly ramified morphology, characterized by the presence of multiple dynamic protrusions that these cells use to scan the brain parenchyma and engulf several neurons per hour (Villani et al. 2019). The mechanisms that allow microglia to use their branches to identify and engulf apoptotic neurons successfully remain unclear. Zebrafish live imaging approaches have shown that microglia, despite having many branches, always select one branch and engulf one neuron at a time (Möller et al. 2022). This branch selection process strongly correlates with the movement of the microglial centrosome that translocates rapidly into one branch toward the forming phagosome. Microglia with two centrosomes—a condition obtained by overexpressing core centrosomal components—engulf more neurons and even remove two neurons simultaneously, indicating that centrosomal migration is a rate-limiting step in microglial neuronal engulfment (Möller et al. 2022). The targeted movement of the microglial centrosome has been shown to involve the PLC/DAG signaling cascade, which also operates in T cells at the immunological synapse, reinforcing the idea of a possible evolutionary link between these two critical cellular interphases (Möller et al. 2022). Besides looking at neuronal microglial interactions during brain development, several studies have also focused on how microglia respond to tumors induced by AKT1 oncogene overexpression in neural cells (Chia et al. 2018, 2019). Interestingly, dynamic interactions between microglia and these AKT1[+] neuronal cells are mediated by ATP signaling that attracts microglia in a *p2ry12*-dependent manner, similar to microglial attraction toward neuronal injuries (Chia et al. 2019). These interactions are not phagocytic but might promote tumor growth as microglial depletion reduces AKT1[+] neuronal cell proliferation (Chia et al. 2019). There has also been considerable interest in the role of microglia in synaptic elimination, a process also known as pruning and first described in mice (Paolicelli et al. 2011; Schafer et al. 2012). Interestingly, although it is established that microglia participate in synaptic pruning, it is an ongoing debate whether microglia do so by actively removing synapses through engulfment (Eyo and Molofsky 2023). Here, the fantastic properties of zebrafish for in vivo live cell imaging of how microglia engage with synapses during circuit remodeling could be used to help resolve these open questions. Furthermore, studies in zebrafish and mice have demonstrated a nonphagocytic role for microglia in the modulation of neuronal activity (Li et al. 2012; Merlini et al. 2021). Indeed, live imaging in zebrafish has revealed that microglial processes contact highly active neurons and that in turn lead to the downregulation of both spontaneous and induced neuronal activity (Li et al. 2012), suggesting an important role for microglia in neuronal modulation.

In conclusion, we now know that microglia are an integral part of the CNS glial pool and that these cells perform a variety of important functions. As we continue to study microglia, their phenotypes, and dynamic state transitions, one clear goal is the development of novel strategies for modulating microglial activities in vivo. The zebrafish model system will remain an invaluable and indispensable resource in this pursuit.

ASTROCYTES

Astrocytes are morphologically complex glial cells that extend dense cellular processes to interact closely with neuronal synapses, brain vasculature, and other glial cells in the CNS. The most numerous cells in the mammalian brain, astrocytes support neuronal activity, maintain homeostasis of the CNS, and are implicated in the control of neural circuit development and function (Clarke and Barres 2013; Nagai et al. 2021; Perez-Catalan et al. 2021). Moreover, many studies suggest that astrocytes play key roles in neurological diseases (Molofsky et al. 2012; Burda and Sofroniew 2014). Despite their importance, compared to our understanding of neuronal development and function, we know very little about how astrocytes develop, what the diverse function of astrocytes might be in different brain regions, and how these properties are regulated.

During development, immature astrocytes derive from radial glial cells. Astrocytes elaborate their cellular processes during postnatal development, coincident with the period of active CNS synaptogenesis, and ultimately form intimate associations with neuronal synapses that are crucial for both cell types (Bushong et al. 2002). How astrocytes establish and maintain their remarkable morphologies is not known. Astrocytes also powerfully control neuronal development. For instance, astrocyte-secreted thrombospondin promotes synapse formation via its neuronally expressed receptor (Christopherson et al. 2005; Eroglu et al. 2009), and additional astrocyte-derived factors are also required for synapse formation and maturation (Kucukdereli et al. 2011; Allen et al. 2012). Based on efforts from several laboratories, it seems highly likely that additional molecules regulating astrocytic process growth, plasticity, and sculpting of neural circuitry await discovery. Finally, astrocytes respond to neurotransmitter release by increasing intracellular calcium levels (Cornell-Bell et al. 1990; Dani et al. 1992), which has been proposed to participate in neural circuit control. For example, norepinephrine powerfully controls astrocyte calcium signaling in mammals (Shigetomi et al. 2016), and a conserved neuromodulatory event (via invertebrate analogs of norepinephrine) regulates neurotransmission changes and behaviors in *Drosophila* (Ma et al. 2016).

Although most of our understanding of astrocyte biology derives from investigation of mouse models, numerous studies suggest striking conservation of astrocyte biology across species (Oikonomou and Shaham 2011; Stork et al. 2014). Curiously, zebrafish had long been proposed to not possess stellate astrocytes until recently, and radial glial cells had been historically proposed to functionally substitute for astrocytes (Grupp et al. 2010; Lyons and Talbot 2015). Recent work, however, has identified astrocytes in zebrafish (Chen et al. 2020), thus positioning zebrafish as a new model to study astrocyte biology in vivo.

Discovery and Characterization of Zebrafish Astrocytes

In mammals, radial glia serve as neural progenitors during early development; by late neurogenesis, most radial glia regress their radial processes from the ventricles and become stellate-like astrocytes (Rowitch and Kriegstein 2010). Similarly, in zebrafish, radial glia have been characterized in various CNS regions during development (Lyons et al. 2003; Johnson et al. 2016). However, in contrast to mammals, zebrafish radial glia persist in most regions of the adult CNS and are thought to be at least in part responsible for the impressive CNS regenerative capacity observed in this species (Kroehne et al. 2011; Than-Trong and Bally-Cuif 2015). Radial glia-like cells in zebrafish are present in the brain and in some regions have elaborated processes near synapses (Lyons and Talbot 2015; Mu et al. 2019), suggesting these cells could perform key functions of astrocytes.

Chen et al. (2020) recently sought to test whether zebrafish radial glia perform necessary astrocytic functions or whether a subset of zebrafish radial glia transform into stellate astrocytes that morphologically and functionally resemble mammalian astrocytes. Previous studies in zebrafish relied on Gfap (glial fibrillary acidic protein) as a marker and which is also expressed in zebrafish radial glia. Instead, Chen et al. focused

on Glast (glutamate aspartate transporter or EAAT1), which is encoded by two orthologs in zebrafish, *slc1a3a* and *slc1a3b*. Transgenic lines and expression constructs were created in which membrane and nuclear markers were expressed under the *slc1a3b* promoter, thus enabling global and single-cell resolution analysis of Glast$^+$ cells. Using these tools, Chen et al. observed radial astrocytes (Mu et al. 2019; see next section) in the hindbrain, Bergmann glia-like cells in the cerebellum, and cells with the appearance of stellate astrocytes in the spinal cord (Fig. 2).

Focusing on these stellate cells, which we hereafter refer to as zebrafish astrocytes, Chen and colleagues demonstrated their genesis from radial glia precursors by time-lapse imaging, showed that these cells express additional astrocyte markers, elaborate fine processes during synapse formation, tile with other astrocytes, exhibit spontaneous microdomain calcium transients with similar kinetics as mouse and *Drosophila* astrocytes, and that these microdomain calcium transients are sensitive to norepinephrine. In all, this work demonstrated that the zebrafish CNS houses a population of astrocytes very similar to those in mammals and *Drosophila*, providing further support for the notion that astrocytes are an ancient, well-conserved CNS cell type. Going forward, zebrafish will represent a powerful tool to study astrocyte development and function in vivo.

Roles for Astrocytes in Neural Circuitry

It is critical to study astrocyte biology in vivo to understand their role in the control of neural circuits. Zebrafish represent an ideal system to do so, as demonstrated in a recent elegant study by Mu and colleagues (2019) who investigated how astrocytes regulate circuit function using whole brain imaging as animals executed simple behaviors. The authors examined the optomotor response in zebrafish larvae, which is a robust reflex enabling animals to maintain position in response to current. In these experiments, head-fixed larvae respond to moving gratings with swim bouts that attempt to match the presented optical flow (Orger et al. 2008). If the visual feedback following fictive swimming bouts is withheld, zebrafish eventually stop responding to the moving grating and become passive, a behavior that has been compared to learned helplessness in mammals (Nagai et al. 2021). Using this behavioral test coupled with whole-brain calcium imaging and cell-specific perturbations, Mu et al. found that zebrafish radial astrocytes are causal in regulating passivity generated by "futile" swim attempts such that radial astrocyte activation increased passivity, while silencing decreased passivity. After accumulated unsuccessful attempts, noradrenergic neurons in the medulla oblongata become active, and the released norepinephrine activates the α1-adenoceptor on radial astrocytes, which then activate GABAergic neurons in the brainstem to trigger behavioral passivity. In all, this work established zebrafish radial astrocytes as an essential player in a circuit that mediates an adaptive behavioral response.

It is interesting to note that spinal cord astrocytes (Chen et al. 2020) and hindbrain radial astrocytes (Mu et al. 2019) exhibit morphological differences, with hindbrain radial astrocytes maintaining a long primary process between the cell body and the dense branches. However, given similarities in molecular markers (both express *glast* and *gfap*) and responses to norepinephrine signaling, it seems likely that spinal cord astrocytes and hindbrain radial astrocytes represent the same cell type or closely related cell types in different CNS areas, whereby surrounding cells or structural constraints might play a role in regulating morphogenesis.

Roles for Astrocytes and Radial Glia in Injury and Disease Models

The injury response of zebrafish radial glia and bridging glia, particularly in the adult CNS, has been extensively discussed (e.g., Lyons and Talbot 2015; Jurisch-Yaksi et al. 2020; Becker and Becker 2022) and we direct the reader to these resources for more information. In the future, it will be important to test whether stellate astrocytes, which are abundant in the larval CNS (Chen et al. 2020), persist in adulthood and how these cells respond to injury and repair. Beyond injury, the study of zebrafish astrocytes can be a powerful contribution to our understanding

of disease models. For example, zebrafish have been used for decades in numerous models of epilepsy (Yaksi et al. 2021), and recent work has uncovered key roles for radial glia in pentylenetetrazole (PTZ)-induced epilepsy. Diaz-Verdugo and colleagues demonstrated that Ca^{2+} signaling in radial glia is highly active and strongly synchronized compared to neurons before seizures began. During seizures, synchronization of radial glia and neural activity increased, and activation of radial glia using optogenetic approaches could strongly modulate neural activity by glutamate and gap junctions (Verdugo et al. 2019).

CONCLUSIONS AND FUTURE DIRECTIONS

The use of zebrafish began as a discovery model to identify genes important for different aspects of development and has since then transitioned toward a highly versatile model organism with which to study glial cell biology by combining genetics, imaging, and physiology of intercellular communication in an intact living animal. With an ever-increasing set of reagents and assays to visualize and manipulate cells of interest, and the continued advancement of microscopy approaches available, we anticipate that intravital imaging will continue to be one of the main strengths that this model provides to longitudinally study glial cell function and their complex interactions in the same animal over time. With the advent of CRISPR/Cas9-mediated genome editing, the direct targeting of specific genes is now highly efficient and allows for the rapid generation of knockins, for example to insert floxed alleles to enable cell-type-specific gene disruption, which has historically been unavailable to the community (Liu et al. 2022). We anticipate that such approaches will also become standard when studying glial cell biology in zebrafish in the next few years. Regardless of these technological considerations, moving the field forward will also necessitate looking beyond understanding the biology of glial cells themselves. It will be very interesting to dissect how glial cells integrate into the multicellular CNS, and how they help regulate formation, function, and dysfunction of the CNS. Possessing all the major classes of glial

Cite this article as *Cold Spring Harb Perspect Biol* doi: 10.1101/cshperspect.a041350

cells as well as a true vasculature, zebrafish represent a powerful tool with which to study glial development, neuron–glial interactions, glial–glial interactions, and glial–vascular interactions, all in intact circuits in a living, behaving vertebrate. Some recent studies have already integrated glial biology into systems neuroscience questions (Mu et al. 2019) and vice versa, that is, integrated circuit approaches into questions that relate to glial biology (Xiao et al. 2022), showing that zebrafish research is excellently suited to move forward toward an integrated understanding of glial cells for nervous system formation, function, and dysfunction.

REFERENCES

Ackerman SD, Monk KR. 2016. The scales and tales of myelination: using zebrafish and mouse to study myelinating glia. *Brain Res* **1641:** 79–91. doi:10.1016/j.brainres.2015.10.011

Allen NJ, Bennett ML, Foo LC, Wang GX, Chakraborty C, Smith SJ, Barres BA. 2012. Astrocyte glypicans 4 and 6 promote formation of excitatory synapses via GluA1 AMPA receptors. *Nature* **486:** 410–414. doi:10.1038/nature11059

Almeida AR, Macklin WB. 2023. Early myelination involves the dynamic and repetitive ensheathment of axons which resolves through a low and consistent stabilization rate. *eLife* **12:** e82111. doi:10.7554/eLife.82111

Almeida RG, Czopka T, ffrench-Constant C, Lyons DA. 2011. Individual axons regulate the myelinating potential of single oligodendrocytes in vivo. *Development* **138:** 4443–4450. doi:10.1242/dev.071001

Almeida RG, Williamson JM, Madden ME, Early JJ, Voas MG, Talbot WS, Bianco IH, Lyons DA. 2021. Myelination induces axonal hotspots of synaptic vesicle fusion that promote sheath growth. *Curr Biol* **31:** 3743–3754.e5. doi:10.1016/j.cub.2021.06.036

Auer F, Vagionitis S, Czopka T. 2018. Evidence for myelin sheath remodeling in the CNS revealed by in vivo imaging. *Curr Biol* **28:** 549–559.e3. doi:10.1016/j.cub.2018.01.017

Auguste YSS, Ferro A, Kahng JA, Xavier AM, Dixon JR, Vrudhula U, Nichitiu AS, Rosado D, Wee TL, Pedmale UV, et al. 2022. Oligodendrocyte precursor cells engulf synapses during circuit remodeling in mice. *Nat Neurosci* **25:** 1273–1278. doi:10.1038/s41593-022-01170-x

Bacmeister CM, Barr HJ, McClain CR, Thornton MA, Nettles D, Welle CG, Hughes EG. 2020. Motor learning promotes remyelination via new and surviving oligodendrocytes. *Nat Neurosci* **23:** 819–831. doi:10.1038/s41593-020-0637-3

Baraban M, Koudelka S, Lyons DA. 2017. Ca^{2+} activity signatures of myelin sheath formation and growth in vivo. *Nat Neurosci* **19:** 1–23. doi:10.1038/s41593-017-0040-x

Becker T, Becker CG. 2022. Regenerative neurogenesis: the integration of developmental, physiological and immune

signals. *Development* **149:** dev199907. doi:10.1242/dev.199907

Braasch I, Salzburger W, Meyer A. 2006. Asymmetric evolution in two fish-specifically duplicated receptor tyrosine kinase paralogons involved in teleost coloration. *Mol Biol Evol* **23:** 1192–1202. doi:10.1093/molbev/msk003

Buchanan J, Elabbady L, Collman F, Jorstad NL, Bakken TE, Ott C, Glatzer J, Bleckert AA, Bodor AL, Brittain D, et al. 2022. Oligodendrocyte precursor cells ingest axons in the mouse neocortex. *Proc Natl Acad Sci* **119:** e2202580119. doi:10.1073/pnas.2202580119

Burda JE, Sofroniew MV. 2014. Reactive gliosis and the multicellular response to CNS damage and disease. *Neuron* **81:** 229–248. doi:10.1016/j.neuron.2013.12.034

Bushong EA, Martone ME, Jones YZ, Ellisman MH. 2002. Protoplasmic astrocytes in CA1 stratum radiatum occupy separate anatomical domains. *J Neurosci* **22:** 183–192. doi:10.1523/JNEUROSCI.22-01-00183.2002

Cahoy JD, Emery B, Kaushal A, Foo LC, Zamanian JL, Christopherson KS, Xing Y, Lubischer JL, Krieg PA, Krupenko SA, et al. 2008. A transcriptome database for astrocytes, neurons, and oligodendrocytes: a new resource for understanding brain development and function. *J Neurosci* **28:** 264–278. doi:10.1523/JNEUROSCI.4178-07.2008

Call CL, Neely SA, Early JJ, James OG, Zoupi L, Williams AC, Chandran S, Lyons DA, Bergles DE. 2022. Oligodendrocytes form paranodal bridges that generate chains of myelin sheaths that are vulnerable to degeneration with age. bioRxiv doi:10.1101/2022.02.16.480718

Casano AM, Albert M, Peri F. 2016. Developmental apoptosis mediates entry and positioning of microglia in the zebrafish brain. *Cell Rep* **16:** 897–906. doi:10.1016/j.celrep.2016.06.033

Chen J, Poskanzer KE, Freeman MR, Monk KR. 2020. Live-imaging of astrocyte morphogenesis and function in zebrafish neural circuits. *Nat Neurosci* **23:** 1297–1306. doi:10.1038/s41593-020-0703-x

Chia K, Mazzolini J, Mione M, Sieger D. 2018. Tumor initiating cells induce Cxcr4-mediated infiltration of pro-tumoral macrophages into the brain. *eLife* **7:** e31918. doi:10.7554/eLife.31918

Chia K, Keatinge M, Mazzolini J, Sieger D. 2019. Brain tumours repurpose endogenous neuron to microglia signalling mechanisms to promote their own proliferation. *eLife* **8:** e46912. doi:10.7554/eLife.46912

Christopherson KS, Ullian EM, Stokes CCA, Mullowney CE, Hell JW, Agah A, Lawler J, Mosher DF, Bornstein P, Barres BA. 2005. Thrombospondins are astrocyte-secreted proteins that promote CNS synaptogenesis. *Cell* **120:** 421–433. doi:10.1016/j.cell.2004.12.020

Clarke LE, Barres BA. 2013. Emerging roles of astrocytes in neural circuit development. *Nat Rev Neurosci* **14:** 311–321. doi:10.1038/nrn3484

Cornell-Bell AH, Finkbeiner SM, Cooper MS, Smith SJ. 1990. Glutamate induces calcium waves in cultured astrocytes: long-range glial signaling. *Science* **247:** 470–473. doi:10.1126/science.1967852

Cunha MI, Su M, Cantuti-Castelvetri L, Müller SA, Schifferer M, Djannatian M, Alexopoulos I, van der Meer F, Winkler A, van Ham TJ, et al. 2020. Pro-inflammatory activation following demyelination is required for myelin clearance and oligodendrogenesis. *J Exp Med* **217**: e20191390. doi:10.1084/jem.20191390

Czopka T. 2016. Insights into mechanisms of central nervous system myelination using zebrafish. *Glia* **64**: 333–349. doi:10.1002/glia.22897

Czopka T, ffrench-Constant C, Lyons DA. 2013. Individual oligodendrocytes have only a few hours in which to generate new myelin sheaths in vivo. *Dev Cell* **25**: 599–609. doi:10.1016/j.devcel.2013.05.013

Dani JW, Chernjavsky A, Smith SJ. 1992. Neuronal activity triggers calcium waves in hippocampal astrocyte networks. *Neuron* **8**: 429–440. doi:10.1016/0896-6273(92)90271-E

Dimou L, Simons M. 2017. Diversity of oligodendrocytes and their progenitors. *Curr Opin Neurobiol* **47**: 73–79. doi:10.1016/j.conb.2017.09.015

Djannatian M, Timmler S, Arends M, Luckner M, Weil M-T, Alexopoulos I, Snaidero N, Schmid B, Misgeld T, Möbius W, et al. 2019. Two adhesive systems cooperatively regulate axon ensheathment and myelin growth in the CNS. *Nat Commun* **10**: 4794. doi:10.1038/s41467-019-12789-z

Djannatian M, Radha S, Weikert U, Safaiyan S, Wrede C, Deichsel C, Kislinger G, Rhomberg A, Ruhwedel T, Campbell DS, et al. 2023. Myelination generates aberrant ultrastructure that is resolved by microglia. *J Cell Biol* **222**: e202204010. doi:10.1083/jcb.202204010

Easley-Neal C, Foreman O, Sharma N, Zarrin AA, Weimer RM. 2019. CSF1R ligands IL-34 and CSF1 are differentially required for microglia development and maintenance in white and gray matter brain regions. *Front Immunol* **10**: 2199. doi:10.3389/fimmu.2019.02199

Elmore MRP, Najafi AR, Koike MA, Dagher NN, Spangenberg EE, Rice RA, Kitazawa M, Matusow B, Nguyen H, West BL, et al. 2014. Colony-stimulating factor 1 receptor signaling is necessary for microglia viability, unmasking a microglia progenitor cell in the adult brain. *Neuron* **82**: 380–397. doi:10.1016/j.neuron.2014.02.040

Erblich B, Zhu L, Etgen AM, Dobrenis K, Pollard JW. 2011. Absence of colony stimulation factor-1 receptor results in loss of microglia, disrupted brain development and olfactory deficits. *PLoS ONE* **6**: e26317. doi:10.1371/journal.pone.0026317

Eroglu C, Allen NJ, Susman MW, O'Rourke NA, Park CY, Özkan E, Chakraborty C, Mulinyawe SB, Annis DS, Huberman AD, et al. 2009. Gabapentin receptor α2δ-1 is a neuronal thrombospondin receptor responsible for excitatory CNS synaptogenesis. *Cell* **139**: 380–392. doi:10.1016/j.cell.2009.09.025

Eyo U, Molofsky AV. 2023. Defining microglial–synapse interactions. *Science* **381**: 1155–1156. doi:10.1126/science.adh7906

Ferrero G, Mahony CB, Dupuis E, Yvernogeau L, Ruggiero ED, Miserocchi M, Caron M, Robin C, Traver D, Bertrand JY, et al. 2018. Embryonic microglia derive from primitive macrophages and are replaced by *cmyb*-dependent definitive microglia in zebrafish. *Cell Rep* **24**: 130–141. doi:10.1016/j.celrep.2018.05.066

Ferrero G, Miserocchi M, Ruggiero ED, Wittamer V. 2020. A *csf1rb* mutation uncouples two waves of microglia development in zebrafish. *Development* **148**: dev194241. doi:10.1242/dev.194241

Foerster S, Hill MFE, Franklin RJM. 2019. Diversity in the oligodendrocyte lineage: plasticity or heterogeneity? *Glia* **25**: 2411.

Foerster S, Neumann B, McClain C, Canio LD, Chen CZ, Reich DS, Simons BD, Franklin RJ. 2020. Proliferation is a requirement for differentiation of oligodendrocyte progenitor cells during CNS remyelination. bioRxiv doi:10.1101/2020.05.21.108373

Franklin RJM, ffrench-Constant C. 2017. Regenerating CNS myelin—from mechanisms to experimental medicines. *Nat Rev Neurosci* **18**: 753–769. doi:10.1038/nrn.2017.136

Green LA, Nebiolo JC, Smith CJ. 2019. Microglia exit the CNS in spinal root avulsion. *PLoS Biol* **17**: e3000159. doi:10.1371/journal.pbio.3000159

Greter M, Lelios I, Pelczar P, Hoeffel G, Price J, Leboeuf M, Kündig TM, Frei K, Ginhoux F, Merad M, et al. 2012. Stroma-derived interleukin-34 controls the development and maintenance of Langerhans cells and the maintenance of microglia. *Immunity* **37**: 1050–1060. doi:10.1016/j.immuni.2012.11.001

Grupp L, Wolburg H, Mack AF. 2010. Astroglial structures in the zebrafish brain. *J Comp Neurol* **518**: 4277–4287. doi:10.1002/cne.22481

Hill RA, Patel KD, Goncalves CM, Grutzendler J, Nishiyama A. 2014. Modulation of oligodendrocyte generation during a critical temporal window after NG2 cell division. *Nat Neurosci* **17**: 1518–1527. doi:10.1038/nn.3815

Hines JH, Ravanelli AM, Schwindt R, Scott EK, Appel B. 2015. Neuronal activity biases axon selection for myelination in vivo. *Nat Neurosci* **18**: 683–689. doi:10.1038/nn.3992

Howe K, Clark MD, Torroja CF, Torrance J, Berthelot C, Muffato M, Collins JE, Humphray S, McLaren K, Matthews L, et al. 2013. The zebrafish reference genome sequence and its relationship to the human genome. *Nature* **496**: 498–503. doi:10.1038/nature12111

Hughes AN, Appel B. 2019. Oligodendrocytes express synaptic proteins that modulate myelin sheath formation. *Nat Commun* **10**: 4125. doi:10.1038/s41467-019-12059-y

Hughes AN, Appel B. 2020. Microglia phagocytose myelin sheaths to modify developmental myelination. *Nat Neurosci* **23**: 1055–1066. doi:10.1038/s41593-020-0654-2

Iyer H, Shen K, Meireles AM, Talbot WS. 2022. A lysosomal regulatory circuit essential for the development and function of microglia. *Sci Adv* **8**: eabp8321. doi:10.1126/sciadv.abp8321

Jaronen M, Wheeler MA, Quintana FJ. 2022. Protocol for inducing inflammation and acute myelin degeneration in larval zebrafish. *Star Protoc* **3**: 101134. doi:10.1016/j.xpro.2022.101134

Johnson K, Barragan J, Bashiruddin S, Smith CJ, Tyrrell C, Parsons MJ, Doris R, Kucenas S, Downes GB, Velez CM, et al. 2016. Gfap-positive radial glial cells are an essential progenitor population for later-born neurons and glia in the zebrafish spinal cord. *Glia* **64**: 1170–1189. doi:10.1002/glia.22990

Jurisch-Yaksi N, Yaksi E, Kizil C. 2020. Radial glia in the zebrafish brain: functional, structural, and physiological comparison with the mammalian glia. *Glia* **68**: 2451–2470. doi:10.1002/glia.23849

Kamen Y, Pivonkova H, Evans KA, Káradóttir RT. 2022. A matter of state: diversity in oligodendrocyte lineage cells. *Neuroscientist* **28:** 144–162. doi:10.1177/1073858420987208

Kana V, Desland FA, Casanova-Acebes M, Ayata P, Badimon A, Nabel E, Yamamuro K, Sneeboer M, Tan I-L, Flanigan ME, et al. 2019. CSF-1 controls cerebellar microglia and is required for motor function and social interaction. *J Exp Med* **216:** 2265–2281. doi:10.1084/jem.20182037

Karttunen MJ, Czopka T, Goedhart M, Early JJ, Lyons DA. 2017. Regeneration of myelin sheaths of normal length and thickness in the zebrafish CNS correlates with growth of axons in caliber. *PLoS ONE* **12:** e0178058. doi:10.1371/journal.pone.0178058

Keren-Shaul H, Spinrad A, Weiner A, Matcovitch-Natan O, Dvir-Szternfeld R, Ulland TK, David E, Baruch K, Lara-Astaiso D, Toth B, et al. 2017. A unique microglia type associated with restricting development of Alzheimer's disease. *Cell* **169:** 1276–1290.e17. doi:10.1016/j.cell.2017.05.018

Klingseisen A, Ristoiu A-M, Kegel L, Sherman DL, Rubio-Brotons M, Almeida RG, Koudelka S, Benito-Kwiecinski SK, Poole RJ, Brophy PJ, et al. 2019. Oligodendrocyte neurofascin independently regulates both myelin targeting and sheath growth in the CNS. *Dev Cell* **51:** 730–744.e6. doi:10.1016/j.devcel.2019.10.016

Koudelka S, Voas MG, Almeida RG, Baraban M, Soetaert J, Meyer MP, Talbot WS, Lyons DA. 2016. Individual neuronal subtypes exhibit diversity in CNS myelination mediated by synaptic vesicle release. *Curr Biol* **26:** 1447–1455. doi:10.1016/j.cub.2016.03.070

Krasnow AM, Ford MC, Valdivia LE, Wilson SW, Attwell D. 2018. Regulation of developing myelin sheath elongation by oligodendrocyte calcium transients in vivo. *Nat Neurosci* **21:** 24–28. doi:10.1038/s41593-017-0031-y

Kroehne V, Freudenreich D, Hans S, Kaslin J, Brand M. 2011. Regeneration of the adult zebrafish brain from neurogenic radial glia-type progenitors. *Development* **138:** 4831–4841. doi:10.1242/dev.072587

Kucukdereli H, Allen NJ, Lee AT, Feng A, Ozlu MI, Conatser LM, Chakraborty C, Workman G, Weaver M, Sage EH, et al. 2011. Control of excitatory CNS synaptogenesis by astrocyte-secreted proteins Hevin and SPARC. *Proc Natl Acad Sci* **108:** E440–E449. doi:10.1073/pnas.1104977108

Li Y, Du X, Liu C, Wen Z, Du J. 2012. Reciprocal regulation between resting microglial dynamics and neuronal activity in vivo. *Dev Cell* **23:** 1189–1202. doi:10.1016/j.devcel.2012.10.027

Li J, Miramontes TG, Czopka T, Monk KR. 2024. Synaptic input and Ca²⁺ activity in zebrafish oligodendrocyte precursor cells contribute to myelin sheath formation. *Nat Neurosci* doi:10.1038/s41593-023-01553-8

Lin H, Lee E, Hestir K, Leo C, Huang M, Bosch E, Halenbeck R, Wu G, Zhou A, Behrens D, et al. 2008. Discovery of a cytokine and its receptor by functional screening of the extracellular proteome. *Science* **320:** 807–811. doi:10.1126/science.1154370

Liu P, Du J, He C. 2013. Developmental pruning of early-stage myelin segments during CNS myelination in vivo. *Cell Res* **23:** 962–964. doi:10.1038/cr.2013.62

Liu F, Kambakam S, Almeida MP, Ming Z, Welker JM, Wierson WA, Schultz-Rogers LE, Ekker SC, Clark KJ, Essner JJ,

McGrail M. 2022. Cre/*lox* regulated conditional rescue and inactivation with zebrafish UFlip alleles generated by CRISPR-Cas9 targeted integration. *Elife* **11:** e71478. doi:10.7554/eLife.71478

Lyons DA, Talbot WS. 2015. Glial cell development and function in zebrafish. *Cold Spring Harb Perspect Biol* **7:** a020586. doi:10.1101/cshperspect.a020586

Lyons DA, Guy AT, Clarke JDW. 2003. Monitoring neural progenitor fate through multiple rounds of division in an intact vertebrate brain. *Development* **130:** 3427–3436. doi:10.1242/dev.00569

Ma Z, Stork T, Bergles DE, Freeman MR. 2016. Neuromodulators signal through astrocytes to alter neural circuit activity and behaviour. *Nature* **539:** 428–432. doi:10.1038/nature20145

Makinodan M, Rosen KM, Ito S, Corfas G. 2012. A critical period for social experience-dependent oligodendrocyte maturation and myelination. *Science* **337:** 1357–1360. doi:10.1126/science.1220845

Maldonado PP, Angulo MC. 2015. Multiple modes of communication between neurons and oligodendrocyte precursor cells. *Neurosci* **21:** 266–276.

Marisca R, Hoche T, Agirre E, Hoodless LJ, Barkey W, Auer F, Castelo-Branco G, Czopka T. 2020. Functionally distinct subgroups of oligodendrocyte precursor cells integrate neural activity and execute myelin formation. *Nat Neurosci* **23:** 363–374. doi:10.1038/s41593-019-0581-2

Marques S, Zeisel A, Codeluppi S, van Bruggen D, Falcão AM, Xiao L, Li H, Häring M, Hochgerner H, Romanov RA, et al. 2016. Oligodendrocyte heterogeneity in the mouse juvenile and adult central nervous system. *Science* **352:** 1326–1329. doi:10.1126/science.aaf6463

Marschallinger J, Iram T, Zardeneta M, Lee SE, Lehallier B, Haney MS, Pluvinage JV, Mathur V, Hahn O, Morgens DW, et al. 2020. Lipid-droplet-accumulating microglia represent a dysfunctional and proinflammatory state in the aging brain. *Nat Neurosci* **23:** 194–208. doi:10.1038/s41593-019-0566-1

Masuda T, Sankowski R, Staszewski O, Prinz M. 2020. Microglia heterogeneity in the single-cell era. *Cell Rep* **30:** 1271–1281. doi:10.1016/j.celrep.2020.01.010

Mazaheri F, Breus O, Durdu S, Haas P, Wittbrodt J, Gilmour D, Peri F. 2014. Distinct roles for BAI1 and TIM-4 in the engulfment of dying neurons by microglia. *Nat Commun* **5:** 4046. doi:10.1038/ncomms5046

McKenzie IA, Ohayon D, Li H, de Faria JP, Emery B, Tohyama K, Richardson WD. 2014. Motor skill learning requires active central myelination. *Science* **346:** 318–322. doi:10.1126/science.1254960

Menassa DA, Gomez-Nicola D. 2018. Microglial dynamics during human brain development. *Front Immunol* **9:** 1014. doi:10.3389/fimmu.2018.01014

Mensch S, Baraban M, Almeida RG, Czopka T, Ausborn J, Manira AE, Lyons DA. 2015. Synaptic vesicle release regulates myelin sheath number of individual oligodendrocytes in vivo. *Nat Neurosci* **18:** 628–630. doi:10.1038/nn.3991

Merlini M, Rafalski VA, Ma K, Kim KY, Bushong EA, Coronado PER, Yan Z, Mendiola AS, Sozmen EG, Ryu JK, et al. 2021. Microglial Gi-dependent dynamics regulate brain network hyperexcitability. *Nat Neurosci* **24:** 19–23. doi:10.1038/s41593-020-00756-7

Möller K, Brambach M, Villani A, Gallo E, Gilmour D, Peri F. 2022. A role for the centrosome in regulating the rate of neuronal efferocytosis by microglia in vivo. *eLife* **11**: e82094. doi:10.7554/eLife.82094

Molofsky AV, Krencik R, Krenick R, Ullian EM, Ullian E, Tsai H, Deneen B, Richardson WD, Barres BA, Rowitch DH. 2012. Astrocytes and disease: a neurodevelopmental perspective. *Gene Dev* **26**: 891–907. doi:10.1101/gad.188326.112

Morris AD, Kucenas S. 2021. A novel lysolecithin model for visualizing damage in vivo in the larval zebrafish spinal cord. *Front Cell Dev Biol* **9**: 654583. doi:10.3389/fcell.2021.654583

Mu Y, Bennett DV, Rubinov M, Narayan S, Yang CT, Tanimoto M, Mensh BD, Looger LL, Ahrens MB. 2019. Glia accumulate evidence that actions are futile and suppress unsuccessful behavior. *Cell* **178**: 27–43.e19. doi:10.1016/j.cell.2019.05.050

Münzel EJ, Becker CG, Becker T, Williams A. 2014. Zebrafish regenerate full thickness optic nerve myelin after demyelination, but this fails with increasing age. *Acta Neuropathol Commun* **2**: 77. doi:10.1186/s40478-014-0077-y

Nagai J, Yu X, Papouin T, Cheong E, Freeman MR, Monk KR, Hastings MH, Haydon PG, Rowitch D, Shaham S, et al. 2021. Behaviorally consequential astrocytic regulation of neural circuits. *Neuron* **109**: 576–596. doi:10.1016/j.neuron.2020.12.008

Nagy C, Maitra M, Tanti A, Suderman M, Théroux J-F, Davoli MA, Perlman K, Yerko V, Wang YC, Tripathy SJ, et al. 2020. Single-nucleus transcriptomics of the prefrontal cortex in major depressive disorder implicates oligodendrocyte precursor cells and excitatory neurons. *Nat Neurosci* **23**: 771–781. doi:10.1038/s41593-020-0621-y

Neely SA, Williamson JM, Klingseisen A, Zoupi L, Early JJ, Williams A, Lyons DA. 2022. New oligodendrocytes exhibit more abundant and accurate myelin regeneration than those that survive demyelination. *Nat Neurosci* **25**: 415–420. doi:10.1038/s41593-021-01009-x

Nugent AA, Lin K, van Lengerich B, Lianoglou S, Przybyla L, Davis SS, Llapashtica C, Wang J, Kim DJ, Xia D, et al. 2020. TREM2 regulates microglial cholesterol metabolism upon chronic phagocytic challenge. *Neuron* **105**: 837–854.e9. doi:10.1016/j.neuron.2019.12.007

Oikonomou G, Shaham S. 2011. The glia of *Caenorhabditis elegans. Glia* **59**: 1253–1263. doi:10.1002/glia.21084

Oosterhof N, Kuil LE, van der Linde HC, Burm SM, Berdowski W, van Ijcken WFJ, van Swieten JC, Hol EM, Verheijen MHG, van Ham TJ. 2018. Colony-stimulating factor 1 receptor (CSF1R) regulates microglia density and distribution, but not microglia differentiation in vivo. *Cell Rep* **24**: 1203–1217.e6. doi:10.1016/j.celrep.2018.06.113

Orger MB, Kampff AR, Severi KE, Bollmann JH, Engert F. 2008. Control of visually guided behavior by distinct populations of spinal projection neurons. *Nat Neurosci* **11**: 327–333. doi:10.1038/nn2048

Paolicelli RC, Bolasco G, Pagani F, Maggi L, Scianni M, Panzanelli P, Giustetto M, Ferreira TA, Guiducci E, Dumas L, et al. 2011. Synaptic pruning by microglia is necessary for normal brain development. *Science* **333**: 1456–1458. doi:10.1126/science.1202529

Park HC, Mehta A, Richardson JS, Appel B. 2002. Olig2 is required for zebrafish primary motor neuron and oligo-

dendrocyte development. *Dev Biol* **248**: 356–368. doi:10.1006/dbio.2002.0738

Perez-Catalan NA, Doe CQ, Ackerman SD. 2021. The role of astrocyte-mediated plasticity in neural circuit development and function. *Neural Dev* **16**: 1. doi:10.1186/s13064-020-00151-9

Peri F, Nüsslein-Volhard C. 2008. Live imaging of neuronal degradation by microglia reveals a role for v0-ATPase a1 in phagosomal fusion in vivo. *Cell* **133**: 916–927. doi:10.1016/j.cell.2008.04.037

Preston MA, Macklin WB. 2015. Zebrafish as a model to investigate CNS myelination. *Glia* **63**: 177–193. doi:10.1002/glia.22755

Pridans C, Raper A, Davis GM, Alves J, Sauter KA, Lefevre L, Regan T, Meek S, Sutherland L, Thomson AJ, et al. 2018. Pleiotropic impacts of macrophage and microglial deficiency on development in rats with targeted mutation of the *Csf1r* locus. *J Immunol* **201**: 2683–2699. doi:10.4049/jimmunol.1701783

Prinz M, Erny D, Hagemeyer N. 2017. Ontogeny and homeostasis of CNS myeloid cells. *Nat Immunol* **18**: 385–392. doi:10.1038/ni.3703

Ranawat N, Masai I. 2021. Mechanisms underlying microglial colonization of developing neural retina in zebrafish. *eLife* **10**: e70550. doi:10.7554/eLife.70550

Ravanelli AM, Appel B. 2015. Motor neurons and oligodendrocytes arise from distinct cell lineages by progenitor recruitment. *Genes Dev* **29**: 2504–2515. doi:10.1101/gad.271312.115

Rojo R, Raper A, Ozdemir DD, Lefevre L, Grabert K, Wollscheid-Lengeling E, Bradford B, Caruso M, Gazova I, Sánchez A, et al. 2019. Deletion of a Csf1r enhancer selectively impacts CSF1R expression and development of tissue macrophage populations. *Nat Commun* **10**: 3215. doi:10.1038/s41467-019-11053-8

Rowitch DH. 2004. Glial specification in the vertebrate neural tube. *Nat Rev Neurosci* **5**: 409–419. doi:10.1038/nrn1389

Rowitch DH, Kriegstein AR. 2010. Developmental genetics of vertebrate glial–cell specification. *Nature* **468**: 214–222. doi:10.1038/nature09611

Schafer DP, Lehrman EK, Kautzman AG, Koyama R, Mardinly AR, Yamasaki R, Ransohoff RM, Greenberg ME, Barres BA, Stevens B. 2012. Microglia sculpt postnatal neural circuits in an activity and complement-dependent manner. *Neuron* **74**: 691–705. doi:10.1016/j.neuron.2012.03.026

Scholz J, Klein MC, Behrens TEJ, Johansen-Berg H. 2009. Training induces changes in white-matter architecture. *Nat Neurosci* **12**: 1370–1371. doi:10.1038/nn.2412

Scott-Hewitt N, Perrucci F, Morini R, Erreni M, Mahoney M, Witkowska A, Carey A, Faggiani E, Schuetz LT, Mason S, et al. 2020. Local externalization of phosphatidylserine mediates developmental synaptic pruning by microglia. *EMBO J* **39**: e105380. doi:10.15252/embj.2020105380

Shen K, Sidik H, Talbot WS. 2016. The Rag-Ragulator complex regulates lysosome function and phagocytic flux in microglia. *Cell Rep* **14**: 547–559. doi:10.1016/j.celrep.2015.12.055

Shigetomi E, Patel S, Khakh BS. 2016. Probing the complexities of astrocyte calcium signaling. *Trends Cell Biol* **26**: 300–312. doi:10.1016/j.tcb.2016.01.003

Cite this article as *Cold Spring Harb Perspect Biol* doi: 10.1101/cshperspect.a041350

Silva NJ, Dorman LC, Vainchtein ID, Horneck NC, Molofsky AV. 2021. In situ and transcriptomic identification of microglia in synapse-rich regions of the developing zebrafish brain. *Nat Commun* **12:** 5916. doi:10.1038/s41467-021-26206-x

Smolders SMT, Kessels S, Vangansewinkel T, Rigo JM, Legendre P, Brône B. 2019. Microglia: brain cells on the move. *Prog Neurobiol* **178:** 101612. doi:10.1016/j.pneurobio.2019.04.001

Spitzer SO, Sitnikov S, Kamen Y, Evans KA, Kronenberg-Versteeg D, Dietmann S, de F O Jr, Agathou S, Káradóttir RT. 2019. Oligodendrocyte progenitor cells become regionally diverse and heterogeneous with age. *Neuron* **101:** 459–471.e5. doi:10.1016/j.neuron.2018.12.020

Squarzoni P, Oller G, Hoeffel G, Pont-Lezica L, Rostaing P, Low D, Bessis A, Ginhoux F, Garel S. 2014. Microglia modulate wiring of the embryonic forebrain. *Cell Rep* **8:** 1271–1279. doi:10.1016/j.celrep.2014.07.042

Stork T, Sheehan A, Tasdemir-Yilmaz OE, Freeman MR. 2014. Neuron-glia interactions through the heartless FGF receptor signaling pathway mediate morphogenesis of *Drosophila* astrocytes. *Neuron* **83:** 388–403. doi:10.1016/j.neuron.2014.06.026

Streisinger G, Walker C, Dower N, Knauber D, Singer F. 1981. Production of clones of homozygous diploid zebra fish (*Brachydanio rerio*). *Nature* **291:** 293–296. doi:10.1038/291293a0

Swinnen N, Smolders S, Avila A, Notelaers K, Paesen R, Ameloot M, Brône B, Legendre P, Rigo J. 2013. Complex invasion pattern of the cerebral cortex by microglial cells during development of the mouse embryo. *Glia* **61:** 150–163. doi:10.1002/glia.22421

Swire M, Kotelevtsev Y, Webb DJ, Lyons DA, ffrench-Constant C. 2019. Endothelin signalling mediates experience-dependent myelination in the CNS. *eLife* **8:** e49493. doi:10.7554/eLife.49493

Tan AM, Zhang W, Levine JM. 2005. NG2: a component of the glial scar that inhibits axon growth. *J Anat* **207:** 717–725. doi:10.1111/j.1469-7580.2005.00452.x

Than-Trong E, Bally-Cuif L. 2015. Radial glia and neural progenitors in the adult zebrafish central nervous system. *Glia* **63:** 1406–1428. doi:10.1002/glia.22856

Thorlakur J, Hreinn S, Stacy S, Ingileif J, Jonsson PV, Snaedal J, Bjornsson S, Huttenlocher J, Levey AI, Lah JJ, et al. 2013. Variant of *TREM2* associated with the risk of Alzheimer's disease. *N Engl J Med* **368:** 107–116. doi:10.1056/NEJMoa1211103

Tomassy GS, Berger DR, Chen HH, Kasthuri N, Hayworth KJ, Vercelli A, Seung HS, Lichtman JW, Arlotta P. 2014. Distinct profiles of myelin distribution along single axons

of pyramidal neurons in the neocortex. *Science* **344:** 319–324. doi:10.1126/science.1249766

Vagionitis S, Auer F, Xiao Y, Almeida RG, Lyons DA, Czopka T. 2022. Clusters of neuronal neurofascin prefigure the position of a subset of nodes of Ranvier along individual central nervous system axons in vivo. *Cell Rep* **38:** 110366. doi:10.1016/j.celrep.2022.110366

Verdugo CD, Myren-Svelstad S, Aydin E, Hoeymissen EV, Deneubourg C, Vanderhaeghe S, Vancraeynest J, Pelgrims R, Cosacak MI, Muto A, et al. 2019. Glia–neuron interactions underlie state transitions to generalized seizures. *Nat Commun* **10:** 3830. doi:10.1038/s41467-019-11739-z

Viganò F, Möbius W, Götz M, Dimou L. 2013. Transplantation reveals regional differences in oligodendrocyte differentiation in the adult brain. *Nat Neurosci* **16:** 1370–1372. doi:10.1038/nn.3503

Villani A, Benjaminsen J, Moritz C, Henke K, Hartmann J, Norlin N, Richter K, Schieber NL, Franke T, Schwab Y, et al. 2019. Clearance by microglia depends on packaging of phagosomes into a unique cellular compartment. *Dev Cell* **49:** 77–88.e7. doi:10.1016/j.devcel.2019.02.014

Wang Y, Szretter KJ, Vermi W, Gilfillan S, Rossini C, Cella M, Barrow AD, Diamond MS, Colonna M. 2012. IL-34 is a tissue-restricted ligand of CSF1R required for the development of Langerhans cells and microglia. *Nat Immunol* **13:** 753–760. doi:10.1038/ni.2360

Wu S, Nguyen LTM, Pan H, Hassan S, Dai Y, Xu J, Wen Z. 2020. Two phenotypically and functionally distinct microglial populations in adult zebrafish. *Sci Adv* **6:** eabd1160. doi:10.1126/sciadv.abd1160

Xiao Y, Petrucco L, Hoodless LJ, Portugues R, Czopka T. 2022. Oligodendrocyte precursor cells sculpt the visual system by regulating axonal remodeling. *Nat Neurosci* **25:** 280–284. doi:10.1038/s41593-022-01023-7

Xu J, Zhu L, He S, Wu Y, Jin W, Yu T, Qu JY, Wen Z. 2015. Temporal-spatial resolution fate mapping reveals distinct origins for embryonic and adult microglia in zebrafish. *Dev Cell* **34:** 632–641. doi:10.1016/j.devcel.2015.08.018

Xu J, Wang T, Wu Y, Jin W, Wen Z. 2016. Microglia colonization of developing zebrafish midbrain is promoted by apoptotic neuron and lysophosphatidylcholine. *Dev Cell* **38:** 214–222. doi:10.1016/j.devcel.2016.06.018

Yaksi E, Jamali A, Verdugo CD, Jurisch-Yaksi N. 2021. Past, present and future of zebrafish in epilepsy research. *FEBS J* **288:** 7243–7255. doi:10.1111/febs.15694

Zeisel A, Muñoz-Manchado AB, Codeluppi S, Lönnerberg P, Manno GL, Juréus A, Marques S, Munguba H, He L, Betsholtz C, et al. 2015. Brain structure. Cell types in the mouse cortex and hippocampus revealed by single-cell RNA-seq. *Science* **347:** 1138–1142. doi:10.1126/science.aaa1934

Generation of Mammalian Astrocyte Functional Heterogeneity

Theresa Bartels,[1] David H. Rowitch,[1] and Omer Ali Bayraktar[2]

[1]Department of Paediatrics and Wellcome-MRC Cambridge Stem Cell Institute, University of Cambridge, Cambridge CB2 0AW, United Kingdom

[2]Wellcome Sanger Institute, Wellcome Genome Campus, Hinxton, Cambridge CB10 1SA, United Kingdom

Correspondence: dhr25@medschl.cam.ac.uk; ob5@sanger.ac.uk

Mammalian astrocytes have regional roles within the brain parenchyma. Indeed, the notion that astrocytes are molecularly heterogeneous could help explain how the central nervous system (CNS) retains embryonic positional information through development into specialized regions into adulthood. A growing body of evidence supports the concept of morphological and molecular differences between astrocytes in different brain regions, which might relate to their derivation from regionally patterned radial glia and/or local neuron inductive cues. Here, we review evidence for regionally encoded functions of astrocytes to provide an integrated concept on lineage origins and heterogeneity to understand regional brain organization, as well as emerging technologies to identify and further investigate novel roles for astrocytes.

Although neurons are a functionally diverse population, astrocytes, the most abundant cell types in the mammalian brain, have traditionally been regarded as relatively homogenous and interchangeable. That view has been irreversibly changed, based on new lines of evidence that have emerged since the publication of Bayraktar et al. (2014). The literature now comprises further studies showing molecular and functional differences among astrocytes for neuron subtype production, neuron subtype survival, circuit formation, and/or maintenance. Increasing evidence suggests that astrocyte heterogeneity is encoded by both cell-intrinsic cues during embryonic pattern formation and cell-extrinsic signals from regional neurons.

A major driver for change has been single-cell transcriptomics, and more recently spatial transcriptomics (ST), enabling the discovery of astrocyte gene expression heterogeneity. These and other studies conclusively show that astrocytes from different regions of the central nervous system (CNS) display molecularly distinct characteristics, supporting the concept of functional diversification.

Profiling by single-cell/-nucleus RNA sequencing can reveal genes differentially expressed in astrocytes by region or level, and both may be biologically significant. Alternatively, activity reporters, such as genetically encoded calcium indicators (GECIs; Broussard et al. 2014; Yu et al. 2020), show a variety of regional readouts. Here, we define diversity at the functional level, which in almost every case is heralded by a regional restricted pattern of astrocyte gene expression, then proven by gain- or loss-of-function studies. Factors satisfying this criterion are few in number so far, but we speculate a lexicon of functionally distinct astrocytes, specifically imbued with a "bell or whistle" that optimizes the activity of a local circuit. Some examples follow in the sections below. In summary, here we review recent progress using new tools to investigate diversified mammalian astrocytes, how functional heterogeneity has been demonstrated among mature astrocyte populations, as well as potentially underlying mechanisms to establish the identity of regionally restricted astrocytic cohorts.

NEW TECHNOLOGIES FOR DISCOVERING ASTROCYTE HETEROGENEITY

Major advances in single-cell/-nucleus and ST technologies provide robust approaches to define astrocyte heterogeneity in the healthy and dis-

eased CNS. These transcriptomic studies can also guide hypothesis generation on diversified astrocyte functions that can be investigated using transgenic perturbations and cell/tissue culture systems. Here, we review these recent technological developments that are expanding our views of astrocyte heterogeneity.

Single-Cell Transcriptomics

Single-cell RNA sequencing (scRNA-seq) profiles the transcriptomes of individual cells in an unbiased manner (Fig. 1A). Compared to bulk methods, scRNA-seq provides a more powerful approach to dissect the cellular composition of complex tissues and discover gene expression heterogeneity within a given cell type. Various scRNA-seq methods offer different advantages: whereas droplet (Klein et al. 2015; Macosko et al. 2015) and split pool barcoding (Rosenberg et al. 2018)-based methods provide high-throughput to routinely profile tens to hundreds

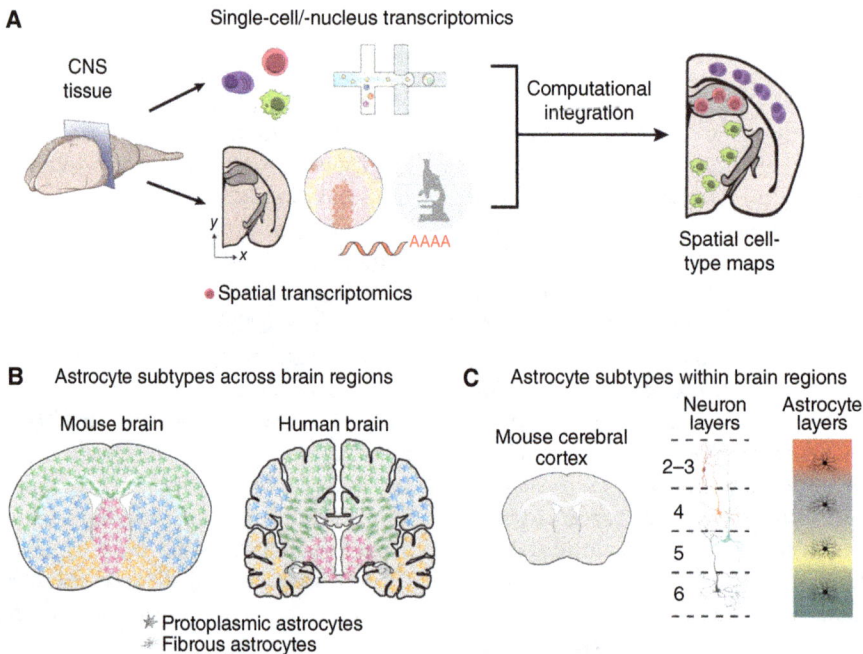

Figure 1. Regional astrocyte heterogeneity mapped using single-cell and spatial transcriptomics (ST). (*A*) Single-cell/-nucleus and ST can be used to uncover the regional molecular heterogeneity of central nervous system (CNS) cell types. (*B*) Mammalian astrocyte heterogeneity across brain regions. (*C*) Astrocyte heterogeneity within a brain region. Astrocyte subtypes across layers of the cerebral cortex are shown.

of thousands of cells, full-length scRNA-seq (Hagemann-Jensen et al. 2020) provides high sensitivity to capture lowly expressed genes. These methods are also applicable to single nuclei extracted from frozen tissue samples (Lake et al. 2016) and are being adapted to fixed archival samples (Vallejo et al. 2022; Janesick et al. 2023), opening the door to profiling a vast array of healthy and diseased tissues. However, because astrocytes express genes at relatively low levels, a caveat to be noted is that single-cell/-nucleus RNA-seq may underreport in this lineage. Indeed, one study used Smart-seq2 to enhance astrocyte lineage capture (Batiuk et al. 2020).

To date, scRNA-seq has been extensively applied to chart cellular diversity in the CNS. To name a few examples, in the developing brain, scRNA-seq aided the characterization of novel human neural stem cell (NSC) populations (Pollen et al. 2015), including new glial progenitors (Huang et al. 2020b), and has been combined with fate tracing to reveal lineage relationships at scale (Bandler et al. 2022). In the adult brain, scRNA-seq was used to characterize the extensive neuronal diversification across layers and regions of the cerebral cortex (Hodge et al. 2019; Yao et al. 2021) and is being extended to regional cellular atlases of the whole brain (Siletti et al. 2023). In neurodegenerative diseases, single-cell methods have also revealed vulnerable cell types and pathological gene expression programs across multiple disorders, including multiple sclerosis (Schirmer et al. 2019, 2021), Parkinson's (Kamath et al. 2022), and Alzheimer's disease (Mathys et al. 2019; Gabitto et al. 2023).

As elaborated in the following sections, scRNA-seq has revealed regional heterogeneity of astrocytes across the mouse (Zeisel et al. 2018; Batiuk et al. 2020; Farhy-Tselnicker et al. 2021) and human brain (Hodge et al. 2019; Siletti et al. 2023). Furthermore, single-cell methods have also identified diverse pathological astrocyte states (Schirmer et al. 2019; Hasel et al. 2021; Burda et al. 2022; Sadick et al. 2022) demonstrating the utility of scRNA-seq for studying astrocyte diversity in health and disease.

Beyond describing cellular heterogeneity, single-cell methods can also predict cell–cell interactions in complex tissues and provide clues on specialized functions of identified cell types. Several computational methods can infer cellular interactions from scRNA-seq data based on the complementary gene expression patterns of cell surface receptors and ligands (Efremova et al. 2020; Jin et al. 2021; Dimitrov et al. 2022). Furthermore, these interactions can be putatively linked to downstream signaling pathways and gene regulatory networks inferred from single-cell gene expression data (Browaeys et al. 2020).

To date, cell interaction analysis has been widely applied to immune cell types with extensively characterized repertoires of surface receptors (Vento-Tormo et al. 2018; Shilts et al. 2022). While we lack high-quality curations of neural receptor–ligands, we believe this is a promising approach to decipher cellular interactions in the CNS and functionally annotate astrocyte subtypes, being already applied to study human astrocyte development (Voss et al. 2023) and microglia heterogeneity in the cerebral cortex (Stogsdill et al. 2022). Perturbation screens incorporating scRNA-seq readouts represent another promising approach to map and functionally investigate neural cell interactions, exemplified by the study of astrocyte–microglial cross talk (Wheeler et al. 2023).

Finally, while transcriptomics is the most developed and widely used single-cell technology, other modalities are increasingly applied to single cells, including chromatin accessibility profiling (Buenrostro et al. 2015; Chen et al. 2019), proteomics (Brunner et al. 2022), and metabolomics (Duncan et al. 2019), and these promise a richer characterization of astrocyte heterogeneity in the future.

Spatial Transcriptomics

The spatial organization of cell types is a fundamental feature of tissue architecture and underlies cell–cell communication, organ function, and pathology. Whereas scRNA-seq can characterize cellular heterogeneity, it requires the dissociation of tissues into cell suspensions and does not capture spatial locations of cell types in intact tissues. ST technologies, developed in the past decade, address this need and offer robust approaches to profile cellular transcriptomes and tissue microenvironments in situ (Fig. 1A; Tian et al. 2023).

ST methods can be broadly classified as sequencing- or imaging-based approaches (extensively reviewed in Rao et al. 2021; Moffitt et al. 2022). Sequencing-based ST technologies enable unbiased profiling of whole transcriptomes in situ. These methods use various approaches to positionally capture RNA (e.g., Visium ST [Ståhl et al. 2016], Slide-seq [Rodrigues et al. 2019], Stereo-seq [Chen et al. 2022], DBiT-seq [Liu et al. 2020]) or cell/nuclei (e.g., XYZeq [Lee et al. 2021], Slide-tags [Russell et al. 2024]) from thin tissue sections, followed by sequencing readouts to spatially profile transcriptomes. While methods differ with regard to their spatial resolution and throughput, the most widely used sequencing-based ST technologies such as Visium and Slide-seq do not provide true single-cell resolution yet can be scaled to many samples for in situ transcriptome mapping (Langlieb et al. 2023).

Alternatively, imaging-based ST technologies enable targeted analysis of transcripts at single-cell or subcellular resolution in situ. The most widely used technologies image transcripts in tissues via probe-based detection approaches based on single-molecule fluorescence in situ hybridization (smFISH) (Rao et al. 2021). Various technologies differ with regards to multiplexing, sensitivity, and throughput. Methods such as RNAscope (Wang et al. 2012) and LaSTmap (Bayraktar et al. 2020) can spatially map the expression of a handful of genes at high sensitivity, whereas methods such as MERFISH (Chen et al. 2015; Moffitt et al. 2018; Xia et al. 2019), in situ sequencing (Ke et al. 2013; Qian et al. 2020), and seqFISH (Lubeck et al. 2014; Shah et al. 2016; Eng et al. 2019) use iterative labeling strategies to commonly profile 100–1000 multiplexed genes. The recent commercialization and automation of highly multiplexed imaging-based ST methods have increased their ease of use and extended their applications (He et al. 2022; Janesick et al. 2023; Yao et al. 2023), while continuing technological developments are moving toward imaging whole transcriptomes at single-cell resolution (Fang et al. 2022) as well as 3D mapping in situ (Wang et al. 2018).

Given their complementary strengths, sequencing-based ST is often used for discovery science, with the aim of identifying spatial locations of resident cell types and regionalized gene expression patterns in tissue samples, whereas imaging-based ST is commonly used for single-cell resolution mapping of targeted cell types and orthogonal validation of scRNA-seq findings in situ (Rao et al. 2021). Computational integration with scRNA-seq can extend insights from ST data (Fig. 1A), enabling deconvolution of individual cell types in sequencing-based ST (Kleshchevnikov et al. 2022) and imputation of whole transcriptomes in imaging-based ST (Ghazanfar et al. 2023; Li et al. 2023b). Importantly, ST provides a robust approach to interrogate tissue microenvironments (Kleshchevnikov et al. 2022) and identify cell–cell interactions between spatially resolved neighboring cell types (Garcia-Alonso et al. 2021; Tanevski et al. 2022; Fischer et al. 2023).

To date, ST has been widely applied to the CNS to chart spatial maps of neuronal and glial cell types across anatomical divisions of the nervous system. Initial ST studies on the spatial organization of individual mouse brain regions (Ståhl et al. 2016; Codeluppi et al. 2018; Moffitt et al. 2018; Wang et al. 2018; Eng et al. 2019; Rodrigues et al. 2019; Bayraktar et al. 2020; Zhang et al. 2021) have been extended to brain-wide spatial cell type maps based on sequencing (Langlieb et al. 2023) and imaging (Borm et al. 2023; Yao et al. 2023; Zhang et al. 2023) ST methods. These studies are creating high-resolution cellular maps of the brain that bridge the classical anatomical definitions of neural cell types with their molecular identities. ST has also been applied to map the spatial organization of the developing (Bhaduri et al. 2021; Braun et al. 2023) and adult human brain (Maynard et al. 2021; Fang et al. 2022), and examine pathological microenvironments in neurodegenerative disorders (Schirmer et al. 2019; Chen et al. 2020). In addition, ST has been coupled with other imaging-based assays of neuronal activity (Moffitt et al. 2018; Sun et al. 2021) and connectivity (Sun et al. 2021), unifying multiple facets of neuronal identity.

As elaborated in the following sections, ST has also helped resolve the regional heterogeneity of astrocytes in health (Zeisel et al. 2018; Batiuk et al. 2020; Bayraktar et al. 2020; Kleshchevnikov et al. 2022; Yao et al. 2023) and disease (Schirmer et al. 2019; Lerma-Martin et al. 2022). Hence, ST has already proven useful for the dissection of neural tissue microenvironments, and for regional mapping of astrocyte subtypes specialized to support local neural circuits or linked to neural pathologies.

Finally, in parallel with the development of single-cell multiomic technologies, spatial methods are being rapidly extended to capture other modalities such as chromatin accessibility (Deng et al. 2022), proteomics (Bhatia et al. 2022), and translational profiling (Zeng et al. 2023).

Transgenic Tools to Identify and Manipulate Astrocyte Subpopulations

Transcriptomic approaches described above have highlighted molecular differences between astrocyte subpopulations. As part of this, transgenic mice have been useful for fluorescence-activated cell sorting purification, one example being the Aldh1L1-GFP reporter mouse generated by the GENSAT program (Tien et al. 2012). Arguably, single-cell/-nucleus RNA transcriptomic approaches obviate the need for purification before RNA sequencing. Nevertheless, mouse transgenic reporter lines remain important tools to validate expression across domains in mammalian CNS, and are critical as cre recombinase drivers for conditional gene manipulation within the astrocyte lineage. Other reviews have highlighted the pluses and minuses of GFAP, Aldh1L1, and other existing drivers for astrocytes and radial glia (Yu et al. 2020) so these will not be covered in detail here. We have used Aldh1L1-cre and GFAP-cre in the ventral spinal cord where expression is reasonably restricted to astroglial lineage (Molofsky et al. 2014), but the same cre drivers lead to ectopic expression in targeting neurons of the dorsal spinal cord and forebrain. Unfortunately, there has not been significant progress in the creation of new useful driver lines for gene manipulation in astrocyte subpopulations in the cortex, with a few notable exceptions (Srinivasan et al. 2016; Miller et al. 2019). Thus, important goals are to develop transgenic mouse lines for conditional gene manipulation that are

drivers: (1) exclusive to the astrocyte lineage, (2) that direct expression at mature or immature stages in brain and/or spinal cord (some of these issues are addressed by using inducible, e.g., creERT2 systems [Srinivasan et al. 2016]), and (3) in a variety of regional subsets of astrocytes. Other useful reporter lines for live cell imaging of calcium and glutamate signaling in astrocytes have provided a vivid picture of regional differences in activity in response to various stimuli, and optogenetic approaches have also been useful. The finding that fish and flies have CNS astrocytes (Tasdemir-Yilmaz and Freeman 2014; Nagai et al. 2021; Neely and Lyons 2021), opens up important possibilities to manipulate astrocyte gene function and activity in nonmammalian models.

HETEROGENEITY OF MAMMALIAN ASTROCYTES

Astrocytes have an extensive repertoire of functions in the mammalian CNS. Recent studies have aimed to address whether this wide spectrum of roles is (1) shared by all astrocytes, (2) divided between functionally heterogeneous astrocyte subtypes intermixed in all CNS domains, or (3) segregated to regional astrocyte domains with specialized local functions in neural circuit support. Here, we review the recent evidence on morphological, molecular, and functional heterogeneity of astrocytes in the adult CNS.

Morphological Heterogeneity of Astrocytes

Astrocytes can be categorized into two broad morphological subtypes found in white versus gray matter compartments (Miller and Raff 1984; Oberheim et al. 2012; Ben Haim and Rowitch 2017; Salmon et al. 2023). Fibrous astrocytes align along white matter tracts, make contact with oligodendrocytes and axons at the nodes of Ranvier, have a regular, star-like appearance, and are characterized by a high density of intermediate filaments. Protoplasmic astrocytes populate the gray matter, make contact with synapses and capillaries, have a rather irregular, bushy-like appearance with thousands of thin processes, and express low levels of intermediate filaments.

Protoplasmic astrocytes organize into distinct, nonoverlapping domains, but couple together as astroglial syncytia via gap junctions; fibrous astrocytes instead form smaller astrocytic networks and their processes overlap (Oberheim et al. 2012).

It is becoming increasingly clear that the morphology of protoplasmic astrocytes varies across and within CNS regions. In the rodent cerebral and cerebellar cortex, protoplasmic astrocytes further segregate into different morphological subtypes in a layer-specific manner (Farmer et al. 2016; Lanjakornsiripan et al. 2018). In the cerebral cortex, astrocytes residing in the upper versus deep layers are distinct in their size, the orientation of their cellular processes across the cortical depth and structural association with synapses (Lanjakornsiripan et al. 2018). In the cerebellar cortex, morphologically distinct Bergmann glia and velate astrocytes reside in the Purkinje and granule cell layers, respectively (Farmer et al. 2016). Astrocyte morphological diversity is equally observed across other mouse brain areas (Emsley and Macklis 2006; Chai et al. 2017; Endo et al. 2022), implying functional specialization for distinct regional neural circuits.

The morphological heterogeneity of astrocytes is also apparent in the human brain. Interlaminar astrocytes are located in layer 1 of the cerebral cortex and possess unusually long cellular processes extending through several cortical layers (Colombo and Reisin 2004; Oberheim et al. 2006, 2009, 2012), with increased morphological complexity in primates compared to other mammals (Falcone et al. 2019). Varicose astrocytes reside in deep cortical layers and white matter, and extend long cellular processes decorated with varicosities (Oberheim et al. 2006, 2009, 2012); these astrocytes appear to be specific to humans and great apes (Falcone et al. 2022).

Molecular Heterogeneity of Astrocytes

We now appreciate that astrocyte gene expression is regionally diversified across the mammalian CNS. To date, a large number of transcriptomic profiling studies have identified substantial gene expression differences between protoplasmic astrocytes from different anatomical divisions of the CNS. These include a wide array of genes involved in neural development, homeostasis, and neurodegeneration, strongly implying that astrocytes are regionally specialized to perform a variety of functions in the healthy and diseased brain.

Early findings from bulk transcriptomic profiling of regional astrocyte populations in the mouse CNS have now been enriched with single-cell and ST, conclusively demonstrating the molecular diversity of astrocytes across the rostrocaudal and dorsoventral divisions of the brain and spinal cord (Zeisel et al. 2015; Farmer et al. 2016; Chai et al. 2017; John Lin et al. 2017; Morel et al. 2017; Boisvert et al. 2018; Clarke et al. 2018; Itoh et al. 2018; Lanjakornsiripan et al. 2018; Zeisel et al. 2018; Miller et al. 2019; Morel et al. 2019; Batiuk et al. 2020; Bayraktar et al. 2020; Endo et al. 2022). Initial studies sampled astrocytes across select brain areas including the olfactory bulb (John Lin et al. 2017; Endo et al. 2022), cerebral cortex (Zeisel et al. 2015; John Lin et al. 2017; Morel et al. 2017; Boisvert et al. 2018; Clarke et al. 2018; Itoh et al. 2018; Lanjakornsiripan et al. 2018; Miller et al. 2019; Morel et al. 2019; Batiuk et al. 2020; Bayraktar et al. 2020; Endo et al. 2022), hippocampus (Chai et al. 2017; Morel et al. 2017; Clarke et al. 2018; Itoh et al. 2018; Batiuk et al. 2020; Endo et al. 2022), striatum (Chai et al. 2017; Morel et al. 2017; Clarke et al. 2018; Endo et al. 2022), thalamus (John Lin et al. 2017; Morel et al. 2017; Endo et al. 2022), hypothalamus (Morel et al. 2017; Boisvert et al. 2018; Endo et al. 2022), midbrain (John Lin et al. 2017; Endo et al. 2022), hindbrain (John Lin et al. 2017; Endo et al. 2022), cerebellum (Farmer et al. 2016; John Lin et al. 2017; Boisvert et al. 2018; Itoh et al. 2018; Endo et al. 2022), and spinal cord (Molofsky et al. 2014; Itoh et al. 2018; Endo et al. 2022). Recent atlassing studies extended transcriptomic profiling of neural cell types, including astrocytes, to effectively cover the entire mouse brain (Saunders et al. 2018; Zeisel et al. 2018; Endo et al. 2022; Yao et al. 2023; Zhang et al. 2023).

Collectively, these molecular profiling studies suggest that astrocyte diversity largely follows the regional ontogeny of the nervous system (Fig. 1B). Protoplasmic astrocytes from developmentally distinct CNS divisions show the most substantial gene expression differences, giving rise to "broad" regional astrocyte subtypes from the olfactory bulb, telencephalon, diencephalon, brainstem, cerebellum, and spinal cord (Zeisel et al. 2018; Yao et al. 2023). The regional astrocyte gene expression domains, resolved using smFISH (Zeisel et al. 2018) and ST (Kleshchevnikov et al. 2022; Yao et al. 2023; Zhang et al. 2023), closely match the anatomical boundaries of brain areas, such as the telencephalon/diencephalon border (Zeisel et al. 2018).

Furthermore, astrocytes are spatially patterned into finer-grained molecular subtypes within each CNS region (Fig. 1C). In the spinal cord (Molofsky et al. 2014) and midbrain (Zeisel et al. 2018), dorsal and ventral astrocytes show distinct gene expression patterns. In the cerebral cortex, astrocytes show rostrocaudal and dorsoventral gene expression gradients across cortical areas (Boisvert et al. 2018; Bayraktar et al. 2020). Within each area, we identified that astrocytes are further diversified across cortical layers, giving rise to molecularly distinct superficial, mid- and deep-layer astrocytes (Bayraktar et al. 2020). Using ST, we observed that the boundaries of cortical astrocyte layers are more coarsely defined than neuronal laminae (Bayraktar et al. 2020), suggesting that the spatial organization of astrocytes can also diverge from that of neurons. Yet, we also identified that astrocytes can show a fine degree of spatial organization, demonstrated by molecularly distinct astrocyte subtypes enriched in the medial habenula (Kleshchevnikov et al. 2022).

What is the extent of astrocyte molecular heterogeneity across and within brain regions? Full-length scRNA-seq of cortical versus hippocampal astrocytes has identified both shared and distinct molecular subtypes across the two brain areas, where ~30% of all astrocyte-expressed genes are shared between subtypes and can be linked to common "core" astrocyte functions, while ~70% are subtype-specific (Batiuk et al. 2020). These subtype-enriched genes are associated with a wide array of astrocyte functions including synaptogenesis, neurotransmission, and ion transport (Batiuk et al. 2020). These observations are consistent with brain-wide bulk tran-

scriptomic profiling studies (Endo et al. 2022) and suggest that a significant component of astrocyte function in the CNS is channeled through specialized subtypes within and across brain regions.

Other studies investigating astrocyte profiles from the dorsal and ventral spinal cord (Molofsky et al. 2014), revealed only about 20 differentially expressed genes out of thousands and would, therefore, suggest that astrocytes are rather "homogeneous." However, when tested, several of these differentially expressed genes (*Sema3a*, *Kcnj10*; see below) were demonstrated to have essential local functions for spinal motor neurons, so such differences in expression can be clear and important clues to region-restricted astrocyte-encoded function.

What are the molecular and cellular pathways differentiating regional astrocytes? Regionally expressed astrocyte genes are involved in a wide array of neural processes, from the regulation of neurite outgrowth and guidance (Molofsky et al. 2014), to synapse development and refinement (Batiuk et al. 2020; Endo et al. 2022) and glutamatergic neurotransmission (Zeisel et al. 2018; Siletti et al. 2023). Many of these regional astrocyte gene expression patterns are apparent throughout development and maintained into adulthood and in aging (Boisvert et al. 2018; Bayraktar et al. 2020). Furthermore, regional astrocytes in the adult brain are also distinguished by the expression of transcription factors that are linked to spatial patterning of the embryonic CNS (Zeisel et al. 2018). As reviewed in the section below, these molecular clues foreshadow the specialized functions of astrocytes in regional neural circuits across the CNS as well as the developmental origins of astrocyte heterogeneity.

The regional gene expression heterogeneity of astrocytes is also apparent in the human brain. In the human cerebral cortex, interlaminar astrocytes are transcriptomically distinct from their protoplasmic and fibrous counterparts (Hodge et al. 2019; Fang et al. 2022; Siletti et al. 2023), while protoplasmic astrocytes show layer-specific gene expression signatures (Bayraktar et al. 2020; Fang et al. 2022). A recent brain-wide cell atlassing study identified substantial regional diversity of human astrocytes (Siletti et al. 2023).

With the most prominent molecular differences observed between telencephalic and nontelencephalic brain areas, protoplasmic and fibrous astrocytes are patterned into molecularly distinct regional subtypes across the human brain (Siletti et al. 2023). While these findings are consistent with astrocyte diversity observed in mouse, there are substantial gene expression differences between human and mouse brain astrocytes (Kelley et al. 2018a; Hodge et al. 2019; Jorstad et al. 2023), and regional astrocyte diversity might show further evolutionary specializations in the human brain.

Finally, we are also beginning to appreciate the molecular diversity of astrocytes in disease. Spatiotemporally distinct astrocyte subtypes and cell states can be identified in neurodegenerative disease and after CNS injury (Schirmer et al. 2019, 2021; Hasel et al. 2021; Burda et al. 2022; Sadick et al. 2022). These pathological astrocyte states are covered in the literature.

Functional Heterogeneity of Astrocytes

While functional studies trail behind the transcriptomic characterization of regional astrocytes, functional astrocyte heterogeneity is no longer a speculative concept. To date, several exciting studies have examined whether the morphological and molecular heterogeneity of astrocytes translates into their functional diversification. As reviewed below, these studies are beginning to demonstrate that astrocytes are regionally specialized across the CNS to support local neuron subtype survival, circuit formation, and maintenance.

We have focused our functional studies of astrocyte diversity in the sensorimotor circuit, which is markedly organized along the dorsoventral axis of the spinal cord. Initially, we found that the genetic ablation of ventral astrocytes in mice perturbs specific synaptic inputs into their neighboring motor neurons (Tsai et al. 2012). Using bulk transcriptomic profiling, we found that dorsal and ventral spinal cord astrocytes differentially express several genes encoding extracellular matrix components or axon and cell migration factors in the CNS, including ventrally enriched expression of semaphorin 3a (Sema3a) (Molof-sky et al. 2014). Using astrocyte-specific conditional knockouts, we identified that ventral astrocyte-derived Sema3a is required for selective survival, correct synaptic input, and axon orientation of α-motor neurons (αMNs) (Molofsky et al. 2014). In the same circuit, we also found that ventral astrocytes support another neuronal population, the fast αMNs, through regionally restricted expression of the inwardly rectifying potassium channel Kir4.1 that selectively regulates fast αMN size and function (Kelley et al. 2018b). Taken together, these findings demonstrate diverse cellular functions for a regional astrocyte subtype.

Other studies have expanded our view of astrocyte functional diversity into the brain. In the lateral habenula, regionally enriched expression of Kir4.1 in astrocytes regulates neuronal bursting activity and the local circuits involved in depression (Cui et al. 2018). In the thalamus and spinal cord, interleukin 33 (IL-33)-expressing regional astrocytes are required for synaptic homeostasis and postnatal circuit development (Vainchtein et al. 2018). Strikingly, astrocyte-derived IL-33 signals primarily to microglia and recruits them to prune synapses, demonstrating that regional astrocyte roles are specialized to cooperate with diverse neural cell types (Vainchtein et al. 2018). In the hippocampus, local astrocytes selectively rely on the activity of the transcription factor nuclear factor I-A (NFIA) to regulate the plasticity of hippocampal neural circuits (Huang et al. 2020a). Striatal and hippocampal astrocytes differ in their morphology as well as their electrophysiological properties, foreshadowing their functional differences (Chai et al. 2017).

Beyond the examples above, transcriptomic profiling studies provide clues about the putative specialized roles of astrocytes in other brain regions. For example, upper cortical layer astrocytes show enriched expression of Chordin-like 1 (Bayraktar et al. 2020), which was shown, in a separate study (Blanco-Suarez et al. 2018), to drive synaptic maturation of cortical circuits. While less is known about the functional heterogeneity of human astrocytes, their molecular heterogeneity is consistent with local neural circuit-adapted astrocyte roles in the human brain.

Collectively, these findings demonstrate how the discovery of regional gene expression patterns in astrocytes can lead the efforts to uncover their functional heterogeneity. Given the extensive molecular heterogeneity of CNS astrocyte populations, we expect that much remains to be characterized about regional astrocyte roles. For future studies, examining cell–cell interactions between astrocytes and other neural cell types in integrated scRNA-seq and ST data sets provide key discovery opportunities. Furthermore, co-culture (Sloan et al. 2017; Li et al. 2023a) or organoid (Szebényi et al. 2021) systems coupled to high-throughput functional perturbations (Leng et al. 2022) can be used to systematically identify regional astrocyte roles.

SPECIFICATION OF ASTROCYTE HETEROGENEITY

How does astrocyte heterogeneity arise in the nervous system? During development, cell-extrinsic signaling cues and cell-intrinsic genetic programs act in combination to specify the generation of diverse neuronal fates in a spatiotemporally regulated manner. Here, we focus on brain and spinal cord development leading up to the diversification of astrocytes, and review emerging evidence on regulation of astrocyte heterogeneity.

Embryonic Patterning Specifies Neural Progenitor Heterogeneity

Regional Heterogeneity of Forebrain Neural Progenitors during Embryonic Development

How are NSCs programmed to produce diversified neural progeny? During embryonic development, radial glial (RG) cells derived from the neuroepithelium are the primary NSCs that produce neurons and glia throughout the brain (Noctor et al. 2002; Anthony et al. 2004; for reviews, see Kriegstein and Alvarez-Buylla 2009; Rowitch and Kriegstein 2010). These cells, like their neuroepithelial precursors, contain apicobasal specializations with their cell bodies retained in the ventricular zone (VZ). Although neuroepithelial cells and RG appear to be ho-

mogenous populations of dividing cells lining the ventricles, they are regionally specialized for producing distinct subtypes of neurons (Campbell 2003; Puelles and Rubenstein 2003). These progenitor domains are established by cell-extrinsic morphogens, such as dorsal bone morphogenetic proteins (BMPs), ventral Sonic hedgehog (Shh), Wnts from the cortical hem, etc., and are further refined in the dorsoventral (DV), anteroposterior (AP), and mediolateral (ML) axes by the expression of numerous transcription factors (Campbell 2003; Sur and Rubenstein 2005). Along the DV axis of the forebrain, major RG domains include the pallium (cortex), subpallial structures (lateral and medial ganglionic eminences, LGE and MGE, respectively), and the septum. Cortical progenitors produce different layers of projection neurons with distinct morphological and functional identities. Conversely, LGE progenitors produce projection neurons of the striatum and olfactory bulb interneurons, while MGE progenitors give rise to the interneurons that populate the different regions of the telencephalon, as well as projection neurons of the pallidum and basal forebrain (Campbell 2003; Wonders and Anderson 2006; Flames et al. 2007). Similar subdivisions are present in the anteroposterior axis of the forebrain (Puelles and Rubenstein 2003; Greig et al. 2013).

Spatial patterning also underlies embryonic spinal cord development (Fig. 2). Early in embryogenesis, gradients of cell-extrinsic organizing signals, such as ventral Shh and dorsal BMPs, lead to the segmentation of the spinal cord neuroepithelium into discrete DV-restricted progenitor domains (Jessell 2000). This is reflected in the segmental expression of homeodomain and basic helix–loop–helix (bHLH) transcription factors that subsequently engage in cross-repressive interactions to refine domain boundaries (Fig. 2A) (e.g., p0, p1, p2, and p3 domains that generate interneuron subtypes) and pMN domain that generates motor neurons. Thus, regional patterning across the CNS confers positional identity to neural progenitors and creates a segmental template that underlies neuronal diversity. The regional heterogeneity of neural progenitors also underlies the development of diversified glial subtypes at later developmental stages

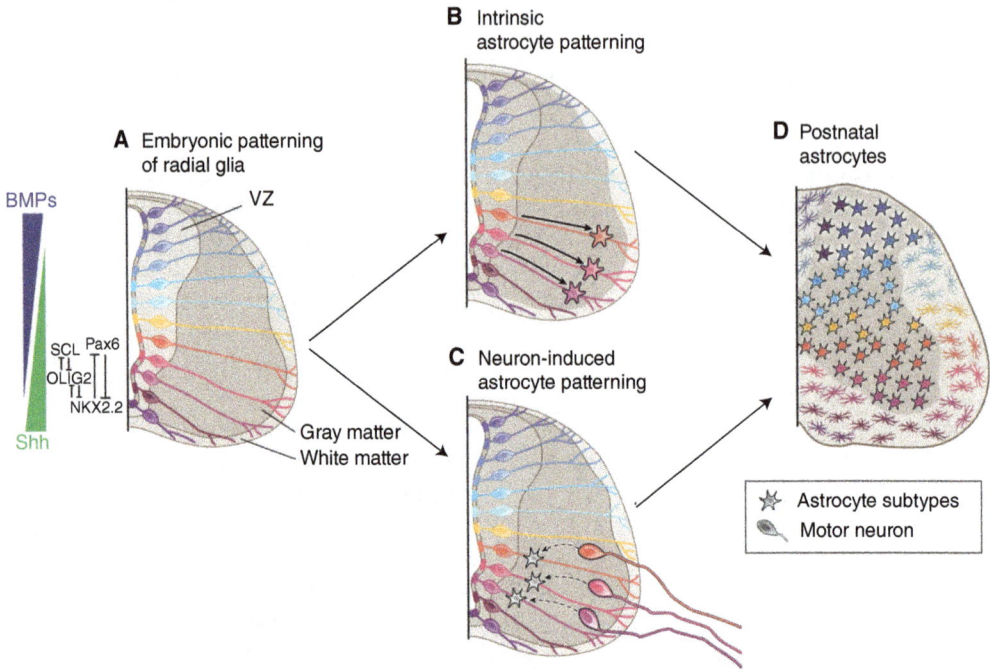

Figure 2. Models of intrinsic versus neuron-induced specification of regional astrocyte heterogeneity in the spinal cord. (A) Embryonic patterning along the dorsoventral axis gives rise to regionally specified radial glia residing in the ventricular zone (VZ). Signaling by Shh and bone morphogenetic proteins (BMPs) regulates the expression of segmental transcription factors, and cross-repressive interactions refine progenitor domains. After neurogenesis, astrocytes are generated in all domains of the spinal cord. They migrate laterally along the radial glia trajectories and do not undergo tangential migration from their domains of origin. (B) Embryonic patterning specifies regionally distinct astrocytes. The segmental code established during early patterning could be used during gliogenesis to specify segmental astrocyte subtypes, and determines the spatial allocation of astrocytes. (C) Extrinsic cues from local neurons might also determine regional astrocyte features (motor neurons are depicted). (D) Segmental heterogeneity of fibrous and protoplasmic astrocytes could give rise to regional astrocyte domains with distinct functions.

of embryogenesis (Rowitch and Kriegstein 2010) as discussed below.

Temporal Heterogeneity of NSCs during Embryonic Development

Temporal factors also contribute to neural diversification from neural progenitors and a number of different studies have demonstrated that the neurogenic potential of invertebrate neuroblasts and mammalian RG changes over time (for reviews, see Kohwi and Doe 2013; El-Danaf et al. 2023). Elegant studies in the *Drosophila* CNS have shown that individual NSCs could change over time to generate distinct neural subtypes; these temporal neural identities are specified by

sequentially expressed transcription factors in neural progenitors (Isshiki et al. 2001; Bayraktar and Doe 2013; Li et al. 2013). In vertebrates, temporal progenitor regulation has been studied in the developing cortex and retina (McConnell 1992; Cepko et al. 1996). More recent studies using scRNA-seq have identified temporal transcription factor series in neural progenitors across the CNS (Rayon et al. 2021; Sagner et al. 2021).

Cortical neurogenesis begins with the production of deeper projection neurons (peaking between embryonic day [E] 12.5 to E13.5 in mouse) and continues with successive generations of superficial projection neurons (E14.5–E15.5) (Cadwell et al. 2019). These periods of neurogenesis are then followed by increased gliogenesis. Astrocytes are first detected around E16 and oligodendrocytes around birth; however, the vast majority of both cell types are produced during the first month of postnatal development (Cadwell et al. 2019). Remarkably, this timing mechanism appears to rely on cell-intrinsic temporal cues since it can be recapitulated in clonal culture (Qian et al. 2000; Shen et al. 2006).

Taken together, it is now clear that the embryonic neuroepithelium is composed of a heterogeneous mix of progenitors that produce different neural subtypes within specific spatiotemporal boundaries. The developmental timing and location of RG, a property tightly linked to their neuroepithelial origin, appear to be the key determinants of the types of neurons generated. As we will describe below, similar, if not identical, spatial and temporal patterning programs appear to diversify the neuronal progeny of adult NSCs and, perhaps, parenchymal astrocytes into functional subtypes.

Development and Specification of Astrocyte Heterogeneity

During development, spatiotemporal patterning programs similar to those discussed above for neurons might also specify diverse astrocyte subtypes from gliogenic progenitors. This notion is consistent with the molecular and functional heterogeneity of regional astrocytes described in the previous section. Yet, regional astrocyte differences can be induced by extrinsic cues from regionally diversified neurons. Which mechanism might apply to an astrocyte cohort of interest? The spinal cord, cerebral cortex, and cerebellum are suitable systems to investigate this issue as the diversified cohort of astrocytes occupy the same domain as the cognate neuron populations and respond to specific signals from them. This section briefly reviews key studies that support two different mechanisms leading to diverse astrocyte end points using the Shh pathway as an exemplar.

Cell-Intrinsic (Developmental) Patterning of Astrocyte Heterogeneity

In the spinal cord, neural tube progenitor domains switch to glial production at about E12.5 in mouse

(Tien et al. 2012). Is astrocyte development in the spinal cord based on a segmental template? The genetic studies of Shh-regulated homeodomain and bHLH transcription factors suggested that astrocytes are specified from their progenitors in a position-dependent manner (Hochstim et al. 2008) and, moreover, that progeny cells stay close register to the domains of their RG progenitors in the spinal cord and brain, with little secondary movement (Fig. 2B; Tsai et al. 2012). The principle of patterning of astroglial progenitor function in terms of progeny output was demonstrated in the DV axis of the forebrain subventricular zone (Merkle et al. 2007). In this and subsequent studies (Molofsky et al. 2014), astroglia can retain positional phenotypes in vitro, suggesting they undergo epigenetic change postpatterning. Consistently, scRNA-seq studies identified that adult astrocyte populations in the mouse (Zeisel et al. 2018) and human brain (Siletti et al. 2023) maintain the regional expression patterns of embryonic patterning transcription factors. Finally, astrocyte heterogeneity within a CNS region can also map back to distinct developmental lineages, exemplified by distinct RG origins of protoplasmic versus fibrous astrocytes in the developing human cortex (Allen et al. 2022).

Our previous study of the astrocyte-encoded function of Sema3a is a compelling example of patterning leading to regionally specialized function. Importantly, the expression pattern of Sema3a in the ventral half of the spinal cord does not correspond to any neuron subtype (Molofsky et al. 2014), indicating that a patterning versus neuron-induced mechanism is required to explain upstream regulation of Sema3a. Despite this, we showed that this astrocyte region–restricted Sema3a function was critical for survival specifically of fast αMNs (Molofsky et al. 2014), suggesting that a regional astrocyte subtype can optimize the function of a particular circuit.

Cell-Extrinsic (Neuronal) Patterning of Astrocyte Heterogeneity

Alternatively, diversified neurons might induce diverse properties of astrocytes via cell-extrinsic cues (Fig. 2C). Indeed, previous studies have shown that neuron-derived signals could regulate mature astrocyte functions and physiology

(Swanson et al. 1997; Barnabé-Heider et al. 2005; Kaneko et al. 2010). In the cerebellum, the manipulation of Shh signaling from Purkinje cells led to a loss of various factors, including Kir4.1, from neighboring Bergmann glia (Farmer et al. 2016). Consistently, the activation of Shh signaling in velate astrocytes induces ectopic Bergmann glia–like gene expression in these astrocytes (Farmer et al. 2016). In the cerebral cortex, we and others showed, through the manipulation of cortical layer neuron identity and positioning, that neuronal cues instruct astrocyte layer identity (Lanjakornsiripan et al. 2018; Bayraktar et al. 2020). At a molecular level, the deletion of Shh from layer 5 neurons or its receptor Smoothened from astrocytes of the cerebral cortex resulted in the loss of region-restricted, deep-layer astrocyte marker expression (Xie et al. 2022).

These findings together suggest that it is likely that both CNS patterning and extrinsic cues from diversified neurons cooperate or act in a combinatorial fashion to determine spatial astrocyte heterogeneity (Fig. 2D). For example, embryonic patterning could establish certain aspects of regional astrocyte identity, and these properties could be refined by neuronal derived cues later during development.

CONCLUSION

While multiple new functions of astrocytes have been reported in recent years, important questions remain. For example, the extent of the functional heterogeneity of region-restricted astrocytes remains unclear. Genetic, transcriptomic, and proteomic tools to characterize astrocyte populations in the developing and mature CNS, combined with gene functional testing, will be critical in making further progress. Such studies are not only important for fundamental knowledge about diversified astrocyte cell–cell regulation of neural circuits; they will also deepen our understanding of pathological mechanisms underlying human neurological disease.

ACKNOWLEDGMENTS

T.B. acknowledges funding from the Wellcome Trust (109142/Z/15/Z) and a Junior Fellowship from the Loulou Foundation. O.A.B. acknowledges funding from the Wellcome Trust (220540/Z/20/A). D.H.R. acknowledges funding from the European Research Council (Advanced Grant 789054) and the Wellcome Trust (Investigator Award 88114). We thank Yeliz Demirci for comments on the manuscript.

REFERENCES

Allen DE, Donohue KC, Cadwell CR, Shin D, Keefe MG, Sohal VS, Nowakowski TJ. 2022. Fate mapping of neural stem cell niches reveals distinct origins of human cortical astrocytes. *Science* **376:** 1441–1446. doi:10.1126/science.abm5224

Anthony TE, Klein C, Fishell G, Heintz N. 2004. Radial glia serve as neuronal progenitors in all regions of the central nervous system. *Neuron* **41:** 881–890. doi:10.1016/S0896-6273(04)00140-0

Bandler RC, Vitali I, Delgado RN, Ho MC, Dvoretskova E, Ibarra Molinas JS, Frazel PW, Mohammadkhani M, Machold R, Maedler S, et al. 2022. Single-cell delineation of lineage and genetic identity in the mouse brain. *Nature* **601:** 404–409. doi:10.1038/s41586-021-04237-0

Barnabé-Heider F, Wasylnka JA, Fernandes KJ, Porsche C, Sendtner M, Kaplan DR, Miller FD. 2005. Evidence that embryonic neurons regulate the onset of cortical gliogenesis via cardiotrophin-1. *Neuron* **48:** 253–265. doi:10.1016/j.neuron.2005.08.037

Batiuk MY, Martirosyan A, Wahis J, de Vin F, Marneffe C, Kusserow C, Koeppen J, Viana JF, Oliveira JF, Voet T, et al. 2020. Identification of region-specific astrocyte subtypes at single cell resolution. *Nat Commun* **11:** 1220. doi:10.1038/s41467-019-14198-8

Bayraktar OA, Doe CQ. 2013. Combinatorial temporal patterning in progenitors expands neural diversity. *Nature* **498:** 449–455. doi:10.1038/nature12266

Bayraktar OA, Fuentealba LC, Alvarez-Buylla A, Rowitch DH. 2014. Astrocyte development and heterogeneity. *Cold Spring Harb Perspect Biol* **7:** a020362. doi:10.1101/cshperspect.a020362

Bayraktar OA, Bartels T, Holmqvist S, Kleshchevnikov V, Martirosyan A, Polioudakis D, Ben Haim L, Young AMH, Batiuk MY, Prakash K, et al. 2020. Astrocyte layers in the mammalian cerebral cortex revealed by a single-cell in situ transcriptomic map. *Nat Neurosci* **23:** 500–509. doi:10.1038/s41593-020-0602-1

Ben Haim L, Rowitch DH. 2017. Functional diversity of astrocytes in neural circuit regulation. *Nat Rev Neurosci* **18:** 31–41. doi:10.1038/nrn.2016.159

Bhaduri A, Sandoval-Espinosa C, Otero-Garcia M, Oh I, Yin R, Eze UC, Nowakowski TJ, Kriegstein AR. 2021. An atlas of cortical arealization identifies dynamic molecular signatures. *Nature* **598:** 200–204. doi:10.1038/s41586-021-03910-8

Bhatia HS, Brunner AD, Öztürk F, Kapoor S, Rong Z, Mai H, Thielert M, Ali M, Al-Maskari R, Paetzold JC, et al. 2022. Spatial proteomics in three-dimensional intact specimens. *Cell* **185:** 5040–5058.e19. doi:10.1016/j.cell.2022.11.021

Cite this article as *Cold Spring Harb Perspect Biol* doi: 10.1101/cshperspect.a041351

Blanco-Suarez E, Liu TF, Kopelevich A, Allen NJ. 2018. Astrocyte-secreted chordin-like 1 drives synapse maturation and limits plasticity by increasing synaptic GluA2 AMPA receptors. *Neuron* **100**: 1116–1132.e13. doi:10.1016/j.neuron.2018.09.043

Boisvert MM, Erikson GA, Shokhirev MN, Allen NJ. 2018. The aging astrocyte transcriptome from multiple regions of the mouse brain. *Cell Rep* **22**: 269–285. doi:10.1016/j.celrep.2017.12.039

Borm LE, Mossi Albiach A, Mannens CCA, Janusauskas J, Özgün C, Fernández-García D, Hodge R, Castillo F, Hedin CRH, Villablanca EJ, et al. 2023. Scalable in situ single-cell profiling by electrophoretic capture of mRNA using EEL FISH. *Nat Biotechnol* **41**: 222–231. doi:10.1038/s41587-022-01455-3

Braun E, Danan-Gotthold M, Borm LE, Lee KW, Vinsland E, Lönnerberg P, Hu L, Li X, He X, Andrusivová Ž, et al. 2023. Comprehensive cell atlas of the first-trimester developing human brain. *Science* **382**: eadf1226. doi:10.1126/science.adf1226

Browaeys R, Saelens W, Saeys Y. 2020. Nichenet: modeling intercellular communication by linking ligands to target genes. *Nat Methods* **17**: 159–162. doi:10.1038/s41592-019-0667-5

Broussard GJ, Liang R, Tian L. 2014. Monitoring activity in neural circuits with genetically encoded indicators. *Front Mol Neurosci* **7**: 97. doi:10.3389/fnmol.2014.00097

Brunner AD, Thielert M, Vasilopoulou C, Ammar C, Coscia F, Mund A, Hoerning OB, Bache N, Apalategui A, Lubeck M, et al. 2022. Ultra-high sensitivity mass spectrometry quantifies single-cell proteome changes upon perturbation. *Mol Syst Biol* **18**: e10798. doi:10.15252/msb.202110798

Buenrostro JD, Wu B, Litzenburger UM, Ruff D, Gonzales ML, Snyder MP, Chang HY, Greenleaf WJ. 2015. Single-cell chromatin accessibility reveals principles of regulatory variation. *Nature* **523**: 486–490. doi:10.1038/nature14590

Burda JE, O'Shea TM, Ao Y, Suresh KB, Wang S, Bernstein AM, Chandra A, Deverasetty S, Kawaguchi R, Kim JH, et al. 2022. Divergent transcriptional regulation of astrocyte reactivity across disorders. *Nature* **606**: 557–564. doi:10.1038/s41586-022-04739-5

Cadwell CR, Bhaduri A, Mostajo-Radji MA, Keefe MG, Nowakowski TJ. 2019. Development and arealization of the cerebral cortex. *Neuron* **103**: 980–1004. doi:10.1016/j.neuron.2019.07.009

Campbell K. 2003. Dorsal-ventral patterning in the mammalian telencephalon. *Curr Opin Neurobiol* **13**: 50–56. doi:10.1016/S0959-4388(03)00009-6

Cepko CL, Austin CP, Yang X, Alexiades M, Ezzeddine D. 1996. Cell fate determination in the vertebrate retina. *Proc Natl Acad Sci* **93**: 589–595. doi:10.1073/pnas.93.2.589

Chai H, Diaz-Castro B, Shigetomi E, Monte E, Octeau JC, Yu X, Cohn W, Rajendran PS, Vondriska TM, Whitelegge JP, et al. 2017. Neural circuit-specialized astrocytes: transcriptomic, proteomic, morphological, and functional evidence. *Neuron* **95**: 531–549.e9. doi:10.1016/j.neuron.2017.06.029

Chen KH, Boettiger AN, Moffitt JR, Wang S, Zhuang X. 2015. RNA imaging. Spatially resolved, highly multiplexed RNA profiling in single cells. *Science* **348**: aaa6090. doi:10.1126/science.aaa6090

Chen S, Lake BB, Zhang K. 2019. High-throughput sequencing of the transcriptome and chromatin accessibility in the same cell. *Nat Biotechnol* **37**: 1452–1457. doi:10.1038/s41587-019-0290-0

Chen WT, Lu A, Craessaerts K, Pavie B, Sala Frigerio C, Corthout N, Qian X, Laláková J, Kühnemund M, et al. 2020. Spatial transcriptomics and in situ sequencing to study Alzheimer's disease. *Cell* **182**: 976–991.e19. doi:10.1016/j.cell.2020.06.038

Chen A, Liao S, Cheng M, Ma K, Wu L, Lai Y, Qiu X, Yang J, Xu J, Hao S, et al. 2022. Spatiotemporal transcriptomic atlas of mouse organogenesis using DNA nanoball-patterned arrays. *Cell* **185**: 1777–1792.e21. doi:10.1016/j.cell.2022.04.003

Clarke LE, Liddelow SA, Chakraborty C, Münch AE, Heiman M, Barres BA. 2018. Normal aging induces A1-like astrocyte reactivity. *Proc Natl Acad Sci* **115**: E1896–E1905. doi:10.1073/pnas.1800165115

Codeluppi S, Borm LE, Zeisel A, La Manno G, van Lunteren JA, Svensson CI, Linnarsson S. 2018. Spatial organization of the somatosensory cortex revealed by osmFISH. *Nat Methods* **15**: 932–935. doi:10.1038/s41592-018-0175-z

Colombo JA, Reisin HD. 2004. Interlaminar astroglia of the cerebral cortex: a marker of the primate brain. *Brain Res* **1006**: 126–131. doi:10.1016/j.brainres.2004.02.003

Cui Y, Yang Y, Ni Z, Dong Y, Cai G, Foncelle A, Ma S, Sang K, Tang S, Li Y, et al. 2018. Astroglial Kir4.1 in the lateral habenula drives neuronal bursts in depression. *Nature* **554**: 323–327. doi:10.1038/nature25752

Deng Y, Bartosovic M, Ma S, Zhang D, Kukanja P, Xiao Y, Su G, Liu Y, Qin X, Rosoklija GB, et al. 2022. Spatial profiling of chromatin accessibility in mouse and human tissues. *Nature* **609**: 375–383. doi:10.1038/s41586-022-05094-1

Dimitrov D, Türei D, Garrido-Rodriguez M, Burmedi PL, Nagai JS, Boys C, Ramirez Flores RO, Kim H, Szalai B, Costa IG, et al. 2022. Comparison of methods and resources for cell–cell communication inference from single-cell RNA-seq data. *Nat Commun* **13**: 3224. doi:10.1038/s41467-022-30755-0

Duncan KD, Fyrestam J, Lanekoff I. 2019. Advances in mass spectrometry based single-cell metabolomics. *Analyst* **144**: 782–793. doi:10.1039/C8AN01581C

Efremova M, Vento-Tormo M, Teichmann SA, Vento-Tormo R. 2020. CellPhoneDB: inferring cell-cell communication from combined expression of multi-subunit ligand-receptor complexes. *Nat Protoc* **15**: 1484–1506. doi:10.1038/s41596-020-0292-x

El-Danaf RN, Rajesh R, Desplan C. 2023. Temporal regulation of neural diversity in *Drosophila* and vertebrates. *Semin Cell Dev Biol* **142**: 13–22. doi:10.1016/j.semcdb.2022.05.011

Emsley JG, Macklis JD. 2006. Astroglial heterogeneity closely reflects the neuronal-defined anatomy of the adult murine CNS. *Neuron Glia Biol* **2**: 175–186. doi:10.1017/S1740925X06000202

Endo F, Kasai A, Soto JS, Yu X, Qu Z, Hashimoto H, Gradinaru V, Kawaguchi R, Khakh BS. 2022. Molecular basis of astrocyte diversity and morphology across the CNS in health and disease. *Science* **378**: eadc9020. doi:10.1126/science.adc9020

Eng CL, Lawson M, Zhu Q, Dries R, Koulena N, Takei Y, Yun J, Cronin C, Karp C, Yuan GC, et al. 2019. Transcriptome-scale super-resolved imaging in tissues by RNA seqFISH+. *Nature* **568:** 235–239. doi:10.1038/s41586-019-1049-y

Falcone C, Wolf-Ochoa M, Amina S, Hong T, Vakilzadeh G, Hopkins WD, Hof PR, Sherwood CC, Manger PR, Noctor SC, et al. 2019. Cortical interlaminar astrocytes across the therian mammal radiation. *J Comp Neurol* **527:** 1654–1674. doi:10.1002/cne.24605

Falcone C, McBride EL, Hopkins WD, Hof PR, Manger PR, Sherwood CC, Noctor SC, Martínez-Cerdeño V. 2022. Redefining varicose projection astrocytes in primates. *Glia* **70:** 145–154. doi:10.1002/glia.24093

Fang R, Xia C, Close JL, Zhang M, He J, Huang Z, Halpern AR, Long B, Miller JA, Lein ES, et al. 2022. Conservation and divergence of cortical cell organization in human and mouse revealed by MERFISH. *Science* **377:** 56–62. doi:10.1126/science.abm1741

Farhy-Tselnicker I, Boisvert MM, Liu H, Dowling C, Erikson GA, Blanco-Suarez E, Farhy C, Shokhirev MN, Ecker JR, Allen NJ. 2021. Activity-dependent modulation of synapse-regulating genes in astrocytes. *eLife* **10:** e70514. doi:10.7554/eLife.70514

Farmer WT, Abrahamsson T, Chierzi S, Lui C, Zaelzer C, Jones EV, Bally BP, Chen GG, Théroux JF, Peng J, et al. 2016. Neurons diversify astrocytes in the adult brain through sonic hedgehog signaling. *Science* **351:** 849–854. doi:10.1126/science.aab3103

Fischer DS, Schaar AC, Theis FJ. 2023. Modeling intercellular communication in tissues using spatial graphs of cells. *Nat Biotechnol* **41:** 332–336. doi:10.1038/s41587-022-01467-z

Flames N, Pla R, Gelman DM, Rubenstein JL, Puelles L, Marín O. 2007. Delineation of multiple subpallial progenitor domains by the combinatorial expression of transcription codes. *J Neurosci* **27:** 9682–9695. doi:10.1523/JNEUROSCI.2750-07.2007

Gabitto MI, Travaglini KJ, Rachleff VM, Kaplan ES, Long B, Ariza J, Ding Y, Mahoney JT, Dee N, Goldy J, et al. 2023. Integrated multimodal cell atlas of Alzheimer's disease. *Res Sq* doi:10.21203/rs.3.rs-2921860/v1

Garcia-Alonso L, Handfield LF, Roberts K, Nikolakopoulou K, Fernando RC, Gardner L, Woodhams B, Arutyunyan A, Polanski K, Hoo R, et al. 2021. Mapping the temporal and spatial dynamics of the human endometrium in vivo and in vitro. *Nat Genet* **53:** 1698–1711. doi:10.1038/s41588-021-00972-2

Ghazanfar S, Guibentif C, Marioni JC. 2023. Stabilized mosaic single-cell data integration using unshared features. *Nat Biotechnol* doi:10.1038/s41587-023-01766-z

Greig LC, Woodworth MB, Galazo MJ, Padmanabhan H, Macklis JD. 2013. Molecular logic of neocortical projection neuron specification, development and diversity. *Nat Rev Neurosci* **14:** 755–769. doi:10.1038/nrn3586

Hagemann-Jensen M, Ziegenhain C, Chen P, Ramsköld D, Hendriks GJ, Larsson AJM, Faridani OR, Sandberg R. 2020. Single-cell RNA counting at allele and isoform resolution using Smart-seq3. *Nat Biotechnol* **38:** 708–714. doi:10.1038/s41587-020-0497-0

Hasel P, Rose IVL, Sadick JS, Kim RD, Liddelow SA. 2021. Neuroinflammatory astrocyte subtypes in the mouse brain. *Nat Neurosci* **24:** 1475–1487. doi:10.1038/s41593-021-00905-6

He S, Bhatt R, Brown C, Brown EA, Buhr DL, Chantranuvatana K, Danaher P, Dunaway D, Garrison RG, Geiss G, et al.

2022. High-plex imaging of RNA and proteins at subcellular resolution in fixed tissue by spatial molecular imaging. *Nat Biotechnol* **40:** 1794–1806. doi:10.1038/s41587-022-01483-z

Hochstim C, Deneen B, Lukaszewicz A, Zhou Q, Anderson DJ. 2008. Identification of positionally distinct astrocyte subtypes whose identities are specified by a homeodomain code. *Cell* **133:** 510–522. doi:10.1016/j.cell.2008.02.046

Hodge RD, Bakken TE, Miller JA, Smith KA, Barkan ER, Graybuck LT, Close JL, Long B, Johansen N, Penn O, et al. 2019. Conserved cell types with divergent features in human versus mouse cortex. *Nature* **573:** 61–68. doi:10.1038/s41586-019-1506-7

Huang AY, Woo J, Sardar D, Lozzi B, Bosquez Huerta NA, Lin CJ, Felice D, Jain A, Paulucci-Holthauzen A, Deneen B. 2020a. Region-specific transcriptional control of astrocyte function oversees local circuit activities. *Neuron* **106:** 992–1008.e9. doi:10.1016/j.neuron.2020.03.025

Huang W, Bhaduri A, Velmeshev D, Wang S, Wang L, Rottkamp CA, Alverez-Buylla A, Rowitch DH, Kriegstein AR. 2020b. Origins and proliferative states of human oligodendrocyte precursor cells. *Cell* **182:** 594–608.e11. doi:10.1016/j.cell.2020.06.027

Isshiki T, Pearson B, Holbrook S, Doe CQ. 2001. *Drosophila* neuroblasts sequentially express transcription factors which specify the temporal identity of their neuronal progeny. *Cell* **106:** 511–521. doi:10.1016/S0092-8674(01)00465-2

Itoh N, Itoh Y, Tassoni A, Ren E, Kaito M, Ohno A, Ao Y, Farkhondeh V, Johnsonbaugh H, Burda J, et al. 2018. Cell-specific and region-specific transcriptomics in the multiple sclerosis model: focus on astrocytes. *Proc Natl Acad Sci* **115:** E302–E309. doi:10.1073/pnas.1716032115

Janesick A, Shelansky R, Gottscho AD, Wagner F, Williams SR, Rouault M, Beliakoff G, Morrison CA, Oliveira MF, Sicherman JT, et al. 2023. High resolution mapping of the tumor microenvironment using integrated single-cell, spatial and in situ analysis. *Nat Commun* **14:** 8353. doi:10.1038/s41467-023-43458-x

Jessell TM. 2000. Neuronal specification in the spinal cord: inductive signals and transcriptional codes. *Nat Rev Genet* **1:** 20–29. doi:10.1038/35049541

Jin S, Guerrero-Juarez CF, Zhang L, Chang I, Ramos R, Kuan CH, Myung P, Plikus MV, Nie Q. 2021. Inference and analysis of cell-cell communication using CellChat. *Nat Commun* **12:** 1088. doi:10.1038/s41467-021-21246-9

John Lin CC, Yu K, Hatcher A, Huang TW, Lee HK, Carlson J, Weston MC, Chen F, Zhang Y, Zhu W, et al. 2017. Identification of diverse astrocyte populations and their malignant analogs. *Nat Neurosci* **20:** 396–405. doi:10.1038/nn.4493

Jorstad NL, Song JHT, Exposito-Alonso D, Suresh H, Castro-Pacheco N, Krienen FM, Yanny AM, Close J, Gelfand E, Long B, et al. 2023. Comparative transcriptomics reveals human-specific cortical features. *Science* **382:** eade9516. doi:10.1126/science.ade9516

Kamath T, Abdulraouf A, Burris SJ, Langlieb J, Gazestani V, Nadaf NM, Balderrama K, Vanderburg C, Macosko EZ. 2022. Single-cell genomic profiling of human dopamine neurons identifies a population that selectively degenerates in Parkinson's disease. *Nat Neurosci* **25:** 588–595. doi:10.1038/s41593-022-01061-1

Kaneko N, Marín O, Koike M, Hirota Y, Uchiyama Y, Wu JY, Lu Q, Tessier-Lavigne M, Alvarez-Buylla A, Okano H, et al. 2010. New neurons clear the path of astrocytic processes for their rapid migration in the adult brain. *Neuron* **67:** 213–223. doi:10.1016/j.neuron.2010.06.018

Ke R, Mignardi M, Pacureanu A, Svedlund J, Botling J, Wählby C, Nilsson M. 2013. In situ sequencing for RNA analysis in preserved tissue and cells. *Nat Methods* **10:** 857–860. doi:10.1038/nmeth.2563

Kelley KW, Nakao-Inoue H, Molofsky AV, Oldham MC. 2018a. Variation among intact tissue samples reveals the core transcriptional features of human CNS cell classes. *Nat Neurosci* **21:** 1171–1184. doi:10.1038/s41593-018-0216-z

Kelley KW, Ben Haim L, Schirmer L, Tyzack GE, Tolman M, Miller JG, Tsai HH, Chang SM, Molofsky AV, Yang Y, et al. 2018b. Kir4.1-dependent astrocyte-fast motor neuron interactions are required for peak strength. *Neuron* **98:** 306–319.e7. doi:10.1016/j.neuron.2018.03.010

Klein AM, Mazutis L, Akartuna I, Tallapragada N, Veres A, Li V, Peshkin L, Weitz DA, Kirschner MW. 2015. Droplet barcoding for single-cell transcriptomics applied to embryonic stem cells. *Cell* **161:** 1187–1201. doi:10.1016/j.cell.2015.04.044

Kleshchevnikov V, Shmatko A, Dann E, Aivazidis A, King HW, Li T, Elmentaite R, Lomakin A, Kedlian V, Gayoso A, et al. 2022. Cell2location maps fine-grained cell types in spatial transcriptomics. *Nat Biotechnol* **40:** 661–671. doi:10.1038/s41587-021-01139-4

Kohwi M, Doe CQ. 2013. Temporal fate specification and neural progenitor competence during development. *Nat Rev Neurosci* **14:** 823–838. doi:10.1038/nrn3618

Kriegstein A, Alvarez-Buylla A. 2009. The glial nature of embryonic and adult neural stem cells. *Annu Rev Neurosci* **32:** 149–184. doi:10.1146/annurev.neuro.051508.135600

Lake BB, Ai R, Kaeser GE, Salathia NS, Yung YC, Liu R, Wildberg A, Gao D, Fung HL, Chen S, et al. 2016. Neuronal subtypes and diversity revealed by single-nucleus RNA sequencing of the human brain. *Science* **352:** 1586–1590. doi:10.1126/science.aaf1204

Langlieb J, Sachdev NS, Balderrama KS, Nadaf NM, Raj M, Murray E, Webber JT, Vanderburg C, Gazestani V, Tward D, et al. 2023. The molecular architecture of the adult mouse brain. *Nature* **624:** 333–342. doi:10.1038/s41586-023-06818-7

Lanjakornsiripan D, Pior BJ, Kawaguchi D, Furutachi S, Tahara T, Katsuyama Y, Suzuki Y, Fukazawa Y, Gotoh Y. 2018. Layer-specific morphological and molecular differences in neocortical astrocytes and their dependence on neuronal layers. *Nat Commun* **9:** 1623. doi:10.1038/s41467-018-03940-3

Lee Y, Bogdanoff D, Wang Y, Hartoularos GC, Woo JM, Mowery CT, Nisonoff HM, Lee DS, Sun Y, Lee J, et al. 2021. XYZeq: spatially resolved single-cell RNA sequencing reveals expression heterogeneity in the tumor microenvironment. *Sci Adv* **7:** eabg4755. doi:10.1126/sciadv.abg4755

Leng K, Rose IVL, Kim H, Xia W, Romero-Fernandez W, Rooney B, Koontz M, Li E, Ao Y, Wang S, et al. 2022. CRISPRi screens in human iPSC-derived astrocytes elucidate regulators of distinct inflammatory reactive states. *Nat Neurosci* **25:** 1528–1542. doi:10.1038/s41593-022-01180-9

Lerma-Martin C, Badia-i-Mompel P, Ramirez Flores RO, Sekol P, Hofmann A, Thäwel T, Riedl CJ, Wünnemann F, Ubarra-Arellano MA, Trobisch T, et al. 2022. Spatial cell type mapping of multiple sclerosis lesions. bioRxiv doi:10.1101/2022.11.03.514906

Li X, Chen Z, Desplan C. 2013. Temporal patterning of neural progenitors in *Drosophila*. *Curr Top Dev Biol* **105:** 69–96. doi:10.1016/B978-0-12-396968-2.00003-8

Li E, Benitez C, Boggess SC, Koontz M, Rose IVL, Draeger N, Teter OM, Samelson AJ, Ullian EM, Kampmann M. 2023a. CRISPRi-based screens in iAssembloids to elucidate neuron–glia interactions. bioRxiv doi:10.1101/2023.04.26.538498

Li T, Horsfall D, Basurto-Lozada D, Roberts K, Prete M, Lawrence JEG, He P, Tuck E, Moore J, Ghazanfar S, et al. 2023b. Webatlas pipeline for integrated single cell and spatial transcriptomic data. bioRxiv doi:10.1101/2023.05.19.541329

Liu Y, Yang M, Deng Y, Su G, Enninful A, Guo CC, Tebaldi T, Zhang D, Kim D, Bai Z, et al. 2020. High-spatial-resolution multi-omics sequencing via deterministic barcoding in tissue. *Cell* **183:** 1665–1681.e18. doi:10.1016/j.cell.2020.10.026

Lubeck E, Coskun AF, Zhiyentayev T, Ahmad M, Cai L. 2014. Single-cell in situ RNA profiling by sequential hybridization. *Nat Methods* **11:** 360–361. doi:10.1038/nmeth.2892

Macosko EZ, Basu A, Satija R, Nemesh J, Shekhar K, Goldman M, Tirosh I, Bialas AR, Kamitaki N, Martersteck EM, et al. 2015. Highly parallel genome-wide expression profiling of individual cells using nanoliter droplets. *Cell* **161:** 1202–1214. doi:10.1016/j.cell.2015.05.002

Mathys H, Davila-Velderrain J, Peng Z, Gao F, Mohammadi S, Young JZ, Menon M, He L, Abdurrob F, Jiang X, et al. 2019. Single-cell transcriptomic analysis of Alzheimer's disease. *Nature* **570:** 332–337. doi:10.1038/s41586-019-1195-2

Maynard KR, Collado-Torres L, Weber LM, Uytingco C, Barry BK, Williams SR, Catallini JL, Tran MN, Besich Z, Tippani M, et al. 2021. Transcriptome-scale spatial gene expression in the human dorsolateral prefrontal cortex. *Nat Neurosci* **24:** 425–436. doi:10.1038/s41593-020-00787-0

McConnell SK. 1992. The control of neuronal identity in the developing cerebral cortex. *Curr Opin Neurobiol* **2:** 23–27. doi:10.1016/0959-4388(92)90156-f

Merkle FT, Mirzadeh Z, Alvarez-Buylla A. 2007. Mosaic organization of neural stem cells in the adult brain. *Science* **317:** 381–384. doi:10.1126/science.1144914

Miller RH, Raff MC. 1984. Fibrous and protoplasmic astrocytes are biochemically and developmentally distinct. *J Neurosci* **4:** 585–592. doi:10.1523/JNEUROSCI.04-02-00585.1984

Miller SJ, Philips T, Kim N, Dastgheyb R, Chen Z, Hsieh YC, Daigle JG, Datta M, Chew J, Vidensky S, et al. 2019. Molecularly defined cortical astroglia subpopulation modulates neurons via secretion of Norrin. *Nat Neurosci* **22:** 741–752. doi:10.1038/s41593-019-0366-7

Moffitt JR, Bambah-Mukku D, Eichhorn SW, Vaughn E, Shekhar K, Perez JD, Rubinstein ND, Hao J, Regev A, Dulac C, et al. 2018. Molecular, spatial, and functional single-cell profiling of the hypothalamic preoptic region. *Science* 362: eaau5324. doi:10.1126/science.aau5324

Moffitt JR, Lundberg E, Heyn H. 2022. The emerging landscape of spatial profiling technologies. *Nat Rev Genet* 23: 741–759. doi:10.1038/s41576-022-00515-3

Molofsky AV, Kelley KW, Tsai HH, Redmond SA, Chang SM, Madireddy L, Chan JR, Baranzini SE, Ullian EM, Rowitch DH. 2014. Astrocyte-encoded positional cues maintain sensorimotor circuit integrity. *Nature* 509: 189–194. doi:10.1038/nature13161

Morel L, Chiang MSR, Higashimori H, Shoneye T, Iyer LK, Yelick J, Tai A, Yang Y. 2017. Molecular and functional properties of regional astrocytes in the adult brain. *J Neurosci* 37: 8706–8717. doi:10.1523/JNEUROSCI.3956-16 .2017

Morel L, Men Y, Chiang MSR, Tian Y, Jin S, Yelick J, Higashimori H, Yang Y. 2019. Intracortical astrocyte subpopulations defined by astrocyte reporter mice in the adult brain. *Glia* 67: 171–181. doi:10.1002/glia.23545

Nagai J, Yu X, Papouin T, Cheong E, Freeman MR, Monk KR, Hastings MH, Haydon PG, Rowitch D, Shaham S, et al. 2021. Behaviorally consequential astrocytic regulation of neural circuits. *Neuron* 109: 576–596. doi:10.1016/j .neuron.2020.12.008

Neely SA, Lyons DA. 2021. Insights into central nervous system glial cell formation and function from zebrafish. *Front Cell Dev Biol* 9: 754606. doi:10.3389/fcell.2021.754606

Noctor SC, Flint AC, Weissman TA, Wong WS, Clinton BK, Kriegstein AR. 2002. Dividing precursor cells of the embryonic cortical ventricular zone have morphological and molecular characteristics of radial glia. *J Neurosci* 22: 3161–3173. doi:10.1523/JNEUROSCI.22-08-03161.2002

Oberheim NA, Wang X, Goldman S, Nedergaard M. 2006. Astrocytic complexity distinguishes the human brain. *Trends Neurosci* 29: 547–553. doi:10.1016/j.tins.2006.08 .004

Oberheim NA, Takano T, Han X, He W, Lin JH, Wang F, Xu Q, Wyatt JD, Pilcher W, Ojemann JG, et al. 2009. Uniquely hominid features of adult human astrocytes. *J Neurosci* 29: 3276–3287. doi:10.1523/JNEUROSCI.4707-08.2009

Oberheim NA, Goldman SA, Nedergaard M. 2012. Heterogeneity of astrocytic form and function. *Methods Mol Biol* 814: 23–45. doi:10.1007/978-1-61779-452-0_3

Pollen AA, Nowakowski TJ, Chen J, Retallack H, Sandoval-Espinosa C, Nicholas CR, Shuga J, Liu SJ, Oldham MC, Diaz A, et al. 2015. Molecular identity of human outer radial glia during cortical development. *Cell* 163: 55–67. doi:10.1016/j.cell.2015.09.004

Puelles L, Rubenstein JLR. 2003. Forebrain gene expression domains and the evolving prosomeric model. *Trends Neurosci* 26: 469–476. doi:10.1016/S0166-2236(03)00234-0

Qian X, Shen Q, Goderie SK, He W, Capela A, Davis AA, Temple S. 2000. Timing of CNS cell generation: a programmed sequence of neuron and glial cell production from isolated murine cortical stem cells. *Neuron* 28: 69–80. doi:10.1016/S0896-6273(00)00086-6

Qian X, Harris KD, Hauling T, Nicoloutsopoulos D, Muñoz-Manchado AB, Skene N, Hjerling-Leffler J, Nilsson M. 2020. Probabilistic cell typing enables fine mapping of closely related cell types in situ. *Nat Methods* 17: 101–106. doi:10.1038/s41592-019-0631-4

Rao A, Barkley D, França GS, Yanai I. 2021. Exploring tissue architecture using spatial transcriptomics. *Nature* 596: 211–220. doi:10.1038/s41586-021-03634-9

Rayon T, Maizels RJ, Barrington C, Briscoe J. 2021. Single-cell transcriptome profiling of the human developing spinal cord reveals a conserved genetic programme with human-specific features. *Development* 148: dev199711. doi:10.1242/dev.199711

Rodriques SG, Stickels RR, Goeva A, Martin CA, Murray E, Vanderburg CR, Welch J, Chen LM, Chen F, Macosko EZ. 2019. Slide-seq: a scalable technology for measuring genome-wide expression at high spatial resolution. *Science* 363: 1463–1467. doi:10.1126/science.aaw1219

Rosenberg AB, Roco CM, Muscat RA, Kuchina A, Sample P, Yao Z, Graybuck LT, Peeler DJ, Mukherjee S, Chen W, et al. 2018. Single-cell profiling of the developing mouse brain and spinal cord with split-pool barcoding. *Science* 360: 176–182. doi:10.1126/science.aam8999

Rowitch DH, Kriegstein AR. 2010. Developmental genetics of vertebrate glial-cell specification. *Nature* 468: 214–222. doi:10.1038/nature09611

Russell AJC, Weir JA, Nadaf NM, Shabet M, Kumar V, Kambhampati S, Raichur R, Marrero GJ, Liu S, Balderrama KS, et al. 2024. Slide-tags enables scalable, single-nucleus barcoding for multi-modal spatial genomics. *Nature* 625: 101–109. doi:10.1038/s41586-023-06837-4

Sadick JS, O'Dea MR, Hasel P, Dykstra T, Faustin A, Liddelow SA. 2022. Astrocytes and oligodendrocytes undergo subtype-specific transcriptional changes in Alzheimer's disease. *Neuron* 110: 1788–1805.e10. doi:10.1016/j.neuron .2022.03.008

Sagner A, Zhang I, Watson T, Lazaro J, Melchionda M, Briscoe J. 2021. A shared transcriptional code orchestrates temporal patterning of the central nervous system. *PLoS Biol* 19: e3001450. doi:10.1371/journal.pbio.3001450

Salmon CK, Syed TA, Kacerovsky JB, Alivodej N, Schober AL, Sloan TFW, Pratte MT, Rosen MP, Green M, Chirgwin-Dasgupta A, et al. 2023. Organizing principles of astrocytic nanoarchitecture in the mouse cerebral cortex. *Curr Biol* 33: 957–972.e5. doi:10.1016/j.cub.2023.01.043

Saunders A, Macosko EZ, Wysoker A, Goldman M, Krienen FM, de Rivera H, Bien E, Baum M, Bortolin L, Wang S, et al. 2018. Molecular diversity and specializations among the cells of the adult mouse brain. *Cell* 174: 1015–1030.e16. doi:10.1016/j.cell.2018.07.028

Schirmer L, Velmeshev D, Holmqvist S, Kaufmann M, Werneburg S, Jung D, Vistnes S, Stockley JH, Young A, Steindel M, et al. 2019. Neuronal vulnerability and multilineage diversity in multiple sclerosis. *Nature* 573: 75–82. doi:10 .1038/s41586-019-1404-z

Schirmer L, Schafer DP, Bartels T, Rowitch DH, Calabresi PA. 2021. Diversity and function of glial cell types in multiple sclerosis. *Trends Immunol* 42: 228–247. doi:10.1016/j.it .2021.01.005

Shah S, Lubeck E, Zhou W, Cai L. 2016. In situ transcription profiling of single cells reveals spatial organization of cells in the mouse hippocampus. *Neuron* 92: 342–357. doi:10 .1016/j.neuron.2016.10.001

Shen Q, Wang Y, Dimos JT, Fasano CA, Phoenix TN, Lemischka IR, Ivanova NB, Stifani S, Morrisey EE, Temple S. 2006. The timing of cortical neurogenesis is encoded within lineages of individual progenitor cells. *Nat Neurosci* **9:** 743–751. doi:10.1038/nn1694

Shilts J, Severin Y, Galaway F, Müller-Sienerth N, Chong ZS, Pritchard S, Teichmann S, Vento-Tormo R, Snijder B, Wright GJ. 2022. A physical wiring diagram for the human immune system. *Nature* **608:** 397–404. doi:10.1038/s41586-022-05028-x

Siletti K, Hodge R, Mossi Albiach A, Hu L, Lee KW, Lonnerberg P, Bakken T, Ding SL, Clark M, Casper T, et al. 2023. Transcriptomic diversity of cell types across the adult human brain. *Science* **382:** eadd7046. doi:10.1126/science.add7046

Sloan SA, Darmanis S, Huber N, Khan TA, Birey F, Caneda C, Reimer R, Quake SR, Barres BA, Paşca SP. 2017. Human astrocyte maturation captured in 3D cerebral cortical spheroids derived from pluripotent stem cells. *Neuron* **95:** 779–790.e6. doi:10.1016/j.neuron.2017.07.035

Srinivasan R, Lu TY, Chai H, Xu J, Huang BS, Golshani P, Coppola G, Khakh BS. 2016. New transgenic mouse lines for selectively targeting astrocytes and studying calcium signals in astrocyte processes in situ and in vivo. *Neuron* **92:** 1181–1195. doi:10.1016/j.neuron.2016.11.030

Ståhl PL, Salmén F, Vickovic S, Lundmark A, Navarro JF, Magnusson J, Giacomello S, Asp M, Westholm JO, Huss M, et al. 2016. Visualization and analysis of gene expression in tissue sections by spatial transcriptomics. *Science* **353:** 78–82. doi:10.1126/science.aaf2403

Stogsdill JA, Kim K, Binan L, Farhi SL, Levin JZ, Arlotta P. 2022. Pyramidal neuron subtype diversity governs microglia states in the neocortex. *Nature* **608:** 750–756. doi:10.1038/s41586-022-05056-7

Sun YC, Chen X, Fischer S, Lu S, Zhan H, Gillis J, Zador AM. 2021. Integrating barcoded neuroanatomy with spatial transcriptional profiling enables identification of gene correlates of projections. *Nat Neurosci* **24:** 873–885. doi:10.1038/s41593-021-00842-4

Sur M, Rubenstein JLR. 2005. Patterning and plasticity of the cerebral cortex. *Science* **310:** 805–810. doi:10.1126/science.1112070

Swanson RA, Liu J, Miller JW, Rothstein JD, Farrell K, Stein BA, Longuemare MC. 1997. Neuronal regulation of glutamate transporter subtype expression in astrocytes. *J Neurosci* **17:** 932–940. doi:10.1523/JNEUROSCI.17-03-00932.1997

Szebényi K, Wenger LMD, Sun Y, Dunn AWE, Limegrover CA, Gibbons GM, Conci E, Paulsen O, Mierau SB, Balmus G, et al. 2021. Human ALS/FTD brain organoid slice cultures display distinct early astrocyte and targetable neuronal pathology. *Nat Neurosci* **24:** 1542–1554. doi:10.1038/s41593-021-00923-4

Tanevski J, Flores ROR, Gabor A, Schapiro D, Saez-Rodriguez J. 2022. Explainable multiview framework for dissecting spatial relationships from highly multiplexed data. *Genome Biol* **23:** 1–31. doi:10.1186/s13059-022-02663-5

Tasdemir-Yilmaz OE, Freeman MR. 2014. Astrocytes engage unique molecular programs to engulf pruned neuronal debris from distinct subsets of neurons. *Genes Dev* **28:** 20–33. doi:10.1101/gad.229518.113

Tian L, Chen F, Macosko EZ. 2023. The expanding vistas of spatial transcriptomics. *Nat Biotechnol* **41:** 773–782. doi:10.1038/s41587-022-01448-2

Tien AC, Tsai HH, Molofsky AV, McMahon M, Foo LC, Kaul A, Dougherty JD, Heintz N, Gutmann DH, Barres BA, et al. 2012. Regulated temporal-spatial astrocyte precursor cell proliferation involves BRAF signalling in mammalian spinal cord. *Development* **139:** 2477–2487. doi:10.1242/dev.077214

Tsai HH, Li H, Fuentealba LC, Molofsky AV, Taveira-Marques R, Zhuang H, Tenney A, Murnen AT, Fancy SP, Merkle F, et al. 2012. Regional astrocyte allocation regulates CNS synaptogenesis and repair. *Science* **337:** 358–362. doi:10.1126/science.1222381

Vainchtein ID, Chin G, Cho FS, Kelley KW, Miller JG, Chien EC, Liddelow SA, Nguyen PT, Nakao-Inoue H, Dorman LC, et al. 2018. Astrocyte-derived interleukin-33 promotes microglial synapse engulfment and neural circuit development. *Science* **359:** 1269–1273. doi:10.1126/science.aal3589

Vallejo AF, Harvey K, Wang T, Wise K, Buler LM, Polo J, Plummer J, Swarbrick A, Martelotto LG. 2022. snPATHO-seq: unlocking the FFPE archives for single nucleus RNA profiling. bioRxiv doi:10.1101/2022.08.23.505054

Vento-Tormo R, Efremova M, Botting RA, Turco MY, Vento-Tormo M, Meyer KB, Park JE, Stephenson E, Polański K, Goncalves A, et al. 2018. Single-cell reconstruction of the early maternal–fetal interface in humans. *Nature* **563:** 347–353. doi:10.1038/s41586-018-0698-6

Voss AJ, Lanjewar SN, Sampson MM, King A, Hill E, Sing A, Sojka C, Bhatia TN, Spangle JM, Sloan SA. 2023. Identification of ligand-receptor pairs that drive human astrocyte development. *Nat Neurosci* **26:** 1339–1351. doi:10.1038/s41593-023-01375-8

Wang F, Flanagan J, Su N, Wang LC, Bui S, Nielson A, Wu X, Vo HT, Ma XJ, Luo Y. 2012. RNAscope: a novel in situ RNA analysis platform for formalin-fixed, paraffin-embedded tissues. *J Mol Diagn* **14:** 22–29. doi:10.1016/j.jmoldx.2011.08.002

Wang X, Allen WE, Wright MA, Sylwestrak EL, Samusik N, Vesuna S, Evans K, Liu C, Ramakrishnan C, Liu J, et al. 2018. Three-dimensional intact-tissue sequencing of single-cell transcriptional states. *Science* **361:** eaat5691. doi:10.1126/science.aat5691

Wheeler MA, Clark IC, Lee HG, Li Z, Linnerbauer M, Rone JM, Blain M, Akl CF, Piester G, Giovannoni F, et al. 2023. Droplet-based forward genetic screening of astrocyte-microglia cross-talk. *Science* **379:** 1023–1030. doi:10.1126/science.abq4822

Wonders CP, Anderson SA. 2006. The origin and specification of cortical interneurons. *Nat Rev Neurosci* **7:** 687–696. doi:10.1038/nrn1954

Xia C, Fan J, Emanuel G, Hao J, Zhuang X. 2019. Spatial transcriptome profiling by MERFISH reveals subcellular RNA compartmentalization and cell cycle-dependent gene expression. *Proc Natl Acad Sci* **116:** 19490–19499. doi:10.1073/pnas.1912459116

Xie Y, Kuan AT, Wang W, Herbert ZT, Mosto O, Olukoya O, Adam M, Vu S, Kim M, Tran D, et al. 2022. Astrocyte-neuron crosstalk through Hedgehog signaling mediates cortical synapse development. *Cell Rep* **38:** 110416. doi:10.1016/j.celrep.2022.110416

Yao Z, van Velthoven CTJ, Nguyen TN, Goldy J, Sedeno-Cortes AE, Baftizadeh F, Bertagnolli D, Casper T,

Chiang M, Crichton K, et al. 2021. A taxonomy of transcriptomic cell types across the isocortex and hippocampal formation. *Cell* **184:** 3222–3241.e26. doi:10.1016/j.cell.2021.04.021

Yao Z, van Velthoven CTJ, Kunst M, Zhang M, McMillen D, Lee C, Jung W, Goldy J, Abdelhak A, Aitken M, et al. 2023. A high-resolution transcriptomic and spatial atlas of cell types in the whole mouse brain. *Nature* **624:** 317–332. doi:10.1038/s41586-023-06812-z

Yu X, Nagai J, Khakh BS. 2020. Improved tools to study astrocytes. *Nat Rev Neurosci* **21:** 121–138. doi:10.1038/s41583-020-0264-8

Zeisel A, Muñoz-Manchado AB, Codeluppi S, Lönnerberg P, La Manno G, Juréus A, Marques S, Munguba H, He L, Betsholtz C, et al. 2015. Brain structure. Cell types in the mouse cortex and hippocampus revealed by single-cell RNA-seq. *Science* **347:** 1138–1142. doi:10.1126/science.aaa1934

Zeisel A, Hochgerner H, Lönnerberg P, Johnsson A, Memic F, van der Zwan J, Häring M, Braun E, Borm LE, La Manno G, et al. 2018. Molecular architecture of the mouse nervous system. *Cell* **174:** 999–1014.e22. doi:10.1016/j.cell.2018.06.021

Zeng H, Huang J, Ren J, Wang CK, Tang Z, Zhou H, Zhou Y, Shi H, Aditham A, Sui X, et al. 2023. Spatially resolved single-cell translatomics at molecular resolution. *Science* **380:** eadd3067. doi:10.1126/science.add3067

Zhang M, Eichhorn SW, Zingg B, Yao Z, Cotter K, Zeng H, Dong H, Zhuang X. 2021. Spatially resolved cell atlas of the mouse primary motor cortex by MERFISH. *Nature* **598:** 137–143. doi:10.1038/s41586-021-03705-x

Zhang M, Pan X, Jung W, Halpern A, Eichhorn SW, Lei Z, Cohen L, Smith KA, Tasic B, Yao Z, et al. 2023. Molecularly defined and spatially resolved cell atlas of the whole mouse brain. *Nature* **624:** 343–354. doi:10.1038/s41586-023-06808-9

Astrocyte Regulation of Synapse Formation, Maturation, and Elimination

Won-Suk Chung,[1,4] Katherine T. Baldwin,[2,4] and Nicola J. Allen[3,4]

[1]Department of Biological Sciences, Korea Advanced Institute of Science and Technology, Yuseong-gu, Daejeon 34141, Korea

[2]Department of Cell Biology and Physiology and UNC Neuroscience Center, University of North Carolina at Chapel Hill, Chapel Hill, North Carolina 27599, USA

[3]Molecular Neurobiology Laboratory, Salk Institute for Biological Studies, La Jolla, California 92037, USA

Correspondence: wonsuk.chung@kaist.ac.kr; ktbaldwin@med.unc.edu; nallen@salk.edu

Astrocytes play an integral role in the development, maturation, and refinement of neuronal circuits. Astrocytes secrete proteins and lipids that instruct the formation of new synapses and induce the maturation of existing synapses. Through contact-mediated signaling, astrocytes can regulate the formation and state of synapses within their domain. Through phagocytosis, astrocytes participate in the elimination of excess synaptic connections. In this work, we will review key findings on the molecular mechanisms of astrocyte–synapse interaction with a focus on astrocyte-secreted factors, contact-mediated mechanisms, and synapse elimination. We will discuss this in the context of typical brain development and maintenance, as well as consider the consequences of dysfunction in these pathways in neurological disorders, highlighting a role for astrocytes in health and disease.

In the adult central nervous system (CNS), astrocytes are closely associated with neuronal synapses, forming the tripartite synapse, a complex of astrocyte processes with presynaptic and postsynaptic structures (Araque et al. 1999).[5] Astrocyte processes contain neurotransmitter receptors, uptake transporters, ion channels, and cell-adhesion molecules that mediate synapse–astrocyte communication (Allen 2014). Due to this structural arrangement, astrocytes can monitor synaptic activity and in turn regulate synaptic transmission (Halassa et al. 2007). Astrocytes occupy nonoverlapping domains (Bushong et al. 2002; Ogata and Kosaka 2002; Halassa et al. 2007), and it has been estimated that in the mouse brain one astrocyte contacts around 100,000 synapses through its fine processes (Fig. 1; Halassa et al. 2007). In addition to their important roles in regulating established synapses in the adult brain, a number of critical findings have highlighted the importance of astrocytes in the establishment and refinement of synaptic connectivity in the CNS.

In the rodent cerebral cortex, the majority of neurons mature and project axons to their targets within a few days after birth; however, by the end

---

[4]These authors contributed equally to this work.

[5]This is an update to a previous article published in *Cold Spring Harbor Perspectives in Biology* [Chung et al. (2015). *Cold Spring Harb Perspect Biol* 7: a020370. doi:10.1101/cshperspect.a020370].

Cite this article as *Cold Spring Harb Perspect Biol* doi: 10.1101/cshperspect.a041352

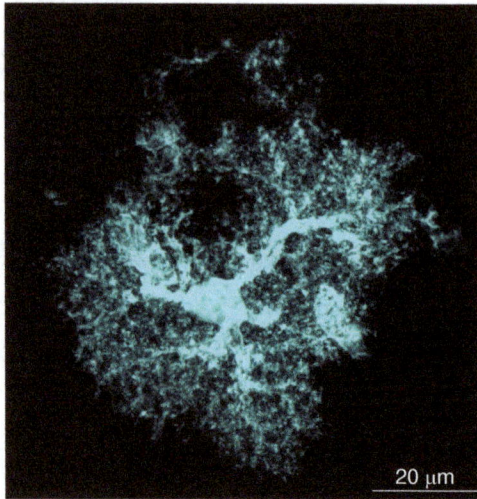

Figure 1. Astrocytes have a complex morphology and interact with synapses through their fine processes. Example image of a single astrocyte from layer V of the visual cortex in a 3-wk-old mouse, visualized by expression of cytosolic mCherry and pseudo-colored cyan. The image shown is a maximum projection image of a 5-μm-thick z-stack, representing approximately 1/10 of the astrocyte in the z dimension. Scale bar, 20 μm. (Image provided by co-author K. Baldwin.)

of the first postnatal week few synapses have formed, and most synaptogenesis occurs in the second and third postnatal weeks (Farhy-Tselnicker and Allen 2018). This period of extensive synapse formation coincides with the differentiation and maturation of astrocytes (Bartels et al. 2023). These observations indicated astrocytes may regulate synaptogenesis, but this was difficult to test as astrocytes are crucial for the survival and health of neurons (Banker 1980). This block was overcome by development of purified cultures of retinal ganglion cell (RGC) neurons (Meyer-Franke et al. 1995), which are isolated from postnatal rodent retinas and cultured in the absence of other cell types with growth factors to support their survival. Under these conditions, RGCs survive and grow neurites but make very few synapses. On the contrary, if RGCs are cultured in the presence of astrocyte feeder layers or fed by culture media previously conditioned by astrocytes (i.e., astrocyte conditioned media [ACM]), then RGCs establish

many synapses and there is an increase in synaptic activity (Pfrieger and Barres 1997; Ullian et al. 2001).

The use of neuron and astrocyte cultures has been a powerful approach to identify molecular mechanisms astrocytes use to regulate synaptic development, with findings validated by in vivo studies in the developing brain. This includes the discovery that astrocytes control distinct stages of excitatory glutamatergic synapse formation via release of different secreted factors. Further, a role for contact-mediated signaling from astrocytes in regulating synaptogenesis was identified, as well as a role for astrocytes in eliminating excess synapses that are formed during development and regulating synapse turnover in the adult brain. In this work, we will review the key findings on astrocytic control of synapse formation, maturation, and elimination to provide the current understanding of astrocytes as active participants in the construction of neural circuits.

ASTROCYTE-SECRETED FACTORS CONTROL SYNAPSE FORMATION AND MATURATION

Astrocyte Synapse-Regulating Factors Are Heterogeneously Expressed and Altered across Development

Since the initial discovery that astrocytes regulate synapse formation through release of secreted factors, many studies have been conducted to identify the molecules and mechanisms underlying these effects. What has emerged is that even for one class of synapse, excitatory glutamatergic synapses, astrocytes produce multiple different factors, each with a specific action. For example, thrombospondin family proteins (TSPs) promote formation of nascent synapses, and glypican family proteins (GPCs) promote synaptic function through recruitment of neurotransmitter receptors (Christopherson et al. 2005; Allen et al. 2012). Why would astrocytes need multiple signals to regulate synapse formation? Transcriptomic studies have revealed that these factors show heterogeneous expression in the developing brain, for example, varying across brain regions or across developmental time points within a brain region (Morel et al. 2017; Farhy-Tselnicker et al. 2021). Further, expression of synapse-regulating factors by astrocytes within a region is regulated by interactions with neurons, including neuronal layer identity, neuronal activity, and release of signaling molecules from neurons such as Shh (Hill et al. 2019; Bayraktar et al. 2020; Farhy-Tselnicker et al. 2021) (for a discussion of astrocyte heterogeneity, see Bartels et al. 2023). What is emerging is a picture whereby expression of each synapse-regulating factor by astrocytes is tuned to the neurons it regulates, and the developmental time frame when it is needed. In this section, we will focus on the identity and mechanism of action of known astrocyte-secreted synapse-regulating factors that in-

duce excitatory glutamatergic synapse formation (Fig. 2). Astrocyte-secreted factors also regulate inhibitory GABAergic synapse formation; however, less is known about the identity of the molecules mediating these effects (Hughes et al. 2010) and whether this occurs in all brain regions (Turko et al. 2019).

Astrocyte-Secreted Factors Induce Synapse Formation

Thrombospondins Induce Formation of Silent Synapses

The first family of astrocyte-secreted proteins to be identified that induce synapse formation were

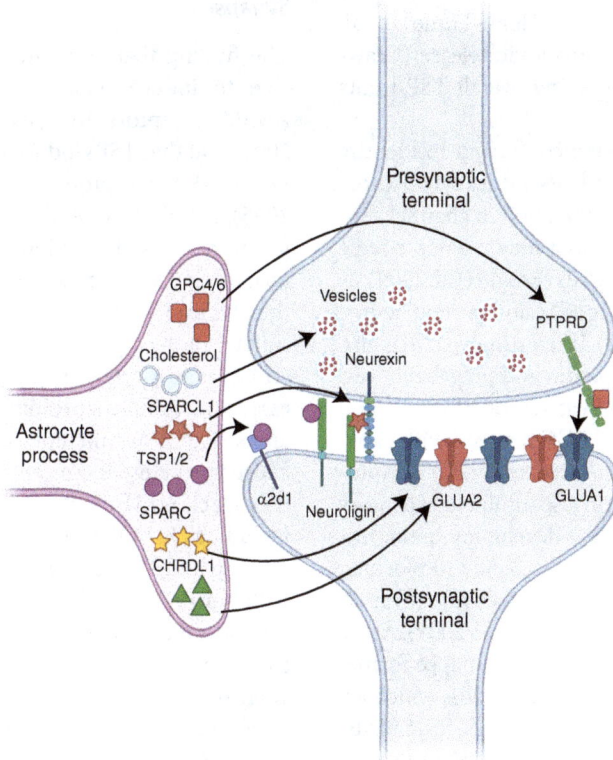

Figure 2. Astrocyte-secreted factors regulate synapse formation and maturation. Overview of astrocyte-secreted factors that regulate synapse formation and maturation, and the receptors in neurons that they signal through. TSP1/2 promotes synapse formation through postsynaptic receptors α2d1 and neuroligin (NL); GPC4/6 induces synapse formation through presynaptic protein tyrosine phosphatase receptor type D (PTPRD) that leads to an increase in postsynaptic GLUA1 AMPA receptors; cholesterol enhances presynaptic function by increasing neurotransmitter vesicle number; SPARCL1 promotes synapse maturation by bridging presynaptic neurexin and postsynaptic NL; SPARC regulates levels of AMPA receptors GLUA1 and GLUA2; and CHRDL1 induces synapse maturation by increasing postsynaptic GLUA2 AMPA receptors. (Created with BioRender.com.)

the extracellular matrix proteins thrombospondins (TSPs; TSP1-5) (Christopherson et al. 2005). Addition of purified TSP protein to cultured neurons increased synapse number, while TSP1/2 double-knockout (KO) mice displayed fewer cortical excitatory synapses during the first week of postnatal development, a time period that corresponds to the initiation of excitatory synapse formation in this region. Interestingly synapses induced by TSPs are postsynaptically silent as they do not contain AMPA glutamate receptors (AMPARs), which was the first indication that astrocytes must make more than one signal to regulate synaptic development. In addition to effects on synapse formation, TSPs increase synaptic glycine receptors and decrease synaptic AMPARs at already formed synapses in spinal cord neurons (Hennekinne et al. 2013) and decrease presynaptic release (Crawford et al. 2012), suggesting overall TSP limits neuronal excitability.

Further studies were conducted to identify the mechanism of how TSPs induce synapse formation, and identified the calcium channel subunit $\alpha2\delta1$ (Cacna2d1) as a major synaptogenic neuronal receptor for TSPs (Eroglu et al. 2009), as well as the postsynaptic adhesion protein neuroligin 1 (Xu et al. 2009a). Interestingly, $\alpha2\delta1$ is also the receptor for gabapentin and pregabalin, two drugs that are used to treat neuropathic pain and epilepsy. Treatment of RGC neurons with gabapentin blocked TSP-induced excitatory synaptogenesis, and gabapentin also inhibited excitatory synapse formation in the developing brain. Importantly, this inhibition is specific to blocking new synapse formation, and does not affect already formed synapses (Eroglu et al. 2009). At the molecular level, TSP promotes synapse formation by binding to postsynaptic $\alpha2\delta1$, which activates the Rho GTPase Rac1 to induce dendritic spine formation (Risher et al. 2018).

In adult mice, the levels of TSPs are low, but can be increased by injury or astrocyte stimulation. Lack of TSPs leads to defects in injury-induced structural plasticity of the developing barrel cortex (Eroglu et al. 2009) and hampers synaptic recovery after stroke, showing an important role for astrocyte-derived TSPs in regulating the formation of new synapses after injury

(Liauw et al. 2008). Further, inhibition of TSP or $\alpha2\delta1$ up-regulation, or blocking their interaction by delivering gabapentin, blocks establishment of pain states and prevents epileptogenic activity (Boroujerdi et al. 2011; Kim et al. 2012; Li et al. 2012). A recent study found that inducing astrocyte calcium signaling in striatal astrocytes through use of DREADDs was sufficient to alter mouse behavior, and that this effect was mediated by an up-regulation of TSP1 and transient formation of new synaptic connections (Nagai et al. 2019). Together, these studies show important roles for astrocyte-derived TSPs beyond development.

Glypicans Induce Formation of Active Synapses

The finding that astrocyte-secreted factors are able to induce synaptic activity and recruit AMPA receptors to synapses (Ullian et al. 2001), but that TSPs induce silent synapses lacking AMPA receptors (Christopherson et al. 2005), led to a biochemical study to identify the factor responsible for inducing active synapse formation (Allen et al. 2012). This identified the glypican (GPC; GPC1-6) family of heparan sulfate proteoglycans, specifically GPC4 and GPC6, as astrocyte-derived proteins that are necessary and sufficient to induce nascent active synapses. GPCs are membrane attached GPI-anchored proteins, and recent work identified that they are cleaved from the membrane by proteases, including ADAM9, to make the soluble form (Huang and Park 2021). Treating RGC neurons with purified GPC4 led to an increase in synapse number and an increase in synaptic activity, measured by electrophysiological recording (Allen et al. 2012). Interestingly, the increase in activity is specifically due to recruitment of GLUA1 containing AMPARs to synapses, a subunit typically associated with immature synapses in the developing cortex (Brill and Huguenard 2008). GPC4/6 had no impact on levels of synaptic GLUA2/3, which is increased as synapses mature, identifying that the role of GPC4/6 is specific to induction of immature but functional synapses and determining that astrocytes must make additional factors that induce synapse maturation. In the developing mouse brain, the peak of expression of GPC4 and GPC6 in the cortex is in the first 2 weeks, coinciding with the peak of synapse formation and the role of GPC4/6 as factors that induce synapse initiation (Farhy-Tselnicker et al. 2021).

The mechanism of GPC4-induced synapse formation was identified as signaling through a protein tyrosine phosphatase receptor (PTPRD) present in the presynaptic terminal (Farhy-Tselnicker et al. 2017). Interaction of GPC4 with PTPRD leads to secretion of the AMPA receptor clustering factor neuronal pentraxin 1 (NP1) from presynaptic terminals, and this NP1 binds to and stabilizes AMPA receptors on nearby postsynaptic dendrites. In this way, it is hypothesized that GPC4 acts as a temporal and spatial cue to axons to instruct them to switch from a growth state to a synaptogenic state when they encounter GPC4 in the extracellular environment. Interestingly, a different pentraxin family member, pentraxin 3 (PTX3), was identified as enriched in astrocytes in the developing brain and sufficient to induce formation of active glutamatergic synapses (Fossati et al. 2019), showing that astrocytes regulate synaptogenesis by both direct (PTX3) and indirect (GPC4) release of pentraxins. Functionally, GPC4 KO mice display altered behavior including juvenile hyperactivity and altered social interaction in adulthood, showing impacts of this factor on neural circuit function (Dowling and Allen 2018). Recently, another GPC family member, GPC3, was identified as up-regulated in glioblastoma and contributing to glioblastoma-induced hyperexcitability by increasing synapse formation in the peritumoral neuropil, showing that additional GPCs are synaptogenic, and highlighting a contribution to pathology (Yu et al. 2020). This suggests the actions of other GPC family members should be the focus of investigation in the healthy brain and in pathology.

Astrocyte-Derived Extracellular Vesicles and Synapse Formation

In addition to identifying individual astrocyte-secreted proteins that regulate synapses, recent work has found that extracellular vesicles released by astrocytes contain cargoes that promote synapse formation (Patel and Weaver 2021). Fibulin 2 was identified as a protein present in vesicles released from cultured astrocytes that was sufficient to induce synapse formation. Interestingly the cargo of astrocyte extracellular vesicles can be altered by extracellular stimuli (Datta Chaudhuri et al. 2020), for example, inflammatory cytokines, suggesting that the vesicles may differ depending on developmental stage and presence of pathology.

Astrocyte-Secreted Factors Regulate Synapse Maturation

Role of SPARC Family Proteins in Synapse Maturation

SPARCL1 was identified as a synaptogenic protein secreted by astrocytes that is sufficient to induce synapse formation between RGC neurons in culture, whose effects are antagonized by the related protein SPARC, also produced by astrocytes (Kucukdereli et al. 2011). Analysis of SPARCL1 KO and SPARC KO mice showed that lack of SPARCL1 significantly impaired the formation and maturation of synapses, whereas lack of SPARC led to accelerated synapse formation (Kucukdereli et al. 2011). As SPARCL1 increases in astrocytes with maturation, a role for SPARCL1 in maturation and refinement of dendritic spines in the cortex was investigated (Risher et al. 2014). This revealed spines in SPARCL1 KO mice maintain multiple innervation into mature ages, with inputs from both intracortical and thalamocortical axons, features that are usually lost during postnatal development. Together with the expression pattern, this strongly suggests an important role for SPARCL1 in synapse maturation in vivo. Indeed, SPARCL1 KO mice show reduced plasticity in the visual cortex, suggesting a disruption of circuit maturation (Singh et al. 2016). The mechanism of SPARCL1-induced synaptogenesis was revealed to be via interacting with synaptogenic cell-adhesion molecules in both the presynaptic (neurexin) and postsynaptic neuroligin (NL) terminal, acting as a bridge to enable normally incompatible isoforms of these proteins to bind (Singh et al. 2016).

In addition to blocking SPARCL1-induced synaptogenesis, SPARC also regulates synaptic accumulation of AMPARs. In development, SPARC decreases AMPA receptors at synapses, fitting with an inhibitory role (Jones et al. 2011). After injury, for example stroke, SPARC is up-regulated and in this context promotes synaptic recruitment of AMPA receptors (Jones et al. 2018). Thus, the impact of SPARC on synapses is context dependent. Treatment of autaptic cholinergic neurons with SPARC results in an enhanced presynaptic release probability and a decrease in the number of vesicles available for release, features indicative of an immature presynaptic terminal (Albrecht et al. 2012), and SPARC can lead to synapse loss (López-Murcia et al. 2015). This shows that SPARC has multiple mechanisms to decrease synaptic communication in the developing brain.

Chordin-Like 1 Induces Synapse Maturation

As astrocytes regulate synaptic recruitment of all AMPA receptor subunits, and GPC4 was identified to only regulate GLUA1 and immature synapse formation (Allen et al. 2012), a biochemical screen was conducted to identify astrocyte-secreted molecules that are sufficient to recruit GLUA2 AMPA receptors and induce synapse maturation (Blanco-Suarez et al. 2018). This identified chordin-like 1 (CHRDL1) as necessary and sufficient to recruit GLUA2 to synapses between RGC neurons in vitro, and to enhance synaptic activity. The known role of CHRDL1 is as a secreted antagonist of BMP signaling; however, in this context, CHRDL1 appears to act in a BMP-independent manner, so its mechanism is yet to be identified. CHRDL1 KO mice show decreased recruitment of GLUA2 AMPA receptors to developing synapses in the cortex, and altered synaptic function, overall displaying features of immaturity. Interestingly, this prolonged synaptic immaturity enables enhanced plasticity in the cortex both during development when plasticity is normally occurring, and in adulthood when plasticity is greatly diminished. Together this shows that in the healthy brain, astrocyte-secreted CHRDL1 induces synapse maturation and inhibits plasticity. Interestingly, after acute injury such as a localized stroke, CHRDL1 expression is up-regulated in astrocytes (Blanco-Suarez and Allen 2022). This up-regulated CHRDL1 contributes to failure of synapses to recover after injury; as in CHRDL1 KO mice spine number is preserved after injury, whereas in wild-type mice spines are greatly decreased in number. This shows that astrocytes, via CHRDL1, are inhibitory to synaptic remodeling both in health and after injury.

Astrocyte-Secreted Lipids Regulate Synaptogenesis

Astrocyte-secreted cholesterol complexed with apolipoprotein E lipoparticles was identified as a positive regulator of glutamatergic presynaptic function, initially in studies of RGC neurons in culture (Mauch et al. 2001). Cholesterol enhances both presynaptic release probability and quantal content, thus increasing the efficacy of synaptic transmission (Mauch et al. 2001; Goritz et al. 2005). Further studies investigated the requirement for astrocyte-derived lipids, including cholesterol, for synapse development and function in vivo. Decreasing function of SREBP in astrocytes, which is required for production of cholesterol, led to immature synapses that contain less synaptic vesicles and showed functional impairments (van Deijk et al. 2017), as well as impaired brain development (Ferris et al. 2017). This shows the importance of lipids derived from astrocytes for correct synaptic development and maturation.

Astrocyte Secreted Factors and Neurodevelopmental Disorders

In addition to important roles in typical development, astrocytes show a dysregulation in release of factors that regulate synapses in genetic neurodevelopmental disorders, and these contribute to pathogenesis (Blanco-Suárez et al. 2017). Individual factors such as SPARCL1, SPARC, and TSP are decreased in Fragile X syndrome (FXS) and Down syndrome (DS) and contribute to stunted neuronal development in culture (Garcia et al. 2010; Cheng et al. 2016; Wallingford et al. 2017). Two studies took an unbiased proteomic approach to ask how the total protein secretion profile of astrocytes is changed in neurodevelopmental disorders;

the first focused on disorders with an excess of dendritic growth and synapse formation (Costello syndrome [CS]; Krencik et al. 2015), and the second on three disorders where astrocytes inhibit dendritic growth and synaptogenesis (FXS, DS, Rett syndrome [RTT]; Caldwell et al. 2022). Both studies identified many proteins up and down-regulated in release, and importantly found differences that would not have been revealed by studying gene expression alone. For the disorders that show stunted neurite outgrowth (RTT, FXS, DS) there was strong overlap in the altered protein profiles, but there was little overlap between these changes and CS, where neuronal growth and synapse formation is enhanced. This demonstrates that astrocytes show disorder-specific changes in their protein secretion profiles that match the phenotype that is induced in the neurons.

CONTACT-MEDIATED MECHANISMS REGULATING SYNAPSE FORMATION AND FUNCTION

In addition to secreted factors, astrocytes regulate synapse formation and function through direct contact. Several synaptogenic astrocyte-expressed cell-adhesion molecules (CAMs) have recently been identified (Fig. 3). These astrocyte CAMs interact with neuronal CAMs to control many aspects of synapse development and regulate the balance between synaptic excitation and inhibition. Notably, many of these interactions are bidirectional and also regulate the development of astrocytes. Here we will discuss the role of astrocyte–neuron contact in regulating synapse formation and function. For additional discussion of the role of astrocyte–neuron contact in astrocyte development, we refer to Ahrens et al. (2023).

Astrocyte–Neuron Contact during Development

Astrocyte–neuron contact is required at multiple developmental stages for proper synapse formation and function. Prior to synapse formation, neurons undergo a developmental switch to become receptive to soluble synaptogenic signals from astrocytes. This switch is induced by direct contact with astrocytes. When RGCs from 17-

day-old embryos (E17 RGCs) are cultured together with postnatal RGCs in the presence of ACM, E17 RGCs fail to receive synapses from postnatal RGCs, whereas their axons establish synapses onto postnatal RGCs. By E19, RGCs start responding to ACM (Barker et al. 2008). This switch in receptivity between E17 and E19 correlates with the appearance of astrocytes in the retina. Indeed, physical contact with astrocytes, but not amacrine cells, in culture was sufficient for E17 RGCs to become receptive to synaptogenic signals secreted from astrocytes. Contact by astrocytes causes the synaptic adhesion molecule neurexin to partition out of the dendrites where it is inhibitory to synapse formation (Barker et al. 2008).

Direct contact between astrocytes and neurons induces the formation of new synapses. This discovery was made using individual rat hippocampal neurons grown as distinct micro-islands in astrocyte-condition media (Hama et al. 2004). Under these conditions, the number of functional and mature excitatory synapses formed by individual neurons increased nearly fourfold upon the addition of an individual astrocyte to the micro-island. Integrin-mediated protein kinase C signaling was proposed to play a critical role in this process (Hama et al. 2004). In the developing spinal cord, homophilic interaction between astrocytic and neuronal γ-protocadherins (γ-Pcdh) controls the formation of both excitatory and inhibitory synapses (Garrett and Weiner 2009). In cultures of interneurons and spinal cord astrocytes, loss of γ-Pcdh from either neurons or astrocytes delays excitatory and inhibitory synapse formation. With astrocytic deletion, excitatory synapse number recovers over time, but remains decreased with neuronal deletion. A similar phenotype is found in the mouse spinal cord, where decreased synapse number eventually recovers with astrocyte-specific γ-Pcdh deletion but persists when γ-Pcdh is deleted from all cells (Garrett and Weiner 2009). In the mouse cortex, loss of γ-Pcdh from astrocytes decreases neuronal dendritic complexity, suggesting that this mechanism may extend beyond the spinal cord (Garrett et al. 2012).

Astrocyte contact-mediated mechanisms regulate the stability of newly formed dendritic pro-

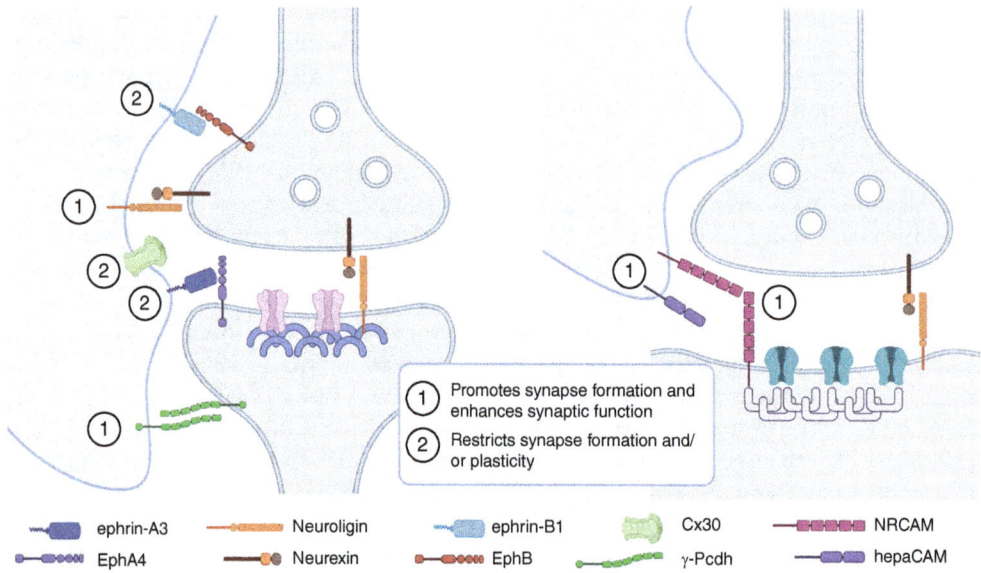

Figure 3. Contact-mediated mechanisms of astrocyte-synapse regulation. A summary of astrocyte-expressed cell-adhesion molecules and their respective neuronal binding partners that mediate direct contact between astrocytes and synapses. At excitatory synapses (*left*), neurexin–neuroligin and γ-protocadherins (γ-Pcdh) promote synapse formation and enhance synaptic function, whereas interactions between ephrins and Eph receptors generally play restrictive roles. Connexin 30 (Cx30) restricts the extent of astrocyte process infiltration into the synaptic cleft. At the inhibitory synapse (*right*), neuronal cell-adhesion molecule (NRCAM) and hepatic and glial cell-adhesion molecule (hepaCAM) promote synapse formation and enhance synaptic inhibition. (Created with BioRender.com.)

trusions and promote the maturation of dendritic spines. In hippocampal slice cultures, dendritic protrusions in contact with perisynaptic astrocyte processes are longer lived and more likely to form spines (Nishida and Okabe 2007). Bidirectional signaling between astrocyte-expressed ephrin-A3 and neuron-expressed EphA4 plays a key role in spine stability and morphology. Perturbing Ephrin/EphA signaling either by delivering soluble Ephrin A3 in hippocampal slice cultures or by transfecting neurons with a kinase-inactive EphA4 results in defects in spine morphology (Murai et al. 2003). Similarly, dendrites in mice lacking ephA4 or ephrin-3A have irregularly shaped and disorganized spines (Carmona et al. 2009; Filosa et al. 2009). Interestingly, activating EphA4 signaling induces spine retraction in acute slices from the adult hippocampus, suggesting that finely tuned and temporally precise signaling is required for proper regulation of spine morphology (Murai et al. 2003).

Astrocytic Cell-Adhesion Molecules Regulating Synapse Formation and Function

In addition to controlling spine morphology, ephrin/Eph interactions play important roles in synaptic function. Both ephrin-A3 and EphA4 KO mice show reduced glutamate levels near the synapse and impaired LTP (Filosa et al. 2009). EphrinA3 KO mice also have impaired contextual memory (Carmona et al. 2009). In development, loss of ephrin-B1 from astrocytes increases excitatory synapse number and activity, reduces inhibitory synapse number and response, and leads to impaired social behavior (Nguyen et al. 2020b). In the adult mouse, deletion of ephrin-B1 from astrocytes increases excitatory synapse number and enhances long-term contextual memory, while ephrin-B1 overexpression decreases spine density and impairs new memory formation (Nguyen et al. 2020a). In vitro evidence suggests that astrocytic ephrin-B1 may mediate its effect by binding to presynaptic

 Cite this article as *Cold Spring Harb Perspect Biol* doi: 10.1101/cshperspect.a041352

EphB (Koeppen et al. 2018). Together, these findings indicate a common role for ephrin/Eph signaling in restricting synapse formation and plasticity. The astrocyte-enriched gap junction protein Connexin 30 (Cx30) also functions to restrict excitatory synaptic plasticity, employing an adhesion-based mechanism to regulate how far astrocyte processes extend into the synaptic cleft (Pannasch et al. 2014).

NLs are cell-adhesion molecules known for their roles in synapse assembly, stabilization, and specification. In neurons, NLs are expressed postsynaptically where they interact with presynaptic neurexins to promote the formation of stable synapses (Craig and Kang 2007). Interestingly, several NL family members, NL1, NL2, and NL3 are expressed at equal or higher levels in astrocytes compared to neurons (Zhang et al. 2014). In astrocytes of the mouse cortex, NL2 controls the extent of local excitatory synapse formation (Stogsdill et al. 2017). Deletion of NL2 from individual astrocytes reduces the number excitatory synapses within the astrocyte's territory but does not affect synapse formation in neighboring wild-type astrocytes. The inhibitory synapse number remains unchanged. Deleting NL2 from a large population of astrocytes in the cortex reduces excitatory synaptic activity and increases inhibitory synapse activity (Stogsdill et al. 2017).

A recent study used proximity-based quantitative proteomics in vivo to uncover numerous proteins present at sites of astrocyte–synapse contact (Takano et al. 2020). Among these, they identified the neuronal cell-adhesion molecule (NRCAM). In the mouse cortex, deletion of NRCAM from astrocytes reduces contact between astrocyte processes and inhibitory synapses, but not excitatory synapses. NRCAM regulates inhibitory synapse organization and function via homophilic interaction between astrocytic and neuronal NRCAM. Deletion of NRCAM from either astrocytes or neurons reduces excitatory synapse formation in the mouse cortex, and deletion of NRCAM from astrocytes reduces inhibitory synaptic activity (Takano et al. 2020). Hepatic and glial cell-adhesion molecule (hepaCAM) is also expressed on astrocyte processes at inhibitory synapses (Baldwin et al. 2021). Loss of hep-

aCAM from astrocytes reduces inhibitory synaptic strength and increases excitatory synaptic strength. HepaCAM is present at inhibitory postsynapses, but whether astrocytic hepaCAM regulates synaptic function via homophilic interaction with neuronal hepaCAM remains to be explored.

Collectively, these recent advancements highlight key roles for direct astrocyte–neuron contact in mediating synapse formation, maturation, and function. These functions are mediated by a growing collection of cell-adhesion molecules, some with distinct and some with overlapping roles. For example, while NL2 deletion reduces excitatory synapse activity and increases inhibitory synapse activity, deletion of hepaCAM does the opposite. Similar to hepaCAM deletion, NRCAM deletion also reduces inhibitory synapse activity, although hepaCAM and NRCAM have distinct functions in astrocyte morphogenesis, and it remains unclear whether they function through common or different mechanisms at inhibitory synapses. As a group, the molecules discussed in this section seem to regulate the balance between synaptic excitation and inhibition, at least in the cortex and hippocampus. Additional studies are needed to determine whether these mechanisms are shared with other brain regions, or whether they show heterogeneity between and within brain regions. At the cellular and molecular level, how astrocytes control the localization of cell-adhesion molecules to their perisynaptic process, and perhaps specific synapses, remains to be explored, as do the neuronal mechanisms through which these interactions effect change at the synapse.

ASTROCYTES CONTROL SYNAPSE ELIMINATION

During development, neurons initially generate excessive projections that make redundant synaptic connections with target cells. Within critical time periods, these excessive synapses are eliminated, whereas the remaining inputs are further strengthened to form mature neural circuit (Shatz 1983; Sanes and Lichtman 1999). Synapse elimination has also been observed during synaptic plasticity in the adult brain (Xu et al. 2009b; Yang et al. 2009; Roberts et al. 2010) during learning and memory formation. Thus, synapse elimination plays a central role in remodeling and reorganization of our nervous system throughout life. Moreover, in many neurodegenerative diseases, such as Alzheimer's disease (AD) and Parkinson's disease (PD), excessive synapse loss has been observed during disease progression (DeKosky and Scheff 1990; Day et al. 2006), highlighting the importance of understanding the mechanisms of synapse elimination.

How is synapse elimination achieved? Since the discovery that microglia mediate developmental synapse pruning by phagocytosing synapses through the classical complement cascade (Stevens et al. 2007; Paolicelli et al. 2011; Schafer et al. 2012), numerous studies have reported critical roles for microglia in mediating developmental synapse elimination as well as abnormal synapse loss in diseased brains (Hong et al. 2016; Lui et al. 2016; Sekar et al. 2016; Gunner et al. 2019). However, a gene expression analysis using purified mouse astrocytes and microarray revealed that astrocytes also express many genes that are implicated in engulfment and phagocytosis (Cahoy et al. 2008). These genes can be categorized into two main phagocytic pathways that begin in parallel and converge into common downstream mechanisms (Fig. 4). The first pathway consists of MEGF10 phagocytic receptor (an ortholog of *Drosophila* Draper and *Caenorhabditis elegans* CED-1) and its associated proteins, such as GULP (an ortholog of *Drosophila* dCed-6 and *C. elegans* CED-6) and ABCA1 (ATP-binding cassette, subfamily A member 1), which participate in cellular debris recognition and engulfment (Zhou et al. 2001; MacDonald et al.

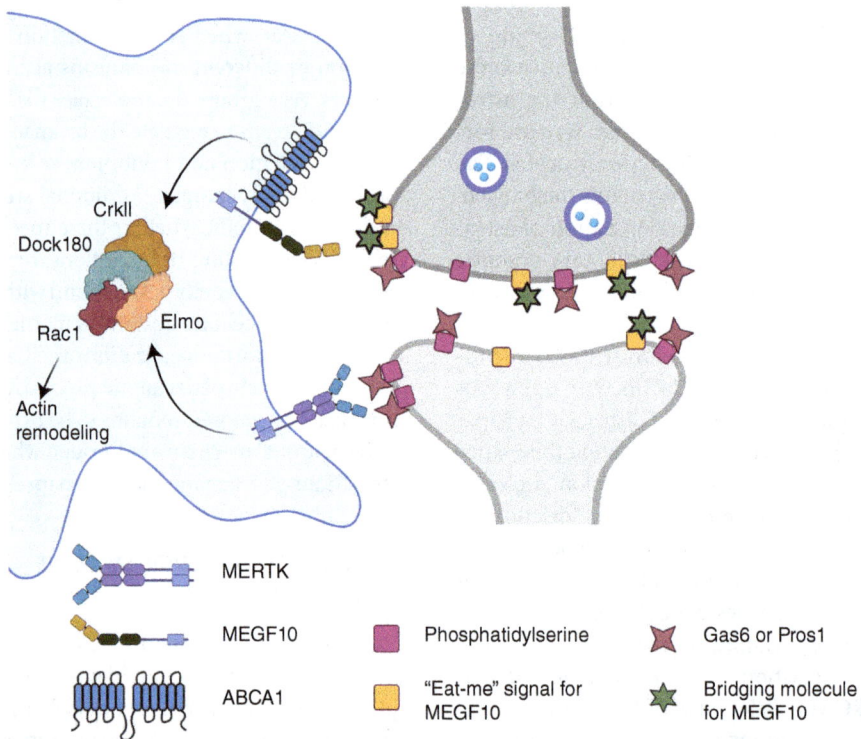

Figure 4. Astrocytes eliminate synapses via phagocytosis. A summary of astrocyte-expressed phagocytic receptors and their bridging molecules that mediate recognition of "eat-me" signals on synaptic surface. Astrocytic MERTK recognizes synaptic phosphatidylserine through Gas6 and Pros1. The identities of bridging molecules and respective synaptic "eat-me" signals for MEGF10 are less clear, although C1q can bind with MEGF10 (Iram et al. 2016). Activation of phagocytic receptors induces remodeling of the actin cytoskeleton through the Crkll/Dock180/Elmo/Rac1 complex to engulf synapses. (Created with BioRender.com.)

2006). The second pathway includes the TAM receptors, TYRO3, AXL, and MERTK (Lemke 2013), which require the soluble mediators Growth arrest-specific 6 (Gas6) or Protein S (Pros1) to recognize "eat-me" signals, such as phosphatidylserine (PtdSer) exposed on the surface of apoptotic cells. Upon recognizing such "eat-me" signals on targets, both MEGF10 and TAM phagocytic receptors induce rearrangement of the actin cytoskeleton, likely through CrKII/DOCK180/ELMO/Rac1 modules (Wu et al. 2005), thereby activating membrane dynamics of phagocytes to enable engulfment of surrounding cellular debris.

Synapse Elimination by Astrocytes during the Normal Synapse Remodeling

Astrocytes indeed use these two phagocytic pathways in directly mediating developmental synapse elimination (Chung et al. 2013). In mice deficient in both MEGF10 and MERTK, developing RGCs fail to normally refine their connections and retain excess functional synapses with postsynaptic neurons in the dorsal lateral geniculate nucleus (dLGN), thus providing evidence that astrocytes actively engulf live synapses rather than simply cleaning up already dead synapses. Blocking spontaneous retinal waves in both eyes significantly reduces astrocyte-mediated phagocytosis of bilateral synaptic inputs, whereas selective blocking of activity in only one eye induces preferential engulfment of the silenced synapses by astrocytes, indicating that astrocytes actively contribute to neural activity-dependent circuit refinement by phagocytosing weakened synapses. *Drosophila* glia, which resemble mammalian astrocytes, also phagocytose synapses during metamorphosis using Draper (an ortholog of MEGF10) and Crk/Mbc/dCed-12 (orthologs of CrKII/DOCK1/ELMO) signaling pathways, indicating that the phagocytic function of astrocytes is evolutionarily conserved (Tasdemir-Yilmaz and Freeman 2014).

Importantly, synapses in the adult brains also undergo constant formation and elimination during learning and memory formation. A recent study developed synapse phagocytosis reporters through tagging different synapse types with a pH-indicator (i.e., mCherry-eGFP) to demonstrate glia-mediated elimination of both excitatory and inhibitory synapses in the normal adult hippocampal CA1 region (Lee et al. 2021). This work revealed that astrocytes and microglia continuously phagocytose excitatory and inhibitory synapses, and that excitatory synapses are the major population that undergo glia-mediated synapse elimination in the adult CA1. Surprisingly, compared to microglia, astrocytes appear to eliminate significantly more synapses during neuronal activity-dependent synaptic turnover of the adult CA1. Deleting *Megf10* specifically in the adult astrocytes reduces astrocytes' ability to eliminate excitatory, but not inhibitory synapses, resulting in a rapid increase in the number of excitatory synapses with the appearance of multisynaptic boutons and multisynaptic spines, a characteristic feature of immature synapses. Electrophysiology and behavior tests showed that *Megf10* cKO animals exhibit enhanced basal synaptic transmission but defective synaptic plasticity (for both LTP and LTD) with defective learning and memory formation. Overall, this work challenges the general consensus that microglia are the primary synapse phagocytes that control synapse number in the brain. In addition, this work shows that in the adult brain regions where synapses undergo rapid turnover, multiple synapses are formed redundantly, and continuous astrocyte-mediated synapse elimination is needed for maintaining proper neural circuit homeostasis.

Another recent study used a pH indicator protein expressed in Purkinje cells (PCs) of the adult mouse cerebellum to demonstrate that synapse elimination is a shared function of astrocytes across diverse brain regions. Bergmann glia (BG), the specialized gray matter astrocytes of the cerebellum ingest PC synapses during sensory learning adaptation (Morizawa et al. 2022). Deleting ABCA1, an associated protein in the MEGF10 phagocytic pathway, reduces the phagocytic capacity of BG, resulting in transient defects in horizontal optokinetic response (HOKR) learning, as well as in the reduction of PC postsynaptic spine volume. In addition to astrocytes and microglia, oligodendrocyte precursor cells (OPCs) were recently shown to engulf synapses in the developing and adult visual cortex. While the exact phagocytic machinery for OPCs remain unclear, this OPC-mediated synapse elimination was particularly prevalent in the visual cortex during periods of sensory-mediated refinement (Auguste et al. 2022).

Thus, the primacy of one glial cell type over the others in phagocytic elimination of synapses is likely determined by various factors, including age, brain region, synapse type, and the presence of specific stimuli. Astrocytes may hold advantages over other glial cells in neuronal activity-dependent synapse remodeling, since astrocytic processes are structurally associated with synapses and constantly monitor changes in neuronal activity. Recent discoveries showing that astrocytes express various adhesion molecules that were once thought to be exclusively expressed by synapses and utilize them in interacting with synapses (Stogsdill et al. 2017), also emphasize that astrocytes could be the first responders in recognizing and eliminating unnecessary synapses in the brain.

Synapse Elimination by Astrocytes in Diseased/Injured Brains

Synapse loss is a common feature of injured and diseased brains. Since astrocytes and microglia undergo reactive gliosis with the increased gene expression profiles related to proinflammatory responses, it has been suggested that reactive glia may cause the abnormal synapse loss found in injured or diseased brains. Compared to microglia-mediated synapse elimination, there has been relatively little exploration of astrocytic synapse elimination in the diseased or injured brains. However, several recent studies have indicated that astrocytes play a critical role in synapse elimination in pathological conditions.

Astrocytes participate in eliminating neuronal debris and synapses after ischemic stroke (Morizawa et al. 2017). Experiments using a transient middle cerebral artery occlusion (MCAO) mouse model revealed that astrocytes in the penumbra region surrounding the ischemic core become phagocytic with enhanced expression of ABCA1 and Galectin-3 and engulf degenerating neuronal cell debris as well as pre- and post-synapses. Knocking out ABCA1 in astrocytes decreased the phagocytic capacity of astrocytes in ischemic brains. Interestingly, microglia phagocytosis showed early onset in the ischemic core regions, while astrocytic phagocytosis showed late onset in the penumbra regions. The distinct roles of astrocytes and microglia in eliminating synapses in different stroke types were also proposed recently (Shi et al. 2021). In this study, both astrocytes and microglia use MEGF10 and MERTK phagocytic receptors in eliminating synapses after ischemic stroke and that reducing phagocytic capacity of astrocytes or microglia preserved synapse loss, improving neurobehavioral outcomes after recovery. However, in hemorrhagic stroke, astrocytes lost their phagocytic ability, so that further knocking out *Megf10* or *Mertk* in astrocytes did not rescue the synapse loss and behavior symptoms. Thus, depending on the specific injury types, astrocytic synapse elimination may be differentially regulated and play distinct roles during the disease progression.

In the case of neurodegenerative diseases, reactive astrocytes have been shown to engulf presynaptic debris in AD model mouse as well as in AD patients (Gomez-Arboledas et al. 2018), suggesting astrocytes may participate in removing damaged neurites and synapses (Hulshof et al. 2022). More recently, it has been shown that astrocytes could be the major population in eliminating excitatory synapses in AD brains (Dejanovic et al. 2022). By measuring the astrocyte–synapse interface association as well as the amounts of engulfed excitatory and inhibitory synapses by glial cells, the authors show that astrocytes preferentially phagocytose excitatory synapses, while microglia preferentially eliminate inhibitory synapses in P301S mutant Tau overexpressing mice. However, throughout the progression of AD, it has been suggested that oligomeric amyloid β may reduce the phagocytic capacity of astrocytes contributing to the accumulation of amyloid β plaque as well as dystrophic synapses (Sanchez-Mico et al. 2021). In addition, reactive astrocytes induced by systemic LPS injection exhibit reduced *Megf10* and *Mertk* mRNA levels with the reduced phagocytic capacity during developmental synapse pruning (Liddelow et al. 2017).

Thus, current data supports the notion that astrocytes continuously participate in synapse

Cite this article as *Cold Spring Harb Perspect Biol* doi: 10.1101/cshperspect.a041352

elimination in adult and aging brains, but this process is likely altered during the initiation and progression of various neurodegenerative diseases. It is worth noting that synapse elimination is the result of bidirectional communication between synapses and glial cells. The expression levels of phagocytic receptors and bridging molecules that bind to the receptors may be changed during the course of the disease, allowing astrocytes to abnormally engage with synapses, and eliminate them. At the same time, alterations in neuronal/synaptic presentation of surface molecules could be an important component that initiates glia-mediated synapse elimination. During AD and PD, neuronal activity can be dynamically changed, often preceding the initiation of abnormal synapse elimination. How specific patterns of neuronal activity induce the expression and presentation of synaptic "eat-me" signals, and whether these synaptic changes interact with glial phagocytic receptors in the context of these diseases are fascinating subjects for future study.

CONCLUSION

Here we have reviewed recent advances in our knowledge of astrocyte–synapse interactions that highlight the important roles of astrocytes in the formation, maturation, and elimination of synapses. We particularly focused on findings from mammalian systems, but many of these glial functions are well conserved in other species, including *C. elegans*, *Drosophila*, and zebrafish. In addition to advances in understanding the role of astrocytes in the healthy brain, numerous studies have highlighted the integral role of astrocytes in mediating synaptic dysfunction in diverse neurological disorders. Moving forward, this suggests astrocytes should be considered important targets for rescuing synaptic dysfunction.

REFERENCES

*Reference is also in this subject collection.

* Ahrens MB, Khakh BS, Poskanzer KE. 2023. Astrocyte calcium signaling: summary and new perspectives. *Cold Spring Harb Perspect Biol* doi:10.1101/cshperspect.a041353

Albrecht D, López-Murcia FJ, Pérez-González AP, Lichtner G, Solsona C, Llobet A. 2012. SPARC prevents maturation of cholinergic presynaptic terminals. *Mol Cell Neurosci* **49**: 364–374. doi:10.1016/j.mcn.2012.01.005

Allen NJ. 2014. Astrocyte regulation of synaptic behavior. *Annu Rev Cell Dev Biol* **30**: 439–463. doi:10.1146/annurev-cellbio-100913-013053

Allen NJ, Bennett ML, Foo LC, Wang GX, Chakraborty C, Smith SJ, Barres BA. 2012. Astrocyte glypicans 4 and 6 promote formation of excitatory synapses via GluA1 AMPA receptors. *Nature* **486**: 410–414. doi:10.1038/nature11059

Araque A, Parpura V, Sanzgiri RP, Haydon PG. 1999. Tripartite synapses: glia, the unacknowledged partner. *Trends Neurosci* **22**: 208–215. doi:10.1016/s01662236(98)01349-6

Auguste YSS, Ferro A, Kahng JA, Xavier AM, Dixon JR, Vrudhula U, Nichitiu AS, Rosado D, Wee TL, Pedmale UV, et al. 2022. Oligodendrocyte precursor cells engulf synapses during circuit remodeling in mice. *Nat Neurosci* **25**: 1273–1278. doi:10.1038/s41593-022-01170-x

Baldwin KT, Tan CX, Strader ST, Jiang C, Savage JT, Elorza-Vidal X, Contreras X, Rülicke T, Hippenmeyer S, Estévez R, et al. 2021. HepaCAM controls astrocyte self-organization and coupling. *Neuron* **109**: 2427–2442.e10. doi:10.1016/j.neuron.2021.05.025

Banker GA. 1980. Trophic interactions between astroglial cells and hippocampal neurons in culture. *Science* **209**: 809–810. doi:10.1126/science.7403847

Barker AJ, Koch SM, Reed J, Barres BA, Ullian EM. 2008. Developmental control of synaptic receptivity. *J Neurosci* **28**: 8150–8160. doi:10.1523/JNEUROSCI.1744-08.2008

* Bartels T, Rowitch DH, Bayraktar OA. 2023. Generation of mammalian astrocyte functional heterogeneity. *Cold Spring Harb Perspect Biol* doi:10.1101/cshperspect.a041351

Bayraktar OA, Bartels T, Holmqvist S, Kleshchevnikov V, Martirosyan A, Polioudakis D, Ben Haim L, Young AMH, Batiuk MY, Prakash K, et al. 2020. Astrocyte layers in the mammalian cerebral cortex revealed by a single-cell in situ transcriptomic map. *Nat Neurosci* **23**: 500–509. doi:10.1038/s41593-020-0602-1

Blanco-Suarez E, Allen NJ. 2022. Astrocyte-secreted chordin-like 1 regulates spine density after ischemic injury. *Sci Rep* **12**: 4176. doi:10.1038/s41598-022-08031-4

Blanco-Suárez E, Caldwell ALM, Allen NJ. 2017. Role of astrocyte-synapse interactions in CNS disorders. *J Physiol* **595**: 1903–1916. doi:10.1113/JP270988

Blanco-Suarez E, Liu TF, Kopelevich A, Allen NJ. 2018. Astrocyte-secreted chordin-like 1 drives synapse maturation and limits plasticity by increasing synaptic GluA2 AMPA receptors. *Neuron* **100**: 1116–1132.e13. doi:10.1016/j.neuron.2018.09.043

Boroujerdi A, Zeng J, Sharp K, Kim D, Steward O, Luo ZD. 2011. Calcium channel α-2-δ-1 protein upregulation in dorsal spinal cord mediates spinal cord injury-induced neuropathic pain states. *Pain* **152**: 649–655. doi:10.1016/j.pain.2010.12.014

Brill J, Huguenard JR. 2008. Sequential changes in AMPA receptor targeting in the developing neocortical excitatory circuit. *J Neurosci* **28**: 13918–13928. doi:10.1523/JNEUROSCI.3229-08.2008

Bushong EA, Martone ME, Jones YZ, Ellisman MH. 2002. Protoplasmic astrocytes in CA1 stratum radiatum occupy separate anatomical domains. *J Neurosci* **22:** 183–192. doi:10.1523/JNEUROSCI.22-01-00183.2002

Cahoy JD, Emery B, Kaushal A, Foo LC, Zamanian JL, Christopherson KS, Xing Y, Lubischer JL, Krieg PA, Krupenko SA, et al. 2008. A transcriptome database for astrocytes, neurons, and oligodendrocytes: a new resource for understanding brain development and function. *J Neurosci* **28:** 264–278. doi:10.1523/JNEUROSCI.4178-07.2008

Caldwell ALM, Sancho L, Deng J, Bosworth A, Miglietta A, Diedrich JK, Shokhirev MN, Allen NJ. 2022. Aberrant astrocyte protein secretion contributes to altered neuronal development in multiple models of neurodevelopmental disorders. *Nat Neurosci* **25:** 1163–1178. doi:10.1038/s41593-022-01150-1

Carmona MA, Murai KK, Wang L, Roberts AJ, Pasquale EB. 2009. Glial ephrin-A3 regulates hippocampal dendritic spine morphology and glutamate transport. *Proc Natl Acad Sci* **106:** 12524–12529. doi:10.1073/pnas.0903328106

Cheng C, Lau SKM, Doering LC. 2016. Astrocyte-secreted thrombospondin-1 modulates synapse and spine defects in the fragile X mouse model. *Mol Brain* **9:** 74. doi:10.1186/s13041-016-0256-9

Christopherson KS, Ullian EM, Stokes CC, Mullowney CE, Hell JW, Agah A, Lawler J, Mosher DF, Bornstein P, Barres BA. 2005. Thrombospondins are astrocyte-secreted proteins that promote CNS synaptogenesis. *Cell* **120:** 421–433. doi:10.1016/j.cell.2004.12.020

Chung WS, Clarke LE, Wang GX, Stafford BK, Sher A, Chakraborty C, Joung J, Foo LC, Thompson A, Chen C, et al. 2013. Astrocytes mediate synapse elimination through MEGF10 and MERTK pathways. *Nature* **504:** 394–400. doi:10.1038/nature12776

Craig AM, Kang Y. 2007. Neurexin–neuroligin signaling in synapse development. *Curr Opin Neurobiol* **17:** 43–52. doi:10.1016/j.conb.2007.01.011

Crawford DC, Jiang X, Taylor A, Mennerick S. 2012. Astrocyte-derived thrombospondins mediate the development of hippocampal presynaptic plasticity in vitro. *J Neurosci* **32:** 13100–13110. doi:10.1523/JNEUROSCI.2604-12.2012

Datta Chaudhuri A, Dasgheyb RM, DeVine LR, Bi H, Cole RN, Haughey NJ. 2020. Stimulus-dependent modifications in astrocyte-derived extracellular vesicle cargo regulate neuronal excitability. *Glia* **68:** 128–144. doi:10.1002/glia.23708

Day M, Wang Z, Ding J, An X, Ingham CA, Shering AF, Wokosin D, Ilijic E, Sun Z, Sampson AR, et al. 2006. Selective elimination of glutamatergic synapses on striatopallidal neurons in Parkinson disease models. *Nat Neurosci* **9:** 251–259. doi:10.1038/nn1632

Dejanovic B, Wu T, Tsai MC, Graykowski D, Gandham VD, Rose CM, Bakalarski CE, Ngu H, Wang Y, Pandey S, et al. 2022. Complement C1q-dependent excitatory and inhibitory synapse elimination by astrocytes and microglia in Alzheimer's disease mouse models. *Nat Aging* **2:** 837–850. doi:10.1038/s43587-022-00281-1

DeKosky ST, Scheff SW. 1990. Synapse loss in frontal cortex biopsies in Alzheimer's disease: correlation with cognitive severity. *Ann Neurol* **27:** 457–464. doi:10.1002/ana.410270502

Dowling C, Allen NJ. 2018. Mice lacking glypican 4 display juvenile hyperactivity and adult social interaction deficits. *Brain Plast* **4:** 197–209. doi:10.3233/BPL-180079

Eroglu C, Allen NJ, Susman MW, O'Rourke NA, Park CY, Özkan E, Chakraborty C, Mulinyawe SB, Annis DS, Huberman AD, et al. 2009. Gabapentin receptor α2δ-1 is a neuronal thrombospondin receptor responsible for excitatory CNS synaptogenesis. *Cell* **139:** 380–392. doi:10.1016/j.cell.2009.09.025

Farhy-Tselnicker I, Allen NJ. 2018. Astrocytes, neurons, synapses: a tripartite view on cortical circuit development. *Neural Dev* **13:** 7. doi:10.1186/s13064-018-0104-y

Farhy-Tselnicker I, van Casteren ACM, Lee A, Chang VT, Aricescu AR, Allen NJ. 2017. Astrocyte-secreted glypican 4 regulates release of neuronal pentraxin 1 from axons to induce functional synapse formation. *Neuron* **96:** 428–445.e13. doi:10.1016/j.neuron.2017.09.053

Farhy-Tselnicker I, Boisvert MM, Liu H, Dowling C, Erikson GA, Blanco-Suarez E, Farhy C, Shokhirev MN, Ecker JR, Allen NJ. 2021. Activity-dependent modulation of synapse regulating genes in astrocytes. *eLife* **10:** e70514. doi:10.7554/eLife.70514

Ferris HA, Perry RJ, Moreira GV, Shulman GI, Horton JD, Kahn CR. 2017. Loss of astrocyte cholesterol synthesis disrupts neuronal function and alters whole-body metabolism. *Proc Natl Acad Sci* **114:** 1189–1194. doi:10.1073/pnas.1620506114

Filosa A, Paixão S, Honsek SD, Carmona MA, Becker L, Feddersen B, Gaitanos L, Rudhard Y, Schoepfer R, Klopstock T, et al. 2009. Neuron-glia communication via EphA4/ephrin-A3 modulates LTP through glial glutamate transport. *Nat Neurosci* **12:** 1285–1292. doi:10.1038/nn.2394

Fossati G, Pozzi D, Canzi A, Mirabella F, Valentino S, Morini R, Ghirardini E, Filipello F, Moretti M, Gotti C, et al. 2019. Pentraxin 3 regulates synaptic function by inducing AMPA receptor clustering via ECM remodeling and β1-integrin. *EMBO J* **38:** e99529. doi:10.15252/embj.201899529

Garcia O, Torres M, Helguera P, Coskun P, Busciglio J. 2010. A role for thrombospondin-1 deficits in astrocyte-mediated spine and synaptic pathology in down's syndrome. *PLoS ONE* **5:** e14200. doi:10.1371/journal.pone.0014200

Garrett AM, Weiner JA. 2009. Control of CNS synapse development by γ-protocadherin mediated astrocyte-neuron contact. *J Neurosci* **29:** 11723–11731. doi:10.1523/JNEUROSCI.2818-09.2009

Garrett AM, Schreiner D, Lobas MA, Weiner JA. 2012. γ-Protocadherins control cortical dendrite arborization by regulating the activity of a FAK/PKC/MARCKS signaling pathway. *Neuron* **74:** 269–276. doi:10.1016/j.neuron.2012.01.028

Gomez-Arboledas A, Davila JC, Sanchez-Mejias E, Navarro V, Nuñez-Diaz C, Sanchez-Varo R, Sanchez-Mico MV, Trujillo-Estrada L, Fernandez-Valenzuela JJ, Vizuete M, et al. 2018. Phagocytic clearance of presynaptic dystrophies by reactive astrocytes in Alzheimer's disease. *Glia* **66:** 637–653. doi:10.1002/glia.23270

Goritz C, Mauch DH, Pfrieger FW. 2005. Multiple mechanisms mediate cholesterol-induced synaptogenesis in a

CNS neuron. *Mol Cell Neurosci* **29:** 190–201. doi:10.1016/j.mcn.2005.02.006

Gunner G, Cheadle L, Johnson KM, Ayata P, Badimon A, Mondo E, Nagy MA, Liu L, Bemiller SM, Kim KW, et al. 2019. Sensory lesioning induces microglial synapse elimination via ADAM10 and fractalkine signaling. *Nat Neurosci* **22:** 1075–1088. doi:10.1038/s41593019-0419-y

Halassa MM, Fellin T, Takano H, Dong JH, Haydon PG. 2007. Synaptic islands defined by the territory of a single astrocyte. *J Neurosci* **27:** 6473–6477. doi:10.1523/JNEUROSCI.1419-07.2007

Hama H, Hara C, Yamaguchi K, Miyawaki A. 2004. PKC signaling mediates global enhancement of excitatory synaptogenesis in neurons triggered by local contact with astrocytes. *Neuron* **41:** 405–415. doi:10.1016/S0896-6273(04)00007-8

Hennekinne L, Colasse S, Triller A, Renner M. 2013. Differential control of thrombospondin over synaptic glycine and AMPA receptors in spinal cord neurons. *J Neurosci* **33:** 11432–11439. doi:10.1523/JNEUROSCI.5247-12.2013

Hill SA, Blaeser AS, Coley AA, Xie Y, Shepard KA, Harwell CC, Gao WJ, Garcia ADR. 2019. Sonic hedgehog signaling in astrocytes mediates cell type-specific synaptic organization. *eLife* **8:** e45545. doi:10.7554/eLife.45545

Hong S, Beja-Glasser VF, Nfonoyim BM, Frouin A, Li S, Ramakrishnan S, Merry KM, Shi Q, Rosenthal A, Barres BA, et al. 2016. Complement and microglia mediate early synapse loss in Alzheimer mouse models. *Science* **352:** 712–716. doi:10.1126/science.aad8373

Huang K, Park S. 2021. Heparan sulfated glypican-4 is released from astrocytes by proteolytic shedding and GPI-anchor cleavage mechanisms. *eNeuro* **8:** ENEURO.0069-21.2021. doi:10.1523/ENEURO.0069-21.2021

Hughes EG, Elmariah SB, Balice-Gordon RJ. 2010. Astrocyte secreted proteins selectively increase hippocampal GABAergic axon length, branching, and synaptogenesis. *Mol Cell Neurosci* **43:** 136–145. doi:10.1016/j.mcn.2009.10.004

Hulshof LA, van Nuijs D, Hol EM, Middeldorp J. 2022. The role of astrocytes in synapse loss in Alzheimer's disease: a systematic review. *Front Cell Neurosci* **16:** 899251. doi:10.3389/fncel.2022.899251

Iram T, Ramirez-Ortiz Z, Byrne MH, Coleman UA, Kingery ND, Means TK, Frenkel D, El Khoury J. 2016. Megf10 is a receptor for C1Q that mediates clearance of apoptotic cells by astrocytes. *J Neurosci* **36:** 5185–5192. doi:10.1523/JNEUROSCI.3850-15.2016

Jones EV, Bernardinelli Y, Tse YC, Chierzi S, Wong TP, Murai KK. 2011. Astrocytes control glutamate receptor levels at developing synapses through SPARC-β-integrin interactions. *J Neurosci* **31:** 4154–4165. doi:10.1523/jneurosci.4757-10.2011

Jones EV, Bernardinelli Y, Zarruk JG, Chierzi S, Murai KK. 2018. SPARC and GluA1-containing AMPA receptors promote neuronal health following CNS injury. *Front Cell Neurosci* **12:** 22. doi:10.3389/fncel.2018.00022

Kim DS, Li KW, Boroujerdi A, Peter Yu Y, Zhou CY, Deng P, Park J, Zhang X, Lee J, Corpe M, et al. 2012. Thrombospondin-4 contributes to spinal sensitization and neuropathic pain states. *J Neurosci* **32:** 8977–8987. doi:10.1523/JNEUROSCI.6494-11.2012

Koeppen J, Nguyen AQ, Nikolakopoulou AM, Garcia M, Hanna S, Woodruff S, Figueroa Z, Obenaus A, Ethell IM. 2018. Functional consequences of synapse remodeling following astrocyte-specific regulation of ephrin-B1 in the adult hippocampus. *J Neurosci* **38:** 5710–5726. doi:10.1523/JNEUROSCI.3618-17.2018

Krencik R, Hokanson KC, Naraya AR, Dvornik J, Rooney GE, Rauen KA, Weiss LA, Rowitch DH, Ullian EM. 2015. Dysregulation of astrocyte extracellular signaling in Costello syndrome. *Sci Transl Med* **7:** 286ra66. doi:10.1126/scitranslmed.aaa5645

Kucukdereli H, Allen NJ, Lee AT, Feng A, Ozlu MI, Conatser LM, Chakraborty C, Workman G, Weaver M, Sage EH, et al. 2011. Control of excitatory CNS synaptogenesis by astrocyte-secreted proteins Hevin and SPARC. *Proc Natl Acad Sci* **108:** E440–E449. doi:10.1073/pnas.1104977108

Lee JH, Kim JY, Noh S, Lee H, Lee SY, Mun JY, Park H, Chung WS. 2021. Astrocytes phagocytose adult hippocampal synapses for circuit homeostasis. *Nature* **590:** 612–617. doi:10.1038/s41586-020-03060-3

Lemke G. 2013. Biology of the TAM receptors. *Cold Spring Harb Perspect Biol* **5:** a009076. doi:10.1101/cshperspect.a009076

Li H, Graber KD, Jin S, McDonald W, Barres BA, Prince DA. 2012. Gabapentin decreases epileptiform discharges in a chronic model of neocortical trauma. *Neurobiol Dis* **48:** 429–438. doi:10.1016/j.nbd.2012.06.019

Liauw J, Hoang S, Choi M, Eroglu C, Choi M, Sun GH, Percy M, Wildman-Tobriner B, Bliss T, Guzman RG, et al. 2008. Thrombospondins 1 and 2 are necessary for synaptic plasticity and functional recovery after stroke. *J Cereb Blood Flow Metab* **28:** 1722–1732. doi:10.1038/jcbfm.2008.65

Liddelow SA, Guttenplan KA, Clarke LE, Bennett FC, Bohlen CJ, Schirmer L, Bennett ML, Münch AE, Chung WS, Peterson TC, et al. 2017. Neurotoxic reactive astrocytes are induced by activated microglia. *Nature* **541:** 481–487. doi:10.1038/nature21029

López-Murcia FJ, Terni B, Llobet A. 2015. SPARC triggers a cell-autonomous program of synapse elimination. *Proc Natl Acad Sci* **112:** 13366–13371. doi:10.1073/pnas.1512202112

Lui H, Zhang J, Makinson SR, Cahill MK, Kelley KW, Huang HY, Shang Y, Oldham MC, Martens LH, Gao F, et al. 2016. Progranulin deficiency promotes circuit-specific synaptic pruning by microglia via complement activation. *Cell* **165:** 921–935. doi:10.1016/j.cell.2016.04.001

MacDonald JM, Beach MG, Porpiglia E, Sheehan AE, Watts RJ, Freeman MR. 2006. The *Drosophila* cell corpse engulfment receptor draper mediates glial clearance of severed axons. *Neuron* **50:** 869–881. doi:10.1016/j.neuron.2006.04.028

Mauch DH, Nägler K, Schumacher S, Göritz C, Müuller EC, Otto A, Pfrieger FW. 2001. CNS synaptogenesis promoted by glia-derived cholesterol. *Science* **294:** 1354–1357. doi:10.1126/science.294.5545.1354

Meyer-Franke A, Kaplan MR, Pfrieger FW, Barres BA. 1995. Characterization of the signaling interactions that promote the survival and growth of developing retinal ganglion cells in culture. *Neuron* **15:** 805–819. doi:10.1016/0896-6273(95)90172-8

Morel L, Chiang MSR, Higashimori H, Shoneye T, Iyer LK, Yelick J, Tai A, Yang Y. 2017. Molecular and functional properties of regional astrocytes in the adult brain. *J Neurosci* **37:** 8706–8717. doi:10.1523/jneurosci.3956-16.2017

Morizawa YM, Hirayama Y, Ohno N, Shibata S, Shigetomi E, Sui Y, Nabekura J, Sato K, Okajima F, Takebayashi H, et al. 2017. Reactive astrocytes function as phagocytes after brain ischemia via ABCA1-mediated pathway. *Nat Commun* **8:** 28. doi:10.1038/s41467017-00037-1

Morizawa YM, Matsumoto M, Nakashima Y, Endo N, Aida T, Ishikane H, Beppu K, Moritoh S, Inada H, Osumi N, et al. 2022. Synaptic pruning through glial synapse engulfment upon motor learning. *Nat Neurosci* **25:** 1458–1469. doi:10.1038/s41593-022-01184-5

Murai KK, Nguyen LN, Irie F, Yamaguchi Y, Pasquale EB. 2003. Control of hippocampal dendritic spine morphology through ephrin-A3/EphA4 signaling. *Nat Neurosci* **6:** 153–160. doi:10.1038/nn994

Nagai J, Rajbhandari AK, Gangwani MR, Hachisuka A, Coppola G, Masmanidis SC, Fanselow MS, Khakh BS. 2019. Hyperactivity with disrupted attention by activation of an astrocyte synaptogenic cue. *Cell* **177:** 1280–1292.e20. doi:10.1016/j.cell.2019.03.019

Nguyen AQ, Koeppen J, Woodruff S, Mina K, Figueroa Z, Ethell IM. 2020a. Astrocytic ephrin-B1 controls synapse formation in the hippocampus during learning and memory. *Front Synaptic Neurosci* **12:** 10. doi:10.3389/fnsyn.2020.00010

Nguyen AQ, Sutley S, Koeppen J, Mina K, Woodruff S, Hanna S, Vengala A, Hickmott PW, Obenaus A, Ethell IM. 2020b. Astrocytic ephrin-B1 controls excitatory-inhibitory balance in developing hippocampus. *J Neurosci* **40:** 6854–6871. doi:10.1523/JNEUROSCI.0413-20.2020

Nishida H, Okabe S. 2007. Direct astrocytic contacts regulate local maturation of dendritic spines. *J Neurosci* **27:** 331–340. doi:10.1523/JNEUROSCI.4466-06.2007

Ogata K, Kosaka T. 2002. Structural and quantitative analysis of astrocytes in the mouse hippocampus. *Neuroscience* **113:** 221–233. doi:10.1016/s0306-4522(02)00041-6

Pannasch U, Freche D, Dallérac G, Ghézali G, Escartin C, Ezan P, Cohen-Salmon M, Benchenane K, Abudara V, Dufour A, et al. 2014. Connexin 30 sets synaptic strength by controlling astroglial synapse invasion. *Nat Neurosci* **17:** 549–558. doi:10.1038/nn.3662

Paolicelli RC, Bolasco G, Pagani F, Maggi L, Scianni M, Panzanelli P, Giustetto M, Ferreira TA, Guiducci E, Dumas L, et al. 2011. Synaptic pruning by microglia is necessary for normal brain development. *Science* **333:** 1456–1458. doi:10.1126/science.1202529

Patel MR, Weaver AM. 2021. Astrocyte-derived small extracellular vesicles promote synapse formation via fibulin-2-mediated TGF-β signaling. *Cell Rep* **34:** 108829. doi:10.1016/j.celrep.2021.108829

Pfrieger FW, Barres BA. 1997. Synaptic efficacy enhanced by glial cells in vitro. *Science* **277:** 1684–1687. doi:10.1126/science.277.5332.1684

Risher WC, Patel S, Kim IH, Uezu A, Bhagat S, Wilton DK, Pilaz LJ, Singh Alvarado J, Calhan OY, Silver DL, et al. 2014. Astrocytes refine cortical connectivity at dendritic spines. *eLife* **3:** e04047. doi:10.7554/eLife.04047

Risher WC, Kim N, Koh S, Choi JE, Mitev P, Spence EF, Pilaz LJ, Wang D, Feng G, Silver DL, et al. 2018. Thrombospondin receptor α2δ-1 promotes synaptogenesis and spino-

genesis via postsynaptic Rac1. *J Cell Biol* **217:** 3747–3765. doi:10.1083/jcb.201802057

Roberts TF, Tschida KA, Klein ME, Mooney R. 2010. Rapid spine stabilization and synaptic enhancement at the onset of behavioural learning. *Nature* **463:** 948–952. doi:10.1038/nature08759

Sanchez-Mico MV, Jimenez S, Gomez-Arboledas A, Muñoz-Castro C, Romero-Molina C, Navarro V, Sanchez-Mejias E, Nuñnez-Diaz C, Sanchez-Varo R, Galea E, et al. 2021. Amyloid-β impairs the phagocytosis of dystrophic synapses by astrocytes in Alzheimer's disease. *Glia* **69:** 997–1011. doi:10.1002/glia.23943

Sanes JR, Lichtman JW. 1999. Development of the vertebrate neuromuscular junction. *Annu Rev Neurosci* **22:** 389–442. doi:10.1146/annurev.neuro.22.1.389

Schafer DP, Lehrman EK, Kautzman AG, Koyama R, Mardinly AR, Yamasaki R, Ransohoff RM, Greenberg ME, Barres BA, Stevens B. 2012. Microglia sculpt postnatal neural circuits in an activity and complement-dependent manner. *Neuron* **74:** 691–705. doi:10.1016/j.neuron.2012.03.026

Sekar A, Bialas AR, de Rivera H, Davis A, Hammond TR, Kamitaki N, Tooley K, Presumey J, Baum M, Van Doren V, et al. 2016. Schizophrenia risk from complex variation of complement component 4. *Nature* **530:** 177–183. doi:10.1038/nature16549

Shatz CJ. 1983. The prenatal development of the cat's retinogeniculate pathway. *J Neurosci* **3:** 482–499. doi:10.1523/JNEUROSCI.03-03-00482.1983

Shi X, Luo L, Wang J, Shen H, Li Y, Mamtilahun M, Liu C, Shi R, Lee JH, Tian H, et al. 2021. Stroke subtype-dependent synapse elimination by reactive gliosis in mice. *Nat Commun* **12:** 6943. doi:10.1038/s41467-021-27248-x

Singh SK, Stogsdill JA, Pulimood NS, Dingsdale H, Kim YH, Pilaz LJ, Kim IH, Manhaes AC, Rodrigues WS Jr, Pamukcu A, et al. 2016. Astrocytes assemble thalamocortical synapses by bridging NRX1α and NL1 via Hevin. *Cell* **164:** 183–196. doi:10.1016/j.cell.2015.11.034

Stevens B, Allen NJ, Vazquez LE, Howell GR, Christopherson KS, Nouri N, Micheva KD, Mehalow AK, Huberman AD, Stafford B, et al. 2007. The classical complement cascade mediates CNS synapse elimination. *Cell* **131:** 1164–1178. doi:10.1016/j.cell.2007.10.036

Stogsdill JA, Ramirez J, Liu D, Kim YH, Baldwin KT, Enustun E, Ejikeme T, Ji RR, Eroglu C. 2017. Astrocytic neuroligins control astrocyte morphogenesis and synaptogenesis. *Nature* **551:** 192–197. doi:10.1038/nature24638

Takano T, Wallace JT, Baldwin KT, Purkey AM, Uezu A, Courtland JL, Soderblom EJ, Shimogori T, Maness PF, Eroglu C, et al. 2020. Chemico-genetic discovery of astrocytic control of inhibition in vivo. *Nature* **588:** 296–302. doi:10.1038/s41586-020-2926-0

Tasdemir-Yilmaz OE, Freeman MR. 2014. Astrocytes engage unique molecular programs to engulf pruned neuronal debris from distinct subsets of neurons. *Genes Dev* **28:** 20–33. doi:10.1101/gad.229518.113

Turko P, Groberman K, Browa F, Cobb S, Vida I. 2019. Differential dependence of GABAergic and glutamatergic neurons on glia for the establishment of synaptic transmission. *Cereb Cortex* **29:** 1230–1243. doi:10.1093/cercor/bhy029

Ullian EM, Sapperstein SK, Christopherson KS, Barres BA. 2001. Control of synapse number by glia. *Science* **291**: 657–661. doi:10.1126/science.291.5504.657

van Deijk ALF, Camargo N, Timmerman J, Heistek T, Brouwers JF, Mogavero F, Mansvelder HD, Smit AB, Verheijen MHG. 2017. Astrocyte lipid metabolism is critical for synapse development and function in vivo. *Glia* **65**: 670–682. doi:10.1002/glia.23120

Wallingford J, Scott AL, Rodrigues K, Doering LC. 2017. Altered developmental expression of the astrocyte-secreted factors Hevin and SPARC in the fragile X mouse model. *Front Mol Neurosci* **10**: 268. doi:10.3389/fnmol.2017.00268

Wu Y, Singh S, Georgescu MM, Birge RB. 2005. A role for Mer tyrosine kinase in αvβ5 integrin-mediated phagocytosis of apoptotic cells. *J Cell Sci* **118**: 539–553. doi:10.1242/jcs.01632

Xu J, Xiao N, Xia J. 2009a. Thrombospondin 1 accelerates synaptogenesis in hippocampal neurons through neuroligin 1. *Nat Neurosci* **13**: 22–24. doi:10.1038/nn.2459

Xu T, Yu X, Perlik AJ, Tobin WF, Zweig JA, Tennant K, Jones T, Zuo Y. 2009b. Rapid formation and selective stabilization of synapses for enduring motor memories. *Nature* **462**: 915–919. doi:10.1038/nature08389

Yang G, Pan F, Gan WB. 2009. Stably maintained dendritic spines are associated with lifelong memories. *Nature* **462**: 920–924. doi:10.1038/nature08577

Yu K, Lin CJ, Hatcher A, Lozzi B, Kong K, Huang-Hobbs E, Cheng YT, Beechar VB, Zhu W, Zhang Y, et al. 2020. PIK3CA variants selectively initiate brain hyperactivity during gliomagenesis. *Nature* **578**: 166–171. doi:10.1038/s41586-020-1952-2

Zhang Y, Chen K, Sloan SA, Bennett ML, Scholze AR, O'Keeffe S, Phatnani HP, Guarnieri P, Caneda C, Ruderisch N, et al. 2014. An RNA-sequencing transcriptome and splicing database of glia, neurons, and vascular cells of the cerebral cortex. *J Neurosci* **34**: 11929–11947. doi:10.1523/JNEUROSCI.1860-14.2014

Zhou Z, Hartwieg E, Horvitz HR. 2001. CED-1 is a transmembrane receptor that mediates cell corpse engulfment in *C. elegans*. *Cell* **104**: 43–56. doi:10.1016/S0092-8674(01)00190-8

Astrocyte Calcium Signaling

Misha B. Ahrens,[1] Baljit S. Khakh,[2] and Kira E. Poskanzer[3]

[1]Janelia Research Campus, Howard Hughes Medical Institute, Ashburn, Virginia 20147, USA

[2]Department of Physiology and Department of Neurobiology, University of California Los Angeles, Los Angeles, California 90095, USA

[3]Department of Biochemistry & Biophysics, University of California, San Francisco, San Francisco, California 94143, USA

Correspondence: ahrensm@janelia.hhmi.org; bkhakh@mednet.ucla.edu; kira.poskanzer@ucsf.edu

Astrocytes are predominant glial cells that tile the central nervous system and participate in well-established functional and morphological interactions with neurons, blood vessels, and other glia. These ubiquitous cells display rich intracellular Ca^{2+} signaling, which has now been studied for over 30 years. In this review, we provide a summary and perspective of recent progress concerning the study of astrocyte intracellular Ca^{2+} signaling as well as discussion of its potential functions. Progress has occurred in the areas of imaging, silencing, activating, and analyzing astrocyte Ca^{2+} signals. These insights have collectively permitted exploration of the relationships of astrocyte Ca^{2+} signals to neural circuit function and behavior in a variety of species. We summarize these aspects along with a framework for mechanistically interpreting behavioral studies to identify directly causal effects. We finish by providing a perspective on new avenues of research concerning astrocyte Ca^{2+} signaling.

Calcium (Ca^{2+}) ions are a ubiquitous inorganic signaling species that play myriad fundamental roles in nearly all aspects of biology across cell types in diverse species from most phyla. One of the first demonstrations of dynamic intracellular Ca^{2+} fluctuations and regulation in astroglia came from the laboratory of Stephen Smith (Cornell-Bell et al. 1990). Using hippocampal explant cultures and organic Ca^{2+} indicator dyes, Smith and other researchers demonstrated that astroglia exhibited spontaneous intracellular Ca^{2+} elevations that propagated in complicated patterns throughout cells as well as developed into Ca^{2+} waves that propagated over long distances and involved many individual cells (Cornell-Bell et al. 1990; Charles et al. 1991; Dani et al. 1992). Further, these investigators found that the activation of neurotransmitter receptors such as those for glutamate increased Ca^{2+} signaling within astroglial cells. However, these studies did not identify all the molecular mechanisms of astrocyte Ca^{2+} signaling, or its physiological functions. Furthermore, these and other early studies mainly assessed astrocyte somata and therefore lacked relevant context in terms of signaling in astrocyte processes and near plasma membrane regions. Consequently, after a necessary and useful period of documenting astrocyte Ca^{2+} signaling in a variety of experimental systems, focus over the past three decades has shifted to analysis of the mechanisms and physiological roles of astrocyte Ca^{2+} fluxes in situ and, more recently, in vivo.

The study of astrocyte Ca^{2+} signaling now represents a large effort by many researchers around the world and it is impossible to summarize all of the studies published in the last 30 years. Instead, we make general points that we believe are accepted broadly in the field. We then provide a perspective on how the field can better explore astrocyte Ca^{2+} and its roles in neural circuits. Our perspective, by necessity for a field still in development and growth, contains some speculative components that we hope at the very least will stimulate discussion and new experiments.

A more detailed historical perspective on astrocyte Ca^{2+} signaling is available in the previous version of this collection (Khakh and McCarthy 2015). Furthermore, several reviews are also available on this topic (Volterra et al. 2014; Bazargani and Attwell 2016; Shigetomi et al. 2016; Kofuji and Araque 2021). We do not discuss gliotransmission—the proposed transmission of chemical information through molecules released from vesicles in a Ca^{2+}- dependent manner from astrocytes to neurons and other glia—because this topic has been extensively reviewed elsewhere (Araque et al. 2014; Fiacco and McCarthy 2018; Savtchouk and Volterra 2018).

PROGRESS OVER THE LAST FEW YEARS

Imaging Astrocyte Calcium Signals

Although much of the early work was performed using organic Ca^{2+} indicator dyes such as Fluo-4 and Fura-2, almost all current studies are performed using genetically encoded Ca^{2+} indicators (GECIs). A variety of these are now available, as they have been iteratively improved over the last 15 years (Zhang and Looger 2023; Zhang et al. 2023). GECIs can be expressed selectively within astrocytes in vivo and allow for straightforward observations of Ca^{2+} dynamics in brain slices and in a variety of awake behaving animals. In addition, GECIs can be targeted to subcellular compartments such as the plasma membrane, mitochondria, and endoplasmic reticulum. Such reliable tools have greatly expanded our appreciation of the richness of astrocyte Ca^{2+} dynamics as highly complex in terms of location within astrocytes, their spatial extent (i.e., local microdomains or more global events encompassing whole astrocytes), and their diverse amplitude and kinetics. In particular, GECI-based astrocyte Ca^{2+} imaging has underscored the largely unknown relationships between spatially separable Ca^{2+} events in different locations even of single cells. Thus, we do not yet understand how the many Ca^{2+} events in a given astrocyte are related to each other, nor whether their emergence as observably separable simply reflects the torturous subcellular morphology of astrocytes from which they emerge (Salmon et al. 2023).

Despite progress, key gaps in our understanding still remain and require more detailed work. First, astrocyte Ca^{2+} dynamics have not been recorded sufficiently in entire 3D astrocytes at the required temporal and spatial scales, and thus the extent of signaling in relevant compartments such as near synapses remains unknown. Second, although much progress has been made in documenting astrocyte Ca^{2+} dynamics, the field still lacks biophysical understanding of astrocyte Ca^{2+} sources, sinks, buffers, and regulators. Thus, mechanism-based mathematical models of astrocyte Ca^{2+} dynamics remain a goal of the future. To be of most relevance, such efforts need to be based on precise knowledge of the molecular mechanisms within astrocytes; some relevant advances have been reviewed (Yu et al. 2020) but much work remains. We also note that some of the latest GECIs have been designed for fast imaging, which is necessary to track fast dynamics of neuronal firing. However, it remains unclear whether astrocytes display similarly fast Ca^{2+} signals. Indeed, most past studies have shown that astrocyte Ca^{2+} signals last between a few hundred milliseconds and tens of seconds. Additionally, the most biologically relevant astrocyte Ca^{2+} concentration range changes may not coincide with those of neurons for which the GECIs have been largely developed. We suggest that key challenges in imaging astrocyte Ca^{2+} signaling are the need for speed, the need to capture large volumes rapidly in 3D so as to simultaneously image entire astrocytes or entire collections

Cite this article as *Cold Spring Harb Perspect Biol* doi: 10.1101/cshperspect.a041353

of coupled astrocytes (Peinado et al. 2021), and the need to cover the appropriate range of Ca^{2+} concentrations. To our knowledge, no study has yet achieved these requirements and, as a result, our understanding of astrocyte Ca^{2+} signaling remains incomplete.

Activation and Silencing of Astrocyte Calcium Signals

The last few years have witnessed excellent progress in the development and use of methods to activate and reduce ("silence") astrocyte Ca^{2+} dynamics. Largely, the key approaches have been borrowed or adapted from their intended purposes to study neurons. In terms of activation, the most reliable tools include designer receptors exclusively activated by designer drugs (DREADDs) (Roth 2016). These, irrespective of type (Gs, Gq, or Gi), elevate astrocyte Ca^{2+} signals when acutely activated in brain slices (Chai et al. 2017; Durkee et al. 2019), which recalls early studies that showed that a variety of GPCR agonists evoked astrocyte Ca^{2+} signaling (Porter and McCarthy 1997). However, in mice, at least some of the DREADDs seem to result in reduced Ca^{2+} signaling (Vaidyanathan et al. 2021; Delepine et al. 2023), which is perhaps due to intracellular Ca^{2+} store depletion that may occur during the minutes of DREADD activation in vivo. Furthermore, the precise relationship between the type of Ca^{2+} signaling triggered by DREADDs, how large the evoked signals are relative to each other, and physiological astrocyte Ca^{2+} signaling is still not clear. Thus, DREADDs are useful tools, but their use comes with caveats in terms of interpreting experimental results and their physiological relevance. Optical activation tools such as opto-α1AR (Airan et al. 2009) and melanopsin (Mederos et al. 2019) have also been developed and used in various species (Mu et al. 2019; Mederos et al. 2021), but these also carry the similar limitations in terms of relevance to physiological signaling.

For silencing, several tools are available and have been validated. IP3R2 knockout mice largely lack intracellular store-mediated Ca^{2+} signaling (Petravicz et al. 2008; Agulhon et al. 2010),

but continue to display additional forms of dynamics that may reflect transmembrane fluxes (Srinivasan et al. 2015). IP3-binding proteins or "sponges" have been validated and can be used (Xie et al. 2010). CalEx is useful to largely attenuate astrocyte Ca^{2+} dynamics and appears useful in this regard (Yu et al. 2018, 2021). However, a caveat with its use is that Ca^{2+} dynamics are silenced broadly irrespective of the source, meaning that precise mechanistic interpretations are not possible in terms of how the relevant signals arose. Recently, iβARK has been used to silence selectively GPCR Ca^{2+} signals mediated by the Gq pathway (Nagai et al. 2021a) but cannot be used to probe other forms of astrocyte Ca^{2+} dynamics. Furthermore, this tool reduced Gq GCPR Ca^{2+} responses by about 80%, implying that in some circumstances the remaining 20% may be sufficient for downstream effects. Thus, a positive result with iβARK can be informative, but a negative result is harder to interpret in settings where we know that some Gq GPCR signaling remains. Although frequently used in the field, we suggest that owing to their relatively low Ca^{2+} permeability and ability to depolarize cells, channelrhodopsin-based approaches are not ideally suited to probe astrocyte Ca^{2+} dynamics (Octeau et al. 2019). In the case of inhibitory opsins, these elevate Ca^{2+} within astrocytes likely owing to effects mediated by transporters (Poskanzer and Yuste 2016). These two examples show aptly how tools designed for neurons can have different, but nonetheless interesting and potentially useful, effects in astrocytes. Overall, there is no doubt that improved tools are needed to genetically interrogate astrocyte Ca^{2+} dynamics in vivo, including for optical suppression of Ca^{2+} levels. In addition, astrocyte-specific RNA-seq data sets could be mined to identify key molecules to target to selectively explore endogenous mechanisms; such data are beginning to emerge (Endo et al. 2022). Indeed, a mechanistic understanding of astrocyte Ca^{2+} dynamics is still an open topic that merits much further work. In this regard, the availability of large multiomic data sets documenting the molecular makeup of astrocytes should be used to design specific experiments that interrogate endogenous mechanisms of astrocyte Ca^{2+} signaling. In species where this is feasible, large-scale genome-wide screens to identify key steps involved in astrocyte Ca^{2+} signaling could be highly informative.

Analyzing Astrocyte Calcium Signals

As GECIs have been more widely used over the last several years, it has become apparent that Ca^{2+} signals in astrocytes are more heterogeneous, irregular, and propagative than those observable in earlier organic dye-based experiments that revealed mostly the somata. The fluorescence dynamics evident using GECIs also made clear that Ca^{2+} signals within astrocytic somata and their fine processes could differ in area, frequency, and other critical physiological metrics. Thus, understanding astrocyte function became more closely tied to the problem of accurately analyzing the Ca^{2+} dynamics themselves. Further, because astrocyte Ca^{2+} signals are qualitatively different from somatic neuronal Ca^{2+} signals widely used to study population-level neuronal activity, common fluorescence analytical tools designed for neuronal Ca^{2+} (Pachitariu et al. 2017; Giovannucci et al. 2019) have proven inadequate when applied to astrocyte Ca^{2+}. To tackle this image analysis problem, several methods have been developed. These include—but are not limited to—region-of-interest (ROI)-based (GECIquant [Srinivasan et al. 2015] and CaSCaDe [Agarwal et al. 2017]), event-based (AQuA [Fig. 1; Wang et al. 2019]), and hybrid (Bojarskaite et al. 2020) approaches. When choosing an analysis method to match a particular in vivo or ex vivo Ca^{2+} imaging experiment, users should consider the overall analytical goals, flexibility, and ease of usability. In general, ROI-based approaches can be simple to implement and may account for specific morphological features of astrocytes, but often cannot accurately reflect many important aspects of astrocyte Ca^{2+}, including the area of the astrocyte or astrocytic network encompassed by Ca^{2+} changes, and features related to Ca^{2+} propagation distance or speed. In contrast, event-based approaches allow users to quantify a greater number of Ca^{2+} features, including those that are now known to differ across behavioral states (Vaidyanathan et al. 2021; Reitman et al. 2023).

Since event-based analyses are not constrained by size or shape, they can be applied across spatiotemporal imaging scales and even for experiments combining astrocyte Ca^{2+} imaging with other fluorescent indicators that are highly dynamic and heterogeneous, such as those sensitive to changes in extracellular neuroactive molecules (Pittolo et al. 2022).

Consequences of Astrocyte Calcium Signals

The combination of the aforementioned approaches has allowed exploration of the consequences of astrocyte Ca^{2+} dynamics for neural circuits and behavior, but we emphasize our view that this area of work is still in its infancy. Such experiments have been performed in a variety of species, because of the finding that core features of astrocytes, including Ca^{2+} signaling, are preserved. There is a vast body of literature on this topic that has been reviewed by several excellent researchers (Nimmerjahn and Bergles 2015; Nagai et al. 2021b; Nimmerjahn and Hirrlinger 2022). The major development on this front has been the realization that astrocyte Ca^{2+} dynamics are engaged in a behaviorally relevant manner, are associated with distinct brain states, and that either activating or silencing astrocyte Ca^{2+} dynamics can lead to changes in behavior and neural circuit dynamics. Instead of restating these general findings that have been reviewed, we instead provide pointers on the interpretation of the data relating to the behaviorally relevant responses ascribed to astrocyte Ca^{2+} signaling. We summarize these as three different types of astrocyte effects: (1) seconds-timescale responses, (2) slower homeostatic functions, and (3) unnatural effects. Some of these ideas were recently reviewed (Nagai et al. 2021b).

1. *Seconds–timescale responses*: These may reflect computation upon input signals and intervention on neuronal activity. In this category are examples suggesting that astrocytes integrate incoming neuronal signals over seconds and then switch to a mode resulting in altered neuronal functions. For example, zebrafish radial astrocytes in a specific subregion of the brainstem temporally integrate neuromodulator-encoded behavioral failures to accumulate evidence of futility over sec-

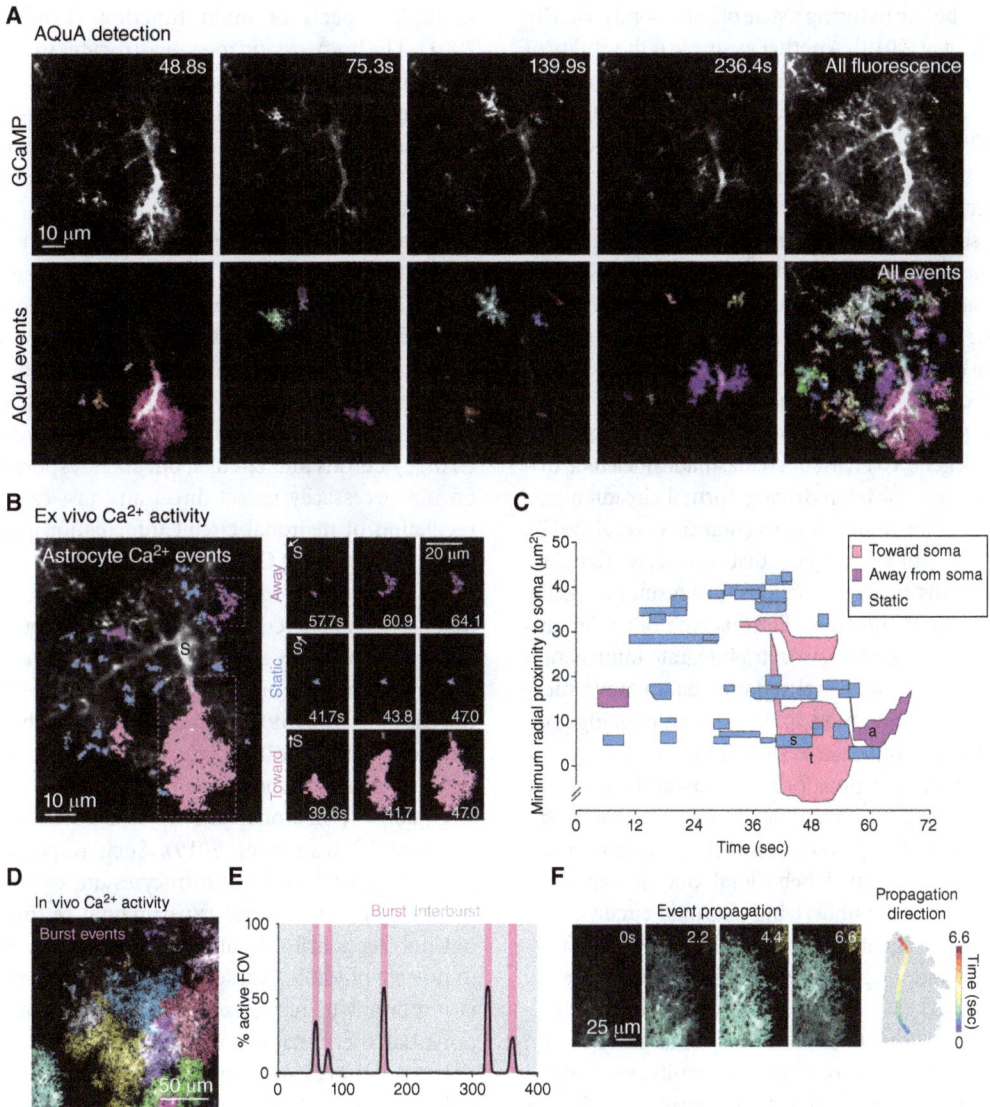

Figure 1. Using AQuA to analyze astrocyte Ca^{2+} activity. (*A*) Individual representative frames from 5-min GCaMP imaging experiment with AQuA-detected events shown *below*. Each color, all of which are random, represents an individual event. *Right* column: average GCaMP fluorescence (*top*) and (*bottom*) all AQuA-detected events. (*B*) Same GCaMP6f ex vivo experiment as in *A*, with AQuA events overlaid from 1 min of the video (*left*). Soma marked with black "s." (*Right*) Image sequence for each propagation direction class (blue, static; pink, toward soma; purple, away from soma). Soma direction marked with S and white arrow. (*C*) Spatiotemporal plot of Ca^{2+} activity from 1 min of video. Each event is represented by a polygon proportional to its area as it changes over its lifetime. (*D*) In vivo GCaMP6f images, showing AQuA-detected events during a burst period at locomotion onset. (*E*) Population Ca^{2+} events represented as percentage of the imaging field active as a function of time. (*F*) (*Left*) Individual Ca^{2+} event from *D*, plotted to demonstrate propagation direction (change of centroid relative to original location). (*Right*) Each frame direction is overlaid on the event (gray). (Figure reprinted from Wang et al. 2019, with permission from the authors.)

onds before inducing a state of behavioral passivity (Mu et al. 2019). Another example is the ability of *Drosophila* astrocytes to regulate sleep, olfaction-driven chemotaxis, and touch-induced startle via ATP/adenosine release (Ma et al. 2016; Blum et al. 2020). In mice, cortical astrocytes have been shown to regulate neuronal synchrony on the time course of seconds (Reitman et al. 2023). Furthermore, genetically impairing Ca^{2+} signaling in mouse striatal astrocytes was sufficient to guide very specific behaviors—highly stereotyped self-grooming (Yu et al. 2018)—also suggestive of seconds–timescale interactions with neuronal circuit activity. On longer timescales of integration, expression of a clock gene, *Cry1*, in superchiasmatic nucleus astrocytes is involved in driving normal circadian patterns of locomotor activity (Brancaccio et al. 2017). These studies suggest that astrocyte signaling performs specific functions that result in specific behavioral outcomes. For this type of action, understanding how astrocytes integrate information at a biophysical level requires quantitative measurements and modeling, and is an ongoing goal. Although informative, there is insufficient mechanistic or computational understanding of the underlying Ca^{2+} dynamics in any of the cases mentioned in the preceding sentences. Furthermore, are the associated behavioral outputs simply a readout of the underlying neuronal circuits, with astrocyte signaling increasing the likelihood of normally rare neuronal responses that can be readily observed over the course of the evaluated timescales? Alternatively, do astrocytes impart new and unique functions on neuronal circuits over the time course of the Ca^{2+} signaling events? The evidence to support this latter possibility is still limited.

2. *Slower homeostatic functions*: Astrocytes are important for brain homeostasis/function and therefore it is natural (perhaps even obvious) that some behaviors will be altered when astrocytes are changed. This type of explanation may be applicable in several cases and may be of particular interest in the context of how astrocytes contribute to disease phenotypes. One example is Kir4.1 up-regulation in the LHb, which, through altered K^+ homeostasis, drives behaviors associated with depression (Cui et al. 2018). Another example is how altered astrocyte-mediated glutamate homeostasis affects

multiple aspects of brain function (Danbolt 2001). The homeostatic roles of astrocytes in metabolism and tissue health (Ioannou et al. 2019) may be particularly meaningful in disease-related behaviors to understand the phenotypes or to modulate them for beneficial effect. Is it possible that astrocyte Ca^{2+} signaling also alters such essential functions to alter neuronal circuits and behavior? We believe this possibility is challenging to definitively rule out or in with currently available tools, but should be considered, as homeostasis is a core function of astrocytes. Thus, mechanisms downstream of Ca^{2+} that affect K^+ and neurotransmitter homeostasis are expected to change neurons and circuits, but such responses do not necessarily reflect direct and fast causal regulation of neuronal circuit information processing by astrocyte Ca^{2+} dynamics.

3. *Potentially unnatural effects*: In some cases, a neural circuit or behavioral alteration may be explained by a coincident effect unrelated to normal astrocyte biology. One example may be the use of channelrhodopsin in astrocytes, which elevates extracellular K^+ levels when activated for seconds or more, which may result in unnatural alterations of neuronal activity and potentially behavior (Octeau et al. 2019). Such responses perhaps best tell us what astrocytes are capable of doing under artificial experimental settings, but not necessarily about what they actually do in normal physiological settings. This distinction is important to recognize. Another example of potential interpretative caveats is the outcome of use of promoters and driver lines that are meant to be astrocyte selective but result in alteration of neurons with resultant coincident changes in circuits and behavior that confound interpretation of astrocytic effects that may or may not occur in parallel. In such cases of mixed astrocytic and neuronal expression, separating cause and effect is highly problematic at behavioral levels of study.

More generally, optogenetic and chemogenetic approaches often lack holistic context for the cellular effects, which can make it problematic to separate direct and secondary astrocyte contributions to behavior. This is especially a concern when long durations of stimulation are used. Furthermore, exogenous stimulation may not faithfully recapitulate endogenous pathways used in vivo, which of course must also apply to opto-/chemogenetic manipulations of neuronal firing. However, in the case of neuronal studies, the most informative experiments attempt to recapitulate endogenous firing patterns by the use of effectors such as channelrhodopsin. At this stage in the field, this is impossible to do for astrocytes because there is no consensus on the relevant physiological astrocytic activity, and the most commonly used DREADD approaches are challenging to regulate over seconds or minutes.

To resolve the interpretative confounds caused by the aforementioned three points, in the future, explorations of astrocyte Ca^{2+} signaling could be aided by suppression of defined types of ongoing physiological activity as a complementary interventional approach. However, this requires significantly improved methods, such as the development of efficient light-activated suppressors of intracellular Ca^{2+} release. It is also a prerequisite for such studies to commence that we have a molecular understanding of the mechanisms that mediate physiological Ca^{2+} signaling so that improved methods can be developed and used appropriately. Furthermore, an equivalent astrocyte response may have different circuit consequences that will be dictated by the biophysical properties of neurons in specific brain regions. Along with the evidence of region-specific astrocyte properties, there are likely to exist astrocyte mechanisms with differential effects between brain areas.

SPECULATION ON ASTROCYTE CALCIUM SIGNALING

In the preceding sections, we summarized foundational astrocyte physiology research and laid out current challenges. In this final, more speculative section, we aim to take recent successes as a launchpad for considering new ways to better understand the role that astrocytes play in neural circuit activity. We are driven by the conviction that the technical and conceptual challenges posed by the study of astrocytes can give us important clues about their functions. For example, considering how astrocytes differ from neurons at the cell biological level, and how they relate to other cells can give us

starting points to look for unexpected functions of astrocytes in neural circuits.

Astrocyte Signaling in Relation to Broader Areas of Biology

In general, astrocyte Ca^{2+} signals appear to operate on much slower timescales than fast electrical events in neurons. Compared to cells in general, neurons appear to be the outlier, having evolved to compute on timescales of milliseconds. Although networks of neurons can exhibit long-timescale phenomena such as evidence accumulation (Pinto et al. 2022) and single neurons can integrate signals over many seconds (Thornquist et al. 2020; Zhang et al. 2021), much subcellular machinery is dedicated to energy-intensive phenomena that occur on the timescale of milliseconds, such as ion channel opening, inactivation, and closing during action potentials. In this realm, therefore, astrocytes may be more like most other cell types in the body and across biology that employ slower signaling. The comparatively lower mechanistic pressure to be very fast and precise may allow astrocytes to function in a larger variety of ways. Indeed, astrocytes have been observed to be involved in a broad array of functions including synaptic plasticity, neural development, neuroprotection, sleep, computation, behavioral states, metabolism of toxic waste from neurons, and phagocytosis of microglial debris. However, at this juncture, it is unclear how many of these responses strictly require Ca^{2+} signaling, although some certainly do.

Among the evident complexity of biology, how do we put astrocytes in a framework by which to best understand them, and propose specific experiments to best uncover these functions? Three broad questions emerge of relevance to the topic of this review. What is the "goal" or function of astrocyte Ca^{2+} signaling? What mechanisms are used to achieve that goal or function? How are these mechanisms implemented by the biology of astrocytes? Since the potential functions and mechanisms of astrocyte–neuron and astrocyte–astrocyte communication in circuits is large, as are the potential contributions of astrocytes to behavior, we wonder whether it is useful to consider the roles that astrocytes *could* play in the implementations of algorithms solving specific goals of an animal. Animal behavior unfolds on multiple timescales, from about 100 msec—the time it takes to startle —to about a second—the time it takes to grasp an object, say a word, take a step, or decide to jump from a branch—to longer timescales for integrating, memorizing, and learning from new information. Information processing on timescales of milliseconds most likely relies on neuronal processing, but as described above, elements of neural algorithms that unfold on timescales of seconds may engage astrocytic processing. For example, we speculate that integrating information in an approximate way that does not require

the precision seen in, for example, eye movement integrators designed to keep one's gaze fixated, could be implemented in astrocytes, which may have advantages in terms of energy or cell number requirements, over implementations in recurrent neural circuits. This has been observed in zebrafish astrocytes, where behavioral state switches occur as a result of many seconds of futile swimming. In this case, the integration of action failures underlying the behavioral switch could not be found in neuronal networks, but was instead observed in astrocytes (Fig. 2; Mu et al. 2019). Since astrocytes respond to local circuit activity and can integrate neurotransmitter and neuromodulatory signals for prolonged pe-

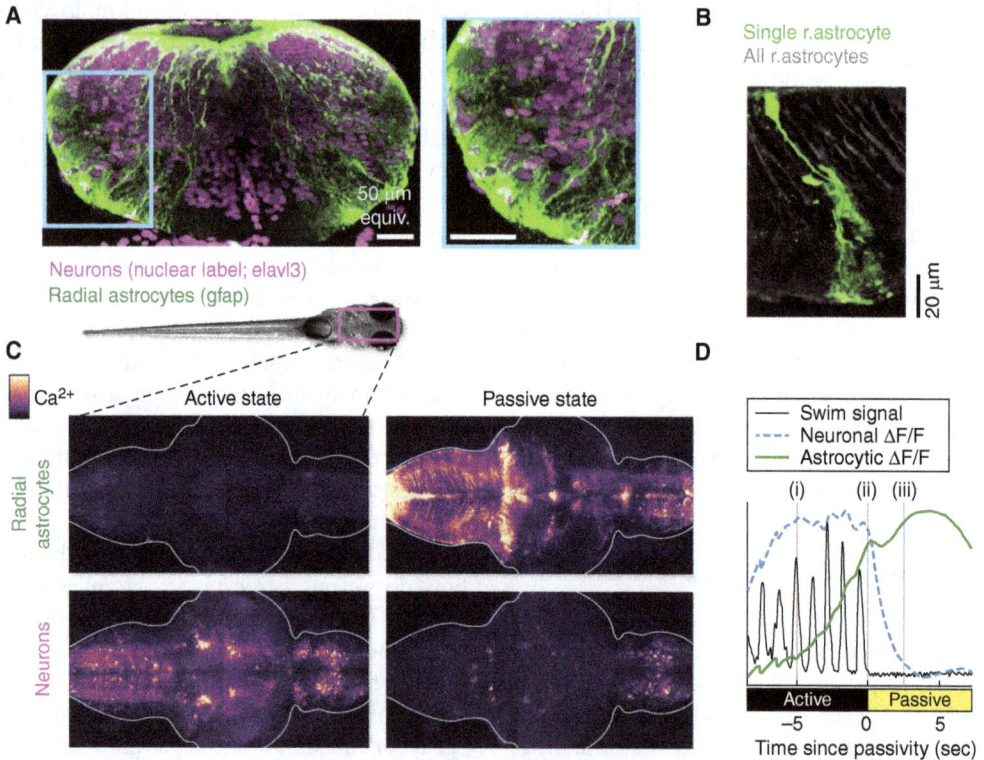

Figure 2. Astrocyte-like glial cells in larval zebrafish. (*A*) Astrocytes and neurons (depicted by a nuclear label) in a coronal hindbrain section, showing the radial processes emanating to neuropil-rich areas containing leafy structures similar to those of mammalian astrocytes. (*B*) Single radial astrocyte labeled with dye injection showing radial-glia-like and astrocyte-like morphology. (*C*) Calcium in radial astrocytes in neurons as animals transition from a behaviorally active to a passive state as a result of experiencing behavioral futility. (*D*) Swimming, average neuronal activity, and average astrocyte activity as animals undergo behavioral-state transitions, showing integration in astrocytes; astrocytes causally trigger the behavioral state transition. (All panels adapted from Mu et al. 2019, with permission from the authors © 2019; work published by Elsevier.)

Cite this article as *Cold Spring Harb Perspect Biol* doi: 10.1101/cshperspect.a041353

riods (Araque et al. 2014; Nagai et al. 2021b), they might contain an "after-image" of circuit activity that can be exploited by proximal neurons at later times. When studying circuits for working memory, attractor dynamics (Major and Tank 2004), behavioral-timescale synaptic plasticity (Bittner et al. 2017), and other operations that require information storage over multiple seconds, it could be useful to investigate astrocytic contributions in addition to those of neurons (Kol et al. 2020).

Although imaging astrocytes using existing sensors for Ca^{2+} and other molecules is now frequently performed, understanding signaling between astrocytes and other cells will likely require novel or improved sensors to monitor the various types of molecules mediating such multicellular communication. Obvious places to start are species such as K^+, lactate, ATP, adenosine, lipids, chemokines, and cytokines. However, designing sensors for such potential extracellular signaling species is nontrivial. For example, in the case of ATP that perhaps has the best-accepted roles in astrocytic responses, the ATP affinity of its native receptors range from nanomolar to about a millimolar in the case of P2Y and P2X7 receptors, respectively (Khakh 2001; Khakh and North 2006, 2012). In this regard, we are aware of no sensors, or strategies to develop them, that can cover binding affinities spanning six orders of magnitude, despite the fact that the biophysics of the natively expressed receptors tells us that this is what occurs physiologically. We suggest that although much progress has been made by borrowing tools from neurons and applying them to astrocytes, in the future additional tools designed with astrocyte biological ground truth and necessity in mind will be critical.

Astrocyte–Astrocyte Networks

To obtain clues about the possible circuit functions of astrocytes (i.e., both the way they interact with neuronal circuits and the way astrocytes interact with each other) as compared to the single-astrocyte level, one can look at their anatomical and molecular properties. With their leafy processes, their surface area-to-volume ratio is much larger than that of neurons, so their ability to sense and influence the local extracellular environment may be greater than that of neurons. One dominant feature of astrocytic networks is that they are richly coupled via gap junctions (Giaume et al. 2021); for example, patch-clamping an astrocyte with a dye-filled pipette almost immediately causes its neighbors to become labeled (Poskanzer and Yuste 2011; Anders et al. 2014). If astrocytes are playing many functional roles at the same time, then coupled networks of these individual cells may also exhibit emergent or unexpected features (Hansson and Rönnbäck 1995). What experiments would allow us to observe or test these ideas? One way to start may be to focus on better understanding propagation and integration of Ca^{2+} signals within astrocytic networks, by which we mean local groups of tens to hundreds of gap junctionally connected astrocytes. These kinds of intra-astrocyte network dynamics are currently almost entirely unknown for Ca^{2+}, and for other important intracellular signals like cAMP, which has been shown to be critical in differentiating intracellular astrocytic responses to external stimuli (Oe et al. 2020). Several fundamental features of Ca^{2+} signals within astrocyte populations remain a mystery, including the basic dynamics of signal spread within astrocyte networks. For example, how does the Ca^{2+} activity, or indeed any other physiologically relevant signal in a single astrocyte, influence the activity of its gap junctionally coupled network neighbors? If this kind of astrocyte network coupling of Ca^{2+} signals occurs, is it dynamic over time, under varying conditions or states? Could molecular signals that are known to activate astrocytes be encoded by this coupled astrocyte network, rather than at the individual cell level? These questions are central to understanding the recent literature of astrocytic control of neurons, as well as how astrocytes may fundamentally interact at the network level.

FUTURE OUTLOOK

Arguably, it is the potentially broad spectrum of functions that makes it both enticing and challenging to study the functions of astrocytes. In neurons, fast timescales have been a blessing (spikes can be measured by simple metal wires) and a challenge (tracking causal interactions across populations can be hard). Astrocytes, on the other hand, pose a different set of challenges. For example, their input–output characteristics (i.e., the steps between ligand binding to an extracellular receptor, activation of intracellular pathways, and the downstream effects for neighboring cells) remain to be determined. The current state of knowledge suggests that astrocytes communicate using multiple molecular signals that overlap and differ from those used for neuronal communication, making it problematic to determine which are relevant in specific circumstances. This calls for the development of technology specifically designed to (1) measure multimodal molecular information inside and outside astrocytes (to measure beyond Ca^{2+} and a handful of other signals); to (2) manipulate the function of astrocytes through controllable, specific, but minimal intervention in endogenous molecular pathways; and to (3) achieve these endpoints in behaving animals whenever possible. Since it is clear that much is still unknown about the function and mechanisms of astrocytes, the development of the most impactful tools ought not to be rigidly guided by what is currently known, assumed, or guessed. Single-cell molecular profiling (e.g., RNA sequencing) can help identify and prioritize the most prevalent pathways worth studying. Similarly, taking the view that much may still be unknown about the ways in which astrocytes influence neurons, and simultaneously measuring multiple molecular signals across populations of astrocytes and neurons may be crucial for determining their functions and mechanisms at tissue and circuit scales.

Much progress has been made in the last 30 years, but studying how astrocytes influence each other and regulate neurons to modify or mediate behavior remain essential goals for the field. We expect that findings resulting from the use of new types of tools, generally outlined in this review, will be informative to obtain comprehensive understanding of astrocyte Ca^{2+} signaling. The next few years hold great promise as new approaches are developed and new and existing biological questions are addressed with rigor and confidence.

REFERENCES

Agarwal A, Wu PH, Hughes EG, Fukaya M, Tischfield MA, Langseth AJ, Wirtz D, Bergles DE. 2017. Transient opening of the mitochondrial permeability transition pore induces microdomain calcium transients in astrocyte processes. *Neuron* 93: 587–605.e7. doi:10.1016/j.neuron.2016 .12.034

Agulhon C, Fiacco TA, McCarthy KD. 2010. Hippocampal short- and long-term plasticity are not modulated by astrocyte Ca^{2+} signaling. *Science* 327: 1250–1254. doi:10 .1126/science.1184821

Airan RD, Thompson KR, Fenno LE, Bernstein H, Deisseroth K. 2009. Temporally precise in vivo control of intracellular signalling. *Nature* 458: 1025–1029. doi:10.1038/na ture07926

Anders S, Minge D, Griemsmann S, Herde MK, Steinhäuser C, Henneberger C. 2014. Spatial properties of astrocyte gap junction coupling in the rat hippocampus. *Philos Trans R Soc Lond B Biol Sci* 369: 20130600. doi:10.1098/rstb.2013 .0600

Araque A, Carmignoto G, Haydon PG, Oliet SH, Robitaille R, Volterra A. 2014. Gliotransmitters travel in time and space. *Neuron* 81: 728–739. doi:10.1016/j.neuron.2014.02.007

Bazargani N, Attwell D. 2016. Astrocyte calcium signaling: the third wave. *Nat Neurosci* 19: 182–189. doi:10.1038/nn .4201

Bittner KC, Milstein AD, Grienberger C, Romani S, Magee JC. 2017. Behavioral time scale synaptic plasticity underlies CA1 place fields. *Science* 357: 1033–1036. doi:10.1126/ science.aan3846

Blum ID, Keleş MF, Baz ES, Han E, Park K, Luu S, Issa H, Brown M, Ho MCW, Tabuchi M, et al. 2020. Astroglial calcium signaling encodes sleep need in *Drosophila*. *Curr Biol* 31: 150–162.e7. doi:10.1016/j.cub.2020 .10.012

Bojarskaite L, Bjørnstad DM, Pettersen KH, Cunen C, Hermansen GH, Åbjørsbråten KS, Chambers AR, Sprengel R, Vervaeke K, Tang W, et al. 2020. Astrocytic Ca^{2+} signaling is reduced during sleep and is involved in the regulation of slow wave sleep. *Nat Commun* 11: 3240. doi:10.1038/ s41467-020-17062-2

Brancaccio M, Patton AP, Chesham JE, Maywood ES, Hastings MH. 2017. Astrocytes control circadian timekeeping in the suprachiasmatic nucleus via glutamatergic signaling. *Neuron* 93: 1420–1435.e5. doi:10.1016/j.neuron.2017 .02.030

Chai H, Diaz-Castro B, Shigetomi E, Monte E, Octeau JC, Yu X, Cohn W, Rajendran PS, Vondriska TM, Whitelegge JP, et al. 2017. Neural circuit-specialized astrocytes: transcriptomic, proteomic, morphological, and functional evidence. *Neuron* 95: 531–549.e9. doi:10.1016/j.neuron .2017.06.029

Charles AC, Merrill JE, Dirksen ER, Sanderson MJ. 1991. Intercellular signaling in glial cells: calcium waves and oscillations in response to mechanical stimulation and glutamate. *Neuron* 6: 983–992. doi:10.1016/0896-6273 (91)90238-u

Cite this article as *Cold Spring Harb Perspect Biol* doi: 10.1101/cshperspect.a041353

Cornell-Bell AH, Finkbeiner SM, Cooper MS, Smith SJ. 1990. Glutamate induces calcium waves in cultured astrocytes: long-range glial signaling. *Science* **247:** 470–473. doi:10.1126/science.1967852

Cui Y, Yang Y, Ni Z, Dong Y, Cai G, Foncelle A, Ma S, Sang K, Tang S, Li Y, et al. 2018. Astroglial Kir4.1 in the lateral habenula drives neuronal bursts in depression. *Nature* **554:** 323–327. doi:10.1038/nature25752

Danbolt NC. 2001. Glutamate uptake. *Prog Neurobiol* **6:** 1–105. doi:10.1016/s0301-0082(00)00067-8

Dani JW, Chernjavsky A, Smith SJ. 1992. Neuronal activity triggers calcium waves in hippocampal astrocyte networks. *Neuron* **8:** 429–440. doi:10.1016/0896-6273(92)90271-e

Delepine C, Shih J, Li K, Gaudeaux P, Sur M. 2023. Differential effects of astrocyte manipulations on learned motor behavior and neuronal ensembles in the motor cortex. *J Neurosci* **43:** 2696–2713. doi:10.1523/JNEUROSCI.1982-22.2023

Durkee CA, Covelo A, Lines J, Kofuji P, Aguilar J, Araque A. 2019. G_{i/o} protein-coupled receptors inhibit neurons but activate astrocytes and stimulate gliotransmission. *Glia* **67:** 1076–1093. doi:10.1002/glia.23589

Endo F, Kasai A, Soto JS, Yu X, Qu Z, Hashimoto H, Gradinaru V, Kawaguchi R, Khakh BS. 2022. Molecular basis of astrocyte diversity and morphology across the CNS in health and disease. *Science* **378:** eadc9020. doi:10.1126/science.adc9020

Fiacco TA, McCarthy KD. 2018. Multiple lines of evidence indicate that gliotransmission does not occur under physiological conditions. *J Neurosci* **38:** 3–13. doi:10.1523/JNEUROSCI.0016-17.2017

Giaume C, Naus CC, Sáez JC, Leybaert L. 2021. Glial connexins and pannexins in the healthy and diseased brain. *Physiol Rev* **101:** 93–145. doi:10.1152/physrev.00043.2018

Giovannucci A, Friedrich J, Gunn P, Kalfon J, Brown BL, Koay SA, Taxidis J, Najafi F, Gauthier JL, Zhou P, et al. 2019. Caiman an open source tool for scalable calcium imaging data analysis. *eLife* **8:** e38173. doi:10.7554/eLife.38173

Hansson E, Rönnbäck L. 1995. Astrocytes in glutamate neurotransmission. *FASEB J* **9:** 343–350. doi:10.1096/fasebj.9.5.7534736

Ioannou MS, Jackson J, Sheu SH, Chang CL, Weigel AV, Liu H, Pasolli HA, Xu CS, Pang S, Matthies D, et al. 2019. Neuron-astrocyte metabolic coupling protects against activity-induced fatty acid toxicity. *Cell* **177:** 1522–1535.e14. doi:10.1016/j.cell.2019.04.001

Khakh BS. 2001. Molecular physiology of P2X receptors and ATP signalling at synapses. *Nat Rev Neurosci* **2:** 165–174. doi:10.1038/35058521

Khakh BS, McCarthy KD. 2015. Astrocyte calcium signaling: from observations to functions and the challenges therein. *Cold Spring Harb Perspect Biol* **7:** a020404. doi:10.1101/cshperspect.a020404

Khakh BS, North RA. 2006. P2X receptors as cell-surface ATP sensors in health and disease. *Nature* **442:** 527–532. doi:10.1038/nature04886

Khakh BS, North RA. 2012. Neuromodulation by extracellular ATP and P2X receptors in the CNS. *Neuron* **76:** 51–69. doi:10.1016/j.neuron.2012.09.024

Kofuji P, Araque A. 2021. G-protein-coupled receptors in astrocyte-neuron communication. *Neuroscience* **456:** 71–84. doi:10.1016/j.neuroscience.2020.03.025

Kol A, Adamsky A, Groysman M, Kreisel T, London M, Goshen I. 2020. Astrocytes contribute to remote memory formation by modulating hippocampal-cortical communication during learning. *Nat Neurosci* **23:** 1229–1239. doi:10.1038/s41593-020-0679-6

Ma Z, Stork T, Bergles DE, Freeman MR. 2016. Neuromodulators signal through astrocytes to alter neural circuit activity and behaviour. *Nature* **539:** 428–432. doi:10.1038/nature20145

Major G, Tank D. 2004. Persistent neural activity: prevalence and mechanisms. *Curr Opin Neurobiol* **14:** 675–684. doi:10.1016/j.conb.2004.10.017

Mederos S, Hernández-Vivanco A, Ramírez-Franco J, Martín-Fernández M, Navarrete M, Yang A, Boyden ES, Perea G. 2019. Melanopsin for precise optogenetic activation of astrocyte-neuron networks. *Glia* **67:** 915–934. doi:10.1002/glia.23580

Mederos S, Sánchez-Puelles C, Esparza J, Valero M, Ponomarenko A, Perea G. 2021. GABAergic signaling to astrocytes in the prefrontal cortex sustains goal-directed behaviors. *Nat Neurosci* **24:** 82–92. doi:10.1038/s41593-020-00752-x

Mu Y, Bennett DV, Rubinov M, Narayan S, Yang CT, Tanimoto M, Mensh BD, Looger LL, Ahrens MB. 2019. Glia accumulate evidence that actions are futile and suppress unsuccessful behavior. *Cell* **178:** 27–43.e19. doi:10.1016/j.cell.2019.05.050

Nagai J, Bellafard A, Qu Z, Yu X, Ollivier M, Gangwani MR, Diaz-Castro B, Coppola G, Schumacher SM, Golshani P, et al. 2021a. Specific and behaviorally consequential astrocyte G_q GPCR signaling attenuation in vivo with iβARK. *Neuron* **109:** 2256–2274.e9. doi:10.1016/j.neuron.2021.05.023

Nagai J, Yu X, Papouin T, Cheong E, Freeman MR, Monk KR, Hastings MH, Haydon PG, Rowitch D, Shaham S, et al. 2021b. Behaviorally consequential astrocytic regulation of neural circuits. *Neuron* **109:** 576–596. doi:10.1016/j.neuron.2020.12.008

Nimmerjahn A, Bergles DE. 2015. Large-scale recording of astrocyte activity. *Curr Opin Neurobiol* **32:** 95–106. doi:10.1016/j.conb.2015.01.015

Nimmerjahn A, Hirrlinger J. 2022. Astrocyte regulation of neural circuit function and animal behavior. *Glia* **70:** 1453–1454. doi:10.1002/glia.24223

Octeau JC, Gangwani MR, Allam SL, Tran D, Huang S, Hoang-Trong TM, Golshani P, Rumbell TH, Kozloski JR, Khakh BS. 2019. Transient, consequential increases in extracellular potassium ions accompany channelrhodopsin2 excitation. *Cell Rep* **27:** 2249–2261.e7. doi:10.1016/j.celrep.2019.04.078

Oe Y, Wang X, Patriarchi T, Konno A, Ozawa K, Yahagi K, Hirai H, Tsuboi T, Kitaguchi T, Tian L, et al. 2020. Distinct temporal integration of noradrenaline signaling by astrocytic second messengers during vigilance. *Nat Commun* **11:** 471. doi:10.1038/s41467-020-14378-x. Erratum in Nat Commun 2020 Jul 7;11(1):3447.

Pachitariu M, Stringer C, Dipoppa M, Schröder S, Rossi LF, Dalgleish H, Carandini M, Harris KD. 2017. Suite2p: beyond 10,000 neurons with standard two-photon micros-copy. bioRxiv doi:10.1101/061507

Peinado A, Bendek E, Yokoyama S, Poskanzer KE. 2021. Deformable mirror-based axial scanning for two-photon mammalian brain imaging. *Neurophotonics* **8:** 015003. doi:10.1117/1.NPh.8.1.015003

Petravicz J, Fiacco TA, McCarthy KD. 2008. Loss of IP3 receptor-dependent Ca^{2+} increases in hippocampal astrocytes does not affect baseline CA1 pyramidal neuron synaptic activity. *J Neurosci* **28:** 4967–4973. doi:10.1523/JNEUROSCI.5572-07.2008

Pinto L, Tank DW, Brody CD. 2022. Multiple timescales of sensory-evidence accumulation across the dorsal cortex. *eLife* **11:** e70263. doi:10.7554/eLife.70263

Pittolo S, Yokoyama S, Willoughby DD, Taylor CR, Reitman ME, Tse V, Wu Z, Etchenique R, Li Y, Poskanzer KE. 2022. Dopamine activates astrocytes in prefrontal cortex via α1-adrenergic receptors. *Cell Rep* **40:** 111426. doi:10.1016/j.celrep.2022.111426

Porter JT, McCarthy KD. 1997. Astrocytic neurotransmitter receptors in situ and in vivo. *Prog Neurobiol* **51:** 439–455. doi:10.1016/s0301-0082(96)00068-8

Poskanzer KE, Yuste R. 2011. Astrocytic regulation of cortical UP states. *Proc Natl Acad Sci* **108:** 18453–18458. doi:10.1073/pnas.1112378108

Poskanzer KE, Yuste R. 2016. Astrocytes regulate cortical state switching in vivo. *Proc Natl Acad Sci* **113:** E2675–E2684. doi:10.1073/pnas.1520759113

Reitman ME, Tse V, Mi X, Willoughby DD, Peinado A, Aivazidis A, Myagmar BE, Simpson PC, Bayraktar OA, Yu G, et al. 2023. Norepinephrine links astrocytic activity to regulation of cortical state. *Nat Neurosci* **26:** 579–593. doi:10.1038/s41593-023-01284-w

Roth BL. 2016. DREADDs for neuroscientists. *Neuron* **89:** 683–694. doi:10.1016/j.neuron.2016.01.040

Salmon CK, Syed TA, Kacerovsky JB, Alivodej N, Schober AL, Sloan TFW, Pratte MT, Rosen MP, Green M, Chirgwin-Dasgupta A, et al. 2023. Organizing principles of astrocytic nanoarchitecture in the mouse cerebral cortex. *Curr Biol* **33:** 957–972.e5. doi:10.1016/j.cub.2023.01.043

Savtchouk I, Volterra A. 2018. Gliotransmission: beyond black-and-white. *J Neurosci* **38:** 14–25. doi:10.1523/JNEUROSCI.0017-17.2017

Shigetomi E, Patel S, Khakh BS. 2016. Probing the complexities of astrocyte calcium signaling. *Trends Cell Biol* **26:** 300–312. doi:10.1016/j.tcb.2016.01.003

Srinivasan R, Huang BS, Venugopal S, Johnston AD, Chai H, Zeng H, Golshani P, Khakh BS. 2015. Ca^{2+} signaling in astrocytes from Ip3r2$^{-/-}$ mice in brain slices and during startle responses in vivo. *Nat Neurosci* **18:** 708–717. doi:10.1038/nn.4001

Thornquist SC, Langer K, Zhang SX, Rogulja D, Crickmore MA. 2020. CaMKII measures the passage of time to coordinate behavior and motivational state. *Neuron* **105:** 334–345.e9. doi:10.1016/j.neuron.2019.10.018

Vaidyanathan TV, Collard M, Yokoyama S, Reitman ME, Poskanzer KE. 2021. Cortical astrocytes independently regulate sleep depth and duration via separate GPCR pathways. *eLife* **10:** e63329. doi:10.7554/eLife.63329

Volterra A, Liaudet N, Savtchouk I. 2014. Astrocyte Ca^{2+} signalling: Sn unexpected complexity. *Nat Rev Neurosci* **15:** 327–335. doi:10.1038/nrn3725

Wang Y, DelRosso NV, Vaidyanathan TV, Cahill MK, Reitman ME, Pittolo S, Mi X, Yu G, Poskanzer KE. 2019. Accurate quantification of astrocyte and neurotransmitter fluorescence dynamics for single-cell and population-level physiology. *Nat Neurosci* **22:** 1936–1944. doi:10.1038/s41593-019-0492-2

Xie Y, Wang T, Sun GY, Ding S. 2010. Specific disruption of astrocytic Ca^{2+} signaling pathway in vivo by adeno-associated viral transduction. *Neuroscience* **170:** 992–1003. doi:10.1016/j.neuroscience.2010.08.034

Yu X, Taylor AMW, Nagai J, Golshani P, Evans CJ, Coppola G, Khakh BS. 2018. Reducing astrocyte calcium signaling in vivo alters striatal microcircuits and causes repetitive behavior. *Neuron* **99:** 1170–1187.e9. doi:10.1016/j.neuron.2018.08.015

Yu X, Nagai J, Khakh BS. 2020. Improved tools to study astrocytes. *Nat Rev Neurosci* **21:** 121–138. doi:10.1038/s41583-020-0264-8

Yu X, Moye SL, Khakh BS. 2021. Local and CNS-wide astrocyte intracellular calcium signaling attenuation in vivo with CalEx^{flox} mice. *J Neurosci* **41:** 4556–4574. doi:10.1523/JNEUROSCI.0085-21.2021

Zhang Y, Looger LL. 2023. Fast and sensitive GCaMP calcium indicators for neuronal imaging. *J Physiol.* doi:10.1113/JP283832

Zhang SX, Lutas A, Yang S, Diaz A, Fluhr H, Nagel G, Gao S, Andermann ML. 2021. Hypothalamic dopamine neurons motivate mating through persistent cAMP signalling. *Nature* **597:** 245–249. doi:10.1038/s41586-021-03845-0

Zhang Y, Rózsa M, Liang Y, Bushey D, Wei Z, Zheng J, Reep D, Broussard GJ, Tsang A, Tsegaye G, et al. 2023. Fast and sensitive GCaMP calcium indicators for imaging neural populations. *Nature* **615:** 884–891. doi:10.1038/s41586-023-05828-9

Cite this article as *Cold Spring Harb Perspect Biol* doi: 10.1101/cshperspect.a041353

Astrocyte Regulation of Cerebral Blood Flow in Health and Disease

Anusha Mishra,[1,5] Grant R. Gordon,[2,5] Brian A. MacVicar,[3] and Eric A. Newman[4]

[1]Department of Neurology, Jungers Center for Neurosciences Research, Oregon Health & Science University, Portland, Oregon 97239, USA

[2]Hotchkiss Brain Institute, Department of Physiology and Pharmacology, Cumming School of Medicine, University of Calgary, Calgary, Alberta T2N 4N1, Canada

[3]Djavad Mowafaghian Centre for Brain Health, Department of Psychiatry, University of British Columbia, Vancouver, British Columbia V6T 1Z3, Canada

[4]Department of Neuroscience, University of Minnesota, Minneapolis, Minnesota 55455, USA

Correspondence: ean@umn.edu

Astrocytes play an important role in controlling microvascular diameter and regulating local cerebral blood flow (CBF) in several physiological and pathological scenarios. Neurotransmitters released from active neurons evoke Ca^{2+} increases in astrocytes, leading to the release of vasoactive metabolites of arachidonic acid (AA) from astrocyte endfeet. Synthesis of prostaglandin E_2 (PGE_2) and epoxyeicosatrienoic acids (EETs) dilate blood vessels while 20-hydroxyeicosatetraenoic acid (20-HETE) constricts vessels. The release of K^+ from astrocyte endfeet also contributes to vasodilation or constriction in a concentration-dependent manner. Whether astrocytes exert a vasodilation or vasoconstriction depends on the local microenvironment, including the metabolic status, the concentration of Ca^{2+} reached in the endfoot, and the resting vascular tone. Astrocytes also contribute to the generation of steady-state vascular tone. Tonic release of both 20-HETE and ATP from astrocytes constricts vascular smooth muscle cells, generating vessel tone, whereas tone-dependent elevations in endfoot Ca^{2+} produce tonic prostaglandin dilators to limit the degree of constriction. Under pathological conditions, including Alzheimer's disease, epilepsy, stroke, and diabetes, disruption of normal astrocyte physiology can compromise the regulation of blood flow, with negative consequences for neurological function.

Astrocyte endfoot processes completely envelop all blood vessels in the brain.[6] Work over the past 20 years has demonstrated the importance of astrocytes in cerebral blood flow (CBF) regulation. However, several other cell types also contribute to this process, which has made untangling the specific contributions of astrocytes an enduring area of research. Indeed, neurovascular coupling (NVC) is accomplished by the coordinated activity of excitatory and inhibitory neurons, vascular endothelium, mural cells, astrocytes, and even microglia (Bisht et al. 2021;

---

[5]These authors contributed equally to this work.
[6]This is an update to a previous article published in *Cold Spring Harbor Perspectives in Biology* [MacVicar and Newman (2015). *Cold Spring Harb Perspect Biol* **8:** a020388. doi: 10.1101/cshperspect.a020388].

Cite this article as *Cold Spring Harb Perspect Biol* doi: 10.1101/cshperspect.a041354

Howarth et al. 2021; Császár et al. 2022). Each cell type uses multiple cellular pathways and diffusible messengers, many of which are recruited in a redundant fashion to create a "fail-safe" system. This parallel processing has likely evolved because matching CBF to the metabolic demands is essential for healthy brain function. The goal of this work is to describe the mechanisms and circumstances under which astrocytes modify CBF.

The maintenance of brain homeostasis and cognitive processing requires substantial energy expenditures relative to the rest of the body. It is estimated that the brain accounts for 20% of total energy consumption although it represents only 2% of body weight (Attwell and Laughlin 2001). The greatest proportion of the brain's energy expenditure is due to excitatory synaptic transmission (Howarth et al. 2012), suggesting that glutamatergic transmission may be preferentially impacted by reduced energy supply. Brain metabolism is almost exclusively due to oxidative phosphorylation, where glucose is the primary energy substrate and O_2 is the final electron acceptor in the mitochondrial electron transport chain (Magistretti et al. 1995). Although some glycogen is stored in granules in astrocytes (Brown and Ransom 2007; Oe et al. 2016; Howarth et al. 2021; Dienel et al. 2023), the brain lacks sufficient energy reserves to maintain function over extended periods greater than a few minutes. Therefore, the moment-to-moment delivery of O_2 and glucose through the blood is fundamental to providing a consistent energy supply to adequately support brain function.

There are at least four important physiological states under which CBF is locally regulated. First, basal CBF, which is remarkably higher than the rest of the body (Magistretti et al. 1995; Raichle 2015), is regulated so that the brain receives an adequate supply of energy substrates at all times. This process, in part, helps feed intrinsic fluctuations in cortical neural activity, even in the absence of overt sensory, motor, or cognitive activities (Raichle 2015). As such, resting CBF oscillates at ~0.1 Hz, corresponding to fluctuations in high frequency neural activity (Mateo et al. 2017). Second, autoregulatory mechanisms limit CBF variability in the face of changes in systemic blood pressure. Here, the vasculature itself is intrinsically sensitive to increases in pressure, which trigger the myogenic response (vasoconstriction) to ensure relatively constant CBF (Schaeffer and Iadecola 2021). Third, changes in blood gasses result in altered CBF to maintain proper balance between energy substrates and byproducts of metabolism (Schaeffer and Iadecola 2021). Fourth, CBF is regulated in response to sudden changes in brain activity triggered by external (sensory inputs) or internal (motor command or neural processing) events. This homeostatic response, named functional hyperemia, increases delivery of glucose and O_2 at times of enhanced metabolic demand (Howarth et al. 2021). Astrocytes regulate CBF under each of these physiological states, although most work has been done on functional hyperemia.

HISTORICAL OVERVIEW OF FUNCTIONAL HYPEREMIA

Functional hyperemia was first hypothesized in the 1880s by Angelo Mosso (1880). In patients with skull defects, which allowed direct observation of the cortical surface, Mosso found that sensory stimulation increased brain volume, representing increased CBF. A decade later, Roy and Sherrington (1890) showed that stimulation of sensory nerves in dogs produced increases in cortical blood flow. They speculated that "the chemical products of cerebral metabolism … can cause variations of the caliber of the cerebral vessels: that in this reaction the brain possesses an intrinsic mechanism by which its vascular supply can be varied locally in correspondence with local variations of functional activity" (Roy and Sherrington 1890).

Around the same time, astrocyte morphology was described by Virchow (1858), Golgi (1894), Ramón y Cajal (1995, first published in 1897), and others. They observed that astrocytes contacted both blood vessels, which are enveloped by astrocyte endfeet, and neurons (Fig. 1). Cajal wrote, "The perivascular neuroglial cells live only in the proximity of the capillaries of the gray matter, to which they send one or more thick appendages inserted in the outer side of the endothelium…. The object of such

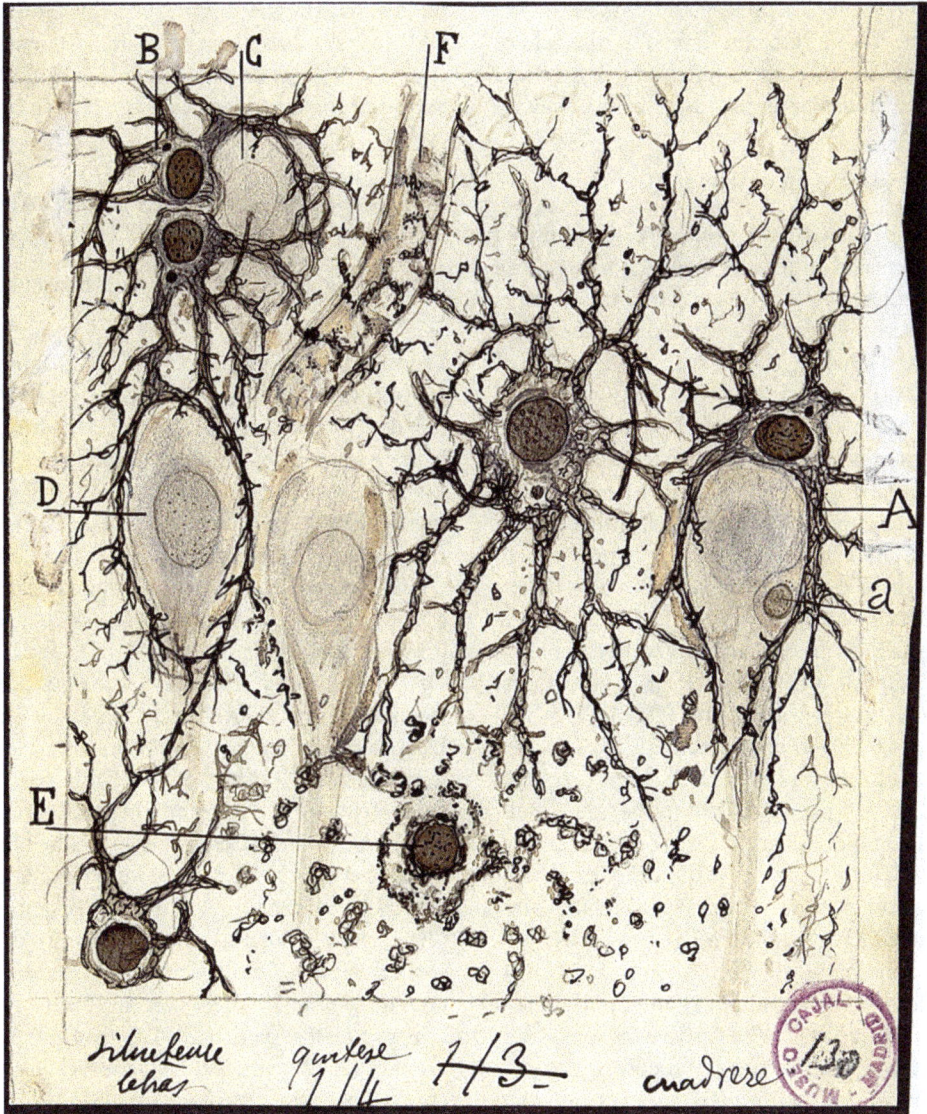

Figure 1. Drawing of brain astrocytes by Santiago Ramon y Cajal. Astrocytes, the darker cells in the drawing (*A*, *B*) contact both neurons, the lighter cells (*a*, *C*, *D*), and a blood vessel (*F*). As suggested by Cajal, astrocytes are ideally situated to mediate signaling from neurons to blood vessels and to regulate cerebral blood flow in response to neuronal activity. Drawing is reprinted with permission from Ricardo Martínez Murillo, Director of the Instituto Cajal.

elements is to evoke, by contraction of the afore-mentioned appendices, local dilations of the vessels" (Cajal 1895). Cajal's suggestion that astrocytes regulate blood flow, albeit by an incorrect mechanism, was prescient.

More recently, Paulson and Newman (1987) proposed that astrocytes mediate functional hy-peremia by a K^+ siphoning mechanism, releasing K^+ onto blood vessels from their endfeet in response to neuronal activity. Later, Harder et al. (1998) found that epoxyeicosatrienoic acids (EETs) derived from arachidonic acid (AA) in astrocytes underlie functional hyperemia. Soon after, Zonta et al. (2003) showed that Ca^{2+}-dependent synthesis of prostaglandins in astrocytes mediates NVC. Since these reports, the role of astrocytes in mediating functional hyperemia has been studied intensely, as detailed below.

NEUROVASCULAR COUPLING

The control of CBF was originally hypothesized to be mediated by negative feedback whereby the metabolites generated by active neurons (such as CO_2) were the signals that caused increased CBF (Roy and Sherrington 1890). While a recent study has brought new light to the CO_2 hypothesis (Hosford et al. 2022), work over the past two decades has shown that brain activity can directly increase CBF in a feedforward manner, independent of negative feedback. Indeed, CBF increases to such an extent that more O_2 is provided to active brain regions than is consumed (Offenhauser et al. 2005; Devor et al. 2011). This oversupply of O_2 is the basis of the blood oxygenation level dependent (BOLD) effect in functional magnetic resonance imaging (fMRI). It is well recognized that neuronal activity is causal for enhancing CBF and the BOLD signal (Lee et al. 2010), with synaptic activity (rather than spiking) being the primary driver (Logothetis et al. 2001; O'Herron et al. 2016). Indeed, direct signaling from neurons to blood vessels contributes to local CBF regulation through several well-described mechanisms (Howarth et al. 2021). Additionally, astrocytes also play an important role in regulating NVC. While astrocytes were originally thought to be the key mediator, directly transducing neural signals into vasodilation (Zonta et al. 2003), more recent discoveries suggest that they modulate, rather than mediate, CBF through several mechanisms depending on the physiological context. These include (1) augmenting CBF increase only when astrocytes are recruited during intense or sustained neural activity (Dunn et al. 2013; Gu et al. 2018; Institoris et al. 2022); (2) preferentially regulating capillaries over arterioles (Mishra et al. 2016); (3) controlling the dynamic range and polarity of the neural-evoked CBF response (Blanco et al. 2008; He et al. 2012); (4) potentially resetting microvascular diameter after functional hyperemia (Gu et al. 2018); (5) setting steady-state microvascular tone to control basal CBF (Kur and Newman 2014; Kim et al. 2015; Rosenegger et al. 2015); and, finally (6) contributing to intrinsic arteriole oscillations (Haidey et al. 2021).

MECHANISMS OF ASTROCYTE-MEDIATED NEUROVASCULAR COUPLING

Arachidonic Acid-Mediated Neurovascular Coupling

Elevation in astrocyte free Ca^{2+} bidirectionally controls vascular diameter (Fig. 2). This link was best demonstrated by uncaging Ca^{2+} in astrocytes in brain slices, retinal explants, and in vivo. These experiments showed unequivocally that Ca^{2+} transients in astrocytes can induce dilations and constrictions in adjacent vasculature. Zonta et al. (2003) showed that arterioles exhibited dilations, as opposed to the constrictions reported by Mulligan and MacVicar (2004), when, in the latter, Ca^{2+} was uncaged in adjacent astrocyte endfeet (see section on Bidirectional Control of Vessel Diameter for potential mechanisms that underlie the difference in polarity of NVC). The vasoconstriction was blocked by inhibiting phospholipase A2 (PLA2), which liberates AA from membrane lipids, and by preventing the conversion of AA to the vasoconstrictive lipid 20-HETE. Takano et al. (2006) showed that Ca^{2+} uncaging in vivo dilates cortical arterioles via a mechanism dependent on cyclooxygenase-1 (COX1), an enzyme expressed in astrocytes that synthesize PGE_2 from AA. Later, He et al. (2012) demonstrated that evoked arteriole dilations were absent or inverted to constrictions in brain slices from mice lacking IP_3 receptor type 2 (IP_3R_2), which gives rise to G_q-coupled receptor-dependent Ca^{2+} signals in astrocytes, or those lacking cytoplasmic PLA2 enzyme. These findings cemented a role for astrocyte Ca^{2+} and AA-derived substances in arteriole NVC.

Accumulating evidence shows that NVC also occurs at the capillary level (Chaigneau et al. 2003; Lacar et al. 2012; Hall et al. 2014; Kornfield and Newman 2014), where even small changes in diameter can give rise to pronounced effects on CBF as a significant fraction of total vascular resistance resides in capillaries (Blinder et al. 2013; Hall et al. 2014). Capillary contractility is

Cite this article as *Cold Spring Harb Perspect Biol* doi: 10.1101/cshperspect.a041354

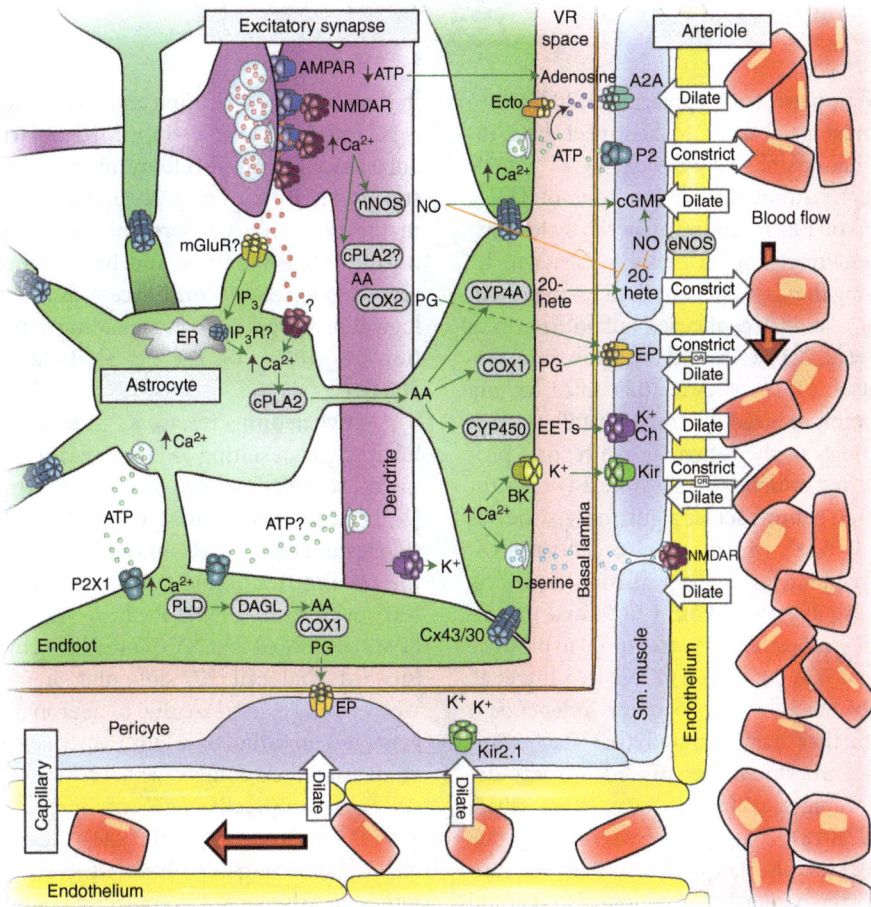

Figure 2. Summary of proposed Ca^{2+}-dependent signaling pathways in neurovascular coupling. Signaling to arterioles and capillaries by glutamatergic synaptic transmission is shown, either through direct neuronal mediators or through astrocyte pathways. For neuronal to vessel signaling subsequent to AMPAR and NMDAR activation, NO from nNOS or NOS1 causes vasodilation either through the canonical cGMP pathway or via inhibition of 20-HETE synthesis by acting on a CYP450 enzyme subtype. Prostaglandin (PG) production in neurons occurs via the enzyme COX2, acting on vascular EP4 receptors. Extracellular K^+ elevation by voltage-gated K^+ channels may trigger K^+-mediated dilation via Kir 2.1 channels on mural cells. For astrocyte to vessel signaling, glutamate release causes an elevation in free Ca^{2+} that occurs through an unknown mechanism. While IP$_3$R2 is a prominent Ca^{2+} source, it may not be involved in functional hyperemia. Activation of cPLA2 liberates AA from the plasma membrane and its metabolism leads to three primary pathways acting on arterioles: (1) the conversion to 20-HETE by CYP4A to cause vasoconstriction, (2) the conversion to PGs by COX1 to cause vasodilation, and (3) the conversion to EETs via a subtype of CYP450 enzyme to cause dilation. Calcium elevation in endfeet can also lead to K^+ efflux via BK channels to cause dilation, and trigger the release of D-serine, which facilitates opening of endothelial NMDARs causing dilation. At the capillary level, astrocyte endfoot Ca^{2+} triggers PLD and DAGL activity to generate AA. COX1 then converts AA to PGs to cause dilation. This mechanism involves ATP release and P2X1 receptor opening, but the source of the ATP is unclear. (A2A) Adenosine A2 receptor, (AA) arachidonic acid, (AMPAR) α-amino-3-hydroxy-5-methyl-4- isoxazolepropionic acid receptor, (ATP) adenosine 5′-triphosphate, (BK) large conductance Ca^{2+} and depolarization-gated K^+ channels, (cGMP) cyclic guanosine monophosphate, (COX1) cyclooxygenase 1, (COX2) cyclooxygenase 2, (Cx43/30), connexin 43 or 30 gap junction channels, (CYP4A) cytochrome P450 4A, (CYP450) cytochrome 450 enzymes, (DAGL) diacylglycerol lipase, (Ecto) ectonucleotidase, (EETs) epoxyeicosatrienoic acid, (eNOS) endothelial nitric oxide synthase, (EP) E-type prostanoid receptor, (20-HETE) 20-hydroxyeicosatetraenoic acid, (IP3) inositol 1,4,5-trisphosphate, (IP3R2) IP3 receptor 2, (Kir) inwardly rectifying K^+ channels, (mGluR) metabotropic glutamate receptor, (NMDAR) N-methyl-D-aspartate receptor, (nNOS), neuronal nitric oxide synthase, (NO) nitric oxide, (P2) purinergic 2 receptors, (P2X1) purinergic 2 receptor X1, (PG) prostaglandin, (PLA2) cytosolic phospholipase A2, (PLD2) phospholipase D2, (VR) Virchow Robin space.

mediated by pericytes (Peppiatt et al. 2006; Puro 2007; Hartmann et al. 2021). Mishra et al. (2016) showed that neural activity-evoked Ca^{2+} signals in astrocytes stimulate the synthesis of PGE_2, which acts on EP4 receptors on pericytes to dilate capillaries. Preventing Ca^{2+} rises in astrocytes by introducing the Ca^{2+} chelator BAPTA into them inhibited dilation of capillaries, but not arterioles (Mishra et al. 2016). Around the same time, Biesecker et al. (2016) showed that mice lacking IP_3R2 exhibit reduced dilation of capillaries but not arterioles in the retina. These reports provided the first evidence for the necessity of astrocyte Ca^{2+} in neurovascular signaling. Astrocytes expressed PLD2, which can also generate AA, and was required for capillary dilation. Astrocyte endfeet also contained both COX1 and prostaglandin E synthase enzymes required to produce PGE_2 from AA, but lacked COX2 (Mishra et al. 2016). Evidence from two other independent groups further demonstrated that Ca^{2+} transients in astrocyte endfeet precede activity-dependent capillary dilation (Otsu et al. 2015; Lind et al. 2018). Together, these results indicate that astrocytes can dilate both arterioles and capillaries via the synthesis and release of PGE_2, and thus regulate CBF (Fig. 3).

More recent work conducted in vivo using optogenetics has further demonstrated the sufficiency of astrocyte activation in controlling CBF. Optically activating ChR2 expressed selectively in cortical astrocytes drove dilation of both penetrating and, more surprisingly, pial surface arterioles over a broad area (Hatakeyama et al. 2021), and drove increases in BOLD signals (Takata et al. 2018).

Potassium and D-Serine

NVC may also be mediated by the glial release of K^+ on to blood vessels. Raising extracellular K^+ concentration ($[K^+]_o$) from a resting level of $\sim$3 mM up to $\sim$15 mM dilates blood vessels. Potassium-induced vasodilation is mediated by an increase in the conductance of inwardly rectifying K^+ channels (Filosa et al. 2006; Haddy et al. 2006) and by activation of the Na^+-K^+ ATPase on the vascular smooth muscle cells (Bunger et al. 1976; Haddy 1983), both resulting

in their hyperpolarization and relaxation. Larger K^+ increases, above $\sim$15 mM, depolarize vascular smooth muscle cells, resulting in vasoconstriction (Girouard et al. 2010). However, $[K^+]_o$ in the brain normally reaches $\sim$12 mM only during intense activity and has not been observed to exceed 15 mM except during pathological processes such as spreading depolarization or stroke (Vyskocil et al. 1972; Somjen 2001).

Active neurons release K^+ into the extracellular space, resulting in an increase in $[K^+]_o$. Light stimulation produces slow, transient $[K^+]_o$ increases of $\sim$1 mM in the cat visual cortex (Singer and Lux 1975; Connors et al. 1979) and in the cat and frog retina (Dick and Miller 1985; Karwoski et al. 1985). Paulson and Newman (1987) proposed that NVC is mediated by a feedforward glial cell K^+ siphoning mechanism, whereby $[K^+]_o$ increase due to neuronal activity generates an influx of K^+ into astrocytes and K^+ efflux from their endfeet, which have a high density of inwardly rectifying K^+ channels (Newman 1984, 1986; Newman et al. 1984). However, the glial K^+ siphoning hypothesis, when tested in the retina, was shown not to contribute significantly to NVC (Metea et al. 2007). Later studies showed that NVC may be mediated by a second K^+-associated mechanism. Neuronal activity-evoked increases in Ca^{2+} and production of EETs in astrocytes activate Ca^{2+}-activated large conductance K^+ (BK) channels expressed in endfeet (Price et al. 2002; Gebremedhin et al. 2003; Filosa et al. 2006; Dunn and Nelson 2010), which result in K^+ efflux onto blood vessels and vessel dilation (Filosa et al. 2006; Girouard et al. 2010).

Another potential astrocyte-derived mediator for NVC is the Ca^{2+}-dependent release of D-serine. This amino acid is an important coagonist for N-methyl-D-aspartate receptors (NMDARs), which are critical for CBF regulation. The canonical view of NMDARs involves the generation of nitric oxide (NO) via neuronal nitric oxide synthase (NOS). However, these receptors are also expressed on vascular endothelial cells (Lu et al. 2017), and D-serine mediates vasodilation in brain slices evoked by direct astrocyte stimulation (Stobart et al. 2013). Furthermore, endothelial knockdown of NMDARs sig-

 Cite this article as *Cold Spring Harb Perspect Biol* doi: 10.1101/cshperspect.a041354

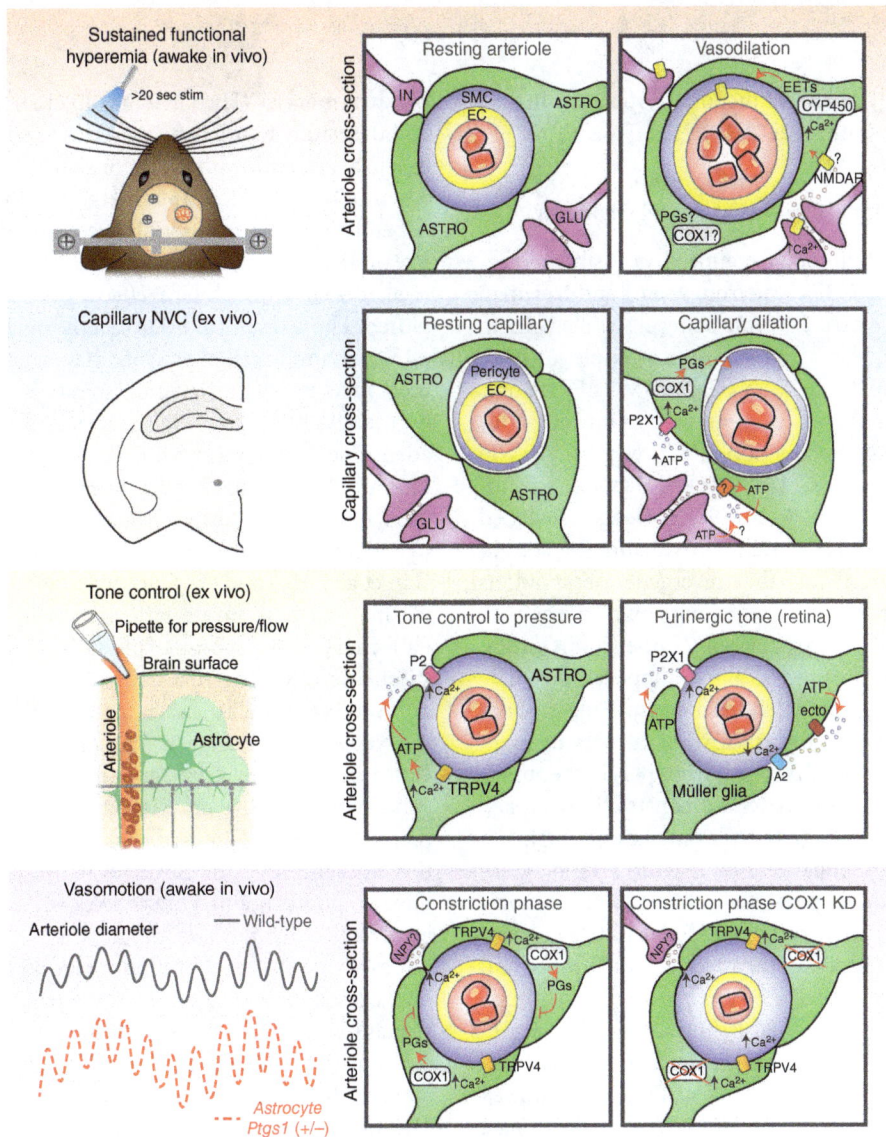

Figure 3. Diagram showing signaling pathways in astrocytes involved in four types of cerebral blood flow (CBF) regulation beyond brief functional hyperemia, which is typically studied. From *top* to *bottom*: (1) Astrocytes help amplify functional hyperemia to sustained neuronal activity through an NMDAR-dependent mechanism. The location of this receptor is unclear. NMDARs activate a CYP450 enzyme to generate EETs, boosting vasodilation that is primarily mediated by neuronal messengers. (2) Astrocytes participate in neurovascular coupling at the capillary level, but have no Ca^{2+}-dependent role at the arteriole level to brief neuronal activity. This involves an ATP- and COX1-dependent mechanism (see Fig. 2 for full pathway). (3) Astrocytes continually act to control steady-state arteriole tone in response to luminal pressure via TRPV4 in the neocortex, or through tonic purinergic signaling in the retina via P2X receptors. (4) Astrocyte endfeet sense vasoconstriction via TRPV4 and act to limit constriction by generating vasodilatory PGs from COX1 activity. Knocking down COX1 (*Ptgs1*) from astrocytes enhances vasoconstriction and oscillatory vasomotion in vivo. This mechanism likely acts more generally to constrain cerebral vasoconstriction. (A2) Adenosine A2 receptor, (ASTRO) astrocyte, (ATP) adenosine 5′-triphosphate, (COX1) cyclooxygenase 1, (CYP450) cytochrome 450 enzymes, (Ecto) ectonucleotidase, (EC) endothelial cell, (EETs) epoxyeicosatrienoic acid, (GLU) glutamatergic excitatory neuron, (IN) interneuron, (NMDAR) *N*-methyl-D-aspartate receptor, (P2) purinergic 2 receptors, (P2X1) purinergic 2 receptor X1, (PGs) prostaglandins, (Ptgs1) prostaglandin-endoperoxide synthase 1, (SMC) smooth muscle cell, (TRPV4) transient receptor potential cation channel subfamily V member 4.

nificantly reduced functional hyperemia in vivo (Hogan-Cann et al. 2019).

Bidirectional Control of Vessel Diameter

The first hints of the complexity of astrocyte regulation of CBF came from Metea and Newman (2006), where Ca^{2+} uncaging in Müller glia in retinal explants triggered both arteriole constrictions via 20-HETE or dilations via EETs. Here, NO was shown to be an important factor for switching the direction of the response: an NO donor converted light-evoked dilations to constrictions, whereas an NO scavenger switched light-evoked constrictions to dilations. A second explanation for the divergent observations that astrocyte Ca^{2+} signaling could trigger both dilations and constrictions came from Gordon et al.'s (2008) work in the hippocampus. The authors reported that the polarity of the vascular response to astrocyte Ca^{2+} transients in brain slices depended on the level of pO_2 in the superfusion solution. Arteriole constrictions evoked by Ca^{2+} uncaging in astrocytes in high O_2 (95%) solutions were reversed to arteriole dilations in more physiological O_2 (20%) solutions. Low pO_2 increased extracellular lactate concentrations, which in turn reduced the uptake and clearance of PGE_2 by the prostaglandin transporter, allowing PGE_2-mediated dilations to dominate. In high pO_2, lower lactate levels facilitated increased clearance of PGE_2 and allowed the conversion of AA to 20-HETE to dominate and cause constrictions (Fig. 3). The increased adenosine tone in low pO_2 solutions further reduced vasoconstriction (Gordon et al. 2008). A similar switch from arteriole dilation to constriction was observed in retinal explants when tissue pO_2 was altered from a low O_2 level to a high level (Mishra et al. 2011). However, this effect of O_2 was not evident in the in vivo retina when animals were made hyperoxic. This could be because hyperoxia-induced vasoconstriction (Mishra et al. 2011) maintains tissue pO_2 within a physiological range in vivo (Yu et al. 1999), allowing vasodilatory mechanisms to prevail (Mishra et al. 2011). High pO_2 has similar constrictive effects on capillaries in the cerebellum (Hall et al. 2014) and cortex, in a 20-HETE-de-

pendent manner (Hirunpattarasilp et al. 2022). Another study found that astrocytes in vivo and in vitro respond with elevations in free Ca^{2+} when pO_2 goes below 15 mmHg, in a manner dependent on mitochondrial O_2 sensing, downstream IP_3 signaling, and ATP release (Angelova et al. 2015), although CBF effects were not explored. These results indicate that the polarity of astrocyte modulation of CBF may reflect the metabolic state of the tissue and may be modified by the levels of extracellular lactate and ATP/adenosine. Intriguingly, a link between the magnitude of CBF changes and lactate levels has also been observed in human subjects. Functional hyperemia correlates with brain lactate levels (Lin et al. 2010) and exogenously increasing the plasma lactate/pyruvate ratio results in larger CBF changes in response to physiological visual stimulation (Mintun et al. 2004).

Another variable dictating the polarity of the vessel's response may be the concentration of Ca^{2+} reached within astrocyte endfeet. Two independent groups have demonstrated this in acute brain slices. First, Girouard et al. (2010) showed that uncaging Ca^{2+} in astrocytes led to bidirectional responses of adjacent arterioles, which could be stratified by the size of the Ca^{2+} increase achieved in endfeet: large Ca^{2+} signals (>500 nM) caused vasoconstrictions while smaller Ca^{2+} signals (<500 nM) evoked vasodilation. Both constrictions and dilations depended on Ca^{2+}-sensitive BK channel opening, corresponding to high or low levels of K^+ efflux from endfeet onto vessels, respectively (Girouard et al. 2010). Separately, Haidey and Gordon (2021) used the whole-cell patch technique to directly clamp free Ca^{2+} within astrocytes to different concentrations and measure the resulting arteriole response. They reported the same general phenomenon: elevating Ca^{2+} to the moderate value of 250 nM caused dilations and elevating it highly to 750 nM caused constrictions. Here, the moderate rise in astrocyte Ca^{2+} increased dilations via PGE_2, while large astrocyte Ca^{2+} increases caused constrictions via 20-HETE. This study did not explore a role for BK channels. Haidey and Gordon's findings also support the O_2 switching hypothesis, showing that the moderate elevation to 250 nM free astrocyte Ca^{2+} during hyperoxia resulted in vasoconstriction.

Cite this article as *Cold Spring Harb Perspect Biol* doi: 10.1101/cshperspect.a041354

Together, these findings indicate that the divergent observations made in early studies—where NVC could result in dilations in some cases while leading to constrictions in others—were likely due to differences in the extent of astrocyte Ca^{2+} increases or ambient conditions in the microenvironment (e.g., tissue concentration of O_2, NO, lactate, etc.).

RESOLVING ASTROCYTE CONTROVERSIES

While astrocyte stimulation is sufficient to induce changes in vessel diameter and CBF both ex vivo and in vivo, the question of whether they are necessary for physiological NVC is still open. It is well established that neuronal activity evokes Ca^{2+} signaling in astrocytes in vivo, but how consistent this response is and whether it precedes changes in CBF has been contentious. Many studies show that astrocyte Ca^{2+} signals develop slowly (>3 sec), and follow vasodilation (Nizar et al. 2013; Bonder and McCarthy 2014; Paukert et al. 2014; Tran et al. 2018), suggesting astrocytes do not initiate functional hyperemia. However, lack of observing fast or high-fidelity astrocyte Ca^{2+} signals could be due to limitations of the Ca^{2+} indicator tools available, ROI based analysis, and/or signal-to-noise problems. Recent studies using unbiased event-based analysis and targeted genetically encoded Ca^{2+} indicators have demonstrated rapid Ca^{2+} signaling in astrocyte processes and endfeet preceding arteriole dilation (Lind et al. 2013, 2018; Otsu et al. 2015; Stobart et al. 2018; Del Franco et al. 2022), but they did not examine a causal role for astrocyte Ca^{2+} signals in functional hyperemia. One prominent pathway to raise astrocyte cytoplasmic $[Ca^{2+}]$ is IP_3R2-dependent release of Ca^{2+} from the endoplasmic reticulum. Multiple groups reported that functional hyperemia (Takata et al. 2013; Bonder and McCarthy 2014; Del Franco et al. 2022) and BOLD fMRI signals (Jego et al. 2014) still occurred in IP_3R2 knockout animals and concluded that astrocyte Ca^{2+} signaling was not involved in vascular regulation. IP_3-dependent Ca^{2+} signaling occurs downstream of Gq-coupled receptors, and recent evidence from chemogenetic and optogenetic Gq activation studies suggest that astrocyte Gq-coupled receptor activation either do not

contribute to NVC (Ozawa et al. 2023) or contribute only to the late phase following sustained neuronal activation (Institoris et al. 2022). However, there are other smaller and/or spatially restricted Ca^{2+} signals that can be detected in astrocytes lacking IP_3R2 (Srinivasan et al. 2015; Rungta et al. 2016; Stobart et al. 2018), which could still contribute to functional hyperemia. A second important consideration is the interpretation of experiments involving genetic knockout of IP_3R2. As discussed earlier, astrocyte-mediated regulation of CBF reflects only one component of several NVC systems (interneuron release of peptides, NO, PGE2 release from neurons etc. [Howarth et al. 2021]) that modify CBF and compensation from other pathways, within astrocytes or other cell types, can occur in knockout mice, even when genes are conditionally deleted in adulthood (Hösli et al. 2022). Therefore, the intact functional hyperemia observed in IP_3R2 knockouts does not definitively prove that this pathway has no role in CBF regulation. In support of astrocyte Ca^{2+}-dependent vascular control, ex vivo experiments show that buffering Ca^{2+} in astrocytes with BAPTA eliminates neuronal activity-dependent cortical capillary dilation without reducing arteriole dilation (Mishra et al. 2016). In awake mice in vivo, clamping and lowering astrocyte-free Ca^{2+} with a high-affinity plasma membrane Ca^{2+} ATPase pump called CalEx, had no effect on arteriole dilation evoked by a short 5-sec whisker stimulation, but reduced the late component of the response when stimulation was prolonged for 30 sec (Institoris et al. 2022). This suggests that astrocyte Ca^{2+} amplifies the late component of functional hyperemia. If astrocyte Ca^{2+} is also involved in the initiation of NVC, this process may only require small-amplitude, kinetically fast increases in Ca^{2+} in spatially restricted regions that astrocyte CalEx is unable to silence (Mishra et al. 2016). Identifying the endogenous trigger(s) and source(s) of the Ca^{2+} signals within astrocytes that underlie NVC remains an ongoing endeavor.

Another point of contention is the glutamate receptor involved in activating astrocytes. Astrocytic mGluR5 was proposed to mediate NVC (Zonta et al. 2003), but this work was performed on young tissues. mGluR5 is expressed in and can mediate astrocyte Ca^{2+} signaling in juvenile but not in adult mice (Sun et al. 2013) or rats (Duffy and Macvicar 1995). While it is now clear that mGluR5 is not responsible for functional hyperemia in adults (Calcinaghi et al. 2011; Mishra et al. 2016), which glutamate receptors, and on which cell types, are responsible for astrocyte-mediated CBF effects is still uncertain. This is important to resolve because synaptic glutamate release is the primary driver of CBF increases. NMDARs are well known to be critical for CBF regulation, and while the existence of functional astrocyte NMDARs are debated, there is some evidence for their ability to help control steady-state arteriole tone (Mehina et al. 2017). Furthermore, work in awake mice shows that NMDAR antagonism with APV dramatically reduces the late component of functional hyperemia when sensory stimulation is prolonged; the same time period in which astrocytes are implicated to enhance the CBF response (Institoris et al. 2022). Thus, selective knockout of astrocyte NMDARs as well as other glutamate receptors will be important to determine whether this is an important astrocyte-mediated pathway in NVC. Besides glutamate, many other neurotransmitters can evoke Ca^{2+} signaling in astrocytes (Schipke and Kettenmann 2004) and could also lead to CBF modulation. While we do not review all the possibilities here, a notable example is the local release of ATP, subsequent to an increase in neuronal activity. ATP is important for astrocyte-mediated NVC at the capillary level via P2X1 receptors (Mishra et al. 2016). However, it is still unclear whether ATP is generated by active neurons themselves or by astrocytes upon stimulation via another messenger from neurons (Figs. 2 and 3). It is also unclear whether P2X1 receptors on astrocytes themselves are involved, or whether the receptors are on another cell type upstream of astrocytes.

OTHER ROLES FOR ASTROCYTES IN CBF REGULATION

Regulation of basal vascular tone ensures a constant supply of O_2 and nutrients to the brain and is mediated by several mechanisms, including extrinsic autonomic innervation, intrinsic subcort-ical innervation, cortical interneurons, and vascular cells themselves (Cohen et al. 1996; Cauli et al. 2004; Hamel 2006; Bekar et al. 2012). Recent evidence indicates that vasoactive agents from astrocytes also contribute to vascular tone. As discussed above, astrocytes can produce the vasoconstrictor 20-HETE (Imig et al. 1996; Zou et al. 1996; Gebremedhin et al. 1998). The CYP450 enzymes that synthesize 20-HETE are inhibited by NO, a potent vasodilator (Oyekan et al. 1999; Hall et al. 2014), suggesting that a balance between 20-HETE and NO is important for setting the basal tone of cerebral blood vessels (Gordon et al. 2007; Metea et al. 2007). This conclusion is supported by observations that blocking NO synthesis constricts vessels via 20-HETE (Zonta et al. 2003; Mulligan and MacVicar 2004) and appears to be an important mechanism in setting basal tone during the very long-lasting (2 h) decrease in CBF following cortical spreading depression (Fig. 2; Fordsmann et al. 2013). Dynamic changes in NO synthesis also contribute to basal tone regulation. Interestingly, mitochondria are enriched in astrocyte endfeet (Busija et al. 2016; Göbel et al. 2020). A recent report showed that NO synthesis from astrocytic mitochondria via a NOS-independent pathway requiring reduction of NO_2^- underlies vasodilation during mild hypoxia (Christie et al. 2023). Other reports have suggested that hypoxic dilation depends, to some extent, on NO production by mitochondrial NOS (Lacza et al. 2001), potentially from astrocyte endfeet.

Astrocytes also generate vascular tone by the tonic release of ATP (Kur and Newman 2014). Hippocampal astrocytes tonically release ATP, resulting in extracellular ATP levels of ~10 μM (Pascual et al. 2005) while astrocytes and retinal Müller cells release ATP in response to cellular Ca^{2+} increases (Newman 2001). Released ATP tonically constricts arterioles by activating P2X1 receptors on vascular smooth muscle cells. Enzymatic degradation of extracellular ATP, blockage of P2X1 receptors, and selective poisoning of glial cells with the toxin fluorocitrate reduced vascular tone (Kur and Newman 2014). Additionally, cortical astrocytes in brain slices tonically release dilatory prostaglandins via COX1 activity (Rosenegger et al. 2015) and this

Cite this article as *Cold Spring Harb Perspect Biol* doi: 10.1101/cshperspect.a041354

mechanism can even be initiated in response to vasoconstriction (Haidey et al. 2021). Chemogenetically induced arteriole constriction triggered a small, but constant elevation of astrocyte-free Ca^{2+} in a TRPV4-dependent manner, which triggered COX1 activity to mitigate vasoconstriction through the production of a vasodilator (Haidey et al. 2021). In other work in brain slices, astrocytes were shown to be required for sustaining luminal pressure-induced vasoconstriction. When astrocytes were filled with BAPTA to buffer their free Ca^{2+} and luminal pressure was applied, steady-state vascular tone was reduced after an equilibration period, suggesting the loss of an astrocyte-derived vasoconstrictor (Kim et al. 2015).

The discoveries of steady-state vascular tone control by astrocytes have opened new avenues to explore how astrocytes contribute to CBF regulation. Notably, arterioles and low-order capillaries are rarely static in vivo and instead their diameter oscillates at ∼0.1 Hz in a process called vasomotion (Mateo et al. 2017). Recent work suggests astrocytes may be important players contributing to this rhythmic vascular activity. In awake mice with an acute cranial window, astrocyte endfoot Ca^{2+} levels oscillate in a manner that is anticorrelated to arteriole diameter: peak vasoconstriction correlated with higher endfoot Ca^{2+}, whereas peak dilation correlated with lower endfoot Ca^{2+} (Haidey et al. 2021). Interestingly, astrocyte-selective knockdown of COX1 using floxed PTGS1 mice amplified vasomotor oscillations in arteriole diameter compared to controls (Fig. 3). This is consistent with astrocyte endfeet limiting the degree of vasoconstriction in a Ca^{2+} and COX1-dependent manner, resulting in larger downward oscillations when COX1 levels are lower.

Finally, astrocytes contribute to CBF regulation through CO_2 sensing. CO_2 has long been recognized as a potent vasodilator and may arise either through brain-generated CO_2 from metabolism or systemic elevations that travel via the blood to the brain. In acute brain slices, local elevation of CO_2 evoked a Ca^{2+} increase in astrocytes and downstream COX1-dependent vasodilation (Howarth et al. 2017). Glutathione was identified as a key mediator in this pathway due to the dependence of microsomal prostaglandin E synthase-1 on this cellular antioxidant. More recently, systemic CO_2 elevation was shown to occlude functional hyperemia using fMRI, and the effects could not be explained by a ceiling effect (Hosford et al. 2022). In the same work, using direct O_2 measurements in the cortex as a proxy, functional hyperemia was blocked by selectively knocking out sodium bicarbonate cotransporters in astrocytes, a protein critical for CO_2 sensing via conversion to bicarbonate. This intriguing mechanism points to a Ca^{2+}-independent form of NVC through astrocytes, and further suggests that multiple known NVC pathways could converge onto CO_2 signaling (Fig. 4).

ASTROCYTES AND BLOOD FLOW DYSREGULATION IN PATHOLOGY

Alzheimer's Disease

Alzheimer's disease (AD) is characterized classically by extracellular accumulation of amyloid β (Aβ) plaques and intracellular inclusions of neurofibrillary tau tangles (Alzheimer 1907; Kosik et al. 1986; Goedert et al. 1988; Graeber and Mehraein 1999). In mouse models of AD, many physiological and phenotypic changes are also observed in the cells surrounding plaques, including abnormal synchronous neuronal hyperactivity, decreased glutamate uptake by astrocytes (Hefendehl et al. 2016), and signs of profound microglia activation (Terry et al. 1991; Wyss-Coray 2006; Venneti et al. 2008; Akiyama et al. 2000). Additionally, astrocytes exhibit spontaneous intercellular Ca^{2+} waves and higher resting Ca^{2+} levels (Kuchibhotla et al. 2009; Lines et al. 2022), while reactive astrocytes around Aβ plaques show enhanced Ca^{2+} transients due to their abnormally high expression of the metabotropic P2Y1 receptors (Delekate et al. 2014). The alterations in both neuronal and astrocyte Ca^{2+} signaling could synergistically lead to impaired vascular control and possibly vasoconstrictions and tissue hypoxia (Mulligan and MacVicar 2004; Gordon et al. 2008; Attwell et al. 2010). Aβ may also disrupt vasoregulation by increasing reactive oxygen species (Park et al. 2004, 2005; Nortley et al. 2019) that could be enhanced during transient hypoxia (Zhang

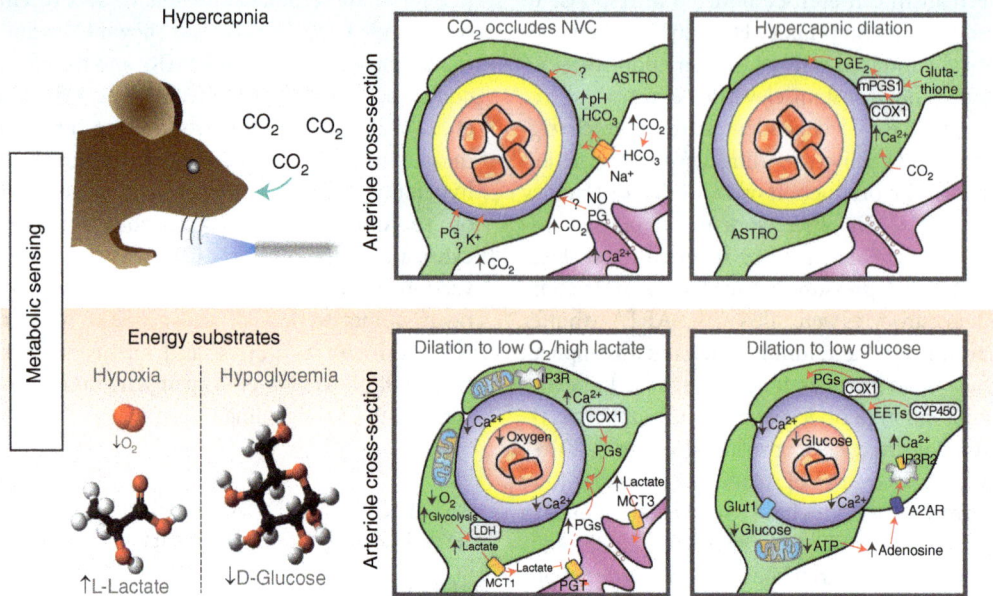

Figure 4. Diagram showing signaling pathways in astrocytes involved in cerebral blood flow (CBF) regulatory mechanisms related to metabolism and metabolites. *Upper* panels show two different mechanisms for how elevated CO_2 (hypercapnia) causes vasodilation through astrocytes. *Top middle* panel depicts a mechanism involving the sodium bicarbonate cotransporter, which is highly expressed by astrocytes. The mechanism mediating vasodilation is unclear. *Top right* panel shows a pathway involving Ca^{2+} elevation by CO_2 and the activation of COX1 and mPGS1 to generate vasodilatory PGE_2. *Lower* panels depict two mechanisms showing how energy substrates control CBF through astrocytes. *Bottom middle* panel shows how low O_2/high lactate (hypoxia) causes dilation by elevating PG signaling, in part by hindering PG uptake. *Bottom right* panel shows how low blood glucose (hypoglycemia) causes vasodilation through astrocytes via adenosine elevation, A2A receptors, and the Ca^{2+}-dependent generation of PGs and EETs. (A2AR) Adenosine A2 receptor, (ASTRO) astrocytes, (ATP) adenosine $5'$-triphosphate, (CO_2) carbon dioxide, (COX1) cyclooxygenase 1, (CYP450) cytochrome 450 enzymes, (EETs) epoxyeicosatrienoic acid, (Glut1) glucose transporter 1, (HCO3) bicarbonate, (IP3R2) inositol 1,4,5-trisphosphate receptor 2, (LDH) lactate dehydrogenase, (MCT1) monocarboxylate transporter 1, (MCT3) monocarboxylate transporter 3, (mPGS1) microsomal prostaglandin E synthase-1, (NO) nitric oxide, (PG) prostaglandins, (PGE_2) prostaglandin E2.

et al. 2014). Neurovascular dysfunction is emerging as an early symptom in AD-related dementia, preceding even Aβ and tau pathologies (Iturria-Medina et al. 2016). Thus, disruptions in neurovascular signaling by astrocytes as well as neurons could be important in the development of Alzheimer's disease and vascular dementia (Iadecola 2013).

Diabetes and Hypoglycemia

Vascular pathology is the most serious manifestation of diabetes (Coucha et al. 2018) and the regulation of CBF is compromised in the disease. Functional hyperemia in the retina is disrupted in patients with diabetic retinopathy, a serious complication of diabetes. In healthy individuals, flickering light dilates retinal arterioles by ~7% (Polak et al. 2002; Garhöfer et al. 2004a). This response is reduced by ~60% in patients with type 1 or type 2 diabetes (Garhöfer et al. 2004b; Nguyen et al. 2009; Pemp et al. 2009) and in an animal model of type 1 diabetes (Mishra and Newman 2010, 2012).

Functional hyperemia may be lost in the diabetic retina due to the disruption of signaling

Cite this article as *Cold Spring Harb Perspect Biol* doi: 10.1101/cshperspect.a041354

from glial cells to blood vessels. Glial-evoked dilation of blood vessels in the retina is reduced by NO (Metea and Newman 2006). In early stages of diabetic retinopathy, there is an up-regulation of iNOS (Du et al. 2002; Mishra and Newman 2010), leading to increased NO levels (Kowluru et al. 2000) and a reduction in glial-evoked vessel dilation (Mishra and Newman 2010). Inhibition of iNOS in diabetic animals to reduce NO production restored both glial-evoked and light-evoked vessel dilation to control levels (Mishra and Newman 2010, 2012).

Hypoglycemia, a reduction in blood glucose concentration, is a serious complication of insulin treatment for diabetes (Johnson-Rabbett and Seaquist 2019). Hypoglycemia induces vessel dilation and a global increase in CBF (Neil et al. 1987), mediated, in part, by the Ca^{2+}-dependent release of PGE_2 and EETs from astrocytes. Astrocyte Ca^{2+} signaling in the mouse somatosensory cortex increases as blood glucose falls and hypoglycemia-induced arteriole dilation is decreased when astrocyte Ca^{2+} signaling is reduced or PGE_2 and EETs synthesis is blocked (Nippert et al. 2022). These findings implicate astrocytes in hypoglycemia-evoked CBF increases (Fig. 4).

Epilepsy

Epileptic events, consisting of aberrant, high-frequency, synchronized neuronal activity, are associated with profound changes in vessel diameter, CBF, and oxygenation (Zhao et al. 2009; Gómez-Gonzalo et al. 2011; Farrell et al. 2016). While many studies have observed hyperemic responses in the seizure focus during ictal activity, pathological drops in CBF have been reported after the termination of long-duration ictal events (Farrell et al. 2016). This causes dangerously low levels of O_2 (<10 mgHg) persisting for ~1 h. Sustained vasoconstriction was also observed in vivo in mice (Tran et al. 2020), with correlated Ca^{2+} signals in astrocytes and vascular smooth muscles. A follow-up study found that generation of AA from breakdown of the endocannabinoid 2-AG and subsequent metabolism by COX2 into PGE_2 caused the vasoconstriction by acting on the EP1 receptor (Farrell et al. 2021). This is an intriguing new pathway for CBF reg-

ulation. While PGE_2 is mostly studied as a vasodilator during functional hyperemia, which it achieves by acting on EP4 receptors at low doses, it can also vasoconstrict at high doses by acting on EP1 receptors (Dabertrand et al. 2013; Czigler et al. 2020). The role of COX2 and 2-AG suggests that neurons are the most likely source of high PGE_2 in epilepsy. However, astrocytes may be involved to limit the degree of vasoconstriction through TRPV4-COX1 feedback dilation pathways (Haidey et al. 2021). An in vivo study of a 4-AP model of epilepsy found that high endfoot Ca^{2+} was associated with reduced vessel diameter, whereas lower endfoot Ca^{2+} was associated with larger vessel diameters during ictal events (Zhang et al. 2019). It is not clear whether these astrocyte Ca^{2+} signals exacerbate the reduction in CBF or attempt to minimize it.

Stroke

Stroke is a vascular disorder caused by either the blockage (ischemic stroke) or rupture (hemorrhagic stroke) of a cerebral blood vessel. In patients surviving ischemic stroke, brain-wide impairments in cerebral autoregulation, hypercapnic hyperemia and NVC have been reported (Krainik et al. 2005; Lin et al. 2011; Salinet et al. 2015). A recent rodent study found that the reduction in NVC after stroke is caused by increased synthesis of 20-HETE, one of the vasoconstrictors produced by astrocytes (Li et al. 2021). Increased 20-HETE is also reported in both ischemic and hemorrhagic stroke patients and predicts worse outcomes (Crago et al. 2011; Donnelly et al. 2015; Yi et al. 2017). Other groups have reported that astrocytes react to ischemic events by exhibiting large Ca^{2+} signals (Ding et al. 2009; Rakers and Petzold 2017), which may increase 20-HETE synthesis (Haidey and Gordon 2021). Observations that attenuating reactive astrogliosis mitigates CBF defects and improves neurological outcomes after ischemic stroke further support this hypothesis (Begum et al. 2018).

Hemorrhagic stroke also disrupts cerebral autoregulation (Koide et al. 2021) and impairs NVC in animal models. Interestingly, after a hemorrhage, activity-dependent responses of arterioles switch from vasodilations to constrictions, resulting in decreased CBF (Balbi et al. 2017). This inversion of NVC has been attributed to abnormally large purinergic-mediated Ca^{2+} signals in astrocytes (Pappas et al. 2015, 2016), which increase BK channel activity and lead to large K^+ efflux from endfeet, causing vessel constriction (Koide et al. 2012).

Cortical spreading depolarization, which is common after both ischemic and hemorrhagic strokes, also results in large Ca^{2+} waves in astrocytes (Chuquet et al. 2007), and is followed by a wave of vasoconstriction and NVC inversion (Chuquet et al. 2007; Major et al. 2017). These effects are partly mediated by an increase in 20-HETE (Fordsmann et al. 2013), although the exact role of astrocytes in this process has yet to be determined.

CONCLUSION

The regulation of CBF is essential for proper brain function. Astrocytes contribute to the regulation of CBF in several ways. Ca^{2+}-dependent synthesis of AA metabolites by astrocytes can bidirectionally modulate CBF: synthesis of PGE_2 and EETs dilates blood vessels while 20-HETE constricts vessels. Ca^{2+}-dependent release of K^+ also contributes to bidirectional vascular regulation. The precise physiological contexts in which astrocytes participate in NVC as well as the mechanisms underlying astrocyte Ca^{2+} signaling related to this process continue to be refined. While astrocytes may help control capillary bed perfusion or amplify sustained elevations in CBF to neuronal activation, new roles for astrocytes outside functional hyperemia have also been proposed. These include vascular tone setting, vasomotion, and metabolic sensing. Under pathological conditions, including Alzheimer's disease, epilepsy, stroke, and diabetic retinopathy, disruption of normal astrocyte physiology can compromise CBF regulation. This can exert a negative impact on tissue health and contribute to neurodegeneration.

ACKNOWLEDGMENTS

The authors' work is supported by Fondation Leducq of France (B.A.M. and E.A.N.), the Canadian Institutes of Health Research (244825, 245760, FDN-148397 to B.A.M., PTJ-173468 to G.R.G.), TCE-117869 in the framework of the ERA-NET NEURON (B.A.M.), a Canada Research Chair (B.A.M.), the National Institutes of Health of the United States (EY004077, EY023216, R01EY026514, and EY026882 to E.A.N. and NS110690 and AG066518 to A.M.), and John and Tami Marick Foundation (A.M.).

REFERENCES

Akiyama H, Barger S, Barnum S, Bradt B, Bauer J, Cole GM, Cooper NR, Eikelenboom P, Emmerling M, Fiebich BL, et al. 2000. Inflammation and Alzheimer's disease. *Neurobiol Aging* **21:** 383–421. doi:10.1016/S0197-4580(00)00124-X

Alzheimer A. 1907. Über eine eigenartige erkrankung der hirnrinde [Concerning a strange illness of the cerebral cortex]. *Allg Z Psychiat* **64:** 146–148.

Angelova PR, Kasymov V, Christie I, Sheikhbahaei S, Turovsky E, Marina N, Korsak A, Zwicker J, Teschemacher AG, Ackland GL, et al. 2015. Functional oxygen sensitivity of astrocytes. *J Neurosci* **35:** 10460–10473. doi:10.1523/JNEUROSCI.0045-15.2015

Attwell D, Laughlin SB. 2001. An energy budget for signaling in the grey matter of the brain. *J Cereb Blood Flow Metab* **21:** 1133–1145. doi:10.1097/00004647-200110000-00001

Attwell D, Buchan AM, Charpak S, Lauritzen M, MacVicar BA, Newman EA. 2010. Glial and neuronal control of brain blood flow. *Nature* **468:** 232–243. doi:10.1038/nature09613

Balbi M, Koide M, Wellman GC, Plesnila N. 2017. Inversion of neurovascular coupling after subarachnoid hemorrhage in vivo. *J Cereb Blood Flow Metab* **37:** 3625–3634. doi:10.1177/0271678X16686595

Begum G, Song S, Wang S, Zhao H, Bhuiyan MIH, Li E, Nepomuceno R, Ye Q, Sun M, Calderon MJ, et al. 2018. Selective knockout of astrocytic Na^+/H^+ exchanger isoform 1 reduces astrogliosis, BBB damage, infarction, and improves neurological function after ischemic stroke. *Glia* **66:** 126–144. doi:10.1002/glia.23232

Bekar LK, Wei HS, Nedergaard M. 2012. The locus coeruleus-norepinephrine network optimizes coupling of cerebral blood volume with oxygen demand. *J Cereb Blood Flow Metab* **32:** 2135–2145. doi:10.1038/jcbfm.2012.115

Biesecker KR, Srienc AI, Shimoda AM, Agarwal A, Bergles DE, Kofuji P, Newman EA. 2016. Glial cell calcium signaling mediates capillary regulation of blood flow in the retina. *J Neurosci* **36:** 9435–9445. doi:10.1523/JNEUROSCI.1782-16.2016

Bisht K, Okojie KA, Sharma K, Lentferink DH, Sun YY, Chen HR, Uweru JO, Amancherla S, Calcuttawala Z, Campos-Salazar AB, et al. 2021. Capillary-associated microglia regulate vascular structure and function through PANX1-P2RY12 coupling in mice. *Nat Commun* **12:** 5289. doi:10.1038/s41467-021-25590-8

Blanco VM, Stern JE, Filosa JA. 2008. Tone-dependent vascular responses to astrocyte-derived signals. *Am J Physiol*

Heart Circ Physiol **294:** H2855–H2863. doi:10.1152/aj pheart.91451.2007

Blinder P, Tsai PS, Kaufhold JP, Knutsen PM, Suhl H, Kleinfeld D. 2013. The cortical angiome: an interconnected vascular network with noncolumnar patterns of blood flow. *Nat Neurosci* **16:** 889–897. doi:10.1038/nn.3426

Bonder DE, McCarthy KD. 2014. Astrocytic Gq-GPCR-linked IP3R-dependent Ca^{2+} signaling does not mediate neurovascular coupling in mouse visual cortex in vivo. *J Neurosci* **34:** 13139–13150. doi:10.1523/JNEUROSCI.2591-14.2014

Brown AM, Ransom BR. 2007. Astrocyte glycogen and brain energy metabolism. *Glia* **55:** 1263–1271. doi:10.1002/glia.20557

Bunger R, Haddy FJ, Querengasser A, Gerlach E. 1976. Studies on potassium induced coronary dilation in the isolated Guinea pig heart. *Pflugers Arch* **363:** 27–31. doi:10.1007/BF00587398

Busija DW, Rutkai I, Dutta S, Katakam PV. 2016. Role of mitochondria in cerebral vascular function: energy production, cellular protection, and regulation of vascular tone. *Compr Physiol* **6:** 1529–1548. doi:10.1002/cphy.c150051

Cajal SR. 1895. Algunas conjeturas sobre el mecanismo anatomico de la ideacion, asociacion y atencion [Some conjectures about the anatomical mechanisms of ideation, association and attention]. *Revista de Medicina y Cirugia Prácticas (Madrid)* **19:** 497–508.

Cajal SR. 1995. *Histology of the nervous system of man and vertebrates* (ed. Swanson N. Swanson LW), 1st Ed. Oxford University Press, New York.

Calcinaghi N, Jolivet R, Wyss MT, Ametamey SM, Gasparini F, Buck A, Weber B. 2011. Metabotropic glutamate receptor mGluR5 is not involved in the early hemodynamic response. *J Cereb Blood Flow Metab* **31:** e1–e10. doi:10.1038/jcbfm.2011.96

Cauli B, Tong XK, Rancillac A, Serluca N, Lambolez B, Rossier J, Hamel E. 2004. Cortical GABA interneurons in neurovascular coupling: relays for subcortical vasoactive pathways. *J Neurosci* **24:** 8940–8949. doi:10.1523/JNEUROSCI.3065-04.2004

Chaigneau E, Oheim M, Audinat E, Charpak S. 2003. Two-photon imaging of capillary blood flow in olfactory bulb glomeruli. *Proc Natl Acad Sci* **100:** 13081–13086. doi:10.1073/pnas.2133652100

Christie IN, Theparambil SM, Doronin M, Hosford PS, Brazhe A, Hobbs A, Semyanov A, Abramov AY, Angelova P, Gourine AV. 2023. Astrocyte mitochondria produce nitric oxide from nitrite to modulate cerebral blood flow during brain hypoxia. *Cell Reports* **42:** 113514. doi:10.1016/j.celrep.2023.113514

Chuquet J, Hollender L, Nimchinsky EA. 2007. High-resolution in vivo imaging of the neurovascular unit during spreading depression. *J Neurosci* **27:** 4036–4044. doi:10.1523/JNEUROSCI.0721-07.2007

Cohen Z, Bonvento G, Lacombe P, Hamel E. 1996. Serotonin in the regulation of brain microcirculation. *Prog Neurobiol* **50:** 335–362. doi:10.1016/S0301-0082(96)00033-0

Connors B, Dray A, Fox P, Hilmy M, Somjen G. 1979. LSD's effect on neuron populations in visual cortex gauged by transient responses of extracellular potassium evoked by optical stimuli. *Neurosci Lett* **13:** 147–150. doi:10.1016/0304-3940(79)90032-6

Coucha M, Abdelsaid M, Ward R, Abdul Y, Ergul A. 2018. Impact of metabolic diseases on cerebral circulation: structural and functional consequences. *Compr Physiol* **8:** 773–799. doi:10.1002/cphy.c170019

Crago EA, Thampatty BP, Sherwood PR, Kuo CWJ, Bender C, Balzer J, Horowitz M, Poloyac SM. 2011. Cerebrospinal fluid 20-HETE is associated with delayed cerebral ischemia and poor outcomes after aneurysmal subarachnoid hemorrhage. *Stroke* **42:** 1872–1877. doi:10.1161/STROKEAHA.110.605816

Császár E, Lénárt N, Cserép C, Környei Z, Fekete R, Pósfai B, Balázsfi D, Hangya B, Schwarcz AD, Szabadits E, et al. 2022. Microglia modulate blood flow, neurovascular coupling, and hypoperfusion via purinergic actions. *J Exp Med* **219:** e20211071. doi:10.1084/jem.20211071

Czigler A, Toth L, Szarka N, Szilágyi K, Kellermayer Z, Harci A, Vecsernyes M, Ungvari Z, Szolics A, Koller A, et al. 2020. Prostaglandin E2, a postulated mediator of neurovascular coupling, at low concentrations dilates whereas at higher concentrations constricts human cerebral parenchymal arterioles. *Prostaglandins Other Lipid Mediat* **146:** 106389. doi:10.1016/j.prostaglandins.2019.106389

Dabertrand F, Hannah RM, Pearson JM, Hill-Eubanks DC, Brayden JE, Nelson MT. 2013. Prostaglandin E2, a postulated astrocyte-derived neurovascular coupling agent, constricts rather than dilates parenchymal arterioles. *J Cereb Blood Flow Metab* **33:** 479–482. doi:10.1038/jcbfm.2013.9

Delekate A, Füchtemeier M, Schumacher T, Ulbrich C, Foddis M, Petzold GC. 2014. Metabotropic P2Y1 receptor signalling mediates astrocytic hyperactivity in vivo in an Alzheimer's disease mouse model. *Nat Commun* **5:** 5422. doi:10.1038/ncomms6422

Del Franco AP, Chiang PP, Newman EA. 2022. Dilation of cortical capillaries is not related to astrocyte calcium signaling. *Glia* **70:** 508–521. doi:10.1002/glia.24119

Devor A, Sakadžić S, Saisan PA, Yaseen MA, Roussakis E, Srinivasan VJ, Vinogradov SA, Rosen BR, Buxton RB, Dale AM, et al. 2011. "Overshoot" of O_2 is required to maintain baseline tissue oxygenation at locations distal to blood vessels. *J Neurosci* **31:** 13676–13681. doi:10.1523/JNEUROSCI.1968-11.2011

Dick E, Miller RF. 1985. Extracellular K^+ activity changes related to electroretinogram components. I: Amphibian (I-type) retinas. *J Gen Physiol* **85:** 885–909. doi:10.1085/jgp.85.6.885

Dienel GA, Gillinder L, McGonigal A, Borges K. 2023. Potential new roles for glycogen in epilepsy. *Epilepsia* **64:** 29–53. doi:10.1111/epi.17412

Ding S, Wang T, Cui W, Haydon PG. 2009. Photothrombosis ischemia stimulates a sustained astrocytic Ca^{2+} signaling in vivo. *Glia* **57:** 767–776. doi:10.1002/glia.20804

Donnelly MK, Crago EA, Conley YP, Balzer JR, Ren D, Ducruet AF, Kochanek PM, Sherwood PR, Poloyac SM. 2015. 20-HETE is associated with unfavorable outcomes in subarachnoid hemorrhage patients. *J Cereb Blood Flow Metab* **35:** 1515–1522. doi:10.1038/jcbfm.2015.75

Du Y, Smith MA, Miller CM, Kern TS. 2002. Diabetes-induced nitrative stress in the retina, and correction by ami-

noguanidine. *J Neurochem* **80:** 771–779. doi:10.1046/j
.0022-3042.2001.00737.x

Duffy S, MacVicar BA. 1995. Adrenergic calcium signaling
in astrocyte networks within the hippocampal slice. *J
Neurosci* **15:** 5535–5550. doi:10.1523/JNEUROSCI.15-
08-05535.1995

Dunn KM, Nelson MT. 2010. Potassium channels and neurovascular coupling. *Circ J* **74:** 608–616. doi:10.1253/circj
.CJ-10-0174

Dunn KM, Hill-Eubanks DC, Liedtke WB, Nelson MT. 2013.
TRPV4 channels stimulate Ca^{2+}-induced Ca^{2+} release in
astrocytic endfeet and amplify neurovascular coupling responses. *Proc Natl Acad Sci* **110:** 6157–6162. doi:10.1073/
pnas.1216514110

Farrell JS, Gaxiola-Valdez I, Wolff MD, David LS, Dika HI,
Geeraert BL, Rachel Wang X, Singh S, Spanswick SC,
Dunn JF, et al. 2016. Postictal behavioural impairments
are due to a severe prolonged hypoperfusion/hypoxia
event that is COX-2 dependent. *eLife* **5:** e19352. doi:10
.7554/eLife.19352

Farrell JS, Colangeli R, Dong A, George AG, Addo-Osafo K,
Kingsley PJ, Morena M, Wolff MD, Dudok B, He K, et al.
2021. In vivo endocannabinoid dynamics at the timescale
of physiological and pathological neural activity. *Neuron*
109: 2398–2403.e4. doi:10.1016/j.neuron.2021.05.026

Filosa JA, Bonev AD, Straub SV, Meredith AL, Wilkerson
MK, Aldrich RW, Nelson MT. 2006. Local potassium signaling couples neuronal activity to vasodilation in the
brain. *Nat Neurosci* **9:** 1397–1403. doi:10.1038/nn1779

Fordsmann JC, Ko RWY, Choi HB, Thomsen K, Witgen BM,
Mathiesen C, Lønstrup M, Piilgaard H, MacVicar BA,
Lauritzen M. 2013. Increased 20-HETE synthesis explains
reduced cerebral blood flow but not impaired neurovascular coupling after cortical spreading depression in rat cerebral cortex. *J Neurosci* **33:** 2562–2570. doi:10.1523/
JNEUROSCI.2308-12.2013

Garhöfer G, Zawinka C, Resch H, Huemer KH, Dorner GT,
Schmetterer L. 2004a. Diffuse luminance flicker increases
blood flow in major retinal arteries and veins. *Vis Res* **44:**
833–838. doi:10.1016/j.visres.2003.11.013

Garhöfer G, Zawinka C, Resch H, Kothy P, Schmetterer L,
Dorner GT. 2004b. Reduced response of retinal vessel di-
ameters to flicker stimulation in patients with diabetes. *Br J
Ophthalmol* **88:** 887–891. doi:10.1136/bjo.2003.033548

Gebremedhin D, Lange AR, Narayanan J, Aebly MR, Jacobs
ER, Harder DR. 1998. Cat cerebral arterial smooth muscle
cells express cytochrome P450 4A2 enzyme and produce
the vasoconstrictor 20-HETE which enhances L-type Ca^{2+}
current. *J Physiol* **507:** 771–781. doi:10.1111/j.1469-7793
.1998.771bs.x

Gebremedhin D, Yamaura K, Zhang C, Bylund J, Koehler RC,
Harder DR. 2003. Metabotropic glutamate receptor activation enhances the activities of two types of Ca^{2+}-activat-
ed K^+ channels in rat hippocampal astrocytes. *J Neurosci*
23: 1678–1687. doi:10.1523/JNEUROSCI.23-05-01678
.2003

Girouard H, Bonev AD, Hannah RM, Meredith A, Aldrich
RW, Nelson MT. 2010. Astrocytic endfoot Ca^{2+} and BK
channels determine both arteriolar dilation and constriction. *Proc Natl Acad Sci* **107:** 3811–3816. doi:10.1073/pnas
.0914722107

Göbel J, Engelhardt E, Pelzer P, Sakthivelu V, Jahn HM, Jevtic
M, Folz-Donahue K, Kukat C, Schauss A, Frese CK, et al.
2020. Mitochondria-endoplasmic reticulum contacts in
reactive astrocytes promote vascular remodeling. *Cell
Metab* **31:** 791–808.e8. doi:10.1016/j.cmet.2020.03.005

Goedert M, Wischik CM, Crowther RA, Walker JE, Klug A.
1988. Cloning and sequencing of the cDNA encoding a
core protein of the paired helical filament of Alzheimer
disease: identification as the microtubule-associated protein tau. *Proc Natl Acad Sci* **85:** 4051–4055. doi:10.1073/
pnas.85.11.4051

Golgi C. 1894. *Untersuchungen über den feineren bau des
zentralen und peripherischen nervensystems* [*Investigations into the finer structure of the central and peripheral
nervous systems*]. Gustav Fischer, Jena, Germany.

Gómez-Gonzalo M, Losi G, Brondi M, Uva L, Sulis-Sato S, De
Curtis M, Ratto GM, Carmignoto G. 2011. Ictal but not
interictal epileptic discharges activate astrocyte endfeet
and elicit cerebral arteriole responses. *Front Cell Neurosci*
5: 8. doi:10.3389/fncel.2011.00008

Gordon GRJ, Mulligan SJ, MacVicar BA. 2007. Astrocyte
control of the cerebrovasculature. *Glia* **55:** 1214–1221.
doi:10.1002/glia.20543

Gordon GRJ, Choi HB, Rungta RL, Ellis-Davies GCR, Mac-
Vicar BA. 2008. Brain metabolism dictates the polarity of
astrocyte control over arterioles. *Nature* **456:** 745–749.
doi:10.1038/nature07525

Graeber MB, Mehraein P. 1999. Reanalysis of the first case of
Alzheimer's disease. *Eur Arch Psychiatry Clin Neurosci*
249: 10–13. doi:10.1007/PL00014167

Gu X, Chen W, Volkow ND, Koretsky AP, Du C, Pan Y. 2018.
Synchronized astrocytic Ca^{2+} responses in neurovascular
coupling during somatosensory stimulation and for the
resting state. *Cell Rep* **23:** 3878–3890. doi:10.1016/j
.celrep.2018.05.091

Haddy FJ. 1983. Potassium effects on contraction in arterial
smooth muscle mediated by Na^+, K^+-ATPase. *FASEB J* **42:**
239–243.

Haddy FJ, Vanhoutte PM, Feletou M. 2006. Role of potassium
in regulating blood flow and blood pressure. *Am J Physiol
Regul Integr Comp Physiol* **290:** R546–R552. doi:10.1152/
ajpregu.00491.2005

Haidey JN, Gordon GR. 2021. Direct deviations in astrocyte
free Ca^{2+} concentration control multiple arteriole tone
states. *Neuroglia* **2:** 48–56. doi:10.3390/neuroglia2010006

Haidey JN, Peringod G, Institoris A, Gorzo KA, Nicola W,
Vandal M, Ito K, Liu S, Fielding C, Visser F, et al. 2021.
Astrocytes regulate ultra-slow arteriole oscillations via
stretch-mediated TRPV4-COX-1 feedback. *Cell Rep* **36:**
109405. doi:10.1016/j.celrep.2021.109405

Hall CN, Reynell C, Gesslein B, Hamilton NB, Mishra A,
Sutherland BA, O'Farrell FM, Buchan AM, Lauritzen M,
Attwell D. 2014. Capillary pericytes regulate cerebral
blood flow in health and disease. *Nature* **508:** 55–60.
doi:10.1038/nature13165

Hamel E. 2006. Perivascular nerves and the regulation of
cerebrovascular tone. *J Appl Physiol* **100:** 1059–1064.
doi:10.1152/japplphysiol.00954.2005

Harder DR, Alkayed NJ, Lange AR, Gebremedhin D, Roman
RJ. 1998. Functional hyperemia in the brain: hypothesis
for astrocyte-derived vasodilator metabolites. *Stroke* **29:**
229–234. doi:10.1161/01.STR.29.1.229

Hartmann DA, Berthiaume AA, Grant RI, Harrill SA, Koski T, Tieu T, McDowell KP, Faino AV, Kelly AL, Shih AY. 2021. Brain capillary pericytes exert a substantial but slow influence on blood flow. *Nat Neurosci* **24:** 633–645. doi:10.1038/s41593-020-00793-2

Hatakeyama N, Unekawa M, Murata J, Tomita Y, Suzuki N, Nakahara J, Takuwa H, Kanno I, Matsui K, Tanaka KF, et al. 2021. Differential pial and penetrating arterial responses examined by optogenetic activation of astrocytes and neurons. *J Cereb Blood Flow Metab* **41:** 2676–2689. doi:10.1177/0271678X211010355

He L, Linden DJ, Sapirstein A. 2012. Astrocyte inositol triphosphate receptor Type 2 and cytosolic phospholipase A$_2$ α regulate arteriole responses in mouse neocortical brain slices. *PLoS ONE* **7:** e42194. doi:10.1371/journal.pone.0042194

Hefendehl JK, LeDue J, Ko RW, Mahler J, Murphy TH, MacVicar BA. 2016. Mapping synaptic glutamate transporter dysfunction in vivo to regions surrounding Aβ plaques by iGluSnFR two-photon imaging. *Nat Commun* **7:** 13441. doi:10.1038/ncomms13441

Hirunpattarasilp C, Barkaway A, Davis H, Pfeiffer T, Sethi H, Attwell D. 2022. Hyperoxia evokes pericyte-mediated capillary constriction. *J Cereb Blood Flow Metab* **42:** 2032–2047. doi:10.1177/0271678X221111598

Hogan-Cann AD, Lu P, Anderson CM. 2019. Endothelial NMDA receptors mediate activity-dependent brain hemodynamic responses in mice. *Proc Natl Acad Sci* **116:** 10229–10231. doi:10.1073/pnas.1902647116

Hosford PS, Wells JA, Nizari S, Christie IN, Theparambil SM, Castro PA, Hadjihambi A, Barros LF, Ruminot I, Lythgoe MF, et al. 2022. CO$_2$ signaling mediates neurovascular coupling in the cerebral cortex. *Nat Commun* **13:** 2125. doi:10.1038/s41467-022-29622-9

Hösli L, Binini N, Ferrari KD, Thieren L, Looser ZJ, Zuend M, Zanker HS, Berry S, Holub M, Möbius W, et al. 2022. Decoupling astrocytes in adult mice impairs synaptic plasticity and spatial learning. *Cell Rep* **38:** 110484. doi:10.1016/j.celrep.2022.110484

Howarth C, Gleeson P, Attwell D. 2012. Updated energy budgets for neural computation in the neocortex and cerebellum. *J Cereb Blood Flow Metab* **32:** 1222–1232. doi:10.1038/jcbfm.2012.35

Howarth C, Sutherland BA, Choi HB, Martin C, Lind BL, Khennouf L, LeDue JM, Pakan JMP, Ko RWY, Ellis-Davies G, et al. 2017. A critical role for astrocytes in hypercapnic vasodilation in brain. *J Neurosci* **37:** 2403–2414. doi:10.1523/JNEUROSCI.0005-16.2016

Howarth C, Mishra A, Hall C. 2021. More than just summed neuronal activity: how multiple cell types shape the BOLD response. *Philos Trans R Soc Lond B Biol Sci* **376:** 20190630. doi:10.1098/rstb.2019.0630

Iadecola C. 2013. The pathobiology of vascular dementia. *Neuron* **80:** 844–866. doi:10.1016/j.neuron.2013.10.008

Imig JD, Zou AP, Stec DE, Harder DR, Falck JR, Roman RJ. 1996. Formation and actions of 20-hydroxyeicosatetraenoic acid in rat renal arterioles. *Am J Physiol* **270:** R217–R227. doi:10.1152/ajpregu.1996.270.1.R217

Institoris A, Vandal M, Peringod G, Catalano C, Tran CH, Yu X, Visser F, Breiteneder C, Molina L, Khakh BS, et al. 2022. Astrocytes amplify neurovascular coupling to sustained activation of neocortex in awake mice. *Nat Commun* **13:** 7872. doi:10.1038/s41467-022-35383-2

Iturria-Medina Y, Sotero RC, Toussaint PJ, Mateos-Pérez JM, Evans AC, Weiner MW, Aisen P, Petersen R, Jack CR, Jagust W, et al. 2016. Early role of vascular dysregulation on late-onset Alzheimer's disease based on multifactorial data-driven analysis. *Nat Commun* **7:** 11934. doi:10.1038/ncomms11934

Jego P, Pacheco-Torres J, Araque A, Canals S. 2014. Functional MRI in mice lacking IP3-dependent calcium signaling in astrocytes. *J Cereb Blood Flow Metab* **34:** 1599–1603. doi:10.1038/jcbfm.2014.144

Johnson-Rabbett B, Seaquist ER. 2019. Hypoglycemia in diabetes: the dark side of diabetes treatment. A patient-centered review. *J Diabetes* **11:** 711–718. doi:10.1111/1753-0407.12933

Karwoski CJ, Frambach DA, Proenza LM. 1985. Laminar profile of resistivity in frog retina. *J Neurophysiol* **54:** 1607–1619. doi:10.1152/jn.1985.54.6.1607

Kim KJ, Iddings JA, Stern JE, Blanco VM, Croom D, Kirov SA, Filosa JA. 2015. Astrocyte contributions to flow/pressure-evoked parenchymal arteriole vasoconstriction. *J Neurosci* **35:** 8245–8257. doi:10.1523/JNEUROSCI.4486-14.2015

Koide M, Bonev AD, Nelson MT, Wellman GC. 2012. Inversion of neurovascular coupling by subarachnoid blood depends on large-conductance Ca^{2+}-activated K$^+$ (BK) channels. *Proc Natl Acad Sci* **109:** E1387–E1395. doi:10.1073/pnas.1121359109

Koide M, Ferris HR, Nelson MT, Wellman GC. 2021. Impaired cerebral autoregulation after subarachnoid hemorrhage: a quantitative assessment using a mouse model. *Front Physiol* **12:** 688468. doi:10.3389/fphys.2021.688468

Kornfield TE, Newman EA. 2014. Regulation of blood flow in the retinal trilaminar vascular network. *J Neurosci* **34:** 11504–11513. doi:10.1523/JNEUROSCI.1971-14.2014

Kosik KS, Joachim CL, Selkoe DJ. 1986. Microtubule-associated protein tau (τ) is a major antigenic component of paired helical filaments in Alzheimer disease. *Proc Natl Acad Sci* **83:** 4044–4048. doi:10.1073/pnas.83.11.4044

Kowluru RA, Engerman RL, Kern TS. 2000. Abnormalities of retinal metabolism in diabetes or experimental galactosemia. VIII: Prevention by aminoguanidine. *Curr Eye Res* **21:** 814–819. doi:10.1076/ceyr.21.4.814.5545

Krainik A, Hund-Georgiadis M, Zysset S, von Cramon DY. 2005. Regional impairment of cerebrovascular reactivity and BOLD signal in adults after stroke. *Stroke* **36:** 1146–1152. doi:10.1161/01.STR.0000166178.40973.a7

Kuchibhotla KV, Lattarulo CR, Hyman BT, Bacskai BJ. 2009. Synchronous hyperactivity and intercellular calcium waves in astrocytes in Alzheimer mice. *Science* **323:** 1211–1215. doi:10.1126/science.1169096

Kur J, Newman EA. 2014. Purinergic control of vascular tone in the retina. *J Physiol* **592:** 491–504. doi:10.1113/jphysiol.2013.267294

Lacar B, Herman P, Platel JC, Kubera C, Hyder F, Bordey A. 2012. Neural progenitor cells regulate capillary blood flow in the postnatal subventricular zone. *J Neurosci* **32:** 16435–16448. doi:10.1523/JNEUROSCI.1457-12.2012

Lacza Z, Puskar M, Figueroa JP, Zhang J, Rajapakse N, Busija DW. 2001. Mitochondrial nitric oxide synthase is constitutively active and is functionally upregulated in hypoxia.

Free Radic Biol Med **31:** 1609–1615. doi:10.1016/S0891-5849(01)00754-7

Lee JH, Durand R, Gradinaru V, Zhang F, Goshen I, Kim DS, Fenno LE, Ramakrishnan C, Deisseroth K. 2010. Global and local fMRI signals driven by neurons defined optogenetically by type and wiring. *Nature* **465:** 788–792. doi:10.1038/nature09108

Li Z, McConnell HL, Stackhouse TL, Pike MM, Zhang W, Mishra A. 2021. Increased 20-HETE signaling suppresses capillary neurovascular coupling after ischemic stroke in regions beyond the infarct. *Front Cell Neurosci* **15:** 762843. doi:10.3389/fncel.2021.762843

Lin AL, Fox PT, Hardies J, Duong TQ, Gao JH. 2010. Nonlinear coupling between cerebral blood flow, oxygen consumption, and ATP production in human visual cortex. *Proc Natl Acad Sci* **107:** 8446–8451. doi:10.1073/pnas.0909711107

Lin WH, Hao Q, Rosengarten B, Leung WH, Wong KS. 2011. Impaired neurovascular coupling in ischaemic stroke patients with large or small vessel disease. *Eur J Neurol* **18:** 731–736. doi:10.1111/j.1468-1331.2010.03262.x

Lind BL, Brazhe AR, Jessen SB, Tan FCC, Lauritzen MJ. 2013. Rapid stimulus-evoked astrocyte Ca^{2+} elevations and hemodynamic responses in mouse somatosensory cortex in vivo. *Proc Natl Acad Sci* **110:** E4678–E4687.

Lind BL, Jessen SB, Lønstrup M, Josephine C, Bonvento G, Lauritzen M. 2018. Fast Ca^{2+} responses in astrocyte endfeet and neurovascular coupling in mice. *Glia* **66:** 348–358. doi:10.1002/glia.23246

Lines J, Baraibar AM, Fang C, Martin ED, Aguilar J, Lee MK, Araque A, Kofuji P. 2022. Astrocyte-neuronal network interplay is disrupted in Alzheimer's disease mice. *Glia* **70:** 368–378. doi:10.1002/glia.24112

Logothetis NK, Pauls J, Augath M, Trinath T, Oeltermann A. 2001. Neurophysiological investigation of the basis of the fMRI signal. *Nature* **412:** 150–157. doi:10.1038/35084005

Lu L, Hogan-Cann AD, Globa AK, Lu P, Nagy JI, Bamji SX, Anderson CM. 2017. Astrocytes drive cortical vasodilatory signaling by activating endothelial NMDA receptors. *J Cereb Blood Flow Metab* **39:** 481–496. doi:10.1177/0271678X17734100

Magistretti PJ, Pellerin L, Martin JL. 1995. Brain energy metabolism: an integrated cellular perspective. In *Psychopharmacology—4th generation of progress.* Raven, New York.

Major S, Petzold GC, Reiffurth C, Windmüller O, Foddis M, Lindauer U, Kang EJ, Dreier JP. 2017. A role of the sodium pump in spreading ischemia in rats. *J Cereb Blood Flow Metab* **37:** 1687–1705. doi:10.1177/0271678X16639059

Mateo C, Knutsen PM, Tsai PS, Shih AY, Kleinfeld D. 2017. Entrainment of arteriole vasomotor fluctuations by neural activity is a basis of blood-oxygenation-level-dependent "resting-state" connectivity. *Neuron* **96:** 936–948.e3. doi:10.1016/j.neuron.2017.10.012

Mehina EMF, Murphy-Royal C, Gordon GR. 2017. Steady-State free Ca^{2+} in astrocytes is decreased by experience and impacts arteriole tone. *J Neurosci* **37:** 8150–8165. doi:10.1523/JNEUROSCI.0239-17.2017

Metea MR, Newman EA. 2006. Glial cells dilate and constrict blood vessels: a mechanism of neurovascular coupling. *J Neurosci* **26:** 2862–2870. doi:10.1523/JNEUROSCI.4048-05.2006

Metea MR, Kofuji P, Newman EA. 2007. Neurovascular coupling is not mediated by potassium siphoning from glial cells. *J Neurosci* **27:** 2468–2471. doi:10.1523/JNEUROSCI.3204-06.2007

Mintun MA, Vlassenko AG, Rundle MM, Raichle ME. 2004. Increased lactate/pyruvate ratio augments blood flow in physiologically activated human brain. *Proc Natl Acad Sci* **101:** 659–664. doi:10.1073/pnas.0307457100

Mishra A, Newman EA. 2010. Inhibition of inducible nitric oxide synthase reverses the loss of functional hyperemia in diabetic retinopathy. *Glia* **58:** 1996–2004. doi:10.1002/glia.21068

Mishra A, Newman EA. 2012. Aminoguanidine reverses the loss of functional hyperemia in a rat model of diabetic retinopathy. *Front Neuroenerg* **3:** 10. doi:10.3389/fnene.2011.00010

Mishra A, Hamid A, Newman EA. 2011. Oxygen modulation of neurovascular coupling in the retina. *Proc Natl Acad Sci* **108:** 17827–17831. doi:10.1073/pnas.1110533108

Mishra A, Reynolds JP, Chen Y, Gourine AV, Rusakov DA, Attwell D. 2016. Astrocytes mediate neurovascular signaling to capillary pericytes but not to arterioles. *Nat Neurosci* **19:** 1619–1627. doi:10.1038/nn.4428

Mosso A. 1880. Sulla circolazione del sangue nel cervello dell'uomo [On the circulation of blood in the human brain]. *R Accad Lincei* **5:** 237–358.

Mulligan SJ, MacVicar BA. 2004. Calcium transients in astrocyte endfeet cause cerebrovascular constrictions. *Nature* **431:** 195–199. doi:10.1038/nature02827

Neil HA, Gale EA, Hamilton SJ, Lopez-Espinoza I, Kaura R, McCarthy ST. 1987. Cerebral blood flow increases during insulin-induced hypoglycaemia in type 1 (insulin-dependent) diabetic patients and control subjects. *Diabetologia* **30:** 305–309. doi:10.1007/BF00299022

Newman EA. 1984. Regional specialization of retinal glial cell membrane. *Nature* **309:** 155–157. doi:10.1038/309155a0

Newman EA. 1986. High potassium conductance in astrocyte endfeet. *Science* **233:** 453–454. doi:10.1126/science.3726539

Newman EA. 2001. Propagation of intercellular calcium waves in retinal astrocytes and Müller cells. *J Neurosci* **21:** 2215–2223. doi:10.1523/JNEUROSCI.21-07-02215.2001

Newman EA, Frambach DA, Odette LL. 1984. Control of extracellular potassium levels by retinal glial cell K$^+$ siphoning. *Science* **225:** 1174–1175. doi:10.1126/science.6474173

Nguyen TT, Kawasaki R, Wang JJ, Kreis AJ, Shaw J, Vilser W, Wong TY. 2009. Flicker light-induced retinal vasodilation in diabetes and diabetic retinopathy. *Diabetes Care* **32:** 2075–2080. doi:10.2337/dc09-0075

Nippert AR, Chiang PP, Del Franco AP, Newman EA. 2022. Astrocyte regulation of cerebral blood flow during hypoglycemia. *J Cereb Blood Flow Metab* **42:** 1534–1546. doi:10.1177/0271678X221089091

Nizar K, Uhlirova H, Tian P, Saisan PA, Cheng Q, Reznichenko L, Weldy KL, Steed TC, Sridhar VB, MacDonald CL, et al. 2013. In vivo stimulus-induced vasodilation occurs without IP$_3$ receptor activation and may precede astrocytic calcium increase. *J Neurosci* **33:** 8411–8422. doi:10.1523/JNEUROSCI.3285-12.2013

Nortley R, Korte N, Izquierdo P, Hirunpattarasilp C, Mishra A, Jaunmuktane Z, Kyrargyri V, Pfeiffer T, Khennouf L, Madry C, et al. 2019. Amyloid β oligomers constrict human capillaries in Alzheimer's disease via signaling to pericytes. *Science* 365: eaav9518. doi:10.1126/science.aav 9518

Oe Y, Baba O, Ashida H, Nakamura KC, Hirase H. 2016. Glycogen distribution in the microwave-fixed mouse brain reveals heterogeneous astrocytic patterns. *Glia* 64: 1532–1545. doi:10.1002/glia.23020

Offenhauser N, Thomsen K, Caesar K, Lauritzen M. 2005. Activity-induced tissue oxygenation changes in rat cerebellar cortex: interplay of postsynaptic activation and blood flow. *J Physiol* 565: 279–294. doi:10.1113/jphysiol .2005.082776

O'Herron P, Chhatbar PY, Levy M, Shen Z, Schramm AE, Lu Z, Kara P. 2016. Neural correlates of single-vessel haemodynamic responses in vivo. *Nature* 534: 378–382. doi:10 .1038/nature17965

Otsu Y, Couchman K, Lyons DG, Collot M, Agarwal A, Mallet JM, Pfrieger FW, Bergles DE, Charpak S. 2015. Calcium dynamics in astrocyte processes during neurovascular coupling. *Nat Neurosci* 18: 210–218. doi:10.1038/nn.3906

Oyekan AO, Youseff T, Fulton D, Quilley J, McGiff JC. 1999. Renal cytochrome P450 ω-hydroxylase and epoxygenase activity are differentially modified by nitric oxide and sodium chloride. *J Clin Invest* 104: 1131–1137. doi:10.1172/ JCI6786

Ozawa K, Nagao M, Konno A, Iwai Y, Vittani M, Kusk P, Mishima T, Hirai H, Nedergaard M, Hirase H. 2023. Astrocytic GPCR-induced Ca^{2+} signaling is not causally related to local cerebral blood flow changes. *Int J Mol Sci* 24: 13590. doi:10.3390/ijms241713590

Pappas AC, Koide M, Wellman GC. 2015. Astrocyte Ca^{2+} signaling drives inversion of neurovascular coupling after subarachnoid hemorrhage. *J Neurosci* 35: 13375. doi:10 .1523/JNEUROSCI.1551-15.2015

Pappas AC, Koide M, Wellman GC. 2016. Purinergic signaling triggers endfoot high-amplitude Ca^{2+} signals and causes inversion of neurovascular coupling after subarachnoid hemorrhage. *J Cereb Blood Flow Metab* 36: 1901–1912. doi:10.1177/0271678X16650911

Park L, Anrather J, Forster C, Kazama K, Carlson GA, Iadecola C. 2004. Abeta-induced vascular oxidative stress and attenuation of functional hyperemia in mouse somatosensory cortex. *J Cereb Blood Flow Metab* 24: 334–342. doi:10 .1097/01.WCB.0000105800.49957.1E

Park L, Anrather J, Zhou P, Frys K, Pitstick R, Younkin S, Carlson GA, Iadecola C. 2005. NADPH-oxidase-derived reactive oxygen species mediate the cerebrovascular dysfunction induced by the amyloid β peptide. *J Neurosci* 25: 1769–1777. doi:10.1523/JNEUROSCI.5207-04.2005

Pascual O, Casper KB, Kubera C, Zhang J, Revilla-Sanchez R, Sul JY, Takano H, Moss SJ, McCarthy K, Haydon PG. 2005. Astrocytic purinergic signaling coordinates synaptic networks. *Science* 310: 113–116. doi:10.1126/science .1116916

Paukert M, Agarwal A, Cha J, Doze VA, Kang JU, Bergles DE. 2014. Norepinephrine controls astroglial responsiveness to local circuit activity. *Neuron* 82: 1263–1270. doi:10 .1016/j.neuron.2014.04.038

Paulson OB, Newman EA. 1987. Does the release of potassium from astrocyte endfeet regulate cerebral blood flow? *Science* 237: 896–898. doi:10.1126/science.3616619

Pemp B, Garhofer G, Weigert G, Karl K, Resch H, Wolzt M, Schmetterer L. 2009. Reduced retinal vessel response to flicker stimulation but not to exogenous nitric oxide in type 1 diabetes. *Invest Ophthalmol Vis Sci* 50: 4029–4032. doi:10.1167/iovs.08-3260

Peppiatt CM, Howarth C, Mobbs P, Attwell D. 2006. Bidirectional control of CNS capillary diameter by pericytes. *Nature* 443: 700–704. doi:10.1038/nature05193

Polak K, Schmetterer L, Riva CE. 2002. Influence of flicker frequency on flicker-induced changes of retinal vessel diameter. *Invest Ophthalmol Vis Sci* 43: 2721–2726.

Price DL, Ludwig JW, Mi H, Schwarz TL, Ellisman MH. 2002. Distribution of rSlo Ca^{2+}-activated K^+ channels in rat astrocyte perivascular endfeet. *Brain Res* 956: 183–193. doi:10.1016/S0006-8993(02)03266-3

Puro DG. 2007. Physiology and pathobiology of the pericyte-containing retinal microvasculature: new developments. *Microcirculation* 14: 1–10. doi:10.1080/10739680601072 099

Raichle ME. 2015. The restless brain: how intrinsic activity organizes brain function. *Philos Trans R Soc B Biol Sci* 370: 20140172. doi:10.1098/rstb.2014.0172

Rakers C, Petzold GC. 2017. Astrocytic calcium release mediates peri-infarct depolarizations in a rodent stroke model. *J Clin Invest* 127: 511–516. doi:10.1172/JCI89354

Rosenegger DG, Tran CHT, Wamsteeker Cusulin JI, Gordon GR. 2015. Tonic local brain blood flow control by astrocytes independent of phasic neurovascular coupling. *J Neurosci* 35: 13463–13474. doi:10.1523/JNEUROSCI .1780-15.2015

Roy CS, Sherrington CS. 1890. On the regulation of the blood-supply of the brain. *J Physiol* 11: 85–108. doi:10.1113/jphy siol.1890.sp000321

Rungta RL, Bernier L-P, Dissing-Olesen L, Groten CJ, LeDue JM, Ko R, Drissler S, MacVicar BA. 2016. Ca^{2+} transients in astrocyte fine processes occur via Ca^{2+} influx in the adult mouse hippocampus. *Glia* 64: 2093–2103. doi:10.1002/ glia.23042

Salinet ASM, Robinson TG, Panerai RB. 2015. Effects of cerebral ischemia on human neurovascular coupling, CO_2 reactivity, and dynamic cerebral autoregulation. *J Appl Physiol* 118: 170–177. doi:10.1152/japplphysiol.00620 .2014

Schaeffer S, Iadecola C. 2021. Revisiting the neurovascular unit. *Nat Neurosci* 24: 1198–1209. doi:10.1038/s41593-021-00904-7

Schipke CG, Kettenmann H. 2004. Astrocyte responses to neuronal activity. *Glia* 47: 226–232. doi:10.1002/glia .20029

Singer W, Lux HD. 1975. Extracellular potassium gradients and visual receptive fields in the cat striate cortex. *Brain Res* 96: 378–383. doi:10.1016/0006-8993(75)90751-9

Somjen GG. 2001. Mechanisms of spreading depression and hypoxic spreading depression-like depolarization. *Physiol Rev* 81: 1065–1096. doi:10.1152/physrev.2001.81.3.1065

Srinivasan R, Huang BS, Venugopal S, Johnston AD, Chai H, Zeng H, Golshani P, Khakh BS. 2015. Ca^{2+} signaling in astrocytes from $Ip3r2^{-/-}$ mice in brain slices and during startle responses in vivo. *Nat Neurosci* **18:** 708–717. doi:10 .1038/nn.4001

Stobart JLL, Lu L, Anderson HDI, Mori H, Anderson CM. 2013. Astrocyte-induced cortical vasodilation is mediated by D-serine and endothelial nitric oxide synthase. *Proc Natl Acad Sci* **110:** 3149–3154. doi:10.1073/pnas.1215 929110

Stobart JL, Ferrari KD, Barrett MJP, Gluck C, Stobart MJ, Zuend M, Weber B. 2018. Cortical circuit activity evokes rapid astrocyte calcium signals on a similar timescale to neurons. *Neuron* **98:** 726–735.e4. doi:10.1016/j.neuron .2018.03.050

Sun W, McConnell E, Pare JF, Xu Q, Chen M, Peng W, Lovatt D, Han X, Smith Y, Nedergaard M. 2013. Glutamate-dependent neuroglial calcium signaling differs between young and adult brain. *Science* **339:** 197–200. doi:10 .1126/science.1226740

Takano T, Tian GF, Peng W, Lou N, Libionka W, Han X, Nedergaard M. 2006. Astrocyte-mediated control of cerebral blood flow. *Nat Neurosci* **9:** 260–267. doi:10.1038/ nn1623

Takata N, Nagai T, Ozawa K, Oe Y, Mikoshiba K, Hirase H. 2013. Cerebral blood flow modulation by basal forebrain or whisker stimulation can occur independently of large cytosolic Ca^{2+} signaling in astrocytes. *PLoS ONE* **8:** e66525. doi:10.1371/journal.pone.0066525

Takata N, Sugiura Y, Yoshida K, Koizumi M, Hiroshi N, Honda K, Yano R, Komaki Y, Matsui K, Suematsu M, et al. 2018. Optogenetic astrocyte activation evokes BOLD fMRI response with oxygen consumption without neuronal activity modulation. *Glia* **66:** 2013–2023. doi:10.1002/ glia.23454

Terry RD, Masliah E, Salmon DP, Butters N, DeTeresa R, Hill R, Hansen LA, Katzman R. 1991. Physical basis of cognitive alterations in Alzheimer's disease: synapse loss is the major correlate of cognitive impairment. *Ann Neurol* **30:** 572–580. doi:10.1002/ana.410300410

Tran CHT, Peringod G, Gordon GR. 2018. Astrocytes integrate behavioral state and vascular signals during functional hyperemia. *Neuron* **100:** 1133–1148.e3. doi:10 .1016/j.neuron.2018.09.045

Tran CHT, George AG, Teskey GC, Gordon GR. 2020. Seizures elevate gliovascular unit Ca^{2+} and cause sustained vasoconstriction. *JCI Insight* **5:** e136469. doi:10.1172/jci .insight.136469

Venneti S, Wang G, Nguyen J, Wiley CA. 2008. The positron emission tomography ligand DAA1106 binds with high affinity to activated microglia in human neurological disorders. *J Neuropathol Exp Neurol* **67:** 1001–1010. doi:10 .1097/NEN.0b013e318188b204

Virchow RL. 1858. *Cellular pathology as based upon physiological and pathological histology.* John Churchill, London.

Vyskočil F, Kříž N, Bureš J. 1972. Potassium-selective microelectrodes used for measuring the extracellular brain potassium during spreading depression and anoxic depolarization in rats. *Brain Res* **39:** 255–259. doi:10.1016/ 0006-8993(72)90802-5

Wyss-Coray T. 2006. Inflammation in Alzheimer disease: driving force, bystander or beneficial response? *Nat Med* **12:** 1005–1015.

Yi X, Lin J, Wang C, Zhou Q. 2017. CYP genetic variants, CYP metabolite levels, and neurologic deterioration in acute ischemic stroke in Chinese population. *J Stroke Cerebrovasc Dis* **26:** 969–978. doi:10.1016/j.jstrokecerebrovasdis .2016.11.004

Yu DY, Cringle SJ, Alder V, Su EN. 1999. Intraretinal oxygen distribution in the rat with graded systemic hyperoxia and hypercapnia. *Invest Ophthalmol Vis Sci* **40:** 2082–2087.

Zhang J, Malik A, Choi HB, Ko RW, Dissing-Olesen L, MacVicar BA. 2014. Microglial CR3 activation triggers long-term synaptic depression in the hippocampus via NADPH oxidase. *Neuron* **82:** 195–207. doi:10.1016/j.neuron.2014 .01.043

Zhang C, Tabatabaei M, Bélanger S, Girouard H, Moeini M, Lu X, Lesage F. 2019. Astrocytic endfoot Ca^{2+} correlates with parenchymal vessel responses during 4-AP induced epilepsy: an in vivo two-photon lifetime microscopy study. *J Cereb Blood Flow Metab* **39:** 260–271. doi:10.1177/ 0271678X17725417

Zhao M, Ma H, Suh M, Schwartz TH. 2009. Spatiotemporal dynamics of perfusion and oximetry during ictal discharges in the rat neocortex. *J Neurosci* **29:** 2814. doi:10 .1523/JNEUROSCI.4667-08.2009

Zonta M, Angulo MC, Gobbo S, Rosengarten B, Hossmann KA, Pozzan T, Carmignoto G. 2003. Neuron-to-astrocyte signaling is central to the dynamic control of brain microcirculation. *Nat Neurosci* **6:** 43–50. doi:10.1038/nn980

Zou AP, Fleming JT, Falck JR, Jacobs ER, Gebremedhin D, Harder DR, Roman RJ. 1996. 20-HETE is an endogenous inhibitor of the large-conductance Ca^{2+}-activated K^+ channel in renal arterioles. *Am J Physiol* **270:** R228–R237.

Cite this article as *Cold Spring Harb Perspect Biol* doi: 10.1101/cshperspect.a041354

The Astrocyte: Metabolic Hub of the Brain

L. Felipe Barros,[1,2] Stefanie Schirmeier,[3] and Bruno Weber[4]

[1]Centro de Estudios Científicos, Valdivia 5110465, Chile

[2]Universidad San Sebastián, Facultad de Medicina y Ciencia, Valdivia 5110693, Chile

[3]Technische Universität Dresden, Department of Biology, 01217 Dresden, Germany

[4]University of Zurich, Institute of Pharmacology and Toxicology, 8057 Zurich, Switzerland

Correspondence: bweber@pharma.uzh.ch

Astrocytic metabolism has taken center stage. Interposed between the neuron and the vasculature, astrocytes exert control over the fluxes of energy and building blocks required for neuronal activity and plasticity. They are also key to local detoxification and waste recycling. Whereas neurons are metabolically rigid, astrocytes can switch between different metabolic profiles according to local demand and the nutritional state of the organism. Their metabolic state even seems to be instructive for peripheral nutrient mobilization and has been implicated in information processing and behavior. Here, we summarize recent progress in our understanding of astrocytic metabolism and its effects on metabolic homeostasis and cognition.

In the first edition of *Glia*, we discussed how excitatory synaptic activity is an avid consumer of metabolic energy, generated by the oxidation of blood-borne glucose.[5] Astrocytes are central players in brain metabolism (Fig. 1), supplying neurons with energy substrates and precursors for biosynthesis, while recycling neurotransmitters, oxidized scavengers, and other waste products (Weber and Barros 2015). Here, we review the energy metabolism of mammalian astrocytes and glial cells, which play similar roles in invertebrates, illuminated by emerging techniques, such as genetically encoded sensors. In a nutshell, astrocytes take the brunt of the metabolic load, subsidizing neurons so these can allocate more resources to information processing.

MORPHOLOGY OF ASTROCYTES AND THEIR ROLE AS INTERFACE CELLS

Neurons, glial cells, and the cerebral vasculature form a tightly coupled ensemble, adeptly described by the recently coined term "neuro-gliavascular unit" (Kugler et al. 2021). This anatomical and functional unit is fundamental to our understanding of brain metabolism and astrocytes are responsible for its cohesion. Astrocytes are complex spongiform cells (Aten et al. 2022) with a central cell body and a dense radial arrangement of processes that follow a branch-branchlet-leaflet scheme, parceling the neuropil into largely nonoverlapping domains. Terminal processes originate from every part of the astrocyte and can also form loop-like structures (Arizono et al.

---

[5]This is an update to a previous article published in *Cold Spring Harbor Perspectives in Biology* [Weber and Barros (2015). *Cold Spring Harb Perspect Biol* **7:** a020396. doi: 10.1101/cshperspect.a020396].

Cite this article as *Cold Spring Harb Perspect Biol* doi: 10.1101/cshperspect.a041355

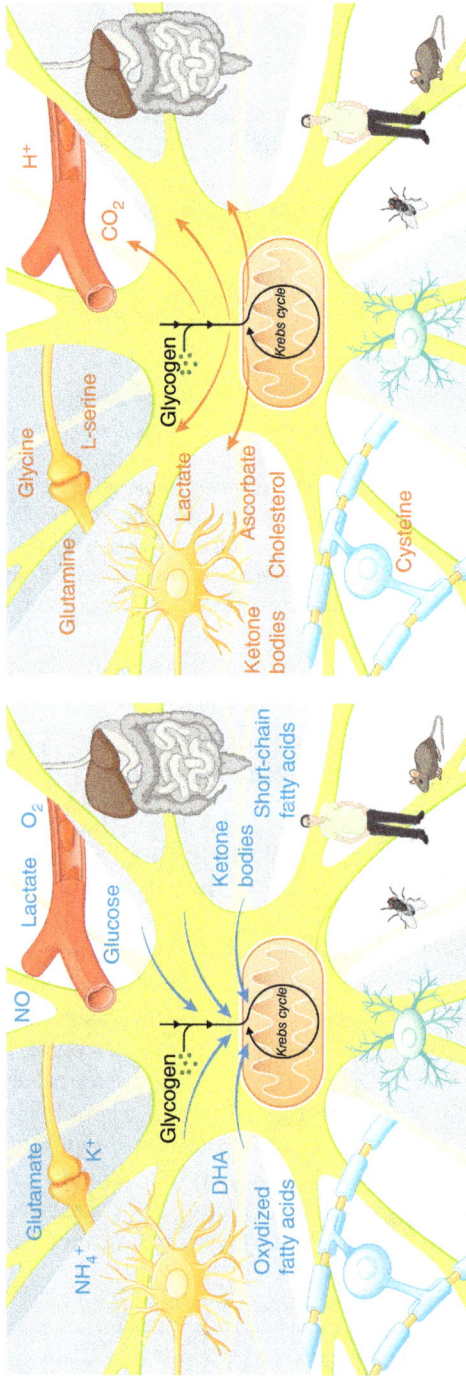

Figure 1. The astrocyte is the metabolic hub of the brain. The backbone of metabolism is the glycolytic pathway plus the Krebs cycle, where fuel and building blocks are generated for neural activity, growth, plasticity, and repair. Astrocytic glycogen is the main energy and carbon store of brain tissue. (*Left* panel) Wedged between blood and the rest of the parenchyma, the astrocyte inputs and integrates substrates, waste products, and regulatory signals, both local and systemic. (*Right* panel) The astrocyte controls the internal milieu of the brain and sustains the function of neurons and other parenchymal cells through the controlled output of energy-rich lactate and other metabolic precursors and signals. Acting locally, the metabolism of astrocytes and equivalent glial cells in invertebrates affects multiple functions of the brain and distant organs. The coupling between astrocytes and neurons has received great deal of attention and has been studied in various model systems, but less is known regarding the metabolic interaction between astrocytes and oligodendrocytes, microglia, smooth muscle, pericytes, and endothelial cells. (DHA) Dehydroascorbate.

Cite this article as *Cold Spring Harb Perspect Biol* doi: 10.1101/cshperspect.a041355

2020; Aten et al. 2022). Each astrocyte is estimated to enwrap about four neuronal somata and 10^5 synapses (Bushong et al. 2002; Halassa et al. 2007; Oberheim et al. 2008). Astrocytes cover virtually the entire basal lamina of the cerebral vasculature (Mathiisen et al. 2010) with delicate processes termed "perivascular astrocytic endfeet" (Reichenbach 1989). Recent data show that every astrocyte has contact with at least one but up to four capillaries (Hösli et al. 2022). Astrocytes also extend peripheral processes (Derouiche and Frotscher 2001) that make close contact with neurons at somata, dendrites, and axons (Aten et al. 2022). The term "tripartite synapse" (Araque et al. 1999) comes from the fact that most of the synapses, pre- and postsynapse, are touched by astrocytic processes (Ventura and Harris 1999). The synapse–astrocyte interface has attracted much attention in relation to neurotransmission (Halassa and Haydon 2010), but it is also of paramount importance for metabolism. The astrocyte removes the neurotransmitter glutamate from the synapse through high-affinity surface transporters, triggering intracellular metabolic events that are described elsewhere in this paper. The spatial relationship between astrocytes and synapses is complex (Bernardinelli et al. 2014a). Coverage of individual synapses by astrocytic processes varies across brain regions, reaching almost 100% in cerebellum (Grosche et al. 1999), 86% in hippocampus (Aten et al. 2022), and only 68% in neocortex (Kikuchi et al. 2020). The degree of coverage is dynamic and dependent on synaptic activity (Bernardinelli et al. 2014b); it modulates local levels of glutamate (Oliet et al. 2001) as well as its cotransmitter D-serine (Panatier et al. 2006). Conceivably, coverage may also modulate the metabolic exchange between astrocytes and neurons.

ANATOMY OF METABOLISM AND METABOLITE TRANSPORT

The presence of the blood–brain barrier (BBB) means that brain tissue is relatively isolated metabolically. The BBB prevents free diffusion of circulating molecules, protecting neural cells from harmful substances while permitting precise regulation of the brain extracellular milieu, which is essential for signaling. But the BBB also taxes endothelial cells with the transport of metabolic substrates and waste (for review, see Weiler et al. 2017). Metabolite flux across the BBB is tightly controlled. Only small gases like O_2 and CO_2 and lipophilic molecules under 450 Da and with a polar surface area of lower than 90 Å^2 can diffuse freely in and out of the nervous system (van de Waterbeemd et al. 1998). Through regulation of metabolite transport, the BBB protects the nervous system from changes in circulating metabolite concentrations, occurring as, for example, the effects of malnutrition (Kumagai et al. 1995; Simpson et al. 1999; Hertenstein et al. 2021).

The main energy source for the brain is glucose (Siesjö 1978). Glucose is transported across the BBB via the endothelial isoform of GLUT1 (55 kDa) (Dick et al. 1984; Gerhart et al. 1989; Sivitz et al. 1989; Harik et al. 1990; Farrell and Pardridge 1991; Maher et al. 1991; Simpson et al. 2001). Endothelial GLUT1 expression is up-regulated during hypoglycemia (Kumagai et al. 1995; Simpson et al. 1999). Furthermore, expression of the sodium-dependent glucose transporters, SGLT1 and SGLT2, is induced in endothelial cells by ischemia (Nishizaki et al. 1995; Nishizaki and Matsuoka 1998; Enerson and Drewes 2006; Vemula et al. 2009). Plus, the transport of other metabolites, like ketone bodies or fatty acids, can be adapted to the nutritional status of the organism (Pifferi et al. 2021; Düking et al. 2022). Such adaptations are likely protecting the brain from metabolic stress and seem to be a conserved mechanism that is also present in insects (Hertenstein et al. 2021).

Glucose is then further transported into astrocytic endfeet that enwrap microvessels and express the astrocytic form of GLUT1 (45 kDa) (Maher et al. 1991, 1994; Mathiisen et al. 2010). Astrocytes maintain a substantial pool of glucose, unlike other cell types (Bittner et al. 2010, 2011; Prebil et al. 2011; Ruminot et al. 2011). Since GLUT1 is a rather low affinity transporter, it can facilitate both the influx and efflux of glucose. The neuronal transporter GLUT3, however, has a higher affinity and is therefore best suited for glucose uptake (Barros and Deitmer 2010). Intuitively, we would expect neurons to consume more glucose than astrocytes in the nervous system. However, it is, in fact, astrocytes that consume most, at least as primary cells and in tissue slices (Bouzier-Sore et al. 2003, 2006; Barros et al. 2009; Jakoby et al. 2014).

Notably, there seems to be a discrepancy between the rate of glycolysis and the rate of mitochondrial oxidative phosphorylation in astrocytes (Hyder et al. 2006), resulting from differential transcriptional and posttranslational regulation of key enzymes (Lovatt et al. 2007; Cahoy et al. 2008; Herrero-Mendez et al. 2009; Halim et al. 2010). A glycolytic rate that outruns the rate of oxidative phosphorylation leads to a net production of pyruvate in astrocytes, which is exported as lactate and used as an efficient fuel for the neuronal tricarboxylic acid (TCA) cycle and oxidative phosphorylation (OXPHOS) (Schurr et al. 1988; Bouzier-Sore et al. 2006; Wyss et al. 2011; Mächler et al. 2016). Their low dependence on OXPHOS as a source of ATP and their greater metabolic flexibility renders astrocytes rather insensitive to OXPHOS inhibition in vitro and in vivo (Bolaños et al. 1994; Almeida et al. 2001; San Martín et al. 2017; Supplie et al. 2017; Fiebig et al. 2019). Inhibition of OXPHOS in astrocytes leads to a compensatory up-regulation of glycolysis (Almeida et al. 2004). Neurons in contrast, are very sensitive to a lack of OXPHOS, as they are unable to up-regulate glycolysis as efficiently (Bolaños et al. 2010). In neurons, the glycolytic rate needs to be kept low, because a significant amount of glucose must be metabolized in the pentose–phosphate pathway (PPP) to produce building blocks and NADPH, an essential cofactor in antioxidant protection of neurons (see Bonvento and Bolaños 2021 for a comprehensive review). Astrocytes and neurons have very different metabolic machinery and thus very different metabolic needs. While the flux of glucose through glycolysis must be kept low in neurons to allow for a sufficient level of antioxidants, high levels of glycolysis are essential for astrocytes, forcing neurons and astrocytes to cooperate metabolically to maintain nervous system function.

This metabolic coupling involves the transfer of lactate from the astrocyte to the neuron; a phenomenon termed the astrocyte neuron lactate shuttle (ANLS) (Pellerin and Magistretti 1994). The vectorial flux of lactate is fostered by differential expression of transporters and enzymes.

Astrocytes express the monocarboxylate transporter 4 (MCT4) lactate-permeable ion channels and lactate dehydrogenase 5 (LDH5) all of which promote lactate export (Pierre and Pellerin 2005; Sotelo-Hitschfeld et al. 2015; Karagiannis et al. 2016; Contreras-Baeza et al. 2019), whereas neurons express MCT2 and LDH1, which promote lactate import (Aubert et al. 2005; Barros and Deitmer 2010). MCT2 expression maps local glucose consumption and the expression of genes involved in local K^+ dynamics across the brain (Medel et al. 2022) and its inhibition disrupts neurovascular coupling (Roumes et al. 2021). As discussed below, glucose uptake, glycolytic rate, and lactate production by astrocytes are all sensitive to neuronal activity and thus the rate of neuronal ATP consumption.

In contrast to synapses, long axons are not in ample contact with astrocytes, and they are often myelinated. On the one hand, myelination permits efficient signal transduction, but on the other hand, it blocks axonal access to the interstitial space and thus to circulation-derived metabolites. Axons and oligodendrocytes have been shown to have a similar metabolic relationship as synapses and astrocytes (Fünfschilling et al. 2012; Lee et al. 2012; Saab et al. 2016).

The metabolic division between glial cells and neurons has long been thought to be an adaptation to the highly complex mammalian nervous system. In recent years, however, increasing evidence suggests that metabolic specialization of glial cells and neurons is a basic mechanism of nervous system function, since it is conserved from insects to man (Volkenhoff et al. 2015; Delgado et al. 2018; González-Gutiérrez et al. 2020; reviewed in Rittschof and Schirmeier 2018). In insects, glial cells are glycolytic and produce lactate and alanine that are shuttled to neurons (Rittschof and Schirmeier 2017; Rabah et al. 2023). Lactate supply is essential for neuronal function as lack of glial glycolysis induces severe neurodegeneration and premature death (Volkenhoff et al. 2015). Remarkably, glycolysis in insect neurons is largely dispensable, even though neurons take up glucose and can likely metabolize it via glycolysis and the phosphogluconate pathway (PPP) (Volkenhoff et al. 2015, 2018).

Cite this article as *Cold Spring Harb Perspect Biol* doi: 10.1101/cshperspect.a041355

COMPARTMENTALIZATION AND ENERGY RESERVOIRS

The BBB provides protection against circulating toxins (Obermeier et al. 2013), behavioral stability in the face of starvation and disease, and the possibility of metabolic specialization, for example, co-option of the amino acid glutamate for the purposes of neurotransmission. A necessary trade-off is that complex and energetically expensive chemical reactions need to be carried out "in-house." Neurons are deficient in several metabolic pathways, which are correspondingly stronger in astrocytes, including the production of building blocks for biosynthesis (Yu et al. 1983; Herrero-Mendez et al. 2009) antioxidation (Schmidt and Dringen 2012), and waste disposal (Bak et al. 2006; Bélanger et al. 2011). The metabolic reactions in astrocytes are also present in other cell types of the body. The uniqueness of astrocytic metabolism stems from its intimate and heavily biased relationship with the super-specialized neuron, in a context of relative insulation from circulation.

Metabolic processes occurring within neurons and astrocytes are distributed between membrane compartments and are undertaken by enzymes and transporters. Enzymes transform molecules while transporters move them between compartments. As the control of flux is distributed throughout multiple nodes of the metabolic network, specific enzymes or transporters are no longer considered to be rate limiting. Mitochondria, the endoplasmic reticulum, and other membrane-bound organelles host specific reactions, integrated with the rest of the metabolic network through exchange with the cytosol. The nucleus is well connected to the cytosol, behaving as a metabolic buffer.

METABOLISM DEPENDENT ON NEURONAL ACTIVITY

Glucose enters the brain parenchyma via endothelial GLUT1, a facilitative transporter whose commanding role is underscored by the neurological manifestations of GLUT1 haploinsufficiency (Wang et al. 2015). What happens to the sugar beyond the endothelium is still unclear.

Electron microscopy of chemically fixed tissue showed that capillaries are fully enwrapped by astrocytic endfeet (Mathiisen et al. 2010), suggesting astrocytes would control the flux of glucose to neurons. Alternatively, cryofixating brain tissue, considered a less invasive method, showed only partial coverage (Korogod et al. 2015), suggesting the brain interstice is a single well-mixed compartment that feeds all parenchymal cells on equal terms.

According to biophysical considerations, neural activity imposes comparable metabolic demands on both astrocytes and neurons, in line with their similar mitochondrial endowment and TCA cycle fluxes (Attwell and Laughlin 2001; Harris et al. 2012; Barros 2022). The unexpected discovery of energy-inefficient lactate production despite oxygen availability (i.e., aerobic glycolysis) showed that activated brain metabolism is not only greater but is different (Fox et al. 1988; Prichard et al. 1991; Hu and Wilson 1997). Astrocytes have been found to play a central role in aerobic glycolysis. Extracellular K^+ is a major mediator between excitatory neuronal activity and astrocytic energy metabolism. This cation is released by postsynaptic neurons, amplifying the presynaptic release of glutamate by a factor of 100. Upon reaching astrocytic processes (Rasmussen et al. 2019; Armbruster et al. 2022), K^+ stimulates their glucose transport and consumption (Bittner et al. 2011; Fernández-Moncada et al. 2021), the latter comediated by the Na^+-bicarbonate transporter NBCe1 and the $\alpha2\beta2$ Na^+/K^+ ATPase (Ruminot et al. 2011, 2019; Köhler et al. 2018). Activation of glycolysis produces an ATP surplus and acute inhibition of astrocytic oxygen consumption, the so-called crabtree effect (Fernández-Moncada et al. 2018). At the same time, astrocytes release lactate through a voltage-sensitive anion channel (Sotelo-Hitschfeld et al. 2015; Zuend et al. 2020), diminishing its tonic hold of glycolysis (Sotelo-Hitschfeld et al. 2012). Glutamate, NH_4^+ and nitric oxide are also capable of modulating astrocytic metabolism in an acute manner. Glutamate activates GLUT1 and has a delayed stimulatory effect on glycolysis that evolves over minutes (Pellerin and Magistretti 1994; Loaiza et al. 2003; Bittner et al. 2011), the occurrence that gave birth to the astrocyte-to-neuron lactate shuttle hypothesis (Pellerin and Magistretti 1994; Pellerin et al. 2007; Magistretti and Allaman 2018). NH_4^+ and nitric oxide veer glycolytic pyruvate away from mitochondria into lactate (Lerchundi et al. 2015; San Martín et al. 2017). The net result of these quick events is that more oxygen and lactate are made available to the active brain area (Zuend et al. 2020; Hosford et al. 2022; Barros et al. 2023). Aerobic glycolysis in astrocytes also promotes the production of D-serine, an NMDA receptor co-agonist that is defective in mice with Alzheimer's disease (Le Douce et al. 2020).

Astrocytes can store glucose in the form of glycogen, which is converted into lactate during memory processing, exercise, hypoglycemia, and ischemia (Dringen et al. 1993; Gibbs et al. 2006; Newman et al. 2011; Suzuki et al. 2011; Oe et al. 2016; Matsui et al. 2017; Waitt et al. 2017). Several neuronal signals might mobilize glycogen, including noradrenaline, adenosine, and vasoactive intestinal peptide (VIP). Extracellular K^+ has been proposed to mobilize glycogen via the soluble adenylyl cyclase (sAC) (Choi et al. 2012), a mechanism that awaits confirmation (Theparambil et al. 2016; Horvat et al. 2021; Jakobsen et al. 2021). A fraction of the glucose captured by astrocytes may go through glycogen before becoming pyruvate and lactate, a phenomenon termed the "glycogen shunt" (Shulman et al. 2001; Walls et al. 2009).

Meanwhile, active neurons increase their energy consumption, chiefly at the $\alpha3\beta1$ Na^+/K^+ ATPase (Harris et al. 2012; Baeza-Lehnert et al. 2019). The identity of the substrate that fuels active neurons (i.e., glucose versus astrocytic lactate) is an ongoing debate (Bak and Walls 2018; Barros and Weber 2018; Magistretti and Allaman 2018; Dienel 2019), informed by multiple technical approaches, including an expanding armamentarium of genetically encoded sensors (Barros et al. 2018; Koveal et al. 2022; San Martín et al. 2022). Genetically encoded indicators have become the gold standard to measure calcium transients (e.g., using GCaMPs). Measuring intracellular metabolite concentrations is far more challenging, because small and slow changes imposed on high baseline levels need to be detected. Whenever possible, intensiometric or fluorescent lifetime signals should be converted into molar metabolite concentrations. Two recent in vitro studies were based on genetically encoded metabolite sensors. The first, explored the initial few seconds after neurotransmission in hippocampal granule cells in acute slices and showed a transient increase in cytosolic $NADH/NAD^+$, pointing to a transient mismatch between glycolysis and mitochondrial metabolism. Still, these cells did not appear to release lactate (Díaz-García et al. 2017). In the second study, pyramidal cells electrically stimulated in culture showed similar degrees of activation of glucose consumption and mitochondrial pyruvate consumption, without apparent changes in the intracellular levels of lactate or pyruvate (Baeza-Lehnert et al. 2019). Whether or not neurons produce lactate and hence contribute to aerobic glycolysis under physiological conditions remains to be clarified. Also unclear is the extent to which neurons are energized by glucose versus lactate, how much of the glucose is diverted through the PPP (Herrero-Mendez et al. 2009), and whether there are different fueling strategies for neuronal subtypes and across brain regions. In this respect, juvenile neurons subjected to memory tasks were found to rely more on glucose than adult neurons, an observation that helps to reconcile ostensibly conflicting observations (Cruz et al. 2022). Comprehensive reviews of the energetics of neurotransmission are available (Magistretti and Allaman 2018; Yellen 2018; Dienel 2019; Bonvento and Bolaños 2021; Barros et al. 2023).

SUPPLY OF BUILDING BLOCKS

The brain responds to developmental and environmental cues with structural changes that underlie performance, ranging from synaptic growth to cell proliferation. The new structures are made of amino acids, sugars, lipids, nucleotides, and cofactors, most of which are generated locally de novo, either from glycolytic intermediates or from TCA cycle intermediates. While acute aerobic glycolysis caters for the urgent energy demands of neurotransmission (see above), persistent aerobic glycolysis defines the conditions for the generation of building blocks for tissue plasticity, akin to the Warburg effect occurring in tumors, inflammation, wound repair, and other proliferative conditions (Warburg 1925; Vander Heiden et al. 2009; Goyal et al. 2014; Russell et al. 2019). Standing brain tissue aerobic glycolysis peaks during early childhood, declines with aging (Goyal et al. 2017), and might be protective against Alzheimer's disease (Goyal et al. 2023). As well as its involvement in D-serine generation for glutamatergic signaling (Le Douce et al. 2020), the phosphorylated pathway that branches off astrocytic glycolysis is required to generate glycine, cysteine, phosphoglycerides, sphingolipids, phosphatidylserine, and methylenetetrahydrofolate for neurons, precursors that may be in short supply when aerobic glycolysis is defective. Building block synthesis requires continuous replenishment of TCA cycle intermediates (i.e., anaplerosis), a process occurring mostly in astrocytes through the carboxylation of pyruvate (Yu et al. 1983) and only to a lesser extent in neurons, either from astrocytic glutamine or directly from extracellular glutamate (Divakaruni et al. 2017). Cholesterol is another component of neurons that is synthetized in astrocytes (Pfrieger and Ungerer 2011; Ferris et al. 2017). A striking example of the importance of glial cell metabolism for neurons is that a single amino acid substitution in the PPP enzyme transketolase, involved in glial lipid synthesis, is the main reason why the frontal neocortex of the modern human brain has more neurons than that of Neanderthals (Pinson et al. 2022).

WASTE RECYCLING

Neuronal function produces waste products that must be recycled to avoid toxicity and/or replenish precursor pools. Mitochondrial respiration generates CO_2 and reactive oxygen species (ROS). Since CO_2 is a small gas, it can diffuse freely into the blood, to be excreted by the lungs. ROS are highly reactive compounds that need to be detoxified fast. Glutathione (GSH) and ascorbate are the most important scavengers of ROS in neurons (Harrison and May 2009; Schmidt and Dringen 2012). Glutathione is oxidized in the detoxification process of ROS, xenobiotics, and endogenous toxins. Most oxidized glutathione is reduced in situ to replenish the GSH pool, but a fraction of this is lost via multidrug resistance transporters and has to be replenished by de

novo synthesis. Neuronal GSH synthesis requires cysteine supplied by astrocytes (Schmidt and Dringen 2012). Dehydroascorbic acid (DHA), the oxidized form of ascorbate is reduced locally by reactions consuming NAD(P) H. However, some DHA, which is toxic, is transported into astrocytes via glucose transporters and recycled to ascorbate through GSH or enzymatic reactions (Nualart et al. 2003). Ascorbate is then shuttled back to neurons in an activity-dependent manner (Siushansian et al. 1996; Harrison and May 2009).

The activity-dependent production of ROS in neurons leads to the formation of peroxidated lipids. These lipids are transported into astrocytes in an ApoE- or ApoD- (in insects) dependent manner and stored within lipid droplets (LDs) to prevent cytotoxicity (Liu et al. 2015, 2017; Ioannou et al. 2019; Smolič et al. 2021; Yin et al. 2021). Astrocytes can then use for energy production via β-oxidation. In a very similar manner, excess lactate derived from glial cells is recycled and used in neurons to produce acetyl-CoA, which in turn allows for production of free fatty acids that are shuttled back to the glial cells, where they can likewise be stored in LDs for use in energy production (Liu et al. 2017). Such toxic fatty acids are mainly produced in hyperactive neurons, in which ROS production is also elevated. Thus fatty acid/lipid transfer from neurons to glial cells is a mechanism that allows neurons to deal with the cytotoxic effects of lipid peroxidation and fatty acid production that occur when neurons are highly activity (Ioannou et al. 2019; Smolič et al. 2021; Yin et al. 2021). It is worth highlighting that several risk genes for Alzheimer's disease have been linked to neuron–glia lipid shuttling and glial lipid metabolism (Moulton et al. 2021). Thus, carefully regulated neural lipid homeostasis seems to play an essential role in preventing neurodegenerative phenotypes and its deregulation contributes to Alzheimer's disease progression and likely the advance of other neurodegenerative diseases (Di Paolo and Kim 2011; Reed 2011; Kunkle et al. 2019; Lin et al. 2019; Chung et al. 2020; Yang et al. 2022).

Other waste products of neuronal activity are ammonia and K^+, which are both recycled via astrocytes. Ammonia (NH_3) is formed during the glutamate and GABA cycles and when glutamine is used for anaplerosis. NH_3 captures a proton in the neuronal cytosol forming ammonium (NH_4^+). As neurons lack glutamine synthase, they cannot process this nitrogen excess, which is shuttled to astrocytes as NH_4^+, NH_3, or amino acids (Bak et al. 2006; Cooper 2012; Rothman et al. 2012). The NH_4^+ that enters astrocytes via K^+ channels and transporters is recycled to glutamine, which is ferried back to neurons (Nagaraja and Brookes 1998; Kelly and Rose 2010). Metabolism also generates other toxins, such as methylglyoxal, a by-product of glycolysis that promotes the formation of advanced glycation end products that leads to slowly progressing cell degeneration. Due to their high glycolytic rates, astrocyte produce rather high amounts of methylglyoxal and thus express a robust glyoxalase system that protects both themselves and neurons against methylglyoxal toxicity (Bélanger et al. 2011).

ASTROCYTES AS METABOLIC SENSORS AND KEY REGULATORS OF SYSTEMIC METABOLISM

As summarized above, astrocytes play multiple roles in provision of metabolites to neurons and in maintaining metabolic homeostasis in the nervous system. In has recently become apparent that astrocytes participate in nutrient sensing, regulating systemic metabolism and behavior (see below). By expressing diverse metabolic and neurotransmitter receptors/transporters, astrocytes are capable of sensing and responding to metabolic and synaptic cues (Perea et al. 2009; García-Cáceres et al. 2019). The role of astrocytic Ca^{2+} responses and gliotransmission in regulating local metabolism and synapse physiology is well known (Araque et al. 2014; Verkhratsky and Nedergaard 2018; Schaeffer and Iadecola 2021). But exploration of astrocyte impact on systemic metabolism is only just beginning.

Astrocytes closely monitor glucose concentrations in the nervous system. By expressing glucose transporters of different affinities, astrocytes are able to monitor a wide range of glucose concentrations (Simpson et al. 2007; Thorens 2015; Koepsell 2020). In the hypothalamus and the hindbrain, the two main areas of central glucose

sensing in the brain, astrocytes have been directly implicated in regulating systemic homeostasis (Donovan and Watts 2014). In the hypothalamus, astrocytic insulin signaling seems to be essential for regulation of systemic glucose handling and transport of glucose into the brain (Guillod-Maximin et al. 2004; García-Cáceres et al. 2016). Elevated glucose levels also induce reduced astrocytic coverage of pro-opiomelanocortin (POMC) neurons that are implicated in feeding behavior. This leads to increased excitatory input onto those neurons (Nuzzaci et al. 2020). In the hindbrain, glucose deprivation has been shown to trigger astrocytic Ca^{2+} responses that precede neuronal Ca^{2+} responses (McDougal et al. 2013; Rogers et al. 2020). Further, purinergic signaling in hindbrain astrocytes has been implicated in regulating the rise in blood glucose levels following hypoglycemia in rats (Rogers et al. 2016, 2018).

In addition to closely monitoring glucose levels and instructing the adaptation of blood glucose levels, glial lipid metabolism acts as a sensor for the body metabolic status, while glial cells seem to influence systemic lipid and carbohydrate mobilization in response to changes in glial lipid metabolism (Varela et al. 2021; McMullen et al. 2023). Astrocytes might be involved in the neuroprotective effects of short-chain fatty acids produced by gut bacteria from dietary fibers (Cuervo-Zanatta et al. 2023).

BRAIN METABOLISM GOVERNING BEHAVIOR

Neural activity and cognition are quickly compromised by hypoxia, hypoglycemia and ischemia. In want of energy, neurotransmission and action potentials cease and the tissue is brought to an electric standstill that helps preserve cell viability. Chronic energy deprivation may also lead to a compensatory deficit in function, as demonstrated by the inhibitory effect of food scarcity on AMPA currents of cortical neurons, which saves ATP at the cost of coding precision in the visual cortex (Padamsey et al. 2022). Less intuitively, causality also works in the opposite direction. Cellular-resolution imaging of energy metabolism revealed that up-regulation of mushroom body energy flux is both necessary and sufficient to drive long-term memory formation in *Drosophila*, with lactate and alanine shuttling between glial cells and neurons mediating different types of memory (Plaçais et al. 2017; Barros 2023; Rabah et al. 2023), whereas inhibition of OXPHOS mediates the aggressive response of honey bees exposed to pheromones and induces aggression in flies (Li-Byarlay et al. 2014). Also, as mentioned above, the anaplerotic function of a metabolic enzyme has been singled out as a major factor in the evolution of our thick frontal neocortex, which is involved in social behavior (Pinson et al. 2022). These experimental findings support the concept that energy metabolism not only has a permissive role but may determine network behavior and cognition under physiological conditions (Vergara et al. 2019).

CONCLUDING REMARKS

The astrocyte structures the microscopic anatomy of the brain and, by expressing highly regulated metabolic enzymes and transporters, it plays central roles in the supply of building blocks for tissue growth and remodeling, waste recycling, and activity-dependent energy homeostasis. There has been substantial progress over the last decade on the mechanistic understanding of astrocytic metabolism and its regulation, aided by insights from studying invertebrates. The emerging picture is that of a highly plastic cell, which adapts quickly to the demands of neurotransmission, but that also uses metabolic signals to control neuronal output. Challenges ahead are to figure out the additional ways in which metabolism is linked to activity, and whether it is used to regulate neural circuit function, as well as to determine the relative contribution of these mechanisms over various spatiotemporal scales, and their response to disease. Across the brain, astrocytes differ in terms of morphology and gene expression (Zeisel et al. 2018; Batiuk et al. 2020), setting the stage for metabolic diversity. Thus, current general conclusions might be less valid for certain brain regions. The development of methods that allow targeting of specific astrocytic subtypes seems imperative.

Brain metabolism has been assumed to be rather rigid. Recent data, however, suggest that glial metabolism adapts to support neuronal function under adverse conditions (Lavrentyev et al. 2004; Schulz et al. 2015; Ioannou et al. 2019; Weightman Potter et al. 2019; White et al. 2020; Hertenstein et al. 2021; Asadollahi et al. 2022; Silva et al. 2022). Changes in brain metabolism observed during disease are usually perceived as pathological. But are they part of the problem or part of the adaptation? A more thorough understanding of this metabolic flexibility would seem essential to further probe the complexities of brain disease.

Energy metabolism continues to be the main area of metabolic research, but metabolism is a far wider-reaching field. Making inroads into neglected metabolic pathways appears increasingly feasible thanks to progress made with single-cell omics technology, targeted gene manipulation, and high-resolution metabolic imaging. Studies using these techniques will likely reveal important insights into metabolic interactions in the future.

ACKNOWLEDGMENTS

We are grateful to our colleagues for their continuing support and discussions. We thank Dr. Karen Everett for critical reading of the manuscript. This work was partly supported by Fondecyt grant 1230145 and the Swiss National Science Foundation.

REFERENCES

Almeida A, Almeida J, Bolaños JP, Moncada S. 2001. Different responses of astrocytes and neurons to nitric oxide: the role of glycolytically generated ATP in astrocyte protection. *Proc Natl Acad Sci* **98**: 15294–15299. doi:10.1073/pnas.261560998

Almeida A, Moncada S, Bolaños JP. 2004. Nitric oxide switches on glycolysis through the AMP protein kinase and 6-phosphofructo-2-kinase pathway. *Nat Cell Biol* **6**: 45–51. doi:10.1038/ncb1080

Araque A, Parpura V, Sanzgiri RP, Haydon PG. 1999. Tripartite synapses: glia, the unacknowledged partner. *Trends Neurosci* **22**: 208–215. doi:10.1016/S0166-2236(98)01349-6

Araque A, Carmignoto G, Haydon PG, Oliet SHR, Robitaille R, Volterra A. 2014. Gliotransmitters travel in time and space. *Neuron* **81**: 728–739. doi:10.1016/j.neuron.2014.02.007

Arizono M, Inavalli V, Panatier A, Pfeiffer T, Angibaud J, Levet F, Ter Veer MJT, Stobart J, Bellocchio L, Mikoshiba K, et al. 2020. Structural basis of astrocytic Ca^{2+} signals at tripartite synapses. *Nat Commun* **11:** 1906. doi:10.1038/s41467-020-15648-4

Armbruster M, Naskar S, Garcia JP, Sommer M, Kim E, Adam Y, Haydon PG, Boyden ES, Cohen AE, Dulla CG. 2022. Neuronal activity drives pathway-specific depolarization of peripheral astrocyte processes. *Nat Neurosci* **25:** 607–616. doi:10.1038/s41593-022-01049-x

Asadollahi E, Trevisiol A, Saab AS, Looser ZJ, Dibaj P, Kusch K, Ruhwedel T, Möbius W, Jahn O, Baes M, et al. 2022. Myelin lipids as nervous system energy reserves. bioRxiv doi:10.1101/2022.02.24.481621

Aten S, Kiyoshi CM, Arzola EP, Patterson JA, Taylor AT, Du Y, Guiher AM, Philip M, Camacho EG, Mediratta D, et al. 2022. Ultrastructural view of astrocyte arborization, astrocyte–astrocyte and astrocyte–synapse contacts, intracellular vesicle-like structures, and mitochondrial network. *Prog Neurobiol* **213:** 102264. doi:10.1016/j.pneurobio.2022.102264

Attwell D, Laughlin SB. 2001. An energy budget for signaling in the grey matter of the brain. *J Cereb Blood Flow Metab* **21:** 1133–1145. doi:10.1097/00004647-200110000-00001

Aubert A, Costalat R, Magistretti PJ, Pellerin L. 2005. Brain lactate kinetics: modeling evidence for neuronal lactate uptake upon activation. *Proc Natl Acad Sci* **102:** 16448–16453. doi:10.1073/pnas.0505427102

Baeza-Lehnert F, Saab AS, Gutiérrez R, Larenas V, Díaz E, Horn M, Vargas M, Hösli L, Stobart J, Hirrlinger J, et al. 2019. Non-canonical control of neuronal energy status by the Na$^+$ pump. *Cell Metab* **29:** 668–680.e4. doi:10.1016/j.cmet.2018.11.005

Bak LK, Walls AB. 2018. Crosstalk opposing view: lack of evidence supporting an astrocyte-to-neuron lactate shuttle coupling neuronal activity to glucose utilisation in the brain. *J Physiol* **596:** 351–353. doi:10.1113/JP274945

Bak LK, Schousboe A, Waagepetersen HS. 2006. The glutamate/GABA-glutamine cycle: aspects of transport, neurotransmitter homeostasis and ammonia transfer. *J Neurochem* **98:** 641–653. doi:10.1111/j.1471-4159.2006.03913.x

Barros LF. 2022. How expensive is the astrocyte? *J Cereb Blood Flow Metab* **42:** 738–745. doi:10.1177/0271678X221077343

Barros LF. 2023. Glial metabolism checkpoints memory. *Nat Metab* **5:** 1852–1853. doi:10.1038/s42255-023-00886-9

Barros LF, Deitmer JW. 2010. Glucose and lactate supply to the synapse. *Brain Res Rev* **63:** 149–159. doi:10.1016/j.brainresrev.2009.10.002

Barros LF, Weber B. 2018. Crosstalk proposal: an important astrocyte-to-neuron lactate shuttle couples neuronal activity to glucose utilisation in the brain. *J Physiol* **596:** 347–350. doi:10.1113/JP274944

Barros LF, Courjaret R, Jakoby P, Loaiza A, Lohr C, Deitmer JW. 2009. Preferential transport and metabolism of glucose in Bergmann glia over Purkinje cells: a multiphoton study of cerebellar slices. *Glia* **57:** 962–970. doi:10.1002/glia.20820

Barros LF, Bolaños JP, Bonvento G, Bouzier-Sore AK, Brown A, Hirrlinger J, Kasparov S, Kirchhoff F, Murphy AN, Pellerin L, et al. 2018. Current technical approaches to brain energy metabolism. *Glia* **66:** 1138–1159. doi:10.1002/glia.23248

Barros LF, Ruminot I, Sotelo-Hitschfeld T, Lerchundi R, Fernández-Moncada I. 2023. Metabolic recruitment in brain

tissue. *Annu Rev Physiol* **85:** 115–135. doi:10.1146/annurev-physiol-021422-091035

Batiuk MY, Martirosyan A, Wahis J, de Vin F, Marneffe C, Kusserow C, Koeppen J, Viana JF, Oliveira JF, Voet T, et al. 2020. Identification of region-specific astrocyte subtypes at single cell resolution. *Nat Commun* **11:** 1220. doi:10.1038/s41467-019-14198-8

Bélanger M, Yang J, Petit JM, Laroche T, Magistretti PJ, Allaman I. 2011. Role of the glyoxalase system in astrocyte-mediated neuroprotection. *J Neurosci* **31:** 18338–18352. doi:10.1523/JNEUROSCI.1249-11.2011

Bernardinelli Y, Muller D, Nikonenko I. 2014a. Astrocyte-synapse structural plasticity. *Neural Plast* **2014:** 232105. doi:10.1155/2014/232105

Bernardinelli Y, Randall J, Janett E, Nikonenko I, König S, Jones EV, Flores CE, Murai KK, Bochet CG, Holtmaat A, et al. 2014b. Activity-dependent structural plasticity of perisynaptic astrocytic domains promotes excitatory synapse stability. *Curr Biol* **24:** 1679–1688. doi:10.1016/j.cub.2014.06.025

Bittner CX, Loaiza A, Ruminot I, Larenas V, Sotelo-Hitschfeld T, Gutierrez R, Cordova A, Valdebenito R, Frommer WB, Barros LF. 2010. High resolution measurement of the glycolytic rate. *Front Neuroenergetics* **2:** 26. doi:10.3389/fnene.2010.00026

Bittner CX, Valdebenito R, Ruminot I, Loaiza A, Larenas V, Sotelo-Hitschfeld T, Moldenhauer H, San Martín A, Gutiérrez R, Zambrano M, et al. 2011. Fast and reversible stimulation of astrocytic glycolysis by K$^+$ and a delayed and persistent effect of glutamate. *J Neurosci* **31:** 4709–4713. doi:10.1523/JNEUROSCI.5311-10.2011

Bolaños JP, Peuchen S, Heales SJ, Land JM, Clark JB. 1994. Nitric oxide-mediated inhibition of the mitochondrial respiratory chain in cultured astrocytes. *J Neurochem* **63:** 910–916. doi:10.1046/j.1471-4159.1994.63030910.x

Bolaños JP, Almeida A, Moncada S. 2010. Glycolysis: a bioenergetic or a survival pathway? *Trends Biochem Sci* **35:** 145–149. doi:10.1016/j.tibs.2009.10.006

Bonvento G, Bolaños JP. 2021. Astrocyte-neuron metabolic cooperation shapes brain activity. *Cell Metab* **33:** 1546–1564. doi:10.1016/j.cmet.2021.07.006

Bouzier-Sore AK, Voisin P, Canioni P, Magistretti PJ, Pellerin L. 2003. Lactate is a preferential oxidative energy substrate over glucose for neurons in culture. *J Cereb Blood Flow Metab* **23:** 1298–1306. doi:10.1097/01.WCB.0000091761.61714.25

Bouzier-Sore AK, Voisin P, Bouchaud V, Bezancon E, Franconi JM, Pellerin L. 2006. Competition between glucose and lactate as oxidative energy substrates in both neurons and astrocytes: a comparative NMR study. *Eur J Neurosci* **24:** 1687–1694. doi:10.1111/j.1460-9568.2006.05056.x

Bushong EA, Martone ME, Jones YZ, Ellisman MH. 2002. Protoplasmic astrocytes in CA1 stratum radiatum occupy separate anatomical domains. *J Neurosci* **22:** 183–192. doi:10.1523/JNEUROSCI.22-01-00183.2002

Cahoy JD, Emery B, Kaushal A, Foo LC, Zamanian JL, Christopherson KS, Xing Y, Lubischer JL, Krieg PA, Krupenko SA, et al. 2008. A transcriptome database for astrocytes, neurons, and oligodendrocytes: a new resource for understanding brain development and function. *J Neurosci* **28:** 264–278. doi:10.1523/JNEUROSCI.4178-07.2008

Choi HB, Gordon GR, Zhou N, Tai C, Rungta RL, Martinez J, Milner TA, Ryu JK, McLarnon JG, Tresguerres M, et al. 2012. Metabolic communication between astrocytes and neurons via bicarbonate-responsive soluble adenylyl cyclase. *Neuron* 75: 1094–1104. doi:10.1016/j.neuron.2012.08.032

Chung HL, Wangler MF, Marcogliese PC, Jo J, Ravenscroft TA, Zuo Z, Duraine L, Sadeghzadeh S, Li-Kroeger D, Schmidt RE, et al. 2020. Loss- or gain-of-function mutations in ACOX1 cause axonal loss via different mechanisms. *Neuron* 106: 589–606.e6. doi:10.1016/j.neuron.2020.02.021

Contreras-Baeza Y, Sandoval PY, Alarcón R, Galaz A, Cortés-Molina F, Alegría K, Baeza-Lehnert F, Arce-Molina R, Guequén A, Flores CA, et al. 2019. Monocarboxylate transporter 4 (MCT4) is a high affinity transporter capable of exporting lactate in high-lactate environments. *J Biol Chem* 294: 20135–20147. doi:10.1074/jbc.RA119.009093

Cooper AJ. 2012. The role of glutamine synthetase and glutamate dehydrogenase in cerebral ammonia homeostasis. *Neurochem Res* 37: 2439–2455. doi:10.1007/s11064-012-0803-4

Cruz E, Bessières B, Magistretti P, Alberini CM. 2022. Differential role of neuronal glucose and PFKFB3 in memory formation during development. *Glia* 70: 2207–2231. doi:10.1002/glia.24248

Cuervo-Zanatta D, Syeda T, Sánchez-Valle V, Irene-Fierro M, Torres-Aguilar P, Torres-Ramos MA, Shibayama-Salas M, Silva-Olivares A, Noriega LG, Torres N, et al. 2023. Dietary fiber modulates the release of gut bacterial products preventing cognitive decline in an Alzheimer's mouse model. *Cell Mol Neurobiol* 43: 1519–1618. doi:10.1007/s10571-022-01268-7

Delgado MG, Oliva C, López E, Ibacache A, Galaz A, Delgado R, Barros LF, Sierralta J. 2018. Chaski, a novel *Drosophila* lactate/pyruvate transporter required in glia cells for survival under nutritional stress. *Sci Rep* 8: 1186. doi:10.1038/s41598-018-19595-5

Derouiche A, Frotscher M. 2001. Peripheral astrocyte processes: monitoring by selective immunostaining for the actin-binding ERM proteins. *Glia* 36: 330–341. doi:10.1002/glia.1120

Díaz-García CM, Mongeon R, Lahmann C, Koveal D, Zucker H, Yellen G. 2017. Neuronal stimulation triggers neuronal glycolysis and not lactate uptake. *Cell Metab* 26: 361–374.e4. doi:10.1016/j.cmet.2017.06.021

Dick AP, Harik SI, Klip A, Walker DM. 1984. Identification and characterization of the glucose transporter of the blood-brain barrier by cytochalasin B binding and immunological reactivity. *Proc Natl Acad Sci* 81: 7233–7237. doi:10.1073/pnas.81.22.7233

Dienel GA. 2019. Brain glucose metabolism: integration of energetics with function. *Physiol Rev* 99: 949–1045. doi:10.1152/physrev.00062.2017

Di Paolo G, Kim TW. 2011. Linking lipids to Alzheimer's disease: cholesterol and beyond. *Nat Rev Neurosci* 12: 284–296. doi:10.1038/nrn3012

Divakaruni AS, Wallace M, Buren C, Martyniuk K, Andreyev AY, Li E, Fields JA, Cordes T, Reynolds IJ, Bloodgood BL, et al. 2017. Inhibition of the mitochondrial pyruvate carrier protects from excitotoxic neuronal death. *J Cell Biol* 216: 1091–1105. doi:10.1083/jcb.201612067

Donovan CM, Watts AG. 2014. Peripheral and central glucose sensing in hypoglycemic detection. *Physiology (Bethesda)* 29: 314–324. doi:10.1152/physiol.00069.2013

Dringen R, Gebhardt R, Hamprecht B. 1993. Glycogen in astrocytes: possible function as lactate supply for neighboring cells. *Brain Res* 623: 208–214. doi:10.1016/0006-8993(93)91429-V

Düking T, Spieth L, Berghoff SA, Piepkorn L, Schmidke AM, Mitkovski M, Kannaiyan N, Hosang L, Scholz P, Shaib AH, et al. 2022. Ketogenic diet uncovers differential metabolic plasticity of brain cells. *Sci Adv* 8: eabo7639–eabo7639. doi:10.1126/sciadv.abo7639

Enerson BE, Drewes LR. 2006. The rat blood—brain barrier transcriptome. *J Cereb Blood Flow Metab* 26: 959–973. doi:10.1038/sj.jcbfm.9600249

Farrell CL, Pardridge WM. 1991. Blood-brain barrier glucose transporter is asymmetrically distributed on brain capillary endothelial lumenal and ablumenal membranes: an electron microscopic immunogold study. *Proc Natl Acad* 88: 5779–5783. doi:10.1073/pnas.88.13.5779

Fernández-Moncada I, Ruminot I, Robles-Maldonado D, Alegría K, Deitmer JW, Barros LF. 2018. Neuronal control of astrocytic respiration through a variant of the Crabtree effect. *Proc Natl Acad Sci* 115: 1623–1628. doi:10.1073/pnas.1716469115

Fernández-Moncada I, Robles-Maldonado D, Castro P, Alegría K, Epp R, Ruminot I, Barros LF. 2021. Bidirectional astrocytic GLUT1 activation by elevated extracellular K⁺. *Glia* 69: 1012–1021. doi:10.1002/glia.23944

Ferris HA, Perry RJ, Moreira GV, Shulman GI, Horton JD, Kahn CR. 2017. Loss of astrocyte cholesterol synthesis disrupts neuronal function and alters whole-body metabolism. *Proc Natl Acad Sci* 114: 1189–1194. doi:10.1073/pnas.1620506114

Fiebig C, Keiner S, Ebert B, Schäffner I, Jagasia R, Lie DC, Beckervordersandforth R. 2019. Mitochondrial dysfunction in astrocytes impairs the generation of reactive astrocytes and enhances neuronal cell death in the cortex upon photothrombotic lesion. *Front Mol Neurosci* 12: 40. doi:10.3389/fnmol.2019.00040

Fox PT, Raichle ME, Mintun MA, Dence C. 1988. Nonoxidative glucose consumption during focal physiologic neural activity. *Science* 241: 462–464. doi:10.1126/science.3260686

Fünfschilling U, Supplie LM, Mahad D, Boretius S, Saab AS, Edgar J, Brinkmann BG, Kassmann CM, Tzvetanova ID, Möbius W, et al. 2012. Glycolytic oligodendrocytes maintain myelin and long-term axonal integrity. *Nature* 485: 517–521. doi:10.1038/nature11007

García-Cáceres C, Quarta C, Varela L, Gao Y, Gruber T, Legutko B, Jastroch M, Johansson P, Ninkovic J, Yi CX, et al. 2016. Astrocytic insulin signaling couples brain glucose uptake with nutrient availability. *Cell* 166: 867–880. doi:10.1016/j.cell.2016.07.028

García-Cáceres C, Balland E, Prevot V, Luquet S, Woods SC, Koch M, Horvath TL, Yi CX, Chowen JA, Verkhratsky A, et al. 2019. Role of astrocytes, microglia, and tanycytes in brain control of systemic metabolism. *Nat Neurosci* 22: 7–14. doi:10.1038/s41593-018-0286-y

Gerhart DZ, LeVasseur RJ, Broderius MA, Drewes LR. 1989. Glucose transporter localization in brain using light and electron immunocytochemistry. *J Neurosci Res* 22: 464–472. doi:10.1002/jnr.490220413

Gibbs ME, Anderson DG, Hertz L. 2006. Inhibition of glycogenolysis in astrocytes interrupts memory consolidation in young chickens. *Glia* 54: 214–222. doi:10.1002/glia.20377

González-Gutiérrez A, Ibacache A, Esparza A, Barros LF, Sierralta J. 2020. Neuronal lactate levels depend on glia-derived lactate during high brain activity in *Drosophila*. *Glia* 68: 1213–1227. doi:10.1002/glia.23772

Goyal MS, Hawrylycz M, Miller JA, Snyder AZ, Raichle ME. 2014. Aerobic glycolysis in the human brain is associated with development and neotenous gene expression. *Cell Metab* 19: 49–57. doi:10.1016/j.cmet.2013.11.020

Goyal MS, Vlassenko AG, Blazey TM, Su Y, Couture LE, Durbin TJ, Bateman RJ, Benzinger TL, Morris JC, Raichle ME. 2017. Loss of brain aerobic glycolysis in normal human aging. *Cell Metab* 26: 353–360.e3. doi:10.1016/j.cmet.2017.07.010

Goyal MS, Blazey T, Metcalf NV, McAvoy MP, Strain J, Rahmani M, Durbin TJ, Xiong C, Bezinger TLS, Morris JC, et al. 2023. Brain aerobic glycolysis and resilience in Alzheimer disease. *Proc Natl Acad Sci* 120: e2212256120. doi:10.1073/pnas.2212256120

Grosche J, Matyash V, Möller T, Verkhratsky A, Reichenbach A, Kettenmann H. 1999. Microdomains for neuron-glia interaction: parallel fiber signaling to Bergmann glial cells. *Nat Neurosci* 2: 139–143. doi:10.1038/5692

Guillod-Maximin E, Lorsignol A, Alquier T, Pénicaud L. 2004. Acute intracarotid glucose injection towards the brain induces specific c-fos activation in hypothalamic nuclei: involvement of astrocytes in cerebral glucose-sensing in rats. *J Neuroendocrinol* 16: 464–471. doi:10.1111/j.1365-2826.2004.01185.x

Halassa MM, Haydon PG. 2010. Integrated brain circuits: astrocytic networks modulate neuronal activity and behavior. *Annu Rev Physiol* 72: 335–355. doi:10.1146/annurev-physiol-021909-135843

Halassa MM, Fellin T, Takano H, Dong JH, Haydon PG. 2007. Synaptic islands defined by the territory of a single astrocyte. *J Neurosci* 27: 6473–6477. doi:10.1523/JNEUROSCI.1419-07.2007

Halim ND, McFate T, Mohyeldin A, Okagaki P, Korotchkina LG, Patel MS, Jeoung NH, Harris RA, Schell MJ, Verma A. 2010. Phosphorylation status of pyruvate dehydrogenase distinguishes metabolic phenotypes of cultured rat brain astrocytes and neurons. *Glia* 58: 1168–1176. doi:10.1002/glia.20996

Harik SI, Kalaria RN, Andersson L, Lundahl P, Perry G. 1990. Immunocytochemical localization of the erythroid glucose transporter: abundance in tissues with barrier functions. *J Neurosci* 10: 3862–3872. doi:10.1523/JNEUROSCI.10-12-03862.1990

Harris JJ, Jolivet R, Attwell D. 2012. Synaptic energy use and supply. *Neuron* 75: 762–777. doi:10.1016/j.neuron.2012.08.019

Harrison FE, May JM. 2009. Vitamin C function in the brain: vital role of the ascorbate transporter SVCT2. *Free Radic Biol Med* 46: 719–730. doi:10.1016/j.freeradbiomed.2008.12.018

Herrero-Mendez A, Almeida A, Fernández E, Maestre C, Moncada S, Bolaños JP. 2009. The bioenergetic and antioxidant status of neurons is controlled by continuous degradation of a key glycolytic enzyme by APC/C–Cdh1. *Nat Cell Biol* 11: 747–752. doi:10.1038/ncb1881

Hertenstein H, McMullen E, Weiler A, Volkenhoff A, Becker HM, Schirmeier S. 2021. Starvation-induced regulation of carbohydrate transport at the blood–brain barrier is TGF-β-signaling dependent. *eLife* 10: e62503–e62503. doi:10.7554/eLife.62503

Horvat A, Muhič M, Smolič T, Begić E, Zorec R, Kreft M, Vardjan N. 2021. Ca^{2+} as the prime trigger of aerobic glycolysis in astrocytes. *Cell Calcium* 95: 102368. doi:10.1016/j.ceca.2021.102368

Hosford PS, Wells JA, Nizari S, Christie IN, Theparambil SM, Castro PA, Hadjihambi A, Barros LF, Ruminot I, Lythgoe MF, Gourine AV. 2022. CO$_2$ signalling mediates neurovascular coupling in the cerebral cortex. *Nat Commun* 13: 2125. doi:10.1038/s41467-022-29622-9

Hösli L, Zuend M, Bredell G, Zanker HS, Porto de Oliveira CE, Saab AS, Weber B. 2022. Direct vascular contact is a hallmark of cerebral astrocytes. *Cell Rep* 39: 110599. doi:10.1016/j.celrep.2022.110599

Hu Y, Wilson GS. 1997. A temporary local energy pool coupled to neuronal activity: fluctuations of extracellular lactate levels in rat brain monitored with rapid-response enzyme-based sensor. *J Neurochem* 69: 1484–1490. doi:10.1046/j.1471-4159.1997.69041484.x

Hyder F, Patel AB, Gjedde A, Rothman DL, Behar KL, Shulman RG. 2006. Neuronal–glial glucose oxidation and glutamatergic–GABAergic function. *J Cereb Blood Flow Metab* 26: 865–877. doi:10.1038/sj.jcbfm.9600263

Ioannou MS, Jackson J, Sheu SH, Chang CL, Weigel AV, Liu H, Pasolli HA, Xu CS, Pang S, Matthies D, et al. 2019. Neuron-astrocyte metabolic coupling protects against activity-induced fatty acid toxicity. *Cell* 177: 1522–1535.e14. doi:10.1016/j.cell.2019.04.001

Jakobsen E, Andersen JV, Christensen SK, Siamka O, Larsen MR, Waagepetersen HS, Aldana BI, Bak LK. 2021. Pharmacological inhibition of mitochondrial soluble adenylyl cyclase in astrocytes causes activation of AMP-activated protein kinase and induces breakdown of glycogen. *Glia* 69: 2828–2844. doi:10.1002/glia.24072

Jakoby P, Schmidt E, Ruminot I, Gutierrez R, Barros LF, Deitmer JW. 2014. Higher transport and metabolism of glucose in astrocytes compared with neurons: a multiphoton study of hippocampal and cerebellar tissue slices. *Cereb Cortex* 24: 222–231. doi:10.1093/cercor/bhs309

Karagiannis A, Sylantyev S, Hadjihambi A, Hosford PS, Kasparov S, Gourine AV. 2016. Hemichannel-mediated release of lactate. *J Cereb Blood Flow Metab* 36: 1202–1211. doi:10.1177/0271678X15611912

Kelly T, Rose CR. 2010. Ammonium influx pathways into astrocytes and neurones of hippocampal slices. *J Neurochem* 115: 1123–1136. doi:10.1111/j.1471-4159.2010.07009.x

Kikuchi T, Gonzalez-Soriano J, Kastanauskaite A, Benavides-Piccione R, Merchan-Perez A, DeFelipe J, Blazquez-Llorca L. 2020. Volume electron microscopy study of the relationship between synapses and astrocytes in the developing rat somatosensory cortex. *Cereb Cortex* 30: 3800–3819. doi:10.1093/cercor/bhz343

Cite this article as *Cold Spring Harb Perspect Biol* doi: 10.1101/cshperspect.a041355

Koepsell H. 2020. Glucose transporters in brain in health and disease. *Pflugers Arch* **472:** 1299–1343. doi:10.1007/s00424-020-02441-x

Köhler S, Winkler U, Sicker M, Hirrlinger J. 2018. NBCe1 mediates the regulation of the NADH/NAD$^+$ redox state in cortical astrocytes by neuronal signals. *Glia* **66:** 2233–2245. doi:10.1002/glia.23504

Korogod N, Petersen CC, Knott GW. 2015. Ultrastructural analysis of adult mouse neocortex comparing aldehyde perfusion with cryo fixation. *eLife* **4:** e05793. doi:10.7554/eLife.05793

Koveal D, Rosen PC, Meyer DJ, Díaz-García CM, Wang Y, Cai LH, Chou PJ, Weitz DA, Yellen G. 2022. A high-throughput multiparameter screen for accelerated development and optimization of soluble genetically encoded fluorescent biosensors. *Nat Commun* **13:** 2919. doi:10.1038/s41467-022-30685-x

Kugler EC, Greenwood J, MacDonald RB. 2021. The "neuroglial-vascular" unit: the role of glia in neurovascular unit formation and dysfunction. *Front Cell Dev Biol* **9:** 732820. doi:10.3389/fcell.2021.732820

Kumagai AK, Kang YS, Boado RJ, Pardridge WM. 1995. Upregulation of blood-brain barrier GLUT1 glucose transporter protein and mRNA in experimental chronic hypoglycemia. *Diabetes* **44:** 1399–1404. doi:10.2337/diab.44.12.1399

Kunkle BW, Grenier-Boley B, Sims R, Bis JC, Damotte V, Naj AC, Boland A, Vronskaya M, van der Lee SJ, Amlie-Wolf A, et al. 2019. Genetic meta-analysis of diagnosed Alzheimer's disease identifies new risk loci and implicates Aβ, tau, immunity and lipid processing. *Nat Genet* **51:** 414–430. doi:10.1038/s41588-019-0358-2

Lavrentyev EN, Matta SG, Cook GA. 2004. Expression of three carnitine palmitoyltransferase-I isoforms in 10 regions of the rat brain during feeding, fasting, and diabetes. *Biochem Biophys Res Commun* **315:** 174–178. doi:10.1016/j.bbrc.2004.01.040

Le Douce J, Maugard M, Veran J, Matos M, Jégo P, Vigneron PA, Faivre E, Toussay X, Vandenberghe M, Balbastre Y, et al. 2020. Impairment of glycolysis-derived l-serine production in astrocytes contributes to cognitive deficits in Alzheimer's disease. *Cell Metab* **31:** 503–517.e8. doi:10.1016/j.cmet.2020.02.004

Lee Y, Morrison BM, Li Y, Lengacher S, Farah MH, Hoffman PN, Liu Y, Tsingalia A, Jin L, Zhang PW, et al. 2012. Oligodendroglia metabolically support axons and contribute to neurodegeneration. *Nature* **487:** 443–448. doi:10.1038/nature11314

Lerchundi R, Fernández-Moncada I, Contreras-Baeza Y, Sotelo-Hitschfeld T, Mächler P, Wyss MT, Stobart J, Baeza-Lehnert F, Alegría K, Weber B, et al. 2015. NH4$^+$ triggers the release of astrocytic lactate via mitochondrial pyruvate shunting. *Proc Natl Acad Sci* **112:** 11090–11095. doi:10.1073/pnas.1508259112

Li-Byarlay H, Rittschof CC, Massey JH, Pittendrigh BR, Robinson GE. 2014. Socially responsive effects of brain oxidative metabolism on aggression. *Proc Natl Acad Sci* **111:** 12533–12537. doi:10.1073/pnas.1412306111

Lin G, Wang L, Marcogliese PC, Bellen HJ. 2019. Sphingolipids in the pathogenesis of Parkinson's disease and Parkinsonism. *Trends Endocrinol Metab* **30:** 106–117. doi:10.1016/j.tem.2018.11.003

Liu L, Zhang K, Sandoval H, Yamamoto S, Jaiswal M, Sanz E, Li Z, Hui J, Graham BH, Quintana A, et al. 2015. Glial lipid droplets and ROS induced by mitochondrial defects promote neurodegeneration. *Cell* **160:** 177–190. doi:10.1016/j.cell.2014.12.019

Liu L, MacKenzie KR, Putluri N, Maletić-Savatić M, Bellen HJ. 2017. The glia-neuron lactate shuttle and elevated ros promote lipid synthesis in neurons and lipid droplet accumulation in glia via APOE/D. *Cell Metab* **26:** 719–737.e6. doi:10.1016/j.cmet.2017.08.024

Loaiza A, Porras OH, Barros LF. 2003. Glutamate triggers rapid glucose transport stimulation in astrocytes as evidenced by real-time confocal microscopy. *J Neurosci* **23:** 7337–7342. doi:10.1523/JNEUROSCI.23-19-07337.2003

Lovatt D, Sonnewald U, Waagepetersen HS, Schousboe A, He W, Lin JH, Han X, Takano T, Wang S, Sim FJ, et al. 2007. The transcriptome and metabolic gene signature of protoplasmic astrocytes in the adult murine cortex. *J Neurosci* **27:** 12255–12266. doi:10.1523/JNEUROSCI.3404-07.2007

Mächler P, Wyss MT, Elsayed M, Stobart J, Gutierrez R, von Faber-Castell A, Kaelin V, Zuend M, San Martin A, Romero-Gómez I, et al. 2016. In vivo evidence for a lactate gradient from astrocytes to neurons. *Cell Metab* **23:** 94–102. doi:10.1016/j.cmet.2015.10.010

Magistretti PJ, Allaman I. 2018. Lactate in the brain: from metabolic end-product to signalling molecule. *Nat Rev Neurosci* **19:** 235–249. doi:10.1038/nrn.2018.19

Maher F, Davies-Hill TM, Lysko PG, Henneberry RC, Simpson IA. 1991. Expression of two glucose transporters, GLUT1 and GLUT3, in cultured cerebellar neurons: evidence for neuron-specific expression of GLUT3. *Mol Cell Neurosci* **2:** 351–360. doi:10.1016/1044-7431(91)90066-W

Maher F, Vannucci SJ, Simpson IA. 1994. Glucose transporter proteins in brain. *FASEB J* **8:** 1003–1011. doi:10.1096/fasebj.8.13.7926364

Mathiisen TM, Lehre KP, Danbolt NC, Ottersen OP. 2010. The perivascular astroglial sheath provides a complete covering of the brain microvessels: an electron microscopic 3D reconstruction. *Glia* **58:** 1094–1103. doi:10.1002/glia.20990

Matsui T, Omuro H, Liu YF, Soya M, Shima T, McEwen BS, Soya H. 2017. Astrocytic glycogen-derived lactate fuels the brain during exhaustive exercise to maintain endurance capacity. *Proc Natl Acad Sci* **114:** 6358–6363. doi:10.1073/pnas.1702739114

McDougal D, Hermann G, Rogers R. 2013. Astrocytes in the nucleus of the solitary tract are activated by low glucose or glucoprivation: evidence for glial involvement in glucose homeostasis. *Front Neurosci* **7:** 249. doi:10.3389/fnins.2013.00249

McMullen E, Hertenstein H, Strassburger K, Deharde L, Brankatschk M, Schirmeier S. 2023. Glycolytically impaired *Drosophila* glial cells fuel neural metabolism via β-oxidation. *Nat Commun* **14:** 2996. doi:10.1038/s41467-023-38813-x

Medel V, Crossley N, Gajardo I, Muller E, Barros LF, Shine JM, Sierralta J. 2022. Whole-brain neuronal MCT2 lactate transporter expression links metabolism to human brain structure and function. *Proc Natl Acad Sci* **119:** e2204619119. doi:10.1073/pnas.2204619119

Moulton MJ, Barish S, Ralhan I, Chang J, Goodman LD, Harland JG, Marcogliese PC, Johansson JO, Ioannou MS, Bellen HJ. 2021. Neuronal ROS-induced glial lipid droplet formation is altered by loss of Alzheimer's disease–associated genes. *Proc Natl Acad Sci* **118:** e2112095118. doi:10.1073/pnas.2112095118

Nagaraja TN, Brookes N. 1998. Intracellular acidification induced by passive and active transport of ammonium ions in astrocytes. *Am J Physiol* **274:** C883–C891. doi:10.1152/ajpcell.1998.274.4.C883

Newman LA, Korol DL, Gold PE. 2011. Lactate produced by glycogenolysis in astrocytes regulates memory processing. *PLoS ONE* **6:** e28427. doi:10.1371/journal.pone.0028427

Nishizaki T, Matsuoka T. 1998. Low glucose enhances Na$^+$/glucose transport in bovine brain artery endothelial cells. *Stroke* **29:** 844–849. doi:10.1161/01.STR.29.4.844

Nishizaki T, Kammesheidt A, Sumikawa K, Asada T, Okada Y. 1995. A sodium- and energy-dependent glucose transporter with similarities to SGLT1-2 is expressed in bovine cortical vessels. *Neurosci Res* **22:** 13–22. doi:10.1016/0168-0102(95)00876-U

Nualart FJ, Rivas CI, Montecinos VP, Godoy AS, Guaiquil VH, Golde DW, Vera JC. 2003. Recycling of vitamin C by a bystander effect. *J Biol Chem* **278:** 10128–10133. doi:10.1074/jbc.M210686200

Nuzzaci D, Cansell C, Liénard F, Nédélec E, Ben Fradj S, Castel J, Foppen E, Denis R, Grouselle D, Laderrière A, et al. 2020. Postprandial hyperglycemia stimulates neuroglial plasticity in hypothalamic POMC neurons after a balanced meal. *Cell Rep* **30:** 3067–3078.e5. doi:10.1016/j.celrep.2020.02.029

Oberheim NA, Tian GF, Han X, Peng W, Takano T, Ransom B, Nedergaard M. 2008. Loss of astrocytic domain organization in the epileptic brain. *J Neurosci* **28:** 3264–3276. doi:10.1523/JNEUROSCI.4980-07.2008

Obermeier B, Daneman R, Ransohoff RM. 2013. Development, maintenance and disruption of the blood-brain barrier. *Nat Med* **19:** 1584–1596. doi:10.1038/nm.3407

Oe Y, Baba O, Ashida H, Nakamura KC, Hirase H. 2016. Glycogen distribution in the microwave-fixed mouse brain reveals heterogeneous astrocytic patterns. *Glia* **64:** 1532–1545. doi:10.1002/glia.23020

Oliet SH, Piet R, Poulain DA. 2001. Control of glutamate clearance and synaptic efficacy by glial coverage of neurons. *Science* **292:** 923–926. doi:10.1126/science.1059162

Padamsey Z, Katsanevaki D, Dupuy N, Rochefort NL. 2022. Neocortex saves energy by reducing coding precision during food scarcity. *Neuron* **110:** 280–296.e10. doi:10.1016/j.neuron.2021.10.024

Panatier A, Theodosis DT, Mothet JP, Touquet B, Pollegioni L, Poulain DA, Oliet SH. 2006. Glia-derived D-serine controls NMDA receptor activity and synaptic memory. *Cell* **125:** 775–784. doi:10.1016/j.cell.2006.02.051

Pellerin L, Magistretti PJ. 1994. Glutamate uptake into astrocytes stimulates aerobic glycolysis: a mechanism coupling neuronal activity to glucose utilization. *Proc Natl Acad Sci* **91:** 10625–10629. doi:10.1073/pnas.91.22.10625

Pellerin L, Bouzier-Sore AK, Aubert A, Serres S, Merle M, Costalat R, Magistretti PJ. 2007. Activity-dependent regulation of energy metabolism by astrocytes: an update. *Glia* **55:** 1251–1262. doi:10.1002/glia.20528

Perea G, Navarrete M, Araque A. 2009. Tripartite synapses: astrocytes process and control synaptic information. *Trends Neurosci* **32:** 421–431. doi:10.1016/j.tins.2009.05.001

Pfrieger FW, Ungerer N. 2011. Cholesterol metabolism in neurons and astrocytes. *Prog Lipid Res* **50:** 357–371. doi:10.1016/j.plipres.2011.06.002

Pierre K, Pellerin L. 2005. Monocarboxylate transporters in the central nervous system: distribution, regulation and function. *J Neurochem* **94:** 1–14. doi:10.1111/j.1471-4159.2005.03168.x

Pifferi F, Laurent B, Plourde M. 2021. Lipid transport and metabolism at the blood-brain interface: implications in health and disease. *Front Physiol* **12:** 645646. doi:10.3389/fphys.2021.645646

Pinson A, Xing L, Namba T, Kalebic N, Peters J, Oegema CE, Traikov S, Reppe K, Riesenberg S, Maricic T, et al. 2022. Human TKTL1 implies greater neurogenesis in frontal neocortex of modern humans than Neanderthals. *Science* **377:** eabl6422. doi:10.1126/science.abl6422

Plaçais PY, de Tredern E, Scheunemann L, Trannoy S, Goguel V, Han KA, Isabel G, Preat T. 2017. Upregulated energy metabolism in the *Drosophila* mushroom body is the trigger for long-term memory. *Nat Commun* **8:** 15510. doi:10.1038/ncomms15510

Prebil M, Vardjan N, Jensen J, Zorec R, Kreft M. 2011. Dynamic monitoring of cytosolic glucose in single astrocytes. *Glia* **59:** 903–913. doi:10.1002/glia.21161

Prichard J, Rothman D, Novotny E, Petroff O, Kuwabara T, Avison M, Howseman A, Hanstock C, Shulman R. 1991. Lactate rise detected by 1H NMR in human visual cortex during physiologic stimulation. *Proc Natl Acad Sci* **88:** 5829–5831. doi:10.1073/pnas.88.13.5829

Rabah Y, Francés R, Minatchy J, Guédon L, Desnous C, Placais PY, Preat T. 2023. Glycolysis-derived alanine from glia fuels neuronal mitochondria for memory in *Drosophila*. *Nat Metab* **5:** 2002–2019. doi:10.1038/s42255-023-00910-y

Rasmussen R, Nicholas E, Petersen NC, Dietz AG, Xu Q, Sun Q, Nedergaard M. 2019. Cortex-wide changes in extracellular potassium ions parallel brain state transitions in awake behaving mice. *Cell Rep* **28:** 1182–1194.e4. doi:10.1016/j.celrep.2019.06.082

Reed TT. 2011. Lipid peroxidation and neurodegenerative disease. *Free Radic Biol Med* **51:** 1302–1319. doi:10.1016/j.freeradbiomed.2011.06.027

Reichenbach A. 1989. Attempt to classify glial cells by means of their process specialization using the rabbit retinal Müller cell as an example of cytotopographic specialization of glial cells. *Glia* **2:** 250–259. doi:10.1002/glia.440020406

Rittschof CC, Schirmeier S. 2018. Insect models of central nervous system energy metabolism and its links to behavior. *Glia* **66:** 1160–1175. doi:10.1002/glia.23235

Rogers RC, Ritter S, Hermann GE. 2016. Hindbrain cytoglucopenia-induced increases in systemic blood glucose levels by 2-deoxyglucose depend on intact astrocytes and adenosine release. *Am J Physiol Regul Integr Comp Physiol* **310:** R1102–R1108. doi:10.1152/ajpregu.00493.2015

Rogers RC, McDougal DH, Ritter S, Qualls-Creekmore E, Hermann GE. 2018. Response of catecholaminergic neurons in the mouse hindbrain to glucoprivic stimuli is astrocyte dependent. *Am J Physiol Regul Integr Comp Physiol* **315:** R153–R164. doi:10.1152/ajpregu.00368.2017

Rogers RC, Burke SJ, Collier JJ, Ritter S, Hermann GE. 2020. Evidence that hindbrain astrocytes in the rat detect low glucose with a glucose transporter 2-phospholipase C-calcium release mechanism. *Am J Physiol Regul Integr Comp Physiol* **318:** R38–R48. doi:10.1152/ajpregu.00133.2019

Rothman DL, De Feyter HM, Maciejewski PK, Behar KL. 2012. Is there in vivo evidence for amino acid shuttles carrying ammonia from neurons to astrocytes? *Neurochem Res* **37:** 2597–2612. doi:10.1007/s11064-012-0898-7

Roumes H, Jollé C, Blanc J, Benkhaled I, Chatain CP, Massot P, Raffard G, Bouchaud V, Biran M, Pythoud C, et al. 2021. Lactate transporters in the rat barrel cortex sustain whisker-dependent BOLD fMRI signal and behavioral performance. *Proc Natl Acad Sci* **118:** e2112466118. doi:10.1073/pnas.2112466118

Ruminot I, Gutiérrez R, Peña-Münzenmayer G, Añazco C, Sotelo-Hitschfeld T, Lerchundi R, Niemeyer MI, Shull GE, Barros LF. 2011. NBCe1 mediates the acute stimulation of astrocytic glycolysis by extracellular K$^+$. *J Neurosci* **31:** 14264–14271. doi:10.1523/JNEUROSCI.2310-11.2011

Ruminot I, Schmälzle J, Leyton B, Barros LF, Deitmer JW. 2019. Tight coupling of astrocyte energy metabolism to synaptic activity revealed by genetically encoded FRET nanosensors in hippocampal tissue. *J Cereb Blood Flow Metab* **39:** 513–523. doi:10.1177/0271678X17737012

Russell DG, Huang L, VanderVen BC. 2019. Immunometabolism at the interface between macrophages and pathogens. *Nat Rev Immunol* **19:** 291–304. doi:10.1038/s41577-019-0124-9

Saab AS, Tzvetavona ID, Trevisiol A, Baltan S, Dibaj P, Kusch K, Möbius W, Goetze B, Jahn HM, Huang W, et al. 2016. Oligodendroglial NMDA receptors regulate glucose import and axonal energy metabolism. *Neuron* **91:** 119–132. doi:10.1016/j.neuron.2016.05.016

San Martín A, Arce-Molina R, Galaz A, Pérez-Guerra G, Barros LF. 2017. Nanomolar nitric oxide concentrations quickly and reversibly modulate astrocytic energy metabolism. *J Biol Chem* **292:** 9432–9438. doi:10.1074/jbc.M117.777243

San Martín A, Arce-Molina R, Aburto C, Baeza-Lehnert F, Barros LF, Contreras-Baeza Y, Pinilla A, Ruminot I, Rauseo D, Sandoval PY. 2022. Visualizing physiological parameters in cells and tissues using genetically encoded indicators for metabolites. *Free Radic Biol Med* **182:** 34–58. doi:10.1016/j.freeradbiomed.2022.02.012

Schaeffer S, Iadecola C. 2021. Revisiting the neurovascular unit. *Nat Neurosci* **24:** 1198–1209. doi:10.1038/s41593-021-00904-7

Schmidt MM, Dringen R. 2012. GSH synthesis and metabolism. In *Advances in neurobiology: neural metabolism in vivo* (ed. Gruetter R, Choi IY), Vol. 4, pp. 1029–1050. Springer, New York.

Schulz JG, Laranjeira A, Van Huffel L, Gärtner A, Vilain S, Bastianen J, Van Veldhoven PP, Dotti CG. 2015. Glial β-oxidation regulates *Drosophila* energy metabolism. *Sci Rep* **5:** 7805. doi:10.1038/srep07805

Schurr A, West CA, Rigor BM. 1988. Lactate-supported synaptic function in the rat hippocampal slice preparation. *Science* **240:** 1326–1328. doi:10.1126/science.3375817

Shulman RG, Hyder F, Rothman DL. 2001. Cerebral energetics and the glycogen shunt: neurochemical basis of functional imaging. *Proc Natl Acad Sci* **98:** 6417–6422. doi:10.1073/pnas.101129298

Siesjö BK. 1978. *Brain energy metabolism.* John Wiley and Sons, New York.

Silva B, Mantha OL, Schor J, Pascual A, Plaçais PY, Pavlowsky A, Preat T. 2022. Glia fuel neurons with locally synthesized ketone bodies to sustain memory under starvation. *Nat Metab* **4:** 213–224. doi:10.1038/s42255-022-00528-6

Simpson IA, Appel NM, Hokari M, Oki J, Holman GD, Maher F, Koehler-Stec EM, Vannucci SJ, Smith QR. 1999. Blood–brain barrier glucose transporter: effects of hypo- and hyperglycemia revisited. *J Neurochem* **72:** 238–247. doi:10.1046/j.1471-4159.1999.0720238.x

Simpson IA, Vannucci SJ, DeJoseph MR, Hawkins RA. 2001. Glucose transporter asymmetries in the bovine blood-brain barrier. *J Biol Chem* **276:** 12725–12729. doi:10.1074/jbc.M010897200

Simpson IA, Carruthers A, Vannucci SJ. 2007. Supply and demand in cerebral energy metabolism: the role of nutrient transporters. *J Cereb Blood Flow Metab* **27:** 1766–1791. doi:10.1038/sj.jcbfm.9600521

Siushansian R, Dixon SJ, Wilson JX. 1996. Osmotic swelling stimulates ascorbate efflux from cerebral astrocytes. *J Neurochem* **66:** 1227–1233. doi:10.1046/j.1471-4159.1996.66031227.x

Sivitz W, DeSautel S, Walker PS, Pessin JE. 1989. Regulation of the glucose transporter in developing rat brain. *Endocrinol* **124:** 1875–1880. doi:10.1210/endo-124-4-1875

Smolič T, Tavčar P, Horvat A, Černe U, Halužan Vasle A, Tratnjek L, Kreft ME, Scholz N, Matis M, Petan T, et al. 2021. Astrocytes in stress accumulate lipid droplets. *Glia* **69:** 1540–1562. doi:10.1002/glia.23978

Sotelo-Hitschfeld T, Fernández-Moncada I, Barros LF. 2012. Acute feedback control of astrocytic glycolysis by lactate. *Glia* **60:** 674–680. doi:10.1002/glia.22304

Sotelo-Hitschfeld T, Niemeyer MI, Mächler P, Ruminot I, Lerchundi R, Wyss MT, Stobart J, Fernández-Moncada I, Valdebenito R, Garrido-Gerter P, et al. 2015. Channel-mediated lactate release by k$^+$-stimulated astrocytes. *J Neurosci* **35:** 4168–4178. doi:10.1523/JNEUROSCI.5036-14.2015

Supplie LM, Düking T, Campbell G, Diaz F, Moraes CT, Götz M, Hamprecht B, Boretius S, Mahad D, Nave KA. 2017. Respiration-deficient astrocytes survive as glycolytic cells in vivo. *J Neurosci* **37:** 4231–4242. doi:10.1523/JNEUROSCI.0756-16.2017

Suzuki A, Stern SA, Bozdagi O, Huntley GW, Walker RH, Magistretti PJ, Alberini CM. 2011. Astrocyte-neuron lactate transport is required for long-term memory formation. *Cell* **144:** 810–823. doi:10.1016/j.cell.2011.02.018

Theparambil SM, Weber T, Schmälzle J, Ruminot I, Deitmer JW. 2016. Proton fall or bicarbonate rise: glycolytic rate in mouse astrocytes is paved by intracellular alkalinization. *J Biol Chem* **291:** 19108–19117. doi:10.1074/jbc.M116.730143

Thorens B. 2015. GLUT2, glucose sensing and glucose homeostasis. *Diabetologia* **58:** 221–232. doi:10.1007/s00125-014-3451-1

Vander Heiden MG, Cantley LC, Thompson CB. 2009. Understanding the Warburg effect: the metabolic requirements of cell proliferation. *Science* **324:** 1029–1033. doi:10.1126/science.1160809

van de Waterbeemd H, Camenisch G, Folkers G, Chretien JR, Raevsky OA. 1998. Estimation of blood-brain barrier crossing of drugs using molecular size and shape, and H-bonding descriptors. *J Drug Target* **6:** 151–165. doi:10.3109/10611869808997889

Varela L, Kim JG, Fernández-Tussy P, Aryal B, Liu ZW, Fernández-Hernando C, Horvath TL. 2021. Astrocytic lipid metabolism determines susceptibility to diet-induced obesity. *Sci Adv* **7:** eabj2814–eabj2814. doi:10.1126/sciadv.abj2814

Vemula S, Roder KE, Yang T, Bhat GJ, Thekkumkara TJ, Abbruscato TJ. 2009. A functional role for sodium-dependent glucose transport across the blood–brain barrier during oxygen glucose deprivation. *J Pharmacol Exp Ther* **328:** 487–495. doi:10.1124/jpet.108.146589

Ventura R, Harris KM. 1999. Three-dimensional relationships between hippocampal synapses and astrocytes. *J Neurosci* **19:** 6897–6906. doi:10.1523/JNEUROSCI.19-16-06897.1999

Vergara RC, Jaramillo-Riveri S, Luarte A, Moënne-Loccoz C, Fuentes R, Couve A, Maldonado PE. 2019. The energy homeostasis principle: neuronal energy regulation drives local network dynamics generating behavior. *Front Comput Neurosci* **13:** 49. doi:10.3389/fncom.2019.00049

Verkhratsky A, Nedergaard M. 2018. Physiology of astroglia. *Physiol Rev* **98:** 239–389. doi:10.1152/physrev.00042.2016

Volkenhoff A, Weiler A, Letzel M, Stehling M, Klämbt C, Schirmeier S. 2015. Glial glycolysis is essential for neuronal survival in *Drosophila*. *Cell Metab* **22:** 437–447. doi:10.1016/j.cmet.2015.07.006

Volkenhoff A, Hirrlinger J, Kappel JM, Klämbt C, Schirmeier S. 2018. Live imaging using a FRET glucose sensor reveals glucose delivery to all cell types in the *Drosophila* brain. *J Insect Physiol* **106:** 55–64. doi:10.1016/j.jinsphys.2017.07.010

Waitt AE, Reed L, Ransom BR, Brown AM. 2017. Emerging roles for glycogen in the CNS. *Front Mol Neurosci* **10:** 73. doi:10.3389/fnmol.2017.00073

Walls AB, Heimbürger CM, Bouman SD, Schousboe A, Waagepetersen HS. 2009. Robust glycogen shunt activity in astrocytes: effects of glutamatergic and adrenergic agents. *Neuroscience* **158:** 284–292. doi:10.1016/j.neuroscience.2008.09.058

Wang D, Pascual JM, DeVivo DC. 2015. Glucose transporter type 1 deficiency syndrome. In *Generereviews* (ed. Pagon RA, Adam MP, Ardinger HH). University of Washington, Seattle.

Warburg O. 1925. The metabolism of carcinoma cells. *J Cancer Res* **9:** 148–163. doi:10.1158/jcr.1925.148

Weber B, Barros LF. 2015. The astrocyte: powerhouse and recycling center. *Cold Spring Harb Perspect Biol* **7:** a020396. doi:10.1101/cshperspect.a020396

Weightman Potter PG, Vlachaki Walker JM, Robb JL, Chilton JK, Williamson R, Randall AD, Ellacott KLJ, Beall C. 2019. Basal fatty acid oxidation increases after recurrent low glucose in human primary astrocytes. *Diabetologia* **62:** 187–198. doi:10.1007/s00125-018-4744-6

Weiler A, Volkenhoff A, Hertenstein H, Schirmeier S. 2017. Metabolite transport across the mammalian and insect brain diffusion barriers. *Neurobiol Dis* **107:** 15–31. doi:10.1016/j.nbd.2017.02.008

White CJ, Lee J, Choi J, Chu T, Scafidi S, Wolfgang MJ. 2020. Determining the bioenergetic capacity for fatty acid oxidation in the mammalian nervous system. *Mol Cell Biol* **40:** e00037. doi:10.1128/MCB.00037-20

Wyss MT, Jolivet R, Buck A, Magistretti PJ, Weber B. 2011. In vivo evidence for lactate as a neuronal energy source. *J Neurosci* **31:** 7477–7485. doi:10.1523/JNEUROSCI.0415-11.2011

Yang D, Wang X, Zhang L, Fang Y, Zheng Q, Liu X, Yu W, Chen S, Ying J, Hua F. 2022. Lipid metabolism and storage in neuroglia: role in brain development and neurodegenerative diseases. *Cell Biosci* **12:** 106–106. doi:10.1186/s13578-022-00828-0

Yellen G. 2018. Fueling thought: management of glycolysis and oxidative phosphorylation in neuronal metabolism. *J Cell Biol* **217:** 2235–2246. doi:10.1083/jcb.201803152

Yin J, Spillman E, Cheng ES, Short J, Chen Y, Lei J, Gibbs M, Rosenthal JS, Sheng C, Chen YX, et al. 2021. Brain-specific lipoprotein receptors interact with astrocyte derived apolipoprotein and mediate neuron-glia lipid shuttling. *Nat Commun* **12:** 2408–2408. doi:10.1038/s41467-021-22751-7

Yu AC, Drejer J, Hertz L, Schousboe A. 1983. Pyruvate carboxylase activity in primary cultures of astrocytes and neurons. *J Neurochem* **41:** 1484–1487. doi:10.1111/j.1471-4159.1983.tb00849.x

Zeisel A, Hochgerner H, Lönnerberg P, Johnsson A, Memic F, van der Zwan J, Häring M, Braun E, Borm LE, La Manno G, et al. 2018. Molecular architecture of the mouse nervous system. *Cell* **174:** 999–1014.e22. doi:10.1016/j.cell.2018.06.021

Zuend M, Saab AS, Wyss MT, Ferrari KD, Höesli L, Looser ZJ, Stobart JL, Duran J, Guinovart JJ, Barros LF, et al. 2020. Arousal-induced cortical activity triggers lactate release from astrocytes. *Nat Metab* **2:** 179–191. doi:10.1038/s42255-020-0170-4

Cite this article as *Cold Spring Harb Perspect Biol* doi: 10.1101/cshperspect.a041355

Reactive Astrocytes and Emerging Roles in Central Nervous System (CNS) Disorders

Shane A. Liddelow,[1,2,3] Michelle L. Olsen,[4] and Michael V. Sofroniew[5]

[1]Neuroscience Institute; [2]Department of Neuroscience and Physiology; [3]Department of Ophthalmology, NYU School of Medicine, New York, New York 10016, USA

[4]School of Neuroscience, Virginia Tech, Blacksburg, Virginia 24061, USA

[5]Department of Neurobiology, David Geffen School of Medicine, University of California, Los Angeles, California 90095, USA

Correspondence: shane.liddelow@nyulangone.org; molsen1@vt.edu; sofroniew@mednet.ucla.edu

In addition to their many functions in the healthy central nervous system (CNS), astrocytes respond to CNS damage and disease through a process called "reactivity." Recent evidence reveals that astrocyte reactivity is a heterogeneous spectrum of potential changes that occur in a context-specific manner. These changes are determined by diverse signaling events and vary not only with the nature and severity of different CNS insults but also with location in the CNS, genetic predispositions, age, and potentially also with "molecular memory" of previous reactivity events. Astrocyte reactivity can be associated with both essential beneficial functions as well as with harmful effects. The available information is rapidly expanding and much has been learned about molecular diversity of astrocyte reactivity. Emerging functional associations point toward central roles for astrocyte reactivity in determining the outcome in CNS disorders.

Astrocytes exert many essential functions in the healthy central nervous system (CNS) as reviewed and discussed in other articles in this collection and elsewhere (Allen and Eroglu 2017; Verkhratsky and Nedergaard 2018; Khakh and Deneen 2019).[6] In addition, astrocytes respond to all forms of CNS damage, infection, and disease with a variety of potential changes in molecular expression, cellular structure, and function—commonly referred to as astrocyte "reactivity." Emerging evidence demonstrates that astrocytes play important roles in most if not all CNS disorders.

The last decade of astrocyte research has led to exciting findings related to the complexity of astrocyte reactivity. The application of transcriptomics in preclinical models and human brain to understand the diverse astrocyte response to insult and new genetic tools to target astrocytes have shepherded in a new era of astrocyte research. Modern culture and organoid systems have further provided for exciting discoveries

---

[6]This is an update to a previous article published in *Cold Spring Harbor Perspectives in Biology* [Sofroniew (2015). *Cold Spring Harb Perspect Biol* **7**: a020420].

Cite this article as *Cold Spring Harb Perspect Biol* doi: 10.1101/cshperspect.a041356

about function and evolutionary conservation of astrocyte reactivity. An amazing array of tools and data are now available to the budding astrocyte biologist. These advances in technology have brought new researchers to the field, seeded and expanded exciting cross-discipline collaborations that have uncovered a great deal about astrocytes in health and disease. Here, we update a previous report in this series (Sofroniew 2015b) and provide an overview of currently available information about the mechanisms, functions, and impact of astrocyte reactivity, with a particular focus on recent advances in defining the heterogeneity and regulation of astrocyte reactivity and their diverse roles in different CNS disorders.

WHAT ARE REACTIVE ASTROCYTES?

Astrocyte reactivity can now be defined as a spectrum of potential molecular, cellular, and functional changes in astrocytes that occur in response to all forms and severities of pathology in surrounding CNS tissue, as evidenced by numerous studies in both experimental animals and human pathological specimens, as well as in vitro cell-based assays. These pathological insults include microbial infection, ischemia, traumatic injury, autoimmune attack, seizure activity, exposure to environmental toxins, peripheral metabolic disorders, foreign bodies including medical implants, neurodegenerative disease, and neoplastic growth (Sofroniew and Vinters 2010; Pekny et al. 2016; Robel and Sontheimer 2016; Liddelow and Barres 2017; Campbell et al. 2020; O'Shea et al. 2020; Sofroniew 2020; Escartin et al. 2021; Han et al. 2021; Krawczyk et al. 2022). Box 1 summarizes the key features of astrocyte reactivity, which are discussed below. Various terms are sometimes used to refer to astrocyte responses to CNS damage or disease, and their usage may vary among authors. In this article, we will use "astrocyte reactivity" and "reactive astrocytes" as general, all-inclusive descriptors of all forms of astrocyte responses to non-cell-autonomous cues associated with CNS damage or disease. As discussed in more detail later, these terms encompass astrocyte responses

---

BOX 1. WORKING DEFINITION AND KEY FEATURES OF ASTROCYTE REACTIVITY

- Astrocyte reactivity occurs in response to all forms of CNS injury, infection, and disease.
- Astrocyte reactivity is a diverse set of changes to normal physiological function. This is largely defined by measurements of changes in gene expression and protein levels, morphology, and other molecular changes to lipids, metabolites, etc.
- Astrocyte reactivity-related changes can alter astrocyte activities through both loss and gain of functions.
- Astrocyte reactivity-related changes are context dependent and may vary with the type and severity of insult, the time span after an insult, the types and combinations of molecular triggers, the CNS location, and may depend on the heterogeneous nature of individual astrocytes. This may occur within a single cell, or across a population of cells in a specific CNS region.
- Astrocyte reactivity-related changes are regulated in a context-specific manner by combinatorial interactions of diverse inter- and intracellular signaling molecules.
- Astrocyte reactivity-related changes can be homeostatic and adaptive and can influence disorder outcomes in beneficial ways.
- Astrocyte reactivity-related changes have the potential to be maladaptive and influence disorder outcomes in detrimental ways.
- Diseased astrocytes should be differentiated from reactive astrocytes. One should refer to either "diseased" astrocytes (caused by genetic mutations), or to "reactive" astrocytes (responding to an external stimuli).

Cite this article as *Cold Spring Harb Perspect Biol* doi: 10.1101/cshperspect.a041356

of considerable diversity and heterogeneity, and considerable effort is now aimed at identifying different forms of astrocyte reactivity.

ASTROCYTES AS A NEXUS FOR MULTICELLULAR RESPONSES TO CNS INSULTS

While this article focuses on astrocyte reactivity, it is essential to note that astrocyte responses occur in the context of coordinated multicellular responses to CNS insults. Different types of glia, including microglia, astrocytes, NG2-positive oligodendrocyte progenitors, respond most robustly. Non-neural cells intrinsic to the CNS, such as endothelia, perivascular fibroblasts, pericytes, and meningeal cells also respond to insults. Additionally, blood-derived cells and molecules such as leukocytes, platelets, and blood serum proteins, fatty acids, and high concentrations of other serum components like glutamate and K^+

enter the CNS and contribute to these responses. Together, these cells produce intercellular signaling molecules that can influence the activities and functions of different cell types, including astrocytes. Thus, the response to CNS insults is a complex mixture of events involving multiple cell type interactions that change over time. Astrocytes take part in these interactions both by receiving and sending instructive signals that influence other cells (Tables 1 and 2).

HETEROGENEITY OF ASTROCYTES AND REACTIVE ASTROCYTES

The concept of astrocyte heterogeneity is not new. Classical neuroanatomists of the late nineteenth and early twentieth centuries described multiple types of astrocytes based on morphologies that were common across mammals, including the well-known bushy astrocytes in gray matter, fibrous astrocytes in white matter, and glia limitans

Table 1. Examples of extracellular triggers of astrocyte reactivity

Categories	Source	Molecules
Blood and serum proteins and molecules	Bloodstream	Albumin, fibrinogen, thrombin, complement, fatty acids, K^+, glutamate
Cytokines and growth factors	Other glia and local nonneural cells, infiltrating leukocytes, tumor cells	CNTF, C1Q, EGF, EDN1, FGFs, IL-1α, IL-1β, IL-6, IL-10, INF-β, INF-γ, LIF, SHH, TGF, TNF
Damage-associated molecular patterns (DAMPs)	Cell damage	ATP, HMGB1, nitric oxide, reactive oxygen species (ROS)
Degeneration-associated proteins	Neurodegenerative disorders	α-Synuclein, β-amyloid, mutant-Huntingtin, prions, Tau, among others
Environmental toxins	External environment	Amphetamine, herbicides, insecticides, methamphetamine, MDMA, MPTP
Foreign bodies	Medical implants, traumatic injuries	Cationic surfaces
Hypoxia and metabolic stress	Ischemia	Oxygen deprivation, glucose deprivation
Inorganic molecules	Systemic metabolic toxicity (liver failure)	Ammonium (NH_4^+)
Mechanical stretch	Acceleration, blast, compression deceleration	Stretch receptors
Oxidative stress	Ischemia, metabolic stress	H_2O_2, Free radicals, NO
Pathogen-associated molecular patterns (PAMPs)	Microbial infections	Bacterial (e.g., lipopolysaccharide [LPS]), prions, toxoplasma, viral (e.g., dsRNA), yeast (e.g., zymosan)
Transmitters	Local neurons	Glutamate, noradrenalin

Abbreviations for growth factors, cytokines, chemokines, receptors, and transcription factors are per standard nomenclature (Human Gene Compendium, GeneCards, www.genecards.org). See main text for literature references.

Table 2. Examples of effector molecular released by reactive astrocytes

Categories	Molecules
Chemokines	CCL2, CCL5, CCL7, CCL8, CXCL1, CXCL9, CXCL10, CXCL12, CXCL16
Cytokines	CNTF, IFN-γ, IL-1β, IL-6, IL-11, IL-15, LIF, TGF, TNF
Extracellular matrix	BCAN, collagens, MMP3, NCAN, SEMA4A, TIMP1
Growth factors and other proteins	BDNF, BMPs, CRYAB, FGF2, GDNF, GPC4, GPC6, NGF, PTX3, THSP1, VEGFA
Small molecules	ATP, lipids, nitric oxide (NO), prostaglandin E (PGE)
Transmitters	Glutamate, kynurenic acid (KYN), D-serine

Abbreviations are as per standard nomenclature (Human Gene Compendium, GeneCards, www.genecards.org). See main text for literature references.

astrocytes along meninges, as well as others (Verkhratsky and Nedergaard 2018). Recent and ongoing studies are expanding and correlating information about structural, genetic, and functional diversity of astrocytes across the healthy CNS (Khakh and Sofroniew 2015; Srinivasan et al. 2016; Zhang et al. 2016; Chai et al. 2017; Haim and Rowitch 2017; John Lin et al. 2017; Wu et al. 2017; Khakh and Deneen 2019; Batiuk et al. 2020; Falcone et al. 2021; Torres-Ceja and Olsen 2022) and are beginning to identify their developmental origins (Molofsky et al. 2014; Clavreul et al. 2019). In addition, there is a steadily growing interest in characterizing the diversity of astrocyte reactivity and understanding how it is regulated.

Historically, astrocyte reactivity has been treated as a uniform entity, but this is not the case. It is now appreciated to be a spectrum of potential changes, ranging from reversible alterations in gene expression and protein levels, to altered morphology, cell proliferation, and tissue rearrangement. Common features exist across different forms and intensities of astrocyte reactivity, but even these occur along a gradient that can vary immensely. Thus, astrocyte reactivity is complex, context-dependent, and multivariate phenomenon that depends on various factors such as the type, severity, and time frame of insult, the CNS location, and the types and combinations of molecular triggers driving the heterogeneity of individual reactive astrocytes.

Information is gradually accumulating that allows for the subdivision of astrocyte reactivity into different categories, which may exhibit common features or definable differences. It remains unclear to what extent specific substates of astrocyte reactivity are programmatically induced and share common molecular and functional features across multiple types of insults. Experimental measures to characterize differences among reactive astrocytes include discriminating proliferative from nonproliferative reactivity, characterizing differences in molecular signatures identified by transcriptomic and proteomic analyses, and identifying differences in physiological and functional changes as discussed in the following sections.

DISCRIMINATING ASTROCYTE DISEASE FROM ASTROCYTE REACTIVITY

There is an emerging concept that diseased astrocytes contribute to neural dysfunction and degeneration. As discussed in more detail below, genetic mutations or polymorphisms can cause cell-autonomous astrocyte dysfunction, leading to neuronal dysfunction and neurodegeneration. This type of disease-precipitated cell-autonomous astrocyte dysfunction should be distinguished from astrocyte reactivity, which is triggered by external non-cell-autonomous signals generated by dysfunctioning and/or degenerating neural tissue (Box 1). Current molecular markers used to study astrocytes in disorders, such as GFAP, do not readily differentiate between the two. Therefore, it is important to avoid conflating non-cell-autonomously triggered astrocyte reactivity with cell-autonomously-precipitated astrocyte disease based simply on the common expression of certain markers.

NONPROLIFERATIVE AND PROLIFERATIVE ASTROCYTE REACTIVITY

In the healthy adult CNS, astrocytes are postmitotic cells that rarely divide (Wanner et al. 2013).

Cite this article as *Cold Spring Harb Perspect Biol* doi: 10.1101/cshperspect.a041356

Nonproliferative astrocyte reactivity typically occurs in neural tissue that is not overtly damaged and retains its basic tissue architecture such that the reactive astrocytes maintain many of the fundamental features of cell structure, cellular interactions, and basic functions that they have in healthy tissue, together with additional context-specific gain- or loss-of-function changes as discussed in more detail below. They exhibit variable degrees of cellular hypertrophy and reorganization of their processes, but generally retain the discrete, nonoverlapping cellular domains exhibited by astrocytes in healthy gray matter (Bushong et al. 2002; Wilhelmsson et al. 2006; Han et al. 2021).

In response to overt tissue damage, such as stroke, severe trauma, infection, foreign bodies, autoimmune inflammation, neoplasm, or severe neurodegeneration, some substates of reactive astrocytes can re-enter the cell cycle (O'Shea et al. 2020; Sofroniew 2020; Krawczyk et al. 2022). Proliferating reactive astrocytes form borders that separate damage, inflamed, and fibrotic tissue from adjacent viable neural tissue (Sofroniew 2020). Emerging evidence indicates that reactive astrocytes that proliferate exhibit markedly different features and functions compared with reactive astrocytes that do not proliferate, or astrocytes in healthy tissue. Newly proliferated reactive astrocytes form borders that separate damaged, inflamed, and fibrotic tissue from adjacent viable neural tissue (see section on traumatic injury and stroke below). During this border formation, newly proliferated astrocytes adopt new cellular interactions with nonneural cells (Sofroniew 2020; Han et al. 2021; Burda et al. 2022). During this border formation, newly proliferated astrocytes interact with many other CNS cells, adopting new interactions (Sofroniew 2020). Although most border-forming reactive astrocytes in narrow zones immediately abutting tissue lesions are newly proliferated, proliferation drops off rapidly with increased distance from lesions, which are surrounded by large areas of nonproliferative reactive astrocytes as well (Wanner et al. 2013). Loss or attenuation of these newly proliferated reactive astrocytes leads to increased spread of inflammation and serum proteins, increased loss of neural tissue, and decreased func-

tional recovery (Bush et al. 1999; Wanner et al. 2013; Sofroniew 2015a; Frik et al. 2018; Williamson et al. 2021; Burda et al. 2022).

Thus, nonproliferative and proliferative astrocyte reactivity represent two broad categories that are easily differentiated and are associated with molecular and functional differences. Notably, these categories should not be regarded as homogenous as each appears comprised of substates that likely have different functional states.

TRANSCRIPTIONAL AND PROTEOMIC PROFILING OF REACTIVE ASTROCYTES

Transcriptomic investigation of astrocytes in the context of infection, injury, and disease have generated much information about diverse molecular expression profiles of reactive astrocytes. These studies have employed bulk analyses of whole tissue samples, pooled purified astrocytes, or single-cell/nucleus analysis (examples include Haumont et al. 1989; Zamanian et al. 2012; Sirko et al. 2015; Anderson et al. 2016; Boisvert et al. 2018; Clarke et al. 2018; Barbar et al. 2020; Carroll et al. 2020; Pan et al. 2020; Zhou et al. 2020; Diaz-Castro et al. 2021; Hasel et al. 2021; Wei et al. 2021; Burda et al. 2022; Sadick et al. 2022). Multiple online databases are available that provide searchable expression profiles of multiple reactive astrocytes subtypes (Table 3) that have led to the identification of molecular markers up-regulated in different forms of reactive astrocytes (i.e., stroke, traumatic injury, inflammation, and neurodegeneration). We provide a short list of reactive astrocyte genes that are commonly up-regulated in astrocytes across multiple disease/injury conditions (Table 4). Notably, many of these genes are not expressed in astrocytes at baseline but become up-regulated in the context of reactivity and are, in many cases, up-regulated across multiple CNS cell types. At the single-cell/nucleus level, transcriptomic heterogeneity has been identified within an individual insult/disease (or disease model), in addition to similarities across certain insults of some, but not all, astrocyte reactivity substates. Examples of this include similarities in inflammatory reactive astrocytes substates identified by transcriptional analyses that are present following systemic in-

Table 3. Online resources and data sets for probing "omics" changes in reactive astrocyte substates

URL	Data type	Species	Data set variables
Development, aging, different brain regions			
www.brainrnaseq.org	RNA-seq (sorted cells and TRAP)	Mouse, human	Normal, aging
igc1.salk.edu:3838/ astrocyte_aging_transcriptome	RNA-seq (TRAP)	Mouse	Aging
dropviz.org	scRNA-seq	Mouse	Different brain regions
Reactivity and disease			
www.gliaseq.com	RNA-seq, sc/snRNA-seq, scATAC-seq, nanostring, proteomics, lipidomics, metabolomics, spatial transcriptomics	Rodent, human, mESC, hiPSC	Normal, different brain regions, inflammation, disease
www.gliaweb.net[a]	All	All	All
astrocyte.rnaseq.sofroniewlab .neurobio.ucla.edu	RNA-seq	Mouse	Spinal cord injury, inflammation
astrocyternaseq.org	RNA-seq	Mouse	Different brain regions, disease models, inflammation
cellxgene.cziscience.com/ collections	scRNA-seq, scATAC-seq, snmC-seq2, spatial transcriptomics	Mouse, human	Normal (different brain regions), disease
seqseek.ninds.nih.gov	scRNA-seq	Mouse	Normal, spinal cord injury
singlecell.broadinstitute.org/ single_cell	sc/snRNA-seq, scATAC-seq	Mouse, human	Different brain regions, disease,
cells.ucsc.edu	scRNA-seq	many	Development, aging, disease
adsn.ddnetbio.com	scRNA-seq	Human	Disease
ki.se/en/mbb/oligointernode	sc/snRNA-seq, scATAC-seq	Mouse, human	Disease
tr.astrocytereactivity.com/home	snRNA-seq, scATAC-seq	Mouse	Disease, trauma
General CNS data sets			
www.proteinatlas.org	scRNA-seq, proteomics, pathology	Mouse, human	Disease
gtexportal.org/home	scRNA-seq, RNA-seq, QTL, eGTEx	Human	CNS regions, peripheral tissues

Note that many data sets have low capture counts for astrocytes. Additional aggregation/meta-analyses are always recommended. Not all data sets have validated reactive astrocyte data, but many have data from astrocyte states with different physiology (e.g., development, aging, or multiple brain regions/species. See individual websites for publication references. This is by no means an exhaustive list, as more resources are being produced.

[a]gliaweb.net is an umbrella repository for all glia-sequencing efforts across the field. It allows for individual laboratories to link and share their independent data sets with ease.

jection of lipopolysaccharide (LPS) as well as mouse models of AD, MS, and acute stab wound trauma (Hasel et al. 2021; Castranio et al. 2023).

Transcriptomic analysis of reactive astrocytes has also revealed diverse gene expression level changes driven by type of insult, the duration from onset, and distinct CNS region. Yet, how these transcriptomic changes correlate with changes in protein, lipid, or metabolite levels and subsequent alterations in astrocyte function are only beginning to be evaluated. For example, recent advances in proteomic technologies, in-

Table 4. Examples of molecular markers used to identify reactive astrocytes in different contexts

Gene ID	Expressed in astrocytes at baseline	Astrocyte enriched[a]	Protein function	Subcellular location
C3	Yes[a,b,c]	No	Activator of the complement system	Cytosol, secreted
Cd44	Yes[b]	No	Cell-surface glycoprotein	Membrane
Cp	No	No	Metalloprotein	Secreted
Cxcl10	No	No	Chemokine	Secreted
Gfap	Yes[a,b]	Yes[a]	Intermediate filament, structure	Cytoskeleton
Hspb1	Yes[b]	No	Molecular chaperone	Cytoplasm, nucleus
Lcn2	No	No	Lipid transport	Secreted
Lgals3	Yes[b]	No[a]	Cell–cell and cell–matrix interactions	Cytoplasm, membrane, extracellular
Osmr	Yes[b]	No	Cytokine receptor	Membrane
S100b	Yes[b]	No	Ca^{2+}-binding protein	Cytoplasm, nucleus
Serpina3n	Yes[a,b]	No	Serine protease inhibitor	Secreted
Steap4	No	No	Metalloreductase	Golgi
Timp1	No	No	Matrix metalloproteinase	Secreted
Vim	Yes[a,b,c]	No[a]	Intermediate filament, structure	Cytoskeleton

Abbreviations are as per standard nomenclature (Human Gene Compendium, GeneCards, www.genecards.org). See main text for literature references.

[a]With caveats.

[b]In astrocyte subpopulations.

[c]Present in astrocytes developmental in CNS stem cell niches.

cluding higher sensitivity and lower protein input requirements, have enabled large scale screening of the astrocyte proteome across contexts. At present, most studies exploring the proteome of reactive or diseased astrocytes are limited to analysis of astrocyte enriched or common astrocyte reactive molecules in bulk tissue (examples include Diaz-Castro et al. 2019; Johnson et al. 2020; Heaven et al. 2022; Bac et al. 2023). These studies have confirmed commonly identified astrocyte dysregulated genes and enriched pathways and provided additional evidence of diverse astrocyte populations. Far fewer studies have evaluated the proteome of isolated astrocyte populations with evaluation limited to cultured astrocytes, including immunopanned rodent astrocytes, human fetal astrocytes, or iPSC-derived astrocytes, in response to cytokines and other inflammatory mediators or disease conditions (Levine et al. 2016; Dozio and Sanchez 2018; Guttenplan et al. 2021; Labib et al. 2022). Ultimately, proteomic approaches that allow for regional and temporal control of cell-type-specific metabolic labeling of

newly synthesized proteins in vivo followed by isolation of tagged proteins using click chemistry (Prabhakar et al. 2023), cell-type-specific biotin labeling (Rayaprolu et al. 2022), or similar evolving technologies will be necessary to correlate transcriptomic and proteomic data.

As the field continues toward characterizing the functional outcomes of reactive astrocyte substates, the number of transcriptomically defined substates that can feed such studies is rich. A current bottleneck at the functional testing level is that in vivo methods provide insufficient fidelity to properly measure individual functions within a heterogeneous population of astrocytes, while some in vitro methods are often unable to properly recapitulate in vivo gene expression. This can be due to inclusion of serum in culture media, which drives a reactive phenotype at baseline (Foo et al. 2011), or omission of other CNS cells required for proper astrocyte gene expression signatures (Hasel et al. 2017). Future endeavors to label or capture astrocytes at the substate-specific level in vivo, such as the newly described FIND-seq approach, which enables capture of small subsets of cells based on nucleic acid detection rather than cell surface markers (Clark et al. 2023), may enable purification of these populations for protein/lipid/metabolomic analysis, further expanding our understanding of the nuances of astrocyte reactivity.

CONTEXT-SPECIFIC REGULATION OF ASTROCYTE REACTIVITY

Astrocyte reactivity represents a broad spectrum of potential changes that occur in context-specific manners as determined by a wide variety of different potential extracellular signals that can change over time and can vary within tissue microdomains. These external signals can derive from a wide variety of sources and activate a broad spectrum of potential intracellular signal transducers and result in the release of a wide variety of potential effector molecules (Tables 1, 2, and 5).

Extracellular signals that can induce or modulate astrocyte reactivity include many molecules that are also used in physiological activities, such as purines, transmitters, steroid hormones, growth factors, and serum proteins, as well as pathological signals such as molecules derived from microbial infections, tissue damage, inflammation, or neurodegeneration-associated events. In addition, mechanical stress can trigger astrocyte reactivity. These diverse instructive signals can derive from many different non-cell-autonomous sources, including (1) local neural and nonneural cells intrinsic to CNS tissue such

as neurons, microglia, oligodendrocyte lineage cells, other astrocytes, endothelia, pericytes, and fibroblasts; (2) nonneural cells that gain entry into the CNS, such as bone marrow–derived leukocytes, fibrocytes, and microbial infectious agents; (3) foreign materials introduced by trauma or medical implants; or (4) tissue stretching or compression caused by trauma or neoplasm (Burda and Sofroniew 2014; Liddelow and Barres 2017; O'Shea et al. 2020; Lantoine et al. 2021). Examples of different types of triggers of astrocyte reactivity and their potential sources are presented in Table 1.

These diverse extracellular signals can induce changes associated with different substates of reactive astrocytes via a wide range of potential intracellular signaling mechanisms. These intracellular mechanisms include signal transducers that act via phosphorylation, acetylation or SUMOylation, Ca^{2+} signaling, chromatin modulators, DNA-binding transcription factors, transcriptional regulators that bind protein or RNA, and microRNAs (Chen et al. 2018; Shigetomi et al. 2019; Bai et al. 2021; Burda et al. 2022). Notably, different intrinsic properties of individual astrocytes associated with regional or local differences (Makarava et al. 2023), aging, molecular memory of previous events, genetic mutations or polymorphisms can cell-autonomously influence intracellular signaling mechanisms and modulate reactivity responses. Examples of different types of intracellular regulators of astrocyte reactivity are presented in Table 5.

Comparisons both within and across disorders are revealing remarkable heterogeneity of

Table 5. Examples of astrocyte intrinsic signaling pathways and transcriptional regulators of astrocyte reactivity

Categories	Molecules
Calcium signaling	Ca^{2+}
Chromatin regulators	DNMT3B, EP300, MECP2, MECOM, MEF3C, SMARCA4, SMARCE1
MicroRNAs	DICER (ribonuclease), miR-21, miR-124, miR-146a, miR-153, miR-155, miR-181a, miR-200a-3p, miR-218, miR-223, miR-330, miR-326, miR-3099
Signal transducers	ADAM8, Camp, ERK, G-proteins, IRAK1, JAK2, MAPK1, MAP3K13 (LZK), mTOR, PKA, PKC
Transcription regulators	ARNT, ATF4, BCL3, BCL6, CEPBA, CLOCK, CREB1, FOS, HIF1A, HTT, IRF1, IRF5, IRF8, IRF9, JUN, KLF4, NFE2L2 (NRF2), NFKB, NOTCH1, NURR1, MYD88, OLIG2, RUNX1, SMARCAa4, SMAD3, SMAD4, SOCS3, SOX9, SP1, SPI1, SREDBF1, STAT2, STAT3, STAT5, TCF4 (TCF7L12), TP53, WT1, YAP1, ZBTB16, and many others

See main text for literature references.

Cite this article as *Cold Spring Harb Perspect Biol* doi: 10.1101/cshperspect.a041356

reactive astrocyte transcriptional signatures, and dissecting the underlying molecular signaling mechanisms indicates this heterogeneity derives from highly combinatorial and context-specific interactions among the many different types of regulators. For example, modulating astrocyte Jak2-Stat3 signaling, which typically occurs via cytokine or growth factor exposure, can have different effects in different contexts. Deletion of Jak2-Stat3 signaling from reactive astrocytes in traumatic injury or infection increased inflammation and tissue loss and worsened outcome (Drögemüller et al. 2008; Wanner et al. 2013), whereas doing so reduced plaque load and improved outcome in the β-amyloidopathy APP/PS1ΔE9 AD model (Ceyzériat et al. 2018; Reichenbach et al. 2019), but had little effect in the 3xTg amyloidopathy AD model (Guillemaud et al. 2020). In Huntington's disease (HD) models, activating astrocyte Jak2-Stat3 signaling increased proteolytic degradation of mutant Huntingtin and slowed disease progression (Abjean et al. 2023).

Some of these differences may be explained by diverse transcriptional responses regulated by *Stat3* in different reactivity contexts. Multiple lines of evidence show that reactivity-associated transcriptional changes are regulated in a highly combinatorial manner. ATAC-seq analysis highlights genes regulated by *Stat3* also have DNA-binding sites for multiple other transcription factors (Burda et al. 2022). In addition, Stat3 can regulate the expression of chromatin modulators and thereby indirectly influence gene expression (Burda et al. 2022). Such combinatorial interactions of multiple regulators can result in different effects on the same functional systems. For example, Stat3 signaling can either promote or inhibit different inflammatory signatures in reactive astrocytes by interacting with interferon or Il6 signaling pathways (Leng et al. 2022). Thus, Stat3 cannot be regarded as a master switch that stereotypically activates a particular signature or substate of astrocyte reactivity. Nor can the effects of modulating *Stat3* be extrapolated directly from one context to another.

Similar observations have been made for other reactivity regulators as well. The induction and maintenance of diverse astrocyte reactivity signatures and states is regulated in a context-dependent manner by the combinatorial and complex interactions of a vast array of mechanisms including intracellular signal transducers, chromatin regulators, transcription regulators, and microRNAs (Table 5). These findings highlight the need to understand the complex and combinatorial signaling regulation of astrocyte reactivity in specific contexts to develop rational therapeutic approaches based on modulating signaling pathways. Importantly, interpreting the effects of manipulating broad transcriptomic regulators that influence multiple reactive substates should now be done with caution. As the field moves forward, there is a need to develop and engage more specific tools, such as FIND-seq (Clark et al. 2023), to disentangle molecular drivers of specific responses associated with individual reactivity substates.

FUNCTIONS AND EFFECTS OF ASTROCYTE REACTIVITY

In response to the many signaling events described above, reactive astrocytes have the potential to release large variety molecules that impact on nearby cells (Table 2), as well as alter their physical contacts with adjacent cells, which are important for physiological functions like synapse stabilization, neurotransmitter reuptake and recycling, and maintenance of the glymphatic space. Reactive astrocyte-secreted molecules can exert many different functions and effects that are substate dependent, and that may be beneficial but can also give rise to maladaptive effects, and that modulation of these opposing effects is under tight temporal and spatial control. In the following sections, we highlight representative examples of major categories of insults that astrocytes respond to.

REACTIVE ASTROCYTES IN TRAUMATIC INJURY AND STROKE

Astrocytes respond to tissue acute damage like trauma and stroke in a graded manner (Sofroniew and Vinters 2010; Gleichman and Carmichael 2014; Burda et al. 2016). Scar border–forming astrocytes exhibit substantial transcriptional reprogramming and proliferate immediately abutting areas of severe tissue damage to wall off damaged and fibrotic areas and protect adjacent neural tissue (Wanner et al. 2013; Gleichman and Carmichael 2014; Sofroniew 2015a; Burda et al. 2022). Such reactive astrocytes were previously thought to be detrimental scars that prevent recovery (Sofroniew 2018), but recent studies have shown that astrocyte borders are necessary for axon regeneration (Anderson et al. 2016). Moreover, substantive axon regeneration through lesions can be achieved by providing growth-stimulating and chemoattractive factors, and this growth is attenuated by disrupting astrocyte borders (Anderson et al. 2016, 2018). In spared neural tissue adjacent to overt lesions, there is a gradient of nonproliferative astrocyte reactivity that diminishes with distance from the lesion (Wanner et al. 2013; Burda et al. 2016). This nonproliferative reactive astrocyte population can influence synaptic plasticity (Overman et al. 2012; Burda et al. 2016; Carmichael et al. 2017; Sozmen et al. 2019; Brennan et al. 2021; Lawal et al. 2022).

REACTIVE ASTROCYTES AND INFLAMMATION

Inflammation is a natural response of the body to protect cells from infection and disease-associated molecules. In the periphery, white blood cells, platelets, and other components mount the response, leading to swelling, heat production, and cytokine release. In the CNS, astrocytes and microglia, along with infiltrating peripheral immune cells, play a key role in the inflammatory response (Han et al. 2021). The interaction between astrocytes and microglia is crucial for appropriate responses and resolution of inflammation. This interaction could be manipulated for developing new therapeutic strategies in neurodegenerative diseases.

Astrocytes have been shown to produce both pro- and anti-inflammatory molecules, and specific molecular regulators can modulate their functions to either increase or limit CNS inflammation. Astrocytes produce a wide range of chemokines not only that open the blood–brain barrier (Argaw et al. 2012) and attract inflammatory cells (Hamby et al. 2012; Zamanian et al. 2012; Liddelow et al. 2017; Hasel et al. 2021) but also molecules that can exert potent suppressive effects on inflammatory cells (Kostianovsky et al. 2008; Hasel et al. 2021). Proliferative reactive astrocyte scar borders can limit the spread of inflammatory cells but may also limit the infiltration of anti-inflammatory modulators (Voskuhl et al. 2009; Wanner et al. 2013; Sofroniew 2015a). Modulating astrocyte functions can either attenuate or exacerbate CNS inflammation (Herrmann et al. 2008; Brambilla et al. 2009; Sofroniew 2015a), opening the door for gain or loss of astrocyte functions to impact CNS inflammation.

Clinical evidence suggests that disrupting astrocyte function worsens outcomes in CNS autoimmune diseases. Neuromyelitis optica (NMO), which specifically affects astrocytes, is a severe CNS autoimmune disease that causes vision loss and paralysis. Autoantibodies to aquaporin-4 (AQP4) on astrocytes drive an autoimmune response leading to complement-mediated astrocyte lysis (Lennon et al. 2005; Roemer et al. 2007), and patients with autoimmune-mediated CNS demyelination caused by AQP4 autoantibodies tend to have a worse outcome (Kitley et al. 2014; Sato et al. 2014), suggesting that AQP4 is not just a passive autoimmune antigen in the CNS, but that disrupting astrocyte function with AQP4 antibodies worsens outcomes. Experimental evidence suggests that astrocytes play a critical role in preventing the spread of inflammation during an autoimmune attack on the CNS (Voskuhl et al. 2009; Haroon et al. 2011), and loss of astrocytes exacerbates the spread of autoimmune tissue loss in NMO (Sofroniew 2015a).

Together, these findings provide compelling clinical and experimental evidence that astrocytes play a crucial role in CNS inflammatory responses, both as regulators and targets of inflammation. The interaction between astrocytes and other CNS resident cells, namely microglia, is vital for an appropriate response to such noxious stimuli. Manipulating this interaction has therapeutic potential for neurodegenerative diseases. Astrocytes have the ability to produce both pro- and anti-inflammatory molecules, and their functions can either exacerbate or attenuate CNS inflammation. Clinical studies have shown that CNS autoimmune diseases that affect astrocytes tend to be more severe, suggesting that disrupting astrocyte function can worsen outcomes. Changes in function in specific substates of reactive astrocytes can exacerbate CNS inflammation and tissue damage. Understanding the role of astrocytes in CNS inflammation is crucial for developing effective therapeutic strategies for neurodegenerative diseases.

REACTIVE ASTROCYTES AND CANCER

In response to brain cancer, reactive astrocytes can play a significant role in promoting tumor growth, angiogenesis, and invasion. Tumor cells themselves, or altered vascular permeability, and infiltrating peripheral immune cells, are likely drivers of these reactive transitions that have been measured by hypertrophy, proliferation, and increased GFAP protein or *Gfap* gene levels. At their most basic, tumor-associated reactive astrocytes form a barrier around the tumor and release various cytokines, chemokines, and growth factors that contribute to tumor progression and pathophysiology (John Lin et al. 2017; Krawczyk et al. 2022; Perelroizen et al. 2022). In other reports, breast-to-brain metastasis cancer cells infiltrate the synaptic cleft and exclude astrocyte processes (Zeng et al. 2019), while other studies report carcinoma–astrocyte gap junctions that further promote metastasis (Chen et al. 2016). However, it remains unclear whether astrocytes in these pathologies are reactive per se, or simply displaced and unable to perform physiological functions. Important recent discoveries suggest that metastatic cancer–astrocyte introductions do induce reactivity, with tumor-associated reactive astrocytes releasing lipocalin 2 (LCN2) to further drive inflammation in the CNS (Adler et al. 2023). Global deletion of *Lcn2* in mice attenuated this neuroinflammation and inhibited the seeding of metastases in this study. Interestingly, studies using complete depletion of proliferating astrocytes (using *Gfap*-TK mice) have observed a regression of glioblastoma and prolonged mouse survival, presumably due to a lack of newly proliferated reactive astrocytes (Perelroizen et al. 2022). Nevertheless, targeting all *Gfap*-positive cells may re-

move several reactive substates of astrocytes that may have opposing functions. Melanoma metastases, among the most aggressively CNS penetrant of peripheral cancers, cleave amyloid precursor protein thereby forming and secreting amyloid β. This mounts a cascade involving prometastatic anti-inflammatory astrocytes, which in turn inhibit microglia phagocytosis of the initial melanoma (Kleffman et al. 2022). Thus, while complicated, at a cell biological level, some reactive astrocytes create a unique microenvironment that may simultaneously promote tumor growth while facilitating immune suppression and resistance to therapeutic interventions, while other reactive astrocytes are driven by microglia-chemotherapy responses to actively kill CNS-derived tumors like oligodendroglioma (Gibson et al. 2019). Targeting specific aspects of reactive astrocyte interactions with cancer cells has been proposed as a promising therapeutic strategy for the treatment of brain cancer, but much remains unknown about the specific substate functional changes in reactive astrocytes in these contexts.

REACTIVE ASTROCYTES AND PERIPHERAL INFECTION

Peripheral infections can result in high levels of circulating cytokines and inflammatory mediators, including microbial pathogen-associated molecular pattern molecules (PAMPs), that can access the CNS and drive astrocyte transcriptome profiles toward a proinflammatory and cytotoxic state. This response is essential to limit the spread of infection, and diversion from this tightly controlled CNS inflammatory response will alter the molecular expression and function of reactive astrocytes (and microglia). Studies have shown a direct correlation between peripheral immune response signals, microglia–astrocyte communication, and neurological diseases. The GLP1R signaling pathway has been found to reduce astrocyte reactivity and mitigate neurodegeneration in part by reducing neurotoxic astrocyte reactivity in a number of in vivo mouse neurodegeneration models (Yun et al. 2018; Sterling et al. 2020, 2023).

Recent research has also begun to uncover the interactions between peripheral immune cells and the microglia–astrocyte metabolic signaling cascade in the CNS. Microbiome-reprogrammed natural killer cells have been found to be sufficient to drive reactive states in microglia and induce neurotoxic reactive astrocytes (Sanmarco et al. 2021). Additionally, mechanisms have been identified that are initiated by altered interferon signaling and regulated by dietary tryptophan metabolites, microglial aryl hydrocarbon receptor signaling, and VEGFb/TGF-α, which tune astrocyte reactivity in CNS autoimmunity (for reviews, see Linnerbauer et al. 2020; Han et al. 2021).

Exposure of astrocytes to PAMPs such as LPS and immune cell- and peripherally derived cytokines can drive astrocytes toward proinflammatory states with the potential for neurotoxicity (Zamanian et al. 2012; Liddelow et al. 2017; Barbar et al. 2020; Guttenplan et al. 2021; Hasel et al. 2021). Although the potential for neurotoxicity by reactive astrocytes was first discussed in the context of SOD1 mutations associated with amyotrophic lateral sclerosis (ALS) (Di Giorgio et al. 2007, 2008; Nagai et al. 2007), the identity of molecular mediators of this neurotoxicity has eluded researchers for decades and has often been attributed to glutamate excitotoxicity or lack of trophic support (Han et al. 2021). Recent investigations using in vitro experiments validated with in vivo acute injury paradigms (Liddelow et al. 2017), and analysis of gene expression, protein, and lipid secretion changes (Guttenplan et al. 2020a) highlight that fully saturated very long chain fatty acids are the most likely driver of this neurotoxicity. These saturated lipids drive PERTK-ATF3-mediated lipo-apoptosis in neurons. Importantly, previous work described that healthy neurons themselves are not susceptible to this astrocyte-mediated cell death, and instead these cells must be susceptible (modeled using axotomy or inflammatory stressors) for lipo-apoptosis to occur (Guttenplan et al. 2020a). Comorbid peripheral infections may alter the molecular expression and function of reactive astrocytes in ways that exacerbate tissue damage and compromise neural repair (Failli et al. 2012; Heintz and Mair 2014; Sanmarco et al. 2021). Further research is needed to explore the potential impact of concurrent low-grade infections or alterations in the microbiome on the induction of reactive astrocytes and consequent interactions with CNS cells driving pathology.

DISEASE-ASSOCIATED GENE MUTATIONS THAT ALTER ASTROCYTE FUNCTION

Genetic mutations, like those in Huntington's disease, familial ALS, and Alexander disease, can cause cell-autonomous astrocyte dysfunction that leads to neuronal dysfunction and neurodegeneration. These "diseased" astrocytes exhibit alterations in transcriptomics, proteomics, or function due to the mutation. This should be differentiated from astrocyte reactivity, which is triggered by external non-cell-autonomous signals from degenerating neural tissue (Fig. 1).

Alexander disease (AxD) is a well-studied example of cell-autonomous disease in astrocytes, caused by a dominant gain of mutation of the GFAP gene. Patients with early-onset type 1 AxD exhibit macrocephaly, motor and cognitive delay, seizures, psychomotor disturbances and premature death, while those with late-onset type II AxD present with bulbar symptoms, autonomic dysfunction, and spinal cord and brainstem atrophy (Brenner et al. 2001; Prust et al. 2011). In murine models of type II AxD, AxD astrocytes down-regulate expression of key homeostatic proteins including those involved in extracellular glutamate (Glt1) and K^+ regulation (Kir4.1) (Minkel et al. 2015), providing a mechanistic link between astrocyte and neuronal dysfunction and disease-associated phenotypes. A recent proteomic study confirmed these functional changes and revealed additional disruption of the typical "mature astrocyte proteome," fatty acid metabolism, and alterations in protein expression associated with other CNS cell populations (Heaven et al. 2022), indicating diseased astrocytes can drive neurological disorders.

Diseased astrocytes may not be reactive at baseline but may react to pathological stimuli and become reactive during disease progression. Often the mutated gene is highly expressed in astrocytes as well as other CNS cell populations, including several neurodegenerative disease-associated mutations (Huntington's disease: HTT; ALS: SOD1; AD: PSEN1, PSEN2, APP; Parkinson's disease (PD): LRRK1, PINK1, CD38). A

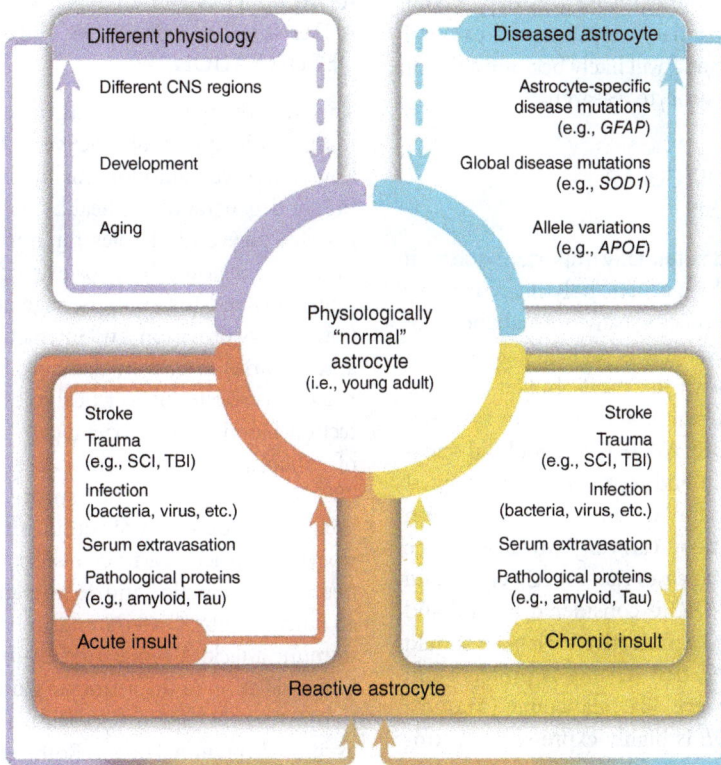

Figure 1. Changes in astrocyte states during development, disease, and reactive responses. Astrocytes have varying gene expression profiles and functions during development, disease, and in response to external stimuli. Aging can cause changes in gene expression, which may not be a true reactive state but rather a change in physiological demands. Diseased astrocytes exhibit altered gene expression due to mutations in astrocyte-specific genes (e.g., GFAP or SOD1). Reactive astrocytes result from external factors such as immune responses, pathogenic proteins, or traumatic injury, and may occur at any stage or in diseased astrocytes. Transitions between states may occur bidirectionally (e.g., acute insult, solid arrows), but some may be irreversible (dashed arrows). (SCI) Spinal cord injury, (TBI) traumatic brain injury.

rich literature now exists indicating diseased astrocytes alter their functional properties in ways that may contribute significantly to neurological symptoms and neurodegeneration. For example, in Huntington's disease, astrocytes as well as neurons accumulate nuclear inclusions of mutant huntingtin protein (mHTT). Astrocytes with mutant huntingtin protein down-regulate Kir4.1, leading to increased extracellular potassium and neuronal excitability. These effects occur at early stages of symptom onset and in the absence of astrocyte changes associated with reactivity. Both genetically mutated and abnormal "physiological" astrocytes likely contribute to the detrimental effects on neuronal function and neurotoxicity (Tong et al. 2014).

The role of diseased astrocytes in neuronal dysfunction and disease progression has been well characterized in ALS, where mutant *SOD1*-expressing astrocytes contribute to neuronal dysfunction and degeneration. Mutant *SOD1* produced in neurons also contributes to neuronal degeneration in the full disease (Lobsiger and Cleveland 2007; Nagai et al. 2007; Yamanaka et al. 2008). Reactive astrocyte formation is drastically reduced by mutant *SOD1*, and gene expression during inflammation is altered (Guttenplan et al. 2020b). Investigation into the interplay between altered baseline astrocyte gene expression/function and secondary responses to external stimuli is ongoing and will likely become a major focus of reactive astrocyte research.

ASTROCYTE MOLECULAR POLYMORPHISMS

An intriguing and potentially important question is the degree to which genetic polymorphisms alter functions of physiologically normal and reactive astrocytes. Available evidence suggests that this is the case, with the most studied example represented by human polymorphisms of apolipoprotein E (*APOE*). APOE is a major lipid transporter in the brain and represents the strongest genetic risk factor for late-onset AD. In Western European populations, carrying two alleles of the *APOE4* isoform increases risk for AD eight- and 12-fold, while *APOE3* is considered neutral and *APOE2* considered a protective variant. It should be noted, however, that *APOE4* is protective in disorders of the eye like glaucoma (Margeta et al. 2020). *APOE* is highly expressed in astrocytes, representing one of the most abundant protein-coding transcripts. Studies using various transgenic mouse models show that astrocyte-secreted APOE4, but not APOE3 or APOE2, is associated with early seeding of amyloid pathology, increased blood–brain barrier leak, reduced astrocyte end-foot coverage of blood vessels, and increased neuronal degeneration (Bell et al. 2012; Liu et al. 2017; Jackson et al. 2022). In contrast, selective reduction of *APOE4* in mice leads to significantly reduced amyloid β plaque load, and reduced overall cortical Gfap levels (Liu et al. 2017; Mahan et al. 2022). APOE4 is also reported to alter astrocyte transcriptional responses to proinflammatory stimuli and result in maladaptive immune and metabolic responses (Lee et al. 2023). As new genetic polymorphisms associated with neurologic disease are identified, the field must consider the notion that these may impact astrocyte function and alter reactive astrocyte substates in disease-relevant ways. For example, a polymorphism of *CD38*, which is highly expressed in astrocytes (Margeta et al. 2020), has recently been identified as a risk factor for PD (Guerreiro et al. 2020; Kia et al. 2021).

ASTROCYTE-MEDIATED NON-CELL-AUTONOMOUS DYSFUNCTION OR DEGENERATION

There are many ways that loss of astrocyte functions or disruption of reactive astrocytes could lead to non-cell-autonomous neuronal dysfunction or degeneration. In healthy neural tissue, astrocytes play critical roles for normal neuronal function including homeostasis of extracellular fluid, ions and transmitters, regulation of blood flow, energy provision, and interactions with synapses (Barres 2008). In damaged neural tissue, reactive astrocytes play critical roles in neuroprotection, blood–brain barrier repair, and regulation of inflammation (Sofroniew 2020; Han et al. 2021). Thus, perhaps not surprisingly, experimentally induced loss of specific functions of astrocytes or reactive astrocyte substates can cause neuronal dysfunction and degeneration and worsen outcome after CNS trauma, ischemia, or autoimmune attack. A pioneering example is that selective deletion of the astrocyte glutamate uptake transporter, GLT1 (*Eaat2*) will lead to seizures and excitotoxic neuronal death (Rothstein et al. 1996). Further examples include (1) astrocyte-selective expression of mutant *SOD* causes neuronal dysfunction and degeneration (Lobsiger and Cleveland 2007; Nagai ct al. 2007; Yamanaka et al. 2008), (2) astrocyte-selective deletion of the endoribonuclease, Dicer, cases severe ataxia, progressive cerebellar degeneration, seizures, and premature death (Tao et al. 2011), and (3) astrocyte-selective deletion of the Wnt signaling molecule adenopolyposis coli causes delayed degeneration of cerebellar Purkinje neurons (Wang et al. 2011). Experimental disruption of various astrocyte signaling molecules can alter reactive astrocytes and cause neuronal dysfunction and tissue degeneration during traumatic injury, ischemia, and autoimmune attack (Okada et al. 2006; Drögemüller et al. 2008; Herrmann et al. 2008; Li et al. 2008; Haroon et al. 2011).

THERAPEUTIC TARGETING OF REACTIVE ASTROCYTES

There is increasing recognition that molecular mechanisms associated with specific functions

Cite this article as *Cold Spring Harb Perspect Biol* doi: 10.1101/cshperspect.a041356

of astrocytes in physiological and pathological/ reactive states may be potential targets for novel therapeutic strategies for CNS disorders. As emphasized throughout this article, reactive astrocytes exert both essential beneficial functions as well as harmful effects in specific contexts as determined by specific signaling events—often by the same cell. Thus, useful therapeutic strategies will need to be specifically targeted, likely to individual functions or substates and not to global drivers of reactivity, so as to preserve or augment beneficial effects of reactive astrocytes while blocking or reducing harmful ones. The once prevalent view that wholesale blockade of reactive astrocytes could be a therapeutic strategy is no longer tenable and would likely do more harm than good. This is well reported in the literature by bulk blockade of reactivity using *Stat3* deletion, which can have both net positive and net negative effects in the same disease models (e.g., in AD; Sadick and Liddelow 2019; Smit et al. 2021). Specific aspects of reactive astrocytes that are being explored as potential targets for therapeutic manipulations include mechanisms that regulate extracellular glutamate and K^+, enzymes that generate or neutralize reactive oxygen species, and production of certain cytokines; however, a clear understanding of which transcriptomically defined substates of reactive astrocytes have alterations in these functions, and whether they are associated with specific regions of dysfunction/pathology remains to be unearthed.

CONCLUDING REMARKS

Astrocyte reactivity is a complex and multifaceted response to various forms and degrees of CNS pathology. The response can range from subtle, reversible changes to the formation of long-lasting scar border. Specific responses are controlled by various extracellular and intracellular signaling mechanisms, which are context-dependent and influenced by factors such as location, genetics, age, and previous reactivity events. Reactive astrocytes can have both beneficial and harmful effects, and changes in their function can cause or contribute to CNS disorders. As such, they are potential targets for novel therapies

aimed at augmenting or attenuating their context-specific functions. Effective therapies will need to be directed at these specific functions.

REFERENCES

Abjean L, Ben Haim L, Riquelme-Perez M, Gipchtein P, Derbois C, Palomares MA, Petit F, Hérard AS, Gaillard MC, Guillermier M, et al. 2023. Reactive astrocytes promote proteostasis in Huntington's disease through the JAK2-STAT3 pathway. *Brain* 146: 149–166. doi:10.1093/brain/awac068

Adler O, Zait Y, Cohen N, Blazquez R, Doron H, Monteran L, Scharff Y, Shami T, Mundhe D, Glehr G, et al. 2023. Reciprocal interactions between innate immune cells and astrocytes facilitate neuroinflammation and brain metastasis via lipocalin-2. *Nat Cancer* 4: 401–418. doi:10.1038/s43018-023-00519-w

Allen NJ, Eroglu C. 2017. Cell biology of astrocyte-synapse interactions. *Neuron* 96: 697–708. doi:10.1016/j.neuron.2017.09.056

Anderson MA, Burda JE, Ren Y, Ao Y, O'Shea TM, Kawaguchi R, Coppola G, Khakh BS, Deming TJ, Sofroniew MV. 2016. Astrocyte scar formation aids central nervous system axon regeneration. *Nature* 532: 195–200. doi:10.1038/nature17623

Anderson MA, O'Shea TM, Burda JE, Ao Y, Barlatey SL, Bernstein AM, Kim JH, James ND, Rogers A, Kato B, et al. 2018. Required growth facilitators propel axon regeneration across complete spinal cord injury. *Nature* 561: 396–400. doi:10.1038/s41586-018-0467-6

Argaw AT, Asp L, Zhang J, Navrazhina K, Pham T, Mariani JN, Mahase S, Dutta DJ, Seto J, Kramer EG, et al. 2012. Astrocyte-derived VEGF-A drives blood-brain barrier disruption in CNS inflammatory disease. *J Clin Invest* 122: 2454–2468. doi:10.1172/JCI60842

Bac B, Hicheri C, Weiss C, Buell A, Vilcek N, Spaeni C, Geula C, Savas JN, Disterhoft JF. 2023. The TgF344-AD rat: behavioral and proteomic changes associated with aging and protein expression in a transgenic rat model of Alzheimer's disease. *Neurobiol Aging* 123: 98–110. doi:10.1016/j.neurobiolaging.2022.12.015

Bai Y, Su X, Piao L, Jin Z, Jin R. 2021. Involvement of astrocytes and microRNA dysregulation in neurodegenerative diseases: from pathogenesis to therapeutic potential. *Front Mol Neurosci* 14: 556215. doi:10.3389/fnmol.2021.556215

Barbar L, Jain T, Zimmer M, Kruglikov I, Sadick JS, Wang M, Kalpana K, Rose IVL, Burstein SR, Rusielewicz T, et al. 2020. CD49f is a novel marker of functional and reactive human iPSC-derived astrocytes. *Neuron* 107: 436–453. e12. doi:10.1016/j.neuron.2020.05.014

Barres BA. 2008. The mystery and magic of glia: a perspective on their roles in health and disease. *Neuron* 60: 430–440. doi:10.1016/j.neuron.2008.10.013

Batiuk MY, Martirosyan A, Wahis J, de Vin F, Marneffe C, Kusserow C, Koeppen J, Viana JF, Oliveira JF, Voet T, et al. 2020. Identification of region-specific astrocyte subtypes at single cell resolution. *Nat Commun* 11: 1220. doi:10.1038/s41467-019-14198-8

Bell RD, Winkler EA, Singh I, Sagare AP, Deane R, Wu Z, Holtzman DM, Betsholtz C, Armulik A, Sallstrom J, et al.

2012. Apolipoprotein E controls cerebrovascular integrity via cyclophilin A. *Nature* **485:** 512–516. doi:10.1038/nature11087

Boisvert MM, Erikson GA, Shokhirev MN, Allen NJ. 2018. The aging astrocyte transcriptome from multiple regions of the mouse brain. *Cell Rep* **22:** 269–285. doi:10.1016/j.celrep.2017.12.039

Brambilla R, Persaud T, Hu X, Karmally S, Shestopalov VI, Dvoriantchikova G, Ivanov D, Nathanson L, Barnum SR, Bethea JR. 2009. Transgenic inhibition of astroglial NF-κB improves functional outcome in experimental autoimmune encephalomyelitis by suppressing chronic central nervous system inflammation. *J Immunol* **182:** 2628–2640. doi:10.4049/jimmunol.0802954

Brennan FH, Noble BT, Wang Y, Guan Z, Davis H, Mo X, Harris C, Eroglu C, Ferguson AR, Popovich PG. 2021. Acute post-injury blockade of α2δ-1 calcium channel subunits prevents pathological autonomic plasticity after spinal cord injury. *Cell Rep* **34:** 108667. doi:10.1016/j.celrep.2020.108667

Brenner M, Johnson AB, Boespflug-Tanguy O, Rodriguez D, Goldman JE, Messing A. 2001. Mutations in GFAP, encoding glial fibrillary acidic protein, are associated with alexander disease. *Nat Genet* **27:** 117–120. doi:10.1038/83679

Burda JE, Sofroniew MV. 2014. Reactive gliosis and the multicellular response to CNS damage and disease. *Neuron* **81:** 229–248. doi:10.1016/j.neuron.2013.12.034

Burda JE, Bernstein AM, Sofroniew MV. 2016. Astrocyte roles in traumatic brain injury. *Exp Neurol* **275:** 305–315. doi:10.1016/j.expneurol.2015.03.020

Burda JE, O'Shea TM, Ao Y, Suresh KB, Wang S, Bernstein AM, Chandra A, Deverasetty S, Kawaguchi R, Kim JH, et al. 2022. Divergent transcriptional regulation of astrocyte reactivity across disorders. *Nature* **606:** 557–564. doi:10.1038/s41586-022-04739-5

Bush TG, Puvanachandra N, Horner CH, Polito A, Ostenfeld T, Svendsen CN, Mucke L, Johnson MH, Sofroniew MV. 1999. Leukocyte infiltration, neuronal degeneration, and neurite outgrowth after ablation of scar-forming, reactive astrocytes in adult transgenic mice. *Neuron* **23:** 297–308. doi:10.1016/S0896-6273(00)80781-3

Bushong EA, Martone MA, Jones YZ, Ellisman MH. 2002. Protoplasmic astrocytes in CA1 stratum radiatum occupy separate anatomical domains. *J Neurosci* **22:** 183–192. doi:10.1523/JNEUROSCI.22-01-00183.2002

Campbell SC, Muñoz-Ballester C, Chaunsali L, Mills WA III, Yang JH, Sontheimer H, Robel S. 2020. Potassium and glutamate transport is impaired in scar-forming tumor-associated astrocytes. *Neurochem Int* **133:** 104628. doi:10.1016/j.neuint.2019.104628

Carmichael ST, Kathirvelu B, Schweppe CA, Nie EH. 2017. Molecular, cellular and functional events in axonal sprouting after stroke. *Exp Neurol* **287:** 384–394. doi:10.1016/j.expneurol.2016.02.007

Carroll JA, Race B, Williams K, Striebel J, Chesebro B. 2020. RNA-seq and network analysis reveal unique glial gene expression signatures during prion infection. *Mol Brain* **13:** 71. doi:10.1186/s13041-020-00610-8

Castranio EL, Hasel P, Haure-Mirande JV, Ramirez Jimenez AV, Hamilton BW, Kim RD, Glabe CG, Wang M, Zhang B, Gandy S, et al. 2023. Microglial *INPP5D* limits plaque formation and glial reactivity in the PSAPP mouse model of Alzheimer's disease. *Alzheimers Dement* **19:** 2239–2252. doi:10.1002/alz.12821

Ceyzériat K, Ben Haim L, Denizot A, Pommier D, Matos M, Guillemaud O, Palomares MA, Abjean L, Petit F, Gipchtein P, et al. 2018. Modulation of astrocyte reactivity improves functional deficits in mouse models of Alzheimer's disease. *Acta Neuropathol Commun* **6:** 104. doi:10.1186/s40478-018-0606-1

Chai H, Diaz-Castro B, Shigetomi E, Monte E, Octeau JC, Yu X, Cohn W, Rajendran PS, Vondriska TM, Whitelegge JP, et al. 2017. Neural circuit-specialized astrocytes: transcriptomic, proteomic, morphological, and functional evidence. *Neuron* **95:** 531–549.e9. doi:10.1016/j.neuron.2017.06.029

Chen Q, Boire A, Jin X, Valiente M, Er EE, Lopez-Soto A, Jacob L, Patwa R, Shah H, Xu K, et al. 2016. Carcinoma-astrocyte gap junctions promote brain metastasis by cGAMP transfer. *Nature* **533:** 493–498. doi:10.1038/nature18268

Chen M, Geoffroy CG, Meves JM, Narang A, Li Y, Nguyen MT, Khai VS, Kong X, Steinke CL, Carolino KI, et al. 2018. Leucine zipper-bearing kinase is a critical regulator of astrocyte reactivity in the adult mammalian CNS. *Cell Rep* **22:** 3587–3597. doi:10.1016/j.celrep.2018.02.102

Clark IC, Wheeler MA, Lee HG, Li Z, Sanmarco LM, Thaploo S, Polonio CM, Shin SW, Scalisi G, Henry AR, et al. 2023. Identification of astrocyte regulators by nucleic acid cytometry. *Nature* **614:** 326–333. doi:10.1038/s41586-022-05613-0

Clarke LE, Liddelow SA, Chakraborty C, Munch AE, Heiman M, Barres BA. 2018. Normal aging induces A1-like astrocyte reactivity. *Proc Natl Acad Sci* **115:** E1896–E1905. doi:10.1073/pnas.1800165115

Clavreul S, Abdeladim L, Hernández Garzón E, Niculescu D, Durand J, Ieng SH, Barry R, Bonvento G, Beaurepaire E, Livet J, et al. 2019. Cortical astrocytes develop in a plastic manner at both clonal and cellular levels. *Nat Commun* **10:** 4884. doi:10.1038/s41467-019-12791-5

Diaz-Castro B, Gangwani MR, Yu X, Coppola G, Khakh BS. 2019. Astrocyte molecular signatures in Huntington's disease. *Sci Transl Med* **11:** eaaw8546. doi:10.1126/scitranslmed.aaw8546

Diaz-Castro B, Bernstein AM, Coppola G, Sofroniew MV, Khakh BS. 2021. Molecular and functional properties of cortical astrocytes during peripherally induced neuroinflammation. *Cell Rep* **36:** 109508. doi:10.1016/j.celrep.2021.109508

Di Giorgio FP, Carrasco MA, Siao MC, Maniatis T, Eggan K. 2007. Non–cell autonomous effect of glia on motor neurons in an embryonic stem cell-based ALS model. *Nat Neurosci* **10:** 608–614. doi:10.1038/nn1885

Di Giorgio FP, Boulting GL, Bobrowicz S, Eggan KC. 2008. Human embryonic stem cell-derived motor neurons are sensitive to the toxic effect of glial cells carrying an ALS-causing mutation. *Cell Stem Cell* **3:** 637–648. doi:10.1016/j.stem.2008.09.017

Dozio V, Sanchez JC. 2018. Profiling the proteomic inflammatory state of human astrocytes using DIA mass spectrometry. *J Neuroinflamm* **15:** 331. doi:10.1186/s12974-018-1371-6

Drögemüller K, Helmuth U, Brunn A, Sakowicz-Burkiewicz M, Gutmann DH, Mueller W, Deckert M, Schlüter D. 2008. Astrocyte gp130 expression is critical for the control of *toxoplasma* encephalitis. *J Immunol* **181**: 2683–2693. doi:10.4049/jimmunol.181.4.2683

Escartin C, Galea E, Lakatos A, O'Callaghan JP, Petzold GC, Serrano-Pozo A, Steinhauser C, Volterra A, Carmignoto G, Agarwal A, et al. 2021. Reactive astrocyte nomenclature, definitions, and future directions. *Nat Neurosci* **24**: 312–325. doi:10.1038/s41593-020-00783-4

Failli V, Kopp MA, Gericke C, Martus P, Klingbeil S, Brommer B, Laginha I, Chen Y, DeVivo MJ, Dirnagl U, et al. 2012. Functional neurological recovery after spinal cord injury is impaired in patients with infections. *Brain* **135**: 3238–3250. doi:10.1093/brain/aws267

Falcone C, Penna E, Hong T, Tarantal AF, Hof PR, Hopkins WD, Sherwood CC, Noctor SC, Martínez-Cerdeño V. 2021. Cortical interlaminar astrocytes are generated prenatally, mature postnatally, and express unique markers in human and nonhuman primates. *Cereb Cortex* **31**: 379–395. doi:10.1093/cercor/bhaa231

Foo LC, Allen NJ, Bushong EA, Ventura PB, Chung WS, Zhou L, Cahoy JD, Daneman R, Zong H, Ellisman MH, et al. 2011. Development of a method for the purification and culture of rodent astrocytes. *Neuron* **71**: 799–811. doi:10.1016/j.neuron.2011.07.022

Frik J, Merl-Pham J, Plesnila N, Mattugini N, Kjell J, Kraska J, Gómez RM, Hauck SM, Sirko S, Götz M. 2018. Cross-talk between monocyte invasion and astrocyte proliferation regulates scarring in brain injury. *EMBO Rep* **19**: e45294. doi:10.15252/embr.201745294

Gibson EM, Nagaraja S, Ocampo A, Tam LT, Wood LS, Pallegar PN, Greene JJ, Geraghty AC, Goldstein AK, Ni L, et al. 2019. Methotrexate chemotherapy induces persistent triglial dysregulation that underlies chemotherapy-related cognitive impairment. *Cell* **176**: 43–55.e13. doi:10.1016/j.cell.2018.10.049

Gleichman AJ, Carmichael ST. 2014. Astrocytic therapies for neuronal repair in stroke. *Neurosci Lett* **565**: 47–52. doi:10.1016/j.neulet.2013.10.055

Guerreiro S, Privat AL, Bressac L, Toulorge D. 2020. CD38 in neurodegeneration and neuroinflammation. *Cells* **9**: 471. doi:10.3390/cells9020471

Guillemaud O, Ceyzériat K, Saint-Georges T, Cambon K, Petit F, Ben Haim L, Carrillo-de Sauvage MA, Guillermier M, Bernier S, Hérard AS, et al. 2020. Complex roles for reactive astrocytes in the triple transgenic mouse model of Alzheimer disease. *Neurobiol Aging* **90**: 135–146. doi:10.1016/j.neurobiolaging.2020.02.010

Guttenplan KA, Stafford BK, El-Danaf RN, Adler DI, Münch AE, Weigel MK, Huberman AD, Liddelow SA. 2020a. Neurotoxic reactive astrocytes drive neuronal death after retinal injury. *Cell Rep* **31**: 107776. doi:10.1016/j.celrep.2020.107776

Guttenplan KA, Weigel MK, Adler DI, Couthouis J, Liddelow SA, Gitler AD, Barres BA. 2020b. Knockout of reactive astrocyte activating factors slows disease progression in an ALS mouse model. *Nat Commun* **11**: 3753. doi:10.1038/s41467-020-17514-9

Guttenplan KA, Weigel MK, Prakash P, Wijewardhane PR, Hasel P, Rufen-Blanchette U, Münch AE, Blum JA, Fine J, Neal MC, et al. 2021. Neurotoxic reactive astrocytes induce

cell death via saturated lipids. *Nature* **599**: 102–107. doi:10.1038/s41586-021-03960-y

Haim LB, Rowitch DH. 2017. Functional diversity of astrocytes in neural circuit regulation. *Nat Rev Neurosci* **18**: 31–41. doi:10.1038/nrn.2016.159

Hamby ME, Coppola G, Ao Y, Geschwind DH, Khakh BS, Sofroniew MV. 2012. Inflammatory mediators alter the astrocyte transcriptome and calcium signaling elicited by multiple g-protein-coupled receptors. *J Neurosci* **32**: 14489–14510. doi:10.1523/JNEUROSCI.1256-12.2012

Han RT, Kim RD, Molofsky AV, Liddelow SA. 2021. Astrocyte-immune cell interactions in physiology and pathology. *Immunity* **54**: 211–224. doi:10.1016/j.immuni.2021.01.013

Haroon F, Drögemüller K, Händel U, Brunn A, Reinhold D, Nishanth G, Mueller W, Trautwein C, Ernst M, Deckert M, et al. 2011. Gp130-dependent astrocytic survival is critical for the control of autoimmune central nervous system inflammation. *J Immunol* **186**: 6521–6531. doi:10.4049/jimmunol.1001135

Hasel P, Dando O, Jiwaji Z, Baxter P, Todd AC, Heron S, Márkus NM, McQueen J, Hampton DW, Torvell M, et al. 2017. Neurons and neuronal activity control gene expression in astrocytes to regulate their development and metabolism. *Nat Commun* **8**: 15132. doi:10.1038/ncomms15132

Hasel P, Rose IVL, Sadick JS, Kim RD, Liddelow SA. 2021. Neuroinflammatory astrocyte subtypes in the mouse brain. *Nat Neurosci* **24**: 1475–1487. doi:10.1038/s41593-021-00905-6

Haumont D, Deckelbaum RJ, Richelle M, Dahlan W, Coussaert E, Bihain BE, Carpentier YA. 1989. Plasma lipid and plasma lipoprotein concentrations in low birth weight infants given parenteral nutrition with twenty or ten percent lipid emulsion. *J Pediatr* **115**: 787–793. doi:10.1016/S0022-3476(89)80663-8

Heaven MR, Herren AW, Flint DL, Pacheco NL, Li J, Tang A, Khan F, Goldman JE, Phinney BS, Olsen ML. 2022. Metabolic enzyme alterations and astrocyte dysfunction in a murine model of alexander disease with severe reactive gliosis. *Mol Cell Proteomics* **21**: 100180. doi:10.1016/j.mcpro.2021.100180

Heintz C, Mair W. 2014. You are what you host: microbiome modulation of the aging process. *Cell* **156**: 408–411. doi:10.1016/j.cell.2014.01.025

Herrmann JE, Imura T, Song B, Qi J, Ao Y, Nguyen TK, Korsak RA, Takeda K, Akira S, Sofroniew MV. 2008. STAT3 is a critical regulator of astrogliosis and scar formation after spinal cord injury. *J Neurosci* **28**: 7231–7243. doi:10.1523/JNEUROSCI.1709-08.2008

Jackson RJ, Meltzer JC, Nguyen H, Commins C, Bennett RE, Hudry E, Hyman BT. 2022. APOE4 derived from astrocytes leads to blood-brain barrier impairment. *Brain* **145**: 3582–3593. doi:10.1093/brain/awab478

John Lin CC, Yu K, Hatcher A, Huang TW, Lee HK, Carlson J, Weston MC, Chen F, Zhang Y, Zhu W, et al. 2017. Identification of diverse astrocyte populations and their malignant analogs. *Nat Neurosci* **20**: 396–405. doi:10.1038/nn.4493

Johnson ECB, Dammer EB, Duong DM, Ping L, Zhou M, Yin L, Higginbotham LA, Guajardo A, White B, Troncoso JC, et al. 2020. Large-scale proteomic analysis of Alzheimer's disease brain and cerebrospinal fluid reveals early changes in energy metabolism associated with microglia and astrocyte activation. *Nat Med* **26:** 769–780. doi:10.1038/s41591-020-0815-6

Khakh BS, Deneen B. 2019. The emerging nature of astrocyte diversity. *Annu Rev Neurosci* **42:** 187–207. doi:10.1146/annurev-neuro-070918-050443

Khakh BS, Sofroniew MV. 2015. Diversity of astrocyte functions and phenotypes in neural circuits. *Nat Neurosci* **18:** 942–952. doi:10.1038/nn.4043

Kia DA, Zhang D, Guelfi S, Manzoni C, Hubbard L, Reynolds RH, Botía J, Ryten M, Ferrari R, Lewis PA, et al. 2021. Identification of candidate Parkinson disease genes by integrating genome-wide association study, expression, and epigenetic data sets. *JAMA Neurol* **78:** 464–472. doi:10.1001/jamaneurol.2020.5257

Kitley J, Waters P, Woodhall M, Leite MI, Murchison A, George J, Küker W, Chandratre S, Vincent A, Palace J. 2014. Neuromyelitis optica spectrum disorders with aquaporin-4 and myelin-oligodendrocyte glycoprotein antibodies: a comparative study. *JAMA Neurol* **71:** 276–283. doi:10.1001/jamaneurol.2013.5857

Kleffman K, Levinson G, Rose IVL, Blumenberg LM, Shadaloey SAA, Dhabaria A, Wong E, Galáan-Echevarría F, Karz A, Argibay D, et al. 2022. Melanoma-secreted amyloid β suppresses neuroinflammation and promotes brain metastasis. *Cancer Discov* **12:** 1314–1335. doi:10.1158/2159-8290.CD-21-1006

Kostianovsky AM, Maier LM, Anderson RC, Bruce JN, Anderson DE. 2008. Astrocytic regulation of human monocytic/microglial activation. *J Immunol* **181:** 5425–5432. doi:10.4049/jimmunol.181.8.5425

Krawczyk MC, Haney JR, Pan L, Caneda C, Khankan RR, Reyes SD, Chang JW, Morselli M, Vinters HV, Wang AC, et al. 2022. Human astrocytes exhibit tumor microenvironment-, age-, and sex-related transcriptomic signatures. *J Neurosci* **42:** 1587–1603. doi:10.1523/JNEUROSCI.0407-21.2021

Labib D, Wang Z, Prakash P, Zimmer M, Smith MD, Frazel PW, Barbar L, Sapar ML, Calabresi PA, Peng J, et al. 2022. Proteomic alterations and novel markers of neurotoxic reactive astrocytes in human induced pluripotent stem cell models. *Front Mol Neurosci* **15:** 870085. doi:10.3389/fnmol.2022.870085

Lantoine J, Procès A, Villers A, Halliez S, Buée L, Ris L, Gabriele S. 2021. Inflammatory molecules released by mechanically injured astrocytes trigger presynaptic loss in cortical neuronal networks. *ACS Chem Neurosci* **12:** 3885–3897. doi:10.1021/acschemneuro.1c00488

Lawal O, Ulloa Severino FP, Eroglu C. 2022. The role of astrocyte structural plasticity in regulating neural circuit function and behavior. *Glia* **70:** 1467–1483. doi:10.1002/glia.24191

Lee S, Williams HC, Gorman AA, Devanney NA, Harrison DA, Walsh AE, Goulding DS, Tuck T, Schwartz JL, Zajac DJ, et al. 2023. APOE4 drives transcriptional heterogeneity and maladaptive immunometabolic responses of astrocytes. bioRxiv doi:10.1101/2023.02.06.527204

Leng K, Rose IVL, Kim H, Xia W, Romero-Fernandez W, Rooney B, Koontz M, Li E, Ao Y, Wang S, et al. 2022. CRISPRi screens in human iPSC-derived astrocytes elucidate regulators of distinct inflammatory reactive states. *Nat Neurosci* **25:** 1528–1542. doi:10.1038/s41593-022-01180-9

Lennon VA, Kryzer TJ, Pittock SJ, Verkman AS, Hinson SR. 2005. Igg marker of optic-spinal multiple sclerosis binds to the aquaporin-4 water channel. *J Exp Med* **202:** 473–477. doi:10.1084/jem.20050304

Levine J, Kwon E, Paez P, Yan W, Czerwieniec G, Loo JA, Sofroniew MV, Wanner IB. 2016. Traumatically injured astrocytes release a proteomic signature modulated by STAT3-dependent cell survival. *Glia* **64:** 668–694. doi:10.1002/glia.22953

Li L, Lundkvist A, Andersson D, Wilhelmsson U, Nagai N, Pardo AC, Nodin C, Ståhlberg A, Aprico K, Larsson K, et al. 2008. Protective role of reactive astrocytes in brain ischemia. *J Cereb Blood Flow Metab* **28:** 468–481. doi:10.1038/sj.jcbfm.9600546

Liddelow SA, Barres BA. 2017. Reactive astrocytes: production, function, and therapeutic potential. *Immunity* **46:** 957–967. doi:10.1016/j.immuni.2017.06.006

Liddelow SA, Guttenplan KA, Clarke LE, Bennett FC, Bohlen CJ, Schirmer L, Bennett ML, Münch AE, Chung WS, Peterson TC, et al. 2017. Neurotoxic reactive astrocytes are induced by activated microglia. *Nature* **541:** 481–487. doi:10.1038/nature21029

Linnerbauer M, Wheeler MA, Quintana FJ. 2020. Astrocyte crosstalk in CNS inflammation. *Neuron* **108:** 608–622. doi:10.1016/j.neuron.2020.08.012

Liu CC, Zhao N, Fu Y, Wang N, Linares C, Tsai CW, Bu G. 2017. Apoe4 accelerates early seeding of amyloid pathology. *Neuron* **96:** 1024–1032.e3. doi:10.1016/j.neuron.2017.11.013

Lobsiger CS, Cleveland DW. 2007. Glial cells as intrinsic components of non-cell-autonomous neurodegenerative disease. *Nat Neurosci* **10:** 1355–1360. doi:10.1038/nn1988

Mahan TE, Wang C, Bao X, Choudhury A, Ulrich JD, Holtzman DM. 2022. Selective reduction of astrocyte apoE3 and apoE4 strongly reduces Aβ accumulation and plaque-related pathology in a mouse model of amyloidosis. *Mol Neurodegener* **17:** 13. doi:10.1186/s13024-022-00516-0

Makarava N, Mychko O, Molesworth K, Chang JC, Henry RJ, Tsymbalyuk N, Gerzanich V, Simard JM, Loane DJ, Baskakov IV. 2023. Region-specific homeostatic identity of astrocytes is essential for defining their response to pathological insults. *Cells* **12:** 2172. doi:10.3390/cells12172172

Margeta MA, Letcher SM, Igo RP Jr, Cooke Bailey JN, Pasquale LR, Haines JL, Butovsky O, Wiggs JL, NEIGHBORHOOD Consortium 2020. Association of *APOE* with primary open-angle glaucoma suggests a protective effect for *APOE ε4*. *Invest Ophthalmol Vis Sci* **61:** 3. doi:10.1167/iovs.61.8.3

Minkel HR, Anwer TZ, Arps KM, Brenner M, Olsen ML. 2015. Elevated GFAP induces astrocyte dysfunction in caudal brain regions: a potential mechanism for hindbrain involved symptoms in type II alexander disease. *Glia* **63:** 2285–2297. doi:10.1002/glia.22893

Molofsky AV, Kelley KW, Tsai HH, Redmond SA, Chang SM, Madireddy L, Chan JR, Baranzini SE, Ullian EM, Rowitch DH. 2014. Astrocyte-encoded positional cues maintain sensorimotor circuit integrity. *Nature* **509:** 189–194. doi:10.1038/nature13161

Nagai M, Re DB, Nagata T, Chalazonitis A, Jessell TM, Wichterle H, Przedborski S. 2007. Astrocytes expressing ALS-linked mutated SOD1 release factors selectively toxic to motor neurons. *Nat Neurosci* **10:** 615–622. doi:10.1038/nn1876

Okada S, Nakamura M, Katoh H, Miyao T, Shimazaki T, Ishii K, Yamane J, Yoshimura A, Iwamoto Y, Toyama Y, et al. 2006. Conditional ablation of Stat3 or Socs3 discloses a dual role for reactive astrocytes after spinal cord injury. *Nature Med* **12:** 829–834. doi:10.1038/nm1425

O'Shea TM, Wollenberg AL, Kim JH, Ao Y, Deming TJ, Sofroniew MV. 2020. Foreign body responses in mouse central nervous system mimic natural wound responses and alter biomaterial functions. *Nat Commun* **11:** 6203. doi:10.1038/s41467-020-19906-3

Overman JJ, Clarkson AN, Wanner IB, Overman WT, Eckstein I, Maguire JL, Dinov ID, Toga AW, Carmichael ST. 2012. A role for ephrin-A5 in axonal sprouting, recovery, and activity-dependent plasticity after stroke. *Proc Natl Acad Sci* **109:** E2230–E2239. doi:10.1073/pnas.1204386109

Pan J, Ma N, Yu B, Zhang W, Wan J. 2020. Transcriptomic profiling of microglia and astrocytes throughout aging. *J Neuroinflamm* **17:** 97. doi:10.1186/s12974-020-01774-9

Pekny M, Pekna M, Messing A, Steinhäuser C, Lee JM, Parpura V, Hol EM, Sofroniew MV, Verkhratsky A. 2016. Astrocytes: a central element in neurological diseases. *Acta Neuropathol* **131:** 323–345. doi:10.1007/s00401-015-1513-1

Perelroizen R, Philosof B, Budick-Harmelin N, Chernobylsky T, Ron A, Katzir R, Shimon D, Tessler A, Adir O, Gaoni-Yogev A, et al. 2022. Astrocyte immunometabolic regulation of the tumour microenvironment drives glioblastoma pathogenicity. *Brain* **145:** 3288–3307. doi:10.1093/brain/awac222

Prabhakar P, Pielot R, Landgraf P, Wissing J, Bayrhammer A, van Ham M, Gundelfinger ED, Jänsch L, Dieterich DC, Müller A. 2023. Monitoring regional astrocyte diversity by cell type-specific proteomic labeling in vivo. *Glia* **71:** 682–703. doi:10.1002/glia.24304

Prust M, Wang J, Morizono H, Messing A, Brenner M, Gordon E, Hartka T, Sokohl A, Schiffmann R, Gordish-Dressman H, et al. 2011. GFAP mutations, age at onset, and clinical subtypes in alexander disease. *Neurology* **77:** 1287–1294. doi:10.1212/WNL.0b013e3182309f72

Rayaprolu S, Bitarafan S, Santiago JV, Betarbet R, Sunna S, Cheng L, Xiao H, Nelson RS, Kumar P, Bagchi P, et al. 2022. Cell type-specific biotin labeling in vivo resolves regional neuronal and astrocyte proteomic differences in mouse brain. *Nat Commun* **13:** 2927. doi:10.1038/s41467-022-30623-x

Reichenbach N, Delekate A, Plescher M, Schmitt F, Krauss S, Blank N, Halle A, Petzold GC. 2019. Inhibition of Stat3-mediated astrogliosis ameliorates pathology in an Alzheimer's disease model. *EMBO Mol Med* **11:** e9665. doi:10.15252/emmm.201809665

Robel S, Sontheimer H. 2016. Glia as drivers of abnormal neuronal activity. *Nat Neurosci* **19:** 28–33. doi:10.1038/nn.4184

Roemer SF, Parisi JE, Lennon VA, Benarroch EE, Lassmann H, Bruck W, Mandler RN, Weinshenker BG, Pittock SJ, Wingerchuk DM, et al. 2007. Pattern-specific loss of aquaporin-4 immunoreactivity distinguishes neuromyelitis

optica from multiple sclerosis. *Brain* **130:** 1194–1205. doi:10.1093/brain/awl371

Rothstein JD, Dykes-Hoberg M, Pardo CA, Bristol LA, Jin L, Kuncl RW, Kanai Y, Hediger MA, Wang Y, Schielke JP, et al. 1996. Knockout of glutamate transporters reveals a major role for astroglial transport in excitotoxicity and clearance of glutamate. *Neuron* **16:** 675–686. doi:10.1016/S0896-6273(00)80086-0

Sadick JS, Liddelow SA. 2019. Don't forget astrocytes when targeting Alzheimer's disease. *Br J Pharmacol* **176:** 3585–3598. doi:10.1111/bph.14568

Sadick JS, O'Dea MR, Hasel P, Dykstra T, Faustin A, Liddelow SA. 2022. Astrocytes and oligodendrocytes undergo subtype-specific transcriptional changes in Alzheimer's disease. *Neuron* **110:** 1788–1805.e10. doi:10.1016/j.neuron.2022.03.008

Sanmarco LM, Wheeler MA, Gutiérrez-Vázquez C, Polonio CM, Linnerbauer M, Pinho-Ribeiro FA, Li Z, Giovannoni F, Batterman KV, Scalisi G, et al. 2021. Gut-licensed IFNγ$^+$ NK cells drive LAMP1$^+$TRAIL$^+$ anti-inflammatory astrocytes. *Nature* **590:** 473–479. doi:10.1038/s41586-020-03116-4

Sato DK, Callegaro D, Lana-Peixoto MA, Waters PJ, de Haidar Jorge FM, Takahashi T, Nakashima I, Apostolos-Pereira SL, Talim N, Simm RF, et al. 2014. Distinction between MOG antibody-positive and AQP4 antibody-positive NMO spectrum disorders. *Neurology* **82:** 474–481. doi:10.1212/WNL.0000000000000101

Shigetomi E, Saito K, Sano F, Koizumi S. 2019. Aberrant calcium signals in reactive astrocytes: a key process in neurological disorders. *Int J Mol Sci* **20:** 996. doi:10.3390/ijms20040996

Sirko S, Irmler M, Gascón S, Bek S, Schneider S, Dimou L, Obermann J, De Souza Paiva A, Poirier F, Beckers J, et al. 2015. Astrocyte reactivity after brain injury—the role of galectins 1 and 3. *Glia* **63:** 2340–2361. doi:10.1002/glia.22898

Smit T, Deshayes NAC, Borchelt DR, Kamphuis W, Middeldorp J, Hol EM. 2021. Reactive astrocytes as treatment targets in Alzheimer's disease—systematic review of studies using the APPswePS1dE9 mouse model. *Glia* **69:** 1852–1881. doi:10.1002/glia.23981

Sofroniew MV. 2015a. Astrocyte barriers to neurotoxic inflammation. *Nat Rev Neurosci* **16:** 249–263. doi:10.1038/nrn3898

Sofroniew MV. 2015b. Astrogliosis. *Cold Spring Harb Perspect Biol* **7:** a020420. doi:10.1101/cshperspect.a020420

Sofroniew MV. 2018. Dissecting spinal cord regeneration. *Nature* **557:** 343–350. doi:10.1038/s41586-018-0068-4

Sofroniew MV. 2020. Astrocyte reactivity: subtypes, states, and functions in CNS innate immunity. *Trends Immunol* **41:** 758–770. doi:10.1016/j.it.2020.07.004

Sofroniew MV, Vinters HV. 2010. Astrocytes: biology and pathology. *Acta Neuropathol* **119:** 7–35. doi:10.1007/s00401-009-0619-8

Sozmen EG, DiTullio DJ, Rosenzweig S, Hinman JD, Bridges SP, Marin MA, Kawaguchi R, Coppola G, Carmichael ST. 2019. White matter stroke induces a unique oligo-astrocyte niche that inhibits recovery. *J Neurosci* **39:** 9343–9359. doi:10.1523/JNEUROSCI.0103-19.2019

Srinivasan R, Lu TY, Chai H, Xu J, Huang BS, Golshani P, Coppola G, Khakh BS. 2016. New transgenic mouse lines for selectively targeting astrocytes and studying calcium signals in astrocyte processes in situ and in vivo. *Neuron* **92:** 1181–1195. doi:10.1016/j.neuron.2016.11.030

Sterling JK, Adetunji MO, Guttha S, Bargoud AR, Uyhazi KE, Ross AG, Dunaief JL, Cui QN. 2020. GLP-1 receptor agonist NLY01 reduces retinal inflammation and neuron death secondary to ocular hypertension. *Cell Rep* **33:** 108271. doi:10.1016/j.celrep.2020.108271

Sterling J, Hua P, Dunaief JL, Cui QN, VanderBeek BL. 2023. Glucagon-like peptide 1 receptor agonist use is associated with reduced risk for glaucoma. *Br J Ophthalmol* **107:** 215–220. doi:10.1136/bjophthalmol-2021-319232

Tao J, Wu H, Lin Q, Wei W, Lu XH, Cantle JP, Ao Y, Olsen RW, Yang XW, Mody I, et al. 2011. Deletion of astroglial dicer causes non-cell-autonomous neuronal dysfunction and degeneration. *J Neurosci* **31:** 8306–8319. doi:10.1523/JNEUROSCI.0567-11.2011

Tong X, Ao Y, Faas GC, Nwaobi SE, Xu J, Haustein MD, Anderson MA, Mody I, Olsen ML, Sofroniew MV, et al. 2014. Astrocyte Kir4.1 ion channel deficits contribute to neuronal dysfunction in Huntington's disease model mice. *Nat Neurosci* **17:** 694–703. doi:10.1038/nn.3691

Torres-Ceja B, Olsen ML. 2022. A closer look at astrocyte morphology: development, heterogeneity, and plasticity at astrocyte leaflets. *Curr Opin Neurobiol* **74:** 102550. doi:10.1016/j.conb.2022.102550

Verkhratsky A, Nedergaard M. 2018. Physiology of astroglia. *Physiol Rev* **98:** 239–389. doi:10.1152/physrev.00042.2016

Voskuhl RR, Peterson RS, Song B, Ao Y, Morales LB, Tiwari-Woodruff S, Sofroniew MV. 2009. Reactive astrocytes form scar-like perivascular barriers to leukocytes during adaptive immune inflammation of the CNS. *J Neurosci* **29:** 11511–11522. doi:10.1523/JNEUROSCI.1514-09.2009

Wang X, Imura T, Sofroniew MV, Fushiki S. 2011. Loss of adenomatous polyposis coli in Bergmann glia disrupts their unique architecture and leads to cell nonautonomous neurodegeneration of cerebellar Purkinje neurons. *Glia* **59:** 857–868. doi:10.1002/glia.21154

Wanner IB, Anderson MA, Song B, Levine J, Fernandez A, Gray-Thompson Z, Ao Y, Sofroniew MV. 2013. Glial scar borders are formed by newly proliferated, elongated astrocytes that interact to corral inflammatory and fibrotic cells via STAT3-dependent mechanisms after spinal cord injury. *J Neurosci* **33:** 12870–12886. doi:10.1523/JNEUROSCI.2121-13.2013

Wei H, Wu X, You Y, Duran RC, Zheng Y, Narayanan KL, Hai B, Li X, Tallapragada N, Prajapati TJ, et al. 2021. Systematic analysis of purified astrocytes after SCI unveils Zeb2os function during astrogliosis. *Cell Rep* **34:** 108721. doi:10.1016/j.celrep.2021.108721

Wilhelmsson U, Bushong EA, Price DL, Smarr BL, Phung V, Terada M, Ellisman MH, Pekny M. 2006. Redefining the concept of reactive astrocytes as cells that remain within their unique domains upon reaction to injury. *Proc Natl Acad Sci* **103:** 17513–17518. doi:10.1073/pnas.0602841103

Williamson MR, Fuertes CJA, Dunn AK, Drew MR, Jones TA. 2021. Reactive astrocytes facilitate vascular repair and remodeling after stroke. *Cell Rep* **35:** 109048. doi:10.1016/j.celrep.2021.109048

Wu YE, Pan L, Zuo Y, Li X, Hong W. 2017. Detecting activated cell populations using single-cell RNA-seq. *Neuron* **96:** 313–329.e6. doi:10.1016/j.neuron.2017.09.026

Yamanaka K, Chun SJ, Boillee S, Fujimori-Tonou N, Yamashita H, Gutmann DH, Takahashi R, Misawa H, Cleveland DW. 2008. Astrocytes as determinants of disease progression in inherited amyotrophic lateral sclerosis. *Nat Neurosci* **11:** 251–253. doi:10.1038/nn2047

Yun SP, Kam TI, Panicker N, Kim S, Oh Y, Park JS, Kwon SH, Park YJ, Karuppagounder SS, Park H, et al. 2018. Block of A1 astrocyte conversion by microglia is neuroprotective in models of Parkinson's disease. *Nat Med* **24:** 931–938. doi:10.1038/s41591-018-0051-5

Zamanian JL, Xu L, Foo LC, Nouri N, Zhou L, Giffard RG, Barres BA. 2012. Genomic analysis of reactive astrogliosis. *J Neurosci* **32:** 6391–6410. doi:10.1523/JNEUROSCI.6221-11.2012

Zeng Q, Michael IP, Zhang P, Saghafinia S, Knott G, Jiao W, McCabe BD, Galván JA, Robinson HPC, Zlobec I, et al. 2019. Synaptic proximity enables NMDAR signalling to promote brain metastasis. *Nature* **573:** 526–531. doi:10.1038/s41586-019-1576-6

Zhang Y, Sloan SA, Clarke LE, Caneda C, Plaza CA, Blumenthal PD, Vogel H, Steinberg GK, Edwards MS, Li G, et al. 2016. Purification and characterization of progenitor and mature human astrocytes reveals transcriptional and functional differences with mouse. *Neuron* **89:** 37–53. doi:10.1016/j.neuron.2015.11.013

Zhou Y, Song WM, Andhey PS, Swain A, Levy T, Miller KR, Poliani PL, Cominelli M, Grover S, Gilfillan S, et al. 2020. Human and mouse single-nucleus transcriptomics reveal TREM2-dependent and TREM2-independent cellular responses in Alzheimer's disease. *Nat Med* **26:** 131–142. doi:10.1038/s41591-019-0695-9

Cite this article as *Cold Spring Harb Perspect Biol* doi: 10.1101/cshperspect.a041356

Features, Fates, and Functions of Oligodendrocyte Precursor Cells

Robert A. Hill,[1] Akiko Nishiyama,[2] and Ethan G. Hughes[3]

[1]Department of Biological Sciences, Dartmouth College, Hanover, New Hampshire 03755, USA

[2]Department of Physiology and Neurobiology, University of Connecticut, Storrs, Connecticut 06269, USA

[3]Department of Cell and Developmental Biology, University of Colorado School of Medicine, Aurora, Colorado 80045, USA

Correspondence: robert.hill@dartmouth.edu

Oligodendrocyte precursor cells (OPCs) are a central nervous system resident population of glia with a distinct molecular identity and an ever-increasing list of functions. OPCs generate oligodendrocytes throughout development and across the life span in most regions of the brain and spinal cord. This process involves a complex coordination of molecular checkpoints and biophysical cues from the environment that initiate the differentiation and integration of new oligodendrocytes that synthesize myelin sheaths on axons. Outside of their progenitor role, OPCs have been proposed to play other functions including the modulation of axonal and synaptic development and the participation in bidirectional signaling with neurons and other glia. Here, we review OPC identity and known functions and discuss recent findings implying other roles for these glial cells in brain physiology and pathology.

In this work, we provide a brief overview of oligodendrocyte precursor cell (OPC) development, fate, and touch on many of their known and proposed functions. Since their discovery, OPCs have been referred to by several names. This list includes small branching cells, type 1 oligodendrocytes, O-2A progenitor cells, NG2 cells, NG2 glia, polydendrocytes, synantocytes, and a few others. In this work, we will refer to these cells as OPCs to indicate their primary function. To be specific, these are the central nervous system (CNS) resident population of highly branched and tiled glial cells that express genes encoding the NG2 chondroitin sulfate proteoglycan (*Cspg4*) and the α receptor for platelet derived growth factor (*Pdgfra*). OPCs can generate oligodendrocytes and may play other roles in the nervous system. Before covering these many features, we first outline the historical context for the discovery, identification, and evolving definition of OPCs.

A BRIEF HISTORY OF THE DISCOVERY OF OPCs

OPCs were first described at the turn of the twentieth century by pioneering anatomists. Remarkably accurate sketches from dog and human brain tissue of cells resembling OPCs were made by the Scottish pathologist William Ford Robertson in 1899. Robertson (1899) called the cells "small branching cells." Subsequently, Río-Hortega (1921) used silver carbonate stain to correctly discriminate between microglia and oligodendrocytes and later classified oligodendrocytes into four types, referring to Robertson's cells as the "first type" (Río-Hortega 1928). This "first type" likely included some OPCs and a subpopulation of oligodendrocytes. Even with such prescient descriptions, these "small branching cells" remained relatively unrecognized and not studied for almost a century.

In the late 1970s to early 1980s, scientists used the A2B5 monoclonal antibody that recognizes a ganglioside to differentially mark two types of GFAP$^+$ astrocytes in cultures taken from rodent optic nerves (Raff et al. 1983a). It was found that A2B5$^+$ GFAP$^-$ cells differentiated into GalC$^+$ oligodendrocytes when maintained in chemically defined serum-free medium, whereas the same cells differentiated into A2B5$^+$ GFAP$^+$ type 2 astrocytes when cultured in the presence of serum (Raff et al. 1983b). This suggested that A2B5$^+$ cells give rise to both astrocytes and oligodendrocytes and were thus called bipotential O-2A (oligodendrocyte-type 2 astrocyte) progenitor cells. Similar studies found that the NG2 chondroitin sulfate proteoglycan, previously identified to label a population of cultured glia, was also expressed by the A2B5$^+$ O-2A progenitor cells and was down-regulated as the cells differentiated into GalC$^+$ oligodendrocytes (Stallcup and Beasley 1987). In the presence of serum, NG2 expression persisted on the type 2 astrocytes. These findings indicated that NG2 is an antigen expressed by O-2A progenitor cells, but a long debate followed as to the in vivo correlates of these cultured cells.

At the end of the 1980s, it was discovered that platelet-derived growth factor A (PDGF-AA) was the predominant mitogen for O-2A progenitor cells, and that its receptor PDGFRA was responsible for mediating the mitogenic effect of PDGF-AA on O-2A progenitor cells (Pringle et al. 1992). These studies also showed the first appearance of *Pdgfra* mRNA$^+$ cells in germinal regions of brain tissue, which was shortly followed by their migration out of the germinal zone and expansion in the parenchyma. *Pdgfra* expression was down-regulated in cells that underwent terminal differentiation into oligoden-

drocytes but persisted on some cells into adulthood. Thus, *Pdgfra* mRNA expression seemed to mark OPCs.

Subsequent studies investigated the coexpression of *Cspg4* and *Pdgfra*, revealing that there was an almost complete overlap between PDGFRA$^+$ and NG2$^+$ cell populations (Nishiyama et al. 1996). The notion that NG2$^+$ cells and *Pdgfra* mRNA$^+$ cells were the same cells and likely to be OPCs became accepted in the late 1990s; however, at that time, available techniques did not allow direct demonstration that OPCs could generate oligodendrocytes, as both NG2 and PDGFRA are lost upon their terminal differentiation into oligodendrocytes. Nonetheless, the establishment of these molecular markers and the characterization of these cells in culture and in tissues established them as the fourth major glial cell population in the CNS.

OPC DEVELOPMENT AND FATE

During mammalian brain development, PDFRA and NG2$^+$ OPCs are generated from distinct progenitor domains within the ganglionic eminences, ventricular zones, and spinal cord. The neural progenitors populating these regions are characterized by the production of specific transcription factors and give rise to OPCs that migrate throughout the brain and spinal cord (Rakic and Zecevic 2003; Cai et al. 2005; Kessaris et al. 2006; Huang et al. 2020).

Direct evidence for the oligodendrocyte fate of OPCs in mice has come mainly from Cre-lox fate mapping showing that NG2$^+$ and/or PDGFRA$^+$ cells are indeed OPCs (Kessaris et al. 2006; Dimou et al. 2008; Rivers et al. 2008; Zhu et al. 2008, 2011; Nishiyama et al. 2009; Kang et al. 2010). In addition to oligodendrocytes, these studies showed that OPCs also generate protoplasmic astrocytes in the gray matter of the ventral forebrain (Zhu et al. 2008). This suggested that some OPCs in prenatal CNS behaved like the culture identified bipotential O-2A cells; however, OPCs never generated fibrous astrocytes in white matter, and the astrocyte fate of OPCs is specifically restricted to early developmental stages (Zhu et al. 2011; Huang et al. 2019). Other studies initially suggested that a small number of neurons in the anterior pyriform cortex were generated from *Pdgfra*-expressing cells (Rivers et al. 2008), but a follow-up study did not support the original conclusion (Clarke et al. 2012). Neuronal fates were also reported in *Plp1*-CreER mice (Guo et al. 2010); however, there is reported nonspecific activation of the *Plp1* promoter in cells other than OPCs (Michalski et al. 2011). Neuronal fates were not observed in *Cspg4*-CreER mice (Zhu et al. 2011), in a different line of *Pdgfra*-CreER mice (Kang et al. 2010), nor in *Olig2*-CreER mice (Dimou et al. 2008). Thus, the current perspective is that, except for some protoplasmic astrocytes generated prenatally, OPCs only generate oligodendrocytes, at least during normal development and throughout life (Young et al. 2013; Tripathi et al. 2017; Hill et al. 2018; Hughes et al. 2018).

OPC RESIDENCY AND MORPHOLOGY

OPCs reside in almost all regions of the CNS. They exhibit a tiled distribution with complex multibranched arborization (Fig. 1). This tiling is established early in development and is found even in regions where no oligodendrocytes or myelin sheaths are generated (Lin et al. 2005; Goebbels et al. 2017). The widespread distribution raises several questions related to how and why this patterning is established and maintained and whether all OPCs are indeed progenitors for oligodendrocytes or if they serve other roles in the brain. For example, it is possible that there are different genetic and/or functional classes of OPCs participating in progenitor and nonprogenitor roles (Dimou and Simons 2017; Marisca et al. 2020; Beiter et al. 2022).

The tiling of OPCs is regulated through a balance of local proliferation, oligodendrocyte differentiation, and programed cell death (Raff et al. 1993; Trapp et al. 1997). A mix of growth factor signals are critical for the developmental establishment of the resident OPC populations including PDGF-AA, as established initially in cultured OPCs but confirmed in knockout and overexpressing mouse models (Noble et al. 1988; Richardson et al. 1988; Pringle et al. 1992; Calver et al. 1998), fibroblast growth factor (McKinnon et al. 1990; Baron et al. 2000), and neurotrophins

(Casaccia-Bonnefil et al. 1996; Cohen et al. 1996) among others (Barres et al. 1992). Cellular sources for these growth factors include neurons, other glia, and signals released from vascular endothelial cells. For example, migrating OPCs use the vasculature as a scaffold to populate the developing nervous system (Tsai et al. 2016; Lepiemme et al. 2022). Contact mediated signaling is also involved in OPC separation after cell division (Huang et al. 2020), highlighting the balance between diffusible environmental cues and OPC cell surface–specific signals. When OPCs do not receive permissive signals for differentiation and integration, they initiate programmed cell death pathways associated with cellular autophagy and apoptosis. When these pathways are disrupted, the balance between OPC self-renewal, death, and oligodendrocyte differentiation is skewed (Meireles et al. 2018; Sun et al. 2018). Thus, the coincidence of sufficient growth factor availability, permitting tissue substrates, and enhancing developmental migration results in the establishment of lifelong residency by OPCs throughout the CNS.

In the adult brain, it is less clear which signals maintain OPC tiling, but it is likely that a combination of diffusible and membrane tethered cell–cell contact signals regulate the local populations. Imaging studies of OPC process dynamics in zebrafish spinal cord and mouse cortex have found evidence for contact repulsion when neighboring OPC processes touch (Kirby et al. 2006; Hughes et al. 2013). This proposed ability for OPCs to sense their neighbors results in rapid OPC replacement via division and local migration when single OPCs die or differentiate into myelinating oligodendrocytes (Kirby et al. 2006; Hughes et al. 2013, 2018; Hill et al. 2017). This process likely also accounts for the rapid replacement of OPCs when widespread OPC-specific genetic-based cell ablation approaches are used (Xing et al. 2023). Moreover, like the developmental role for PDGF-AA, adult OPC population maintenance is dependent on signaling through PDGFRA (Đặng et al. 2019). This means that sustained environmental cues from neurons, other glial cells, and the neurovascular unit contribute to the tiling behavior of OPCs in the adult.

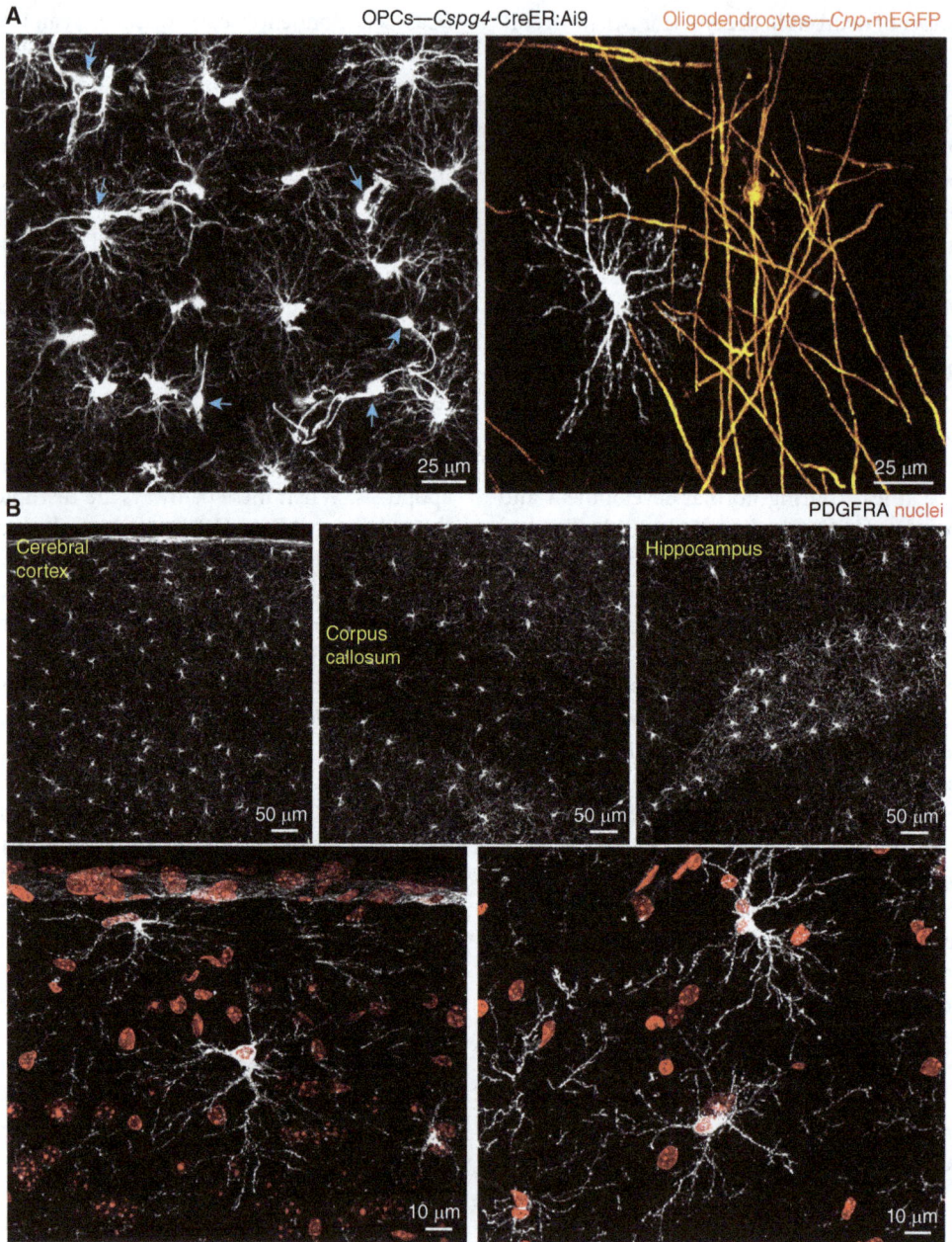

Figure 1. Oligodendrocyte precursor cell (OPC) morphology and residency in the brain. (*A*) OPCs imaged in the cerebral cortex of a living transgenic mouse with tdTomato fluorescent protein labeling (Ai9 [Madisen et al. 2010]) in all cells with *Cspg4* activity at the time of Cre recombination [(Cspg4-CreER [Zhu et al. 2011]). Both OPCs and vascular mural cells (blue arrows) are labeled, but OPCs have distinct multibranched and ramified morphologies. This mouse model allows titration of cellular labeling such that single OPCs (shown in gray) can be visualized as in the image on the *right*. Sparse labeling allows for analysis of OPC morphology in contrast to a myelinating oligodendrocyte (shown in orange) labeled with membrane tethered enhanced green fluorescent protein (EGFP) (Cnp-mEGFP [Deng et al. 2014]). (*B*) Immunohistochemistry for OPC-specific α receptor for platelet derived growth factor (PDGFRA) in various brain regions of the mouse forebrain shows widespread tiled distribution and ramified morphology for individual cells across the tissue.

Once OPCs are resident, they exhibit diverse morphologies dependent on the brain region and physiological context. For example, at baseline, gray and white matter OPCs differ in their process arborization, complexity, and size (Osorio et al. 2023). Similar differences are found between OPCs in neuron soma–rich versus axon-rich regions of the developing zebrafish (Marisca et al. 2020). The different morphologies observed in the zebrafish were associated with different fates and cellular activity, suggesting a connection between OPC form and function. Genetic underpinnings for this diversity are not clear but an association with cell division history was found as the emergence of a new morphological phenotype occurred almost exclusively after a cell division event instead of a single cell directly changing morphology without first dividing (Marisca et al. 2020). Without definitive genetic markers, connecting OPC morphological and functional heterogeneity has been challenging, and whether similar connections between morphology and function are present in developing and adult mammals is not clear. Future intravital imaging approaches allowing longitudinal investigations of OPC shape and function could reveal how closely these features are linked in other settings.

After acute injury, in neurodegenerative contexts, and even some psychiatric conditions, OPCs display morphological transformations often characterized by hypertrophy and increased branching (Ong and Levine 1999; Vanzulli et al. 2020; Yu et al. 2022; Chapman et al. 2023). These shape changes are generally considered to be a reactive phenotype contributing to the glial scar; however, a direct role for OPCs in this injury response is not clear. The best studied OPC morphological injury response is in the context of spinal cord injury and neocortical focal injury via mechanical or laser-mediated lesions (Hughes et al. 2013; von Streitberg et al. 2021). These studies indicate that OPC processes polarize toward the lesion within hours followed by cell soma migration and cell division. This response is thought to contribute to the barrier that is established with astrocytes and microglia, both by the presence of OPC processes and via the deposition of extracellular matrix. Inhibition of injury induced OPC proliferation leads to deficits in wound closure suggesting that this OPC-specific response is beneficial for injury containment and tissue regeneration (von Streitberg et al. 2021). Overall, OPCs can quickly alter their morphology in response to cellular and tissue damage, hinting at another link between OPC morphology and function.

INTRINSIC AND ADAPTIVE GENERATION OF OLIGODENDROCYTES

The primary function of OPCs is the generation of oligodendrocytes, a process that can occur in various contexts depending on the demands of the tissue. Many recent articles have extensively covered this topic highlighting that the signals that induce oligodendrocyte generation are multifaceted (Almeida and Lyons 2017; Bechler et al. 2018; Monje 2018; Chapman and Hill 2020; Xin and Chan 2020). These signals range from biophysical cues such as axon caliber to activity-dependent and/or sensory-dependent release of signals from neurons and other cells in the brain (Fig. 2A; Liu et al. 2012; Makinodan et al. 2012; Gibson et al. 2014; Hill et al. 2014; Bechler et al. 2015; Hughes et al. 2018; Mayoral et al. 2018; Mitew et al. 2018). Whether there are distinct programs that are initiated for developmental, intrinsic, and/or activity-modulated generation of oligodendrocytes is an active area of investigation in the field.

The generation of new oligodendrocytes has been shown to be important for specific motor, learning, and memory tasks (McKenzie et al. 2014; Xiao et al. 2016; Pan et al. 2020; Steadman et al. 2020; Wang et al. 2020). These experiments have all relied on a genetic trick to block new oligodendrocyte generation via the inducible OPC-specific deletion of transcription factors such as *Myrf* and *Olig2*, which are required for proper differentiation in adult animals. Although powerful, additional methods to explore the precise requirement for new myelin from the generation of new oligodendrocytes and the specific neural circuits that they modify will help further our understanding of the necessity for adaptive myelination in these and other learning paradigms. There is extensive literature demonstrating that OPCs are

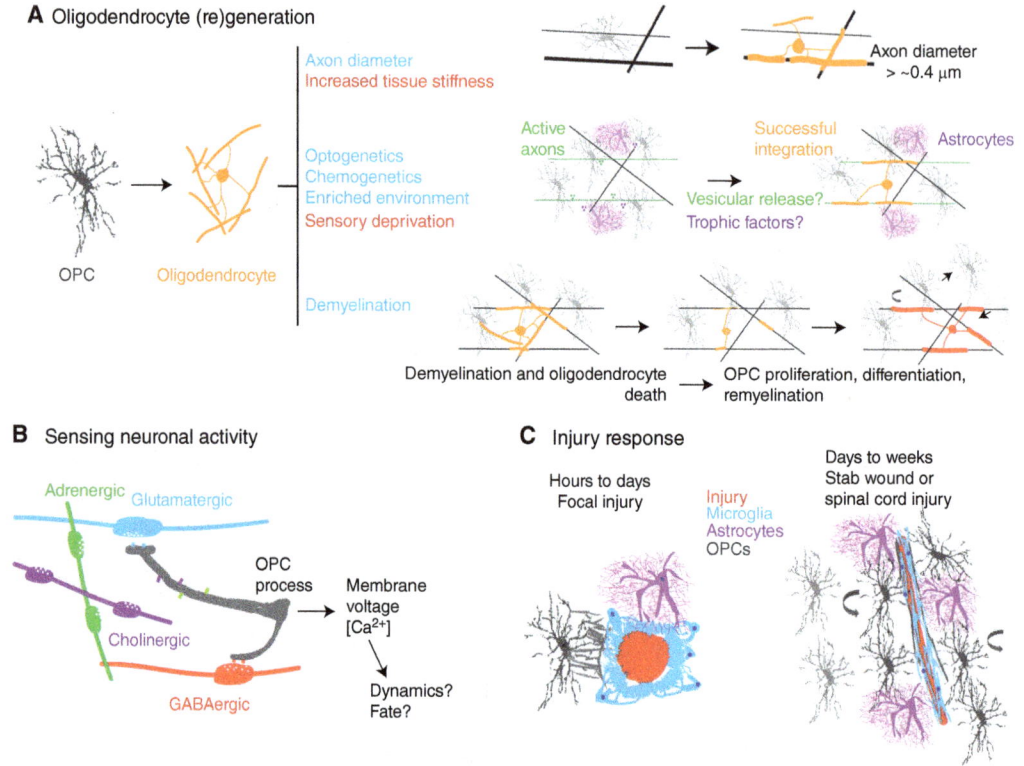

Figure 2. Established functions for oligodendrocyte precursor cells (OPCs). (*A*) The primary function of OPCs is to generate oligodendrocytes across the life span. This can be modulated by various environmental factors ranging from biophysical cues to neuron activity–dependent signaling. Factors that enhance oligodendrocyte generation are indicated in cyan and factors that decrease oligodendrocyte generation are shown in red. Similar processes are initiated in response to oligodendrocyte death and demyelination, in which OPCs serve as the main source of oligodendrocytes that are generated after myelin loss. (*B*) OPCs have many neurotransmitter receptors and receive direct synaptic input from glutamatergic and GABAergic neurons. These inputs initiate distinct patterns of OPC membrane depolarization and rises in intracellular calcium; however, the primary functional outcome of these inputs is not resolved. (*C*) After injury, OPCs respond within hours with process rearrangement and contribution to the glial scar with microglia and astrocytes. Depending on the scale of the damage, OPCs continue to respond via cell migration and proliferation.

the major source of remyelinating oligodendrocytes in demyelinating and neurodegenerative contexts. The signals inducing OPC differentiation after demyelination are likely a combination of the intrinsic and adaptive programs used in development and the adult coupled with an injury response and reactive OPC phenotypes. Remyelination by OPCs is discussed below.

SYNAPTIC INPUT TO OPCs

In addition to generating oligodendrocytes that myelinate axons, OPCs themselves have exten-

sive interactions with regions of axons that are unmyelinated. This includes direct neuronal synaptic input allowing OPCs to monitor neuronal activity via neurotransmitter receptor-mediated signaling (Bergles et al. 2000, 2010). All postnatal OPCs are thought to receive this form of synaptic input, which is lost once they differentiate into myelinating oligodendrocytes (De Biase et al. 2010; Kukley et al. 2010). The reason for such specific and specialized signaling is not clear, but the ability to sense patterns of neuronal activity is potentially linked to OPC fate decisions and successful integration during oligodendro-

Cite this article as *Cold Spring Harb Perspect Biol* doi: 10.1101/cshperspect.a041425

cyte generation (Fig. 2). Although neurotransmitter exposure and pharmacological manipulations can change OPC behavior and fate (Pende et al. 1994; Gallo et al. 1996; Li et al. 2013; Lundgaard et al. 2013; Zonouzi et al. 2015), direct evidence for how physiological neurotransmitter receptor activation is linked to OPC behavior and fate is inconsistent. Thus far, cell-type-specific in vivo manipulations have mainly involved glutamatergic receptor subtypes *N*-methyl-D-aspartate (NMDA) and α-amino-3-hydroxy-5-methyl-4-isoxazolepropionic acid (AMPA) and GABAergic receptor subtypes.

Deletion of NMDA receptors in OPCs shows no effects on OPC fate (De Biase et al. 2011; Guo et al. 2012) but instead causes altered axonal metabolism followed by delayed myelin degeneration (Saab et al. 2016), suggesting a more prominent role for the NMDA receptor in myelinating oligodendrocytes. Manipulations of AMPA receptors in OPCs have variable outcomes. One study demonstrated that OPC deletion of *Gria2*, *Gria3*, and *Gria4* genes that encode AMPA receptor subunits results in decreased oligodendrocyte survival and integration during differentiation without impacting OPC proliferation (Kougioumtzidou et al. 2017). *Gria2* regulates channel conductance and limits calcium permeability for AMPA receptors; therefore, other studies have focused specifically on altering the functionality or expression of *Gria2* to link AMPA mediated changes in intracellular calcium with OPC behavior (Chen et al. 2018; Khawaja et al. 2021). OPC-specific *Gria2* overexpression showed no effects on the generation of oligodendrocytes during development but increased OPC differentiation after injury (Khawaja et al. 2021). This finding is consistent with work showing that nonspecific AMPA receptor antagonism increases oligodendrocyte generation after white matter injury (Gautier et al. 2015). Viral mediated overexpression of modified *Gria2* causes changes in OPC proliferation with some minor decreases in oligodendrocyte generation (Chen et al. 2018). Therefore, common outcomes of these manipulations point toward *Gria2* modulating OPC proliferation in adult animals with minimal or inconsistent outcomes on oligodendrocyte differentiation

(Kukley 2023). Studies in zebrafish have shown that *gria4a* (an orthlog of *Gria4*) decreases OPC migration and the number of myelin sheaths made by mature oligodendrocytes (Piller et al. 2021); however, it is difficult to make a direct link between this result and the *Gria2* manipulations in the mouse given the differences in the subunits and experimental readouts used. The connection between direct glutamatergic synaptic input signaling through NMDA and/or AMPA receptors and OPC fate outcomes is yet to be resolved. Altogether, how synaptic glutamate release impacts OPC fate requires more detailed investigations using consistent methodologies, molecular manipulations, and animal models.

OPCs also express a variety of genes encoding GABA receptors including ionotropic GABAA and metabotropic GABAB receptors with a variety of subunit compositions (Habermacher et al. 2019). Moreover, OPCs receive direct synaptic input from interneurons in the developing hippocampus and cerebral cortex (Lin and Bergles 2004; Vélez-Fort et al. 2010; Balia et al. 2015). In fact, OPCs and parvalbumin (PV)- positive fast-spiking interneurons arise from similar germinal niches, and this developmental source predicts PV synaptic input to lineage related OPCs (Orduz et al. 2019). Deletion of GABAB receptors in OPCs results in decreased oligodendrocyte differentiation (Fang et al. 2022). However, it is unclear whether this is because of the lack of OPCs differentiating or that there are also fewer PV axons to myelinate because, intriguingly, loss of GABABR in OPCs also results in decreased survival of PV-positive neurons. In contrast, conditional OPC GABA γ2 subunit deletion does not cause a change in OPC proliferation or oligodendrocyte generation (Balia et al. 2017). However, γ2 deletion in OPCs does result in a change in myelin patterning and targeting on PV positive axons (Benamer et al. 2020), suggesting a more subtle but important contribution of GABAergic signaling through GABA receptors containing the γ2 subunit in OPCs. Like the story with glutamatergic signaling, additional experiments are needed to more clearly define how synaptic input from GABAergic neurons impacts OPC behavior.

Separate from direct synaptic input from glutamatergic and GABAergic neurons, OPCs also possess many other neurotransmitter receptors including cholinergic, adrenergic, dopaminergic, and purinergic, among others (Káradóttir and Attwell 2007; Akay et al. 2021). For example, muscarinic acetylcholine receptors have recently emerged as drivers of OPC fate for remyelination therapy as discussed below (Deshmukh et al. 2013; Mei et al. 2014; Green et al. 2017). Recent experiments have also shown the norepinephrine signaling onto OPCs can regulate local calcium signals and OPC fate in vivo (Fiore et al. 2022; Lu et al. 2022). Given the number of receptor subunits expressed by OPCs, deletion of one or two may not be sufficient to fully block the downstream signaling. A recent zebrafish study, also linking synaptic release to local calcium signals, shows that disruption of major postsynaptic organizers (PSD-95 or Gephyrin) impairs OPC differentiation and oligodendrocyte myelination (Li et al. 2023). As is the case of growth factor signaling in modulating OPC behavior and fate, the same is true for these neurotransmitter signals that have many roles and specific contexts for how and when these neuromodulators impact OPC behavior.

OPCs IN AXON PLASTICITY, PHAGOCYTOSIS, AND IMMUNE SIGNALING

OPCs also exhibit roles for modulating axonal growth, plasticity, and regeneration after injury. There is extensive literature suggesting that the OPC response to tissue damage, like spinal cord injury, contributes to the glial scar and an inhibitory environment for regeneration and repair (Levine 2016; Bradbury and Burnside 2019). This idea primarily comes from the increased production of extracellular matrix molecules in damaged tissue, some of which are thought to derive from OPCs (Asher et al. 2002; Garwood et al. 2004). These findings initially suggested that OPCs could be a repelling source for axons; however, as was just discussed, OPCs make extensive synaptic contacts with axons and also have been shown to attract axons in culture and some injury contexts (Yang et al. 2006; Busch

et al. 2010; Filous et al. 2014). Thus, OPCs might limit axon regeneration because of their adhesion with growing axons instead of their repulsion via production of NG2 and/or other secreted chondroitin sulfate proteoglycans known to limit axon regeneration (Duncan et al. 2020). Further understanding of the mechanisms impacting synapse formation between OPCs and growing axons and how reactive OPCs vary from homeostatic OPCs in their axon interactions will help resolve this question.

During development, several studies show that OPCs can modulate axon and synapse plasticity. Taking advantage of the zebrafish optic tectum where OPCs are resident without a local myelinating oligodendrocyte population, OPC ablation was found to increase axonal arborization and complexity, resulting in a behavioral change in prey capture (Xiao et al. 2022). Another recent study discovered that OPCs contain a significant amount of axon-derived debris identified via serial electron microscopy, suggesting that these cells are involved in the engulfment/phagocytosis of axons (Buchanan et al. 2022). Similarly, immunohistochemistry in the developing mouse cortex provides some evidence for synaptic debris in OPCs proposing a role for these cells in developmental synaptic pruning (Auguste et al. 2022). Finally, intravital imaging of OPCs adjacent to neuronal cell death events revealed a targeted rearrangement of OPC processes surrounding the dying neuron (Damisah et al. 2020). Altogether, these studies suggest a potential role for OPCs in modulating axonal and synaptic structure and participating in cell debris processing through paracrine or phagocytic mechanisms independent of myelination (Fig. 3A,B). Many questions remain for whether these observations are connected and the molecular signaling pathways involved. For example, there was no evidence for OPC-specific axonal engulfment in the zebrafish study (Xiao et al. 2022) and whether the engulfment of the axonal debris in the electron microscopy study was passive or active could not be determined (Buchanan et al. 2022). There is some evidence that OPCs express genes that encode phagocytic receptors such as *Mertk*, *Ptprj*, and *Lrp1* (Buchanan et al. 2022); however, a direct connection between

Cite this article as *Cold Spring Harb Perspect Biol* doi: 10.1101/cshperspect.a041425

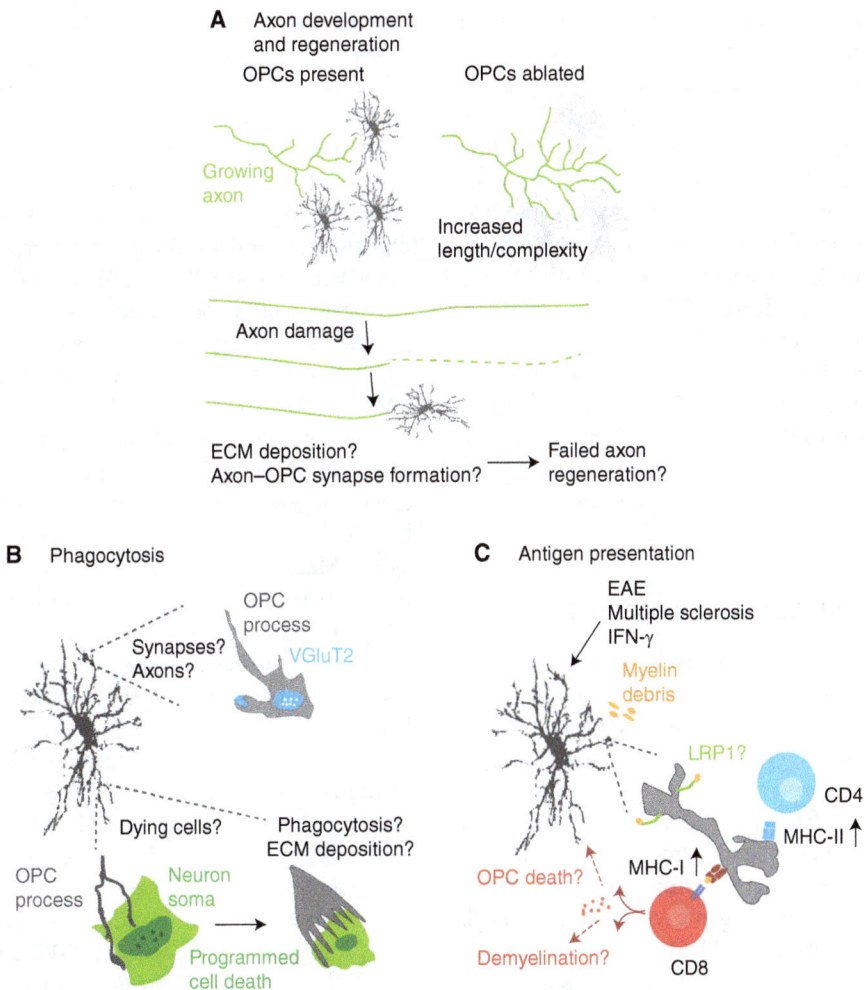

Figure 3. Alternative functions for oligodendrocyte precursor cells (OPCs). (*A*) OPCs might play a role in axon growth and regeneration. During development, OPC ablation can result in altered axonal arborizations and, after injury, OPCs might inhibit axonal regeneration either via deposition of extracellular matrix (ECM) or via synaptic connectivity with regenerating axons. (*B*) OPCs exhibit phagocytic-like behavior with some evidence showing synaptic and axonal debris in OPCs and directed OPC process rearrangement when neighboring neurons die. (*C*) OPCs up-regulate major histocompatibility complex (MHC)-encoding transcripts in various diseases and experimental contexts and engage in antigen presentation and activation of CD8 T cells, potentially initiating and/or exacerbating disease. (ECM) Extracellular matrix, (EAE) experimental autoimmune encephalomyelitis, (IFN-γ) interferon γ.

these genes and OPC phagocytosis and/or engulfment of debris is lacking.

Potentially downstream of phagocytosis, other work has discovered that OPCs can participate in immune signaling through specific activation and antigen presentation (Falcão et al. 2018; Kirby et al. 2019; Fernández-Castañeda et al. 2020; Meijer et al. 2022; Harrington et al. 2023). These observations initially came from single-cell RNA sequencing in experimental autoimmune encephalomyelitis and multiple sclerosis (MS) tissues and demonstrated that a subpopulation of oligodendrocyte lineage cells up-regulate transcripts associated with antigen presentation including those encoding MHC-I and MHC-II (Falcão et al. 2018; Jäkel et al.

2019). Similar signals were detected in cultured OPCs exposed to interferon γ, suggesting that this is a cell autonomous response by some OPCs to cytokine exposure (Jäkel et al. 2019; Kirby et al. 2019). The upstream signal leading to the recognition, engulfment, and MHC-I antigen presentation is not well defined in OPCs, but one study demonstrated a role for OPC production of the phagocytic receptor LRP1 in this process (Fernández-Castañeda et al. 2020). A particularly detrimental outcome of this response and increased MHC-I production by OPCs is the activation of CD8 T cells and resulting cytotoxic OPC (and oligodendrocyte) death, potentially depleting the pool available for remyelination and exacerbating autoimmune-mediated demyelination (Fig. 3C).

OPCs IN NEUROPATHOLOGY AND AGING

In diseases such as MS, myelin-producing oligodendrocytes undergo cell death resulting in demyelination. Following this demyelinating injury, a spontaneous regenerative response can lead to the production of new oligodendrocytes and myelin. Genetic fate mapping has conclusively shown that these new oligodendrocytes are generated from OPCs both in the surrounding parenchyma as well as from neurogenic zones (Tripathi et al. 2010; Samanta et al. 2015). Remyelination can restore action potential propagation and protect from axonal injury (Smith et al. 1979; Mei et al. 2016). Therefore, several screens for compounds that promote differentiation of OPCs were conducted to identify regenerative approaches to treat demyelinating injuries (Deshmukh et al. 2013; Mei et al. 2014; Najm et al. 2015; Early et al. 2018; Hubler et al. 2018). Some of these candidate drugs had positive results in animal models of demyelination and moved to human clinical trials where they continue to be evaluated for promoting remyelination (for review, see Lubetzki et al. 2020). Continued validation of electrophysiological and diagnostic imaging biomarkers of remyelination will aid in determining the efficacy of therapies that target restoration of function elicited via OPC differentiation to drive myelin repair (Caverzasi et al. 2023). Patients with MS show varying levels of oligodendrocyte generation

(Yeung et al. 2019), suggesting that therapies focused on remyelination via harnessing the regenerative capacity of OPCs may be essential for promoting functional recovery.

Although the survival rates of myelinating oligodendrocytes are remarkably high across life span (Tripathi et al. 2017), gradual oligodendrocyte death and myelin degeneration are associated with aging (Hill et al. 2018). Furthermore, the capacity of oligodendrocyte and myelin regeneration also declines with age (Shields et al. 1999; Psachoulia et al. 2009; Chapman et al. 2023) mediated in part by age-associated changes in OPCs (Sim et al. 2002; de la Fuente et al. 2020). During development, OPCs start out as a homogeneous population but become functionally heterogeneous across brain regions and aging (Marques et al. 2016; Spitzer et al. 2019). Age-associated transcriptomic and electrophysiological changes of OPCs lead to a decreased potential for differentiation and generation of myelinating oligodendrocytes (Shen et al. 2008; Spitzer et al. 2019). Changes to extrinsic factors such as tissue stiffness in the aged microenvironment may limit the capacity of OPCs to generate new oligodendrocytes (Segel et al. 2019). Recent studies show blood- and CSF-derived factors from young animals promote OPC proliferation and differentiation via youthful monocytes and FGF17, respectively (Ruckh et al. 2012; Iram et al. 2022). In addition, intrinsic factors such as TET1-mediated DNA hydroxymethylation result in an age-dependent decline in myelin repair (Moyon et al. 2021), suggesting multiple factors lead to age-related decline in OPC functions. Recent studies sought to identify therapeutics that increase the capacity for oligodendrogenesis and myelin regeneration in the aged nervous system. The small molecule fasting mimetic, metformin, which modulates the AMPK pathway, restores the differentiation and regeneration capacity of aged OPCs (Neumann et al. 2019). Genetically or pharmacologically enhancing oligodendrogenesis and myelination can reverse age-related memory decline and neurodegeneration (Wang et al. 2020; Chen et al. 2021). Future work continuing to focus on the effects of aging on OPCs and exploration of interventions that facilitate OPC function will help to counteract age-related decline as well as treatment of neurodegenerative diseases.

CONCLUDING REMARKS

OPCs are well recognized as a separate glial cell population in the brain, and we know a great deal about their role as the precursors for oligodendrocytes. Many molecular signals regulating this process have been discovered and several are now being applied in clinical settings to enhance oligodendrocyte regeneration in neurodegenerative contexts. Even with this extensive and rich literature, many questions remain regarding when, where, and how these molecular signaling cascades result in the initiation, differentiation, and successful integration of a mature myelinating oligodendrocyte in the intact nervous system (Hughes and Stockton 2021). Precise molecular and cellular manipulation coupled with in vivo assays will continue to reveal the signals involved in these OPC fate decisions.

Other remaining questions have been highlighted throughout this review. These include heterogeneity within the OPC population along with further investigations into the functions played by OPCs beyond their progenitor role (Dimou and Simons 2017). When considering heterogeneity, it is important to be clear whether the heterogeneity arises from genetic diversity within the OPC population due to developmental source or other intrinsic predetermined genetic programs. There is little evidence for source-dependent or genetic heterogeneity within OPCs. The different transcriptome subtypes of oligodendrocytes revealed via RNA sequencing primarily indicates the continuum of differentiation states from OPC to premyelinating oligodendrocytes to fully mature myelinating oligodendrocytes, with more evidence suggesting diversity in the myelinating stage compared to the OPC stage also somewhat complicated by the cell cycle in OPCs (Marques et al. 2016). Even without clear genetic diversity, there is more evidence for OPC functional heterogeneity likely representing plasticity of OPC states in response to cues from the local microenvironment (Kamen et al. 2022). These include differences in OPC properties and functional responses by

brain region (Hill et al. 2013; Viganò et al. 2013; Marisca et al. 2020; Sherafat et al. 2021) and across different ages (Shen et al. 2008; Neumann et al. 2019; Spitzer et al. 2019). Contributions from the local microenvironment, whether they be from physiological signaling from neighboring neurons or glia or activation of OPCs in response to pathological conditions, likely drive these heterogenous states. This is not to say that these microenvironmental impacts cannot have long-term consequences for OPC function, even when these cells are placed in a new environment. It just asks the question of whether OPC heterogeneity is genetically predetermined during development or if these different states emerge through environmental influences. Future work is sure to further clarify the duration, functional implications, and reversibility of these OPC functional states.

As we have highlighted throughout, another major remaining question is: What are other functions (beyond their progenitor role) played by OPCs in nervous system physiology and pathology? These range from engaging in bidirectional signaling with almost every other cell type in the brain and potentially even the peripheral immune system, to responding to damage and disease with distinct behaviors. Overall, recent work has established that OPCs are multifunctional glia that likely contribute significantly toward nervous system development, plasticity, and neurodegeneration on top of their primary role of making myelinating cells.

ACKNOWLEDGMENTS

This work was supported by the following funding sources: National Institutes of Health R01 NS122800 and the Esther A. and Joseph Klingenstein Fund and Simons Foundation to R.A.H.; National Institutes of Health R01 NS073425 and NS116182 to A.N.; and National Institutes of Health R01 NS115975 and NS125230 to E.G.H.

REFERENCES

Akay LA, Effenberger AH, Tsai LH. 2021. Cell of all trades: oligodendrocyte precursor cells in synaptic, vascular, and immune function. *Genes Dev* **35**: 180–198. doi:10.1101/gad.344218.120

Almeida RG, Lyons DA. 2017. On myelinated axon plasticity and neuronal circuit formation and function. *J Neurosci* **37**: 10023–10034. doi:10.1523/JNEUROSCI.3185-16.2017

Asher RA, Morgenstern DA, Shearer MC, Adcock KH, Pesheva P, Fawcett JW. 2002. Versican is upregulated in CNS injury and is a product of oligodendrocyte lineage cells. *J Neurosci* **22**: 2225–2236. doi:10.1523/JNEUROSCI.22-06-02225.2002

Auguste YSS, Ferro A, Kahng JA, Xavier AM, Dixon JR, Vrudhula U, Nichitiu A-S, Rosado D, Wee TL, Pedmale UV, et al. 2022. Oligodendrocyte precursor cells engulf synapses during circuit remodeling in mice. *Nat Neurosci* **25**: 1273–1278. doi:10.1038/s41593-022-01170-x

Balia M, Vélez-Fort M, Passlick S, Schäfer C, Audinat E, Steinhäuser C, Seifert G, Angulo MC. 2015. Postnatal down-regulation of the GABAA receptor γ2 subunit in neocortical NG2 cells accompanies synaptic-to-extrasynaptic switch in the GABAergic transmission mode. *Cereb Cortex* **25**: 1114–1123. doi:10.1093/cercor/bht309

Balia M, Benamer N, Angulo MC. 2017. A specific GABAergic synapse onto oligodendrocyte precursors does not regulate cortical oligodendrogenesis. *Glia* **65**: 1821–1832. doi:10.1002/glia.23197

Baron W, Metz B, Bansal R, Hoekstra D, de Vries H. 2000. PDGF and FGF-2 signaling in oligodendrocyte progenitor cells: regulation of proliferation and differentiation by multiple intracellular signaling pathways. *Mol Cell Neurosci* **15**: 314–329. doi:10.1006/mcne.1999.0827

Barres BA, Hart IK, Coles HS, Burne JF, Voyvodic JT, Richardson WD, Raff MC. 1992. Cell death and control of cell survival in the oligodendrocyte lineage. *Cell* **70**: 31–46. doi:10.1016/0092-8674(92)90531-G

Bechler ME, Byrne L, Ffrench-Constant C. 2015. CNS myelin sheath lengths are an intrinsic property of oligodendrocytes. *Curr Biol* **25**: 2411–2416. doi:10.1016/j.cub.2015.07.056

Bechler ME, Swire M, Ffrench-Constant C. 2018. Intrinsic and adaptive myelination-A sequential mechanism for smart wiring in the brain. *Dev Neurobiol* **78**: 68–79. doi:10.1002/dneu.22518

Beiter RM, Rivet-Noor C, Merchak AR, Bai R, Johanson DM, Slogar E, Sol-Church K, Overall CC, Gaultier A. 2022. Evidence for oligodendrocyte progenitor cell heterogeneity in the adult mouse brain. *Sci Rep* **12**: 12921. doi:10.1038/s41598-022-17081-7

Benamer N, Vidal M, Balia M, Angulo MC. 2020. Myelination of parvalbumin interneurons shapes the function of cortical sensory inhibitory circuits. *Nat Commun* **11**: 5151. doi:10.1038/s41467-020-18984-7

Bergles DE, Roberts JD, Somogyi P, Jahr CE. 2000. Glutamatergic synapses on oligodendrocyte precursor cells in the hippocampus. *Nature* **405**: 187–191. doi:10.1038/35012083

Bergles DE, Jabs R, Steinhäuser C. 2010. Neuron–glia synapses in the brain. *Brain Res Rev* **63**: 130–137. doi:10.1016/j.brainresrev.2009.12.003

Bradbury EJ, Burnside ER. 2019. Moving beyond the glial scar for spinal cord repair. *Nat Commun* **10**: 3879. doi:10.1038/s41467-019-11707-7

Buchanan J, Elabbady L, Collman F, Jorstad NL, Bakken TE, Ott C, Glatzer J, Bleckert AA, Bodor AL, Brittain D, et al. 2022. Oligodendrocyte precursor cells ingest axons in the mouse neocortex. *Proc Natl Acad Sci* **119**: e2202580119. doi:10.1073/pnas.2202580119

Busch SA, Horn KP, Cuascut FX, Hawthorne AL, Bai L, Miller RH, Silver J. 2010. Adult NG2$^+$ cells are permissive to neurite outgrowth and stabilize sensory axons during macrophage-induced axonal dieback after spinal cord injury. *J Neurosci* **30**: 255–265. doi:10.1523/JNEUROSCI.3705-09.2010

Cai J, Qi Y, Hu X, Tan M, Liu Z, Zhang J, Li Q, Sander M, Qiu M. 2005. Generation of oligodendrocyte precursor cells from mouse dorsal spinal cord independent of Nkx6 regulation and Shh signaling. *Neuron* **45**: 41–53. doi:10.1016/j.neuron.2004.12.028

Calver AR, Hall AC, Yu WP, Walsh FS, Heath JK, Betsholtz C, Richardson WD. 1998. Oligodendrocyte population dynamics and the role of PDGF in vivo. *Neuron* **20**: 869–882. doi:10.1016/S0896-6273(00)80469-9

Casaccia-Bonnefil P, Carter BD, Dobrowsky RT, Chao MV. 1996. Death of oligodendrocytes mediated by the interaction of nerve growth factor with its receptor p75. *Nature* **383**: 716–719. doi:10.1038/383716a0

Caverzasi E, Papinutto N, Cordano C, Kirkish G, Gundel TJ, Zhu A, Akula AV, Boscardin WJ, Neeb H, Henry RG, et al. 2023. MWF of the corpus callosum is a robust measure of remyelination: results from the ReBUILD trial. *Proc Natl Acad Sci* **120**: e2217635120. doi:10.1073/pnas.2217635120

Chapman TW, Hill RA. 2020. Myelin plasticity in adulthood and aging. *Neurosci Lett* **715**: 134645. doi:10.1016/j.neulet.2019.134645

Chapman TW, Olveda GE, Bame X, Pereira E, Hill RA. 2023. Oligodendrocyte death initiates synchronous remyelination to restore cortical myelin patterns in mice. *Nat Neurosci* **26**: 555–569. doi:10.1038/s41593-023-01271-1

Chen TJ, Kula B, Nagy B, Barzan R, Gall A, Ehrlich I, Kukley M. 2018. In vivo regulation of oligodendrocyte precursor cell proliferation and differentiation by the AMPA-receptor subunit GluA2. *Cell Rep* **25**: 852–861.e7. doi:10.1016/j.celrep.2018.09.066

Chen JF, Liu K, Hu B, Li RR, Xin W, Chen H, Wang F, Chen L, Li RX, Ren SY, et al. 2021. Enhancing myelin renewal reverses cognitive dysfunction in a murine model of Alzheimer's disease. *Neuron* **109**: 2292–2307.e5. doi:10.1016/j.neuron.2021.05.012

Clarke LE, Young KM, Hamilton NB, Li H, Richardson WD, Attwell D. 2012. Properties and fate of oligodendrocyte progenitor cells in the corpus callosum, motor cortex, and piriform cortex of the mouse. *J Neurosci* **32**: 8173–8185. doi:10.1523/JNEUROSCI.0928-12.2012

Cohen RI, Marmur R, Norton WT, Mehler MF, Kessler JA. 1996. Nerve growth factor and neurotrophin-3 differentially regulate the proliferation and survival of developing rat brain oligodendrocytes. *J Neurosci* **16**: 6433–6442. doi:10.1523/JNEUROSCI.16-20-06433.1996

Damisah EC, Hill RA, Rai A, Chen F, Rothlin CV, Ghosh S, Grutzendler J. 2020. Astrocytes and microglia play orchestrated roles and respect phagocytic territories during neuronal corpse removal in vivo. *Sci Adv* **6**: eaba3239. doi:10.1126/sciadv.aba3239

Đặng TC, Ishii Y, Nguyen VD, Yamamoto S, Hamashima T, Okuno N, Nguyen QL, Sang Y, Ohkawa N, Saitoh Y, et al. 2019. Powerful homeostatic control of oligodendroglial lineage by PDGFRα in adult brain. *Cell Rep* **27**: 1073–1089.e5. doi:10.1016/j.celrep.2019.03.084

De Biase LM, Nishiyama A, Bergles DE. 2010. Excitability and synaptic communication within the oligodendrocyte lineage. *J Neurosci* **30**: 3600–3611. doi:10.1523/JNEUROSCI.6000-09.2010

De Biase LM, Kang SH, Baxi EG, Fukaya M, Pucak ML, Mishina M, Calabresi PA, Bergles DE. 2011. NMDA receptor signaling in oligodendrocyte progenitors is not required for oligodendrogenesis and myelination. *J Neurosci* **31**: 12650–12662. doi:10.1523/JNEUROSCI.2455-11.2011

de la Fuente AG, Queiroz RML, Ghosh T, McMurran CE, Cubillos JF, Bergles DE, Fitzgerald DC, Jones CA, Lilley KS, Glover CP, et al. 2020. Changes in the oligodendrocyte progenitor cell proteome with ageing. *Mol Cell Proteomics* **19**: 1281–1302. doi:10.1074/mcp.RA120.002102

Deng Y, Kim B, He X, Kim S, Lu C, Wang H, Cho SG, Hou Y, Li J, Zhao X, et al. 2014. Direct visualization of membrane architecture of myelinating cells in transgenic mice expressing membrane-anchored EGFP. *Genesis* **52**: 341–349. doi:10.1002/dvg.22751

Deshmukh VA, Tardif V, Lyssiotis CA, Green CC, Kerman B, Kim HJ, Padmanabhan K, Swoboda JG, Ahmad I, Kondo T, et al. 2013. A regenerative approach to the treatment of multiple sclerosis. *Nature* **502**: 327–332. doi:10.1038/nature12647

Dimou L, Simons M. 2017. Diversity of oligodendrocytes and their progenitors. *Curr Opin Neurobiol* **47**: 73–79. doi:10.1016/j.conb.2017.09.015

Dimou L, Simon C, Kirchhoff F, Takebayashi H, Götz M. 2008. Progeny of Olig2-expressing progenitors in the gray and white matter of the adult mouse cerebral cortex. *J Neurosci* **28**: 10434–10442. doi:10.1523/JNEUROSCI.2831-08.2008

Duncan GJ, Manesh SB, Hilton BJ, Assinck P, Plemel JR, Tetzlaff W. 2020. The fate and function of oligodendrocyte progenitor cells after traumatic spinal cord injury. *Glia* **68**: 227–245. doi:10.1002/glia.23706

Early JJ, Cole KL, Williamson JM, Swire M, Muskavitch M, Lyons DA. 2018. An automated high-resolution in vivo screen in zebrafish to identify chemical regulators of myelination. *eLife* **7**: e35136. doi:10.7554/eLife.35136

Falcão AM, van Bruggen D, Marques S, Meijer M, Jäkel S, Agirre E, Samudyata FE, Vanichkina DP, Ffrench-Constant C, et al. 2018. Disease-specific oligodendrocyte lineage cells arise in multiple sclerosis. *Nat Med* **24**: 1837–1844. doi:10.1038/s41591-018-0236-y

Fang LP, Zhao N, Caudal LC, Chang HF, Zhao R, Lin CH, Hainz N, Meier C, Bettler B, Huang W, et al. 2022. Impaired bidirectional communication between interneurons and oligodendrocyte precursor cells affects social cognitive behavior. *Nat Commun* **13**: 1394. doi:10.1038/s41467-022-29020-1

Fernández-Castañeda A, Chappell MS, Rosen DA, Seki SM, Beiter RM, Johanson DM, Liskey D, Farber E, Onengut-Gumuscu S, Overall CC, et al. 2020. The active contribution of OPCs to neuroinflammation is mediated by LRP1.

Acta Neuropathol **139**: 365–382. doi:10.1007/s00401-019-02073-1

Filous AR, Tran A, Howell CJ, Busch SA, Evans TA, Stallcup WB, Kang SH, Bergles DE, Lee S, Levine JM, et al. 2014. Entrapment via synaptic-like connections between NG2 proteoglycan+ cells and dystrophic axons in the lesion plays a role in regeneration failure after spinal cord injury. *J Neurosci* **34**: 16369–16384. doi:10.1523/JNEUROSCI.1309-14.2014

Fiore F, Dereddi RR, Alhalaseh K, Coban I, Harb A, Agarwal A. 2022. Norepinephrine regulates Ca²⁺ signals and fate of oligodendrocyte progenitor cells in the cortex. bioRxiv doi:10.1101/2022.08.31.505555

Gallo V, Zhou JM, McBain CJ, Wright P, Knutson PL, Armstrong RC. 1996. Oligodendrocyte progenitor cell proliferation and lineage progression are regulated by glutamate receptor-mediated K+ channel block. *J Neurosci* **16**: 2659–2670. doi:10.1523/JNEUROSCI.16-08-02659.1996

Garwood J, Garcion E, Dobbertin A, Heck N, Calco V, Ffrench-Constant C, Faissner A. 2004. The extracellular matrix glycoprotein Tenascin-C is expressed by oligodendrocyte precursor cells and required for the regulation of maturation rate, survival and responsiveness to platelet-derived growth factor. *Eur J Neurosci* **20**: 2524–2540. doi:10.1111/j.1460-9568.2004.03727.x

Gautier HOB, Evans KA, Volbracht K, James R, Sitnikov S, Lundgaard I, James F, Lao-Peregrin C, Reynolds R, Franklin RJM, et al. 2015. Neuronal activity regulates remyelination via glutamate signalling to oligodendrocyte progenitors. *Nat Commun* **6**: 8518. doi:10.1038/ncomms9518

Gibson EM, Purger D, Mount CW, Goldstein AK, Lin GL, Wood LS, Inema I, Miller SE, Bieri G, Zuchero JB, et al. 2014. Neuronal activity promotes oligodendrogenesis and adaptive myelination in the mammalian brain. *Science* **344**: 1252304. doi:10.1126/science.1252304

Goebbels S, Wieser GL, Pieper A, Spitzer S, Weege B, Yan K, Edgar JM, Yagensky O, Wichert SP, Agarwal A, et al. 2017. A neuronal PI(3,4,5)P3-dependent program of oligodendrocyte precursor recruitment and myelination. *Nat Neurosci* **20**: 10–15. doi:10.1038/nn.4425

Green AJ, Gelfand JM, Cree BA, Bevan C, Boscardin WJ, Mei F, Inman J, Arnow S, Devereux M, Abounasr A, et al. 2017. Clemastine fumarate as a remyelinating therapy for multiple sclerosis (ReBUILD): a randomised, controlled, double-blind, crossover trial. *Lancet* **390**: 2481–2489. doi:10.1016/S0140-6736(17)32346-2

Guo F, Maeda Y, Ma J, Xu J, Horiuchi M, Miers L, Vaccarino F, Pleasure D. 2010. Pyramidal neurons are generated from oligodendroglial progenitor cells in adult piriform cortex. *J Neurosci* **30**: 12036–12049. doi:10.1523/JNEUROSCI.1360-10.2010

Guo F, Maeda Y, Ko EM, Delgado M, Horiuchi M, Soulika A, Miers L, Burns T, Itoh T, Shen H, et al. 2012. Disruption of NMDA receptors in oligodendroglial lineage cells does not alter their susceptibility to experimental autoimmune encephalomyelitis or their normal development. *J Neurosci* **32**: 639–645. doi:10.1523/JNEUROSCI.4073-11.2012

Habermacher C, Angulo MC, Benamer N. 2019. Glutamate versus GABA in neuron–oligodendroglia communication. *Glia* **67**: 2092–2106. doi:10.1002/glia.23618

Harrington EP, Catenacci RB, Smith MD, Heo D, Miller CE, Meyers KR, Glatzer J, Bergles DE, Calabresi PA. 2023. MHC class I and MHC class II reporter mice enable analysis of immune oligodendroglia in mouse models of multiple sclerosis. *eLife* **12:** e82938. doi:10.7554/eLife.82938

Hill RA, Patel KD, Medved J, Reiss AM, Nishiyama A. 2013. NG2 cells in white matter but not gray matter proliferate in response to PDGF. *J Neurosci* **33:** 14558–14566. doi:10.1523/JNEUROSCI.2001-12.2013

Hill RA, Patel KD, Goncalves CM, Grutzendler J, Nishiyama A. 2014. Modulation of oligodendrocyte generation during a critical temporal window after NG2 cell division. *Nat Neurosci* **17:** 1518–1527. doi:10.1038/nn.3815

Hill RA, Damisah EC, Chen F, Kwan AC, Grutzendler J. 2017. Targeted two-photon chemical apoptotic ablation of defined cell types in vivo. *Nat Commun* **8:** 15837. doi:10.1038/ncomms15837

Hill RA, Li AM, Grutzendler J. 2018. Lifelong cortical myelin plasticity and age-related degeneration in the live mammalian brain. *Nat Neurosci* **21:** 683–695. doi:10.1038/s41593-018-0120-6

Huang W, Guo Q, Bai X, Scheller A, Kirchhoff F. 2019. Early embryonic NG2 glia are exclusively gliogenic and do not generate neurons in the brain. *Glia* **67:** 1094–1103. doi:10.1002/glia.23590

Huang W, Bhaduri A, Velmeshev D, Wang S, Wang L, Rottkamp CA, Alvarez-Buylla A, Rowitch DH, Kriegstein AR. 2020. Origins and proliferative states of human oligodendrocyte precursor cells. *Cell* **182:** 594–608.e11. doi:10.1016/j.cell.2020.06.027

Hubler Z, Allimuthu D, Bederman I, Elitt MS, Madhavan M, Allan KC, Shick HE, Garrison E, T. Karl M, Factor DC, et al. 2018. Accumulation of 8,9-unsaturated sterols drives oligodendrocyte formation and remyelination. *Nature* **560:** 372–376. doi:10.1038/s41586-018-0360-3

Hughes EG, Stockton ME. 2021. Premyelinating oligodendrocytes: mechanisms underlying cell survival and integration. *Front Cell Dev Biol* **9:** 714169. doi:10.3389/fcell.2021.714169

Hughes EG, Kang SH, Fukaya M, Bergles DE. 2013. Oligodendrocyte progenitors balance growth with self-repulsion to achieve homeostasis in the adult brain. *Nat Neurosci* **16:** 668–676. doi:10.1038/nn.3390

Hughes EG, Orthmann-Murphy JL, Langseth AJ, Bergles DE. 2018. Myelin remodeling through experience-dependent oligodendrogenesis in the adult somatosensory cortex. *Nat Neurosci* **21:** 696–706. doi:10.1038/s41593-018-0121-5

Iram T, Kern F, Kaur A, Myneni S, Morningstar AR, Shin H, Garcia MA, Yerra L, Palovics R, Yang AC, et al. 2022. Young CSF restores oligodendrogenesis and memory in aged mice via Fgf17. *Nature* **605:** 509–515. doi:10.1038/s41586-022-04722-0

Jäkel S, Agirre E, Mendanha Falcão A, van Bruggen D, Lee KW, Knuesel I, Malhotra D, Ffrench-Constant C, Williams A, Castelo-Branco G. 2019. Altered human oligodendrocyte heterogeneity in multiple sclerosis. *Nature* **566:** 543–547. doi:10.1038/s41586-019-0903-2

Kamen Y, Pivonkova H, Evans KA, Káradóttir RT. 2022. A matter of state: diversity in oligodendrocyte lineage cells. *Neuroscientist* **28:** 144–162. doi:10.1177/10738584 20987208

Kang SH, Fukaya M, Yang JK, Rothstein JD, Bergles DE. 2010. NG2$^+$ CNS glial progenitors remain committed to the oligodendrocyte lineage in postnatal life and following neurodegeneration. *Neuron* **68:** 668–681. doi:10.1016/j.neuron.2010.09.009

Káradóttir R, Attwell D. 2007. Neurotransmitter receptors in the life and death of oligodendrocytes. *Neuroscience* **145:** 1426–1438. doi:10.1016/j.neuroscience.2006.08.070

Kessaris N, Fogarty M, Iannarelli P, Grist M, Wegner M, Richardson WD. 2006. Competing waves of oligodendrocytes in the forebrain and postnatal elimination of an embryonic lineage. *Nat Neurosci* **9:** 173–179. doi:10.1038/nn1620

Khawaja RR, Agarwal A, Fukaya M, Jeong H-K, Gross S, Gonzalez-Fernandez E, Soboloff J, Bergles DE, Kang SH. 2021. Glua2 overexpression in oligodendrocyte progenitors promotes postinjury oligodendrocyte regeneration. *Cell Rep* **35:** 109147. doi:10.1016/j.celrep.2021.109147

Kirby BB, Takada N, Latimer AJ, Shin J, Carney TJ, Kelsh RN, Appel B. 2006. In vivo time-lapse imaging shows dynamic oligodendrocyte progenitor behavior during zebrafish development. *Nat Neurosci* **9:** 1506–1511. doi:10.1038/nn1803

Kirby L, Jin J, Cardona JG, Smith MD, Martin KA, Wang J, Strasburger H, Herbst L, Alexis M, Karnell J, et al. 2019. Oligodendrocyte precursor cells present antigen and are cytotoxic targets in inflammatory demyelination. *Nat Commun* **10:** 3887. doi:10.1038/s41467-019-11638-3

Kougioumtzidou E, Shimizu T, Hamilton NB, Tohyama K, Sprengel R, Monyer H, Attwell D, Richardson WD. 2017. Signalling through AMPA receptors on oligodendrocyte precursors promotes myelination by enhancing oligodendrocyte survival. *eLife* **6:** e28080. doi:10.7554/eLife.28080

Kukley M. 2023. Recent insights into the functional role of AMPA receptors in the oligodendrocyte lineage cells in vivo. *Int J Mol Sci* **24:** 4138. doi:10.3390/ijms24044138

Kukley M, Nishiyama A, Dietrich D. 2010. The fate of synaptic input to NG2 glial cells: neurons specifically downregulate transmitter release onto differentiating oligodendroglial cells. *J Neurosci* **30:** 8320–8331. doi:10.1523/JNEUROSCI.0854-10.2010

Lepiemme F, Stoufflet J, Javier-Torrent M, Mazzucchelli G, Silva CG, Nguyen L. 2022. Oligodendrocyte precursors guide interneuron migration by unidirectional contact repulsion. *Science* **376:** eabn6204. doi:10.1126/science.abn6204

Levine J. 2016. The reactions and role of NG2 glia in spinal cord injury. *Brain Res* **1638:** 199–208. doi:10.1016/j.brainres.2015.07.026

Li C, Xiao L, Liu X, Yang W, Shen W, Hu C, Yang G, He C. 2013. A functional role of NMDA receptor in regulating the differentiation of oligodendrocyte precursor cells and remyelination. *Glia* **61:** 732–749. doi:10.1002/glia.22469

Li J, Miramontes T, Czopka T, Monk K. 2022. Synapses and Ca^{2+} activity in oligodendrocyte precursor cells predict where myelin sheaths form. bioRxiv doi:10.1101/2022.03.18.484955

Lin S, Bergles DE. 2004. Synaptic signaling between GABAergic interneurons and oligodendrocyte precursor cells in the hippocampus. *Nat Neurosci* **7:** 24–32. doi:10.1038/nn1162

Cite this article as *Cold Spring Harb Perspect Biol* doi: 10.1101/cshperspect.a041425

Lin SC, Huck JHJ, Roberts JDB, Macklin WB, Somogyi P, Bergles DE. 2005. Climbing fiber innervation of NG2-expressing glia in the mammalian cerebellum. *Neuron* **46:** 773–785. doi:10.1016/j.neuron.2005.04.025

Liu J, Dietz K, DeLoyht JM, Pedre X, Kelkar D, Kaur J, Vialou V, Lobo MK, Dietz DM, Nestler EJ, et al. 2012. Impaired adult myelination in the prefrontal cortex of socially isolated mice. *Nat Neurosci* **15:** 1621–1623. doi:10.1038/nn.3263

Lu TY, Hanumaihgari P, Hsu ET, Agarwal A, Kawaguchi R, Calabresi PA, Bergles DE. 2023. Norepinephrine modulates calcium dynamics in cortical oligodendrocyte precursor cells promoting proliferation during arousal in mice. *Nat Neurosci* **26:** 1739–1750. doi:10.1038/s41593-023-01426-0

Lubetzki C, Zalc B, Williams A, Stadelmann C, Stankoff B. 2020. Remyelination in multiple sclerosis: from basic science to clinical translation. *Lancet Neurol* **19:** 678–688. doi:10.1016/S1474-4422(20)30140-X

Lundgaard I, Luzhynskaya A, Stockley JH, Wang Z, Evans KA, Swire M, Volbracht K, Gautier HOB, Franklin RJM, Ffrench-Constant C, et al. 2013. Neuregulin and BDNF induce a switch to NMDA receptor–dependent myelination by oligodendrocytes. *PLoS Biol* **11:** e1001743. doi:10.1371/journal.pbio.1001743

Madisen L, Zwingman TA, Sunkin SM, Oh SW, Zariwala HA, Gu H, Ng LL, Palmiter RD, Hawrylycz MJ, Jones AR, et al. 2010. A robust and high-throughput Cre reporting and characterization system for the whole mouse brain. *Nat Neurosci* **13:** 133–140. doi:10.1038/nn.2467

Makinodan M, Rosen KM, Ito S, Corfas G. 2012. A critical period for social experience–dependent oligodendrocyte maturation and myelination. *Science* **337:** 1357–1360. doi:10.1126/science.1220845

Marisca R, Hoche T, Agirre E, Hoodless LJ, Barkey W, Auer F, Castelo-Branco G, Czopka T. 2020. Functionally distinct subgroups of oligodendrocyte precursor cells integrate neural activity and execute myelin formation. *Nat Neurosci* **23:** 363–374. doi:10.1038/s41593-019-0581-2

Marques S, Zeisel A, Codeluppi S, van Bruggen D, Mendanha Falcão A, Xiao L, Li H, Häring M, Hochgerner H, Romanov RA, et al. 2016. Oligodendrocyte heterogeneity in the mouse juvenile and adult central nervous system. *Science* **352:** 1326–1329. doi:10.1126/science.aaf6463

Mayoral SR, Etxeberria A, Shen YAA, Chan JR. 2018. Initiation of CNS myelination in the optic nerve is dependent on axon caliber. *Cell Rep* **25:** 544–550.e3. doi:10.1016/j.celrep.2018.09.052

McKenzie IA, Ohayon D, Li H, de Faria JP, Emery B, Tohyama K, Richardson WD. 2014. Motor skill learning requires active central myelination. *Science* **346:** 318–322. doi:10.1126/science.1254960

McKinnon RD, Matsui T, Dubois-Dalcq M, Aaronson SA. 1990. FGF modulates the PDGF-driven pathway of oligodendrocyte development. *Neuron* **5:** 603–614. doi:10.1016/0896-6273(90)90215-2

Mei F, Fancy SPJ, Shen YAA, Niu J, Zhao C, Presley B, Miao E, Lee S, Mayoral SR, Redmond SA, et al. 2014. Micropillar arrays as a high-throughput screening platform for therapeutics in multiple sclerosis. *Nat Med* **20:** 954–960. doi:10.1038/nm.3618

Mei F, Lehmann-Horn K, Shen YAA, Rankin KA, Stebbins KJ, Lorrain DS, Pekarek K, A Sagan S, Xiao L, Teuscher C, et al. 2016. Accelerated remyelination during inflammatory demyelination prevents axonal loss and improves functional recovery. *eLife* **5:** e18246. doi:10.7554/eLife.18246

Meijer M, Agirre E, Kabbe M, van Tuijn CA, Heskol A, Zheng C, Mendanha Falcão A, Bartosovic M, Kirby L, Calini D, et al. 2022. Epigenomic priming of immune genes implicates oligodendroglia in multiple sclerosis susceptibility. *Neuron* **110:** 1193–1210.e13. doi:10.1016/j.neuron.2021.12.034

Meireles AM, Shen K, Zoupi L, Iyer H, Bouchard EL, Williams A, Talbot WS. 2018. The lysosomal transcription factor TFEB represses myelination downstream of the rag-ragulator complex. *Dev Cell* **47:** 319–330.e5. doi:10.1016/j.devcel.2018.10.003

Michalski JP, Anderson C, Beauvais A, De Repentigny Y, Kothary R. 2011. The proteolipid protein promoter drives expression outside of the oligodendrocyte lineage during embryonic and early postnatal development. *PLoS ONE* **6:** e19772. doi:10.1371/journal.pone.0019772

Mitew S, Gobius I, Fenlon LR, McDougall SJ, Hawkes D, Xing YL, Bujalka H, Gundlach AL, Richards LJ, Kilpatrick TJ, et al. 2018. Pharmacogenetic stimulation of neuronal activity increases myelination in an axon-specific manner. *Nat Commun* **9:** 306. doi:10.1038/s41467-017-02719-2

Monje M. 2018. Myelin plasticity and nervous system function. *Annu Rev Neurosci* **41:** 61–76. doi:10.1146/annurev-neuro-080317-061853

Moyon S, Frawley R, Marechal D, Huang D, Marshall-Phelps KLH, Kegel L, Bøstrand SMK, Sadowski B, Jiang YH, Lyons DA, et al. 2021. TET1-mediated DNA hydroxymethylation regulates adult remyelination in mice. *Nat Commun* **12:** 3359. doi:10.1038/s41467-021-23735-3

Najm FJ, Madhavan M, Zaremba A, Shick E, Karl RT, Factor DC, Miller TE, Nevin ZS, Kantor C, Sargent A, et al. 2015. Drug-based modulation of endogenous stem cells promotes functional remyelination in vivo. *Nature* **522:** 216–220. doi:10.1038/nature14335

Neumann B, Baror R, Zhao C, Segel M, Dietmann S, Rawji KS, Foerster S, McClain CR, Chalut K, van Wijngaarden P, et al. 2019. Metformin restores CNS remyelination capacity by rejuvenating aged stem cells. *Cell Stem Cell* **25:** 473–485.e8. doi:10.1016/j.stem.2019.08.015

Nishiyama A, Lin XH, Giese N, Heldin CH, Stallcup WB. 1996. Co-localization of NG2 proteoglycan and PDGF α-receptor on O2A progenitor cells in the developing rat brain. *J Neurosci Res* **43:** 299–314. doi:10.1002/(SICI)1097-4547(19960201)43:3<299::AID-JNR5>3.0.CO;2-E

Nishiyama A, Komitova M, Suzuki R, Zhu X. 2009. Polydendrocytes (NG2 cells): multifunctional cells with lineage plasticity. *Nat Rev Neurosci* **10:** 9–22. doi:10.1038/nrn2495

Noble M, Murray K, Stroobant P, Waterfield MD, Riddle P. 1988. Platelet-derived growth factor promotes division and motility and inhibits premature differentiation of the oligodendrocyte/type-2 astrocyte progenitor cell. *Nature* **333:** 560–562. doi:10.1038/333560a0

Ong WY, Levine JM. 1999. A light and electron microscopic study of NG2 chondroitin sulfate proteoglycan-positive oligodendrocyte precursor cells in the normal and kainate-lesioned rat hippocampus. *Neuroscience* **92:** 83–95. doi:10.1016/S0306-4522(98)00751-9

Orduz D, Benamer N, Ortolani D, Coppola E, Vigier L, Pierani A, Angulo MC. 2019. Developmental cell death regulates lineage-related interneuron-oligodendroglia functional clusters and oligodendrocyte homeostasis. *Nat Commun* **10:** 4249. doi:10.1038/s41467-019-11904-4

Osorio MJ, Mariani JN, Zou L, Schanz SJ, Heffernan K, Cornwell A, Goldman SA. 2023. Glial progenitor cells of the adult human white and grey matter are contextually distinct. *Glia* **71:** 524–540. doi:10.1002/glia.24291

Pan S, Mayoral SR, Choi HS, Chan JR, Kheirbek MA. 2020. Preservation of a remote fear memory requires new myelin formation. *Nat Neurosci* **23:** 487–499. doi:10.1038/s41593-019-0582-1

Pende M, Holtzclaw LA, Curtis JL, Russell JT, Gallo V. 1994. Glutamate regulates intracellular calcium and gene expression in oligodendrocyte progenitors through the activation of DL-alpha-amino-3-hydroxy-5-methyl-4-isoxazolepropionic acid receptors. *Proc Natl Acad Sci* **91:** 3215–3219. doi:10.1073/pnas.91.8.3215

Piller M, Werkman IL, Brown EA, Latimer AJ, Kucenas S. 2021. Glutamate signaling via the AMPAR subunit GluR4 regulates oligodendrocyte progenitor cell migration in the developing spinal cord. *J Neurosci* **41:** 5353–5371. doi:10.1523/JNEUROSCI.2562-20.2021

Pringle NP, Mudhar HS, Collarini EJ, Richardson WD. 1992. PDGF receptors in the rat CNS: during late neurogenesis, PDGF alpha-receptor expression appears to be restricted to glial cells of the oligodendrocyte lineage. *Development* **115:** 535–551. doi:10.1242/dev.115.2.535

Psachoulia K, Jamen F, Young KM, Richardson WD. 2009. Cell cycle dynamics of NG2 cells in the postnatal and ageing brain. *Neuron Glia Biol* **5:** 57–67. doi:10.1017/S1740925X09990354

Raff MC, Abney ER, Cohen J, Lindsay R, Noble M. 1983a. Two types of astrocytes in cultures of developing rat white matter: differences in morphology, surface gangliosides, and growth characteristics. *J Neurosci* **3:** 1289–1300. doi:10.1523/JNEUROSCI.03-06-01289.1983

Raff MC, Miller RH, Noble M. 1983b. A glial progenitor cell that develops in vitro into an astrocyte or an oligodendrocyte depending on culture medium. *Nature* **303:** 390–396. doi:10.1038/303390a0

Raff MC, Barres BA, Burne JF, Coles HS, Ishizaki Y, Jacobson MD. 1993. Programmed cell death and the control of cell survival: lessons from the nervous system. *Science* **262:** 695–700. doi:10.1126/science.8235590

Rakic S, Zecevic N. 2003. Early oligodendrocyte progenitor cells in the human fetal telencephalon. *Glia* **41:** 117–127. doi:10.1002/glia.10140

Richardson WD, Pringle N, Mosley MJ, Westermark B, Dubois-Dalcq M. 1988. A role for platelet-derived growth factor in normal gliogenesis in the central nervous system. *Cell* **53:** 309–319. doi:10.1016/0092-8674(88)90392-3

Río-Hortega P. 1921. Estudios sobre la neurología. La glía de escasas radiaciones (oligodendroglía) [Studies on neurology. Glia of few processes (oligodendroglia)]. *Bol Real Sociedad Española Hist Nat* **21:** 63–92.

Río-Hortega P. 1928. Tercera aportación al conocimiento morfológico e interpretación functional de la oligoden-

droglía [Third contribution to the morphological knowledge and functional interpretation of the oligodendroglia]. *Memor Real Sociedad Española Hist Nat* **14:** 5–122.

Rivers LE, Young KM, Rizzi M, Jamen F, Psachoulia K, Wade A, Kessaris N, Richardson WD. 2008. PDGFRA/NG2 glia generate myelinating oligodendrocytes and piriform projection neurons in adult mice. *Nat Neurosci* **11:** 1392–1401. doi:10.1038/nn.2220

Robertson WF. 1899. On a new method of obtaining a black reaction in certain tissue-elements of the central nervous system (platinum method). *Scottish Med Surg* **4:** 23–30.

Ruckh JM, Zhao JW, Shadrach JL, van Wijngaarden P, Rao TN, Wagers AJ, Franklin RJM. 2012. Rejuvenation of regeneration in the aging central nervous system. *Cell Stem Cell* **10:** 96–103. doi:10.1016/j.stem.2011.11.019

Saab AS, Tzvetavona ID, Trevisiol A, Baltan S, Dibaj P, Kusch K, Möbius W, Goetze B, Jahn HM, Huang W, et al. 2016. Oligodendroglial NMDA receptors regulate glucose import and axonal energy metabolism. *Neuron* **91:** 119–132. doi:10.1016/j.neuron.2016.05.016

Samanta J, Grund EM, Silva HM, Lafaille JJ, Fishell G, Salzer JL. 2015. Inhibition of Gli1 mobilizes endogenous neural stem cells for remyelination. *Nature* **526:** 448–452. doi:10.1038/nature14957

Segel M, Neumann B, Hill MFE, Weber IP, Viscomi C, Zhao C, Young A, Agley CC, Thompson AJ, Gonzalez GA, et al. 2019. Niche stiffness underlies the ageing of central nervous system progenitor cells. *Nature* **573:** 130–134. doi:10.1038/s41586-019-1484-9

Shen S, Sandoval J, Swiss VA, Li J, Dupree J, Franklin RJM, Casaccia-Bonnefil P. 2008. Age-dependent epigenetic control of differentiation inhibitors is critical for remyelination efficiency. *Nat Neurosci* **11:** 1024–1034. doi:10.1038/nn.2172

Sherafat A, Pfeiffer F, Reiss AM, Wood WM, Nishiyama A. 2021. Microglial neuropilin-1 promotes oligodendrocyte expansion during development and remyelination by *trans*-activating platelet-derived growth factor receptor. *Nat Commun* **12:** 2265. doi:10.1038/s41467-021-22532-2

Shields SA, Gilson JM, Blakemore WF, Franklin RJ. 1999. Remyelination occurs as extensively but more slowly in old rats compared to young rats following gliotoxin-induced CNS demyelination. *Glia* **28:** 77–83. doi:10.1002/(SICI)1098-1136(199910)28:1<77::AID-GLIA9>3.0.CO;2-F

Sim FJ, Zhao C, Penderis J, Franklin RJM. 2002. The age-related decrease in CNS remyelination efficiency is attributable to an impairment of both oligodendrocyte progenitor recruitment and differentiation. *J Neurosci* **22:** 2451–2459. doi:10.1523/JNEUROSCI.22-07-02451.2002

Smith KJ, Blakemore WF, Mcdonald WI. 1979. Central remyelination restores secure conduction. *Nature* **280:** 395–396. doi:10.1038/280395a0

Spitzer SO, Sitnikov S, Kamen Y, Evans KA, Kronenberg-Versteeg D, Dietmann S, de Faria O, Agathou S, Káradóttir RT. 2019. Oligodendrocyte progenitor cells become regionally diverse and heterogeneous with age. *Neuron* **101:** 459–471.e5. doi:10.1016/j.neuron.2018.12.020

Stallcup WB, Beasley L. 1987. Bipotential glial precursor cells of the optic nerve express the NG2 proteoglycan. *J Neurosci* **7:** 2737–2744. doi:10.1523/JNEUROSCI.07-09-02737.1987

Cite this article as *Cold Spring Harb Perspect Biol* doi: 10.1101/cshperspect.a041425

Steadman PE, Xia F, Ahmed M, Mocle AJ, Penning ARA, Geraghty AC, Steenland HW, Monje M, Josselyn SA, Frankland PW. 2020. Disruption of oligodendrogenesis impairs memory consolidation in adult mice. *Neuron* **105:** 150–164.e6. doi:10.1016/j.neuron.2019.10.013

Sun LO, Mulinyawe SB, Collins HY, Ibrahim A, Li Q, Simon DJ, Tessier-Lavigne M, Barres BA. 2018. Spatiotemporal control of CNS myelination by oligodendrocyte programmed cell death through the TFEB–PUMA axis. *Cell* **175:** 1811–1826.e21. doi:10.1016/j.cell.2018.10.044

Trapp BD, Nishiyama A, Cheng D, Macklin W. 1997. Differentiation and death of premyelinating oligodendrocytes in developing rodent brain. *J Cell Biol* **137:** 459–468. doi:10.1083/jcb.137.2.459

Tripathi RB, Rivers LE, Young KM, Jamen F, Richardson WD. 2010. NG2 glia generate new oligodendrocytes but few astrocytes in a murine experimental autoimmune encephalomyelitis model of demyelinating disease. *J Neurosci* **30:** 16383–16390. doi:10.1523/JNEUROSCI.3411-10.2010

Tripathi RB, Jackiewicz M, McKenzie IA, Kougioumtzidou E, Grist M, Richardson WD. 2017. Remarkable stability of myelinating oligodendrocytes in mice. *Cell Rep* **21:** 316–323. doi:10.1016/j.celrep.2017.09.050

Tsai HH, Niu J, Munji R, Davalos D, Chang J, Zhang H, Tien AC, Kuo CJ, Chan JR, Daneman R, et al. 2016. Oligodendrocyte precursors migrate along vasculature in the developing nervous system. *Science* **351:** 379–384. doi:10.1126/science.aad3839

Vanzulli I, Papanikolaou M, De-La-Rocha IC, Pieropan F, Rivera AD, Gomez-Nicola D, Verkhratsky A, Rodríguez JJ, Butt AM. 2020. Disruption of oligodendrocyte progenitor cells is an early sign of pathology in the triple transgenic mouse model of Alzheimer's disease. *Neurobiol Aging* **94:** 130–139. doi:10.1016/j.neurobiolaging.2020.05.016

Vélez-Fort M, Maldonado PP, Butt AM, Audinat E, Angulo MC. 2010. Postnatal switch from synaptic to extrasynaptic transmission between interneurons and NG2 cells. *J Neurosci* **30:** 6921–6929. doi:10.1523/JNEUROSCI.0238-10.2010

Viganò F, Möbius W, Götz M, Dimou L. 2013. Transplantation reveals regional differences in oligodendrocyte differentiation in the adult brain. *Nat Neurosci* **16:** 1370–1372. doi:10.1038/nn.3503

von Streitberg A, Jäkel S, Eugenin von Bernhardi J, Straube C, Buggenthin F, Marr C, Dimou L. 2021. NG2-glia transiently overcome their homeostatic network and contribute to wound closure after brain injury. *Front Cell Dev Biol* **9:** 662056. doi:10.3389/fcell.2021.662056

Wang F, Ren SY, Chen JF, Liu K, Li RX, Li ZF, Hu B, Niu JQ, Xiao L, Chan JR, et al. 2020. Myelin degeneration and diminished myelin renewal contribute to age-related def-

icits in memory. *Nat Neurosci* **23:** 481–486. doi:10.1038/s41593-020-0588-8

Xiao L, Ohayon D, McKenzie IA, Sinclair-Wilson A, Wright JL, Fudge AD, Emery B, Li H, Richardson WD. 2016. Rapid production of new oligodendrocytes is required in the earliest stages of motor-skill learning. *Nat Neurosci* **19:** 1210–1217. doi:10.1038/nn.4351

Xiao Y, Petrucco L, Hoodless LJ, Portugues R, Czopka T. 2022. Oligodendrocyte precursor cells sculpt the visual system by regulating axonal remodeling. *Nat Neurosci* **25:** 280–284. doi:10.1038/s41593-022-01023-7

Xin W, Chan JR. 2020. Myelin plasticity: sculpting circuits in learning and memory. *Nat Rev Neurosci* **21:** 682–694. doi:10.1038/s41583-020-00379-8

Xing YL, Poh J, Chuang BHA, Moradi K, Mitew S, Richardson WD, Kilpatrick TJ, Osanai Y, Merson TD. 2023. High-efficiency pharmacogenetic ablation of oligodendrocyte progenitor cells in the adult mouse CNS. *Cell Rep Methods* **3:** 100414. doi:10.1016/j.crmeth.2023.100414

Yang Z, Suzuki R, Daniels SB, Brunquell CB, Sala CJ, Nishiyama A. 2006. NG2 glial cells provide a favorable substrate for growing axons. *J Neurosci* **26:** 3829–3839. doi:10.1523/JNEUROSCI.4247-05.2006

Yeung MSY, Djelloul M, Steiner E, Bernard S, Salehpour M, Possnert G, Brundin L, Frisén J. 2019. Dynamics of oligodendrocyte generation in multiple sclerosis. *Nature* **566:** 538–542. doi:10.1038/s41586-018-0842-3

Young KM, Psachoulia K, Tripathi RB, Dunn SJ, Cossell L, Attwell D, Tohyama K, Richardson WD. 2013. Oligodendrocyte dynamics in the healthy adult CNS: evidence for myelin remodeling. *Neuron* **77:** 873–885. doi:10.1016/j.neuron.2013.01.006

Yu G, Su Y, Guo C, Yi C, Yu B, Chen H, Cui Y, Wang X, Wang Y, Chen X, et al. 2022. Pathological oligodendrocyte precursor cells revealed in human schizophrenic brains and trigger schizophrenia-like behaviors and synaptic defects in genetic animal model. *Mol Psychiatry* **27:** 5154–5166. doi:10.1038/s41380-022-01777-3

Zhu X, Bergles DE, Nishiyama A. 2008. NG2 cells generate both oligodendrocytes and gray matter astrocytes. *Development* **135:** 145–157. doi:10.1242/dev.004895

Zhu X, Hill RA, Dietrich D, Komitova M, Suzuki R, Nishiyama A. 2011. Age-dependent fate and lineage restriction of single NG2 cells. *Development* **138:** 745–753. doi:10.1242/dev.047951

Zonouzi M, Scafidi J, Li P, McEllin B, Edwards J, Dupree JL, Harvey L, Sun D, Hübner CA, Cull-Candy SG, et al. 2015. GABAergic regulation of cerebellar NG2 cell development is altered in perinatal white matter injury. *Nat Neurosci* **18:** 674–682. doi:10.1038/nn.3990

Regulators of Oligodendrocyte Differentiation

Ben Emery[1] and Teresa L. Wood[2]

[1]Jungers Center for Neurosciences Research, Department of Neurology, Oregon Health & Science University, Portland, Oregon 97239, USA

[2]Department of Pharmacology, Physiology and Neuroscience, New Jersey Medical School, Rutgers University, Newark, New Jersey 07103, USA

Correspondence: terri.wood@rutgers.edu

Myelination has evolved as a mechanism to ensure fast and efficient propagation of nerve impulses along axons. Within the central nervous system (CNS), myelination is carried out by highly specialized glial cells, oligodendrocytes. The formation of myelin is a prolonged aspect of CNS development that occurs well into adulthood in humans, continuing throughout life in response to injury or as a component of neuroplasticity. The timing of myelination is tightly tied to the generation of oligodendrocytes through the differentiation of their committed progenitors, oligodendrocyte precursor cells (OPCs), which reside throughout the developing and adult CNS. In this article, we summarize our current understanding of some of the signals and pathways that regulate the differentiation of OPCs, and thus the myelination of CNS axons.

Oligodendrocytes are the myelinating cells of the central nervous system (CNS). They arise from the differentiation of oligodendrocyte precursor cells (OPCs) (Fig. 1), a population that is present throughout the late embryonic to adult CNS. Although OPCs are present from late embryonic ages, their subsequent exit of the cell cycle and differentiation into oligodendrocytes that myelinate the CNS is a remarkably protracted process that largely occurs during postnatal development. Neuroimaging studies indicate that myelination largely occurs over the first several decades of life in humans and the first several months in mice (Lebel et al. 2012; Hammelrath et al. 2016). Nevertheless, substantial evidence indicates that some degree of ongoing oligodendrocyte differentiation occurs throughout adult life, either during repair of CNS damage or as part of continual remodeling of myelination within circuits (Kang et al. 2010; Young et al. 2013; Hill et al. 2014, 2018; Yeung et al. 2014; Hughes et al. 2018). Recent studies have provided evidence that mature oligodendrocytes may be capable of reinitiating the myelination program and generating new myelin sheaths in the context of injury (Duncan et al. 2018; Kirby et al. 2019; Yeung et al. 2019). In spite of these tantalizing findings, the emerging consensus is that mature oligodendrocytes are extremely inefficient at generating new myelin segments relative to actively differentiating cells (Bacmeister et al. 2022; Neely et al. 2022). As such, the fate decision of an OPC to initiate the differentiation process into a postmitotic oligodendrocyte represents a critical checkpoint in the myelination process. OPCs themselves and the biology of the mature, myeli-

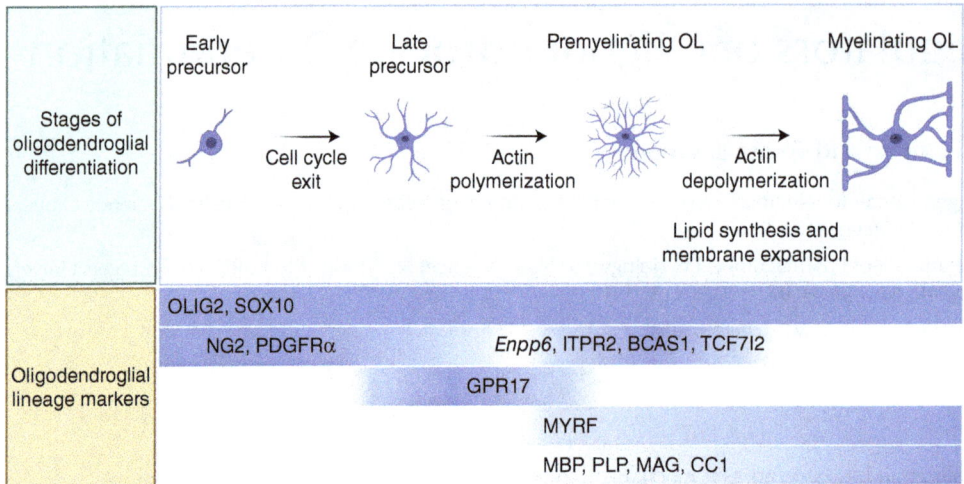

Figure 1. Commonly used markers to determine stages of oligodendrocyte differentiation. As an oligodendrocyte precursor cell (OPC) exits the cell cycle and terminally differentiates into a premyelinating oligodendrocyte and finally a myelinating oligodendrocyte, a series of markers can be used to determine its differentiation status. The transcription factors OLIG2 and SOX10 will identify the entire oligodendrocyte lineage and can be used in conjunction with NG2 and PDGFRα to identify OPCs. As they initiate differentiation, OPCs will transiently express GPR17 before transiently expressing makers such as ENPP6, ITPR2, BCAS1, and TCF7L2 as premyelinating oligodendrocytes. Mature oligodendrocytes can be identified by the expression of the antigen for CC1 as well as mature myelin proteins such as myelin basic protein (MBP) and proteolipid protein 1 (PLP1). (Figure generated with BioRender, https://www.biorender.com.)

nating oligodendrocyte are described in detail in accompanying articles in this collection (Hill et al. 2023; Simons et al. 2024). In this work, we will focus largely on the transition from an OPC to an oligodendrocyte state and how this step is regulated. Most of the data presented reference the wealth of information from rodent studies with findings from chicken, zebrafish, and humans specifically noted.

DEVELOPMENT OF THE OLIGODENDROCYTE LINEAGE

Specification and Proliferation of OPCs

OPCs first arise from radial glial cells of the ventricular germinal zones of the developing spinal cord and brain, being identifiable by embryonic day 12.5 in mice, embryonic day 14 in rats, embryonic day 6 in chicks, 48 h postfertilization in zebrafish, and gestational week 6.5 in humans (Pringle and Richardson 1993; Ono et al. 1995; Hajihosseini et al. 1996; Calver et al. 1998; Park

et al. 2002). The initial specification of neural progenitor cells to the oligodendrocyte lineage occurs within ventral domains of the spinal cord, mediated by gradients of ventrally secreted sonic hedgehog (SHH), dorsally secreted bone morphogenic proteins (BMPs), and wingless-related integration site proteins (WNTs) (Orentas and Miller 1996; Pringle et al. 1996; Orentas et al. 1999). Acting via the GLI family of transcription factors, SHH signaling induces expression of the oligodendrocyte transcription factor (OLIG) 2 within the motor neural progenitor (pMN) domain (Persson et al. 2002). In turn, OLIG2 promotes the specification of both OPCs and motor neurons from the pMN and is critically required for the specification of OPCs within many regions of the CNS (Novitch et al. 2001; Lu et al. 2002; Park et al. 2002; Zhou and Anderson 2002). Once specified, these ventrally derived OPCs proliferate and disperse widely through the developing spinal cord. Although these ventrally derived OPCs will colonize much of the spinal cord, there is also a secondary wave of OPC specification from radial glial cells in the dorsal spinal cord. Genetic fate mapping studies in mice indicate that these dorsally derived cells will largely replace the ventrally derived OPCs within dorsal white matter tracts (Tripathi et al. 2011). A broadly similar pattern of OPC specification occurs within the brain; at earlier developmental time points, OPCs are specified within the ventral ventricular zone of the medial ganglionic eminence. Subsequent populations of OPCs arise from the lateral and caudal ganglionic eminences, with a late population arising from the postnatal cortex (Kessaris et al. 2006). As with the spinal cord, over time these dorsally derived cells will largely replace the earlier-born, ventrally specified OPCs. Interestingly, in spite of this replacement, the sequential populations of OPCs appear to be relatively redundant; when any one population of OPCs is genetically ablated, the remaining OPCs will compensate to colonize the brain with a normal complement of OPCs and oligodendrocytes (Kessaris et al. 2006).

Once generated, OPCs both proliferate and migrate away from the germinal zones, ultimately achieving a remarkably uniform distribution throughout both the white and gray matter (Nishiyama et al. 2002). During their migration, OPCs use the vasculature as a scaffold. Tsai et al. (2016) showed that disrupting the developing vasculature through deletion of the adhesion G-protein-coupled receptor (GPR) A2 caused OPCs to stall close to their germinal zones, failing to disperse. This dependence on the vasculature for migration is thought to extend to the adult CNS in the context of injury (Niu et al. 2019). The proliferation of OPCs is highly dependent on mitogens, most notably platelet-derived growth factor (PDGF) derived from both astrocytes and neurons (Raff et al. 1988; Richardson et al. 1988; Fruttiger et al. 2000). Accordingly, deletion or overexpression of PDGF-A causes a depletion or hyperproliferation of OPCs, respectively (Calver et al. 1998). Within the adult CNS, OPCs will balance their proliferation with the migration, death, and differentiation of their neighbors to maintain a homeostatic density throughout life (Hughes et al. 2013). Indeed, even in the adult CNS, OPCs show a remarkable ability to regain their normal density when the vast majority of

them are experimentally ablated (Robins et al. 2013; Đặng et al. 2019b; Xing et al. 2023). Recent studies indicate that this process is likely regulated at least in part by self-repulsive interactions mediated by protocadherin 15, which mediates the repulsion of recently divided daughter cells and acts to inhibit OPC proliferation (Huang et al. 2020; Zhen et al. 2022).

Cell Cycle Exit

As discussed above, the early OPC that arises from the neural tube is highly proliferative. Although it may seem obvious that exit from the cell cycle is necessary for terminal OPC differentiation, whether or not cell cycle exit is sufficient to induce OPC differentiation and how it is coordinated with the onset of the differentiation program has been the focus of studies in several laboratories. Early in vitro experiments demonstrated that OPCs exposed to either PDGF or FGF-2 alone divide a defined number of times and then differentiate (Noble et al. 1988; Bögler et al. 1990). In contrast, the combination of PDGF and FGF-2 maintains OPCs in the cell cycle and prevents their differentiation (Bögler et al. 1990) due at least in part to FGF-2 sustaining expression of PDGFRα in the OPCs (McKinnon et al. 1990a). Subsequent studies revealed the importance of simultaneous stimulation of the insulin-like growth factor receptor (IGF1R) in OPC cultures through micromolar insulin concentrations in the media and demonstrated that IGF-1/IGF1R activation increases the mitogenic actions of PDGF and FGF (Jiang et al. 2001). OPCs cultured in the absence of PDGF and FGF-2 exit the cell cycle and begin to differentiate (Raff et al. 1984). Consistent with the in vitro data, global loss of PDGFRα in adult mice results in a rapid differentiation of adult OPCs in the CNS (Đặng et al. 2019a), and inhibition of FGFR1 in OPCs in vivo during postnatal developmental myelination promotes oligodendrocyte differentiation (Zhou et al. 2006). Moreover, genetic deletion of the cyclin-dependent kinase (CDK) inhibitor p27[kip1] in mice inhibits OPC differentiation in vivo (Casaccia-Bonnefil et al. 1997). Taken together, these findings led to the conclusion that cell cycle exit is necessary for OPC differentiation.

Data from several laboratories provided the basis for our current understanding that while necessary, cell cycle exit is by itself insufficient to promote OPC differentiation. Overexpression of either p27Kip1 or of a dominant-negative CDK-2 in OPCs in vitro successfully blocks proliferation but fails to induce differentiation of the cells (Tang et al. 1999; Belachew et al. 2002). In contrast, OPCs with a genetic deletion of the CDK inhibitor p21^{Cip1} exit the cell cycle but show deficits in differentiation in vitro and in vivo (Zezula et al. 2001).

The conclusion that cell cycle exit is necessary but insufficient to promote OPC differentiation then leads to obvious questions of how cell cycle exit is coordinated with differentiation and how the onset of differentiation is regulated. The first step to consider to be able to answer these questions is to address what constitutes the onset of differentiation. In the next sections, we introduce the myriad of cellular programs that must be induced in the early stages of oligodendrocyte differentiation including those involved in morphological differentiation, myelin gene expression, and membrane expansion (Fig. 1) before we discuss the intrinsic and extrinsic factors that regulate these programs.

Morphological Differentiation

Differentiation from OPCs to premyelinating oligodendrocytes involves a progression of cell morphological changes (Fig. 1; for reviews, see Michalski and Kothary 2015; Snaidero and Simons 2017; Brown and Macklin 2020). The earliest outward evidence of OPC differentiation is the transition of the bipolar OPC into a multibranched late OPC. The branching pattern then becomes more complex as the cells progress into a premyelinating oligodendrocyte that eventually makes contact with axons to initiate myelin wrapping. These dynamic morphological changes require extensive rearrangement of the actin cytoskeleton. New information and models for how cytoskeletal changes are orchestrated in premyelinating to mature oligodendrocytes were published by several groups in 2014–2015 (Snai-

dero et al. 2014; Nawaz et al. 2015; Zuchero et al. 2015). In these studies, the authors proposed that the initial phase of cytoskeletal changes occurs through actin polymerization whereby the oligodendrocytes extend multiple processes that eventually flatten as they prepare to contact axons and initiate ensheathment (Fig. 1). The second phase of cytoskeletal changes then involves actin depolymerization that is required for active axon wrapping. The exception to this is in the inner tongue that retains actin filaments that are hypothesized to provide the force necessary for wrapping to proceed (Simons et al. 2024).

Myelin Gene Expression

Two essential components of mature oligodendrocyte myelin are myelin proteins and lipids (Figs. 1 and 2). The initiation of myelin protein gene expression is a hallmark of OPC transition into terminal differentiation following cell cycle exit. Major myelin proteins include 2′,3′-cyclic nucleotide 3′-phosphodiesterase (CNP), MBP, PLP1, myelin-associated glycoprotein (MAG), as well as myelin oligodendrocyte glycoprotein (MOG). Chronologically, MAG and MBP are the earliest of the myelin proteins to appear, whereas MOG is the latest to appear in developing oligodendrocytes (Dubois-Dalcq et al. 1986; Monge et al. 1986; Trapp 1990; Solly et al. 1996). Initiation of gene expression sometimes precedes the appearance of the myelin protein such as in the case of PLP1. *Plp1* mRNA is one if the earliest detectable myelin protein mRNAs along with *Mbp* mRNA. However, unlike MBP protein, PLP1 protein in differentiating oligodendrocytes does not appear until several days following detection of the mRNA and after the appearance of MBP protein (Dubois-Dalcq et al. 1986; Monge et al. 1986; Sorg et al. 1987). Although the *Plp1* promoter is commonly used for transgene expression in mature oligodendrocytes (Mallon et al. 2002; Doerflinger et al. 2003), its expression is initiated in premyelinating oligodendrocytes as well as transiently in early spinal cord glial and neuronal precursors (Harlow et al. 2014). This early expression of *Plp1* mRNA has led to some difficulties in deciphering some transgenic or Cre driver lines based on the *Plp1* promoter.

Cite this article as *Cold Spring Harb Perspect Biol* doi: 10.1101/cshperspect.a041358

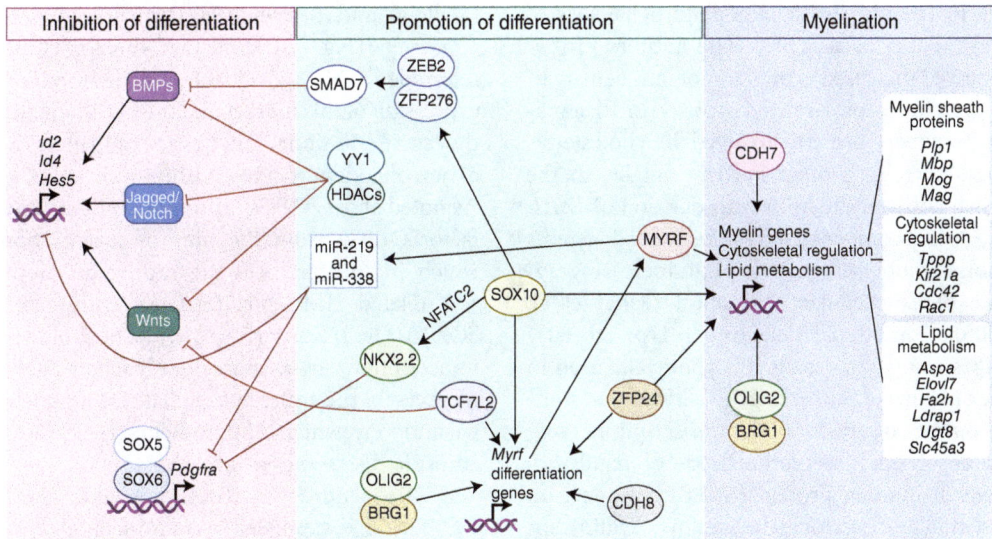

Figure 2. Stage-specific transcriptional circuits during oligodendrocyte differentiation. In oligodendrocyte precursor cells (OPCs) (*left* panel), a series of transcription factors including ID2, ID4, and HES5 are expressed downstream from Wnt, Bmp, and Notch signaling. Together with SoxD proteins SOX5 and SOX6, these serve to inhibit differentiation. During differentiation (*middle* panel), the coordinated activity of SOX10, OLIG2, TCF7L2, ZFP24, and YY1 will serve to feed back and inhibit these OPC transcription factors, while inducing additional prodifferentiation factors such as ZEB2 and MYRF. In mature oligodendrocytes, SOX10, MYRFf, and ZFP24 will continue to regulate several hundred genes involved in myelin biogenesis (*right* panel). (Figure generated with BioRender, https://www.biorender.com.)

Membrane Expansion

Along with increased branching and the onset of myelin protein synthesis, the transition from a premyelinating to a myelinating oligodendrocyte requires a massive expansion in lipid and fatty acid production to produce the cell membrane required for both increased cellular branching complexity and axon wrapping (Fig. 1). Unlike regular cellular membranes that have relatively equivalent proportions of proteins and lipids, myelin membranes contain ~70% lipids, the vast majority (46%) of which is cholesterol (for reviews, see Saher and Stumpf 2015; Poitelon et al. 2020). Cholesterol is important for the structure and fluidity of membranes (Ohvo-Rekilä et al. 2002), and in oligodendrocytes, cholesterol is necessary for myelin membrane expansion (Saher et al. 2005; Saher and Stumpf 2015). Cholesterol also promotes PLP1 incorporation into the myelin membrane (Saher et al. 2012).

Since cholesterol normally does not cross the blood–brain barrier, cells in the CNS must syn-

thesize their own cholesterol or, alternatively, acquire it from neighboring cells (Morell and Jurevics 1996; Camargo et al. 2017). Saher et al. (2005) disrupted cholesterol biosynthesis in myelinating glia through conditional deletion of the gene for squalene synthase (farnesyl-pyrophosphate:farnesyl-pyrophosphate farnesyltransferase, *Fdft1*), a rate-limiting enzyme in the cholesterol biosynthesis pathway. Oligodendroglial loss of *Fdft1* results in thinner myelin and reduced numbers of myelinated axons in white matter in both the brain and spinal cord, but the impact was greater in the spinal cord. The mice also develop tremors by 2–3 wk of age. Similarly, deletion of the gene encoding 3-hydroxy-3-methylglutaryl coenzyme A synthase (*hmgcs1*) in zebrafish compromises axon wrapping (Mathews et al. 2014). These findings support the conclusion that oligodendroglia rely on endogenous cholesterol biosynthesis for normal developmental myelination. However, it should be noted that the number of myelinated axons and the behavioral phenotype both improve with age in the mice with oligodendroglial loss of *Fdft1* (Saher et al. 2005). The authors further observed increased expression of low-density lipoprotein receptor-related protein (LRP), a major transport protein involved in cholesterol-containing lipoprotein particle uptake, in the spinal cords of the mice with deletion of *Fdft1*, suggesting compensation for loss of endogenous cholesterol biosynthesis through increasing uptake of extracellular cholesterol (Saher et al. 2005). Conditional deletion of *Lrp1* in early OPCs using also results in hypomyelination in the optic nerve (Lin et al. 2017). Moreover, studies on the coactivator of the transcriptional regulator of lipid biosynthesis, sterol regulatory element-binding protein (SREBP) support an essential function for astrocytes in providing lipids for myelination. Oligodendroglial-specific deletion of *Scap* (SREBP cleavage-activating protein) leads to delayed CNS myelination but recovery by 3 mo of age. However, simultaneous deletion of *Scap* from astrocytes, or deletion of astrocytes during postnatal development, leads to persistent hypomyelination (Camargo et al. 2017). Taken together, these data support the conclusion that oligodendrocytes rely on both endogenous cholesterol synthesis as well as uptake of extracellular cholesterol from neighboring cells to accomplish normal myelination. Finally, in addition to cholesterol, oligodendrocytes also require fatty acid synthesis to successfully accomplish myelination, since it is required for production of all noncholesterol myelin lipids. Oligodendroglia-specific deletion of the gene encoding fatty acid synthase (*Fasn*) results in hypomyelination in both the brain and spinal cord (Dimas et al. 2019).

MARKERS FOR STAGES OF THE OLIGODENDROCYTE LINEAGE

OPC Markers

Both developmentally and in the adult CNS, OPCs can be identified through their expression of PDGFRα (which as discussed above is the receptor for their dominant mitogen, PDGF-AA) (Calver et al. 1998) and the chondroitin sulfate proteoglycan neuronal/glial antigen 2, or NG2

(Stallcup and Beasley 1987), encoded by the *Cspg4* gene (Fig. 1). Indeed, OPCs are sometimes referred to as NG2-glia to reflect their prominent expression of this marker and the likelihood that they serve additional functions to the generation of new oligodendrocytes (Nishiyama et al. 2009). As noted above, OPCs express the transcription factor OLIG2 from the time of specification, which in turn induces the expression of the SRY-related HMG-box (Sox) family member SOX10 (Fig. 1; Kuspert et al. 2011). Both of these transcription factors subsequently serve as stable markers for the entire lineage. The O4 monoclonal antibody, which binds to sulfated galactocerebroside, also serves as a marker for both OPCs and oligodendrocytes (Reynolds and Hardy 1997). NG2 is expressed by additional cell types such as pericytes and O4, SOX10 and OLIG2 are also expressed by both OPCs and mature oligodendrocytes. As such, a combination of markers is often required to unequivocally identify OPCs.

Markers of Cell Cycle Exit and Oligodendrocyte Differentiation

As OPCs exit the cell cycle and begin their differentiation into premyelinating oligodendrocytes, they down-regulate NG2 and PDGFRα and up-regulate many of the proteins associated with the myelin sheath such as MBP and PLP1 (Fig. 1). Historically, premyelinating or newly generated oligodendrocytes have been identified in both human and rodent tissue on the basis of their morphology—a halo of fine processes that contact nearby axons—as well as their relatively low protein levels of markers such as MBP and the DM20 isoform of PLP1 (Trapp et al. 1997; Chang et al. 2002). More recently, several markers have been identified that are transiently expressed by newly formed oligodendrocytes (NFOLs), allowing more confident identification at either the protein or the mRNA level (Fig. 1). These include ectonucleotide pyrophosphatase/phosphodiesterase 6 (*Ennp6*), which has primarily been assessed at the level of mRNA expression to identify newly generated oligodendrocytes (Xiao et al. 2016) as well as inositol 1,4,5 trisphosphate receptor type 2 (ITPR2) and brain-enriched myelin-associated protein 1 (BCAS1),

Cite this article as *Cold Spring Harb Perspect Biol* doi: 10.1101/cshperspect.a041358

both of which have been used at the protein level to identify newly differentiated and actively myelinating oligodendrocytes (Marques et al. 2016; Fard et al. 2017). Together, these markers have provided an important advance, allowing for the distinction between actively differentiating or newly generated oligodendrocytes and their more established counterparts in contexts such as myelin repair and plasticity. They also provide an important complementary technique to genetic fate mapping or labeling dividing OPCs and their progeny with thymidine analogs, both approaches that have proven highly useful in tracking OPCs and identifying newly generated oligodendrocytes in rodent tissue (Rivers et al. 2008; Kang et al. 2010; Zhu et al. 2011; Young et al. 2013). Other markers that are transiently expressed during the differentiation process include the GPR 17 (Chen et al. 2009b) in OPCs as they commit to differentiating and the transcription factor 7-like 2 (TCF7L2) (Fancy et al. 2009).

Markers of Myelinating Oligodendrocytes

Mature, myelinating oligodendrocytes highly express proteins associated with the myelin sheath, including MBP, PLP1, and MAG (Fig. 1). Although these proteins are highly useful in labeling the mature myelin sheaths, it can be remarkably difficult to unequivocally identify the oligodendrocyte cell body with them, especially in densely myelinated white matter tracks. For this reason, the CC1 monoclonal antibody, which identifies oligodendrocyte soma with relatively little staining of the processes or other CNS-resident cells, has emerged as a highly utilized marker for mature oligodendrocytes (Bhat et al. 1996). Although CC1's antigen was initially proposed to be the adenomatous polyposis coli (APC) protein, it has more recently been demonstrated to instead bind to quaking 7, an RNA-binding protein (Bin et al. 2016). This is consistent with the known role of quaking proteins in the regulation of *Mbp* mRNA transport and myelination (Li et al. 2000). In addition to CC1, antibodies to aspartoacylase (ASPA) and myelin regulatory factor (MYRF) can both be used to identify postmitotic oligodendrocytes (Moffett et al. 2011; Hornig et al. 2013).

Single-Cell Approaches to Study Oligodendrocyte Differentiation and Heterogeneity

The advent of single-cell (sc) and single-nuclei (sn) RNA-sequencing approaches has provided a highly granular picture of the dynamic gene expression changes accompanying oligodendrocyte differentiation. Based on gene expression, the lineage can be broken down into OPCs (expressing markers such as *Pdgfra*, *Ptprz1*, *Pcdh15*), differentiation-committed OPCs (COPs, distinguished by decreased cell cycle markers and elevated expression of *Gpr17*, *Bmp4*, *Nkx2-2*), NFOLs (identified by markers such as *Tcf7l2*, *Itpr2*, *Casr*), myelin-forming oligodendrocytes (defined by their up-regulation of many myelin genes such as *Mag*, *Mog*, *Mbp*, *Plp1*) as well as several populations of mature oligodendrocytes (defined by later markers such as *Klk6* and *Apod*) (Marques et al. 2016). Although research is ongoing as to whether these transcriptionally distinct populations of mature oligodendrocytes represent a continuum of plastic states verses stable populations, they do show distinct spatial distributions between different white and gray matter tracts within the CNS. This has led to the suggestion that they may represent functionally distinct classes of oligodendrocytes specialized for distinct forms of myelination (Floriddia et al. 2020). The application of scRNA-seq and snRNA-seq technology to a range of developmental and disease states has documented active differentiation of OPCs and also the presence of disease-specific oligodendrocyte states, revealing unexpected roles such as the expression of adaptive immunity and antigen presentation proteins by OPCs and oligodendrocytes (Falcão et al. 2018; Lee et al. 2021; Pandey et al. 2022).

INTRINSIC AND EXTRINSIC REGULATORS OF OLIGODENDROCYTE DIFFERENTIATION

Understanding the hallmarks of differentiating oligodendrocytes then leads to important questions of how these events are initiated and regulated during development. Each step of oligodendrocyte differentiation is highly regulated to orchestrate cytoskeletal reorganization, expression of the myelin proteins, and membrane expansion including increased expression of lipid synthesis enzymes at the appropriate stages. Initiation and progression through the differentiation stages depend on the regulation of gene expression and cell signaling within the differentiating oligodendroglia and on extrinsic signals coming from neighboring cells and the local environment. In the next sections, we provide an overview of cell-intrinsic and -extrinsic regulators of oligodendrocyte differentiation. Wherever possible with the current state of knowledge, we have endeavored to connect these regulators of differentiation with the hallmarks outlined in the previous sections including cytoskeletal reorganization, myelin gene expression, and lipid metabolism.

Intrinsic Regulators

The transcriptional and epigenetic regulation of oligodendroglial specification and differentiation has been perhaps more thoroughly mapped out than for any other CNS lineage, and a full discussion of the current state of knowledge is beyond the scope of this article. For recent comprehensive reviews, see Elbaz and Popko (2019), Sock and Wegner (2019), Parras et al. (2020), and Selcen et al. (2023). Much of the study of transcriptional and epigenetic regulation of oligodendrocyte development has been carried out in the context of early development, often within the spinal cord (e.g., Zhou et al. 2001; Fu et al. 2002; Stolt et al. 2002). It is generally assumed, however, that with minor exceptions, similar transcriptional programs will mediate oligodendrocyte differentiation and myelination throughout the CNS and into adulthood.

TRANSCRIPTIONAL CONTROL OF OLIGODENDROCYTE DEVELOPMENT AND MYELINATION

Transcription Factors Maintaining OPCs in the Precursor State

From the time of their specification, OPCs can be identified by the expression of a core set of tran-scription factors including OLIG2, NKX2.2, and SOX10. Depending on the CNS region in which OPCs are specified, the bHLH transcription factor OLIG2 is present either just before or soon after specification. Within ventral regions, OLIG2 is already present at the time of specification (Lu et al. 2002; Zhou and Anderson 2002). Within more dorsal regions, it is induced shortly following OPC specification (Cai et al. 2005). OLIG2 subsequently induces both NKX2.2 and SOX10 (Zhou et al. 2001; Liu et al. 2007), directly binding to conserved regulatory elements within the *Sox10* gene (Kuspert et al. 2011; Yu et al. 2013). There is reciprocity in this arrangement, with SOX10 in turn positively regulating the expression of *Olig2* via an upstream enhancer (Weider et al. 2015). This ensures stable expression of these two core transcription factors throughout the development of the lineage.

A series of transcription factors act to maintain OPCs in their undifferentiated, proliferative state (Fig. 2, left panel). These include HES5, ID2, and ID4, which are induced downstream from Jagged/Notch and BMP signaling, respectively (see below). Levels of nuclear ID2 and ID4 are strongly associated with the maintenance of OPCs; overexpression of either factor prevents OPC differentiation and enhances proliferation while their knockout or knockdown leads to precocious oligodendrocyte differentiation (Kondo and Raff 2000a; Wang et al. 2001; Marin-Husstege et al. 2006). In a similar manner, deletion of *Hes5* leads to precocious expression of myelin genes and proteins (Liu et al. 2006). The exact molecular mechanisms by which HES5, ID2, and ID4 act to inhibit oligodendrocyte differentiation are somewhat unclear and may be mixed—on one hand they appear to directly bind the promoters of genes including *Mbp* (Liu et al. 2006). On the other hand, they also appear to physically interact with transcription factors such as OLIG1, OLIG2, and SOX10 that would otherwise serve to promote differentiation, presumably sequestering them or suppressing their activity (Samanta and Kessler 2004; Liu et al. 2006).

A number of members of the Sox family of transcription factors are highly expressed in OPCs and have a critical role in regulating their proliferation and survival. The SoxD proteins

SOX5 and SOX6 serve to antagonize oligodendrocyte differentiation; deletion of both *Sox5* and *Sox6* in the oligodendrocyte lineage leads to precocious differentiation of oligodendrocytes in the spinal cord and forebrain (Stolt et al. 2006; Baroti et al. 2016). This inhibition of differentiation was initially proposed to occur via competition with SOX10 at the regulatory elements of myelin genes (Stolt et al. 2006), but a later study indicated that they additionally interact with the promoter of the *Pdgfra* gene, encoding the receptor for the major mitogen for OPCs (Baroti et al. 2016). In a somewhat analogous manner, SoxE proteins SOX9 and SOX10 are also required for normal OPC migration and survival via their promotion of *Pdgfra* expression (Finzsch et al. 2008).

Transcription Factors Promoting Differentiation and Myelination

Oligodendrocyte differentiation requires either down-regulation or antagonism of the above genetic pathways, along with the coordinated activity of a large number of broadly prodifferentiation factors, often acting through or in concert with SOX10. As noted above, the expression of SOX10 is induced early in OPC specification, and SOX10 is active in OPCs in promoting the expression of OPC genes such as *Pdgfra* (Finzsch et al. 2008; Kuspert et al. 2011). Subsequently, SOX10 continues to be expressed and to play an important role in regulating differentiation and myelin gene expression—the most striking consequence of deletion of *Sox10* is a robust reduction in oligodendrocyte differentiation (Stolt et al. 2002). Nevertheless, SOX10's presence in the oligodendrocyte lineage from the time of specification (Kuhlbrodt et al. 1998) indicates the requirement for additional regulators to either enhance SOX10's prodifferentiation activity, or counter the above transcriptional regulators and pathways that would otherwise serve to maintain OPCs in their undifferentiated state (Fig. 2, middle panel).

One such example of a feedback mechanism during oligodendrocyte differentiation is the OLIG2 target ZEB2, which antagonizes BMP signaling and ID2 expression, at least in part through induction of SMAD7 (Weng et al. 2012). The activity of ZEB2 is augmented by ZFP276, which is induced by SOX10 during differentiation and subsequently acts with ZEB2 to inhibit the expression of OPC genes including *Id2*, *Cspg4*, *Wnt7a*, and *Tgfb2* (Aberle et al. 2022). In a somewhat analogous manner, the YY1 transcription factor recruits Hdac1 to the *Id1* and *Tcf7l2* genes, inhibiting their expression (He et al. 2007). Accordingly, deletion of either *Yy1* or *Zeb2* in the oligodendrocyte lineage causes a stall in oligodendrocyte differentiation and block in CNS myelination (He et al. 2007; Weng et al. 2012). A somewhat more complex regulation of oligodendrocyte differentiation is provided by TCF7L2, which is transiently expressed during oligodendrocyte differentiation (Fancy et al. 2009). As with YY1 and ZEB2, loss of *Tcf7l2* causes a block in oligodendrocyte differentiation (Ye et al. 2009a; Hammond et al. 2015). Generally considered an effector of the Wnt pathway, TCF7L2 initially complexes with β-catenin to inhibit differentiation (Ye et al. 2009a). At later time points, TCF7L2 instead binds histone deacetylases (HDACs) and KAISO to inhibit Wnt/β-catenin signaling and then finally acts with SOX10 at the regulatory elements of genes promoting differentiation (Ye et al. 2009a; Zhao et al. 2016). An additional factor is NKX2.2, which is induced early in development both during dorsoventral patterning and downstream from SOX10 within OPCs (Liu et al. 2007). OLIG2 and NKX2.2 usually antagonize the expression of each other (Sun et al. 2003). In response to dephosphorylation and activation of the NFATC2 protein, NFATC2 and SOX10 overcome this cross repression to elevate the expression of NKX2.2 (Qi et al. 2001; Weider et al. 2018). Once induced, NKX2.2 can interact with a series of transcriptional repressors to modulate differentiation via targets such as *Pdgfra* (Zhu et al. 2014; Zhang et al. 2020).

The above examples predominantly serve as repressors of genes that would otherwise serve to maintain OPCs in a proliferative state (i.e., "inhibiting the inhibitors"). In contrast, several transcriptional regulators will positively regulate the differentiation and myelination program. The role of ZFP24 (also known as ZFP191) in myelination was initially identified when a spontaneous mouse mutation associated with tremors, seizures, and dysmyelination was mapped to the *Zfp24* gene (Howng et al. 2010). ZFP24 is broadly expressed throughout the oligodendrocyte lineage but is dephosphorylated and activated during oligodendrocyte differentiation, allowing it to further promote differentiation and myelination (Elbaz et al. 2018). Unlike ZFP24, the expression of *Myrf* mRNA is induced early during oligodendrocyte differentiation (Emery et al. 2009) and is largely dependent on the activity of SOX10, which binds directly to the regulatory elements in the first intron of the *Myrf* gene in conjunction with OLIG2 and ZFP24 (Hornig et al. 2013; Yu et al. 2013; Lopez-Anido et al. 2015; Elbaz et al. 2018). Once induced, MYRF and SOX10 bind to an overlapping set of regulatory elements of many genes involved in myelin and membrane lipid regulation, with MYRF enhancing SOX10's activity at promyelination genes and repressing its activity at OPC-specific SOX10 targets (Bujalka et al. 2013; Hornig et al. 2013; Aprato et al. 2020). MYRF and SOX10 also interact with the thyroid hormone-induced KLF9 and its close relative KLF13 (Bernhardt et al. 2022), providing a potential mechanism for the finding that thyroid hormone strongly promotes the differentiation of OPCs into mature oligodendrocytes (Kondo and Raff 2000a).

With the advent of genome-wide approaches such as RNA-seq and ChIP-seq, assessment of the roles of promyelination transcription factors including OLIG2, SOX10, and MYRF shifted from a focus on the regulatory elements of defined candidate targets such as *Mbp* to an unbiased assessment of their targets in myelinating oligodendrocytes. These approaches have shown that OLIG2, SOX10, and MYRF bind an overlapping set of regulatory elements for hundreds of genes in myelinating oligodendrocytes, including those encoding lipid regulatory proteins (*Aspa, Elovl7, Fa2h, Ldrap1, Ugt8, Slc45a3*), cytoskeletal regulators (*Tppp, Kif21a, Cdc42, Rac1*), and many of the protein components of the myelin sheaths (*Plp1, Mbp, Mog, Mag*) (Bujalka et al. 2013; Yu et al. 2013; Lopez-Anido et al. 2015; Darr et al. 2017). OLIG2 appears to play a prepatterning role in myelin gene expression (see below); once the oligodendrocytes are fully differentiated its expression is no longer strictly necessary (Mei et al. 2013; Yu et al. 2013; Wang et al. 2022). In contrast, the continued activity of MYRF and SOX10 at myelin genes is necessary throughout the life of the oligodendrocyte, as conditional ablation of either *Myrf* or *Sox10* (along with the closely related and compensatory *Sox8*) in mature oligodendrocytes causes a severe disruption to myelin gene expression and ultimately breakdown of the myelin sheaths (Koenning et al. 2012; Turnescu et al. 2018).

Epigenetic Control of Oligodendrocyte Development and Myelination

Many of the known OLIGs also pair with epigenetic regulators to modify the accessibility of the regulatory elements of target genes. In this section, we briefly describe some of the major epigenetic pathways currently understood to influence oligodendrocyte differentiation.

Histone-Modifying HDACs and HATs

One of the best-understood mechanisms regulating gene expression is the modification of permissive acetylation markers on lysine residues in histone tails, controlled by the balance of activities between histone acetyltransferases (HATs) and HDACs. The overall chromatin structure of OPCs is euchromatic, characterized by an overall open chromatin structure and permissive marks such as H3K9 and H3K14 histone acetylation, contrasting with the overall compacted and inaccessible chromatin of differentiated oligodendrocytes (Liu et al. 2012). During their differentiation, the overall levels of acetylated histones decrease in oligodendrocytes, suggesting a shift to poised or inactive chromatin states (Marin-Husstege et al. 2002). Correspondingly, pharmacological inhibition of HDACs in the developing rat brain or remyelinating mouse brain strongly inhibits oligodendrocyte maturation (Marin-Husstege et al. 2002; Shen et al. 2005; Shen et al. 2008), confirming the broad importance of histone deacetylation during oligodendrocyte differentiation. At the genetic level, ablation of both *Hdac1* and *Hdac2* blocks oligodendrocyte differentiation both in the mouse and in culture, associated with a stabilization of β-CATENIN and activation of WNT signaling (Ye et al. 2009a). The activity of HDACs is guided by the YY1 transcription factor during oligodendrocyte differentiation, with YY1 recruiting HDAC1 to the *Id4* and *Tcf7l2* promoters to repress their expression and allow differentiation (He et al. 2007).

Histone Methylation

Another important regulator of chromatin accessibility is the presence of repressive histone marks such as H3K27 and H3K9 trimethylation. Overall levels of H3K9me3 increase during oligodendrocyte lineage progression, with the activity of histone methyltransferase activity strongly induced by differentiation signals such as T3 (Liu et al. 2015). Both H3K9me3 and H3K27me3 markers show a largely nonoverlapping set of target loci in OPCs compared to postmitotic oligodendrocytes, indicating a highly dynamic program of methylation-mediated gene repression (Liu et al. 2015). Pharmacological inhibition of H3K9 histone methyltransferases or siRNA-mediated knockdown of the methyltransferase *Ezh2* reduces oligodendrocyte differentiation in vitro, with the cells showing aberrant expression of neuronal genes and electrical excitability (Liu et al. 2015). In vivo, disruption of the PRC2 complex that mediates H3K27 trimethylation through conditional knockout of *Ezh2* or *Eed* strongly disrupts CNS myelination, with OPCs instead showing elevation of NOTCH, WNT, and BMP pathways and inappropriate expression of neuronal and astrocyte genes (Wang et al. 2020a,b).

Chromatin-Remodeling Complexes

Additional regulation of DNA accessibility is provided by the SWI/SFN/BAF complex of ATP-dependent, chromatin-remodeling enzymes. Taking a genome-wide approach (Yu et al. 2013) demonstrated that the BRG1 ATPase component of the SWI/SFN/BAF family is a transcriptional target of OLIG2. Conditional ablation of *Brg1* using an *Olig1*-Cre driver did not

alter OPC numbers but caused a severe CNS hypomyelination and block in oligodendrocyte differentiation (Yu et al. 2013). BRG1 occupancy at the regulatory elements of oligodendrocyte genes strongly overlaps with the occupancy of OLIG2, suggesting that OLIG2 may act to recruit BRG1-containing, chromatin-remodeling complexes in much the same way that SOX10 does within Schwann cells (Weider et al. 2012; Yu et al. 2013). It should be noted that a subsequent study in which *Brg1* was deleted using the *Cnp1*-Cre or *Cspg4*-Cre drivers (Bischof et al. 2015) found a considerably more modest reduction in oligodendrocyte differentiation than Yu et al. found with the *Olig1*-Cre driver. Whether these differences reflect the comparatively earlier timing of *Brg1* inactivation seen with the *Olig1*-Cre line, compensation by other family members, or a loss of SOX10 expression in the *Olig1*-Cre; *Brg1*[Floxed] mice remains unclear (Yu et al. 2013; Bischof et al. 2015).

In contrast, the chromodomain helicase DNA-binding (CDH) family has clear roles during differentiation. At earlier stages of differentiation, CDH8 shows a strong co-occupancy with OLIG2 and SOX10 at oligodendrocyte-related genes (Zhao et al. 2018). Conditional ablation of *Cdh8* results in reduced chromatin accessibility and reduced levels of the permissive H3K4me3 markers at the promoters of key regulators of differentiation including *Zeb2*, *Myrf*, and *Nkx2.2*, as well as at myelin genes such as *Mbp* and *Ugt8a* (Zhao et al. 2018). The reduced H3K4 methylation at these sites suggests that CDH8 recruits the MLL1/KMT2A complex, responsible for the trimethylation at lysine 4 of histone H3. Consistent with this possibility, Zhao et al. (2018) found that CDH8 coimmunoprecipitated with members of this complex. In a similar manner, CDH7 is induced by OLIG2 and BRG1 and is guided by SOX10 to the promoters of genes involved in myelination, cytoskeletal organization, and lipid homeostasis (He et al. 2016).

Noncoding RNAs

In addition to the above epigenetic changes to chromatin structure, a variety of RNA-mediated epigenetic changes are implicated in CNS myelination. MicroRNAs are short noncoding RNAs that regulate gene expression. Showing their general importance in the lineage, conditional ablation of the DICER enzyme (required to generate mature microRNAs from their full-length precursors; Bartel 2004) strongly disrupts OPC differentiation into postmitotic oligodendrocytes (Dugas et al. 2010; Zhao et al. 2010). Several approaches have been taken to identify the individual microRNAs that mediate this effect and their target genes. Both Dugas et al. (2010) and Wang et al. (2017) took a profiling approach, identifying a cohort of microRNAs, most notably *miR-219* and *miR-338*, which were strongly induced during differentiation. The predicted targets of *miR-219* and *miR-338* included OPC genes such as *Hes5*, *Pdgfra*, and *Sox6*, indicating that the induction of these microRNAs serves as a feedback loop to the OPC maintenance program (Dugas et al. 2010; Wang et al. 2017). Subsequently, Wang and colleagues confirmed that genetic ablation of the two genes encoding *miR-219* (*miR-219-1* and *miR-219-2*) in mice caused hypomyelination, exacerbated by deletion of *miR-338*, also identifying the myelin inhibitor *Lingo1* as an additional *miR-219* target (Wang et al. 2017). There is a strong interaction between these microRNAs and the transcriptional program of oligodendrocyte differentiation. For example, during differentiation, SOX10 mediates the induction of *miR-204*, which in turn targets *Ccnd1* and *Sox4* to reduce OPC proliferation (Wittstatt et al. 2020). In much the same way, *miR-219* is a direct transcriptional target of the promyelination factor MYRF (Bujalka et al. 2013). Together, the induction of these microRNAs provides a feedback mechanism to ensure that the OPC maintenance program is shut down as the differentiation process is initiated.

If *miR-219* and *miR-338* are induced to down-regulate the OPC maintenance program, their activity is countered by several microRNAs that act to inhibit or temper the differentiation program. *miR-145-5p* is highly expressed in OPCs and is down-regulated during differentiation (Letzen et al. 2010). It acts to maintain OPCs in a proliferative state, with *miR-145* knockdown promoting differentiation (Kornfeld et al. 2021). One of *miR-145*'s direct targets is *Myrf*, so inhi-

bition of *miR-145* in OPCs leads to an increase in *Myrf* transcripts as well as MYRF target gene levels (Kornfeld et al. 2021). The *mir-17-92* family serves to promote OPC proliferation, mediated at least in part through targeting of *Pten* and activation of Akt signaling (Budde et al. 2010). *MiR-125a-3p* also serves to restrict the differentiation of OPCs to oligodendrocytes, likely through inhibition of targets such as *Fyn* and *Nrg1* (Lecca et al. 2016).

In addition to microRNAs, recent work has shown that long noncoding RNAs (lncRNAs) also regulate oligodendrocyte differentiation. lncRNAs are noncoding RNAs of >200 nt that can modulate gene expression via a variety of mechanisms such as binding protein epigenetic regulators (Srinivas et al. 2023). He and colleagues found that a subset of these lncRNAs show dynamic expression during oligodendrocyte differentiation, with some being relatively selective to the oligodendrocyte lineage and with many showing genomic occupancy by SOX10 at their transcription start sites (Lopez-Anido et al. 2015; He et al. 2017). Selecting one of these lncRNAs (dubbed *lncOL1*) for study, they showed that its ablation in mice strongly delayed developmental myelination. Mechanistically, *lncOL1* physically interacts with the SUZ12/PRC2 complex, suggesting it contributes to targeting OPC genes for H3K27me3-mediated repression (He et al. 2017). The role of additional lncRNAs in the oligodendrocyte lineage remains to be determined but given the large number that are regulated during oligodendrocyte development and their induction by SOX10 and OLIG2 (He et al. 2017; Wei et al. 2021), they likely represent an important element of interplay between the transcriptional and epigenetic control of myelination.

EXTRINSIC TO INTRINSIC: LIGANDS, RECEPTORS, AND INTRACELLULAR SIGNALING PATHWAYS

There are numerous extracellular modulators that regulate the progression of the bipolar OPC into a myelinating oligodendrocyte. Growth factors secreted by neurons or other glial cells, and sometimes by differentiating OPCs themselves (McMorris et al. 1986; Carson et al. 1993;

McKinnon et al. 1993; Du et al. 2003; Wang et al. 2007), release of neurotransmitters and other molecules from axons (Demerens et al. 1996; Bongarzone et al. 1998; Bozzali and Wrabetz 2004; Agresti et al. 2005; Cavaliere et al. 2012; Fannon et al. 2015), systemic hormones such as thyroid hormone (Bhat et al. 1979; Almazan et al. 1985; Lee and Petratos 2016), and extracellular matrix molecules (Colognato et al. 2007; Colognato and Tzvetanova 2011; Eyermann et al. 2012; Kang and Yao 2022; Yamada et al. 2022) are among the factors that inhibit or promote differentiation of the OPC to a myelinating oligodendrocyte. For purposes of this discussion, we focus on a subset of factors with cell-autonomous actions on developing oligodendroglia through cell surface receptors and known signaling cascades.

Ligand–Receptors Inhibiting Oligodendrocyte Differentiation and Myelination

A number of intracellular signaling pathways are known that regulate oligodendrocyte differentiation and myelination. Some of these are particular ligand–receptor interactions, which then directly impact nuclear transcription through different mechanisms. The BMP, NOTCH, and WNT ligands act primarily to inhibit OPC differentiation through binding cell surface receptors and mediating downstream actions largely through nuclear regulation of transcription. BMP receptors (BMPRs)-Ia, -Ib, and -II are expressed by oligodendroglia in the adult brain (Miyagi et al. 2012) and in developing OPCs (Zhang et al. 2014; Marques et al. 2016; Ornelas et al. 2020). Studies on the loss of *Bmpr1a/1b* in the embryonic neural tube support a role for these receptors in inhibiting oligodendrocyte differentiation and myelination (See et al. 2007). BMP2 and BMP4 have been the most studied BMP ligands in the lineage, both acting to inhibit OPC differentiation (Mabie et al. 1999; Grinspan et al. 2000; See and Grinspan 2009). RNA-seq and scRNA-seq databases indicate high expression of BMP1, 4, and 7 in OPCs suggesting autocrine–paracrine regulation of differentiation (Zhang et al. 2014; Marques et al. 2016). BMP stimulation of BMPRs results in phosphorylation and activation of receptor-associated SMAD proteins. The SMADs can then as-

sociate in the nucleus with p300 and HDACs in a transcription regulatory complex to modulate gene expression such as inducing expression of ID2 and ID4, which inhibit oligodendrocyte differentiation (Kondo and Raff 2000b; Gomes et al. 2003; Samanta and Kessler 2004; Cheng et al. 2007; Emery and Lu 2015; Grinspan 2015).

The NOTCH signaling pathway is another well-known pathway that inhibits OPC differentiation through nuclear actions. JAGGED-1 and DELTA-1 ligand binding to NOTCH receptors causes proteolytic cleavage and release of the NOTCH intracellular domain, which then translocates to the nucleus to regulate gene transcription. Deletion of *Notch1* in OPCs enhances the appearance of PLP1[+] and MAG[+] oligodendrocytes in the spinal cord by late embryonic stages without altering the number of OPCs (Genoud et al. 2002). Heterozygous deletion of *Notch1* globally also results in increased MBP and PLP and premature myelination in the brain (Givogri et al. 2002). Heterozygous loss of *Notch1* also decreases expression of the NOTCH target HES5, which can compete with SOX10 to inhibit transcription of *Mbp* (Liu et al. 2006; Xiao et al. 2020).

Canonical WNT signaling directly stabilizes β-CATENIN protein, another transcriptional regulator of oligodendroglial progression (Rim et al. 2022). Wnt ligands were initially described in oligodendrocyte development as factors in the dorsal spinal cord that inhibit OPC development (Shimizu et al. 2005). WNT ligand binding to the seven-transmembrane domain frizzled receptors of the GPCR family and to the LRP5/6 coreceptors causes stabilization of β-CATENIN through inactivating the cytoplasmic β-CATENIN destruction complex (Rim et al. 2022). β-CATENIN then translocates to the nucleus to regulate transcription. Early in vitro studies demonstrated that treatment of mouse embryonic spinal cord whole-mount preparations with a WNT antagonist increases the number of O4[+] cells (Shimizu et al. 2005). Conversely, constitutively activating WNT/β-CATENIN signaling by preventing β-CATENIN degradation in oligodendroglia delays expression of PLP1 and MBP and myelination in postnatal mouse spinal cords (Feigenson et al. 2009). Similarly, overexpression of activated β-CATENIN in the oligodendrocyte lineage delays OPC differentiation and results in hypomyelination in the spinal cord (Fancy et al. 2009). Interestingly, subsequent studies revealed that WNT pathway inhibition of oligodendrocyte differentiation functions at least in part through the BMP pathway. Inhibiting BMP signaling in rat OPCs in vitro with the BMP antagonist noggin prevents WNT-mediated inhibition of differentiation (Feigenson et al. 2011). Similarly, mouse OPCs lacking BMPR1a and BMPR1b differentiate normally when exposed to WNT ligands in vitro (Feigenson et al. 2011). Subsequent studies where canonical WNT signaling was disrupted in oligodendroglia through deletion of *Ctnnb1* showed that loss of WNT signaling increases the number of PDGFRα$^+$ OPCs at embryonic stages but delays the appearance of PLP1$^+$ oligodendrocytes at postnatal ages (Dai et al. 2014b), suggesting stage-specific effects of WNT signaling during oligodendrocyte specification and differentiation. The complexity of β-CATENIN actions on oligodendrocyte differentiation can be explained in part by its interactions with other transcriptional regulators at different stages including TCF7L2 (Ye et al. 2009b; Guo et al. 2015; Rim et al. 2022). Numerous frizzled receptors are expressed by OPCs and oligodendrocytes (Marques et al. 2016), but their exact functions have not been investigated in this lineage with the exception of frizzled 8a, which when knocked down in zebrafish, results in reduced OPC maturation (Kim et al. 2008).

Several GPRs in addition to the frizzled receptors discussed above have been identified that regulate oligodendrocyte differentiation and myelination (Mogha et al. 2016). GPR17 acts as a negative regulator of OPC differentiation and is down-regulated as the cells mature (Chen et al. 2009a; Boda et al. 2011; Fumagalli et al. 2011). Global loss of *Gpr17* results in precocious myelination, whereas overexpression of *Gpr17* in the oligodendrocyte lineage inhibits myelination (Chen et al. 2009a). GPR17 is activated in OPCs by uracil nucleotides or cysteinyl leukotrienes causing inhibition of intracellular cAMP generation (Ciana et al. 2006; Fumagalli et al. 2011). GPR17 was identified as a transcriptional target

of OLIG1, which represses *Gpr17* expression; however, the subsequent activation of GPR17 in turn induces nuclear translocation of ID2/4, which binds to and represses OLIG1/2 activity (Chen et al. 2009a).

GPR37 and GPR56 are both negative regulators of developmental myelination acting at different stages in the lineage. GPR37 is expressed in the later stages of differentiating oligodendroglia (Yang et al. 2016). Systemic deletion of *Gpr37* results in hypermyelination, and GPR37-deficient OPCs cocultured with dorsal root ganglion neurons differentiate more rapidly into late-stage myelinating oligodendrocytes (Yang et al. 2016). These studies also revealed that the accelerated differentiation seen with loss of GPR37 is mediated at least in part through the activation of mitogen-activated protein kinase (MAPK)/ ERK1/2 signaling. GPR56 or adhesion GPCR G1 (ADGRG1) promotes OPC proliferation, and loss of this receptor early in the lineage results in hypomyelination in both zebrafish and mice (Ackerman et al. 2015; Giera et al. 2015; Chiou et al. 2021). GPR56 can be activated by microglia-derived transglutaminase-2 in a laminin-dependent manner (Giera et al. 2018).

Ligand–Receptors Promoting Oligodendrocyte Differentiation and Myelination

The receptor tyrosine kinases TRKB, FGF receptor (FGFR)2/3, and IGF1R all have essential oligodendrocyte cell-autonomous functions through their respective ligands as positive mediators of oligodendrocyte differentiation and/or developmental myelination (Ye et al. 1995a, 2002; Bhat and Zhang 1996; Yim et al. 2001; Bansal et al. 2003; Oh et al. 2003; Zeger et al. 2007; Van't Veer et al. 2009; Furusho et al. 2012; Wong et al. 2013; Furusho et al. 2017).

Brain-derived neurotrophic factor (BDNF) secreted by neurons and astrocytes binds to TRKB on oligodendroglia to enhance oligodendrocyte development at multiple stages. Exposure of primary rat OPCs to BDNF in vitro results in increased numbers of MBP$^+$ oligodendrocytes, enhances their maturation, and promotes axon myelination (Du et al. 2003; Xiao et al.

2010). Heterozygous loss of *Bdnf* in mice systemically delays myelination in both the corpus callosum and spinal cord (Xiao et al. 2010). Similarly, the deletion of *Trkb* in premyelinating oligodendrocytes delays myelination in both the corpus callosum and spinal cord (Wong et al. 2013).

The IGF ligands IGF-1 and IGF-2 are circulating peptide hormones expressed locally by a number of cell types in the CNS during developmental myelination, and oligodendroglia at all stages express IGF1R (Zhang et al. 2014; Marques et al. 2016). Successful in vitro culturing of oligodendroglia requires stimulation of IGF1R generally through the addition of micromolar levels of insulin in the culture medium (Barres et al. 1992). Global overexpression or deletion of *Igf1* in mice results in hypermyelination and hypomyelination, respectively, but are also accompanied by significant alterations in overall body size and brain weights (Carson et al. 1993; Powell-Braxton et al. 1993; Beck et al. 1995; Ye et al. 1995b). Oligodendroglial-specific deletion of the *Igf1r* reduces MBP and the number of mature CC1$^+$ oligodendrocytes due at least in part to decreased OPC proliferation and increased apoptosis (Zeger et al. 2007). These findings and the need for IGF1R stimulation for oligodendroglia survival over time in vitro have made it difficult to define specific functions for IGF1R signaling in the differentiation of oligodendrocytes. However, there is evidence that IGF-1 mediates commitment and differentiation of oligodendrocytes from OPC bipotential glial precursors and adult neural progenitor cells (McMorris and Dubois-Dalcq 1988; Hsieh et al. 2004).

Several of the 22 known FGF ligands are known to bind to three of the four FGF receptors (FGFR1–3) expressed by oligodendroglia (Fortin et al. 2005; Zhang et al. 2014; Marques et al. 2016). FGF-2 acting through the FGFR1 enhances proliferation and prevents differentiation of OPCs (McKinnon et al. 1990b; Reimers et al. 2001; Fortin et al. 2005; Zhou et al. 2006). However, *Cnp-Cre* deletion of *Fgfr1* and *Fgfr2* results in hypomyelination in the spinal cord predominantly by regulating later stages of oligodendrocyte maturation involving myelin production (Furusho et al. 2012). A subsequent study re-

vealed that FGFR2 is the main mediator of oligodendrocyte maturation and myelin growth through activating the MAPK/ERK1/2 pathway and increasing *Myrf* (Furusho et al. 2017).

INTRACELLULAR SIGNALING: MAPK AND PI3K/AKT/MTOR PATHWAYS

Well-studied intracellular signaling pathways activated downstream from multiple cell surface receptors in oligodendroglia are the MAPK/ERK, p38 MAPK, and phosphoinositide 3-kinase (PI3K)/Akt/mechanistic target of rapamycin (mTOR) pathways. BDNF/TRKB and FGF/FGFRs preferentially activate the MAPK/ERK1/2 pathway (Du et al. 2006; Van't Veer et al. 2009; Furusho et al. 2012, 2017), whereas IGF-1/IGF1R predominantly activates the PI3K/AKT/mTOR pathway in OPCs (Ness and Wood 2002; Ness et al. 2002; Romanelli et al. 2007, 2009). Activation of these signaling pathways has complex outcomes regulating molecules in the cytoplasm and nucleus through phosphorylation cascades. As will be seen from the data presented below, it is not always simple to separate the impact of these pathways on morphological maturation and lineage progression from their actions on myelination since this developmental progression is intricately linked.

MAPK Signaling

Activation of several cell surface receptors critical for oligodendrocyte differentiation including BDNF and FGFs activate RAS GTPases resulting in a phosphorylation cascade of RAF followed by MEK1/2, which are the upstream kinases of serine–threonine extracellular signaling-regulated kinases (ERK)1/2 (Wortzel and Seger 2011). Studies on the function of the MAPK/ERK pathway in oligodendrocyte differentiation have yielded mixed findings. Inhibiting ERK1/2 signaling in vitro through the use of a MEK1/2 inhibitor reduces the progression of late OPCs to immature and mature oligodendrocytes (Galabova-Kovacs et al. 2008; Fyffe-Maricich et al. 2011). However, expressing a constitutively active MEK1 in OPCs resulting in hyperactivation of ERK1/2 promotes myelination in cocultures of

OPCs with dorsal root ganglion neurons without altering in cell survival, proliferation, or differentiation (Xiao et al. 2012). Consistent with a prominent effect of ERK1/2 signaling in myelination, mice carrying a global *Erk1* deletion combined with an oligodendrocyte lineage-specific *Erk2* deletion are hypomyelinated, particularly in the spinal cord, but have normal OPC proliferation and differentiation (Ishii et al. 2012). Consistent with these findings, in vivo expression of a constitutively active MEK1 construct in developing oligodendroglia causes CNS hypermyelination with a transient increase in OPC proliferation but no effect on oligodendrocyte differentiation (Ishii et al. 2013). These studies support the conclusion that ERK1/2 signaling regulates myelin thickness independently of oligodendrocyte differentiation during development in vivo.

A second MAPK pathway, the p38 MAPK pathway, has also been studied in the context of oligodendrocyte lineage progression. The p38 MAPK family consists of four isoforms, α, β, δ, and γ. Early studies using pharmacological inhibitors of p38 MAPK on OPCs in vitro reported inhibition of OPC differentiation through preventing expression of myelin gene expression (Fragoso et al. 2007; Chew et al. 2010; Haines et al. 2015). Specific deletion of *p38α* in the oligodendrocyte lineage inhibited MBP expression in differentiating OPCs in vitro and delayed OPC differentiation and myelination in the corpus callosum in vivo (Chung et al. 2015). Activation of this pathway regulates oligodendrocyte differentiation in part through the downstream effector mitogen-activated protein kinase-activated protein kinase 2 (Haines et al. 2010). More recent studies on the γ isoform of p38 MAPK support an opposing role to the α/β isoforms in oligodendrocyte differentiation. The timing of p38γ expression correlates with developmental myelination such that it is highest in PDGFRα⁺ OPCs and is down-regulated as the cells mature both in vitro and in vivo (Marziali et al. 2023). Cre-lox deletion of *p38γ* in early postnatal NG2⁺ OPCs accelerates OPC differentiation and transiently enhances myelination in the brain, an effect that may be partially mediated by p38α/β activity (Marziali et al. 2023). Thus, p38γ signaling may

serve as a brake through as yet unknown mechanisms for the myelination-promoting actions of p38α/β.

PI3K/AKT/mTOR Signaling

Activation of PI3K results in phosphorylation of PtdIns(4,5)2 (PIP_2) to generate PIP_3, which then phosphorylates the serine–threonine kinase AKT leading to activation of mTOR (Gaesser and Fyffe-Maricich 2016; Giguère 2018). Constitutive activation of AKT through aspartate substitution at Ser 473 (referred to as Akt-DD) in differentiating oligodendroglia enhances myelin thickness (Flores et al. 2008; Narayanan et al. 2009). The phosphatase and tensin homolog (PTEN) normally acts as a negative regulator of the PI3K/AKT pathway by dephosphorylating PIP_3 to PIP_2 (Worby and Dixon 2014). Similar to constitutive activation of AKT, deleting *Pten* in differentiating oligodendroglia results in CNS hypermyelination (Goebbels et al. 2010).

mTOR, a serine/threonine kinase, is a downstream mediator of the PI3K/AKT pathway regulating cell growth, proliferation, and survival in many cell types from flies to mammals (Giguère 2018; Liu and Sabatini 2020). Receptor activation of PI3K activates the raptor-associated mTOR complex (mTORC) 1 through phosphoinositide-dependent kinase (PDK)1, which phosphorylates AKT on Thr^{308}, leading to inhibition of the tuberous sclerosis complex (TSC), thereby inhibiting RAS homolog enriched in brain (RHEB) (Fig. 3). Activation of mTORC2 also occurs through PI3K activation (Gan et al. 2011); however, several studies indicate that mTORC2 can be activated by TSC (Huang et al. 2008; Ikenoue et al. 2008). Downstream, mTORC2 phosphorylates AKT on Ser^{473} resulting in its full activation (Sarbassov et al. 2005).

Initial in vitro studies on the function of mTOR in the oligodendrocyte lineage demonstrated that treating primary rat OPCs with the mTOR inhibitor rapamycin blocks differentiation of the late-stage O4⁺/GalC⁻ precursor into a GalC⁺ immature oligodendrocyte and prevents expression of PLP1 and MBP (Tyler et al. 2009, 2011). Subsequent studies confirmed that inhibiting mTOR pharmacologically in OPCs cul-

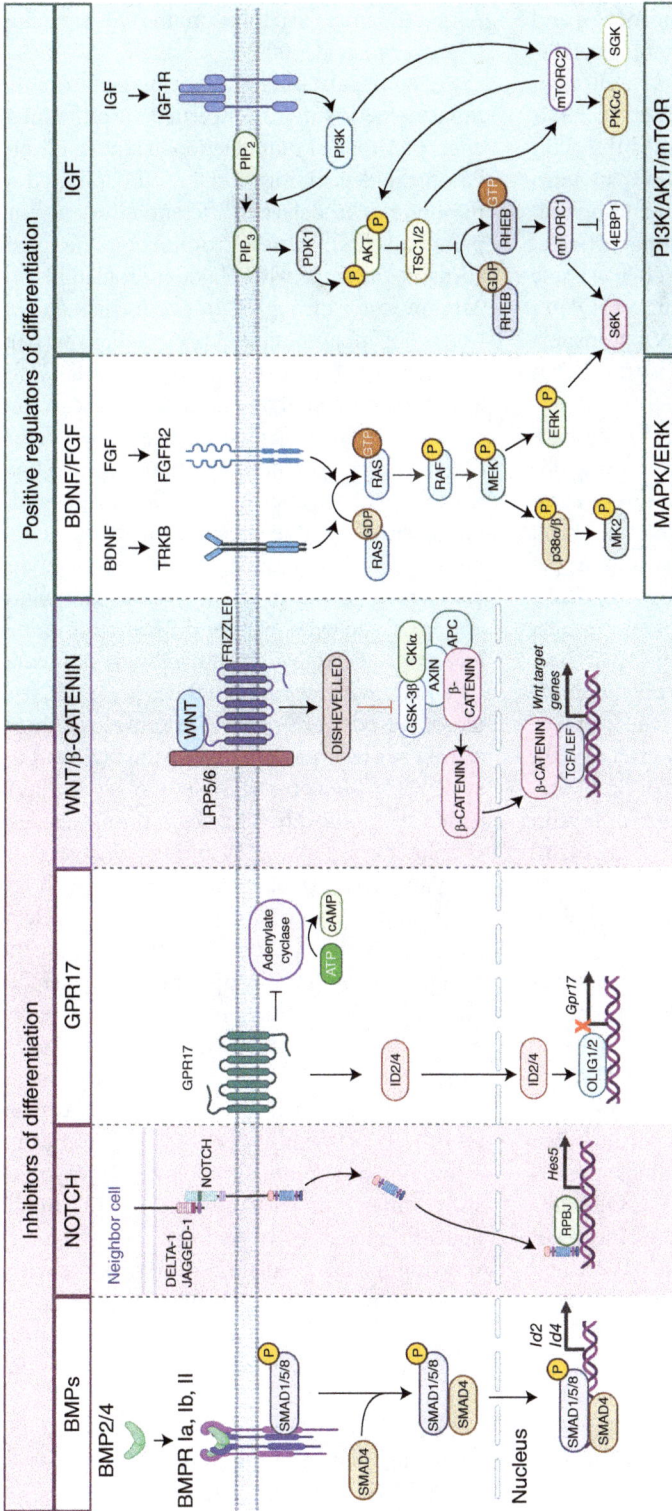

Figure 3. Extrinsic to intrinsic regulators of oligodendrocyte differentiation. A number of extracellular signals interact with cell surface receptors to inhibit or promote oligodendrocyte differentiation. The cell surface receptors connect to intracellular signaling molecules that can directly translocate to the nucleus to regulate gene transcription or can be intermediates in a phosphorylation cascade targeting cytoplasmic and nuclear targets to regulate oligodendrocyte differentiation processes including morphology, metabolism, and gene transcription. Bone morphogenic protein (BMP), Notch, and G-protein–coupled receptor (GPR)17 signaling negatively regulate oligodendrocyte differentiation during developmental myelination. The mitogen-activated protein kinase (MAPK) and phosphoinositide 3-kinase (PI3K)/AKT/ mechanistic target of rapamycin (mTOR) intracellular signaling pathways act as positive mediators of oligodendrocyte differentiation downstream from a number of extracellular ligands and cell surface receptors. The Wnt signaling pathway has both negative and positive functions in developmental oligodendrocyte differentiation and myelination. (Figure generated with BioRender, https:// biorender.com.)

tured in differentiation media prevents the progression to MBP$^+$ mature oligodendrocytes and compromises branching and morphological differentiation of rat OPCs in vitro as well as in zebrafish in vivo (Guardiola-Diaz et al. 2012; Musah et al. 2020). mTOR regulates the actin cytoskeleton in differentiating OPCs in vitro through multiple mechanisms including decreasing proteins involved in actin polymerization and branching (Musah et al. 2020). These studies also showed that inhibiting mTOR in OPCs in vitro disrupts *Mbp* mRNA transport to the distal processes for protein translation (Musah et al. 2020).

Genetic studies in zebrafish and mice have confirmed the role for mTOR in promoting oligodendrocyte differentiation and myelination (Bercury et al. 2014; Lebrun-Julien et al. 2014; Wahl et al. 2014; Fedder-Semmes and Appel 2021; Khandker et al. 2022; Dahl et al. 2023). Oligodendroglia deletion of *Mtor* in mice results in the accumulation of PDGFRα$^+$ and NG2$^+$ precursors and reduces O4$^+$ immature oligodendrocytes in the spinal cord during the peak of postnatal differentiation (Ornelas et al. 2020), resulting in the delayed appearance of CC1$^+$ mature oligodendrocytes and expression of myelin proteins (Wahl et al. 2014; Musah et al. 2020). Mice with oligodendroglial deletion of either *Mtor* or *Rptor* are hypomyelinated in the spinal cord into adulthood despite near complete recovery in the number of mature oligodendrocytes and myelinated axons (Bercury et al. 2014; Lebrun-Julien et al. 2014; Wahl et al. 2014; Musah et al. 2020). Deletion of both *Rptor* and *Rictor* in oligodendroglia revealed a similar phenotype to that of *Rptor* alone (Lebrun-Julien et al. 2014). In vitro and in vivo findings support the conclusion that mTOR promotes OPC differentiation in part through suppressing BMP signaling both at the level of the BMPR and through disrupting the expression of *Id2* (Ornelas et al. 2020). In addition to regulating BMP signaling, mTOR promotes actin polymerization and branching in differentiating oligodendrocytes in the spinal cord recapitulating in vitro findings (Musah et al. 2020). Consistent with the gene deletion studies showing loss of *Mtor* results in hypomyelination, the hypermyelina-

tion resulting from activation of Akt in the AKT-DD mice depends on mTOR signaling (Narayanan et al. 2009).

A significant finding that emerged from the mouse genetic studies concerned the differential effect of *Mtor* or *Rptor* deletion in oligodendroglia on the developing spinal cord compared to the brain. The delayed differentiation and impaired myelin thickness prominent in the spinal cords in the mice with oligodendroglial loss of *Mtor* or *Rptor* through *Cnp-Cre*-mediated deletion is not apparent in the corpus callosum (Bercury et al. 2014; Wahl et al. 2014). Recently, Dahl and colleagues analyzed a mouse line carrying deletion of *Rictor* using a noninducible *Pdgfra-Cre* and observed a more pronounced phenotype in the brain than reported from *Cnp-Cre* deletion of *Rictor*, likely due to earlier deletion in the lineage (Dahl et al. 2023). ScRNA-seq analyses comparing gene expression in immature oligodendroglia in wild-type and *Mtor* conditional knockout mice during postnatal developmental myelination uncovered a number of mTOR-regulated genes involved in cholesterol biosynthesis as well as in cytoskeletal remodeling (Khandker et al. 2022). Many of the cholesterol biosynthesis gene alterations were validated at the protein level (Khandker et al. 2022), and the changes seen in cytoskeletal targets were consistent with the prior studies on mTOR-regulated cytoskeletal changes discussed above. Not surprisingly, the mTOR-regulated transcriptome is considerably more extensive in the spinal cord compared to brain oligodendroglia (Khandker et al. 2022), consistent with the more pronounced hypomyelination in the spinal cord with oligodendroglia loss of mTOR (Wahl et al. 2014). The finding that expression of a number of cholesterol biosynthesis enzymes is downstream from mTOR in oligodendroglia is consistent with prior studies showing mTOR regulation of lipid and cholesterol enzymes in rapamycin-treated OPCs differentiating in vitro (Tyler et al. 2011) and in mice with oligodendroglial deletion of *Rptor* (Lebrun-Julien et al. 2014).

Surprisingly, the scRNA-seq study by Khandker et al. (2022) also revealed differences between wild-type brain and spinal cord oligodendroglia: Spinal cord oligodendroglia have higher expression of cholesterol biosynthesis genes than oligodendroglia from the brain. These findings help explain why *Mtor* deletion in this lineage has a more profound effect on myelination in the spinal cord compared to the brain.

The SREBPs are master transcriptional regulators of lipid metabolism, with different family members controlling fatty acid or cholesterol synthesis genes or both in the liver (Horton et al. 2002). mTORC1 activates the SREBPs in other cell types including Schwann cells (Norrmén et al. 2014; Eid et al. 2017). Deletion of *Rptor* in oligodendroglia reduced expression of SREBP2, which primarily regulates cholesterol biosynthesis in the developing spinal cord (Lebrun-Julien et al. 2014). No changes were seen in SREBP1 proteins that control fatty acid synthesis; however, the SREBP1 target, FASN, was reduced in the spinal cord with oligodendroglial loss of *Rptor*. Similarly, FASN was reduced by inhibition of mTOR in OPCs in vitro (Tyler et al. 2011) and in spinal cord oligodendroglia with *Mtor* deletion (Khandker et al. 2022). It should also be noted that while the mouse genetic studies place mTOR upstream of cholesterol biosynthesis, studies in zebrafish additionally place mTOR signaling downstream where it mediates cholesterol-dependent myelin gene expression (Mathews and Appel 2016).

Integrating Signaling Pathways and Oligodendrocyte Differentiation during Developmental Myelination

The findings on intracellular signaling pathway regulation of oligodendrocyte differentiation and myelination reveal that although the different pathways have some distinct functions, particularly in how they regulate early stages of OPC differentiation, a common theme that has emerged is that they all modulate myelin wrapping. This raises interesting questions such as (1) do these pathways cross talk, and (2) what are their immediate molecular targets? Several studies have addressed cross talk between the MAPK/ERK and PI3K/AKT/mTOR pathways in oligodendroglia. During OPC differentiation in vitro, inhibiting any point in the PI3K/AKT/mTOR pathway increases activation of ERK1/2, but inhibiting ERK1/2 has no impact on phosphorylation of either AKT or mTOR (Dai et al. 2014a); inhibiting either pathway reduces MBP expression. Similarly, intraperitoneal injection of rapamycin into postnatal pups increases pERK1/2 in differentiating oligodendroglia in the corpus callosum (Dai et al. 2014a). In genetic mouse models, hypomyelination resulting from ERK1/2 deletion can be partially or completely rescued by oligodendroglial expression of constitutively active AKT or PI3K, respectively (Ishii et al. 2019). In contrast, hyperactivation of ERK1/2 fails to rescue the myelination deficits in the oligodendroglial *mTOR* knockout mice (Ishii et al. 2019). These data suggest that ERK1/2 signaling regulates myelination in part through the PI3K/Akt/mTOR pathway.

Defining the signaling cascades leading to the induction of myelin genes, cytoskeletal reorganization, and lipid biosynthesis in differentiating oligodendroglia will be a major challenge in the years ahead. Although activation of ERK1/2 has been linked to MYRF expression (Furusho et al. 2017), and some of the cytoskeletal and cholesterol biosynthesis targets have been identified downstream from mTOR (Khandker et al. 2022), in both cases, these proteins are likely downstream from a kinase cascade that ultimately mediates expression at the RNA and/or protein level. Two well-studied immediate targets of mTORC1 kinase activity are p70S6K1/2, which is activated by mTORC1 phosphorylation, and 4EBP1, which is inhibited by mTORC1 phosphorylation causing derepression of the eukaryotic translation initiation factor (EIF)4E. Collectively, these targets have a major function in regulating cap-dependent RNA translation (Kim et al. 2002; Inoki et al. 2005). Initial analyses of mice carrying germline deletions of *p70s6k1/2* or of *Eeif4ebp1/2* indicated no myelination defects (Lebrun-Julien et al. 2014). However, a recent study demonstrated that pharmacological inhibition of p70S6K1 in differentiating oligodendroglia reduces MBP expression in rat OPCs in vitro and the number of myelin internodes formed in zebrafish in vivo (Benardais et al. 2022). As discussed previously, a primary mTORC2 direct target is AKT-Ser473, but mTORC2 can also regulate cell polarization and the actin cytoskeleton through directly phosphorylating protein kinase C (PKC) and through regulating the Rho family of small GTPases in other cell types (Jacinto et al. 2004; Sarbassov et al. 2004; Inoki et al. 2005; Facchinetti et al. 2008; Ikenoue et al. 2008; Angliker and Rüegg 2013; Giguère 2018). The recent study by Dahl and colleagues supports a role for RICTOR in phosphorylating AKT-Ser473 and PKCα/β and in regulating cytoskeletal targets involved in myelin wrapping in differentiating brain OPCs (Dahl et al. 2023).

FUTURE PERSPECTIVES

Although the work we review here on the regulation of oligodendroglial differentiation has been extensive and contributed to by a large number of investigators, there are still many open questions for future studies. An area of considerable current interest is the connection between neuronal activity and oligodendrocyte differentiation and myelination, although a number of investigations have made this connection and are reviewed in accompanying articles in this collection (Hill et al. 2023). The studies on signaling pathways are only beginning to connect these intracellular pathways from the cell surface to the numerous molecules known to regulate myelin protein expression, lipid biosynthesis, and cytoskeletal reorganization both in the cytoplasm and nucleus. Fully mapping how these pathways mediate their ultimate effects will be an exciting challenge for the future. As we explore the regulation of oligodendrocyte development, there is an ongoing question of how the insights we gain relate to the impaired differentiation seen in the context of adult CNS injury and remyelination. There is still much to be understood about how the mechanisms underlying developmental myelination overlap or are distinct from adult myelin maintenance and remyelination.

ACKNOWLEDGMENTS

The authors thank Marie L. Mather for designing the figures and critical reading of the manuscript. This work was supported by the following funding sources: National Institutes of Health R01 NS120981 (B.E.) and R37 NS082203 (T.L.W.), an endowment from the Warren family (B.E.), and National Multiple Sclerosis Society RG170728557 (T.L.W.).

REFERENCES

*Reference is also in this subject collection.

Aberle T, Piefke S, Hillgärtner S, Tamm ER, Wegner M, Küspert M. 2022. Transcription factor Zfp276 drives oligodendroglial differentiation and myelination by switching off the progenitor cell program. *Nucleic Acids Res* **50:** 1951–1968. doi:10.1093/nar/gkac042

Ackerman SD, Garcia C, Piao X, Gutmann DH, Monk KR. 2015. The adhesion GPCR Gpr56 regulates oligodendrocyte development via interactions with Gα12/13 and RhoA. *Nat Commun* **6:** 6122. doi:10.1038/ncomms7122

Agresti C, Meomartini ME, Amadio S, Ambrosini E, Volonté C, Aloisi F, Visentin S. 2005. ATP regulates oligodendrocyte progenitor migration, proliferation, and differentiation: involvement of metabotropic P2 receptors. *Brain Res Rev* **48:** 157–165. doi:10.1016/j.brainresrev.2004.12.005

Almazan G, Honegger P, Matthieu JM. 1985. Triiodothyronine stimulation of oligodendroglial differentiation and myelination. A developmental study. *Dev Neurosci* **7:** 45–54. doi:10.1159/000112275

Angliker N, Rüegg MA. 2013. In vivo evidence for mTORC2-mediated actin cytoskeleton rearrangement in neurons. *Bioarchitecture* **3:** 113–118. doi:10.4161/bioa.26497

Aprato J, Sock E, Weider M, Elsesser O, Fröb F, Wegner M. 2020. Myrf guides target gene selection of transcription factor Sox10 during oligodendroglial development. *Nucleic Acids Res* **48:** 1254–1270. doi:10.1093/nar/gkz1158

Bacmeister CM, Huang R, Osso LA, Thornton MA, Conant L, Chavez AR, Poleg-Polsky A, Hughes EG. 2022. Motor learning drives dynamic patterns of intermittent myelination on learning-activated axons. *Nat Neurosci* **25:** 1300–1313. doi:10.1038/s41593-022-01169-4

Bansal R, Magge S, Winkler S. 2003. Specific inhibitor of FGF receptor signaling: FGF-2-mediated effects on proliferation, differentiation, and MAPK activation are inhibited by PD173074 in oligodendrocyte-lineage cells. *J Neurosci Res* **74:** 486–493. doi:10.1002/jnr.10773

Baroti T, Zimmermann Y, Schillinger A, Liu L, Lommes P, Wegner M, Stolt CC. 2016. Transcription factors Sox5 and Sox6 exert direct and indirect influences on oligodendroglial migration in spinal cord and forebrain. *Glia* **64:** 122–138. doi:10.1002/glia.22919

Barres BA, Hart IK, Coles HS, Burne JF, Voyvodic JT, Richardson WD, Raff MC. 1992. Cell death in the oligodendrocyte lineage. *J Neurobiol* **23:** 1221–1230. doi:10.1002/neu.480230912

Bartel DP. 2004. MicroRNAs: genomics, biogenesis, mechanism, and function. *Cell* **116:** 281–297. doi:10.1016/s0092-8674(04)00045-5

Beck KD, Powell-Braxton L, Widmer HR, Valverde J, Hefti F. 1995. Igf1 gene disruption results in reduced brain size, CNS hypomyelination, and loss of hippocampal granule and striatal parvalbumin-containing neurons. *Neuron* **14:** 717–730. doi:10.1016/0896-6273(95)90216-3

Belachew S, Aguirre AA, Wang H, Vautier F, Yuan X, Anderson S, Kirby M, Gallo V. 2002. Cyclin-dependent kinase-2 controls oligodendrocyte progenitor cell cycle progression and is downregulated in adult oligodendrocyte progenitors. *J Neurosci* 22: 8553–8562. doi:10.1523/JNEUROSCI.22-19-08553.2002

Benardais K, Ornelas IM, Fauveau M, Brown TL, Finseth LT, Panic R, Deboux C, Macklin WB, Wood TL, Nait-Oumesmar B. 2022. p70S6 kinase regulates oligodendrocyte differentiation and is active in remyelinating lesions. *Brain Commun* 4: fcac025. doi:10.1093/braincomms/fcac025

Bercury KK, Dai J, Sachs HH, Ahrendsen JT, Wood TL, Macklin WB. 2014. Conditional ablation of raptor or rictor has differential impact on oligodendrocyte differentiation and CNS myelination. *J Neurosci* 34: 4466–4480. doi:10.1523/JNEUROSCI.4314-13.2014

Bernhardt C, Sock E, Fröb F, Hillgärtner S, Nemer M, Wegner M. 2022. KLF9 and KLF13 transcription factors boost myelin gene expression in oligodendrocytes as partners of SOX10 and MYRF. *Nucleic Acids Res* 50: 11509–11528. doi:10.1093/nar/gkac953

Bhat NR, Zhang P. 1996. Activation of mitogen-activated protein kinases in oligodendrocytes. *J Neurochem* 66: 1986–1994. doi:10.1046/j.1471-4159.1996.66051986.x

Bhat NR, Sarlieve LL, Rao GS, Pieringer RA. 1979. Investigations on myelination in vitro. Regulation by thyroid hormone in cultures of dissociated brain cells from embryonic mice. *J Biol Chem* 254: 9342–9344. doi:10.1016/S0021-9258(19)83519-6

Bhat RV, Axt KJ, Fosnaugh JS, Smith KJ, Johnson KA, Hill DE, Kinzler KW, Baraban JM. 1996. Expression of the APC tumor suppressor protein in oligodendroglia. *Glia* 17: 169–174. doi:10.1002/(SICI)1098-1136(199606)17:2<169::AID-GLIA8>3.0.CO;2-Y

Bin JM, Harris SN, Kennedy TE. 2016. The oligodendrocyte-specific antibody "CC1" binds Quaking 7. *J Neurochem* 139: 181–186. doi:10.1111/jnc.13745

Bischof M, Weider M, Kuspert M, Nave KA, Wegner M. 2015. Brg1-dependent chromatin remodelling is not essentially required during oligodendroglial differentiation. *J Neurosci* 35: 21–35. doi:10.1523/JNEUROSCI.1468-14.2015

Boda E, Viganò F, Rosa P, Fumagalli M, Labat-Gest V, Tempia F, Abbracchio MP, Dimou L, Buffo A. 2011. The GPR17 receptor in NG2 expressing cells: focus on in vivo cell maturation and participation in acute trauma and chronic damage. *Glia* 59: 1958–1973. doi:10.1002/glia.21237

Bögler O, Wren D, Barnett SC, Land H, Noble M. 1990. Cooperation between two growth factors promotes extended self-renewal and inhibits differentiation of oligodendrocyte-type-2 astrocyte (O-2A) progenitor cells. *Proc Natl Acad Sci* 87: 6368–6372. doi:10.1073/pnas.87.16.6368

Bongarzone ER, Howard SG, Schonmann V, Campagnoni AT. 1998. Identification of the dopamine D3 receptor in oligodendrocyte precursors: potential role in regulating differentiation and myelin formation. *J Neurosci* 18: 5344–5353. doi:10.1523/JNEUROSCI.18-14-05344.1998

Bozzali M, Wrabetz L. 2004. Axonal signals and oligodendrocyte differentiation. *Neurochem Res* 29: 979–988. doi:10.1023/B:NERE.0000021242.12455.75

Brown TL, Macklin WB. 2020. The actin cytoskeleton in myelinating cells. *Neurochem Res* 45: 684–693. doi:10.1007/s11064-019-02753-0

Budde H, Schmitt S, Fitzner D, Opitz L, Salinas-Riester G, Simons M. 2010. Control of oligodendroglial cell number by the miR-17-92 cluster. *Development* 137: 2127–2132. doi:10.1242/dev.050633

Bujalka H, Koenning M, Jackson S, Perreau VM, Pope B, Hay CM, Mitew S, Hill AF, Lu QR, Wegner M, et al. 2013. MYRF is a membrane-associated transcription factor that autoproteolytically cleaves to directly activate myelin genes. *PLoS Biol* 11: e1001625. doi:10.1371/journal.pbio.1001625

Cai J, Qi Y, Hu X, Tan M, Liu Z, Zhang J, Li Q, Sander M, Qiu M. 2005. Generation of oligodendrocyte precursor cells from mouse dorsal spinal cord independent of Nkx6 regulation and Shh signaling. *Neuron* 45: 41–53. doi:10.1016/j.neuron.2004.12.028

Calver AR, Hall AC, Yu WP, Walsh FS, Heath JK, Betsholtz C, Richardson WD. 1998. Oligodendrocyte population dynamics and the role of PDGF in vivo. *Neuron* 20: 869–882. doi:10.1016/S0896-6273(00)80469-9

Camargo N, Goudriaan A, van Deijk AF, Otte WM, Brouwers JF, Lodder H, Gutmann DH, Nave KA, Dijkhuizen RM, Mansvelder HD, et al. 2017. Oligodendroglial myelination requires astrocyte-derived lipids. *PLoS Biol* 15: e1002605. doi:10.1371/journal.pbio.1002605

Carson MJ, Behringer RR, Brinster RL, McMorris FA. 1993. Insulin-like growth factor I increases brain growth and central nervous system myelination in transgenic mice. *Neuron* 10: 729–740. doi:10.1016/0896-6273(93)90173-O

Casaccia-Bonnefil P, Tikoo R, Kiyokawa H, Friedman V Jr, Chao MV, Koff A. 1997. Oligodendrocyte precursor differentiation is perturbed in the absence of the cyclin-dependent kinase inhibitor p27[Kip1]. *Genes Dev* 11: 2335–2346. doi:10.1101/gad.11.18.2335

Cavaliere F, Urra O, Alberdi E, Matute C. 2012. Oligodendrocyte differentiation from adult multipotent stem cells is modulated by glutamate. *Cell Death Dis* 3: e268. doi:10.1038/cddis.2011.144

Chang A, Tourtellotte WW, Rudick R, Trapp BD. 2002. Premyelinating oligodendrocytes in chronic lesions of multiple sclerosis. *N Engl J Med* 346: 165–173. doi:10.1056/NEJMoa010994

Chen Y, Wu H, Wang S, Koito H, Li J, Ye F, Hoang J, Escobar SS, Gow A, Arnett HA, et al. 2009a. The oligodendrocyte-specific G protein-coupled receptor GPR17 is a cell-intrinsic timer of myelination. *Nat Neurosci* 12: 1398–1406. doi:10.1038/nn.2410

Chen Y, Wu H, Wang S, Koito H, Li J, Ye F, Hoang J, Escobar SS, Gow A, Arnett HA, et al. 2009b. The oligodendrocyte-specific G protein-coupled receptor GPR17 is a cell-intrinsic timer of myelination. *Nat Neurosci* 12: 1398–1406. doi:10.1038/nn.2410

Cheng X, Wang Y, He Q, Qiu M, Whittemore SR, Cao Q. 2007. Bone morphogenetic protein signaling and olig1/2 interact to regulate the differentiation and maturation of adult oligodendrocyte precursor cells. *Stem Cells* 25: 3204–3214. doi:10.1634/stemcells.2007-0284

Chew LJ, Coley W, Cheng Y, Gallo V. 2010. Mechanisms of regulation of oligodendrocyte development by p38 mitogen-activated protein kinase. *J Neurosci* **30:** 11011–11027. doi:10.1523/JNEUROSCI.2546-10.2010

Chiou B, Gao C, Giera S, Folts CJ, Kishore P, Yu D, Oak HC, Jiang R, Piao X. 2021. Cell type-specific evaluation of ADGRG1/GPR56 function in developmental central nervous system myelination. *Glia* **69:** 413–423. doi:10.1002/glia.23906

Chung SH, Biswas S, Selvaraj V, Liu XB, Sohn J, Jiang P, Chen C, Chmilewsky F, Marzban H, Horiuchi M, et al. 2015. The p38α mitogen-activated protein kinase is a key regulator of myelination and remyelination in the CNS. *Cell Death Dis* **6:** e1748. doi:10.1038/cddis.2015.119

Ciana P, Fumagalli M, Trincavelli ML, Verderio C, Rosa P, Lecca D, Ferrario S, Parravicini C, Capra V, Gelosa P, et al. 2006. The orphan receptor GPR17 identified as a new dual uracil nucleotides/cysteinyl-leukotrienes receptor. *EMBO J* **25:** 4615–4627. doi:10.1038/sj.emboj.7601341

Colognato H, Tzvetanova ID. 2011. Glia unglued: how signals from the extracellular matrix regulate the development of myelinating glia. *Dev Neurobiol* **71:** 924–955. doi:10.1002/dneu.20966

Colognato H, Galvin J, Wang Z, Relucio J, Nguyen T, Harrison D, Yurchenco PD, Ffrench-Constant C. 2007. Identification of dystroglycan as a second laminin receptor in oligodendrocytes, with a role in myelination. *Development* **134:** 1723–1736. doi:10.1242/dev.02819

Dahl KD, Almeida AR, Hathaway HA, Bourne J, Brown TL, Finseth LT, Wood TL, Macklin WB. 2023. mTORC2 loss in oligodendrocyte progenitor cells results in regional hypomyelination in the central nervous system. *J Neurosci* **43:** 540–558. doi:10.1523/JNEUROSCI.0010-22.2022

Dai J, Bercury KK, Macklin WB. 2014a. Interaction of mTOR and Erk1/2 signaling to regulate oligodendrocyte differentiation. *Glia* **62:** 2096–2109. doi:10.1002/glia.22729

Dai ZM, Sun S, Wang C, Huang H, Hu X, Zhang Z, Lu QR, Qiu M. 2014b. Stage-specific regulation of oligodendrocyte development by Wnt/β-catenin signaling. *J Neurosci* **34:** 8467–8473. doi:10.1523/JNEUROSCI.0311-14.2014

Đặng TC, Ishii Y, Nguyen V, Yamamoto S, Hamashima T, Okuno N, Nguyen QL, Sang Y, Ohkawa N, Saitoh Y, et al. 2019a. Powerful homeostatic control of oligodendroglial lineage by PDGFRα in adult brain. *Cell Rep* **27:** 1073–1089. e5. doi:10.1016/j.celrep.2019.03.084

Đặng TC, Ishii Y, Nguyen VD, Yamamoto S, Hamashima T, Okuno N, Nguyen QL, Sang Y, Ohkawa N, Saitoh Y, et al. 2019b. Powerful homeostatic control of oligodendroglial lineage by PDGFRα in adult brain. *Cell Rep* **27:** 1073–1089. e5. doi:10.1016/j.celrep.2019.03.084

Darr AJ, Danzi MC, Brady L, Emig-Agius D, Hackett A, Golshani R, Warner N, Lee J, Lemmon VP, Tsoulfas P. 2017. Identification of genome-wide targets of Olig2 in the adult mouse spinal cord using ChIP-Seq. *PLoS ONE* **12:** e0186091. doi:10.1371/journal.pone.0186091

Demerens C, Stankoff B, Logak M, Anglade P, Allinquant B, Couraud F, Zalc B, Lubetzki C. 1996. Induction of myelination in the central nervous system by electrical activity. *Proc Natl Acad Sci* **93:** 9887–9892. doi:10.1073/pnas.93.18.9887

Dimas P, Montani L, Pereira JA, Moreno D, Trötzmüller M, Gerber J, Semenkovich CF, Köfeler HC, Suter U. 2019. CNS myelination and remyelination depend on fatty acid synthesis by oligodendrocytes. *eLife* **8:** e44702. doi:10.7554/eLife.44702

Doerflinger NH, Macklin WB, Popko B. 2003. Inducible site-specific recombination in myelinating cells. *Genesis* **35:** 63–72. doi:10.1002/gene.10154

Du Y, Fischer TZ, Lee LN, Lercher LD, Dreyfus CF. 2003. Regionally specific effects of BDNF on oligodendrocytes. *Dev Neurosci* **25:** 116–126. doi:10.1159/000072261

Du Y, Lercher LD, Zhou R, Dreyfus CF. 2006. Mitogen-activated protein kinase pathway mediates effects of brain-derived neurotrophic factor on differentiation of basal forebrain oligodendrocytes. *J Neurosci Res* **84:** 1692–1702. doi:10.1002/jnr.21080

Dubois-Dalcq M, Behar T, Hudson L, Lazzarini RA. 1986. Emergence of three myelin proteins in oligodendrocytes cultured without neurons. *J Cell Biol* **102:** 384–392. doi:10.1083/jcb.102.2.384

Dugas JC, Cuellar TL, Scholze A, Ason B, Ibrahim A, Emery B, Zamanian JL, Foo LC, McManus MT, Barres BA. 2010. Dicer1 and miR-219 are required for normal oligodendrocyte differentiation and myelination. *Neuron* **65:** 597–611. doi:10.1016/j.neuron.2010.01.027

Duncan ID, Radcliff AB, Heidari M, Kidd G, August BK, Wierenga LA. 2018. The adult oligodendrocyte can participate in remyelination. *Proc Natl Acad Sci* **115:** 201808064. doi:10.1073/pnas.1808064115

Eid W, Dauner K, Courtney KC, Gagnon A, Parks RJ, Sorisky A, Zha X. 2017. mTORC1 activates SREBP-2 by suppressing cholesterol trafficking to lysosomes in mammalian cells. *Proc Natl Acad Sci* **114:** 7999–8004. doi:10.1073/pnas.1705304114

Elbaz B, Popko B. 2019. Molecular control of oligodendrocyte development. *Trends Neurosci* **42:** 263–277. doi:10.1016/j.tins.2019.01.002

Elbaz B, Aaker JD, Isaac S, Kolarzyk A, Brugarolas P, Eden A, Popko B. 2018. Phosphorylation state of ZFP24 controls oligodendrocyte differentiation. *Cell Rep* **23:** 2254–2263. doi:10.1016/j.celrep.2018.04.089

Emery B, Lu QR. 2015. Transcriptional and epigenetic regulation of oligodendrocyte development and myelination in the central nervous system. *Cold Spring Harb Perspect Biol* **7:** a020461. doi:10.1101/cshperspect.a020461

Emery B, Agalliu D, Cahoy JD, Watkins TA, Dugas JC, Mulinyawe SB, Ibrahim A, Ligon KL, Rowitch DH, Barres BA. 2009. Myelin gene regulatory factor is a critical transcriptional regulator required for CNS myelination. *Cell* **138:** 172–185. doi:10.1016/j.cell.2009.04.031

Eyermann C, Czaplinski K, Colognato H. 2012. Dystroglycan promotes filopodial formation and process branching in differentiating oligodendroglia. *J Neurochem* **120:** 928–947. doi:10.1111/j.1471-4159.2011.07600.x

Facchinetti V, Ouyang W, Wei H, Soto N, Lazorchak A, Gould C, Lowry C, Newton AC, Mao Y, Miao RQ, et al. 2008. The mammalian target of rapamycin complex 2 controls folding and stability of Akt and protein kinase C. *EMBO J* **27:** 1932–1943. doi:10.1038/emboj.2008.120

Falcão AM, van Bruggen D, Marques S, Meijer M, Jäkel S, Agirre E, Samudyata, Floriddia EM, Vanichkina DP, Ffrench-Constant C, et al. 2018. Disease-specific oligodendrocyte lineage cells arise in multiple sclerosis. *Nat Med* **24:** 1837–1844. doi:10.1038/s41591-018-0236-y

Fancy SP, Baranzini SE, Zhao C, Yuk DI, Irvine KA, Kaing S, Sanai N, Franklin RJ, Rowitch DH. 2009. Dysregulation of the Wnt pathway inhibits timely myelination and remyelination in the mammalian CNS. *Genes Dev* **23**: 1571–1585. doi:10.1101/gad.1806309

Fannon J, Tarmier W, Fulton D. 2015. Neuronal activity and AMPA-type glutamate receptor activation regulates the morphological development of oligodendrocyte precursor cells. *Glia* **63**: 1021–1035. doi:10.1002/glia.22799

Fard MK, Meer F, Sánchez P, Cantuti-Castelvetri L, Mandad S, Jäkel S, Fornasiero EF, Schmitt S, Ehrlich M, Starost L, et al. 2017. BCAS1 expression defines a population of early myelinating oligodendrocytes in multiple sclerosis lesions. *Sci Transl Med* **9**: eaam7816. doi:10.1126/scitranslmed.aam7816

Fedder-Semmes KN, Appel B. 2021. The Akt-mTOR pathway drives myelin sheath growth by regulating cap-dependent translation. *J Neurosci* **41**: 8532–8544. doi:10.1523/JNEUROSCI.0783-21.2021

Feigenson K, Reid M, See J, Crenshaw EB 3rd, Grinspan JB. 2009. Wnt signaling is sufficient to perturb oligodendrocyte maturation. *Mol Cell Neurosci* **42**: 255–265. doi:10.1016/j.mcn.2009.07.010

Feigenson K, Reid M, See J, Crenshaw IE, Grinspan JB. 2011. Canonical Wnt signalling requires the BMP pathway to inhibit oligodendrocyte maturation. *ASN Neuro* **3**: e00061. doi:10.1042/AN20110004

Finzsch M, Stolt CC, Lommes P, Wegner M. 2008. Sox9 and Sox10 influence survival and migration of oligodendrocyte precursors in the spinal cord by regulating PDGF receptor α expression. *Development* **135**: 637–646. doi:10.1242/dev.010454

Flores AI, Narayanan SP, Morse EN, Shick HE, Yin X, Kidd G, Avila RL, Kirschner DA, Macklin WB. 2008. Constitutively active Akt induces enhanced myelination in the CNS. *J Neurosci* **28**: 7174–7183. doi:10.1523/JNEUROSCI.0150-08.2008

Floriddia EM, Lourenço T, Zhang S, van Bruggen D, Hilscher MM, Kukanja P, Gonçalves dos Santos JP, Altınkök M, Yokota C, Llorens-Bobadilla E, et al. 2020. Distinct oligodendrocyte populations have spatial preference and different responses to spinal cord injury. *Nat Commun* **11**: 5860. doi:10.1038/s41467-020-19453-x

Fortin D, Rom E, Sun H, Yayon A, Bansal R. 2005. Distinct fibroblast growth factor (FGF)/FGF receptor signaling pairs initiate diverse cellular responses in the oligodendrocyte lineage. *J Neurosci* **25**: 7470–7479. doi:10.1523/JNEUROSCI.2120-05.2005

Fragoso G, Haines JD, Roberston J, Pedraza L, Mushynski WE, Almazan G. 2007. P38 mitogen-activated protein kinase is required for central nervous system myelination. *Glia* **55**: 1531–1541. doi:10.1002/glia.20567

Fruttiger M, Calver AR, Richardson WD. 2000. Platelet-derived growth factor is constitutively secreted from neuronal cell bodies but not from axons. *Curr Biol* **10**: 1283–1286. doi:10.1016/S0960-9822(00)00757-0

Fu H, Qi Y, Tan M, Cai J, Takebayashi H, Nakafuku M, Richardson W, Qiu M. 2002. Dual origin of spinal oligodendrocyte progenitors and evidence for the cooperative role of *Olig2* and *Nkx2.2* in the control of oligodendrocyte differentiation. *Development* **129**: 681–693. doi:10.1242/dev.129.3.681

Fumagalli M, Daniele S, Lecca D, Lee PR, Parravicini C, Fields RD, Rosa P, Antonucci F, Verderio C, Trincavelli ML, et al. 2011. Phenotypic changes, signaling pathway, and functional correlates of GPR17-expressing neural precursor cells during oligodendrocyte differentiation. *J Biol Chem* **286**: 10593–10604. doi:10.1074/jbc.M110.162867

Furusho M, Dupree JL, Nave KA, Bansal R. 2012. Fibroblast growth factor receptor signaling in oligodendrocytes regulates myelin sheath thickness. *J Neurosci* **32**: 6631–6641. doi:10.1523/JNEUROSCI.6005-11.2012

Furusho M, Ishii A, Bansal R. 2017. Signaling by FGF receptor 2, not FGF receptor 1, regulates myelin thickness through activation of ERK1/2-MAPK, which promotes mTORC1 activity in an Akt-independent manner. *J Neurosci* **37**: 2931–2946. doi:10.1523/JNEUROSCI.3316-16.2017

Fyffe-Maricich SL, Karlo JC, Landreth GE, Miller RH. 2011. The ERK2 mitogen-activated protein kinase regulates the timing of oligodendrocyte differentiation. *J Neurosci* **31**: 843–850. doi:10.1523/JNEUROSCI.3239-10.2011

Gaesser JM, Fyffe-Maricich SL. 2016. Intracellular signaling pathway regulation of myelination and remyelination in the CNS. *Exp Neurol* **283**: 501–511. doi:10.1016/j.expneurol.2016.03.008

Galabova-Kovacs G, Catalanotti F, Matzen D, Reyes GX, Zezula J, Herbst R, Silva A, Walter I, Baccarini M. 2008. Essential role of B-Raf in oligodendrocyte maturation and myelination during postnatal central nervous system development. *J Cell Biol* **180**: 947–955. doi:10.1083/jcb.200709069

Gan X, Wang J, Su B, Wu D. 2011. Evidence for direct activation of mTORC2 kinase activity by phosphatidylinositol 3,4,5-trisphosphate. *J Biol Chem* **286**: 10998–11002. doi:10.1074/jbc.M110.195016

Genoud S, Lappe-Siefke C, Goebbels S, Radtke F, Aguet M, Scherer SS, Suter U, Nave KA, Mantei N. 2002. Notch1 control of oligodendrocyte differentiation in the spinal cord. *J Cell Biol* **158**: 709–718. doi:10.1083/jcb.200202002

Giera S, Deng Y, Luo R, Ackerman SD, Mogha A, Monk KR, Ying Y, Jeong SJ, Makinodan M, Bialas AR, et al. 2015. The adhesion G protein-coupled receptor GPR56 is a cell-autonomous regulator of oligodendrocyte development. *Nat Commun* **6**: 6121. doi:10.1038/ncomms7121

Giera S, Luo R, Ying Y, Ackerman SD, Jeong SJ, Stoveken HM, Folts CJ, Welsh CA, Tall GG, Stevens B, et al. 2018. Microglial transglutaminase-2 drives myelination and myelin repair via GPR56/ADGRG1 in oligodendrocyte precursor cells. *eLife* **7**: e33385. doi:10.7554/eLife.33385

Giguère V. 2018. Canonical signaling and nuclear activity of mTOR—a teamwork effort to regulate metabolism and cell growth. *FEBS J* **285**: 1572–1588. doi:10.1111/febs.14384

Givogri MI, Costa RM, Schonmann V, Silva AJ, Campagnoni AT, Bongarzone ER. 2002. Central nervous system myelination in mice with deficient expression of Notch1 receptor. *J Neurosci Res* **67**: 309–320. doi:10.1002/jnr.10128

Goebbels S, Oltrogge JH, Kemper R, Heilmann I, Bormuth I, Wolfer S, Wichert SP, Mobius W, Liu X, Lappe-Siefke C, et al. 2010. Elevated phosphatidylinositol 3,4,5-trisphosphate in glia triggers cell-autonomous membrane wrapping and myelination. *J Neurosci* **30**: 8953–8964. doi:10.1523/JNEUROSCI.0219-10.2010

Gomes WA, Mehler MF, Kessler JA. 2003. Transgenic over-expression of BMP4 increases astroglial and decreases oligodendroglial lineage commitment. *Dev Biol* 255: 164–177. doi:10.1016/S0012-1606(02)00037-4

Grinspan JB. 2015. Bone morphogenetic proteins: inhibitors of myelination in development and disease. *Vitam Horm* 99: 195–222. doi:10.1016/bs.vh.2015.05.005

Grinspan JB, Edell E, Carpio DF, Beesley JS, Lavy L, Pleasure D, Golden JA. 2000. Stage-specific effects of bone morphogenetic proteins on the oligodendrocyte lineage. *J Neurobiol* 43: 1–17. doi:10.1002/(SICI)1097-4695(200004)43:1<1::AID-NEU1>3.0.CO;2-0

Guardiola-Diaz HM, Ishii A, Bansal R. 2012. Erk1/2 MAPK and mTOR signaling sequentially regulates progression through distinct stages of oligodendrocyte differentiation. *Glia* 60: 476–486. doi:10.1002/glia.22281

Guo F, Lang J, Sohn J, Hammond E, Chang M, Pleasure D. 2015. Canonical Wnt signaling in the oligodendroglial lineage-puzzles remain. *Glia* 63: 1671–1693. doi:10.1002/glia.22813

Haines JD, Fang J, Mushynski WE, Almazan G. 2010. Mitogen-activated protein kinase activated protein kinase 2 (MK2) participates in p38 MAPK regulated control of oligodendrocyte differentiation. *Glia* 58: 1384–1393. doi:10.1002/glia.21014

Haines JD, Fulton DL, Richard S, Almazan G. 2015. P38 mitogen-activated protein kinase pathway regulates genes during proliferation and differentiation in oligodendrocytes. *PLoS ONE* 10: e0145843. doi:10.1371/journal.pone.0145843

Hajihosseini M, Tham TN, Dubois-Dalcq M. 1996. Origin of oligodendrocytes within the human spinal cord. *J Neurosci* 16: 7981–7994. doi:10.1523/JNEUROSCI.16-24-07981.1996

Hammelrath L, Škokić S, Khmelinskii A, Hess A, van der Knaap N, Staring M, Lelieveldt BPF, Wiedermann D, Hoehn M. 2016. Morphological maturation of the mouse brain: an in vivo MRI and histology investigation. *Neuroimage* 125: 144–152. doi:10.1016/j.neuroimage.2015.10.009

Hammond E, Lang J, Maeda Y, Pleasure D, Angus-Hill M, Xu J, Horiuchi M, Deng W, Guo F. 2015. The Wnt effector transcription factor 7-Like 2 positively regulates oligodendrocyte differentiation in a manner independent of Wnt/β-catenin signaling. *J Neurosci* 35: 5007–5022. doi:10.1523/JNEUROSCI.4787-14.2015

Harlow DE, Saul KE, Culp CM, Vesely EM, Macklin WB. 2014. Expression of proteolipid protein gene in spinal cord stem cells and early oligodendrocyte progenitor cells is dispensable for normal cell migration and myelination. *J Neurosci* 34: 1333–1343. doi:10.1523/JNEUROSCI.2477-13.2014

He Y, Dupree J, Wang J, Sandoval J, Li J, Liu H, Shi Y, Nave K-A, Casaccia-Bonnefil P. 2007. The transcription factor Yin Yang 1 is essential for oligodendrocyte progenitor differentiation. *Neuron* 55: 217–230. doi:10.1016/j.neuron.2007.06.029

He D, Marie C, Zhao C, Kim B, Wang J, Deng Y, Clavairoly A, Frah M, Wang H, He X, et al. 2016. Chd7 cooperates with Sox10 and regulates the onset of CNS myelination and remyelination. *Nat Neurosci* 19: 678–689. doi:10.1038/nn.4258

He D, Wang J, Lu Y, Deng Y, Zhao C, Xu L, Chen Y, Hu Y-C, Zhou W, Lu QR. 2017. lncRNA functional networks in oligodendrocytes reveal stage-specific myelination control by an lncOL1/Suz12 complex in the CNS. *Neuron* 93: 362–378. doi:10.1016/j.neuron.2016.11.044

Hill RA, Patel KD, Goncalves CM, Grutzendler J, Nishiyama A. 2014. Modulation of oligodendrocyte generation during a critical temporal window after NG2 cell division. *Nat Neurosci* 17: 1518–1527. doi:10.1038/nn.3815

Hill RA, Li AM, Grutzendler J. 2018. Lifelong cortical myelin plasticity and age-related degeneration in the live mammalian brain. *Nat Neurosci* 21: 683–695. doi:10.1038/s41593-018-0120-6

* Hill RA, Nishiyama A, Hughes EG. 2023. Features, fates, and functions of oligodendrocyte precursor cells. *Cold Spring Harb Perspect Biol* doi:10.1101/cshperspect.a041425

Hornig J, Fröb F, Vogl MR, Hermans-Borgmeyer I, Tamm ER, Wegner M. 2013. The transcription factors Sox10 and Myrf define an essential regulatory network module in differentiating oligodendrocytes. *PLoS Genet* 9: e1003907. doi:10.1371/journal.pgen.1003907

Horton JD, Goldstein JL, Brown MS. 2002. SREBPs: transcriptional mediators of lipid homeostasis. *Cold Spring Harb Symp Quant Biol* 67: 491–498. doi:10.1101/sqb.2002.67.491

Howng SYB, Avila RL, Emery B, Traka M, Lin W, Watkins T, Cook S, Bronson R, Davisson M, Barres BA, et al. 2010. ZFP191 is required by oligodendrocytes for CNS myelination. *Genes Dev* 24: 301–311. doi:10.1101/gad.1864510

Hsieh J, Aimone JB, Kaspar BK, Kuwabara T, Nakashima K, Gage FH. 2004. IGF-I instructs multipotent adult neural progenitor cells to become oligodendrocytes. *J Cell Biol* 164: 111–122. doi:10.1083/jcb.200308101

Huang J, Dibble CC, Matsuzaki M, Manning BD. 2008. The TSC1-TSC2 complex is required for proper activation of mTOR complex 2. *Mol Cell Biol* 28: 4104–4115. doi:10.1128/MCB.00289-08

Huang W, Bhaduri A, Velmeshev D, Wang S, Wang L, Rottkamp CA, Alvarez-Buylla A, Rowitch DH, Kriegstein AR. 2020. Origins and proliferative states of human oligodendrocyte precursor cells. *Cell* 182: 594–608.e11. doi:10.1016/j.cell.2020.06.027

Hughes EG, Kang SH, Fukaya M, Bergles DE. 2013. Oligodendrocyte progenitors balance growth with self-repulsion to achieve homeostasis in the adult brain. *Nat Neurosci* 16: 668–676. doi:10.1038/nn.3390

Hughes EG, Orthmann-Murphy JL, Langseth AJ, Bergles DE. 2018. Myelin remodeling through experience-dependent oligodendrogenesis in the adult somatosensory cortex. *Nat Neurosci* 21: 696–706. doi:10.1038/s41593-018-0121-5

Ikenoue T, Inoki K, Yang Q, Zhou X, Guan KL. 2008. Essential function of TORC2 in PKC and Akt turn motif phosphorylation, maturation and signalling. *EMBO J* 27: 1919–1931. doi:10.1038/emboj.2008.119

Inoki K, Ouyang H, Li Y, Guan KL. 2005. Signaling by target of rapamycin proteins in cell growth control. *Microbiol Mol Biol Rev* 69: 79–100. doi:10.1128/MMBR.69.1.79-100.2005

Ishii A, Fyffe-Maricich SL, Furusho M, Miller RH, Bansal R. 2012. ERK1/ERK2 MAPK signaling is required to increase myelin thickness independent of oligodendrocyte differentiation and initiation of myelination. *J Neurosci* **32:** 8855–8864. doi:10.1523/JNEUROSCI.0137-12.2012

Ishii A, Furusho M, Bansal R. 2013. Sustained activation of ERK1/2 MAPK in oligodendrocytes and Schwann cells enhances myelin growth and stimulates oligodendrocyte progenitor expansion. *J Neurosci* **33:** 175–186. doi:10.1523/JNEUROSCI.4403-12.2013

Ishii A, Furusho M, Macklin W, Bansal R. 2019. Independent and cooperative roles of the Mek/ERK1/2-MAPK and PI3K/Akt/mTOR pathways during developmental myelination and in adulthood. *Glia* **67:** 1277–1295. doi:10.1002/glia.23602

Jacinto E, Loewith R, Schmidt A, Lin S, Rüegg MA, Hall A, Hall MN. 2004. Mammalian TOR complex 2 controls the actin cytoskeleton and is rapamycin insensitive. *Nat Cell Biol* **6:** 1122–1128. doi:10.1038/ncb1183

Jiang F, Frederick TJ, Wood TL. 2001. IGF-I synergizes with FGF-2 to stimulate oligodendrocyte progenitor entry into the cell cycle. *Dev Biol* **232:** 414–423. doi:10.1006/dbio.2001.0208

Kang M, Yao Y. 2022. Laminin regulates oligodendrocyte development and myelination. *Glia* **70:** 414–429. doi:10.1002/glia.24117

Kang SH, Fukaya M, Yang JK, Rothstein JD, Bergles DE. 2010. NG2+ CNS glial progenitors remain committed to the oligodendrocyte lineage in postnatal life and following neurodegeneration. *Neuron* **68:** 668–681. doi:10.1016/j.neuron.2010.09.009

Kessaris N, Fogarty M, Iannarelli P, Grist M, Wegner M, Richardson WD. 2006. Competing waves of oligodendrocytes in the forebrain and postnatal elimination of an embryonic lineage. *Nat Neurosci* **9:** 173–179. doi:10.1038/nn1620

Khandker L, Jeffries MA, Chang YJ, Mather ML, Evangelou AV, Bourne JN, Tafreshi AK, Ornelas IM, Bozdagi-Gunal O, Macklin WB, et al. 2022. Cholesterol biosynthesis defines oligodendrocyte precursor heterogeneity between brain and spinal cord. *Cell Rep* **38:** 110423. doi:10.1016/j.celrep.2022.110423

Kim DH, Sarbassov DD, Ali SM, King JE, Latek RR, Erdjument-Bromage H, Tempst P, Sabatini DM. 2002. mTOR interacts with raptor to form a nutrient-sensitive complex that signals to the cell growth machinery. *Cell* **110:** 163–175. doi:10.1016/S0092-8674(02)00808-5

Kim S, Kim SH, Kim H, Chung AY, Cha YI, Kim CH, Huh TL, Park HC. 2008. Frizzled 8a function is required for oligodendrocyte development in the zebrafish spinal cord. *Dev Dyn* **237:** 3324–3331. doi:10.1002/dvdy.21739

Kirby L, Jin J, Cardona JG, Smith MD, Martin KA, Wang J, Strasburger H, Herbst L, Alexis M, Karnell J, et al. 2019. Oligodendrocyte precursor cells present antigen and are cytotoxic targets in inflammatory demyelination. *Nat Commun* **10:** 3887. doi:10.1038/s41467-019-11638-3

Koenning M, Jackson S, Hay CM, Faux C, Kilpatrick TJ, Willingham M, Emery B. 2012. Myelin gene regulatory factor is required for maintenance of myelin and mature oligodendrocyte identity in the adult CNS. *J Neurosci* **32:** 12528–12542. doi:10.1523/JNEUROSCI.1069-12.2012

Kondo T, Raff M. 2000a. Basic helix-loop-helix proteins and the timing of oligodendrocyte differentiation. *Development* **127:** 2989–2998. doi:10.1242/dev.127.14.2989

Kondo T, Raff M. 2000b. The Id4 HLH protein and the timing of oligodendrocyte differentiation. *EMBO J* **19:** 1998–2007. doi:10.1093/emboj/19.9.1998

Kornfeld SF, Cummings SE, Fathi S, Bonin SR, Kothary R. 2021. MiRNA-145-5p prevents differentiation of oligodendrocyte progenitor cells by regulating expression of myelin gene regulatory factor. *J Cell Physiol* **236:** 997–1012. doi:10.1002/jcp.29910

Kuhlbrodt K, Herbarth B, Sock E, Hermans-Borgmeyer I, Wegner M. 1998. Sox10, a novel transcriptional modulator in glial cells. *J Neurosci* **18:** 237–250. doi:10.1523/JNEUROSCI.18-01-00237.1998

Kuspert M, Hammer A, Bosl MR, Wegner M. 2011. Olig2 regulates Sox10 expression in oligodendrocyte precursors through an evolutionary conserved distal enhancer. *Nucleic Acids Res* **39:** 1280–1293. doi:10.1093/nar/gkq951

Lebel C, Gee M, Camicioli R, Wieler M, Martin W, Beaulieu C. 2012. Diffusion tensor imaging of white matter tract evolution over the lifespan. *Neuroimage* **60:** 340–352. doi:10.1016/j.neuroimage.2011.11.094

Lebrun-Julien F, Bachmann L, Norrmen C, Trotzmuller M, Kofeler H, Ruegg MA, Hall MN, Suter U. 2014. Balanced mTORC1 activity in oligodendrocytes is required for accurate CNS myelination. *J Neurosci* **34:** 8432–8448. doi:10.1523/JNEUROSCI.1105-14.2014

Lecca D, Marangon D, Coppolino GT, Méndez AM, Finardi A, Costa GD, Martinelli V, Furlan R, Abbracchio MP. 2016. MiR-125a-3p timely inhibits oligodendroglial maturation and is pathologically up-regulated in human multiple sclerosis. *Sci Rep* **6:** 34503. doi:10.1038/srep34503

Lee JY, Petratos S. 2016. Thyroid hormone signaling in oligodendrocytes: from extracellular transport to intracellular signal. *Mol Neurobiol* **53:** 6568–6583. doi:10.1007/s12035-016-0013-1

Lee SH, Rezzonico MG, Friedman BA, Huntley MH, Meilandt WJ, Pandey S, Chen YJJ, Easton A, Modrusan Z, Hansen DV, et al. 2021. TREM2-independent oligodendrocyte, astrocyte, and T cell responses to tau and amyloid pathology in mouse models of Alzheimer disease. *Cell Rep* **37:** 110158. doi:10.1016/j.celrep.2021.110158

Letzen BS, Liu C, Thakor NV, Gearhart JD, All AH, Kerr CL. 2010. MicroRNA expression profiling of oligodendrocyte differentiation from human embryonic stem cells. *PLoS ONE* **5:** e10480. doi:10.1371/journal.pone.0010480

Li Z, Zhang Y, Li D, Feng Y. 2000. Destabilization and mislocalization of myelin basic protein mRNAs in *quaking* dysmyelination lacking the QKI RNA-binding proteins. *J Neurosci* **20:** 4944–4953. doi:10.1523/JNEUROSCI.20-13-04944.2000

Lin JP, Mironova YA, Shrager P, Giger RJ. 2017. LRP1 regulates peroxisome biogenesis and cholesterol homeostasis in oligodendrocytes and is required for proper CNS myelin development and repair. *eLife* **6:** e30498. doi:10.7554/eLife.30498

Liu GY, Sabatini DM. 2020. mTOR at the nexus of nutrition, growth, ageing and disease. *Nat Rev Mol Cell Biol* **21:** 183–203. doi:10.1038/s41580-019-0199-y

Liu A, Li J, Marin-Husstege M, Kageyama R, Fan Y, Gelinas C, Casaccia-Bonnefil P. 2006. A molecular insight of Hes5-dependent inhibition of myelin gene expression: old partners and new players. *EMBO J* **25:** 4833–4842. doi:10.1038/sj.emboj.7601352

Liu Z, Hu X, Cai J, Liu B, Peng X, Wegner M, Qiu M. 2007. Induction of oligodendrocyte differentiation by Olig2 and Sox10: evidence for reciprocal interactions and dosage-dependent mechanisms. *Dev Biol* 302: 683–693. doi:10 .1016/j.ydbio.2006.10.007

Liu J, Dietz K, DeLoyht JM, Pedre X, Kelkar D, Kaur J, Vialou V, Lobo MK, Dietz DM, Nestler EJ, et al. 2012. Impaired adult myelination in the prefrontal cortex of socially isolated mice. *Nat Neurosci* 15: 1621–1623. doi:10.1038/nn .3263

Liu J, Magri L, Zhang F, Marsh NO, Albrecht S, Huynh JL, Kaur J, Kuhlmann T, Zhang W, Slesinger PA, et al. 2015. Chromatin landscape defined by repressive histone methylation during oligodendrocyte differentiation. *J Neurosci* 35: 352–365. doi:10.1523/JNEUROSCI.2606-14.2015

Lopez-Anido C, Sun G, Koenning M, Srinivasan R, Hung HA, Emery B, Keles S, Svaren J. 2015. Differential Sox10 genomic occupancy in myelinating glia. *Glia* 63: 1897–1914. doi:10.1002/glia.22855

Lu QR, Sun T, Zhu Z, Ma N, Garcia M, Stiles CD, Rowitch DH. 2002. Common developmental requirement for Olig function indicates a motor neuron/oligodendrocyte connection. *Cell* 109: 75–86. doi:10.1016/S0092-8674(02) 00678-5

Mabie PC, Mehler MF, Kessler JA. 1999. Multiple roles of bone morphogenetic protein signaling in the regulation of cortical cell number and phenotype. *J Neurosci* 19: 7077–7088. doi:10.1523/JNEUROSCI.19-16-07077.1999

Mallon BS, Shick HE, Kidd GJ, Macklin WB. 2002. Proteolipid promoter activity distinguishes two populations of NG2-positive cells throughout neonatal cortical development. *J Neurosci* 22: 876–885. doi:10.1523/JNEUROSCI .22-03-00876.2002

Marin-Husstege M, Muggironi M, Liu A, Casaccia-Bonnefil P. 2002. Histone deacetylase activity is necessary for oligodendrocyte lineage progression. *J Neurosci* 22: 10333–10345. doi:10.1523/JNEUROSCI.22-23-10333.2002

Marin-Husstege M, He Y, Li J, Kondo T, Sablitzky F, Casaccia-Bonnefil P. 2006. Multiple roles of Id4 in developmental myelination: predicted outcomes and unexpected findings. *Glia* 54: 285–296. doi:10.1002/glia.20385

Marques S, Zeisel A, Codeluppi S, van Bruggen D, Mendanha Falcão A, Xiao L, Li H, Häring M, Hochgerner H, Romanov RA, et al. 2016. Oligodendrocyte heterogeneity in the mouse juvenile and adult central nervous system. *Science* 352: 1326–1329. doi:10.1126/science.aaf6463

Marziali LN, Hwang Y, Palmisano M, Cuenda A, Sim FJ, Gonzalez A, Volsko C, Dutta R, Trapp BD, Wrabetz L, Feltri M. 2023. p38γ MAPK delays myelination and remyelination and is abundant in multiple sclerosis lesions. *Brain* doi:10.1093/brain/awad421

Mathews ES, Appel B. 2016. Cholesterol biosynthesis supports myelin gene expression and axon ensheathment through modulation of P13K/Akt/mTor signaling. *J Neurosci* 36: 7628–7639. doi:10.1523/JNEUROSCI.0726-16 .2016

Mathews ES, Mawdsley DJ, Walker M, Hines JH, Pozzoli M, Appel B. 2014. Mutation of 3-hydroxy-3-methylglutaryl CoA synthase I reveals requirements for isoprenoid and cholesterol synthesis in oligodendrocyte migration arrest, axon wrapping, and myelin gene expression. *J Neurosci* 34: 3402–3412. doi:10.1523/JNEUROSCI.4587-13.2014

McKinnon R, Matsui T, Dubois-Dalcq M, Aaronson SA. 1990a. FGF modulates the PDGF-driven pathway of oligodendrocyte development. *Neuron* 5: 603–614. doi:10 .1016/0896-6273(90)90215-2

McKinnon RD, Matsui T, Dubois-Dalcq M, Aaronson SA. 1990b. FGF modulates the PDGF-driven pathway of oligodendrocyte development. *Neuron* 5: 603–614. doi:10 .1016/0896-6273(90)90215-2

McKinnon R, Smith C, Behar T, Smith T, Dubois-Dalcq M. 1993. Distinct effects of bFGF and PDGF on oligodendrocyte progenitor cells. *Glia* 7: 245–254. doi:10.1002/glia .440070308

McMorris FA, Dubois-Dalcq M. 1988. Insulin-like growth factor I promotes cell proliferation and oligodendroglial commitment in rat glial progenitor cells developing in vitro. *J Neurosci Res* 21: 199–209. doi:10.1002/jnr.490 210212

McMorris F, Smith T, DeSalvo S, Furlanetto R. 1986. Insulin-like growth factor I/somatomedin C: a potent inducer of oligodendrocyte development. *Proc Natl Acad Sci* 83: 822–826. doi:10.1073/pnas.83.3.822

Mei F, Wang H, Liu S, Niu J, Wang L, He Y, Etxeberria A, Chan JR, Xiao L. 2013. Stage-specific deletion of Olig2 conveys opposing functions on differentiation and maturation of oligodendrocytes. *J Neurosci* 33: 8454–8462. doi:10.1523/JNEUROSCI.2453-12.2013

Michalski JP, Kothary R. 2015. Oligodendrocytes in a nutshell. *Front Cell Neurosci* 9: 340. doi:10.3389/fncel.2015 .00340

Miyagi M, Mikawa S, Sato T, Hasegawa T, Kobayashi S, Matsuyama Y, Sato K. 2012. BMP2, BMP4, noggin, BMPRIA, BMPRIB, and BMPRII are differentially expressed in the adult rat spinal cord. *Neuroscience* 203: 12–26. doi:10.1016/j.neuroscience.2011.12.022

Moffett JR, Arun P, Ariyannur PS, Garbern JY, Jacobowitz DM, Namboodiri AMA. 2011. Extensive aspartoacylase expression in the rat central nervous system. *Glia* 59: 1414–1434. doi:10.1002/glia.21186

Mogha A, D'Rozario M, Monk KR. 2016. G protein-coupled receptors in myelinating glia. *Trends Pharmacol Sci* 37: 977–987. doi:10.1016/j.tips.2016.09.002

Monge M, Kadiiski D, Jacque CM, Zalc B. 1986. Oligodendroglial expression and deposition of four major myelin constituents in the myelin sheath during development. An in vivo study. *Dev Neurosci* 8: 222–235. doi:10.1159/ 000112255

Morell P, Jurevics H. 1996. Origin of cholesterol in myelin. *Neurochem Res* 21: 463–470. doi:10.1007/BF02527711

Musah AS, Brown TL, Jeffries MA, Shang Q, Hashimoto H, Evangelou AV, Kowalski A, Batish M, Macklin WB, Wood TL. 2020. Mechanistic target of rapamycin regulates the oligodendrocyte cytoskeleton during myelination. *J Neurosci* 40: 2993–3007. doi:10.1523/JNEUROSCI.1434-18 .2020

Narayanan SP, Flores AI, Wang F, Macklin WB. 2009. Akt signals through the mammalian target of rapamycin pathway to regulate CNS myelination. *J Neurosci* 29: 6860–6870. doi:10.1523/JNEUROSCI.0232-09.2009

Nawaz S, Sánchez P, Schmitt S, Snaidero N, Mitkovski M, Velte C, Brückner BR, Alexopoulos I, Czopka T, Jung SY, et al. 2015. Actin filament turnover drives leading edge growth during myelin sheath formation in the central nervous system. *Dev Cell* **34**: 139–151. doi:10.1016/j .devcel.2015.05.013

Neely SA, Williamson JM, Klingseisen A, Zoupi L, Early JJ, Williams A, Lyons DA. 2022. New oligodendrocytes exhibit more abundant and accurate myelin regeneration than those that survive demyelination. *Nat Neurosci* **25**: 415–420. doi:10.1038/s41593-021-01009-x

Ness JK, Wood TL. 2002. Insulin-like growth factor I, but not neurotrophin-3, sustains Akt activation and provides long-term protection of immature oligodendrocytes from glutamate-mediated apoptosis. *Mol Cell Neurosci* **20**: 476–488. doi:10.1006/mcne.2002.1149

Ness JK, Mitchell NE, Wood TL. 2002. IGF-I and NT-3 signaling pathways in developing oligodendrocytes: differential regulation and activation of receptors and the downstream effector Akt. *Dev Neurosci* **24**: 437–445. doi:10 .1159/000069050

Nishiyama A, Watanabe M, Yang Z, Bu J. 2002. Identity, distribution, and development of polydendrocytes: NG2-expressing glial cells. *J Neurocytol* **31**: 437–455. doi:10.1023/A:1025783412651

Nishiyama A, Komitova M, Suzuki R, Zhu X. 2009. Polydendrocytes (NG2 cells): multifunctional cells with lineage plasticity. *Nat Rev Neurosci* **10**: 9–22. doi:10.1038/nrn 2495

Niu J, Tsai HH, Hoi KK, Huang N, Yu G, Kim K, Baranzini SE, Xiao L, Chan JR, Fancy SPJ. 2019. Aberrant oligodendroglial–vascular interactions disrupt the blood–brain barrier, triggering CNS inflammation. *Nat Neurosci* **22**: 709–718. doi:10.1038/s41593-019-0369-4

Noble M, Murray K, Stroobant P, Waterfield MD, Riddle P. 1988. Platelet-derived growth factor promotes division and motility and inhibits premature differentiation of the oligodendrocyte/type-2 astrocyte progenitor cell. *Nature* **333**: 560–562. doi:10.1038/333560a0

Norrmén C, Figlia G, Lebrun-Julien F, Pereira JA, Trötzmüller M, Köfeler HC, Rantanen V, Wessig C, van Deijk AL, Smit AB, et al. 2014. mTORC1 controls PNS myelination along the mTORC1-RXRγ-SREBP-lipid biosynthesis axis in Schwann cells. *Cell Rep* **9**: 646–660. doi:10.1016/j.celrep .2014.09.001

Novitch BG, Chen AI, Jessell TM. 2001. Coordinate regulation of motor neuron subtype identity and pan-neuronal properties by the bHLH repressor Olig2. *Neuron* **31**: 773–789. doi:10.1016/S0896-6273(01)00407-X

Oh LYS, Denninger A, Colvin JS, Vyas A, Tole S, Ornitz DM, Bansal R. 2003. Fibroblast growth factor receptor 3 signaling regulates the onset of oligodendrocyte terminal differentiation. *J Neurosci* **23**: 883–894. doi:10.1523/JNEURO SCI.23-03-00883.2003

Ohvo-Rekilä H, Ramstedt B, Leppimäki P, Slotte JP. 2002. Cholesterol interactions with phospholipids in membranes. *Prog Lipid Res* **41**: 66–97. doi:10.1016/S0163-7827(01)00020-0

Ono K, Bansal R, Payne J, Rutishauser U, Miller RH. 1995. Early development and dispersal of oligodendrocyte precursors in the embryonic chick spinal cord. *Development* **121**: 1743–1754. doi:10.1242/dev.121.6.1743

Orentas DM, Miller RH. 1996. The origin of spinal cord oligodendrocytes is dependent on local influences from the notochord. *Dev Biol* **177**: 43–53. doi:10.1006/dbio .1996.0143

Orentas DM, Hayes JE, Dyer KL, Miller RH. 1999. Sonic hedgehog signaling is required during the appearance of spinal cord oligodendrocyte precursors. *Development* **126**: 2419–2429. doi:10.1242/dev.126.11.2419

Ornelas IM, Khandker L, Wahl SE, Hashimoto H, Macklin WB, Wood TL. 2020. The mechanistic target of rapamycin pathway downregulates bone morphogenetic protein signaling to promote oligodendrocyte differentiation. *Glia* **68**: 1274–1290. doi:10.1002/glia.23776

Pandey S, Shen K, Lee SH, Shen YAA, Wang Y, Otero-García M, Kotova N, Vito ST, Laufer BI, Newton DF, et al. 2022. Disease-associated oligodendrocyte responses across neurodegenerative diseases. *Cell Rep* **40**: 111189. doi:10.1016/j .celrep.2022.111189

Park HC, Mehta A, Richardson JS, Appel B. 2002. Olig2 is required for zebrafish primary motor neuron and oligodendrocyte development. *Dev Biol* **248**: 356–368. doi:10 .1006/dbio.2002.0738

Parras C, Marie C, Zhao C, Lu QR. 2020. Chromatin remodelers in oligodendroglia. *Glia* **68**: 1604–1618. doi:10.1002/ glia.23837

Persson M, Stamataki D, te Welscher P, Andersson E, Böse J, Rüther U, Ericson J, Briscoe J. 2002. Dorsal-ventral patterning of the spinal cord requires Gli3 transcriptional repressor activity. *Gene Dev* **16**: 2865–2878. doi:10.1101/ gad.243402

Poitelon Y, Kopec AM, Belin S. 2020. Myelin fat facts: an overview of lipids and fatty acid metabolism. *Cells* **9**: 812. doi:10.3390/cells9040812

Powell-Braxton L, Hollingshead P, Warburton C, Dowd M, Pitts-Meek S, Dalton D, Gillett N, Stewart TA. 1993. IGF-I is required for normal embryonic growth in mice. *Genes Dev* **7**: 2609–2617. doi:10.1101/gad.7.12b.2609

Pringle NP, Richardson WD. 1993. A singularity of PDGF α-receptor expression in the dorsoventral axis of the neural tube may define the origin of the oligodendrocyte lineage. *Development* **117**: 525–533. doi:10.1242/dev.117.2.525

Pringle NP, Yu W-P, Guthrie S, Roelink H, Lumsden A, Peterson AC, Richardson WD. 1996. Determination of neuroepithelial cell fate: induction of the oligodendrocyte lineage by ventral midline cells and sonic hedgehog. *Dev Biol* **177**: 30–42. doi:10.1006/dbio.1996.0142

Qi Y, Cai J, Wu Y, Wu R, Lee J, Fu H, Rao M, Sussel L, Rubenstein J, Qiu M. 2001. Control of oligodendrocyte differentiation by the *Nkx2.2* homeodomain transcription factor. *Development* **128**: 2723–2733. doi:10.1242/dev.128 .14.2723

Raff MC, Williams BP, Miller RH. 1984. The in vitro differentiation of a bipotential glial progenitor cell. *EMBO J* **3**: 1857–1864. doi:10.1002/j.1460-2075.1984.tb02059.x

Raff MC, Lillien LE, Richardson WD, Burne JF, Noble MD. 1988. Platelet-derived growth factor from astrocytes drives the clock that times oligodendrocyte development in culture. *Nature* **333**: 562–565. doi:10.1038/333562a0

Reimers D, López-Toledano MA, Mason I, Cuevas P, Redondo C, Herranz AS, Lobo MV, Bazán E. 2001. Developmental expression of fibroblast growth factor (FGF) receptors in neural stem cell progeny. Modulation of neuronal and glial lineages by basic FGF treatment. *Neurol Res* **23**: 612–621. doi:10.1179/016164101101199090

Reynolds R, Hardy R. 1997. Oligodendroglial progenitors labeled with the O4 antibody persist in the adult rat cerebral cortex in vivo. *J Neurosci Res* **47**: 455–470. doi:10 .1002/(SICI)1097-4547(19970301)47:5<455::AID-JNR1> 3.0.CO;2-G

Richardson WD, Pringle N, Mosley MJ, Westermark B, Dubois-Dalcg M. 1988. A role for platelet-derived growth factor in normal gliogenesis in the central nervous system. *Cell* **53**: 309–319. doi:10.1016/0092-8674(88)90392-3

Rim EY, Clevers H, Nusse R. 2022. The Wnt pathway: from signaling mechanisms to synthetic modulators. *Annu Rev Biochem* **91**: 571–598. doi:10.1146/annurev-biochem-040320-103615

Rivers LE, Young KM, Rizzi M, Jamen F, Psachoulia K, Wade A, Kessaris N, Richardson WD. 2008. PDGFRA/NG2 glia generate myelinating oligodendrocytes and piriform projection neurons in adult mice. *Nat Neurosci* **11**: 1392–1401. doi:10.1038/nn.2220

Robins SC, Villemain A, Liu X, Djogo T, Kryzskaya D, Storch KF, Kokoeva MV. 2013. Extensive regenerative plasticity among adult NG2-glia populations is exclusively based on self-renewal. *Glia* **61**: 1735–1747. doi:10.1002/glia.22554

Romanelli RJ, Lebeau AP, Fulmer CG, Lazzarino DA, Hochberg A, Wood TL. 2007. Insulin-like growth factor type-I receptor internalization and recycling mediate the sustained phosphorylation of AKT. *J Biol Chem* **282**: 22513–22524. doi:10.1074/jbc.M704309200

Romanelli RJ, Mahajan KR, Fulmer CG, Wood TL. 2009. Insulin-like growth factor-I-stimulated Akt phosphorylation and oligodendrocyte progenitor cell survival require cholesterol-enriched membranes. *J Neurosci Res* **87**: 3369–3377. doi:10.1002/jnr.22099

Saher G, Stumpf SK. 2015. Cholesterol in myelin biogenesis and hypomyelinating disorders. *Biochim Biophys Acta* **1851**: 1083–1094. doi:10.1016/j.bbalip.2015.02.010

Saher G, Brügger B, Lappe-Siefke C, Möbius W, Tozawa R, Wehr MC, Wieland F, Ishibashi S, Nave KA. 2005. High cholesterol level is essential for myelin membrane growth. *Nat Neurosci* **8**: 468–475. doi:10.1038/nn1426

Saher G, Rudolphi F, Corthals K, Ruhwedel T, Schmidt KF, Löwel S, Dibaj P, Barrette B, Möbius W, Nave KA. 2012. Therapy of Pelizaeus–Merzbacher disease in mice by feeding a cholesterol-enriched diet. *Nat Med* **18**: 1130–1135. doi:10.1038/nm.2833

Samanta J, Kessler JA. 2004. Interactions between ID and OLIG proteins mediate the inhibitory effects of BMP4 on oligodendroglial differentiation. *Development* **131**: 4131–4142. doi:10.1242/dev.01273

Sarbassov DD, Ali SM, Kim DH, Guertin DA, Latek RR, Erdjument-Bromage H, Tempst P, Sabatini DM. 2004. Rictor, a novel binding partner of mTOR, defines a rapamycin-insensitive and raptor-independent pathway that regulates the cytoskeleton. *Curr Biol* **14**: 1296–1302. doi:10 .1016/j.cub.2004.06.054

Sarbassov DD, Guertin DA, Ali SM, Sabatini DM. 2005. Phosphorylation and regulation of Akt/PKB by the rictor-mTOR complex. *Science* **307**: 1098–1101. doi:10 .1126/science.1106148

See JM, Grinspan JB. 2009. Sending mixed signals: bone morphogenetic protein in myelination and demyelination. *J Neuropathol Exp Neurol* **68**: 595–604. doi:10.1097/NEN .0b013e3181a66ad9

See J, Mamontov P, Ahn K, Wine-Lee L, Crenshaw EB 3rd, Grinspan JB. 2007. BMP signaling mutant mice exhibit glial cell maturation defects. *Mol Cell Neurosci* **35**: 171–182. doi:10.1016/j.mcn.2007.02.012

Selcen I, Prentice E, Casaccia P. 2023. The epigenetic landscape of oligodendrocyte lineage cells. *Ann NY Acad Sci* **1522**: 24–41. doi:10.1111/nyas.14959

Shen S, Li J, Casaccia-Bonnefil P. 2005. Histone modifications affect timing of oligodendrocyte progenitor differentiation in the developing rat brain. *J Cell Biol* **169**: 577–589. doi:10 .1083/jcb.200412101

Shen S, Sandoval J, Swiss VA, Li J, Dupree J, Franklin RJM, Casaccia-Bonnefil P. 2008. Age-dependent epigenetic control of differentiation inhibitors is critical for remyelination efficiency. *Nat Neurosci* **11**: 1024–1034. doi:10 .1038/nn.2172

Shimizu T, Kagawa T, Wada T, Muroyama Y, Takada S, Ikenaka K. 2005. Wnt signaling controls the timing of oligodendrocyte development in the spinal cord. *Dev Biol* **282**: 397–410. doi:10.1016/j.ydbio.2005.03.020

* Simons M, Gibson EM, Nave KA. 2024. Oligodendrocytes: myelination, plasticity, and axonal support. *Cold Spring Harb Perspect Biol* doi:10.1101/cshperspect.a041359

Snaidero N, Simons M. 2017. The logistics of myelin biogenesis in the central nervous system. *Glia* **65**: 1021–1031. doi:10.1002/glia.23116

Snaidero N, Möbius W, Czopka T, Hekking LH, Mathisen C, Verkleij D, Goebbels S, Edgar J, Merkler D, Lyons DA, et al. 2014. Myelin membrane wrapping of CNS axons by PI (3,4,5)P3-dependent polarized growth at the inner tongue. *Cell* **156**: 277–290. doi:10.1016/j.cell.2013.11.044

Sock E, Wegner M. 2019. Transcriptional control of myelination and remyelination. *Glia* **67**: 2153–2165. doi:10 .1002/glia.23636

Solly SK, Thomas JL, Monge M, Demerens C, Lubetzki C, Gardinier MV, Matthieu JM, Zalc B. 1996. Myelin/oligodendrocyte glycoprotein (MOG) expression is associated with myelin deposition. *Glia* **18**: 39–48. doi:10.1002/ (SICI)1098-1136(199609)18:1<39::AID-GLIA4>3.0.CO; 2-Z

Sorg BA, Smith MM, Campagnoni AT. 1987. Developmental expression of the myelin proteolipid protein and basic protein mRNAs in normal and dysmyelinating mutant mice. *J Neurochem* **49**: 1146–1154. doi:10.1111/j.1471-4159.1987.tb10005.x

Srinivas T, Mathias C, Oliveira-Mateos C, Guil S. 2023. Roles for lncRNAs in brain development and pathogenesis: emerging therapeutic opportunities. *Mol Ther* **31**: 1550–1561. doi:10.1016/j.ymthe.2023.02.008

Stallcup WB, Beasley L. 1987. Bipotential glial precursor cells of the optic nerve express the NG2 proteoglycan. *J Neurosci* **7**: 2737–2744. doi:10.1523/JNEUROSCI.07-09-02737 .1987

Stolt CC, Rehberg S, Ader M, Lommes P, Riethmacher D, Schachner M, Bartsch U, Wegner M. 2002. Terminal differentiation of myelin-forming oligodendrocytes depends on the transcription factor Sox10. *Gene Dev* **16**: 165–170. doi:10.1101/gad.215802

Stolt CC, Schlierf A, Lommes P, Hillgärtner S, Werner T, Kosian T, Sock E, Kessaris N, Richardson WD, Lefebvre V, et al. 2006. Soxd proteins influence multiple stages of oligodendrocyte development and modulate SoxE protein function. *Devl Cell* **11**: 697–709. doi:10.1016/j.devcel.2006 .08.011

Cite this article as *Cold Spring Harb Perspect Biol* doi: 10.1101/cshperspect.a041358

Sun T, Dong H, Wu L, Kane M, Rowitch DH, Stiles CD. 2003. Cross-repressive interaction of the Olig2 and Nkx2.2 transcription factors in developing neural tube associated with formation of a specific physical complex. *J Neurosci* **23**: 9547–9556. doi:10.1523/JNEUROSCI.23-29-09547.2003

Tang XM, Beesley JS, Grinspan JB, Seth P, Kamholz J, Cambi F. 1999. Cell cycle arrest induced by ectopic expression of p27 is not sufficient to promote oligodendrocyte differentiation. *J Cell Biochem* **76**: 270–279. doi:10.1002/(SICI) 1097-4644(20000201)76:2<270::AID-JCB10>3.0.CO;2-6

Trapp BD. 1990. Myelin-associated glycoprotein. Location and potential functions. *Ann NY Acad Sci* **605**: 29–43. doi:10.1111/j.1749-6632.1990.tb42378.x

Trapp BD, Nishiyama A, Cheng D, Macklin W. 1997. Differentiation and death of premyelinating oligodendrocytes in developing rodent brain. *J Cell Biol* **137**: 459–468. doi:10 .1083/jcb.137.2.459

Tripathi RB, Clarke LE, Burzomato V, Kessaris N, Anderson PN, Attwell D, Richardson WD. 2011. Dorsally and ventrally derived oligodendrocytes have similar electrical properties but myelinate preferred tracts. *J Neurosci* **31**: 6809–6819. doi:10.1523/JNEUROSCI.6474-10.2011

Tsai HH, Niu J, Munji R, Davalos D, Chang J, Zhang H, Tien AC, Kuo CJ, Chan JR, Daneman R, et al. 2016. Oligodendrocyte precursors migrate along vasculature in the developing nervous system. *Science* **351**: 379–384. doi:10.1126/ science.aad3839

Turnescu T, Arter J, Reiprich S, Tamm ER, Waisman A, Wegner M. 2018. Sox8 and Sox10 jointly maintain myelin gene expression in oligodendrocytes. *Glia* **66**: 279–294. doi:10.1002/glia.23242

Tyler WA, Gangoli N, Gokina P, Kim HA, Covey M, Levison SW, Wood TL. 2009. Activation of the mammalian target of rapamycin (mTOR) is essential for oligodendrocyte differentiation. *J Neurosci* **29**: 6367–6378. doi:10.1523/ JNEUROSCI.0234-09.2009

Tyler WA, Jain MR, Cifelli SE, Li Q, Ku L, Feng Y, Li H, Wood TL. 2011. Proteomic identification of novel targets regulated by the mammalian target of rapamycin pathway during oligodendrocyte differentiation. *Glia* **59**: 1754–1769. doi:10.1002/glia.21221

Van't Veer A, Du Y, Fischer TZ, Boetig DR, Wood MR, Dreyfus CF. 2009. Brain-derived neurotrophic factor effects on oligodendrocyte progenitors of the basal forebrain are mediated through trkB and the MAP kinase pathway. *J Neurosci Res* **87**: 69–78. doi:10.1002/jnr.21841

Wahl SE, McLane LE, Bercury KK, Macklin WB, Wood TL. 2014. Mammalian target of rapamycin promotes oligodendrocyte differentiation, initiation and extent of CNS myelination. *J Neurosci* **34**: 4453–4465. doi:10.1523/ JNEUROSCI.4311-13.2014

Wang S, Sdrulla A, Johnson JE, Yokota Y, Barres BA. 2001. A role for the helix-loop-helix protein Id2 in the control of oligodendrocyte development. *Neuron* **29**: 603–614. doi:10.1016/S0896-6273(01)00237-9

Wang Z, Colognato H, Ffrench-Constant C. 2007. Contrasting effects of mitogenic growth factors on myelination in neuron-oligodendrocyte co-cultures. *Glia* **55**: 537–545. doi:10.1002/glia.20480

Wang H, Moyano AL, Ma Z, Deng Y, Lin Y, Zhao C, Zhang L, Jiang M, He X, Ma Z, et al. 2017. miR-219 cooperates with miR-338 in myelination and promotes myelin repair in the

CNS. *Dev Cell* **40**: 566–582. doi:10.1016/j.devcel.2017.03 .001

Wang J, Yang L, Dong C, Wang J, Xu L, Qiu Y, Weng Q, Zhao C, Xin M, Lu QR. 2020a. EED-mediated histone methylation is critical for CNS myelination and remyelination by inhibiting WNT, BMP, and senescence pathways. *Sci Adv* **6**: eaaz6477. doi:10.1126/sciadv.aaz6477

Wang W, Cho H, Kim D, Park Y, Moon JH, Lim SJ, Yoon SM, McCane M, Aicher SA, Kim S, et al. 2020b. PRC2 acts as a critical timer that drives oligodendrocyte fate over astrocyte identity by repressing the notch pathway. *Cell Rep* **32**: 108147. doi:10.1016/j.celrep.2020.108147

Wang J, Yang L, Jiang M, Zhao C, Liu X, Berry K, Waisman A, Langseth AJ, Novitch BG, Bergles DE, et al. 2022. *Olig2* ablation in immature oligodendrocytes does not enhance CNS myelination and remyelination. *J Neurosci* **42**: 8542–8555. doi:10.1523/JNEUROSCI.0237-22.2022

Wei H, Dong X, You Y, Hai B, Duran RC-D, Wu X, Kharas N, Wu JQ. 2021. OLIG2 regulates lncRNAs and its own expression during oligodendrocyte lineage formation. *Bmc Biol* **19**: 132. doi:10.1186/s12915-021-01057-6

Weider M, Küspert M, Bischof M, Vogl Michael R, Hornig J, Loy K, Kosian T, Müller J, Hillgärtner S, Tamm Ernst R, et al. 2012. Chromatin-remodeling factor Brg1 is required for Schwann cell differentiation and myelination. *Dev Cell* **23**: 193–201. doi:10.1016/j.devcel.2012.05.017

Weider M, Wegener A, Schmitt C, Küspert M, Hillgärtner S, Bösl MR, Hermans-Borgmeyer I, Nait-Oumesmar B, Wegner M. 2015. Elevated in vivo levels of a single transcription factor directly convert satellite glia into oligodendrocyte-like cells. *PLoS Genet* **11**: e1005008. doi:10.1371/ journal.pgen.1005008

Weider M, Starost LJ, Groll K, Küspert M, Sock E, Wedel M, Fröb F, Schmitt C, Baroti T, Hartwig AC, et al. 2018. Nfat/ calcineurin signaling promotes oligodendrocyte differentiation and myelination by transcription factor network tuning. *Nat Commun* **9**: 899. doi:10.1038/s41467-018-03336-3

Weng Q, Chen Y, Wang H, Xu X, Yang B, He Q, Shou W, Chen Y, Higashi Y, Berghe V, et al. 2012. Dual-mode modulation of Smad signaling by Smad-interacting protein Sip1 is required for myelination in the central nervous system. *Neuron* **73**: 713–728. doi:10.1016/j.neuron.2011 .12.021

Wittstatt J, Weider M, Wegner M, Reiprich S. 2020. MicroRNA miR-204 regulates proliferation and differentiation of oligodendroglia in culture. *Glia* **68**: 2015–2027. doi:10 .1002/glia.23821

Wong AW, Xiao J, Kemper D, Kilpatrick TJ, Murray SS. 2013. Oligodendroglial expression of TrkB independently regulates myelination and progenitor cell proliferation. *J Neurosci* **33**: 4947–4957. doi:10.1523/JNEUROSCI.3990-12 .2013

Worby CA, Dixon JE. 2014. PTEN. *Annu Rev Biochem* **83**: 641–669. doi:10.1146/annurev-biochem-082411-113907

Wortzel I, Seger R. 2011. The ERK cascade: distinct functions within various subcellular organelles. *Genes Cancer* **2**: 195–209. doi:10.1177/1947601911407328

Xiao J, Wong AW, Willingham MM, van den Buuse M, Kilpatrick TJ, Murray SS. 2010. Brain-derived neurotrophic factor promotes central nervous system myelination via a direct effect upon oligodendrocytes. *Neurosignals* **18**: 186–202. doi:10.1159/000323170

Xiao J, Ferner AH, Wong AW, Denham M, Kilpatrick TJ, Murray SS. 2012. Extracellular signal-regulated kinase 1/2 signaling promotes oligodendrocyte myelination in vitro. *J Neurochem* **122**: 1167–1180. doi:10.1111/j.1471-4159.2012.07871.x

Xiao L, Ohayon D, McKenzie IA, Sinclair-Wilson A, Wright JL, Fudge AD, Emery B, Li H, Richardson WD. 2016. Rapid production of new oligodendrocytes is required in the earliest stages of motor-skill learning. *Nat Neurosci* **19**: 1210–1217. doi:10.1038/nn.4351

Xiao G, Du J, Wu H, Ge X, Xu X, Yang A, Zhu Y, Hu X, Zheng K, Zhu Q, et al. 2020. Differential inhibition of Sox10 functions by Notch-Hes pathway. *Cell Mol Neurobiol* **40**: 653–662. doi:10.1007/s10571-019-00764-7

Xing YL, Poh J, Chuang BHA, Moradi K, Mitew S, Richardson WD, Kilpatrick TJ, Osanai Y, Merson TD. 2023. High-efficiency pharmacogenetic ablation of oligodendrocyte progenitor cells in the adult mouse CNS. *Cell Rep Methods* **3**: 100414. doi:10.1016/j.crmeth.2023.100414

Yamada M, Iwase M, Sasaki B, Suzuki N. 2022. The molecular regulation of oligodendrocyte development and CNS myelination by ECM proteins. *Front Cell Dev Biol* **10**: 952135. doi:10.3389/fcell.2022.952135

Yang HJ, Vainshtein A, Maik-Rachline G, Peles E. 2016. G protein-coupled receptor 37 is a negative regulator of oligodendrocyte differentiation and myelination. *Nat Commun* **7**: 10884. doi:10.1038/ncomms10884

Ye P, Carson J, D'Ercole AJ. 1995a. In vivo actions of insulin-like growth factor-I (IGF-I) on brain myelination: studies of IGF-I and IGF binding protein (IGFBP-1) transgenic mice. *J Neurosci* **15**: 7344–7356. doi:10.1523/JNEUROSCI.15-11-07344.1995

Ye P, Carson J, D'Ercole AJ. 1995b. Insulin-like growth factor-I influences the initiation of myelination: studies of the anterior commissure of transgenic mice. *Neurosci Lett* **201**: 235–238. doi:10.1016/0304-3940(95)12194-3

Ye P, Li L, Richards G, DiAugustine RP, D'Ercole AJ. 2002. Myelination is altered in insulin-like growth factor-I null mutant mice. *J Neurosci* **22**: 6041–6051. doi:10.1523/JNEUROSCI.22-14-06041.2002

Ye F, Chen Y, Hoang T, Montgomery RL, Zhao XH, Bu H, Hu T, Taketo MM, van Es JH, Clevers H, et al. 2009a. HDAC1 and HDAC2 regulate oligodendrocyte differentiation by disrupting the β-catenin–TCF interaction. *Nat Neurosci* **12**: 829–838. doi:10.1038/nn.2333

Ye F, Chen Y, Hoang T, Montgomery RL, Zhao XH, Bu H, Hu T, Taketo MM, van Es JH, Clevers H, et al. 2009b. HDAC1 and HDAC2 regulate oligodendrocyte differentiation by disrupting the β-catenin–TCF interaction. *Nat Neurosci* **12**: 829–838. doi:10.1038/nn.2333

Yeung MSY, Zdunek S, Bergmann O, Bernard S, Salehpour M, Alkass K, Perl S, Tisdale J, Possnert G, Brundin L, et al. 2014. Dynamics of oligodendrocyte generation and myelination in the human brain. *Cell* **159**: 766–774. doi:10.1016/j.cell.2014.10.011

Yeung MSY, Djelloul M, Steiner E, Bernard S, Salehpour M, Possnert G, Brundin L, Frisén J. 2019. Dynamics of oligodendrocyte generation in multiple sclerosis. *Nature* **566**: 538–542. doi:10.1038/s41586-018-0842-3

Yim SH, Hammer JA, Quarles RH. 2001. Differences in signal transduction pathways by which platelet-derived and fibroblast growth factors activate extracellular signal-regulated kinase in differentiating oligodendrocytes. *J Neurochem* **76**: 1925–1934. doi:10.1046/j.1471-4159.2001.00199.x

Young KM, Psachoulia K, Tripathi RB, Dunn S-J, Cossell L, Attwell D, Tohyama K, Richardson WD. 2013. Oligodendrocyte dynamics in the healthy adult CNS: evidence for myelin remodeling. *Neuron* **77**: 873–885. doi:10.1016/j.neuron.2013.01.006

Yu Y, Chen Y, Kim B, Wang H, Zhao C, He X, Liu L, Liu W, Wu LMN, Mao M, et al. 2013. Olig2 targets chromatin remodelers to enhancers to initiate oligodendrocyte differentiation. *Cell* **152**: 248–261. doi:10.1016/j.cell.2012.12.006

Zeger M, Popken G, Zhang J, Xuan S, Lu QR, Schwab MH, Nave KA, Rowitch D, D'Ercole AJ, Ye P. 2007. Insulin-like growth factor type 1 receptor signaling in the cells of oligodendrocyte lineage is required for normal *in vivo* oligodendrocyte development and myelination. *Glia* **55**: 400–411. doi:10.1002/glia.20469

Zezula J, Casaccia-Bonnefil P, Ezhevsky SA, Osterhout DJ, Levine JM, Dowdy SF, Chao MV, Koff A. 2001. P21cip1 is required for the differentiation of oligodendrocytes independently of cell cycle withdrawal. *EMBO Rep* **2**: 27–34. doi:10.1093/embo-reports/kve008

Zhang Y, Chen K, Sloan SA, Bennett ML, Scholze AR, O'Keeffe S, Phatnani HP, Guarnieri P, Caneda C, Ruderisch N, et al. 2014. An RNA-sequencing transcriptome and splicing database of glia, neurons, and vascular cells of the cerebral cortex. *J Neurosci* **34**: 11929–11947. doi:10.1523/JNEUROSCI.1860-14.2014

Zhang C, Huang H, Chen Z, Zhang Z, Lu W, Qiu M. 2020. The transcription factor NKX2-2 regulates oligodendrocyte differentiation through domain-specific interactions with transcriptional corepressors. *J Biol Chem* **295**: 1879–1888. doi:10.1074/jbc.RA119.011163

Zhao X, He X, Han X, Yu Y, Ye F, Chen Y, Hoang T, Xu X, Mi Q-S, Xin M, et al. 2010. MicroRNA-mediated control of oligodendrocyte differentiation. *Neuron* **65**: 612–626. doi:10.1016/j.neuron.2010.02.018

Zhao C, Deng Y, Liu L, Yu K, Zhang L, Wang H, He X, Wang J, Lu C, Wu LN, et al. 2016. Dual regulatory switch through interactions of Tcf7l2/Tcf4 with stage-specific partners propels oligodendroglial maturation. *Nat Commun* **7**: 10883. doi:10.1038/ncomms10883

Zhao C, Dong C, Frah M, Deng Y, Marie C, Zhang F, Xu L, Ma Z, Dong X, Lin Y, et al. 2018. Dual requirement of CHD8 for chromatin landscape establishment and histone methyltransferase recruitment to promote CNS myelination and repair. *Dev Cell* **45**: 753–768.e8. doi:10.1016/j.devcel.2018.05.022

Zhen Y, Cullen CL, Ricci R, Summers BS, Rehman S, Ahmed ZM, Foster AY, Emery B, Gasperini R, Young KM. 2022. Protocadherin 15 suppresses oligodendrocyte progenitor cell proliferation and promotes motility through distinct signalling pathways. *Commun Biol* **5**: 511. doi:10.1038/s42003-022-03470-1

Zhou Q, Anderson DJ. 2002. The bHLH transcription factors OLIG2 and OLIG1 couple neuronal and glial subtype specification. *Cell* **109:** 61–73. doi:10.1016/S0092-8674(02)00677-3

Zhou Q, Choi G, Anderson DJ. 2001. The bHLH transcription factor Olig2 promotes oligodendrocyte differentiation in collaboration with Nkx2.2. *Neuron* **31:** 791–807. doi:10.1016/S0896-6273(01)00414-7

Zhou YX, Flint NC, Murtie JC, Le TQ, Armstrong RC. 2006. Retroviral lineage analysis of fibroblast growth factor receptor signaling in FGF2 inhibition of oligodendrocyte progenitor differentiation. *Glia* **54:** 578–590. doi:10.1002/glia.20410

Zhu X, Hill RA, Dietrich D, Komitova M, Suzuki R, Nishiyama A. 2011. Age-dependent fate and lineage restriction of single NG2 cells. *Development* **138:** 745–753. doi:10.1242/dev.047951

Zhu Q, Zhao X, Zheng K, Li H, Huang H, Zhang Z, Mastracci T, Wegner M, Chen Y, Sussel L, et al. 2014. Genetic evidence that *Nkx2.2* and *Pdgfra* are major determinants of the timing of oligodendrocyte differentiation in the developing CNS. *Development* **141:** 548–555. doi:10.1242/dev.095323

Zuchero JB, Fu MM, Sloan SA, Ibrahim A, Olson A, Zaremba A, Dugas JC, Wienbar S, Caprariello AV, Kantor C, et al. 2015. CNS myelin wrapping is driven by actin disassembly. *Dev Cell* **34:** 152–167. doi:10.1016/j.devcel.2015.06.011

Oligodendrocytes: Myelination, Plasticity, and Axonal Support

Mikael Simons,[1,2] Erin M. Gibson,[3] and Klaus-Armin Nave[4]

[1]Institute of Neuronal Cell Biology, Technical University Munich, Munich 80802, Germany

[2]German Center for Neurodegenerative Diseases, Munich Cluster of Systems Neurology (SyNergy), Institute for Stroke and Dementia Research, Munich 81377, Germany

[3]Department of Psychiatry and Behavioral Sciences, Stanford University School of Medicine, Stanford 94305, California, USA

[4]Department of Neurogenetics, Max Planck Institute for Multidisciplinary Sciences, Göttingen 37075, Germany

Correspondence: mikael.simons@dzne.de; egibson1@stanford.edu; nave@mpinat.mpg.de

The myelination of axons has evolved to enable fast and efficient transduction of electrical signals in the vertebrate nervous system. Acting as an electric insulator, the myelin sheath is a multilamellar membrane structure around axonal segments generated by the spiral wrapping and subsequent compaction of oligodendroglial plasma membranes. These oligodendrocytes are metabolically active and remain functionally connected to the subjacent axon via cytoplasmic-rich myelinic channels for movement of metabolites and macromolecules to and from the internodal periaxonal space under the myelin sheath. Increasing evidence indicates that oligodendrocyte numbers, specifically in the forebrain, and myelin as a dynamic cellular compartment can both respond to physiological demands, collectively referred to as adaptive myelination. This review summarizes our current understanding of how myelin is generated, how its function is dynamically regulated, and how oligodendrocytes support the long-term integrity of myelinated axons.

Information processing in complex organisms requires fast nerve transmission, which is accomplished in vertebrates by the ensheathment of axons with myelin.[5] In humans, the majority of the myelin sheaths are formed during the first decades of life when oligodendrocyte precursor cells (OPCs) proliferate and differentiate. Myelinating oligodendrocytes have enormous biosynthetic capacity, as they are able to generate numerous myelinated segments of multilamellar membrane on different axons. In mammals, myelination is not confined to early development but extends into adult life where it contributes to forebrain plasticity by being modifiable by experience and various environmental factors. At the same time, oligodendroglia provide metabolic support to their associated axons, an evolutionarily conserved function that preceded the invention of myelin in the vertebrate lineage. Here, we review the intrinsic and adaptive mechanisms of

---

[5]This is an update to a previous article published in *Cold Spring Harbor Perspectives in Biology* [Simons and Nave (2015). *Cold Spring Harb Perspect Biol* **8**: a020479. doi:10.1101/cshperspect.a020479].

Cite this article as *Cold Spring Harb Perspect Biol* doi: 10.1101/cshperspect.a041359

myelin formation and the function of myelin for axonal support.

MYELIN STRUCTURE

Myelin became subject of intense investigations with the refinement of electron microscopy (EM) and the application of biochemistry (later proteomics), when it was found that myelin can be physically purified from most other brain membranes (Norton and Poduslo 1973). Mouse genetics represented an early link between these two levels of analysis by providing molecular insight into the myelin architecture. By EM analysis, myelin is a multilayered stack of uniformly thick membranes with a characteristic periodic structure of alternating electron-dense and light

layers, termed the "major dense line" and the "intraperiod line," respectively (Fig. 1). These are adhesion zones and represent the closely apposed cytoplasmic and extracellular myelin membrane bilayers that appear in repeating patterns of ~12 nm ("periodicity"). Compacted myelin provides the high electrical resistance and low capacitance that is essential for saltatory impulse propagation. At their edges, myelin lamellae are tightly connected by autotypic tight junctions between the membranes of the outer leaflet consisting of Claudin-11 (Gow et al. 1999; Morita et al. 1999). These interlamellar strands run radially through myelin forming the so-called "radial components." Most of what we know about myelin ultrastructure is based on EM studies performed on glutaraldehyde fixed and dehy-

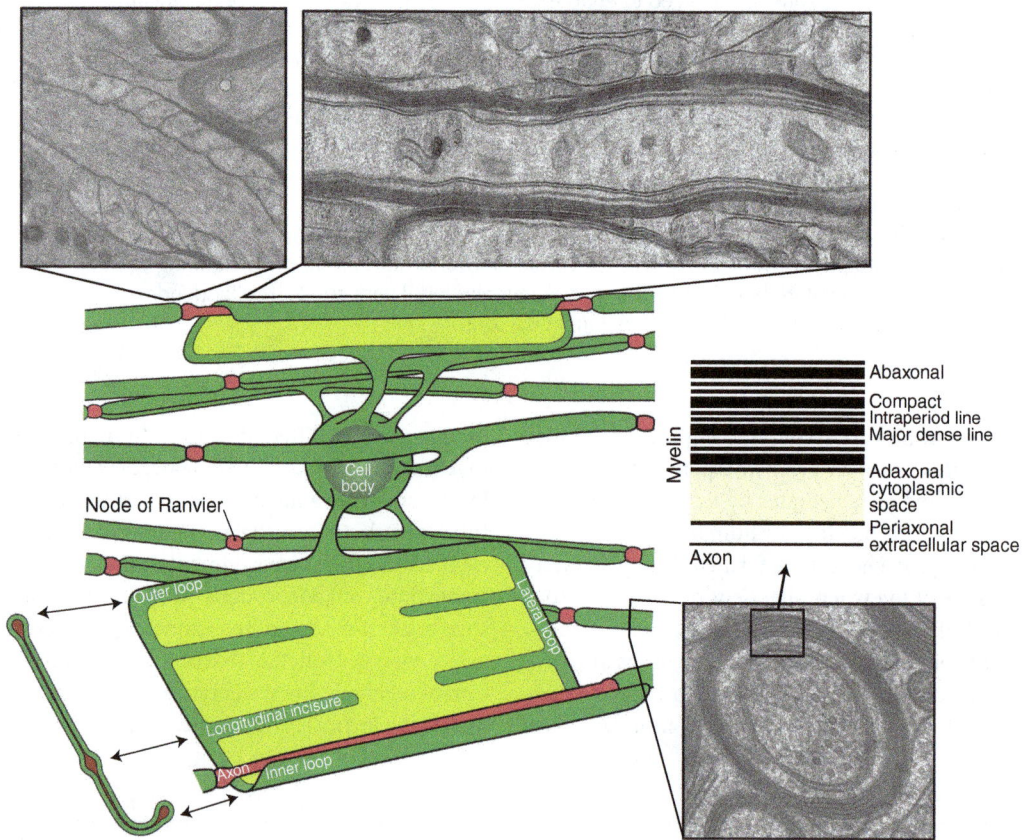

Figure 1. Schematic view of a myelinating oligodendrocytes in its wrapped and conceptually "unrolled" state. Electron micrographs show (*left*) paranodal loops, a longitudinal (*right*) and a cross-section view (*bottom*) of myelin. (This figure is adapted and modified, with permission, from Aggarwal et al. 2011a.)

Cite this article as *Cold Spring Harb Perspect Biol* doi: 10.1101/cshperspect.a041359

drated tissue, often associated with shrinkage and collapse of intracellular spaces. High-pressure freezing EM with an enhanced preservation of tissue architecture has enabled the visualization of the cytoplasmic spaces within myelin (Möbius et al. 2016). These noncompacted regions comprise the outer and inner periaxonal "tongues" (or "lips") of myelin membranes, the "paranodal loops," and the "Schmidt–Lanterman incisures" (the latter in the peripheral nervous system [PNS]). The periaxonal cytosolic space also contributes to ionic flow underlying saltatory impulse propagation (Cohen et al. 2020). Cytoplasmic channels can also be found in developing myelin sheaths of the central nervous system (CNS), but largely disappear when myelination is completed (Snaidero et al. 2014). Another method to visualize these cytoplasmic channels within myelin (termed "myelinic channels" going forward) is to inject diffusible dyes into oligodendrocytes or transgenically express fluorescent proteins. Using the fluorescent dye Lucifer Yellow an extensive network of interconnected cytoplasmic pockets (~1.9 pockets per 10 μm sheath length) was observed in slices of acutely isolated spinal cord (Velumian et al. 2011). During development, myelinic channels are likely to serve as tracks for the delivery of the newly synthesized membrane to the growing tip (see below). In the adult CNS, they may provide a functional connection between the oligodendroglial soma and the periaxonal space, allowing the distribution of glial metabolites to the axonal compartment. Considering that oligodendrocytes are also interconnected through gap junctions with each other and with astrocytes, glial cells may provide a functional "syncytium" within white matter tracts (Abrams and Scherer 2012; Nualart-Marti et al. 2013).

The myelin sheath is also firmly attached to the axon at each end of an internode where the myelin membrane never obtains a compacted structure but terminates in the paranodal loops. These form septate junctions with the axonal surface, held together by the adhesion proteins Caspr and Contactin on the axonal side and Neurofascin-155 on the glial (Rasband and Peles 2016, 2023). In the intermodal region (i.e., between the two paranodal domains) the innermost lamellae

of the myelin sheath remain noncompacted and form a cytoplasmic compartment, the adaxonal part of the myelinic channel system. It faces the periaxonal space along its length and is ideally positioned to communicate to the neuron (Edgar et al. 2021).

Apart from its unique ultrastructure, there are several other exceptional features of myelin. One is certainly its molecular composition. Myelin contains 70%–80% lipids (by dry weight) and there is only a small set of proteins that reside within compacted myelin of which myelin basic protein (MBP) and the proteolipid proteins (PLPs) are most abundant. Even if recent proteome studies have uncovered many more proteins within purified myelin fractions (Ishii et al. 2009; Jahn et al. 2009; Dhaunchak et al. 2010; Xu et al. 2011; Manrique-Hoyos et al. 2012; de Monasterio-Schrader et al. 2013), it still holds true that the proteome of compact myelin is dominated by a few major, predominantly low molecular weight, and mostly myelin-specific proteins.

Another defining feature of myelin is its extraordinary stability. This was compellingly illustrated when 5000-year-old myelin with almost intact ultrastructure was dissected from a Tyrolean Ice Man (Hess et al. 1998). The underlying molecular basis for the stability of myelin is likely its lipid composition with high levels of saturated, long chain fatty acids, together with an enrichment of glycosphingolipids (~20% molar percentage of total lipids) and cholesterol (~40% of molar percentage of total lipids) (O'Brien 1965; Coetzee et al. 1996). In addition, myelin comprises a high proportion of plasmalogens (etherlipids) with saturated long-chain fatty acids (Chrast et al. 2011). In fact, ~20% of the fatty acids in myelin have hydrocarbon chains longer than 18 carbon atoms (~1% in the gray matter) and only ~6% of the fatty acids are polyunsaturated (~20% in gray matter). Van der Waals dispersion forces that are generated by the interaction between methylene groups of these long and saturated hydrocarbon chains provide the major forces that hold these molecules together. In addition, more than 50% of galactosylceramide and sulfatide are hydroxylated at the 2-C atom by fatty acid 2-hydroxylase (FA2H), thereby providing additional hydroxyl groups for hydrogen bonding and, thus, increasing the packing density of lipids in myelin (Eckhardt et al. 2005).

Another peculiarity of the myelin membrane is its metabolic stability. Myelin membrane components have a half-life on the order of several weeks to months (Fard et al. 2017). The half-life of cholesterol in myelin was estimated to be >7–8 mo (Smith and Eng 1965). Using in vivo pulse-chase labeling of proteins synthesis with ^{15}N isotopes, followed by mass spectrometry at various ages, myelin proteins together with histones, nucleoporins, and laminins were found to be among the most long-lived proteins in a mouse (Toyama et al. 2013). Thus, myelin is a stable system in contrast to most membranes, which are highly dynamic, far from equilibrium steady state (Aggarwal et al. 2011a). However, myelin is compartmentalized into structurally and biochemically distinct domains. It is likely that the noncompacted regions are much more dynamic and metabolically active than the tightly compacted regions that lack direct access to the membrane trafficking machinery of oligodendrocytes.

MYELIN FORMATION

Most oligodendrocytes generate between 20 and 60 myelinating segments, with internodal lengths of ~20 µm to 200 µm and up to 100 membrane turns (Matthews and Duncan 1971; Hildebrand et al. 1993; Chong et al. 2012). When estimating the "surface" area that one oligodendrocyte would have (in terms of myelin membrane), it adds up to $5–50 \times 10^3$ µm^2, a number not reached by any other cell in our body (Pfeiffer et al. 1993; Baron and Hoekstra 2010; Stadelmann et al. 2019). Myelin wrapping of single axonal segments takes only a few hours, which appears demanding on the machinery of protein and lipid synthesis (Czopka et al. 2013; Almeida and Macklin 2023).

Following oligodendrocyte maturation, myelination is a complex sequence of events that is best understood when conceptualized as separate steps, including (1) recognition of target axons and axon-glia signaling, (2) membrane outgrowth and axonal wrapping, (3) trafficking of membrane components, (4) myelin compac-

tion, and (5) node formation. We will discuss some aspects here, but refer to comprehensive reviews for a more detailed description (Nave and Werner 2014; Stadelmann et al. 2019; Emery and Wood 2023; Rasband and Peles 2023).

We know little about the axonal factors that determine whether they will be myelinated or not (Simons and Lyons 2013). In general, axons are fully covered with myelin with only small gaps of segments at the node of Ranvier that are left unmyelinated. However, in particular within the gray matter, many axons are incompletely myelinated with myelinated segments interspersed with long, unmyelinated tracts and other axons lacking myelin entirely (Tomassy et al. 2014). One would assume that specific adhesion molecules must be expressed on the surface of axons to determine which segments require myelination. However, searches for single instructive signaling molecules on axons that can drive CNS myelination (similar to NRG1 in the PNS) have not been successful, and most identified factors only modulate ongoing myelination (Demerens et al. 1996; Benninger et al. 2006; Micu et al. 2006; Brinkmann et al. 2008; White et al. 2008; Câmara et al. 2009; Laursen et al. 2009; De Biase et al. 2011; Wake et al. 2011; Lundgaard et al. 2013).

Evidence has been provided that target size alone can determine the initiation of myelination (Rosenberg et al. 2008; Lee et al. 2012a; Bechler et al. 2015). When OPCs were cultured in the presence of inert polystyrene fibers to model axons they were able to ensheath such "axons" above a critical diameter. Thus, similar to the situation in the PNS (Fledrich et al. 2019), axonal caliber appears critical for myelination as central axons below 400 nm are rarely ensheathed. It is possible that due to physical constraints the ensheathing processes of oligodendrocytes are unable to navigate around axons below a specific threshold caliber. Although larger axons are preferentially myelinated in vivo, axons as small as 200–300 nm can be myelinated in some regions of the CNS. In addition, there is an intermediate range of axon diameter, between 0.2 µm and 0.8 µm, where CNS axons can be myelinated or not. Experimentally increasing axon size above 0.2 µm (by targeting PTEN and elevating neuronal PI3K/Akt/mTOR signaling) causes the de novo myelination of cerebellar granule cells (Goebbels et al. 2017) and is associated with numerous changes of neuronal gene expression, some of which may possibly sense diameter.

On the other hand, not all "thick" cellular processes such as dendrites and glial extensions are myelinated. Thus, axon diameter is not the only defining factor. The current model of CNS myelination is based on the combination of promyelinating axonal cues with the loss of negative repulsive factors. Most cellular surfaces are covered by a negatively charged glycocalyx that mediates electrostatic repulsion, thereby preventing unspecific contact. In addition, there are a growing number of inhibitory molecules including PSA-NCAM on immature neurons (Charles et al. 2000), JAM2 on the somatodendritic compartment (Redmond et al. 2016), Lsamp on neurons of the limbic system (Sharma et al. 2015), and class 3 semaphorins that help oligodendrocytes (Piaton et al. 2011; Syed et al. 2011) to define which structures to select for myelination. When OPCs and newly differentiated oligodendrocytes, which constantly sample their environment with highly motile processes, encounter such repulsive molecules they rapidly retract, mediated by elevation of intracellular Ca^{2+} levels (Kirby et al. 2006; Czopka et al. 2013; Hughes et al. 2013; Baraban et al. 2018; Krasnow et al. 2018). How specificity is provided to axon selection by oligodendrocytes is unknown.

Once a particular axon has been engaged by a myelinating oligodendrocyte process, dramatic changes in plasma membrane architecture are induced. How the processes are converted into flat sheets that spread and wind along the axons to generate a multilayered stack of membranes has been difficult to address. By light microscopy, a coil of an average periodicity of 5.7 to 7 μm along the internodal dimension has been observed (Butt and Berry 2000; Pedraza et al. 2009; Sobottka et al. 2011; Ioannidou et al. 2012). This has led to the suggestion that either myelin extends as a coil twisting around the axon in a corkscrew motion (Pedraza et al. 2009) or that myelin thickening is achieved by forming new layers on top of the inner ones in a "croissant-like" manner (Sobottka et al. 2011). Using

EM of high-pressure frozen samples and three-dimensional reconstructions based on serial block face EM, snapshots of the myelin ultrastructure have been obtained during development of the optic nerve (Snaidero et al. 2014). This analysis revealed that the innermost layers of myelin have the shortest lateral extensions, whereas the layers toward the top of the myelin sheath build up in size with each consecutive wrap. Together, the data suggest that myelin grows by two distinct but coordinated motions: the wrapping of the leading edge at the inner tongue around the axon (i.e., underneath the previously deposited membrane) and the lateral extension of myelin membrane layers toward the nodal regions (Fig. 2). Thus, the lateral cytoplasmic-rich edges of each myelin layer always stay in close contact with the axonal surface and move in a continuous helical manner toward the future node where they align and form the paranodal loops. While the lateral edges move toward the future node, they provide a shelter for the movement of the inner tongue. This might be necessary as the leading edge of the growing myelin sheath needs to wrap around the axon with low adhesiveness (Nawaz et al. 2015; Zuchero et al. 2015). Thus, one function of the paranodal adhesion molecules could be to "clamp" the leading edge to the axon preventing it from targeting surfaces unselectively (Djannatian et al. 2019; Elazar et al. 2019). This model implies that newly synthesized membrane material has to be transported all the way through the developing myelin sheath toward the innermost layer facing the axon. In fact, as discussed above, there is an elaborated system of cytoplasmic "myelinic channels" within compacted myelin (Snaidero et al. 2017; Edgar et al. 2021) that provide a helical path for the transport of vesicles to the growth zone at the leading edge. Recent data provide evidence that vesicular SNARE proteins VAMP2/3-mediated exocytosis drive membrane expansion within myelin sheaths to initiate wrapping and power sheath elongation (Lam et al. 2022; Pan et al. 2023). To wrap myelin around the axons, a mechanism is required to overcome the adhesive forces that are generated when the inner tongue crawls underneath the forming myelin sheath. A highly dynamic actin cytoskeleton appears to

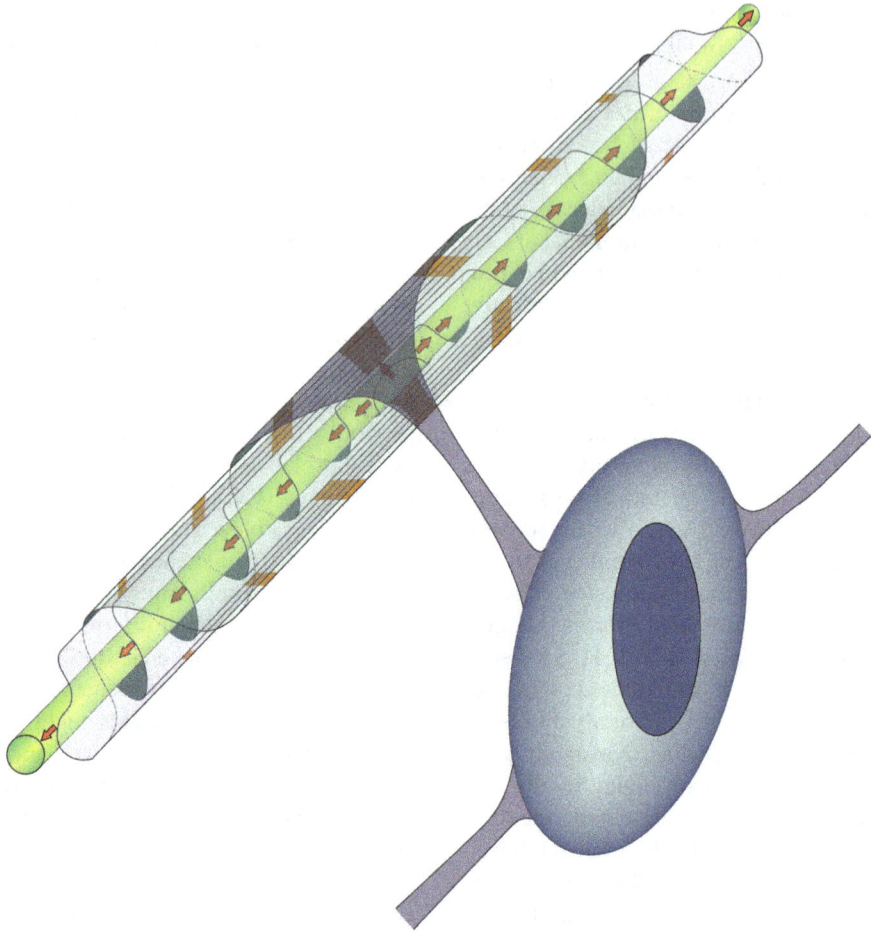

Figure 2. Model showing the direction of how myelin wraps around the axon. There are two motions: the wrapping of the leading edge at the inner tongue around the axon underneath the previously deposited membrane and the lateral extension of myelin membrane layers toward the nodal regions. (For details see Snaidero et al. 2014.)

play a crucial role in this process (Nawaz et al. 2015; Zuchero et al. 2015). Iterative cycles of polymerization/depolymerization at the leading edge may drive its protrusion by both polymerizing/nucleating factors (such as members of the Wiskott–Aldrich syndrome protein family), which regulate the Arp2/3 (actin-related proteins) complex, and depolymerizing factors (ADF/cofilin family members), which break down actin behind the front and free actin monomers for reassembly (Nawaz et al. 2015; Zuchero et al. 2015). In addition, hydrostatic pressure built up by MBP could represent another force driving the leading edge around the axon (Zu-

chero et al. 2015). Lengthening of the myelin sheath is controlled by the rate of Ca^{2+} transients, which may regulate the activity of cytoskeletal dynamics and/or growth promoting signaling such as the PI3K-AKT-mTOR pathway (Zonouzi et al. 2011; Baraban et al. 2018; Krasnow et al. 2018).

The different myelin components are synthesized in oligodendrocytes at several subcellular localizations and are transported by various mechanisms to the growing myelin sheath. Early pulse labeling experiments in Schwann cells suggested that newly synthesized lipids and integral myelin proteins first reach the outer myelin membranes and then diffuse laterally, at least after the initial wraps have been made (Gould and Dawson 1976; Gould 1977). In addition to membrane transport via the biosynthetic secretory pathway, the membrane-associated protein MBP needs to be delivered to myelin, which in development occurs by the transport of its mRNA within cytoplasmic granules (Colman et al. 1982; Ainger et al. 1993). Subsequently, the local translation of MBP is initiated at the tip of the glial process that is in contact to the axon in vivo (or in the outermost edge of the oligodendrocyte in vitro) (White et al. 2008; Laursen et al. 2009; Wake et al. 2011). Whereas MBP is thus synthesized in the distal process (i.e., closest to the axon), myelin compaction mediated by MBP progresses from the outside to the inside of the sheath (i.e., proximal to distal) (Snaidero et al. 2014). Loss of MBP in mature oligodendrocytes, triggered by tamoxifen-induced Cre recombination, causes progressive myelin thinning in brain and spinal cord, indicating that white matter integrity requires continuous myelin synthesis and turnover (Meschkat et al. 2022). It is obvious that if the sites of MBP translation and compaction are spatially separated, there must be a mechanism in place that prevents membrane compaction at the site of MBP synthesis. The myelin protein CNP1 appears to be such a factor that delays myelin compaction. In mice lacking CNP1, myelin compaction proceeds faster and extends to the innermost layers of the myelin sheath, whereas transgenic overexpression of CNP1 results in areas of myelin that lack compaction (Snaidero et al. 2017). Thus, the equilibrium of CNP1 and MBP levels seems to regulate the speed of compaction in early development and the preservation of myelinic channels (see also below). The periaxonal cytoplasmic space is also stabilized by protein interactions involving septin filaments and anillin (Patzig et al. 2016; Erwig et al. 2019).

Once MBP is bound to two adjacent cytoplasmic membrane surfaces, it appears to "polymerize" by lateral interactions with previously deposited MBP "monomers," thereby driving membrane zippering at the cytoplasmic surfaces of the myelin bilayer (Aggarwal et al. 2013). The self-association of high-order aggregates is triggered by membrane binding of MBP, which results in charge neutralization and major conformational changes of the protein (Musse et al. 2009; Aggarwal et al. 2013; Bakhti et al. 2013). One consequence of the formation of MBP polymers is the depletion of most peripheral and membrane-associated proteins from compacted myelin. This occurs by MBP-mediated protein extrusion and generation of a diffusion barrier, which limits the entry of proteins with larger cytoplasmic domains (Aggarwal et al. 2011b).

Compact myelin becomes a protein-poor membrane, lacking major glycoproteins at the extracellular leaflet, which is likely to uncover weak (generic) forces that promote the association of two bilayers at their extracellular surface (Bakhti et al. 2013). In addition, with the maturation of oligodendrocytes, the plasma membrane undergoes major transformations of its structure. Whereas OPCs are covered by a dense layer of large and negatively charged self-repulsive oligosaccharides, the compacted myelin of fully matured oligodendrocytes lacks most of these glycoprotein and complex glycolipids. Such a conversion may contribute to the transformation of oligodendrocyte lineage cells from the self-avoiding and "repulsive" OPCs that tile the entire CNS to "sticky" oligodendrocytes that ensheath axons with a multilayered stack of self-associating membranes.

OLIGODENDROGLIAL SUPPORT OF AXON FUNCTION AND INTEGRITY

Once myelination is completed, the developmentally important cytosolic channels decrease in size, but remain a significant compartment of noncompacted myelin that can be visualized by dye injection, the expression of fluorescent proteins, or high-pressure freezing EM (Stassart et al. 2018; Edgar et al. 2021). This cytosolic space connects the mature oligodendroglial soma to the innermost tip of the oligodendrocyte process. As a "myelinic channel," the tube-like cytosolic compartment runs in parallel to the thin extracellular periaxonal space where it faces the axonal internode along its length ("inner lip"). It is filled with a plethora of soluble and membrane-associated myelin proteins, a molecular complexity that exceeds the proteome of compacted myelin (Jahn et al. 2009; Gargareta et al. 2022). Moreover, the detection of microtubules and multivesicular bodies within the myelinic channel by EM (mostly in the paranodal loops and the Schmitt–Lanterman incisures of myelin in the PNS) suggests that motor-driven transport processes in this compartment are maintained throughout adult life. This likely reflects continued myelin growth and turnover, which is essential for metabolic coupling (see below) and myelin maintenance (Meschkat et al. 2022).

Well-myelinated mice that carry mutations in oligodendroglial genes can exhibit late-onset axonal degeneration and even premature death, the first demonstration that oligodendrocytes support axonal integrity independent of myelination (Griffiths et al. 1998; Lappe-Siefke et al. 2003; Edgar et al. 2009; Buscham et al.

2022). Widespread axonal swellings followed by Wallerian degeneration in the presence of merely subtle morphological myelin defects was surprising and contrasted with normal axonal viability in other, severely dysmyelinated mice, such as MBP-deficient *shiverer* (Rosenbluth 1980; Griffiths et al. 1998) despite the severe neurological symptoms. We suggest that "no myelin" is better than "bad myelin" and that the axon supportive functions of oligodendrocytes are most relevant for myelinated axons, presumably because these are effectively shielded by myelin itself from trophic support provided by the extracellular milieu (Nave 2010).

According to this model, oligodendrocytes use myelinic channels to deliver metabolites directly to the periaxonal space in which the axons reside (Fig. 3). CNP1, which delineates the cytosolic channels and prevents excessive myelin compaction (see above), keeps these channel

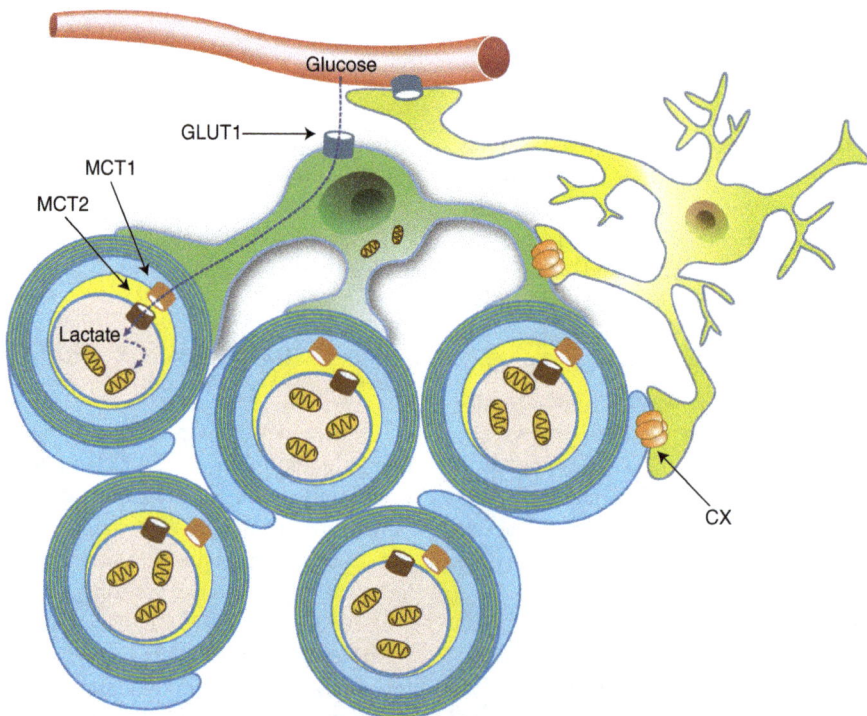

Figure 3. Schematic depiction of an oligodendrocyte that takes up blood-derived glucose and delivers glycolysis products (pyruvate/lactate) via monocarboxylate transporters (MCT1 and MCT2) to myelinated axons. Oligodendrocytes and myelin membranes are also coupled by gap junctions to astrocytes and thus indirectly to the blood–brain barrier. (This figure is adapted and modified, with permission, from Saab et al. 2013.)

structures open (Snaidero et al. 2017). Its absence does not prevent developmental myelination but causes a local collapse of myelinic channels. The resulting physical perturbation of metabolite diffusion and transport processes along tubular tracks most likely affects the oligodendroglial support of axons including the required turnover of monocarboxylate transporters (MCT1, see below), but also the continued growth of myelin and turnover of compacted myelin sheaths (Meschkat et al. 2022). The latter may be relevant to the CNP mutant phenotype that includes axonal swellings and degeneration, but also secondary neuroinflammation (Lappe-Siefke et al. 2003; Edgar et al. 2009), which is an unspecific sign of myelin dysfunction. Even 50% reduced CNP levels in heterozygous mice cause an abnormal white matter aging phenotype and specific behavioral abnormalities (Hagemeyer et al. 2012). These include executive function defects with catatonia-like episodes, which are indeed caused by the secondary neuroinflammation, as they can be rescued by microglia depletion (Janova et al. 2018).

Experiments in mice lacking COX10 from oligodendrocytes and thus mitochondrial respiration demonstrated that an important mechanism of oligodendroglial support is the direct delivery of glycolysis products (pyruvate and/or lactate) from the glial to the axonal compartment (Fünfschilling et al. 2012; Lee et al. 2012b). While oligodendrocytes in development use glucose and lactate (Rinholm et al. 2011) for lipid synthesis and rapid myelination, once myelination is complete, oligodendrocytes lacking respiration survive without signs of demyelination or degeneration. Instead, oligodendroglial pyruvate/lactate is efficiently taken up by axons for mitochondrial ATP production. Apparently, the energy metabolism of developing oligodendrocytes switches to aerobic glycolysis, which is reminiscent of the Warburg effect (Fünfschilling et al. 2012), but the underlying mechanisms are not yet well understood (Rao et al. 2017; Marangon et al. 2022). An unexpected low level of lactate dehydrogenase (LDH) in mature oligodendrocytes (E Spaete et al., unpubl. observation) suggests pyruvate is their major glycolysis product. However, oligodendrocytes coupled

to LDH-positive astrocytes can also shuttle lactate to LDH-positive axons. This would protect lactate from being metabolized in the myelinic channels.

Independent evidence for glycolytic support came from mice with reduced Slc16a1 gene expression, encoding MCT1 (the glial transporter for pyruvate and lactate). Here, 50% reduced gene dosage in heterozygous mouse mutants interfered with long-term maintenance of myelinated axons (Lee et al. 2012b). The late-onset axon swellings and degeneration profiles were indeed similar to that of PLP1-deficient myelin mutant mice (Griffiths et al. 1998). The latter exhibit a reduced axonal energy balance at rest, which does not further decrease following axonal stimulation. However, mutant myelin contains more GLUT1 and MCT1, suggesting some developmental compensation (Trevisiol et al. 2020). Indeed, when myelinated wild-type optic nerves were studied ex vivo, pharmacologically blocking MCT1 only affected axons that rapidly fire (Trevisiol et al. 2017). In brain slices, stimulation of callosal axons revealed that oligodendrocytes can also transfer glucose to axons (Meyer et al. 2018). Interestingly, the constitutive loss of MCT1 expression in the oligodendrocyte lineage was better tolerated than targeting MCT1 expression only later in life (Philips et al. 2021). This suggests developmental compensations of MCT1 deficiency during development, which could also explain a merely mild neurological phenotype of human patients with MCT1 deficiency (van Hasselt et al. 2014). The reported loss of MCT1 in the spinal cord of human ALS patients is presumably also of late onset and therefore relevant to the natural course of this disease (Lee et al. 2012b). The compensating transporters that shuttle and export metabolites in and out of oligodendrocytes are not known, but gap junction hemi-channels (Niu et al. 2016) are good candidates. Trafficking of all these proteins would be affected by physical injury to the myelinic channel system and is thus highly relevant to immune-mediated myelin injury in patients with multiple sclerosis (Schäffner et al. 2023). Also advanced age perturbs the structural integrity of myelin (Peters 2002), and when modeled in myelin mutant mice, such myelin dysfunction drives the neuronal deposition of amyloid in Alzheimer mouse models (Depp et al. 2023).

The oligodendroglial release of glycolysis products must be quantitatively matched to the axonal energy needs because the latter differs widely with respect to spiking frequencies. Any prolonged excess of lactate might cause local acidosis and damage. One regulatory mechanism is the stimulation of myelin-associated NMDA receptors (Saab et al. 2016) by spiking axons that release glutamate (Micu et al. 2016). According to this model, within minutes of NMDAR stimulation the glucose transporter GLUT1 traffics outward and is functionally incorporated into the oligodendroglial/myelin membrane. This elevates the glycolytic flux as a function of average spiking frequency (Saab et al. 2016). In turn, it is possible that some clinical effects (negative symptoms) of anti-NR1 autoantibodies in human patients with so-called "NMDA receptor encephalitis" involve the down-regulation of NMDAR with reduced metabolic support of white matter tracts (Matute et al. 2020; Arinrad et al. 2022).

In the cortex, GABAergic interneurons are frequently myelinated (Stedehouder et al. 2017; Dubey et al. 2022). How their axons would signal to oligodendrocytes is not well understood, although a major function may be in providing metabolic support (Kole et al. 2022). It is possible that here the axon–myelin unit relies merely on signaling by periaxonal potassium transients, which is much faster and has been recently identified as a second mechanism stimulating oligodendroglial pyruvate/lactate export (Losser et al. 2024).

In addition to glycolytic support, the lipid turnover and fatty acid metabolism of myelinated fibers can also contribute to axonal ATP generation. This was demonstrated by ex vivo recordings and metabolic imaging of ATP in optic nerve axons at low glucose conditions. Here, stable compound action potentials and ATP levels require fatty acid β-oxidation including peroxisomal β-oxidation in the oligodendrocyte/myelin compartment (Asadollahi et al. 2022). Such a mechanism could be relevant in prolonged hypoglycemia and in neurological

diseases, in which axonal "starvation" is a risk for irreversible damage.

It is important to note that in the mammalian PNS, axons also receive metabolic support from Schwann cells (Deck et al. 2022), and that this marks a fundamental function of axon-associated glial cells in vertebrates and nonmyelinating invertebrates (Volkenhoff et al. 2015). Moreover, axonal energy metabolism is an important, but not the only aspect of axonal support. In the CNS, oligodendrocytes also help detoxify iron by secreting the ferritin heavy chain (FHC) as a chelator, a function that appears preserved in nervous system evolution (Mukherjee et al. 2020). Oligodendrocyte-to-axon transfer via exosomes (Frühbeis et al. 2020) also includes the protein SIRT-2, an NAD^+-dependent protein deacetylase. SIRT-2 is abundantly expressed by oligodendrocytes (Werner et al. 2007) and was shown to activate, after being taken up by axonal mitochondria, the adenine nucleotide translocases and thus ATP production (Chamberlain et al. 2021). Since exosome release is at least in part stimulated by axonal glutamatergic signals (Frühbeis et al. 2013), the release of SIRT-2 participates in matching oligodendroglial support to axonal energy demands. We envision that there will be additional aspects of axon function and long-term health, specifically when deprived by myelin from the extracellular milieu, where direct oligodendroglial support is critical.

MYELIN PLASTICITY

It has become clear that myelin can undergo significant structural changes not only during development but also in adulthood (Young et al. 2013; Almeida et al. 2018). Myelin can be dynamically regulated by experience both during development and in adult life; thus, the extent of myelin sheath formation may also serve as a form of plasticity to adapt brain function in response to environmental stimuli (Fields 2008; Li et al. 2010; Liu et al. 2012; Makinodan et al. 2012; Mangin et al. 2012; Young et al. 2013). Adaptive myelination refers to dynamic events in oligodendroglia driven by extrinsic factors such as experience or neuronal activity, which subsequently induces changes in circuit structure and function (Gibson et al. 2014). These adaptive myelin changes can alter the temporal or spatial components of circuit behavior through multiple mechanisms, including (1) changing myelin microstructure to fine-tune signal convergence and/or transduction velocity (Etxeberria et al. 2016; Kato et al. 2020), (2) altering myelin dynamics to support the metabolic demands of neurons based on activity level (Saab et al. 2016), and (3) "locking" in circuit structure to limit the potential of changes in neural connections (Stedehouder et al. 2018). Myelin plasticity can be mediated at multiple levels of oligodendroglial lineage cells from changes in OPC proliferation or differentiation to interactions between oligodendrocytes and axons. Oligodendrocytes are generally stable throughout life in mice (Tripathi et al. 2017) and humans (Yeung et al. 2014), and adult-born oligodendrocytes contribute to myelin dynamics (Hughes et al. 2018), including changes in myelin internodes (Hill et al. 2018; Bacmeister et al. 2020; Yang et al. 2020). Understanding how these adaptive changes in neuron–oligodendroglia interactions impact brain function remains a pressing question for the field.

Experience-Dependent Myelin Plasticity

Early work related to myelin plasticity derived from studies showing that life experiences altered myelin and myelin-forming cells. Rodents exposed to enriched environments involving both sensory and motor enrichment or learning during adulthood exhibit variable increases in OPC proliferation, differentiation, and myelin integration (Zhao et al. 2011; Hughes et al. 2018; Zheng et al. 2019; Kato et al. 2020). Compound learning through exposure to the complex wheel (McKenzie et al. 2014; Xiao et al. 2016), water maze (Steadman et al. 2020), or contextual fear memory (Pan et al. 2020) tasks increases OPC proliferation, oligodendrogenesis, and myelination in mice. Blocking production of new oligodendrocytes through knockdown of *Myrf*, a transcription factor essential for oligodendrocyte differentiation and myelination, from OPCs in adulthood impairs learning in these complex tasks. Importantly, pharmacologically enhanc-

ing oligodendrocyte differentiation improves memory in contexts of disrupted oligodendrogenesis and myelinogenesis (Pan et al. 2020; Wang et al. 2020). Sensory enrichment also increases new oligodendrocyte integration but does not change existing internode length in a myelin intact microenvironment (Hughes et al. 2018). However, during partial remyelination of a demyelinated region, enhanced motor learning increases oligodendrogenesis and, to a more limited extent, generation of new myelin sheaths from existing oligodendrocytes, which recovers behavioral function (Bacmeister et al. 2020).

Experiences involving deprivation states during adulthood, such as social isolation, change both transcriptional and ultrastructural components of oligodendrocytes in the prefrontal cortex (PFC), subsequently leading to deficits in myelination and behavior that can be normalized following social reintroduction (Liu et al. 2012). More transient social isolation results in chromatin and myelin changes but does not induce consequent behavioral alterations, which suggests putative acute cellular mechanisms associated with experience-dependent plasticity that may not involve more complex behavioral changes. Remarkably, when mice undergo a similar social isolation paradigm during early life development, they similarly exhibit deficits in PFC function and myelination, but these deficiencies do not recover with social reintroduction. This implicates a critical period for social deprivation effects on myelin dynamics (Makinodan et al. 2012), which may depend on vascular endothelin signaling (Swire et al. 2019). Some studies involving sensory deprivation report increases in OPC proliferation in the barrel cortex (Mangin et al. 2012), but this hyperproliferation appears transient and in response to oligodendrocyte loss (Hill et al. 2014). In contrast, while long-term monocular deprivation increases oligodendrogenesis, this sensory deficit is associated with both decreases in internode length and conduction velocity (Etxeberria et al. 2016) and increased myelin remodeling of parvalbumin[+] (PV[+]) interneurons (Yang et al. 2020). These data indicate that experience-dependent changes in myelin dynamics may not be directionally consistent with increased activity always leading to increased myelin and may depend on not only the age, brain region, and cell type studied, but also the specific myelin structural change assessed. Collectively, these data suggest that myelin plasticity may be driven by extrinsic cues impacting both new and existing oligodendroglial lineage cells to modulate myelin parameters and subsequent circuit function. Outside of regulating signal transduction, the role of experience-dependent adaptive myelination in regulating metabolic support of axons, restriction of neural circuit outgrowth, or intrinsic mechanisms such as axon caliber (Sinclair et al. 2017) require further investigation.

Neural Activity-Dependent Myelin Plasticity

It has become increasingly evident that experience-dependent myelin plasticity is driven specifically by changes in neuronal activity within task-associated circuitry. Activity driven by electrical (Barres and Raff 1993; Demerens et al. 1996; Li et al. 2010), pharmacological (Barres and Raff 1993; Wake et al. 2011; Gautier et al. 2015; Mensch et al. 2015; Marisca et al. 2020), optogenetic (Gibson et al. 2014; Geraghty et al. 2019; Ortiz et al. 2019; Yamazaki et al. 2021), or chemogenetic (Mitew et al. 2018; Stedehouder et al. 2018) strategies have begun to elucidate precisely how circuit-specific neural activity regulates myelin structure and function. The first indication that electrical activity could influence oligodendroglial dynamics came from studies showing that blocking neuronal activity by tetrodotoxin or optic nerve transection decreased OPC proliferation (Barres and Raff 1993) and electrically stimulating axons increased OPC proliferation (Li et al. 2010). Since then, numerous groups have shown that selectively activating or silencing neuronal activity alters oligodendroglia–neuron interactions. Local synaptic neurotransmitter release along an axon not only affects the number of OPCs and oligodendrocytes associated with that axon (Mensch et al. 2015) and local synthesis of myelin proteins (Wake et al. 2011), but also drives preferential selection of active axons for myelination over the ensheathment of electrically silenced neighboring axons (Hines et al. 2015). Approximately a quar-

ter of oligodendrocytes that ensheath axons during zebrafish development will be stabilized without requiring neuronal activity. However, when neurons are silenced, most oligodendrocytes that initiate wrapping will retract processes, suggesting that excessive myelin ensheathment may occur during development that is further culled by neuronal activity throughout life (Hines et al. 2015). Taken together, these data implicate that electrical activity may provide a signal to oligodendrocytes that enables them to stabilize their transient contacts and continue with myelin wrapping.

This activity-associated selectivity extends to OPCs as well. Pharmacological stimulation of zebrafish spinal cord using voltage-gated potassium channel blockers identified two subgroups of OPCs; one group exhibits diminished differentiation potential and increased Ca^{2+} signaling while the second group has increased differentiation potential but reduced Ca^{2+} signaling (Marisca et al. 2020). These data could suggest a putative division of OPCs that either respond to neuronal activity through maintenance of synaptic integration or that respond by initiating differentiation. Altogether, this evidence further supports the conclusion that neuron–oligodendroglia interactions are heterogeneous in nature and this diversity in response to neuronal activity may differentially impact subpopulations of oligodendroglia to alter integration of oligodendroglial lineage cells along the axon. Heterogeneity in these processes may, thus, be directly related to brain area function. For example, it could be speculated that OPCs derived to populate the frontal cortex may be more plastic and thus more attuned to myelin dynamics due to the functions of that brain region (e.g., executive functioning, learning, memory, higher-order cognition) versus OPCs that occupy the brainstem or spinal cord, which are established earlier in development and generally less mutable (Gibson et al. 2014).

Oligodendrocytes and myelination have been implicated in solidifying neural circuit structure through synapse formation and stabilization in the cortex. In support of this, activation of PV^+ interneurons in adult mice increased the total length of myelination within interneurons. This enhancement in myelin content was due to an increase in axonal arborization and internode number not internode length, suggesting that the increase in neuronal activity-mediated myelination within these short-range neurons is facilitated by an increase in axonal morphological complexity (Stedehouder et al. 2018). These myelin patterns of interneurons may be critical in the regulation of inhibitory circuitry during adulthood (Benamer et al. 2020). While the role of neuronal activity in short-range neurons may be more associated with maintenance of circuit structure by setting axonal arborization and excitation–inhibition balance, most research to date has focused on the role of myelin plasticity in tuning signal convergence or transduction velocity of long-range axons to enhance circuit function. Optogenetically stimulated neuronal activity increases OPC proliferation, differentiation, and myelination within the premotor cortex but not the motor corticofugal tract (Gibson et al. 2014), which suggests that neuronal activity–induced plasticity may preferentially impact brain regions that remain incompletely myelinated compared to more fully myelinated tracts. Similarly, chemogenetic stimulation of mouse somatosensory cortex increases OPC proliferation and myelination probability; blocking this neuronal activity reduces myelination (Mitew et al. 2018). While it is not known whether this increase in myelination within these brain regions and subsequent enhancement of brain function are driven by existing oligodendrocytes or integration of new oligodendrocytes, it is known that the increase in OPC proliferation and subsequent oligodendrogenesis is partially due to neuronal brain–derived neurotrophic factor (BDNF) signaling to OPCs (Geraghty et al. 2019). It has been suggested that a neuregulin/BDNF switch may occur between intrinsic, developmental myelination and activity-dependent myelination (Lundgaard et al. 2013), but further work investigating exactly how new myelin is integrated into electrically active circuits throughout life is needed, especially regarding the role of oligodendroglia in supporting the changing metabolic needs of activated axons.

Could myelin plasticity be harnessed to therapeutically target demyelinating disorders or hi-jacked to aid in disease progression? OPCs that are recruited to demyelinating lesions generate new synapses with electrically active neurons and blocking neuronal activity or vesicular release within these lesions reduces remyelination by minimizing the differentiation potential of OPCs, implicating a putative role for OPC synapses during the remyelination process (Gautier et al. 2015). Along similar lines, driving neuronal activity within a demyelinated lesion using optogenetic stimulation-enhanced OPC proliferation, differentiation, and remyelination within the injury site. This enhanced remyelination potential accelerated functional recovery within the circuit by increasing conduction velocity of compound action potentials within the lesion (Ortiz et al. 2019). While neuronal activity may induce adaptive myelin responses that aid in the recovery of myelin loss, maladaptive myelination may also contribute to progression of some disorders. Myelin dynamics that occur during adulthood are associated with adaptive, mutable myelin. Deficits in myelin dynamics that occur in aging or brain disorders, such as the neurological disorder associated with cancer therapy (Geraghty et al. 2019), may thus be associated with maladaptive or aplastic myelination. Recently, maladaptive myelination has been characterized in a mouse model of epilepsy to reinforce highly active circuits associated with the production of seizures (Knowles et al. 2022). These myelin dynamics, thus, may "lock in" neural circuitry to reinforce the function of that circuit in a helpful (adaptive) or harmful (maladaptive) manner. This could include establishing internodes to limit the initiation of new connections made between neurons (Kalil and Dent 2014; Stedehouder et al. 2018; Benamer et al. 2020) or changing the transduction velocity within the activated circuit. A better understanding of how myelin dynamics are mediated not only between neuron–glia interactions but also by intracellular downstream signaling events are necessary to completely establish the role of myelin changes in brain plasticity throughout life, aging, and disease.

CONCLUDING REMARKS

The formation of myelin is one of the most complex transformations of a plasma membrane that comprises timely coordinated cell–cell interactions with reciprocal intercellular signaling events, the wrapping and extension of the glial plasma membrane around the axon and the extrusion of cytoplasm from newly generated multilamellar membrane stacks. Each of these events represents a fascinating area of cellular neuroscience with relevance not only for our understanding of normal physiology, but also for various white matter diseases. Whereas the myelin sheath has been regarded for a long time as an inert insulating structure, it has now become clear that myelin is "alive" and metabolically active with cytoplasmic-rich pathways, myelinic channels, for movement of macromolecules into the periaxonal space. The myelin sheath and its subjacent axon need to be regarded as one functional unit, which are not only morphological but also metabolically coupled. It would be plausible to consider axonal support by myelinating glial cells, an adaptation to axonal length in vertebrates. However, the discovery that the same mechanisms operate in *Drosophila* identifies this symbiotic axon–glia relationship as very ancestral in nervous system evolution (Volkenhoff et al. 2015; Mukherjee et al. 2020). In the future, we need to understand what other molecules are exchanged between neurons and glia. It has become clear that myelin can undergo significant structural changes in adulthood (Young et al. 2013). Myelin seems to be dynamically regulated by experience both during development and in adult life; thus, the extent of myelin sheath formation may also serve as a form of plasticity to adapt brain function to environmental stimuli (Fields 2008; Liu et al. 2012; Makinodan et al. 2012; Mangin et al. 2012; Young et al. 2013; McKenzie et al. 2014; Xiao et al. 2016; Pan et al. 2020; Steadman et al. 2020). Clearly, identifying the molecular mechanisms that regulate myelin formation and metabolic pathways within myelin are likely to be key areas for further study.

ACKNOWLEDGMENTS

This work was supported by the Miriam and Sheldon Adelson Foundation (K.A.N. and M.S.), the National Institutes of Health R01NS126610 (E.M.G.), and the Department of Defense W81XWH-21-1-0846 (E.M.G.).

REFERENCES

*Reference is also in this subject collection.

Abrams CK, Scherer SS. 2012. Gap junctions in inherited human disorders of the central nervous system. *Biochim Biophys Acta* **1818:** 2030–2047. doi:10.1016/j.bbamem.2011.08.015

Aggarwal S, Yurlova L, Simons M. 2011a. Central nervous system myelin: structure, synthesis and assembly. *Trends Cell Biol* **21:** 585–593. doi:10.1016/j.tcb.2011.06.004

Aggarwal S, Yurlova L, Snaidero N, Reetz C, Frey S, Zimmermann J, Pähler G, Janshoff A, Friedrichs J, Müller DJ, et al. 2011b. A size barrier limits protein diffusion at the cell surface to generate lipid-rich myelin-membrane sheets. *Dev Cell* **21:** 445–456. doi:10.1016/j.devcel.2011.08.001

Aggarwal S, Snaidero N, Pähler G, Frey S, Sánchez P, Zweckstetter M, Janshoff A, Schneider A, Weil MT, Schaap IA, et al. 2013. Myelin membrane assembly is driven by a phase transition of myelin basic proteins into a cohesive protein meshwork. *PLoS Biol* **11:** e1001577. doi:10.1371/journal.pbio.1001577

Ainger K, Avossa D, Morgan F, Hill SJ, Barry C, Barbarese E, Carson JH. 1993. Transport and localization of exogenous myelin basic protein mRNA microinjected into oligodendrocytes. *J Cell Biol* **123:** 431–441. doi:10.1083/jcb.123.2.431

Almeida AR, Macklin WB. 2023. Early myelination involves the dynamic and repetitive ensheathment of axons which resolves through a low and consistent stabilization rate. *eLife* **12:** e82111. doi:10.7554/eLife.82111

Almeida RG, Pan S, Cole KLH, Williamson JM, Early JJ, Czopka T, Klingseisen A, Chan JR, Lyons DA. 2018. Myelination of neuronal cell bodies when myelin supply exceeds axonal demand. *Curr Biol* **28:** 1296–1305.e5. doi:10.1016/j.cub.2018.02.068

Arinrad S, Wilke JBH, Seelbach A, Doeren J, Hindermann M, Butt UJ, Steixner-Kumar AA, Spieth L, Ronnenberg A, Pan H, et al. 2022. NMDAR1 autoantibodies amplify behavioral phenotypes of genetic white matter inflammation: a mild encephalitis model with neuropsychiatric relevance. *Mol Psychiatry* **27:** 4974–4983. doi:10.1038/s41380-021-01392-8

Asadollahi E, Trevisiol A, Saab AS, Looser ZJ, Dibaj P, Kusch K, Ruhwedel T, Möbius W, Jahn O, Baes M, et al. 2022. Myelin lipids as nervous system energy reserves. bioRxiv doi:10.1101/2022.02.24.481621

Bacmeister CM, Barr HJ, McClain CR, Thornton MA, Nettles D, Welle CG, Hughes EG. 2020. Motor learning promotes remyelination via new and surviving oligodendrocytes. *Nat Neurosci* **23:** 819–831. doi:10.1038/s41593-020-0637-3

Bakhti M, Snaidero N, Schneider D, Aggarwal S, Möbius W, Janshoff A, Eckhardt M, Nave KA, Simons M. 2013. Loss of electrostatic cell-surface repulsion mediates myelin membrane adhesion and compaction in the central nervous system. *Proc Natl Acad Sci* **110:** 3143–3148. doi:10.1073/pnas.1220104110

Baraban M, Koudelka S, Lyons DA. 2018. Ca^{2+} activity sig-natures of myelin sheath formation and growth in vivo. *Nat Neurosci* **21**: 19–23. doi:10.1038/s41593-017-0040-x

Baron W, Hoekstra D. 2010. On the biogenesis of myelin membranes: sorting, trafficking and cell polarity. *FEBS Lett* **584**: 1760–1770. doi:10.1016/j.febslet.2009.10.085

Barres BA, Raff MC. 1993. Proliferation of oligodendrocyte precursor cells depends on electrical activity in axons. *Nature* **361**: 258–260. doi:10.1038/361258a0

Bechler ME, Byrne L, Ffrench-Constant C. 2015. CNS myelin sheath lengths are an intrinsic property of oligodendro-cytes. *Curr Biol* **25**: 2411–2416. doi:10.1016/j.cub.2015.07.056

Benamer N, Vidal M, Balia M, Angulo MC. 2020. Myelina-tion of parvalbumin interneurons shapes the function of cortical sensory inhibitory circuits. *Nat Commun* **11**: 5151. doi:10.1038/s41467-020-18984-7

Benninger Y, Colognato H, Thurnherr T, Franklin RJ, Leone DP, Atanasoski S, Nave KA, Ffrench-Constant C, Suter U, Relvas JB. 2006. Beta1-integrin signaling mediates pre-myelinating oligodendrocyte survival but is not required for CNS myelination and remyelination. *J Neurosci* **26**: 7665–7673. doi:10.1523/JNEUROSCI.0444-06.2006

Brinkmann BG, Agarwal A, Sereda MW, Garratt AN, Müller T, Wende H, Stassart RM, Nawaz S, Humml C, Velanac V, et al. 2008. Neuregulin-1/ErbB signaling serves distinct functions in myelination of the peripheral and central ner-vous system. *Neuron* **59**: 581–595. doi:10.1016/j.neuron.2008.06.028

Buscham TJ, Eichel-Vogel MA, Steyer AM, Jahn O, Strenzke N, Dardawal R, Memhave TR, Siems SB, Müller C, Mesch-kat M, et al. 2022. Progressive axonopathy when oligoden-drocytes lack the myelin protein CMTM5. *eLife* **11**: e75523. doi:10.7554/eLife.75523

Butt AM, Berry M. 2000. Oligodendrocytes and the control of myelination in vivo: new insights from the rat anterior medullary velum. *J Neurosci Res* **59**: 477–488. doi:10.1002/(SICI)1097-4547(20000215)59:4<477::AID-JNR2>3.0.CO;2-J

Câmara J, Wang Z, Nunes-Fonseca C, Friedman HC, Grove M, Sherman DL, Komiyama NH, Grant SG, Brophy PJ, Peterson A, et al. 2009. Integrin-mediated axoglial inter-actions initiate myelination in the central nervous system. *J Cell Biol* **185**: 699–712. doi:10.1083/jcb.200807010

Chamberlain KA, Huang N, Xie Y, LiCausi F, Li S, Li Y, Sheng ZH. 2021. Oligodendrocytes enhance axonal energy me-tabolism by deacetylation of mitochondrial proteins through transcellular delivery of SIRT2. *Neuron* **109**: 3456–3472.e8. doi:10.1016/j.neuron.2021.08.011

Charles P, Hernandez MP, Stankoff B, Aigrot MS, Colin C, Rougon G, Zalc B, Lubetzki C. 2000. Negative regulation of central nervous system myelination by polysialylated-neural cell adhesion molecule. *Proc Natl Acad Sci* **97**: 7585–7590. doi:10.1073/pnas.100076197

Chong SY, Rosenberg SS, Fancy SP, Zhao C, Shen YA, Hahn AT, McGee AW, Xu X, Zheng B, Zhang LI, et al. 2012. Neurite outgrowth inhibitor Nogo-A establishes spatial segregation and extent of oligodendrocyte myelination. *Proc Natl Acad Sci* **109**: 1299–1304. doi:10.1073/pnas.1113540109

Chrast R, Saher G, Nave KA, Verheijen MH. 2011. Lipid metabolism in myelinating glial cells: lessons from human inherited disorders and mouse models. *J Lipid Res* **52**: 419–434. doi:10.1194/jlr.R009761

Coetzee T, Fujita N, Dupree J, Shi R, Blight A, Suzuki K, Suzuki K, Popko B. 1996. Myelination in the absence of galactocerebroside and sulfatide: normal structure with abnormal function and regional instability. *Cell* **86**: 209–219. doi:10.1016/S0092-8674(00)80093-8

Cohen CCH, Popovic MA, Klooster J, Weil MT, Möbius W, Nave KA, Kole MHP. 2020. Saltatory conduction along myelinated axons involves a periaxonal nanocircuit. *Cell* **180**: 311–322.e15. doi:10.1016/j.cell.2019.11.039

Colman DR, Kreibich G, Frey AB, Sabatini DD. 1982. Syn-thesis and incorporation of myelin polypeptides into CNS myelin. *J Cell Biol* **95**: 598–608. doi:10.1083/jcb.95.2.598

Czopka T, Ffrench-Constant C, Lyons DA. 2013. Individual oligodendrocytes have only a few hours in which to gen-erate new myelin sheaths in vivo. *Dev Cell* **25**: 599–609. doi:10.1016/j.devcel.2013.05.013

De Biase LM, Kang SH, Baxi EG, Fukaya M, Pucak ML, Mishina M, Calabresi PA, Bergles DE. 2011. NMDA recep-tor signaling in oligodendrocyte progenitors is not required for oligodendrogenesis and myelination. *J Neurosci* **31**: 12650–12662. doi:10.1523/JNEUROSCI.2455-11.2011

Deck M, Van Hameren G, Campbell G, Bernard-Marissal N, Devaux J, Berthelot J, Lattard A, Médard JJ, Gautier B, Guelfi S, et al. 2022. Physiology of PNS axons relies on glycolytic metabolism in myelinating Schwann cells. *PLoS ONE* **17**: e0272097. doi:10.1371/journal.pone.0272097

Demerens C, Stankoff B, Logak M, Anglade P, Allinquant B, Couraud F, Zalc B, Lubetzki C. 1996. Induction of myeli-nation in the central nervous system by electrical activity. *Proc Natl Acad Sci* **93**: 9887–9892. doi:10.1073/pnas.93.18.9887

de Monasterio-Schrader P, Patzig J, Möbius W, Barrette B, Wagner TL, Kusch K, Edgar JM, Brophy PJ, Werner HB. 2013. Uncoupling of neuroinflammation from axonal de-generation in mice lacking the myelin protein tetraspanin-2. *Glia* **61**: 1832–1847. doi:10.1002/glia.22561

Depp C, Sun T, Sasmita AO, Spieth L, Berghoff SA, Naza-renko T, Overhoff K, Steixner-Kumar AA, Subramanian S, Arinrad S, et al. 2023. Myelin dysfunction drives amyloid deposition in models of Alzheimer's disease. *Nature* **618**: 349–357. doi:10.1038/s41586-023-06120-6

Dhaunchak AS, Huang JK, De Faria Junior O, Roth AD, Pedraza L, Antel JP, Bar-Or A, Colman DR. 2010. A pro-teome map of axoglial specializations isolated and purified from human central nervous system. *Glia* **58**: 1949–1960. doi:10.1002/glia.21064

Djannatian M, Timmler S, Arends M, Luckner M, Weil MT, Alexopoulos I, Snaidero N, Schmid B, Misgeld T, Möbius W, et al. 2019. Two adhesive systems cooperatively regulate axon ensheathment and myelin growth in the CNS. *Nat Commun* **10**: 4794. doi:10.1038/s41467-019-12789-z

Dubey M, Pascual-Garcia M, Helmes K, Wever DD, Hamada MS, Kushner SA, Kole MHP. 2022. Myelination synchro-nizes cortical oscillations by consolidating parvalbumin-mediated phasic inhibition. *eLife* **11**: e73827. doi:10.7554/eLife.73827

Eckhardt M, Yaghootfam A, Fewou SN, Zöller I, Gieselmann V. 2005. A mammalian fatty acid hydroxylase responsible for the formation of α-hydroxylated galactosylceramide in myelin. *Biochem J* **388:** 245–254. doi:10.1042/BJ20041451

Edgar JM, McLaughlin M, Werner HB, McCulloch MC, Barrie JA, Brown A, Faichney AB, Snaidero N, Nave KA, Griffiths IR. 2009. Early ultrastructural defects of axons and axon-glia junctions in mice lacking expression of *Cnp1*. *Glia* **57:** 1815–1824. doi:10.1002/glia.20893

Edgar JM, McGowan E, Chapple KJ, Möbius W, Lemgruber L, Insall RH, Nave KA, Boullerne A. 2021. Rio-Hortega's drawings revisited with fluorescent protein defines a cytoplasm-filled channel system of CNS myelin. *J Anat* **239:** 1241–1255. doi:10.1111/joa.13577

Elazar N, Vainshtein A, Golan N, Vijayaragavan B, Schaeren-Wiemers N, Eshed-Eisenbach Y, Peles E. 2019. Axoglial adhesion by Cadm4 regulates CNS myelination. *Neuron* **101:** 224–231.e5. doi:10.1016/j.neuron.2018.11.032

* Emery B, Wood TL. 2023. Regulators of oligodendrocyte differentiation. *Cold Spring Harb Perspect Biol* doi:10.1101/cshperspect.a041358

Erwig MS, Patzig J, Steyer AM, Dibaj P, Heilmann M, Heilmann I, Jung RB, Kusch K, Möbius W, Jahn O, et al. 2019. Anillin facilitates septin assembly to prevent pathological outfoldings of central nervous system myelin. *eLife* **8:** e43888. doi:10.7554/eLife.43888

Etxeberria A, Hokanson KC, Dao DQ, Mayoral SR, Mei F, Redmond SA, Ullian EM, Chan JR. 2016. Dynamic modulation of myelination in response to visual stimuli alters optic nerve conduction velocity. *J Neurosci* **36:** 6937–6948. doi:10.1523/JNEUROSCI.0908-16.2016

Fard MK, van der Meer F, Sánchez P, Cantuti-Castelvetri L, Mandad S, Jäkel S, Fornasiero EF, Schmitt S, Ehrlich M, Starost L, et al. 2017. BCAS1 expression defines a population of early myelinating oligodendrocytes in multiple sclerosis lesions. *Sci Transl Med* **9:** eaam7816. doi:10.1126/scitranslmed.aam7816

Fields RD. 2008. White matter in learning, cognition and psychiatric disorders. *Trends Neurosci* **31:** 361–370. doi:10.1016/j.tins.2008.04.001

Fledrich R, Kungl T, Nave KA, Stassart RM. 2019. Axo-glial interdependence in peripheral nerve development. *Development* **146:** dev151704. doi:10.1242/dev.151704

Frühbeis C, Fröhlich D, Kuo WP, Amphornrat J, Thilemann S, Saab AS, Kirchhoff F, Möbius W, Goebbels S, Nave KA, et al. 2013. Neurotransmitter-triggered transfer of exosomes mediates oligodendrocyte-neuron communication. *PLoS Biol* **11:** e1001604. doi:10.1371/journal.pbio.1001604

Frühbeis C, Kuo-Elsner WP, Müller C, Barth K, Peris L, Tenzer S, Möbius W, Werner HB, Nave KA, Fröhlich D, et al. 2020. Oligodendrocytes support axonal transport and maintenance via exosome secretion. *PLoS Biol* **18:** e3000621. doi:10.1371/journal.pbio.3000621

Fünfschilling U, Supplie LM, Mahad D, Boretius S, Saab AS, Edgar J, Brinkmann BG, Kassmann CM, Tzvetanova ID, Möbius W, et al. 2012. Glycolytic oligodendrocytes maintain myelin and long-term axonal integrity. *Nature* **485:** 517–521. doi:10.1038/nature11007

Gargareta VI, Reuschenbach J, Siems SB, Sun T, Piepkorn L, Mangana C, Späte E, Goebbels S, Huitinga I, Möbius W, et al. 2022. Conservation and divergence of myelin proteome and oligodendrocyte transcriptome profiles between humans and mice. *eLife* **11:** e77019. doi:10.7554/eLife.77019

Gautier HO, Evans KA, Volbracht K, James R, Sitnikov S, Lundgaard I, James F, Lao-Peregrin C, Reynolds R, Franklin RJ, et al. 2015. Neuronal activity regulates remyelination via glutamate signalling to oligodendrocyte progenitors. *Nat Commun* **6:** 8518. doi:10.1038/ncomms9518

Geraghty AC, Gibson EM, Ghanem RA, Greene JJ, Ocampo A, Goldstein AK, Ni L, Yang T, Marton RM, Paşca SP, et al. 2019. Loss of adaptive myelination contributes to methotrexate chemotherapy-related cognitive impairment. *Neuron* **103:** 250–265.e8. doi:10.1016/j.neuron.2019.04.032

Gibson EM, Purger D, Mount CW, Goldstein AK, Lin GL, Wood LS, Inema I, Miller SE, Bieri G, Zuchero JB, et al. 2014. Neuronal activity promotes oligodendrogenesis and adaptive myelination in the mammalian brain. *Science* **344:** 1252304. doi:10.1126/science.1252304

Goebbels S, Wieser GL, Pieper A, Spitzer S, Weege B, Yan K, Edgar JM, Yagensky O, Wichert SP, Agarwal A, et al. 2017. A neuronal PI(3,4,5)P(3)-dependent program of oligodendrocyte precursor recruitment and myelination. *Nat Neurosci* **20:** 10–15. doi:10.1038/nn.4425

Gould RM. 1977. Incorporation of glycoproteins into peripheral nerve myelin. *J Cell Biol* **75:** 326–338. doi:10.1083/jcb.75.2.326

Gould RM, Dawson RM. 1976. Incorporation of newly formed lecithin into peripheral nerve myelin. *J Cell Biol* **68:** 480–496. doi:10.1083/jcb.68.3.480

Gow A, Southwood CM, Li JS, Pariali M, Riordan GP, Brodie SE, Danias J, Bronstein JM, Kachar B, Lazzarini RA. 1999. CNS myelin and Sertoli cell tight junction strands are absent in Osp/claudin-11 null mice. *Cell* **99:** 649–659. doi:10.1016/S0092-8674(00)81553-6

Griffiths I, Klugmann M, Anderson T, Yool D, Thomson C, Schwab MH, Schneider A, Zimmermann F, McCulloch M, Nadon N, et al. 1998. Axonal swellings and degeneration in mice lacking the major proteolipid of myelin. *Science* **280:** 1610–1613. doi:10.1126/science.280.5369.1610

Hagemeyer N, Goebbels S, Papiol S, Kästner A, Hofer S, Begemann M, Gerwig UC, Boretius S, Wieser GL, Ronnenberg A, et al. 2012. A myelin gene causative of a catatonia-depression syndrome upon aging. *EMBO Mol Med* **4:** 528–539. doi:10.1002/emmm.201200230

Hess MW, Kirschning E, Pfaller K, Debbage PL, Hohenberg H, Klima G. 1998. 5000-year-old myelin: uniquely intact in molecular configuration and fine structure. *Curr Biol* **8:** R512–R513. doi:10.1016/S0960-9822(07)00334-X

Hildebrand C, Remahl S, Persson H, Bjartmar C. 1993. Myelinated nerve fibres in the CNS. *Prog Neurobiol* **40:** 319–384. doi:10.1016/0301-0082(93)90015-K

Hill RA, Patel KD, Goncalves CM, Grutzendler J, Nishiyama A. 2014. Modulation of oligodendrocyte generation during a critical temporal window after NG2 cell division. *Nat Neurosci* **17:** 1518–1527. doi:10.1038/nn.3815

Hill RA, Li AM, Grutzendler J. 2018. Lifelong cortical myelin plasticity and age-related degeneration in the live mammalian brain. *Nat Neurosci* **21:** 683–695. doi:10.1038/s41593-018-0120-6

Hines JH, Ravanelli AM, Schwindt R, Scott EK, Appel B. 2015. Neuronal activity biases axon selection for myelination in vivo. *Nat Neurosci* **18:** 683–689. doi:10.1038/nn.3992

Hughes EG, Kang SH, Fukaya M, Bergles DE. 2013. Oligodendrocyte progenitors balance growth with self-repulsion to achieve homeostasis in the adult brain. *Nat Neurosci* 16: 668–676. doi:10.1038/nn.3390

Hughes EG, Orthmann-Murphy JL, Langseth AJ, Bergles DE. 2018. Myelin remodeling through experience-dependent oligodendrogenesis in the adult somatosensory cortex. *Nat Neurosci* 21: 696–706. doi:10.1038/s41593-018-0121-5

Ioannidou K, Anderson KI, Strachan D, Edgar JM, Barnett SC. 2012. Time-lapse imaging of the dynamics of CNS glial-axonal interactions in vitro and ex vivo. *PLoS ONE* 7: e30775. doi:10.1371/journal.pone.0030775

Ishii A, Dutta R, Wark GM, Hwang SI, Han DK, Trapp BD, Pfeiffer SE, Bansal R. 2009. Human myelin proteome and comparative analysis with mouse myelin. *Proc Natl Acad Sci* 106: 14605–14610. doi:10.1073/pnas.0905936106

Jahn O, Tenzer S, Werner HB. 2009. Myelin proteomics: molecular anatomy of an insulating sheath. *Mol Neurobiol* 40: 55–72. doi:10.1007/s12035-009-8071-2

Janova H, Arinrad S, Balmuth E, Mitjans M, Hertel J, Habes M, Bittner RA, Pan H, Goebbels S, Begemann M, et al. 2018. Microglia ablation alleviates myelin-associated catatonic signs in mice. *J Clin Invest* 128: 734–745. doi:10.1172/JCI97032

Kalil K, Dent EW. 2014. Branch management: mechanisms of axon branching in the developing vertebrate CNS. *Nat Rev Neurosci* 15: 7–18. doi:10.1038/nrn3650

Kato D, Wake H, Lee PR, Tachibana Y, Ono R, Sugio S, Tsuji Y, Tanaka YH, Tanaka YR, Masamizu Y, et al. 2020. Motor learning requires myelination to reduce asynchrony and spontaneity in neural activity. *Glia* 68: 193–210. doi:10.1002/glia.23713

Kirby BB, Takada N, Latimer AJ, Shin J, Carney TJ, Kelsh RN, Appel B. 2006. In vivo time-lapse imaging shows dynamic oligodendrocyte progenitor behavior during zebrafish development. *Nat Neurosci* 9: 1506–1511. doi:10.1038/nn1803

Knowles JK, Xu H, Soane C, Batra A, Saucedo T, Frost E, Tam LT, Fraga D, Ni L, Villar K, et al. 2022. Maladaptive myelination promotes generalized epilepsy progression. *Nat Neurosci* 25: 596–606. doi:10.1038/s41593-022-01052-2

Kole K, Voesenek BJB, Brinia ME, Petersen N, Kole MHP. 2022. Parvalbumin basket cell myelination accumulates axonal mitochondria to internodes. *Nat Commun* 13: 7598. doi:10.1038/s41467-022-35350-x

Krasnow AM, Ford MC, Valdivia LE, Wilson SW, Attwell D. 2018. Regulation of developing myelin sheath elongation by oligodendrocyte calcium transients in vivo. *Nat Neurosci* 21: 24–28. doi:10.1038/s41593-017-0031-y

Lam M, Takeo K, Almeida RG, Cooper MH, Wu K, Iyer M, Kantarci H, Zuchero JB. 2022. CNS myelination requires VAMP2/3-mediated membrane expansion in oligodendrocytes. *Nat Commun* 13: 5583. doi:10.1038/s41467-022-33200-4

Lappe-Siefke C, Goebbels S, Gravel M, Nicksch E, Lee J, Braun PE, Griffiths IR, Nave KA. 2003. Disruption of Cnp1 uncouples oligodendroglial functions in axonal support and myelination. *Nat Genet* 33: 366–374. doi:10.1038/ng1095

Laursen LS, Chan CW, Ffrench-Constant C. 2009. An integrin-contactin complex regulates CNS myelination by differential Fyn phosphorylation. *J Neurosci* 29: 9174–9185. doi:10.1523/JNEUROSCI.5942-08.2009

Lee S, Leach MK, Redmond SA, Chong SY, Mellon SH, Tuck SJ, Feng ZQ, Corey JM, Chan JR. 2012a. A culture system to study oligodendrocyte myelination processes using engineered nanofibers. *Nat Methods* 9: 917–922. doi:10.1038/nmeth.2105

Lee Y, Morrison BM, Li Y, Lengacher S, Farah MH, Hoffman PN, Liu Y, Tsingalia A, Jin L, Zhang PW, et al. 2012b. Oligodendroglia metabolically support axons and contribute to neurodegeneration. *Nature* 487: 443–448. doi:10.1038/nature11314

Li Q, Brus-Ramer M, Martin JH, McDonald JW. 2010. Electrical stimulation of the medullary pyramid promotes proliferation and differentiation of oligodendrocyte progenitor cells in the corticospinal tract of the adult rat. *Neurosci Lett* 479: 128–133. doi:10.1016/j.neulet.2010.05.043

Liu J, Dietz K, Deloyht JM, Pedre X, Kelkar D, Kaur J, Vialou V, Lobo MK, Dietz DM, Nestler EJ, et al. 2012. Impaired adult myelination in the prefrontal cortex of socially isolated mice. *Nat Neurosci* 15: 1621–1623. doi:10.1038/nn.3263

Losser J, Ravotto L, Jung RB, Werner HB, Ruhwedel T, Möbius W, Bergles DE, Barros F, Nave KA, Weber B, et al. 2024. Oligodendrocyte-axon metabolic coupling is mediated by extracellular K+ and maintains axonal health. *Nat Neurosci* (in press).

Lundgaard I, Luzhynskaya A, Stockley JH, Wang Z, Evans KA, Swire M, Volbracht K, Gautier HO, Franklin RJ, ffrench-Charles C, et al. 2013. Neuregulin and BDNF induce a switch to NMDA receptor-dependent myelination by oligodendrocytes. *PLoS Biol* 11: e1001743. doi:10.1371/journal.pbio.1001743

Makinodan M, Rosen KM, Ito S, Corfas G. 2012. A critical period for social experience-dependent oligodendrocyte maturation and myelination. *Science* 337: 1357–1360. doi:10.1126/science.1220845

Mangin JM, Li P, Scafidi J, Gallo V. 2012. Experience-dependent regulation of NG2 progenitors in the developing barrel cortex. *Nat Neurosci* 15: 1192–1194. doi:10.1038/nn.3190

Manrique-Hoyos N, Jürgens T, Grønborg M, Kreutzfeldt M, Schedensack M, Kuhlmann T, Schrick C, Brück W, Urlaub H, Simons M, et al. 2012. Late motor decline after accomplished remyelination: impact for progressive multiple sclerosis. *Ann Neurol* 71: 227–244. doi:10.1002/ana.22681

Marangon D, Audano M, Pedretti S, Fumagalli M, Mitro N, Lecca D, Caruso D, Abbracchio MP. 2022. Rewiring of glucose and lipid metabolism induced by G protein-coupled receptor 17 silencing enables the transition of oligodendrocyte progenitors to myelinating cells. *Cells* 11: 2369. doi:10.3390/cells11152369

Marisca R, Hoche T, Agirre E, Hoodless LJ, Barkey W, Auer F, Castelo-Branco G, Czopka T. 2020. Functionally distinct subgroups of oligodendrocyte precursor cells integrate neural activity and execute myelin formation. *Nat Neurosci* 23: 363–374. doi:10.1038/s41593-019-0581-2

Matthews MA, Duncan D. 1971. A quantitative study of morphological changes accompanying the initiation and progress of myelin production in the dorsal funiculus of the rat spinal cord. *J Comp Neurol* 142: 1–22. doi:10.1002/cne.901420102

Matute C, Palma A, Serrano-Regal MP, Maudes E, Barman S, Sánchez-Gómez MV, Domercq M, Goebels N, Dalmau J.

2020. N-Methyl-D-Aspartate receptor antibodies in autoimmune encephalopathy alter oligodendrocyte function. *Ann Neurol* **87:** 670–676. doi:10.1002/ana.25699

McKenzie IA, Ohayon D, Li H, de Faria JP, Emery B, Tohyama K, Richardson WD. 2014. Motor skill learning requires active central myelination. *Science* **346:** 318–322. doi:10.1126/science.1254960

Mensch S, Baraban M, Almeida R, Czopka T, Ausborn J, El Manira A, Lyons DA. 2015. Synaptic vesicle release regulates myelin sheath number of individual oligodendrocytes in vivo. *Nat Neurosci* **18:** 628–630. doi:10.1038/nn.3991

Meschkat M, Steyer AM, Weil MT, Kusch K, Jahn O, Piepkorn L, Agüi-Gonzalez P, Phan NTN, Ruhwedel T, Sadowski B, et al. 2022. White matter integrity in mice requires continuous myelin synthesis at the inner tongue. *Nat Commun* **13:** 1163. doi:10.1038/s41467-022-28720-y

Meyer N, Richter N, Fan Z, Siemonsmeier G, Pivneva T, Jordan P, Steinhäuser C, Semtner M, Nolte C, Kettenmann H. 2018. Oligodendrocytes in the mouse corpus callosum maintain axonal function by delivery of glucose. *Cell Rep* **22:** 2383–2394. doi:10.1016/j.celrep.2018.02.022

Micu I, Jiang Q, Coderre E, Ridsdale A, Zhang L, Woulfe J, Yin X, Trapp BD, McRory JE, Rehak R, et al. 2006. NMDA receptors mediate calcium accumulation in myelin during chemical ischaemia. *Nature* **439:** 988–992. doi:10.1038/nature04474

Micu I, Plemel JR, Lachance C, Proft J, Jansen AJ, Cummins K, van Minnen J, Stys PK. 2016. The molecular physiology of the axo-myelinic synapse. *Exp Neurol* **276:** 41–50. doi:10.1016/j.expneurol.2015.10.006

Mitew S, Gobius I, Fenlon LR, McDougall SJ, Hawkes D, Xing YL, Bujalka H, Gundlach AL, Richards LJ, Kilpatrick TJ, et al. 2018. Pharmacogenetic stimulation of neuronal activity increases myelination in an axon-specific manner. *Nat Commun* **9:** 306. doi:10.1038/s41467-017-02719-2

Möbius W, Nave KA, Werner HB. 2016. Electron microscopy of myelin: structure preservation by high-pressure freezing. *Brain Res* **1641:** 92–100. doi:10.1016/j.brainres.2016.02.027

Morita K, Sasaki H, Fujimoto K, Furuse M, Tsukita S. 1999. Claudin-11/OSP-based tight junctions of myelin sheaths in brain and Sertoli cells in testis. *J Cell Biol* **145:** 579–588. doi:10.1083/jcb.145.3.579

Mukherjee C, Kling T, Russo B, Miebach K, Kess E, Schifferer M, Pedro LD, Weikert U, Fard MK, Kannaiyan N, et al. 2020. Oligodendrocytes provide antioxidant defense function for neurons by secreting ferritin heavy chain. *Cell Metab* **32:** 259–272.e10. doi:10.1016/j.cmet.2020.05.019

Musse AA, Gao W, Rangaraj G, Boggs JM, Harauz G. 2009. Myelin basic protein co-distributes with other PI(4,5)P2-sequestering proteins in Triton X-100 detergent-resistant membrane microdomains. *Neurosci Lett* **450:** 32–36. doi:10.1016/j.neulet.2008.11.022

Nave KA. 2010. Myelination and support of axonal integrity by glia. *Nature* **468:** 244–252. doi:10.1038/nature09614

Nave KA, Werner HB. 2014. Myelination of the nervous system: mechanisms and functions. *Annu Rev Cell Dev Biol* **30:** 503–533. doi:10.1146/annurev-cellbio-100913-013101

Nawaz S, Sánchez P, Schmitt S, Snaidero N, Mitkovski M, Velte C, Brückner BR, Alexopoulos I, Czopka T, Jung

SY, et al. 2015. Actin filament turnover drives leading edge growth during myelin sheath formation in the central nervous system. *Dev Cell* **34:** 139–151. doi:10.1016/j.devcel.2015.05.013

Niu J, Li T, Yi C, Huang N, Koulakoff A, Weng C, Li C, Zhao CJ, Giaume C, Xiao L. 2016. Connexin-based channels contribute to metabolic pathways in the oligodendroglial lineage. *J Cell Sci* **129:** 1902–1914. doi:10.1242/jcs.178731

Norton WT, Poduslo SE. 1973. Myelination in rat brain: method of myelin isolation. *J Neurochem* **21:** 749–757. doi:10.1111/j.1471-4159.1973.tb07519.x

Nualart-Marti A, Solsona C, Fields RD. 2013. Gap junction communication in myelinating glia. *Biochim Biophys Acta* **1828:** 69–78. doi:10.1016/j.bbamem.2012.01.024

O'Brien JS. 1965. Stability of the myelin membrane. *Science* **147:** 1099–1107. doi:10.1126/science.147.3662.1099

Ortiz FC, Habermacher C, Graciarena M, Houry PY, Nishiyama A, Nait Oumesmar B, Angulo MC. 2019. Neuronal activity in vivo enhances functional myelin repair. *JCI Insight* **5:** e123434. doi:10.1172/jci.insight.123434

Pan S, Mayoral SR, Choi HS, Chan JR, Kheirbek MA. 2020. Preservation of a remote fear memory requires new myelin formation. *Nat Neurosci* **23:** 287–499. doi:10.1038/s41593-019-0582-1

Pan L, Trimarco A, Zhang AJ, Fujimori K, Urade Y, Sun LO, Taveggia C, Zhang Y. 2023. Oligodendrocyte-lineage cell exocytosis and L-type prostaglandin D synthase promote oligodendrocyte development and myelination. *eLife* **12:** e77441. doi:10.7554/eLife.77441

Patzig J, Erwig MS, Tenzer S, Kusch K, Dibaj P, Möbius W, Goebbels S, Schaeren-Wiemers N, Nave KA, Werner HB. 2016. Septin/anillin filaments scaffold central nervous system myelin to accelerate nerve conduction. *eLife* **5:** e17119. doi:10.7554/eLife.17119

Pedraza L, Huang JK, Colman D. 2009. Disposition of axonal caspr with respect to glial cell membranes: implications for the process of myelination. *J Neurosci Res* **87:** 3480–3491. doi:10.1002/jnr.22004

Peters A. 2002. The effects of normal aging on myelin and nerve fibers: a review. *J Neurocytol* **31:** 581–593.

Pfeiffer SE, Warrington AE, Bansal R. 1993. The oligodendrocyte and its many cellular processes. *Trends Cell Biol* **3:** 191–197. doi:10.1016/0962-8924(93)90213-K

Philips T, Mironova YA, Jouroukhin Y, Chew J, Vidensky S, Farah MH, Pletnikov MV, Bergles DE, Morrison BM, Rothstein JD. 2021. MCT1 deletion in oligodendrocyte lineage cells causes late-onset hypomyelination and axonal degeneration. *Cell Rep* **34:** 108610. doi:10.1016/j.celrep.2020.108610

Piaton G, Aigrot MS, Williams A, Moyon S, Tepavcevic V, Moutkine I, Gras J, Matho KS, Schmitt A, Soellner H, et al. 2011. Class 3 semaphorins influence oligodendrocyte precursor recruitment and remyelination in adult central nervous system. *Brain* **134:** 1156–1167. doi:10.1093/brain/awr022

Rao VTS, Khan D, Cui QL, Fuh SC, Hossain S, Almazan G, Multhaup G, Healy LM, Kennedy TE, Antel JP. 2017. Distinct age and differentiation-state dependent metabolic profiles of oligodendrocytes under optimal and stress conditions. *PLoS ONE* **12:** e0182372. doi:10.1371/journal.pone.0182372

Rasband MN, Peles E. 2016. The nodes of Ranvier: molecular assembly and maintenance. *Cold Spring Harb Perspect Biol* **8:** a020495. doi: 10.1101/cshperspect.a020495

* Rasband MN, Peles E. 2023. The nodes of Ranvier and the organization of myelinated axons. *Cold Spring Harb Perspect Biol* doi: 10.1101/cshperspect.a41361

Redmond SA, Mei F, Eshed-Eisenbach Y, Osso LA, Leshkowitz D, Shen YA, Kay JN, Aurrand-Lions M, Lyons DA, Peles E, et al. 2016. Somatodendritic expression of JAM2 inhibits oligodendrocyte myelination. *Neuron* **91:** 824–836. doi:10.1016/j.neuron.2016.07.021

Rinholm JE, Hamilton NB, Kessaris N, Richardson WD, Bergersen LH, Attwell D. 2011. Regulation of oligodendrocyte development and myelination by glucose and lactate. *J Neurosci* **31:** 538–548. doi:10.1523/JNEUROSCI.3516-10.2011

Rosenberg SS, Kelland EE, Tokar E, De la Torre AR, Chan JR. 2008. The geometric and spatial constraints of the microenvironment induce oligodendrocyte differentiation. *Proc Natl Acad Sci* **23:** 14662–14667. doi:10.1073/pnas.0805640105

Rosenbluth J. 1980. Central myelin in the mouse mutant shiverer. *J Comp Neurol* **194:** 639–648. doi:10.1002/cne.901940310

Saab AS, Tzvetanova ID, Nave KA. 2013. The role of myelin and oligodendrocytes in axonal energy metabolism. *Curr Opin Neurobiol* **23:** 1065–1072. doi: 10.1016/j.conb.2013.09.008

Saab AS, Tzvetavona ID, Trevisiol A, Baltan S, Dibaj P, Kusch K, Möbius W, Goetze B, Jahn HM, Huang W, et al. 2016. Oligodendroglial NMDA receptors regulate glucose import and axonal energy metabolism. *Neuron* **91:** 119–132. doi:10.1016/j.neuron.2016.05.016

Schäffner E, Bosch-Queralt M, Edgar J, Lehning M, Strauß J, Fleischer N, Kungl T, Wieghofer P, Berghoff SA, Reinert T, et al. 2023. Myelin insulation as a risk factor for axonal degeneration in autoimmune demyelinating disease. *Nat Neurosci* **26:** 1218–1228. doi:10.1038/s41593-023-01366-9

Sharma K, Schmitt S, Bergner CG, Tyanova S, Kannaiyan N, Manrique-Hoyos N, Kongi K, Cantuti L, Hanisch UK, Philips MA, et al. 2015. Cell type- and brain region-resolved mouse brain proteome. *Nat Neurosci* **18:** 1819–1831. doi:10.1038/nn.4160

Simons M, Lyons DA. 2013. Axonal selection and myelin sheath generation in the central nervous system. *Curr Opin Cell Biol* **25:** 512–519. doi:10.1016/j.ceb.2013.04.007

Sinclair JL, Fischl MJ, Alexandrova O, Heß M, Grothe B, Leibold C, Kopp-Scheinpflug C. 2017. Sound-Evoked activity influences myelination of brainstem axons in the trapezoid body. *J Neurosci* **37:** 8239–8255. doi:10.1523/JNEUROSCI.3728-16.2017

Smith ME, Eng LF. 1965. The turnover of the lipid components of myelin. *J Am Oil Chem Soc* **42:** 1013–1018. doi:10.1007/BF02636894

Snaidero N, Möbius W, Czopka T, Hekking LH, Mathisen C, Verkleij D, Goebbels S, Edgar J, Merkler D, Lyons DA, et al. 2014. Myelin membrane wrapping of CNS axons by PI (3,4,5)P3-dependent polarized growth at the inner tongue. *Cell* **156:** 277–290. doi:10.1016/j.cell.2013.11.044

Snaidero N, Velte C, Myllykoski M, Raasakka A, Ignatev A, Werner HB, Erwig MS, Möbius W, Kursula P, Nave KA, et

al. 2017. Antagonistic functions of MBP and CNP establish cytosolic channels in CNS myelin. *Cell Rep* **18:** 314–323. doi:10.1016/j.celrep.2016.12.053

Sobottka B, Ziegler U, Kaech A, Becher B, Goebels N. 2011. CNS live imaging reveals a new mechanism of myelination: the liquid croissant model. *Glia* **59:** 1841–1849. doi:10.1002/glia.21228

Stadelmann C, Timmler S, Barrantes-Freer A, Simons M. 2019. Myelin in the central nervous system: structure, function, and pathology. *Physiol Rev* **99:** 1381–1431. doi:10.1152/physrev.00031.2018

Stassart RM, Möbius W, Nave KA, Edgar JM. 2018. The axon-myelin unit in development and degenerative disease. *Front Neurosci* **12:** 467. doi:10.3389/fnins.2018.00467

Steadman PE, Xia F, Ahmed M, Mocle AJ, Penning ARA, Geraghty AC, Steenland HW, Monje M, Josselyn SA, Frankland PW. 2020. Disruption of oligodendrogenesis impairs memory consolidation in adult mice. *Neuron* **105:** 150–164.e6. doi:10.1016/j.neuron.2019.10.013

Stedehouder J, Couey JJ, Brizee D, Hosseini B, Slotman JA, Dirven CMF, Shpak G, Houtsmuller AB, Kushner SA. 2017. Fast-spiking parvalbumin interneurons are frequently myelinated in the cerebral cortex of mice and humans. *Cereb Cortex* **27:** 5001–5013. doi:10.1093/cercor/bhx203

Stedehouder J, Brizee D, Shpak G, Kushner SA. 2018. Activity-dependent myelination of parvalbumin interneurons mediated by axonal morphological plasticity. *J Neurosci* **38:** 3631–3642. doi:10.1523/JNEUROSCI.0074-18.2018

Swire M, Kotelevtsev Y, Webb DJ, Lyons DA, Ffrench-Constant C. 2019. Endothelin signalling mediates experience-dependent myelination in the CNS. *eLife* **8:** e49493. doi:10.7554/eLife.49493

Syed YA, Hand E, Möbius W, Zhao C, Hofer M, Nave KA, Kotter MR. 2011. Inhibition of CNS remyelination by the presence of semaphorin 3A. *J Neurosci* **31:** 3719–3728. doi:10.1523/JNEUROSCI.4930-10.2011

Tomassy GS, Berger DR, Chen HH, Kasthuri N, Hayworth KJ, Vercelli A, Seung HS, Lichtman JW, Arlotta P. 2014. Distinct profiles of myelin distribution along single axons of pyramidal neurons in the neocortex. *Science* **344:** 319–324. doi:10.1126/science.1249766

Toyama BH, Savas JN, Park SK, Harris MS, Ingolia NT, Yates JR 3rd, Hetzer MW. 2013. Identification of long-lived proteins reveals exceptional stability of essential cellular structures. *Cell* **154:** 971–982. doi:10.1016/j.cell.2013.07.037

Trevisiol A, Saab AS, Winkler U, Marx G, Imamura H, Möbius W, Kusch K, Nave KA, Hirrlinger J. 2017. Monitoring ATP dynamics in electrically active white matter tracts. *eLife* **6:** e24241. doi:10.7554/eLife.24241

Trevisiol A, Kusch K, Steyer AM, Gregor I, Nardis C, Winkler U, Köhler S, Restrepo A, Möbius W, Werner HB, et al. 2020. Structural myelin defects are associated with low axonal ATP levels but rapid recovery from energy deprivation in a mouse model of spastic paraplegia. *PLoS Biol* **18:** e3000943. doi:10.1371/journal.pbio.3000943

Tripathi RB, Jackiewicz M, McKenzie IA, Kougioumtzidou E, Grist M, Richardson WD. 2017. Remarkable stability of myelinating oligodendrocytes in mice. *Cell Rep* **21:** 316–323. doi:10.1016/j.celrep.2017.09.050

van Hasselt PM, Ferdinandusse S, Monroe GR, Ruiter JP, Turkenburg M, Geerlings MJ, Duran K, Harakalova M, van der Zwaag B, Monavari AA, et al. 2014. Monocarboxylate transporter 1 deficiency and ketone utilization. *N Engl J Med* **371:** 1900–1907. doi:10.1056/NEJMoa 1407778

Velumian AA, Samoilova M, Fehlings MG. 2011. Visualization of cytoplasmic diffusion within living myelin sheaths of CNS white matter axons using microinjection of the fluorescent dye Lucifer yellow. *Neuroimage* **56:** 27–34. doi:10.1016/j.neuroimage.2010.11.022

Volkenhoff A, Weiler A, Letzel M, Stehling M, Klämbt C, Schirmeier S. 2015. Glial glycolysis is essential for neuronal survival in *Drosophila*. *Cell Metab* **22:** 437–447. doi:10 .1016/j.cmet.2015.07.006

Wake H, Lee PR, Fields RD. 2011. Control of local protein synthesis and initial events in myelination by action potentials. *Science* **333:** 1647–1651. doi:10.1126/science .1206998

Wang F, Ren SY, Chen JF, Liu K, Li RX, Li ZF, Hu B, Niu JQ, Xiao L, Chan JR, et al. 2020. Myelin degeneration and diminished myelin renewal contribute to age-related deficits in memory. *Nat Neurosci* **23:** 481–486. doi:10.1038/ s41593-020-0588-8

Werner HB, Kuhlmann K, Shen S, Uecker M, Schardt A, Dimova K, Orfaniotou F, Dhaunchak A, Brinkmann BG, Möbius W, et al. 2007. Proteolipid protein is required for transport of sirtuin 2 into CNS myelin. *J Neurosci* **27:** 7717–7730. doi:10.1523/JNEUROSCI.1254-07.2007

White R, Gonsior C, Krämer-Albers EM, Stöhr N, Hüttelmaier S, Trotter J. 2008. Activation of oligodendroglial Fyn kinase enhances translation of mRNAs transported in hnRNP A2-dependent RNA granules. *J Cell Biol* **181:** 579–586. doi:10.1083/jcb.200706164

Xiao L, Ohayon D, McKenzie IA, Sinclair-Wilson A, Wright JL, Fudge AD, Emery B, Li H, Richardson WD. 2016. Rapid production of new oligodendrocytes is required in the earliest stages of motor-skill learning. *Nat Neurosci* **19:** 1210–1217. doi:10.1038/nn.4351

Xu CJ, Wang Y, Liao M. 2011. Effect of central myelin on the proliferation and differentiation into O4$^+$ oligodendrocytes of GFP-NSCs. *Mol Cell Biochem* **358:** 173–178. doi:10.1007/s11010-011-0932-0

Yamazaki Y, Abe Y, Fujii S, Tanaka KF. 2021. Oligodendrocytic Na$^+$–K$^+$–Cl– co-transporter 1 activity facilitates axonal conduction and restores plasticity in the adult mouse brain. *Nat Commun* **12:** 5146. doi: 10.1038/s41467-021-25488-5

Yang SM, Michel K, Jokhi V, Nedivi E, Arlotta P. 2020. Neuron class-specific responses govern adaptive myelin remodeling in the neocortex. *Science* **370:** eabd2109. doi:10.1126/science.abd2109

Yeung MS, Zdunek S, Bergmann O, Bernard S, Salehpour M, Alkass K, Perl S, Tisdale J, Possnert G, Brundin L, et al. 2014. Dynamics of oligodendrocyte generation and myelination in the human brain. *Cell* **159:** 766–774. doi:10 .1016/j.cell.2014.10.011

Young KM, Psachoulia K, Tripathi RB, Dunn SJ, Cossell L, Attwell D, Tohyama K, Richardson WD. 2013. Oligodendrocyte dynamics in the healthy adult CNS: evidence for myelin remodeling. *Neuron* **77:** 873–885. doi:10.1016/j .neuron.2013.01.006

Zhao YY, Shi XY, Qiu X, Zhang L, Lu W, Yang S, Li C, Cheng GH, Yang ZW, Tang Y. 2011. Enriched environment increases the total number of CNPase positive cells in the corpus callosum of middle-aged rats. *Acta Neurobiol Exp (Wars)* **71:** 322–330.

Zheng J, Sun X, Ma C, Li BM, Luo F. 2019. Voluntary wheel running promotes myelination in the motor cortex through Wnt signaling in mice. *Mol Brain* **12:** 85. doi:10 .1186/s13041-019-0506-8

Zonouzi M, Renzi M, Farrant M, Cull-Candy SG. 2011. Bidirectional plasticity of calcium-permeable AMPA receptors in oligodendrocyte lineage cells. *Nat Neurosci* **14:** 1430–1438. doi:10.1038/nn.2942

Zuchero JB, Fu MM, Sloan SA, Ibrahim A, Olson A, Zaremba A, Dugas JC, Wienbar S, Caprariello AV, Kantor C, et al. 2015. CNS myelin wrapping is driven by actin disassembly. *Dev Cell* **34:** 152–167. doi:10.1016/j.devcel.2015.06.011

Cite this article as *Cold Spring Harb Perspect Biol* doi: 10.1101/cshperspect.a041359

Schwann Cell Development and Myelination

James Salzer,[1] M. Laura Feltri,[2,3,4] and Claire Jacob[5]

[1]Neuroscience Institute, New York University Grossman School of Medicine, New York, New York 10016, USA

[2]Institute for Myelin and Glia Exploration, Jacobs School of Medicine and Biomedical Sciences, State University of New York at Buffalo, Buffalo, New York 14203, USA

[3]IRCCS Neurological Institute Carlo Besta, Milano 20133, Italy

[4]Department of Biotechnology and Translational Sciences, Universita' Degli Studi di Milano, Milano 20133, Italy

[5]Faculty of Biology, Institute of Developmental Biology and Neurobiology, Johannes Gutenberg University Mainz, Mainz 55128, Germany

Correspondence: James.Salzer@nyulangone.org

Glial cells in the peripheral nervous system (PNS), which arise from the neural crest, include axon-associated Schwann cells (SCs) in nerves, synapse-associated SCs at the neuromuscular junction, enteric glia, perikaryon-associated satellite cells in ganglia, and boundary cap cells at the border between the central nervous system (CNS) and the PNS. Here, we focus on axon-associated SCs. These SCs progress through a series of formative stages, which culminate in the generation of myelinating SCs that wrap large-caliber axons and of nonmyelinating (Remak) SCs that enclose multiple, small-caliber axons. In this work, we describe SC development, extrinsic signals from the axon and extracellular matrix (ECM) and the intracellular signaling pathways they activate that regulate SC development, and the morphogenesis and organization of myelinating SCs and the myelin sheath. We review the impact of SCs on the biology and integrity of axons and their emerging role in regulating peripheral nerve architecture. Finally, we explain how transcription and epigenetic factors control and fine-tune SC development and myelination.

STAGES OF SCHWANN CELL DEVELOPMENT

Schwann cells (SCs) derive from neural crest cells, a multipotent group of cells that arises after closure of the neural tube at around embryonic day (E)9.5–E11.5 in the mouse.[6] Neural crest cells originate at the cranial, trunk, cardiac, and sacral levels of the neural tube, delaminate and migrate throughout the body following three main streams. Depending on the location and migrating stream, neural crest cells give rise to various peripheral nervous system (PNS) neurons, glia, endoneurial fibroblasts (EFs), and other cells including melanocytes and cranial chondrocytes (Le Douarin and Teillet 1974; Joseph et al. 2004).

The formative stages of SC differentiation from neural crest to precursors and then to

---

[6]This is an update to a previous article published in *Cold Spring Harbor Perspectives in Biology* [Salzer (2015). *Cold Spring Harb Perspect Biol* **8**: a020529. doi:10.1101/cshperspect.a020529].

Cite this article as *Cold Spring Harb Perspect Biol* doi: 10.1101/cshperspect.a041360

mature cells (Fig. 1) have been extensively characterized (see Jessen and Mirsky 2019; Muppirala et al. 2021 for recent reviews). Axon-associated SCs originate from trunk neural crest cells that migrate ventrally and rapidly differentiate into SC precursors (Weston 1963; Jessen et al. 1994), which associate with extending axons and surround their growth cones (Wanner et al. 2006). SC precursors retain significant pluripotency and can also give rise to melanocytes, neurons, and chondrocytes (Kastriti et al. 2022; Taveggia and Feltri 2022). In developing nerve trunks, SC precursors interact with axon bundles and generate immature SCs, whose survival depends on the Neuregulin 1 (NRG1) growth factor expressed on axons (Jessen et al. 1994; Meyer and Birchmeier 1995). At this stage, axons of varying caliber are tightly packed into bundles with SCs present typically on the outside. Interaction with axons and NRG1 stimulates the proliferation of SC precursors, generating a sufficient number of cells to surround all axon bundles and deposit a basal lamina.

SCs that surround an axon bundle and are contained within the same basal lamina are described as an SC "family" (Peters et al. 1991). Basal lamina deposition (see below) contributes to SC precursor proliferation and survival and coincides with their differentiation into immature SCs. With proliferation, a daughter SC detaches from the "family," bringing along its associated axon (Webster et al. 1973). When such SCs establish a 1:1 relationship with an axon, they are referred to as "promyelinating" SCs (Fig. 1). These SCs typically ensheath axons with a diameter ≥ 1 μm that will be myelinated. The recognition and isolation of large-caliber axons destined to be myelinated, away from small-caliber axons that will remain unmyelinated, is referred to as radial sorting (Peters and Muir 1959). As promyelinating SCs transition to myelinating SCs, they turn on Egr2 (also known as Krox20), the master transcription factor of SC myelination that drives expression of myelin protein and lipid biosynthetic genes. The small axons that remain in bundles after radial sorting

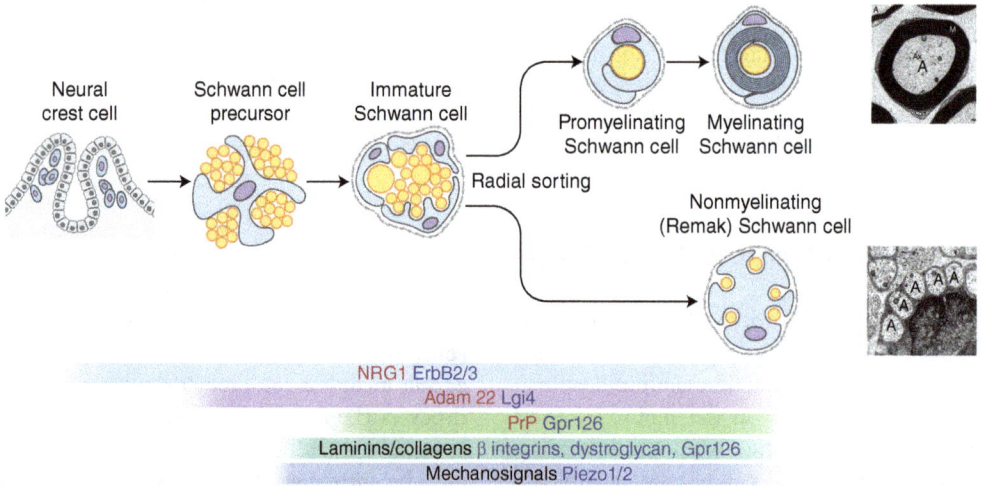

Figure 1. Schematic of the formative stages in the Schwann cell (SC) lineage. At the *right* are electron micrographs showing cross-sections through a myelinating SC and a portion of a Remak SC; axon profiles are labeled A. Key extrinsic signals and their SC receptors (in blue text) are listed *below*. Axonal signals (red) include type III NRG1, which activates ErbB2/3; Adam 22, which interacts with secreted Lgi4; and prion protein (PrP), which signals via Gpr126. Extracellular matrix components including laminins and collagens in the basal lamina, which signal via the abaxonal SC receptors (black) β integrins, dystroglycan, and Gpr126. Mechanical forces signal through Piezo1 and 2. The basal lamina is shown surrounding SCs at the immature and all later stages. (This figure is modified from schematics in Jessen and Mirsky 2005 and Fledrich et al. 2019.)

 Cite this article as *Cold Spring Harb Perspect Biol* doi: 10.1101/cshperspect.a041360

will be sorted into separate pockets of a single SC, which will differentiate into a mature Remak (nonmyelinating) SC. Remak bundles differ from the axon bundles in developing nerves as all axons are small in caliber and are properly ensheathed by SC processes (for review, see Feltri et al. 2016).

These two alternative modes of axon ensheathment by SCs (i.e., myelination vs. Remak ensheathment) (Peters et al. 1991), are responsible for distinct modes of action potential propagation: saltatory conduction versus continuous propagation, respectively. Accordingly, their associated fibers have distinct physiological functions. Axons that are myelinated include large-caliber (13–20 μm), rapidly conducting type A skeletomotor and proprioceptor fibers and mid-caliber (6–12 μm), fast-conducting type B mechanoreceptors, and preganglionic autonomic efferent fibers. Axons enclosed by Remak SCs are small (typically <1 μm diameter), slowly conducting (0.5–4 m/sec) type C fibers. The latter include efferent, postganglionic autonomic nerve fibers and afferent nerve fibers from the skin that carry pain and temperature information.

Myelinating and Remak SCs are transcriptionally quite distinct and accordingly express different repertoires of proteins. For example, the transcription factor Egr1 is present in Remak SCs, whereas, as noted, myelinating SCs express Egr2 (discussed further below). Myelinating and Remak SCs also differ in the cell-adhesion molecules and receptors that mediate their interactions with the axon and the basal lamina (Harty and Monk 2017; Bosch-Queralt et al. 2023). The most striking difference is the selective and robust expression of myelin proteins by myelinating SCs. In contrast, the P75NTR receptor, L1CAM, and GAP43 are characteristic of Remak SCs. For detailed information on markers present in SCs at different stages of differentiation, including Remak SCs, see Jessen et al. (2015).

Recent single-cell transcriptomic data further delineate markers that are differentially expressed during SC development and between myelin-forming versus Remak SCs (Gerber et al. 2021; Tasdemir-Yilmaz et al. 2021). These studies also highlight heterogeneity within mature SC populations, notably of subgroups of my-

elin-forming SCs (mSCs) with unique patterns of gene expression (Yim et al. 2022). One such group of mSCs expresses PMP2; these mSCs preferentially ensheath motor axons and are selectively reduced in amyotrophic lateral sclerosis nerve samples (Yim et al. 2022). These results suggest myelin-forming SC subtypes exhibit different characteristics, either due to variations in local, extrinsic signals, and/or possibly distinct developmental origins, that may be relevant to nerve pathology. Future studies, that manipulate or genetically fate-map these SC subtypes based on RNA-seq data will be useful to further elucidate their functional significance.

EXTRINSIC SIGNALS THAT REGULATE SCHWANN CELL DEVELOPMENT

The major extrinsic signals that drive SC differentiation are provided by the axon and the extracellular matrix (ECM)/basal lamina (Salzer 2015; Fledrich et al. 2019; Muppirala et al. 2021; Wilson et al. 2021), as discussed below.

Axonal Signals

NRG1 on the axon controls virtually every aspect of SC development including proliferation, survival, and differentiation of immature SCs, radial sorting, and their binary choice to become Remak versus myelinating SCs (Newbern and Birchmeier 2010). In particular, levels of NRG1 type III, an abundant membrane-tethered NRG isoform, correlate with the ensheathment fate of axons. Axons destined to be myelinated present much higher levels of NRG1 type III than do axons in Remak fibers (Taveggia et al. 2005). Haploinsufficiency or full genetic inactivation of NRG1 type III result in significant PNS hypomyelination (Michailov et al. 2004; Taveggia et al. 2005; Brinkmann et al. 2008), a reduced percentage of myelinated axons, and aberrant axon sorting including in Remak fibers (Taveggia et al. 2005; Raphael et al. 2011). Conversely, overexpression of NRG1 type III results in significant hypermyelination evident as increased myelin thickness (Michailov et al. 2004). Thus, threshold levels of NRG1 type III trigger SC myelination, and above that threshold, the amount of compact myelin that forms is graded to the levels of NRG1 (Nave and Salzer 2006). As NRG1 type III also controls SC proliferation and survival, the elevated levels of NRG1 type III of larger axons drive the generation of the additional SCs required for myelination—thereby coordinating SC numbers to their myelinating fate (Taveggia et al. 2005). Once myelination is complete, ongoing signaling from axonal NRG1 or its SC receptors is not required to maintain the myelin sheath (Atanasoski et al. 2006; Fricker et al. 2011).

NRG1 type III is a juxtacrine signal that activates the ErbB2/3 coreceptors on SCs. ErbB receptors are classical tyrosine kinases whose downstream effectors include the mitogen-activated protein kinase (MAPK) ERK, phosphatidylinositol-4,5-bisphosphate 3-kinase (PI3K), and phospholipase Cγ (PLCγ) pathways, each of which has been reported to be essential for proper myelination (Newbern and Birchmeier 2010; Fledrich et al. 2019). Expression of constitutively active ERK1/2 is able to rescue myelination in NRG1-deficient mice, in part via an increase in protein translation (Sheean et al. 2014).

An additional key axonal signal required for myelination is Adam22 (a disintegrin and metalloproteinase domain-containing protein 22) (Özkaynak et al. 2010; Kegel et al. 2014). Adam22 interacts with leucine-rich glioma-inactivated 4 (Lgi4), a SC-secreted protein that is mutated in the *claw paw* hypomyelinated mutant mouse (Bermingham et al. 2006). Pan-knockouts of Adam22 (Sagane et al. 2005), neuron-specific deletion of *Adam22*, and SC-specific deletion of *Lgi4* (Özkaynak et al. 2010) all result in major defects in myelination, arresting SCs at the promyelinating stage (Kegel et al. 2013). It is unclear presently how binding of Lgi4 to Adam22 on the axon signals back to SCs to regulate ensheathment and myelination. Interestingly, Lgi4–Adam22 interactions are also required for PNS gliogenesis during early development, including for proliferation of glial-restricted progenitors (Nishino et al. 2010).

Recent studies suggest that axon size, independent of its content of signaling molecules, also regulates SC development and myelination (Park et al. 2022). PIEZO1 and PIEZO2 are among the most abundant mechanosensitive ion channels present in SCs and may regulate myelination via activation of Yes-associated protein 1 (YAP1) and WW domain–containing transcription regulator protein 1 (WWTR1 or TAZ) (Acheta et al. 2022), as discussed further below.

Basal Lamina Signals

Axons direct SC assembly of a basal lamina, which in turn has an essential, autocrine role in SC ensheathment and myelination of axons (Madrid et al. 1975; Bunge et al. 1986; Feltri and Wrabetz 2005; Chernousov et al. 2008). The SC basal lamina is a thin, dense mixture of ECM components, including laminins and collagens, that is closely opposed to the abaxonal SC membrane (Chernousov et al. 2008). The basal lamina is synthesized by SCs with contributions from EFs (Obremski et al. 1993). Conditional knockout of laminin (*LAMC1*) in SCs results in aberrant basal lamina formation, major defects of axon sorting, and hypomyelination in vivo, underscoring laminin's key role in these events (Chen and Strickland 2003). Both the scaffold-forming and cell-adhesion activities of laminin are required to promote SC ensheathment (McKee et al. 2012). Other components of the matrix, notably the collagens, also regulate SC myelination (Chernousov et al. 2008).

SC laminin receptors include dystroglycan and the α6β1 and α6β4 integrins. These receptors activate downstream signaling pathways (e.g., ILK, FAK, Rac1, cdc42, and PKA) to further induce the formation of the lamellipodia-like processes that interdigitate, contact, and separate axons (Benninger et al. 2007; Nodari et al. 2007; Pereira et al. 2009; Pellegatta et al. 2013; Grove and Brophy 2014). In addition, both collagen and laminin bind to and activate Gpr126/Adgrg6 on SCs, which is required for ensheathment and myelination (Monk et al. 2009). Gpr126 mediates mechano-signals from the ECM (collagen IV/laminin 211) (Paavola et al. 2014; Petersen et al. 2015; Mitgau et al. 2022) to activate adenylate cyclase via Gαs (Mogha et al. 2013), elevating cAMP levels that activate PKA and promote myelination (Howe and McCarthy 2000). Inter-

Cite this article as *Cold Spring Harb Perspect Biol* doi: 10.1101/cshperspect.a041360

estingly, the prion protein (PrP) expressed on axons has been implicated as an agonistic ligand for Gpr126 to sustain cAMP levels and myelination in the adult (Küffer et al. 2016). Thus, Gpr126 appears to function as a signal on both the abaxonal (outer) and adaxonal (inner) SC membranes.

MECHANISMS OF SCHWANN CELL ENSHEATHMENT AND RADIAL SORTING

The combined activity of these extrinsic signals drives expansion of the SC plasma membrane and promotes radial sorting and myelination of axons. All of these morphogenetic changes in SCs are driven by remodeling of the actin cytoskeleton. Thus, inhibitors of actin assembly (Fernandez-Valle et al. 1997) and of myosin II contractility (Wang et al. 2008) block axon sorting and myelination in cocultures. Among the master regulators of actin assembly and its organization in cells are Rho, Rac1, and cdc42, foundational members of the Rho GTPase family of proteins (Hall 2012). Each of these have been implicated in regulating radial sorting and SC myelination. Thus, conditional inactivation of Rac1 or Cdc42 during SC development results in impaired axon sorting, delayed myelination, and significant hypomyelination (Benninger et al. 2007; Nodari et al. 2007; Guo et al. 2013). Rho and its key effector Rho kinase (ROCK) promote radial sorting (Pereira et al. 2009) and the coordinated progression of the inner turn of the myelin sheath around the axon at the onset of myelination (Melendez-Vasquez et al. 2004).

Extrinsic signals from the basal lamina and axon, in turn, are known to activate these Rho family GTPases. Rac1 is activated downstream of laminin/β1-integrin signaling (Benninger et al. 2007; Nodari et al. 2007) via lymphoid cell kinase (Lck) (Ness et al. 2013). Indeed, overexpression of Rac1 partially rescues the phenotype of SC conditional *β1-integrin* nulls (Nodari et al. 2007). In addition, increased activity of the PI3K/Akt pathway, which is downstream of NRG1 signaling (Maurel and Salzer 2000), drives exuberant SC wrapping of axons in Remak and myelinated fibers, and even of collagen fibers, via mTOR signaling and Rac1 activity (Goebbels et al. 2010;

Doménech-Estevez et al. 2016). RhoA activation is downstream of Gpr56, which is required for radial sorting (Ackerman et al. 2018) and of laminin signaling via ILK (Montani et al. 2014). Cdc42 is activated in SCs by soluble NRG1 in vitro (Benninger et al. 2007); however, the importance of NRG1 as an activator of Cdc42 in vivo is not yet established.

Effectors of these GTPases include N-WASP (the neuronal Wiskott–Aldrich syndrome protein), which is activated downstream of Rac1 and cdc42 and binds to the Arp2/3 complex to nucleate new branched arrays of actin filaments (Rohatgi et al. 1999; Burianek and Soderling 2013). N-Wasp localizes to the leading edge of SC processes. Pharmacological inhibition of N-WASP impairs myelination in cocultures (Bacon et al. 2007) and conditional inactivation markedly delays sorting and profoundly impairs myelin wrapping, resulting in shorter internodes and thinner sheaths (Jin et al. 2011; Novak et al. 2011). These latter results indicate that elongation and circumferential membrane wrapping share similar actin regulators. Finally, the Rho/ROCK pathway promotes lamellipodia by MLCK (myosin light chain kinase) and activation of Myosin II (Melendez-Vasquez et al. 2004) and, separately, via the actin binding protein Profilin1 (PFN1) (Montani et al. 2014). In the latter case, SC-specific gene ablation of *Pfn1* in mice results in profound radial sorting and myelination defects (Montani et al. 2014).

In addition to proteins that promote actin assembly and branching, depolymerization of the actin cytoskeleton is also actively regulated, including by the actin-severing protein Cofilin. NRG1 activates cofilin via the cofilin-phosphatase Slingshot-1, which is activated by translocation to and association with F-actin at the leading edge of lamellipodia (Nagata-Ohashi et al. 2004). In agreement, NRG1 promotes translocation of nonphosphorylated Cofilin to the leading edge of cultured SCs (Sparrow et al. 2012). Further, knockdown of *Cofilin1* in cultured SCs blocks their engagement or alignment along axons, their assembly of the basal lamina, and their ability to myelinate (Sparrow et al. 2012).

Taken together, these findings suggest that actin is dynamically and obligately regulated during axon segregation and myelination. Based on ligand activation and localization studies, NRG1/ErbB signaling activates Cofilin and Cdc42 at the leading edge of the adaxonal membrane, leading to a cycle of actin depolymerization and N-WASP-dependent assembly, respectively. These results are consistent with actin treadmilling known to underlie lamellipodial extension in most cells (Pollard and Borisy 2003). In addition, Rac1 and Rho/ROCK act in the abaxonal compartment downstream of laminin/β1-integrin to drive actinomyosin activity required for radial sorting.

MORPHOGENESIS OF THE SCHWANN CELL MYELIN SHEATH

Early electron microscopic studies established that the myelin sheath forms as the result of spiral wrapping of the SC plasma membrane around the axon (Geren 1954; Robertson 1955). The myelin sheath is initially loosely spiraled for the first few turns, then compacts with upregulation of myelin proteins, notably P0 (Kidd et al. 2013). In a classic study, Bunge et al. (1989) provided compelling evidence that spiral wrapping results from circumnavigation of the inner myelin turn around the axon rather than by the countervailing movement of the outer SC compartment that contains the SC nucleus. This model was also suggested by earlier, live imaging studies of developing nerves in tadpole tails (Speidel 1964). Growth of the inner membrane requires it to intercalate between the existing inner membrane and the axon (Fig. 1), disrupting existing interactions between the axon and the SC. Whether movement of the inner turn is primarily guided by interactions with the axon (heterotypic interactions), by interactions with the glial membrane (homotypic interactions), or by both is not yet known.

In addition to wrapping of the inner turn, the myelin sheath also expands at its lateral margins. The innermost turns are initially narrower than the full length of the internode, as suggested by older electron microscopy (EM) studies (Webster 1971) and expands laterally as the sheath matures. This interpretation is strongly supported by changes in the pattern of Caspr staining, which delineates the leading and lateral edges of the SC and transitions from a loose spiral around the axon to its subsequent localization at the paranodes (Pedraza et al. 2001). Thus, morphogenesis of the SC myelin sheath mirrors that of oligodendrocyte myelin (Snaidero et al. 2014; Simons et al. 2023).

New components of the membrane are added in the abaxonal and perinuclear region (Gould 1977), and potentially within the paranodes (Gould and Mattingly 1990). Inserted components—in particular membrane lipids—then likely diffuse throughout the forming myelin sheath. Even after compact myelin has formed, the sheath rapidly expands and its internal circumference continues to increase to accommodate radial expansion of the axon (Webster 1971). Together, these results suggest the myelin sheath conforms to the classical model of a fluid mosaic membrane (Singer and Nicolson 1972; Nicolson and Ferreira de Mattos 2023), with diffusion potentially enhanced by its high lipid content.

REGULATION OF MYELIN SHEATH THICKNESS AND INTERNODE LENGTH

Myelination typically commences around axons that are >1 µm in diameter (Peters et al. 1991), in agreement with theoretical models that suggest myelination enhances conduction velocity in PNS axons of this diameter (Rushton 1951). Together with axon diameter, key determinants of nerve conduction velocity include the thickness of the myelin sheath, its length (i.e., internode length), and the width of the node of Ranvier (Moore et al. 1978; Wu et al. 2012; Arancibia-Cárcamo et al. 2017). These parameters are optimized for their role in saltatory conduction (Waxman 1980).

Accordingly, myelin sheath thickness is tightly regulated by and correlates to the diameter of the axon it surrounds. The relationship between sheath thickness and axon diameter is conventionally expressed as the g-ratio (i.e., axon diameter/total fiber diameter) and is typically ∼0.67 in the PNS. The tight correlation of myelin sheath thickness to axon diameter results from a balance of promyelinating signals with other signals that inhibit/terminate myelination (Salzer

Cite this article as *Cold Spring Harb Perspect Biol* doi: 10.1101/cshperspect.a041360

2015; Bolino 2021). As discussed above, the major (promyelinating) determinant of myelin sheath thickness is the level of NRG1 on the axon, which is strongly correlated to axon diameter (Michailov et al. 2004; Taveggia et al. 2005). Recent studies also implicate mechanical signals that are sensitive to axonal diameter and act through ACTL6a, a component of the SWI/SNF chromatin remodeling complex, to determine myelin thickness (Park et al. 2022).

Mechanisms that limit sheath thickness include down-regulation of promyelinating signals (e.g., the NRG1/ErbB signaling pathway). NRG1 activity is itself subject to negative regulation, for example via TACE cleavage (La Marca et al. 2011). In addition, SC expression of *ErbB2* (Cohen et al. 1992; Jin et al. 1993) and *Akt* (Heller et al. 2014; Sheean et al. 2014) is markedly down-regulated with myelination. Furthermore, negative regulators of signaling, for example of the PI3K/Akt/mTOR pathway, also control myelin thickness (Cotter et al. 2010; Noseda et al. 2013; Norrmén et al. 2014). Paradoxically, these inhibitors include components of the basal lamina, which are required for initial SC differentiation, radial sorting, and myelination, but also constrain the amount of myelin that forms, preventing hypermyelination. Laminin interactions in the abaxonal compartment activate Sgk1 (Heller et al. 2014) and inhibit PKA, which is activated by NRG1 (Heller et al. 2014; Ghidinelli et al. 2017), to limit myelination of small fibers. Collagen VI in the basal lamina also limits hypermyelination by regulating FAK, Akt, and Erk signaling (Michailov et al. 2004; Chen et al. 2014).

Internode length is regulated independently of sheath thickness (Tricaud 2018) and NRG1 levels (Michailov et al. 2004). Much of the expansion of the internode during development results from axon elongation dictated by growth of the developing limbs (Hildebrand et al. 1994). Thus, experimental manipulations that elongate limbs increase internode length (Abe et al. 2004; Simpson et al. 2013), whereas those that constrain limb growth during development reduce internode length (Hildebrand et al. 1989; Jacobs and Myers 1993). This linkage may reflect adhesive interactions that bind SCs to axons and mechanotransduction signals that are activated during limb growth (Tricaud 2018). Internode length improves conduction speed until a plateau is reached (Simpson et al. 2013). Conversely, the short internodes characteristic of remyelinated segments (Scherer and Salzer 2001) reduce nerve conduction velocities. Such short, remyelinated segments may result from myelin formation around adult axons that are no longer actively elongating.

Formation of the myelin sheath represents a dramatic expansion of the SC plasma membrane. By some estimates, myelinating SCs may generate up to 20 mm^2 of membrane around the largest axons—some 2000 times that of a typical epithelial cell (Kidd et al. 2013). The rapid membrane production that occurs during myelination requires coordination of gene transcription, protein translation, directed trafficking and insertion of myelin components to sites of membrane addition, and mechanisms to establish the proper stoichiometry of these myelin components. The mechanisms that coordinate these various cell biological processes during myelination are poorly understood. Production of this huge expanse of membrane imposes significant stresses upon the SCs, including in the ER, rendering them especially vulnerable to mutations that affect the stoichiometry, the folding and/or the trafficking of myelin proteins (Scherer and Svaren 2023).

ORGANIZATION OF THE MYELINATING SCHWANN CELL AND ITS MYELIN SHEATH

Myelinating SCs are radially and longitudinally polarized cells (Salzer 2003; Özçelik et al. 2010; Pereira et al. 2012; Tricaud 2018). With myelination, SCs organize into distinct membrane domains, each with a unique array of proteins, and a communicating set of cytoplasmic compartments (Fig. 2). Longitudinal (axial) polarity is evident by the overall organization of the myelinating SCs and axons into nodal, paranodal, juxtaparanodal, and internodal domains (Rasband and Peles 2023).

Radial polarity is characterized by the distinct compositions of the adaxonal and abaxonal membranes, which lie on either side of the SC—with the compacted membranes of the myelin sheath in between. The adaxonal membrane, which is separated from the axolemma in the

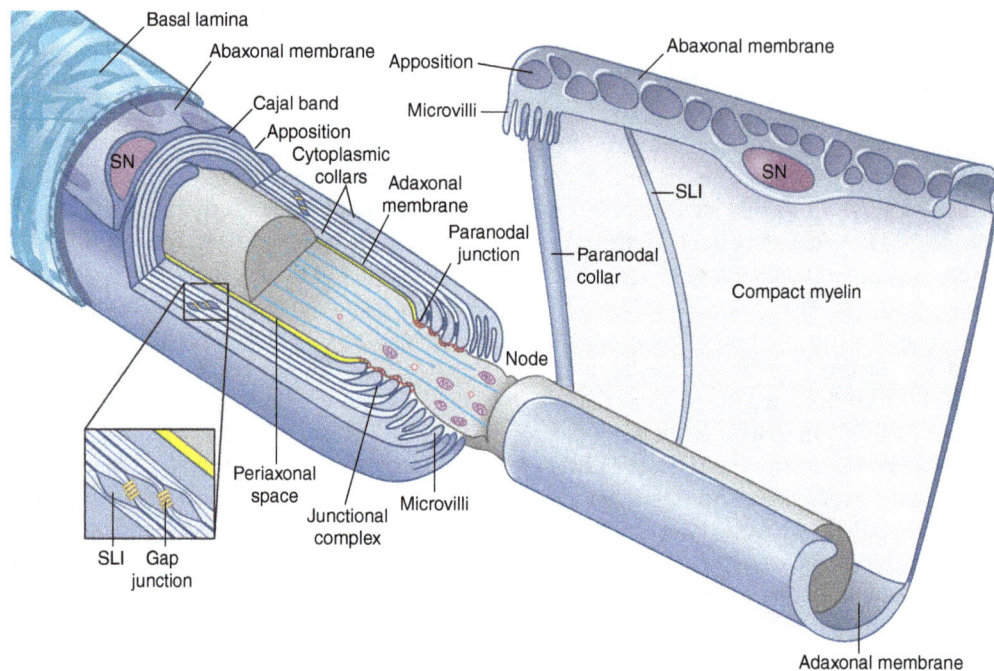

Figure 2. Organization of myelinating Schwann cells (SCs). Schematic organization of myelinating SCs (blue) surrounding an axon (gray); the *left* cell is shown in longitudinal cross-section and the *right* cell is shown unwrapped. Myelinating SCs are surrounded by a basal lamina (shown only on the *left*), which is in direct contact with the abaxonal membrane. The abaxonal compartment contains the SC nucleus (SN) and is divided into Cajal bands and periodic appositions that form between the abaxonal membrane and outer turn of compact myelin. The SC adaxonal membrane is separated from the axonal membrane by the periaxonal space (shown in yellow). Compact myelin is interrupted by Schmidt Lanterman incisures (SLIs), which retain cytoplasm and are enriched in gap and other junctions; a similar autotypic junctional complex of adherens, tight and gap junctions form between the apposed membranes of the paranodal loops. Also shown are the paranodal loops and junctions (red) and the SC microvilli contacting the axon at the node. The axon diameter is reduced in the region of the node and paranodes. (This figure legend is reprinted from Salzer 2015 © *Cold Spring Harbor Perspectives in Biology*; the *left* side of the image is adapted, with permission, from Salzer 2003; the *right* side is adapted, with permission, from Nave 2010.)

internode by a gap of ~15 nm (the periaxonal space), is enriched in adhesion molecules— the Myelin-associated glycoprotein (MAG) and members of the CADM (Necl) family of proteins (Maurel et al. 2007; Spiegel et al. 2007); the latter are tethered to a 4.1G-based cytoskeleton in the SC. These cell adhesion molecules (CAMs) in the inner SC membrane likely maintain spacing and interactions with the axon during wrapping of promyelinating and myelinating SCs. CAMs that mediate interactions with other domains are discussed in detail elsewhere (Salzer et al. 2008; Rasband and Peles 2023). The adaxonal SC membrane also contains solute carriers that maintain the periaxonal space (Marshall-Phelps et al. 2020) and monocarboxylate transporters that contribute to axo-glial metabolic exchange (Doménech-Estevez et al. 2015). SC receptors for axonal signals (e.g., NRG1 and PrP) are also likely to be expressed in the adaxonal membrane, although this remains to be convincingly demonstrated.

The outer, abaxonal membrane lies adjacent to and interacts with laminin and other components of the basal lamina via integrins (initially via α6β1 and later via α6β4), dystroglycan, and Gpr126. Unlike the inner cytoplasmic compartment, which is uniformly spaced, the outer cytoplasmic compartment is interrupted by periodic appositions between the abaxonal membrane and the outer wrap of the compact myelin sheath (Fig. 2). These appositions are enriched in a dystroglycan complex linked via dystrophin-related protein 2 (Drp2) to periaxin (Sherman et al. 2012); they delineate a network of anastomosing, cytoplasmic channels termed Cajal bands (Court et al. 2004) that depend on expression of periaxin and dystroglycan and signaling from Gpr56 (Court et al. 2004; Ackerman et al. 2018). These cytoplasmic channels provide a conduit for centrifugal transport of RNA and proteins originating in the cell soma en route to the paranodal collar (Gould and Mattingly 1990; Court et al. 2004). Consistent with their proposed role in transport, Cajal bands are enriched in large arrays of cytoskeletal proteins, including microtubules, intermediate filaments, utrophin family members (Court et al. 2009, 2011), and an actin/spectrin complex (Susuki et al. 2011; Kidd et al. 2013; Walko et al. 2013). They also represent sites of localized protein synthesis (Gould and Mattingly 1990), intracellular signaling from laminin receptors (Heller et al. 2014), and based on their enrichment in caveolae (Mugnaini et al. 1977; Köling 1985; Mikol et al. 1999), may function in the uptake of metabolites from the extracellular space.

In between these two membranes is the multilamellar myelin sheath—the result of compaction of the circumferentially wrapped, SC plasma membrane. Myelin provides a high resistance, low capacitance sheath essential for impulse propagation by saltatory conduction. The myelin sheath itself is comprised of 40 or more lamellae (Peters et al. 1991). Electron micrographs of compact myelin reveal interperiod lines, which represent the appositions of the extracellular leaflets, alternating with major dense lines, representing the tight apposition of the cytoplasmic leaflets (Fig. 3).

Myelin has a unique composition among plasma membrane equivalents. It has an unusually high lipid content (~70%), which includes galactosphingolipids, certain phospholipids (correspondingly named sphingomyelin), saturated long-chain fatty acids, and, in particular, cholesterol—which is required for myelin sheath assembly (Saher and Simons 2010). In addition, myelin is highly enriched in relatively few proteins (Fig. 3). Like central nervous system (CNS) myelin proteins (Toyama et al. 2013), these are likely to be quite stable in the mature sheath. The most abundant SC myelin protein by far is P0, which is encoded by the Myelin protein zero gene *MPZ*. P0 is a transmembrane adhesion molecule of the immunoglobulin superfamily and constitutes nearly half of the protein mass of PNS myelin (Siems et al. 2020). P0 promotes apposition of the extracellular leaflets via homophilic adhesion (Fig. 3; Filbin et al. 1990; Shapiro et al. 1996). Another key set of proteins are isoforms of myelin basic protein (MBP), which constitute ~20% of myelin protein mass. MBP is a peripheral membrane protein that neutralizes the charges of membrane phospholipids, in particular of phosphatidylserines (Kidd et al. 2013). MBP and the cytoplasmic segment of P0 together promote apposition of the intracellular leaflets to form the major dense line (Martini et al. 1995). Other compact myelin proteins include several tetraspanins, notably PMP22, and periaxin, which are frequently mutated in inherited neuropathies (Scherer and Svaren 2023). Mass spectroscopy of myelin-enriched fractions has identified many additional proteins that are present at much lower levels (Siems et al. 2020). Characterization of these proteins is, as yet, incomplete, including whether they are localized to compact myelin or to other sites in the SC. A number of these have been implicated in the etiology of inherited neuropathies (Siems et al. 2020).

Compact myelin is interrupted by Schmidt–Lanterman incisures (SLIs) (Fig. 2). These are interspersed along the internode and are most abundant in heavily myelinated, large-diameter fibers. SLIs (or clefts) provide a conduit for communication between the inner and outer collars of cytoplasm via the presence of gap junctions, which are arrayed between adjacent membranes (Balice-Gordon et al. 1998). Adjacent membranes in the SLIs also harbor autotypic adherens junctions (Fannon et al. 1995), tight junctions, and a series of PDZ-domain-containing proteins, including the MAGUKs, that are linked to the 4.1G cytoskeleton (Terada et al. 2019) and F-actin (Trapp et al. 1989). SLIs also show enrichment of

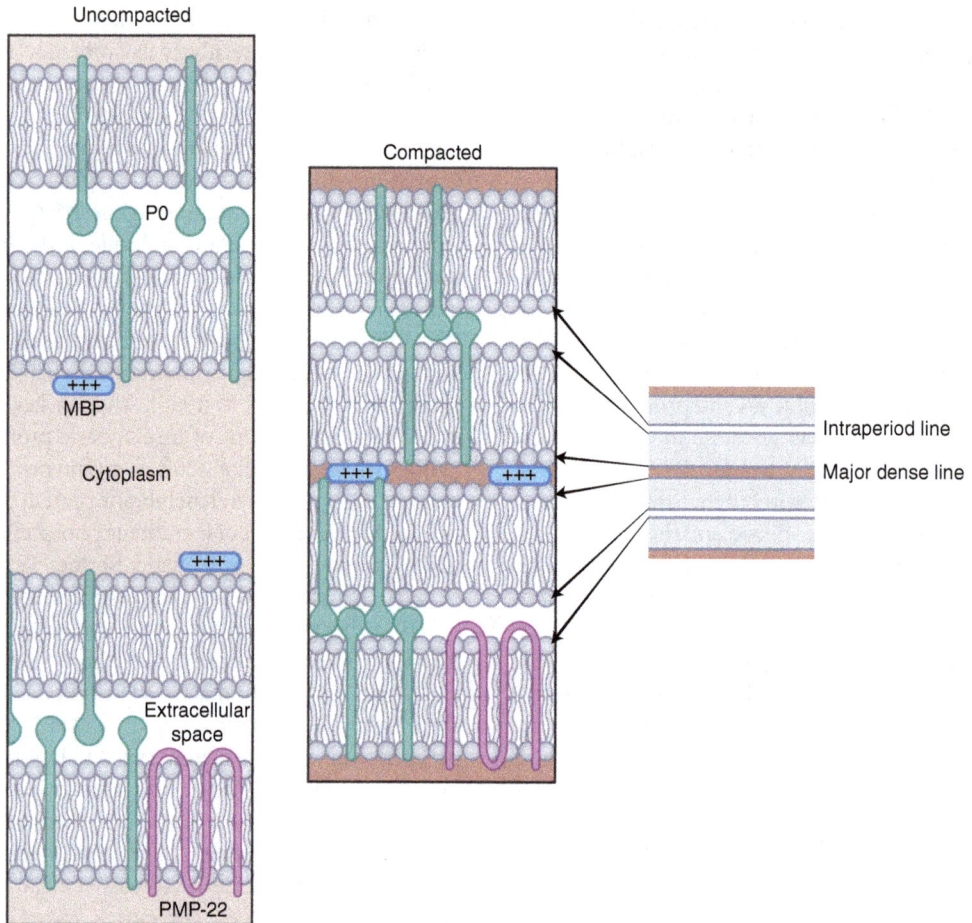

Figure 3. Schematic organization of the compact myelin sheath. Myelin forms, initially, as loose wraps of uncompacted membrane. With the onset of myelin transcription, myelin proteins are up-regulated including P0, myelin basic protein (MBP), and PMP22, which are shown diagrammatically. Compaction is mediated by extracellular interactions of P0 tetramers on one membrane interacting with P0 tetramers on the opposing membrane, shown here only as dimer/dimer interactions for simplicity. Compaction of the cytoplasmic leaflets is mediated by electrostatic interactions of MBP with the phospholipid bilayer supplemented by the interactions with the cytoplasmic tail of P0. Compact myelin appears in electron microscopy (EM) as a major dense line (MDL) (representing tight apposition of the cytoplasmic leaflets) alternating with the intraperiod lines (representing the two apposed extracellular leaflets) as shown. (This figure and figure legend is reprinted from Salzer 2015 © *Cold Spring Harbor Perspectives in Biology*.)

the tyrosine kinase Src (Terada et al. 2013) and are likely to be a site of autocrine signaling (Heller et al. 2014), consistent with their rich array of cell junctions. Many of the adhesion molecules present in the adaxonal membrane (i.e., MAG and the Necls [Cadms]) are also enriched in the SLIs (Maurel et al. 2007; Spiegel et al. 2007) where they likely contribute to the integrity of these structures. As these clefts develop after the myelin

sheath has substantially formed (Small et al. 1987), they are not required for myelin formation, but rather may assist in its maintenance (Gillespie et al. 2000).

In keeping with its polarized organization, components of classical polarity complexes are enriched at distinct sites of myelinating SCs (Masaki 2012; Pereira et al. 2012; Tricaud 2018). The mechanisms that drive the targeting of these complexes to their respective intracellular sites, the regulation of their activity, and their precise roles during myelination in vivo are incompletely understood.

SCHWANN CELLS REGULATE AXON SURVIVAL, DIAMETER, AND TRANSPORT

The interactions between axons and SCs are bidirectional, and SCs have a profound impact on the biology, function, and integrity of axons. As noted, SCs support distinct modes of axon ensheathment and axon conduction. The separation of axons by Remak SCs serves to minimize epiphatic conduction. Myelinating SCs enable saltatory conduction via their elaboration of the myelin sheath and also by their dramatic reorganization of the content and distribution of axonal surface proteins, including the accumulation of ion channels at nodes of Ranvier (Rasband and Peles 2023). SCs also coordinate the injury response by transdifferentiating into repair SCs that promote axonal regeneration in the PNS (Stassart et al. 2023). Here, we briefly consider the role of SCs in supporting axon integrity and regulating axonal diameter and transport.

Both myelin-forming and Remak SCs have an essential role in supporting axon integrity and neuronal survival during early development, postnatally, and in the adult (Harty and Monk 2017; Bosch-Queralt et al. 2023). A fundamental role of SCs is to ensheath axons—a highly conserved function of glia throughout evolution (Nave and Werner 2021). Among the key roles of glial ensheathment is the metabolic support of axons (Simons et al. 2023). Because axons extend a significant distance from their cell bodies—up to 1 meter in humans—they depend on axonal transport, an energy-dependent process, to survive. This renders axons vulnerable, so that axon degeneration is an early event in many neurological diseases. SCs are evenly distributed along the entire length of the axon and are therefore ideally poised to provide local support to axons, both structurally and metabolically. This support is independent from myelination and is believed to rely on the intimate apposition of the SC and axon membranes at various sites. Indeed, similar to oligodendrocytes in the CNS that transfer lac-

tate and pyruvate to axons to support their metabolic needs (Nave and Werner 2021), SCs utilize monocarboxylate transporters to transfer lactate and pyruvate to PNS axons (Doménech-Estevez et al. 2015; Bouçanova and Chrast 2020; Bouçanova et al. 2021; Jia et al. 2021; Deck et al. 2022). Presumably, SCs in close contact with axons sense and respond to their metabolic status. Although the molecular mechanisms that mediate this sensing and energy delivery are still largely unknown, they are the subject of intense investigation.

A striking example of the dependence of axons on SCs is the massive loss of axons and neurons during development that results from depletion of immature SCs due to disruption of the NRG1/ErbB signaling axis (Meyer and Birchmeier 1995; Riethmacher et al. 1997). In addition to loss of metabolic support, these effects may result from loss of trophic signals from immature SCs, which are a rich source of growth factors including neurotrophins (Davies 1998; Scherer and Salzer 2001). Loss of SCs may also impair the targeting and peripheral innervation of axons thereby depriving them of access to their normal, target-derived neurotrophic factors (Birchmeier 2009; Bosch-Queralt et al. 2023).

In addition to prospective metabolic support of axons, myelinating SCs have long been known to promote radial expansion of the underlying axon (Aguayo et al. 1979; Windebank et al. 1985; de Waegh et al. 1992), which in turn increases nerve conduction velocity (Gasser and Grundfest 1939). Radial expansion is locally regulated and is most evident along the internode of large PNS fibers. The local nature of this expansion is evident in the reduced diameter of the axon—underneath regions of noncompact myelin (e.g., SLIs [Price et al. 1993] and in the node and paranodes) where the diameter may be as little as 20% of that of the internode of large axons (Berthold 1996). Reduced diameter of the axon at the node may enhance conduction velocity by reducing the surface area and thereby the capacitance of the nodal axolemma (Halter and Clark 1993; Johnson et al. 2015).

Radial expansion is driven in part by increasing the spacing of the neurofilament (NF) cytoskeleton via increased numbers and phosphorylation of NF subunits (Yuan et al. 2017; see also

Dale and Garcia 2012 for an alternate view). In agreement, NF numbers and phosphorylation are increased along the myelinated internode (de Waegh et al. 1992; Cole et al. 1994) and are reduced, less phosphorylated, and more tightly packed at the node (Price et al. 1993; Hsieh et al. 1994). Axon diameter and NF phosphorylation are significantly reduced in dysmyelinating mouse mutants (de Waegh and Brady 1990, 1991; de Waegh et al. 1992; Kirkpatrick and Brady 1994; Brady et al. 1999; Martini 2001). It is not yet known whether myelinating SCs also impact other mechanisms that regulate axon diameter for example, via the submembranous actin cytoskeleton (Leite et al. 2016) and myosin II contractility (Fan et al. 2017; Costa et al. 2018).

Signals on SCs that regulate axon diameter include MAG and CMTM6 (chemokine-like factor-like MARVEL-transmembrane domain-containing family member-6); it is likely that other signals remain to be identified (Nave and Werner 2021). MAG expression modestly increases NF phosphorylation and axon caliber by activating axonal Cdk5 (Yin et al. 1998; Eichel et al. 2020). Another adaxonal SC protein, CMTM6 is up-regulated with myelination and interacts with MAG on the adaxonal membrane. However, CMTM6 constrains the expansion of axons during myelination and is required to prevent overexpansion of axons in adult nerves (Eichel et al. 2020). Surprisingly, NF packing density and phosphorylation are unaltered in the axons of CMTM6 knockouts, suggesting alternate mechanisms drive their expanded axon diameters. Enlarged axons in the Remak fibers of the *Cmtm6* knockouts indicate that nonmyelinating SCs also regulate axon diameter.

In addition, SC myelination regulates axonal transport. The rates of slow and fast axonal transport are reduced in the nodal and paranodal regions (de Waegh et al. 1992; Zimmermann 1996; Salzer 2003). Indeed, more than 90% of the membranous organelles of the entire axon accumulate in the nodes and paranodes (Berthold et al. 1993). Reduced transport rates in the nodal region were evident by live imaging organelle transport in myelinated axons of *Xenopus laevis* (Cooper and Smith 1974) and labeled (fluorescent) NFs in myelinating cocultures (Monsma et al. 2014). It is also consistent with the accumulation of meta-bolically labeled, axonal glycoproteins at this site (Armstrong et al. 1987). The mechanism(s) responsible for the reduction of transport in the nodal/paranodal region are not known but may result from the constricted axon diameter at this site (Zimmermann 1996) and potentially from alterations of microtubule–motor interactions and organization. Axon–SC interactions at the paranodal junctions are also known to regulate local axonal transport (Zhang et al. 2020; Ishibashi and Baba 2022), although the mechanisms involved remain to be elucidated.

SCHWANN CELLS REGULATE THE ARCHITECTURE OF PERIPHERAL NERVES

SCs are the most abundant cells of the PNS, corresponding to ~70% of all cells (Stierli et al. 2018). SCs interact with many of the cellular components of peripheral nerves to regulate their development and pathology (see Taveggia and Feltri 2022; Bosch-Queralt et al. 2023 for recent reviews). This includes regulating their organization into various compartments (e.g., the endoneurium, the perineurium, and the epineurium) (Fig. 4). These compartments, with their distinct cellular and ECM compositions, provide critical mechanical and metabolic support for nerves. The cellular components of the endoneurial compartment include bundles of axon–SC units with interspersed EFs and a collagenous ECM (Richard et al. 2012). Remak bundles and myelinated axons form fascicles in the endoneurium that are enclosed by the perineurium, which contributes to the formation of the blood–nerve barrier (BNB) (Rechthand and Rapoport 1987; Iwanaga et al. 2022).

SC secretion of the morphogen Desert hedgehog (Dhh) is required during development to form the perineurium and its BNB, the epineurium, and for accumulation of endoneurial collagen (Parmantier et al. 1999; Jessen and Mirsky 2019). Dhh also suppresses the postnatal formation of mini-fascicles that are otherwise generated by EFs (Parmantier et al. 1999; Zotter et al. 2022). The hedgehog-activated transcription factor Gli1 is present in epineurial and perineurial cells, EFs, and pericytes, and regulates endoneurial development (Zotter et al. 2022). Sur-

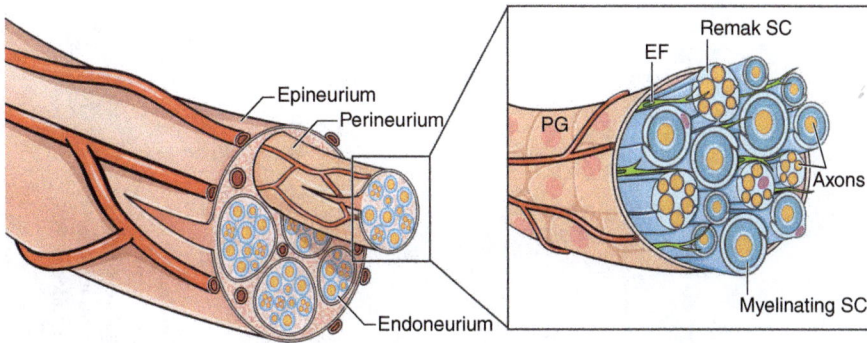

Figure 4. Architecture of peripheral nerves. Schematic of a peripheral nerve highlighting the three cellular compartments: the epineurium, perineurium, and endoneurium. The perineurium is comprised of layers of perineurial glia (PNG) that surround the endoneurium, which, as shown on the *right*, contain large axons (orange) that are myelinated (myelin sheath is dark blue), smaller axons (orange) that are ensheathed by Remak Schwann cells (SCs), and scattered endoneurial fibroblasts (EFs) (green). These fascicles are joined together by the epineurium. The vasculature, including small arterioles of the vasa nervorum, is also shown (red). (This figure is adapted from Zotter et al. 2022 and reprinted with permission from the author(s) © 2022.)

prisingly, myelinating SCs regulate *Gli1* expression independent of Dhh, suggesting other SC signals cooperate with Dhh to control postnatal architecture. Finally, myelinating SCs regulate the formation of the peripheral nerve vasculature via both positive (Mukouyama et al. 2005) and negative (Taïb et al. 2022) signals.

TRANSCRIPTIONAL AND EPIGENETIC REGULATION OF SCHWANN CELL DEVELOPMENT AND MYELINATION

Each step of SC development, from the specification of neural crest cells into SC precursors to their progression through the SC lineage culminating in myelination, is controlled by transcription factors and epigenetic regulators (Fig. 5; Jacob et al. 2011a; Jacob 2015, 2017; Ma and Svaren 2018; Sock and Wegner 2019; Duman et al. 2020a). Epigenetic regulators comprise nucleosome-remodeling complexes, DNA methylation enzymes, histone modifiers, and noncoding RNAs.

Transcriptional Control of SC Specification and Progression to the Immature SC Stage

Many transcription factors are sequentially expressed at different stages of the SC lineage to control the progression of SC development.

However, the transcription factor Sox10 is required for each developmental step in the SC lineage (Kuhlbrodt et al. 1998; Britsch et al. 2001; Paratore et al. 2001; Schreiner et al. 2007; Finzsch et al. 2010; Fröb et al. 2012; Weider and Wegner 2017). Therefore, Sox10 can be considered as one of the main drivers of SC identity and differentiation. Although Sox10 is essential for the specification of the SC lineage (Britsch et al. 2001; Paratore et al. 2001), it is present in all neural crest cells, thus additional factors are required. Indeed, the two highly homologous class I histone deacetylases HDAC1 and HDAC2 (HDAC1/2) interact with Sox10 to promote the activation of the *Pax3* promoter (Jacob et al. 2014). Pax3 is a transcription factor that acts in synergy with Sox10 to activate the expression of several genes including *Sox10* itself (Werner et al. 2007; Wahlbuhl et al. 2012) and the early markers of the SC lineage *Fabp7* and *P0*. In the absence of HDAC1/2 or of Pax3, *Fabp7* and *P0* are not expressed, *Sox10* expression is strongly reduced, and SC precursors are not formed (Jacob et al. 2014). HDAC1/2 and Pax3 are thus essential for the specification of neural crest cells into the SC lineage. In addition to Pax3, Fabp7 and P0, Sox10 also activates the expression of the NRG1 receptor *ErbB3* in neural crest cells, which promotes the specification into SC precursors

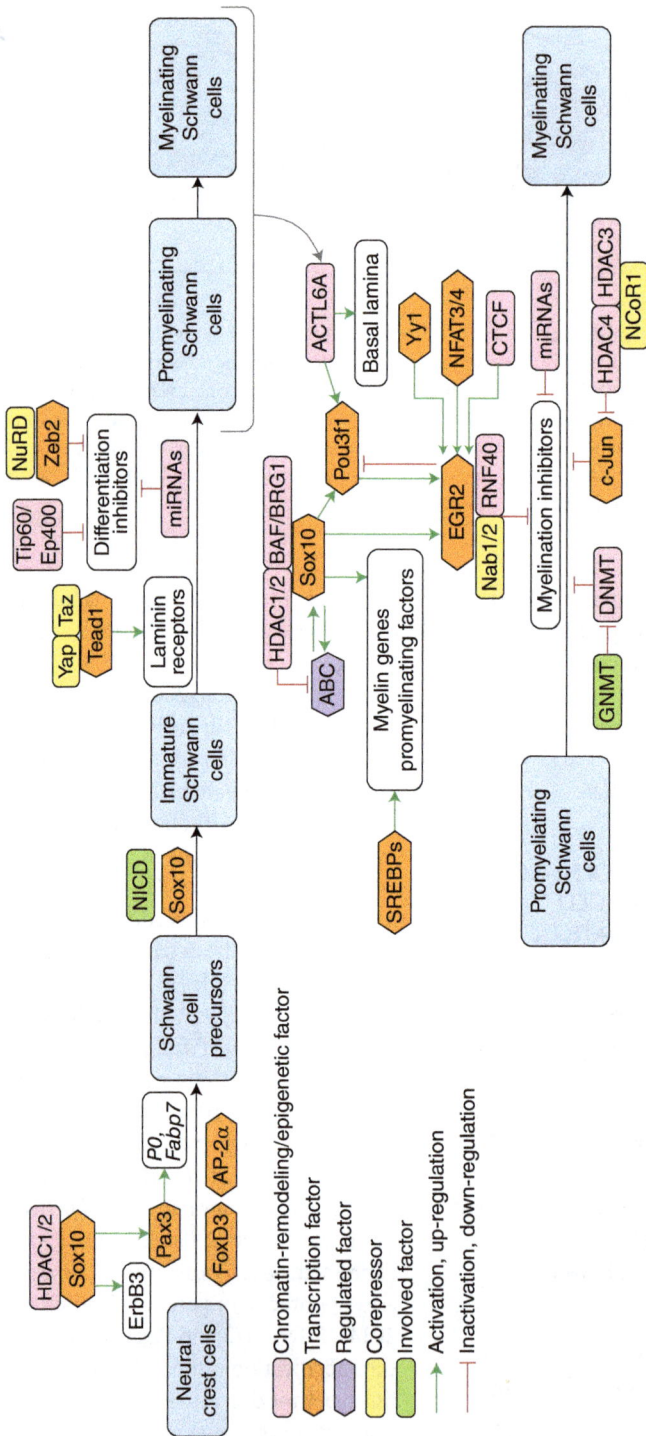

Figure 5. Transcriptional and epigenetic regulation of Schwann cell (SC) development. Schematic representation of the transcriptional (orange hexagons) and epigenetic (pink ovals) regulation of the different stages of SC development from neural crest cells to myelinating SCs is shown. Positive (green arrows) and negative/inhibitory (red arrows) regulatory effects are indicated. Additional regulatory factors, including corepressors (yellow) are also shown. The differentiation into non-myelinating (Remak) SCs and the maintenance of the myelinating SC transcriptional state in adults are not included in this figure for reasons of simplicity.

(Britsch et al. 2001; Adameyko et al. 2009; Prasad et al. 2011).

As noted, SC precursors are directly generated from multipotent neural crest cells. They can also originate from an intermediate progenitor, which is already PNS-committed, such as boundary cap cells. In this case, specification into SC precursors occurs in cells where the expression of the transcription factor *FoxD3* is maintained (Nitzan et al. 2013). The manipulation of other factors such as the down-regulation of the homeodomain transcription factor Hmx1 or the overexpression of constitutively active Notch intracellular domain (NICD) in neural crest cells leads to reduced neurogenesis in favor of SC precursors (Morrison et al. 2000; Adameyko et al. 2009), suggesting a potential involvement of these factors in regulating the generation of SC precursors. While the transcription factors Sox10, Pax3, FoxD3, and AP-2α continue to be expressed by neural crest cells as they specify into SC precursors, other key neural crest transcription factors such as Sox9, Snail, and Slug are down-regulated (Del Barrio and Nieto 2002; Spokony et al. 2002). Before delaminating from the dorsal neural tube and migrating to their final destination to give rise to different cell types, neural crest cells first need to undergo epithelial-to-mesenchymal transition (EMT) (Theveneau and Mayor 2012). Sox9, Snail, and/or Slug are necessary or synergize with other factors to induce EMT (Locascio et al. 2002; Cheung et al. 2005) but need to be down-regulated in migrating neural crest cells giving rise to most derived lineages including SC precursors (Del Barrio and Nieto 2002; Spokony et al. 2002). Delamination from the dorsal neural tube is also controlled by transcription factors: Slug, Sox9, Sox10, and FoxD3 cooperate to induce a transient cadherin switch from N-cadherin and cadherin 6B to cadherins 7 and 11, which is necessary to trigger delamination (Theveneau and Mayor 2012).

SC precursors retain a high level of multipotency; they can further differentiate into immature SCs, but also into several other lineages (for details see Kastriti et al. 2022). Woodhoo et al. (2009) found high levels of NICD in rat peripheral nerves from the embryonic SC precursor stage until the immature SC stage at birth. In this study, the authors show that canonical Notch signaling is not required for the specification into SC precursors but is critical for differentiation of SC precursors into immature SCs; however, Notch effectors prevent further differentiation of the lineage (Woodhoo et al. 2009). In contrast, AP-2α, which persists as neural crest cells progress to SC precursors, is down-regulated in immature SCs and has been shown to repress the progression of the lineage into immature SCs (Stewart et al. 2001).

Gene Regulatory Networks Control the Transition to Promyelinating and Myelinating Stages

The last steps of the differentiation process, namely, differentiation into promyelinating SCs and then into myelinating and nonmyelinating Remak SCs are controlled by several transcription factors, among which Sox10, Egr2, Pou3f1, and Brn-2 hold key functions (Topilko et al. 1994; Bermingham et al. 1996; Jaegle et al. 1996, 2003). These transcription factors are interconnected: Sox10 activates the transcription of Pou3f1 and Brn-2 (Jalagur et al. 2011), which in turn synergize with Sox10 to activate the transcription of Egr2, the master transcription factor of myelination (Murphy et al. 1996; Ghislain and Charnay 2006; Reiprich et al. 2010). Egr2 then synergizes with Sox10 to activate the transcription of myelin protein genes, lipid biosynthesis genes, and genes involved in the formation of the node of Ranvier (Topilko et al. 1994; Leblanc et al. 2005; Saur et al. 2021). Together, these feedforward transcriptional mechanisms promote radial sorting and are essential for SC myelination. Other factors are also prominently involved in the regulation of these processes. Indeed, upon mechanical stimuli, the Hippo pathway effectors— the transcription factor Tead and its coactivators Yap and Taz—induce expression of laminin receptors, which are necessary for radial sorting and myelination (Poitelon et al. 2016; Deng et al. 2017). Simultaneously, the nuclear actin-related protein ACTL6A, which is part of several chromatin-remodeling complexes such as the SWI/SNF complex, acts as a sensor for large-caliber axons and ensures their radial sorting and myelination by inducing the derepression of ECM-related genes to produce SC basal lamina and of promyelinating factors such as Pou3f1 (Park et al. 2022). In addition, the transcription factors Yy1, Nfatc3, and Nfatc4 and the effectors Yap and Taz contribute to the induction of *Egr2* expression (Kao et al. 2009; He et al. 2010; Grove et al. 2017; Weintraub et al. 2017).

As noted, a striking feature of the transcriptional phenotype of SCs is that they can adopt alternate fates as a myelinating versus Remak SC depending on extrinsic signals. This binary distinction is reinforced via transcriptional cross-repression in which Sox2 and c-Jun, two major inhibitors of myelination, inhibit *Egr2* expression and vice versa (Le et al. 2005; Parkinson et al. 2008). Maintenance of the myelinating phenotype requires ongoing expression of both Egr2 and Sox10; conditional inactivation of either results in dedifferentiation, which is particularly rapid in the case of Egr2 and is associated with Sox2 up-regulation (Decker et al. 2006; Bremer et al. 2011). These results indicate that Sox10 and Egr2 are subject to ongoing turnover in mature myelinating SCs and provide the framework for their differentiation into repair SCs following injury (Stassart et al. 2023).

Regulation of the Transcriptional Program of Myelination by Epigenetic and Posttranslational Modifications: HDAC-Dependent Mechanisms

Epigenetic regulation of transcription plays a crucial role in PNS myelination (see Fig. 5 for summary; Pereira et al. 2012). Among the best-characterized modifiers are the HDACs. HDAC1 and HDAC2 are highly homologous and can compensate for the loss of each other; however, they have primary functions when both are expressed. HDAC2 interacts with Sox10 to protect it from degradation and maintain it on its target genes (Duman et al. 2020b) and recruits KDM3A and JMJD2C, two histone H3K9 demethylases, to remove repressive H3K9 methylation marks on Sox10-target genes (Brügger et al. 2017). HDAC1 prevents the activation of β-catenin in early postnatal SCs to protect SCs from apoptosis. Once SCs have entered the myelinating stage, HDAC1 is down-regulated and levels of active β-catenin (ABC) increase, up-regulating Sox10 and Egr2. In turn, Sox10 increases the levels of ABC

and Egr2. This feedforward loop enhances the myelination process (Jacob et al. 2011b) and is likely necessary for timely myelination (Lewallen et al. 2011; Tawk et al. 2011). Sox10 also recruits the nucleosome remodeling complex BAF/BRG1 and interacts with HDAC1/2 to activate the transcription of Pou3f1 and Egr2 and of myelin proteins such as P0 (Jacob et al. 2011b; Weider et al. 2012). In addition, Sox10 interacts with HDAC1/2 to activate its own expression (Jacob et al. 2011b). NF-κB has also been shown to promote Pou3f1 expression (Nickols et al. 2003) and to recruit the BAF/BRG1 complex and HDAC1/2 on the *Sox10* promoter to activate *Sox10* expression (Chen et al. 2011; Limpert et al. 2013); however, ablation of NF-κB in SCs leads to a minimal phenotype (Morton et al. 2013). Therefore, this NF-κB-dependent mechanism is not essential for myelination. In adult SCs, HDAC1/2, presumably through their interaction with Sox10, are necessary to maintain high levels of P0, the integrity of paranodes and nodes of Ranvier, and optimal myelination (Brügger et al. 2015). Two other HDACs, HDAC3 (class I) and HDAC4 (class II), are also involved: HDAC4 has been shown to recruit the repressor complex NcoR1 together with HDAC3 upon cAMP signaling to silence the *Jun* promoter and thereby favor myelination (Gomis-Coloma et al. 2018). However, two other studies on the functions of HDAC3 in SCs show a hypermyelination phenotype in the absence of HDAC3, which can appear somewhat contradictory to a function of HDAC3 in silencing the *Jun* promoter. One of these two studies proposes that HDAC3 replaces HDAC2 at the transition between developmental myelination and myelination maintenance to slow down the activation of the *P0* promoter and the production of myelin while still maintaining myelin homeostasis (Rosenberg et al. 2018). The replacement of HDAC2 by HDAC3 may, however, only be partial since HDAC1/2 are necessary in adult SCs for the maintenance of appropriate levels of P0 (Brügger et al. 2015).

Mechanisms Related to Other Epigenetic Factors

Similar to Sox10, Egr2 recruits histone-modifying enzymes to promote activation of its target genes. Indeed, Egr2-mediated recruitment of the Rnf40-containing E3 ligase leads to histone H2B monoubiquitination of Egr2 target genes. This results in high expression levels of myelin and lipid biosynthesis genes and repression of inhibitors of myelination, which appears essential for the myelination process (Wüst et al. 2020). In addition, the CCCTC-binding factor (CTCF) that mediates chromatin looping contributes to the induction of Egr2 expression and thereby to the transition from the promyelinating to the myelinating stage (Wang et al. 2020).

The NuRD remodeling complex, which comprises the ATPases CHD3 or CHD4 and recruits HDAC1/2, is also required for radial sorting and myelination. The transcription factor Zeb2 recruits the NuRD complex at the immature SC stage to repress the transcription of the inhibitors of differentiation *Sox2*, *Hey2*, and *Ednrb* and thereby favors differentiation into the promyelinating and then to the myelinating stages (Quintes et al. 2016; Wu et al. 2016). In addition, at late developmental myelination stages, Egr2 interacts with the transcriptional repressors NAB1/2, which recruit the NuRD complex to silence the expression of other inhibitors of myelination including *Id2*, *Id4*, and *Jun* (Parkinson et al. 2004; Srinivasan et al. 2006; Mager et al. 2008; Hung et al. 2012). Pou3f1 and Sox2 are also known to be repressed by Egr2 (Zorick et al. 1999; Le et al. 2005; Parkinson et al. 2008), a mechanism likely occurring through the interaction of Egr2 with NAB1/2/NuRD (Srinivasan et al. 2006; Mager et al. 2008). Indeed, Pou3f1 is an intermediate activator of myelination; it is first required to induce the up-regulation of Egr2 but needs to be down-regulated afterward for myelination to proceed (Ryu et al. 2007). In addition to the Egr2-mediated down-regulation of Pou3f1 and Sox2, the chromatin-remodeling complex Tip60/Ep400, which replaces the histones H2A.Z and H3.1 by H2A and H3.3, respectively, is necessary for the timely down-regulation of Pou3f1, Sox2, and other immature SC markers such as AP-2α, Pax3, Sox1, and Sox3. Tip60/Ep400 is also required for terminal differentiation of myelinating and nonmyelinating SCs (Fröb et al. 2019).

To achieve optimal myelination, the regulation of DNA methylation in specific genes is re-quired. During the myelination process, DNA methyltransferases and the methyl donor SAMe are down-regulated, whereas Gadd45a, Gadd45b, and Apobec1, which promote DNA demethylation, and glycine *N*-methyltransferase, which transforms SAMe into *S*-adenosylhomocysteine and *N*-methylglycine, are up-regulated (Varela-Rey et al. 2014). These regulations are necessary for appropriate expression of myelin-related genes and myelin thickness (Varela-Rey et al. 2014).

Finally, microRNAs (miRNAs) have been implicated in regulating SC biology. miRNAs are single-stranded, small noncoding RNAs that can regulate gene expression by inducing the degradation of their target mRNAs. They are processed from hairpin-containing primary transcripts into precursor miRNAs by the microprocessor proteins Drosha and DGCR8, and then into miRNA duplexes by Dicer (Suster and Feng 2021). miRNAs have a critical function in the regulation of radial sorting and myelination and in the maintenance of myelin integrity (Bremer et al. 2010; Pereira et al. 2010; Verrier et al. 2010; Yun et al. 2010; Gokey et al. 2012; Gökbuget et al. 2015, 2018; Lin et al. 2015). Indeed, these studies show that miRNAs down-regulate immature SC factors and prevent inappropriate expression of injury-induced genes in developing and mature SCs.

In addition to promyelinating transcription mechanisms, there are also inhibitory mechanisms that ensure optimal myelin sheath thickness and prevent pathological hypermyelination. Repressive H3K27 trimethylation marks (H3K27me3) have been implicated in these mechanisms. For instance, a nuclear variant of ErbB3 binds to the DNA to induce enrichment of H3K27me3 at the promoter of myelin-related genes, probably through activation of the histone methyltransferase EZH2 (Ness et al. 2016). In addition, the PRC2 complex, which contains EZH2, represses the promoter of *Igfbp2* and thereby prevents Igfbp2-induced activation of Akt, which is an activator of the myelination process (Ma et al. 2015).

Wrapping Up and Looking Forward

There has been substantial progress in the identification of axonal, basal lamina, and mechanotransduction signals that converge on SCs to regulate their development and differentiation. The axon–SC unit functions within the context of the much larger ecosystem of the PNS. There is increasing appreciation that SCs help to coordinate and direct peripheral nerve development. Factors from axons drive SC differentiation, which in turn signal to other cells to direct the development of nerve compartments and vasculature. The best studied of such axon-regulated SC signals is Dhh, but myelinating SCs likely produce other signals that drive morphological changes during differentiation, including Gli1 activation in EFs (Zotter et al. 2022). SCs also contribute to the development and function of other structures and organs, such as the skin, synapses, and the vascular, enteric, and immune systems. Thus, in addition to the myriad of primary peripheral neuropathies, SC pathology is also relevant to cancer, pain, enteric, and dermatological diseases (Taveggia and Feltri 2022).

The extent of reciprocal signaling from other peripheral nerve cell types and compartments back to the axon–SC unit is less well understood. Recent transcriptomic data indicate that a significant number of candidate ligands and their cognate receptor mRNAs are reciprocally expressed by EFs and SCs under homeostatic conditions (Peng et al. 2020; Toma et al. 2020). These include well-characterized growth factors such as nerve growth factor, hepatocyte growth factor, and NRG1 in fibroblasts versus PDGFA and EGF in SCs—in addition to many novel ligands with unknown roles in the PNS. Two candidate factors secreted by fibroblasts that may impinge directly on SCs include the vasodilator peptide adrenomedullin, which was previously shown to up-regulate cAMP expression in SCs (Dumont et al. 2002), and BMP7, previously shown to inhibit myelin gene expression in SCs (Liu et al. 2016). These data highlight the prospective richness of intercellular signaling that may coordinate peripheral nerve development.

The trophic and metabolic support of axons by immature and mature SCs is well known. However, the precise mechanisms involved remain to be established. Given that this trophic support of axons may be relevant to the pathology in primary axonal diseases and to neuropa-thies secondary to demyelination, elucidating the mechanisms involved has significant clinical implications.

Recent transcriptomic data has identified unexpected heterogeneity within SCs that may underlie the different sensitivities of SCs and the resultant clinical presentations of peripheral nerve disease. Understanding the functional significance of SC diversity revealed by recent scRNA-seq studies is an emerging area with important implications for PNS functions in health and disease.

Finally, more work is needed to elucidate how transcription and epigenetic factors are regulated and to better understand the mechanisms these factors control. The integration of these mechanisms into dynamic regulatory networks promises to greatly diversify the possibilities of therapeutic interventions for developmental disorders, for the maintenance of peripheral nerve integrity in adults during aging, and to promote regeneration after injury and pathology.

IN MEMORIUM: LAURA FELTRI

As this article was going to press, Laura Feltri, our co-author and beloved colleague, passed away after a long illness. Trained as a neurologist, Laura focused her career on and made seminal contributions to elucidating the biology and pathobiology of Schwann cells and peripheral nerves. Her many transformative scientific contributions will of necessity be described elsewhere. Together with Larry Wrabetz, her husband and frequent collaborator, she established major centers of PNS research in Milan and Buffalo where they trained many current and next-generation leaders in the field. Despite her illness, Laura courageously assumed key leadership positions, authored a series of major reviews, and completed some of the most important research of her career. We are impoverished by her loss. We dedicate this article to her memory.

ACKNOWLEDGMENTS

The authors gratefully acknowledge their laboratory colleagues, past and present, for their contributions to the studies cited here and Jill Gregory for outstanding artwork. M.L.F. thanks Lawrence Wrabetz for his scientific contributions and making her career possible. Work from the laboratory of J.L.S. has been supported by the NIH and the National Multiple Sclerosis Society. Work in the M.L.F. laboratory has been funded by Telethon, Italy, NIH-NINDS and NICHD, the US Army Department of Defense, the National Multiple Sclerosis Society, the Charcot Marie Tooth Association, the Legacy of Angels foundation, and the European community. Work from the laboratory of C.J. has been supported by the Deutsche Forschungsgemeinschaft, the Swiss National Science Foundation, the International Foundation for Research in Parapegia, the Swiss Multiple Sclerosis Society, the Olga Mayenfisch Foundation, the "Pool de Recherche" of the University of Fribourg, and the Novartis Foundation. The authors regret any omissions in citing relevant publications of other colleagues in this review.

REFERENCES

Reference is also in this subject collection.

Abe I, Ochiai N, Ichimura H, Tsujino A, Sun J, Hara Y. 2004. Internodes can nearly double in length with gradual elongation of the adult rat sciatic nerve. *J Orthop Res* **22**: 571–577. doi:10.1016/j.orthres.2003.08.019

Acheta J, Bhatia U, Haley J, Hong J, Rich K, Close R, Bechler ME, Belin S, Poitelon Y. 2022. Piezo channels contribute to the regulation of myelination in Schwann cells. *Glia* **70**: 2276–2289. doi:10.1002/glia.24251

Ackerman SD, Luo R, Poitelon Y, Mogha A, Harty BL, D'Rozario M, Sanchez NE, Lakkaraju AKK, Gamble P, Li J, et al. 2018. GPR56/ADGRG1 regulates development and maintenance of peripheral myelin. *J Exp Med* **215**: 941–961. doi:10.1084/jem.20161714

Adameyko I, Lallemend F, Aquino JB, Pereira JA, Topilko P, Müller T, Fritz N, Beljajeva A, Mochii M, Liste I, et al. 2009. Schwann cell precursors from nerve innervation are a cellular origin of melanocytes in skin. *Cell* **139**: 366–379. doi:10.1016/j.cell.2009.07.049

Aguayo AJ, Bray GM, Perkins SC. 1979. Axon–Schwann cell relationships in neuropathies of mutant mice. *Ann NY Acad Sci* **317**: 512–531.

Arancibia-Cárcamo IL, Ford MC, Cossell L, Ishida K, Tohyama K, Attwell D. 2017. Node of Ranvier length as a potential regulator of myelinated axon conduction speed. *eLife* **6**: e23329. doi:10.7554/eLife.23329

Armstrong R, Toews AD, Morell P. 1987. Axonal transport through nodes of Ranvier. *Brain Res* **412**: 196–199. doi:10.1016/0006-8993(87)91461-2

Atanasoski S, Scherer SS, Sirkowski E, Leone D, Garratt AN, Birchmeier C, Suter U. 2006. ErbB2 signaling in Schwann cells is mostly dispensable for maintenance of myelinated peripheral nerves and proliferation of adult Schwann cells after injury. *J Neurosci* **26**: 2124–2131. doi:10.1523/JNEUROSCI.4594-05.2006

Bacon C, Lakics V, Machesky L, Rumsby M. 2007. N-WASP regulates extension of filopodia and processes by oligodendrocyte progenitors, oligodendrocytes, and Schwann cells —implications for axon ensheathment at myelination. *Glia* **55**: 844–858. doi:10.1002/glia.20505

Balice-Gordon RJ, Bone LJ, Scherer SS. 1998. Functional gap junctions in the Schwann cell myelin sheath. *J Cell Biol* **142**: 1095–1104. doi:10.1083/jcb.142.4.1095

Benninger Y, Thurnherr T, Pereira JA, Krause S, Wu X, Chrostek-Grashoff A, Herzog D, Nave KA, Franklin RJ, Meijer D, et al. 2007. Essential and distinct roles for cdc42 and rac1 in the regulation of Schwann cell biology during peripheral nervous system development. *J Cell Biol* **177**: 1051–1061. doi:10.1083/jcb.200610108

Bermingham JR, Scherer SS, O'Connell S, Arroyo E, Kalla KA, Powell FL, Rosenfeld MG. 1996. Tst-1/Oct-6/SCIP regulates a unique step in peripheral myelination and is required for normal respiration. *Genes Dev* **10**: 1751–1762. doi:10.1101/gad.10.14.1751

Bermingham JR Jr, Shearin H, Pennington J, O'Moore J, Jaegle M, Driegen S, van Zon A, Darbas A, Ozkaynak E, Ryu EJ, et al. 2006. The claw paw mutation reveals a role for Lgi4 in peripheral nerve development. *Nat Neurosci* **9**: 76–84. doi:10.1038/nn1598

Berthold CH. 1996. Development of nodes of Ranvier in feline nerves: an ultrastructural presentation. *Microsc Res Tech* **34**: 399–421. doi:10.1002/1097-0029(19960801)34:5<399::AID-JEMT1070340502>3.0.CO;2-N

Berthold CH, Fabricius C, Rydmark M, Andersén B. 1993. Axoplasmic organelles at nodes of Ranvier. I: Occurrence and distribution in large myelinated spinal root axons of the adult cat. *J Neurocytol* **22**: 925–940. doi:10.1007/BF01218351

Birchmeier C. 2009. ErbB receptors and the development of the nervous system. *Exp Cell Res* **315**: 611–618. doi:10.1016/j.yexcr.2008.10.035

Bolino A. 2021. Myelin biology. *Neurotherapeutics* **18**: 2169–2184. doi:10.1007/s13311-021-01083-w

Bosch-Queralt M, Fledrich R, Stassart RM. 2023. Schwann cell functions in peripheral nerve development and repair. *Neurobiol Dis* **176**: 105952. doi:10.1016/j.nbd.2022.105952

Bouçanova F, Chrast R. 2020. Metabolic interaction between Schwann cells and axons under physiological and disease conditions. *Front Cell Neurosci* **14**: 148. doi:10.3389/fncel.2020.00148

Bouçanova F, Pollmeier G, Sandor K, Morado Urbina C, Nijssen J, Médard JJ, Bartesaghi L, Pellerin L, Svensson CI, Hedlund E, et al. 2021. Disrupted function of lactate transporter MCT1, but not MCT4, in Schwann cells affects the maintenance of motor end-plate innervation. *Glia* **69**: 124–136. doi:10.1002/glia.23889

Brady ST, Witt AS, Kirkpatrick LL, de Waegh SM, Readhead C, Tu PH, Lee VM. 1999. Formation of compact myelin is required for maturation of the axonal cytoskeleton. *J Neu-*

rosci **19:** 7278–7288. doi:10.1523/JNEUROSCI.19-17-07278.1999

Bremer J, O'Connor T, Tiberi C, Rehrauer H, Weis J, Aguzzi A. 2010. Ablation of Dicer from murine Schwann cells increases their proliferation while blocking myelination. *PLoS ONE* **5:** e12450. doi:10.1371/journal.pone.0012450

Bremer M, Fröb F, Kichko T, Reeh P, Tamm ER, Suter U, Wegner M. 2011. Sox10 is required for Schwann-cell homeostasis and myelin maintenance in the adult peripheral nerve. *Glia* **59:** 1022–1032. doi:10.1002/glia.21173

Brinkmann BG, Agarwal A, Sereda MW, Garratt AN, Müller T, Wende H, Stassart RM, Nawaz S, Humml C, Velanac V, et al. 2008. Neuregulin-1/ErbB signaling serves distinct functions in myelination of the peripheral and central nervous system. *Neuron* **59:** 581–595. doi:10.1016/j.neuron.2008.06.028

Britsch S, Goerich DE, Riethmacher D, Peirano RI, Rossner M, Nave KA, Birchmeier C, Wegner M. 2001. The transcription factor Sox10 is a key regulator of peripheral glial development. *Genes Dev* **15:** 66–78. doi:10.1101/gad.186601

Brügger V, Engler S, Pereira JA, Ruff S, Horn M, Welzl H, Münger E, Vaquié A, Sidiropoulos PN, Egger B, et al. 2015. HDAC1/2-dependent P0 expression maintains paranodal and nodal integrity independently of myelin stability through interactions with neurofascins. *PLoS Biol* **13:** e1002258. doi:10.1371/journal.pbio.1002258

Brügger V, Duman M, Bochud M, Münger E, Heller M, Ruff S, Jacob C. 2017. Delaying histone deacetylase response to injury accelerates conversion into repair Schwann cells and nerve regeneration. *Nat Commun* **8:** 14272. doi:10.1038/ncomms14272

Bunge RP, Bunge MB, Eldridge CF. 1986. Linkage between axonal ensheathment and basal lamina production by Schwann cells. *Annu Rev Neurosci* **9:** 305–328. doi:10.1146/annurev.ne.09.030186.001513

Bunge RP, Bunge MB, Bates M. 1989. Movements of the Schwann cell nucleus implicate progression of the inner (axon-related) Schwann cell process during myelination. *J Cell Biol* **109:** 273–284. doi:10.1083/jcb.109.1.273

Burianek LE, Soderling SH. 2013. Under lock and key: spatiotemporal regulation of WASP family proteins coordinates separate dynamic cellular processes. *Semin Cell Dev Biol* **24:** 258–266. doi:10.1016/j.semcdb.2012.12.005

Chen ZL, Strickland S. 2003. Laminin γ1 is critical for Schwann cell differentiation, axon myelination, and regeneration in the peripheral nerve. *J Cell Biol* **163:** 889–899. doi:10.1083/jcb.200307068

Chen Y, Wang H, Yoon SO, Xu X, Hottiger MO, Svaren J, Nave KA, Kim HA, Olson EN, Lu QR. 2011. HDAC-mediated deacetylation of NF-κB is critical for Schwann cell myelination. *Nat Neurosci* **14:** 437–441. doi:10.1038/nn.2780

Chen P, Cescon M, Megighian A, Bonaldo P. 2014. Collagen VI regulates peripheral nerve myelination and function. *FASEB J* **28:** 1145–1156. doi:10.1096/fj.13-239533

Chernousov MA, Yu WM, Chen ZL, Carey DJ, Strickland S. 2008. Regulation of Schwann cell function by the extracellular matrix. *Glia* **56:** 1498–1507. doi:10.1002/glia.20740

Cheung M, Chaboissier MC, Mynett A, Hirst E, Schedl A, Briscoe J. 2005. The transcriptional control of trunk neural crest induction, survival, and delamination. *Dev Cell* **8:** 179–192. doi:10.1016/j.devcel.2004.12.010

Cohen JA, Yachnis AT, Arai M, Davis JG, Scherer SS. 1992. Expression of the neu proto-oncogene by Schwann cells during peripheral nerve development and Wallerian degeneration. *J Neurosci Res* **31:** 622–634. doi:10.1002/jnr.490310406

Cole JS, Messing A, Trojanowski JQ, Lee VM. 1994. Modulation of axon diameter and neurofilaments by hypomyelinating Schwann cells in transgenic mice. *J Neurosci* **14:** 6956–6966. doi:10.1523/JNEUROSCI.14-11-06956.1994

Cooper PD, Smith RS. 1974. The movement of optically detectable organelles in myelinated axons of *Xenopus laevis*. *J Physiol* **242:** 77–97. doi:10.1113/jphysiol.1974.sp010695

Costa AR, Pinto-Costa R, Sousa SC, Sousa MM. 2018. The regulation of axon diameter: from axonal circumferential contractility to activity-dependent axon swelling. *Front Mol Neurosci* **11:** 319. doi:10.3389/fnmol.2018.00319

Cotter L, Özçelik M, Jacob C, Pereira JA, Locher V, Baumann R, Relvas JB, Suter U, Tricaud N. 2010. Dlg1-PTEN interaction regulates myelin thickness to prevent damaging peripheral nerve overmyelination. *Science* **328:** 1415–1418. doi:10.1126/science.1187735

Court FA, Sherman DL, Pratt T, Garry EM, Ribchester RR, Cottrell DF, Fleetwood-Walker SM, Brophy PJ. 2004. Restricted growth of Schwann cells lacking Cajal bands slows conduction in myelinated nerves. *Nature* **431:** 191–195. doi:10.1038/nature02841

Court FA, Hewitt JE, Davies K, Patton BL, Uncini A, Wrabetz L, Feltri ML. 2009. A laminin-2, dystroglycan, utrophin axis is required for compartmentalization and elongation of myelin segments. *J Neurosci* **29:** 3908–3919. doi:10.1523/JNEUROSCI.5672-08.2009

Court FA, Zambroni D, Pavoni E, Colombelli C, Baragli C, Figlia G, Sorokin L, Ching W, Salzer JL, Wrabetz L, et al. 2011. MMP2-9 cleavage of dystroglycan alters the size and molecular composition of Schwann cell domains. *J Neurosci* **31:** 12208–12217. doi:10.1523/JNEUROSCI.0141-11.2011

Dale JM, Garcia ML. 2012. Neurofilament phosphorylation during development and disease: which came first, the phosphorylation or the accumulation? *J Amino Acids* **2012:** 382107.

Davies AM. 1998. Neuronal survival: early dependence on Schwann cells. *Curr Biol* **8:** R15–18. doi:10.1016/S0960-9822(98)70009-0

Deck M, Van Hameren G, Campbell G, Bernard-Marissal N, Devaux J, Berthelot J, Lattard A, Médard JJ, Gautier B, Guelfi S, et al. 2022. Physiology of PNS axons relies on glycolytic metabolism in myelinating Schwann cells. *PLoS ONE* **17:** e0272097. doi:10.1371/journal.pone.0272097

Decker L, Desmarquet-Trin-Dinh C, Taillebourg E, Ghislain J, Vallat JM, Charnay P. 2006. Peripheral myelin maintenance is a dynamic process requiring constant Krox20 expression. *J Neurosci* **26:** 9771–9779. doi:10.1523/JNEUROSCI.0716-06.2006

Del Barrio MG, Nieto MA. 2002. Overexpression of snail family members highlights their ability to promote chick neural crest formation. *Development* **129:** 1583–1593. doi:10.1242/dev.129.7.1583

Cite this article as *Cold Spring Harb Perspect Biol* doi: 10.1101/cshperspect.a041360

Deng Y, Wu LMN, Bai S, Zhao C, Wang H, Wang J, Xu L, Sakabe M, Zhou W, Xin M, et al. 2017. A reciprocal regulatory loop between TAZ/YAP and G-protein Gαs regulates Schwann cell proliferation and myelination. *Nat Commun* **8:** 15161. doi:10.1038/ncomms15161

de Waegh S, Brady ST. 1990. Altered slow axonal transport and regeneration in a myelin-deficient mutant mouse: the trembler as an in vivo model for Schwann cell-axon interactions. *J Neurosci* **10:** 1855–1865. doi:10.1523/JNEUROSCI.10-06-01855.1990

de Waegh SM, Brady ST. 1991. Local control of axonal properties by Schwann cells: neurofilaments and axonal transport in homologous and heterologous nerve grafts. *J Neurosci Res* **30:** 201–212. doi:10.1002/jnr.490300121

de Waegh SM, Lee VM, Brady ST. 1992. Local modulation of neurofilament phosphorylation, axonal caliber, and slow axonal transport by myelinating Schwann cells. *Cell* **68:** 451–463. doi:10.1016/0092-8674(92)90183-D

Domènech-Estevez E, Baloui H, Repond C, Rosafio K, Medard JJ, Tricaud N, Pellerin L, Chrast R. 2015. Distribution of monocarboxylate transporters in the peripheral nervous system suggests putative roles in lactate shuttling and myelination. *J Neurosci* **35:** 4151–4156. doi:10.1523/JNEUROSCI.3534-14.2015

Domènech-Estevez E, Baloui H, Meng X, Zhang Y, Deinhardt K, Dupree JL, Einheber S, Chrast R, Salzer JL. 2016. Akt regulates axon wrapping and myelin sheath thickness in the PNS. *J Neurosci* **36:** 4506–4521. doi:10.1523/JNEUROSCI.3521-15.2016

Duman M, Martinez-Moreno M, Jacob C, Tapinos N. 2020a. Functions of histone modifications and histone modifiers in Schwann cells. *Glia* **68:** 1584–1595. doi:10.1002/glia.23795

Duman M, Vaquié A, Nocera G, Heller M, Stumpe M, Siva Sankar D, Dengjel J, Meijer D, Yamaguchi T, Matthias P, et al. 2020b. EEF1A1 deacetylation enables transcriptional activation of remyelination. *Nat Commun* **11:** 3420. doi:10.1038/s41467-020-17243-z

Dumont CE, Muff R, Flühmann B, Fischer JA, Born W. 2002. Paracrine/autocrine function of adrenomedullin in peripheral nerves of rats. *Brain Res* **955:** 64–71. doi:10.1016/S0006-8993(02)03365-6

Eichel MA, Gargareta VI, D'Este E, Fledrich R, Kungl T, Buscham TJ, Lüders KA, Miracle C, Jung RB, Distler U, et al. 2020. CMTM6 expressed on the adaxonal Schwann cell surface restricts axonal diameters in peripheral nerves. *Nat Commun* **11:** 4514. doi:10.1038/s41467-020-18172-7

Fan A, Tofangchi A, Kandel M, Popescu G, Saif T. 2017. Coupled circumferential and axial tension driven by actin and myosin influences in vivo axon diameter. *Sci Rep* **7:** 14188. doi:10.1038/s41598-017-13830-1

Fannon AM, Sherman DL, Ilyina-Gragerova G, Brophy PJ, Friedrich VL Jr, Colman DR. 1995. Novel E-cadherin-mediated adhesion in peripheral nerve: Schwann cell architecture is stabilized by autotypic adherens junctions. *J Cell Biol* **129:** 189–202. doi:10.1083/jcb.129.1.189

Feltri ML, Wrabetz L. 2005. Laminins and their receptors in Schwann cells and hereditary neuropathies. *J Peripher Nerv Syst* **10:** 128–143. doi:10.1111/j.1085-9489.2005.0010204.x

Feltri ML, Poitelon Y, Previtali SC. 2016. How Schwann cells sort axons: new concepts. *Neuroscientist* **22:** 252–265. doi:10.1177/1073858415572361

Fernandez-Valle C, Gorman D, Gomez AM, Bunge MB. 1997. Actin plays a role in both changes in cell shape and gene-expression associated with Schwann cell myelination. *J Neurosci* **17:** 241–250. doi:10.1523/JNEUROSCI.17-01-00241.1997

Filbin MT, Walsh FS, Trapp BD, Pizzey JA, Tennekoon GI. 1990. Role of myelin Po protein as a homophilic adhesion molecule. *Nature* **344:** 871–872. doi:10.1038/344871a0

Finzsch M, Schreiner S, Kichko T, Reeh P, Tamm ER, Bösl MR, Meijer D, Wegner M. 2010. Sox10 is required for SC identity and progression beyond the immature SC stage. *J Cell Biol* **189:** 701–712. doi:10.1083/jcb.200912142

Fledrich R, Kungl T, Nave KA, Stassart RM. 2019. Axo-glial interdependence in peripheral nerve development. *Development* **146:** dev151704. doi:10.1242/dev.151704

Fricker FR, Lago N, Balarajah S, Tsantoulas C, Tanna S, Zhu N, Fageiry SK, Jenkins M, Garratt AN, Birchmeier C, et al. 2011. Axonally derived neuregulin-1 is required for remyelination and regeneration after nerve injury in adulthood. *J Neurosci* **31:** 3225–3233. doi:10.1523/JNEUROSCI.2568-10.2011

Fröb F, Bremer M, Finzsch M, Kichko T, Reeh P, Tamm ER, Charnay P, Wegner M. 2012. Establishment of myelinating Schwann cells and barrier integrity between central and peripheral nervous systems depend on Sox10. *Glia* **60:** 806–819. doi:10.1002/glia.22310

Fröb F, Sock E, Tamm ER, Saur AL, Hillgärtner S, Williams TJ, Fujii T, Fukunaga R, Wegner M. 2019. Ep400 deficiency in Schwann cells causes persistent expression of early developmental regulators and peripheral neuropathy. *Nat Commun* **10:** 2361. doi:10.1038/s41467-019-10287-w

Gasser HS, Grundfest H. 1939. Axon diameters in relation to the spike dimensions and the conduction velocity in mammalian A fibers. *Am J Physiol* **127:** 393–414. doi:10.1152/ajplegacy.1939.127.2.393

Gerber D, Pereira JA, Gerber J, Tan G, Dimitrieva S, Yánguez E, Suter U. 2021. Transcriptional profiling of mouse peripheral nerves to the single-cell level to build a sciatic nerve ATlas (SNAT). *eLife* **10:** e58591. doi:10.7554/eLife.58591

Geren BB. 1954. The formation from the Schwann cell surface of myelin in the peripheral nerves of chick embryos. *Exp Cell Res* **7:** 558–562. doi:10.1016/S0014-4827(54)80098-X

Ghidinelli M, Poitelon Y, Shin YK, Ameroso D, Williamson C, Ferri C, Pellegatta M, Espino K, Mogha A, Monk K, et al. 2017. Laminin 211 inhibits protein kinase A in Schwann cells to modulate neuregulin 1 type III-driven myelination. *PLoS Biol* **15:** e2001408. doi:10.1371/journal.pbio.2001408

Ghislain J, Charnay P. 2006. Control of myelination in Schwann cells: a Krox20 cis-regulatory element integrates Oct6, Brn2 and Sox10 activities. *EMBO Rep* **7:** 52–58. doi:10.1038/sj.embor.7400573

Gillespie CS, Sherman DL, Fleetwood-Walker SM, Cottrell DF, Tait S, Garry EM, Wallace VC, Ure J, Griffiths IR, Smith A, et al. 2000. Peripheral demyelination and neuropathic pain behavior in periaxin-deficient mice. *Neuron* **26:** 523–531. doi:10.1016/S0896-6273(00)81184-8

Goebbels S, Oltrogge JH, Kemper R, Heilmann I, Bormuth I, Wolfer S, Wichert SP, Mobius W, Liu X, Lappe-Siefke C, et al. 2010. Elevated phosphatidylinositol 3,4,5-trisphosphate in glia triggers cell-autonomous membrane wrapping and myelination. *J Neurosci* 30: 8953–8964. doi:10 .1523/JNEUROSCI.0219-10.2010

Gökbuget D, Pereira JA, Bachofner S, Marchais A, Ciaudo C, Stoffel M, Schulte JH, Suter U. 2015. The Lin28/let-7 axis is critical for myelination in the peripheral nervous system. *Nat Commun* 6: 8584. doi:10.1038/ncomms9584

Gökbuget D, Pereira JA, Opitz L, Christe D, Kessler T, Marchais A, Suter U. 2018. The miRNA biogenesis pathway prevents inappropriate expression of injury response genes in developing and adult Schwann cells. *Glia* 66: 2632–2644. doi:10.1002/glia.23516

Gokey NG, Srinivasan R, Lopez-Anido C, Krueger C, Svaren J. 2012. Developmental regulation of microRNA expression in Schwann cells. *Mol Cell Biol* 32: 558–568. doi:10 .1128/MCB.06270-11

Gomis-Coloma C, Velasco-Aviles S, Gomez-Sanchez JA, Casillas-Bajo A, Backs J, Cabedo H. 2018. Class IIa histone deacetylases link cAMP signaling to the myelin transcriptional program of Schwann cells. *J Cell Biol* 217: 1249–1268. doi:10.1083/jcb.201611150

Gould RM. 1977. Incorporation of glycoproteins into peripheral nerve myelin. *J Cell Biol* 75: 326–338. doi:10.1083/jcb .75.2.326

Gould RM, Mattingly G. 1990. Regional localization of RNA and protein metabolism in Schwann cells in vivo. *J Neurocytol* 19: 285–301. doi:10.1007/BF01188399

Grove M, Brophy PJ. 2014. FAK is required for Schwann cell spreading on immature basal lamina to coordinate the radial sorting of peripheral axons with myelination. *J Neurosci* 34: 13422–13434. doi:10.1523/JNEUROSCI.1764-14.2014

Grove M, Kim H, Santerre M, Krupka AJ, Han SB, Zhai J, Cho JY, Park R, Harris M, Kim S, et al. 2017. YAP/TAZ initiate and maintain Schwann cell myelination. *eLife* 6: e20982. doi:10.7554/eLife.20982

Guo L, Moon C, Zheng Y, Ratner N. 2013. Cdc42 regulates Schwann cell radial sorting and myelin sheath folding through NF2/merlin-dependent and independent signaling. *Glia* 61: 1906–1921. doi:10.1002/glia.22567

Hall A. 2012. Rho family GTPases. *Biochem Soc Trans* 40: 1378–1382. doi:10.1042/BST20120103

Halter JA, Clark JW Jr. 1993. The influence of nodal constriction on conduction velocity in myelinated nerve fibers. *Neuroreport* 4: 89–92. doi:10.1097/00001756-199 301000-00023

Harty BL, Monk KR. 2017. Unwrapping the unappreciated: recent progress in Remak Schwann cell biology. *Curr Opin Neurobiol* 47: 131–137. doi:10.1016/j.conb.2017.10.003

He Y, Kim JY, Dupree J, Tewari A, Melendez-Vasquez C, Svaren J, Casaccia P. 2010. Yy1 as a molecular link between neuregulin and transcriptional modulation of peripheral myelination. *Nat Neurosci* 13: 1472–1480. doi:10.1038/nn .2686

Heller BA, Ghidinelli M, Voelkl J, Einheber S, Smith R, Grund E, Morahan G, Chandler D, Kalaydjieva L, Giancotti F, et al. 2014. Functionally distinct PI 3-kinase pathways regulate myelination in the peripheral nervous system. *J Cell Biol* 204: 1219–1236. doi:10.1083/jcb.201307057

Hildebrand C, Westerberg M, Mustafa GY. 1989. Influence of an experimental hindlimb maldevelopment on axon number and nodal spacing in the rat sciatic nerve. *Brain Res Dev Brain Res* 50: 169–175. doi:10.1016/0165-3806 (89)90192-2

Hildebrand C, Bowe CM, Remahl IN. 1994. Myelination and myelin sheath remodelling in normal and pathological PNS nerve fibres. *Prog Neurobiol* 43: 85–141. doi:10 .1016/0301-0082(94)90010-8

Howe DG, McCarthy KD. 2000. Retroviral inhibition of cAMP-dependent protein kinase inhibits myelination but not Schwann cell mitosis stimulated by interaction with neurons. *J Neurosci* 20: 3513–3521. doi:10.1523/ JNEUROSCI.20-10-03513.2000

Hsieh ST, Kidd GJ, Crawford TO, Xu Z, Lin WM, Trapp BD, Cleveland DW, Griffin JW. 1994. Regional modulation of neurofilament organization by myelination in normal axons. *J Neurosci* 14: 6392–6401. doi:10.1523/JNEUROSCI .14-11-06392.1994

Hung H, Kohnken R, Svaren J. 2012. The nucleosome remodeling and deacetylase chromatin remodeling (NuRD) complex is required for peripheral nerve myelination. *J Neurosci* 32: 1517–1527. doi:10.1523/JNEUROSCI.2895-11.2012

Ishibashi T, Baba H. 2022. Paranodal axoglial junctions, an essential component in axonal homeostasis. *Front Cell Dev Biol* 10: 951809. doi:10.3389/fcell.2022.951809

Iwanaga T, Takahashi-Iwanaga H, Nio-Kobayashi J, Ebara S. 2022. Structure and barrier functions of the perineurium and its relationship with associated sensory corpuscles: a review. *Biomed Res* 43: 145–159. doi:10.2220/biomedres .43.145

Jacob C. 2015. Transcriptional control of neural crest specification into peripheral glia. *Glia* 63: 1883–1896. doi:10 .1002/glia.22816

Jacob C. 2017. Chromatin-remodeling enzymes in control of Schwann cell development, maintenance and plasticity. *Curr Opin Neurobiol* 47: 24–30. doi:10.1016/j.conb.2017 .08.007

Jacob C, Lebrun-Julien F, Suter U. 2011a. How histone deacetylases control myelination. *Mol Neurobiol* 44: 303–312. doi:10.1007/s12035-011-8198-9

Jacob C, Christen CN, Pereira JA, Somandin C, Baggiolini A, Lötscher P, Özçelik M, Tricaud N, Meijer D, Yamaguchi T, et al. 2011b. HDAC1 and HDAC2 control the transcriptional program of myelination and the survival of Schwann cells. *Nat Neurosci* 14: 429–436. doi:10.1038/ nn.2762

Jacob C, Lötscher P, Engler S, Baggiolini A, Varum Tavares S, Brügger V, John N, Büchmann-Møller S, Snider PL, Conway SJ, et al. 2014. HDAC1 and HDAC2 control the specification of neural crest cells into peripheral glia. *J Neurosci* 34: 6112–6122. doi:10.1523/JNEUROSCI.5212-13.2014

Jacobs JM, Myers R. 1993. Adaptation of tibial nerve myelinated fibre parameters to reduced limb length caused by irradiation. *Neuropathol Appl Neurobiol* 19: 52–56. doi:10.1111/j.1365-2990.1993.tb00404.x

Jaegle M, Mandemakers W, Broos L, Zwart R, Karis A, Visser P, Grosveld F, Meijer D. 1996. The POU factor Oct-6 and Schwann cell differentiation. *Science* 273: 507–510. doi:10 .1126/science.273.5274.507

Jaegle M, Ghazvini M, Mandemakers W, Piirsoo M, Driegen S, Levavasseur F, Raghoenath S, Grosveld F, Meijer D. 2003. The POU proteins Brn-2 and Oct-6 share important functions in Schwann cell development. *Genes Dev* **17:** 1380–1391. doi:10.1101/gad.258203

Jagalur NB, Ghazvini M, Mandemakers W, Driegen S, Maas A, Jones EA, Jaegle M, Grosveld F, Svaren J, Meijer D. 2011. Functional dissection of the Oct6 Schwann cell enhancer reveals an essential role for dimeric Sox10 binding. *J Neurosci* **31:** 8585–8594. doi:10.1523/JNEUROSCI.0659-11.2011

Jessen KR, Mirsky R. 2005. The origin and development of glial cells in peripheral nerves. *Nat Rev Neurosci* **6:** 671–682. doi:10.1038/nrn1746

Jessen KR, Mirsky R. 2019. Schwann cell precursors; multipotent glial cells in embryonic nerves. *Front Mol Neurosci* **12:** 69. doi:10.3389/fnmol.2019.00069

Jessen KR, Brennan A, Morgan L, Mirsky R, Kent A, Hashimoto Y, Gavrilovic J. 1994. The Schwann cell precursor and its fate: a study of cell death and differentiation during gliogenesis in rat embryonic nerves. *Neuron* **12:** 509–527. doi:10.1016/0896-6273(94)90209-7

Jessen KR, Mirsky R, Lloyd AC. 2015. Schwann cells: development and role in nerve repair. *Cold Spring Harb Perspect Biol* **7:** a020487. doi:10.1101/cshperspect.a020487

Jia L, Liao M, Mou A, Zheng Q, Yang W, Yu Z, Cui Y, Xia X, Qin Y, Chen M, et al. 2021. Rheb-regulated mitochondrial pyruvate metabolism of Schwann cells linked to axon stability. *Dev Cell* **56:** 2980–2994.e6. doi:10.1016/j.devcel.2021.09.013

Jin JJ, Nikitin AY, Rajewsky MF. 1993. Schwann cell lineage-specific neu (erbB-2) gene expression in the developing rat nervous system. *Cell Growth Differ* **4:** 227–237.

Jin F, Dong B, Georgiou J, Jiang Q, Zhang J, Bharioke A, Qiu F, Lommel S, Feltri ML, Wrabetz L, et al. 2011. N-WASp is required for Schwann cell cytoskeletal dynamics, normal myelin gene expression and peripheral nerve myelination. *Development* **138:** 1329–1337. doi:10.1242/dev.058677

Johnson C, Holmes WR, Brown A, Jung P. 2015. Minimizing the caliber of myelinated axons by means of nodal constrictions. *J Neurophysiol* **114:** 1874–1884. doi:10.1152/jn.00338.2015

Joseph NM, Mukouyama YS, Mosher JT, Jaegle M, Crone SA, Dormand EL, Lee KF, Meijer D, Anderson DJ, Morrison SJ. 2004. Neural crest stem cells undergo multilineage differentiation in developing peripheral nerves to generate endoneurial fibroblasts in addition to Schwann cells. *Development* **131:** 5599–5612. doi:10.1242/dev.01429

Kao SC, Wu H, Xie J, Chang CP, Ranish JA, Graef IA, Crabtree GR. 2009. Calcineurin/NFAT signaling is required for neuregulin-regulated Schwann cell differentiation. *Science* **323:** 651–654. doi:10.1126/science.1166562

Kastriti ME, Faure L, Von Ahsen D, Bouderlique TG, Bostrom J, Solovieva T, Jackson C, Bronner M, Meijer D, Hadjab S, et al. 2022. Schwann cell precursors represent a neural crest-like state with biased multipotency. *EMBO J* **41:** e108780. doi:10.15252/embj.2021108780

Kegel L, Aunin E, Meijer D, Bermingham JR. 2013. LGI proteins in the nervous system. *ASN Neuro* **5:** 167–181. doi:10.1042/AN20120095

Kegel L, Jaegle M, Driegen S, Aunin E, Leslie K, Fukata Y, Watanabe M, Fukata M, Meijer D. 2014. Functional phy-

logenetic analysis of LGI proteins identifies an interaction motif crucial for myelination. *Development* **141:** 1749–1756. doi:10.1242/dev.107995

Kidd GJ, Ohno N, Trapp BD. 2013. Biology of Schwann cells. *Handb Clin Neurol* **115:** 55–79. doi:10.1016/B978-0-444-52902-2.00005-9

Kirkpatrick LL, Brady ST. 1994. Modulation of the axonal microtubule cytoskeleton by myelinating Schwann cells. *J Neurosci* **14:** 7440–7450. doi:10.1523/JNEUROSCI.14-12-07440.1994

Köling A. 1985. Membrane architecture of myelinated nerve fibres in the human dental pulp studied by freeze-fracturing. *Arch Oral Biol* **30:** 121–128. doi:10.1016/0003-9969(85)90103-7

Küffer A, Lakkaraju AK, Mogha A, Petersen SC, Airich K, Doucerain C, Marpakwar R, Bakirci P, Senatore A, Monnard A, et al. 2016. The prion protein is an agonistic ligand of the G protein-coupled receptor Adgrg6. *Nature* **536:** 464–468. doi:10.1038/nature19312

Kuhlbrodt K, Herbarth B, Sock E, Hermans-Borgmeyer I, Wegner M. 1998. Sox10, a novel transcriptional modulator in glial cells. *J Neurosci* **18:** 237–250. doi:10.1523/JNEUROSCI.18-01-00237.1998

La Marca R, Cerri F, Horiuchi K, Bachi A, Feltri ML, Wrabetz L, Blobel CP, Quattrini A, Salzer JL, Taveggia C. 2011. TACE (ADAM17) inhibits Schwann cell myelination. *Nat Neurosci* **14:** 857–865. doi:10.1038/nn.2849

Le N, Nagarajan R, Wang JY, Araki T, Schmidt RE, Milbrandt J. 2005. Analysis of congenital hypomyelinating Egr2Lo/Lo nerves identifies Sox2 as an inhibitor of Schwann cell differentiation and myelination. *Proc Natl Acad Sci* **102:** 2596–2601. doi:10.1073/pnas.0407836102

Leblanc SE, Srinivasan R, Ferri C, Mager GM, Gillian-Daniel AL, Wrabetz L, Svaren J. 2005. Regulation of cholesterol/lipid biosynthetic genes by Egr2/Krox20 during peripheral nerve myelination. *J Neurochem* **93:** 737–748. doi:10.1111/j.1471-4159.2005.03056.x

Le Douarin NM, Teillet MA. 1974. Experimental analysis of the migration and differentiation of neuroblasts of the autonomic nervous system and of neurectodermal mesenchymal derivatives, using a biological cell marking technique. *Dev Biol* **41:** 162–184. doi:10.1016/0012-1606(74)90291-7

Leite SC, Sampaio P, Sousa VF, Nogueira-Rodrigues J, Pinto-Costa R, Peters LL, Brites P, Sousa MM. 2016. The actin-binding protein α-Adducin is required for maintaining axon diameter. *Cell Rep* **15:** 490–498. doi:10.1016/j.celrep.2016.03.047

Lewallen KA, Shen YA, De la Torre AR, Ng BK, Meijer D, Chan JR. 2011. Assessing the role of the cadherin/catenin complex at the Schwann cell–axon interface and in the initiation of myelination. *J Neurosci* **31:** 3032–3043. doi:10.1523/JNEUROSCI.4345-10.2011

Limpert AS, Bai S, Narayan M, Wu J, Yoon SO, Carter BD, Lu QR. 2013. NF-κB forms a complex with the chromatin remodeler BRG1 to regulate Schwann cell differentiation. *J Neurosci* **33:** 2388–2397. doi:10.1523/JNEUROSCI.3223-12.2013

Lin HP, Oksuz I, Hurley E, Wrabetz L, Awatramani R. 2015. Microprocessor complex subunit DiGeorge syndrome critical region gene 8 (Dgcr8) is required for Schwann cell myelination and myelin maintenance. *J Biol Chem* **290:** 24294–24307. doi:10.1074/jbc.M115.636407

Liu X, Zhao Y, Peng S, Zhang S, Wang M, Chen Y, Zhang S, Yang Y, Sun C. 2016. BMP7 retards peripheral myelination by activating p38 MAPK in Schwann cells. *Sci Rep* **6:** 31049. doi:10.1038/srep31049

Locascio A, Manzanares M, Blanco MJ, Nieto MA. 2002. Modularity and reshuffling of snail and slug expression during vertebrate evolution. *Proc Natl Acad Sci* **99:** 16841–16846. doi:10.1073/pnas.262525399

Ma KH, Svaren J. 2018. Epigenetic control of Schwann cells. *Neuroscientist* **24:** 627–638. doi:10.1177/10738584177 51112

Ma KH, Hung HA, Srinivasan R, Xie H, Orkin SH, Svaren J. 2015. Regulation of peripheral nerve myelin maintenance by gene repression through Polycomb Repressive Complex 2. *J Neurosci* **35:** 8640–8652. doi:10.1523/JNEURO SCI.2257-14.2015

Madrid RE, Jaros E, Cullen MJ, Bradley WG. 1975. Genetically determined defect of Schwann cell basement membrane in dystrophic mouse. *Nature* **257:** 319–321. doi:10 .1038/257319a0

Mager GM, Ward RM, Srinivasan R, Jang SW, Wrabetz L, Svaren J. 2008. Active gene repression by the Egr2.NAB complex during peripheral nerve myelination. *J Biol Chem* **283:** 18187–18197. doi:10.1074/jbc.M803330200

Marshall-Phelps KLH, Kegel L, Baraban M, Ruhwedel T, Almeida RG, Rubio-Brotons M, Klingseisen A, Benito-Kwiecinski SK, Early JJ, Bin JM, et al. 2020. Neuronal activity disrupts myelinated axon integrity in the absence of NKCC1b. *J Cell Biol* **219:** e201909022. doi:10.1083/jcb .201909022

Martini R. 2001. The effect of myelinating Schwann cells on axons. *Muscle Nerve* **24:** 456–466. doi:10.1002/mus.1027

Martini R, Mohajeri MH, Kasper S, Giese KP, Schachner M. 1995. Mice doubly deficient in the genes for P0 and myelin basic protein show that both proteins contribute to the formation of the major dense line in peripheral nerve myelin. *J Neurosci* **15:** 4488–4495. doi:10.1523/JNEUROSCI .15-06-04488.1995

Masaki T. 2012. Polarization and myelination in myelinating glia. *ISRN Neurol* **2012:** 769412. doi:10.5402/2012/769412

Maurel P, Salzer JL. 2000. Axonal regulation of Schwann cell proliferation and survival and the initial events of myelination requires PI 3-kinase activity. *J Neurosci* **20:** 4635–4645. doi:10.1523/JNEUROSCI.20-12-04635.2000

Maurel P, Einheber S, Galinska J, Thaker P, Lam I, Rubin MB, Scherer SS, Murakami Y, Gutmann DH, Salzer JL. 2007. Nectin-like proteins mediate axon–Schwann cell interactions along the internode and are essential for myelination. *J Cell Biol* **178:** 861–874. doi:10.1083/jcb.200705132

McKee KK, Yang DH, Patel R, Chen ZL, Strickland S, Takagi J, Sekiguchi K, Yurchenco PD. 2012. Schwann cell myelination requires integration of laminin activities. *J Cell Sci* **125:** 4609–4619.

Melendez-Vasquez CV, Einheber S, Salzer JL. 2004. Rho kinase regulates Schwann cell myelination and formation of associated axonal domains. *J Neurosci* **24:** 3953–3963. doi:10.1523/JNEUROSCI.4920-03.2004

Meyer D, Birchmeier C. 1995. Multiple essential functions of neuregulin in development. *Nature* **378:** 386–390. doi:10 .1038/378386a0

Michailov GV, Sereda MW, Brinkmann BG, Fischer TM, Haug B, Birchmeier C, Role L, Lai C, Schwab MH, Nave KA. 2004. Axonal neuregulin-1 regulates myelin sheath thickness. *Science* **304:** 700–703. doi:10.1126/science .1095862

Mikol DD, Hong HL, Cheng HL, Feldman EL. 1999. Caveolin-1 expression in Schwann cells. *Glia* **27:** 39–52. doi:10 .1002/(SICI)1098-1136(199907)27:1<39::AID-GLIA5>3 .0.CO;2-#

Mitgau J, Franke J, Schinner C, Stephan G, Berndt S, Placantonakis DG, Kalwa H, Spindler V, Wilde C, Liebscher I. 2022. The N terminus of adhesion G protein-coupled receptor GPR126/ADGRG6 as allosteric force integrator. *Front Cell Dev Biol* **10:** 873278. doi:10.3389/fcell.2022 .873278

Mogha A, Benesh AE, Patra C, Engel FB, Schöneberg T, Liebscher I, Monk KR. 2013. Gpr126 functions in Schwann cells to control differentiation and myelination via G-protein activation. *J Neurosci* **33:** 17976–17985. doi:10.1523/JNEUROSCI.1809-13.2013

Monk KR, Naylor SG, Glenn TD, Mercurio S, Perlin JR, Dominguez C, Moens CB, Talbot WS. 2009. A G protein-coupled receptor is essential for Schwann cells to initiate myelination. *Science* **325:** 1402–1405. doi:10.1126/ science.1173474

Monsma PC, Li Y, Fenn JD, Jung P, Brown A. 2014. Local regulation of neurofilament transport by myelinating cells. *J Neurosci* **34:** 2979–2988. doi:10.1523/JNEURO SCI.4502-13.2014

Montani L, Buerki-Thurnherr T, de Faria JP, Pereira JA, Dias NG, Fernandes R, Gonçalves AF, Braun A, Benninger Y, Böttcher RT, et al. 2014. Profilin 1 is required for peripheral nervous system myelination. *Development* **141:** 1553–1561. doi:10.1242/dev.101840

Moore JW, Joyner RW, Brill MH, Waxman SD, Najar-Joa M. 1978. Simulations of conduction in uniform myelinated fibers. Relative sensitivity to changes in nodal and internodal parameters. *Biophys J* **21:** 147–160. doi:10.1016/ S0006-3495(78)85515-5

Morrison SJ, Perez S, Verdi JM, Hicks C, Weinmaster G, Anderson DJ. 2000. Transient Notch activation initiates an irreversible switch from neurogenesis to gliogenesis by neural crest stem cells. *Cell* **101:** 499–510. doi:10.1016/ S0092-8674(00)80860-0

Morton PD, Dellarole A, Theus MH, Walters WM, Berge SS, Bethea JR. 2013. Activation of NF-κB in SCs is dispensable for myelination in vivo. *J Neurosci* **33:** 9932–9936. doi:10 .1523/JNEUROSCI.2483-12.2013

Mugnaini E, Osen KK, Schnapp B, Friedrich VL Jr. 1977. Distribution of Schwann cell cytoplasm and plasmalemmal vesicles (caveolae) in peripheral myelin sheaths. An electron microscopic study with thin sections and freeze-fracturing. *J Neurocytol* **6:** 647–668. doi:10.1007/BF011 76378

Mukouyama YS, Gerber HP, Ferrara N, Gu C, Anderson DJ. 2005. Peripheral nerve-derived VEGF promotes arterial differentiation via neuropilin 1-mediated positive feedback. *Development* **132:** 941–952. doi:10.1242/dev.01675

Muppirala AN, Limbach LE, Bradford EF, Petersen SC. 2021. Schwann cell development: from neural crest to myelin sheath. *Wiley Interdisc Rev Dev Biol* **10:** e398. doi:10.1002/wdev.398

Murphy P, Topilko P, Schneider-Maunoury S, Seitanidou T, Baron-Van Evercooren A, Charnay P. 1996. The regulation of Krox-20 expression reveals important steps in the control of peripheral glial cell development. *Development* **122:** 2847–2857. doi:10.1242/dev.122.9.2847

Nagata-Ohashi K, Ohta Y, Goto K, Chiba S, Mori R, Nishita M, Ohashi K, Kousaka K, Iwamatsu A, Niwa R, et al. 2004. A pathway of neuregulin-induced activation of cofilin-phosphatase Slingshot and cofilin in lamellipodia. *J Cell Biol* **165:** 465–471. doi:10.1083/jcb.200401136

Nave KA. 2010. Myelination and the trophic support of long axons. *Nat Rev Neurosci* **11:** 275–283. doi:10.1038/nrn2797

Nave KA, Salzer JL. 2006. Axonal regulation of myelination by neuregulin 1. *Curr Opin Neurobiol* **16:** 492–500. doi:10.1016/j.conb.2006.08.008

Nave KA, Werner HB. 2021. Ensheathment and myelination of axons: evolution of glial functions. *Annu Rev Neurosci* **44:** 197–219. doi:10.1146/annurev-neuro-100120-122621

Ness JK, Snyder KM, Tapinos N. 2013. Lck tyrosine kinase mediates β1-integrin signalling to regulate Schwann cell migration and myelination. *Nat Commun* **4:** 1912. doi:10.1038/ncomms2928

Ness JK, Skiles AA, Yap EH, Fajardo EJ, Fiser A, Tapinos N. 2016. Nuc-ErbB3 regulates H3K27me3 levels and HMT activity to establish epigenetic repression during peripheral myelination. *Glia* **64:** 977–992. doi:10.1002/glia.22977

Newbern J, Birchmeier C. 2010. Nrg1/ErbB signaling networks in Schwann cell development and myelination. *Semin Cell Dev Biol* **21:** 922–928. doi:10.1016/j.semcdb.2010.08.008

Nickols JC, Valentine W, Kanwal S, Carter BD. 2003. Activation of the transcription factor NF-κB in Schwann cells is required for peripheral myelin formation. *Nat Neurosci* **6:** 161–167. doi:10.1038/nn995

Nicolson GL, Ferreira de Mattos G. 2023. The fluid-mosaic model of cell membranes: a brief introduction, historical features, some general principles, and its adaptation to current information. *Biochim Biophys Acta Biomembr* **1865:** 184135. doi:10.1016/j.bbamem.2023.184135

Nishino J, Saunders TL, Sagane K, Morrison SJ. 2010. Lgi4 promotes the proliferation and differentiation of glial lineage cells throughout the developing peripheral nervous system. *J Neurosci* **30:** 15228–15240. doi:10.1523/JNEUROSCI.2286-10.2010

Nitzan E, Pfaltzgraff ER, Labosky PA, Kalcheim C. 2013. Neural crest and Schwann cell progenitor-derived melanocytes are two spatially segregated populations similarly regulated by Foxd3. *Proc Natl Acad Sci* **110:** 12709–12714. doi:10.1073/pnas.1306287110

Nodari A, Zambroni D, Quattrini A, Court FA, D'Urso A, Recchia A, Tybulewicz VL, Wrabetz L, Feltri ML. 2007. β1 integrin activates Rac1 in Schwann cells to generate radial lamellae during axonal sorting and myelination. *J Cell Biol* **177:** 1063–1075. doi:10.1083/jcb.200610014

Norrmén C, Figlia G, Lebrun-Julien F, Pereira JA, Trötzmüller M, Köfeler HC, Rantanen V, Wessig C, van Deijk AL,

Smit AB, et al. 2014. mTORC1 controls PNS myelination along the mTORC1-RXRγ-SREBP-lipid biosynthesis axis in Schwann cells. *Cell Rep* **9:** 646–660. doi:10.1016/j.celrep.2014.09.001

Noseda R, Belin S, Piguet F, Vaccari I, Scarlino S, Brambilla P, Martinelli Boneschi F, Feltri ML, Wrabetz L, Quattrini A, et al. 2013. DDIT4/REDD1/RTP801 is a novel negative regulator of Schwann cell myelination. *J Neurosci* **33:** 15295–15305. doi:10.1523/JNEUROSCI.2408-13.2013

Novak N, Bar V, Sabanay H, Frechter S, Jaegle M, Snapper SB, Meijer D, Peles E. 2011. N-WASP is required for membrane wrapping and myelination by Schwann cells. *J Cell Biol* **192:** 243–250. doi:10.1083/jcb.201010013

Obremski VJ, Johnson MI, Bunge MB. 1993. Fibroblasts are required for Schwann cell basal lamina deposition and ensheathment of unmyelinated sympathetic neurites in culture. *J Neurocytol* **22:** 102–117. doi:10.1007/BF01181574

Özçelik M, Cotter L, Jacob C, Pereira JA, Relvas JB, Suter U, Tricaud N. 2010. Pals1 is a major regulator of the epithelial-like polarization and the extension of the myelin sheath in peripheral nerves. *J Neurosci* **30:** 4120–4131. doi:10.1523/JNEUROSCI.5185-09.2010

Özkaynak E, Abello G, Jaegle M, van Berge L, Hamer D, Kegel L, Driegen S, Sagane K, Bermingham JR Jr, Meijer D. 2010. Adam22 is a major neuronal receptor for Lgi4-mediated Schwann cell signaling. *J Neurosci* **30:** 3857–3864. doi:10.1523/JNEUROSCI.6287-09.2010

Paavola KJ, Sidik H, Zuchero JB, Eckart M, Talbot WS. 2014. Type IV collagen is an activating ligand for the adhesion G protein-coupled receptor GPR126. *Sci Signal* **7:** ra76. doi:10.1126/scisignal.2005347

Paratore C, Goerich DE, Suter U, Wegner M, Sommer L. 2001. Survival and glial fate acquisition of neural crest cells are regulated by an interplay between the transcription factor Sox10 and extrinsic combinatorial signaling. *Development* **128:** 3949–3961. doi:10.1242/dev.128.20.3949

Park HJ, Tsai E, Huang D, Weaver M, Frick L, Alcantara A, Moran JJ, Patzig J, Melendez-Vasquez CV, Crabtree GR, et al. 2022. ACTL6a coordinates axonal caliber recognition and myelination in the peripheral nerve. *iScience* **25:** 104132. doi:10.1016/j.isci.2022.104132

Parkinson DB, Bhaskaran A, Droggiti A, Dickinson S, D'Antonio M, Mirsky R, Jessen KR. 2004. Krox-20 inhibits Jun-NH2-terminal kinase/c-Jun to control Schwann cell proliferation and death. *J Cell Biol* **164:** 385–394. doi:10.1083/jcb.200307132

Parkinson DB, Bhaskaran A, Arthur-Farraj P, Noon LA, Woodhoo A, Lloyd AC, Feltri ML, Wrabetz L, Behrens A, Mirsky R, et al. 2008. c-Jun is a negative regulator of myelination. *J Cell Biol* **181:** 625–637. doi:10.1083/jcb.200803013

Parmantier E, Lynn B, Lawson D, Turmaine M, Namini SS, Chakrabarti L, McMahon AP, Jessen KR, Mirsky R. 1999. Schwann cell–derived desert hedgehog controls the development of peripheral nerve sheaths. *Neuron* **23:** 713–724. doi:10.1016/s0896-6273(01)80030-1

Pedraza L, Huang JK, Colman DR. 2001. Organizing principles of the axoglial apparatus. *Neuron* **30:** 335–344. doi:10.1016/S0896-6273(01)00306-3

Pellegatta M, De Arcangelis A, D'Urso A, Nodari A, Zambroni D, Ghidinelli M, Matafora V, Williamson C,

Georges-Labouesse E, Kreidberg J, et al. 2013. α6β1 and α7β1 integrins are required in Schwann cells to sort axons. *J Neurosci* **33:** 17995–18007. doi:10.1523/JNEUROSCI .3179-13.2013

Peng K, Sant D, Andersen N, Silvera R, Camarena V, Piñero G, Graham R, Khan A, Xu XM, Wang G, et al. 2020. Magnetic separation of peripheral nerve-resident cells underscores key molecular features of human Schwann cells and fibroblasts: An immunochemical and transcriptomics approach. *Sci Rep* **10:** 18433. doi:10.1038/s41598-020-74128-3

Pereira JA, Benninger Y, Baumann R, Gonçalves AF, Özçelik M, Thurnherr T, Tricaud N, Meijer D, Fässler R, Suter U, et al. 2009. Integrin-linked kinase is required for radial sorting of axons and Schwann cell remyelination in the peripheral nervous system. *J Cell Biol* **185:** 147–161. doi:10 .1083/jcb.200809008

Pereira JA, Baumann R, Norrmén C, Somandin C, Miehe M, Jacob C, Lühmann T, Hall-Bozic H, Mantei N, Meijer D, et al. 2010. Dicer in Schwann cells is required for myelination and axonal integrity. *J Neurosci* **30:** 6763–6775. doi:10 .1523/JNEUROSCI.0801-10.2010

Pereira JA, Lebrun-Julien F, Suter U. 2012. Molecular mechanisms regulating myelination in the peripheral nervous system. *Trends Neurosci* **35:** 123–134. doi:10.1016/j.tins .2011.11.006

Peters A, Muir AR. 1959. The relationship between axons and Schwann cells during development of peripheral nerves in the rat. *Q J Exp Physiol Cogn Med Sci* **44:** 117–130.

Peters A, Palay SL, Webster HD. 1991. *The fine structure of the nervous system.* Oxford University Press, New York.

Petersen SC, Luo R, Liebscher I, Giera S, Jeong SJ, Mogha A, Ghidinelli M, Feltri ML, Schöneberg T, Piao X, et al. 2015. The adhesion GPCR GPR126 has distinct, domain-dependent functions in Schwann cell development mediated by interaction with laminin-211. *Neuron* **85:** 755–769. doi:10 .1016/j.neuron.2014.12.057

Poitelon Y, Lopez-Anido C, Catignas K, Berti C, Palmisano M, Williamson C, Ameroso D, Abiko K, Hwang Y, Gregorieff A, et al. 2016. YAP and TAZ control peripheral myelination and the expression of laminin receptors in Schwann cells. *Nat Neurosci* **19:** 879–887. doi:10.1038/ nn.4316

Pollard TD, Borisy GG. 2003. Cellular motility driven by assembly and disassembly of actin filaments. *Cell* **112:** 453–465. doi:10.1016/S0092-8674(03)00120-X

Prasad MK, Reed X, Gorkin DU, Cronin JC, McAdow AR, Chain K, Hodonsky CJ, Jones EA, Svaren J, Antonellis A, et al. 2011. SOX10 directly modulates ERBB3 transcription via an intronic neural crest enhancer. *BMC Dev Biol* **11:** 40. doi:10.1186/1471-213X-11-40

Price RL, Lasek RJ, Katz MJ. 1993. Neurofilaments assume a less random architecture at nodes and in other regions of axonal compression. *Brain Res* **607:** 125–133. doi:10.1016/ 0006-8993(93)91497-G

Quintes S, Brinkmann BG, Ebert M, Fröb F, Kungl T, Arlt FA, Tarabykin V, Huylebroeck D, Meijer D, Suter U, et al. 2016. Zeb2 is essential for SC differentiation, myelination and nerve repair. *Nature Neuroscience* **19:** 1050–1059. doi:10.1038/nn.4321

Raphael AR, Lyons DA, Talbot WS. 2011. ErbB signaling has a role in radial sorting independent of Schwann cell number. *Glia* **59:** 1047–1055. doi:10.1002/glia.21175

* Rasband MN, Peles E. 2023. The nodes of Ranvier and the organization of myelinated axons. *Cold Spring Harb Perspect Biol* doi:10.1101/cshperspect.a41361

Rechthand E, Rapoport SI. 1987. Regulation of the microenvironment of peripheral nerve: role of the blood-nerve barrier. *Prog Neurobiol* **28:** 303–343. doi:10.1016/0301-0082(87)90006-2

Reiprich S, Kriesch J, Schreiner S, Wegner M. 2010. Activation of Krox20 gene expression by Sox10 in myelinating SCs. *J Neurochem* **112:** 744–754. doi:10.1111/j.1471-4159 .2009.06498.x

Richard L, Topilko P, Magy L, Decouvelaere AV, Charnay P, Funalot B, Vallat JM. 2012. Endoneurial fibroblast-like cells. *J Neuropathol Exp Neurol* **71:** 938–947. doi:10 .1097/NEN.0b013e318270a941

Riethmacher D, Sonnenberg-Riethmacher E, Brinkmann V, Yamaai T, Lewin GR, Birchmeier C. 1997. Severe neuropathies in mice with targeted mutations in the erbB3 receptor. *Nature* **389:** 725–730. doi:10.1038/39593

Robertson JD. 1955. The ultrastructure of adult vertebrate peripheral myelinated nerve fibers in relation to myelinogenesis. *J Biophys Biochem Cytol* **1:** 271–278. doi:10.1083/ jcb.1.4.271

Rohatgi R, Ma L, Miki H, Lopez M, Kirchhausen T, Takenawa T, Kirschner MW. 1999. The interaction between N-WASP and the Arp2/3 complex links Cdc42-dependent signals to actin assembly. *Cell* **97:** 221–231. doi:10.1016/ S0092-8674(00)80732-1

Rosenberg LH, Cattin AL, Fontana X, Harford-Wright E, Burden JJ, White IJ, Smith JG, Napoli I, Quereda V, Policarpi C, et al. 2018. HDAC3 regulates the transition to the homeostatic myelinating Schwann cell state. *Cell Rep* **25:** 2755–2765.e5. doi:10.1016/j.celrep.2018.11.045

Rushton WAH. 1951. A theory of the effects of fibre size in medullated nerve. *J Physiol* **115:** 101–122. doi:10.1113/ jphysiol.1951.sp004655

Ryu EJ, Wang JY, Le N, Baloh RH, Gustin JA, Schmidt RE, Milbrandt J. 2007. Misexpression of Pou3f1 results in peripheral nerve hypomyelination and axonal loss. *J Neurosci* **27:** 11552–11559. doi:10.1523/JNEUROSCI.5497-06 .2007

Sagane K, Hayakawa K, Kai J, Hirohashi T, Takahashi E, Miyamoto N, Ino M, Oki T, Yamazaki K, Nagasu T. 2005. Ataxia and peripheral nerve hypomyelination in ADAM22-deficient mice. *BMC Neurosci* **6:** 33. doi:10 .1186/1471-2202-6-33

Saher G, Simons M. 2010. Cholesterol and myelin biogenesis. *Subcell Biochem* **51:** 489–508. doi:10.1007/978-90-481-8622-8_18

Salzer JL. 2003. Polarized domains of myelinated axons. *Neuron* **40:** 297–318. doi:10.1016/S0896-6273(03)00628-7

Salzer JL. 2015. Schwann cell myelination. *Cold Spring Harb Perspect Biol* **7:** a020529. doi:10.1101/cshperspect .a020529

Salzer JL, Brophy PJ, Peles E. 2008. Molecular domains of myelinated axons in the peripheral nervous system. *Glia* **56:** 1532–1540. doi:10.1002/glia.20750

Saur AL, Fröb F, Weider M, Wegner M. 2021. Formation of the node of Ranvier by Schwann cells is under control of transcription factor Sox10. *Glia* **69:** 1464–1477. doi:10.1002/glia.23973

Scherer S, Salzer J. 2001. Axon–Schwann cell interactions during peripheral nerve degeneration and regeneration. In *Glial cell development* (ed. Jessen KR, Richardson WD), pp. 299–330. Oxford University Press, London.

* Scherer SS, Svaren J. 2023. Peripheral nervous system (PNS) myelin diseases. *Cold Spring Harb Perspect Biol* doi:10.1101/cshperspect.a41376

Schreiner S, Cossais F, Fischer K, Scholz S, Bösl MR, Holtmann B, Sendtner M, Wegner M. 2007. Hypomorphic Sox10 alleles reveal novel protein functions and unravel developmental differences in glial lineages. *Development* **134:** 3271–3281. doi:10.1242/dev.003350

Shapiro L, Doyle JP, Hensley P, Colman DR, Hendrickson WA. 1996. Crystal structure of the extracellular domain from P0, the major structural protein of peripheral nerve myelin. *Neuron* **17:** 435–449. doi:10.1016/S0896-6273(00)80176-2

Sheean ME, McShane E, Cheret C, Walcher J, Müller T, Wulf-Goldenberg A, Hoelper S, Garratt AN, Krüger M, Rajewsky K, et al. 2014. Activation of MAPK overrides the termination of myelin growth and replaces Nrg1/ErbB3 signals during Schwann cell development and myelination. *Genes Dev* **28:** 290–303. doi:10.1101/gad.230045.113

Sherman DL, Wu LM, Grove M, Gillespie CS, Brophy PJ. 2012. Drp2 and periaxin form Cajal bands with dystroglycan but have distinct roles in Schwann cell growth. *J Neurosci* **32:** 9419–9428. doi:10.1523/JNEUROSCI.1220-12.2012

Siems SB, Jahn O, Eichel MA, Kannaiyan N, Wu LMN, Sherman DL, Kusch K, Hesse D, Jung RB, Fledrich R, et al. 2020. Proteome profile of peripheral myelin in healthy mice and in a neuropathy model. *eLife* **9:** e51406. doi:10.7554/eLife.51406

* Simons M, Gibson EM, Nave KA. 2023. Oligodendrocytes: myelination, plasticity, and axonal support. *Cold Spring Harb Perspect Biol* doi:10.1101/cshperspect.a41359

Simpson AH, Gillingwater TH, Anderson H, Cottrell D, Sherman DL, Ribchester RR, Brophy PJ. 2013. Effect of limb lengthening on internodal length and conduction velocity of peripheral nerve. *J Neurosci* **33:** 4536–4539. doi:10.1523/JNEUROSCI.4176-12.2013

Singer SJ, Nicolson GL. 1972. The fluid mosaic model of the structure of cell membranes. *Science* **175:** 720–731. doi:10.1126/science.175.4023.720

Small JR, Ghabriel MN, Allt G. 1987. The development of Schmidt-Lanterman incisures: an electron microscope study. *J Anat* **150:** 277–286.

Snaidero N, Möbius W, Czopka T, Hekking LH, Mathisen C, Verkleij D, Goebbels S, Edgar J, Merkler D, Lyons DA, et al. 2014. Myelin membrane wrapping of CNS axons by PI(3,4,5)P3-dependent polarized growth at the inner tongue. *Cell* **156:** 277–290. doi:10.1016/j.cell.2013.11.044

Sock E, Wegner M. 2019. Transcriptional control of myelination and remyelination. *Glia* **67:** 2153–2165. doi:10.1002/glia.23636

Sparrow N, Manetti ME, Bott M, Fabianac T, Petrilli A, Bates ML, Bunge MB, Lambert S, Fernandez-Valle C. 2012. The actin-severing protein cofilin is downstream of neuregulin

signaling and is essential for Schwann cell myelination. *J Neurosci* **32:** 5284–5297. doi:10.1523/JNEUROSCI.6207-11.2012

Speidel C. 1964. In vitro studies of myelinated nerve fibers. *Int Rev Cytol* **16:** 173–231. doi:10.1016/S0074-7696(08)60297-1

Spiegel I, Adamsky K, Eshed Y, Milo R, Sabanay H, Sarig-Nadir O, Horresh I, Scherer SS, Rasband MN, Peles E. 2007. A central role for Necl4 (SynCAM4) in Schwann cell-axon interaction and myelination. *Nat Neurosci* **10:** 861–869. doi:10.1038/nn1915

Spokony RF, Aoki Y, Saint-Germain N, Magner-Fink E, Saint-Jeannet JP. 2002. The transcription factor Sox9 is required for cranial neural crest development in *Xenopus*. *Development* **129:** 421–432. doi:10.1242/dev.129.2.421

Srinivasan R, Mager GM, Ward RM, Mayer J, Svaren J. 2006. NAB2 represses transcription by interacting with the CHD4 subunit of the nucleosome remodeling and deacetylase (NuRD) complex. *J Biol Chem* **281:** 15129–15137. doi:10.1074/jbc.M600775200

* Stassart RM, Gomez-Sanchez JA, Lloyd AC. 2023. Schwann cells as orchestrators of nerve repair; implications for tissue regeneration and pathologies. *Cold Spring Harb Perspect Biol* doi:10.1101/cshperspect.a41363

Stewart HJ, Brennan A, Rahman M, Zoidl G, Mitchell PJ, Jessen KR, Mirsky R. 2001. Developmental regulation and overexpression of the transcription factor AP-2, a potential regulator of the timing of Schwann cell generation. *Eur J Neurosci* **14:** 363–372. doi:10.1046/j.0953-816x.2001.01650.x

Stierli S, Napoli I, White IJ, Cattin AL, Monteza Cabrejos A, Garcia Calavia N, Malong L, Ribeiro S, Nihouarn J, Williams R, et al. 2018. The regulation of the homeostasis and regeneration of peripheral nerve is distinct from the CNS and independent of a stem cell population. *Development* **145:** dev170316. doi:10.1242/dev.170316

Suster I, Feng Y. 2021. Multifaceted regulation of microRNA biogenesis: essential roles and functional integration in neuronal and glial development. *Int J Mol Sci* **22:** 6765. doi:10.3390/ijms22136765

Susuki K, Raphael AR, Ogawa Y, Stankewich MC, Peles E, Talbot WS, Rasband MN. 2011. Schwann cell spectrins modulate peripheral nerve myelination. *Proc Natl Acad Sci* **108:** 8009–8014. doi:10.1073/pnas.1019600108

Taïb S, Lamandé N, Martin S, Coulpier F, Topilko P, Brunet I. 2022. Myelinating Schwann cells and Netrin-1 control intra-nervous vascularization of the developing mouse sciatic nerve. *eLife* **11:** e64773. doi:10.7554/eLife.64773

Tasdemir-Yilmaz OE, Druckenbrod NR, Olukoya OO, Dong W, Yung AR, Bastille I, Pazyra-Murphy MF, Sitko AA, Hale EB, Vigneau S, et al. 2021. Diversity of developing peripheral glia revealed by single-cell RNA sequencing. *Dev Cell* **56:** 2516–2535.e8. doi:10.1016/j.devcel.2021.08.005

Taveggia C, Feltri ML. 2022. Beyond wrapping: canonical and noncanonical functions of Schwann cells. *Annu Rev Neurosci* **45:** 561–580. doi:10.1146/annurev-neuro-110920-030610

Taveggia C, Zanazzi G, Petrylak A, Yano H, Rosenbluth J, Einheber S, Xu X, Esper RM, Loeb JA, Shrager P, et al. 2005. Neuregulin-1 type III determines the ensheathment fate of axons. *Neuron* **47:** 681–694. doi:10.1016/j.neuron.2005.08.017

Tawk M, Makoukji J, Belle M, Fonte C, Trousson A, Hawkins T, Li H, Ghandour S, Schumacher M, Massaad C. 2011. Wnt/β-catenin signaling is an essential and direct driver of myelin gene expression and myelinogenesis. *J Neurosci* **31:** 3729–3742. doi:10.1523/JNEUROSCI.4270-10.2011

Terada N, Saitoh Y, Ohno N, Komada M, Yamauchi J, Ohno S. 2013. Involvement of Src in the membrane skeletal complex, MPP6-4.1G, in Schmidt-Lanterman incisures of mouse myelinated nerve fibers in PNS. *Histochem Cell Biol* **140:** 213–222. doi:10.1007/s00418-012-1073-6

Terada N, Saitoh Y, Kamijo A, Yamauchi J, Ohno N, Sakamoto T. 2019. Structures and molecular composition of Schmidt-Lanterman incisures. *Adv Exp Med Biol* **1190:** 181–198. doi:10.1007/978-981-32-9636-7_12

Theveneau E, Mayor R. 2012. Neural crest delamination and migration: from epithelium-to-mesenchyme transition to collective cell migration. *Dev Biol* **366:** 34–54. doi:10.1016/j.ydbio.2011.12.041

Toma JS, Karamboulas K, Carr MJ, Kolaj A, Yuzwa SA, Mahmud N, Storer MA, Kaplan DR, Miller FD. 2020. Peripheral nerve single-cell analysis identifies mesenchymal ligands that promote axonal growth. *eNeuro* **7:** ENEURO.0066-20.2020. doi:10.1523/ENEURO.0066-20.2020

Topilko P, Schneider-Maunoury S, Levi G, Baron-Van Evercooren A, Babinet C, Charnay P. 1994. Krox-20 controls myelination in the peripheral nervous system. *Nature* **371:** 796–799. doi:10.1038/371796a0

Toyama BH, Savas JN, Park SK, Harris MS, Ingolia NT, Yates JR III, Hetzer MW. 2013. Identification of long-lived proteins reveals exceptional stability of essential cellular structures. *Cell* **154:** 971–982. doi:10.1016/j.cell.2013.07.037

Trapp BD, Andrews SB, Wong A, O'Connell M, Griffin JW. 1989. Co-localization of the myelin-associated glycoprotein and the microfilament components, F-actin and spectrin, in Schwann cells of myelinated nerve fibres. *J Neurocytol* **18:** 47–60. doi:10.1007/BF01188423

Tricaud N. 2018. Myelinating Schwann cell polarity and mechanically-driven myelin sheath elongation. *Front Cell Neurosci* **11:** 414. doi:10.3389/fncel.2017.00414

Varela-Rey M, Iruarrizaga-Lejarreta M, Lozano JJ, Aransay AM, Fernandez AF, Lavin JL, Mósen-Ansorena D, Berdasco M, Turmaine M, Luka Z, et al. 2014. S-adenosylmethionine levels regulate the Schwann cell DNA methylome. *Neuron* **81:** 1024–1039. doi:10.1016/j.neuron.2014.01.037

Verrier JD, Semple-Rowland S, Madorsky I, Papin JE, Notterpek L. 2010. Reduction of Dicer impairs SC differentiation and myelination. *J Neurosci Res* **88:** 2558–2568.

Wahlbuhl M, Reiprich S, Vogl MR, Bösl MR, Wegner M. 2012. Transcription factor Sox10 orchestrates activity of a neural crest-specific enhancer in the vicinity of its gene. *Nucleic Acids Res* **40:** 88–101. doi:10.1093/nar/gkr734

Walko G, Wogenstein KL, Winter L, Fischer I, Feltri ML, Wiche G. 2013. Stabilization of the dystroglycan complex in Cajal bands of myelinating Schwann cells through plectin-mediated anchorage to vimentin filaments. *Glia* **61:** 1274–1287. doi:10.1002/glia.22514

Wang H, Tewari A, Einheber S, Salzer JL, Melendez-Vasquez CV. 2008. Myosin II has distinct functions in PNS and CNS myelin sheath formation. *J Cell Biol* **182:** 1171–1184. doi:10.1083/jcb.200802091

Wang J, Wang J, Yang L, Zhao C, Wu LN, Xu L, Zhang F, Weng Q, Wegner M, Lu QR. 2020. CTCF-mediated chromatin looping in EGR2 regulation and SUZ12 recruitment critical for peripheral myelination and repair. *Nat Commun* **11:** 4133. doi:10.1038/s41467-020-17955-2

Wanner IB, Mahoney J, Jessen KR, Wood PM, Bates M, Bunge MB. 2006. Invariant mantling of growth cones by Schwann cell precursors characterize growing peripheral nerve fronts. *Glia* **54:** 424–438. doi:10.1002/glia.20389

Waxman SG. 1980. Determinants of conduction velocity in myelinated nerve fibers. *Muscle Nerve* **3:** 141–150. doi:10.1002/mus.880030207

Webster HD. 1971. The geometry of peripheral myelin sheaths during their formation and growth in rat sciatic nerves. *J Cell Biol* **48:** 348–367. doi:10.1083/jcb.48.2.348

Webster HD, Martin R, O'Connell MF. 1973. The relationships between interphase Schwann cells and axons before myelination: a quantitative electron microscopic study. *Dev Biol* **32:** 401–416. doi:10.1016/0012-1606(73)90250-9

Weider M, Wegner M. 2017. SoxE factors: transcriptional regulators of neural differentiation and nervous system development. *Semin Cell Dev Biol* **63:** 35–42. doi:10.1016/j.semcdb.2016.08.013

Weider M, Küspert M, Bischof M, Vogl MR, Hornig J, Loy K, Kosian T, Müller J, Hillgärtner S, Tamm ER, et al. 2012. Chromatin-remodeling factor Brg1 is required for Schwann cell differentiation and myelination. *Dev Cell* **23:** 193–201. doi:10.1016/j.devcel.2012.05.017

Weintraub AS, Li CH, Zamudio AV, Sigova AA, Hannett NM, Day DS, Young RA. 2017. YY1 is a structural regulator of enhancer-promoter loops. *Cell* **171:** 1573–1588.e28. doi:10.1016/j.cell.2017.11.008

Werner T, Hammer A, Wahlbuhl M, Bösl MR, Wegner M. 2007. Multiple conserved regulatory elements with overlapping functions determine Sox10 expression in mouse embryogenesis. *Nucleic Acids Res* **35:** 6526–6538. doi:10.1093/nar/gkm727

Weston JA. 1963. A radioautographic analysis of the migration and localization of trunk neural crest cells in the chick. *Dev Biol* **6:** 279–310. doi:10.1016/0012-1606(63)90016-2

Wilson ER, Della-Flora Nunes G, Weaver MR, Frick LR, Feltri ML. 2021. Schwann cell interactions during the development of the peripheral nervous system. *Dev Neurobiol* **81:** 464–489. doi:10.1002/dneu.22744

Windebank AJ, Wood P, Bunge RP, Dyck PJ. 1985. Myelination determines the caliber of dorsal root ganglion neurons in culture. *J Neurosci* **5:** 1563–1569. doi:10.1523/JNEUROSCI.05-06-01563.1985

Woodhoo A, Alonso MB, Droggiti A, Turmaine M, D'Antonio M, Parkinson DB, Wilton DK, Al-Shawi R, Simons P, Shen J, et al. 2009. Notch controls embryonic Schwann cell differentiation, postnatal myelination and adult plasticity. *Nat Neurosci* **12:** 839–847. doi:10.1038/nn.2323

Wu LM, Williams A, Delaney A, Sherman DL, Brophy PJ. 2012. Increasing internodal distance in myelinated nerves accelerates nerve conduction to a flat maximum. *Curr Biol* **22:** 1957–1961. doi:10.1016/j.cub.2012.08.025

Wu LM, Wang J, Conidi A, Zhao C, Wang H, Ford Z, Zhang L, Zweier C, Ayee BG, Maurel P, et al. 2016. Zeb2 recruits

HDAC-NuRD to inhibit Notch and controls Schwann cell differentiation and remyelination. *Nat Neurosci* **19:** 1060–1072. doi:10.1038/nn.4322

Wüst HM, Wegener A, Fröb F, Hartwig AC, Wegwitz F, Kari V, Schimmel M, Tamm ER, Johnsen SA, Wegner M, et al. 2020. Egr2-guided histone H2B monoubiquitination is required for peripheral nervous system myelination. *Nucleic Acids Res* **48:** 8959–8976. doi:10.1093/nar/gkaa606

Yim AKY, Wang PL, Bermingham JR Jr, Hackett A, Strickland A, Miller TM, Ly C, Mitra RD, Milbrandt J. 2022. Disentangling glial diversity in peripheral nerves at single-nuclei resolution. *Nat Neurosci* **25:** 238–251. doi:10.1038/s41593-021-01005-1

Yin X, Crawford TO, Griffin JW, Tu PH, Lee VMY, Li C, Roder J, Trapp BD. 1998. Myelin-associated glycoprotein is a myelin signal that modulates the caliber of myelinated axons. *J Neurosci* **18:** 1953–1962. doi:10.1523/JNEUROSCI.18-06-01953.1998

Yuan A, Rao MV, Veeranna NR. 2017. Neurofilaments and neurofilament proteins in health and disease. *Cold Spring Harb Perspect Biol* **9:** a018309. doi:10.1101/cshperspect.a018309

Yun B, Anderegg A, Menichella D, Wrabetz L, Feltri ML, Awatramani R. 2010. MicroRNA-deficient Schwann cells display congenital hypomyelination. *J Neurosci* **30:** 7722–7728. doi:10.1523/JNEUROSCI.0876-10.2010

Zhang Y, Yuen S, Peles E, Salzer JL. 2020. Accumulation of neurofascin at nodes of Ranvier is regulated by a paranodal switch. *J Neurosci* **40:** 5709–5723. doi:10.1523/JNEUROSCI.0830-19.2020

Zimmermann H. 1996. Accumulation of synaptic vesicle proteins and cytoskeletal specializations at the peripheral node of Ranvier. *Microsc Res Tech* **34:** 462–473. doi:10.1002/(SICI)1097-0029(19960801)34:5<462::AID-JEMT6>3.0.CO;2-O

Zorick TS, Syroid DE, Brown A, Gridley T, Lemke G. 1999. Krox-20 controls SCIP expression, cell cycle exit and susceptibility to apoptosis in developing myelinating Schwann cells. *Development* **126:** 1397–1406. doi:10.1242/dev.126.7.1397

Zotter B, Dagan O, Brady J, Baloui H, Samanta J, Salzer JL. 2022. Gli1 regulates the postnatal acquisition of peripheral nerve architecture. *J Neurosci* **42:** 183–201. doi:10.1523/JNEUROSCI.3096-20.2021

The Nodes of Ranvier: Mechanisms of Assembly and Maintenance

Matthew N. Rasband[1] and Elior Peles[2]

[1]Department of Neuroscience, Baylor College of Medicine, Houston, Texas 77030, USA

[2]Department of Molecular Cell Biology, The Weizmann Institute of Science, Rehovot 76100, Israel

Correspondence: rasband@bcm.edu; peles@weizmann.ac.il

Action potential propagation along myelinated axons requires clustered voltage-gated sodium and potassium channels. These channels must be restricted to nodes of Ranvier where the action potential is regenerated. Several mechanisms have evolved to facilitate and ensure the correct assembly and stabilization of these essential axonal domains. This review highlights the current understanding of the axon-intrinsic and glial-extrinsic mechanisms that control the formation and maintenance of the nodes of Ranvier in both the peripheral (PNS) and central (CNS) nervous systems.

Axons conduct electrical signals, called action potentials (APs), between neurons in a circuit, in response to sensory input, and between motor neurons and muscles.[3] In mammals and other vertebrates, many axons are myelinated. Myelin, made by Schwann cells and oligodendrocytes in the peripheral nervous system (PNS) and central nervous system (CNS), respectively, is a multilamellar sheet of glial membrane that wraps around axons to increase transmembrane resistance and decrease membrane capacitance. Although myelin is traditionally viewed as a passive contributor to nervous system function, it is now recognized that myelinating glia also play many active roles including regulation of axon diameter, axonal energy metabolism, and the clustering of ion channels at gaps in the myelin sheath called nodes of Ranvier. Together, the active and passive properties conferred on axons by myelin result in axons with high AP conduction velocities, low metabolic demands, and reduced space requirements as compared to unmyelinated axons. Thus, myelin and the clustering of ion channels in axons permitted the evolution of the complex nervous systems found in vertebrates. This review highlights the current understanding of the axonal-intrinsic and glial-extrinsic mechanisms that control the formation and maintenance of the nodes of Ranvier in both the PNS and CNS.

THE ORGANIZATION OF MYELINATED AXONS

Morphology

Myelinated axons are divided into distinct domains including nodes of Ranvier, paranodal ax-

---

[3]This is an update to a previous article published in *Cold Spring Harbor Perspectives in Biology* [Rasband and Peles (2016). *Cold Spring Harb Perspect Biol* **8:** a020495. doi:10.1101/cshperspect.a020495].

Cite this article as *Cold Spring Harb Perspect Biol* doi: 10.1101/cshperspect.a041361

oglial junctions (PNJs), juxtaparanodes (JXPs), and internodes (INDs) (Fig. 1). At PNS nodes of Ranvier, microvilli emanate from the outer aspect of the Schwann cells that myelinate the flanking INDs to contact the nodal axolemma (Berthold et al. 1983; Ichimura and Ellisman 1991). In contrast, myelinating oligodendrocytes in the CNS do not form nodal microvilli. Instead, many nodes are contacted by perinodal astrocyte or oligodendrocyte progenitor cell (OPC) processes (Black and Waxman 1988; Butt et al. 1994, 1999). Whereas microvilli in the PNS contribute to Na^+ channel clustering during development, the functions of perinodal astrocytes and OPCs remain obscure. Another important difference between myelinated PNS and CNS axons is that the latter lacks a basal lamina; PNS myelinated axons have a basal lamina that completely surrounds individual Schwann cell–axon units

and even extends through the nodal gap (Court et al. 2006). In the nodal gap, glial processes are embedded in extracellular matrix (ECM)-rich material that participates in node formation and is thought to serve as a cationic pool through the negative charges of sulfated proteoglycan chains (Fig. 2A; Bekku et al. 2010; Susuki et al. 2013). Nodes of Ranvier are bordered by the PNJ, a specialized axoglial contact formed between the axolemma and the paranodal loops of the myelinating cells. Here, myelin lamellae split into a series of cytoplasmic loops that are closely apposed to the axon, being separated by a gap of only 2.5–3 nm. These loops spiral around the axon to form septate-like junctions with the axon; these junctions constitute the largest known vertebrate intercellular junction. In electron micrographs of longitudinal sections through the paranodal region, the junctions appear as a series of ladder-

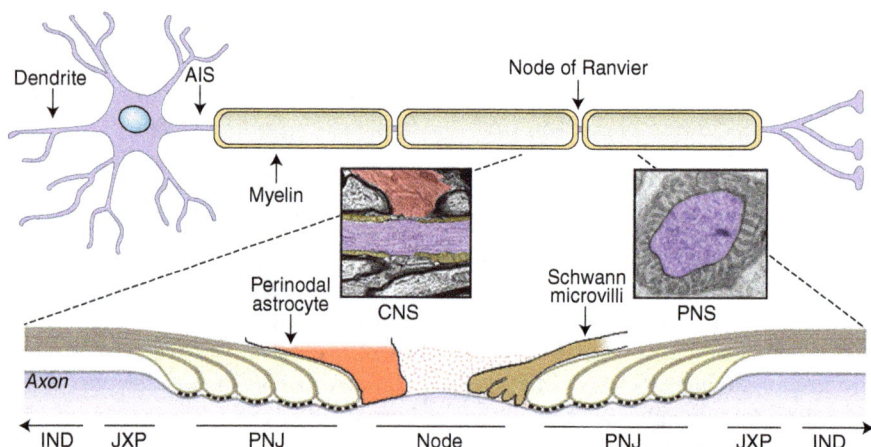

Figure 1. Axonal domains along the myelinated axons. A neuron containing the soma, branched dendrites, and a myelinated axon is shown. Axons are myelinated by Schwann cells in the peripheral nervous system (PNS) and oligodendrocytes in the central nervous system (CNS). Action potentials generated at the axon initial segment (AIS), travel down the axon, and are regenerated at the nodes of Ranvier until reaching the nerve terminals. The axonal membrane is divided into distinct domains: The internodes (INDs; shown partially in the *lower* cartoon) comprise the majority of the axon and are located beneath the compact myelin sheath. The juxtaparanodes (JXPs) are located at the end of the INDs. Near the nodes of Ranvier, the myelin sheath ends with a series of cytoplasmic loops (e.g., paranodal loops) that generate a specialized junction with the axon (paranodal junction [PNJ], often referred to as the axoglial junction). Bordered by the PNJ are the nodes of Ranvier, which are gaps between myelin segments. In the PNS, the nodal axolemma is contacted by microvilli that originate from the outer aspect of the myelinating Schwann cells, whereas in the CNS, many nodes are contacted by a process from a perinodal astrocyte or oligodendrocyte progenitor cell (OPC). The nodal gap is filled with highly charged extracellular matrix material (dots). *Insets* show electron microscopy (EM) images of a longitudinal section through the node of Ranvier in the CNS (perinodal astrocyte in red, paranodes in brown, and axon in purple) and the PNS (axon in purple).

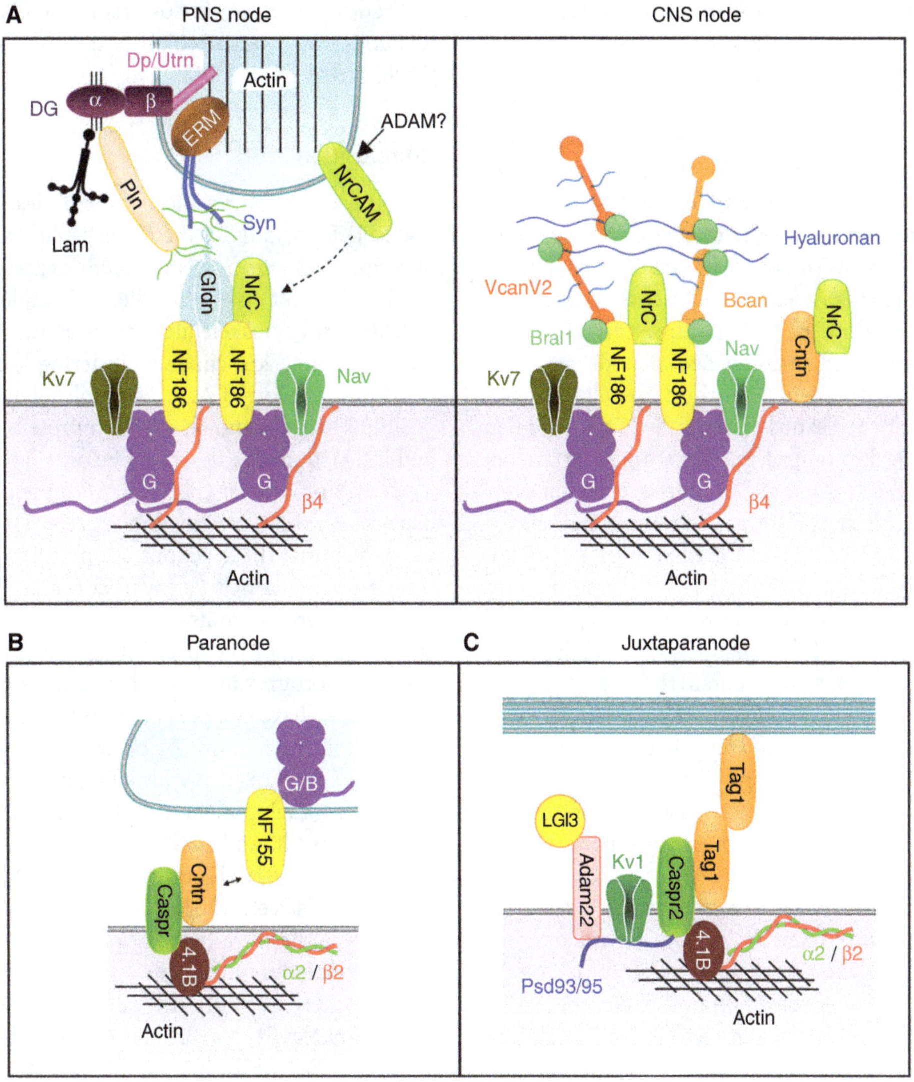

Figure 2. Molecular organization of axonal subdomains. Simplified illustration showing some of the molecules and interactions involved in (*A*) nodal, (*B*) paranodal, and (*C*) juxtaparanodal domains. Peripheral nervous system (PNS) nodes are contacted by microvilli of Schwann cells. Dystroglycan (DG) is autoclaved into α and β chains, which remain associated. β-DG interacts with dystrophin (Dp) and utrophin (Utrn). A transmembrane form of NrCAM is present at the microvilli. Interaction of laminins (Lam) and perlecan (Pln) with α-DG requires proper glycosylation of α-DG (thin black lines). Pln and sydecans3/4 (Syn) are modified by heparan sulfate side chains (green lines). The furin-shed gliomedin (Gldn) trimerizes and is associated with heparan sulfate through its amino-terminal region and collagen-like domain and interacts with NrCAM and NF186 through its olfactomedin domain. G and β4 represent AnkG and β4 spectrin, respectively. In the central nervous system (CNS), the nodal extracellular matrix (ECM) is enriched with shed NrCAM (NrC), but its cellular source is unknown. VcanV2 and Bcan interact with Bral1 and hyaluronan through their G1 globular domains and with NF186 through the G3 domains. The intervening regions of VcanV2 and Bcan are modified by chondroitin sulfate side chains (light blue lines). Contactin (Cntn) was found at CNS nodes, but only weakly at a few PNS nodes. The cytoplasmic partners of NF155 at the paranodal junction are ankG in oligodendrocytes and AnkB in Schwann cells. The ligand of Adam22 at the juxtaparanode is LGI3. (Figure is adapted from Chang et al. 2014 with permission from the authors.)

like densities (i.e., transverse bands or septa) that arise from the axon and contact the glial membranes (Rosenbluth 2009). Paranodal junctions (PNJs) have diverse functions that include attachment of the myelin sheath to the axon, separating the electrical activity of nodal axolemma from internodal axolemma, and functioning as a boundary to limit the lateral diffusion of axonal membrane proteins (more will be said of this boundary function in later sections) (Rosenbluth 2009). As detailed below, the PNJ plays essential roles in the formation and maintenance of nodal and JXP membrane domains. The third specialized region in myelinated axons, the so-called JXP, is located beneath the compact myelin at the interface between the PNJ and IND. This region is characterized by clusters of intramembranous particles in freeze-fracture replicas (Stolinski et al. 1981; Tao-Cheng and Rosenbluth 1984), thought to correspond to delayed rectifier K^+ channels (Chiu and Ritchie 1980). These channels may stabilize conduction and help to maintain the internodal resting potential, especially during myelination and remyelination (Chiu and Ritchie 1984; Wang et al. 1993; Vabnick and Shrager 1998; Rasband 2010). As described below, the formation of this domain depends on the presence of an intact PNJ. Finally, the internodal axolemma located beneath the compact myelin is also considered a unique domain since there are distinct membrane proteins and structures that comprise this region. In the PNS, the organization of the internodal axolemma is dictated by the overlying myelin sheath. Here, the membrane contains two linear rows of juxtaparanodal-type intramembranous particles that flank a paranodal-type aggregate at the inner mesaxon and under the Schmidt–Lanterman incisures (Miller and Pinto da Silva 1977; Stolinski et al. 1985). However, in contrast to the PNJ, no distinct junctional specialization (like transverse bands) is present between the axolemma and the adaxonal Schwann cell membrane at the inner mesaxon. Similar to the JXP, the aggregates found at the juxta-mesaxonal (JXM) and juxta-incisural lines correspond to delayed rectifier K^+ channels (Arroyo et al. 1999). Hence, the radial organization of paranodal and juxtaparanodal components along the INDs corresponds to the longitudinal (i.e., axial) polarity of these proteins near

the nodes of Ranvier. In contrast to the PNS, no juxtamesaxonal organization is detected in the CNS (Arroyo et al. 2001).

Composition

The molecular composition of the nodes, paranodes, JXPs, and INDs is a remarkable example of reciprocal neuron–glia interactions mediated by ECM proteins, cell adhesion molecules (CAMs), and cytoskeletal scaffolds all functioning to facilitate ion channel clustering and the rapid propagation of APs. The molecular compositions of each domain and their protein–protein interactions are described below (Fig. 2).

Nodes of Ranvier (Fig. 2A): AP propagation depends on the rapid de- and repolarization of the axolemma; these actions are performed by Na^+ and K^+ channels clustered at nodes. However, nodes are not uniform in their ion channel composition. Like synapses with different ionotropic neurotransmitter receptors, nodes of Ranvier can also have one or more different kinds of voltage-gated Na^+ and K^+ channels, although this diversity is relatively limited. For example, nodes can include Nav1.1, Nav1.2, Nav1.6, Nav1.7, Nav1.8, and Nav1.9 (Fjell et al. 2000; Boiko et al. 2001; Henry et al. 2005; Duflocq et al. 2008; Black et al. 2012). These Na^+ channels function as the pore-forming protein subunits that mediate ion flux across the membrane. Na^+ channels also interact with the accessory β-subunits Navβ1, Navβ2, and Navβ4 (Chen et al. 2002, 2004; Buffington and Rasband 2013). The β-subunits are covalently linked to Na^+ channels through an extracellular disulfide bond (Chen et al. 2012; Buffington and Rasband 2013) to promote the surface expression of Na^+ channels and change their biophysical properties. In addition to Na^+ channels, nodes of Ranvier are also enriched in K^+ channels (Cooper 2011). These include Kv3.1b, Kv7.2, and Kv7.3 (Devaux et al. 2003, 2004; Stevens et al. 2021), which together regulate neuronal excitability (Battefeld et al. 2014; King et al. 2014). These channels have unique physiologies and together with the diversity of Na^+ channels and their accessory subunits emphasize that not all nodes are created equal.

Cite this article as *Cold Spring Harb Perspect Biol* doi: 10.1101/cshperspect.a041361

Besides ion channels, nodes of Ranvier are also enriched with the cytoskeletal and scaffolding proteins ankyrin G (ankG) and β4 spectrin that together link the Na$^+$ and K$^+$ channels to the underlying cytoskeleton (Kordeli et al. 1995; Berghs et al. 2000). These scaffolds also link ion channels to extrinsic (glial) interactions through CAMs: the 186-kDa isoform of neurofascin (NF186) and NrCAM (Davis et al. 1996). Nodal CAMs interact with a diverse and rich ECM. In the PNS, nodal ECM proteins include gliomedin, syndecans, laminins, NG2, and versican. A specialized ECM also exists in the CNS and consists of the chondroitin sulfate proteoglycans brevican, versican, neurocan, and phosphacan, as well as tenascin-R, Bral1, and shed NrCAM (Susuki et al. 2013). Gliomedin, shed NrCAM, versican, brevican, and Bral1 are all direct binding partners of NF186 (Eshed et al. 2005; Susuki et al. 2013). The aforementioned nodal microvilli in the PNS are also specifically enriched with unique proteins including ezrin, radixin, moesin, EBP50, dystrophin, and utrophin (Occhi et al. 2005). In contrast, no specific proteins have been identified at the terminals of perinodal glial processes in the CNS.

Paranodal junctions (Fig. 2B): PNJs flank nodes of Ranvier and consist of a heterotrimeric CAM complex consisting of the glial, 155-kDa isoform of neurofascin (NF155) and an axonal complex of the glycosylphosphatidyl-inositol (GPI) anchored contactin and Caspr (Charles et al. 2002). These CAMs are all essential for the generation of the characteristic septate-like junction at the paranodes (Bhat et al. 2001; Boyle et al. 2001; Gollan et al. 2002; Rosenbluth 2009). A specialized cytoskeleton is also found at paranodes on both glial and axonal sides of the paranode. Specifically, NF155 interacts with β2 spectrin through ankyrinB (ankB) in Schwann cells and ankG in oligodendrocytes (Chang et al. 2014; Susuki et al. 2018). On the axonal side, Caspr is linked to the αII and β2 spectrin-based cytoskeleton by protein 4.1B (Ogawa et al. 2006). The assembly of the paranodal axonal cytoskeleton is driven by axon–glia interactions between NF155 and Caspr. The axonal paranodal cytoskeletal complex is also the molecular basis of the barrier functions of paranodes (see below)

(Zhang et al. 2013). The significance of proper PNJs in human health is emphasized by the identification of human frameshift mutations in Caspr. These mutations cause severe arthrogryposis multiplex congenita, characterized by congenital joint contractures. These patients also have significantly reduced motor nerve conduction velocities (Laquerriere et al. 2014).

Juxtaparanodes (Fig. 2C): JXPs are characterized mainly by the high-density clustering of Kv1 K$^+$ channels consisting of Kv1.1, Kv1.2, Kv1.4, and KVβ2 subunits (Rasband et al. 2001). Intriguingly, at JXP there is also a CAM interaction homologous to that found at paranodes: GPI-anchored TAG-1 (also known as contactin-2) and Caspr2 form a complex and both are required for proper clustering of JXP Kv1 channels (Poliak et al. 2003). Proteomics of JXP Kv1 channels also revealed ADAM22 (a disintegrin and metalloprotease domain 22) at JXP (Ogawa et al. 2010). Caspr2, Kv1 channels, and ADAM22 all have canonical PDZ-binding motifs. Consistent with this fact, the PDZ-domain proteins PSD95 and PSD93 are also enriched at JXPs. However, these proteins are dispensable for the assembly of the Caspr2/TAG-1/Kv1 channel complex (Horresh et al. 2008). On the other hand, the loss of ADAM22 blocks the clustering of PSD95/93 without affecting Kv1 channel clustering (Ogawa et al. 2010). Recently, human mutations in the gene encoding ADAM22-binding extracellular protein LGI3 were reported, and LGI3 was shown to also be enriched at the JXP. These patients exhibited myokymia and a mislocalized Kv1 channel protein complex (Marafi et al. 2022). Thus, LGI3, Caspr2, and TAG-1 appear to be essential for the proper clustering of JXP proteins. Nevertheless, how axon–glia interactions regulate the localization of LGI3 remains unknown.

Internodes: as described above, the PNS IND has a thin line of alternating JXP–PNJ–JXP proteins that follows the spiral of the inner mesaxon along the IND (hence termed the internodal mesaxonal line). These contact sites between the myelin sheath and the axon share molecular features of the JXP and PNJ. Furthermore, the PNS IND is enriched in CAMs of the Cadm (also known as Necl and SynCAM)-family of proteins (Maurel et al. 2007; Spiegel et al. 2007). Specifically, the internodal axolemma is enriched in Cadm3 and Cadm2 that binds to internodal Cadm4 found on the adaxonal surface of Schwann cells and oligodendrocytes. Axon–glia interactions mediated by these proteins regulate both CNS and PNS myelination (Elazar et al. 2019a,b). In the PNS, an axon–glial interaction mediated by glial Cadm4 and axonal Cadm3 is required for the organization of the axonal membrane, the accurate positioning of Caspr at the paranodal area, as well as the clustering of Kv1 channels at the JXP and the internodal mesaxonal line (Golan et al. 2013). The presence of Kv1 channels at the mesaxonal line also depends on the presence of the cytoskeletal adaptor protein 4.1G, which associates with Cadm4 in Schwann cells (Ivanovic et al. 2012).

Taken together, a picture emerges of a highly organized cytoskeleton and membrane protein distribution found along myelinated axons that uses common kinds of protein complexes important for domain assembly and/or function (for a more detailed list of the molecular composition of nodes of Ranvier see Table 1 in Rasband and Peles 2021). In the next section, we will discuss how these different types of proteins interact to organize the myelinated axon into discrete domains.

ASSEMBLY OF THE NODES OF RANVIER

Formation of the nodes of Ranvier depends on both axonal-intrinsic and glial-extrinsic mechanisms. The assembly of nodes is orchestrated through the placement of several cytoskeletal and tethering proteins along the axolemma at locations that are determined by the contacting oligodendrocytes and myelinating Schwann cells.

Intrinsic Regulators of Node Formation

In neurons, ankG is required for Na^+ channel clustering at the axon initial segment (AIS), which is structurally and functionally similar to nodes, although its assembly is independent of extrinsic mechanisms (Rasband 2010). Since nodal Na^+ and K^+ channels, NF186, NrCAM,

and β4 spectrin all have ankyrin-binding motifs (Stevens and Rasband 2022), ankG has been assumed to function at the center of node assembly as a master "protein accumulator." What evidence supports this conclusion? Mutation of the ankG-binding domain in NF186 blocks its ability to cluster at nodes (Dzhashiashvili et al. 2007; Susuki et al. 2013), and the ankG-binding domain in Na^+ channels is both necessary and sufficient for channel localization; chimeras between the Na^+ channel's ankG-binding motif and an unrelated transmembrane protein can cause it to cluster at nodes (Garrido et al. 2003; Lemaillet et al. 2003; Gasser et al. 2012). The affinity of ankG for Na^+ channels increases 1000-fold following phosphorylation of the ankG-binding motif by the protein kinase CK2 (also enriched at nodes) (Bréchet et al. 2008). Remarkably, a nearly identical ankyrin-binding motif is found in Kv7.2/3 K^+ channels and likely evolved by convergent molecular evolution to facilitate efficient node of Ranvier function (Hill et al. 2008). Clustering of β4 spectrin at nodes depends on its 15th spectrin repeat, which functions as its ankG-binding domain (Yang et al. 2007), and mutation or loss of β4 spectrin leads to disrupted nodes, ataxia, and premature death (Komada and Soriano 2002; Yang et al. 2004; Wang et al. 2018; Liu et al. 2020). Silencing expression of *Ank3* (the gene encoding ankG) in myelinating DRG-neuron/Schwann cell cocultures was reported to block the clustering of Na^+ channels (Dzhashiashvili et al. 2007).

Despite extensive evidence supporting the notion that ankG is required for nodal ion channel clustering, Ho et al. (2014) showed that Na^+ and K^+ channels are still clustered at nodes of Ranvier in PNS and CNS axons after conditional deletion of *Ank3*. This very surprising result revealed redundancy in the ankyrins and spectrins that can function to assemble nodes of Ranvier. Specifically, Ho et al. (2014) found that in the absence of either ankG or β4 spectrin, the orthologues ankR and β1 spectrin can substitute as a "backup" clustering mechanism. This secondary clustering mechanism also participates in the early developmental assembly of nodes of Ranvier. The higher affinities of ankG for NF186 and β4 spectrin ensure that at adult nodes,

Cite this article as *Cold Spring Harb Perspect Biol* doi: 10.1101/cshperspect.a041361

ankR and β1 spectrin are not the main mediators of ion channel clustering and maintenance. Importantly, the genetic removal of both ankG and ankR completely blocks the clustering of nodal ion channels (Ho et al. 2014). In contrast, genetic deletion of both β1 and β4 spectrin results in the destabilization of the ankyrin–ion channel protein complex and the eventual loss of ion channels from nodes (Liu et al. 2020). Thus, ankyrins are responsible for the initial clustering of nodal ion channels, while spectrins are required to maintain the nodal ion channel protein complex. Together, these observations emphasize the importance of intrinsic cytoskeletal and scaffolding protein interactions for proper node assembly and function.

Extrinsic Regulators of Node Formation

Freeze-fracture electron microscopy (EM) studies suggested developmental and cell contact-mediated changes in the axolemma (Tao-Cheng and Rosenbluth 1983). Subsequent studies revealed dynamic changes in the pattern of both Na^+ and K^+ channel clustering during development and after demyelination (Dugandzija-Novakovic et al. 1995; Rasband et al. 1998; Vabnick et al. 1999). During PNS nerve development, different nodal domains follow a stereotypical sequence of events: Na^+ channels are first clustered at the nodes, followed by the generation of the PNJ, and only then by the clustering of K^+ channels at the juxtaparanodal region (Vabnick et al. 1996, 1999; Schafer et al. 2006). In the CNS, Na^+ channel clustering follows the formation of paranodes (Rasband et al. 1999). In both the CNS and the PNS, Na^+ channels cluster initially at sites adjacent to the edges of processes extended by oligodendrocytes (Rasband et al. 1999; Susuki et al. 2013) and myelinating Schwann cells (Vabnick et al. 1996; Ching et al. 1999). Further longitudinal growth of these processes causes displacement of the clusters until ultimately two neighboring clusters appear to fuse, thus forming a new node of Ranvier. The results indicate that these Na^+ channel clusters are positioned by glial cell contact. In agreement, Na^+ channels are diffuse along retinal ganglion cell axons until they cross the lamina cribrosa and become myelinated, after which they are clustered at nodes (Boiko et al. 2001). Na^+ channels are not clustered after ablation of oligodendrocytes (Mathis et al. 2001) or Schwann cells (Vabnick et al. 1997) and are dispersed after demyelination (Dugandzija-Novakovic et al. 1995). Furthermore, nodal Na^+ channels are associated with the edges of myelinating Schwann cells in nerves that display shorter INDs as a result of remyelination (Dugandzija-Novakovic et al. 1995) or genetically impaired myelination (Koszowski et al. 1998). These results strongly support the idea that axoglial interactions induce the clustering of Na^+ channels at the node of Ranvier. As detailed below, two distinct axoglial contact sites at the nodes and the paranodes control the molecular assembly of the nodes of Ranvier.

Nodal Axoglial Interactions

PNS nodes (Fig. 2A): During PNS myelination, heminodal clustering of Na^+ channels (i.e., their accumulation at the edges of a growing myelin segment) depends on the interaction of axonal NF186 with glial gliomedin and NrCAM (Fig. 3A; Eshed et al. 2005; Feinberg et al. 2010). Gliomedin binds to NrCAM present on the microvillar membrane and also associates with other nodal ECM components to create high avidity CAM-binding multimolecular complexes that drive the accumulation of NF186 in the underlying axolemma (Eshed et al. 2007; Feinberg et al. 2010). The clustering activity of gliomedin is negatively regulated by bone morphogenetic protein-1 (BMP1)/Tolloid-like (Tll) proteinases, ensuring the correct spatial and temporal assembly of PNS nodes of Ranvier (Eshed-Eisenbach et al. 2020). Genetic deletion of *Bmp1/Tll* in Schwann cells results in robust nodal clustering along the axon at sites that are not associated with myelin segments. Notably, similar transient clusters were detected in wild-type myelinating cultures, where they preceded the formation of heminodes (Eshed-Eisenbach et al. 2020; Malavasi et al. 2021). In the zebrafish spinal cord, some clusters of Nfasc were also found to be stable, suggesting that a subset may preposition the nodes of Ranvier (Vagionitis et al. 2022). As

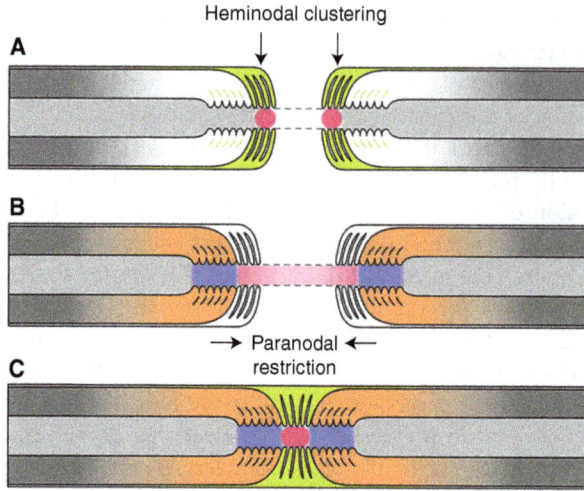

Figure 3. The formation of peripheral nervous system (PNS) nodes of Ranvier. (*A*) Na$^+$ channels (red circle) are trapped at heminodes that are contacted by Schwann cell microvilli (MV; orange). Axon–glia interaction at this site is mediated by the binding of gliomedin and glial NrCAM to axonal NF186. A transmembrane form of glial NrCAM traps gliomedin on Schwann cell MV and enhances its binding to axonal NF186. (*B*) The distribution of Na$^+$ channels is restricted between two forming myelin segments by the paranodal cytoskeleton (blue). Three cell adhesion molecules (CAMs), NF155 present at the glial paranodal loops and an axonal complex of Caspr and contactin mediate axon–glia interaction and the formation of the paranodal axoglial junction (PNJ) and assembly of the paranodal cytoskeleton. (*C*) These two extrinsic cooperating mechanisms provide reciprocal backup systems and ensure that Na$^+$ channels are found at high density at the nodes. Finally, the entire nodal complex requires stabilization and interaction with cytoskeletal and scaffolding proteins. (Figure reprinted from Feinberg et al. 2010 with permission from Elsevier Inc. © 2010.)

a binding partner for NF186, ankG is recruited to the NF186 complexes, followed by β4 spectrin and Na$^+$ channels (Lambert et al. 1997; Eshed et al. 2005; Dzhashiashvili et al. 2007; Yang et al. 2007). The initial accumulation of NF186 at heminodes as well as its internodal clearance requires its extracellular domain, while its stabilization in mature nodes requires its cytoplasmic ankG-binding domain (Zhang et al. 2012). NF186 binds both ankG (Garver et al. 1997) and ECM proteins (Eshed et al. 2005; Susuki et al. 2013), and thus plays a key role during node formation in both CNS and PNS (Sherman et al. 2005; Zonta et al. 2008; Thaxton et al. 2011). In the PNS, secretion of gliomedin into the nodal gap and its accumulation on microvilli by binding to NrCAM initiates the molecular assembly of the nodes of Ranvier (Feinberg et al. 2010). Different binding sites for gliomedin and NrCAM on NF186 allow the multimolecular glial complexes to cluster axonal NF186 at heminodes rapidly and

efficiently (Labasque et al. 2011). By genetically eliminating the expression of different combinations of nodal and paranodal CAMs, it was shown that Schwann cells govern the assembly of nodes of Ranvier by two distinct contact-dependent mechanisms: (1) active clustering of Na$^+$ channels at heminodes (Fig. 3A), and (2) restricting Na$^+$ channel distribution to the nodal gap (Fig. 3B; Feinberg et al. 2010; Amor et al. 2017). The first mechanism of heminodal clustering requires gliomedin, the glial form of NrCAM, and axonal NF186. The second cooperating mechanism that allows the accumulation of ion channels at mature nodes occurs independently of these axonodal CAMs and depends on the formation of the PNJ (see below); thus, mice lacking NF186 still have clustered nodal ankG and ion channels (Amor et al. 2017). These two cooperating processes provide reciprocal backup systems to ensure that Na$^+$ channels are clustered at nodes in the PNS (Feinberg et al. 2010; Rasband and Peles 2021).

How do different axonodal proteins reach the PNS nodes of Ranvier? In general, targeting proteins to specific membrane domains is achieved either by selective transport or random transport followed by specific retention in the correct domain or removal of mislocalized proteins. Retention can be accomplished by anchoring to large molecular scaffolds or by restricting lateral diffusion (Jensen et al. 2011; Lasiecka and Winckler 2011). At PNS nodes during development, CAMs are clustered from a cell surface pool, whereas the accumulation of Na^+ channels and ankG requires vesicular transport from the cell soma (Zhang et al. 2012). These results are consistent with the model that Na^+ channel targeting to the axolemma is ankG-dependent, but gliomedin-induced CAM clustering requires a mobile surface pool of CAMs.

CNS nodes (Fig. 2A): As in the PNS, specific glial-extrinsic mechanisms contribute to the CNS node of Ranvier formation. One of the first studies to investigate how Na^+ channels are clustered in CNS axons concluded that a soluble factor secreted by oligodendrocytes promotes Na^+ channel clustering (Kaplan et al. 1997). For GABAergic neurons, such clustering was found to be induced by contactin-1 and phosphacan or tenascin-R secreted by oligodendrocytes (Freeman et al. 2015; Dubessy et al. 2019). However, other evidence suggested that contact by oligodendrocytes initiates node formation along axons in the optic nerve (Rasband et al. 1999). In contrast to the PNS where contact-dependent mechanisms are necessary to assemble nodes (see above), a combination of both contact-dependent interactions and secreted factors participate in CNS node formation (Susuki et al. 2013). Furthermore, the sequence of contact-dependent events in the CNS is distinct from that in the PNS. First, myelination begins in the CNS together with the formation of the PNJ (Rasband et al. 1999; Susuki et al. 2013). As described above, the paranodal cytoskeleton functions as a barrier to restrict membrane proteins between the adjacent myelin segments. Thus, the initiating event in CNS node formation is the axoglial interactions that occur at the PNJ; these interactions subsequently recruit and assemble the paranodal β2 spectrin-based cytoskeleton. After PNJ assembly,

Na^+ channels accumulate at the edges of the nascent myelin sheath together with NF186, ankG, and β4 spectrin (Susuki et al. 2013). The nodal ECM proteins, although sufficient to induce clustering of nodal proteins alone, then bind to NF186 and likely stabilize the entire nodal protein complex rather than initiate its assembly. Together, although some of the ECM proteins are unique to the CNS, these observations reveal two extrinsic, glia-directed mechanisms for ion channel clustering similar to the PNS: the primary paranodal cytoskeleton mechanism and a secondary NF186–ECM mechanism. These two mechanisms work in concert with ankG-β4 spectrin to cluster ion channels. The idea that multiple independent, yet overlapping, processes contribute to the CNS node of Ranvier formation has been tested by generating mice where two of these overlapping mechanisms are disrupted simultaneously (Susuki et al. 2013; Amor et al. 2017). Loss of only a single mechanism (e.g., PNJ cytoskeleton or ECM) results in mild impairments of channel clustering rather than overt loss, but disruption of both profoundly impairs nodal ion channel clustering (Susuki et al. 2013). Taken together, these observations support the conclusion that two glia-directed mechanisms converge on the clustering of ankG-β4 spectrin to assemble CNS nodes of Ranvier.

CAM-Mediated Formation of a Cytoskeletal Barrier at the Paranodal Junction

The ECM and cytoskeletal mechanisms described above work through ankG. This is consistent with the observation that the ankG-binding domain in Na^+ channels is both necessary and sufficient for nodal Na^+ channel clustering (Gasser et al. 2012). But how do PNJs function to restrict axonal membrane proteins to distinct domains (Fig. 3B)? Loss of Caspr, contactin, or NF155 all result in disrupted paranodes with the absence of transverse bands. Although these paranodal mutants still have clustered Na^+ channels (due to the redundant ECM/NF186 and cytoskeletal mechanisms), they fail to exclude JXP proteins from PNJ domains. Thus, paranodal CAMs are essential for the barrier function of the PNJ. How do these CAMs restrict JXP proteins from paranodes? Two mechanisms have been proposed. First, as the PNJ develops, the CAMs that define this domain acquire biochemical properties consistent with lipid raft–associated proteins (Schafer et al. 2004). Thus, the lipid membrane environments flanking each node of Ranvier are unique from those found at nodes and JXP domains, and the unique biochemical properties of these PNJ lipid domains can influence how membrane proteins partition among nodal, PNJ, and JXP membrane domains. Indeed, mutants lacking myelin lipids have paranodal abnormalities and impaired paranodal barriers with Kv1 K$^+$ channels found in paranodal domains (Dupree et al. 1999; Ishibashi et al. 2002). Second, the paranodal CAMs assemble a specialized paranodal cytoskeleton consisting of protein 4.1B, α2 spectrin, and β2 spectrin (Ogawa et al. 2006; Horresh et al. 2010). Previous studies of how AISs form revealed the existence of an intraxonal boundary, or barrier, consisting of α2 spectrin, β2 spectrin, and ankB that functioned to restrict ankG and β4 spectrin to the proximal axon (Galiano et al. 2012). Loss of the boundary permitted ankG to extend into the distal axon. Since paranodes share many of these same proteins, it was proposed that the paranodal cytoskeleton may also function as a repeating boundary to restrict membrane proteins (i.e., ion channels) to JXP and nodes. This possibility was directly tested using mice that lacked paranodal β2 spectrin or paranodal 4.1B. In *Sptbn1* (the gene encoding β2 spectrin) mutant mice, although the paranodal CAMs remained at paranodes, and PNJs still had transverse bands, Kv1 K$^+$ channels were no longer restricted to JXP domains (Zhang et al. 2013). Thus, paranodal CAMs, their specialized lipid environments, and transverse bands are not sufficient to form paranodal barriers. Instead, the paranodal submembranous cytoskeleton constitutes the paranodal cytoskeletal barrier. In further support of this, the interaction between Caspr and protein 4.1B at paranodes has been shown to be important for the generation of a robust paranodal membrane barrier (Horresh et al. 2010).

To directly test whether the paranodal cytoskeleton functions as one of two mechanisms to cluster ankG-β4 spectrin and ion channels at

nodes of Ranvier, we generated mice lacking both NF186 and β2 spectrin. Thus, although these animals have an intact PNJ, they do not have a paranodal cytoskeleton and they do not have an ECM–NF186 mechanism to recruit ankG. Consistent with the model that these two mechanisms can function independently, mice lacking both NF186 and β2 spectrin in axons fail to cluster ion channels at nodes of Ranvier (Amor et al. 2017). These experiments are the most definitive to date in demonstrating the two overlapping mechanisms of nodal ion channel clustering.

MAINTAINING EXCITABLE DOMAINS IN MYELINATED AXONS

The interactions of the nodal complex with cytoskeletal and ECM components, as well as the formation of the paranodal barrier, all play a role in both assembly and maintenance of the nodes. NF186 is dispensable for the assembly of the AIS and nodes (Hedstrom et al. 2007; Zonta et al. 2008; Feinberg et al. 2010), but is critical for the maintenance of the AIS (Zonta et al. 2011), as well as of nodes in the PNS (Zhang et al. 2012) and the CNS (Desmazieres et al. 2014). In the PNS, the removal of gliomedin and NrCAM, and hence the glial clustering signal operating through NF186, resulted in the gradual loss of Na$^+$ channels and other axonal components that form the nodes (Amor et al. 2014). These results revealed that continuous axon–glia interaction mediated by these molecules is required for long-term maintenance of Na$^+$ channels at nodes of Ranvier (Amor et al. 2014). In the CNS, perinodal ECM components accumulate after Na$^+$ channels are clustered, and thus likely play a stabilizing role (Susuki et al. 2013). This highly variable and redundant ECM always contains BralI, which is a brain-specific link protein that stabilizes the binding of lecticans and hyaluronic acid (Cicanic et al. 2012). In *BralI* knockout mice, changes in the nodal ECM (i.e., brevican, versicanV2, and hyaluronan are missing) do not result in changes in Na$^+$ channel clustering at the node. However, *BralI* mutant mice exhibit slower AP conduction (Bekku et al. 2010), probably since the BralI-associated ECM serves as an extracellular ion pool that facilitates nodal APs.

Thus, the nodal ECM, at least in the CNS, is crucial for maintaining functional nodes. Given the analogy between CNS and PNS nodes, it is likely that in addition to gliomedin and NrCM, other ECM components that bind NF186 are involved in the stabilization and maintenance of PNS nodes. The involvement of other glial-derived ligands in preserving the nodal complex in the PNS is an attractive idea in light of the observation that Na^+ channels were still present in about two-thirds of the nodes in adult mice lacking both gliomedin and NrCAM (Amor et al. 2014), or NF186 (Desmazieres et al. 2014). The nodal gap in the PNS contains several proteins that could cooperate with gliomedin and NrCAM in maintaining the composition of the nodal axolemma (Eshed-Eisenbach and Peles 2013). Some candidates include the adhesion receptor dystroglycan (Saito et al. 2003; Occhi et al. 2005); the heparan sulfate proteoglycans (HSPGs) syndecan 3, syndecan 4 (Goutebroze et al. 2003; Melendez-Vasquez et al. 2005), and perlecan (Bangratz et al. 2012); the ECM components collagen XXVIII (Grimal et al. 2010) and collagen V (Melendez-Vasquez et al. 2005); and the gliomedin-related protein myocilin (Kwon et al. 2013). Gliomedin binds HSPGs (Eshed et al. 2007) and this interaction could further contribute to nodal stabilization, similar to the role HSPGs play in CNS nodes (Dours-Zimmerman et al. 2009; Bekku et al. 2010; Susuki et al. 2013).

Whereas nodal ankyrins are responsible for the initial clustering of ion channels, their maintenance at the nodes of Ranvier requires an intact spectrin cytoskeleton. Since β4 and β1 spectrin are each sufficient to stabilize nodes of Ranvier (Ho et al. 2014), Liu et al. (2020) generated mice lacking both spectrins. Although ankG and ion channels were initially clustered, over time they were gradually lost from nodes, and mice developed tremors and severe ataxia. Similarly, since all neuronal spectrins require α2 spectrin as an obligate component, mice lacking neuronal α2 spectrin also had significantly impaired nodal ion channel clustering in addition to axon degeneration (Huang et al. 2017).

Given their significant role in node assembly, it is not surprising that the PNJs are also required for node maintenance. Thus, in mutant mice with paranodal abnormalities such as mice lacking Caspr (Rios et al. 2003), the galactolipids galactocerebroside (GalC) and sulfatide (Dupree et al. 1998; Rosenbluth et al. 2003), or NF155 (Pillai et al. 2009), nodes gradually become larger and more irregular in shape. The initial accumulation of NF186 at heminodes in the PNS requires its extracellular domain. However, NF186's cytoplasmic ankG-binding domain mediates its transport and stabilization in mature nodes, suggesting that additional mechanisms, likely involving ankG, regulate node maintenance and stability (Zhang et al. 2012). In addition, axoglial interactions at nodes also control nodal stability and the organization by preventing the intrusion of the adjacent paranodal loops onto the nodal axolemma (Thaxton et al. 2011; Amor et al. 2014; Desmazieres et al. 2014). Maintaining the exquisite molecular organization of myelinated axons and particularly the nodal environ is of interest given the presence of autoantibodies to nodal CAMs in Guillain–Barre syndrome (GBS), chronic inflammatory demyelinating polyneuropathy (CIDP), and multiple sclerosis (Mathey et al. 2007; Devaux 2012; Devaux et al. 2012; Ng et al. 2012). The presence of such antibodies in animal models of these neuropathies leads to the disorganization of the nodes of Ranvier, formation of some binary nodes, lengthening of the nodal gap, and reduced nerve conduction (Lonigro and Devaux 2009; Gao et al. 2021). In addition to the direct disruption of nodes by autoimmune attack described above, loss of the myelin sheath (e.g., stroke and spinal cord injury), dysmyelination (e.g., Charcot–Marie–Tooth disease), or hypomyelination (e.g., Pelizaeus–Merzbacher disease) all include as a common sequela the aberrant clustering of Na^+ and K^+ channels. Furthermore, genetic studies have identified mutations in ion channels (e.g., Kv7.2, Nav1.1, and Nav1.6), scaffolding proteins (e.g., ankG, α2, and β4 spectrin), and CAMs (e.g., Caspr, contactin-1, and Caspr2) that are found at and near nodes of Ranvier (Faivre-Sarrailh and Devaux 2013; Susuki 2013; Wang et al. 2018). Together, these observations emphasize the necessity of nodes for human health, and suggest the proper organization of nodes of Ranvier and their associated domains is an essential consideration for any therapeutic effort aimed at nervous system repair or preservation after disease or injury.

CONCLUDING REMARKS

The cellular and molecular mechanisms that evolved to facilitate the assembly of membrane domains along myelinated axons provide exquisite examples of how myelinating glia shape and influence the function and efficiency of neurons. Although much is known about both the extrinsic and intrinsic mechanisms regulating the assembly of nodes, paranodes, and JXPs, more work remains to be done, especially as we consider how to both preserve and repair these domains after injury. One important question for future studies will be to determine how these important domains are maintained over the lifetime of an animal. For example, it is clear that ion channels must be continually replaced at nodes of Ranvier. The developmental mechanisms for the recruitment of preexisting pools of ion channels from membrane sources are not used to replace and replenish existing nodes. How are ion channels targeted to preexisting nodes? What molecular motors are used to transport these nodal proteins and how is a node recognized as a site for cargo delivery? These and many other questions remain to be answered for a complete understanding of how the nodes of Ranvier are both assembled and maintained. We believe any therapeutic approach aimed at the repair or restoration of nervous system function will require a detailed understanding of these mechanisms.

ACKNOWLEDGMENTS

We thank Dr. Yael Eshed-Eisenbach for the discussion and comments. Work done in the authors' laboratories is supported by the National Institute of Health, NINDS (NS122073 MNR), the Israel Science Foundation, the United State-Israel Binational Science Foundation, and the Dr. Miriam and Sheldon G. Adelson Medical Research Foundation. E.P. is the Incumbent of the Hanna Hertz Professorial Chair for Multiple Sclerosis and Neuroscience.

REFERENCES

Amor V, Feinberg K, Eshed-Eisenbach Y, Vainshtein A, Frechter S, Grumet M, Rosenbluth J, Peles E. 2014. Long-term maintenance of Na$^+$ channels at nodes of Ranvier depends on glial contact mediated by gliomedin and NrCAM. *J Neurosci* **34:** 5089–5098. doi:10.1523/JNEUROSCI.4752-13.2014

Amor V, Zhang C, Vainshtein A, Zhang A, Zollinger DR, Eshed-Eisenbach Y, Brophy PJ, Rasband MN, Peles E. 2017. The paranodal cytoskeleton clusters Na$^+$ channels at nodes of Ranvier. *eLife* **6:** e21392. doi:10.7554/eLife.21392

Arroyo EJ, Xu YT, Zhou L, Messing A, Peles E, Chiu SY, Scherer SS. 1999. Myelinating Schwann cells determine the internodal localization of Kv1.1, Kv1.2, Kvβ2, and Caspr. *J Neurocytol* **28:** 333–347. doi:10.1023/A:1007009613484

Arroyo EJ, Xu T, Poliak S, Watson M, Peles E, Scherer SS. 2001. Internodal specializations of myelinated axons in the central nervous system. *Cell Tissue Res* **305:** 53–66. doi:10.1007/s004410100403

Bangratz M, Sarrazin N, Devaux J, Zambroni D, Echaniz-Laguna A, René F, Boërio D, Davoine CS, Fontaine B, Feltri ML, et al. 2012. A mouse model of Schwartz–Jampel syndrome reveals myelinating Schwann cell dysfunction with persistent axonal depolarization in vitro and distal peripheral nerve hyperexcitability when perlecan is lacking. *Am J Pathol* **180:** 2040–2055. doi:10.1016/j.ajpath.2012.01.035

Battefeld A, Tran BT, Gavrilis J, Cooper EC, Kole MH. 2014. Heteromeric Kv7.2/7.3 channels differentially regulate action potential initiation and conduction in neocortical myelinated axons. *J Neurosci* **34:** 3719–3732. doi:10.1523/JNEUROSCI.4206-13.2014

Bekku Y, Vargová L, Goto Y, Vorísek I, Dmytrenko L, Narasaki M, Ohtsuka A, Fässler R, Ninomiya Y, Syková E, et al. 2010. Bral1: its role in diffusion barrier formation and conduction velocity in the CNS. *J Neurosci* **30:** 3113–3123. doi:10.1523/JNEUROSCI.5598-09.2010

Berghs S, Aggujaro D, Dirkx R, Maksimova E, Stabach P, Hermel JM, Zhang JP, Philbrick W, Slepnev V, Ort T, et al. 2000. βIV spectrin, a new spectrin localized at axon initial segments and nodes of Ranvier in the central and peripheral nervous system. *J Cell Biol* **151:** 985–1002. doi:10.1083/jcb.151.5.985

Berthold CH, Nordborg C, Hildebrand C, Conradi S, Sourander P, Lugnegard H. 1983. Sural nerve biopsies from workers with a history of chronic exposure to organic solvents and from normal control cases. Morphometric and ultrastructural studies. *Acta Neuropathol* **62:** 73–86. doi:10.1007/BF00684923

Bhat MA, Rios JC, Lu Y, Garcia-Fresco GP, Ching W, St Martin M, Li J, Einheber S, Chesler M, Rosenbluth J, et al. 2001. Axon-glia interactions and the domain organization of myelinated axons requires neurexin IV/Caspr/Paranodin. *Neuron* **30:** 369–383. doi:10.1016/S0896-6273(01)00294-X

Black JA, Waxman SG. 1988. The perinodal astrocyte. *Glia* **1:** 169–183. doi:10.1002/glia.440010302

Black JA, Frézel N, Dib-Hajj SD, Waxman SG. 2012. Expression of Nav1.7 in DRG neurons extends from peripheral terminals in the skin to central preterminal branches and terminals in the dorsal horn. *Mol Pain* **8:** 82. doi:10.1186/1744-8069-8-82

Boiko T, Rasband MN, Levinson SR, Caldwell JH, Mandel G, Trimmer JS, Matthews G. 2001. Compact myelin dictates the differential targeting of two sodium channel isoforms in the same axon. *Neuron* **30:** 91–104. doi:10.1016/S0896-6273(01)00265-3

Boyle ME, Berglund EO, Murai KK, Weber L, Peles E, Ranscht B. 2001. Contactin orchestrates assembly of the septate-like junctions at the paranode in myelinated peripheral nerve. *Neuron* **30:** 385–397. doi:10.1016/S0896-6273(01)00296-3

Bréchet A, Fache MP, Brachet A, Ferracci G, Baude A, Irondelle M, Pereira S, Leterrier C, Dargent B. 2008. Protein kinase CK2 contributes to the organization of sodium channels in axonal membranes by regulating their interactions with ankyrin G. *J Cell Biol* **183:** 1101–1114. doi:10.1083/jcb.200805169

Buffington SA, Rasband MN. 2013. Na$^+$ channel-dependent recruitment of Na$_v$β4 to axon initial segments and nodes of Ranvier. *J Neurosci* **33:** 6191–6202. doi:10.1523/JNEUROSCI.4051-12.2013

Butt AM, Duncan A, Berry M. 1994. Astrocyte associations with nodes of Ranvier: ultrastructural analysis of HRP-filled astrocytes in the mouse optic nerve. *J Neurocytol* **23:** 486–499. doi:10.1007/BF01184072

Butt AM, Duncan A, Hornby MF, Kirvell SL, Hunter A, Levine JM, Berry M. 1999. Cells expressing the NG2 antigen contact nodes of Ranvier in adult CNS white matter. *Glia* **26:** 84–91. doi:10.1002/(SICI)1098-1136(199903)26:1<84::AID-GLIA9>3.0.CO;2-L

Chang KJ, Zollinger DR, Susuki K, Sherman DL, Makara MA, Brophy PJ, Cooper EC, Bennett V, Mohler PJ, Rasband MN. 2014. Glial ankyrins facilitate paranodal axoglial junction assembly. *Nat Neurosci* **17:** 1673–1681. doi:10.1038/nn.3858

Charles P, Tait S, Faivre-Sarrailh C, Barbin G, Gunn-Moore F, Denisenko-Nehrbass N, Guennoc AM, Girault JA, Brophy PJ, Lubetzki C. 2002. Neurofascin is a glial receptor for the paranodin/Caspr-contactin axonal complex at the axoglial junction. *Curr Biol* **12:** 217–220. doi:10.1016/S0960-9822(01)00680-7

Chen C, Bharucha V, Chen Y, Westenbroek RE, Brown A, Malhotra JD, Jones D, Avery C, Gillespie PJ, Kazen-Gillespie KA, et al. 2002. Reduced sodium channel density, altered voltage dependence of inactivation, and increased susceptibility to seizures in mice lacking sodium channel β2-subunits. *Proc Natl Acad Sci* **99:** 17072–17077. doi:10.1073/pnas.212638099

Chen C, Westenbroek RE, Xu X, Edwards CA, Sorenson DR, Chen Y, McEwen DP, O'Malley HA, Bharucha V, Meadows LS, et al. 2004. Mice lacking sodium channel β1 subunits display defects in neuronal excitability, sodium channel expression, and nodal architecture. *J Neurosci* **24:** 4030–4042. doi:10.1523/JNEUROSCI.4139-03.2004

Chen C, Calhoun JD, Zhang Y, Lopez-Santiago L, Zhou N, Davis TH, Salzer JL, Isom LL. 2012. Identification of the cysteine residue responsible for disulfide linkage of Na$^+$ channel α and β2 subunits. *J Biol Chem* **287:** 39061–39069. doi:10.1074/jbc.M112.397646

Ching W, Zanazzi G, Levinson SR, Salzer JL. 1999. Clustering of neuronal sodium channels requires contact with myelinating Schwann cells. *J Neurocytol* **28:** 295–301. doi:10.1023/A:1007053411667

Chiu SY, Ritchie JM. 1980. Potassium channels in nodal and internodal axonal membrane of mammalian myelinated fibres. *Nature* **284:** 170–171. doi:10.1038/284170a0

Chiu SY, Ritchie JM. 1984. On the physiological role of internodal potassium channels and the security of conduction in myelinated nerve fibres. *Proc R Soc Lond B Biol Sci* **220:** 415–422. doi:10.1098/rspb.1984.0010

Cicanic M, Sykova E, Vargova L. 2012. Bral1: "Superglue" for the extracellular matrix in the brain white matter. *Int J Biochem Cell Biol* **44:** 596–599. doi:10.1016/j.biocel.2012.01.009

Cooper EC. 2011. Made for "anchorin": Kv7.2/7.3 (KCNQ2/KCNQ3) channels and the modulation of neuronal excitability in vertebrate axons. *Semin Cell Dev Biol* **22:** 185–192. doi:10.1016/j.semcdb.2010.10.001

Court FA, Wrabetz L, Feltri ML. 2006. Basal lamina: Schwann cells wrap to the rhythm of space-time. *Curr Opin Neurobiol* **16:** 501–507. doi:10.1016/j.conb.2006.08.005

Davis JQ, Lambert S, Bennett V. 1996. Molecular composition of the node of Ranvier: identification of ankyrin-binding cell adhesion molecules neurofascin (mucin$^+$/third FNIII domain$^-$) and NrCAM at nodal axon segments. *J Cell Biol* **135:** 1355–1367. doi:10.1083/jcb.135.5.1355

Desmazieres A, Zonta B, Zhang A, Wu LM, Sherman DL, Brophy PJ. 2014. Differential stability of PNS and CNS nodal complexes when neuronal neurofascin is lost. *J Neurosci* **34:** 5083–5088. doi:10.1523/JNEUROSCI.4662-13.2014

Devaux JJ. 2012. Antibodies to gliomedin cause peripheral demyelinating neuropathy and the dismantling of the nodes of Ranvier. *Am J Pathol* **181:** 1402–1413. doi:10.1016/j.ajpath.2012.06.034

Devaux J, Alcaraz G, Grinspan J, Bennett V, Joho R, Crest M, Scherer SS. 2003. Kv3.1b is a novel component of CNS nodes. *J Neurosci* **23:** 4509–4518. doi:10.1523/JNEUROSCI.23-11-04509.2003

Devaux JJ, Kleopa KA, Cooper EC, Scherer SS. 2004. KCNQ2 is a nodal K$^+$ channel. *J Neurosci* **24:** 1236–1244. doi:10.1523/JNEUROSCI.4512-03.2004

Devaux JJ, Odaka M, Yuki N. 2012. Nodal proteins are target antigens in Guillain-Barré syndrome. *J Peripher Nerv Syst* **17:** 62–71. doi:10.1111/j.1529-8027.2012.00372.x

Dours-Zimmerman MT, Maurer K, Rauch U, Stoffel W, Fassler R, Zimmermann DR. 2009. Versican V2 assembles the extracellular matrix surrounding the nodes of Ranvier in the CNS. *J Neurosci* **29:** 7731–7742. doi:10.1523/JNEUROSCI.4158-08.2009

Dubessy AL, Mazuir E, Rappeneau Q, Ou S, Abi Ghanem C, Piquand K, Aigrot MS, Thétiot M, Desmazières A, Chan E, et al. 2019. Role of a Contactin multi-molecular complex secreted by oligodendrocytes in nodal protein clustering in the CNS. *Glia* **67:** 2248–2263. doi:10.1002/glia.23681

Duflocq A, Le Bras B, Bullier E, Couraud F, Davenne M. 2008. Nav1.1 is predominantly expressed in nodes of Ranvier and axon initial segments. *Mol Cell Neurosci* **39**: 180–192. doi:10.1016/j.mcn.2008.06.008

Dugandzija-Novakovic S, Koszowski AG, Levinson SR, Shrager P. 1995. Clustering of Na$^+$ channels and node of Ranvier formation in remyelinating axons. *J Neurosci* **15**: 492–503. doi:10.1523/JNEUROSCI.15-01-00492.1995

Dupree JL, Coetzee T, Blight A, Suzuki K, Popko B. 1998. Myelin galactolipids are essential for proper node of Ranvier formation in the CNS. *J Neurosci* **18**: 1642–1649. doi:10.1523/JNEUROSCI.18-05-01642.1998

Dupree JL, Girault JA, Popko B. 1999. Axo-glial interactions regulate the localization of axonal paranodal proteins. *J Cell Biol* **147**: 1145–1152. doi:10.1083/jcb.147.6.1145

Dzhashiashvili Y, Zhang Y, Galinska J, Lam I, Grumet M, Salzer JL. 2007. Nodes of Ranvier and axon initial segments are ankyrin G-dependent domains that assemble by distinct mechanisms. *J Cell Biol* **177**: 857–870. doi:10.1083/jcb.200612012

Elazar N, Vainshtein A, Golan N, Vijayaragavan B, Schaeren-Wiemers N, Eshed-Eisenbach Y, Peles E. 2019a. Axoglial adhesion by Cadm4 regulates CNS myelination. *Neuron* **101**: 224–231.e5. doi:10.1016/j.neuron.2018.11.032

Elazar N, Vainshtein A, Rechav K, Tsoory M, Eshed-Eisenbach Y, Peles E. 2019b. Coordinated internodal and paranodal adhesion controls accurate myelination by oligodendrocytes. *J Cell Biol* **218**: 2887–2895. doi:10.1083/jcb.201906099

Eshed Y, Feinberg K, Poliak S, Sabanay H, Sarig-Nadir O, Spiegel I, Bermingham JR Jr, Peles E. 2005. Gliomedin mediates Schwann cell-axon interaction and the molecular assembly of the nodes of Ranvier. *Neuron* **47**: 215–229. doi:10.1016/j.neuron.2005.06.026

Eshed Y, Feinberg K, Carey DJ, Peles E. 2007. Secreted gliomedin is a perinodal matrix component of peripheral nerves. *J Cell Biol* **177**: 551–562. doi:10.1083/jcb.200612139

Eshed-Eisenbach Y, Peles E. 2013. The making of a node: a co-production of neurons and glia. *Curr Opin Neurobiol* **23**: 1049–1056. doi:10.1016/j.conb.2013.06.003

Eshed-Eisenbach Y, Devaux J, Vainshtein A, Golani O, Lee SJ, Feinberg K, Sukhanov N, Greenspan DS, Susuki K, Rasband MN, et al. 2020. Precise spatiotemporal control of Nodal Na$^+$ channel clustering by bone morphogenetic protein-1/Tolloid-like proteinases. *Neuron* **106**: 806–815. e6. doi:10.1016/j.neuron.2020.03.001

Faivre-Sarrailh C, Devaux JJ. 2013. Neuro–glial interactions at the nodes of Ranvier: implication in health and diseases. *Front Cell Neurosci* **7**: 196. doi:10.3389/fncel.2013.00196

Feinberg K, Eshed-Eisenbach Y, Frechter S, Amor V, Salomon D, Sabanay H, Dupree JL, Grumet M, Brophy PJ, Shrager P, et al. 2010. A glial signal consisting of gliomedin and NrCAM clusters axonal Na$^+$ channels during the formation of nodes of Ranvier. *Neuron* **65**: 490–502. doi:10.1016/j.neuron.2010.02.004

Fjell J, Hjelmström P, Hormuzdiar W, Milenkovic M, Aglieco F, Tyrrell L, Dib-Hajj S, Waxman SG, Black JA. 2000. Localization of the tetrodotoxin-resistant sodium channel NaN in nociceptors. *Neuroreport* **11**: 199–202. doi:10.1097/00001756-200001170-00039

Freeman SA, Desmazières A, Simonnet J, Gatta M, Pfeiffer F, Aigrot MS, Rappeneau Q, Guerreiro S, Michel PP, Yanagawa Y, et al. 2015. Acceleration of conduction velocity linked to clustering of nodal components precedes myelination. *Proc Natl Acad Sci* **112**: E321–E328. doi:10.1073/pnas.1419099112

Galiano MR, Jha S, Ho TS, Zhang C, Ogawa Y, Chang KJ, Stankewich MC, Mohler PJ, Rasband MN. 2012. A distal axonal cytoskeleton forms an intra-axonal boundary that controls axon initial segment assembly. *Cell* **149**: 1125–1139. doi:10.1016/j.cell.2012.03.039

Gao Y, Kong L, Liu S, Liu K, Zhu J. 2021. Impact of Neurofascin on chronic inflammatory demyelinating polyneuropathy via changing the node of Ranvier function: a review. *Front Mol Neurosci* **14**: 779385. doi:10.3389/fnmol.2021.779385

Garrido JJ, Giraud P, Carlier E, Fernandes F, Moussif A, Fache MP, Debanne D, Dargent B. 2003. A targeting motif involved in sodium channel clustering at the axonal initial segment. *Science* **300**: 2091–2094. doi:10.1126/science.1085167

Garver TD, Ren Q, Tuvia S, Bennett V. 1997. Tyrosine phosphorylation at a site highly conserved in the L1 family of cell adhesion molecules abolishes ankyrin binding and increases lateral mobility of neurofascin. *J Cell Biol* **137**: 703–714. doi:10.1083/jcb.137.3.703

Gasser A, Ho TSY, Cheng X, Chang KJ, Waxman SG, Rasband MN, Dib-Hajj S. 2012. An ankyrinG-binding motif is necessary and sufficient for targeting Nav1.6 sodium channels to axon initial segments and nodes of Ranvier. *J Neurosci* **32**: 7232–7243. doi:10.1523/JNEUROSCI.5434-11.2012

Golan N, Kartvelishvily E, Spiegel I, Salomon D, Sabanay H, Rechav K, Vainshtein A, Frechter S, Maik-Rachline G, Eshed-Eisenbach Y, et al. 2013. Genetic deletion of Cadm4 results in myelin abnormalities resembling Charcot-Marie-Tooth neuropathy. *J Neurosci* **33**: 10950–10961. doi:10.1523/JNEUROSCI.0571-13.2013

Gollan L, Sabanay H, Poliak S, Berglund EO, Ranscht B, Peles E. 2002. Retention of a cell adhesion complex at the paranodal junction requires the cytoplasmic region of Caspr. *J Cell Biol* **157**: 1247–1256. doi:10.1083/jcb.200203050

Goutebroze L, Carnaud M, Denisenko N, Boutterin MC, Girault JA. 2003. Syndecan-3 and syndecan-4 are enriched in Schwann cell perinodal processes. *BMC Neurosci* **4**: 29. doi:10.1186/1471-2202-4-29

Grimal S, Puech S, Wagener R, Ventéo S, Carroll P, Fichard-Carroll A. 2010. Collagen XXVIII is a distinctive component of the peripheral nervous system nodes of Ranvier and surrounds nonmyelinating glial cells. *Glia* **58**: 1977–1987. doi:10.1002/glia.21066

Hedstrom KL, Xu X, Ogawa Y, Frischknecht R, Seidenbecher CI, Shrager P, Rasband MN. 2007. Neurofascin assembles a specialized extracellular matrix at the axon initial segment. *J Cell Biol* **178**: 875–886. doi:10.1083/jcb.200705119

Henry MA, Sorensen HJ, Johnson LR, Levinson SR. 2005. Localization of the Nav1.8 sodium channel isoform at nodes of Ranvier in normal human radicular tooth pulp. *Neurosci Lett* **380**: 32–36. doi:10.1016/j.neulet.2005.01.017

Hill AS, Nishino A, Nakajo K, Zhang G, Fineman JR, Selzer ME, Okamura Y, Cooper EC. 2008. Ion channel clustering at the axon initial segment and node of Ranvier evolved sequentially in early chordates. *PLoS Genet* **4**: e1000317. doi:10.1371/journal.pgen.1000317

Ho TS, Zollinger DR, Chang KJ, Xu M, Cooper EC, Stankewich MC, Bennett V, Rasband MN. 2014. A hierarchy of ankyrin-spectrin complexes clusters sodium channels at nodes of Ranvier. *Nat Neurosci* **17**: 1664–1672. doi:10.1038/nn.3859

Horresh I, Poliak S, Grant S, Bredt D, Rasband MN, Peles E. 2008. Multiple molecular interactions determine the clustering of Caspr2 and Kv1 channels in myelinated axons. *J Neurosci* **28**: 14213–14222. doi:10.1523/JNEUROSCI.3398-08.2008

Horresh I, Bar V, Kissil JL, Peles E. 2010. Organization of myelinated axons by Caspr and Caspr2 requires the cytoskeletal adapter protein 4.1B. *J Neurosci* **30**: 2480–2489. doi:10.1523/JNEUROSCI.5225-09.2010

Huang CY, Zhang C, Zollinger DR, Leterrier C, Rasband MN. 2017. An αII spectrin-based cytoskeleton protects large-diameter myelinated axons from degeneration. *J Neurosci* **37**: 11323–11334. doi:10.1523/JNEUROSCI.2113-17.2017

Ichimura T, Ellisman MH. 1991. Three-dimensional fine structure of cytoskeletal-membrane interactions at nodes of Ranvier. *J Neurocytol* **20**: 667–681. doi:10.1007/BF01187068

Ishibashi T, Dupree JL, Ikenaka K, Hirahara Y, Honke K, Peles E, Popko B, Suzuki K, Nishino H, Baba H. 2002. A myelin galactolipid, sulfatide, is essential for maintenance of ion channels on myelinated axon but not essential for initial cluster formation. *J Neurosci* **22**: 6507–6514. doi:10.1523/JNEUROSCI.22-15-06507.2002

Ivanovic A, Horresh I, Golan N, Spiegel I, Sabanay H, Frechter S, Ohno S, Terada N, Möbius W, Rosenbluth J, et al. 2012. The cytoskeletal adapter protein 4.1G organizes the internodes in peripheral myelinated nerves. *J Cell Biol* **196**: 337–344. doi:10.1083/jcb.201111127

Jensen CS, Rasmussen HB, Misonou H. 2011. Neuronal trafficking of voltage-gated potassium channels. *Mol Cell Neurosci* **48**: 288–297. doi:10.1016/j.mcn.2011.05.007

Kaplan MR, Meyer-Franke A, Lambert S, Bennett V, Duncan ID, Levinson SR, Barres BA. 1997. Induction of sodium channel clustering by oligodendrocytes. *Nature* **386**: 724–728. doi:10.1038/386724a0

King CH, Lancaster E, Salomon D, Peles E, Scherer SS. 2014. Kv7.2 regulates the function of peripheral sensory neurons. *J Comp Neurol* **522**: 3262–3280. doi:10.1002/cne.23595

Komada M, Soriano P. 2002. βIV-spectrin regulates sodium channel clustering through ankyrin-G at axon initial segments and nodes of Ranvier. *J Cell Biol* **156**: 337–348. doi:10.1083/jcb.200110003

Kordeli E, Lambert S, Bennett V. 1995. Ankyring. A new ankyrin gene with neural-specific isoforms localized at the axonal initial segment and node of Ranvier. *J Biol Chem* **270**: 2352–2359. doi:10.1074/jbc.270.5.2352

Koszowski AG, Owens GC, Levinson SR. 1998. The effect of the mouse mutation *claw paw* on myelination and nodal frequency in sciatic nerves. *J Neurosci* **18**: 5859–5868. doi:10.1523/JNEUROSCI.18-15-05859.1998

Kwon HS, Johnson TV, Joe MK, Abu-Asab M, Zhang J, Chan CC, Tomarev SI. 2013. Myocilin mediates myelination in the peripheral nervous system through ErbB2/3 signaling. *J Biol Chem* **288**: 26357–26371. doi:10.1074/jbc.M112.446138

Labasque M, Devaux JJ, Lévêque C, Faivre-Sarrailh C. 2011. Fibronectin type III-like domains of neurofascin-186 protein mediate gliomedin binding and its clustering at the developing nodes of Ranvier. *J Biol Chem* **286**: 42426–42434. doi:10.1074/jbc.M111.266353

Lambert S, Davis JQ, Bennett V. 1997. Morphogenesis of the node of Ranvier: co-clusters of ankyrin and ankyrin-binding integral proteins define early developmental intermediates. *J Neurosci* **17**: 7025–7036. doi:10.1523/JNEUROSCI.17-18-07025.1997

Laquerriere A, Maluenda J, Camus A, Fontenas L, Dieterich K, Nolent F, Zhou J, Monnier N, Latour P, Gentil D, et al. 2014. Mutations in CNTNAP1 and ADCY6 are responsible for severe arthrogryposis multiplex congenita with axoglial defects. *Hum Mol Genet* **23**: 2279–2289. doi:10.1093/hmg/ddt618

Lasiecka ZM, Winckler B. 2011. Mechanisms of polarized membrane trafficking in neurons—focusing in on endosomes. *Mol Cell Neurosci* **48**: 278–287. doi:10.1016/j.mcn.2011.06.013

Lemaillet G, Walker B, Lambert S. 2003. Identification of a conserved ankyrin-binding motif in the family of sodium channel α subunits. *J Biol Chem* **278**: 27333–27339. doi:10.1074/jbc.M303327200

Liu CH, Stevens SR, Teliska LH, Stankewich M, Mohler PJ, Hund TJ, Rasband MN. 2020. Nodal β spectrins are required to maintain Na⁺ channel clustering and axon integrity. *eLife* **9**: e52378. doi:10.7554/eLife.52378

Lonigro A, Devaux JJ. 2009. Disruption of neurofascin and gliomedin at nodes of Ranvier precedes demyelination in experimental allergic neuritis. *Brain* **132**: 260–273. doi:10.1093/brain/awn281

Malavasi EL, Ghosh A, Booth DG, Zagnoni M, Sherman DL, Brophy PJ. 2021. Dynamic early clusters of nodal proteins contribute to node of Ranvier assembly during myelination of peripheral neurons. *eLife* **10**: e68089. doi:10.7554/eLife.68089

Marafi D, Kozar N, Duan R, Bradley S, Yokochi K, Al Mutairi F, Saadi NW, Whalen S, Brunet T, Kotzaeridou U, et al. 2022. A reverse genetics and genomics approach to gene paralog function and disease: myokymia and the juxtaparanode. *Am J Hum Genet* **109**: 1713–1723. doi:10.1016/j.ajhg.2022.07.006

Mathey EK, Derfuss T, Storch MK, Williams KR, Hales K, Woolley DR, Al-Hayani A, Davies SN, Rasband MN, Olsson T, et al. 2007. Neurofascin as a novel target for autoantibody-mediated axonal injury. *J Exp Med* **204**: 2363–2372. doi:10.1084/jem.20071053

Mathis C, Denisenko-Nehrbass N, Girault JA, Borrelli E. 2001. Essential role of oligodendrocytes in the formation and maintenance of central nervous system nodal regions. *Development* **128**: 4881–4890. doi:10.1242/dev.128.23.4881

Maurel P, Einheber S, Galinska J, Thaker P, Lam I, Rubin MB, Scherer SS, Murakami Y, Gutmann DH, Salzer JL. 2007. Nectin-like proteins mediate axon Schwann cell interactions along the internode and are essential for myelination. *J Cell Biol* **178**: 861–874. doi:10.1083/jcb .200705132

Melendez-Vasquez C, Carey DJ, Zanazzi G, Reizes O, Maurel P, Salzer JL. 2005. Differential expression of proteoglycans at central and peripheral nodes of Ranvier. *Glia* **52**: 301–308. doi:10.1002/glia.20245

Miller RG, Pinto da Silva P. 1977. Particle rosettes in the periaxonal Schwann cell membrane and particle clusters in the axolemma of rat sciatic nerve. *Brain Res* **130**: 135–141. doi:10.1016/0006-8993(77)90848-4

Ng JK, Malotka J, Kawakami N, Derfuss T, Khademi M, Olsson T, Linington C, Odaka M, Tackenberg B, Prüss H, et al. 2012. Neurofascin as a target for autoantibodies in peripheral neuropathies. *Neurology* **79**: 2241–2248. doi:10.1212/WNL.0b013e31827689ad

Occhi S, Zambroni D, Del Carro U, Amadio S, Sirkowski EE, Scherer SS, Campbell KP, Moore SA, Chen ZL, Strickland S, et al. 2005. Both laminin and Schwann cell dystroglycan are necessary for proper clustering of sodium channels at nodes of Ranvier. *J Neurosci* **25**: 9418–9427. doi:10.1523/JNEUROSCI.2068-05.2005

Ogawa Y, Schafer DP, Horresh I, Bar V, Hales K, Yang Y, Susuki K, Peles E, Stankewich MC, Rasband MN. 2006. Spectrins and ankyrinB constitute a specialized paranodal cytoskeleton. *J Neurosci* **26**: 5230–5239. doi:10.1523/JNEUROSCI.0425-06.2006

Ogawa Y, Oses-Prieto J, Kim MY, Horresh I, Peles E, Burlingame AL, Trimmer JS, Meijer D, Rasband MN. 2010. ADAM22, a Kv1 channel-interacting protein, recruits membrane-associated guanylate kinases to juxtaparanodes of myelinated axons. *J Neurosci* **30**: 1038–1048. doi:10 .1523/JNEUROSCI.4661-09.2010

Pillai AM, Thaxton C, Pribisko AL, Cheng JG, Dupree JL, Bhat MA. 2009. Spatiotemporal ablation of myelinating glia-specific *neurofascin* (*Nfasc NF155*) in mice reveals gradual loss of paranodal axoglial junctions and concomitant disorganization of axonal domains. *J Neurosci Res* **87**: 1773–1793. doi:10.1002/jnr.22015

Poliak S, Salomon D, Elhanany H, Sabanay H, Kiernan B, Pevny L, Stewart CL, Xu X, Chiu SY, Shrager P, et al. 2003. Juxtaparanodal clustering of *Shaker*-like K+ channels in myelinated axons depends on Caspr2 and TAG-1. *J Cell Biol* **162**: 1149–1160. doi:10.1083/jcb.200305018

Rasband MN. 2010. The axon initial segment and the maintenance of neuronal polarity. *Nat Rev Neurosci* **11**: 552–562. doi:10.1038/nrn2852

Rasband MN, Peles E. 2021. Mechanisms of node of Ranvier assembly. *Nat Rev Neurosci* **22**: 7–20. doi:10.1038/ s41583-020-00406-8

Rasband MN, Trimmer JS, Schwarz TL, Levinson SR, Ellisman MH, Schachner M, Shrager P. 1998. Potassium channel distribution, clustering, and function in remyelinating rat axons. *J Neurosci* **18**: 36–47. doi:10.1523/JNEUROSCI .18-01-00036.1998

Rasband MN, Peles E, Trimmer JS, Levinson SR, Lux SE, Shrager P. 1999. Dependence of nodal sodium channel clustering on paranodal axoglial contact in the developing CNS. *J Neurosci* **19**: 7516–7528. doi:10.1523/JNEURO SCI.19-17-07516.1999

Rasband MN, Park EW, Vanderah TW, Lai J, Porreca F, Trimmer JS. 2001. Distinct potassium channels on

pain-sensing neurons. *Proc Natl Acad Sci* **98**: 13373–13378. doi:10.1073/pnas.231376298

Rios JC, Rubin M, St Martin M, Downey RT, Einheber S, Rosenbluth J, Levinson SR, Bhat M, Salzer JL. 2003. Paranodal interactions regulate expression of sodium channel subtypes and provide a diffusion barrier for the node of Ranvier. *J Neurosci* **23**: 7001–7011. doi:10.1523/JNEURO SCI.23-18-07001.2003

Rosenbluth J. 2009. Multiple functions of the paranodal junction of myelinated nerve fibers. *J Neurosci Res* **87**: 3250–3258. doi:10.1002/jnr.22013

Rosenbluth J, Dupree JL, Popko B. 2003. Nodal sodium channel domain integrity depends on the conformation of the paranodal junction, not on the presence of transverse bands. *Glia* **41**: 318–325. doi:10.1002/glia.10179

Saito F, Moore SA, Barresi R, Henry MD, Messing A, Ross-Barta SE, Cohn RD, Williamson RA, Sluka KA, Sherman DL, et al. 2003. Unique role of dystroglycan in peripheral nerve myelination, nodal structure, and sodium channel stabilization. *Neuron* **38**: 747–758. doi:10.1016/S0896-6273(03)00301-5

Schafer DP, Bansal R, Hedstrom KL, Pfeiffer SE, Rasband MN. 2004. Does paranode formation and maintenance require partitioning of neurofascin 155 into lipid rafts? *J Neurosci* **24**: 3176–3185. doi:10.1523/JNEUROSCI.5427-03.2004

Schafer DP, Custer AW, Shrager P, Rasband MN. 2006. Early events in node of Ranvier formation during myelination and remyelination in the PNS. *Neuron Glia Biol* **2**: 69–79. doi:10.1017/S1740925X06000093

Sherman DL, Tait S, Melrose S, Johnson R, Zonta B, Court FA, Macklin WB, Meek S, Smith AJ, Cottrell DF, et al. 2005. Neurofascins are required to establish axonal domains for saltatory conduction. *Neuron* **48**: 737–742. doi:10.1016/j.neuron.2005.10.019

Spiegel I, Adamsky K, Eshed Y, Milo R, Sabanay H, Sarig-Nadir O, Horresh I, Scherer SS, Rasband MN, Peles E. 2007. A central role for Necl4 (SynCAM4) in Schwann cell-axon interaction and myelination. *Nat Neurosci* **10**: 861–869. doi:10.1038/nn1915

Stevens SR, Rasband MN. 2022. Pleiotropic ankyrins: scaffolds for ion channels and transporters. *Channels (Austin)* **16**: 216–229. doi:10.1080/19336950.2022.2120467

Stevens SR, Longley CM, Ogawa Y, Teliska LH, Arumanayagam AS, Nair S, Oses-Prieto JA, Burlingame AL, Cykowski MD, Xue M, et al. 2021. Ankyrin-R regulates fast-spiking interneuron excitability through perineuronal nets and Kv3.1b K+ channels. *eLife* **10**: e6649. doi:10 .7554/eLife.66491

Stolinski C, Breathnach AS, Martin B, Thomas PK, King RH, Gabriel G. 1981. Associated particle aggregates in juxtaparanodal axolemma and adaxonal Schwann cell membrane of rat peripheral nerve. *J Neurocytol* **10**: 679–691. doi:10.1007/BF01262597

Stolinski C, Breathnach AS, Thomas PK, Gabriel G, King RH. 1985. Distribution of particle aggregates in the internodal axolemma and adaxonal Schwann cell membrane of rodent peripheral nerve. *J Neurol Sci* **67**: 213–222. doi:10.1016/0022-510X(85)90117-0

Susuki K. 2013. Node of Ranvier disruption as a cause of neurological diseases. *ASN Neuro* **5**: 209–219. doi:10 .1042/AN20130025

Cite this article as *Cold Spring Harb Perspect Biol* doi: 10.1101/cshperspect.a041361

Susuki K, Chang KJ, Zollinger DR, Liu Y, Ogawa Y, Eshed-Eisenbach Y, Dours-Zimmermann MT, Oses-Prieto JA, Burlingame AL, Seidenbecher CI, et al. 2013. Three mechanisms assemble central nervous system nodes of Ranvier. *Neuron* **78:** 469–482. doi:10.1016/j.neuron.2013.03.005

Susuki K, Zollinger DR, Chang KJ, Zhang C, Huang CY, Tsai CR, Galiano MR, Liu Y, Benusa SD, Yermakov LM, et al. 2018. Glial βII spectrin contributes to paranode formation and maintenance. *J Neurosci* **38:** 6063–6075. doi:10.1523/JNEUROSCI.3647-17.2018

Tao-Cheng JH, Rosenbluth J. 1983. Axolemmal differentiation in myelinated fibers of rat peripheral nerves. *Brain Res* **285:** 251–263. doi:10.1016/0165-3806(83)90023-8

Tao-Cheng JH, Rosenbluth J. 1984. Extranodal particle accumulations in the axolemma of myelinated frog optic axons. *Brain Res* **308:** 289–300. doi:10.1016/0006-8993(84)91068-0

Thaxton C, Pillai AM, Pribisko AL, Dupree JL, Bhat MA. 2011. Nodes of Ranvier act as barriers to restrict invasion of flanking paranodal domains in myelinated axons. *Neuron* **69:** 244–257. doi:10.1016/j.neuron.2010.12.016

Vabnick I, Shrager P. 1998. Ion channel redistribution and function during development of the myelinated axon. *J Neurobiol* **37:** 80–96. doi:10.1002/(SICI)1097-4695(19 9810)37:1<80::AID-NEU7>3.0.CO;2-4

Vabnick I, Novaković SD, Levinson SR, Schachner M, Shrager P. 1996. The clustering of axonal sodium channels during development of the peripheral nervous system. *J Neurosci* **16:** 4914–4922. doi:10.1523/JNEUROSCI.16-16-04914.1996

Vabnick I, Messing A, Chiu SY, Levinson SR, Schachner M, Roder J, Li C, Novakovic S, Shrager P. 1997. Sodium channel distribution in axons of hypomyelinated and MAG null mutant mice. *J Neurosci Res* **50:** 321–336. doi:10.1002/(SICI)1097-4547(19971015)50:2<321::AID-JNR20>3.0.CO;2-9

Vabnick I, Trimmer JS, Schwarz TL, Levinson SR, Risal D, Shrager P. 1999. Dynamic potassium channel distributions during axonal development prevent aberrant firing patterns. *J Neurosci* **19:** 747–758. doi:10.1523/JNEUROSCI.19-02-00747.1999

Vagionitis S, Auer F, Xiao Y, Almeida RG, Lyons DA, Czopka T. 2022. Clusters of neuronal neurofascin prefigure the position of a subset of nodes of Ranvier along individual central nervous system axons in vivo. *Cell Rep* **38:** 110366. doi:10.1016/j.celrep.2022.110366

Wang H, Kunkel DD, Martin TM, Schwartzkroin PA, Tempel BL. 1993. Heteromultimeric K⁺ channels in terminal and juxtaparanodal regions of neurons. *Nature* **365:** 75–79. doi:10.1038/365075a0

Wang CC, Ortiz-González XR, Yum SW, Gill SM, White A, Kelter E, Seaver LH, Lee S, Wiley G, Gaffney PM, et al. 2018. βIV spectrinopathies cause profound intellectual disability, congenital hypotonia, and motor axonal neuropathy. *Am J Hum Genet* **102:** 1158–1168. doi:10.1016/j.ajhg.2018.04.012

Yang Y, Lacas-Gervais S, Morest DK, Solimena M, Rasband MN. 2004. βIV spectrins are essential for membrane stability and the molecular organization of nodes of Ranvier. *J Neurosci* **24:** 7230–7240. doi:10.1523/JNEUROSCI.2125-04.2004

Yang Y, Ogawa Y, Hedstrom KL, Rasband MN. 2007. βIV spectrin is recruited to axon initial segments and nodes of Ranvier by ankyrinG. *J Cell Biol* **176:** 509–519. doi:10.1083/jcb.200610128

Zhang Y, Bekku Y, Dzhashiashvili Y, Armenti S, Meng X, Sasaki Y, Milbrandt J, Salzer JL. 2012. Assembly and maintenance of nodes of Ranvier rely on distinct sources of proteins and targeting mechanisms. *Neuron* **73:** 92–107. doi:10.1016/j.neuron.2011.10.016

Zhang C, Susuki K, Zollinger DR, Dupree JL, Rasband MN. 2013. Membrane domain organization of myelinated axons requires βII spectrin. *J Cell Biol* **203:** 437–443. doi:10.1083/jcb.201308116

Zonta B, Tait S, Melrose S, Anderson H, Harroch S, Higginson J, Sherman DL, Brophy PJ. 2008. Glial and neuronal isoforms of neurofascin have distinct roles in the assembly of nodes of Ranvier in the central nervous system. *J Cell Biol* **181:** 1169–1177. doi:10.1083/jcb.200712154

Zonta B, Desmazieres A, Rinaldi A, Tait S, Sherman DL, Nolan MF, Brophy PJ. 2011. A critical role for neurofascin in regulating action potential initiation through maintenance of the axon initial segment. *Neuron* **69:** 945–956. doi:10.1016/j.neuron.2011.02.021

Perisynaptic Schwann Cells: Guardians of Neuromuscular Junction Integrity and Function in Health and Disease

Thomas W. Gould,[1] Chien-Ping Ko,[2] Hugh Willison,[3] and Richard Robitaille[4,5]

[1]Department of Physiology and Cell Biology, University of Nevada, Reno School of Medicine, Reno, Nevada 89557, USA

[2]Section of Neurobiology, Department of Biological Sciences, University of Southern California, Los Angeles, California 90089-2520, USA

[3]School of Infection and Immunity, University of Glasgow, Glasgow G12 8TA, Scotland

[4]Département de neurosciences, Université de Montréal, Montréal, Québec H3C 3J7, Canada

[5]Centre Interdisciplinaire de Recherche sur le Cerveau et l'apprentissage, Université de Montréal, Montréal, Québec H3C 3J7, Canada

Correspondence: richard.robitaille@umontreal.ca

The neuromuscular junction (NMJ) is a highly reliable synapse to carry the control of the motor commands of the nervous system over the muscles. Its development, organization, and synaptic properties are highly structured and regulated to support such reliability and efficacy. Yet, the NMJ is also highly plastic, able to react to injury, and able to adapt to changes. This balance between structural stability and synaptic efficacy on one hand and structural plasticity and repair on another hand is made possible by perisynaptic Schwann cells (PSCs), glial cells at this synapse. They regulate synaptic efficacy and structural plasticity of the NMJ in a dynamic, bidirectional manner owing to their ability to decode synaptic transmission and by their interactions with trophic-related factors. Alteration of these fundamental roles of PSCs is also important in the maladapted response of NMJs in various diseases and in aging.

The vertebrate neuromuscular junction (NMJ), arguably the best-characterized synapse in the peripheral nervous system (PNS), is composed of three closely associated cellular components: the presynaptic nerve terminal, the postsynaptic specialization, and nonmyelinating Schwann cells.[6] This synapse is fully integrated with the complex functions required to properly control motor functions (Davis et al. 2022). The synapse-associated glial cells are called perisynaptic Schwann cells (PSCs) or terminal Schwann cells (for reviews, see Todd and Robitaille 2006; Feng and Ko 2007; Griffin and Thompson 2008; Sugiura and Lin 2011; Santosa et al. 2018; Fuertes-Alvarez and Izeta 2021). PSCs are a distinct group (Castro et al. 2020)

---

[6]This is an update to a previous article published in *Cold Spring Harbor Perspectives in Biology* [Ko and Robitalle (2015). *Cold Spring Harb Perspect Biol* **7**: a020503. doi:10.1101/cshperspect.a020503].

Cite this article as *Cold Spring Harb Perspect Biol* doi: 10.1101/cshperspect.a041362

among a diverse peripheral glial population (Reed et al. 2022). Multiple roles of PSCs have gained great appreciation since the 1990s and, along with the novel roles of astrocytes in central synapses, have led to the concept of the "tripartite" synapse (Araque et al. 1999, 2014; Volterra et al. 2002; Auld and Robitaille 2003; Kettenmann and Ransom 2013).

Thus, to fully understand synaptic formation and function, it is critical to also consider the active and essential roles of synapse-associated glial cells. We will discuss evidence supporting the existence of a synapse–glia–synapse regulatory loop that helps maintain and restore synaptic efficacy at the NMJ. We will also explore the multiple functions that PSCs exert, functions that are adapted to a given situation at the NMJ (e.g., synapse formation, stability, and reinnervation). This will highlight the great adaptability and plasticity of the morphological and functional properties of PSCs. In this review, we will focus on the multiple roles PSCs play in synaptic formation, maintenance, remodeling, and regeneration, as well as synaptic function and plasticity. Recent advances also highlighted the impact on NMJ structure and functions in disease when PSCs properties are maladapted. Based on the evidence presented, we propose a model in which PSCs, through specific receptor activation, play a prominent role in a continuum of synaptic efficacy, stability, and plasticity at the NMJ. These synaptic-regulated functions allow PSCs to orchestrate the stability and plasticity of the NMJ and, hence, are important for maintaining and adapting synaptic efficacy.

THE TRIPARTITE ORGANIZATION OF THE VERTEBRATE NEUROMUSCULAR JUNCTION

At the vertebrate NMJ, the motor nerve endings are capped by nonmyelinating Schwann cells, in contrast to the motor axons, which are wrapped around by myelinating Schwann cells (Corfas et al. 2004). The existence of PSCs was first suggested by Ranvier (1878), who reported clusters of "arborization nuclei," which were distinct from muscle fiber nuclei and were later identified as nuclei of "teloglia" or terminal Schwann cells at the NMJ (Couteaux 1938, 1960; Tello 1944; Boeke 1949; Ko

et al. 2007; Griffin and Thompson 2008). The identity and intimate contacts of Schwann cells with the nerve terminals were further confirmed with transmission, scanning, and freeze-fracture electron microscopies (Heuser et al. 1976; Desaki and Uehara 1981; Ko 1981). With the advance of immunofluorescence microscopy and the availability of fluorescent probes for PSCs, the tripartite nature of the vertebrate NMJ is further appreciated (Fig. 1). For amphibian muscles, two vital probes for PSCs, peanut agglutinin (PNA) (Ko 1987) and the monoclonal antibody (mAb) 2A12 (Astrow et al. 1998), have been particularly useful to reveal the tripartite organization of the NMJ and the dynamic relationship between PSCs and nerve terminals (see below). Figure 1A–D shows an example of a frog NMJ multiple labeled with mAb 2A12 for PSC somata (asterisks) and processes, with antineurofilament antibody for axons and antisynapsin I antibody for nerve terminals, and with α-bungarotoxin (α-BTX) for acetylcholine receptors (AChRs) on muscle fibers. The merged fluorescent image (Fig. 1D) further reveals the tripartite arrangement, which can also be shown in the electron micrograph of a cross section of the frog NMJ (Fig. 1I). Unfortunately, neither PNA nor mAb2A12 labels mammalian NMJs.

For mammalian muscles, an antibody to the Ca^{2+}-binding protein S100 (Reynolds and Woolf 1992) has been most commonly used for probing mammalian PSCs. Another very useful approach to label mammalian PSCs is the use of transgenic mice that express variants of the green fluorescent protein (GFP) family in axons and Schwann cells to view the dynamic behavior of axons and PSCs in living animals (Kang et al. 2003; Zuo et al. 2004; Li and Thompson 2011; Heredia et al. 2018, 2020). Interestingly, Castro et al. (2020) reported that a combination of labeling with S100 with the neuron–glia antigen 2 (NG2) is a unique molecular marker of S100b+ PSCs in skeletal muscle. This appears to be the first approach that would allow one to selectively label and identify PSCs. Interestingly, Jablonka-Shariff et al. (2021) isolated PSCs from the sternomastoid muscles of S100-GFP mice. Using microarray analysis of the PSC-enriched NMJ samples they found that the T-box transcription factor 21 was expressed in PSCs, thus providing an additional approach through which PSCs can be selectively studied and manipulated. An overview of approaches and tools has been recently reviewed by Negro et al. (2022b). Figure 1 shows an NMJ labeled with α-BTX for AChRs in a mouse that expresses GFP under the control of the S100b promoter in PSC somata (asterisks) and processes (Fig. 1E–H), and cyan fluorescent protein (CFP) in nerve terminal and the preterminal axon (arrow in Fig. 1F). The tripartite organization of the mouse NMJ is further shown in the merged image (Fig. 1H).

A variety of proteins have been demonstrated to be selectively expressed by mammalian PSCs, including LNX-1, an E3 ubiquitin ligase (Young et al. 2005), NaV1.6 (Musarella et al. 2006), TrkC in intact muscles (Hess et al. 2007), GAP-43 (Woolf et al. 1992), p75 neurotrophin receptor (Hassan et al. 1994), nestin (Kang et al. 2007), and the transcription factor zinc-finger proliferation 1 (Ellerton et al. 2008) in denervated muscles. More recently, Procacci et al. (2022) reported the specific presence of Kir4.1, a member of an inwardly rectifying subtype of K^+ channel, in PSCs of adult mice. In the BAC transgenic Kir4.1-CreERT2 mice line that was generated for this study, administration of tamoxifen at specific developmental time points allowed the authors to conclude that Kir4.1 was present in immature Schwann cells, but that its expression is then down-regulated selectively in myelin-forming Schwann cells. This suggests that Kir4.1 may play a functional role maintained exclusively in nonmyelinating Schwann cells including L1 cell-adhesion molecule-positive ($L1^+$) Remak SCs Schwann cells of peripheral nerve and PSCs of the NMJ.

More recently, tissue and single-cell transcriptomic approaches have been used to identify PSC-specific transcripts. For example, Jablonka-Shariff et al. (2022) isolated PSCs from the sternomastoid muscles of S100b-GFP mice. Using microarray analysis of the PSC-enriched NMJ samples, they found that PSCs express a unique combination of genes producing proteins involved in the structural and functional regulation of NMJs, such Erb2 receptor tyrosine kinase (synaptogenesis and pruning), GRB2-associated protein 1 (synaptic pruning), agrin (AChR aggregation), and P2YR (PSCs regulation). Seaberg

et al. (2022) confirmed the presence of these genes in PSCs using the RNAscope approach, but also observed that a number were expressed in other cell types (e.g., muscle spindles and along intramuscular nerves). Similarly, in single-cell transcriptomic analyses of peripheral nerve, several of these genes were enriched in Remak Schwann cell clusters (Wolbert et al. 2020; Gerber et al. 2021). Hence, future studies may benefit from a careful assessment of the extent to which putative PSC marker genes are also expressed in $L1^+$ Remak Schwann cells of peripheral nerves, similar to the study of Procacci et al. (2022). An overview of the approaches and tools used to identify Schwann cell subtypes has been recently reviewed by Negro et al. (2022b).

It has been shown that there are three to five PSC somata in both frog and mammalian mature NMJs, and that the number of PSCs is correlated with the endplate size (Herrera et al. 1990; Love and Thompson 1998; Lubischer and Bebinger 1999; Jordan and Williams 2001). Although animal models have been essential for our understanding of NMJ functions and PSCs roles at this synapse, the human NMJ remains understudied. Recent progress was made by taking advantage of the state-of-the-art labeling of NMJs, and PSCs in particular (Fig. 1J,K; Jones et al. 2017; Aubertin-Leheudre et al. 2020; Alhindi et al. 2021). As expected, human NMJs were shown to be morphologically different from rodent NMJs (Jones et al. 2017; Aubertin-Leheudre et al. 2020) and with a distinct gene profile (Jones et al. 2017). The size principle (i.e., number of PSCs as a function of the NMJ area) appears maintained as human NMJs are overall smaller than mouse NMJs and an average of only two PSCs were observed per NMJ (Aubertin-Leheudre et al. 2020; Alhindi et al. 2021). While Jones et al. (2017) and Alhindi et al. (2021) took advantage of surgical interventions on patients with various chronic diseases, Aubertin-Leheudre et al. (2020) developed a new approach using needle biopsies with enhanced success rate. This allows one to obtain NMJ samples from healthy subjects.

It is not clear why PSCs are nonmyelinating even though they can be labeled with antibodies to some myelinating glial markers, such as myelin protein zero (P0), myelin-associated glycoprotein

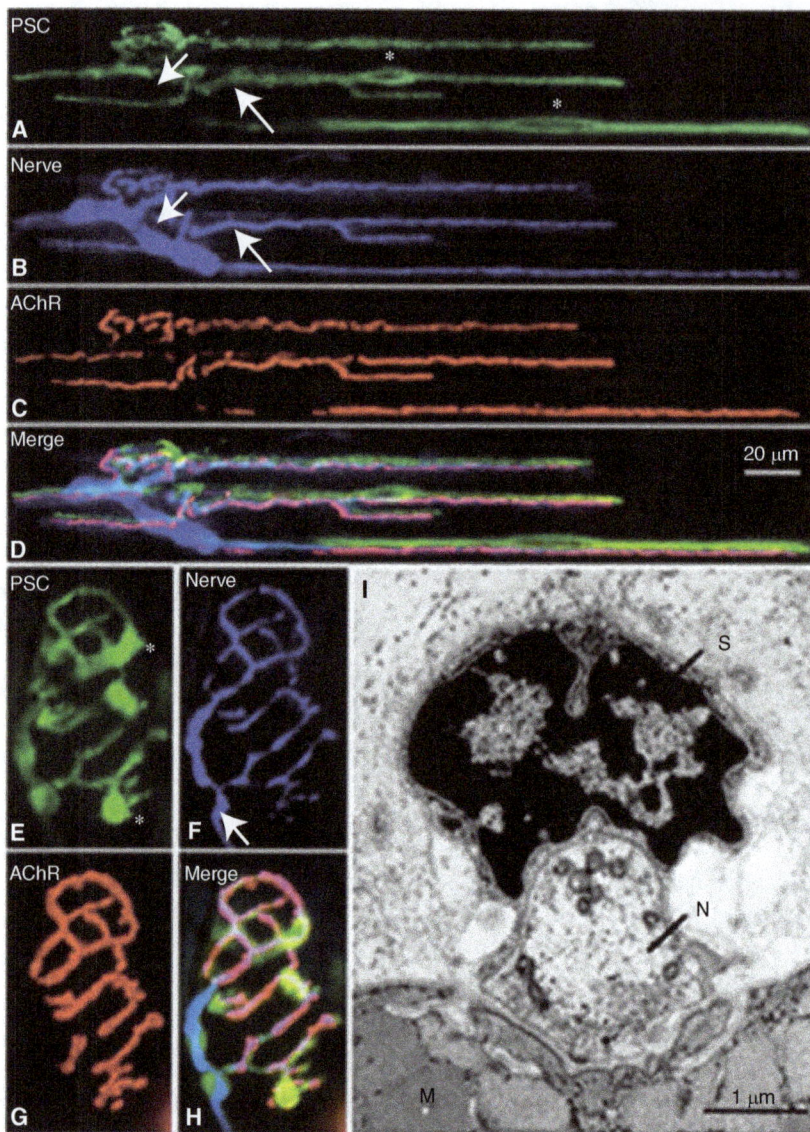

Figure 1. The tripartite organization at the neuromuscular junction (NMJ). (*A–D*) A frog NMJ fluorescently labeled with a monoclonal antibody (2A12) for perisynaptic Schwann cells (PSCs) (green), antineurofilament and antisynapsin I antibodies for nerve fibers and nerve terminals (*B*, blue); and α-bungarotoxin (α-BTX) for acetylcholine receptors (AChRs) (*C*, red). The merged picture (*D*) further shows the tripartite arrangement of the frog NMJ. PSC somata (asterisks in *A*) and processes (arrow in *A*) can be labeled with 2A12 antibody, which does not label Schwann cells along the axon (arrowheads in *A* and *B*). Scale bar in *D* applies to *A–D*. (*E–H*) An NMJ in a transgenic mouse that expresses green fluorescent protein (GFP) in Schwann cells (*E*, green) and cyan fluorescent protein (CFP) in nerve terminals (*F*, blue), labeled with α-BTX for AChRs (*G*, red). Similar to frog NMJ shown in *D*, the merged picture (*H*) further illustrates these three closely associated elements at the mammalian NMJ. PSC somata (asterisk in *E*) and processes, including those associated with the preterminal axon (arrow in *F*) all express GFP. (*I*) Electron micrograph of a frog NMJ in cross section further shows the tripartite arrangement with the PSC (S) capping the nerve terminal (N), which is in apposition with postjunctional folds on the muscle fiber (M). (*Figure and legend continued on following page.*)

(MAG), galactocerebroside, and 2′, 3′-cyclic nucleotide 3′-phosphodiesterase (CNPase) (Georgiou and Charlton 1999). Meehan et al. (2018) reported that the labeling of sulfated GalC stopped at the heminode and did not include PSCs. The genes encoding some of these myelin molecules were also detected by Castro et al. (2020) in their RNA sequencing (RNA-seq) analysis. It is also not well understood why PSCs cap, but do not enclose entirely, the motor nerve terminal. An otherwise

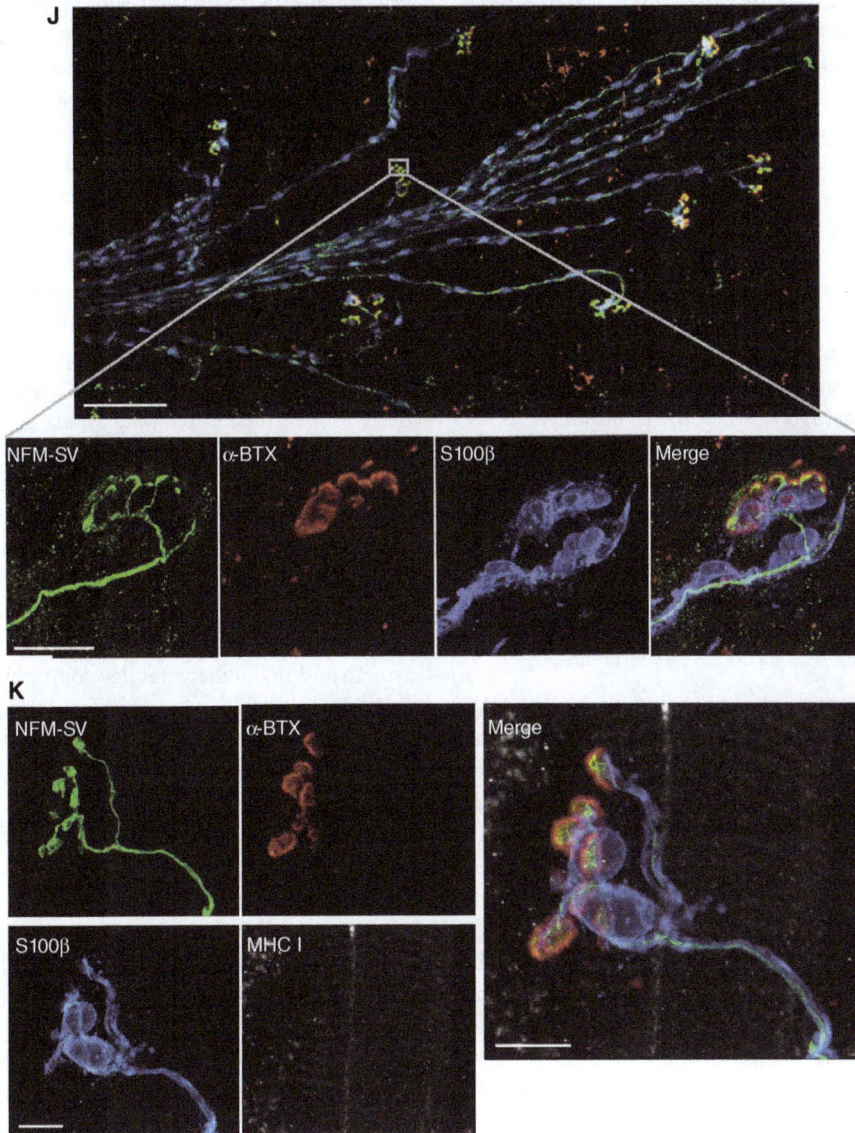

Figure 1. (*Continued*) (*J,K*) False color images of human NMJs obtained using the adapted needle biopsy. The presynaptic terminal (neurofilament and synaptic vesicles, SV2/NF-M, green), postsynaptic apparatus (nicotinic AChRs, α-BTX, red), and glia (PSCs, S100B, blue) were labeled. MHC type I fiber was also labeled (type I fiber, MHCI, gray). Scale bar, 1 μm. Scale bars, 50 μm (*A–H*), close-up 5 μm (*J*); 10 μm (*K*). (*A–H*, Reproduced from Ko and Thompson 2003, Cover Picture, © Springer, with kind permission of Springer Science + Business Media; *I*, reprinted, with permission, from Reddy et al. 2003, Elsevier © 2003.)

complete enclosure of the nerve terminal would obviously severely compromise synaptic function. Frog PSCs, however, project finger-like processes, which contain L-type calcium channels (Robitaille et al. 1996), into the synaptic cleft and interdigitate with active zones—sites of transmitter release. In contrast, mammalian PSC "fingers" are usually excluded from the cleft, which may be attributed to laminin 11 (a5b2g1) in the synaptic cleft (Patton et al. 1998). Besides the synaptic cleft and the muscle surface, a basal lamina also covers PSCs (Saito and Zacks 1969; Engel 1994). However, the extracellular matrix molecules associated with PSC basal lamina are distinct from those in the synaptic cleft and the extrasynaptic muscle surface (Ko 1987; Astrow et al. 1997; Patton et al. 1997; for review, see Patton 2003). It has been suggested that the PSC-associated extracellular matrix may play a role in guiding nerve terminal sprouts at the frog NMJ (Chen and Ko 1994; Ko and Chen 1996; see below). Interestingly, fibroblast-like cells (kranocytes) capping the NMJ have also been shown (Connor and McMahan 1987; Court et al. 2008). Although their role is yet elusive, it was suggested that they could facilitate synaptic regeneration.

ROLE OF PSCs IN SYNAPTOGENESIS

The intimate arrangement of the tripartite NMJ raises a question as to whether PSCs participate in synaptogenesis. To address this question, one needs to know first if Schwann cells are necessary during the initial navigation of axons to their target muscles (Keynes 1987). It has been shown that motor axons can reach their target muscles and even form the initial nerve–muscle contacts, albeit only transiently, in mice genetically lacking the expression of several of the receptors for the epidermal growth family (EGF) family member Neuregulin 1 (mouse *ErbB2*: Morris et al. 1999; Woldeyesus et al. 1999; Lin et al. 2000; *ErbB3*: Riethmacher et al. 1997), as well as in *Splotch* (Grim et al. 1992) mutant mice exhibiting a point mutation in PAX3 mutants, all of which lack all Schwann cells in the developing peripheral nerves. Furthermore, functional nerve–muscle contacts can be formed in cultures without Schwann cells (Kullberg et al. 1977; Chow and Poo 1985). These studies suggest that (1) Schwann cells depend on Neuregulin 1 signaling during embryonic development, and (2) Schwann cells are not necessary for axonal pathfinding and the initial formation of nerve–muscle contacts.

Although Schwann cells are dispensable for the initial stages of NMJ formation, they play a critical role in promoting subsequent synaptic growth, maturation, and maintenance at developing NMJs. Indeed, the fact that synaptic contacts form in erbB mutants, but do not persist, suggests that Schwann cells are required for the maintenance of NMJs (Gould et al. 2019; Liu et al. 2019). Additionally, in frog muscles, PSCs appear shortly after the earliest discernible nerve–muscle contacts in tadpoles, and PSCs then quickly extend processes beyond nerve terminals (Herrera et al. 2000). The subsequent growth of nerve terminals appears to follow along the preceding PSC sprouts as shown with repeated in vivo observations of identified developing NMJs in tadpoles (Reddy et al. 2003). Combining repeated in vivo observations with an ablation technique that takes advantage of mAb2A12 and complement-mediated lysis to selectively ablate PSCs in vivo, Reddy et al. (2003) revealed major perturbations in NMJ structure and establishment, which further shows the critical role of PSCs in promoting synaptic growth and maintenance at developing amphibian NMJs in vivo.

A number of mechanisms have been identified as potential regulators of NMJ formation and maturation. Indeed, mice with conditional inactivation of neurofibromin 1 (Nf1) and phosphatase and tensin homolog (Pten), specifically in Schwann cells, resulted in delayed NMJ maturation (Li et al. 2019). In complementary experiments, the same group showed that Schwann cell–specific Pten inactivation and epidermal growth factor receptor (EGFR) overexpression such as *Neuregulin 1* (*Nrg1*) resulted in NMJ malformation and, including structural disorganization of AChRs (Zhang et al. 2019). Therefore, changes in the amount of neuregulin signaling appear to affect the development of NMJs both directly and indirectly through SC. Supporting this idea, Wang et al. (2017) observed that soluble Neuregulin 1 (NRG1) also regulates AChR cluster density and size at the earliest stage before nerve–

Cite this article as *Cold Spring Harb Perspect Biol* doi: 10.1101/cshperspect.a041362

AChR cluster contact. Finally, NRG1 regulates presynaptic specialization and the positioning of PSCs.

The dependence of PSC survival on NRG1 signaling extends into postnatal developmental periods (Griffin and Thompson 2008). Trachtenberg and Thompson (1996) showed that denervation in neonate, but not in adult, leads to rapid apoptosis of PSCs at rat NMJs, which can be prevented by treatment with exogenous NRG1. This indicates that peripheral nerve axons are the source of the NRG1 that synapse maintenance (in neonates) is required for erbB2/3-expressing PSC. In addition, PSC morphology and NMJ structure can be altered by applications of NRG1 or Schwann cell transplants to mammalian muscles (Trachtenberg and Thompson 1997). The observation that partial denervation in neonatal but not in adult rat muscles results in apoptosis of PSCs together with the absence of nerve terminal sprouting in denervated neonatal muscles further confirms the importance of PSCs in promoting synaptic growth (Lubischer and Thompson 1999). Taken together, these studies suggest that PSCs are essential for the growth and maintenance of developing motor nerve terminals at both amphibian and mammalian muscles.

The molecular mechanisms of how PSCs participate in synaptic growth and maintenance at embryonic and neonatal NMJs are becoming clarified. Recent studies showed that genetic blockade of neuromuscular activity prevents the loss of motor neurons induced by the absence of embryonic Schwann cells in erbB mutants, suggesting that activity-regulated, muscle-derived factors may trigger the retrograde degeneration of motor neurons (Gould et al. 2019; Liu et al. 2019). One such molecule that mediates a portion of these effects is the coagulation factor thrombin (Gould et al. 2019). Additional pathways may involve regulation of the protein Agrin, which is also expressed by Schwann cells in frog, and which enhances AChR aggregation in muscle culture (Yang et al. 2001). Interestingly, metalloproteinases (MMPs) are important regulators of Agrin as they are involved in its cleavage in the synaptic cleft (Chan et al. 2020), suggesting that modulation of Schwann cell–derived Agrin by MMPs may contribute to NMJ formation

and/or maintenance. Additional evidence supporting a role in synaptogenesis for Schwann cells comes from studies of *Xenopus* nerve–muscle cultures (Peng et al. 2003). In particular, Feng and Ko (2008) have shown that Schwann cell–conditioned medium contains transforming growth factor (TGF)-β1. TGF-β1 plays a necessary and sufficient role in promoting NMJ formation and synaptic growth in *Drosophila* (Fuentes-Medel et al. 2012). It has also been shown that *Xenopus* Schwann cell–conditioned medium can acutely enhance transmitter release in developing NMJs in culture (Cao and Ko 2007). However, the in vivo role of TGF-β or other Schwann cell–derived factors in synaptogenesis at NMJs remains to be examined.

One hallmark of mammalian NMJ formation is the innervation of multiple nerve terminals at a single NMJ (polyneuronal innervation) and the subsequent removal of all but one of the nerve endings (synapse elimination) by the second week after birth (Sanes and Lichtman 1999). The potential role of PSCs in pruning excess nerve terminals at multiply innervated NMJs in postnatal muscles has been suggested (Griffin and Thompson 2008). For example, the retraction of nerve terminals and Schwann cell processes from the sites of synapse elimination occurs at a similar time course (Culican et al. 1998). Interestingly, retracting nerve terminals shed numerous membrane-bound remnants called axosomes, which are engulfed by Schwann cells during synapse elimination (Bishop et al. 2004), a phenomenon also seen following nerve injury (Cunningham et al. 2020). Using time-lapse imaging of labeled single PSCs, Brill et al. (2011) have revealed that young PSCs intermingle dynamically in contrast to the static tile patterns seen in adult NMJs. A study using serial electron microscopy has further shown that PSCs participate in synapse elimination by phagocytosis of nerve terminals, although the process involves all axons and PSCs do not seem to select the winner of the competing developing nerve terminals (Smith et al. 2013). Jung et al. (2019) developed a model of PSCs and synapse elimination to further examine the impacts of PSCs in synaptic competition. Their model reproduced synapse elimination and indicated that it should be accelerated by enhanced synaptic activity of one axon by increased areas of vacancies and PSCs. The enhanced synaptic activity of one axon observed in this model relates to the importance of synaptic strength in the regulation of synaptic competition (Darabid et al. 2014). Interestingly, using simultaneous Ca^{2+} imaging of PSCs and synaptic recording of dually innervated mouse NMJs, Darabid et al. (2013) showed that the activity of single PSCs reflects the synaptic strength of each competing nerve terminal and the state of synaptic competition. Hence, PSCs decode synaptic transmission at a later stage of synaptic competition, allowing them to identify the strongest nerve terminal, which is likely to remain and ultimately form the monosynaptic NMJ innervation. This is consistent with the maintained competence of PSCs and Schwann cells in general to detect transmitter release during development (Heredia et al. 2018a; Perez-Gonzalez et al. 2022). The same group (Darabid et al. 2018) went on to show that PSCs biased the outcome of synaptic competition by preferentially enhancing synaptic performances of the stronger nerve terminal via the activation of their P2Y1 receptors (Darabid et al. 2018). In addition, PSC's activation via their P2Y1 receptors also regulates postsynaptic properties by providing resistance to fatigue by regulating perisynaptic K^+ (Heredia et al. 2018b). This illustrates the integrated influence of PSCs on the NMJ as a tripartite synapse, interacting with pre- and postsynaptic elements of the synapse.

ROLE OF PSCs IN SYNAPTIC MAINTENANCE, REMODELING, AND REGENERATION AT ADULT NMJs

Maintenance

It is remarkable that the tripartite organization is maintained at the adult vertebrate NMJ despite the continual mechanical disruptions by muscle contractions throughout the animal's life span. To address the question of whether PSCs play a role in synaptic maintenance, Reddy et al. (2003) took advantage of the selective labeling of PSCs with mAb2A12 and combined this with complement-mediated cell lysis to selectively ablate PSCs from frog NMJs in vivo. They observed no significant changes in synaptic structures and function shortly after PSC ablation (within

5 h). At mouse NMJs, Halstead et al. (2004, 2005) have shown that both NMJ morphology and synaptic transmission are also not acutely affected after selective PSC ablation with an autoantibody against disialosyl epitopes of gangliosides seen in Miller Fisher syndrome. This lack of effect may be caused by the ability of PSCs to both decrease and increase synaptic efficacy (see below; Robitaille 1998; Castonguay and Robitaille 2001; Todd et al. 2010), hence resulting in no net change in the synaptic output at the NMJ. However, partial or total retraction of some nerve terminals and a 50% reduction in transmitter release were seen 1 week after PSC ablation. Hastings et al. (2020) used mice expressing diphtheria toxin receptor (DTR) preferentially in PSCs and local application of diphtheria toxin (DTX) to kill PSCs. They observed that motor axons did not fully degenerate. PSCs returned within 3 weeks and reestablished their coverage at the NMJ. They also observed increased clusters of AChRs and nerve terminal varicosities. Furthermore, PSC ablation after maturation of the mammalian NMJ (P30) also resulted in functional defects whereby NMJ fragmentation and alteration of synaptic transmission revealed by reduced miniature endplate potential amplitude and frequency occurred 6 days after ablation (Barik et al. 2016). These observations suggest that, although NMJ innervation is not completely lost, PSCs are essential for proper morphological innervation and function in the long-term maintenance of NMJs.

Remodeling

Although the tripartite arrangement is maintained at adult NMJs, nerve terminals at frog NMJs undergo extension and/or retraction throughout adult life (Wernig et al. 1980; Herrera et al. 1990; Chen et al. 1991). To address whether PSCs also undergo similar dynamic remodeling, repeated in vivo observations of identified frog NMJs double labeled with a vital florescent dye for nerve terminals and PNA for PSCs have been shown (Chen et al. 1991; Chen and Ko 1994; Ko and Chen 1996). These studies revealed that PSCs and associated extracellular matrix often lead the nerve terminal sprouts. The dynamic

Cite this article as *Cold Spring Harb Perspect Biol* doi: 10.1101/cshperspect.a041362

relationship between PSCs and nerve terminals has also been confirmed using direct injection of fluorescent dyes into adult frog PSCs and nerve terminals (Macleod et al. 2001; Dickens et al. 2003). These findings suggest that the dynamic behavior of PSCs may contribute to the constant remodeling of nerve terminals seen at amphibian NMJs. In contrast, mammalian NMJs are relatively stable (Lichtman et al. 1987; Wigston 1989; Martineau et al. 2018). However, there are minor nerve terminal filopodia and lamellipodia adjacent to PSCs (Robbins and Polak 1988), which also protrude short and unstable processes beyond AChR clusters at mammalian NMJs (Zuo et al. 2004). It is still unclear why adult mammalian NMJs show reduced morphological remodeling compared with amphibian NMJs. This appears an evolutionary characteristic as the human NMJ appears to show less variability than the rodent one (Jones et al. 2017; Aubertin-Leheudre et al. 2020). However, this stability is challenged in diseases such as amyotrophic lateral sclerosis (ALS) and in aging (Bruneteau et al. 2013; Martineau et al. 2018, 2020; Aubertin-Leheudre et al. 2020).

Degeneration and Regeneration

After nerve injury, nerve terminals degenerate, and PSCs become phagocytic to remove debris of degenerating nerve terminals at denervated NMJs (Birks et al. 1960). Several injury signals emerging from the injured axons and axon terminals have been identified to trigger PSCs injury responses. In particular, the group of Montecucco and Rigoni has shown that PSCs are activated by mitochondrial alarmins from axon degenerating nerve terminals (Duregotti et al. 2015), by mitochondrial hydrogen peroxide as part of an autoimmune attack in an animal model of Miller Fisher syndrome (Rodella et al. 2016) and ATP (Negro et al. 2016). PSCs present a complex set of responses to these injury cues. One of the first reactions of PSCs is to the engulf the injured terminals (Lee et al. 2017; Martineau et al. 2020), accompanied by the presence of phagocytotic markers such as Galectin 3 (Martineau et al. 2020; Perez-Gonzalez et al. 2022). Perez-Gonzalez et al. (2022) further showed that the phagocytic response was

up-regulated by the reduced activation of PSC-derived muscarinic AChRs due to reduced ACh release of degenerating nerve terminals. In a recent work, Piovesana et al. (2024) showed that PSCs CB1 receptors facilitate denervation and re-innervation following nerve injury. Furthermore, as originally showed for myelination by Schwann cells (Monk et al. 2011), Jablonka-Shariff et al. (2020) showed that the adhesion G protein–coupled receptor (Gpr126/Adgrg6), regulated PSCs upon injury and was central for their proper response toward NMJ reinnervation.

One seminal study that stimulated our current framework of PSC function was the discovery of profuse sprouting of PSC processes shortly after denervation at the mammalian NMJ (Reynolds and Woolf 1992; Astrow et al. 1994; Son and Thompson 1995a,b). It is interesting to note that Schwann cells at denervated NMJs can release acetylcholine (Dennis and Miledi 1974), although the functional significance of this response is unknown. Furthermore, during reinnervation, nerve terminals grow along PSC "bridges" formed with PSC sprouts originating from adjacent denervated junctions and form the so-called "escaped fibers" that innervate the adjacent denervated endplates (Fig. 2A,Ba–d). A similar role of PSC "bridges" in guiding nerve terminal sprouts has also been shown after partial denervation at the adult NMJ (Fig. 2A,B; Son and Thompson 1995a,b; Love and Thompson 1999). This process was shown to be in part responsible for the changes observed in the pattern of muscle innervation following peripheral nerve injuries, leading to grouping of motor units (Kang et al. 2019).

Bermedo-García et al. (2022) proposed that the functional regeneration of the mouse NMJ relies on long-lasting morphological adaptations. The dynamic relationship between PSCs and regenerating nerve terminals after nerve injury at NMJs has been examined with repeated in vivo observations of the same NMJs labeled with vital dyes (O'Malley et al. 1999; Koirala et al. 2000) or in transgenic mice that express GFP in Schwann cells and CYP in axons (Kang et al. 2003). These in vivo studies further confirm that PSC sprouts guide regenerating nerve terminals following nerve injury. It has been suggested

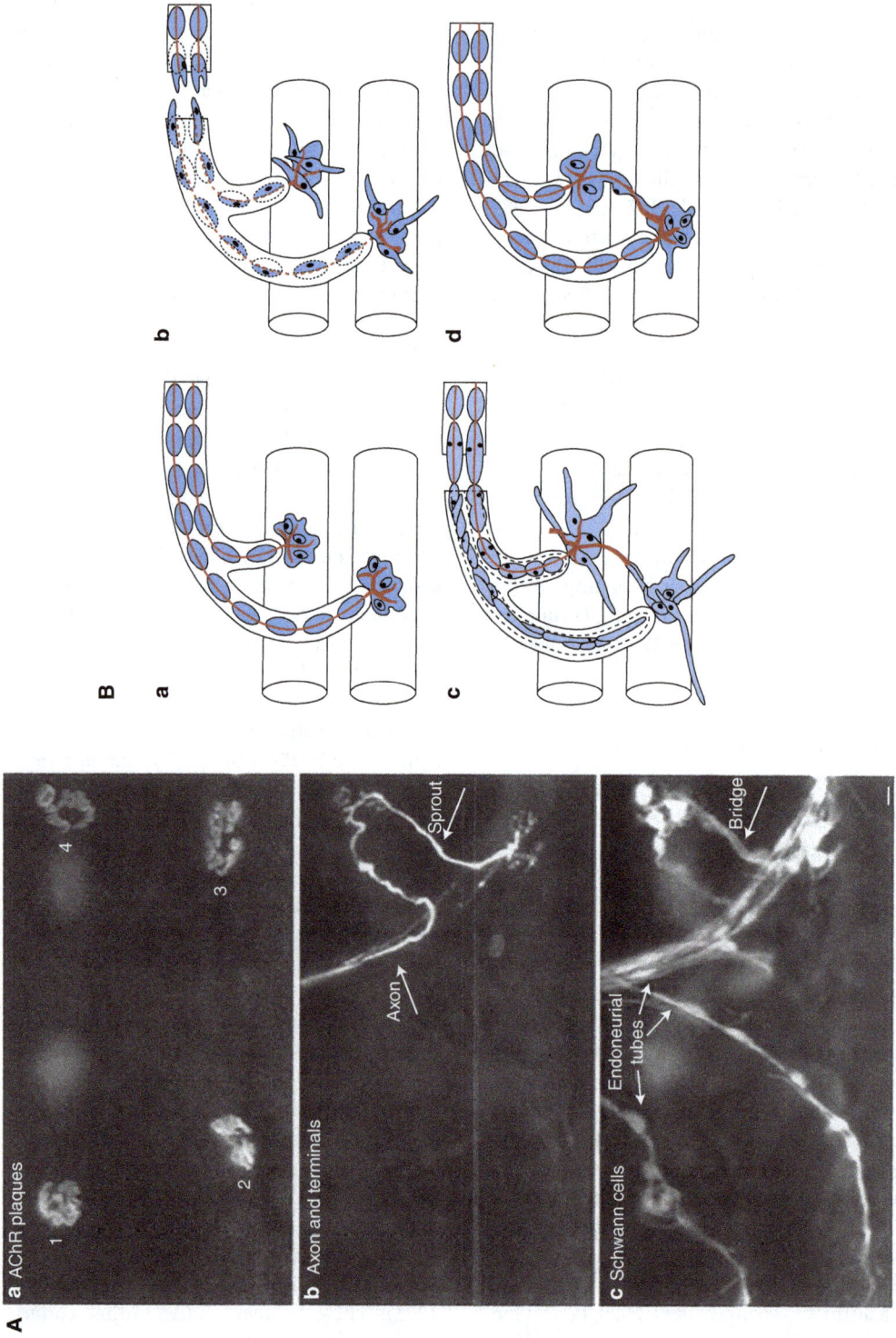

Figure 2. (*See following page for legend.*)

Cite this article as *Cold Spring Harb Perspect Biol* doi: 10.1101/cshperspect.a041362

that NRG1–ErbB signaling is involved in PSC sprouting, as the exogenous application of NRG1 to neonatal muscles or expression of constitutively activated ErbB2 receptors in PSCs induces sprouting and migration of PSCs away from endplate sites (Trachtenberg and Thompson 1997; Hayworth et al. 2006; Moody et al. 2006).

The essential role of PSCs in synaptic repair has also been shown by the absence of nerve terminal sprouting following partial denervation when PSC bridge formation is blocked by direct stimulation or exercise of muscles (Love and Thompson 1999; Love et al. 2003; Tam and Gordon 2003). The importance of PSC sprouts has further been implicated in mdx mice (a model for Duchenne muscular dystrophy) (Personius and Sawyer 2005; Marques et al. 2006), in which presynaptic expression of neuronal nitric oxide (NO) synthase is decreased and formation of PSC "bridges" is impaired, suggesting that these defects may contribute to the less effective reinnervation and muscle weakness in these mutant muscles.

PSCs associated with NMJs of fast-fatigable muscle fibers, which exhibit impaired nerve sprouting after nerve injury and enhanced loss of innervation in animal models of ALS, express the chemorepellent Semaphorin 3A (Sema3A) after nerve injury (De Winter et al. 2006), suggesting that PSCs may negatively regulate rein-nervation after injury or synaptic maintenance in ALS. On the other hand, a recent report suggests that the timing of Sema3A activation within PSCs after nerve injury is important for the repair of NMJs (Daneshvar and Anderson 2022). Enhanced presence of a cell-surface glycoprotein, CD44, in PSCs in an ALS mouse model further suggests a potential role of PSCs in motor neuron disease (Gorlewicz et al. 2009). Impaired PSC sprouting seen in aged muscles may also explain the poor reinnervation after nerve injury during aging (Kawabuchi et al. 2001). Besides guiding presynaptic nerve terminals, PSCs are thought to play a role in clustering postsynaptic AChRs by releasing neuronal isoforms of Agrin at the frog NMJ (Yang et al. 2001). Furthermore, PSCs may play a role in the synthesis of AChRs by expressing Neuregulin-2 at the mammalian NMJ (Rimer et al. 2004). Together, these results suggest that PSCs play an important role in synaptic repair following regeneration and disease at the NMJ. PSCs dynamically facilitate reinnervation through direct response to one of the axonal injury signals (Negro et al. 2022a). Indeed, these authors showed hydrogen peroxide promotes axon regeneration and NMJ repair via connective tissue growth factor. They showed that sequestration of connective tissue growth factor or the inactivation of H_2O_2 delayed NMJ recovery by affecting Schwann cell migration leading to impaired axon regrowth (Negro et al. 2022a).

Figure 2. Perisynaptic Schwann cells (PSCs) regulate neuromuscular junction (NMJ) repair and remodeling. (A) Sprouting after partial denervation. Four endplates are depicted in a rat soleus muscle 3 d after partial denervation triple-labeled with Cy5 conjugated α-bungarotoxin (α-BTX) for acetylcholine receptors (AChRs). (a) Antibodies to neurofilament and synaptic vesicle protein (with a FITC-conjugated secondary antibody) for axons and nerve terminals. (b,c) Antibody to S100 (with a rhodamine-conjugated secondary antibody) for PSCs and Schwann cells associated with the endoneurial tubes. Following partial denervation, Schwann cell processes extend profusely beyond the original endplate sites rich in AChR clusters (compare a and c). While endplates 1 and 2 remain denervated, endplate 3 becomes innervated by a nerve sprout growing along a Schwann cell "bridge" linked to endplate 4, which is innervated (compare a and b). The role of PSCs in guiding nerve terminal sprouting is further depicted in a cartoon in B. (A is reprinted from Love and Thompson 1999 under the terms of the Creative Commons Attribution 4.0 International License [CC-BY].) (B) Schematic diagram summarizing the role of PSCs in reinnervation after nerve injury (a,b) and in sprouting after partial denervation (c,d) at mammalian NMJs. (a) Normal muscle fibers with intact NMJs (nerves in red, Schwann cells in blue). (b) PSCs sprout after nerve injury. (c) Regenerating nerve fibers grow along the endoneurial tubes and reinnervate synaptic sites (the middle muscle fiber). In addition, PSCs protrude processes further to form "bridges" connecting neighboring synaptic sites. (d) The PSC "bridges" guide regenerating nerve terminals to innervate adjacent endplates. The regenerating nerve fibers can continue to grow in a retrograde direction along other endoneurial tubes to innervate more endplates. (A is reprinted, with permission, from Son et al. 1996, Elsevier Science Ltd. ©1996.)

In addition, the same research group showed that CXCL12 from PSCs promoted regeneration of injured motor axon terminals through binding with the neuronal chemokine receptor type 4 (CXCR4) (Negro et al. 2017). As a whole, these observations reveal the extended mechanisms by which PSCs are regulated by nerve injury that result in an intricate and complex response designed to facilitate NMJ reinnervation. A detailed review of the different mechanisms leading to NMJ repair was recently published (Rigoni and Negro 2020). These repair mechanisms represent an interesting set of potential therapeutic targets in various diseases.

As a whole, PSCs regulate several aspects of an NMJ's response to injury and its repair. Interestingly, axonal Schwann cells play very similar roles in the context of axon injury (see Stassart et al. 2024). For instance, both PSCs and axonal Schwann cells contribute to the cleaning of debris following injury through phagocytic mechanisms. Although PSCs extend processes between denervated and innervated NMJs, axonal Schwann cells form a bridge at the injury site to circumvent the inhibitory environment on axonal growth that the scar creates. And in both cases, these glial Schwann cell extensions are essential for axonal and nerve terminal regrowth. Hence, despite different cellular environments and molecular signaling, both cells serve the same fundamental roles in handling the response to injury and the ensuing processes for nerve and NMJ repair.

PHYSIOLOGICAL ROLES OF PSCs IN SYNAPTIC FUNCTION

The roles of PSCs in the regulation of maintenance and morphological plasticity of the NMJ underline a large degree of plasticity in PSCs as they must adjust their properties in various synaptic contexts. Furthermore, these properties imply that PSCs must analyze the synaptic situation to adjust to the changing synaptic environment. To this end, PSCs decode synaptic properties of the NMJ by the detection of synaptic transmission and attune to the fine changes that can take place. Hence, PSCs detect synaptic communication, decode the message and, in re-

turn, modulate synaptic properties in an intricate way adapted to the synaptic context.

PSCs DETECT SYNAPTIC TRANSMISSION

The development of fluorescent probes to detect free intracellular Ca^{2+} (Tsien 1981) has been a major advance in the study of the dynamic properties of glial cells and PSCs in particular. Indeed, the excitability of PSCs, like other glial cells, does not rely on electrical properties like neurons but rather on a biochemical excitability that largely relies on Ca^{2+}-dependent mechanisms (Auld and Robitaille 2003; Araque et al. 2014). Observations at the vertebrate NMJ were among the first to show that glial cells associated with intact chemical synapses detected synaptic transmission via G protein–coupled receptors (GPCRs) that controlled internal stores of Ca^{2+} (Fig. 3A,C; Jahromi et al. 1992; Reist and Smith 1992). PSCs at other vertebrate NMJs were also shown to detect the release of neurotransmitters on stimulation of the motor nerve (Fig. 3A,B; Rochon et al. 2001; Lin and Bennett 2006; Todd et al. 2007, 2010).

PSCs at mature amphibian and mouse NMJs possess subsets of receptors, consistent with their presence at a cholinergic synapse. They mainly possess different types of GPCRs, in particular muscarinic and purinergic receptors that regulate the release of Ca^{2+} from internal stores (Fig. 3A; Robitaille 1995; Robitaille et al. 1997; Castonguay and Robitaille 2001; Rochon et al. 2001). At adult NMJs, detection of synaptic transmission by PSCs is mediated by muscarinic receptors (M1, M3, or M5) (Wright et al. 2009) and by purinergic receptors—in particular, adenosine A1 receptors (Rochon et al. 2001). Although the characterization of the muscarinic receptor system follows a clear nomenclature, the properties of the purinergic receptor systems still elude a clear classification (Robitaille et al. 1997; Rochon et al. 2001; Rousse et al. 2010). In addition to these GPCRs, PSCs also possess α7 nicotinic AChRs (Noronha-Matos et al. 2020). Activation of these receptors triggers the release of adenosine by PSCs via ENT1 transporters to regulate tetanic-induced ACh spillover from the NMJ

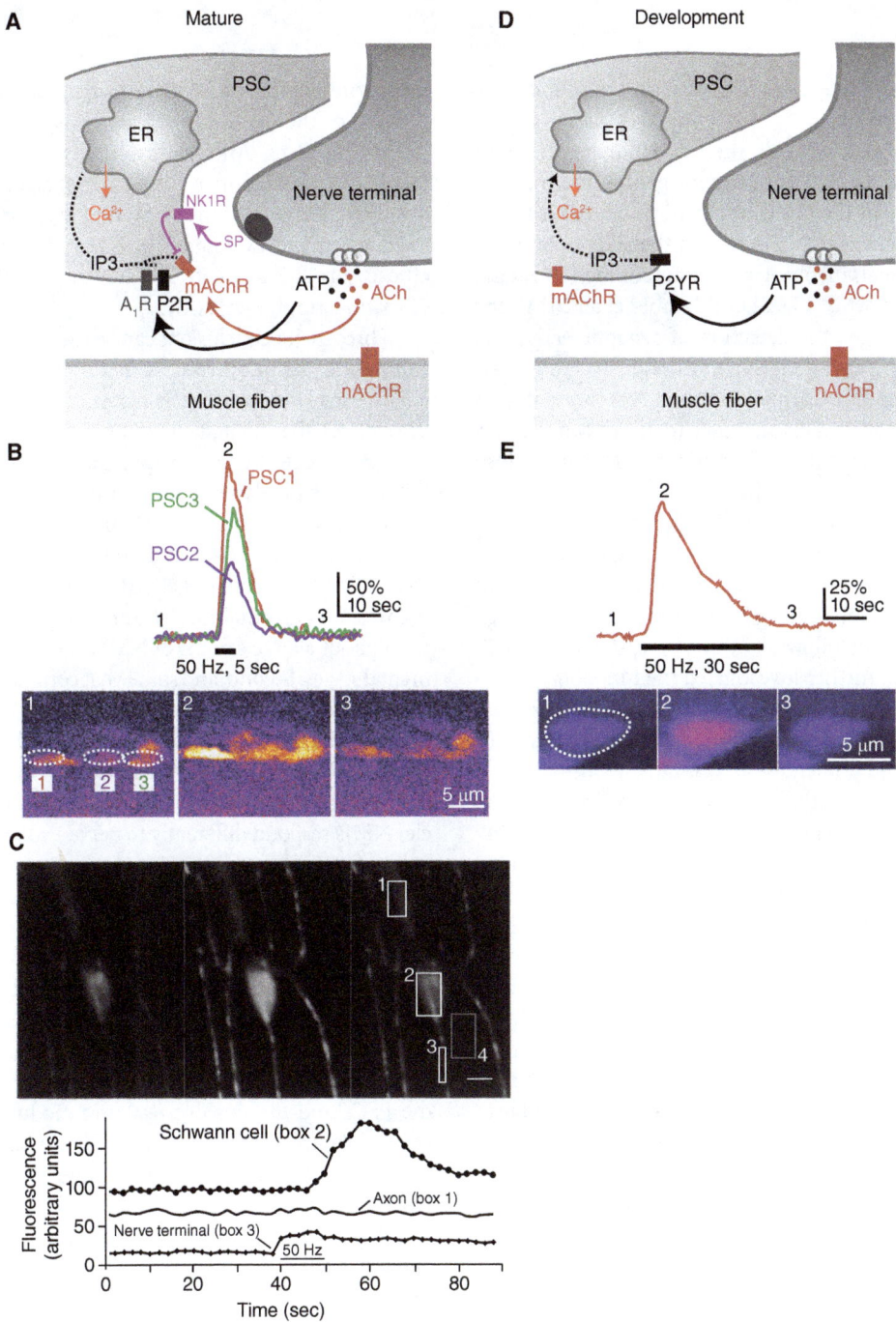

Figure 3. Perisynaptic Schwann cells (PSCs) detect synaptic transmission. (*A*) Diagram depicting the receptors and their actions by which PSCs detect synaptic transmission at mature neuromuscular junction (NMJ) and the main regulatory mechanisms. (*B*) (*Top*) Changes in fluorescence of a Ca^{2+} indicator in PSCs of a mature mouse NMJ before, during, and after motor nerve stimulation. (*Bottom*) False color confocal images of the PSCs loaded with a Ca^{2+} indicator and from which the traces have been measured. (*C*) Images of an amphibian neuromuscular preparation showing the changes in fluorescence observed in the axonal compartment (1), the soma of a PSC (2), and the presynaptic terminal area (3) before, during, and after motor nerve stimulation (bar). (*D*) Diagram depicting the receptors and their actions by which PSCs detect synaptic transmission at developing NMJ. (*E*) (*Top*) Changes in fluorescence of a Ca^{2+} indicator in a PSC of an immature (P7) mouse NMJ before, during, and after motor nerve stimulation. (*Bottom*) False color confocal images of the PSCs loaded with a Ca^{2+} indicator and from which the traces have been measured. (*C*, Reprinted from Reist and Smith 1992 under the terms of the Creative Commons Attribution 4.0 International License [CC-BY].)

through activation of presynaptic A1 inhibitory receptors.

Consistent with their dynamic involvement in the regulation of the formation and maintenance of the NMJ, PSCs at immature NMJs (at postnatal day 7) also detect the activity of nerve terminals involved in synaptic competition at the mouse NMJ (Fig. 3D,E; Darabid et al. 2013). Interestingly, the detection of synaptic activity is solely dependent on purinergic receptors, although muscarinic receptors are present and functional. This appears to be dependent on the localization of the purinergic receptors close to active zones, whereas muscarinic receptors appear more evenly distributed over the PSCs (Darabid et al. 2013). One strong characteristic that defines PSCs is their great plasticity and adaptability during NMJ formation in development and following injury. Perez-Gonzalez et al. (2022) further investigated the functional adaptation of PSCs following injury. They observed that PSC properties regulating their activation adapted a phenotype seen during postnatal development—that is, a reduced muscarinic activation and enhanced purinergic regulation. Interestingly, they described a form of "glial injury memory," whereby PSC properties remained altered for 60 d after injury. This property challenges the assumption that PSC properties are restored rapidly after NMJ reinnervation and gives new perspectives on PSCs implications for a complete recovery. This may also allow them to be more sensitive or more rapidly adaptive to subsequent injury and amenable to adjustments to the general motor recovery. This could potentially promote the persistence of PSCs in a repair state, in which they release neurotrophic factors to enhance the stability of the reinnervated NMJ. This might also render them more receptive and quickly adaptable to future injuries, facilitating adjustments in overall motor recovery (Perez-Gonzalez et al. 2022).

Interestingly, the biochemical excitability of PSCs, which allows them to detect neurotransmitter release, can be regulated. Indeed, Bourque and Robitaille (1998) showed that the peptide substance P released during sustained and intense synaptic activity at the mature amphibian NMJ caused a reduction in the sensitivity of the

muscarinic detection, leading to a reduction in the size of the nerve-evoked Ca^{2+} responses in PSCs (Fig. 3A). Another molecule, NO, acts in an autocrine manner. Descarries et al. (1998) observed that the synthesizing enzymes for NO are present in PSCs and that NO reduced the efficacy of ATP to elicit Ca^{2+} elevation in PSCs of mature amphibian NMJ.

Three major conclusions can be reached when comparing the properties of PSCs at different NMJs. First, the basic mechanisms are common throughout the different types of NMJs studied. Indeed, the detection of synaptic transmission by PSCs at adult NMJs is mainly carried by muscarinic and/or purinergic receptors (Robitaille 1995; Robitaille et al. 1997; Rochon et al. 2001; Colomar and Robitaille 2004; Darabid et al. 2013, 2018), indicating that fundamental mechanisms are preserved throughout a large sample of NMJs and developmental stages. Second, and somewhat contradictorily, PSC properties are also tuned with the properties of the NMJ they are associated with. For instance, PSCs of weaker (e.g., soleus muscle) and stronger (e.g., levator auris longus [LAL] muscle) NMJs respond differently to nerve-evoked release of neurotransmitters in which the weaker synapses systematically evoked smaller Ca^{2+} elevation in PSCs. These differences are largely the result of the different intrinsic properties of the PSCs at the different synapses (Rousse et al. 2010). Third, the excitability of PSCs can be dynamically modulated either by presynaptic signaling or in an autocrine manner, indicating that the properties of the PSCs and the possible resulting modulation can be adapted (Bélair et al. 2010). Hence, similar to the neuronal elements at the synapse, there are basic, fundamental mechanisms that drive PSCs excitability and responsiveness to synaptic activity and are in tune with the properties of the synapse they are associated with. This implies that PSCs are adapted to a given synaptic environment and, hence, can participate to the regulation of NMJ properties in a precise and adapted manner.

PSCs DECODE SYNAPTIC PROPERTIES AND ACTIVITY

Adjustments of PSCs according to the different synaptic contexts suggested that the Ca^{2+}-dependent biochemical excitability of PSCs allowed them to decode synaptic activity. Furthermore, owing to the impacts on cell activity of cytoplasmic changes of Ca^{2+} and the importance of the amplitude and kinetics of Ca^{2+} transients, one could argue that such changes represent a code that reflects the level and type of synaptic activity (Todd et al. 2010). PSCs' ability to decode the nature of the synaptic properties was also unraveled during synaptic competition that occurs postnatally at the NMJ (Fig. 4D–G; Darabid et al. 2013, 2018), whereby Ca^{2+} elevations were quite variable and were dependent on the synaptic strength (amount of transmitter release) of each nerve terminal (Fig. 4E,F). Importantly, PSC Ca^{2+} responses were unaltered in the presence of the K^+ channel blocker tetraethylammonium (TEA), which increased transmitter release without directly affecting PSC excitability (Rousse et al. 2010; Darabid et al. 2013). This indicates that differences in Ca^{2+} kinetics elicited by the two nerve terminals were also determined by the intrinsic properties of PSCs. Hence, PSCs not only detect the two terminals but also decode the ongoing competition.

As a whole, these observations highlight that PSCs, through dynamic Ca^{2+} regulation, decode synaptic communication in a given situation. This is particularly important when considering the PSCs as synaptic partners because their properties should be adapted to a given synaptic environment.

PSCs MODULATE SYNAPTIC ACTIVITY AND PLASTICITY

The ability of PSCs to be in tune with the properties of the synapse and decode the pattern of synaptic activity at adult NMJs and ongoing synaptic competition at developing NMJs are strong indicators that PSCs should be able, in return, to talk back to the pre- and postsynaptic elements and modulate the properties of the synaptic communication. The first observation of synaptic activity modulation by PSCs was made by Robitaille (1998), using the amphibian NMJ. He showed that injection of molecules that increased G-protein activity specifically in PSCs reduced the amount of transmitter release (Fig. 5A).

More importantly, he showed that blocking G-protein activation prevented a large portion of synaptic depression, a short-term synaptic plasticity that occurs at this synapse (Fig. 5B,C). Hence, this was one of the first examples of direct evidence that glial cells at an intact vertebrate synapse were controlling transmitter release and modulating synaptic plasticity. This piece of evidence was a key observation from which the concept of the "tripartite synapse" originated (Araque et al. 1999; Auld and Robitaille 2003). This provided a direct demonstration of the dynamic, bidirectional neuron–glia interactions that occur at the NMJ and further emphasizes that PSCs are active and competent synaptic partners at this synapse.

Synaptic plasticity at any synapse is often a balance of reduction (depression) and increase (potentiation) of synaptic efficacy. Hence, it was hypothesized that, if glial cells are indeed competent partners, they would also have the ability to increase synaptic efficacy. This was observed by Castonguay and Robitaille (2001), who showed that selectively chelating Ca^{2+} in PSCs of amphibian NMJs resulted in an increase in synaptic depression, suggesting that a potentiation event was perturbed. These results indicate that PSCs have the ability to both decrease and increase synaptic efficacy, hence, fine tuning the net output at the NMJ regulating muscle functions. However, these results did not indicate that PSCs used this ability to simultaneously regulate synaptic efficacy in a given synaptic context. This was unraveled when studying the decoding ability of PSCs. As indicated above, Todd et al. (2010) observed that different patterns of synaptic activity elicited Ca^{2+} responses with different kinetics and amplitudes (Figs. 4 and 5D). Concomitantly, these different patterns of motor nerve stimulation-induced different forms of synaptic plasticity, such that the continuous stimulation produced a long-lasting potentiation, whereas the bursting pattern generated a long-lasting depression (Fig. 5D,E). Using selective blockade of Ca^{2+} elevation in PSCs through photo-activation of caged Ca^{2+}, Todd et al. (2010) showed that different Ca^{2+} signaling modalities in PSCs were responsible for the two forms of synaptic plasticity. This differential modulation was

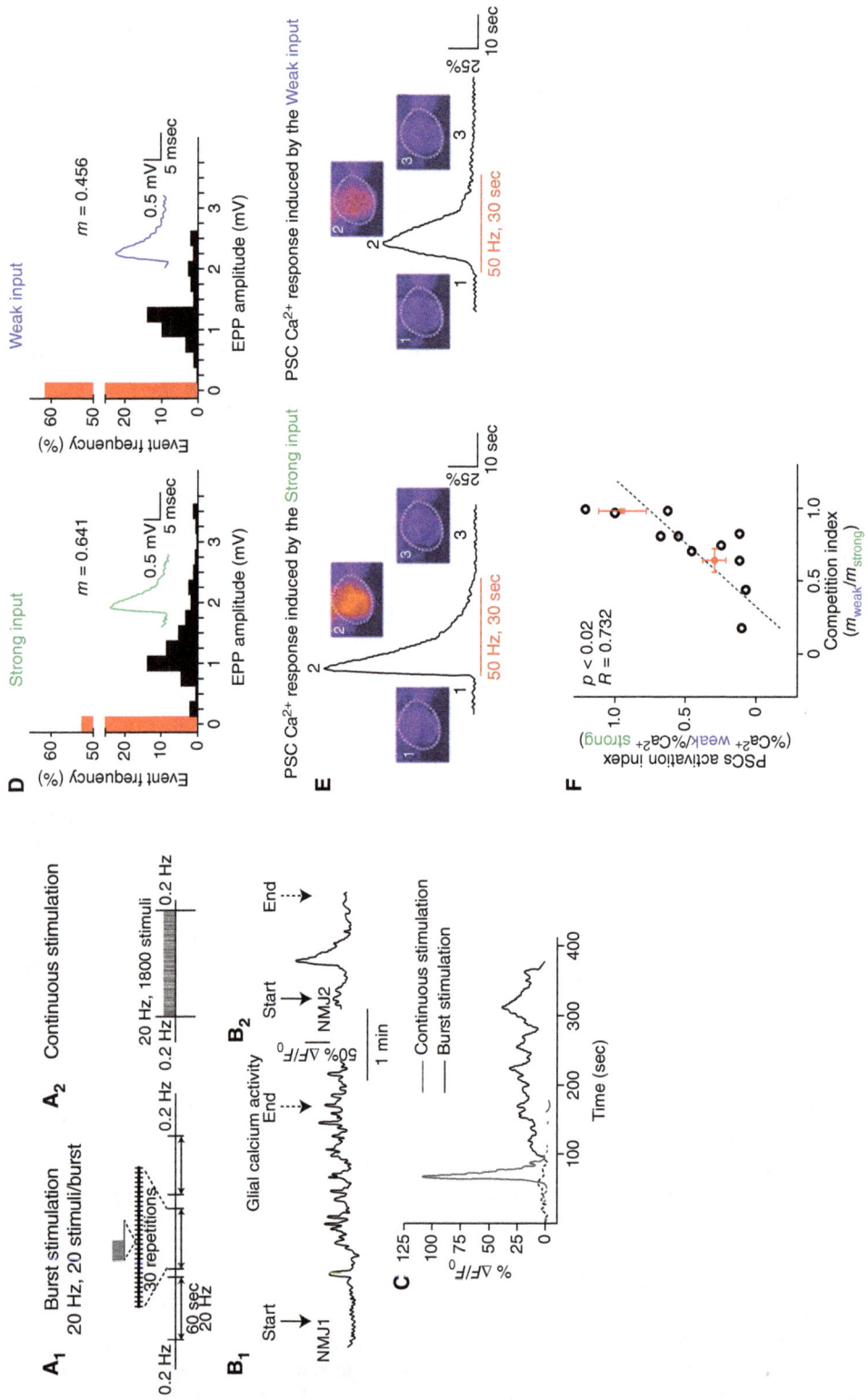

Figure 4. (*See following page for legend.*)

caused by the activation of complementary adenosine receptors (A1 receptors causing depression, A2A receptors causing the potentiation) on hydrolysis of ATP following its release by PSCs (Fig. 5D). Hence, not only do PSCs decode the ongoing synaptic transmission but, as a result, they also react to produce distinctive and adapted modulation. This modulation further illustrates that PSCs, much like other glial cells, release neuromodulatory substances identified as gliotransmitters (Araque et al. 2014). In addition to the involvement of ATP and adenosine in the modulation of synaptic transmission, observations from amphibian, lizard, and mouse NMJs indicated that PSCs may also produce and release other potential neuromodulatory substances, such as glutamate, prostaglandins, and NO (Descarries et al. 1998; Pinard et al. 2003; Pinard and Robitaille 2008; Lindgren et al. 2013). However, it is unclear whether PSCs combine any of these gliotransmitters and whether the same PSC can release them.

PLASTICITY OF PSC PROPERTIES

The results discussed above highlight the fine and efficient regulation of transmitter release by PSCs. In addition, acute modulations in the excitability of PSCs have been discussed, providing evidence that these cells are intrinsically plastic, capable of adapting to a changing synaptic environment. It further raises the possibility that

PSCs themselves could undergo long-term changes in their properties, allowing them to adjust to the changes in the synaptic properties themselves. Bélair et al. (2005, 2010) showed that PSC properties also underwent a long-term plasticity of their properties during long-term synaptic changes. These changes resulted in the alteration of the PSCs decoding ability and possibly of the outcome of their modulation of synaptic transmission. Importantly, these observations reveal that, similar to neurons, glial cells also undergo plastic changes in their properties. Long-term plasticity of PSC properties originates from the study of their properties during synapse development where a switch of the type of signaling mechanisms during the maturation of the synapse occurs (from purinergic to muscarinic). Hence, not only do PSCs interact dynamically and in a bidirectional manner with the pre- and postsynaptic elements of the NMJ, but also these interactions are highly plastic indicating that PSC regulation of the NMJ properties can also be adaptive.

MALADAPTED PROPERTIES OF PSCs IN DISEASES AND POTENTIAL THERAPEUTIC TARGETS

As discussed above, PSCs play essential roles in the regulation of NMJ innervation, its maintenance, and its repair upon injury. These functions are regulated by intricate, interrelated

Figure 4. Perisynaptic Schwann cells (PSCs) decode synaptic information. (A) Diagram depicting the Ca^{2+} responses in PSCs and the mechanisms involved when motor nerve activity is induced using two different patterns of stimulation (continuous or bursting activity) at mature mouse neuromuscular junctions (NMJs). (B) The bursting pattern consists of 30 repetitions of 20 pulses at 20 Hz repeated every 2 sec and a continuous pattern of stimulation at 20 Hz for 90 sec. (C) Typical Ca^{2+} responses elicited by the bursting and the continuous motor nerve stimulation illustrated in B. Note the difference in the kinetics of the Ca^{2+} responses revealing the ability of PSCs to decode the pattern of synaptic activity. (D) Diagram depicting the Ca^{2+} responses in PSCs elicited by the independent activity of competing nerve terminals (weak and strong) at NMJs during synapse formation. (E) Quantal analysis based on the failure rates of two competing inputs at an immature NMJ. Note the larger percentage of failures of the weak nerve terminal. (F) Independent Ca^{2+} responses in the PSC that cover the same two terminals (weak and strong) as in E. Note the difference in the amplitude of the two responses, the stronger terminal eliciting a larger Ca^{2+} response. A PSC activation index as a function of the synaptic strength index shows a continuum in the amplitude of Ca^{2+} responses as a function of the relative strength of competing nerve terminals. These results indicate that a single PSC can decipher the strength of nerve terminals competing for the territory at the same NMJ. (A_1, A_2, B_1, B_2, and C, Reprinted, with permission, from Todd et al. 2010, the authors. D–F, Reprinted from Darabid et al. 2013 under the terms of the Creative Commons Attribution 4.0 International License [CC-BY].)

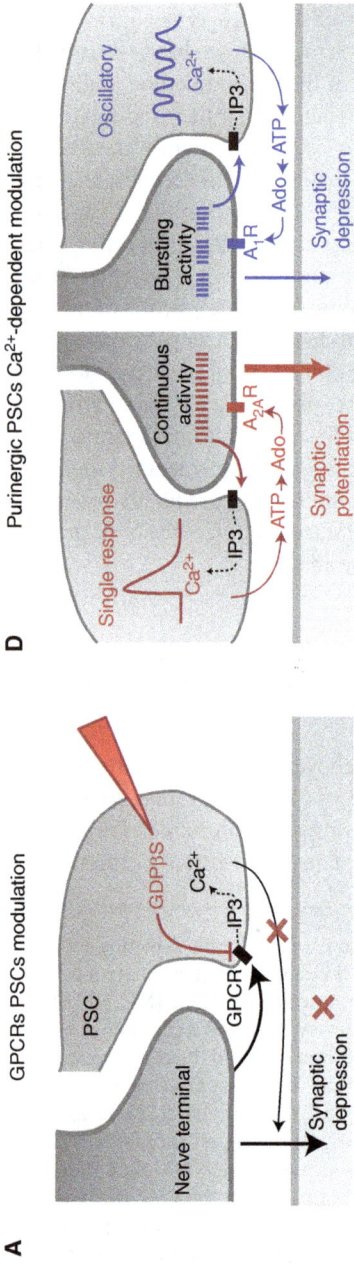

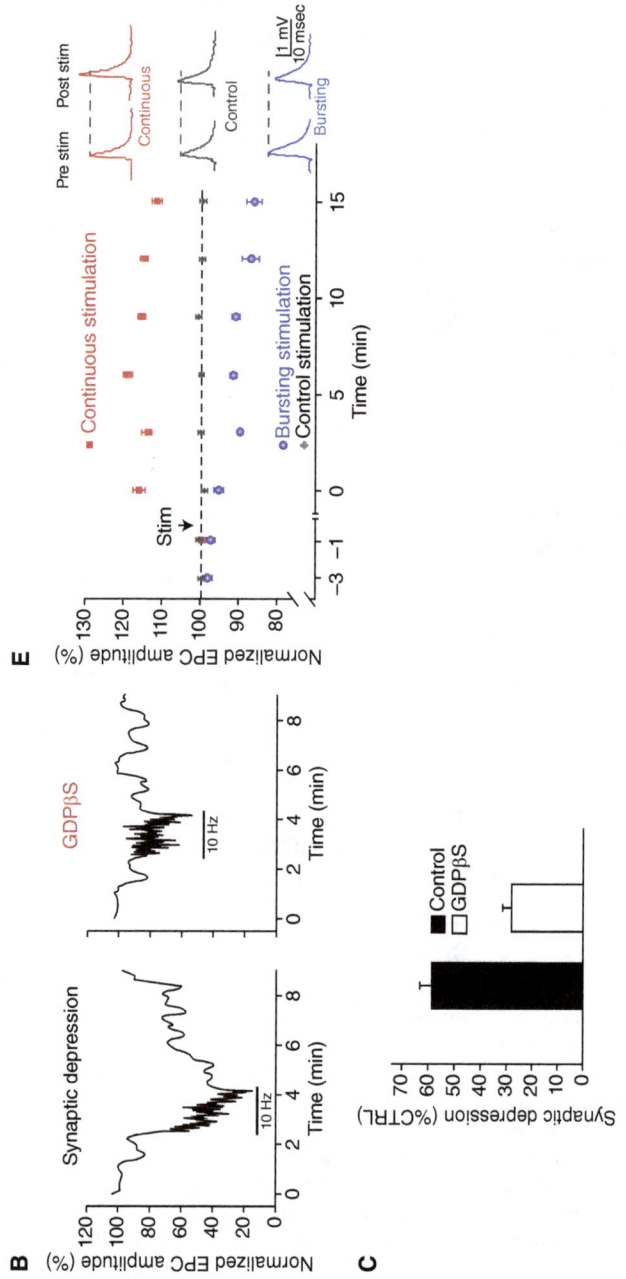

Figure 5. (*See following page for legend.*)

mechanisms that are tuned according to the status of the NMJ. Hence, owing to these critical roles and their fine regulation, several groups have proposed that PSCs may represent a relevant clinical target in several neurodegenerative and neuromuscular diseases, in particular ALS (Arbour et al. 2017; Santosa et al. 2018; Alhindi et al. 2022; Verma et al. 2022) as well as autoimmune neuropathic diseases such as Guillain–Barré syndrome (McGonigal et al. 2022).

Several essential PSC properties and functions are indeed altered in various pathologies. In a mouse model of ALS, Arbour et al. (2015) and Martineau et al. (2020) reported a chaotic extension of PSC processes that did not support properly sprouting of nerve terminals. Interestingly, Carrasco et al. (2016) reported that PSC morphology alterations preceded denervation in a mouse model of ALS.

Another function that appears particularly altered in the disease is PSCs' phagocytosis capacity. Indeed, Martineau et al. (2020) showed that PSCs have a paradoxical expression of the phagocytosis marker Galectin-3 where it was present in PSCs at innervated NMJs, rather than the denervated ones. Although PSCs appear to be the main element responsible for clearance of nerve terminal debris (Cunningham et al. 2020), and not immune cells, in Guillain–Barré syndrome, there is evidence that the immune response can be important for debris clearance in ALS (Van Dyke et al. 2016).

Another possible alteration of PSCs that can contribute to pathological conditions of NMJs is their coverage of the nerve terminal, particularly due to a reduced number. For instance, although pathological NMJs are not denervated in two mouse models of spinal bulbar muscular atrophy (Poort et al. 2016), Lee et al. (2011) and Neve et al. (2016) have found that the PSCs are reduced in number and incompletely cover the endplate site in a mutant mouse model of spinal muscular atrophy (SMA). The PSC defects may contribute to the abnormal and delayed maturation of NMJs in this neuromuscular disease. Altered coverage by PSCs is also a phenomenon observed in aging (Fuertes-Alvarez and Izeta 2021; Hastings et al. 2023), in which NMJs show sign of instability and synaptic transmission is altered (Taetzsch and Valdez 2018). Furthermore, transcriptome analysis revealed that PSCs acquired a profile of enhanced phagocytosis and proinflammatory activity (Hastings et al. 2023).

Functional properties of PSCs were also characterized in ALS models. It was found that the muscarinic contribution to PSCs activation was greatly enhanced in a mouse model of ALS (Arbour et al. 2015; Martineau et al. 2020) regardless of the status of innervation (Perez-Gonzalez et al. 2022). Hence, PSCs do not respond to denervation by reducing their muscarinic regulation as observed during injury in healthy mice. This lack of adjustment in PSC muscarinic activation was interpreted as a major constraint that

---

Figure 5. Perisynaptic Schwann cells (PSCs) modulate synaptic transmission and plasticity. (*A*) Diagram of the glial mechanisms involved in synaptic regulation at the mature amphibian neuromuscular junction (NMJ) upon manipulation of GTP-binding proteins. (*B*) (*Left*) Relative endplate potential amplitude at the frog NMJ before (0.2 Hz), during (10 Hz, 90 sec), and after (0.2 Hz) high-frequency motor nerve stimulation. Note the occurrence of synaptic depression during the high-frequency stimulation. (*Right*) Same protocol was performed on the same NMJ as in the *left* panel but following the injection of GDPβS in a PSC to block G protein activity. (*C*) Histogram illustrating the average depression in control and after injection of GDPβS in PSCs. Note that synaptic depression was significantly reduced following the selective G protein blockade in PSCs. (*D*) Diagram of the glial mechanisms involved in synaptic regulation at the mature mouse NMJ due to the differential activation of PSCs by the two different patterns of stimulation. (*E*) Changes in EPP amplitude at mouse NMJ evoked by the two patterns of stimulation are illustrated in Figure 4B. The continuous stimulation induced a long-lasting potentiation that was caused by the phasic and rapid Ca^{2+} elevation in PSC (*inset*), while the bursting pattern of stimulation induced a long-lasting depression that was caused by the small and sustained changes in Ca^{2+} in PSCs. Both forms of plasticity were altered when selectively blocking Ca^{2+} elevation in PSCs. These results indicate that, based on their decoding of synaptic activity, PSCs regulate synaptic efficacy and plasticity. (*B–C*, Reprinted, with permission, from Robitaille 1998, Elsevier © 1998. *E*, Reprinted from Todd et al. 2010 under the terms of the Creative Commons Attribution 4.0 International License [CC-BY].)

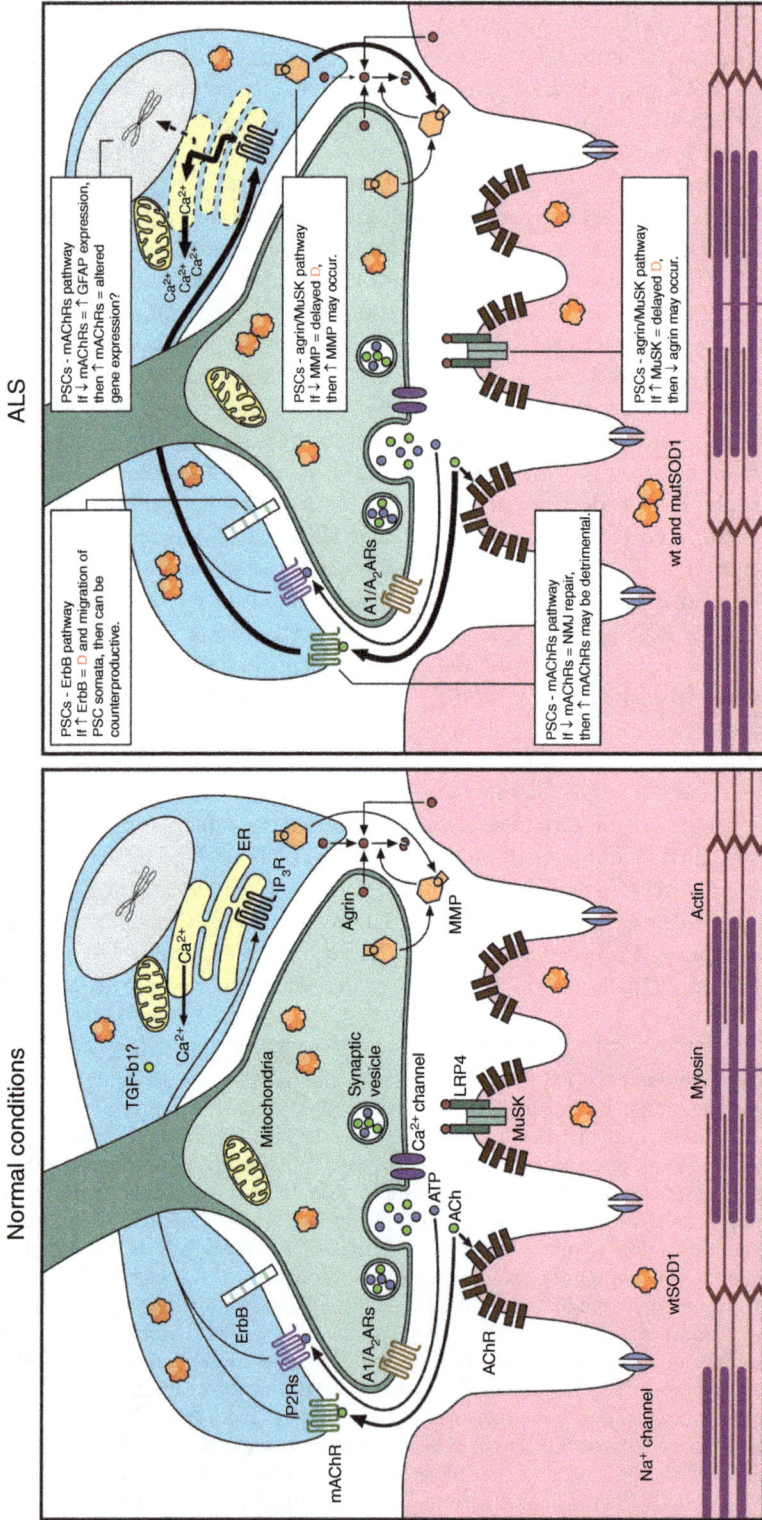

Figure 6. (*See following page for legend.*)

Cite this article as *Cold Spring Harb Perspect Biol* doi: 10.1101/cshperspect.a041362

prevents them from entering the repair mode that would allow them to support NMJ denervation and reinnervation (Arbour et al. 2015, 2017; Martineau et al. 2020; Perez-Gonzalez et al. 2022), thus potentially contributing to the maladapted PSC properties and their inability to perform their response to NMJ denervation occurring in ALS. Potential mechanisms that can be altered and contribute to NMJ malfunction and instability in ALS are described in Figure 6.

PSCs INTEGRATE SYNAPTIC ACTIVITY TO ESTABLISH SYNAPTIC PROPERTIES

The two main roles of PSCs at the NMJ (i.e., morphological stability/plasticity and synaptic regulation) may appear to represent two independent functions. However, a number of observations indicate that in fact both functions are tightly linked and are essential for the balance between synaptic efficacy and stability and synaptic plasticity and repair. Indeed, the same receptor systems that PSCs use to detect and decode synaptic transmission are also used to regulate a number of genes involved in their reaction to injury (Georgiou et al. 1994, 1999). In fact, using the amphibian NMJ model, Georgiou et al. (1994, 1999) have shown that interruption of synaptic communication is sufficient to trigger an injury-like response in PSCs. This was mediated specifically by the muscarinic receptors, not purinergic, and appears not to depend directly on Ca^{2+}, but rather on CREB-like regulation pathways. Interestingly, Wright et al.

(2009) observed that the blockade of muscarinic receptors in vivo induced injury-related changes in PSCs, in particular, an abundant level of PSC process sprouting that is normally observed after axonal injury (Son and Thompson 1995b). This suggests that the muscarinic receptor system is particularly important in regulating the PSCs in a mode of maintenance and regulation of synaptic efficacy. Consistent with the data at mature NMJs, it is remarkable that in conditions when important changes occur at the NMJ such as developing NMJs during synaptic competition, only purinergic receptors, and not muscarinic, are actively recruited by synaptic transmission. Hence, it appears that the contribution of muscarinic receptors is much reduced in situations in which major morphological and functional rearrangement of the NMJs are required (synapse formation or after injury). However, it is unclear whether these changes in receptor activation are caused by receptor levels, the type of receptors, and/or the cellular mechanisms they control. Furthermore, the regulation of PSC excitability by trophic factors, such as neurotrophin-3 (NT-3), brain-derived neurotrophic factor (BDNF) or nerve growth factor (NGF) (Todd et al. 2007), that regulate NMJ formation and stabilization also points to the possibility that PSCs' two main functions are interdependent.

We propose a model to integrate the different functions and properties of PSCs according to the functional states of the NMJ. We propose that activation of PSCs by GPCRs determines the balance between synaptic efficacy/mainte-

---

Figure 6. Potential molecular interactions at neuromuscular junctions (NMJs) in amyotrophic lateral sclerosis (ALS). (*Left*) Normal conditions (presynaptic nerve terminal in green, postsynaptic muscle fiber in pink, and perisynaptic Schwann cells [PSCs] in blue): Ach, coreleased with ATP binds to nAChRs on the muscle fiber and to mAChRs on PSCs. Activation of mAChRs will trigger an increase in intracellular Ca^{2+} via the activation of the IP3R of the ER, whereas ATP will activate PSCs P2Y G protein–coupled receptors. In return, PSCs detection of neurotransmission will regulate synaptic activity and factors involved in NMJ structure and maintenance. (*Right*) Under ALS pathological conditions, these different PSC signaling pathways can be altered to promote NMJ denervation (D). Overactivation of PSC muscarinic pathway leads to greater intracellular Ca^{2+} responses and altered gene expression and, hence, influences NMJ repair. Furthermore, the activation of the ErbB pathway can be implicated in alterations in PSC position and morphology as well as synaptic loss. Finally, the PSC agrin/MuSK pathway can be altered such that metalloproteinase (MMP) release by PSCs can be up-regulated and released agrin can be reduced, leading to NMJ instability. Boxes indicate the hypothesis and proposed mechanisms. Dotted lines indicate pathways that are yet to be confirmed. Line thickness illustrates the relative increase or decrease of the pathway in comparison to the normal condition. (Figure reprinted, with permission, from Arbour et al. 2017, John Wiley and Sons © 2017.)

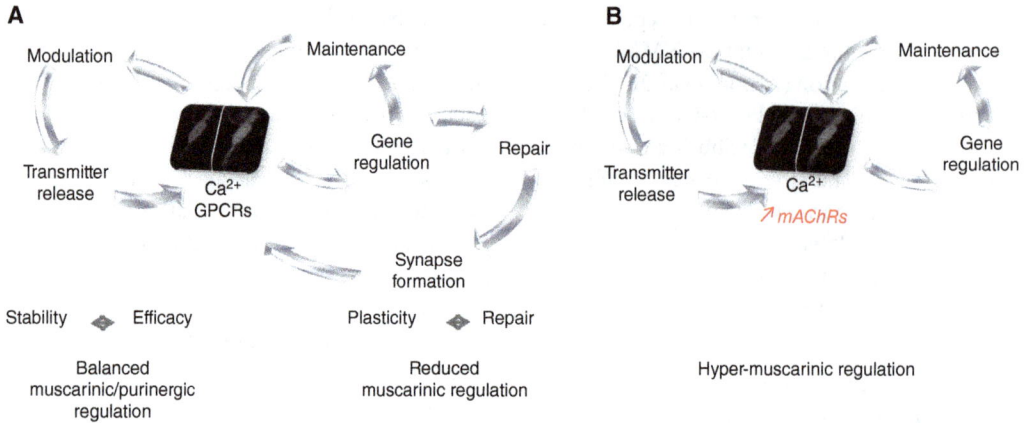

Figure 7. Model of perisynaptic Schwann cells (PSCs) balanced regulation of neuromuscular junction (NMJ) stability and plasticity. (*A*) Normal balanced response of PSCs: PSCs (illustrated as a responding cell) detect synaptic activation through activation of G protein–coupled receptors leading to the activation of Ca^{2+}-dependent events that lead to the modulation of synaptic transmission and plasticity (modulation, *left* loop). The same receptor activation also leads to the regulation of the expression of several genes that promote PSCs activity to sustain the maintenance and stability of the NMJ (maintenance, *right* loop). However, upon dysregulation of transmitter release or following injury, the signaling in PSCs is perturbed, leading to a change in the gene regulation and a switch of PSC phenotype from maintenance to repair (repair, far *right* loop). This repair mode includes the phagocytic response, leading to the removal of remnants of injured nerve terminals, and PSC bridging processes to facilitate nerve terminal sprouting toward denervated endplates. Hence, PSCs can integrate both the efficacy and the plasticity of the NMJ to establish the appropriate response according to the state of the NMJ. (*B*) Maladapted PSC response in amyotrophic lateral sclerosis (ALS). In ALS, because of the hyper-muscarinic activation, PSCs fail to detect the changes in synaptic properties, thus preventing them from accessing their program for NMJ repair following denervation occurring during the disease. Hence, the repair mode would become inaccessible and prevent PSCs from contributing to NMJ repair and reinnervation.

nance and remodeling/repair (Fig. 7). At adult NMJs, normal synaptic activity would be detected by a set of muscarinic and purinergic receptors that regulate the feedback modulation to synaptic functions (modulation, left loop). However, the same receptors also impose a regulation of the expression of a number of genes that allow PSCs to ensure the maintenance and efficacy of the NMJ (maintenance, right loop). Upon injury or in diseases, the balance between muscarinic and purinergic receptor activation would be impaired, altering the gene regulation, thus allowing a change in PSCs phenotype that would allow them to enter into a repair mode (repair, far right loop). Ultimately, this would allow the NMJ to be repaired and synaptic communication re-established. At this point, PSCs would regain their normal functions (left and right loops). However, in ALS, we argue that this passage between the maintenance mode and the repair would not be possible because of the hyper-muscarinic activation, thus preventing proper management of NMJ denervation/reinnervation processes by PSCs.

CONCLUDING REMARKS

PSCs control synapse stability where they can guide growing nerve terminals and contribute to synaptic growth and maintenance at developing NMJs. Similar to developing NMJs, PSCs in adult muscles guide nerve terminal growth during synaptic sprouting and repair after nerve injury. They also control synapse efficacy and the maintenance of the NMJ. This surveillance allows them to alter their properties to allow for synapse repair after injury or other weakening of the synapse. Molecular mechanisms underlying these functions are now being unraveled and ma-

Cite this article as *Cold Spring Harb Perspect Biol* doi: 10.1101/cshperspect.a041362

jor advances are made toward the understanding of human NMJs. These developments will lead to a better understanding of the functions of PSCs and further facilitate the development of therapeutic targets aiming at treating neuromuscular disorders and support motor functions.

ACKNOWLEDGMENTS

This work was supported by grants from the National Institutes of Health (NIH) to C.-P.K. R.R. was supported by the Canadian Institutes for Health Research grants, the Natural Science and Engineering Research Council, ALS Canada, and ALS Association.

REFERENCES

*Reference is also in this subject collection.

Alhindi A, Boehm I, Forsythe RO, Miller J, Skipworth RJE, Simpson H, Jones RA, Gillingwater TH. 2021. Terminal Schwann cells at the human neuromuscular junction. *Brain Commun* **3:** fcab081. doi:10.1093/braincomms/fcab081

Alhindi A, Boehm I, Chaytow H. 2022. Small junction, big problems: neuromuscular junction pathology in mouse models of amyotrophic lateral sclerosis (ALS). *J Anat* **241:** 1089–1107. doi:10.1111/joa.13463

Araque A, Parpura V, Sanzgiri RP, Haydon PG. 1999. Tripartite synapses: glia, the unacknowledged partner. *Trends Neurosci* **22:** 208–215. doi:10.1016/S0166-2236 (98)01349-6

Araque A, Carmignoto G, Haydon PG, Oliet SHR, Robitaille R, Volterra A. 2014. Gliotransmitters travel in time and space. *Neuron* **81:** 728–739. doi:10.1016/j.neuron.2014.02 .007

Arbour D, Tremblay E, Martineau É, Julien JP, Robitaille R. 2015. Early and persistent abnormal decoding by glial cells at the neuromuscular junction in an ALS model. *J Neurosci* **35:** 688–706. doi:10.1523/JNEUROSCI.1379-14.2015

Arbour D, Vande Velde C, Robitaille R. 2017. New perspectives on amyotrophic lateral sclerosis: the role of glial cells at the neuromuscular junction. *J Physiol* **595:** 647–661. doi:10.1113/JP270213

Astrow SH, Son YJ, Thompson WJ. 1994. Differential neural regulation of a neuromuscular junction-associated antigen in muscle fibers and Schwann cells. *J Neurobiol* **25:** 937–952.

Astrow SH, Tyner TR, Nguyen MT, Ko CP. 1997. A Schwann cell matrix component of neuromuscular junctions and peripheral nerves. *J Neurocytol* **26:** 63–75. doi:10.1023/A :1018515526035

Astrow SH, Qiang H, Ko CP. 1998. Perisynaptic Schwann cells at neuromuscular junctions revealed by a novel monoclonal antibody. *J Neurocytol* **27:** 667–681. doi:10 .1023/A:1006916232627

Aubertin-Leheudre M, Pion CH, Vallée J, Marchand S, Morais JA, Bélanger M, Robitaille R. 2020. Improved human muscle biopsy method to study neuromuscular junction structure and functions with aging. *J Gerontol A Biol Sci Med Sci* **75:** 2098–2102. doi:10.1093/gerona/glz292

Auld DS, Robitaille R. 2003. Perisynaptic Schwann cells at the neuromuscular junction: nerve- and activity-dependent contributions to synaptic efficacy, plasticity, and reinnervation. *Neuroscientist* **9:** 144–157. doi:10.1177/ 1073858403252229

Barik A, Li L, Sathyamurthy A, Xiong WC, Mei L. 2016. Schwann cells in neuromuscular junction formation and maintenance. *J Neurosci* **36:** 9770–9781. doi:10.1523/ JNEUROSCI.0174-16.2016

Bélair E-L, Vallée J, Robitaille R. 2005. Long-term in vivo modulation of synaptic efficacy at the neuromuscular junction of *Rana pipiens* frogs. *J Physiol* **569:** 163–178. doi:10.1113/jphysiol.2005.094805

Bélair E-L, Vallée J, Robitaille R. 2010. Bidirectional plasticity of glial cells induced by chronic treatments in vivo. *J Physiol* **588:** 1039–1056.

Bermedo-García F, Zelada D, Martínez E, Tabares L, Henríquez JP. 2022. Functional regeneration of the murine neuromuscular synapse relies on long-lasting morphological adaptations. *BMC Biol* **20:** 158. doi:10.1186/s12915-022-01358-4

Birks R, Katz B, Miledi R. 1960. Physiological and structural changes at the amphibian myoneural junction, in the course of nerve degeneration. *J Physiol* **150:** 145–168. doi:10.1113/jphysiol.1960.sp006379

Bishop DL, Misgeld T, Walsh MK, Gan WB, Lichtman JW. 2004. Axon branch removal at developing synapses by axosome shedding. *Neuron* **44:** 651–661. doi:10.1016/j .neuron.2004.10.026

Boeke J. 1949. The sympathetic end formation, its synaptology, the interstitial cells, the periterminal network, and its bearing on the neurone theory. Discussion and critique. *Acta Anat (Basel)* **8:** 18–61. doi:10.1159/000140398

Bourque MJ, Robitaille R. 1998. Endogenous peptidergic modulation of perisynaptic Schwann cells at the frog neuromuscular junction. *J Physiol* **512:** 197–209. doi:10.1111/j .1469-7793.1998.197bf.x

Brill MS, Lichtman JW, Thompson W, Zuo Y, Misgeld T. 2011. Spatial constraints dictate glial territories at murine neuromuscular junctions. *J Cell Biol* **195:** 293–305. doi:10 .1083/jcb.201108005

Bruneteau G, Simonet T, Bauché S, Mandjee N, Malfatti E, Girard E, Tanguy ML, Behin A, Khiami F, Sariali E, et al. 2013. Muscle histone deacetylase 4 upregulation in amyotrophic lateral sclerosis: potential role in reinnervation ability and disease progression. *Brain* **136:** 2359–2368. doi:10.1093/brain/awt164

Cao G, Ko CP. 2007. Schwann cell-derived factors modulate synaptic activities at developing neuromuscular synapses. *J Neurosci* **27:** 6712–6722. doi:10.1523/JNEUROSCI .1329-07.2007

Carrasco DI, Seburn KL, Pinter MJ. 2016. Altered terminal Schwann cell morphology precedes denervation in SOD1 mice. *Exp Neurol* **275:** 172–181. doi:10.1016/j.expneurol .2015.09.014

Castonguay A, Robitaille R. 2001. Differential regulation of transmitter release by presynaptic and glial Ca^{2+} internal stores at the neuromuscular synapse. *J Neurosci* **21**: 1911–1922. doi:10.1523/JNEUROSCI.21-06-01911.2001

Castro R, Taetzsch T, Vaughan SK, Godbe K, Chappell J, Settlage RE, Valdez G. 2020. Specific labeling of synaptic Schwann cells reveals unique cellular and molecular features. *eLife* **9**: e56935. doi:10.7554/eLife.56935

Chan ZC, Oentaryo MJ, Lee CW. 2020. MMP-mediated modulation of ECM environment during axonal growth and NMJ development. *Neurosci Lett* **724**: 134822. doi:10.1016/j.neulet.2020.134822

Chen L, Ko CP. 1994. Extension of synaptic extracellular matrix during nerve terminal sprouting in living frog neuromuscular junctions. *J Neurosci* **14**: 796–808. doi:10.1523/JNEUROSCI.14-02-00796.1994

Chen LL, Folsom DB, Ko CP. 1991. The remodeling of synaptic extracellular matrix and its dynamic relationship with nerve terminals at living frog neuromuscular junctions. *J Neurosci* **11**: 2920–2930. doi:10.1523/JNEUROSCI.11-09-02920.1991

Chow I, Poo MM. 1985. Release of acetylcholine from embryonic neurons upon contact with muscle cell. *J Neurosci* **5**: 1076–1082. doi:10.1523/JNEUROSCI.05-04-01076.1985

Colomar A, Robitaille R. 2004. Glial modulation of synaptic transmission at the neuromuscular junction. *Glia* **47**: 284–289. doi:10.1002/glia.20086

Connor EA, McMahan UJ. 1987. Cell accumulation in the junctional region of denervated muscle. *J Cell Biol* **104**: 109–120. doi:10.1083/jcb.104.1.109

Corfas G, Velardez MO, Ko CP, Ratner N, Peles E. 2004. Mechanisms and roles of axon–Schwann cell interactions. *J Neurosci* **24**: 9250–9260. doi:10.1523/JNEUROSCI.3649-04.2004

Court FA, Gillingwater TH, Melrose S, Sherman DL, Greenshields KN, Morton AJ, Harris JB, Willison HJ, Ribchester RR. 2008. Identity, developmental restriction and reactivity of extralaminal cells capping mammalian neuromuscular junctions. *J Cell Sci* **121**: 3901–3911. doi:10.1242/jcs.031047

Couteaux R. 1938. Sur l'origine de la sole des plaques motrices. *C R Soc Biol* **127**: 218–221.

Couteaux R. 1960. Motor end-plate structure. In *The structure and function of muscle* (ed. Bourne GH), pp. 337–380. Academic, New York.

Culican SM, Nelson CC, Lichtman JW. 1998. Axon withdrawal during synapse elimination at the neuromuscular junction is accompanied by disassembly of the postsynaptic specialization and withdrawal of Schwann cell processes. *J Neurosci* **18**: 4953–4965. doi:10.1523/JNEUROSCI.18-13-04953.1998

Cunningham ME, Meehan GR, Robinson S, Yao D, McGonigal R, Willison HJ. 2020. Perisynaptic Schwann cells phagocytose nerve terminal debris in a mouse model of Guillain-Barré syndrome. *J Peripher Nerv Syst* **25**: 143–151. doi:10.1111/jns.12373

Daneshvar N, Anderson JE. 2022. Preliminary study of S100B and Sema3A expression patterns in regenerating muscle implicates P75-expressing terminal Schwann cells and muscle satellite cells in neuromuscular junction restoration. *Front Cell Dev Biol* **10**: 874756. doi:10.3389/fcell.2022.874756

Darabid H, Arbour D, Robitaille R. 2013. Glial cells decipher synaptic competition at the mammalian neuromuscular junction. *J Neurosci* **33**: 1297–1313. doi:10.1523/JNEUROSCI.2935-12.2013

Darabid H, Perez-Gonzalez AP, Robitaille R. 2014. Neuromuscular synaptogenesis: coordinating partners with multiple functions. *Nat Rev Neurosci* **15**: 703–718. doi:10.1038/nrn3821

Darabid H, St-Pierre-See A, Robitaille R. 2018. Purinergic-dependent glial regulation of synaptic plasticity of competing terminals and synapse elimination at the neuromuscular junction. *Cell Rep* **25**: 2070–2082.e6. doi:10.1016/j.celrep.2018.10.075

Davis LA, Fogarty MJ, Brown A, Sieck GC. 2022. Structure and function of the mammalian neuromuscular junction. *Compr Physiol* **12**: 3731–3766. doi:10.1002/cphy.c210022

Dennis MJ, Miledi R. 1974. Electrically induced release of acetylcholine from denervated Schwann cells. *J Physiol* **237**: 431–452. doi:10.1113/jphysiol.1974.sp010490

Desaki J, Uehara Y. 1981. The overall morphology of neuromuscular junctions as revealed by scanning electron microscopy. *J Neurocytol* **10**: 101–110. doi:10.1007/BF01181747

Descarries LM, Cai S, Robitaille R. 1998. Localization and characterization of nitric oxide synthase at the frog neuromuscular junction. *J Neurocytol* **27**: 829–840. doi:10.1023/A:1006907531778

De Winter F, Vo T, Stam FJ, Wisman LA, Bär PR, Niclou SP, van Muiswinkel FL, Verhaagen J. 2006. The expression of the chemorepellent Semaphorin 3A is selectively induced in terminal Schwann cells of a subset of neuromuscular synapses that display limited anatomical plasticity and enhanced vulnerability in motor neuron disease. *Mol Cell Neurosci* **32**: 102–117. doi:10.1016/j.mcn.2006.03.002

Dickens P, Hill P, Bennett MR. 2003. Schwann cell dynamics with respect to newly formed motor—nerve terminal branches on mature (*Bufo marinus*) muscle fibers. *J Neurocytol* **32**: 381–392. doi:10.1023/B:NEUR.0000011332.96472.b2

Duregotti E, Negro S, Scorzeto M, Zornetta I, Dickinson BC, Chang CJ, Montecucco C, Rigoni M. 2015. Mitochondrial alarmins released by degenerating motor axon terminals activate perisynaptic Schwann cells. *Proc Natl Acad Sci* **112**: E497–E505. doi:10.1073/pnas.1417108112

Ellerton EL, Thompson WJ, Rimer M. 2008. Induction of zinc-finger proliferation 1 expression in non-myelinating Schwann cells after denervation. *Neuroscience* **153**: 975–985. doi:10.1016/j.neuroscience.2008.02.078

Engel AG. 1994. The neuromuscular junction. In *Myology* (ed. Engel AG, Franzini-Armstrong C). McGraw-Hill, New York.

Feng Z, Ko CP. 2007. Neuronal glia interactions at the vertebrate neuromuscular junction. *Curr Opin Pharmacol* **7**: 316–324. doi:10.1016/j.coph.2006.12.003

Feng Z, Ko CP. 2008. Schwann cells promote synaptogenesis at the neuromuscular junction via transforming growth factor-β1. *J Neurosci* **28**: 9599–9609. doi:10.1523/JNEUROSCI.2589-08.2008

Fuentes-Medel Y, Ashley J, Barria R, Maloney R, Freeman M, Budnik V. 2012. Integration of a retrograde signal during synapse formation by glia-secreted TGF-β ligand. *Curr Biol* **22**: 1831–1838. doi:10.1016/j.cub.2012.07.063

Fuertes-Alvarez S, Izeta A. 2021. Terminal Schwann cell aging: implications for age-associated neuromuscular dysfunction. *Aging Dis* 12: 494–514. doi:10.14336/AD.2020.0708

Georgiou J, Charlton MP. 1999. Non-myelin-forming perisynaptic Schwann cells express protein zero and myelin associated glycoprotein. *Glia* 27: 101–109. doi:10.1002/(SICI)1098-1136(199908)27:2<101::AID-GLIA1>3.0.CO;2-H

Georgiou J, Robitaille R, Trimble WS, Charlton MP. 1994. Synaptic regulation of glial protein expression in vivo. *Neuron* 12: 443–455. doi:10.1016/0896-6273(94)90284-4

Georgiou J, Robitaille R, Charlton MP. 1999. Muscarinic control of cytoskeleton in perisynaptic glia. *J Neurosci* 19: 3836–3846. doi:10.1523/JNEUROSCI.19-10-03836.1999

Gerber D, Pereira JA, Gerber J, Tan G, Dimitrieva S, Yánguez E, Suter U. 2021. Transcriptional profiling of mouse peripheral nerves to the single-cell level to build a sciatic nerve ATlas (SNAT). *eLife* 10: e58591. doi:10.7554/eLife.58591

Gorlewicz A, Wlodarczyk J, Wilczek E, Gawlak M, Cabaj A, Majczynski H, Nestorowicz K, Herbik MA, Grieb P, Slawinska U, et al. 2009. CD44 is expressed in non-myelinating Schwann cells of the adult rat, and may play a role in neurodegeneration-induced glial plasticity at the neuromuscular junction. *Neurobiol Dis* 34: 245–258. doi:10.1016/j.nbd.2009.01.011

Gould TW, Dominguez B, de Winter F, Yeo GW, Liu P, Sundararaman B, Stark T, Vu A, Degen JL, Lin W, Lee KF. 2019. Glial cells maintain synapses by inhibiting an activity-dependent retrograde protease signal. *PLoS Genet* 15: e1007948. doi:10.1371/journal.pgen.1007948

Griffin JW, Thompson WJ. 2008. Biology and pathology of nonmyelinating Schwann cells. *Glia* 56: 1518–1531. doi:10.1002/glia.20778

Grim M, Halata Z, Franz T. 1992. Schwann cells are not required for guidance of motor nerves in the hindlimb in Splotch mutant mouse embryos. *Anat Embryol (Berl)* 186: 311–318. doi:10.1007/BF00185979

Halstead SK, O'Hanlon GM, Humphreys PD, Morrison DB, Morgan BP, Todd AJ, Plomp JJ, Willison HJ. 2004. Anti-disialoside antibodies kill perisynaptic Schwann cells and damage motor nerve terminals via membrane attack complex in a murine model of neuropathy. *Brain* 127: 2109–2123. doi:10.1093/brain/awh231

Halstead SK, Morrison I, O'Hanlon GM, Humphreys PD, Goodfellow JA, Plomp JJ, Willison HJ. 2005. Anti-disialosyl antibodies mediate selective neuronal or Schwann cell injury at mouse neuromuscular junctions. *Glia* 52: 177–189. doi:10.1002/glia.20228

Hassan SM, Jennekens FG, Veldman H, Oestreicher BA. 1994. GAP-43 and p75NGFR immunoreactivity in presynaptic cells following neuromuscular blockade by botulinum toxin in rat. *J Neurocytol* 23: 354–363. doi:10.1007/BF01666525

Hastings RL, Mikesh M, Lee YI, Thompson WJ. 2020. Morphological remodeling during recovery of the neuromuscular junction from terminal Schwann cell ablation in adult mice. *Sci Rep* 10: 11132. doi:10.1038/s41598-020-67630-1

Hastings RL, Flordelys Avila M, Suneby E, Juros D, O'Young A, Peres da Silva J, Valdez G. 2023. Cellular and molecular evidence that synaptic Schwann cells contribute to aging of mouse neuromuscular junctions. *Aging Cell* 22: e13981. doi:10.1111/acel.13981

Hayworth CR, Moody SE, Chodosh LA, Krieg P, Rimer M, Thompson WJ. 2006. Induction of neuregulin signaling in mouse Schwann cells in vivo mimics responses to denervation. *J Neurosci* 26: 6873–6884. doi:10.1523/JNEUROSCI.1086-06.2006

Heredia DJ, Hennig GW, Gould TW. 2018. Ex vivo imaging of cell-specific calcium signaling at the tripartite synapse of the mouse diaphragm. *J Vis Exp* 140: 58347. doi:10.3791/58347

Heredia DJ, Feng CY, Agarwal A, Nennecker K, Hennig GW, Gould TW. 2018a. Postnatal restriction of activity-induced Ca^{2+} responses to Schwann cells at the neuromuscular junction are caused by the proximo-distal loss of axonal synaptic vesicles during development. *J Neurosci* 38: 8650–8665. doi:10.1523/JNEUROSCI.0956-18.2018

Heredia DJ, Feng CY, Hennig GW, Renden RB, Gould TW. 2018b. Activity-induced Ca^{2+} signaling in perisynaptic Schwann cells of the early postnatal mouse is mediated by $P2Y_1$ receptors and regulates muscle fatigue. *eLife* 7: e30839. doi:10.7554/eLife.30839

Heredia DJ, De Angeli C, Fedi C, Gould TW. 2020. Calcium signaling in Schwann cells. *Neurosci Lett* 729: 134959. doi:10.1016/j.neulet.2020.134959

Herrera AA, Banner LR, Nagaya N. 1990. Repeated, in vivo observation of frog neuromuscular junctions: remodelling involves concurrent growth and retraction. *J Neurocytol* 19: 85–99. doi:10.1007/BF01188441

Herrera AA, Qiang H, Ko CP. 2000. The role of perisynaptic Schwann cells in development of neuromuscular junctions in the frog (*Xenopus laevis*). *J Neurobiol* 45: 237–254. doi:10.1002/1097-4695(200012)45:4<237::AID-NEU5>3.0.CO;2-J

Hess DM, Scott MO, Potluri S, Pitts EV, Cisterni C, Balice-Gordon RJ. 2007. Localization of TrkC to Schwann cells and effects of neurotrophin-3 signaling at neuromuscular synapses. *J Comp Neurol* 501: 465–482. doi:10.1002/cne.21163

Heuser JE, Reese TS, Landis DM. 1976. Preservation of synaptic structure by rapid freezing. *Cold Spring Harb Symp Quant Biol* 40: 17–24. doi:10.1101/SQB.1976.040.01.004

Jablonka-Shariff A, Lu CY, Campbell K, Monk KR, Snyder-Warwick AK. 2020. Gpr126/Adgrg6 contributes to the terminal Schwann cell response at the neuromuscular junction following peripheral nerve injury. *Glia* 68: 1182–1200. doi:10.1002/glia.23769

Jablonka-Shariff A, Broberg C, Rios R, Snyder-Warwick AK. 2021. T-box transcription factor 21 is expressed in terminal Schwann cells at the neuromuscular junction. *Muscle Nerve* 64: 109–115. doi:10.1002/mus.27257

Jahromi BS, Robitaille R, Charlton MP. 1992. Transmitter release increases intracellular calcium in perisynaptic Schwann cells in situ. *Neuron* 8: 1069–1077. doi:10.1016/0896-6273(92)90128-Z

Jones RA, Harrison C, Eaton SL, Llavero Hurtado M, Graham LC, Alkhammash L, Oladiran OA, Gale A, Lamont DJ, Simpson H, et al. 2017. Cellular and molecular anatomy of the human neuromuscular junction. *Cell Rep* 21: 2348–2356. doi:10.1016/j.celrep.2017.11.008

Jordan CL, Williams TJ. 2001. Testosterone regulates terminal Schwann cell number and junctional size during developmental synapse elimination. *Dev Neurosci* 23: 441–451. doi:10.1159/000048731

Jung JH, Smith I, Mikesh M. 2019. Terminal Schwann cell and vacant site mediated synapse elimination at developing neuromuscular junctions. *Sci Rep* 9: 18594. doi:10.1038/s41598-019-55017-w

Kang H, Tian L, Thompson W. 2003. Terminal Schwann cells guide the reinnervation of muscle after nerve injury. *J Neurocytol* 32: 975–985. doi:10.1023/B:NEUR.0000020636.27222.2d

Kang H, Tian L, Son YJ, Zuo Y, Procaccino D, Love F, Hayworth C, Trachtenberg J, Mikesh M, Sutton L, et al. 2007. Regulation of the intermediate filament protein nestin at rodent neuromuscular junctions by innervation and activity. *J Neurosci* 27: 5948–5957. doi:10.1523/JNEUROSCI.0621-07.2007

Kang H, Tian L, Thompson WJ. 2019. Schwann cell guidance of nerve growth between synaptic sites explains changes in the pattern of muscle innervation and remodeling of synaptic sites following peripheral nerve injuries. *J Comp Neurol* 527: 1388–1400. doi:10.1002/cne.24625

Kawabuchi M, Zhou CJ, Wang S, Nakamura K, Liu WT, Hirata K. 2001. The spatiotemporal relationship among Schwann cells, axons and postsynaptic acetylcholine receptor regions during muscle reinnervation in aged rats. *Anat Rec* 264: 183–202. doi:10.1002/ar.1159

Kettenmann H, Ransom BR. 2013. *Neuroglia*. Oxford University Press, New York.

Keynes RJ. 1987. Schwann cells during neural development and regeneration: leaders or followers? *Trends Neurosci* 10: 137–139. doi:10.1016/0166-2236(87)90037-3

Ko CP. 1981. Electrophysiological and freeze-fracture studies of changes following denervation at frog neuromuscular junctions. *J Physiol* 321: 627–639. doi:10.1113/jphysiol.1981.sp014007

Ko CP. 1987. A lectin, peanut agglutinin, as a probe for the extracellular matrix in living neuromuscular junctions. *J Neurocytol* 16: 567–576. doi:10.1007/BF01668509

Ko CP, Chen L. 1996. Synaptic remodeling revealed by repeated in vivo observations and electron microscopy of identified frog neuromuscular junctions. *J Neurosci* 16: 1780–1790. doi:10.1523/JNEUROSCI.16-05-01780.1996

Ko CP, Thompson W. 2003. Special issue—the neuromuscular junction. *J Neurocytol* 32: 423–1037.

Ko CP, Sugiura Y, Feng Z. 2007. The biology of perisynaptic (terminal) Schwann cells. In *Biology of Schwann cells* (ed. Armati PJ), pp. 72–99. Cambridge University Press, New York.

Koirala S, Qiang H, Ko CP. 2000. Reciprocal interactions between perisynaptic Schwann cells and regenerating nerve terminals at the frog neuromuscular junction. *J Neurobiol* 44: 343–360. doi:10.1002/1097-4695(20000905)44:3<343::AID-NEU5>3.0.CO;2-O

Kullberg RW, Lentz TL, Cohen MW. 1977. Development of the myotomal neuromuscular junction in *Xenopus laevis*: an electrophysiological and fine-structural study. *Dev Biol* 60: 101–129. doi:10.1016/0012-1606(77)90113-0

Lee YI, Mikesh M, Smith I, Rimer M, Thompson W. 2011. Muscles in a mouse model of spinal muscular atrophy show profound defects in neuromuscular development even in the absence of failure in neuromuscular transmission or loss of motor neurons. *Dev Biol* 356: 432–444. doi:10.1016/j.ydbio.2011.05.667

Lee YI, Thompson WJ, Harlow ML. 2017. Schwann cells participate in synapse elimination at the developing neuromuscular junction. *Curr Opin Neurobiol* 47: 176–181. doi:10.1016/j.conb.2017.10.010

Li Y, Thompson WJ. 2011. Nerve terminal growth remodels neuromuscular synapses in mice following regeneration of the postsynaptic muscle fiber. *J Neurosci* 31: 13191–13203. doi:10.1523/JNEUROSCI.2953-11.2011

Li XX, Zhang SJ, Chiu AP, Lo LH, To JC, Cui HN, Rowlands DK, Keng VW. 2019. Conditional inactivation of Nf1 and Pten in Schwann cells results in abnormal neuromuscular junction maturation. *G3 (Bethesda)* 9: 297–303. doi:10.1534/g3.118.200795

Lichtman JW, Magrassi L, Purves D. 1987. Visualization of neuromuscular junctions over periods of several months in living mice. *J Neurosci* 7: 1215–1222. doi:10.1523/JNEUROSCI.07-04-01215.1987

Lin YQ, Bennett MR. 2006. Schwann cells in rat vascular autonomic nerves activated via purinergic receptors. *Neuroreport* 17: 531–535. doi:10.1097/01.wnr.0000209001.09987.77

Lin W, Sanchez HB, Deerinck T, Morris JK, Ellisman M, Lee KF. 2000. Aberrant development of motor axons and neuromuscular synapses in erbB2-deficient mice. *Proc Natl Acad Sci* 97: 1299–1304. doi:10.1073/pnas.97.3.1299

Lindgren CA, Newman ZL, Morford JJ, Ryan SB, Battani KA, Su Z. 2013. Cyclooxygenase-2, prostaglandin E2 glycerol ester and nitric oxide are involved in muscarine-induced presynaptic enhancement at the vertebrate neuromuscular junction. *J Physiol* 591: 4749–4764. doi:10.1113/jphysiol.2013.256727

Liu Y, Sugiura Y, Chen F, Lee KF, Ye Q, Lin W. 2019. Blocking skeletal muscle DHPRs/Ryr1 prevents neuromuscular synapse loss in mutant mice deficient in type III Neuregulin 1 (CRD-Nrg1). *PLoS Genet* 15: e1007857. doi:10.1371/journal.pgen.1007857

Love FM, Thompson WJ. 1998. Schwann cells proliferate at rat neuromuscular junctions during development and regeneration. *J Neurosci* 18: 9376–9385. doi:10.1523/JNEUROSCI.18-22-09376.1998

Love FM, Thompson WJ. 1999. Glial cells promote muscle reinnervation by responding to activity-dependent postsynaptic signals. *J Neurosci* 19: 10390–10396. doi:10.1523/JNEUROSCI.19-23-10390.1999

Love FM, Son YJ, Thompson WJ. 2003. Activity alters muscle reinnervation and terminal sprouting by reducing the number of Schwann cell pathways that grow to link synaptic sites. *J Neurobiol* 54: 566–576. doi:10.1002/neu.10191

Lubischer JL, Bebinger DM. 1999. Regulation of terminal Schwann cell number at the adult neuromuscular junction. *J Neurosci* 19: RC46. doi:10.1523/JNEUROSCI.19-24-j0004.1999

Lubischer JL, Thompson WJ. 1999. Neonatal partial denervation results in nodal but not terminal sprouting and a decrease in efficacy of remaining neuromuscular junctions

Cite this article as *Cold Spring Harb Perspect Biol* doi: 10.1101/cshperspect.a041362

in rat soleus muscle. *J Neurosci* **19**: 8931–8944. doi:10 .1523/JNEUROSCI.19-20-08931.1999

Macleod GT, Dickens PA, Bennett MR. 2001. Formation and function of synapses with respect to Schwann cells at the end of motor nerve terminal branches on mature amphibian (*Bufo marinus*) muscle. *J Neurosci* **21**: 2380–2392. doi:10.1523/JNEUROSCI.21-07-02380.2001

Marques MJ, Pereira EC, Minatel E, Neto HS. 2006. Nerve terminal and Schwann-cell response after nerve injury in the absence of nitric oxide. *Muscle Nerve* **34**: 225–231. doi:10.1002/mus.20576

Martineau É, Di Polo A, Vande Velde C, Robitaille R. 2018. Dynamic neuromuscular remodeling precedes motor-unit loss in a mouse model of ALS. *eLife* **7**: e41973. doi:10.7554/eLife.41973

Martineau É, Arbour D, Vallée J, Robitaille R. 2020. Properties of glial cell at the neuromuscular junction are incompatible with synaptic repair in the $SOD1^{G37R}$ ALS mouse model. *J Neurosci* **40**: 7759–7777. doi:10.1523/JNEUROSCI.1748-18.2020

McGonigal R, Campbell CI, Barrie JA, Yao D, Cunningham ME, Crawford CL, Rinaldi S, Rowan EG, Willison HJ. 2022. Schwann cell nodal membrane disruption triggers bystander axonal degeneration in a Guillain Barré syndrome mouse model. *J Clin Invest* **132**: e158524. doi:10 .1172/JCI158524

Meehan GR, McGonigal R, Cunningham ME, Wang Y, Barrie JA, Halstead SK, Gourlay D, Yao D, Willison HJ. 2018. Differential binding patterns of anti-sulfatide antibodies to glial membranes. *J Neuroimmunol* **323**: 28–35. doi:10 .1016/j.jneuroim.2018.07.004

Monk KR, Oshima K, Jörs S, Heller S, Talbot WS. 2011. Gpr126 is essential for peripheral nerve development and myelination in mammals. *Development* **138**: 2673–2680. doi:10.1242/dev.062224

Moody SE, Chodosh LA, Krieg P, Rimer M, Thompson WJ. 2006. Induction of neuregulin signaling in mouse Schwann cells in vivo mimics responses to denervation. *J Neurosci* **26**: 6873–6884. doi:10.1523/JNEUROSCI.1086-06.2006

Morris JK, Lin W, Hauser C, Marchuk Y, Getman D, Lee KF. 1999. Rescue of the cardiac defect in ErbB2 mutant mice reveals essential roles of ErbB2 in peripheral nervous system development. *Neuron* **23**: 273–283. doi:10.1016/ S0896-6273(00)80779-5

Musarella M, Alcaraz G, Caillol G, Boudier JL, Couraud F, Autillo-Touati A. 2006. Expression of Nav1.6 sodium channels by Schwann cells at neuromuscular junctions: role in the motor endplate disease phenotype. *Glia* **53**: 13–23. doi:10.1002/glia.20252

Negro S, Bergamin E, Rodella U, Duregotti E, Scorzeto M, Jalink K, Montecucco C, Rigoni M. 2016. ATP released by injured neurons activates Schwann cells. *Front Cell Neurosci* **10**: 134. doi:10.3389/fncel.2016.00134

Negro S, Lessi F, Duregotti E, Aretini P, La Ferla M, Franceschi S, Menicagli M, Bergamin E, Radice E, Thelen M, et al. 2017. CXCL12α/SDF-1 from perisynaptic Schwann cells promotes regeneration of injured motor axon terminals. *EMBO Mol Med* **9**: 1000–1010. doi:10.15252/emmm .201607257

Negro S, Lauria F, Stazi M, Tebaldi T, D'Este G, Pirazzini M, Megighian A, Lessi F, Mazzanti CM, Sales G, et al. 2022a.

Hydrogen peroxide induced by nerve injury promotes axon regeneration via connective tissue growth factor. *Acta Neuropathol Commun* **10**: 189. doi:10.1186/ s40478-022-01495-5

Negro S, Pirazzini M, Rigoni M. 2022b. Models and methods to study Schwann cells. *J Anat* **241**: 1235–1258. doi:10 .1111/joa.13606

Neve A, Trüb J, Saxena S, Schümperli D. 2016. Central and peripheral defects in motor units of the diaphragm of spinal muscular atrophy mice. *Mol Cell Neurosci* **70**: 30–41. doi:10.1016/j.mcn.2015.11.007

Noronha-Matos JB, Oliveira L, Peixoto AR, Almeida L, Castellão-Santana LM, Ambiel CR, Alves-do Prado W, Correia-de-Sá P. 2020. Nicotinic α7 receptor-induced adenosine release from perisynaptic Schwann cells controls acetylcholine spillover from motor endplates. *J Neurochem* **154**: 263–283. doi:10.1111/jnc.14975

O'Malley JP, Waran MT, Balice-Gordon RJ. 1999. In vivo observations of terminal Schwann cells at normal, denervated, and reinnervated mouse neuromuscular junctions. *J Neurobiol* **38**: 270–286. doi:10.1002/(SICI)1097-4695 (19990205)38:2<270::AID-NEU9>3.0.CO;2-F

Patton BL. 2003. Basal lamina and the organization of neuromuscular synapses. *J Neurocytol* **32**: 883–903. doi:10 .1023/B:NEUR.0000020630.74955.19

Patton BL, Miner JH, Chiu AY, Sanes JR. 1997. Distribution and function of laminins in the neuromuscular system of developing, adult, and mutant mice. *J Cell Biol* **139**: 1507–1521. doi:10.1083/jcb.139.6.1507

Patton BL, Chiu AY, Sanes JR. 1998. Synaptic laminin prevents glial entry into the synaptic cleft. *Nature* **393**: 698–701. doi:10.1038/31502

Peng HB, Yang JF, Dai Z, Lee CW, Hung HW, Feng ZH, Ko CP. 2003. Differential effects of neurotrophins and Schwann cell-derived signals on neuronal survival/growth and synaptogenesis. *J Neurosci* **23**: 5050–5060. doi:10 .1523/JNEUROSCI.23-12-05050.2003

Perez-Gonzalez AP, Provost F, Rousse I, Piovesana R, Benzina O, Darabid H, Lamoureux B, Wang YS, Arbour D, Robitaille R. 2022. Functional adaptation of glial cells at neuromuscular junctions in response to injury. *Glia* **70**: 1605–1629. doi:10.1002/glia.24184

Personius KE, Sawyer RP. 2005. Terminal Schwann cell structure is altered in diaphragm of mdx mice. *Muscle Nerve* **32**: 656–663. doi:10.1002/mus.20405

Pinard A, Robitaille R. 2008. Postsynaptic nitrinergic modulation underlies glutamate-induced synaptic depression at a vertebrate neuromuscular junction. *Eur J Neurosci* **28**: 577–587.

Pinard A, Lévesque S, Vallée J, Robitaille R. 2003. Glutamatergic modulation of synaptic plasticity at a PNS vertebrate cholinergic synapse. *Eur J Neurosci* **18**: 3241–3250. doi:10 .1111/j.1460-9568.2003.03028.x

Piovesana R, Charron S, Arbour D, Marsicano M, Bellocchio L, Robitaille R. 2024. Cannabinoid type-1 (CB1) receptors in glial cells promote neuromuscular junction repair following nerve injury. bioRxiv doi:10.1101/2024.01.12 .575382

Poort JE, Rheuben MB, Breedlove SM, Jordan CL. 2016. Neuromuscular junctions are pathological but not denervated in two mouse models of spinal bulbar muscular atrophy.

Hum Mol Genet **25:** 3768–3783. doi:10.1093/hmg/ddw222

Procacci NM, Hastings RL, Aziz AA, Christiansen NM, Zhao J, DeAngeli C, LeBlanc N, Notterpek L, Valdez G, Gould TW. 2022. Kir4.1 is specifically expressed and active in non-myelinating Schwann cells. *Glia* **71:** 926–944. doi:10.1002/glia.24315

Ranvier L. 1878. *Leçons sur l'histologie du système nerveux.* F. Savy, Paris.

Reddy LV, Koirala S, Sugiura Y, Herrera AA, Ko CP. 2003. Glial cells maintain synaptic structure and function and promote development of the neuromuscular junction in vivo. *Neuron* **40:** 563–580. doi:10.1016/S0896-6273(03)00682-2

Reed CB, Feltri ML, Wilson ER. 2022. Peripheral glia diversity. *J Anat* **241:** 1219–1234. doi:10.1111/joa.13484

Reist NE, Smith SJ. 1992. Neurally evoked calcium transients in terminal Schwann cells at the neuromuscular junction. *Proc Natl Acad Sci* **89:** 7625–7629. doi:10.1073/pnas.89.16.7625

Reynolds ML, Woolf CJ. 1992. Terminal Schwann cells elaborate extensive processes following denervation of the motor endplate. *J Neurocytol* **21:** 50–66. doi:10.1007/BF01206897

Riethmacher D, Sonnenberg-Riethmacher E, Brinkmann V, Yamaai T, Lewin GR, Birchmeier C. 1997. Severe neuropathies in mice with targeted mutations in the ErbB3 receptor. *Nature* **389:** 725–730. doi:10.1038/39593

Rigoni M, Negro S. 2020. Signals orchestrating peripheral nerve repair. *Cells* **9:** 1768. doi:10.3390/cells9081768

Rimer M, Prieto AL, Weber JL, Colasante C, Ponomareva O, Fromm L, Schwab MH, Lai C, Burden SJ. 2004. Neuregulin-2 is synthesized by motor neurons and terminal Schwann cells and activates acetylcholine receptor transcription in muscle cells expressing ErbB4. *Mol Cell Neurosci* **26:** 271–281. doi:10.1016/j.mcn.2004.02.002

Robbins N, Polak J. 1988. Filopodia, lamellipodia and retractions at mouse neuromuscular junctions. *J Neurocytol* **17:** 545–561. doi:10.1007/BF01189809

Robitaille R. 1995. Purinergic receptors and their activation by endogenous purines at perisynaptic glial cells of the frog neuromuscular junction. *J Neurosci* **15:** 7121–7131. doi:10.1523/JNEUROSCI.15-11-07121.1995

Robitaille R. 1998. Modulation of synaptic efficacy and synaptic depression by glial cells at the frog neuromuscular junction. *Neuron* **21:** 847–855. doi:10.1016/S0896-6273(00)80600-5

Robitaille R, Bourque MJ, Vandaele S. 1996. Localization of L-type Ca²⁺ channels at perisynaptic glial cells of the frog neuromuscular junction. *J Neurosci* **16:** 148–158. doi:10.1523/JNEUROSCI.16-01-00148.1996

Robitaille R, Jahromi BS, Charlton MP. 1997. Muscarinic Ca²⁺ responses resistant to muscarinic antagonists at perisynaptic Schwann cells of the frog neuromuscular junction. *J Physiol* **504:** 337–347. doi:10.1111/j.1469-7793.1997.337be.x

Rochon D, Rousse I, Robitaille R. 2001. Synapse–glia interactions at the mammalian neuromuscular junction. *J Neurosci* **21:** 3819–3829. doi:10.1523/JNEUROSCI.21-11-03819.2001

Rodella U, Scorzeto M, Duregotti E, Negro S, Dickinson BC, Chang CJ, Yuki N, Rigoni M, Montecucco C. 2016. An animal model of Miller Fisher syndrome: mitochondrial hydrogen peroxide is produced by the autoimmune attack of nerve terminals and activates Schwann cells. *Neurobiol Dis* **96:** 95–104. doi:10.1016/j.nbd.2016.09.005

Rousse I, St-Amour A, Darabid H, Robitaille R. 2010. Synapse–glia interactions are governed by synaptic and intrinsic glial properties. *Neuroscience* **167:** 621–32. doi:10.1016/j.neuroscience.2010.02.036

Saito A, Zacks SI. 1969. Ultrastructure of Schwann and perineural sheaths at the mouse neuromuscular junction. *Anat Rec* **164:** 379–390. doi:10.1002/ar.1091640401

Sanes JR, Lichtman JW. 1999. Development of the vertebrate neuromuscular junction. *Annu Rev Neurosci* **22:** 389–442. doi:10.1146/annurev.neuro.22.1.389

Santosa KB, Keane AM, Jablonka-Shariff A, Vannucci B, Snyder-Warwick AK. 2018. Clinical relevance of terminal Schwann cells: an overlooked component of the neuromuscular junction. *J Neurosci Res* **96:** 1125–1135. doi:10.1002/jnr.24231

Seaberg BL, Purao S, Rimer M. 2022. Validation of terminal Schwann cell gene marker expression by fluorescent in situ hybridization using RNAscope. *Neurosci Lett* **771:** 136468. doi:10.1016/j.neulet.2022.136468

Smith IW, Mikesh M, Lee Y, Thompson WJ. 2013. Terminal Schwann cells participate in the competition underlying neuromuscular synapse elimination. *J Neurosci* **33:** 17724–17736. doi:10.1523/JNEUROSCI.3339-13.2013

Son YJ, Thompson WJ. 1995a. Nerve sprouting in muscle is induced and guided by processes extended by Schwann cells. *Neuron* **14:** 133–141. doi:10.1016/0896-6273(95)90247-3

Son YJ, Thompson WJ. 1995b. Schwann cell processes guide regeneration of peripheral axons. *Neuron* **14:** 125–132. doi:10.1016/0896-6273(95)90246-5

Son YJ, Trachtenberg JT, Thompson WJ. 1996. Schwann cells induce and guide sprouting and reinnervation of neuromuscular junctions. *Trends Neurosci* **19:** 280–285. doi:10.1016/S0166-2236(96)10032-1

* Stassart RM, Gomez-Sanchez JA, Lloyd AC. 2024. Schwann cells as orchestrators of nerve repair; implications for tissue regeneration and pathologies. *Cold Spring Harb Perspect Biol* doi:10.1101/cshperspect.a041363

Sugiura Y, Lin W. 2011. Neuron-glia interactions: the roles of Schwann cells in neuromuscular synapse formation and function. *Biosci Rep* **31:** 295–302. doi:10.1042/BSR20100107

Taetzsch T, Valdez G. 2018. NMJ maintenance and repair in aging. *Curr Opin Physiol* **4:** 57–64. doi:10.1016/j.cophys.2018.05.007

Tam SL, Gordon T. 2003. Neuromuscular activity impairs axonal sprouting in partially denervated muscles by inhibiting bridge formation of perisynaptic Schwann cells. *J Neurobiol* **57:** 221–234. doi:10.1002/neu.10276

Tello JF. 1944. Sobre una vaina que envuelve toda la ramificacion del axon en las terminaciones motrices de los musculos estriados. *Trabajos Inst Cajal Invest Biol (Madrid)* **36:** 1–59.

Todd KJ, Robitaille R. 2006. Neuron-glia interactions at the neuromuscular synapse. *Novartis Found Symp* **276:** 222–

229; discussion 229–237, 275–281. doi:10.1002/97804700 32244.ch17

Todd KJ, Auld DS, Robitaille R. 2007. Differential acute neurotrophin signalling to synaptic glia at the mouse neuromuscular junction. *Eur J Neurosci* **25:** 1287–1296.

Todd KJ, Darabid H, Robitaille R. 2010. Perisynaptic glia discriminate patterns of motor nerve activity and influence plasticity at the neuromuscular junction. *J Neurosci* **30:** 11870–11882. doi:10.1523/JNEUROSCI.3165-10.2010

Trachtenberg JT, Thompson WJ. 1996. Schwann cell apoptosis at developing neuromuscular junctions is regulated by glial growth factor. *Nature* **379:** 174–177. doi:10.1038/379174a0

Trachtenberg JT, Thompson WJ. 1997. Nerve terminal withdrawal from rat neuromuscular junctions induced by neuregulin and Schwann cells. *J Neurosci* **17:** 6243–6255. doi:10.1523/JNEUROSCI.17-16-06243.1997

Tsien RY. 1981. A non-disruptive technique for loading calcium buffers and indicators into cells. *Nature* **290:** 527–528. doi:10.1038/290527a0

Van Dyke JM, Smit-Oistad IM, Macrander C, Krakora D, Meyer MG, Suzuki M. 2016. Macrophage-mediated inflammation and glial response in the skeletal muscle of a rat model of familial amyotrophic lateral sclerosis (ALS). *Exp Neurol* **277:** 275–282. doi:10.1016/j.expneurol.2016 .01.008

Verma S, Khurana S, Vats A, Sahu B, Ganguly NK, Chakraborti P, Gourie-Devi M, Taneja V. 2022. Neuromuscular junction dysfunction in amyotrophic lateral sclerosis. *Mol Neurobiol* **59:** 1502–1527. doi:10.1007/s12035-021-02658-6

Volterra A, Magistretti PJ, Haydon PG. 2002. *The tripartite synapse: glia in synaptic transmission.* Oxford University Press, New York.

Wang J, Song F, Loeb JA. 2017. Neuregulin1 fine-tunes pre-, post-, and perisynaptic neuromuscular junction development. *Dev Dyn* **246:** 368–380. doi:10.1002/dvdy.24494

Wernig A, Pecot-Dechavassine M, Stover H. 1980. Sprouting and regression of the nerve at the frog neuromuscular junction in normal conditions and after prolonged paralysis with curare. *J Neurocytol* **9:** 277–303. doi:10.1007/BF01181538

Wigston DJ. 1989. Remodeling of neuromuscular junctions in adult mouse soleus. *J Neurosci* **9:** 639–647. doi:10.1523/JNEUROSCI.09-02-00639.1989

Wolbert J, Li X, Heming M, Mausberg AK, Akkermann D, Frydrychowicz C, Fledrich R, Groeneweg L, Schulz C, Stettner M, et al. 2020. Redefining the heterogeneity of peripheral nerve cells in health and autoimmunity. *Proc Natl Acad Sci* **117:** 9466–9476. doi:10.1073/pnas.19121 39117

Woldeyesus MT, Britsch S, Riethmacher D, Xu L, Sonnenberg-Riethmacher E, Abou-Rebyeh F, Harvey R, Caroni P, Birchmeier C. 1999. Peripheral nervous system defects in erbB2 mutants following genetic rescue of heart development. *Genes Dev* **13:** 2538–2548. doi:10.1101/gad.13.19 .2538

Woolf CJ, Reynolds ML, Chong MS, Emson P, Irwin N, Benowitz LI. 1992. Denervation of the motor endplate results in the rapid expression by terminal Schwann cells of the growth-associated protein GAP-43. *J Neurosci* **12:** 3999–4010. doi:10.1523/JNEUROSCI.12-10-03999.1992

Wright MC, Potluri S, Wang X, Dentcheva E, Gautam D, Tessler A, Wess J, Rich MM, Son YJ. 2009. Distinct muscarinic acetylcholine receptor subtypes contribute to stability and growth, but not compensatory plasticity, of neuromuscular synapses. *J Neurosci* **29:** 14942–14955. doi:10.1523/JNEUROSCI.2276-09.2009

Yang JF, Cao G, Koirala S, Reddy LV, Ko CP. 2001. Schwann cells express active agrin and enhance aggregation of acetylcholine receptors on muscle fibers. *J Neurosci* **21:** 9572–9584. doi:10.1523/JNEUROSCI.21-24-09572 .2001

Young P, Nie J, Wang X, McGlade CJ, Rich MM, Feng G. 2005. LNX1 is a perisynaptic Schwann cell specific E3 ubiquitin ligase that interacts with ErbB2. *Mol Cell Neurosci* **30:** 238–248. doi:10.1016/j.mcn.2005.07.015

Zhang SJ, Li XX, Yu Y, Chiu AP, Lo LH, To JC, Rowlands DK, Keng VW. 2019. Schwann cell-specific PTEN and EGFR dysfunctions affect neuromuscular junction development by impairing Agrin signaling and autophagy. *Biochem Biophys Res Commun* **515:** 50–56. doi:10.1016/j.bbrc .2019.05.014

Zuo Y, Lubischer JL, Kang H, Tian L, Mikesh M, Marks A, Scofield VL, Maika S, Newman C, Krieg P, et al. 2004. Fluorescent proteins expressed in mouse transgenic lines mark subsets of glia, neurons, macrophages, and dendritic cells for vital examination. *J Neurosci* **24:** 10999–11009. doi:10.1523/JNEUROSCI.3934-04.2004

Schwann Cells as Orchestrators of Nerve Repair: Implications for Tissue Regeneration and Pathologies

Ruth M. Stassart,[1] Jose A. Gomez-Sanchez,[2,3] and Alison C. Lloyd[4]

[1]Paul-Flechsig-Institute of Neuropathology, University Clinic Leipzig, Leipzig 04103, Germany

[2]Instituto de Investigación Sanitaria y Biomédica de Alicante (ISABIAL), Alicante 03010, Spain

[3]Instituto de Neurociencias CSIC-UMH, Sant Joan de Alicante 03550, Spain

[4]UCL Laboratory for Molecular Cell Biology, University College London, London WC1E 6BT, United Kingdom

Correspondence: alison.lloyd@ucl.ac.uk

Peripheral nerves exist in a stable state in adulthood providing a rapid bidirectional signaling system to control tissue structure and function. However, following injury, peripheral nerves can regenerate much more effectively than those of the central nervous system (CNS). This multicellular process is coordinated by peripheral glia, in particular Schwann cells, which have multiple roles in stimulating and nurturing the regrowth of damaged axons back to their targets. Aside from the repair of damaged nerves themselves, nerve regenerative processes have been linked to the repair of other tissues and de novo innervation appears important in establishing an environment conducive for the development and spread of tumors. In contrast, defects in these processes are linked to neuropathies, aging, and pain. In this review, we focus on the role of peripheral glia, especially Schwann cells, in multiple aspects of nerve regeneration and discuss how these findings may be relevant for pathologies associated with these processes.

The repair of injured tissue is a fundamental biological process that has evolved to secure survival (Poss 2010). However, the regenerative capacity varies substantially between different tissues and strongly depends on the context and type of injury (Goldman and Poss 2020). In mammals, the peripheral nervous system (PNS) constitutes a prime example of a tissue with a strong regenerative potential. Successful nerve repair relies on peripheral nerve glial cells, the Schwann cells (SCs), which orchestrate and facilitate the repair process. An increasing body of work indicates that aspects of nerve regeneration have important roles in the repair of other tissues, including the digit tip, the mandibular bone and skin wounds in mammals and, remarkably, are important for the regeneration of entire limbs in the newt (Kumar et al. 2007; Kumar and Brockes 2012; Johnston et al. 2016; Parfejevs et al. 2018; Jones et al. 2019; Stierli et al. 2019). Furthermore, de novo innervation of tumors as well as pathologies including neuropathies and pain represent aberrations of this regenerative capacity, offering new strategies for the treatment of these common disorders (Monje et al. 2020; Winkler et al. 2023).

In this review, we will first discuss our evolving understanding of the role of peripheral glial cells in nerve regeneration and then describe the role of these processes in the repair of other tissues. Finally, we will discuss recent evidence implicating the glial repair response as well as de novo and aberrant innervation in disorders such as neuropathies, pain, aging, and cancer.

NERVE REGENERATION

Historically, two main types of acute injury to peripheral nerves have been studied: nerve cut (neurotmesis) and nerve crush (axonotmesis) (Seddon 1943). In both scenarios, the injury triggers a paradigmatic process called Wallerian degeneration (Waller 1850), which comprises the rapid degeneration of the axons distal to the injury site and the clearance of the resulting axonal debris (Arthur-Farraj and Coleman 2021). Hence, the main goal of nerve regeneration is for axons of the surviving neurons to regrow back to their original targets. This process is remarkably efficient following a crush injury, in which the structure of the nerve mostly remains intact, but much less accurate following a cut, when both the basal lamina tubes surrounding the SC/axonal units and the surrounding connective and perineurial tissue are disrupted (Nguyen et al. 2002).

The severing of the axons during the injury process stimulates a transcriptional program in the cell bodies that drives the polarized growth of the damaged axons (Allodi et al. 2012; Mahar and Cavalli 2018; Smith et al. 2020). This requires regrowing axons to travel through distinct cellular environments, often at great distances from their respective cell bodies (Brosius Lutz and Barres 2014; Cattin and Lloyd 2016). This exceptional example of polarized cell biogenesis (Lloyd 2013; Cattin and Lloyd 2016) is orchestrated by peripheral glia cells that nurture and guide the regrowing neuron and promote a conducive environment for directed migration. The role of peripheral glia can be subdivided into regional requirements (Fig. 1): (1) satellite glial cells that surround the cell bodies of the injured axons and are thought to contribute to the survival and reprogramming of neurons toward regrowth; (2) SCs that direct axons across the wound site and orchestrate the remodeling of the distal stump to provide an environment that promotes axonal regrowth over substantial distances; and (3) terminal or perisynaptic SCs, the specialized SCs associated with nerve terminals, and which guide the final reinnervation process (see Gould et al. 2023). The site-specific roles of glia in coordinating the regenerative response is remarkable both in its complexity and efficacy, with important implications for how tissues can repair themselves in the adult in the absence of many developmental cues.

Satellite Glial Cells

Satellite glial cells are neural crest-derived cells that surround the cell bodies of sensory and autonomic neurons and are thought to have a role in protecting and regulating neuronal function (Fig. 1; Hanani and Spray 2020). The cell bodies of individual neurons are completely enveloped by satellite glia cells connected by gap junctions, with an external basal lamina separating the cell body/satellite glial cell complexes from the surrounding microenvironment (Haberberger et al. 2023). Communication between the cell types is indicated by their closeness (20 nm) and the interdigitations of their membrane, and these structures have been termed neuron–glial units. Somewhat surprisingly, little is known of the role of satellite glia cells in homeostasis, whereas they have been considered as important mediators in the neuronal response to nerve injury and how this may contribute to pain (Hanani and Spray 2020). Following nerve injury, satellite glial cells appear "activated" as determined by morphological changes, and this coincides with changes in transcription (Avraham et al. 2020; Jager et al. 2020; Zhao et al. 2023) as well as with an increase in both satellite glial cell coupling and satellite/neuronal coupling, which has been speculated to be responsible for pain, perhaps by increasing the hyperexcitability of neurons (Hanani and Spray 2020; Bosch-Queralt et al. 2023). However, how satellite glial cells become activated and the role of this process remains unclear.

Cite this article as *Cold Spring Harb Perspect Biol* doi: 10.1101/cshperspect.a041363

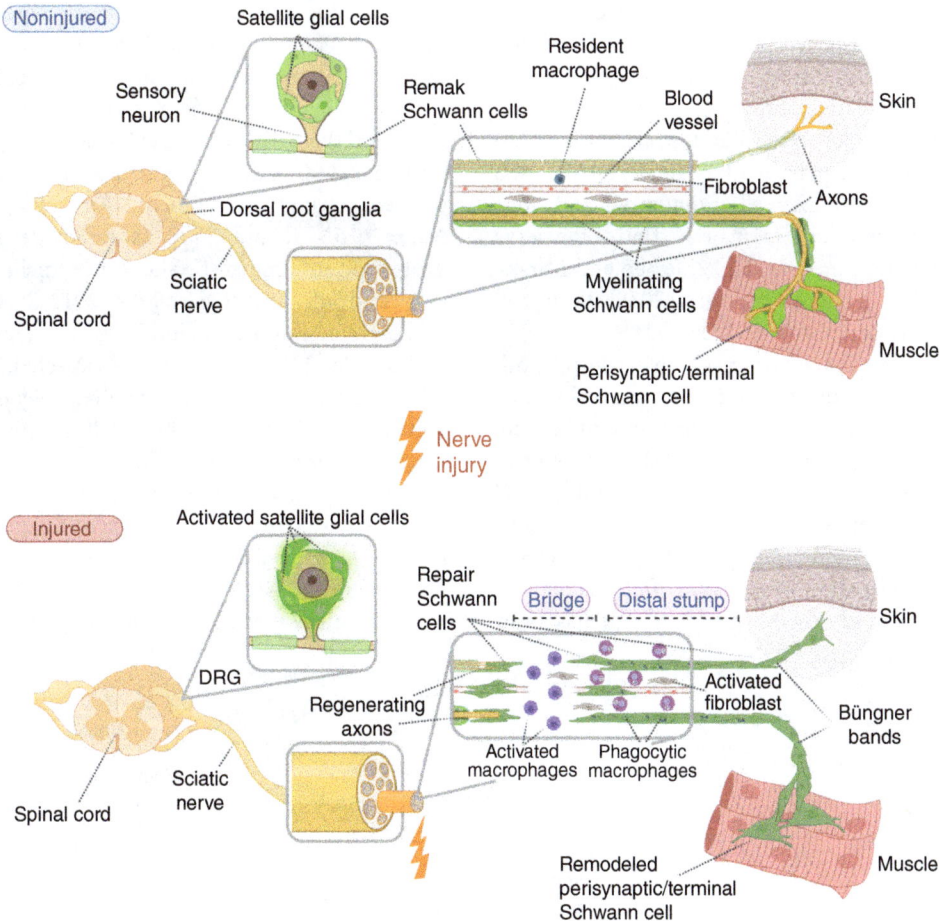

Figure 1. Structure of peripheral nerves in homeostasis and following injury. *Upper* panel shows the structure of peripheral nerves that innervate tissues throughout the body such as muscle and skin. *Lower* panel shows the regenerative response following a transection injury. This process involves complex cellular interactions between neurons, glia, and immune cells along the length of the nerve. Schwann cells (SCs) play a crucial role in this process. In response to axonal injury, SCs undergo reprogramming into repair SCs. Repair SCs, migrating as cellular cords along a newly formed vasculature, promote axonal regrowth across the wound site (the bridge). Within the distal stump, SCs coordinate the remodeling of the cellular environment conducive for axons to regrow back to their targets. This involves the recruitment of macrophages, which, together with SCs, clear axonal and myelin debris. Subsequently, the repair SCs proliferate and align within the original basal lamina tubes, to form the bands of Büngner that act as physical guides for regrowing axons. Upon reaching their targets, remodeled perisynaptic Schwann cells guide the axons back to their targets. Additionally, within the dorsal root ganglia (DRG), satellite glial cells, which surround the cell bodies of sensory neurons, are thought to have a role in protecting and regulating the function of neurons following nerve injury.

Schwann Cells

In the adult nerve, SCs exist as either myelinating (mSCs) or nonmyelinating (nmSCs) subtypes, responsible for the myelination of larger axons or grouping together of smaller axons (Remak bundles), respectively (Fig. 1). In the adult, the turnover of these cells is extremely low, with no detectable turnover of mSCs and only occasional turnover of nmSCs (Stierli et al. 2018), and their role is thought to both nurture the axons and provide the structures that permit saltatory conduction (Harty and Monk 2017; Fledrich et al. 2019a; Taveggia and Feltri 2022). These cells are not postmitotic however, in that following an injury, they are reprogrammed en masse to a progenitor-like state (Stierli et al. 2018; Jessen and Arthur-Farraj 2019). These "repair" SCs then have multiple cell-autonomous and non-cell-autonomous roles in orchestrating the nerve regeneration process (Fig. 2; Jessen and Mirsky 2016; Jessen and Arthur-Farraj 2019; Stierli et al. 2019; Stassart and Woodhoo 2021).

The regulatory processes that control both the homeostatic state and the remarkably efficient reprogramming to a progenitor-like state remain poorly understood and the damage signal from the degenerating axon that initiates SC reprogramming is unknown. However, the subsequent reprogramming of mSCs is associated with a dramatic up-regulation of ERK signaling, which persists for several days and has been

shown to be sufficient to drive this switch in cell state (Harrisingh et al. 2004; Napoli et al. 2012). This involves the inactivation of the Krox-20 (Egr2)-dependent myelination program and the induction of the "repair" cell phenotype including the up-regulation of genes expressed during development (e.g., *cJun*, *L1*, *Ncam*, *p75NTR*, and *Gfap*), as well as de novo expression of injury-specific genes including *Olig1* and *Shh* (Arthur-Farraj et al. 2012; Stassart et al. 2013; Jessen and Arthur-Farraj 2019; Jessen and Mirsky 2019). As would be expected, this dramatic switch in cell state involves epigenetic regulation and chromatin remodeling with distinct complexes associated with either state (Arthur-Farraj et al. 2017; Ma and Svaren 2018; Sock and Wegner 2019; Nocera and Jacob 2020; Gomez-Sanchez et al. 2022). In many ways, this resembles a classical epithelial-to-mesenchymal switch, but the mechanistic controls in SCs re-

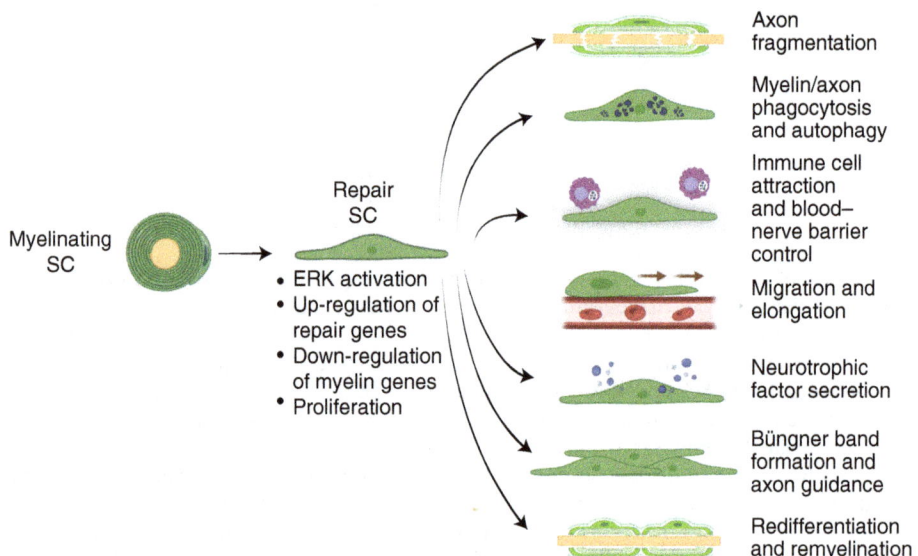

Figure 2. The role of repair Schwann cells (SCs) in nerve regeneration. After an injury, myelinating SCs undergo reprogramming to repair SCs. This highly efficient reprogramming process is driven by the activation of ERK signaling, resulting in a switch in cell state which involves the down-regulation of myelin genes and the up-regulation of repair genes, as well as SC proliferation. The repair SCs have multiple roles in the repair process. They play a crucial role in reorganization of the microenvironment by engaging in axon fragmentation as well as in subsequent autophagy and phagocytosis of myelin and axon debris. They orchestrate the multicellular regeneration process by attracting immune cells, controlling the blood–nerve barrier, and secreting neurotrophic factors that control neuronal survival and biogenesis. They guide the regrowing axons, both across the wound site and within the distal stump by forming the Büngner bands. Finally, the restoration of nerve function is achieved through SC redifferentiation and remyelination.

    Cite this article as *Cold Spring Harb Perspect Biol* doi: 10.1101/cshperspect.a041363

main relatively poorly understood. What is clear is that classes of genes switched on in the repair SCs have multiple, distinct roles in controlling the regeneration process (Fig. 2). These roles differ between the wound site and the distal stump and reflective of these contrasting roles, the transcriptome associated with either region is geared to their local function (Clements et al. 2017).

The Bridge/Wound Site

Following the transection of a nerve, the proximal and distal stumps re-find each other and become "bridged" by new tissue composed of the inflammatory cells and matrix common to many wound sites (Fig. 1; Jurecka et al. 1975; Cattin et al. 2015). This dense, seemingly hostile, nondirectional environment provides a significant barrier to axonal regrowth. This is solved by a complex multicellular response in which cords of migrating SCs guide axons along the surface of newly formed, polarized blood vessels, which provide a pathway toward the distal stump (Cattin and Lloyd 2016). The process is initiated by macrophages, the majority cell type within the wound, which sense the hypoxia of the bridge and secrete VEGF, a potent endothelial cell chemotactic factor. This stimulates the vascularization of the new tissue between the nerve stumps, which is polarized to the subsequent direction of axonal regrowth (Cattin et al. 2015). Reprogrammed SCs emerge from both stumps, migrating as cellular cords as a result of a synergy between changes to the SCs and the environment in which they find themselves (Parrinello et al. 2010; Clements et al. 2017; Stierli et al. 2019). Reprogrammed SCs up-regulate genes including the *EphB2* receptor, the stem cell factor *Sox2*, and *N-cadherin*, that respond to *Ephrin B*–expressing fibroblasts within the wound site (Parrinello et al. 2010). This results in a change in SC behavior from cells that normally repulse each other, in a process known as contact inhibition of locomotion, to "sticky" cells, due to EphB/EphB2/Sox2-dependent relocalization of N-cadherin to the SC surface. Furthermore, the collective migration of SCs is reinforced by TGF-β in the wound environment, which maintains SCs in a more mesenchymal, migratory state and enhances the

EphB response, presumably to ensure that SCs retain their migratory capabilities until the injury has resolved (Clements et al. 2017).

The SC cords provide a conducive surface essential for axonal regrowth across the wound site. However, the directionality of the cords is supplied by polarized blood vessels, because neither SCs nor axons can migrate through the 3D matrix of the wound site, but SCs can migrate along the surface of blood vessels (Cattin et al. 2015). This selective migratory capability, although poorly understood, provides the means for directional migration, which is important for nerve repair and may also allow innervation of new tissue (such as during tumor formation; see below). It may also provide new clinical approaches to improve nerve repair, as an alternative to nerve grafts, when injuries are severe (Barnes et al. 2022).

The Distal Nerve Stump

Once the axons have successfully crossed the wound site, they enter into the distal nerve stump. Here, they encounter a remodeled environment conducive to axonal regrowth, in large part engineered by repair SCs (Fig. 1). Whether following a transection or crush injury, repair SCs orchestrate a multicellular regenerative response along the full length of the distal stump that remodels the environment and provides signals that support the growth of axons back to their targets (Fig. 2). As axonal regrowth is relatively slow (around 1 mm/day) and nerves can be very long (>1 m in humans), this requires a sustained response from the SCs, requiring long-term stability of SCs in the absence of axonal contact.

The role of SCs within the distal stump can be divided into two broad themes:

Remodeling of the microenvironment. This requires the efficient degradation of axons, the clearance of axonal and myelin debris, which is inhibitory to axonal regrowth, and the coordination of the inflammatory response. The onset of axonal degeneration appears to be primarily due to an intrinsic axon specific program, nevertheless SCs play a collaborative role (Jung et al. 2011; Catenaccio et al. 2017; Vaquié et al.

2019; Babetto et al. 2020; Arthur-Farraj and Coleman 2021). SCs form constricting actin-myosin spheres along degenerating axons, which promotes axonal disintegration, a process which is induced in part by a PlGF-mediated activation of VEGFR1 receptors in SCs (Vaquié et al. 2019). Furthermore, repair SCs, together with resident and infiltrating macrophages, remove the myelin and axonal debris and remodel the microenvironment (Perry et al. 1995; Gomez-Sanchez et al. 2015; Jang et al. 2016; Brosius Lutz et al. 2017; Kalinski et al. 2020; Li et al. 2022). SCs undergo a form of autophagy termed myelinophagy (Gomez-Sanchez et al. 2015; Jang et al. 2016; Reed et al. 2020), internalizing and degrading their own myelin sheath, while smaller fragments are phagocytosed by repair SCs or macrophages. This involves the up-regulation of TAM phagocytic receptors (Axl and Mertk) (Brosius Lutz et al. 2017) and the MLKL pseudokinase that facilitates myelin breakdown (Ying et al. 2018). The influx of inflammatory cells, including macrophages, that contribute to the remodeling of the microenvironment, is controlled by repair SCs, which secrete cytokines and chemokines and facilitate blood–nerve barrier permeability (Barrette et al. 2008; Napoli et al. 2012; Cervellini et al. 2018; Malong et al. 2023).

Nurturing and guiding of regrowing axons. Following debris removal, repair SCs play a critical role in providing a conducive surface permitting the migration of regrowing axons. SCs in the distal stump remain within their basal lamina tubes and extend along the surface to form longitudinal cell columns referred to as the bands of Büngner, which serve as physical guidance cues for regrowing axons (Büngner 1891; Arthur-Farraj et al. 2012; Gomez-Sanchez et al. 2017). In addition, neurotrophic factors such as Artemin, GDNF, BDNF, NGF, and others are induced in repair SCs and their secretion contributes to the survival and regrowth of axons, although this remains partially understood (Boyd and Gordon 2003; Gordon 2009; Arthur-Farraj et al. 2012; Fontana et al. 2012). Finally, transfer of materials from SCs to the axons has been shown after injury, including exosomes and ribosomes (Lopez-Verrilli et al. 2013; Müller et al. 2018; Canclini et al. 2020; López-Leal et al. 2020) that may contribute to axonal regeneration via local translation within the axonal compartment (Koley et al. 2019; Dalla Costa et al. 2021). In general, however, our knowledge on the molecular signals and SC-derived factors that specifically support axonal regrowth in the injured distal stump remains relatively poor. Likewise, it remains to be resolved whether and how repair SCs trophically support the regrowing axon during nerve repair (Morrison et al. 2015; Jha et al. 2021).

Terminal Glia

Terminal Glia are specialized nonmyelinating SCs that are tightly engaged at innervation sites, for example at the neuromuscular junction (also referred to as perisynaptic SCs) (see Gould et al. 2023) or around cutaneous sensory end organs (Fig. 1). There is a vast morphological variety of terminal SCs (tSCs), and presumably function, between and within different tissues, for example between the different sense organs of the skin (Li and Ginty 2014; Reed et al. 2021; Suazo et al. 2022; Bosch-Queralt et al. 2023). However, most remain poorly characterized but are likely to have important roles in the maintenance of nerve terminals throughout the body and their repair following injury. In the motor system, perisynaptic SCs, probably the best-studied tSCs, participate in the development, functional maintenance, and repair of the NMJ (Ko and Robitaille 2015). In response to acute nerve injury, perisynaptic SCs become activated and undergo drastic morphological changes (Miledi and Slater 1970; Reynolds and Woolf 1992; Kang et al. 2003; Ko and Robitaille 2015). They extend long processes into the synaptic cleft, most likely phagocytose the degenerating presynaptic axon terminals, and furthermore contribute to activating the immune response (Birks et al. 1960; Miledi and Slater 1970; Ko and Robitaille 2015; Jablonka-Shariff et al. 2020). Upon axonal regeneration, perisynaptic SCs emanate long sprouts that form bridges between neighboring endplates, which act as guidance cues for branching axons (Son and Thompson 1995a,b; O'Malley et al. 1999; Kang et al. 2003, 2014; Negro et al. 2017; Hastings et al. 2020). Interestingly, most perisynaptic SC bridges form connections between innervated and denervated NMJs, and several studies suggest that glial bridging is regulated by postsynaptic signals and muscle activity (Love and Thompson 1999; Castonguay and Robitaille 2001; Rochon et al. 2001; Love et al. 2003; Tam and Gordon 2003; Wright et al. 2009; Todd et al. 2010; Ko and Robitaille 2015; Gordon 2020; Perez-Gonzalez et al. 2022). Notably, perisynaptic SC bridging not only helps in the reinnervation of an NMJ by its original nerve fiber, but also guides fibers that innervate adjacent endplates to their denervated neighbors, which leads to muscle fibers becoming polyneuronally reinnervated (Rich and Lichtman 1989; Son and Thompson 1995b; Ko and Robitaille 2015; Kang et al. 2019). This process likely compensates for failed reinnervation by the original nerve fiber and ultimately results in motor unit remodeling and neurogenic muscle fiber retyping as also observed in neuropathy (Ko and Robitaille 2015; Kang et al. 2019; Gordon 2020). In the skin, signaling between tSCs and sensory end-organs has been shown to be important for the formation of these structures (Li and Ginty 2014; Reed et al. 2021; Rinwa et al. 2021; Schwaller et al. 2021; Meltzer et al. 2022; Bosch-Queralt et al. 2023). In contrast, our knowledge of other tSCs is less, for example, our understanding on the functional contribution of tSCs to the reinnervation of the skin and its sensory end-organs remains poor (Terenghi 1995; Dubový and Aldskogius 1996; Griffin et al. 2010; Li and Ginty 2014; Handler and Ginty 2021). Moreover, the role of tSCs in other sensory structures and tissues is mostly poorly characterized but is likely to be important for maintaining function throughout life, repair following injury, and in pathologies such as pain. For example, the pattern of cutaneous end-organ reinnervation by different fiber types and a takeover of denervated territories by sprouts of nociceptive fibers has been shown to contribute to the development of neuropathic pain (De Logu et al. 2017, 2022; Abdo et al. 2019; Rinwa et al. 2021; Gangadharan et al. 2022).

Restoration of Nerve Function

The regrowth of axons and the reinnervation of their respective target tissues is accompanied by a progressive remodeling and restoration of nerve morphology as well as by a resolution of the inflammatory response as function is restored (Fawcett and Keynes 1990; Navarro et al. 2007; Gao et al. 2013; Stierli et al. 2019). This includes the separation of axons into Remak bundles or the remyelination of large caliber fibers by SCs. Nevertheless, the regenerated nerve is clearly different from a noninjured nerve, characterized by increased axonal sprouting, resulting in a greater number of smaller caliber axons, increased cellularity, and higher levels of extracellular matrix (Salonen et al. 1988; Fawcett and Keynes 1990; Stierli et al. 2018, 2019). Furthermore, remyelinated internodes remain significantly thinner and shorter, contributing to an incomplete restoration of nerve function (Schröder 1972). Another consideration is that remyelination is not an exact recapitulation of myelination during development. Indeed, while signaling factors implicated in SC myelination, such as GPR126/ADGRG6, also contribute to remyelination after injury (Woodhoo et al. 2009; Stassart et al. 2013; Mogha et al. 2016; Jessen and Mirsky 2019; Pellegatta and Taveggia 2019; Nocera and Jacob 2020; Gomez-Sanchez et al. 2022), axonal Neuregulin-1 signaling, a key determinant of both the decision to myelinate and the final thickness of the sheath during development (Michailov et al. 2004; Fledrich et al. 2019a; Salzer et al. 2023), appears to be dispensable for remyelination (Fricker et al. 2013; Stassart and Woodhoo 2021). In addition, longitudinal myelin growth is associated with a concomitant body growth during development, which is absent following repair and may be at least in part responsible for the shorter cells (Fernando et al. 2016; Tricaud 2018; Feltri et al. 2021). Remyelination efficiency may furthermore depend on the permissiveness of the microenvironment, and SCs have been shown to contribute to the regulation of fibrinolysis after injury (Pellegatta et al. 2022). Notably, even less is known about the recovery of Remak bundle structures after injury and the role of nmSCs for the restoration of nerve function (Harty and Monk 2017; Stierli et al. 2018; Ulrichsen et al. 2022). In general, the restoration of the nerve structure represents a complex process in which many events, such as SC remyelination and the resolution of inflammation, proceed in parallel, consistent with repair SCs controlling this process along the length of the nerve as the axons regrow. In this regard, remyelinating SCs have been suggested to contribute to macrophage activity and efflux of the nerve (Fry et al. 2007; Forese et al. 2020); however, our understanding of the role of SCs in shaping the tissue microenvironment during the late phase of peripheral nerve repair remains limited.

THE ROLE OF INNERVATION AND SCs ON TISSUE HOMEOSTASIS AND REGENERATION

All tissues are innervated, permitting bidirectional signaling that can control tissue behavior and feed back sensory information, as a means of homeostatic tissue control and function. What is becoming increasingly clear is that innervation has important roles in regulating tissue niches, particularly stem cell compartments, and signals from nerves have been shown to be involved in controlling stemness and providing a microenvironment that promotes stem cell expansion (Johnston and Miller 2022). These findings have important implications for the homeostatic state, the stimulated state, and repair processes following injury (Fig. 3). While many of these signals involve neuronal control of tissue function, SCs appear to have important roles in both homeostatic and repair processes. These studies are at an early stage and to date are limited to a few tissues, but it is likely that this unappreciated aspect of the stem cell compartment microenvironment has parallels in other tissues with important implications for tissue biology and pathology.

Limb Regeneration

Work indicating the importance of innervation in tissue regeneration first came to prominence from studies of the Salamander (Urodela) family (Kumar and Brockes 2012). Salamanders have the impressive ability to regenerate limbs and tails following loss after injury. This involves the formation of a blastema at the site of injury, composed of proliferating mesenchymal-derived, stem-like cells that retain positional and patterning information permitting the reformation of the limb irrespective of the site of injury

(Tanaka 2016). It had been known for centuries that innervation was required for this process (Todd 1823) but more recently it became clear that repair SCs, which form in response to damaged nerves, have a major role in the efficacy of the regeneration process. For successful regeneration, blastema cell proliferation needs to be sustained and a number of nerve-dependent factors have been identified, but it appears that an SC-derived PROD1 ligand, the newt anterior gradient 2 (NAG), is sufficient to rescue an innervation defect and acts by maintaining blastema cell proliferation via the PROD1 receptor (Kumar et al. 2007).

Tissue Regeneration

The digit tip in adult mammals can regenerate with mechanisms that mirror limb regeneration in urodeles involving the formation of a blastema with the ability to form replacement tissue. Similarly to limb regeneration, digit tip regeneration is also at least partially dependent on nerve innervation (Rinkevich et al. 2011; Takeo et al. 2013; Dolan et al. 2022) and SCs have been shown to play an important role (Johnston et al. 2016; Carr et al. 2019). Repair SCs, formed in response to nerves damaged by the amputation, migrate into and around the blastema, where they secrete ligands (PDGF-AA and oncostatin M), which can rescue the deficits caused by denervation by stimulating the proliferation and differentiation of the blastema cells (Johnston et al. 2016). In a similar manner, SCs also appear to have a role in the repair of bone. In a mouse mandibular denervation model, the same SC-derived paracrine factors important for digit tip repair appear to be required for proper skeletal stem cell function during the repair process (Jones et al. 2019). Denervation caused a decrease in SC numbers, through an unknown mechanism, and this loss of SCs was shown to be responsible for the failure to repair via loss of their paracrine signals—first identified in in vitro coculture experiments and then confirmed by the ability to rescue in vivo. Furthermore, skin is a tissue capable of substantial repair following injury, requiring wound closure and often the formation of a scar. Skin is densely innervated,

Cite this article as *Cold Spring Harb Perspect Biol* doi: 10.1101/cshperspect.a041363

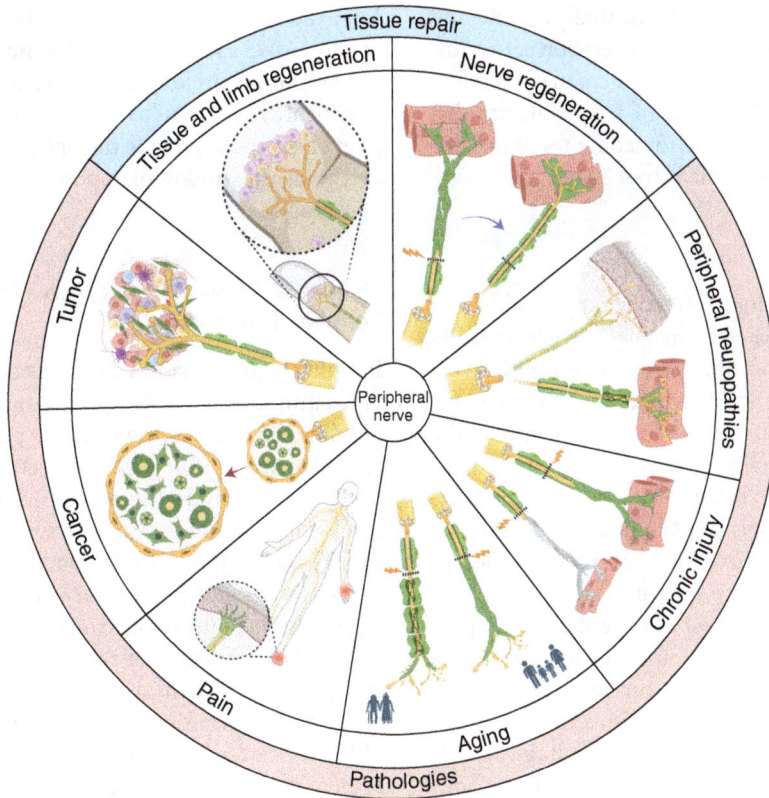

Figure 3. Role of Schwann cells (SCs) in tissue repair and pathologies. SCs play an important role in repair processes involved in tissue regeneration, as well as in various pathologies such as peripheral neuropathies, cancer, and pain.

and injured nerves, and in particular SCs, appear to play a role in repair. Repair SCs, resulting from damage by the wound, secrete as-yet unidentified factors that promote the proliferation of mesenchymal precursor cells required to repair the wound (Johnston et al. 2016; Parfejevs et al. 2018). In addition to the repair of tissue, SCs have been suggested to be implicated in the function of the hematopoietic stem cell (HSC) niche. HSCs within the bone marrow are in close contact with various other cell types, including sympathetic nerve fibers and nmSCs (Mendelson and Frenette 2014; Veiga-Fernandes and Mucida 2016). These SCs have been suggested to influence the maintenance of HSC quiescence via TGF-β signaling, in a study where a loss of HSCs upon experimental denervation of sympathetic fibers was observed (Yamazaki et al. 2011; Yamazaki and Nakauchi 2014). However, future studies are

required to further untangle the specific functions of nmSCs versus the role of axonal signals in controlling HSC behavior (Katayama et al. 2006; Yamazaki et al. 2011; Lucas et al. 2013; Arranz et al. 2014; Yamazaki and Nakauchi 2014; Fielding et al. 2022).

PATHOLOGIES

Despite the efficiency of nerve regeneration, many pathologies result from aberrant repair due to the severity of injury, chronic damage, and failures in aging animals, which can result in loss of function and pain (Fig. 3). Moreover, consistent with a regenerative process, SC-derived tumors that develop in patients with the genetic predisposition syndrome neurofibromatosis type 1 (NF1) mimic the injury response, and innervation is emerging as a critical component of the tumor microenvironment. A better understanding of the mechanisms that promote and restrict functional nerve regeneration across different peripheral nerve disorders may therefore contribute to the design of new strategies that would improve clinical outcome for these diseases (see Scherer and Svaren 2023).

Chronic Nerve Injury

Chronic nerve injury describes any condition in which nerve regeneration is delayed or eventually fails, which constitutes a common and severe problem in clinical practice, for example, nerve transection injuries with long gaps between the nerve stumps are relatively common in humans (Terenghi 1995; Höke 2006; Sulaiman and Gordon 2013; Palispis and Gupta 2017). In general, the time that is required for reinnervation is thought to constitute a key parameter of successful nerve repair, which, in the distal stump, is largely determined by the rate of axonal regrowth, which is slow (1 mm/day), so long human nerves can take more than a year to regrow (Höke and Brushart 2010). Interestingly, cross-suture experiments demonstrated that functional recovery after nerve injury can be enhanced if a chronically axotomized nerve is connected to a freshly denervated distal nerve stump, pointing to the relevance of the distal nerve environment for effective nerve regeneration (Holmes and Young 1942; Fu and Gordon 1995a,b; Sulaiman and Gordon 2000). In the chronically injured distal nerve compartment, SC numbers as well as the preservation of the molecular SC repair phenotype are potential critical variables. SC numbers drop during chronic denervation, which may add to the failure of nerve regeneration, although their numbers still remain higher compared to nonlesioned controls (Salonen et al. 1988; Siironen et al. 1994; Kim et al. 2000; Atanasoski et al. 2001; Serhan et al. 2008; Yang et al. 2008; Jonsson et al. 2013; Benito et al. 2017; Jessen and Mirsky 2019). However, a gradual fading of the SC repair phenotype with time after injury may be the key determinant for a failure of regeneration in chronically injured nerves (Boyd and Gordon 2003; Eggers et al. 2010; Arthur-Farraj et al. 2012; Jessen and Mirsky 2019), which can

be somewhat restored by the overexpression of c-Jun in SCs (Wagstaff et al. 2021). The inability of SCs to sustain a repair-permissive molecular signature during prolonged denervation suggests that identifying means to maintain the SC repair phenotype could provide therapeutic benefit by extending the window of opportunity for the repair process.

Similarities between Acute Nerve Injury and Peripheral Neuropathies

Peripheral neuropathies are classically subdivided into axonal and demyelinating forms, depending on whether the primary pathology is caused by acquired or genetic defects in neurons or SCs, respectively (Suter and Scherer 2003; Dyck and Thomas 2005; Pareyson et al. 2006; Bilbao and Schmidt 2015). Both forms share features with acute nerve injury including overlapping mechanisms of axonal degeneration, the variable presence of regeneration and remyelination, as well as the (aberrant) activation of the SC repair program (Bilbao and Schmidt 2015; Stassart and Woodhoo 2021). The most evident similarities between acute nerve injury and peripheral neuropathies constitute the occurrence of Wallerian degeneration-like mechanisms, which have been implicated as part of "dying back" axonal degeneration in different neuropathies (Coleman 2005; Conforti et al. 2014; Coleman and Höke 2020; Arthur-Farraj and Coleman 2021; Stassart and Woodhoo 2021). Wallerian degeneration in neuropathy is likely to induce a repair response in SCs, but whether failure(s) in the initiation or maintenance of the SC repair response determines the success of axonal regrowth in neuropathies remains largely unknown (King 1999; Bilbao and Schmidt 2015). Interestingly, SCs in demyelinating peripheral neuropathies such as in Charcot–Marie–Tooth Disease 1A and 1B (CMT1A and CMT1B) exhibit molecular phenotypes similar to repair SCs in acute nerve injury (Hanemann et al. 1997; Hutton et al. 2011; Patzkó et al. 2012; D'Antonio et al. 2013; Martini et al. 2013; Fledrich et al. 2014; Hantke et al. 2014; Klein et al. 2014; Fledrich et al. 2019b). However, while the acquisition of a repair state of SCs

Cite this article as *Cold Spring Harb Perspect Biol* doi: 10.1101/cshperspect.a041363

after acute nerve injury is a response to axonal breakdown, in neuropathy, SCs express repair-associated genes while continuing to express myelin genes. Indeed, the molecular repair-like phenotype can be observed in SCs that myelinate morphologically intact–appearing axons (Hutton et al. 2011; Fledrich et al. 2014; Klein et al. 2014). Moreover, aberrant expression patterns and timing of SC repair have been shown to contribute to disease pathogenesis in neuropathy—for example, the persistent expression of the repair-associated growth factor Neuregulin-1 in SCs leads to the development of onion bulb formations in CMT1A disease (Fledrich et al. 2019b). Hence, the repair-like phenotype in neuropathic SCs can exert either protective or detrimental effects on nerve integrity and function, which is dependent on the disease characteristics and disease course dynamics in different forms of peripheral neuropathies (Hantke et al. 2014; Florio et al. 2018; Fledrich et al. 2019b; Stassart and Woodhoo 2021). Notably, shortened internodal length and thinner myelin sheaths are a feature of both remyelination in acute nerve injury and chronic neuropathies (Bilbao and Schmidt 2015) and although not experimentally proven so far, the SC repair response is presumably required for successful remyelination in demyelinating neuropathies (Stassart and Woodhoo 2021). Together these findings show that while the SC repair response is a shared phenomenon between traumatic nerve injuries and peripheral neuropathies, untangling the multifaceted functions and dynamics of the repair cell state may be required to provide novel therapeutic targets for these largely untreatable diseases.

Pain

Failed nerve repair is associated with pain, with both SCs and satellite glial cells implicated in the pain response. For instance, glial cells that become activated by chronic nerve injury, have been reported to induce changes in the levels of ion channels or pain-inducing factors such as TNF-a in sensory neurons. Furthermore, over the longer term, this process can also alter how pain signals are transmitted to the central nervous system (CNS) (Campana 2007; Robel and Sontheimer 2016; Wei et al. 2019; Hanani and Spray 2020). In an extremely common example of chronic nerve injury, the formation of painful neuromas is a frequent complication of even a minor injury to a nerve, in which an imperfect repair process results in a nodule of regenerating axons that fail to innervate. This creates an environment rich in growing axons, activated SCs, and neuroinflammation resulting in aberrant activation of sensory nerves and pain (Neumeister and Winters 2020). Recently, a specialized type of SC was described in the skin (Abdo et al. 2019), which associates with nociceptive axons to form a mesh-like network between the dermis and epidermis responsible for perceiving changes in temperature and mechanical stimuli (Abdo et al. 2019; Rinwa et al. 2021). Using optogenetic methods, it was shown that the SCs themselves are mechanosensitive and provide nociceptive information to the neurons indicating a sensing role for SCs in pain communication. In line with this, SC transient receptor potential ankyrin 1 (TRAP1) channels have been found to contribute to allodynia and neuroinflammation in neuropathic pain (De Logu et al. 2017), and SCs have recently been demonstrated to induce periorbital mechanical allodynia via a calcitonin gene-related peptide signaling response (De Logu et al. 2022). The ablation of nociceptive SCs in mice resulted in nerve retraction and mechanical, cold, and heat hyperalgesia, indicating an interactive relationship between the axons and SCs within the sensory mesh that senses painful stimuli. Further research into this new sensory structure will be important to ascertain its role in multiple forms of pathological pain. Altogether, our current understanding of the roles of peripheral glia in pain remains limited, but represents an exciting, emerging area of research offering new therapeutic targets to potentially relieve chronic pain.

Aging

Peripheral nerves undergo structural and functional changes with age, with reported structural changes within nerve trunks and terminals suspected to account for age-related functional decline (Ceballos et al. 1999; Yuan et al. 2018; Pannese 2021). Similarly, the injury response becomes impaired with advancing age (Thomas et al. 1980; Verdú et al. 1995, 2000; Scheib and Höke 2016; Painter 2017). However, recent research suggests that this occurs independently of the ability of the axon to regrow; rather, a defective ability of SCs and macrophages to remodel the environment appears to be primarily responsible (Kang and Lichtman 2013; Painter et al. 2014; Yuan et al. 2018). In particular, it is well documented that myelin clearance becomes defective in aged animals, and this accumulated debris inhibits axonal regrowth (Tanaka and Webster 1991; Vaughan 1992). This was beautifully demonstrated in live-imaging studies, which showed the speed of axonal regrowth was unchanged but migration stalled as they attempted to maneuver around accumulated debris (Kang and Lichtman 2013; Painter et al. 2014). Moreover, synapse reinnervation occurred efficiently once axons navigated through the debris and reached the neuromuscular junction (Kang and Lichtman 2013; Painter et al. 2014). The mechanisms underlying these phenomena appear to be due to a less robust activation of the SC repair phenotype caused by a defect in the c-Jun-dependent transcriptional response (Painter et al. 2014; Wagstaff et al. 2021) with a resultant decrease in macrophage recruitment, further contributing to the phenotype. Consistent with a recruitment defect, young mice were unable to recover the regenerating capacity of aged nerves in parabiosis studies (Tanaka et al. 1992; Painter et al. 2014). Therapeutically, finding mechanisms to enhance the SC repair state offers new avenues to improve nerve repair in aging populations.

The Role of Innervation in Cancer

The tumor microenvironment is critical for determining whether cancer cells will form a viable tumor and spread. While much recent research has focused on stromal and immune cells, the importance of tumor innervation in both the initiation and spread of tumors is rapidly accumulating (Anastasaki et al. 2023; Winkler et al. 2023). Studies from tumors as diverse as prostate, skin, pancreatic, and breast have found that signals from local nerves are important for the initiation and/or metastatic spread of tumors (Sloan et al. 2010; Magnon et al. 2013; Peterson et al. 2015; Saloman et al. 2016; Renz et al. 2018). Moreover, surrounding nerves can provide a conducive and protected environment for tumor spread in a process known as perineurial invasion (Bapat et al. 2011; Deborde et al. 2022; Winkler et al. 2023). As such, targeting the innervation of tumors offers potential new approaches for developing new classes of therapeutics. In contrast to the CNS, where compelling evidence indicates brain tumors integrate and are sustained by the electrical activity and plasticity intrinsic to the normal functioning brain (Baker et al. 2023; Winkler et al. 2023), the role of peripheral nerve innervation in the development and spread of tumors rather mimics regenerative processes (Boilly et al. 2017). While poorly understood, the most likely scenario involves the processes involved in nerve repair, orchestrated by SCs. The multicellular environment that permits the migration of the SC cords that direct axonal outgrowth across new tissue at the wound site (a hypoxic environment, de novo vascularization) are all present within a developing tumor, making this a feasible mechanism (Cattin et al. 2015; Cattin and Lloyd 2016). In contrast, other, mainly in vitro studies, have shown that tumor cells secrete neurotrophic factors that can attract regrowing axons, providing an alternative or possibly complementary mechanism (Faulkner et al. 2019). Further in vivo studies, preferably at the earliest stages of tumor formation, are required to distinguish these mechanisms. Also unclear are which signals from nerves promote tumor development and the role of SCs, if any, in the cancer microenvironment. Work from the regeneration field, showing signals from neurons and SCs regulate the stem cell environment, provides a potential model consistent with the importance of cancer stem cells to maintain a tumor. But to date, most work has implicated direct signaling by neurons, particularly sympathetic/parasympathetic neurons, in the tumorigenic process (Winkler et al. 2023).

Cite this article as *Cold Spring Harb Perspect Biol* doi: 10.1101/cshperspect.a041363

Schwann Cell–Derived Tumors

Cancers are often referred to as unrepaired wounds and SC-derived tumors that develop in patients with the genetic predisposition syndrome neurofibromatosis type 1 (NF1) provide a classic example of the resemblance of tumors (neurofibromas) to a damaged, regenerating nerve (Parrinello and Lloyd 2009; Fletcher et al. 2019; Hua and Bergers 2019; MacCarthy-Morrogh and Martin 2020). In these tumors, $Nf1^{-/-}$ SCs proliferate and dissociate from axons in a mixed cell environment composed of axons, macrophages, blood vessels, and other immune and stromal components. In this way, these tumors resemble neuromas, painful structures resulting from aberrant nerve repair; one difference is that neurofibromas have additional proliferative capacity and can often become quite large. The gene responsible for NF1 is the Ras-GAP *Neurofibromin*, a negative regulator of Ras-signaling, which leads to the aberrant activation of ERK-signaling pathways that drive the dedifferentiation and proliferation of SCs during an injury response (Martin et al. 1990; Viskochil et al. 1990; Wallace et al. 1990; Napoli et al. 2012). A fascinating aspect of the development of these tumors, derived from studies in mouse models, is that while *Nf1* loss appears to be the sole genetic change required for their formation, it is not sufficient, in that the microenvironment of the nerve plays a critical role (Zhu et al. 2002; Yang et al. 2008; Parrinello and Lloyd 2009; Liao et al. 2018). Several studies have shown that loss of *Nf1* in SCs during development results in few initial changes to SC differentiation or behavior, with nerves appearing mostly normal. Instead, tumors form much later and appear to be initiated by disruption of normal SC/axonal interactions and can be greatly amplified by injury to the nerve (Zhu et al. 2002; Radomska et al. 2019). Moreover, if loss of *Nf1* is triggered in adult SCs, tumor formation is fully dependent on nerve injury—with the specific environment of the wound site, rather than the degenerating/regenerating environment of the distal stump, conducive for tumor formation (Ribeiro et al. 2013). This opposing behavior of $Nf1^{-/-}$ SCs to either form tumors at the wound site or behave as wild-type cells in response to regrowing axons within the distal stump implies that "normalization" of the environment, perhaps by mimicking the pro-differentiating axonal signals identified in regeneration studies, could induce $Nf1^{-/-}$ SCs to return to normality and provide new therapeutic approaches for the treatment of SC-derived tumors.

CONCLUSION

The regenerative capacity of peripheral nerves is astonishing and requires the coordinated collaboration of multiple cell types to ensure the regrowth of damaged axons over substantial distances and their subsequent reinnervation. SCs, themselves remarkably plastic cells, are the orchestrators of this multicellular response and ensuring their repair capacity has implications for developing new therapies for nerve injury, neuropathies, and the decline of regeneration with age. Exciting new fields of study implicate nerve repair processes in normal tissue repair and in pathological conditions such as cancer. These findings offer new avenues to develop novel classes of therapeutics to both improve tissue repair and cancer treatment.

COMPETING INTEREST STATEMENT

The authors declare no competing interests.

ACKNOWLEDGMENTS

This work was supported by an ERC starting grant (948857, AxoMyoGlia) and DFG grant STA 1728/1-1 to R.M.S., a Miguel Servet Fellowship from the Spanish Health Institute Carlos III (CP22/00078) and Plan Nacional 2022 from the Agencia Estatal de Investigación (PID2022-143269OB-I00) to J.A.G.-S. and a program grant from Cancer Research UK to A.C.L. (C378/A4308). Although we have made efforts to present a well-rounded overview of the role of peripheral glial cells in nerve injury and regeneration, we recognize that this review may not cover all aspects of this fascinating area of research. Therefore, we would like to express our apologies to fellow researchers whose valuable contributions have not been included or emphasized in this review, as we were constrained by limitations of space and scope. All figures were created with BioRender.com.

REFERENCES

*Reference is also in this subject collection.

Abdo H, Calvo-Enrique L, Lopez JM, Song J, Zhang MD, Usoskin D, El Manira A, Adameyko I, Hjerling-Leffler J, Ernfors P. 2019. Specialized cutaneous Schwann cells initiate pain sensation. *Science* 365: 695–699. doi:10.1126/science.aax6452

Allodi I, Udina E, Navarro X. 2012. Specificity of peripheral nerve regeneration: interactions at the axon level. *Prog Neurobiol* 98: 16–37. doi:10.1016/j.pneurobio.2012.05.005

Anastasaki C, Gao Y, Gutmann DH. 2023. Neurons as stromal drivers of nervous system cancer formation and progression. *Dev Cell* 58: 81–93. doi:10.1016/j.devcel.2022.12.011

Arranz L, Sánchez-Aguilera A, Martín-Pérez D, Isern J, Langa X, Tzankov A, Lundberg P, Muntión S, Tzeng YS, Lai DM, et al. 2014. Neuropathy of haematopoietic stem cell niche is essential for myeloproliferative neoplasms. *Nature* 512: 78–81. doi:10.1038/nature13383

Arthur-Farraj P, Coleman MP. 2021. Lessons from injury: how nerve injury studies reveal basic biological mechanisms and therapeutic opportunities for peripheral nerve diseases. *Neurotherapeutics* 18: 2200–2221. doi:10.1007/s13311-021-01125-3

Arthur-Farraj PJ, Latouche M, Wilton DK, Quintes S, Chabrol E, Banerjee A, Woodhoo A, Jenkins B, Rahman M, Turmaine M, et al. 2012. c-Jun reprograms Schwann cells of injured nerves to generate a repair cell essential for regeneration. *Neuron* 75: 633–647. doi:10.1016/j.neuron.2012.06.021

Arthur-Farraj PJ, Morgan CC, Adamowicz M, Gomez-Sanchez JA, Fazal SV, Beucher A, Razzaghi B, Mirsky R, Jessen KR, Aitman TJ. 2017. Changes in the coding and noncoding transcriptome and DNA methylome that define the Schwann cell repair phenotype after nerve injury. *Cell Rep* 20: 2719–2734. doi:10.1016/j.celrep.2017.08.064

Atanasoski S, Shumas S, Dickson C, Scherer SS, Suter U. 2001. Differential cyclin D1 requirements of proliferating Schwann cells during development and after injury. *Mol Cell Neurosci* 18: 581–592. doi:10.1006/mcne.2001.1055

Avraham O, Deng PY, Jones S, Kuruvilla R, Semenkovich CF, Klyachko VA, Cavalli V. 2020. Satellite glial cells promote regenerative growth in sensory neurons. *Nat Commun* 11: 4891. doi:10.1038/s41467-020-18642-y

Babetto E, Wong KM, Beirowski B. 2020. A glycolytic shift in Schwann cells supports injured axons. *Nat Neurosci* 23: 1215–1228. doi:10.1038/s41593-020-0689-4

* Baker S, Zong H, Monje M. 2023. Glial malignancies. *Cold Spring Harb Perspect Biol* doi:10.1101/cshperspect.a041373

Bapat AA, Hostetter G, von Hoff DD, Han H. 2011. Perineural invasion and associated pain in pancreatic cancer. *Nat Rev Cancer* 11: 695–707. doi:10.1038/nrc3131

Barnes SL, Miller TA, Simon NG. 2022. Traumatic peripheral nerve injuries: diagnosis and management. *Curr Opin Neurol* 35: 718–727. doi:10.1097/WCO.0000000000001116

Barrette B, Hébert M-A, Filali M, Lafortune K, Vallières N, Gowing G, Julien JP, Lacroix S. 2008. Requirement of myeloid cells for axon regeneration. *J Neurosci* 28: 9363–9376. doi:10.1523/JNEUROSCI.1447-08.2008

Benito C, Davis CM, Gomez-Sanchez JA, Turmaine M, Meijer D, Poli V, Mirsky R, Jessen KR. 2017. STAT3 controls the long-term survival and phenotype of repair Schwann cells during nerve regeneration. *J Neurosci* 37: 4255–4269. doi:10.1523/JNEUROSCI.3481-16.2017

Bilbao JM, Schmidt RE. 2015. *Biopsy diagnosis of peripheral neuropathy*. Springer, New York.

Birks R, Katz B, Miledi R. 1960. Physiological and structural changes at the amphibian myoneural junction, in the course of nerve degeneration. *J Physiol* 150: 145–168. doi:10.1113/jphysiol.1960.sp006379

Boilly B, Faulkner S, Jobling P, Hondermarck H. 2017. Nerve dependence: from regeneration to cancer. *Cancer Cell* 31: 342–354. doi:10.1016/j.ccell.2017.02.005

Bosch-Queralt M, Fledrich R, Stassart RM. 2023. Schwann cell functions in peripheral nerve development and repair. *Neurobiol Dis* 176: 105952. doi:10.1016/j.nbd.2022.105952

Boyd JG, Gordon T. 2003. Neurotrophic factors and their receptors in axonal regeneration and functional recovery after peripheral nerve injury. *Mol Neurobiol* 27: 277–324. doi:10.1385/MN:27:3:277

Brosius Lutz A, Barres BA. 2014. Contrasting the glial response to axon injury in the central and peripheral nervous systems. *Dev Cell* 28: 7–17. doi:10.1016/j.devcel.2013.12.002

Brosius Lutz A, Chung WS, Sloan SA, Carson GA, Zhou L, Lovelett E, Posada S, Zuchero JB, Barres BA. 2017. Schwann cells use TAM receptor-mediated phagocytosis in addition to autophagy to clear myelin in a mouse model of nerve injury. *Proc Natl Acad Sci* 114: E8072–E8080. doi:10.1073/pnas.1710566114

Büngner OV. 1891. Über die degenerations-und regenerationsvorgänge an nerven nach verletzungen. *Beitr Path Anat* 10: 321–390.

Campana WM. 2007. Schwann cells: activated peripheral glia and their role in neuropathic pain. *Brain Behav Immun* 21: 522–527. doi:10.1016/j.bbi.2006.12.008

Canclini L, Farias J, Di Paolo A, Sotelo-Silveira JR, Folle G, Kun A, Sotelo JR. 2020. Association of microtubules and axonal RNA transferred from myelinating Schwann cells in rat sciatic nerve. *PLoS ONE* 15: e0233651. doi:10.1371/journal.pone.0233651

Carr MJ, Toma JS, Johnston APW, Steadman PE, Yuzwa SA, Mahmud N, Frankland PW, Kaplan DR, Miller FD. 2019. Mesenchymal precursor cells in adult nerves contribute to mammalian tissue repair and regeneration. *Cell Stem Cell* 24: 240–256.e9. doi:10.1016/j.stem.2018.10.024

Castonguay A, Robitaille R. 2001. Differential regulation of transmitter release by presynaptic and glial Ca^{2+} internal stores at the neuromuscular synapse. *J Neurosci* 21: 1911–1922. doi:10.1523/JNEUROSCI.21-06-01911.2001

Catenaccio A, Llavero Hurtado M, Diaz P, Lamont DJ, Wishart TM, Court FA. 2017. Molecular analysis of axonal-intrinsic and glial-associated co-regulation of axon degeneration. *Cell Death Dis* **8:** e3166. doi:10.1038/cddis .2017.489

Cattin AL, Lloyd AC. 2016. The multicellular complexity of peripheral nerve regeneration. *Curr Opin Neurobiol* **39:** 38–46. doi:10.1016/j.conb.2016.04.005

Cattin AL, Burden JJ, van Emmenis L, Mackenzie FE, Hoving JJA, Garcia Calavia N, Guo Y, McLaughlin M, Rosenberg LH, Quereda V, et al. 2015. Macrophage-induced blood vessels guide Schwann cell-mediated regeneration of peripheral nerves. *Cell* **162:** 1127–1139. doi:10.1016/j.cell .2015.07.021

Ceballos D, Cuadras J, Verdú E, Navarro X. 1999. Morphometric and ultrastructural changes with ageing in mouse peripheral nerve. *J Anat* **195:** 563–576. doi:10.1046/j .1469-7580.1999.19540563.x

Cervellini I, Galino J, Zhu N, Allen S, Birchmeier C, Bennett DL. 2018. Sustained MAPK/ERK activation in adult Schwann cells impairs nerve repair. *J Neurosci* **38:** 679–690. doi:10.1523/JNEUROSCI.2255-17.2017

Clements MP, Byrne E, Camarillo Guerrero LF, Cattin AL, Zakka L, Ashraf A, Burden JJ, Khadayate S, Lloyd AC, Marguerat S, et al. 2017. The wound microenvironment reprograms Schwann cells to invasive mesenchymal-like cells to drive peripheral nerve regeneration. *Neuron* **96:** 98–114.e7. doi:10.1016/j.neuron.2017.09.008

Coleman M. 2005. Axon degeneration mechanisms: commonality amid diversity. *Nat Rev Neurosci* **6:** 889–898. doi:10.1038/nrn1788

Coleman MP, Höke A. 2020. Programmed axon degeneration: from mouse to mechanism to medicine. *Nat Rev Neurosci* **21:** 183–196. doi:10.1038/s41583-020-0269-3

Conforti L, Gilley J, Coleman MP. 2014. Wallerian degeneration: an emerging axon death pathway linking injury and disease. *Nat Rev Neurosci* **15:** 394–409. doi:10.1038/ nrn3680

Dalla Costa I, Buchanan CN, Zdradzinski MD, Sahoo PK, Smith TP, Thames E, Kar AN, Twiss JL. 2021. The functional organization of axonal mRNA transport and translation. *Nat Rev Neurosci* **22:** 77–91. doi:10.1038/s41583-020-00407-7

D'Antonio M, Musner N, Scapin C, Ungaro D, Del Carro U, Ron D, Feltri ML, Wrabetz L. 2013. Resetting translational homeostasis restores myelination in Charcot–Marie–Tooth disease type 1B mice. *J Exp Med* **210:** 821–838. doi:10.1084/jem.20122005

Deborde S, Gusain L, Powers A, Marcadis A, Yu Y, Chen CH, Frants A, Kao E, Tang LH, Vakiani E, et al. 2022. Reprogrammed Schwann cells organize into dynamic tracks that promote pancreatic cancer invasion. *Cancer Discov* **12:** 2454–2473. doi:10.1158/2159-8290.CD-21-1690

De Logu F, Nassini R, Materazzi S, Carvalho Gonçalves M, Nosi D, Rossi Degl'Innocenti D, Marone IM, Ferreira J, Li Puma S, Benemei S, et al. 2017. Schwann cell TRPA1 mediates neuroinflammation that sustains macrophage-dependent neuropathic pain in mice. *Nat Commun* **8:** 1887. doi:10.1038/s41467-017-01739-2

De Logu F, Nassini R, Hegron A, Landini L, Jensen DD, Latorre R, Ding J, Marini M, Souza Monteiro de Araujo D, Ramírez-Garcia P, et al. 2022. Schwann cell endosome

CGRP signals elicit periorbital mechanical allodynia in mice. *Nat Commun* **13:** 646. doi:10.1038/s41467-022-28204-z

Dolan CP, Imholt F, Yan M, Yang TJ, Gregory J, Qureshi O, Zimmel K, Sherman KM, Smith HM, Falck A, et al. 2022. Digit specific denervation does not inhibit mouse digit tip regeneration. *Dev Biol* **486:** 71–80. doi:10.1016/j.ydbio .2022.03.007

Dubový P, Aldskogius H. 1996. Degeneration and regeneration of cutaneous sensory nerve formations. *Microsc Res Tech* **34:** 362–375. doi:10.1002/(SICI)1097-0029(199607 01)34:4<362::AID-JEMT7>3.0.CO;2-Q

Dyck PJ, Thomas PK. 2005. *Peripheral neuropathy.* Elsevier, New York.

Eggers R, Tannemaat MR, Ehlert EM, Verhaagen J. 2010. A spatio-temporal analysis of motoneuron survival, axonal regeneration and neurotrophic factor expression after lumbar ventral root avulsion and implantation. *Exp Neurol* **223:** 207–220. doi:10.1016/j.expneurol.2009.07.021

Faulkner S, Jobling P, March B, Jiang CC, Hondermarck H. 2019. Tumor neurobiology and the war of nerves in cancer. *Cancer Discov* **9:** 702–710. doi:10.1158/2159-8290.CD-18-1398

Fawcett JW, Keynes RJ. 1990. Peripheral nerve regeneration. *Annu Rev Neurosci* **13:** 43–60. doi:10.1146/annurev.ne.13 .030190.000355

Feltri ML, Weaver MR, Belin S, Poitelon Y. 2021. The Hippo pathway: horizons for innovative treatments of peripheral nerve diseases. *J Peripher Nerv Syst* **26:** 4–16. doi:10.1111/ jns.12431

Fernando RN, Cotter L, Perrin-Tricaud C, Berthelot J, Bartolami S, Pereira JA, Gonzalez S, Suter U, Tricaud N. 2016. Optimal myelin elongation relies on YAP activation by axonal growth and inhibition by Crb3/Hippo pathway. *Nat Commun* **7:** 12186. doi:10.1038/ncomms12186

Fielding C, García-García A, Korn C, Gadomski S, Fang Z, Reguera JL, Pérez-Simón JA, Göttgens B, Méndez-Ferrer S. 2022. Cholinergic signals preserve haematopoietic stem cell quiescence during regenerative haematopoiesis. *Nat Commun* **13:** 543. doi:10.1038/s41467-022-28175-1

Fledrich R, Stassart RM, Klink A, Rasch LM, Prukop T, Haag L, Czesnik D, Kungl T, Abdelaal TAM, Keric N, et al. 2014. Soluble neuregulin-1 modulates disease pathogenesis in rodent models of Charcot–Marie–Tooth disease 1A. *Nat Med* **20:** 1055–1061. doi:10.1038/nm.3664

Fledrich R, Kungl T, Nave KA, Stassart RM. 2019a. Axo-glial interdependence in peripheral nerve development. *Development* **146:** dev151704. doi:10.1242/dev.151704

Fledrich R, Akkermann D, Schütza V, Abdelaal TA, Hermes D, Schäffner E, Soto-Bernardini MC, Götze T, Klink A, Kusch K, et al. 2019b. NRG1 type I dependent autoparacrine stimulation of Schwann cells in onion bulbs of peripheral neuropathies. *Nat Commun* **10:** 1467. doi:10 .1038/s41467-019-09385-6

Fletcher JS, Pundavela J, Ratner N. 2019. After Nf1 loss in Schwann cells, inflammation drives neurofibroma formation. *Neurooncol Adv* **22**(Suppl 1): i23–i32. doi:10.1093/ noajnl/vdz045

Florio F, Ferri C, Scapin C, Feltri ML, Wrabetz L, D'Antonio M. 2018. Sustained expression of negative regulators of myelination protects Schwann cells from dysmyelination in a Charcot–Marie–Tooth 1B mouse model. *J Neurosci* **38:** 4275–4287. doi:10.1523/JNEUROSCI.0201-18.2018

Fontana X, Hristova M, Da Costa C, Patodia S, Thei L, Makwana M, Spencer-Dene B, Latouche M, Mirsky R, Jessen KR, et al. 2012. c-Jun in Schwann cells promotes axonal regeneration and motoneuron survival via paracrine signaling. *J Cell Biol* **198:** 127–141. doi:10.1083/jcb.2012 05025

Forese MG, Pellegatta M, Canevazzi P, Gullotta GS, Podini P, Rivellini C, Previtali SC, Bacigaluppi M, Quattrini A, Taveggia C. 2020. Prostaglandin D2 synthase modulates macrophage activity and accumulation in injured peripheral nerves. *Glia* **68:** 95–110. doi:10.1002/glia.23705

Fricker FR, Antunes-Martins A, Galino J, Paramsothy R, La Russa F, Perkins J, Goldberg R, Brelstaff J, Zhu N, McMahon SB, et al. 2013. Axonal neuregulin 1 is a rate limiting but not essential factor for nerve remyelination. *Brain* **136:** 2279–2297. doi:10.1093/brain/awt148

Fry EJ, Ho C, David S. 2007. A role for Nogo receptor in macrophage clearance from injured peripheral nerve. *Neuron* **53:** 649–662. doi:10.1016/j.neuron.2007.02.009

Fu SY, Gordon T. 1995a. Contributing factors to poor functional recovery after delayed nerve repair: prolonged axotomy. *J Neurosci* **15:** 3876–3885. doi:10.1523/JNEURO SCI.15-05-03876.1995

Fu SY, Gordon T. 1995b. Contributing factors to poor functional recovery after delayed nerve repair: prolonged denervation. *J Neurosci* **15:** 3886–3895. doi:10.1523/JNEUR OSCI.15-05-03886.1995

Gangadharan V, Zheng H, Taberner FJ, Landry J, Nees TA, Pistolic J, Agarwal N, Männich D, Benes V, Helmstaedter M, et al. 2022. Neuropathic pain caused by miswiring and abnormal end organ targeting. *Nature* **606:** 137–145. doi:10.1038/s41586-022-04777-z

Gao X, Wang Y, Chen J, Peng J. 2013. The role of peripheral nerve ECM components in the tissue engineering nerve construction. *Rev Neurosci* **24:** 443–453. doi:10.1515/re vneuro-2013-0022

Goldman JA, Poss KD. 2020. Gene regulatory programmes of tissue regeneration. *Nat Rev Genet* **21:** 511–525. doi:10 .1038/s41576-020-0239-7

Gomez-Sanchez JA, Carty L, Iruarrizaga-Lejarreta M, Palomo-Irigoyen M, Varela-Rey M, Griffith M, Hantke J, Macias-Camara N, Azkargorta M, Aurrekoetxea I, et al. 2015. Schwann cell autophagy, myelinophagy, initiates myelin clearance from injured nerves. *J Cell Biol* **210:** 153–168. doi:10.1083/jcb.201503019

Gomez-Sanchez JA, Pilch KS, van der Lans M, Fazal SV, Benito C, Wagstaff LJ, Mirsky R, Jessen KR. 2017. After nerve injury, lineage tracing shows that myelin and Remak Schwann cells elongate extensively and branch to form repair Schwann cells, which shorten radically on remyelination. *J Neurosci* **37:** 9086–9099. doi:10.1523/JNEURO SCI.1453-17.2017

Gomez-Sanchez JA, Patel N, Martirena F, Fazal SV, Mutschler C, Cabedo H. 2022. Emerging role of HDACs in regeneration and ageing in the peripheral nervous system: repair Schwann cells as pivotal targets. *Int J Mol Sci* **23:** 2996. doi:10.3390/ijms23062996

Gordon T. 2009. The role of neurotrophic factors in nerve regeneration. *Neurosurg Focus* **26:** E3. doi:10.3171/FOC .2009.26.2.E3

Gordon T. 2020. Peripheral nerve regeneration and muscle reinnervation. *Int J Mol Sci* **21:** 8652. doi:10.3390/ijms 21228652

* Gould TW, Ko CP, Willison H, Robitaille R. 2023. Perisynaptic Schwann cells at the neuromuscular synapse. *Cold Spring Harb Perspect Biol* doi:10.1101/cshperspect .a041362

Griffin JW, Pan B, Polley MA, Hoffman PN, Farah MH. 2010. Measuring nerve regeneration in the mouse. *Exp Neurol* **223:** 60–71. doi:10.1016/j.expneurol.2009.12.033

Haberberger RV, Kuramatilake J, Barry CM, Matusica D. 2023. Ultrastructure of dorsal root ganglia. *Cell Tissue Res* **393:** 17–36. doi:10.1007/s00441-023-03770-w

Hanani M, Spray DC. 2020. Emerging importance of satellite glia in nervous system function and dysfunction. *Nat Rev Neurosci* **21:** 485–498. doi:10.1038/s41583-020-0333-z

Handler A, Ginty DD. 2021. The mechanosensory neurons of touch and their mechanisms of activation. *Nat Rev Neurosci* **22:** 521–537. doi:10.1038/s41583-021-00489-x

Hanemann CO, Gabreëls-Festen AA, Stoll G, Müller HW. 1997. Schwann cell differentiation in Charcot-Marie-Tooth disease type 1A (CMT1A): normal number of myelinating Schwann cells in young CMT1A patients and neural cell adhesion molecule expression in onion bulbs. *Acta Neuropathol* **94:** 310–315. doi:10.1007/s004010050712

Hantke J, Carty L, Wagstaff LJ, Turmaine M, Wilton DK, Quintes S, Koltzenburg M, Baas F, Mirsky R, Jessen KR. 2014. c-Jun activation in Schwann cells protects against loss of sensory axons in inherited neuropathy. *Brain* **137:** 2922–2937. doi:10.1093/brain/awu257

Harrisingh MC, Perez-Nadales E, Parkinson DB, Malcolm DS, Mudge AW, Lloyd AC. 2004. The Ras/Raf/ERK signalling pathway drives Schwann cell dedifferentiation. *EMBO J* **23:** 3061–3071. doi:10.1038/sj.emboj.7600309

Harty BL, Monk KR. 2017. Unwrapping the unappreciated: recent progress in Remak Schwann cell biology. *Curr Opin Neurobiol* **47:** 131–137. doi:10.1016/j.conb.2017.10.003

Hastings RL, Mikesh M, Lee YI, Thompson WJ. 2020. Morphological remodeling during recovery of the neuromuscular junction from terminal Schwann cell ablation in adult mice. *Sci Rep* **10:** 11132. doi:10.1038/s41598-020-67630-1

Höke A. 2006. Mechanisms of disease: what factors limit the success of peripheral nerve regeneration in humans? *Nat Clin Pract Neurol* **2:** 448–454. doi:10.1038/ncpneuro0262

Höke A, Brushart T. 2010. Introduction to special issue: challenges and opportunities for regeneration in the peripheral nervous system. *Exp Neurol* **223:** 1–4. doi:10.1016/j .expneurol.2009.12.001

Holmes W, Young JZ. 1942. Nerve regeneration after immediate and delayed suture. *J Anat* **77**(Pt 1): 63–96.10. doi:10 .1093/oxfordjournals.bmb.a070225

Hua Y, Bergers G. 2019. Tumors vs. chronic wounds: an immune cell's perspective. *Front Immunol* **10:** 2178. doi:10.3389/fimmu.2019.02178

Hutton EJ, Carty L, Laurá M, Houlden H, Lunn MPT, Brandner S, Mirsky R, Jessen K, Reilly MM. 2011. c-Jun expression in human neuropathies: a pilot study. *J Peripher Nerv Syst* **16:** 295–303. doi:10.1111/j.1529-8027.2011.00360.x

Jablonka-Shariff A, Lu CY, Campbell K, Monk KR, Snyder-Warwick AK. 2020. Gpr126/Adgrg6 contributes to the

Cite this article as *Cold Spring Harb Perspect Biol* doi: 10.1101/cshperspect.a041363

terminal Schwann cell response at the neuromuscular junction following peripheral nerve injury. *Glia* **68**: 1182–1200. doi:10.1002/glia.23769

Jager SE, Pallesen LT, Richner M, Harley P, Hore Z, McMahon S, Denk F, Vægter CB. 2020. Changes in the transcriptional fingerprint of satellite glial cells following peripheral nerve injury. *Glia* **68**: 1375–1395. doi:10.1002/glia.23785

Jang SY, Shin YK, Park SY, Park JY, Lee HJ, Yoo YH, Kim JK, Park HT. 2016. Autophagic myelin destruction by Schwann cells during Wallerian degeneration and segmental demyelination. *Glia* **64**: 730–742. doi:10.1002/glia.22957

Jessen KR, Arthur-Farraj P. 2019. Repair Schwann cell update: adaptive reprogramming, EMT, and stemness in regenerating nerves. *Glia* **67**: 421–437. doi:10.1002/glia.23532

Jessen KR, Mirsky R. 2016. The repair Schwann cell and its function in regenerating nerves. *J Physiol* **594**: 3521–3531. doi:10.1113/JP270874

Jessen KR, Mirsky R. 2019. The success and failure of the Schwann cell response to nerve injury. *Front Cell Neurosci* **13**: 33. doi:10.3389/fncel.2019.00033

Jha MK, Passero JV, Rawat A, Ament XH, Yang F, Vidensky S, Collins SL, Horton MR, Hoke A, Rutter GA, et al. 2021. Macrophage monocarboxylate transporter 1 promotes peripheral nerve regeneration after injury in mice. *J Clin Invest* **131**: e141964. doi:10.1172/JCI141964

Johnston APW, Miller FD. 2022. The contribution of innervation to tissue repair and regeneration. *Cold Spring Harb Perspect Biol* **14**: a041233. doi:10.1101/cshperspect.a041233

Johnston APW, Yuzwa SA, Carr MJ, Mahmud N, Storer MA, Krause MP, Jones K, Paul S, Kaplan DR, Miller FD. 2016. Dedifferentiated Schwann cell precursors secreting paracrine factors are required for regeneration of the mammalian digit tip. *Cell Stem Cell* **19**: 433–448. doi:10.1016/j.stem.2016.06.002

Jones RE, Salhotra A, Robertson KS, Ransom RC, Foster DS, Shah HN, Quarto N, Wan DC, Longaker MT. 2019. Skeletal stem cell-Schwann cell circuitry in mandibular repair. *Cell Rep* **28**: 2757–2766.e5. doi:10.1016/j.celrep.2019.08.021

Jonsson S, Wiberg R, McGrath AM, Novikov LN, Wiberg M, Novikova LN, Kingham PJ. 2013. Effect of delayed peripheral nerve repair on nerve regeneration, Schwann cell function and target muscle recovery. *PLoS ONE* **8**: e56484. doi:10.1371/journal.pone.0056484

Jung J, Cai W, Lee HK, Pellegatta M, Shin YK, Jang SY, Suh DJ, Wrabetz L, Feltri ML, Park HT. 2011. Actin polymerization is essential for myelin sheath fragmentation during Wallerian degeneration. *J Neurosci* **31**: 2009–2015. doi:10.1523/JNEUROSCI.4537-10.2011

Jurecka W, Ammerer HP, Lassmann H. 1975. Regeneration of a transected peripheral nerve. An autoradiographic and electron microscopic study. *Acta Neuropathol* **32**: 299–312. doi:10.1007/BF00696792

Kalinski AL, Yoon C, Huffman LD, Duncker PC, Kohen R, Passino R, Hafner H, Johnson C, Kawaguchi R, Carbajal KS, et al. 2020. Analysis of the immune response to sciatic nerve injury identifies efferocytosis as a key mechanism of nerve debridement. *eLife* **9**: e60223. doi:10.7554/eLife.60223

Kang H, Lichtman JW. 2013. Motor axon regeneration and muscle reinnervation in young adult and aged animals. *J Neurosci* **33**: 19480–19491. doi:10.1523/JNEUROSCI.4067-13.2013

Kang H, Tian L, Thompson W. 2003. Terminal Schwann cells guide the reinnervation of muscle after nerve injury. *J Neurocytol* **32**: 975–985. doi:10.1023/B:NEUR.0000020636.27222.2d

Kang H, Tian L, Mikesh M, Lichtman JW, Thompson WJ. 2014. Terminal Schwann cells participate in neuromuscular synapse remodeling during reinnervation following nerve injury. *J Neurosci* **34**: 6323–6333. doi:10.1523/JNEUROSCI.4673-13.2014

Kang H, Tian L, Thompson WJ. 2019. Schwann cell guidance of nerve growth between synaptic sites explains changes in the pattern of muscle innervation and remodeling of synaptic sites following peripheral nerve injuries. *J Comp Neurol* **527**: 1388–1400. doi:10.1002/cne.24625

Katayama Y, Battista M, Kao WM, Hidalgo A, Peired AJ, Thomas SA, Frenette PS. 2006. Signals from the sympathetic nervous system regulate hematopoietic stem cell egress from bone marrow. *Cell* **124**: 407–421. doi:10.1016/j.cell.2005.10.041

Kim HA, Pomeroy SL, Whoriskey W, Pawlitzky I, Benowitz LI, Sicinski P, Stiles CD, Roberts TM. 2000. A developmentally regulated switch directs regenerative growth of Schwann cells through cyclin D1. *Neuron* **26**: 405–416. doi:10.1016/S0896-6273(00)81173-3

King RHM. 1999. *Atlas of peripheral nerve pathology*. Arnold, London.

Klein D, Groh J, Wettmarshausen J, Martini R. 2014. Non-uniform molecular features of myelinating Schwann cells in models for CMT1: distinct disease patterns are associated with NCAM and c-Jun upregulation. *Glia* **62**: 736–750. doi:10.1002/glia.22638

Ko CP, Robitaille R. 2015. Perisynaptic Schwann cells at the neuromuscular synapse: adaptable, multitasking glial cells. *Cold Spring Harb Perspect Biol* **7**: a020503. doi:10.1101/cshperspect.a020503

Koley S, Rozenbaum M, Fainzilber M, Terenzio M. 2019. Translating regeneration: local protein synthesis in the neuronal injury response. *Neurosci Res* **139**: 26–36. doi:10.1016/j.neures.2018.10.003

Kumar A, Brockes JP. 2012. Nerve dependence in tissue, organ, and appendage regeneration. *Trends Neurosci* **35**: 691–699. doi:10.1016/j.tins.2012.08.003

Kumar A, Godwin JW, Gates PB, Garza-Garcia AA, Brockes JP. 2007. Molecular basis for the nerve dependence of limb regeneration in an adult vertebrate. *Science* **318**: 772–777. doi:10.1126/science.1147710

Li L, Ginty DD. 2014. The structure and organization of lanceolate mechanosensory complexes at mouse hair follicles. *eLife* **3**: e01901. doi:10.7554/eLife.01901

Li Y, Kang S, Halawani D, Wang Y, Junqueira Alves C, Ramakrishnan A, Estill M, Shen L, Li F, He X, et al. 2022. Macrophages facilitate peripheral nerve regeneration by organizing regeneration tracks through Plexin-B2. *Genes Dev* **36**: 133–148. doi:10.1101/gad.349063.121

Liao CP, Booker RC, Brosseau JP, Chen Z, Mo J, Tchegnon E, Wang Y, Clapp DW, Le LQ. 2018. Contributions of inflammation and tumor microenvironment to neurofibroma tumorigenesis. *J Clin Invest* **128:** 2848–2861. doi:10 .1172/JCI99424

Lloyd AC. 2013. The regulation of cell size. *Cell* **154:** 1194–1205. doi:10.1016/j.cell.2013.08.053

López-Leal R, Díaz-Viraqué F, Catalán RJ, Saquel C, Enright A, Iraola G, Court FA. 2020. Schwann cell reprogramming into repair cells increases miRNA-21 expression in exosomes promoting axonal growth. *J Cell Sci* **133:** jcs239004. doi:10.1242/jcs.239004

Lopez-Verrilli MA, Picou F, Court FA. 2013. Schwann cell-derived exosomes enhance axonal regeneration in the peripheral nervous system. *Glia* **61:** 1795–1806. doi:10.1002/glia.22558

Love FM, Thompson WJ. 1999. Glial cells promote muscle reinnervation by responding to activity-dependent postsynaptic signals. *J Neurosci* **19:** 10390–10396. doi:10.1523/JNEUROSCI.19-23-10390.1999

Love FM, Son YJ, Thompson WJ. 2003. Activity alters muscle reinnervation and terminal sprouting by reducing the number of Schwann cell pathways that grow to link synaptic sites. *J Neurobiol* **54:** 566–576. doi:10.1002/neu .10191

Lucas D, Scheiermann C, Chow A, Kunisaki Y, Bruns I, Barrick C, Tessarollo L, Frenette PS. 2013. Chemotherapy-induced bone marrow nerve injury impairs hematopoietic regeneration. *Nat Med* **19:** 695–703. doi:10.1038/nm.3155

Ma KH, Svaren J. 2018. Epigenetic control of Schwann cells. *Neuroscientist* **24:** 627–638. doi:10.1177/10738584177 51112

MacCarthy-Morrogh L, Martin P. 2020. The hallmarks of cancer are also the hallmarks of wound healing. *Sci Signal* **13:** eaay8690. doi:10.1126/scisignal.aay8690

Magnon C, Hall SJ, Lin J, Xue X, Gerber L, Freedland SJ, Frenette PS. 2013. Autonomic nerve development contributes to prostate cancer progression. *Science* **341:** 1236361. doi:10.1126/science1236361

Mahar M, Cavalli V. 2018. Intrinsic mechanisms of neuronal axon regeneration. *Nat Rev Neurosci* **19:** 323–337. doi:10 .1038/s41583-018-0001-8

Malong L, Napoli I, Casal G, White IJ, Stierli S, Vaughan A, Cattin AL, Burden JJ, Hong KI, Bossio A, et al. 2023. Characterization of the structure and control of the blood-nerve barrier identifies avenues for therapeutic delivery. *Dev Cell* **58:** 174–191.e8. doi:10.1016/j.devcel.2023.01.002

Martin GA, Viskochil D, Bollag G, McCabe PC, Crosier WJ, Haubruck H, Conroy L, Clark R, O'Connell P, Cawthon RM, et al. 1990. The GAP-related domain of the neurofibromatosis type 1 gene product interacts with ras p21. *Cell* **63:** 843–849. doi:10.1016/0092-8674(90)90150-D

Martini R, Klein D, Groh J. 2013. Similarities between inherited demyelinating neuropathies and Wallerian degeneration: An old repair program may cause myelin and axon perturbation under nonlesion conditions. *Am J Pathol* **183:** 655–660. doi:10.1016/j.ajpath.2013.06.002

Meltzer S, Boulanger KC, Osei-Asante E, Handler A, Zhang Q, Sano C, Itohara S, Ginty DD. 2022. A role for axon-glial interactions and Netrin-G1 signaling in the formation of low-threshold mechanoreceptor end organs. *Proc Natl Acad Sci* **119:** e2210421119. doi:10.1073/pnas.221042 1119

Mendelson A, Frenette PS. 2014. Hematopoietic stem cell niche maintenance during homeostasis and regeneration. *Nat Med* **20:** 833–846. doi:10.1038/nm.3647

Michailov GV, Sereda MW, Brinkmann BG, Fischer TM, Haug B, Birchmeier C, Role L, Lai C, Schwab MH, Nave KA. 2004. Axonal neuregulin-1 regulates myelin sheath thickness. *Science* **304:** 700–703. doi:10.1126/science .1095862

Miledi R, Slater CR. 1970. On the degeneration of rat neuromuscular junctions after nerve section. *J Physiol* **207:** 507–528. doi:10.1113/jphysiol.1970.sp009076

Mogha A, Harty BL, Carlin D, Joseph J, Sanchez NE, Suter U, Piao X, Cavalli V, Monk KR. 2016. Gpr126/Adgrg6 has Schwann cell autonomous and nonautonomous functions in peripheral nerve injury and repair. *J Neurosci* **36:** 12351–12367. doi:10.1523/JNEUROSCI.3854-15.2016

Monje M, Borniger JC, D'Silva NJ, Deneen B, Dirks PB, Fattahi F, Frenette PS, Garzia L, Gutmann DH, Hanahan D, et al. 2020. Roadmap for the emerging field of cancer neuroscience. *Cell* **181:** 219–222. doi:10.1016/j.cell.2020.03.034

Morrison BM, Tsingalia A, Vidensky S, Lee Y, Jin L, Farah MH, Lengacher S, Magistretti PJ, Pellerin L, Rothstein JD. 2015. Deficiency in monocarboxylate transporter 1 (MCT1) in mice delays regeneration of peripheral nerves following sciatic nerve crush. *Exp Neurol* **263:** 325–338. doi:10.1016/j.expneurol.2014.10.018

Müller K, Schnatz A, Schillner M, Woertge S, Müller C, von Graevenitz I, Waisman A, van Minnen J, Vogelaar CF. 2018. A predominantly glial origin of axonal ribosomes after nerve injury. *Glia* **66:** 1591–1610. doi:10.1002/glia .23327

Napoli I, Noon LA, Ribeiro S, Kerai AP, Parrinello S, Rosenberg LH, Collins MJ, Harrisingh MC, White IJ, Woodhoo A, et al. 2012. A central role for the ERK-signaling pathway in controlling Schwann cell plasticity and peripheral nerve regeneration in vivo. *Neuron* **73:** 729–742. doi:10.1016/j .neuron.2011.11.031

Navarro X, Vivó M, Valero-Cabré A. 2007. Neural plasticity after peripheral nerve injury and regeneration. *Prog Neurobiol* **82:** 163–201. doi:10.1016/j.pneurobio.2007.06.005

Negro S, Lessi F, Duregotti E, Aretini P, La Ferla M, Franceschi S, Menicagli M, Bergamin E, Radice E, Thelen M, et al. 2017. CXCL12α/SDF-1 from perisynaptic Schwann cells promotes regeneration of injured motor axon terminals. *EMBO Mol Med* **9:** 1000–1010. doi:10.15252/emmm .201607257

Neumeister MW, Winters JN. 2020. Neuroma. *Clin Plast Surg* **47:** 279–283. doi:10.1016/j.cps.2019.12.008

Nguyen QT, Sanes JR, Lichtman JW. 2002. Pre-existing pathways promote precise projection patterns. *Nat Neurosci* **5:** 861–867. doi:10.1038/nn905

Nocera G, Jacob C. 2020. Mechanisms of Schwann cell plasticity involved in peripheral nerve repair after injury. *Cell Mol Life Sci* **77:** 3977–3989. doi:10.1007/s00018-020-03516-9

O'Malley JP, Waran MT, Balice-Gordon RJ. 1999. In vivo observations of terminal Schwann cells at normal, denervated, and reinnervated mouse neuromuscular junctions. *J Neurobiol* **38:** 270–286. doi:10.1002/(SICI)1097-4695 (19990205)38:2<270::AID-NEU9>3.0.CO;2-F

Painter MW. 2017. Aging Schwann cells: mechanisms, implications, future directions. *Curr Opin Neurobiol* **47**: 203–208. doi:10.1016/j.conb.2017.10.022

Painter MW, Brosius Lutz A, Cheng YC, Latremoliere A, Duong K, Miller CM, Posada S, Cobos EJ, Zhang AX, Wagers AJ, et al. 2014. Diminished Schwann cell repair responses underlie age-associated impaired axonal regeneration. *Neuron* **83**: 331–343. doi:10.1016/j.neuron.2014.06.016

Palispis WA, Gupta R. 2017. Surgical repair in humans after traumatic nerve injury provides limited functional neural regeneration in adults. *Exp Neurol* **290**: 106–114. doi:10.1016/j.expneurol.2017.01.009

Pannese E. 2021. Quantitative, structural and molecular changes in neuroglia of aging mammals: a review. *Eur J Histochem* **65**(s1): 3249. doi:10.4081/ejh.2021.3249

Pareyson D, Scaioli V, Laurà M. 2006. Clinical and electrophysiological aspects of Charcot–Marie–Tooth disease. *Neuromolecular Med* **8**: 3–22. doi:10.1385/NMM:8:1-2:3

Parfejevs V, Debbache J, Shakhova O, Schaefer SM, Glausch M, Wegner M, Suter U, Riekstina U, Werner S, Sommer L. 2018. Injury-activated glial cells promote wound healing of the adult skin in mice. *Nat Commun* **9**: 236. doi:10.1038/s41467-017-01488-2

Parrinello S, Lloyd AC. 2009. Neurofibroma development in NF1—insights into tumour initiation. *Trends Cell Biol* **19**: 395–403. doi:10.1016/j.tcb.2009.05.003

Parrinello S, Napoli I, Ribeiro S, Wingfield Digby P, Fedorova M, Parkinson DB, Doddrell RDS, Nakayama M, Adams RH, Lloyd AC. 2010. Ephb signaling directs peripheral nerve regeneration through Sox2-dependent Schwann cell sorting. *Cell* **143**: 145–155. doi:10.1016/j.cell.2010.08.039

Patzkó A, Bai Y, Saporta MA, Katona I, Wu X, Vizzuso D, Feltri ML, Wang S, Dillon LM, Kamholz J, et al. 2012. Curcumin derivatives promote Schwann cell differentiation and improve neuropathy in R98C CMT1B mice. *Brain* **135**: 3551–3566. doi:10.1093/brain/aws299

Pellegatta M, Taveggia C. 2019. The complex work of proteases and secretases in Wallerian degeneration: beyond neuregulin-1. *Front Cell Neurosci* **13**: 93. doi:10.3389/fncel.2019.00093

Pellegatta M, Canevazzi P, Forese MG, Podini P, Valenzano S, Del Carro U, Quattrini A, Taveggia C. 2022. ADAM17 regulates p75NTR-mediated fibrinolysis and nerve remyelination. *J Neurosci* **42**: 2433–2447. doi:10.1523/JNEUROSCI.1341-21.2022

Perez-Gonzalez AP, Provost F, Rousse I, Piovesana R, Benzina O, Darabid H, Lamoureux B, Wang YS, Arbour D, Robitaille R. 2022. Functional adaptation of glial cells at neuromuscular junctions in response to injury. *Glia* **70**: 1605–1629. doi:10.1002/glia.24184

Perry VH, Tsao JW, Fearn S, Brown MC. 1995. Radiation-induced reductions in macrophage recruitment have only slight effects on myelin degeneration in sectioned peripheral nerves of mice. *Eur J Neurosci* **7**: 271–280. doi:10.1111/j.1460-9568.1995.tb01063.x

Peterson SC, Eberl M, Vagnozzi AN, Belkadi A, Veniaminova NA, Verhaegen ME, Bichakjian CK, Ward NL, Dlugosz AA, Wong SY. 2015. Basal cell carcinoma preferentially arises from stem cells within hair follicle and mechano-

sensory niches. *Cell Stem Cell* **16**: 400–412. doi:10.1016/j.stem.2015.02.006

Poss KD. 2010. Advances in understanding tissue regenerative capacity and mechanisms in animals. *Nat Rev Genet* **11**: 710–722. doi:10.1038/nrg2879

Radomska KJ, Coulpier F, Gresset A, Schmitt A, Debbiche A, Lemoine S, Wolkenstein P, Vallat JM, Charnay P, Topilko P. 2019. Cellular origin, tumor progression, and pathogenic mechanisms of cutaneous neurofibromas revealed by mice with Nf1 knockout in boundary cap cells. *Cancer Discov* **9**: 130–147. doi:10.1158/2159-8290.CD-18-0156

Reed CB, Frick LR, Weaver A, Sidoli M, Schlant E, Feltri ML, Wrabetz L. 2020. Deletion of calcineurin in Schwann cells does not affect developmental myelination, but reduces autophagy and delays myelin clearance after peripheral nerve injury. *J Neurosci* **40**: 6165–6176. doi:10.1523/JNEUROSCI.0951-20.2020

Reed CB, Feltri ML, Wilson ER. 2021. Peripheral glia diversity. *J Anat* **241**: 1219–1234. doi:10.1111/joa.13484

Renz BW, Tanaka T, Sunagawa M, Takahashi R, Jiang Z, Macchini M, Dantes Z, Valenti G, White RA, Middelhoff MA, et al. 2018. Cholinergic signaling via muscarinic receptors directly and indirectly suppresses pancreatic tumorigenesis and cancer stemness. *Cancer Discov* **8**: 1458–1473. doi:10.1158/2159-8290.CD-18-0046

Reynolds ML, Woolf CJ. 1992. Terminal Schwann cells elaborate extensive processes following denervation of the motor endplate. *J Neurocytol* **21**: 50–66. doi:10.1007/BF01206897

Ribeiro S, Napoli I, White IJ, Parrinello S, Flanagan AM, Suter U, Parada LF, Lloyd AC. 2013. Injury signals cooperate with Nf1 loss to relieve the tumor-suppressive environment of adult peripheral nerve. *Cell Rep* **5**: 126–136. doi:10.1016/j.celrep.2013.08.033

Rich MM, Lichtman JW. 1989. In vivo visualization of pre- and postsynaptic changes during synapse elimination in reinnervated mouse muscle. *J Neurosci* **9**: 1781–1805. doi:10.1523/JNEUROSCI.09-05-01781.1989

Rinkevich Y, Lindau P, Ueno H, Longaker MT, Weissman IL. 2011. Germ-layer and lineage-restricted stem/progenitors regenerate the mouse digit tip. *Nature* **476**: 409–413. doi:10.1038/nature10346

Rinwa P, Calvo-Enrique L, Zhang MD, Nyengaard JR, Karlsson P, Ernfors P. 2021. Demise of nociceptive Schwann cells causes nerve retraction and pain hyperalgesia. *Pain* **162**: 1816–1827. doi:10.1097/j.pain.0000000000002169

Robel S, Sontheimer H. 2016. Glia as drivers of abnormal neuronal activity. *Nat Neurosci* **19**: 28–33. doi:10.1038/nn.4184

Rochon D, Rousse I, Robitaille R. 2001. Synapse-glia interactions at the mammalian neuromuscular junction. *J Neurosci* **21**: 3819–3382. doi:10.1523/JNEUROSCI.21-11-03819.2001

Saloman JL, Albers KM, Li D, Hartman DJ, Crawford HC, Muha EA, Rhim AD, Davis BM. 2016. Ablation of sensory neurons in a genetic model of pancreatic ductal adenocarcinoma slows initiation and progression of cancer. *Proc Natl Acad Sci* **113**: 3078–3308. doi:10.1073/pnas.1512603113

Salonen V, Aho H, Röyttä M, Peltonen J. 1988. Quantitation of Schwann cells and endoneurial fibroblast-like cells after experimental nerve trauma. *Acta Neuropathol* **75**: 331–336. doi:10.1007/BF00687785

* Salzer J, Feltri ML, Jacob C. 2023. Schwann cell development and myelination. *Cold Spring Harb Perspect Biol* doi:10.1101/cshperspect.a041360

Scheib J, Höke A. 2016. Impaired regeneration in aged nerves: clearing out the old to make way for the new. *Exp Neurol* **284**: 79–83. doi:10.1016/j.expneurol.2016.07.010

* Scherer SS, Svaren J. 2023. Peripheral nervous system (PNS) myelin diseases. *Cold Spring Harb Perspect Biol* doi:10.1101/cshperspect.a41376

Schröder JM. 1972. Altered ratio between axon diameter and myelin sheath thickness in regenerated nerve fibers. *Brain Res* **45**: 49–65. doi:10.1016/0006-8993(72)90215-6

Schwaller F, Bégay V, García-García G, Taberner FJ, Moshourab R, McDonald B, Docter T, Kühnemund J, Ojeda-Alonso J, Paricio-Montesinos R, et al. 2021. USH2A is a Meissner's corpuscle protein necessary for normal vibration sensing in mice and humans. *Nat Neurosci* **24**: 74–81. doi:10.1038/s41593-020-00751-y

Seddon HJ. 1943. Peripheral nerve injuries. *Glasgow Med J* **139**: 61–75. doi:10.1136/bmj.2.4207.286-a

Serhan CN, Yacoubian S, Yang R. 2008. Anti-inflammatory and proresolving lipid mediators. *Annu Rev Pathol* **3**: 279–312. doi:10.1146/annurev.pathmechdis.3.121806.151409

Siironen J, Collan Y, Röyttä M. 1994. Axonal reinnervation does not influence Schwann cell proliferation after rat sciatic nerve transection. *Brain Res* **654**: 303–311. doi:10.1016/0006-8993(94)90492-8

Sloan EK, Priceman SJ, Cox BF, Yu S, Pimentel MA, Tangkanangnukul V, Arevalo JMG, Morizono K, Karanikolas BDW, Wu L, et al. 2010. The sympathetic nervous system induces a metastatic switch in primary breast cancer. *Cancer Res* **70**: 7042–7052. doi:10.1158/0008-5472.CAN-10-0522

Smith TP, Sahoo PK, Kar AN, Twiss JL. 2020. Intra-axonal mechanisms driving axon regeneration. *Brain Res* **1740**: 146864. doi:10.1016/j.brainres.2020.146864

Sock E, Wegner M. 2019. Transcriptional control of myelination and remyelination. *Glia* **67**: 2153–2165. doi:10.1002/glia.23636

Son YJ, Thompson WJ. 1995a. Nerve sprouting in muscle is induced and guided by processes extended by Schwann cells. *Neuron* **14**: 133–141. doi:10.1016/0896-6273(95)90247-3

Son YJ, Thompson WJ. 1995b. Schwann cell processes guide regeneration of peripheral axons. *Neuron* **14**: 125–132. doi:10.1016/0896-6273(95)90246-5

Stassart RM, Woodhoo A. 2021. Axo-glial interaction in the injured PNS. *Dev Neurobiol* **81**: 490–506. doi:10.1002/dneu.22771

Stassart RM, Fledrich R, Velanac V, Brinkmann BG, Schwab MH, Meijer D, Sereda MW, Nave KA. 2013. A role for Schwann cell-derived neuregulin-1 in remyelination. *Nat Neurosci* **16**: 48–54. doi:10.1038/nn.3281

Stierli S, Napoli I, White IJ, Cattin AL, Monteza Cabrejos A, Garcia Calavia N, Malong L, Ribeiro S, Nihouarn J, Williams R, et al. 2018. The regulation of the homeostasis and regeneration of peripheral nerve is distinct from the CNS and independent of a stem cell population. *Development* **145**: dev170316. doi:10.1242/dev.170316

Stierli S, Imperatore V, Lloyd AC. 2019. Schwann cell plasticity-roles in tissue homeostasis, regeneration, and disease. *Glia* **67**: 2203–2215. doi:10.1002/glia.23643

Suazo I, Vega JA, García-Mesa Y, García-Piqueras J, García-Suárez O, Cobo T. 2022. The lamellar cells of vertebrate Meissner and Pacinian corpuscles: development, characterization, and functions. *Front Neurosci* **16**: 790130. doi:10.3389/fnins.2022.790130

Sulaiman OA, Gordon T. 2000. Effects of short- and long-term Schwann cell denervation on peripheral nerve regeneration, myelination, and size. *Glia* **32**: 234–246. doi:10.1002/1098-1136(200012)32:3<234::AID-GLIA40>3.0.CO;2-3

Sulaiman W, Gordon T. 2013. Neurobiology of peripheral nerve injury, regeneration, and functional recovery: from bench top research to bedside application. *Ochsner J* **13**: 100–108.

Suter U, Scherer SS. 2003. Disease mechanisms in inherited neuropathies. *Nat Rev Neurosci* **4**: 714–726. doi:10.1038/nrn1196

Takeo M, Chou WC, Sun Q, Lee W, Rabbani P, Loomis C, Taketo MM, Ito M. 2013. Wnt activation in nail epithelium couples nail growth to digit regeneration. *Nature* **499**: 228–232. doi:10.1038/nature12214

Tam SL, Gordon T. 2003. Neuromuscular activity impairs axonal sprouting in partially denervated muscles by inhibiting bridge formation of perisynaptic Schwann cells. *J Neurobiol* **57**: 221–234. doi:10.1002/neu.10276

Tanaka EM. 2016. The molecular and cellular choreography of appendage regeneration. *Cell* **165**: 1598–1608. doi:10.1016/j.cell.2016.05.038

Tanaka K, Webster HD. 1991. Myelinated fiber regeneration after crush injury is retarded in sciatic nerves of aging mice. *J Comp Neurol* **308**: 180–187. doi:10.1002/cne.903080205

Tanaka K, Zhang QL, Webster HD. 1992. Myelinated fiber regeneration after sciatic nerve crush: morphometric observations in young adult and aging mice and the effects of macrophage suppression and conditioning lesions. *Exp Neurol* **118**: 53–61. doi:10.1016/0014-4886(92)90022-I

Taveggia C, Feltri ML. 2022. Beyond wrapping: canonical and noncanonical functions of Schwann cells. *Annu Rev Neurosci* **45**: 561–580. doi:10.1146/annurev-neuro-110920-030610

Terenghi G. 1995. Peripheral nerve injury and regeneration. *Histol Histopathol* **10**: 709–718.

Thomas PK, King RHM, Sharma AK. 1980. Changes with age in the peripheral nerves of the rat. An ultrastructural study. *Acta Neuropathol* **52**: 1–6. doi:10.1007/BF00687222

Todd T. 1823. On the process of reproduction of the members of the aquatic salamander. *Quart J Sci Arts Lib* **16**: 84–86.

Todd KJ, Darabid H, Robitaille R. 2010. Perisynaptic glia discriminate patterns of motor nerve activity and influence plasticity at the neuromuscular junction. *J Neurosci* **30**: 11870–11882. doi:10.1523/JNEUROSCI.3165-10.2010

Tricaud N. 2018. Myelinating Schwann cell polarity and mechanically driven myelin sheath elongation. *Front Cell Neurosci* **11**: 1–12. doi:10.3389/fncel.2017.00414

Ulrichsen M, Gonçalves NP, Mohseni S, Hjæresen S, Lisle TL, Molgaard S, Madsen NK, Andersen OM, Svenningsen ÅF, Glerup S, et al. 2022. Sortilin modulates Schwann cell signaling and Remak bundle regeneration following nerve injury. *Front Cell Neurosci* **16:** 856734. doi:10.3389/fncel.2022.856734

Vaquié A, Sauvain A, Duman M, Nocera G, Egger B, Meyenhofer F, Falquet L, Bartesaghi L, Chrast R, Lamy CM, et al. 2019. Injured axons instruct Schwann cells to build constricting actin spheres to accelerate axonal disintegration. *Cell Rep* **27:** 3152–3166.e7. doi:10.1016/j.celrep.2019.05.060

Vaughan DW. 1992. Effects of advancing age on peripheral nerve regeneration. *J Comp Neurol* **323:** 219–237. doi:10.1002/cne.903230207

Veiga-Fernandes H, Mucida D. 2016. Neuro-immune interactions at barrier surfaces. *Cell* **165:** 801–811. doi:10.1016/j.cell.2016.04.041

Verdú E, Butí M, Navarro X. 1995. The effect of aging on efferent nerve fibers regeneration in mice. *Brain Res* **696:** 76–82. doi:10.1016/0006-8993(95)00762-F

Verdú E, Ceballos D, Vilches JJ, Navarro X. 2000. Influence of aging on peripheral nerve function and regeneration. *J Peripher Nerv Syst* **5:** 191–208. doi:10.1111/j.1529-8027.2000.00026.x

Viskochil D, Buchberg AM, Xu G, Cawthon RM, Stevens J, Wolff RK, Culver M, Carey JC, Copeland NG, Jenkins NA, et al. 1990. Deletions and a translocation interrupt a cloned gene at the neurofibromatosis type 1 locus. *Cell* **62:** 187–192. doi:10.1016/0092-8674(90)90252-A

Wagstaff LJ, Gomez-Sanchez JA, Fazal SV, Otto GW, Kilpatrick AM, Michael K, Wong LYN, Ma KH, Turmaine M, Svaren J, et al. 2021. Failures of nerve regeneration caused by aging or chronic denervation are rescued by restoring Schwann cell c-Jun. *eLife* **10:** e62232. doi:10.7554/eLife.62232

Wallace MR, Marchuk DA, Andersen LB, Letcher R, Odeh HM, Saulino AM, Fountain JW, Brereton A, Nicholson J, Mitchell AL, et al. 1990. Type 1 neurofibromatosis gene: identification of a large transcript disrupted in three NF1 patients. *Science* **249:** 181–186. doi:10.1126/science.2134734

Waller AV. 1850. XX. Experiments on the section of the glossopharyngeal and hypoglossal nerves of the frog, and observations of the alterations produced thereby in the structure of their primitive fibres. *Phil Trans R Soc* **140:** 423–429. doi:10.1098/rstl.1850.0021

Wei Z, Fei Y, Su W, Chen G. 2019. Emerging role of Schwann cells in neuropathic pain: receptors, glial mediators and myelination. *Front Cell Neurosci* **13:** 116. doi:10.3389/fncel.2019.00116

Winkler F, Venkatesh HS, Amit M, Batchelor T, Demir IE, Deneen B, Gutmann DH, Hervey-Jumper S, Kuner T, Mabbott D, et al. 2023. Cancer neuroscience: state of the field, emerging directions. *Cell* **186:** 1689–1707. doi:10.1016/j.cell.2023.02.002

Woodhoo A, Alonso MBD, Droggiti A, Turmaine M, D'Antonio M, Parkinson DB, Wilton DK, Al-Shawi R, Simons P, Shen J, et al. 2009. Notch controls embryonic Schwann cell differentiation, postnatal myelination and adult plasticity. *Nat Neurosci* **12:** 839–847. doi:10.1038/nn.2323

Wright MC, Potluri S, Wang X, Dentcheva E, Gautam D, Tessler A, Wess J, Rich MM, Son YJ. 2009. Distinct muscarinic acetylcholine receptor subtypes contribute to stability and growth, but not compensatory plasticity, of neuromuscular synapses. *J Neurosci* **29:** 14942–14955. doi:10.1523/JNEUROSCI.2276-09.2009

Yamazaki S, Nakauchi H. 2014. Bone marrow Schwann cells induce hematopoietic stem cell hibernation. *Int J Hematol* **99:** 695–698. doi:10.1007/s12185-014-1588-9

Yamazaki S, Ema H, Karlsson G, Yamaguchi T, Miyoshi H, Shioda S, Taketo MM, Karlsson S, Iwama A, Nakauchi H. 2011. Nonmyelinating Schwann cells maintain hematopoietic stem cell hibernation in the bone marrow niche. *Cell* **147:** 1146–1158. doi:10.1016/j.cell.2011.09.053

Yang DP, Zhang DP, Mak KS, Bonder DE, Pomeroy SL, Kim HA. 2008. Schwann cell proliferation during Wallerian degeneration is not necessary for regeneration and remyelination of the peripheral nerves: axon-dependent removal of newly generated Schwann cells by apoptosis. *Mol Cell Neurosci* **38:** 80–88. doi:10.1016/j.mcn.2008.01.017

Ying Z, Pan C, Shao T, Liu L, Li L, Guo D, Zhang S, Yuan T, Cao R, Jiang Z, et al. 2018. Mixed lineage kinase domain-like protein MLKL breaks down myelin following nerve injury. *Mol Cell* **72:** 457–468.e5. doi:10.1016/j.molcel.2018.09.011

Yuan X, Klein D, Kerscher S, West BL, Weis J, Katona I, Martini R. 2018. Macrophage depletion ameliorates peripheral neuropathy in aging mice. *J Neurosci* **38:** 4610–4620. doi:10.1523/JNEUROSCI.3030-17.2018

Zhao L, Huang W, Yi S. 2023. Cellular complexity of the peripheral nervous system: insights from single-cell resolution. *Front Neurosci* **17:** 1098612. doi:10.3389/fnins.2023.1098612

Zhu Y, Ghosh P, Charnay P, Burns DK, Parada LF. 2002. Neurofibromas in NF1: Schwann cell origin and role of tumor environment. *Science* **296:** 920–922. doi:10.1126/science.1068452

Microglia in Health and Diseases: Integrative Hubs of the Central Nervous System (CNS)

Amanda Sierra,[1,2,3] Veronique E. Miron,[4,5,6] Rosa C. Paolicelli,[7] and Richard M. Ransohoff[8]

[1]Achucarro Basque Center for Neuroscience, Glial Cell Biology Laboratory, Science Park of UPV/EHU, E-48940 Leioa, Bizkaia, Spain

[2]Department of Biochemistry and Molecular Biology, University of the Basque Country EHU/UPV, 48940 Leioa, Spain

[3]Ikerbasque Foundation, Bilbao 48009, Spain

[4]BARLO Multiple Sclerosis Centre, Keenan Research Centre for Biomedical Science at St. Michael's Hospital, Toronto M5B 1T8, Canada

[5]Department of Immunology, University of Toronto, Toronto M5S 1A8, Canada

[6]UK Dementia Research Institute at the University of Edinburgh, Edinburgh BioQuarter, Edinburgh EH16 4TJ, United Kingdom

[7]Department of Biomedical Sciences, Faculty of Biology and Medicine, University of Lausanne, CH-1005 Lausanne, Switzerland

[8]Third Rock Ventures, Boston, Massachusetts 02215, USA

Correspondence: amanda.sierra@achucarro.org; rransohoff@thirdrockventures.com

Microglia are usually referred to as "the innate immune cells of the brain," "the resident macrophages of the central nervous system" (CNS), or "CNS parenchymal macrophages." These labels allude to their inherent immune function, related to their macrophage lineage. However, beyond their classic innate immune responses, microglia also play physiological roles crucial for proper brain development and maintenance of adult brain homeostasis. Microglia sense both external and local stimuli through a variety of surface receptors. Thus, they might serve as integrative hubs at the interface between the external environment and the CNS, able to decode, filter, and buffer cues from outside, with the aim of preserving and maintaining brain homeostasis. In this perspective, we will cast a critical look at how these multiple microglial functions are acquired and coordinated, and we will speculate on their impact on human brain physiology and pathology.

MICROGLIA ACROSS THE LIFE SPAN

Understanding microglia requires becoming acquainted with their ontogeny, for which we rely on a concordant (where compared) set of observations in humans and mice. As with other tissue macrophages, such as those in skin and liver (Langerhans and Kupffer cells; Márquez-Ropero et al. 2020), microglia are not derived from bone marrow myeloid progenitors that give rise to circulating monocytes, but from yolk sac primitive macrophages that express RUNX and c-KIT but not MYB or CD45 (Nayak et al. 2014). Microglial progenitors migrate into the neuroepithelium around embryonic day E9.5 in rodents (Ginhoux et al. 2010), or as early as the fourth gestational week in the fetal human brain (Menassa et al. 2022). Upon their arrival, immature microglial cells start to proliferate and colonize the brain parenchyma, while increasing their morphological complexity. This maturation process results from the integration of intrinsic genetic programs and extrinsic instructional cues provided by the environment, affecting the epigenetic landscape and transcriptional profile of both lineage-determining and signal-dependent transcription factors (Gosselin et al. 2014; Lavin et al. 2014; Matcovitch-Natan et al. 2016). Key transcription factors in microglial development are PU.1, a master regulator of the macrophage lineage (Olson et al. 1995); RUNX, controlling microglial proliferation (Zusso et al. 2012); and CSF1R (colony-stimulating factor receptor 1), a tyrosine kinase receptor critical for microglial survival (Ginhoux et al. 2010). Prominent among the environmental factors controlling microglial development are the CSF1R ligands CSF1 and interleukin (IL)-34 (Greter et al. 2012; Wang et al. 2012) and transforming growth factor β (TGF-β) (Butovsky et al. 2014), which are produced by neurons, microglia themselves, and other cell types. In spite of the identification of these critical factors, very little is known about the brain-specific mechanisms and cell types that drive the acquisition of mature microglial functionality. Recent findings suggest that brain and microglial development are interconnected at least in the neocortex, where developing cortical pyramidal neurons influence microglial transcriptional states (Stogsdill et al. 2022). Therefore, neurons are key players controlling microglial adaptation to the central nervous system (CNS) environment.

Once established in the CNS parenchyma, mature microglia are maintained through adulthood by self-renewal (proliferation and apoptosis) (Askew et al. 2017; Tay et al. 2017) without any contribution from bone marrow–derived cells in physiological conditions (Ajami et al.

2007). In adult mice, microglia are largely quiescent but their proliferation program seems to be reactivated during injury, as RUNX expression is up-regulated (Zusso et al. 2012) and injury induces microglial proliferation (Wirenfeldt et al. 2007). Both in mice and humans, microglia are long-lived cells: some live up to 15 months in the mouse cortex (Füger et al. 2017) and up to several decades in the human brain (Réu et al. 2017). Across their life span, microglia express a large variety of surface receptors, termed the "sensome" (Hickman et al. 2013), that allows sensing of cells, soluble ligands, extracellular matrix, and damage- and pathogen-associated molecular patterns (DAMPs and PAMPs, respectively), all of which are used to detect the activity of the surrounding neuronal network or the presence of pathogens, among other stimuli.

Their longevity implies that microglia could sense and keep track of many of the events experienced by the CNS, integrating the impact of several external stimuli (such as infection, stress, or diet) over time, via epigenetic mechanisms (Allis and Jenuwein 2016). This cellular memory may enable them to rapidly respond to previously encountered pathogens or stimuli, perhaps in a similar manner to the "trained immunity" that macrophages experience upon exposure to challenges (Netea et al. 2016). Just like microglia, many tissue-resident macrophages are long-lived cells (Parihar et al. 2010) and thus susceptible to learning and training. For example, inflammatory challenges impinge on the macrophage metabolism and epigenetic landscape, leading to long-term enhanced responses (Netea et al. 2016). Epigenetic changes, such as DNA methylation and histone modifications, modify DNA accessibility and chromatin structure, thus leading to different patterns of gene expression (Goldberg et al. 2007). While these processes are well documented in macrophages, the microglial epigenetic memory and its functional implications are still poorly understood.

In mice, bacterial lipopolysaccharide (LPS), a ligand for Toll-like receptors (TLRs), and genotoxicity, which induce both tolerance and hyperresponsiveness, can alter epigenetic profiles and transcriptional signatures in microglia, providing vulnerability to develop neuropathology later

Cite this article as *Cold Spring Harb Perspect Biol* doi: 10.1101/cshperspect.a041366

in life (Wendeln et al. 2018; Zhang et al. 2022). Although inflammatory cytokines produced in response to LPS return to basal levels within a few days as a result of a complex cross talk between microglia and other immune cells (Shemer et al. 2020), long-lasting modifications remain on the chromatin landscape and modulate subsequent TLR activation (Lauterbach et al. 2019).

Metabolic pathways are also key in producing chromatin-modifying metabolites, which can shape and influence cellular responses over time (Dai et al. 2020; Diskin et al. 2021). Whether derived from adaptation to nutrient availability and diet, or driven by the metabolomic products of the microbiota, metabolism-induced epigenetics is emerging as an important process in microglial function and dysfunction. For instance, in a mouse model of Alzheimer's disease (AD), lactate-dependent histone modifications have been recently described in plaque-associated microglia, to be responsible for a positive feedback loop that is proposed to increase glycolysis and exacerbate microglial dysfunction (Pan et al. 2022). Gut microbiota significantly influences microglial metabolic and functional states (Thion et al. 2018), with evidence for short-chain fatty acids, such as acetate, to modulate phagocytosis via epigenetic mechanisms (Erny et al. 2021).

Finally, stress, and in particular early life stress, have been linked with long-lasting phenotypic adaptations, and microglia—to some extent—have been implicated in mediating such outcomes via epigenetics (Rahman and Mc-Gowan 2022). Therefore, the individual's life history, including chronic exposure to infectious agents, diet regimens, and stressful events, could be recorded through changes on microglial chromatin, thus inducing long-term consequences on microglial function (Fig. 1) via modulation of immune tolerance or hyperresponsiveness.

MICROGLIA AS INTEGRATIVE HUBS FOR EXTERNAL AND LOCAL SIGNALS

What makes microglia a brain cell? During their maturation process and along their lifetime, microglia must learn to read brain-specific signals and execute brain-specific functions. How this is achieved is not well understood, but it is possibly related to the acquisition of their sensome (Hickman et al. 2013). In response to their sensome activation, microglia react by altering their transcriptional, proteomic, metabolic, and morphological profiles (Paolicelli et al. 2022). These responses lead to a number of microglial states, such as disease-associated microglia (DAM), characterized by a specific transcriptional signature identified via single-cell RNA sequencing; or dark microglia, characterized by an acidophilic cytoplasm visualized by transmission electron microscopy (TEM), among many others. Our current understanding is that these states are dynamic and transient but we do not know whether and how they can become permanent (Paolicelli et al. 2022). Another critical question that the field will need to answer in the next few years is to what extent and through which mechanisms an individual microglia cell can transit across different states and return to baseline, a decisive point to understand resolution of inflammatory or damage challenges. Finally, one last big question is how these states translate into specific microglial functions.

The coordination of microglial functions and how they emerge from altered states is a main goal of microglial research, akin to finding the Rosetta stone to decipher how microglia work. The use of the word "functions" in plural is not trivial, as microglia have been involved in participating in a strikingly large number of tasks. They are endowed with a basic macrophage repertoire of actions: process motility and surveillance, release of inflammatory mediators and other soluble factors, and phagocytosis. And they use this repertoire to support the functioning of neuronal circuits in physiological conditions, including monitoring brain synapses (Tremblay et al. 2010; Paolicelli et al. 2011; Schafer et al. 2012), regulating neuronal excitability (Badimon et al. 2020), removing cell debris (Sierra et al. 2010), or regulating brain blood flow (Császár et al. 2022), among others. We envision that microglia may be integrative hubs of stimuli that produce coordinated responses aimed at maintaining brain homeostasis.

But how can a single cell type integrate such a plethora of stimuli and execute such a variety of

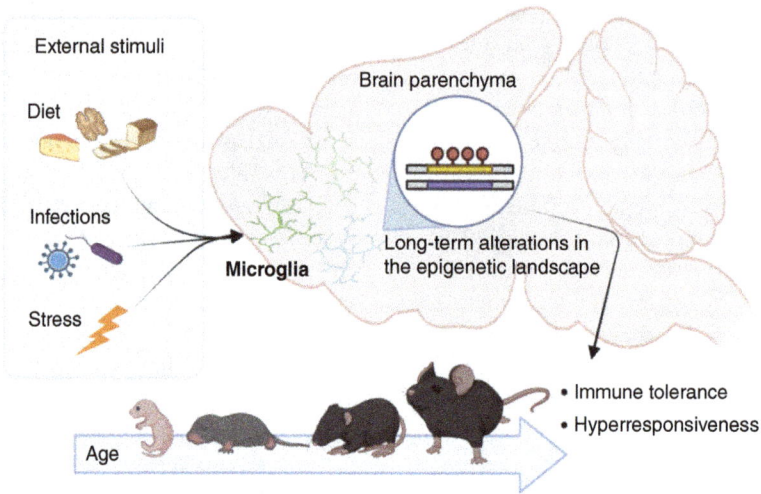

Figure 1. Microglia at the interface between external stimuli and the brain parenchyma. External stimuli, such as infections, diet, and stress, can induce long-term alterations in the microglial epigenetic landscape. The individual's life history is likely to be recorded through chromatin changes, and microglia in particular could play a key role as an interface between the external environment and the brain parenchyma. (Created with BioRender.com.)

functions? Are they pleiotropic at population or at single-cell level? The most straightforward scenario is that the pleiotropy of the population is sustained by specialized groups of microglia in precise neuroanatomical niches, each with a specialized sensome and execution machinery, tailored to a particular function (Fig. 2A,B). For instance, in the adult mouse hippocampus, microglia perform different functions: in CA1 they control glutamatergic synapses and neuronal excitability (Peng et al. 2019; Umpierre and Wu 2021; Basilico et al. 2022). In the neurogenic niche of the subgranular zone of the dentate gyrus (DG) they phagocytose the excess of newborn cells and support neurogenesis (Sierra et al. 2010; Diaz-Aparicio et al. 2020). In the DG, they also remodel the synapses of engram cells to mediate forgetting of remote memories (Wang et al. 2020), and those microglia associated to capillaries regulate blood flow (Bisht et al. 2021). It could be speculated that to sustain hippocampal connectivity, all these functions need to be carried out in coordination: spines need to be remodeled based on the connectivity between neurons, debris needs to be removed before toxic elements are released and affect neighboring cells, and blood flow needs to support ongoing neuronal

activity. Therefore, to achieve an integrated homeostatic support, communication between microglia expressing distinct functions may be expected, perhaps using similar mechanisms as those used for microglial–neuron communication.

An alternative scenario is that individual microglia are in fact pleiotropic and each acts as an autonomous unit with a full sensome and execution machinery, capable of performing different functions while adapting to the particular circumstances of their local environment (Fig. 2C, D). In the hippocampal example above, it is conceivable that capillary-associated microglia encounter an apoptotic cell, or that CA1 microglia sense at the same time the activity of nearby neurons and blood vessels. In these cases, a microglia cell would be expected to integrate all input signals and produce a coherent functional output. But in the presence of competing signals, what are the rules of decision-making? Microglia may simply react stronger and faster to signals coming from closer sources or at higher concentrations, or they may be programmed to prioritize particularly critical tasks. For example, one key signal for microglia is ATP, which is sensed through several purinergic receptors (Calovi et al. 2019).

Cite this article as *Cold Spring Harb Perspect Biol* doi: 10.1101/cshperspect.a041366

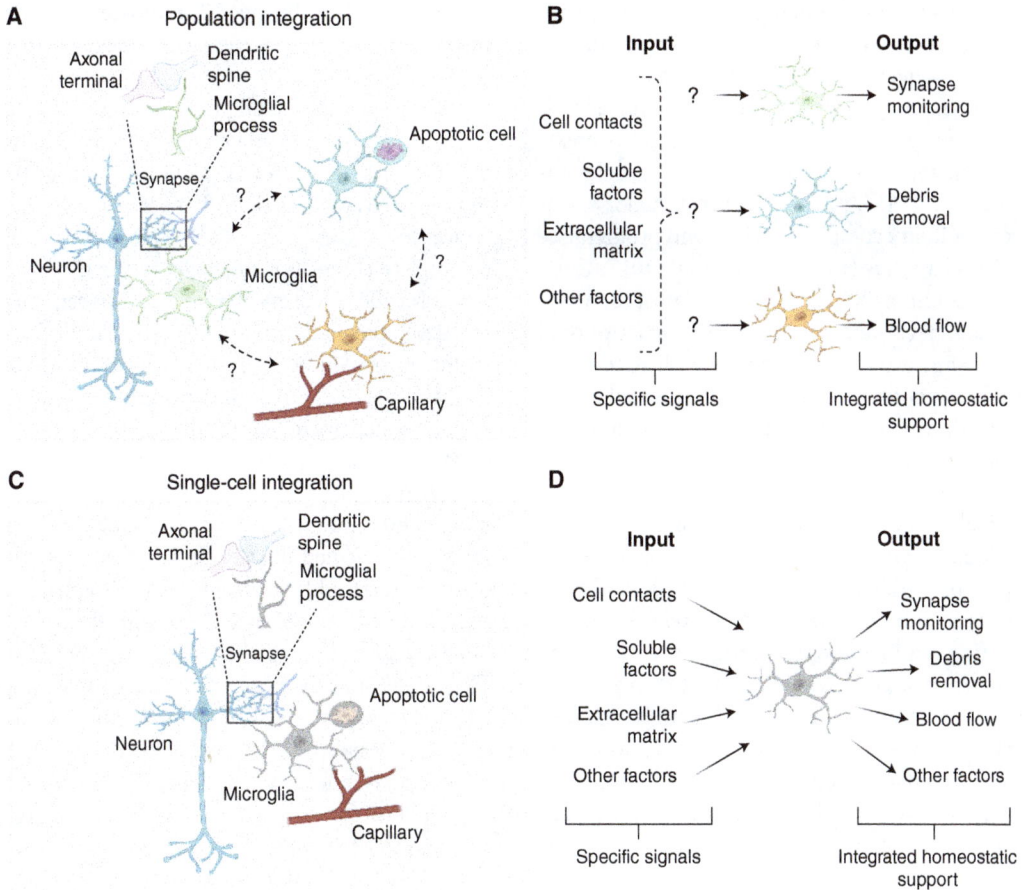

Figure 2. Are microglia integrative hubs of local signals? (*A*) Specialized subpopulations of microglia may be in charge of executing context-tailored functions, such as synapse monitoring (blue microglia), phagocytosis of apoptotic cells (pink microglia), or capillary blood flow regulation (yellow microglia). To achieve a coordinated homeostatic control, microglia–microglia communication (broken arrows) would be necessary. (*B*) The scenario depicted in *A* implies that microglia integrate input signals at the population level, with each subpopulation specialized in sensing specific signals that result in specific output functions, possibly leading to an integrated homeostatic support. (*C*) Alternatively, a single microglia cell may be capable of performing several functions in the same spatiotemporal context. (*D*) The scenario depicted in *B* implies that an integrated homeostatic support requires a single microglia cell to integrate multiple signals and produce a unique functional output. (Created with BioRender.com.)

ATP is released by apoptotic cells as a "find-me" signal to trigger microglial process chemotaxis and engulfment, making microglial phagocytosis very efficient in physiological conditions (Sierra et al. 2010). However, during epileptic seizures, ATP is widely released by hyperactive neuronal networks (Dale and Frenguelli 2009). These signals compete, turning microglia "blind" to apoptotic cells and hindering their detection and removal (Abiega et al. 2016). ATP/ADP re-leased upon neuronal activation by neurons and astrocytes in different models of neuronal excitation recruit microglial processes (Eyo et al. 2014) and drive negative feedback control of neuronal activity by microglia (Badimon et al. 2020). To understanding how decisions are made, it is important to understand the microglial responses in pathological settings, because it is possible that those rules are overridden in pathology and aging.

Evidence of population or single-cell pleiotropism is largely lacking in the literature. Recent scRNA-seq studies have described a large number of microglial transcriptional states associated with different physiopathological contexts in both mice and humans. Some examples include the DAM, interferon (IFN), major histocompatibility complex (MHC), and proliferative microglia states (Keren-Shaul et al. 2017; Chen and Colonna 2021). However, most of these studies have failed to link the transcriptional state with a specific function, with exceptions such as proliferative-region-associated microglia (PAM), which was related to phagocytosis of developing oligodendrocytes (Li et al. 2019a). Assigning specific function(s) to a particular state is a challenging task because some states may lack unique markers and may most likely be associated with simultaneous functions. In addition, states may be transient and not related to cells sharing spatiotemporal coordinates or local environments. Thus, we anticipate that in vivo imaging experiments combined with single-cell transcriptomics/proteomics will help to determine specific microglial functions of given microglial states by addressing the behavior and expression patterns of localized microglial populations.

WHICH MICROGLIAL FUNCTIONS ARE NEEDED FOR HEALTHY BRAIN FUNCTION?

As with most cells in most tissues, microglia do not act alone, but in collaboration with other cell types. For example, blood flow is directly controlled by astrocytic endfeet and contractile pericytes (Attwell et al. 2016) with whom capillary-associated microglia may have to work together (Bisht et al. 2021). Functions strongly associated with microglia, such as phagocytosis of apoptotic cells and synapse remodeling, can be executed by other cells, such as astrocytes, when microglia are functionally impaired (Konishi et al. 2020). In contrast to other tissue macrophages located in barrier tissues, such as the gut or the skin, the CNS is not routinely exposed to pathogenic microbes and the rigid skull along with the shock absorber provided by the cerebrospinal fluid protects the brain from many forms of physical trauma, limiting the need for innate immune responses. Which microglial functions might be dispensable?

One way of addressing this question is to ask, "What happens when microglia are lacking?" For instance, use of a recently developed transgenic mouse, in which deletion of the FIRE superenhancer in the *Csf1r* locus (*Csf1r*-FIRE$^{\Delta/\Delta}$) leads to a complete absence of microglia in the CNS, allowed for the discovery that they are dispensable for developmental myelination yet required to maintain myelin integrity in adulthood (McNamara et al. 2022). Another way is to ask, "What happens when microglia are not functioning properly?" Mendelian genetic associations to human disease may provide useful insights. One paramount example is the Nasu Hakola disease (otherwise termed polycystic lipomembranous osteodysplasia with sclerosing leukoencephalopathy [PLOSL]) caused by homozygous deficiency of either TREM2 (OMIM #618193) or DAP12/TYROBP (transmembrane immune signaling adaptor; OMIM #221770). Other relevant diseases are adult-onset leukoencephalopathy (ALSP) (with axonal spheroids; OMIM #221820), caused by haploinsufficiency of CSF1R. Homozygous CSF1R deficiency causes brain abnormalities, neurodegeneration, and dysosteosclerosis (BANDDOS) (OMIM #618476), in which microglia are absent (Oosterhof et al. 2019; Guo and Ikegawa 2021). In the human CNS parenchyma, expression of TREM2, DAP12, and CSF1R are restricted to microglia (www.brainrnaseq.org) (Zhang et al. 2014). Within the CNS, expression of these microglial genes by border-associated macrophages (BAMs) (www.brainimmuneatlas.org/index.php) (Van Hove et al. 2019) is also observed. In the human periphery, all three genes are expressed by diverse other immune cells (rstats.immgen.org/Skyline/skyline.html). There are several lines of evidence that microglial dysfunction is central for the pathogenesis of these diseases:

1. Disease phenotype selectively involves the CNS with lesser involvement reflecting dysfunction of bone osteoclasts.

2. Expression of all three genes is restricted to microglia, among parenchymal brain cells.

Cite this article as *Cold Spring Harb Perspect Biol* doi: 10.1101/cshperspect.a041366

For these reasons, the disorders are classed as microgliopathies among the leukodystrophies (Ferrer 2022). It remains plausible that the peripheral immune compartment plays some role in the pathogenesis of these complex, poorly understood disorders, an interpretation challenged by the lack of efficacy of peripheral immune modulation (Konno et al. 2018). In ALSP, promising early results of allogeneic hematopoietic stem cell transfer after myeloablation have been attributed to engraftment of the recipient CNS with donor CSF1R-sufficient microglia-like cells (Tipton et al. 2021). Despite these speculations by the clinical investigators, effects of autologous hematopoietic stem cell transplantation (aHSCT) could doubtless in part reflect replacement of a CSF1R-sufficient peripheral immune compartment. However, PLOSL, ALSP, and BANDDOS are rare diseases and remain largely unstudied. Insofar as limited case material allows, their pathophysiological features, including neuropathological, cerebrospinal fluid (CSF), and neuroradiological findings, have been well-characterized. PLOSL and ALSP present nonspecifically in mid-life (typically age 40–50), with leukoencephalopathy (comprising a distinctive regional abnormality of white matter as visualized by MRI brain scans) and progressive cognitive and motor impairment. These features compromise quality of life and also shorten life span. Many patients also exhibit cystic bone lesions, due to abnormal osteoclasts. Homozygous deficiency of TREM2 can also produce a frontotemporal dementia-like syndrome (Guerreiro et al. 2013b).

How do these uncommon experiments of nature inform our understanding of the roles of microglia in the human brain? The "usual suspects" have been cleared: no changes in antimicrobial host defense or impaired removal either of apoptotic cells or cellular debris have been described. Despite widespread, ingenious, and sustained effort, the precise dysfunction in microglial cells of PLOSL and ALSP patients remains obscure (Galimberti et al. 2019; Dash et al. 2020; Berdowski et al. 2022; Bianchin and Snow 2022; Chitu et al. 2022; Ferrer 2022; Stables et al. 2022). The neuropathology of ALSP indicates that the microgliopathy affects most brain cell types: the eponymous axonal spheroids are numerous and strikingly enlarged, carrying markers that suggest impaired axonal transport; bizarre astrocytes are observed; and, in affected regions, white matter is vacuolated—whether because of secondary degeneration of the myelination of abnormal axons or due to primary disturbance of oligodendrocyte function—remains uncertain (Papapetropoulos et al. 2022).

Adding to the conundrum, genetic models have only been partially helpful in understanding the physiological basis of the human phenotypes. For example, *Trem2* knockout (KO) mice show defects in microglial synapse engulfment and altered functional connectivity together with behavioral defects, but do not display the full phenotype of PLOSL patients (Filipello et al. 2018). *Csf1r*-FIRE$^{\Delta/\Delta}$ mice, whose lack of an intronic superenhancer impairs *Csf1r* expression and results in the absence of microglia and other peripheral macrophages, were initially described to have no major neurological or developmental defects (Rojo et al. 2019), although a more recent analysis showed spontaneous demyelination (McNamara et al. 2022) and, when bred to transgenic models of AD, enhanced vascular deposition of amyloid β and premature lethality (Shabestari et al. 2022). Homozygous *Csf1r* KO rodents and zebrafish show alterations in myelination, hippocampal neurogenesis, and glia (Erblich et al. 2011; Patkar et al. 2021; Berdowski et al. 2022) but do not display neuropathological or behavioral phenotypes associated with ALSP. Each of these microglia-lacking mouse models have their own caveats, as has been reviewed elsewhere (Green et al. 2020). For example, *Csf1r* KO mice have deficits in all tissue-resident brain macrophages (Dai et al. 2002), whereas Csf1r-FIRE$^{\Delta/\Delta}$ mice lack intraventricular macrophages (Munro et al. 2020). Thus, the phenotype of these mice cannot be ascribed exclusively to their lack of microglia.

Additional support for the importance of healthy microglia comes from associations with common neurodegenerative disorders, most prominently AD, through genome-wide association studies (GWAS) with supportive follow-up studies. Multiple gene variants associated with increased risk to develop AD involve transcripts largely restricted to microglia within the CNS, suggesting that microglial dysfunction plays a substantial role in susceptibility to AD (Wightman et al. 2021). This line of research has been sufficiently productive that the microglial role in AD is no longer considered secondary, epiphenomenological, or reactive but, rather, primary.

Similarly, in other neurodegenerative diseases, microglia are no longer considered passive responders to disease conditions but now disease modifiers or even causative players in disease progression or initiation (Fig. 3). One example is TREM2, for which the causative role in PLOSL is described above. Heterozygous structural polymorphisms in *TREM2*, associated with germline single-nucleotide polymorphism (SNP) variants, show strong effects on AD susceptibility (Jonsson et al. 2013; Guerreiro et al. 2013a). Subsequent research showed that TREM2 controls the ability of microglia to provide protective barriers around the fibrillar structures of amyloid plaques in human AD brain (Yuan et al. 2016). Other candidate genes for mediating the microglial contribution to AD pathogenesis and whose expression is either restricted to, or enriched in, microglia within the CNS are *BIN1* (bridging integrator 1), whose most significant risk allele is

located in a microglia-specific enhancer (Nott et al. 2019); *SPI1* (Huang et al. 2017); inositol polyphosphate-5-phosphatase (*INPP5D*) (Tsai et al. 2021); and *CD33* (a sialic acid-binding immunoglobulin-like lectin) (Griciuc et al. 2013). Combined, these findings strongly support a role for microglial dysfunction in the initiation or progression of AD, but the direct causal link is, not surprisingly, complex and extensive research has not yet been successfully translated into clinical benefit for patients.

HOW TO STUDY HUMAN MICROGLIA?

Given the emerging importance of microglia in the maintenance or disruption of human CNS function, it is important to specifically study human microglia. Although microglial core functions, gene/protein signatures, and some microglia states are considered to be largely conserved across species, human microglia do show differential responses in turnover, phagocytosis, transcriptomic heterogeneity, and gene expression compared to their rodent counterparts (Miron and Priller 2020). It is challenging, however, to untangle whether these differences reflect different brain regions being compared, intrinsic

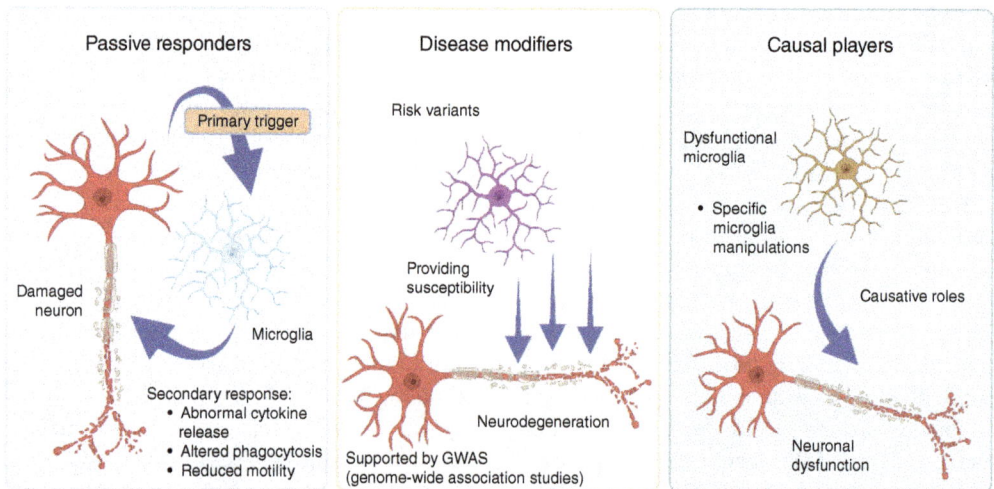

Figure 3. Microglia as disease modifiers: cumulative evidence in recent years has led to a paradigm shift in microglial research. Mostly considered mere bystanders in the past (passive responder to neuronal injuries or infections), microglia are now emerging as cellular disease modifiers, able to influence the onset and progression of neurodegenerative disorders, through the expression of genetic risk variants. (Created with BioRender.com.)

Cite this article as *Cold Spring Harb Perspect Biol* doi: 10.1101/cshperspect.a041366

properties, and/or influences such as history of infection or trauma, microbiome, life span, or other comorbidities. Nevertheless, investigating human microglia behavior and function is an essential component of elucidating the contribution of these cells to CNS development, homeostasis, disease, and repair.

How can we study human microglia? Several complementary techniques have been recently adopted. Postmortem human brain tissue has unveiled the dynamics of microglia numbers with age (Menassa et al. 2022) and transcriptomic and proteomic heterogeneity of microglial states in development, homeostasis, and neurodegenerative disease (Bottcher et al. 2019; Masuda et al. 2019; Mathys et al. 2019; Sankowski et al. 2019). In addition, single-cell RNA sequencing has recently allowed the comparison of shifts in microglia states during development in human versus mouse (Li et al. 2022). Although there has in the past been concern relating to postmortem delay from death to tissue processing leading to RNA degradation and false positive results in RNA se-

quencing, only subtle changes in microglia gene expression have been documented in both mouse and human (Heng et al. 2021). Transcriptomics is typically assessed using single-nuclei or single-cell RNA sequencing (snRNA-seq and scRNA-seq, respectively), both with their own advantages and disadvantages (Box 1).

Ultimately, we need to study the function of human microglia. Primary microglia can be derived from developmental sources or biopsied adult material, and show similar transcriptomic profiles to postmortem microglia (Heng et al. 2021), although they have limited cell yield and availability. Alternatively, microglia-like cells can be generated from human embryonic stem cells (Mancuso et al. 2019) or induced pluripotent stem cells (Muffat et al. 2016; Abud et al. 2017; Haenseler et al. 2017; Pandya et al. 2017), whose function is regulated by genetic variants or pathological insults associated with disease (Bassil et al. 2021; Trudler et al. 2021; Tcw et al. 2022). However powerful, stem cell–based approaches also have limitations (Box 2).

BOX 1. LIMITATIONS OF SINGLE-CELL TRANSCRIPTOMICS

The decision to use scRNA-seq or snRNA-seq is a strategic one (Ding et al. 2020). scRNA-seq relies on the generation of fresh samples of living cells, and is prone to artifacts resulting from tissue processing (Marsh et al. 2022). snRNA-seq allows using fixed cells, reducing gene expression artifacts resulting from dissociation, and making use of existing tissue sections. This approach, however, is confounded by the poor representation of microglia among annotated nuclei (1%–3%) due to relatively lower cell abundance, which can be overcome by exclusion of neuronal and oligodendrocyte nuclei markers or positive selection of microglial nuclei markers such as PU.1 and IRF8 (Nott et al. 2019; van der Poel et al. 2019; Smith et al. 2022). A common issue for both techniques is the identification of transcripts within microglia associated with other cell types—raising the question as to whether these indicate phagocytosis of cell-type components, or expression of these genes by microglia themselves. In addition, the offset between confidence in microglial cluster identification versus depth of sequencing can create challenges in gene pathway analysis and functional interpretations of state-associated roles; this may be addressed in the future by integration of multiple 'omics approaches to fully elucidate microglial state functions, such as single-cell mass cytometry/proteomics/phosphoproteomics, and epigenetics. Another consideration is the inherent variability in neuropathological and clinical presentation of human disease. A solution to this issue is to associate these individual readouts with specific cell changes, for instance as done when associating cell cluster enrichment with Alzheimer's disease neuropathological presentation (Gerrits et al. 2021; Smith et al. 2022) or cognitive dysfunction (Masuda et al. 2019). In the future, we can strengthen our understanding of the range of mechanisms that could underpin disease heterogeneity by further integration of these findings with other potential determinants of disease progression such allelic risk variants (Lopes et al. 2022; Smith et al. 2022) and neuropathological data using digital spatial profiling (as applied to mouse models) (Chen et al. 2020).

BOX 2. LIMITATIONS OF IPSC-DERIVED APPROACHES TO STUDY MICROGLIA

A study assessing microglia derived from induced pluripotent stem cells (iPSCs) generated from individuals harboring the APOE4 variant (conferring higher susceptibility to Alzheimer's disease) surprisingly revealed that individual genetic background has a greater influence on cellular responses in vitro than allelic variant, genetic editing, or culture environment (Tcw et al. 2022). This data points to a common issue in both iPSC and ex vivo primary microglia studies relating to interindividual variability (Kosoy et al. 2022; Tcw et al. 2022). Increasing the number of subjects from which iPSCs are generated, rather than increasing the number of isogenic clones, will therefore likely give greater power in identifying genotype-driven effects (Tcw et al. 2022). Replicability among iPSC cultures may be addressed to some extent by fully automated liquid handling, imaging, and analysis platforms (Bassil et al. 2021). Another issue is that iPSC reprogramming removes the age-associated epigenetic signatures that may contribute to neurodegenerative disease progression (Studer et al. 2015), and it is unclear to what extent the reprogramming influences pathological phenotypes. In addition, iPSC-derived microglia-like cells are greatly influenced by the culture environment, as being maintained in isolation is associated with an immature transcriptomic signature—which can however be improved by 2D coculture with other cell types (Haenseler et al. 2017) or by exposing them to brain substrates (Dolan et al. 2023). Indeed, a human PSC-derived triculture system with microglia, astrocytes, and neurons harboring mutations in amyloid precursor protein (APP) has recently been used to identify cellular cross talk, which may be relevant for Alzheimer's disease (Guttikonda et al. 2021).

The homeostatic signature of iPSC-induced microglia is most comparable to in vivo microglia when in a 3D CNS environment (Guttikonda et al. 2021). One approach is to use microglia-containing organoids; current protocols include generation of cortical, midbrain, and ocular organoids (Popova et al. 2021; Sabate-Soler et al. 2022; Shiraki et al. 2022). These provide an excellent environment for either transplant of microglia-like cells or primitive macrophage progenitors (Popova et al. 2021; Xu et al. 2021), or for genetic or small molecule manipulation to induce endogenous microglia generation (Abud et al. 2017; Ormel et al. 2018; Cakir et al. 2022). However, one must consider the consequence of the absence of vasculature and associated cells, the lack of incoming peripheral immune cells, and the biological relevance of timing of introduction of transplanted cells relative to organoid maturation.

Finally, chimeric models with xenotransplantation into rodent systems provide an intact environment in which to investigate human microglia function in disease models (Hasselmann et al. 2019). In vivo transplantation can lead to up to 80% chimerism in a relatively short period of time (Fattorelli et al. 2021) and closely recapitulates primary microglia signatures (Popova et al. 2021). Interestingly, human iPSC-derived microglia can show opposite responses when grown in isolation in vitro versus following transplant into a humanized mouse brain in vivo. As one example, lipid droplet accumulation in cells harboring a TREM2 mutation was decreased when R47H variant iPSCs were the source of microglia-like cells used for xenotransplantation, in contrast to observations obtained from in vitro culture experiments (Claes et al. 2021). Of interest, comparing human iPSC-derived microglial responses following transplant into mouse versus human brain environments has revealed intrinsic versus context-dependent microglial states (Popova et al. 2021). There are limitations to xenotransplantation of microglia, however, including the requirement for human CSF1 to maintain microglial survival, an immune-deficient host, the variation in transplantation efficiency, and region-specific distribution. Altogether, these systems provide useful platforms through which to elucidate the functions of human microglia, complementing the wealth of rodent-based experimental systems (Table 1).

CONCLUDING REMARKS

Microglia are increasingly recognized as highly complex and multifaceted cells, implicated in a

Table 1. Tools to visualize and manipulate microglia

Category	Tool	Advantages	Disadvantages	References
Cre-lines	Cx3cr1-CreERT2	High efficiency Monocyte turnover avoids targeting	Potential leakiness in neurons May target BAMs Not for short-term/development timeline Heterozygosity to avoid haploinsufficiency	Goldmann et al. 2013
	Cd11b-CreERT2 LysM-CreERT2	High efficiency	Targets BAMs and monocytes but the microglial targeting is controversial	Füger et al. 2017
	Cd11c-Cre	Targets microglia subset	40% efficiency Targets BAMs, monocytes, and DCs	Wlodarczyk et al. 2017
	Crybb1-Cre	High efficiency in embryonic brain macrophages	Targets a subset of BAMs, neurons, and oligodendrocyte precursors	Brioschi et al. 2023
	Sall1-CreERT2	High efficiency Avoids monocyte targeting	Targets other neural cells	Buttgereit et al. 2016
	Tmem119-CreERT2	Targets homeostatic microglia Activity maintained with aging No leakiness	50%–66% efficiency May target monocytes in specific contexts May have reduced activity in chronic injury Not appropriate for <embryonic day 17 Lessened recombination efficiency for inter-LoxP distances of >2 kb or in heterozygous mice	Kaiser and Feng 2019; Faust et al. 2023
	P2ry12-CreERT2	Targets homeostatic microglia 100% efficiency at prenatal time points Activity maintained with injury	May target monocytes in specific contexts Targets small number of CP macrophages	McKinsey et al. 2020
	Hexb-CreERT2	Maintained activity with chronic injury High efficiency	30%–85% efficiency Targets small number of PVMs	Masuda et al. 2020; Faust et al. 2023
	Sall1-Cre:Cx3Cr1-Cre	Microglia specific	Not inducible	Kim et al. 2021

Continued

Table 1. *Continued*

Category	Tool	Advantages	Disadvantages	References
Microglia tracking in vivo	Tmem119-tdt or eGFP	High efficiency; Activity maintained with injury	Targets small number of BAMs and MDMs	Ruan et al. 2020; Kaiser and Feng 2019
	Hexb-tdT	High efficiency; Maintained activity in chronic injury	Targets small number of BAMs	Masuda et al. 2020
	Fms-eGFP or eCFP	Optimal for live imaging	eCFP strength dependent on region; Targets BAMs and monocytes	Sasmono et al. 2003
	Microfetti	Assess clonal expansion and self-renewal	Unequal efficiency of each reporter	Tay et al. 2017
	Flow cytometry (CD11b+ CD45lo)	Distinguish microglia from monocyte-derived cells in some contexts	Cannot distinguish BAMs; Microglia can up-regulate CD45 in specific contexts	Zhang et al. 2002; Lloyd et al. 2019
	TSPO PET ligand	Live assessment	Nonspecific for microglia; May reflect density rather than activation in humans	Venneti et al. 2008; Owen et al. 2017
	AAV-cMG	>80% efficiency of transduction in vitro and in vivo, restricted to microglia; Can be used for rapid targeting of microglia with single injection	Not yet commercially available	Lin et al. 2022
Microglia depletion in vivo	CSF1R inhibitors (e.g., PLX)	99% depletion in 7 d; Allows long-term depletion; Removal of inhibitor allows repopulation	Potential off-target receptor inhibition; Targets all macrophages	Elmore et al. 2014
	Clodronate liposomes	~80% depletion in 3 d	Requires CNS injection; Targets all macrophages	Miron et al. 2013
	Antibody-saporin conjugates	~50% depletion in 1 d	Induces monocyte recruitment; Targets all macrophages	Cantaut-Belarif et al. 2017
	Diphtheria toxin receptor transgenic	>95% depletion in 7 d; Specificity driven by Cre line	Can cause cytokine storm, astrogliosis, and BBB leakage	Bruttger et al. 2015
	Csf1r floxed	~50% depletion in 5 d; Specificity driven by Cre line	Repopulation occurs naturally	Cronk et al. 2018
	Global knockouts (Tgfb1, Csf1r, Sfp1)	Constitutive absence of microglia	Targets all macrophages; Developmental abnormalities and death; Astrogliosis and inflammation	Li et al. 2019b
	Csf1r-FIRE$^{\Delta/\Delta}$	Constitutive absence of microglia; Most BAMs preserved; Survive to adulthood	Lack intracerebroventricular BAMs; Some macrophages lack CSF1R	Rojo et al. 2019

Continued

Cite this article as *Cold Spring Harb Perspect Biol* doi: 10.1101/cshperspect.a041366

Table 1. *Continued*

Category	Tool	Advantages	Disadvantages	References
Microglia cell culture	Rodent primary cells	Mature microglia; Controlled environment	Adult yields low; Reduced homeostatic signature	Gosselin et al. 2017
	Rodent brain explants	Cellular interactions maintained	Reaction to dissection; Lack vasculature and immune system	Masuch and Biber 2019
	Human primary (epilepsy or tumor biopsies)	Mature microglia; Controlled environment	Low cell yield; Often younger cohort (epilepsy) or aged cohort (tumor)	Durafourt et al. 2013
	Mixed species coculture	Simultaneous profiling of multiple cell types	Species-specific responses need to be controlled for	Qiu et al. 2018
	iPSC-derived cells	Large cell yield; Patient-derived genetic modification; Xenotransplantation	Requires coculture with neural cells for ramification and homeostatic signature; Reproducibility issues	Aktories et al. 2022
	ESC-derived cells			
	Organoids	Cellular diversity improved over 2D cultures	Lack vasculature and immune system; Most require microglia transplant	Sabate-Soler et al. 2022
Omics	RABID-seq	Assesses cell–cell communication	Requires transgenic expression of viral receptor	Clark et al. 2021
	scRNA-seq	Allows enrichment of myeloid cells	Requires fresh tissue	Marsh et al. 2022
	snRNA-seq	Avoids batched processing	Low sensitivity	Grubman et al. 2021
	Ribo-seq	Avoids cell isolation artifacts; Indications on proteome	Requires transgenic reporter	Haimon et al. 2018
	CYTOF	Protein assessment	Limited to ~60 targets, including lineage markers for identification	Sankowski et al. 2019
	Imaging mass spectrometry	Multiplexed protein assessment on tissue	Limited number of proteins assessed	Ramaglia et al. 2019

Currently available tools to specifically investigate microglia biology, their advantages and disadvantages. (BAMs) Border-associated macrophages, (Cre) Cre recombinase, (DCs) dendritic cells, (CP) choroid plexus, (PVMs) perivascular macrophages, (MDMs) monocyte-derived macrophages, (eCFP) enhanced cyan fluorescent protein, (AAV) adeno-associated virus, (TSPO) translocator protein, (PET) positron emission tomography, (BBB) blood–brain barrer, (CSF1R) colony stimulating factor receptor 1, (RABID-seq) rabies 1D sequencing, (scRNA-seq) single-cell RNA sequencing, (snRNA-seq) single-nuclei RNA sequencing, (iPSC) induced pluripotent stem cell, (ESC) embryonic stem cell.

variety of biological process across the entire life span of the CNS. Despite recent advances in our knowledge about their identity and function, however, major open questions remain. Here, we propose that microglia may act as integrative hubs of external stimuli and local cues that encode the physiological status of the individual. Understanding the rules of microglial decision-making (at the population or single-cell level) will provide insights into the molecular and cellular basis of microglia-mediated homeostatic control. Whether microglial epigenetic "memory" of life events plays a role in maintaining brain homeostasis is also an interesting and plausible scenario that certainly deserves further investigation.

In light of recent findings from large human genetic studies, microglial are emerging as key disease modifiers, able to influence the onset and course of many neurodegenerative disorders. However, the causal mechanisms that link microglial risk variants to pathological outcomes remains in the shadows. Further studies are urgently required to dissect the precise contribution of microglial dysfunction to specific diseases. Extending these findings to human models will be imperative for establishing relevant links with human diseases and designing novel microglia-targeted therapies.

ACKNOWLEDGMENTS

This work was supported by grants from the Synapsis Foundation - Alzheimer Research Switzerland ARS, Swiss National Science Foundation (SNSF 310030_197940) and European Research Council (ERC StGrant REMIND 804949) to R.C.P.; the Spanish Ministry of Science and Innovation Competitiveness MCIN/AEI/10.13039/501100011033 and FEDER "A way to make Europe" (RTI2018-099267-B-I00, PID2022-136698OB-I00), and a Basque Government Department of Education project (PIBA 2020_1_0030) to A.S.; the John David Eaton Chair from the Multiple Sclerosis Catalyst Fund/BARLO MS Centre, and a Medical Research Council Senior Non-Clinical Fellowship (MRC/V031260/1) to V.E.M.

REFERENCES

Abiega O, Beccari S, Diaz-Aparicio I, Nadjar A, Layé S, Leyrolle Q, Gómez-Nicola D, Domercq M, Pérez-Samartín A, Sánchez-Zafra V, et al. 2016. Neuronal hyperactivity disturbs ATP microgradients, impairs microglial motility, and reduces phagocytic receptor expression triggering apoptosis/microglial phagocytosis uncoupling. *PLoS Biol* 14: e1002466. doi:10.1371/journal.pbio.1002466

Abud EM, Ramirez RN, Martinez ES, Healy LM, Nguyen CHH, Newman SA, Yeromin AV, Scarfone VM, Marsh SE, Fimbres C, et al. 2017. iPSC-derived human microglia-like cells to study neurological diseases. *Neuron* 94: 278–293.e9. doi:10.1016/j.neuron.2017.03.042

Ajami B, Bennett JL, Krieger C, Tetzlaff W, Rossi FM. 2007. Local self-renewal can sustain CNS microglia maintenance and function throughout adult life. *Nat Neurosci* 10: 1538–1543. doi:10.1038/nn2014

Aktories P, Petry P, Kierdorf K. 2022. Microglia in a dish—which techniques are on the menu for functional studies? *Front Cell Neurosci* 16: 908315. doi:10.3389/fncel.2022.908315

Allis CD, Jenuwein T. 2016. The molecular hallmarks of epigenetic control. *Nat Rev Genet* 17: 487–500. doi:10.1038/nrg.2016.59

Askew K, Li K, Olmos-Alonso A, Garcia-Moreno F, Liang Y, Richardson P, Tipton T, Chapman MA, Riecken K, Beccari S, et al. 2017. Coupled proliferation and apoptosis maintain the rapid turnover of microglia in the adult brain. *Cell Rep* 18: 391–405. doi:10.1016/j.celrep.2016.12.041

Attwell D, Mishra A, Hall CN, O'Farrell FM, Dalkara T. 2016. What is a pericyte? *J Cereb Blood Flow Metab* 36: 451–455. doi:10.1177/0271678X15610340

Badimon A, Strasburger HJ, Ayata P, Chen X, Nair A, Ikegami A, Hwang P, Chan AT, Graves SM, Uweru JO, et al. 2020. Negative feedback control of neuronal activity by microglia. *Nature* 586: 417–423. doi:10.1038/s41586-020-2777-8

Basilico B, Ferrucci L, Ratano P, Golia MT, Grimaldi A, Rosito M, Ferretti V, Reverte I, Sanchini C, Marrone MC, et al. 2022. Microglia control glutamatergic synapses in the adult mouse hippocampus. *Glia* 70: 173–195. doi:10.1002/glia.24101

Bassil R, Shields K, Granger K, Zein I, Ng S, Chih B. 2021. Improved modeling of human AD with an automated culturing platform for iPSC neurons, astrocytes and microglia. *Nat Commun* 12: 5220. doi:10.1038/s41467-021-25344-6

Berdowski WM, van der Linde HC, Breur M, Oosterhof N, Beerepoot S, Sanderson L, Wijnands LI, de Jong P, Tsai-Meu-Chong E, de Valk W, et al. 2022. Dominant-acting CSF1R variants cause microglial depletion and altered astrocytic phenotype in zebrafish and adult-onset leukodystrophy. *Acta Neuropathol* 144: 211–239. doi:10.1007/s00401-022-02440-5

Bianchin MM, Snow Z. 2022. Primary microglia dysfunction or microgliopathy: a cause of dementias and other neurological or psychiatric disorders. *Neuroscience* 497: 324–339. doi:10.1016/j.neuroscience.2022.06.032

Bisht K, Okojie KA, Sharma K, Lentferink DH, Sun YY, Chen HR, Uweru JO, Amancherla S, Calcuttawala Z, Campos-Salazar AB, et al. 2021. Capillary-associated microglia regulate vascular structure and function through PANX1-P2RY12 coupling in mice. *Nat Commun* **12:** 5289. doi:10.1038/s41467-021-25590-8

Bottcher C, Schlickeiser S, Sneeboer MAM, Kunkel D, Knop A, Paza E, Fidzinski P, Kraus L, Snijders GJL, Kahn RS, et al. 2019. Human microglia regional heterogeneity and phenotypes determined by multiplexed single-cell mass cytometry. *Nat Neurosci* **22:** 78–90. doi:10.1038/s41593-018-0290-2

Brioschi S, Belk JA, Peng V, Molgora M, Fernandes Rodrigues P, Nguyen KM, Wang S, Du S, Wang WL, Grajales-Reyes GE, et al. 2023. A Cre-deleter specific for embryo-derived brain macrophages reveals distinct features of microglia and border macrophages. *Immunity* **56:** 1027–1045.e8. doi:10.1016/j.immuni.2023.01.028

Bruttger J, Karram K, Wörtge S, Regen T, Marini F, Hoppmann N, Klein M, Blank T, Yona S, Wolf Y, et al. 2015. Genetic cell ablation reveals clusters of local self-renewing microglia in the mammalian central nervous system. *Immunity* **43:** 92–106. doi:10.1016/j.immuni.2015.06.012

Butovsky O, Jedrychowski MP, Moore CS, Cialic R, Lanser AJ, Gabriely G, Koeglsperger T, Dake B, Wu PM, Doykan CE, et al. 2014. Identification of a unique TGF-β–dependent molecular and functional signature in microglia. *Nat Neurosci* **17:** 131–143. doi:10.1038/nn.3599

Buttgereit A, Lelios I, Yu X, Vrohlings M, Krakoski NR, Gautier EL, Nishinakamura R, Becher B, Greter M. 2016. Sall1 is a transcriptional regulator defining microglia identity and function. *Nat Immunol* **17:** 1397–1406. doi:10.1038/ni.3585

Cakir B, Tanaka Y, Kiral FR, Xiang Y, Dagliyan O, Wang J, Lee M, Greaney AM, Yang WS, duBoulay C, et al. 2022. Expression of the transcription factor PU.1 induces the generation of microglia-like cells in human cortical organoids. *Nat Commun* **13:** 430. doi:10.1038/s41467-022-28043-y

Calovi S, Mut-Arbona P, Sperlágh B. 2019. Microglia and the purinergic signaling system. *Neuroscience* **405:** 137–147. doi:10.1016/j.neuroscience.2018.12.021

Cantaut-Belarif Y, Antri M, Pizzarelli R, Colasse S, Vaccari I, Soares S, Renner M, Dallel R, Triller A, Bessis A. 2017. Microglia control the glycinergic but not the GABAergic synapses via prostaglandin E2 in the spinal cord. *J Cell Biol* **216:** 2979–2989. doi:10.1083/jcb.201607048

Chen Y, Colonna M. 2021. Microglia in Alzheimer's disease at single-cell level. Are there common patterns in humans and mice? *J Exp Med* **218:** e20202717. doi:10.1084/jem.20202717

Chen WT, Lu A, Craessaerts K, Pavie B, Frigerio CS, Corthout N, Qian X, Laláková J, Kuhnemund M, Voytyuk I, et al. 2020. Spatial transcriptomics and in situ sequencing to study Alzheimer's disease. *Cell* **182:** 976–991.e19. doi:10.1016/j.cell.2020.06.038

Chitu V, Gökhan S, Stanley ER. 2022. Modeling CSF-1 receptor deficiency diseases—how close are we? *FEBS J* **289:** 5049–5073. doi:10.1111/febs.16085

Claes C, Danhash EP, Hasselmann J, Chadarevian JP, Shabestari SK, England WE, Lim TE, Hidalgo JLS, Spitale RC, Davtyan H, et al. 2021. Plaque-associated human microglia accumulate lipid droplets in a chimeric model of Alzheimer's disease. *Mol Neurodegener* **16:** 50. doi:10.1186/s13024-021-00473-0

Clark IC, Gutiérrez-Vázquez C, Wheeler MA, Li Z, Rothhammer V, Linnerbauer M, Sanmarco LM, Guo L, Blain M, Zandee SEJ, et al. 2021. Barcoded viral tracing of single-cell interactions in central nervous system inflammation. *Science* **372:** eabf1230. doi:10.1126/science.abf1230

Cronk JC, Filiano AJ, Louveau A, Marin I, Marsh R, Ji E, Goldman DH, Smirnov I, Geraci N, Acton S, et al. 2018. Peripherally derived macrophages can engraft the brain independent of irradiation and maintain an identity distinct from microglia. *J Exp Med* **215:** 1627–1647. doi:10.1084/jem.20180247

Császár E, Lénárt N, Cserép C, Környei Z, Fekete R, Pósfai B, Balázsfi D, Hangya B, Schwarcz AD, Szabadits E, et al. 2022. Microglia modulate blood flow, neurovascular coupling, and hypoperfusion via purinergic actions. *J Exp Med* **219:** e20211071. doi:10.1084/jem.20211071

Dai XM, Ryan GR, Hapel AJ, Dominguez MG, Russell RG, Kapp S, Sylvestre V, Stanley ER. 2002. Targeted disruption of the mouse colony-stimulating factor 1 receptor gene results in osteopetrosis, mononuclear phagocyte deficiency, increased primitive progenitor cell frequencies, and reproductive defects. *Blood* **99:** 111–120. doi:10.1182/blood.V99.1.111

Dai Z, Ramesh V, Locasale JW. 2020. The evolving metabolic landscape of chromatin biology and epigenetics. *Nat Rev Genet* **21:** 737–753. doi:10.1038/s41576-020-0270-8

Dale N, Frenguelli BG. 2009. Release of adenosine and ATP during ischemia and epilepsy. *Curr Neuropharmacol* **7:** 160–179. doi:10.2174/157015909789152146

Dash R, Choi HJ, Moon IS. 2020. Mechanistic insights into the deleterious roles of Nasu–Hakola disease associated TREM2 variants. *Sci Rep* **10:** 3663. doi:10.1038/s41598-020-60561-x

Diaz-Aparicio I, Paris I, Sierra-Torre V, Plaza-Zabala A, Rodríguez-Iglesias N, Márquez-Ropero M, Beccari S, Huguet P, Abiega O, Alberdi E, et al. 2020. Microglia actively remodel adult hippocampal neurogenesis through the phagocytosis secretome. *J Neurosci* **40:** 1453–1482. doi:10.1523/JNEUROSCI.0993-19.2019

Ding J, Adiconis X, Simmons SK, Kowalczyk MS, Hession CC, Marjanovic ND, Hughes TK, Wadsworth MH, Burks T, Nguyen LT, et al. 2020. Systematic comparison of single-cell and single-nucleus RNA-sequencing methods. *Nat Biotechnol* **38:** 737–746. doi:10.1038/s41587-020-0465-8

Diskin C, Ryan TAJ, O'Neill LAJ. 2021. Modification of proteins by metabolites in immunity. *Immunity* **54:** 19–31. doi:10.1016/j.immuni.2020.09.014

Dolan MJ, Therrien M, Jereb S, Kamath T, Gazestani V, Atkeson T, Marsh SE, Goeva A, Lojek NM, Murphy S, et al. 2023. Exposure of iPSC-derived human microglia to brain substrates enables the generation and manipulation of diverse transcriptional states in vitro. *Nat Immunol* **24:** 1382–1390. doi:10.1038/s41590-023-01558-2

Durafourt BA, Moore CS, Blain M, Antel JP. 2013. Isolating, culturing, and polarizing primary human adult and fetal microglia. *Methods Mol Biol* **1041:** 199–211. doi:10.1007/978-1-62703-520-0_19

Elmore MR, Najafi AR, Koike MA, Dagher NN, Spangenberg EE, Rice RA, Kitazawa M, Matusow B, Nguyen H, West BL, et al. 2014. Colony-stimulating factor 1 receptor signaling is necessary for microglia viability, unmasking a microglia progenitor cell in the adult brain. *Neuron* **82:** 380–397. doi:10.1016/j.neuron.2014.02.040

Erblich B, Zhu L, Etgen AM, Dobrenis K, Pollard JW. 2011. Absence of colony stimulation factor-1 receptor results in loss of microglia, disrupted brain development and olfactory deficits. *PLoS ONE* **6:** e26317. doi:10.1371/journal.pone.0026317

Erny D, Dokalis N, Mezö C, Castoldi A, Mossad O, Staszewski O, Frosch M, Villa M, Fuchs V, Mayer A, et al. 2021. Microbiota-derived acetate enables the metabolic fitness of the brain innate immune system during health and disease. *Cell Metab* **33:** 2260–2276.e7. doi:10.1016/j.cmet.2021.10.010

Eyo UB, Peng J, Swiatkowski P, Mukherjee A, Bispo A, Wu LJ. 2014. Neuronal hyperactivity recruits microglial processes via neuronal NMDA receptors and microglial P2Y12 receptors after status epilepticus. *J Neurosci* **34:** 10528–10540. doi:10.1523/JNEUROSCI.0416-14.2014

Fattorelli N, Martinez-Muriana A, Wolfs L, Geric I, De Strooper B, Mancuso R. 2021. Stem-cell-derived human microglia transplanted into mouse brain to study human disease. *Nat Protoc* **16:** 1013–1033. doi:10.1038/s41596-020-00447-4

Faust TE, Feinberg PA, O'Connor C, Kawaguchi R, Chan A, Strasburger H, Frosch M, Boyle MA, Masuda T, Amann L, et al. 2023. A comparative analysis of microglial inducible Cre lines. *Cell Rep* **42:** 113031. doi:10.1016/j.celrep.2023.113031

Ferrer I. 2022. The primary microglial leukodystrophies: a review. *Int J Mol Sci* **23:** 6341. doi:10.3390/ijms23116341

Filipello F, Morini R, Corradini I, Zerbi V, Canzi A, Michalski B, Erreni M, Markicevic M, Starvggi-Cucuzza C, Otero K, et al. 2018. The microglial innate immune receptor TREM2 is required for synapse elimination and normal brain connectivity. *Immunity* **48:** 979–991.e8. doi:10.1016/j.immuni.2018.04.016

Füger P, Hefendehl JK, Veeraraghavalu K, Wendeln AC, Schlosser C, Obermüller U, Wegenast-Braun BM, Neher JJ, Martus P, Kohsaka S, et al. 2017. Microglia turnover with aging and in an Alzheimer's model via long-term in vivo single-cell imaging. *Nat Neurosci* **20:** 1371–1376. doi:10.1038/nn.4631

Galimberti D, Fenoglio C, Ghezzi L, Serpente M, Arcaro M, D'Anca M, De Riz M, Arighi A, Fumagalli GG, Pietroboni AM, et al. 2019. Inflammatory expression profile in peripheral blood mononuclear cells from patients with Nasu-Hakola disease. *Cytokine* **116:** 115–119. doi:10.1016/j.cyto.2018.12.024

Gerrits E, Brouwer N, Kooistra SM, Woodbury ME, Vermeiren Y, Lambourne M, Mulder J, Kummer M, Möller T, Biber K, et al. 2021. Distinct amyloid-β and tau-associated microglia profiles in Alzheimer's disease. *Acta Neuropathol* **141:** 681–696. doi:10.1007/s00401-021-02263-w

Ginhoux F, Greter M, Leboeuf M, Nandi S, See P, Gokhan S, Mehler MF, Conway SJ, Ng LG, Stanley ER, et al. 2010. Fate mapping analysis reveals that adult microglia derive from primitive macrophages. *Science* **330:** 841–845. doi:10.1126/science.1194637

Goldberg AD, Allis CD, Bernstein E. 2007. Epigenetics: a landscape takes shape. *Cell* **128:** 635–638. doi:10.1016/j.cell.2007.02.006

Goldmann T, Wieghofer P, Müller PF, Wolf Y, Varol D, Yona S, Brendecke SM, Kierdorf K, Staszewski O, Datta M, et al. 2013. A new type of microglia gene targeting shows TAK1 to be pivotal in CNS autoimmune inflammation. *Nat Neurosci* **16:** 1618–1626. doi:10.1038/nn.3531

Gosselin D, Link VM, Romanoski CE, Fonseca GJ, Eichenfield DZ, Spann NJ, Stender JD, Chun HB, Garner H, Geissmann F, et al. 2014. Environment drives selection and function of enhancers controlling tissue-specific macrophage identities. *Cell* **159:** 1327–1340. doi:10.1016/j.cell.2014.11.023

Gosselin D, Skola D, Coufal NG, Holtman IR, Schlachetzki JCM, Sajti E, Jaeger BN, O'Connor C, Fitzpatrick C, Pasillas MP, et al. 2017. An environment-dependent transcriptional network specifies human microglia identity. *Science* **356:** eaal3222. doi:10.1126/science.aal3222

Green KN, Crapser JD, Hoshfield LA. 2020. To kill a microglia: a case for CSF1R inhibitors. *Trends Immunol* **41:** 771–784. doi:10.1016/j.it.2020.07.001

Greter M, Lelios I, Pelczar P, Hoeffel G, Price J, Leboeuf M, Kündig TM, Frei K, Ginhoux F, Merad M, et al. 2012. Stroma-derived interleukin-34 controls the development and maintenance of langerhans cells and the maintenance of microglia. *Immunity* **37:** 1050–1060. doi:10.1016/j.immuni.2012.11.001

Griciuc A, Serrano-Pozo A, Parrado AR, Lesinski AN, Asselin CN, Mullin K, Hooli B, Choi SH, Hyman BT, Tanzi RE. 2013. Alzheimer's disease risk gene CD33 inhibits microglial uptake of amyloid β. *Neuron* **78:** 631–643. doi:10.1016/j.neuron.2013.04.014

Grubman A, Choo XY, Chew G, Ouyang JF, Sun G, Croft NP, Rossello FJ, Simmons R, Buckberry S, Landin DV, et al. 2021. Transcriptional signature in microglia associated with Aβ plaque phagocytosis. *Nat Commun* **12:** 3015. doi:10.1038/s41467-021-23111-1

Guerreiro R, Wojtas A, Bras J, Carrasquillo M, Rogaeva E, Majounie E, Cruchaga C, Sassi C, Kauwe JS, Younkin S, et al. 2013a. TREM2 variants in Alzheimer's disease. *N Engl J Med* **368:** 117–127. doi:10.1056/NEJMoa1211851

Guerreiro RJ, Lohmann E, Brás JM, Gibbs JR, Rohrer JD, Gurunlian N, Dursun B, Bilgic B, Hanagasi H, Gurvit H, et al. 2013b. Using exome sequencing to reveal mutations in TREM2 presenting as a frontotemporal dementia-like syndrome without bone involvement. *JAMA Neurol* **70:** 78–84. doi:10.1001/jamaneurol.2013.579

Guo L, Ikegawa S. 2021. From HDLS to BANDDOS: fast-expanding phenotypic spectrum of disorders caused by mutations in CSF1R. *J Hum Genet* **66:** 1139–1144. doi:10.1038/s10038-021-00942-w

Guttikonda SR, Sikkema L, Tchieu J, Saurat N, Walsh RM, Harschnitz O, Ciceri G, Sneeboer M, Mazutis L, Setty M, et al. 2021. Fully defined human pluripotent stem cell-derived microglia and tri-culture system model C3 production in Alzheimer's disease. *Nat Neurosci* **24:** 343–354. doi:10.1038/s41593-020-00796-z

Haenseler W, Sansom SN, Buchrieser J, Newey SE, Moore CS, Nicholls FJ, Chintawar S, Schnell C, Antel JP, Allen ND, et al. 2017. A highly efficient human pluripotent stem cell microglia model displays a neuronal-co-culture-specific expression profile and inflammatory response. *Stem Cell Rep* **8:** 1727–1742. doi:10.1016/j.stemcr.2017.05.017

Haimon Z, Volaski A, Orthgiess J, Boura-Halfon S, Varol D, Shemer A, Yona S, Zuckerman B, David E, Chappell-Maor L, et al. 2018. Re-evaluating microglia expression profiles using RiboTag and cell isolation strategies. *Nat Immunol* **19:** 636–644. doi:10.1038/s41590-018-0110-6

Hasselmann J, Coburn MA, England W, Figueroa Velez DX, Shebestari SK, Tu CH, McQuade A, Kolahdouzan M, Echeverria K, Claes C, et al. 2019. Development of a chimeric model to study and manipulate human microglia in vivo. *Neuron* **103:** 1016–1033.e10. doi:10.1016/j.neuron.2019.07.002

Heng Y, Dubbelaar ML, Marie SKN, Boddeke E, Eggen BJL. 2021. The effects of postmortem delay on mouse and human microglia gene expression. *Glia* **69:** 1053–1060. doi:10.1002/glia.23948

Hickman SE, Kingery ND, Ohsumi TK, Borowsky ML, Wang LC, Means TK, El Khoury J. 2013. The microglial sensome revealed by direct RNA sequencing. *Nat Neurosci* **16:** 1896–1905. doi:10.1038/nn.3554

Huang KL, Marcora E, Pimenova AA, Di Narzo AF, Kapoor M, Jin SC, Harari O, Bertelsen S, Fairfax BP, Czajkowski J, et al. 2017. A common haplotype lowers PU.1 expression in myeloid cells and delays onset of Alzheimer's disease. *Nat Neurosci* **20:** 1052–1061. doi:10.1038/nn.4587

Jonsson T, Stefansson H, Steinberg S, Jonsdottir I, Jonsson PV, Snaedal J, Bjornsson S, Huttenlocher J, Levey AI, Lah JJ, et al. 2013. Variant of *TREM2* associated with the risk of Alzheimer's disease. *N Engl J Med* **368:** 107–116. doi:10.1056/NEJMoa1211103

Kaiser T, Feng G. 2019. Tmem119-EGFP and Tmem119-CreERT2 transgenic mice for labeling and manipulating microglia. *eNeuro* **6:** ENEURO.0448-18.2019. doi:10.1523/ENEURO.0448-18.2019

Keren-Shaul H, Spinrad A, Weiner A, Matcovitch-Natan O, Dvir-Szternfeld R, Ulland TK, David E, Baruch K, Lara-Astaiso D, Toth B, et al. 2017. A unique microglia type associated with restricting development of Alzheimer's disease. *Cell* **169:** 1276–1290.e17. doi:10.1016/j.cell.2017.05.018

Kim JS, Kolesnikov M, Peled-Hajaj S, Scheyltjens I, Xia Y, Trzebanski S, Haimon Z, Shemer A, Lubart A, Van Hove H, et al. 2021. A binary Cre transgenic approach dissects microglia and CNS border-associated macrophages. *Immunity* **54:** 176–190.e7. doi:10.1016/j.immuni.2020.11.007

Konishi H, Okamoto T, Hara Y, Komine O, Tamada H, Maeda M, Osako F, Kobayashi M, Nishiyama A, Kataoka Y, et al. 2020. Astrocytic phagocytosis is a compensatory mechanism for microglial dysfunction. *EMBO J* **39:** e104464. doi:10.15252/embj.2020104464

Konno T, Kasanuki K, Ikeuchi T, Dickson DW, Wszolek ZK. 2018. *CSF1R*-related leukoencephalopathy: a major player in primary microgliopathies. *Neurology* **91:** 1092–1104. doi:10.1212/WNL.0000000000006642

Kosoy R, Fullard JF, Zeng B, Bendl J, Dong P, Rahman S, Kleopoulos SP, Shao Z, Girdhar K, Humphrey J, et al. 2022. Genetics of the human microglia regulome refines Alzheimer's disease risk loci. *Nat Genet* **54:** 1145–1154. doi:10.1038/s41588-022-01149-1

Lauterbach MA, Hanke JE, Serefidou M, Mangan MSJ, Kolbe CC, Hess T, Rothe M, Kaiser R, Hoss F, Gehlen J, et al. 2019. Toll-like receptor signaling rewires macrophage metabo-lism and promotes histone acetylation via ATP-citrate lyase. *Immunity* **51:** 997–1011.e7. doi:10.1016/j.immuni.2019.11.009

Lavin Y, Winter D, Blecher-Gonen R, David E, Keren-Shaul H, Merad M, Jung S, Amit I. 2014. Tissue-resident macrophage enhancer landscapes are shaped by the local microenvironment. *Cell* **159:** 1312–1326. doi:10.1016/j.cell.2014.11.018

Li F, Jiang D, Samuel MA. 2019a. Microglia in the developing retina. *Neural Dev* **14:** 12. doi:10.1186/s13064-019-0137-x

Li Q, Cheng Z, Zhou L, Darmanis S, Neff NF, Okamoto J, Gulati G, Bennett ML, Sun LO, Clarke LE, et al. 2019b. Developmental heterogeneity of microglia and brain myeloid cells revealed by deep single-cell RNA sequencing. *Neuron* **101:** 207–223.e10. doi:10.1016/j.neuron.2018.12.006

Li Y, Li Z, Yang M, Wang F, Zhang Y, Li R, Li Q, Gong Y, Wang B, Fan B, et al. 2022. Decoding the temporal and regional specification of microglia in the developing human brain. *Cell Stem Cell* **29:** 620–634.e6. doi:10.1016/j.stem.2022.02.004

Lin R, Zhou Y, Yan T, Wang R, Li H, Wu Z, Zhang X, Zhou X, Zhao F, Zhang L, et al. 2022. Directed evolution of adeno-associated virus for efficient gene delivery to microglia. *Nat Methods* **19:** 976–985. doi:10.1038/s41592-022-01547-7

Lloyd AF, Davies CL, Holloway RK, Labrak Y, Ireland G, Carradori D, Dillenburg A, Borger E, Soong D, Richardson JC, et al. 2019. Central nervous system regeneration is driven by microglia necroptosis and repopulation. *Nat Neurosci* **22:** 1046–1052. doi:10.1038/s41593-019-0418-z

Lopes KP, Snijders GJL, Humphrey J, Allan A, Sneeboer MAM, Navarro E, Schilder BM, Vialle RA, Parks M, Missall R, et al. 2022. Genetic analysis of the human microglial transcriptome across brain regions, aging and disease pathologies. *Nat Genet* **54:** 4–17. doi:10.1038/s41588-021-00976-y

Mancuso R, Van Den Daele J, Fattorelli N, Wolfs L, Balusu S, Burton O, Liston A, Sierksma A, Fourne Y, Poovathingal S, et al. 2019. Stem-cell-derived human microglia transplanted in mouse brain to study human disease. *Nat Neurosci* **22:** 2111–2116. doi:10.1038/s41593-019-0525-x

Márquez-Ropero M, Benito E, Plaza-Zabala A, Sierra A. 2020. Microglial corpse clearance: lessons from macrophages. *Front Immunol* **11:** 506. doi:10.3389/fimmu.2020.00506

Marsh SE, Walker AJ, Kamath T, Dissing-Olesen L, Hammond TR, Yvanka de Soysa T, Young AMH, Murphy S, Abdulraouf A, Nadaf N, et al. 2022. Dissection of artifactual and confounding glial signatures by single-cell sequencing of mouse and human brain. *Nat Neurosci* **25:** 306–316. doi:10.1038/s41593-022-01022-8

Masuch A, Biber K. 2019. Replenishment of organotypic hippocampal slice cultures with neonatal or adult microglia. *Methods Mol Biol* **2034:** 127–147. doi:10.1007/978-1-4939-9658-2_10

Masuda T, Sankowski R, Staszewski O, Böttcher C, Amann L, Sagar, Scheiwe C, Nessler S, Kunz P, van Loo G, et al. 2019. Spatial and temporal heterogeneity of mouse and human microglia at single-cell resolution. *Nature* **566:** 388–392. doi:10.1038/s41586-019-0924-x

Masuda T, Amann L, Sankowski R, Staszewski O, Lenz M, Errico PD, Snaidero N, Costa Jordao MJ, Bottcher C, Kierdorf K, et al. 2020. Novel Hexb-based tools for studying

microglia in the CNS. *Nat Immunol* **21**: 802–815. doi:10
.1038/s41590-020-0707-4

Matcovitch-Natan O, Winter DR, Giladi A, Vargas Aguilar S,
Spinrad A, Sarrazin S, Ben-Yehuda H, David E, Gonzalez
FZ, Perrin P, et al. 2016. Microglia development follows a
stepwise program to regulate brain homeostasis. *Science*
353: aad8670. doi:10.1126/science.aad8670

Mathys H, Davila-Velderrain J, Peng Z, Gao F, Mohammadi
S, Young JZ, Menon M, He L, Abdurrob F, Jiang X, et al.
2019. Single-cell transcriptomic analysis of Alzheimer's
disease. *Nature* **570**: 332–337. doi:10.1038/s41586-019-
1195-2

McKinsey GL, Lizama CO, Keown-Lang AE, Niu A, Santander N, Larpthaveesarp A, Chee E, Gonzalez FF, Arnold TD.
2020. A new genetic strategy for targeting microglia in
development and disease. *eLife* **9**: e54590. doi:10.7554/eL
ife.54590

McNamara NB, Munro DAD, Bestard-Cuche N, Uyeda A,
Bogie JFJ, Hoffmann A, Holloway RK, Molina-Gonzalez I,
Askew KE, Mitchell S, et al. 2023. Microglia regulate central
nervous system myelin growth and integrity. *Nature* **613**:
120–129. doi:10.1038/s41586-022-05534-y

Menassa DA, Muntslag TAO, Martin-Estebané M, Barry-
Carroll L, Chapman MA, Adorjan I, Tyler T, Turnbull B,
Rose-Zerilli MJJ, Nicoll JAR, et al. 2022. The spatiotempo-
ral dynamics of microglia across the human lifespan. *Dev
Cell* **57**: 2127–2139.e6. doi:10.1016/j.devcel.2022.07.015

Miron VE, Priller J. 2020. Investigating microglia in health
and disease: challenges and opportunities. *Trends Immunol* **41**: 785–793. doi:10.1016/j.it.2020.07.002

Miron VE, Boyd A, Zhao JW, Yuen TJ, Ruckh JM, Shadrach
JL, van Wijngaarden P, Wagers AJ, Williams A, Franklin
RJM, et al. 2013. M2 microglia and macrophages drive
oligodendrocyte differentiation during CNS remyelination. *Nat Neurosci* **16**: 1211–1218. doi:10.1038/nn.3469

Munro DAD, Bradford BM, Mariani SA, Hampton DW,
Vink CS, Chandran S, Hume DA, Pridans C, Priller J.
2020. CNS macrophages differentially rely on an intronic
Csf1r enhancer for their development. *Development* **147**:
dev194449. doi:10.1242/dev.194449

Muffat J, Li Y, Yuan B, Mitalipova M, Omer A, Corcoran S,
Bakiasi G, Tsai LH, Aubourg P, Ransohoff RM, et al. 2016.
Efficient derivation of microglia-like cells from human
pluripotent stem cells. *Nat Med* **22**: 1358–1367. doi:10
.1038/nm.4189

Nayak D, Roth TL, McGavern DB. 2014. Microglia development and function. *Annu Rev Immunol* **32**: 367–402.
doi:10.1146/annurev-immunol-032713-120240

Netea MG, Joosten LA, Latz E, Mills KH, Natoli G, Stunnen-
berg HG, O'Neill LA, Xavier RJ. 2016. Trained immunity: a
program of innate immune memory in health and disease.
Science **352**: aaf1098. doi:10.1126/science.aaf1098

Nott A, Holtman IR, Coufal NG, Schlachetzki JCM, Yu M, Hu
R, Han CZ, Pena M, Ziao J, Wu Y, et al. 2019. Brain cell
type–specific enhancer–promoter interactome maps and
disease-risk association. *Science* **366**: 1134–1139. doi:10
.1126/science.aay0793

Olson MC, Scott EW, Hack AA, Su GH, Tenen DG, Singh H,
Simon MC. 1995. PU.1 is not essential for early myeloid
gene expression but is required for terminal myeloid differentiation. *Immunity* **3**: 703–714. doi:10.1016/1074-
7613(95)90060-8

Oosterhof N, Chang IJ, Karimiani EG, Kuil LE, Jensen DM,
Daza R, Young E, Astle L, van der Linde HC, Shivaram GM,
et al. 2019. Homozygous mutations in CSF1R cause a pediatric-onset leukoencephalopathy and can result in congenital absence of microglia. *Am J Hum Genet* **104**: 936–
947. doi:10.1016/j.ajhg.2019.03.010

Ormel PR, Vieira de Sá R, van Bodegraven EJ, Karst H,
Harschnitz O, Sneeboer MAM, Johansen LE, van Dijk
RE, Scheefhals N, Berdenis van Berlekom A, et al. 2018.
Microglia innately develop within cerebral organoids. *Nat
Commun* **9**: 4167. doi:10.1038/s41467-018-06684-2

Owen DR, Narayan N, Wells L, Healy L, Smyth E, Rabiner EA,
Galloway D, Williams JB, Lehr J, Mandhair H, et al. 2017.
Pro-inflammatory activation of primary microglia and
macrophages increases 18 kDa translocator protein ex-
pression in rodents but not humans. *J Cereb Blood Flow
Metab* **37**: 2679–2690. doi:10.1177/0271678X17710182

Pan RY, He L, Zhang J, Liu X, Liao Y, Gao J, Liao Y, Yan Y, Li Q,
Zhou X, et al. 2022. Positive feedback regulation of microglial glucose metabolism by histone H4 lysine 12 lactylation
in Alzheimer's disease. *Cell Metab* **34**: 634–648.e6. doi:10
.1016/j.cmet.2022.02.013

Pandya H, Shen MJ, Ichikawa DM, Sedlock AB, Choi Y, Johnson KR, Kim G, Brown MA, Elkahloun AG, Maric D, et al.
2017. Differentiation of human and murine induced pluripotent stem cells to microglia-like cells. *Nat Neurosci* **20**:
753–759. doi:10.1038/nn.4534

Paolicelli RC, Bolasco G, Pagani F, Maggi L, Scianni M, Panzanelli P, Giustetto M, Ferreira TA, Guiducci E, Dumas L,
et al. 2011. Synaptic pruning by microglia is necessary for
normal brain development. *Science* **333**: 1456–1458.
doi:10.1126/science.1202529

Paolicelli R, Sierra A, Stevens B, Tremblay ME, Aguzzi A,
Ajami B, Amit I, Audinat E, Bechmann I, Bennett M,
et al. 2022. Microglia states and nomenclature: a field at
its crossroads. *Neuron* **110**: 3458–3483. doi:10.1016/j
.neuron.2022.10.020

Papapetropoulos S, Pontius A, Finger E, Karrenbauer V,
Lynch DS, Brennan M, Zappia S, Koehler W, Schoels L,
Hayer SN, et al. 2022. Adult-onset leukoencephalopathy
with axonal spheroids and pigmented glia: review of clinical manifestations as foundations for therapeutic development. *Front Neurol* **12**: 788168. doi:10.3389/fneur.2021
.788168

Parihar A, Eubank TD, Doseff AI. 2010. Monocytes and macrophages regulate immunity through dynamic networks of
survival and cell death. *J Innate Immun* **2**: 204–215. doi:10
.1159/000296507

Patkar OL, Caruso M, Teakle N, Keshvari S, Bush SJ, Pridans
C, Belmer A, Summers KM, Irvine KM, Hume DA. 2021.
Analysis of homozygous and heterozygous Csf1r knockout in the rat as a model for understanding microglial
function in brain development and the impacts of human
CSF1R mutations. *Neurobiol Dis* **151**: 105268. doi:10
.1016/j.nbd.2021.105268

Peng J, Liu Y, Umpierre AD, Xie M, Tian DS, Richardson JR,
Wu LJ. 2019. Microglial P2Y12 receptor regulates ventral
hippocampal CA1 neuronal excitability and innate fear in
mice. *Mol Brain* **12**: 71. doi:10.1186/s13041-019-0492-x

Popova G, Soliman SS, Kim CN, Keefe MG, Hennick KM, Jain
S, Li T, Tejera D, Shin D, Chhun BB, et al. 2021. Human
microglia states are conserved across experimental models

and regulate neural stem cell responses in chimeric organoids. *Cell Stem Cell* **28:** 2153–2166.e6. doi:10.1016/j.stem.2021.08.015

Qiu J, Dando O, Baxter PS, Hasel P, Heron S, Simpson TI, Hardingham GE. 2018. Mixed-species RNA-seq for elucidation of non-cell-autonomous control of gene transcription. *Nat Protoc* **13:** 2176–2199. doi:10.1038/s41596-018-0029-2

Rahman MF, McGowan PO. 2022. Cell-type-specific epigenetic effects of early life stress on the brain. *Transl Psychiatry* **12:** 326. doi:10.1038/s41398-022-02076-9

Ramaglia V, Sheikh-Mohamed S, Legg K, Park C, Rojas OL, Zandee S, Fu F, Ornatsky O, Swanson EC, Pitt D, et al. 2019. Multiplexed imaging of immune cells in staged multiple sclerosis lesions by mass cytometry. *eLife* **8:** e48051. doi:10.7554/eLife.48051

Réu P, Khosravi A, Bernard S, Mold JE, Salehpour M, Alkass K, Perl S, Tisdale J, Possnert G, Druid H, et al. 2017. The lifespan and turnover of microglia in the human brain. *Cell Rep* **20:** 779–784. doi:10.1016/j.celrep.2017.07.004

Rojo R, Raper A, Ozdemir DD, Lefevre L, Grabert K, Wollscheid-Lengeling E, Bradford B, Caruso M, Gazova I, Sánchez S, et al. 2019. Deletion of a Csf1r enhancer selectively impacts CSF1R expression and development of tissue macrophage populations. *Nat Commun* **10:** 3215. doi:10.1038/s41467-019-11053-8

Ruan C, Sun L, Kroshilina A, Beckers L, De Jager P, Bradshaw EM, Hasson SA, Yang G, Elyaman W. 2020. A novel Tmem119-tdTomato reporter mouse model for studying microglia in the central nervous system. *Brain Behav Immun* **83:** 180–191. doi:10.1016/j.bbi.2019.10.009

Sabate-Soler S, Nickels SL, Saraiva C, Berger E, Dubonyte U, Barmpa K, Lan YJ, Kouno T, Jarazo J, Robertson G, et al. 2022. Microglia integration into human midbrain organoids leads to increased neuronal maturation and functionality. *Glia* **70:** 1267–1288. doi:10.1002/glia.24167

Sankowski R, Böttcher C, Masuda T, Geirsdottir L, Sagar, Sindram E, Seredenina T, Muhs A, Scheiwe C, Shah MJ, et al. 2019. Mapping microglia states in the human brain through the integration of high-dimensional techniques. *Nat Neurosci* **22:** 2098–2110. doi:10.1038/s41593-019-0532-y

Sasmono RT, Oceandy D, Pollard JW, Tong W, Pavli P, Wainwright BJ, Ostrowski MC, Himes SR, Hume DA. 2003. A macrophage colony-stimulating factor receptor-green fluorescent protein transgene is expressed throughout the mononuclear phagocyte system of the mouse. *Blood* **101:** 1155–1163. doi:10.1182/blood-2002-02-0569

Schafer DP, Lehrman EK, Kautzman AG, Koyama R, Mardinly AR, Yamasaki R, Ransohoff RM, Greenberg ME, Barres BA, Stevens B. 2012. Microglia sculpt postnatal neural circuits in an activity and complement-dependent manner. *Neuron* **74:** 691–705. doi:10.1016/j.neuron.2012.03.026

Shabestari SK, Morabito S, Danhash EP, McQuade A, Ramirez Sanchez J, Miyoshi E, Chadarevian JP, Claes C, Coburn MA, Hasselmann J, et al. 2022. Absence of microglia promotes diverse pathologies and early lethality in Alzheimer's disease mice. *Cell Rep* **39:** 110961. doi:10.1016/j.celrep.2022.110961

Shemer A, Scheyltjens I, Frumer GR, Kim JS, Grozovski J, Ayanaw S, Dassa B, Van Hove H, Chappell-Maor L,

Boura-Halfon S, et al. 2020. Interleukin-10 prevents pathological microglia hyperactivation following peripheral endotoxin challenge. *Immunity* **53:** 1033–1049.e7. doi:10.1016/j.immuni.2020.09.018

Shiraki N, Maruyama K, Hayashi R, Oguchi A, Murakawa Y, Katayama T, Takigawa T, Sakimoto S, Quantock AJ, Tsujikawa M, et al. 2022. PAX6-positive microglia evolve locally in hiPSC-derived ocular organoids. *Stem Cell Rep* **17:** 221–230. doi:10.1016/j.stemcr.2021.12.009

Sierra A, Encinas JM, Deudero JJ, Chancey JH, Enikolopov G, Overstreet-Wadiche LS, Tsirka SE, Maletic-Savatic M. 2010. Microglia shape adult hippocampal neurogenesis through apoptosis-coupled phagocytosis. *Cell Stem Cell* **7:** 483–495. doi:10.1016/j.stem.2010.08.014

Smith AM, Davey K, Tsartsalis S, Khozoie C, Fancy N, Tang SS, Liaptsi E, Weinert M, McGarry A, Muirhead RCJ, et al. 2022. Diverse human astrocyte and microglial transcriptional responses to Alzheimer's pathology. *Acta Neuropathol* **143:** 75–91. doi:10.1007/s00401-021-02372-6

Stables J, Green EK, Sehgal A, Patkar OL, Keshvari S, Taylor I, Ashcroft ME, Grabert K, Wollscheid-Lengeling E, Szymkowiak S, et al. 2022. A kinase-dead *Csf1r* mutation associated with adult-onset leukoencephalopathy has a dominant inhibitory impact on CSF1R signalling. *Development* **149:** dev200237. doi:10.1242/dev.200237

Stogsdill JA, Kim K, Binan L, Farhi SL, Levin JZ, Arlotta P. 2022. Pyramidal neuron subtype diversity governs microglia states in the neocortex. *Nature* **608:** 750–756. doi:10.1038/s41586-022-05056-7

Studer L, Vera E, Cornacchia D. 2015. Programming and reprogramming cellular age in the era of induced pluripotency. *Cell Stem Cell* **16:** 591–600. doi:10.1016/j.stem.2015.05.004

Tay TL, Mai D, Dautzenberg J, Fernández-Klett F, Lin G, Sagar, Datta M, Drougard A, Stempfl T, Ardura-Fabregat A, et al. 2017. A new fate mapping system reveals context-dependent random or clonal expansion of microglia. *Nat Neurosci* **20:** 793–803. doi:10.1038/nn.4547

Tcw J, Qian L, Pipalia NH, Chao MJ, Liang SA, Shi Y, Jain BR, Bertelsen SE, Kapoor M, Marcora E, et al. 2022. Cholesterol and matrisome pathways dysregulated in astrocytes and microglia. *Cell* **185:** 2213–2233.e25. doi:10.1016/j.cell.2022.05.017

Thion MS, Low D, Silvin A, Chen J, Grisel P, Schulte-Schrepping J, Blecher R, Ulas T, Squarzoni P, Helffel G, et al. 2018. Microbiome influences prenatal and adult microglia in a sex-specific manner. *Cell* **172:** 500–516.e16. doi:10.1016/j.cell.2017.11.042

Tipton PW, Kenney-Jung D, Rush BK, Middlebrooks EH, Nascene D, Singh B, Holtan S, Ayala E, Broderick DF, Lund T, et al. 2021. Treatment of *CSF1R*-related leukoencephalopathy: breaking new ground. *Mov Disord* **36:** 2901–2909. doi:10.1002/mds.28734

Tremblay ME, Lowery RL, Majewska AK. 2010. Microglial interactions with synapses are modulated by visual experience. *PLoS Biol* **8:** e1000527. doi:10.1371/journal.pbio.1000527

Trudler D, Nazor KL, Eisele YS, Grabauskas T, Dolatabadi N, Parker J, Sultan A, Zhong Z, Goodwin MS, Levites Y, et al. 2021. Soluble α–synuclein-antibody complexes activate the NLRP3 inflammasome in hiPSC-derived microglia.

Proc Natl Acad Sci **118:** e2025847118. doi:10.1073/pnas
.2025847118

Tsai AP, Lin PB, Dong C, Moutinho M, Casali BT, Liu Y, Lamb
BT, Landreth GE, Oblak AL, Nho K. 2021. INPP5D expression is associated with risk for Alzheimer's disease and
induced by plaque-associated microglia. *Neurobiol Dis*
153: 105303. doi:10.1016/j.nbd.2021.105303

Umpierre AD, Wu LJ. 2021. How microglia sense and regulate neuronal activity. *Glia* **69:** 1637–1653. doi:10.1002/
glia.23961

van der Poel M, Ulas T, Mizee MR, Hsiao CC, Miedema SSM,
Adelia, Schuurman KG, Helder B, Tas SW, Schultze JL,
et al. 2019. Transcriptional profiling of human microglia
reveals grey-white matter heterogeneity and multiple scle-
rosis-associated changes. *Nat Commun* **10:** 1139. doi:10
.1038/s41467-019-08976-7

Van Hove H, Martens L, Scheyltjens I, De Vlaminck K,
Pombo Antunes AR, De Prijck S, Vandamme N, De Schepper S, Van Isterdael G, Scott CL, et al. 2019. A single-cell
atlas of mouse brain macrophages reveals unique transcriptional identities shaped by ontogeny and tissue environment. *Nat Neurosci* **22:** 1021–1035. doi:10.1038/
s41593-019-0393-4

Venneti S, Wang G, Nguyen J, Wiley CA. 2008. The positron
emission tomography ligand DAA1106 binds with high
affinity to activated microglia in human neurological disorders. *J Neuropathol Exp Neurol* **67:** 1001–1010. doi:10
.1097/NEN.0b013e318188b204

Wang Y, Szretter KJ, Vermi W, Gilfillan S, Rossini C, Cella M,
Barrow AD, Diamond MS, Colonna M. 2012. IL-34 is a
tissue-restricted ligand of CSF1R required for the development of Langerhans cells and microglia. *Nat Immunol* **13:**
753–760. doi:10.1038/ni.2360

Wang C, Yue H, Hu Z, Shen Y, Ma J, Li J, Wang XD, Wang L,
Sun B, Shi P, et al. 2020. Microglia mediate forgetting via
complement-dependent synaptic elimination. *Science*
367: 688–694. doi:10.1126/science.aaz2288

Wendeln AC, Degenhardt K, Kaurani L, Gertig M, Ulas T, Jain
G, Wagner J, Häsler LM, Wild K, Skodras A, et al. 2018.
Innate immune memory in the brain shapes neurological
disease hallmarks. *Nature* **556:** 332–338. doi:10.1038/
s41586-018-0023-4

Wightman DP, Jansen IE, Savage JE, Shadrin AA, Bahrami
S, Holland D, Rongve A, Børte S, Winsvold BS, Drange
OK, et al. 2021. A genome-wide association study
with 1,126,563 individuals identifies new risk loci for
Alzheimer's disease. *Nat Genet* **53:** 1276–1282. doi:10
.1038/s41588-021-00921-z

Wirenfeldt M, Dissing-Olesen L, Anne Babcock A, Nielsen
M, Meldgaard M, Zimmer J, Azcoitia I, Leslie RG, Dag-
naes-Hansen F, Finsen B. 2007. Population control of resident and immigrant microglia by mitosis and apoptosis.
Am J Pathol **171:** 617–631. doi:10.2353/ajpath.2007
.061044

Wlodarczyk A, Holtman IR, Krueger M, Yogev N, Bruttger J,
Khorooshi R, Benmamar-Badel A, de Boer-Bergsma JJ,
Martin NA, Karram K, et al. 2017. A novel microglial sub-
set plays a key role in myelinogenesis in developing brain.
EMBO J **36:** 3292–3308. doi:10.15252/embj.201696056

Xu R, Boreland AJ, Li X, Erickson C, Jin M, Atkins C, Pang ZP,
Daniels BP, Jiang P. 2021. Developing human pluripotent
stem cell-based cerebral organoids with a controllable microglia ratio for modeling brain development and pathology. *Stem Cell Rep* **16:** 1923–1937. doi:10.1016/j.stemcr
.2021.06.011

Yuan P, Condello C, Keene CD, Wang Y, Bird TD, Paul SM,
Luo W, Colonna M, Baddeley D, Grutzendler J. 2016.
TREM2 haplodeficiency in mice and humans impairs
the microglia barrier function leading to decreased amyloid compaction and severe axonal dystrophy. *Neuron* **90:**
724–739. doi:10.1016/j.neuron.2016.05.003

Zhang GX, Li J, Ventura E, Rostami A. 2002. Parenchymal
microglia of naïve adult C57BL/6J mice express high levels of B7.1, B7.2, and MHC class II. *Exp Mol Pathol* **73:** 35–
45. doi:10.1006/exmp.2002.2441

Zhang Y, Chen K, Sloan SA, Bennett ML, Scholze AR,
O'Keeffe S, Phatnani HP, Guarnieri P, Caneda C, Ruderisch N, et al. 2014. An RNA-sequencing transcriptome and
splicing database of glia, neurons, and vascular cells of the
cerebral cortex. *J Neurosci* **34:** 11929–11947. doi:10.1523/
JNEUROSCI.1860-14.2014

Zhang X, Kracht L, Lerario AM, Dubbelaar ML, Brouwer N,
Wesseling EM, Boddeke E, Eggen BJL, Kooistra SM. 2022.
Epigenetic regulation of innate immune memory in microglia. *J Neuroinflammation* **19:** 111. doi:10.1186/s12974-
022-02463-5

Zusso M, Methot L, Lo R, Greenhalgh AD, David S, Stifani S.
2012. Regulation of postnatal forebrain amoeboid microglial cell proliferation and development by the transcription factor Runx1. *J Neurosci* **32:** 11285–11298.
doi:10.1523/JNEUROSCI.6182-11.2012

Role of Microglia in Central Nervous System Development and Plasticity

Dorothy P. Schafer,[1] Beth Stevens,[2] Mariko L. Bennett,[3,4] and Frederick C. Bennett[4,5]

[1]Department of Neurobiology, Brudnick Neuropsychiatric Research Institute, University of Massachusetts Chan Medical School, Worcester, Massachusetts 01605, USA

[2]Department of Neurology, F.M. Kirby Neurobiology Center, Boston Children's Hospital, Harvard Medical School, Boston, Massachusetts 02115, USA; Broad Institute of MIT and Harvard, Cambridge, Massachusetts 02142, USA; Howard Hughes Medical Institute, New York, New York 10032, USA

[3]Department of Neurology, Perelman School of Medicine, University of Pennsylvania, Philadelphia, Pennsylvania 19104, USA

[4]Division of Neurology, Children's Hospital of Philadelphia, University of Pennsylvania, Philadelphia, Pennsylvania 19104, USA

[5]Department of Psychiatry, Perelman School of Medicine, University of Pennsylvania, Philadelphia, Pennsylvania 19104, USA

Correspondence: Beth.Stevens@childrens.harvard.edu

The nervous system comprises a remarkably diverse and complex network of cell types, which must communicate with one another with speed, reliability, and precision. Thus, the developmental patterning and maintenance of these cell populations and their connections with one another pose a rather formidable task. Emerging data implicate microglia, the resident myeloid-derived cells of the central nervous system (CNS), in spatial patterning and synaptic wiring throughout the healthy, developing, and adult CNS. Importantly, new tools to specifically manipulate microglia function have revealed that these cellular functions translate, on a systems level, to effects on overall behavior. In this review, we give a historical perspective of work to identify microglia function in the healthy CNS, and highlight exciting new discoveries about their contributions to CNS development, maintenance, and plasticity.

Microglia, long enigmatic and understudied brain resident macrophages, are now recognized to play important roles in brain development and disease.[6] In addition to classical macrophage functions such as responding to damage, microglia are essential to the formation and regulation of neural circuits (Li and Barres 2018). Enthusiasm for microglia was galvanized by pioneering imaging studies showing that the processes of "resting" microglia are highly motile, surveying the entire brain parenchyma within a matter of hours (Davalos et al. 2005; Nimmerjahn et al. 2005). These findings led to an explosion of interest in the functional roles of microglia in the developing central nervous system (CNS), enabled by the development of new

---

genetic, cell transplantation, pharmacological, and culture tools. Herein we review research on microglia function in CNS development and plasticity. Our goal is to give a comprehensive and critical perspective of this expanding field of research, its relevance to neuropsychiatric disease, and to introduce the latest strategies to manipulate microglia function.

MICROGLIA ONTOGENY

The distinct ontogeny of microglia underpins their important roles in development. Unlike neurons and other glia, microglia belong to the hematopoietic system. Although most blood cells differentiate from hematopoietic stem cells (HSCs), microglia precede their existence, arising from progenitors generated during primitive hematopoiesis, an embryonic process that generates macrophages and nucleated red blood cells, first in the yolk sac (Dzierzak and Bigas 2018). In rodents, yolk sac–derived microglial progenitors migrate to the fetal head early in the second embryonic week (Ginhoux et al. 2010), just as the neural tube begins to form, and progressively specialize to the brain environment (Matcovitch-Natan et al. 2016), the product of bidirectional signals with developing neurons and glia (Han et al. 2023). This is the beginning of a lifelong relationship, because unlike macrophages in most other tissues, embryonic-origin microglia persist throughout the life span, typically with minimal replacement by infiltrating, monocyte-derived macrophages of bone marrow origin (Ginhoux and Guilliams 2016; Liu et al. 2019b). Instead, microglia are long-lived (up to 20 years in humans [Réu et al. 2017]) and sustain their numbers by self-renewal (Ajami et al. 2007). The distinct ontogeny of microglia means that they are present, responsive to, and influenced by neural cells throughout the course of development, as much brain as blood cells.

Precise details of microglial origin and the extent of blood monocyte infiltration over the life span, typically delineated using genetic fate mapping in murine models, are rapidly evolving, controversial, and do not apply to all species, as evidenced by the case of zebrafish (Ferrero et al. 2018). Excellent reviews address these details (Mass 2018; Mass et al. 2023) and controversies (Hume 2023) in the broader context of tissue-resident macrophage biology. Even since their publication our understanding has significantly evolved. For example, in a landmark human study, some cases of clonal hematopoiesis of indeterminate potential (CHIP), the age-associated formation of genetically distinct clonal subpopulations of blood cell progenitors, may protect against Alzheimer's disease, associated with a degree of brain monocyte infiltration (Bouzid et al. 2023) far greater than estimated by fate mapping studies. Another topic important to CNS development is the relationship between microglia and so-called "border-associated macrophages" (BAMs) of the meninges and perivascular spaces, where a recent study argues that microglia and meningeal macrophages share a common progenitor, while perivascular macrophages differentiate perinatally from the latter (Masuda et al. 2022). In summary, although many details are incompletely resolved, the unique ontogeny of microglia, characterized by embryonic origin, self-renewal, and persistence throughout the life span are critically important to understanding their impacts on brain development (Thion et al. 2018).

FUNCTIONS, STATES, AND AN EXPLOSION OF TOOLS

Microglia in the Single-Cell Era

Resident macrophages, including microglia, specialize to the cellular and molecular features of their environment (Okabe and Medzhitov 2016). It is increasingly clear that even within an organ, macrophages adopt context and localization-specific states to meet the demands of their tissues (Park et al. 2022). A major challenge to understanding the breadth of microglial functions, therefore, was an absence of methods to deeply phenotype their heterogeneity. Over the past decade, this has changed dramatically; with the arrival of single-cell techniques, particularly single-cell RNA sequencing, spatial transcriptomics, and high-dimensional flow cytometry, it is now possible to systematically catalog the spectrum of microglial states present in development, health, and disease. Moreover, protocols have been developed and optimized to minimize ex vivo artifacts of microglia isolation, improving the integration of diverse data sets and cross-species comparisons of microglia states in different brain regions and contexts (Marsh et al. 2022; Paolicelli et al. 2022). We are now able to conceptualize and, paired with new tools to isolate and manipulate them, interrogate microglia not only as a cell type, but at the level of their functional specializations (Fig. 1). This is particularly relevant during development, where microglia perform diverse and widely different functions across time and space. Importantly, since pathologic mechanisms of brain disease often recapitulate aspects of development, the field benefits from a rapidly expanding wealth of single-cell data across species, disease, and age that often share common features. For example, "disease-associated microglia," found using single-cell sequencing in a

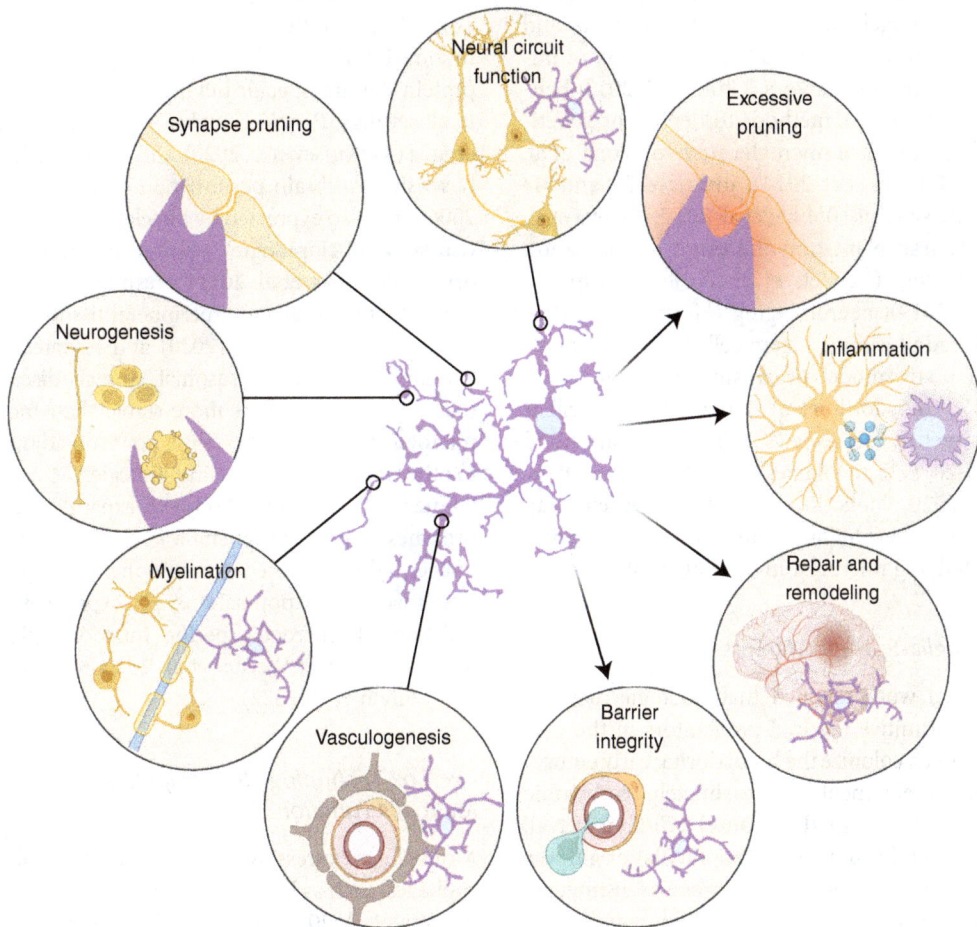

Figure 1. Microglia functional states. Microglia dynamically adopt varied states to perform diverse functions, including related to neural circuit regulation, immunoregulation, myelination, tissue repair/remodeling, barrier integrity, and vasculogenesis. It is unclear whether the same microglial cell can move in and out of these different states, or whether different populations of microglia are more dynamic than others (Paolicelli et al. 2022). Regardless, these functions can have beneficial and harmful effects on normal development and disease pathogenesis, often in a context-dependent manner. For example, appropriate synapse pruning facilitates normal circuit development, but exuberant pruning may underlie the pathogenesis of some neuropsychiatric diseases.

model of brain amyloid pathology (Keren-Shaul et al. 2017), overlap with microglia found in developing white matter tracts (Hammond et al. 2019), raising the possibility that shared functional demands induce their formation. Critical new tools and approaches greatly enhance the study of microglia in the single-cell era, setting the stage for increasingly sophisticated linking of transcriptional states to function.

New Tools and Insights

Paradigm shifts in our understanding of microglia owe largely to new customized tools and approaches. These include the discovery of microglia-specific markers (Chiu et al. 2013; Bennett et al. 2016), methods to deplete or genetically manipulate microglia in vivo (Yona et al. 2013; Elmore et al. 2014), advanced 2D and 3D culture systems (Bohlen et al. 2017; Popova et al. 2021), transplantation and xenotransplantation approaches (Bennett et al. 2018; Hasselmann et al. 2019), neuroimaging (Horti et al. 2019), induced pluripotent stem cell (iPSC) microglia, and many model organisms (Gosselin et al. 2017; Geirsdottir et al. 2019). With excellent comprehensive reviews available (Hammond et al. 2018; Li and Barres 2018; Bennett and Bennett 2020; Paolicelli et al. 2022; Warden et al. 2023), we here highlight three discoveries critical to studying microglia in development.

Microglia-Specific Markers

Seminal work revealed that microglia develop from primitive myeloid progenitors in the yolk sac, which colonize the brain during early embryonic development (approximately embryonic day 9.5 [E9.5] in the mouse) (Ginhoux et al. 2010). As this unique developmental origin was recognized, so too were markers to distinguish microglia from other macrophage populations. The most common way to label microglia in mammalian tissue is by immunostaining with anti-Iba1 antibodies, which recognize a calcium-binding protein expressed by many tissue macrophages including microglia. Another popular marker is the fractalkine receptor, *CX3CR1*, which is robustly expressed by microglia (Jung

et al. 2000; Mizutani et al. 2012). Although also expressed by brain BAMs and Ly6C^{lo} circulating monocytes, CX3CR1-based tools have been key to fundamental discoveries related to microglial roles in synapse pruning during development (Paolicelli et al. 2011; Schafer et al. 2012) and remain widely accepted and used. Other microglia-specific marker genes include *Tmem119*, *P2ry12*, and *Hexb*. *Tmem119* is highly expressed by microglia and with high-quality commercial antibodies widely available, effectively distinguishes true parenchymal microglia from brain BAMs and circulating monocytes (Bennett et al. 2016). Importantly, *Tmem119* is down-regulated in some injuries and diseases, making TMEM119 protein staining specific but imperfectly sensitive in all settings. *P2ry12* is highly expressed by microglia (McKinsey et al. 2020), although it is highly sensitive to brain perturbation (Haynes et al. 2006) and also expressed by platelets, as the protein target of clopidogrel, a powerful antiplatelet drug (O'Connor et al. 2011). Hexb is highly expressed by microglia and peripheral tissue macrophages (Masuda et al. 2020) and mutated in Sandhoff disease, a lysosomal storage disease (Xiao et al. 2022). It is more stable than most microglia markers and can be a powerful adjunct to other more specific labeling strategies. Microglia markers, in combination with expanding approaches to label peripheral and border macrophages, allow nuanced immunophenotyping of CNS macrophage populations, and experimental isolation of microglia-specific functions (Saederup et al. 2010; Werner et al. 2020; Kim et al. 2021; Silvin et al. 2022).

Cre/Lox Technology to Manipulate Microglia Function

Mouse lines expressing Cre recombinase in macrophages, such as Iba1^{Cre} (Pozner et al. 2015; Nakayama et al. 2018) and CD11b^{Cre} (Ferron and Vacher 2005), allow manipulation of microglia, but also target other myeloid-derived cells throughout the body. Microglia-specific gene expression analyses have since enabled the development of more specific genetic tools to manipulate microglia function in vivo, starting with Cx3cr1^{CreER} knockin mice (Yona et al. 2013). Al-

 Cite this article as *Cold Spring Harb Perspect Biol* doi: 10.1101/cshperspect.a041810

though CX3CR1 is widely expressed by macrophages and some monocytes, most turn over within a few weeks due to myelopoiesis, while microglia are long-lived. As a consequence, only microglia and other long-lived macrophage populations demonstrate recombination after ~3 weeks. More recently, microglia-specific markers spurred the development of Hexb[CreER] (Masuda et al. 2020), P2ry12[CreER] (McKinsey et al. 2020), and Tmem119[CreER] (Kaiser and Feng 2019) mice. While the recombination is less efficient in these mice compared to Cx3cr1[CreER] mice, they are more specific to microglia and there is less spurious recombination in the absence of tamoxifen (Faust et al. 2023; Bedolla et al. 2024).

To increase microglia-specificity, a clever combinatorial technique is the use of a "split" or "binary" Cre in which the two domains of Cre recombinase are encoded by two separate promoters. Such a strategy, relying on coexpression of Sall1 and Cx3cr1, very specifically targets microglia (Kim et al. 2021). These mice offer an additional approach, albeit with reduced efficiency, to achieve microglia-specific genetic manipulation without the need for tamoxifen.

Microglia Depletion

Donor-derived macrophages appear in the brain parenchyma after hematopoietic stem cell transplantation (HSCT) suggesting that microglia may be ablated and replaced by infiltrating macrophages (Hoogerbrugge et al. 1988; Unger et al. 1993). Indeed, direct head irradiation or high-dose busulfan chemotherapy led to widespread donor cell engraftment in direct proportion to microglial loss (Mildner et al. 2007; Capotondo et al. 2012). While donor-derived macrophages share most features with endogenous microglia, there exist important transcriptional differences (Bennett et al. 2018). These early observations foreshadowed an explosion in methods to ablate microglia (Han et al. 2019) chemically, pharmacologically, or with mouse transgenic systems.

One striking demonstration of depletion involves targeting CSF1R, a receptor required for microglial survival upon binding to the ligands CSF1 or IL-34 (Stanley and Chitu 2014). Oral treatment with brain-penetrant CSF1R inhibi-

tors (e.g., PLX3397 or 5622) depletes more than 90% of endogenous microglia (Elmore et al. 2014; Huang et al. 2018). Upon treatment cessation, however, remaining microglia rapidly proliferate and return to normal density and morphology within 2 weeks. These findings demonstrate key features of induced microglia ablation—it is never complete and microglia rapidly repopulate from residual cells after inhibitor removal (Basilico et al. 2022).

SPATIAL PATTERNING IN THE DEVELOPING AND MATURE CNS

Being early CNS residents (Ginhoux et al. 2010), microglia are poised to shape the developing CNS. As professional phagocytes, they efficiently clear apoptotic cells and debris, key to successful nervous system development (Wyss-Coray and Mucke 2002; Napoli and Neumann 2009; Ransohoff and Perry 2009; Ransohoff and Cardona 2010; Kettenmann et al. 2011; Fourgeaud et al. 2016). Microglia also produce factors that trigger neuronal apoptosis, including tumor necrosis factor α (TNF-α), reactive oxygen species (ROS), and glutamate (Bessis et al. 2007), and they can spur astrocytes to release neurotoxic saturated lipids (Guttenplan et al. 2021). These data suggest that microglia play a more active role in regulating developmental apoptosis. Recent work in multiple disease states (glioma, brain ischemia, Alzheimer's disease, etc.) has shown that microglia phagocytose live cells, including viable but stressed neurons, via multiple phagocytic pathways (Brown and Neher 2014). In sum, microglia play diverse roles in patterning and maintaining the cell populations that comprise the CNS. The following section reviews studies demonstrating a role for microglia in regulating CNS cell numbers during development and in adulthood.

Microglia Engulfment Mechanisms: Regulation of Cell Death

Essential to the development and spatial patterning of all organ systems, and highly conserved across vertebrate and invertebrate species, is programmed cell death (PCD) (Sulston and Horvitz

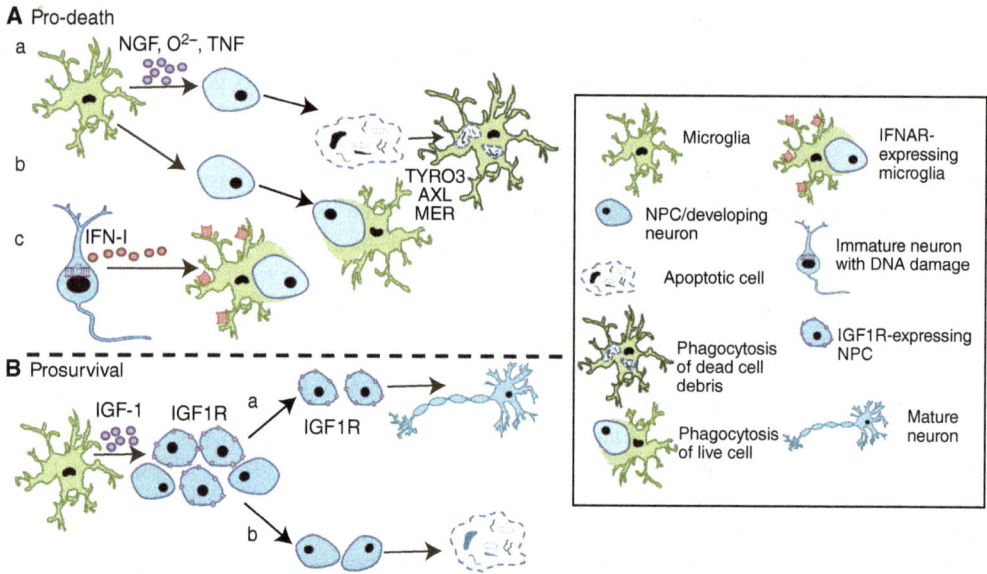

Figure 2. Mechanisms involved in microglia-mediated spatial patterning of neurons. (*A*) The role of microglia in programmed cell death. (*a,b*) Microglia can initiate the cell death program (*a*) or render neurons vulnerable (*b*) through a variety of soluble (e.g., nerve growth factor [NGF], O2⁻, and tumor necrosis factor [TNF]) and membrane-bound (e.g., CD11b and DAP12) factors followed by phagocytosis, commonly using TAM (TYRO2, AXL, MER) family receptors. (*c*) Another example of "phagoptosis" has been described in which a subset of immature neurons in the developing brain have DNA damage that induces type I interferon (IFN) production, and microglia subsequently clear these damaged neurons downstream from interferon receptor (IFNAR) signaling. (*B*) Microglia can also promote the survival of neural progenitor cells (NPCs) in the developing central nervous system (CNS). One mechanism identified was insulin-like growth factor (IGF)-1, a soluble factor made and released by microglia, which is believed to bind IGF-1 receptors expressed by a subset of NPCs. It is speculated that those NPCs expressing the receptor survive (*a*), whereas those that do not express the receptor undergo apoptosis (*b*).

1977; Abrams et al. 1993; Vaux and Korsmeyer 1999; Bangs and White 2000). During vertebrate development, 50% of neurons born undergo PCD and are cleared, a massive and necessary task to maintain CNS homeostasis.

The best-characterized role for microglia in PCD is their large capacity to phagocytose dead cell corpses during development (Fig. 2Aa). One of the first descriptions of microglia in this role was in the postnatal rat cerebral cortex, where, during the first week of rat postnatal development, a large amount of neuronal cell death occurs, particularly in the subplate and layers II/III (Ferrer et al. 1990). During this time, globular, vacuolated phagocytes, which are likely resident microglial cells, were found in increased numbers and had engulfed dead cells throughout the cortical and subplate regions. Since these origi-

nal observations, microglia have been shown to phagocytose dead cells throughout the healthy, developing CNS (Bessis et al. 2007; Sierra et al. 2013). It has been further identified that Tyro3, Axl, and Mer (TAM) receptor signaling is a critical mechanism regulating this clearance of dead and dying cells by microglia in the developing neurogenic niches of the rodent brain. For example, in the subventricular zone (SVZ) and rostral migratory stream (RMS), it was shown that mice deficient in TAM receptors Axl and Mertk ($Axl^{-/-}$; $Mertk^{-/-}$ mice) had defects in clearance of cleaved caspase 3–positive cells in neurogenic niches (Fourgeaud et al. 2016). Interestingly, although PCD and microglial clearance of neuronal corpses occur beyond these neurogenic niches in the CNS (Cunningham et al. 2013), PCD and clearance of corpses in these CNS regions were not TAM receptor-dependent. Clues into other non-TAM receptor-mediated mechanisms could be garnered from invertebrates and teleosts where glial clearance of corpses has been elegantly defined. For example, in *Drosophila melanogaster* and *Caenorhabditis elegans*, dead cell corpses are removed by phagocytic cells through a Draper/dCED-6-dependent signaling pathway (Zhou et al. 2001; Freeman et al. 2003; Tasdemir-Yilmaz and Freeman 2014). In addition, microglia engulf apoptotic neurons in zebrafish via phosphatidylserine receptors BaI1 and TIM-4 (Mazaheri et al. 2014). While microglia deficient in these receptors in zebrafish can recognize their targets, they fail to properly form phagosomes around their targets with BaI1 deficiency and they fail to stabilize the phagosomes with TIM-4 deficiency.

Besides playing an important role in the removal of large amounts of cellular debris produced by PCD, recent work suggests a more active role by which microglia initiate and/or propagate the cell death program before phagocytosis (Fig. 2Ac). The concept of phagocytes initiating cell death before phagocytosis or "phagoptosis" was first conceived in the late 1800s by Elie Metchnikoff (Aliprantis et al. 1996). In the nervous system, it has been shown in the developing rodent retina that genetic depletion of macrophages resulted in a persistence of ocular tissues that are normally developmentally transient, the hyaloid vasculature and pupillary membrane (Lang and Bishop 1993). Similarly, in the chick retina, optic cups grown in the absence of microglia resulted in reduced retinal cell death (Frade and Barde 1998). When purified microglia were added back to these chick optic cup cultures, retinal cell death was increased. The molecular mechanism underlying this effect was identified as microglial-derived nerve growth factor (NGF). Using a similar in vitro approach, microglia were also shown to initiate the PCD of Purkinje neurons in the developing mouse cerebellum and motoneurons in the developing rat spinal cord (Marín-Teva et al. 2004; Sedel et al. 2004). In this context, the microglia-derived cytotoxic factor was identified as superoxide ions (O_2^-) produced during respiratory bursts. It was further shown that

Purkinje neurons expressed molecules associated with damage/PCD (i.e., activated caspase 3) before engulfment. These data suggest a role for microglia in driving the cell death program in neurons that are already rendered vulnerable, a concept that has been suggested in other model organisms (*C. elegans* and *Drosophila*) where dying, but not yet dead, cells are engulfed by neighboring cells, including glia (Logan and Freeman 2007; Kinet and Shaham 2014). However, there is also evidence that microglia can render neurons more competent to die, too. For example, using cultured spinal cord explants from developing rat embryos, microglial ablation, or blocking microglia-derived TNF-α signaling resulted in a decrease in motoneuron PCD (Sedel et al. 2004). In this study, microglial TNF-α expression was transient and down-regulated before the cell death, data suggesting that microglial-derived TNF-α may not directly induce cell death but rather render motoneurons more competent to die.

These initial studies suggested a more active role for microglia in PCD, but they were all performed in vitro or ex vivo. To better address this question under more physiological conditions, other studies have taken advantage of strategies to genetically and/or pharmacologically manipulate microglia in vivo. One of the first of these in vivo studies assessed PCD in the neonatal hippocampus, where microglia were found to associate with neurons undergoing apoptosis (Wakselman et al. 2008). To determine an underlying mechanism, the investigators assessed mice deficient in CD11b (CD11b2/2), a surface receptor integrin, and mice with loss of function in DAP12 (DAP12KI), a transmembrane signal transduction adaptor molecule for the triggering receptor expressed on myeloid cells 2 (TREM2). These two molecules are expressed by peripheral immune cells, including macrophages, where they carry out effector functions such as cell adhesion and phagocytosis (Mócsai et al. 2006). Importantly, in the context of normal development, these receptors are almost exclusively expressed by microglia in the CNS. In CD11b2/2 and DAP12KI mice, there was a significant, although not complete, reduction in neuronal apoptosis. Intriguingly, the investigators provided evidence that CD11b and DAP12 regulate the production of O_2^- by microglia, which was shown earlier to regulate PCD via microglia in cultured cerebellar slices (Marín-Teva et al. 2004). Similarly, another study showed that administration of broad-spectrum tetracyclines that typically dampen the immune response (minocycline or doxycycline) or depletion of microglia by injection of liposomal clodronate into the ventricles resulted in an increase in neural progenitor cells (NPCs) in the brain later in development (Cunningham et al. 2013). Conversely, elevating the immune response with lipopolysaccharide (LPS) injection resulted in a decreased NPC number. These data suggest that microglia may actively regulate NPC numbers in the developing brain by inducing apoptosis of NPCs followed by phagocytosis (Fig. 2Aa). The possibility still exists that this could be a neurogenic effect, which is consistent with another study in which doxycycline in adult mice induced an increase in hippocampal neurogenesis (Sultan et al. 2013). Another important aspect to consider is that the pharmacological agents used are relatively nonspecific to microglia and affect several other cell types outside and inside the CNS (Smilack 1999; Hagberg and Mallard 2005; Bilbo et al. 2006; Buller et al. 2009; Meyer et al. 2009). To address these caveats, a recent study identified I interferon (IFN)-responsive microglia in the developing mouse somatosensory cortex that engulf whole, intact neurons (Fig. 2Ac). They further showed that neurons with nuclear DNA damage accumulate in the setting of microglia-specific interferon receptor (IFNAR) loss (Escoubas et al. 2024). Overall, these studies demonstrate that microglia are critical for initiating and executing the cell death program in a vulnerable population of neurons with DNA damage in the developing brain.

Beyond neurons, microglia can control PCD of other glial cells. For example, pharmacological ablation of microglia in vivo resulted in decreased ability of astrocytes to undergo developmental PCD in the retina (Puñal et al. 2019). Interestingly, similar to neurons, astrocyte PCD was reduced, but still observed upon microglia ablation. Thus, although microglia may

play a role in the initiation of cell death in some cells, they may also be involved in facilitating the progression of cell death in another subset that are already rendered vulnerable by other mechanisms. In fact, when microglia are pharmacologically ablated with a CSF1R inhibitor, astrocytes can then take over the job of clearing dead cell corpses (Konishi et al. 2020).

In addition to development, microglia have now been implicated in PCD that occurs throughout the life of an organism. This work has centered around the hippocampus, where neurogenesis occurs in the subgranular zone of the dentate gyrus during adulthood. Although a small subset of these newly born neurons incorporates into the circuitry, the remaining cells undergo PCD (Song et al. 2012). Using a combination of immunohistochemistry, transgenic green fluorescent protein (GFP)-expressing mice, and bromodeoxyuridine labeling, it was shown that most newly born neurons in the subgranular zone of 1-month-old mice die within the first 4 days of birth (56% of newborn cells; 400 cells/h) and phenotypically "resting" microglia appear to efficiently phagocytose these apoptotic cell bodies (90% of apoptotic cells in the subgranular zone) within this 4-day period (Sierra et al. 2010). Furthermore, microglia were shown to regulate their phagocytic capacity in response to changes in neurogenesis and cell death induced by age (decrease) or LPS-mediated inflammation (increase).

Although this study demonstrated a role for microglia in the elimination of apoptotic NPCs, another study has shown that NPCs themselves can phagocytose dying cells during adult hippocampal neurogenesis (Lu et al. 2011). Thus, it appears that microglia and NPCs may be working together to eliminate apoptotic NPCs. In addition, unlike the developing CNS, it remains unclear whether microglia and/or neighboring NPCs are involved in inducing apoptosis in these newly born neurons in the adult hippocampus. Evidence that apoptotic neurons exist in the absence of microglial or NPC contact suggests that this may be a neuron-autonomous event followed by phagocytosis. However, there is some evidence in the adult mouse dentate gyrus that microglia infiltrate small cytoplasmic openings of granule cells before cell death and that these microglial processes closely oppose the nuclei of granule cells, data suggesting a more active role in initiating or contributing to progression of the death program (Ribak et al. 2009). Similarly, there is evidence that microglia engulf live newborn neurons in the dentate gyrus of the hippocampus following seizures (Luo et al. 2016).

Regulation of Cell Survival, Proliferation, and Differentiation

Microglia also contribute to spatial patterning of the developing and adult CNS by promoting cell survival, proliferation, and/or differentiation (Fig. 2B). Some of the first lines of evidence for this were in vitro studies where microglia-conditioned media was added to neuronal cultures, resulting in enhanced progenitor cell proliferation and enhanced neuron survival and/or maturation (Nagata et al. 1993; Chamak et al. 1994; Morgan et al. 2004). However, other recent work has shown that culturing NPCs isolated from cortices that lack microglia (PU.1/2) had no effect on neuron survival or neurogenesis but rather resulted in decreases in NPC proliferation and astrogenesis, effects that were rescued by adding wild-type (WT) microglia back to the cultures (Antony et al. 2011). Conversely, another study cultured NPCs from adult hippocampus and observed a reduction in progenitor cell proliferation in the presence of microglia (Gebara et al. 2013). In the same study, the investigators used models of enhanced (exercise) or impaired (aging) hippocampal neurogenesis to show an inverse correlation between the number of microglia and the number of progenitor cells in fixed tissue. There are several explanations for these disparate results. The most plausible explanations are differences in culture conditions, ages of mice used for the study, and/or region-specific effects. Furthermore, although it is clear from all of these studies that microglia have the capacity to regulate progenitor cell numbers, in vivo strategies and careful developmental analyses are necessary.

Other work has addressed whether microglia promote neuronal survival in the developing mouse cortex (Fig. 2B; Ueno et al. 2013). Between P3 and P5, "inactivating" microglia with minocycline or ablating them by injecting diphtheria toxin into mice expressing the diphtheria toxin receptor under the control of the CD11b promoter, resulted in an increase in the number of apoptotic neurons specifically in layer V of the cerebral cortex. These manipulations resulted in an increase in the number of apoptotic neurons specifically in layer V of the cerebral cortex. Furthermore, microglia still appeared to engulf apoptotic debris after minocycline treatment, suggesting that microglial phagocytic capacity is not inhibited. Thus, the data suggest that microglia provide trophic support for neurons in the early postnatal CNS. Consistent with this idea, the investigators show that mice deficient in the fractalkine receptor (CX3CR1), which is expressed almost exclusively by microglia in the context of the postnatal CNS, have a similar reduction in layer V neurons. Furthermore, CX3CR1 knockout mice had a reduction in free insulin-like growth factor 1 (IGF-1) and an increase in IGF-1-binding proteins, data suggesting that IGF-1 may be a factor downstream from CX3CR1 that promotes neuron survival. When in vitro and in vivo approaches were used to block IGF signaling (pharmacology and small interfering RNA), there was a significant increase in cortical neuron cell death. However, none of the strategies were completely specific to microglia. Using a more microglia-specific CreER driver Cx3cr1CreER/+ another group more recently showed that microglia-derived brain-derived neurotrophic factor (BDNF) promoted adult neurogenesis in the hippocampus (Harley et al. 2021). Finally, it is important to consider putative cell adhesion mechanisms that could regulate neuronal survival via microglia. This is supported by a recent study that showed that the cell adhesion and axon guidance-related molecule netrin-G1 expressed by corticospinal tract (CST) neurons is critical for microglia to accumulate along axons of CST layer V neurons during development (Fujita et al. 2020).

Genetic ablation of netrin-G1 in layer V neurons resulted in a failure of postnatal microglia to accumulate around layer V neuron axons and increased layer V neuron death was observed. The authors further showed that microglia deficient in the netrin-G ligand-1 (NGL1) failed to accumulate around axons in vitro. Therefore, in addition to the mechanisms described above, cell adhesion between neurons and microglia is likely another mechanism to promote neuronal survival. Future work using microglia-specific Cre/LoxP technology will be critical to gain more insight into the in vivo significance of this and other described microglial mechanisms to promote cell survival.

Taken together, these data demonstrate a role for microglia in the spatial patterning of nervous system tissue by clearing apoptotic cells (neurons and neural progenitors). In addition, experiments also suggest a role for microglia in either initiating the cell death program or progressing cell death in a neuron previously rendered vulnerable. At the same time, there is also new evidence that microglia can also promote neuron and NPC survival and/or proliferation. Given that these seemingly disparate functions (proapoptotic and prosurvival/proliferation) occur during very similar time frames, it remains unclear how microglia carry out these functions simultaneously and raises the intriguing possibility that microglia are a heterogeneous cell population with several different subtypes of cells performing vastly different functions. In addition, although microglia ablation during specific developmental time points has shown some modest effects on cell numbers, ablation of microglia in the adult brain has shown no significant effect (Parkhurst et al. 2013; Elmore et al. 2014). Several different explanations could account for this, such as the importance of context/timing; the developing CNS may be particularly sensitive to these manipulations. In addition, there may be overlapping functions with astrocytes, which can also phagocytose cells, and phagocytosis by astrocytes can compensate for microglia (Konishi et al. 2020; Zhou et al. 2022). Future work using more specific in vivo strategies to manipulate microglia is necessary. Pharmacological strategies (LPS, minocycline, and doxycycline) are nonspecific, with effects on many cell types. Furthermore, although data are promising in mice deficient in CD11b, DAP12, CX3CR1, netrin-G1

ligand, and IGF-1 signaling, other cell types also express these molecules and are, thus, affected by genetic targeting. Using approaches to ablate genes specifically in microglia, such as $Tmem119^{CreER}$, $Hexb^{CreER}$, $Sall1^{ncre}:Cx3cr1^{ccre}$, and $P2ry12^{CreER}$ mouse lines, should provide more concrete answers.

Role of Microglia in White Matter

The intimate relationship between microglia and developmental myelination has long been appreciated, with Penfield (1925) noting that "the three areas where microglia" are found early in development are in "areas in which myelinization is in progress" (Penfield 1925). Appearing just before myelination in mice, these axon tract-associated microglia are transcriptionally distinct from mature microglia (Hammond et al. 2019; Li et al. 2019). These CD11c[+], Clec7a-expressing microglia secrete IGF-1 to support myelination (Wlodarczyk et al. 2017), and engulf oligodendrocyte precursor cells (OPCs) (Nemes-Baran et al. 2020; Irfan et al. 2022) and newly formed oligodendrocytes (Hammond et al. 2019; Li et al. 2019). Slightly later in development, expression of microglial transglutaminase 2 (TGM2) engages OPC G-protein-coupled receptor 56 (GPR56) to support their proliferation and myelin formation (Giera et al. 2018). In zebrafish, microglia remodel myelin sheaths during development, with abnormal sheath formation after microglia depletion (Hughes and Appel 2020). The role of microglia in developmental myelination is further supported by a case report of a child lacking microglia due to compound heterozygous CSF1R mutations with evidence of periventricular white matter injury and agenesis of the corpus callosum (Oosterhof et al. 2019). Lack of microglia either due to CSF1R inhibition (BLZ945) or genetic knockout (Rojo et al. 2019; McNamara et al. 2023; Barclay et al. 2024) results in decreased OPC and mature oligodendrocyte numbers (Hagemeyer et al. 2017), although, of note, other inhibitors and genetic models showed either no effect of oligodendrocyte or OPC number (Liu et al. 2019a), or distinct regional differences (Nandi et al. 2012). In a complementary strategy, mice lacking microglia due to deletion of the

FIRE enhancer from intron 2 of the CSF1R locus show normal developmental myelin ensheathment but problems with myelin integrity and proper growth (McNamara et al. 2023). Together, these findings suggest there is more to discover about the relationship of microglia with oligodendrocyte lineage cells and myelin development.

Microglial support of myelin formation and integrity is lifelong. Microglial processes constantly survey mature myelin sheaths at nodes of Ranvier (Wu et al. 2022). Further, myelin contents are found within microglia located at paranodal regions (Hildebrand et al. 1993). Age-related defects in microglia-mediated myelin fragments clearance suggest a continuous process of non-cell-autonomous myelin curation (Safaiyan et al. 2016). Many clues to the importance of microglia in maintaining existing and supporting new myelin can be found in studies of injury and disease, for example, in the setting of remyelination as discussed in Sierra et al. (2024). How and when microglia support OPC and oligodendrogenesis, maintain myelin integrity, and whether microglia contribute to adaptive myelination are largely unknown, and represent exciting avenues for new discovery.

Role in Regulating Vasculature

There is clear data supporting a role for microglia at the vasculature and regulation of the blood–brain barrier during disease (Zhao et al. 2018; Borjini et al. 2019; Ronaldson and Davis 2020). Still, a large percentage of microglia are also associated with the vasculature in the healthy CNS. These vascular-associated microglia (VAMs) have also been called juxtavascular microglia and capillary-associated microglia (CAM) (Grossmann et al. 2002; Bisht et al. 2021). For simplicity, we will refer to them as VAMs. During development, almost half of all microglia are associated with the brain and retinal vasculature, particularly capillaries, during embryonic and postnatal development across rodents and human (Checchin et al. 2006; Monier et al. 2007; Mondo et al. 2020; Hattori 2022). In developing rats, it was further shown that microglia and neural precursor cells (NPCs) are highly associated with the endothelial cell processes in the ventricular zone (Penna et al. 2021). Live imaging has further revealed that these VAMs in the developing CNS migrate along the vasculature, which is enhanced after an acute traumatic injury (Grossmann et al. 2002; Smolders et al. 2017; Mondo et al. 2020). VAMs move along vessels in the developing cortex with a migratory trajectory and the timing of this migratory pattern aligns with synapse development (Mondo et al. 2020). It is, therefore, intriguing to consider that VAMs use the vasculature to migrate and colonize the developing brain at the appropriate time to regulate synapse development and pruning. Evidence further supports a tight association with vascular pericytes during embryogenesis (Hattori et al. 2022). In turn, pericytes appear to support the proliferation of microglia during embryogenesis as depletion of pericytes in embryos resulted in a decrease in microglia proliferation (Hattori et al. 2022). Together, these studies have sparked interest in understanding what specific functions microglia are performing to regulate vascular development.

Toward a functional role for VAMs in vascular development, when microglia are depleted from P1 rat retinas with clodronate liposomes, there is a reduction in vascular growth and density between P5 and P8 (Checchin et al. 2006; Kubota et al. 2009). This effect of microglia depletion on vascular growth was rescued when microglia were injected intraocularly. Microglia have further been shown to associate with endothelial tip cells at the sprouting edge of developing rodent retinal vasculature (Kubota et al. 2009; Fantin et al. 2010; Rymo et al. 2011). Similar to earlier work with clodronate-mediated depletion (Checchin et al. 2006), mice that lack all macrophages due to genetic or pharmacological disruption of mCSF or CSF1R, there is a defect in vascular branching and anastomosis in the retina. Another recent study showed that microglia can also promote developmental retinal angiogenesis through the production of transforming growth factor β1 (TGF-β1) and this TGF-β1 production is regulated through microglial detection of changing extracellular matrix (ECM) stiffness in the developing retina (Dudiki et al. 2020). Interestingly, one of these studies showed that defects in retinal vascular development due to microglia are largely resolved after 1 month of development, suggesting a transient effect of microglia/macrophages on vascular growth and complexity (Kubota et al. 2009).

Besides development, ~20% of total microglia have also been shown to associate with the vasculature in the healthy adult brain (Mondo et al. 2020; Bisht et al. 2021). These VAMs in the adult CNS are enriched at capillaries, they frequently contact the basal lamina generated by endothelial cells and pericytes, and their cell bodies reside at areas along the vasculature that are void of astrocyte endfeet (Mondo et al. 2020; Bisht et al. 2021; Császár et al. 2022). Emerging evidence also supports a functional role for microglia in the regulation of blood flow and neovascular coupling in the adult brain. For example, when microglia are pharmacologically depleted with a CSF1R inhibitor or in mice lacking the microglial purinergic receptor P2RY12, there is an increase in capillary dilation, reduced vasodilation, and increased blood flow in the adult CNS (Bisht et al. 2021; Császár et al. 2022). Further, pharmacologic blockade or genetic deletion of P2RY12 reduced increased blood flow induced by whisker stimulation in mice (Császár et al. 2022). Specifically, whisker stimulation induced increased blood flow to the corresponding rodent barrel cortex that was reduced upon blockade or deletion of P2RY12, as well as chemogenetic activation of Gq signaling in microglia with designer receptor exclusively activated by designer drugs (DREADD) hM3Dq. Although all genetic and pharmacological manipulations are not specific to VAMs and will affect all CNS microglia, P2RY12 deficiency or activation of hM3Dq did decrease the amount of microglial contact with the vasculature (Bisht et al. 2021; Császár et al. 2022). It is, therefore, intriguing to consider this is a VAM-specific effect. Identification of molecules that specifically regulate VAM association with the vasculature and/or genes that are specific to VAMs versus other resident microglia will be critical to further dissect VAM function.

SYNAPTIC WIRING IN THE CNS

Similar to the excess of neurons born early in development, an excess of axonal and synaptic contacts are formed initially, many of which are later refined and pruned away. While neuron-intrinsic mechanisms and other glial cells can regulate this process (Faust et al. 2021), phagocytic microglia are key during synapse elimination, a process critical for precise synaptic connectivity (Fig. 3; Purves and Lichtman 1980; Lichtman and Colman 2000). In addition, new research suggests that microglia can also influence brain wiring by regulating the number, maturation, and plasticity of synapses in development and throughout the life span. The following section reviews these findings.

Axon Outgrowth and Synaptogenesis

As microglia colonize the developing embryonic brain, they are first enriched in deeper cortical layers and in axonal tracts, including in the corpus callosum and the subpallium (Cunningham et al. 2013; Squarzoni et al. 2014). Mounting evidence supports that microglia are playing a role in the developmental fasciculation and guidance of axon tracts. For example, in mice harboring a loss-of-function DAP12, a key microglia-enriched signaling molecule, there was defasciculation of axons during embryogenesis in dorsal callosal axons (Pont-Lezica et al. 2014). In the same study, similar effects were observed in mice that lacked microglia and all other myeloid cells (Pu.1$^{-/-}$ mice) or embryos that were exposed to maternal inflammation at embryonic day 15.5 (E15.5). Later work further supports a role for DAP12 signaling in axon development as DAP12$^{-/-}$ mouse embryos displayed exuberant outgrowth of dopaminergic axons in the subpallium of the embryonic forebrain (Squarzoni et al. 2014).

A similar phenotype was also observed in microglia depletion via CSF1R blockade as well as mice deficient in CX3CR1 or complement receptor 3 (CR3). In contrast, embryos exposed to maternal immune activation (MIA) showed reduced dopaminergic axon outgrowth. This induced a reduction in dopaminergic axon outgrowth into the subpallium. Mechanistically, it remains unknown how microglia are precisely regulating axon fasciculation and outgrowth during embryogenesis. One study provided

Cite this article as *Cold Spring Harb Perspect Biol* doi: 10.1101/cshperspect.a041810

data that microglia are in direct contact and contain engulfed fragments of dopaminergic axons within their cytoplasm (Squarzoni et al. 2014). This suggests a role in microglia preventing the exuberant outgrowth of axons through engulfment mechanisms. Of the molecules tested in this study, CR3 is the most plausible mechanism as it is an engulfment receptor. Similar to netrin-G1-dependent accumulation of microglia along layer V neuron axons (Fujita et al. 2020), it is also possible that cell adhesion mechanisms between microglia and axons promote fasciculation and axonal outgrowth trajectory. Another interesting possibility is the secretion of ECM by microglia to assist in axon guidance as recently shown in the developing spinal cord following injury (Li et al. 2020b).

After axons reach their targets, there is a process of synaptogenesis that must occur to form a functional circuit. In addition to axon development, microglia have been implicated in promoting the generation of new synapses as well as in restricting the generation of too many synapses. The role of microglia in synaptogenesis was elegantly demonstrating first in layer II/III of the postnatal somatosensory cortex. By two-photon live imaging in mice, it was shown that microglia frequently contacted developing cortical dendrites with their processes. Immediately following this contact there was an increase in local dendritic calcium (Ca^{2+}) and actin accumulation at the point of microglial contact, which was followed by filipodia formation within ~10 min (Miyamoto et al. 2016). Strikingly, when microglia were genetically ablated with diphtheria toxin, there was a decrease in dendritic spines and functional synapses in the postnatal somatosensory cortex.

The mechanistic underpinnings of this microglia-induced synaptogenesis remained an open question after this study. Similarly, microglia within the postnatal somatosensory cortex were also shown by static imaging to frequently contact GABAergic axo-axonic chandelier synapses along axon initial segments and ablation of microglia reduced the number of these chandelier synapses (Gallo et al. 2022). One of the biggest questions arising from these studies is the mechanism by which microglia are

promoting synapses. In culture, interleukin 10 (IL-10) secreted from microglia was found to have a synaptogenic effect on hippocampal neurons (Lim et al. 2013). In vivo, BDNF has emerged as a putative microglia-mediated synaptogenic factor. This is based on work performed in juvenile P30 mice, which demonstrated by two-photon live imaging that microglia more frequently contacted dendrites in the motor cortex during motor learning-induced synaptogenesis (Parkhurst et al. 2013). Genetic ablation of microglia with diphtheria toxin or with microglia-specific BDNF deletion using mice expressing CreER under the control of the CX3CR1 resulted in a reduction in motor learning-induced synaptogenesis. As microglia express low levels of BDNF compared to other resident CNS cell types (Zhang et al. 2014), it remains unclear precisely how this microglia-derived BDNF exerts its effects. It could be by very local release or through other mechanisms downstream from BDNF that are elicited in the microglia.

Synaptic Refinement and Pruning

One of the first clues that suggested a role for microglia in synaptic pruning was the observation that process-bearing, phagocytic microglia were enriched in brain regions undergoing active synaptic remodeling, including the cerebellum, hippocampus, and the visual system (Perry et al. 1985; Milligan et al. 1991; Dalmau et al. 1998; Maślińska et al. 1998; Wierzba-Bobrowicz et al. 1998; Schafer et al. 2012). Since this time, a number of studies have further described microglia as engulfing synaptic material, and ultimately, genetic studies have shown a functional role for microglia in synaptic pruning.

Three seminal studies first revealed that microglia engulf and prune synapses (Paolicelli et al. 2011; Schafer et al. 2012). Tremblay and colleagues, combing high-resolution 3D serial electron microscopy (EM) with two-photon live imaging during the critical period in the juvenile mouse primary visual cortex (V1) (Tremblay et al. 2010), found that microglia frequently contacted pre- and postsynaptic elements (one synaptic element per 40 min). Interestingly, many of the dendritic spines contacted

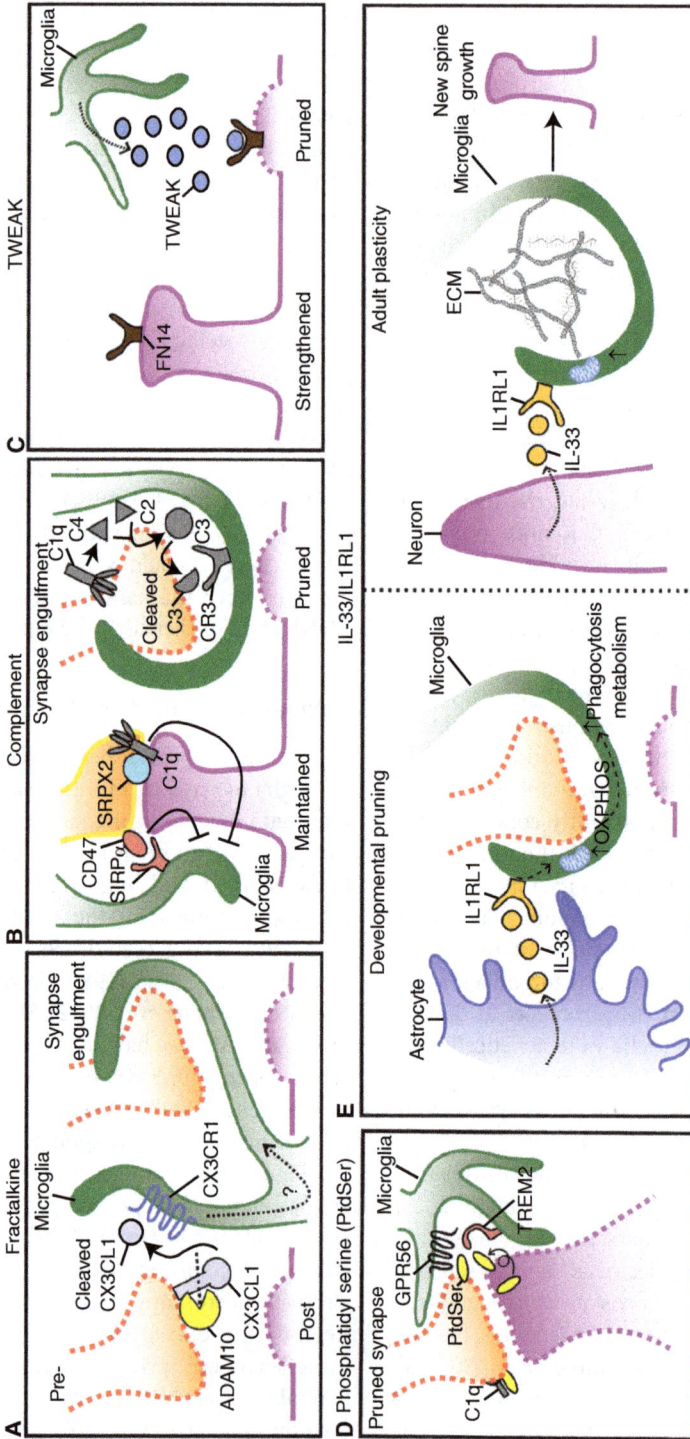

Figure 3. (*See following page for legend.*)

by microglia tended to be smaller and were often no longer present at later imaging sessions, data suggesting microglia are playing a role in the elimination of these spines. The investigators further assessed microglia–synapse interactions following a dark adaptation paradigm known to elicit robust synaptic remodeling (Mower et al. 1983; Philpot et al. 2001; Viegi et al. 2002). Using this paradigm, in dark-adapted mice or dark-adapted mice reexposed to light, live imaging revealed that microglia changed their motility and increased their interactions with spines as compared with light-reared control animals. Furthermore, EM analysis revealed that microglia appeared to more frequently contact synaptic elements and had more phagocytic inclusions as compared with light-reared control animals. These phagocytic inclusions were speculated to be synaptic elements, suggesting a role for microglia in experience-dependent synaptic remodeling in the mature brain. Similar phagocytic inclusions containing synaptic material were shown by Paolicelli et al. (2011) in the developing hippocampus. This group also showed that mice deficient in CX3CR1 had a

delay in synapse maturation in the hippocampus, which seems to be true in the developing mouse barrel cortex as well (Hoshiko et al. 2012). In a later study, CX3CR1 was shown to directly regulate microglial engulfment of synapses in response to sensory deprivation in the rodent somatosensory cortex (Gunner et al. 2019). As CX3CR1 is a chemokine receptor and not an engulfment receptor, it remains unknown precisely how microglial CX3CR1 signaling modulates synapse maturation and engulfment of synapses.

To further test the hypothesis that microglia engulf excess or exuberant synaptic connections during postnatal development, Schafer et al. (2012) took advantage of the mouse retinogeniculate system, a visual pathway and classic model system in which to study activity-dependent synapse elimination. In this system, postsynaptic thalamic relay neurons within the dorsal lateral geniculate nucleus (dLGN) are initially innervated by multiple weak retinal ganglion cell inputs originating from the retina. A subset of these presynaptic inputs are later eliminated, while the remaining inputs are maintained and

Figure 3. Multiple microglial engulfment mechanisms contribute to synaptic pruning. Established mechanisms of axonal and synaptic pruning include: (*A*) CX3CR1–CX3CL1–ADAM10 signaling. Neuron-expressed CX3CL1 is cleaved from the membrane by the protease ADAM10, which initiates signaling to its microglial receptor CX3CR1 (Gunner et al. 2019). CX3CR1 signaling instructs microglial synapse engulfment by a yet-to-be-identified mechanism (Gunner et al. 2019). (*B*) Complement. C1q induces the formation of the C3 convertase through complement factors C2 and C4. C3 convertase then cleaves and activates C3, which then directs microglia to engulf synapses via CR3 (Stevens et al. 2007; Schafer et al. 2012) while signals such as CD47–SIRPα signaling and the C1q inhibitor SRPX2 inhibit excessive pruning by microglia (Lehrman et al. 2018; Cong et al. 2020). (*C*) Cytokine signaling. Microglial TWEAK signals through neuronal Fn14 to weaken synapses to be pruned (Cheadle et al. 2020). (*D*) Phosphatidylserine. TREM2, GPR56, and MERTK (not shown) bind externalized PtdSer at synapses, leading to engulfment (Li et al. 2020a; Scott-Hewitt et al. 2020; Park et al. 2021). (*E*) IL-33. During development (*left*), astrocytic IL-33 signals to microglia via IL1RL1 to up-regulate metabolic pathways to support phagocytosis. In adulthood (*right*), neuronal IL-33 instructs microglia to engulf the ECM, clearing space for new synapses to form (Nguyen et al. 2020). (CD47) Cluster of differentiation 47, (SIPRα) signal-regulatory protein α, (SRPX2) sushi repeat-containing protein X-linked 2, (TGF-β) transforming growth factor β, (MEGF10) multiple epidermal growth factor-like domains protein 10, (MerTK) Mer proto-oncognene tyrosine kinase, (PtdSer) phosphatidylserine, (ADAM10) a disintegrin and metalloproteinase domain-containing protein 10, (CX3CL1) C-X3-C motif chemokine ligand 1, (CX3CR1) C-X3-C motif chemokine receptor 1, (IL-33) interleukin 33, (IL1RL1) interleukin 1 receptor like 1, (CR3) complement receptor 3, (C3) complement component 3, (C1q) complement component 1q, (C4) complement component 4, (C2) complement component 2, (ECM) extracellular matrix, (TREM2) triggering receptor expressed on myeloid cells, (TWEAK, aka TNFSF12) TNF-related WEAK inducer of apoptosis, (Fn14, aka TNFRSF12A) FGF-inducible 14, (GPR56, aka ADGRG1) G-protein couple receptor 56. (Panels *A* and *B* reprinted, with permission, from Faust et al. 2021, Springer Nature BV © 2021.)

strengthened (Chen and Regehr 2000; Hooks and Chen 2006; Huberman 2007; Huberman et al. 2008). It was previously shown in this circuit that classical complement cascade proteins C1q and C3 are localized to synaptic compartments and mediate synaptic pruning in the developing retinogeniculate system (Stevens et al. 2007). In the innate immune system, C1q and/or C3 bind cellular material, inducing its removal by several mechanisms, including phagocytic pathways (Lambris and Tsokos 1986; Gasque 2004). Together, these findings raised the hypothesis that complement C1q and C3 target synapses for elimination by microglia, which, in the context of the healthy postnatal brain, are the primary cell type that expresses the high-affinity receptor for C3, CR3 (Cd11b) (Schafer et al. 2012). Using this retinogeniculate circuit, it was then identified that microglia contained presynaptic inputs within their processes and within lysosomal compartments (Schafer et al. 2012). If C3 or CR3 was ablated, microglia engulfed fewer synapses and synaptic pruning was defective as evidenced by significantly more retinogeniculate synapses in the P30 LGN. By manipulating activity in the eyes, it was further shown that microglia preferentially engulf less active retinogeniculate synapses. Together these data reveal that microglia, indeed, engulf synapses during developmental synaptic pruning in an activity-dependent manner. These data now raise questions about the underlying mechanisms. How do complement and microglia target some synapses but not others? Are synapses, indeed, intact at the time of engulfment (Fig. 3)? Toward the latter question, ex vivo live imaging in the mouse hippocampus has suggested that microglia preferentially engulf or "nibble" small pieces of presynaptic material in a process called trogocytosis, while postsynaptic material remains intact (Weinhard et al. 2018). Similarly, complement regulates microglial trogocytosis of axons in the developing *Xenopus laevis* retinotectal circuit (Lim and Ruthazer 2021). It remains to be determined whether preferential presynaptic trogocytosis by microglia occurs in other brain regions in different contexts (e.g., inflammation).

Since this initial work, a flurry of studies has demonstrated roles for microglia in synapse pruning and remodeling. It is important to note there are also microglia-independent roles in synaptic remodeling and understanding the mechanisms that drive microglia to remodel some circuits and not others will be important going forward (Faust et al. 2021). One question raised by these studies is how secreted complement and other "eat me" signals localize to axonal inputs and synapses. This is an area of active investigation. For example, complement component C1q can bind phosphatidyl serine (PS) (Païdassi et al. 2008), which is normally restricted to the inner leaflet of the plasma membrane but it becomes externalized on the surface of dead or dying cells to promote their clearance. In C1q$^{-/-}$ mice, which have a known defect in retinogeniculate pruning, PS accumulates at presynaptic structures and microglia reduce their engulfment of synapses in the postnatal LGN (Stevens et al. 2007; Scott-Hewitt et al. 2020). These data suggest that local PS exposure regulates which synapses are eliminated via complement-dependent engulfment by microglia. In addition, there appears to be negative regulation preventing excessive engulfment of synapses by microglia. For example, a newly identified, neuronally expressed complement inhibitor called sushi repeat-containing protein X-linked 2 (SRPX2) was shown to bind C1q at the synaptic membrane, thereby preventing excessive C1q-mediated synapse engulfment and pruning (Cong et al. 2020). Although not directly regulating complement, the molecule CD47 inhibits microglia from exuberantly engulfing synapses via complement in the developing retinogeniculate circuit during pruning via its receptor signal-regulatory protein α (SIRPα) expressed by microglia (Lehrman et al. 2018). Intriguingly, in the retina neuron-derived SIRPα appears to counteract microglial SIRPα to facilitate pruning (Jiang et al. 2022). However, this mechanism has not been investigated outside the retinal compartment.

Beyond complement, other molecules known to bind PS, including Adhesion GPR56 and TREM2, have now been shown to regulate synapse engulfment by microglia. First, GPR56 was found to bind PS through its GPCR autoproteolysis inducing (GAIN) domain in a do-

Cite this article as *Cold Spring Harb Perspect Biol* doi: 10.1101/cshperspect.a041810

main-specific manner and global or microglia-specific deletion of Gpr56 leads to a defect in microglial engulfment and pruning of retinogeniculate synapses. Similarly, TREM2 also binds PS and TREM2$^{-/-}$ microglia showed a decreased capacity to engulf and prune synapses in the postanal hippocampus, but not the cortex. Retinogeniculate pruning has not been tested in TREM2$^{-/-}$ mice. Interestingly, another study has suggested that TREM2 in microglia is not directly modulating the microglial of synapses, but rather it induces microglia to limit synapse pruning by astrocytes in the developing hippocampus (Jay et al. 2019). This interplay between microglia and astrocytes to coordinate synapse pruning is, indeed, an emerging theme as both cell types have been identified to prune developing synapses. Most recently, this interplay has been shown to involve IL33-IL1R1 signaling. Specifically, in the postnatal rodent sensorimotor circuit, astrocyte-derived IL-33 can direct microglia to increase the engulfment of synapses via IL1RL1 (also known as ST2) expressed by microglia. It has been further shown that IL-33 binding to microglial IL1RL1 induces a PI3K-dependent increase in glucose metabolism and oxidative phosphorylation. This metabolic adaptation is coupled to elevated phagocytosis in microglia. Intriguingly, IL-33 has also been shown to promote plasticity in adulthood in the hippocampus through a different intercellular signaling mechanism. That is, while astrocytes express the *Il33a* isoform, neurons express the *Il33b* isoform. Neuronal *Il33b* expression was shown to increase in the adult mouse hippocampus exposed to an enriched environment, which promotes circuit remodeling and neurogenesis. It was then shown that neuron-derived IL-33 during enriched environment exposure promotes microglia to engulf, not synapses, but ECM through IL1R1 expressed by microglia. Evidence suggests that this engulfment of ECM is critical to accommodate the formation of new synapses during environmental enrichment. It remains unclear how microglia distinguish IL-33-derived signals from astrocytes and neurons to engulf different cellular substrates. Microglia are also key ECM regulators during embryogenesis, likely through

phagocytosis and ECM remodeling. An elegant study found that a subset of microglia accumulates at the boundaries of the embryonic cortex, striatum, and amygdala, which was necessary to prevent tissue tearing as the brain grows (Lawrence et al. 2024). SPP1 was identified as a regulator of this process and phagocytosis of the ECM and modulation of subsequent ECM secretion during brain growth are likely mediators.

In addition to phagocytic mechanisms, it is important to note that microglia can influence the elimination and maturation of neural circuits by other mechanisms. For example, at later stages of the developing retinogeniculate circuit, microglia-derived TWEAK (a TNF family cytokine also known as TNFSF12) binds neuronal Fn14 (also known as TNFRSF12A) to facilitate the strengthening of synapses in the retinogeniculate circuit (Cheadle et al. 2020). In the cerebellum, ablation of microglia by conditional ablation of CSF1R results in impaired climbing fiber elimination (Nakayama et al. 2018). However, it is not through engulfment mechanisms but rather microglia-mediated promotion of GABAergic inhibition on Purkinje neurons, which, in turn, facilitates the elimination of climbing fibers during developmental cerebellar pruning.

It is now clear that multiple immune mechanisms regulate synaptic refinement and pruning, which converge on microglia as a final executor of the process. Yet, do all these mechanisms cooperate together to drive pruning or do specific mechanisms regulate pruning depending on the circuit and age of the animal? In some cases, evidence suggests the latter as, for example, CX3CR1 regulates pruning in the hippocampus and barrel cortex but not in the visual cortex, cerebellum, or retinogeniculate circuit (Lowery et al. 2017; Schecter et al. 2017; Kaiser et al. 2020). Likewise, complement regulates pruning in the retinogeniculate circuit but not plasticity in the visual cortex (Welsh et al. 2020). This is consistent with a more recent study that showed depletion of microglia had no effect on visual cortex plasticity (Brown et al. 2024). In contrast, loss of microglial P2RY12 signaling resulted in defective visual cortex plasticity (also discussed in the next section) (Sipe et al. 2016). The latter could be explained if P2RY12-deficient microglia adopt a more inflammatory state in the absence of this gene important for homeostasis, impacting plasticity. This remains speculative but could have implications for brain plasticity under neuroinflammatory conditions. It also remains to be determined how these different mechanisms drive pruning in specific circuits at specific times. One possibility is that the immune molecules driving this process, which have receptors on microglia, are genetically predetermined in specific neuron subtypes. Another possibility is that different activity patterns drive a specific immune repertoire in neurons and/or surrounding glial cells. Future work to assess the diversity of immune signaling between cell types and subsequently manipulating activity in specific neuron populations to assess how this changes immune signaling could offer powerful new insight.

Microglial Modulation of Synaptic Function

Whether by engulfment or other mechanisms, many lines of evidence demonstrate that microglia are a functional part of the neural circuit by "listening" to activity and modulating synapse function and, ultimately, behavior. This began with seminal live imaging studies showing that microglia modify their motility in response to changes in neural activity in vivo (Davalos et al. 2005; Nimmerjahn et al. 2005). There was further supported by electrophysiology studies demonstrating that mice deficient in or mutant for microglia-enriched molecules such as CX3CR1 or DAP12, which is the coreceptor for TREM2, had defects or delays in functional synapse development (Roumier et al. 2004; Paolicelli et al. 2011; Hoshiko et al. 2012; Zhan et al. 2014; Filipello et al. 2018). In addition, many cytokines, which are highly enriched in microglia compared with other CNS cell types, can serve as modulators to impact neural excitability and functional synaptic plasticity (for review, see Zipp et al. 2023). Microglia can directly respond to changes in neural activity to, in turn, impact the function of synapses in development and into adulthood. First suggested by transcriptomic studies and in vitro work that microglia express functional neurotransmitter re-

ceptors (Pocock and Kettenmann 2007), it is now being realized that these receptors function in vivo. For example, a recent study demonstrated that GABAb receptor-expressing microglia interact with and potentially engulf inhibitory synapses (Favuzzi et al. 2021). Deletion of GABAb receptor in microglia led to increased inhibitory synapse number, and inhibitory postsynaptic currents, without affecting the excitatory system. This was associated with a reduction in synapse remodeling gene expression. Recent studies also used single-cell and spatial profiling to cement that microglia adopt subspecialized states across cortical layers (Stogsdill et al. 2022), and between brain regions (Ayata et al. 2018), potentially driven by neuronal activity. Additionally, it was shown by several groups that microglia can detect and serve to dampen neural activity during seizures (Badimon et al. 2020; Eyo et al. 2021; Merlini et al. 2021; Feinberg et al. 2022; Gibbs-Shelton et al. 2023). One elegant demonstration of this showed that microglia can sense excessive ATP release from hyperactive neurons (Badimon et al. 2020). In turn, they hydrolyze this ATP into adenosine through CD73, which subsequently quiets hyperexcited neurons. In addition to neurotransmitter receptor expression and manipulation of neurotransmitter sensing in microglia, the sensing of neural activity by microglia to impact synapse function has further been shown by calcium imaging in vivo. For example, it was shown that microglia can flux calcium in response to changes in neural activity in vivo (Umpierre et al. 2020). Since this time, it was shown that microglial calcium flux can be induced through the purinergic receptor P2Y6 during seizures to increase phagocytic and inflammatory signaling in microglia (Umpierre et al. 2024). There is also evidence that microglia can regulate sleep and one study showed that this is through norepinephrine-induced increased fluxes in microglial calcium (Ma et al. 2024). Taken together, these findings demonstrate that microglia cannot only impact the structure of a circuit, but they have the ability to sense activity to, in turn, modulate the function of synapses in development and into adulthood.

ROLES IN MODULATING BEHAVIOR

Decades of studies report an association between abnormal microglial morphology, protein or gene expression and neuropsychiatric diagnoses in human postmortem brain tissue, including in neurodevelopmental syndromes like autism (Liao et al. 2020) and schizophrenia (Trépanier et al. 2016). The contributions of microglia to synapse and circuit regulation increasingly suggest that these observations reflect a causal contribution by microglia to the regulation of behavior. Below, we review emerging work that directly interrogates the role of microglia in behavior regulation, focusing on two categories: mechanistic studies showing that developmental alterations to microglia can regulate behavior and those showing that microglia indeed do impact behavior in neuropsychiatric diseases of development.

Evidence that Microglia Regulate Behavior

Multiple lines of evidence show that microglial dysfunction during development can impact behavior. Transient developmental depletion of microglia with clodronate liposomes causes persistent changes in social, affective, and locomotor behavior (VanRyzin et al. 2016). Also, Cx3cr1-deficient mice have deficits in social interaction and increased repetitive behaviors, associated with deficits in synaptic pruning (Paolicelli et al. 2011; Zhan et al. 2014), and IL-33 deficient mice have similar behavior deficits (Dohi et al. 2017). Another body of work focuses on Hoxb8, a transcription factor broadly important for nervous system patterning, but suggested to be specifically expressed by a subset of microglial progenitors before their engraftment in the brain. Deletion of Hoxb8 leads to an "OCD-like" phenotype of overgrooming behavior (Chen et al. 2010). Although it remains to be verified with tightly controlled fate mapping, one study suggests that so-called "Hoxb8" microglia represent an ontogenically distinct sublineage that may have a particular effect on compulsive behaviors (De et al. 2018).

Recent work further shows that many developmental perturbations affecting microglia produce sex-specific behavioral abnormalities. For example, developmental minocycline treatment induces feminine behaviors in males by inhibiting PGE2 production, and PGE2 is in turn sufficient to masculinizes female behavior (Lenz et al. 2013), while environmental stress induces male-specific social deficits (Block et al. 2022). These provocative behavioral differences in response to developmental challenges likely reflect the striking sexual dimorphism of microglia. A large body of work demonstrates clear differences between male and female microglia numbers, gene expression, response to the microbiome, and function (Weinhard et al. 2018; VanRyzin et al. 2020; Lynch 2022). Supporting a causal effect of microglia on sex-associated behaviors, two studies demonstrate that blocking complement-dependent phagocytosis alters male social behaviors (Kopec et al. 2018; VanRyzin et al. 2019). These sex differences in microglia likely reflect both exposures to sex hormones, but also intrinsic chromosomal sex differences. In sum, this body of evidence argues that microglia contribute to establishing sex-associated behavioral differences, an important future direction for the field.

Evidence that Microglia Contribute to Neuropsychiatric Symptoms and Disorders

Hundreds of human neuropathology studies report abnormalities in microglial phagocytosis and cytokine secretion in patients with developmental CNS pathologies. This literature provides strong circumstantial evidence that microglia impact behavior.

Perhaps most strikingly, homozygous germline loss-of-function mutations in CSF1R lead to microglial absence, profound developmental delay, epilepsy, and early mortality in humans (Oosterhof et al. 2019). Mutations in NRROS, a key regulator of TGF-β signaling, similarly lead to severe epilepsy, neurodegeneration, and death (Dong et al. 2020; Smith et al. 2020). These monogenic disorders exemplify a broader emerging category of "microgliopathies," genetic diseases whose CNS symptoms are thought to arise primarily from microglial dysfunction. Other emerging microgliopathies include neurological disorders due to mutations in TREM2 (Guerreiro et al. 2013; Jonsson et al. 2013),

BRAF (Mass et al. 2017), DAP12 (Paloneva et al. 2000), and USP18 (Meuwissen et al. 2016).

A longstanding hypothesis in schizophrenia is that symptoms result from an overpruning of synapses (Feinberg 1982). Landmark studies (Sekar et al. 2016; Yilmaz et al. 2021) provided human genetic and mouse mechanistic evidence to support a role for microglial synapse elimination in the pathogenesis of schizophrenia, by (1) attributing a large portion of genetic risk associated with the MHC locus to the C4 gene, (2) showing that schizophrenia patients have increased C4 expression, and (3) in mouse, verifying that loss of endogenous C4 or overexpression of schizophrenia risk allele human C4A alter microglial synaptic pruning. Although further validation is ongoing, these data implicate exuberant synapse engulfment in schizophrenia at the population level.

Finally, a major risk factor for developmental disorders is MIA or infection midgestation (Estes and McAllister 2016). Rodent MIA models are widely used and exhibit behavioral abnormalities in offspring often described as "autism-like" or "schizophrenia-like." Accompanying these behavioral differences are abnormal microglial morphology, function, and gene expression. Treatment of MIA-exposed pups with minocycline, a nonspecific inhibitor of some microglial responses, prevents many behavioral phenotypes (Mattei et al. 2017). Two recent studies demonstrate a direct relationship between MIA, microglial state changes, and circuit function. In one, microglial loss of GPR56, a receptor involved in synaptic refinement via PS recognition, phenocopies the reduced PV interneuron abundance observed following MIA (Yu et al. 2022). In another, MIA caused durable changes in microglial chromatin structure and gene expression that could be rescued, along with abnormal dopaminergic signaling, by de- and repopulation of microglia (Hayes et al. 2022). This mirrors an exciting finding demonstrating that methotrexate-induced microglial dysfunction could also be rescued by depletion and repopulation (Gibson et al. 2019), suggesting that developmental insults to microglia can permanently change their function, and in turn alter circuit function.

CONCLUDING REMARKS

Until recently, the microglia field faced a pervasive challenge—tantalizing hypotheses about microglia in the development, refinement, and function of neural circuits greatly outnumbered the tools available to test them. Landmark advances in our ability to identify, profile, and manipulate microglia have dramatically changed this landscape. With the discovery of microglia-specific markers, a deeper understanding of the transcriptional states that microglia adopt, and the development of associated methods to specifically target microglia, it is clear that microglia actively participate in critical aspects of neurodevelopment and circuit function. This includes the active clearance of cells and synapses, the regulation of vascular structure and function, and the remodeling and maintenance of oligodendrocyte lineage cells and myelination. These findings, paired with a rapidly growing understanding that, in performing these functions, microglia can impact behavior, suggest that microglia may contribute to the pathogenesis of neuropsychiatric disorders. Together, these layers of molecular, cellular, and circuit-wide insight into the roles of microglia suggest many new avenues for the treatment of neurologic and psychiatric disorders. In sum, we are now entering a very exciting time for microglia research, where a combination of tools and increasing understanding will allow us to discover new microglial functions, but also understand their specific place in brain development, circuit function, and disease.

ACKNOWLEDGMENTS

All work performed by authors is supported by grants from the Smith Family Foundation (B.S.), the Klingenstein-Simons Neuroscience Fellowship (F.C.B.), the Paul Allen Frontiers Group (F.C.B.), the Whitehall Foundation (F.C.B.), HHMI (B.S.), the National Institute of Neurological Disorders and Stroke (R01NS120960 to F.C.B; R01NS117533 to D.P.S.; R25NS065745 to M.L.B.), the National Institute of Mental Health (R01MH113743 to D.P.S.), the National Institute of Aging (RF1AG068281 to D.P.S.),

Cite this article as *Cold Spring Harb Perspect Biol* doi: 10.1101/cshperspect.a041810

the NIH Common Fund (DP5-OD036159 to M.L.B), and Dr. Miriam and Sheldon G. Adelson Medical Research Foundation (D.P.S.).

REFERENCES

*Reference is also in this subject collection.

Abrams JM, White K, Fessler LI, Steller H. 1993. Programmed cell death during *Drosophila* embryogenesis. *Development* 117: 29–43. doi:10.1242/dev.117.1.29

Ajami B, Bennett JL, Krieger C, Tetzlaff W, Rossi FMV. 2007. Local self-renewal can sustain CNS microglia maintenance and function throughout adult life. *Nat Neurosci* 10: 1538–1543. doi:10.1038/nn2014

Aliprantis AO, Diez-Roux G, Mulder LC, Zychlinsky A, Lang RA. 1996. Do macrophages kill through apoptosis? *Immunol Today* 17: 573–576. doi:10.1016/S0167-5699 (96)10071-2

Antony JM, Paquin A, Nutt SL, Kaplan DR, Miller FD. 2011. Endogenous microglia regulate development of embryonic cortical precursor cells. *J Neurosci Res* 89: 286–298. doi:10.1002/jnr.22533

Ayata P, Badimon A, Strasburger HJ, Duff MK, Montgomery SE, Loh YE, Ebert A, Pimenova AA, Ramirez BR, Chan AT, et al. 2018. Epigenetic regulation of brain region-specific microglia clearance activity. *Nat Neurosci* 21: 1049–1060. doi:10.1038/s41593-018-0192-3

Badimon A, Strasburger HJ, Ayata P, Chen X, Nair A, Ikegami A, Hwang P, Chan AT, Graves SM, Uweru JO, et al. 2020. Negative feedback control of neuronal activity by microglia. *Nature* 586: 417–423. doi:10.1038/s41586-020-2777-8

Bangs P, White K. 2000. Regulation and execution of apoptosis during *Drosophila* development. *Dev Dyn* 218: 68–79. doi:10.1002/(SICI)1097-0177(200005)218 :1<68::AID-DVDY6>3.0.CO;2-9

Barclay KM, Abduljawad N, Cheng Z, Kim MW, Zhou L, Yang J, Rustenhoven J, Mazzitelli JA, Smyth LCD, Kapadia D, et al. 2024. An inducible genetic tool to track and manipulate specific microglial states reveals their plasticity and roles in remyelination. *Immunity* 57: 1394–1412.e8.

Basilico B, Ferrucci L, Khan A, Di Angelantonio S, Ragozzinò D, Reverte I. 2022. What microglia depletion approaches tell us about the role of microglia on synaptic function and behavior. *Front Cell Neurosci* 16: 1022431. doi:10.3389/fncel.2022.1022431

Bedolla AM, McKinsey GL, Ware K, Santander N, Arnold TD, Luo Y. 2024. A comparative evaluation of the strengths and potential caveats of the microglial inducible CreER mouse models. *Cell Rep* 43: 113660. doi:10.1016/j .celrep.2023.113660

Bennett ML, Bennett FC. 2020. The influence of environment and origin on brain resident macrophages and implications for therapy. *Nat Neurosci* 23: 157–166. doi:10 .1038/s41593-019-0545-6

Bennett ML, Bennett FC, Liddelow SA, Ajami B, Zamanian JL, Fernhoff NB, Mulinyawe SB, Bohlen CJ, Adil A, Tucker A, et al. 2016. New tools for studying microglia in the mouse and human CNS. *Proc Natl Acad Sci* 113: E1738–E1746. doi:10.1073/pnas.1525528113

Bennett FC, Bennett ML, Yaqoob F, Mulinyawe SB, Grant GA, Hayden Gephart M, Plowey ED, Barres BA. 2018. A combination of ontogeny and CNS environment establishes microglial identity. *Neuron* 98: 1170–1183.e8. doi:10.1016/j.neuron.2018.05.014

Bessis A, Béchade C, Bernard D, Roumier A. 2007. Microglial control of neuronal death and synaptic properties. *Glia* 55: 233–238. doi:10.1002/glia.20459

Bilbo SD, Rudy JW, Watkins LR, Maier SF. 2006. A behavioural characterization of neonatal infection-facilitated memory impairment in adult rats. *Behav Brain Res* 169: 39–47. doi:10.1016/j.bbr.2005.12.002

Bisht K, Okojie KA, Sharma K, Lentferink DH, Sun YY, Chen HR, Uweru JO, Amancherla S, Calcuttawala Z, Campos-Salazar AB, et al. 2021. Capillary-associated microglia regulate vascular structure and function through PANX1-P2RY12 coupling in mice. *Nat Commun* 12: 5289. doi:10.1038/s41467-021-25590-8

Block CL, Eroglu O, Mague SD, Smith CJ, Ceasrine AM, Sriworarat C, Blount C, Beben KA, Malacon KE, Ndubuizu N, et al. 2022. Prenatal environmental stressors impair postnatal microglia function and adult behavior in males. *Cell Rep* 40: 111161. doi:10.1016/j.celrep.2022.111161

Bohlen CJ, Bennett FC, Tucker AF, Collins HY, Mulinyawe SB, Barres BA. 2017. Diverse requirements for microglial survival, specification, and function revealed by defined-medium cultures. *Neuron* 94: 759–773.e8. doi:10.1016/j .neuron.2017.04.043

Borjini N, Sivilia S, Giuliani A, Fernandez M, Giardino L, Facchinetti F, Calzà L. 2019. Potential biomarkers for neuroinflammation and neurodegeneration at short and long term after neonatal hypoxic-ischemic insult in rat. *J Neuroinflammation* 16: 194. doi:10.1186/s12974-019-1595-0

Bouzid H, Belk JA, Jan M, Qi Y, Sarnowski C, Wirth S, Ma L, Chrostek MR, Ahmad H, Nachun D, et al. 2023. Clonal hematopoiesis is associated with protection from Alzheimer's disease. *Nat Med* 29: 1662–1670. doi:10.1038/ s41591-023-02397-2

Brown GC, Neher JJ. 2014. Microglial phagocytosis of live neurons. *Nat Rev Neurosci* 15: 209–216. doi:10.1038/ nrn3710

Brown TC, Crouse EC, Attaway CA, Oakes DK, Minton SW, Borghuis BG, McGee AW. 2024. Microglia are dispensable for experience-dependent refinement of mouse visual circuitry. *Nat Neurosci* 27: 1462–1467. doi:10.1038/ s41593-024-01706-3

Buller KM, Carty ML, Reinebrant HE, Wixey JA. 2009. Minocycline: a neuroprotective agent for hypoxic-ischemic brain injury in the neonate? *J Neurosci Res* 87: 599–608. doi:10.1002/jnr.21890

Capotondo A, Milazzo R, Politi LS, Quattrini A, Palini A, Plati T, Merella S, Nonis A, di Serio C, Montini E, et al. 2012. Brain conditioning is instrumental for successful microglia reconstitution following hematopoietic stem cell transplantation. *Proc Natl Acad Sci* 109: 15018–15023. doi:10.1073/pnas.1205858109

Chamak B, Morandi V, Mallat M. 1994. Brain macrophages stimulate neurite growth and regeneration by secreting thrombospondin. *J Neurosci Res* 38: 221–233. doi:10 .1002/jnr.490380213

Cheadle L, Rivera SA, Phelps JS, Ennis KA, Stevens B, Burkly LC, Lee WA, Greenberg ME. 2020. Sensory experience engages microglia to shape neural connectivity through a non-phagocytic mechanism. *Neuron* 108: 451–468.e9. doi:10.1016/j.neuron.2020.08.002

Checchin D, Sennlaub F, Levavasseur E, Leduc M, Chemtob S. 2006. Potential role of microglia in retinal blood vessel formation. *Invest Ophthalmol Vis Sci* 47: 3595–3602. doi:10.1167/iovs.05-1522

Chen C, Regehr WG. 2000. Developmental remodeling of the retinogeniculate synapse. *Neuron* 28: 955–966. doi:10.1016/S0896-6273(00)00166-5

Chen SK, Tvrdik P, Peden E, Cho S, Wu S, Spangrude G, Capecchi MR. 2010. Hematopoietic origin of pathological grooming in Hoxb8 mutant mice. *Cell* 141: 775–785. doi:10.1016/j.cell.2010.03.055

Chiu IM, Morimoto ET, Goodarzi H, Liao JT, O'Keeffe S, Phatnani HP, Muratet M, Carroll MC, Levy S, Tavazoie S, et al. 2013. A neurodegeneration-specific gene-expression signature of acutely isolated microglia from an amyotrophic lateral sclerosis mouse model. *Cell Rep* 4: 385–401. doi:10.1016/j.celrep.2013.06.018

Cong Q, Soteros BM, Wollet M, Kim JH, Sia GM. 2020. The endogenous neuronal complement inhibitor SRPX2 protects against complement-mediated synapse elimination during development. *Nat Neurosci* 23: 1067–1078. doi:10.1038/s41593-020-0672-0

Császár E, Lénárt N, Cserép C, Környei Z, Fekete R, Pósfai B, Balázsfi D, Hangya B, Schwarcz AD, Szabadits E, et al. 2022. Microglia modulate blood flow, neurovascular coupling, and hypoperfusion via purinergic actions. *J Exp Med* 219: e20211071. doi:10.1084/jem.20211071

Cunningham CL, Martínez-Cerdeño V, Noctor SC. 2013. Microglia regulate the number of neural precursor cells in the developing cerebral cortex. *J Neurosci* 33: 4216–4233. doi:10.1523/JNEUROSCI.3441-12.2013

Dalmau I, Finsen B, Zimmer J, González B, Castellano B. 1998. Development of microglia in the postnatal rat hippocampus. *Hippocampus* 8: 458–474. doi:10.1002/(SICI)1098-1063(1998)8:5<458::AID-HIPO6>3.0.CO;2-N

Davalos D, Grutzendler J, Yang G, Kim JV, Zuo Y, Jung S, Littman DR, Dustin ML, Gan WB. 2005. ATP mediates rapid microglial response to local brain injury in vivo. *Nat Neurosci* 8: 752–758. doi:10.1038/nn1472

De S, Van Deren D, Peden E, Hockin M, Boulet A, Titen S, Capecchi MR. 2018. Two distinct ontogenies confer heterogeneity to mouse brain microglia. *Development* 145: dev152306. doi:10.1242/dev.152306

Dohi E, Choi EY, Rose IVL, Murata AS, Chow S, Niwa M, Kano SI. 2017. Behavioral changes in mice lacking interleukin-33. *eNeuro* 4: ENEURO.0147-17.2017. doi:10.1523/ENEURO.0147-17.2017

Dong X, Tan NB, Howell KB, Barresi S, Freeman JL, Vecchio D, Piccione M, Radio FC, Calame D, Zong S, et al. 2020. Bi-allelic LoF NRROS variants impairing active TGF-β1 delivery cause a severe infantile-onset neurodegenerative condition with intracranial calcification. *Am J Hum Genet* 106: 559–569. doi:10.1016/j.ajhg.2020.02.014

Dudiki T, Meller J, Mahajan G, Liu H, Zhevlakova I, Stefl S, Witherow C, Podrez E, Kothapalli CR, Byzova TV. 2020. Microglia control vascular architecture via a TGFβ1 dependent paracrine mechanism linked to tissue mechanics. *Nat Commun* 11: 986. doi:10.1038/s41467-020-14787-y

Dzierzak E, Bigas A. 2018. Blood development: hematopoietic stem cell dependence and independence. *Cell Stem Cell* 22: 639–651. doi:10.1016/j.stem.2018.04.015

Elmore MR, Najafi AR, Koike MA, Dagher NN, Spangenberg EE, Rice RA, Kitazawa M, Matusow B, Nguyen H, West BL, et al. 2014. Colony-stimulating factor 1 receptor signaling is necessary for microglia viability, unmasking a microglia progenitor cell in the adult brain. *Neuron* 82: 380–397. doi:10.1016/j.neuron.2014.02.040

Escoubas CC, Dorman LC, Nguyen PT, Lagares-Linares C, Nakajo H, Anderson SR, Barron JJ, Wade SD, Cuevas B, Vainchtein ID, et al. 2024. Type-I-interferon-responsive microglia shape cortical development and behavior. *Cell* 187: 1936–1954.e24. doi:10.1016/j.cell.2024.02.020

Estes ML, McAllister AK. 2016. Maternal immune activation: implications for neuropsychiatric disorders. *Science* 353: 772–777. doi:10.1126/science.aag3194

Eyo UB, Haruwaka K, Mo M, Campos-Salazar AB, Wang L, Speros XS IV, Sabu S, Xu P, Wu LJ. 2021. Microglia provide structural resolution to injured dendrites after severe seizures. *Cell Rep* 35: 109080. doi:10.1016/j.celrep.2021.109080

Fantin A, Vieira JM, Gestri G, Denti L, Schwarz Q, Prykhozhij S, Peri F, Wilson SW, Ruhrberg C. 2010. Tissue macrophages act as cellular chaperones for vascular anastomosis downstream of VEGF-mediated endothelial tip cell induction. *Blood* 116: 829–840. doi:10.1182/blood-2009-12-257832

Faust TE, Gunner G, Schafer DP. 2021. Mechanisms governing activity-dependent synaptic pruning in the developing mammalian CNS. *Nat Rev Neurosci* 22: 657–673. doi:10.1038/s41583-021-00507-y

Faust TE, Feinberg PA, O'Connor C, Kawaguchi R, Chan A, Strasburger H, Frosch M, Boyle MA, Masuda T, Amann L, et al. 2023. A comparative analysis of microglial inducible Cre lines. *Cell Rep* 42: 113031. doi:10.1016/j.celrep.2023.113031

Favuzzi E, Huang S, Saldi GA, Binan L, Ibrahim LA, Fernández-Otero M, Cao Y, Zeine A, Sefah A, Zheng K, et al. 2021. GABA-receptive microglia selectively sculpt developing inhibitory circuits. *Cell* 184: 4048–4063.e32. doi:10.1016/j.cell.2021.06.018

Feinberg I. 1982. Schizophrenia: caused by a fault in programmed synaptic elimination during adolescence? *J Psychiatr Res* 17: 319–334. doi:10.1016/0022-3956(82)90038-3

Feinberg PA, Becker SC, Chung L, Ferrari L, Stellwagen D, Anaclet C, Durán-Laforet V, Faust TE, Sumbria RK, Schafer DP. 2022. Elevated TNF-α leads to neural circuit instability in the absence of interferon regulatory factor 8. *J Neurosci* 42: 6171–6185. doi:10.1523/JNEUROSCI.0601-22.2022

Ferrer I, Bernet E, Soriano E, del Rio T, Fonseca M. 1990. Naturally occurring cell death in the cerebral cortex of the rat and removal of dead cells by transitory phagocytes. *Neuroscience* 39: 451–458. doi:10.1016/0306-4522(90)90281-8

Ferrero G, Mahony CB, Dupuis E, Yvernogeau L, Di Ruggiero E, Miserocchi M, Caron M, Robin C, Traver D, Bertrand JY, et al. 2018. Embryonic microglia derive from primitive macrophages and are replaced by cmyb-dependent definitive microglia in zebrafish. *Cell Rep* **24:** 130–141. doi:10.1016/j.celrep.2018.05.066

Ferron M, Vacher J. 2005. Targeted expression of Cre recombinase in macrophages and osteoclasts in transgenic mice. *Genesis* **41:** 138–145. doi:10.1002/gene.20108

Filipello F, Morini R, Corradini I, Zerbi V, Canzi A, Michalski B, Erreni M, Markicevic M, Starvaggi-Cucuzza C, Otero K, et al. 2018. The microglial innate immune receptor TREM2 is required for synapse elimination and normal brain connectivity. *Immunity* **48:** 979–991.e8. doi:10.1016/j.immuni.2018.04.016

Fourgeaud L, Través PG, Tufail Y, Leal-Bailey H, Lew ED, Burrola PG, Callaway P, Zagórska A, Rothlin CV, Nimmerjahn A, et al. 2016. TAM receptors regulate multiple features of microglial physiology. *Nature* **532:** 240–244. doi:10.1038/nature17630

Frade JM, Barde YA. 1998. Microglia-derived nerve growth factor causes cell death in the developing retina. *Neuron* **20:** 35–41. doi:10.1016/S0896-6273(00)80432-8

Freeman MR, Delrow J, Kim J, Johnson E, Doe CQ. 2003. Unwrapping glial biology: Gcm target genes regulating glial development, diversification, and function. *Neuron* **38:** 567–580. doi:10.1016/S0896-6273(03)00289-7

Fujita Y, Nakanishi T, Ueno M, Itohara S, Yamashita T. 2020. Netrin-G1 regulates microglial accumulation along axons and supports the survival of layer V neurons in the postnatal mouse brain. *Cell Rep* **31:** 107580. doi:10.1016/j.celrep.2020.107580

Gallo NB, Berisha A, Van Aelst L. 2022. Microglia regulate chandelier cell axo-axonic synaptogenesis. *Proc Natl Acad Sci* **119:** e2114476119. doi:10.1073/pnas.2114476119

Gasque P. 2004. Complement: a unique innate immune sensor for danger signals. *Mol Immunol* **41:** 1089–1098. doi:10.1016/j.molimm.2004.06.011

Gebara E, Sultan S, Kocher-Braissant J, Toni N. 2013. Adult hippocampal neurogenesis inversely correlates with microglia in conditions of voluntary running and aging. *Front Neurosci* **7:** 145. doi:10.3389/fnins.2013.00145

Geirsdottir L, David E, Keren-Shaul H, Weiner A, Bohlen SC, Neuber J, Balic A, Giladi A, Sheban F, Dutertre CA, et al. 2019. Cross-species single-cell analysis reveals divergence of the primate microglia program. *Cell* **179:** 1609–1622.e16. doi:10.1016/j.cell.2019.11.010

Gibbs-Shelton S, Benderoth J, Gaykema RP, Straub J, Okojie KA, Uweru JO, Lentferink DH, Rajbanshi B, Cowan MN, Patel B, et al. 2023. Microglia play beneficial roles in multiple experimental seizure models. *Glia* **71:** 1699–1714. doi:10.1002/glia.24364

Gibson EM, Nagaraja S, Ocampo A, Tam LT, Wood LS, Pallegar PN, Greene JJ, Geraghty AC, Goldstein AK, Ni L, et al. 2019. Methotrexate chemotherapy induces persistent tri-glial dysregulation that underlies chemotherapy-related cognitive impairment. *Cell* **176:** 43–55.e13. doi:10.1016/j.cell.2018.10.049

Giera S, Luo R, Ying Y, Ackerman SD, Jeong SJ, Stoveken HM, Folts CJ, Welsh CA, Tall GG, Stevens B, et al. 2018. Microglial transglutaminase-2 drives myelination and myelin repair via GPR56/ADGRG1 in oligodendrocyte precursor cells. *eLife* **7:** e33385. doi:10.7554/eLife.33385

Ginhoux F, Guilliams M. 2016. Tissue-resident macrophage ontogeny and homeostasis. *Immunity* **44:** 439–449. doi:10.1016/j.immuni.2016.02.024

Ginhoux F, Greter M, Leboeuf M, Nandi S, See P, Gokhan S, Mehler MF, Conway SJ, Ng LG, Stanley ER, et al. 2010. Fate mapping analysis reveals that adult microglia derive from primitive macrophages. *Science* **330:** 841–845. doi:10.1126/science.1194637

Gosselin D, Skola D, Coufal NG, Holtman IR, Schlachetzki JCM, Sajti E, Jaeger BN, O'Connor C, Fitzpatrick C, Pasillas MP, et al. 2017. An environment-dependent transcriptional network specifies human microglia identity. *Science* **356:** eaal3222. doi:10.1126/science.aal3222

Grossmann R, Stence N, Carr J, Fuller L, Waite M, Dailey ME. 2002. Juxtavascular microglia migrate along brain microvessels following activation during early postnatal development. *Glia* **37:** 229–240. doi:10.1002/glia.10031

Guerreiro R, Wojtas A, Bras J, Carrasquillo M, Rogaeva E, Majounie E, Cruchaga C, Sassi C, Kauwe JS, Younkin S, et al. 2013. TREM2 variants in Alzheimer's disease. *N Engl J Med* **368:** 117–127. doi:10.1056/NEJMoa1211851

Gunner G, Cheadle L, Johnson KM, Ayata P, Badimon A, Mondo E, Nagy MA, Liu L, Bemiller SM, Kim KW, et al. 2019. Sensory lesioning induces microglial synapse elimination via ADAM10 and fractalkine signaling. *Nat Neurosci* **22:** 1075–1088. doi:10.1038/s41593-019-0419-y

Guttenplan KA, Weigel MK, Prakash P, Wijewardhane PR, Hasel P, Rufen-Blanchette U, Münch AE, Blum JA, Fine J, Neal MC, et al. 2021. Neurotoxic reactive astrocytes induce cell death via saturated lipids. *Nature* **599:** 102–107. doi:10.1038/s41586-021-03960-y

Hagberg H, Mallard C. 2005. Effect of inflammation on central nervous system development and vulnerability. *Curr Opin Neurol* **18:** 117–123. doi:10.1097/01.wco.0000162851.44897.8f

Hagemeyer N, Hanft KM, Akriditou MA, Unger N, Park ES, Stanley ER, Staszewski O, Dimou L, Prinz M. 2017. Microglia contribute to normal myelinogenesis and to oligodendrocyte progenitor maintenance during adulthood. *Acta Neuropathol* **134:** 441–458. doi:10.1007/s00401-017-1747-1

Hammond TR, Robinton D, Stevens B. 2018. Microglia and the brain: complementary partners in development and disease. *Annu Rev Cell Dev Biol* **34:** 523–544. doi:10.1146/annurev-cellbio-100616-060509

Hammond TR, Dufort C, Dissing-Olesen L, Giera S, Young A, Wysoker A, Walker AJ, Gergits F, Segel M, Nemesh J, et al. 2019. Single-cell RNA sequencing of microglia throughout the mouse lifespan and in the injured brain reveals complex cell-state changes. *Immunity* **50:** 253–271.e6. doi:10.1016/j.immuni.2018.11.004

Han J, Zhu K, Zhang XM, Harris RA. 2019. Enforced microglial depletion and repopulation as a promising strategy for the treatment of neurological disorders. *Glia* **67:** 217–231. doi:10.1002/glia.23529

Han CZ, Li RZ, Hansen E, Trescott S, Fixsen BR, Nguyen CT, Mora CM, Spann NJ, Bennett HR, Poirion O, et al. 2023. Human microglia maturation is underpinned by specific gene regulatory networks. *Immunity* **56:** 2152–2171.e13. doi:10.1016/j.immuni.2023.07.016

Harley SBR, Willis EF, Shaikh SN, Blackmore DG, Sah P, Ruitenberg MJ, Bartlett PF, Vukovic J. 2021. Selective ablation of BDNF from microglia reveals novel roles in self-renewal and hippocampal neurogenesis. *J Neurosci* **41:** 4172–4186. doi:10.1523/JNEUROSCI.2539-20.2021

Hasselmann J, Coburn MA, England W, Figueroa Velez DX, Kiani Shabestari S, Tu CH, McQuade A, Kolahdouzan M, Echeverria K, Claes C, et al. 2019. Development of a chimeric model to study and manipulate human microglia in vivo. *Neuron* **103**: 1016–1033.e10. doi:10.1016/j.neuron.2019.07.002

Hattori Y. 2022. The multiple roles of pericytes in vascular formation and microglial functions in the brain. *Life (Basel)* **12**: 1835. doi:10.3390/life12111835

Hattori Y, Itoh H, Tsugawa Y, Nishida Y, Kurata K, Uemura A, Miyata T. 2022. Embryonic pericytes promote microglial homeostasis and their effects on neural progenitors in the developing cerebral cortex. *J Neurosci* **42**: 362–376. doi:10.1523/JNEUROSCI.1201-21.2021

Haynes SE, Hollopeter G, Yang G, Kurpius D, Dailey ME, Gan WB, Julius D. 2006. The P2Y12 receptor regulates microglial activation by extracellular nucleotides. *Nat Neurosci* **9**: 1512–1519. doi:10.1038/nn1805

Hayes LN, An K, Carloni E, Li F, Vincent E, Trippaers C, Paranjpe M, Dölen G, Goff LA, Ramos A, et al. 2022. Prenatal immune stress blunts microglia reactivity, impairing neurocircuitry. *Nature* **610**: 327–334. doi:10.1038/s41586-022-05274-z

Hildebrand C, Remahl S, Persson H, Bjartmar C. 1993. Myelinated nerve fibres in the CNS. *Prog Neurobiol* **40**: 319–384. doi:10.1016/0301-0082(93)90015-K

Hoogerbrugge PM, Suzuki K, Suzuki K, Poorthuis BJ, Kobayashi T, Wagemaker G, van Bekkum DW. 1988. Donor-derived cells in the central nervous system of twitcher mice after bone marrow transplantation. *Science* **239**: 1035–1038. doi:10.1126/science.3278379

Hooks BM, Chen C. 2006. Distinct roles for spontaneous and visual activity in remodeling of the retinogeniculate synapse. *Neuron* **52**: 281–291. doi:10.1016/j.neuron.2006.07.007

Horti AG, Naik R, Foss CA, Minn I, Misheneva V, Du Y, Wang Y, Mathews WB, Wu Y, Hall A, et al. 2019. PET imaging of microglia by targeting macrophage colony-stimulating factor 1 receptor (CSF1R). *Proc Natl Acad Sci* **116**: 1686–1691. doi:10.1073/pnas.1812155116

Hoshiko M, Arnoux I, Avignone E, Yamamoto N, Audinat E. 2012. Deficiency of the microglial receptor CX3CR1 impairs postnatal functional development of thalamocortical synapses in the barrel cortex. *J Neurosci* **32**: 15106–15111. doi:10.1523/JNEUROSCI.1167-12.2012

Huang Y, Xu Z, Xiong S, Sun F, Qin G, Hu G, Wang J, Zhao L, Liang YX, Wu T, et al. 2018. Repopulated microglia are solely derived from the proliferation of residual microglia after acute depletion. *Nat Neurosci* **21**: 530–540. doi:10.1038/s41593-018-0090-8

Huberman AD. 2007. Mechanisms of eye-specific visual circuit development. *Curr Opin Neurobiol* **17**: 73–80. doi:10.1016/j.conb.2007.01.005

Huberman AD, Manu M, Koch SM, Susman MW, Lutz AB, Ullian EM, Baccus SA, Barres BA. 2008. Architecture and activity-mediated refinement of axonal projections from a mosaic of genetically identified retinal ganglion cells. *Neuron* **59**: 425–438. doi:10.1016/j.neuron.2008.07.018

Hughes AN, Appel B. 2020. Microglia phagocytose myelin sheaths to modify developmental myelination. *Nat Neurosci* **23**: 1055–1066. doi:10.1038/s41593-020-0654-2

Hume DA. 2023. Fate-mapping studies in inbred mice: a model for understanding macrophage development and homeostasis? *Eur J Immunol* **53**: e2250242. doi:10.1002/eji.202250242

Irfan M, Evonuk KS, DeSilva TM. 2022. Microglia phagocytose oligodendrocyte progenitor cells and synapses during early postnatal development: implications for white versus gray matter maturation. *FEBS J* **289**: 2110–2127. doi:10.1111/febs.16190

Jay TR, von Saucken VE, Muñoz B, Codocedo JF, Atwood BK, Lamb BT, Landreth GE. 2019. TREM2 is required for microglial instruction of astrocytic synaptic engulfment in neurodevelopment. *Glia* **67**: 1873–1892. doi: 10.1002/glia.23664

Jiang D, Burger CA, Akhanov V, Liang JH, Mackin RD, Albrecht NE, Andrade P, Schafer DP, Samuel MA. 2022. Neuronal signal-regulatory protein α drives microglial phagocytosis by limiting microglial interaction with CD47 in the retina. *Immunity* **55**: 2318–2335.e7. doi:10.1016/j.immuni.2022.10.018

Jonsson T, Stefansson H, Steinberg S, Jonsdottir I, Jonsson PV, Snaedal J, Bjornsson S, Huttenlocher J, Levey AI, Lah JJ, et al. 2013. Variant of *TREM2* associated with the risk of Alzheimer's disease. *N Engl J Med* **368**: 107–116. doi:10.1056/NEJMoa1211103

Jung S, Aliberti J, Graemmel P, Sunshine MJ, Kreutzberg GW, Sher A, Littman DR. 2000. Analysis of fractalkine receptor CX3CR1 function by targeted deletion and green fluorescent protein reporter gene insertion. *Mol Cell Biol* **20**: 4106–4114. doi:10.1128/MCB.20.11.4106-4114.2000

Kaiser T, Feng G. 2019. Tmem119-EGFP and Tmem119-CreERT2 transgenic mice for labeling and manipulating microglia. *eNeuro* **6**: ENEURO.0448-18.2019. doi:10.1523/ENEURO.0448-18.2019

Kaiser N, Pätz C, Brachtendorf S, Eilers J, Bechmann I. 2020. Undisturbed climbing fiber pruning in the cerebellar cortex of CX3CR1-deficient mice. *Glia* **68**: 2316–2329. doi:10.1002/glia.23842

Keren-Shaul H, Spinrad A, Weiner A, Matcovitch-Natan O, Dvir-Szternfeld R, Ulland TK, David E, Baruch K, Lara-Astaiso D, Toth B, et al. 2017. A unique microglia type associated with restricting development of Alzheimer's disease. *Cell* **169**: 1276–1290.e17. doi:10.1016/j.cell.2017.05.018

Kettenmann H, Hanisch UK, Noda M, Verkhratsky A. 2011. Physiology of microglia. *Physiol Rev* **91**: 461–553. doi:10.1152/physrev.00011.2010

Kim JS, Kolesnikov M, Peled-Hajaj S, Scheyltjens I, Xia Y, Trzebanski S, Haimon Z, Shemer A, Lubart A, Van Hove H, et al. 2021. A binary Cre transgenic approach dissects microglia and CNS border-associated macrophages. *Immunity* **54**: 176–190.e7. doi:10.1016/j.immuni.2020.11.007

Kinet MJ, Shaham S. 2014. Noncanonical cell death in the nematode *Caenorhabditis elegans*. *Methods Enzymol* **545**: 157–180. doi:10.1016/B978-0-12-801430-1.00007-X

Konishi H, Okamoto T, Hara Y, Komine O, Tamada H, Maeda M, Osako F, Kobayashi M, Nishiyama A, Kataoka Y, et al. 2020. Astrocytic phagocytosis is a compensatory mechanism for microglial dysfunction. *EMBO J* **39**: e104464. doi:10.15252/embj.2020104464

Cite this article as *Cold Spring Harb Perspect Biol* doi: 10.1101/cshperspect.a041810

Kopec AM, Smith CJ, Ayre NR, Sweat SC, Bilbo SD. 2018. Microglial dopamine receptor elimination defines sex-specific nucleus accumbens development and social behavior in adolescent rats. *Nat Commun* **9:** 3769. doi:10.1038/s41467-018-06118-z

Kubota Y, Takubo K, Shimizu T, Ohno H, Kishi K, Shibuya M, Saya H, Suda T. 2009. M-CSF inhibition selectively targets pathological angiogenesis and lymphangiogenesis. *J Exp Med* **206:** 1089–1102. doi:10.1084/jem.20081605

Lambris JD, Tsokos GC. 1986. The biology and pathophysiology of complement receptors. *Anticancer Res* **6:** 515–523.

Lang RA, Bishop JM. 1993. Macrophages are required for cell death and tissue remodeling in the developing mouse eye. *Cell* **74:** 453–462. doi:10.1016/0092-8674(93)80047-I

Lawrence AR, Canzi A, Bridlance C, Olivié N, Lansonneur C, Catale C, Pizzamiglio L, Kloeckner B, Silvin A, Munro DAD, et al. 2024. Microglia maintain structural integrity during fetal brain morphogenesis. *Cell* **187:** 962–980.e19. doi:10.1016/j.cell.2024.01.012

Lehrman EK, Wilton DK, Litvina EY, Welsh CA, Chang ST, Frouin A, Walker AJ, Heller MD, Umemori H, Chen C, et al. 2018. CD47 protects synapses from excess microglia-mediated pruning during development. *Neuron* **100:** 120–134.e6. doi:10.1016/j.neuron.2018.09.017

Lenz KM, Nugent BM, Haliyur R, McCarthy MM. 2013. Microglia are essential to masculinization of brain and behavior. *J Neurosci* **33:** 2761–2772. doi:10.1523/JNEUROSCI.1268-12.2013

Li Q, Barres BA. 2018. Microglia and macrophages in brain homeostasis and disease. *Nat Rev Immunol* **18:** 225–242. doi:10.1038/nri.2017.125

Li Q, Cheng Z, Zhou L, Darmanis S, Neff NF, Okamoto J, Gulati G, Bennett ML, Sun LO, Clarke LE, et al. 2019. Developmental heterogeneity of microglia and brain myeloid cells revealed by deep single-cell RNA sequencing. *Neuron* **101:** 207–223.e10. doi:10.1016/j.neuron.2018.12.006

Li T, Chiou B, Gilman CK, Luo R, Koshi T, Yu D, Oak HC, Giera S, Johnson-Venkatesh E, Muthukumar AK, et al. 2020a. A splicing isoform of GPR56 mediates microglial synaptic refinement via phosphatidylserine binding. *EMBO J* **39:** e104136. doi:10.15252/embj.2019104136

Li Y, He X, Kawaguchi R, Zhang Y, Wang Q, Monavarfeshani A, Yang Z, Chen B, Shi Z, Meng H, et al. 2020b. Microglia-organized scar-free spinal cord repair in neonatal mice. *Nature* **587:** 613–618. doi:10.1038/s41586-020-2795-6

Liao X, Liu Y, Fu X, Li Y. 2020. Postmortem studies of neuroinflammation in autism spectrum disorder: a systematic review. *Mol Neurobiol* **57:** 3424–3438. doi:10.1007/s12035-020-01976-5

Lichtman JW, Colman H. 2000. Synapse elimination and indelible memory. *Neuron* **25:** 269–278. doi:10.1016/S0896-6273(00)80893-4

Lim TK, Ruthazer ES. 2021. Microglial trogocytosis and the complement system regulate axonal pruning in vivo. *eLife* **10:** e62167. doi:10.7554/eLife.62167

Lim SH, Park E, You B, Jung Y, Park AR, Park SG, Lee JR. 2013. Neuronal synapse formation induced by microglia and interleukin 10. *PLoS ONE* **8:** e81218. doi:10.1371/journal.pone.0081218

Liu Y, Given KS, Dickson EL, Owens GP, Macklin WB, Bennett JL. 2019a. Concentration-dependent effects of CSF1R inhibitors on oligodendrocyte progenitor cells ex vivo and in vivo. *Exp Neurol* **318:** 32–41. doi:10.1016/j.expneurol.2019.04.011

Liu Z, Gu Y, Chakarov S, Bleriot C, Kwok I, Chen X, Shin A, Huang W, Dress RJ, Dutertre CA, et al. 2019b. Fate mapping via Ms4a3-expression history traces monocyte-derived cells. *Cell* **178:** 1509–1525.e19. doi:10.1016/j.cell.2019.08.009

Logan MA, Freeman MR. 2007. The scoop on the fly brain: glial engulfment functions in *Drosophila*. *Neuron Glia Biol* **3:** 63–74. doi:10.1017/S1740925X0700049X

Lowery RL, Tremblay ME, Hopkins BE, Majewska AK. 2017. The microglial fractalkine receptor is not required for activity-dependent plasticity in the mouse visual system. *Glia* **65:** 1744–1761. doi:10.1002/glia.23192

Lu Z, Elliott MR, Chen Y, Walsh JT, Klibanov AL, Ravichandran KS, Kipnis J. 2011. Phagocytic activity of neuronal progenitors regulates adult neurogenesis. *Nat Cell Biol* **13:** 1076–1083. doi:10.1038/ncb2299

Luo C, Koyama R, Ikegaya Y. 2016. Microglia engulf viable newborn cells in the epileptic dentate gyrus. *Glia* **64:** 1508–1517. doi:10.1002/glia.23018

Lynch MA. 2022. Exploring sex-related differences in microglia may be a game-changer in precision medicine. *Front Aging Neurosci* **14:** 868448. doi:10.3389/fnagi.2022.868448

Ma C, Li B, Silverman D, Ding X, Li A, Xiao C, Huang G, Worden K, Muroy S, Chen W, et al. 2024. Microglia regulate sleep through calcium-dependent modulation of norepinephrine transmission. *Nat Neurosci* **27:** 249–258. doi:10.1038/s41593-023-01548-5

Marín-Teva JL, Dusart I, Colin C, Gervais A, van Rooijen N, Mallat M. 2004. Microglia promote the death of developing Purkinje cells. *Neuron* **41:** 535–547. doi:10.1016/S0896-6273(04)00069-8

Marsh SE, Walker AJ, Kamath T, Dissing-Olesen L, Hammond TR, de Soysa TY, Young AMH, Murphy S, Abdulraouf A, Nadaf N, et al. 2022. Dissection of artifactual and confounding glial signatures by single-cell sequencing of mouse and human brain. *Nat Neurosci* **25:** 306–316. doi:10.1038/s41593-022-01022-8

Maślińska D, Laure-Kamionowska M, Kaliszek A. 1998. Morphological forms and localization of microglial cells in the developing human cerebellum. *Folia Neuropathol* **36:** 145–151.

Mass E. 2018. Delineating the origins, developmental programs and homeostatic functions of tissue-resident macrophages. *Int Immunol* **30:** 493–501. doi:10.1093/intimm/dxy044

Mass E, Jacome-Galarza CE, Blank T, Lazarov T, Durham BH, Ozkaya N, Pastore A, Schwabenland M, Chung YR, Rosenblum MK, et al. 2017. A somatic mutation in erythro-myeloid progenitors causes neurodegenerative disease. *Nature* **549:** 389–393. doi:10.1038/nature23672

Mass E, Nimmerjahn F, Kierdorf K, Schlitzer A. 2023. Tissue-specific macrophages: how they develop and choreograph tissue biology. *Nat Rev Immunol* **23:** 563–579. doi:10.1038/s41577-023-00848-y

Masuda T, Amann L, Sankowski R, Staszewski O, Lenz M, D Errico P, Snaidero N, Costa Jordão MJ, Böttcher C, Kierdorf K, et al. 2020. Novel Hexb-based tools for studying microglia in the CNS. *Nat Immunol* **21:** 802–815. doi:10.1038/s41590-020-0707-4

Masuda T, Amann L, Monaco G, Sankowski R, Staszewski O, Krueger M, Del Gaudio F, He L, Paterson N, Nent E, et al. 2022. Specification of CNS macrophage subsets occurs postnatally in defined niches. *Nature* **604:** 740–748. doi:10.1038/s41586-022-04596-2

Matcovitch-Natan O, Winter DR, Giladi A, Vargas Aguilar S, Spinrad A, Sarrazin S, Ben-Yehuda H, David E, Zelada González F, Perrin P, et al. 2016. Microglia development follows a stepwise program to regulate brain homeostasis. *Science* **353:** aad8670. doi:10.1126/science.aad8670

Mattei D, Ivanov A, Ferrai C, Jordan P, Guneykaya D, Buonfiglioli A, Schaafsma W, Przanowski P, Deuther-Conrad W, Brust P, et al. 2017. Maternal immune activation results in complex microglial transcriptome signature in the adult offspring that is reversed by minocycline treatment. *Transl Psychiatry* **7:** e1120. doi:10.1038/tp.2017.80

Mazaheri F, Breus O, Durdu S, Haas P, Wittbrodt J, Gilmour D, Peri F. 2014. Distinct roles for BAI1 and TIM-4 in the engulfment of dying neurons by microglia. *Nat Commun* **5:** 4046. doi:10.1038/ncomms5046

McKinsey GL, Lizama CO, Keown-Lang AE, Niu A, Santander N, Larpthaveesarp A, Chee E, Gonzalez FF, Arnold TD. 2020. A new genetic strategy for targeting microglia in development and disease. *eLife* **9:** e54590. doi:10.7554/eLife.54590

McNamara NB, Munro DAD, Bestard-Cuche N, Uyeda A, Bogie JFJ, Hoffmann A, Holloway RK, Molina-Gonzalez I, Askew KE, Mitchell S, et al. 2023. Microglia regulate central nervous system myelin growth and integrity. *Nature* **613:** 120–129. doi:10.1038/s41586-022-05534-y

Merlini M, Rafalski VA, Ma K, Kim KY, Bushong EA, Rios Coronado PE, Yan Z, Mendiola AS, Sozmen EG, Ryu JK, et al. 2021. Microglial Gi-dependent dynamics regulate brain network hyperexcitability. *Nat Neurosci* **24:** 19–23. doi:10.1038/s41593-020-00756-7

Meuwissen ME, Schot R, Buta S, Oudesluijs G, Tinschert S, Speer SD, Li Z, van Unen L, Heijsman D, Goldmann T, et al. 2016. Human USP18 deficiency underlies type 1 interferonopathy leading to severe pseudo-TORCH syndrome. *J Exp Med* **213:** 1163–1174. doi:10.1084/jem.20151529

Meyer U, Feldon J, Fatemi SH. 2009. In-vivo rodent models for the experimental investigation of prenatal immune activation effects in neurodevelopmental brain disorders. *Neurosci Biobehav Rev* **33:** 1061–1079. doi:10.1016/j.neubiorev.2009.05.001

Mildner A, Schmidt H, Nitsche M, Merkler D, Hanisch UK, Mack M, Heikenwalder M, Brück W, Priller J, Prinz M. 2007. Microglia in the adult brain arise from Ly-6C^{hi}CCR2^{+} monocytes only under defined host conditions. *Nat Neurosci* **10:** 1544–1553. doi:10.1038/nn2015

Milligan CE, Levitt P, Cunningham TJ. 1991. Brain macrophages and microglia respond differently to lesions of the developing and adult visual system. *J Comp Neurol* **314:** 136–146. doi:10.1002/cne.903140113

Miyamoto A, Wake H, Ishikawa AW, Eto K, Shibata K, Murakoshi H, Koizumi S, Moorhouse AJ, Yoshimura Y, Nabekura J. 2016. Microglia contact induces synapse formation in developing somatosensory cortex. *Nat Commun* **7:** 12540. doi:10.1038/ncomms12540

Mizutani M, Pino PA, Saederup N, Charo IF, Ransohoff RM, Cardona AE. 2012. The fractalkine receptor but not CCR2 is present on microglia from embryonic development throughout adulthood. *J Immunol* **188:** 29–36. doi:10.4049/jimmunol.1100421

Mócsai A, Abram CL, Jakus Z, Hu Y, Lanier LL, Lowell CA. 2006. Integrin signaling in neutrophils and macrophages uses adaptors containing immunoreceptor tyrosine-based activation motifs. *Nat Immunol* **7:** 1326–1333. doi:10.1038/ni1407

Mondo E, Becker SC, Kautzman AG, Schifferer M, Baer CE, Chen J, Huang EJ, Simons M, Schafer DP. 2020. A developmental analysis of juxtavascular microglia dynamics and interactions with the vasculature. *J Neurosci* **40:** 6503–6521. doi:10.1523/JNEUROSCI.3006-19.2020

Monier A, Adle-Biassette H, Delezoide AL, Evrard P, Gressens P, Verney C. 2007. Entry and distribution of microglial cells in human embryonic and fetal cerebral cortex. *J Neuropathol Exp Neurol* **66:** 372–382. doi:10.1097/nen.0b013e3180517b46

Morgan SC, Taylor DL, Pocock JM. 2004. Microglia release activators of neuronal proliferation mediated by activation of mitogen-activated protein kinase, phosphatidylinositol-3-kinase/Akt and delta-Notch signalling cascades. *J Neurochem* **90:** 89–101. doi:10.1111/j.1471-4159.2004.02461.x

Mower GD, Christen WG, Caplan CJ. 1983. Very brief visual experience eliminates plasticity in the cat visual cortex. *Science* **221:** 178–180. doi:10.1126/science.6857278

Nagata K, Nakajima K, Takemoto N, Saito H, Kohsaka S. 1993. Microglia-derived plasminogen enhances neurite outgrowth from explant cultures of rat brain. *Int J Dev Neurosci* **11:** 227–237. doi:10.1016/0736-5748(93)90081-N

Nakayama H, Abe M, Morimoto C, Iida T, Okabe S, Sakimura K, Hashimoto K. 2018. Microglia permit climbing fiber elimination by promoting GABAergic inhibition in the developing cerebellum. *Nat Commun* **9:** 2830. doi:10.1038/s41467-018-05100-z

Nandi S, Gokhan S, Dai XM, Wei S, Enikolopov G, Lin H, Mehler MF, Stanley ER. 2012. The CSF-1 receptor ligands IL-34 and CSF-1 exhibit distinct developmental brain expression patterns and regulate neural progenitor cell maintenance and maturation. *Dev Biol* **367:** 100–113. doi:10.1016/j.ydbio.2012.03.026

Napoli I, Neumann H. 2009. Microglial clearance function in health and disease. *Neuroscience* **158:** 1030–1038. doi:10.1016/j.neuroscience.2008.06.046

Nemes-Baran AD, White DR, DeSilva TM. 2020. Fractalkine-dependent microglial pruning of viable oligodendrocyte progenitor cells regulates myelination. *Cell Rep* **32:** 108047. doi:10.1016/j.celrep.2020.108047

Nguyen PT, Dorman LC, Pan S, Vainchtein ID, Han RT, Nakao-Inoue H, Taloma SE, Barron JJ, Molofsky AB, Kheirbek MA, et al. 2020. Microglial remodeling of the extracellular matrix promotes synapse plasticity. *Cell* **182:** 388–403.e15. doi:10.1016/j.cell.2020.05.050

Nimmerjahn A, Kirchhoff F, Helmchen F. 2005. Resting microglial cells are highly dynamic surveillants of brain parenchyma in vivo. *Science* **308:** 1314–1318. doi:10.1126/science.1110647

O'Connor S, Montalescot G, Collet JP. 2011. The P2Y12 receptor as a target of antithrombotic drugs. *Purinergic Signal* **7**: 325–332. doi:10.1007/s11302-011-9241-z

Okabe Y, Medzhitov R. 2016. Tissue biology perspective on macrophages. *Nat Immunol* **17**: 9–17. doi:10.1038/ni .3320

Oosterhof N, Chang IJ, Karimiani EG, Kuil LE, Jensen DM, Daza R, Young E, Astle L, van der Linde HC, Shivaram GM, et al. 2019. Homozygous mutations in CSF1R cause a pediatric-onset leukoencephalopathy and can result in congenital absence of microglia. *Am J Hum Genet* **104**: 936–947. doi:10.1016/j.ajhg.2019.03.010

Païdassi H, Tacnet-Delorme P, Garlatti V, Darnault C, Ghebrehiwet B, Gaboriaud C, Arlaud GJ, Frachet P. 2008. C1q binds phosphatidylserine and likely acts as a multiligand-bridging molecule in apoptotic cell recognition. *J Immunol* **180**: 2329–2338. doi:10.4049/jimmunol.180.4.2329

Paloneva J, Kestilä M, Wu J, Salminen A, Böhling T, Ruotsalainen V, Hakola P, Bakker AB, Phillips JH, Pekkarinen P, et al. 2000. Loss-of-function mutations in TYROBP (DAP12) result in a presenile dementia with bone cysts. *Nat Genet* **25**: 357–361. doi:10.1038/77153

Paolicelli RC, Bolasco G, Pagani F, Maggi L, Scianni M, Panzanelli P, Giustetto M, Ferreira TA, Guiducci E, Dumas L, et al. 2011. Synaptic pruning by microglia is necessary for normal brain development. *Science* **333**: 1456–1458. doi:10.1126/science.1202529

Paolicelli RC, Sierra A, Stevens B, Tremblay ME, Aguzzi A, Ajami B, Amit I, Audinat E, Bechmann I, Bennett M, et al. 2022. Microglia states and nomenclature: a field at its crossroads. *Neuron* **110**: 3458–3483. doi:10.1016/j .neuron.2022.10.020

Park J, Choi Y, Jung E, Lee SH, Sohn JW, Chung WS. 2021. Microglial MERTK eliminates phosphatidylserine-displaying inhibitory post-synapses. *EMBO J* **40**: e107121. doi:10.15252/embj.2020107121

Park MD, Silvin A, Ginhoux F, Merad M. 2022. Macrophages in health and disease. *Cell* **185**: 4259–4279. doi:10.1016/j.cell.2022.10.007

Parkhurst CN, Yang G, Ninan I, Savas JN, Yates JR III, Lafaille JJ, Hempstead BL, Littman DR, Gan WB. 2013. Microglia promote learning-dependent synapse formation through brain-derived neurotrophic factor. *Cell* **155**: 1596–1609. doi:10.1016/j.cell.2013.11.030

Penfield W. 1925. Microglia and the process of phagocytosis in gliomas. *Am J Pathol* **1**: 77–90.15.

Penna E, Mangum JM, Shepherd H, Martínez-Cerdeño V, Noctor SC. 2021. Development of the neuro-immune-vascular plexus in the ventricular zone of the prenatal rat neocortex. *Cereb Cortex* **31**: 2139–2155. doi:10.1093/cercor/bhaa351

Perry VH, Hume DA, Gordon S. 1985. Immunohistochemical localization of macrophages and microglia in the adult and developing mouse brain. *Neuroscience* **15**: 313–326. doi:10.1016/0306-4522(85)90215-5

Philpot BD, Sekhar AK, Shouval HZ, Bear MF. 2001. Visual experience and deprivation bidirectionally modify the composition and function of NMDA receptors in visual cortex. *Neuron* **29**: 157–169. doi:10.1016/S0896-6273(01)00187-8

Pocock JM, Kettenmann H. 2007. Neurotransmitter receptors on microglia. *Trends Neurosci* **30**: 527–535. doi:10 .1016/j.tins.2007.07.007

Pont-Lezica L, Beumer W, Colasse S, Drexhage H, Versnel M, Bessis A. 2014. Microglia shape corpus callosum axon tract fasciculation: functional impact of prenatal inflammation. *Eur J Neurosci* **39**: 1551–1557. doi:10.1111/ejn .12508

Popova G, Soliman SS, Kim CN, Keefe MG, Hennick KM, Jain S, Li T, Tejera D, Shin D, Chhun BB, et al. 2021. Human microglia states are conserved across experimental models and regulate neural stem cell responses in chimeric organoids. *Cell Stem Cell* **28**: 2153–2166.e6. doi:10 .1016/j.stem.2021.08.015

Pozner A, Xu B, Palumbos S, Gee JM, Tvrdik P, Capecchi MR. 2015. Intracellular calcium dynamics in cortical microglia responding to focal laser injury in the PC::G5-tdT reporter mouse. *Front Mol Neurosci* **8**: 12. doi:10.3389/fnmol.2015.00012

Puñal VM, Paisley CE, Brecha FS, Lee MA, Perelli RM, Wang J, O'Koren EG, Ackley CR, Saban DR, Reese BE, et al. 2019. Large-scale death of retinal astrocytes during normal development is non-apoptotic and implemented by microglia. *PLoS Biol* **17**: e3000492. doi:10.1371/journal .pbio.3000492

Purves D, Lichtman JW. 1980. Elimination of synapses in the developing nervous system. *Science* **210**: 153–157. doi:10 .1126/science.7414326

Ransohoff RM, Cardona AE. 2010. The myeloid cells of the central nervous system parenchyma. *Nature* **468**: 253–262. doi:10.1038/nature09615

Ransohoff RM, Perry VH. 2009. Microglial physiology: unique stimuli, specialized responses. *Annu Rev Immunol* **27**: 119–145. doi:10.1146/annurev.immunol.021908.132528

Réu P, Khosravi A, Bernard S, Mold JE, Salehpour M, Alkass K, Perl S, Tisdale J, Possnert G, Druid H, et al. 2017. The lifespan and turnover of microglia in the human brain. *Cell Rep* **20**: 779–784. doi:10.1016/j.celrep.2017.07.004

Ribak CE, Shapiro LA, Perez ZD, Spigelman I. 2009. Microglia-associated granule cell death in the normal adult dentate gyrus. *Brain Struct Funct* **214**: 25–35. doi:10 .1007/s00429-009-0231-7

Rojo R, Raper A, Ozdemir DD, Lefevre L, Grabert K, Wollscheid-Lengeling E, Bradford B, Caruso M, Gazova I, Sánchez A, et al. 2019. Deletion of a Csf1r enhancer selectively impacts CSF1R expression and development of tissue macrophage populations. *Nat Commun* **10**: 3215. doi:10.1038/s41467-019-11053-8

Ronaldson PT, Davis TP. 2020. Regulation of blood–brain barrier integrity by microglia in health and disease: a therapeutic opportunity. *J Cereb Blood Flow Metab* **40**: S6–S24. doi:10.1177/0271678X20951995

Roumier A, Béchade C, Poncer JC, Smalla KH, Tomasello E, Vivier E, Gundelfinger ED, Triller A, Bessis A. 2004. Impaired synaptic function in the microglial KARAP/DAP12-deficient mouse. *J Neurosci* **24**: 11421–11428. doi:10.1523/JNEUROSCI.2251-04.2004

Rymo SF, Gerhardt H, Wolfhagen Sand F, Lang R, Uv A, Betsholtz C. 2011. A two-way communication between microglial cells and angiogenic sprouts regulates angiogenesis in aortic ring cultures. *PLoS ONE* **6**: e15846. doi:10.1371/journal.pone.0015846

Saederup N, Cardona AE, Croft K, Mizutani M, Cotleur AC, Tsou CL, Ransohoff RM, Charo IF. 2010. Selective chemokine receptor usage by central nervous system myeloid cells in CCR2-red fluorescent protein knock-in mice. *PLoS ONE* **5:** e13693. doi:10.1371/journal.pone.0013693

Safaiyan S, Kannaiyan N, Snaidero N, Brioschi S, Biber K, Yona S, Edinger AL, Jung S, Rossner MJ, Simons M. 2016. Age-related myelin degradation burdens the clearance function of microglia during aging. *Nat Neurosci* **19:** 995–998. doi:10.1038/nn.4325

Schafer DP, Lehrman EK, Kautzman AG, Koyama R, Mardinly AR, Yamasaki R, Ransohoff RM, Greenberg ME, Barres BA, Stevens B. 2012. Microglia sculpt postnatal neural circuits in an activity and complement-dependent manner. *Neuron* **74:** 691–705. doi:10.1016/j.neuron.2012.03.026

Schecter RW, Maher EE, Welsh CA, Stevens B, Erisir A, Bear MF. 2017. Experience-dependent synaptic plasticity in V1 occurs without microglial CX3CR1. *J Neurosci* **37:** 10541–10553. doi:10.1523/JNEUROSCI.2679-16.2017

Scott-Hewitt N, Perrucci F, Morini R, Erreni M, Mahoney M, Witkowska A, Carey A, Faggiani E, Schuetz LT, Mason S, et al. 2020. Local externalization of phosphatidylserine mediates developmental synaptic pruning by microglia. *EMBO J* **39:** e105380. doi:10.15252/embj.2020105380

Sedel F, Béchade C, Vyas S, Triller A. 2004. Macrophage-derived tumor necrosis factor α, an early developmental signal for motoneuron death. *J Neurosci* **24:** 2236–2246. doi:10.1523/JNEUROSCI.4464-03.2004

Sekar A, Bialas AR, de Rivera H, Davis A, Hammond TR, Kamitaki N, Tooley K, Presumey J, Baum M, Van Doren V, et al. 2016. Schizophrenia risk from complex variation of complement component 4. *Nature* **530:** 177–183. doi:10.1038/nature16549

Sierra A, Encinas JM, Deudero JJ, Chancey JH, Enikolopov G, Overstreet-Wadiche LS, Tsirka SE, Maletic-Savatic M. 2010. Microglia shape adult hippocampal neurogenesis through apoptosis-coupled phagocytosis. *Cell Stem Cell* **7:** 483–495. doi:10.1016/j.stem.2010.08.014

Sierra A, Abiega O, Shahraz A, Neumann H. 2013. Janus-faced microglia: beneficial and detrimental consequences of microglial phagocytosis. *Front Cell Neurosci* **7:** 6. doi:10.3389/fncel.2013.00006

* Sierra A, Miron VE, Paolicelli RC, Ransohoff RM. 2024. Microglial in health and diseases: integrative hubs of the central nervous system (CNS). *Cold Spring Harb Perspect Biol* **16:** a041366. doi: 10.1101/cshperspect.a041366

Silvin A, Uderhardt S, Piot C, Da Mesquita S, Yang K, Geirsdottir L, Mulder K, Eyal D, Liu Z, Bridlance C, et al. 2022. Dual ontogeny of disease-associated microglia and disease inflammatory macrophages in aging and neurodegeneration. *Immunity* **55:** 1448–1465.e6. doi:10.1016/j.immuni.2022.07.004

Sipe GO, Lowery RL, Tremblay MÈ, Kelly EA, Lamantia CE, Majewska AK. 2016. Microglial P2Y12 is necessary for synaptic plasticity in mouse visual cortex. *Nat Commun* **7:** 10905. doi:10.1038/ncomms10905

Smilack JD. 1999. The tetracyclines. *Mayo Clin Proc* **74:** 727–729. doi:10.4065/74.7.727

Smith C, McColl BW, Patir A, Barrington J, Armishaw J, Clarke A, Eaton J, Hobbs V, Mansour S, Nolan M, et al. 2020. Biallelic mutations in NRROS cause an early onset lethal microgliopathy. *Acta Neuropathol* **139:** 947–951. doi:10.1007/s00401-020-02137-7

Smolders SM, Swinnen N, Kessels S, Arnauts K, Smolders S, Le Bras B, Rigo JM, Legendre P, Brône B. 2017. Age-specific function of α5β1 integrin in microglial migration during early colonization of the developing mouse cortex. *Glia* **65:** 1072–1088. doi:10.1002/glia.23145

Song J, Zhong C, Bonaguidi MA, Sun GJ, Hsu D, Gu Y, Meletis K, Huang ZJ, Ge S, Enikolopov G, et al. 2012. Neuronal circuitry mechanism regulating adult quiescent neural stem-cell fate decision. *Nature* **489:** 150–154. doi:10.1038/nature11306

Squarzoni P, Oller G, Hoeffel G, Pont-Lezica L, Rostaing P, Low D, Bessis A, Ginhoux F, Garel S. 2014. Microglia modulate wiring of the embryonic forebrain. *Cell Rep* **8:** 1271–1279. doi:10.1016/j.celrep.2014.07.042

Stanley ER, Chitu V. 2014. CSF-1 receptor signaling in myeloid cells. *Cold Spring Harb Perspect Biol* **6:** a021857. doi:10.1101/cshperspect.a021857

Stevens B, Allen NJ, Vazquez LE, Howell GR, Christopherson KS, Nouri N, Micheva KD, Mehalow AK, Huberman AD, Stafford B, et al. 2007. The classical complement cascade mediates CNS synapse elimination. *Cell* **131:** 1164–1178. doi:10.1016/j.cell.2007.10.036

Stogsdill JA, Kim K, Binan L, Farhi SL, Levin JZ, Arlotta P. 2022. Pyramidal neuron subtype diversity governs microglia states in the neocortex. *Nature* **608:** 750–756. doi:10.1038/s41586-022-05056-7

Sulston JE, Horvitz HR. 1977. Post-embryonic cell lineages of the nematode, *Caenorhabditis elegans*. *Dev Biol* **56:** 110–156. doi:10.1016/0012-1606(77)90158-0

Sultan S, Gebara E, Toni N. 2013. Doxycycline increases neurogenesis and reduces microglia in the adult hippocampus. *Front Neurosci* **7:** 131. doi:10.3389/fnins.2013.00131

Tasdemir-Yilmaz OE, Freeman MR. 2014. Astrocytes engage unique molecular programs to engulf pruned neuronal debris from distinct subsets of neurons. *Genes Dev* **28:** 20–33. doi:10.1101/gad.229518.113

Thion MS, Ginhoux F, Garel S. 2018. Microglia and early brain development: an intimate journey. *Science* **362:** 185–189. doi:10.1126/science.aat0474

Tremblay MÈ, Lowery RL, Majewska AK. 2010. Microglial interactions with synapses are modulated by visual experience. *PLoS Biol* **8:** e1000527. doi:10.1371/journal.pbio.1000527

Trépanier MO, Hopperton KE, Mizrahi R, Mechawar N, Bazinet RP. 2016. Postmortem evidence of cerebral inflammation in schizophrenia: a systematic review. *Mol Psychiatry* **21:** 1009–1026. doi:10.1038/mp.2016.90

Ueno M, Fujita Y, Tanaka T, Nakamura Y, Kikuta J, Ishii M, Yamashita T. 2013. Layer V cortical neurons require microglial support for survival during postnatal development. *Nat Neurosci* **16:** 543–551. doi:10.1038/nn.3358

Umpierre AD, Bystrom LL, Ying Y, Liu YU, Worrell G, Wu LJ. 2020. Microglial calcium signaling is attuned to neuronal activity in awake mice. *eLife* **9:** e56502. doi:10.7554/eLife.56502

Umpierre AD, Li B, Ayasoufi K, Simon WL, Zhao S, Xie M, Thyen G, Hur B, Zheng J, Liang Y, et al. 2024. Microglial P2Y6 calcium signaling promotes phagocytosis and shapes neuroimmune responses in epileptogenesis. *Neuron* **112:** 1959–1977.e10. doi:10.1016/j.neuron.2024.03.017

Unger ER, Sung JH, Manivel JC, Chenggis ML, Blazar BR, Krivit W. 1993. Male donor-derived cells in the brains of female sex-mismatched bone marrow transplant recipients: a Y-chromosome specific in situ hybridization study. *J Neuropathol Exp Neurol* **52:** 460–470. doi:10.1097/00005072-199309000-00004

VanRyzin JW, Yu SJ, Perez-Pouchoulen M, McCarthy MM. 2016. Temporary depletion of microglia during the early postnatal period induces lasting sex-dependent and sex-independent effects on behavior in rats. *eNeuro* **3:** ENEURO.0297-16.2016. doi:10.1523/ENEURO.0297-16.2016

VanRyzin JW, Marquardt AE, Argue KJ, Vecchiarelli HA, Ashton SE, Arambula SE, Hill MN, McCarthy MM. 2019. Microglial phagocytosis of newborn cells is induced by endocannabinoids and sculpts sex differences in juvenile rat social play. *Neuron* **102:** 435–449.e6. doi:10.1016/j.neuron.2019.02.006

VanRyzin JW, Marquardt AE, Pickett LA, McCarthy MM. 2020. Microglia and sexual differentiation of the developing brain: a focus on extrinsic factors. *Glia* **68:** 1100–1113. doi:10.1002/glia.23740

Vaux DL, Korsmeyer SJ. 1999. Cell death in development. *Cell* **96:** 245–254. doi:10.1016/S0092-8674(00)80564-4

Viegi A, Cotrufo T, Berardi N, Mascia L, Maffei L. 2002. Effects of dark rearing on phosphorylation of neurotrophin Trk receptors. *Eur J Neurosci* **16:** 1925–1930. doi:10.1046/j.1460-9568.2002.02270.x

Wakselman S, Béchade C, Roumier A, Bernard D, Triller A, Bessis A. 2008. Developmental neuronal death in hippocampus requires the microglial CD11b integrin and DAP12 immunoreceptor. *J Neurosci* **28:** 8138–8143. doi:10.1523/JNEUROSCI.1006-08.2008

Warden AS, Han C, Hansen E, Trescott S, Nguyen C, Kim R, Schafer D, Johnson A, Wright M, Ramirez G, et al. 2023. Tools for studying human microglia: in vitro and in vivo strategies. *Brain Behav Immun* **107:** 369–382. doi:10.1016/j.bbi.2022.10.008

Weinhard L, Neniskyte U, Vadisiute A, di Bartolomei G, Aygün N, Riviere L, Zonfrillo F, Dymecki S, Gross C. 2018. Sexual dimorphism of microglia and synapses during mouse postnatal development. *Dev Neurobiol* **78:** 618–626. doi:10.1002/dneu.22568

Welsh CA, Stephany CÉ, Sapp RW, Stevens B. 2020. Ocular dominance plasticity in binocular primary visual cortex does not require C1q. *J Neurosci* **40:** 769–783. doi:10.1523/JNEUROSCI.1011-19.2019

Werner Y, Mass E, Ashok Kumar P, Ulas T, Händler K, Horne A, Klee K, Lupp A, Schütz D, Saaber F, et al. 2020. Cxcr4 distinguishes HSC-derived monocytes from microglia and reveals monocyte immune responses to experimental stroke. *Nat Neurosci* **23:** 351–362. doi:10.1038/s41593-020-0585-y

Wierzba-Bobrowicz T, Kosno-Kruszewska E, Gwiazda E, Lechowicz W. 1998. The comparison of microglia maturation in different structures of the human nervous system. *Folia Neuropathol* **36:** 152–160.

Wlodarczyk A, Holtman IR, Krueger M, Yogev N, Bruttger J, Khorooshi R, Benmamar-Badel A, de Boer-Bergsma JJ, Martin NA, Karram K, et al. 2017. A novel microglial subset plays a key role in myelinogenesis in developing brain. *EMBO J* **36:** 3292–3308. doi:10.15252/embj.201696056

Wu W, He S, Wu J, Chen C, Li X, Liu K, Qu JY. 2022. Long-term in vivo imaging of mouse spinal cord through an optically cleared intervertebral window. *Nat Commun* **13:** 1959. doi:10.1038/s41467-022-29496-x

Wyss-Coray T, Mucke L. 2002. Inflammation in neurodegenerative disease—a double-edged sword. *Neuron* **35:** 419–432. doi:10.1016/S0896-6273(02)00794-8

Xiao C, Tifft C, Toro C. 2022. Sandhoff disease. In *Genereviews* (ed. Adam MP, et al.). University of Washington, Seattle, WA.

Yilmaz M, Yalcin E, Presumey J, Aw E, Ma M, Whelan CW, Stevens B, McCarroll SA, Carroll MC. 2021. Overexpression of schizophrenia susceptibility factor human complement C4A promotes excessive synaptic loss and behavioral changes in mice. *Nat Neurosci* **24:** 214–224. doi:10.1038/s41593-020-00763-8

Yona S, Kim KW, Wolf Y, Mildner A, Varol D, Breker M, Strauss-Ayali D, Viukov S, Guilliams M, Misharin A, et al. 2013. Fate mapping reveals origins and dynamics of monocytes and tissue macrophages under homeostasis. *Immunity* **38:** 79–91. doi:10.1016/j.immuni.2012.12.001

Yu D, Li T, Delpech JC, Zhu B, Kishore P, Koshi T, Luo R, Pratt KJB, Popova G, Nowakowski TJ, et al. 2022. Microglial GPR56 is the molecular target of maternal immune activation-induced parvalbumin-positive interneuron deficits. *Sci Adv* **8:** eabm2545. doi:10.1126/sciadv.abm2545

Zhan Y, Paolicelli RC, Sforazzini F, Weinhard L, Bolasco G, Pagani F, Vyssotski AL, Bifone A, Gozzi A, Ragozzino D, et al. 2014. Deficient neuron-microglia signaling results in impaired functional brain connectivity and social behavior. *Nat Neurosci* **17:** 400–406. doi:10.1038/nn.3641

Zhang Y, Chen K, Sloan SA, Bennett ML, Scholze AR, O'Keeffe S, Phatnani HP, Guarnieri P, Caneda C, Ruderisch N, et al. 2014. An RNA-sequencing transcriptome and splicing database of glia, neurons, and vascular cells of the cerebral cortex. *J Neurosci* **34:** 11929–11947. doi:10.1523/JNEUROSCI.1860-14.2014

Zhao X, Eyo UB, Murugan M, Wu LJ. 2018. Microglial interactions with the neurovascular system in physiology and pathology. *Dev Neurobiol* **78:** 604–617. doi:10.1002/dneu.22576

Zhou Z, Hartwieg E, Horvitz HR. 2001. CED-1 is a transmembrane receptor that mediates cell corpse engulfment in *C. elegans*. *Cell* **104:** 43–56. doi:10.1016/S0092-8674(01)00190-8

Zhou T, Li Y, Li X, Zeng F, Rao Y, He Y, Wang Y, Liu M, Li D, Xu Z, et al. 2022. Microglial debris is cleared by astrocytes via C4b-facilitated phagocytosis and degraded via RUBICON-dependent noncanonical autophagy in mice. *Nat Commun* **13:** 6233. doi:10.1038/s41467-022-33932-3

Zipp F, Bittner S, Schafer DP. 2023. Cytokines as emerging regulators of central nervous system synapses. *Immunity* **56:** 914–925. doi:10.1016/j.immuni.2023.04.011fig

The Blood–Brain Barrier: Composition, Properties, and Roles in Brain Health

Baptiste Lacoste,[1,2,3] Alexandre Prat,[4] Moises Freitas-Andrade,[1] and Chenghua Gu[5]

[1]Ottawa Hospital Research Institute, Neuroscience Program, Ottawa, Ontario K1H 8M5, Canada

[2]Department of Cellular and Molecular Medicine, University of Ottawa, Ottawa, Ontario K1H 8M5, Canada

[3]University of Ottawa Brain and Mind Research Institute, Ottawa, Ontario K1H 8M5, Canada

[4]Department of Neuroscience, Université de Montréal, Montréal, Québec H2X 0A9, Canada

[5]Department of Neurobiology, Howard Hughes Medical Institute, Harvard Medical School, Boston, Massachusetts 02115, USA

Correspondence: blacoste@uottawa.ca; Chenghua_Gu@hms.harvard.edu

Blood vessels are critical to deliver oxygen and nutrients to tissues and organs throughout the body. The blood vessels that vascularize the central nervous system (CNS) possess unique properties, termed the blood–brain barrier (BBB), which allow these vessels to tightly regulate the movement of ions, molecules, and cells between the blood and the brain. This precise control of CNS homeostasis allows for proper neuronal function and protects the neural tissue from toxins and pathogens, and alterations of this barrier are important components of the pathogenesis and progression of various neurological diseases. The physiological barrier is coordinated by a series of physical, transport, and metabolic properties possessed by the brain endothelial cells (ECs) that form the walls of the blood vessels. These properties are regulated by interactions between different vascular, perivascular, immune, and neural cells. Understanding how these cell populations interact to regulate barrier properties is essential for understanding how the brain functions in both health and disease contexts.

Blood vessels convey blood from the heart throughout the body, which is essential to deliver oxygen and nutrients to the tissues, to remove carbon dioxide and metabolic waste, to convey hormonal signaling, and to mediate the interactions with the peripheral immune system.[6]

The vascular tree is comprised of arteries and arterioles, followed by the capillary bed, which is essential for gas and nutrient exchange and venules and veins draining blood from tissues. Each segment has different properties depending on where they are in the vascular tree as well as which organ they vascularize. In particular, the microvasculature, made up of capillaries and postcapillary venules, has different properties to meet the unique requirements of the organ it vascularizes. There are three main structural classes of capillaries. Continuous nonfenestrated capillaries of the skin and lung are joined togeth-

---

[6]This is an update to a previous article published in *Cold Spring Harbor Perspectives in Biology* [Daneman and Prat (2015). *Cold Spring Harb Perspect Biol* 7: a020412. doi:10.1101/cshperspect.a020412].

Cite this article as *Cold Spring Harb Perspect Biol* doi: 10.1101/cshperspect.a041422

er by cellular junctions, have a complete basement membrane (BM), and lack fenestra (pores) in their plasma membrane. Continuous fenestrated vessels of the intestinal villi and endocrine glands have a similar continuous structure but contain diaphragmed fenestra throughout their membrane. Discontinuous capillaries in the liver have large gaps throughout the cell and an incomplete BM. These classes of capillaries differ greatly in their regulation of the movement of solutes between the blood and the tissues, with continuous nonfenestrated capillaries being the most restrictive, and discontinuous being the least restrictive (Aird 2007a,b).

The term blood–brain barrier (BBB) is used to describe the unique properties of the microvasculature of the central nervous system (CNS). CNS vessels are continuous nonfenestrated vessels, but also contain a series of additional properties that allow them to tightly regulate the movement of molecules, ions, and cells between the blood and the CNS (Zlokovic 2008; Daneman 2012). This heavily restricting barrier capacity allows BBB endothelial cells (ECs) to control CNS homeostasis, which is critical to allow for proper neuronal function, as well as protect the CNS from toxins, pathogens, inflammation, and injury. The restrictive nature of the BBB provides an obstacle for drug delivery to the CNS, and, thus, major efforts have been made to generate methods to modulate or bypass the BBB for the delivery of therapeutics (Larsen et al. 2014). Loss of some, or most, of these barrier properties during neurological diseases, including stroke, multiple sclerosis (MS), brain traumas, and neurodevelopmental and neurodegenerative disorders, is a major component of the pathology and progression of these diseases (Zlokovic 2008; Daneman 2012; Ouellette and Lacoste 2021).

While much of what is known about the BBB was discovered largely through animal/rodent studies, recent technical advancements have facilitated the use of human-derived brain vascular cells, expanding our understanding of the BBB (Nishihara et al. 2022; Matsuo et al. 2023; Spitzer et al. 2023). Organoid systems (Bergmann et al. 2018; Koh and Hagiwara 2024) and microfluidic devices (Park et al. 2019; Hajal et al. 2022) provide 3D models to investigate permeability characteristics of human BBB and therapeutic strategies for drug delivery. Advances in human-induced pluripotent stem cell (hiPSC) technology offer an opportunity to understand BBB contributions to neuropathology, as well as reveal similarities/differences between human and animal models of disease (Katt et al. 2019; Nishihara et al. 2022). For instance, BBB dysfunction in MS, is generally understood to be a consequence of neuroinflammation. However, using human MS–derived brain microvascular endothelial-like cells, as an in vitro model of the BBB, Nishihara et al. (2022) demonstrated that BBB impairment is an intrinsic characteristic of ECs in MS. Together, with the advent of single-cell transcriptomic profiling, of both human and mouse brain ECs, a clearer picture of conserved and divergent mechanisms is beginning to surface (Vanlandewijck et al. 2018; Munji et al. 2019; Kalucka et al. 2020; Song et al. 2020; Dion-Albert et al. 2022; Yang et al. 2022). While these new approaches are of value, important limitations have been uncovered (Sabbagh and Nathans 2020; Lu et al. 2021). For instance, BBB models derived from hiPSC were shown to lack functional attributes of ECs, since they are deficient in vascular lineage genes (Lu et al. 2021). In addition, there is evidence that cultured ECs quickly lose their endogenous BBB features (Sabbagh and Nathans 2020). Importantly, as with many nascent technologies, interpretation of the data requires diligent and critical analysis.

CELLS OF THE BBB

Blood vessels are made up of two main cell types: ECs that form the walls of the blood vessels, and mural cells that sit on the abluminal surface of the EC layer. The properties of the BBB are largely manifested within the ECs but are induced and maintained by critical interactions with mural cells, immune cells, glial cells, and neural cells, which interact in the neurovascular unit (NVU) (Fig. 1).

Endothelial Cells

ECs are mesodermal-derived, modified simple squamous epithelial cells lining the blood vessel wall. The diameter of large arteries and veins can

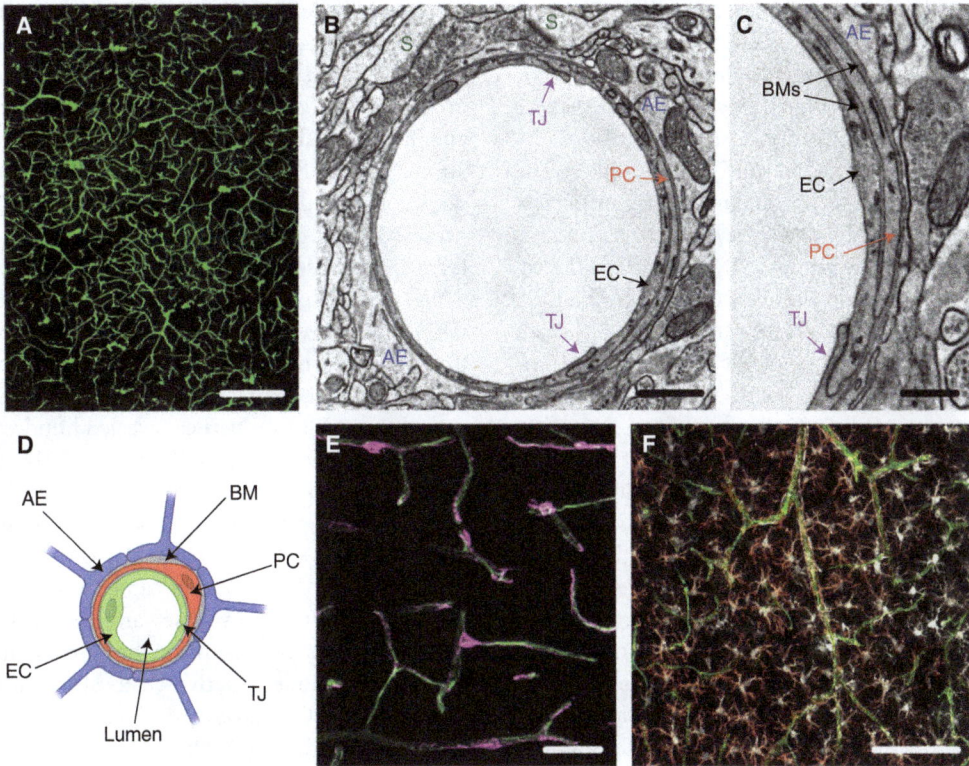

Figure 1. Components of the blood–brain barrier (BBB). (*A*) Immunostained endothelial networks of mouse cerebral cortex showing the density of the central nervous system (CNS) vasculature. (*B*) Electron micrograph (EM) of a cross section of a CNS capillary depicts the close relationship between endothelial cells (ECs), pericytes (PCs), and astrocytic endfeet (AE). (S) Synapses, (TJs) tight junctions. (*C*) Higher magnification of EM in *B* depicts the relationship among EC, PC, basement membranes (BMs), and AE. (*D*) Schematic representation of these elements within the context of the neurovascular unit. (*E*) The immunofluorescence micrograph depicts the relationship between PC (magenta) and EC (green). (*F*) The immunofluorescence micrograph depicts the relationship of astrocytes (doubly labeled with GFAP, red, and Aldh1l1-GFP, white) with blood vessels (green). Astrocytes extend processes that ensheath the blood vessels (courtesy of Pavel Kotchetkov and Julie Ouellette). Scale bars, 200 μm (*A*); 1 μm (*B*); 500 nm (*C*); 100 μm (*E*); 200 μm (*F*).

be made up of dozens of ECs, whereas the smallest capillary is formed by a single EC folding onto itself to form the lumen of the vessel (Conway and Carmeliet 2004; Aird 2007a,b). These CNS microvascular ECs are extremely thin, 39% thinner than muscle ECs, with less than a quarter of a micron separating the luminal from the parenchymal surface (Fig. 1B,C; Coomber and Stewart 1985).

CNS ECs have unique properties compared with ECs in other tissues. CNS ECs are held together by tight junctions (TJs), which greatly limit the paracellular flux of solutes (Reese and Kar-

novsky 1967; Brightman and Reese 1969; Westergaard and Brightman 1973). CNS ECs undergo extremely low rates of transcytosis as compared with peripheral ECs, which greatly restricts the vesicle-mediated transcellular movement of solutes (Coomber and Stewart 1985; Ben-Zvi et al. 2014). This tight regulation of paracellular and transcellular trafficking creates a polarized cell with distinct luminal and abluminal membrane compartments, such that movement between the blood and the brain can be tightly controlled through polarized cellular transport properties (Betz and Goldstein 1978; Betz et al. 1980).

There are two main categories of transporters expressed by CNS ECs. The first are efflux transporters that transport a wide variety of lipophilic molecules that could otherwise diffuse across the cell membrane (Cordon-Cardo et al. 1989; Thiebaut et al. 1989; Löscher and Potschka 2005). The second are highly specific nutrient transporters that facilitate the transport of specific nutrients across the BBB into the CNS, as well as the removal of specific waste products from the CNS into the blood (Mittapalli et al. 2010). CNS ECs contain higher amounts of mitochondria compared to other ECs (Oldendorf et al. 1977), which are thought to be critical to generate ATP to drive the ion gradients critical for transport functions. CNS ECs also express an extremely low level of leukocyte adhesion molecules (LAMs), as compared with ECs in other tissues greatly limiting the number of immune cells that enter the CNS (Henninger et al. 1997; Aird 2007a; Daneman et al. 2010a). In addition, a recent study of single-cell transcriptome atlas of murine ECs from many organs including the brain revealed differential vascular metabolism in CNS ECs (Kalucka et al. 2020). For example, focusing on multiple genes within a single metabolic pathway, they observed that ECs from the brain up-regulated the expression of transporters that were involved in the transport of glucose (*Slc2a1*), amino acids (*Slc3a2*, *Slc7a5*), and fatty acids (*Bsg*). This unbiased approach demonstrated that the heterogeneity of metabolic gene expression signatures of ECs in vascular beds across tissues contributes to the tissue-grouping phenomenon of ECs. The combination of TJs, low transcytosis, low LAMs, efflux transporters, specific metabolism, as well as specific transporters to deliver required nutrients, allows the CNS ECs to tightly regulate brain function.

Heterogeneity of permeability across brain regions has been recently demonstrated. Emerging evidence shows that different brain regions show different levels of impermeability. For example, the circumventricular organs (CVOs), seven specialized regions that include the median eminence (ME), are naturally leaky to perform their functions (Kaur and Ling 2017) despite being directly adjacent to regions with a sealed BBB. CVO neurons sense signaling compounds and secrete brain-derived hormones in the systemic circulation to facilitate rapid communication between the brain and the periphery and regulate processes like feeding, cardiovascular function, and thirst (Ufnal and Skrzypecki 2014; Jiang et al. 2020). The capillaries of these CVOs are continuous fenestrated vessels, and these capillary ECs express Plvap. A recent study established a platform of unbiased, high-throughput single-cell RNA sequencing (scRNA-seq) of one of the CVOs, the ME, and a size-matched region of the somatosensory cortex (cortex) in the mouse. Comparison of these two small brain regions with distinct BBB properties revealed hundreds of regional enriched genes (Pfau et al. 2021), including the lack of canonical Wnt signaling in CVO capillary ECs (Benz et al. 2019; Wang et al. 2019). However, the molecular and subcellular structure differences in capillary ECs between these CVO's leaky vessels and BBB vessels are not the only contributors to the permeability heterogeneity; indeed, perivascular cells, including pericytes (PCs) and astrocytes, as well as their interactions with neighboring ECs, also underlie the permeability heterogeneity (Pfau et al. 2021).

Mural Cells

Mural cells include vascular smooth muscle cells (VSMCs) that surround the large vessels, and PCs that incompletely cover the endothelial wall of the microvasculature. PCs are cells that sit on the abluminal surface of the microvascular endothelial tube and are embedded within the vascular BM (Sims 1986). A challenge in studying PCs is the lack of a specific marker that is expressed uniquely by PCs, and, thus, these cells are often confused with other cells that sit in the perivascular space (Armulik et al. 2011). Currently, the most widely accepted molecular identifier of CNS PCs is positive reactivity to both PDGFR-β and NG2; but other markers, including Anpep (CD13), Desmin, Rgs5, Abcc9, Kcnj8, Dlk, and Zic1, have all been used to identify PCs, with none being perfect identifiers of this cell type (Armulik et al. 2011). Without a PC-specific mark, it has been difficult to assess whether all of the different functions attributed to PCs are performed by all of the same cells, by different subsets of PCs, or even by non-PC cells that sit adjacent to the vasculature. A recent report on identifying ATP13A5 as a marker for CNS PCs in mice may shed new light on solving this issue (Guo et al. 2021). The *Atp13a5* Cre mouse line will be a valuable tool for studying CNS PC-specific functions.

PCs extend long cellular processes along the abluminal surface of the endothelium that can often span several EC bodies. These cells contain contractile proteins and have the ability to constrict to control the diameter of capillaries (Peppiatt et al. 2006; Hall et al. 2014; Hartmann et al. 2021). Although these cells line the endothelial tube, most of the cell body and processes do not touch the endothelium but are separated by the BM they are embedded within. The processes do form cellular adhesions with the endothelium at discrete points, known as peg-and-socket junctions, mediated by the adhesion molecule N-cadherin (Gerhardt et al. 2000). In addition, other PC–endothelial cellular adhesions have been identified including adhesion plaques, gap junctions, and TJs (Courtoy and Boyles 1983; Cuevas et al. 1984; Larson et al. 1987; Diaz-Flores et al. 2009).

CNS PCs possess unique properties compared to PCs in other tissues. CNS PCs are derived from the neural crest, in contrast with PCs in many peripheral tissues, which are derived from the mesoderm (Majesky 2007). In addition, the CNS microvasculature has the highest CNS PCs coverage of any tissue, with an endothelial: PC ratio estimated between 1:1 and 3:1, whereas the muscle has a ratio of 100:1 (Shepro and Morel 1993). PCs play important roles in regulating angiogenesis, deposition of extracellular matrix (ECM), wound healing, regulating immune cell infiltration, and regulation of blood flow in response to neural activity, and reports suggest that they also can be multipotent stem cells of the CNS (Armulik et al. 2011; Karow et al. 2018). In addition, these cells are important for regulating the formation of the BBB during development, as well as maintaining its function in adulthood and aging (Armulik et al. 2010; Daneman et al. 2010b).

While brain PCs show transcriptomic differences from PCs in the periphery (Vanlandewijck et al. 2018), recent studies have revealed the heterogeneity of PCs within the brain (Pfau et al. 2021). Using spatial transcriptome analysis by comparing PCs associated with capillary ECs from the cortex versus CVOs, more than 100 regional enriched genes have been identified. Consistent with the molecular differences, PCs from the two regions also exhibit clear differences in morphology as well as their interactions and coverage of neighboring ECs. Additionally, while cortex PCs generally interacted with a single capillary, some ME PCs contacted more than one capillary resulting in a decreased PC-to-EC ratio. Interestingly, PCs associated with the leaky vessels in the CVO regions show more transcriptional similarity to mural cells in peripheral tissues than cortex PCs.

PC heterogeneity also exists across species. In contrast to one major type of PC in the mouse brain (Vanlandewijck et al. 2018), recent human brain vascular single-nucleus transcriptome studies have discovered two subtypes of human PCs, marked by solute transport and ECM organization (Yang et al. 2022). Moreover, the selective vulnerability of ECM-maintaining PCs and gene expression patterns that implicate dysregulated blood flow were observed in Alzheimer's disease (AD) samples. Further comparison with mouse PC transcriptome and validation in both species will be needed to understand PC molecular and functional heterogeneity across species.

Astrocytes

Astrocytes are a major glial cell type, which extend numerous polarized cellular processes known as peripheral astrocytic processes (Felix et al. 2021) that ensheath either neuronal synapsis, a complex referred to as the tripartite synapse, and blood vessels (Abbott et al. 2006). The endfeet of the basal process almost completely ensheath the cerebral blood vessels; they contribute to the composition of the BM and contain a discrete array of proteins, including dystroglycan, dystrophin, large-conductance Ca^{2+}-dependent K^+ channels, inward-rectifying K^+ channels, connexins, megalencephalic leukoencephalopathy with subcortical cysts 1 (MCL1), prostaglandins and phospholipases, as well as aquaporin 4. The dystroglycan–dystrophin complex is important to link the endfeet cytoskeleton to the BM by binding agrin (Noell et al. 2011; Wolburg et al. 2011). This linkage coordinates aquaporin 4 into orthogonal arrays of particles, which is critical for regulating water homeostasis in the CNS. Astrocytes provide a cellular link between the neuronal circuitry and blood vessels. This gliovascular coupling enables astrocytes to relay signals that regulate blood flow in response to neuronal activity (Attwell et al. 2010; Gordon et al. 2011). This includes regulating the contraction/dilation of VSMCs surrounding arterioles as well as PCs surrounding capillaries. Astrocytes have been identified as important mediators of BBB formation and function because of the ability of purified astrocytes to induce barrier properties in non-CNS blood vessels in transplantation studies (Janzer and Raff 1987), as well as induce barrier properties in cultured ECs in in vitro coculture paradigms (Abbott et al. 2006). One issue with these studies is that the astrocytes are often cultured from neonatal rodent brains and go through many rounds of cell division, suggesting that these studies are analyzing progenitor cells as opposed to mature astrocytes. Recent data analyzing the BBB in dissected rodent embryos suggest that the BBB is formed before astrocyte generation and ensheathment of the vasculature (Daneman et al. 2010b), and, thus, these cells do not play a role in the initial induction of the BBB. The identification of astrocyte-secreted factors that do regulate BBB function suggests that mature astrocytes modulate and maintain the barrier once it is formed. With the advent of cell-type-specific gene databases and improved cell culturing methods to isolate mature astrocytes, future research will better define the signaling pathways that govern the development and maintenance of endfoot–BBB interactions (Cahoy et al. 2008; Guttenplan and Liddelow 2019; Yosef et al. 2020; Freitas-Andrade et al. 2023).

Immune Cells

CNS blood vessels interact with different immune cell populations both in the blood as well as within the CNS. The two main cell populations within the CNS are perivascular macrophages and microglial cells. Perivascular macrophages are monocyte lineage cells that sit on the abluminal side of the vascular tube commonly found in the Virchow–Robin space, a small fluid-filled canal that lines the abluminal surface of the veins and arteries that enter/leave the CNS (Hickey and Kimura 1988; Polfliet et al. 2001). These cells are derived from blood-borne progenitors, and chimera experiments suggest that they are able to cross the BBB and can be 80% replaced within 3 months (Unger et al. 1993; Vass et al. 1993; Williams et al. 2001). These cells provide a first line of innate immunity by phagocytosing cellular debris. Microglial cells are resident CNS parenchymal immune cells that are derived from progenitors in the yolk sac and enter the brain during embryonic development (Ginhoux et al. 2010). These cells are involved in regulating neuronal development, innate immune response, and wound healing, and can act as antigen-presenting cells in adaptive immunity (Streit et al. 2005; Ajami et al. 2007). In addition, different blood-borne immune cell populations, including neutrophils, T cells, and macrophages, can interact with CNS vessels when activated and are thought to regulate BBB properties in response to infection, injury, and disease by releasing reactive oxygen species that can increase vascular permeability (Persidsky et al. 1999; Hudson et al. 2005). Identifying the mechanisms by which both the immune cells and the BBB become "activated" to interact may be important in deciphering the mechanisms by which the BBB is disrupted during different pathological states. Significant gaps in knowledge include how inflammatory signals are relayed from the brain across the BBB to circulating lymphocytes, how ECs recruit T cells to areas of inflammation, and where along the vascular tree this inflammatory signaling is occurring. For detailed reviews on immune cell trafficking at the BBB, see other published reviews on this topic (Ransohoff and Engelhardt 2012; Galea 2021; Sierra et al. 2024).

Basement Membrane

The vascular tube is surrounded by two BMs, the inner vascular BM and the outer parenchymal

Cite this article as *Cold Spring Harb Perspect Biol* doi: 10.1101/cshperspect.a041422

BM, also called the vascular glia limitans perivascularis (Del Zoppo et al. 2006; Sorokin 2010). The vascular BM is an ECM secreted by the ECs and PCs, whereas the parenchymal BM is primarily secreted by astrocytic processes that extend toward the vasculature. These BMs are comprised of different secreted molecules, including type IV collagens, laminin, nidogen, heparin sulfate proteoglycans, and other glycoproteins. The vascular and parenchymal BMs have different compositions; for instance, the former is made up of laminins a4 and a5, whereas the latter contains laminins a1 and a2 (Wu et al. 2009; Sorokin 2010; Nirwane and Yao 2022). These BMs provide an anchor for many signaling processes at the vasculature. For example, integrin-mediated and ECM-driven regulation of Wnt signaling (Astudillo and Larraín 2014) modulates barrier properties. Deletion of β1 integrin in ECs impairs vascular endothelial (VE)-cadherin signaling, which causes loss of TJs resulting in a leaky barrier (Yamamoto et al. 2015). Moreover, integrin α5 in the CNS ECs acts as a receptor for vitronectin, an ECM protein secreted by neighboring PCs, to regulate blood–CNS barrier function. Genetic ablation of vitronectin or mutating vitronectin to prevent integrin binding as well as endothelial-specific deletion of integrin α5 causes barrier leakage in mice (Ayloo et al. 2022). These BMs also provide an additional barrier for molecules and cells to cross before accessing the neural tissue. Disruption of these BMs by matrix metalloproteinases (MMPs) is an important component of BBB dysfunction and leukocyte infiltration that is observed in many different neurological disorders.

FEATURES AND MOLECULES OF THE BBB

The discovery of molecules expressed by CNS ECs has led to the identification of important structural and transport components of the BBB (Fig. 2; Li et al. 2001, 2002; Enerson and Drewes 2006; Cayrol et al. 2008; Daneman et al. 2009, 2010a; Ohtsuki et al. 2014). Recently, the use of large-scale genomic and proteomic experimental approaches has provided greater detail and understanding of the molecular biology of the BBB. The use of acutely purified microvascular fragments, acutely purified ECs, and cultured

ECs combined with microarray technology, RNA sequencing, and mass spectrometry proteomic analysis have enabled large-scale gene expression comparisons of CNS ECs with neural cells as well as ECs from other tissues. In particular, a comparison of the molecular differences between CNS ECs and ECs from nonneural tissues has provided an understanding of the unique molecular composition of the BBB (Munji et al. 2019).

Tight Junctions

CNS ECs are held together by TJs, which create a high-resistance paracellular barrier to molecules and ions, polarizing the luminal and abluminal compartments. Most of what is known about TJs is from work on ECs, which have identified that these cellular adhesions are formed on the apical part of the lateral membrane by homotypic and heterotypic interactions of transmembrane molecules that are linked to the cytoskeleton through interactions with cytoplasmic adaptors. The strength of the junctions varies greatly depending on the tissue in which they are found, and work in cell culture suggests that they have a size-selective permeability to uncharged molecules of up to 4 nm, and then low permeability to larger molecules (Van Itallie and Anderson 2006; Van Itallie et al. 2008). This suggests that the TJs form a 4-nm pore and that larger molecules would pass through discontinuities in the junctions.

The transmembrane molecules include claudins, occludins, and junctional adhesion molecules (JAMs) (Furuse 2010). Claudins are a class of more than 25 different family members that are tetraspanins characterized by a W-GLW-C-C domain in the first extracellular loop (Gupta and Ryan 2010). Evidence in vitro suggests that claudins are essential for the paracellular barrier formation. Expression of claudins is sufficient to form TJ strands in fibroblasts, and disruption of claudins decreases the paracellular barrier properties of canine kidney cells (Furuse et al. 1998, 2001; Van Itallie et al. 2001; Amasheh et al. 2005; Hou et al. 2006). Work with chimeric claudins has shown that amino acid residues in the first extracellular loop define the size and charge se-

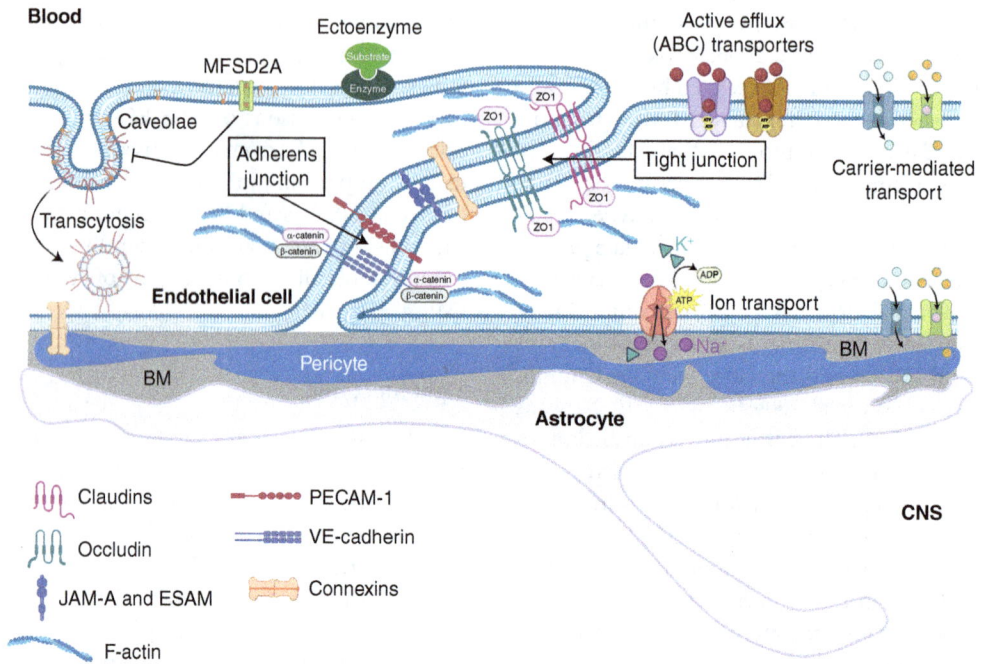

Figure 2. Schematic representation of molecules of the blood–brain barrier (BBB). A diverse array of transporters expressed at the luminal (blood) and abluminal (central nervous system [CNS]) regions of endothelial cells (ECs) facilitates the movement of specific substrates into the brain, while removing metabolic byproducts and toxins into the blood. Tight junctions form a physical barrier that inhibits paracellular diffusion into the CNS, and MFSD2A plays a central role in limiting the transcytotic activity in ECs. Together with ectoenzymes, these molecules form a dynamic barrier at the EC level and help maintain brain homeostasis.

lectivity of the pore within the cellular junction, and, thus, the composition of the claudins within a given cell can determine the permeability of the paracellular barrier (Colegio et al. 2003). Different claudin family members are expressed by different epithelial barriers in different tissues, and mouse knockouts (KOs) have shown that specific family members are essential for specific barriers (cldn1, epidermal barrier, cldn16 kidney epithelia, cldn11, CNS myelin, cldn19, peripheral myelin), many of which have been associated with human disease (Gow et al. 1999; Furuse et al. 2002; Hadj-Rabia et al. 2004; Knohl and Scheinman 2004; Miyamoto et al. 2005; Hampson et al. 2008). Claudin-5 has been shown to be highly expressed by CNS ECs, and mice that lack claudin-5 have a size-selective leak of the BBB (Morita et al. 1999; Nitta et al. 2003). Although earlier studies claimed that CLDN1, CLDN3, and CLDN12 are also expressed at the BBB, recent studies have shown that mRNA for these claudins are either undetectable or at extremely low levels in brain ECs, and that these claudins are dispensable for barrier function (Pfeiffer et al. 2011; Castro Dias et al. 2019a,b).

Occludin is a tetraspanin expressed by epithelial cells and CNS ECs; an in vitro culture experiment disrupting occludin homotypic interactions suggests that it is important for the resistance of the barrier (Balda et al. 1996; McCarthy et al. 1996; Wong and Gumbiner 1997). Occludin is highly enriched in CNS ECs compared with ECs in nonneural tissues, indicating that it may be an important component of the barrier. Occludin-deficient mice, however, are shown to have a normal high-resistance epithelial barrier and a functioning BBB. These mice do have calcification of the CNS suggesting that perhaps occludin specifically regulates calcium flux across the BBB (Saitou et al. 2000). JAMs are immunoglobulin superfamily members that form homotypic interactions at TJs in epithelial cells and ECs. JAMs have been shown to regulate leukocyte extravasation as well as paracellular permeability (Martìn-Padura et al. 1998; Johnson-Legér et al. 2002; Ludwig et al. 2005). In particular, JAM4 has been identified at the BBB in mice (Daneman et al. 2010a). Recently, it has been shown that unique molecular components are required to seal the paracellular barrier at the contact points of three cells. These tricellular junctional complexes are made up of lipolysis-stimulated receptor (LSR), which is required to localize marveld2 to tricellular adhesions (Masuda et al. 2011). It remains unclear what the nature of the size/charge-selective pore is formed by the specific composition of TJ proteins expressed by CNS ECs, and whether the permeability of this pore is static or whether it is dynamically altered in response to neuronal activity.

The transmembrane adhesion complexes are linked to the cytoskeleton through a series of cytoplasmic adaptors including ZO-1, ZO-2, Cingulin, Jacop, MAGIs, and MPPs (Van Itallie and Anderson 2013). In addition, the TJs interact with basal adherens junctions (AJs), which connect all ECs, are made up of VE-cadherin and platelet EC adhesion molecules (PECAM)1, and are linked to the cytoskeleton by catenins. Interestingly, many of the TJ proteins identified, including cldn5, cldn12, ZO-1, and ZO2, appear to be expressed by ECs in all tissues. Thus, a major question is why only CNS ECs form this tight barrier and not ECs in other tissues. Transcriptional analysis comparing CNS ECs with peripheral ECs suggests that several cytoplasmic adaptors, including jacop and MPP7, as well as tricellular TJ molecules, LSR, and marveld2 (Daneman et al. 2010a), are enriched at the BBB suggesting that these molecules may be critical for this barrier formation.

Transporters

CNS ECs are highly polarized cells that have distinct luminal and abluminal compartments. The low permeability of the paracellular junctions allows the transport properties of the cells to control the movement of ions and molecules between the blood and the brain. There are two main types of transporters expressed by CNS ECs: efflux transporters and nutrient transporters. Current work to elucidate the full array of transporters and their substrates is highly sought after both to understand the external requirements for brain metabolism and function, but also to identify targets to aid in drug delivery across the BBB.

Efflux transporters, including Mdr1, BCRP, and MRPs, use the hydrolysis of ATP to transport their substrates up their concentration gradient (Ha et al. 2007). Many of these transporters are localized to the luminal surface and transport a wide array of substrates into the blood compartment. This wide substrate diversity allows these transporters to provide a barrier to many small lipophilic molecules, which would otherwise passively diffuse through the EC membrane. Mdr1, also called P-glycoprotein, has been widely studied in this context, and KO mice show an increase in a wide variety of small lipophilic drugs entering the brain, as well as endogenous molecules (Schinkel et al. 1994, 1995, 1996). Up-regulation of Mdr1 has also been associated with drug-resistant epilepsy and tumors (Potschka et al. 2001; Abbott et al. 2002). An important avenue of research uses structural modeling to predict substrates of these efflux transporters to develop therapeutics that can avoid efflux and, thus, gain entry to the CNS. In addition, developing inhibitors of these efflux transporters is an ongoing research avenue to aid in the delivery of small molecule compounds to the CNS. Interestingly, not much is known about the endogenous molecules that are effluxed by these transporters, how this is important to regulate brain function, and whether inhibitors would alter the tissue distribution of important endogenous molecules.

Nutrient transporters facilitate the movement of specific nutrients down their concentration gradient. CNS ECs express a wide variety of these transporters to deliver very specific nutrients across the physical barrier of the CNS ECs into the CNS parenchyma. Many of these belong to the solute carrier class of facilitated transporters, including slc2a1 (glucose), slc16a1 (lactate, pyruvate), slc7a1 (cationic amino acids), and slc7a5 (neutral amino acids, L-DOPA) (Zlokovic

2008; Daneman 2012). Slc2a1, also called glut1, has been largely studied for its role in providing the CNS with glucose. Expression of this transporter is highly enriched in CNS ECs compared with ECs in nonneural tissues, and it facilitates the transport of glucose down its concentration gradient from the blood into the brain (Cornford et al. 1994). In humans, Glut1 deficiency leads to an epileptic syndrome that is treated by being fed a high-ketone diet (De Vivo et al. 1991, 2002). In addition, CNS ECs express a variety of different receptor-mediated transport systems, including the transferrin receptor (transferrin/iron), Ager (amyloid), and low-density receptor-related lipoprotein (LRP)1/LRP8. Many of these transport systems are being targeted as Trojan horses to aid in drug delivery to the CNS. Although most of these transporters provide nutrients from the blood to the brain (slc2a1, slc16a1, slc7a5, Tfr), several are also important for removing waste products from the brain (Ager). A complete characterization of BBB transporters, their substrates, and their direction of transport is critical to determine the external nutrient requirements of the CNS and how the BBB mediates the interaction between the blood and the CNS. Recently, systemic proteins have been implicated in regulating neurogenesis differently in youth and during aging (Villeda et al. 2011, 2014; Katsimpardi et al. 2014); however, it remains unclear whether this is because of specific transport, localized permeability of the BBB, or nonspecific passive movement of small amounts of systemic factors.

Transcytosis

In CNS ECs, the rate of transcytosis is dramatically lower than in ECs in nonneural tissues but is up-regulated as a major component of BBB dysfunction during injury and disease. Transcytosis through ECs is mediated through caveolin-based vesicle trafficking. Caveolin-1 is expressed by all ECs and is up-regulated at the BBB following traumatic brain injury (TBI) (Liu et al. 2010; Zhao et al. 2011; Gu et al. 2012; Badaut et al. 2015). Plasmalemma vesicle-associated protein (PLVAP) expression is enriched in peripheral ECs compared with CNS ECs, and has been implicated in vesicle trafficking, formation of fenes-

tra, and leukocyte extravasation in these "leaky" vascular beds. This molecule is also up-regulated in CNS ECs in a variety of diseases in which there is BBB leakage (Shue et al. 2008; Keuschnigg et al. 2009). Therefore, the lack of PLVAP in healthy CNS ECs appears to be important for limiting permeability.

The mechanisms of CNS EC's ability to maintain a low rate of nonspecific transcytosis (other than having designated transporters to deliver specific nutrients to the brain) were largely unknown until recent work demonstrated that these low rates result from an active inhibition of transcytosis in CNS ECs. Recent studies established Mfsd2a functions as a transcytosis inhibitor to regulate the BBB (Ben-Zvi et al. 2014; Andreone et al. 2017; Chow and Gu 2017). Mfsd2a is a lipid transporter (Nguyen et al. 2014) enriched in CNS ECs, and $Mfsd2a^{-/-}$ mice and zebrafish have increased caveolae-mediated vesicular trafficking causing leaky barriers without affecting TJs (Ben-Zvi et al. 2014; Andreone et al. 2017; O'Brown et al. 2019). Moreover, Andreone et al. established Mfsd2a's mechanism of action: Lipids translocated by Mfsd2a establish a unique lipid composition in the CNS EC plasma membrane that inhibits caveolae vesicle formation, thereby suppressing transcytosis. This lipid transport-dependent transcytosis inhibition mechanism is further supported by the cryo-electron microscopy structure of mouse MFSD2A (Cater et al. 2021; Wood et al. 2021). Mfsd2a is down-regulated after stroke and in aging when the BBB is compromised (Yang et al. 2020). The rate of transcytosis is dynamically modulated during development and disease (Chow and Gu 2017; Sadeghian et al. 2018; O'Brown et al. 2019). These findings imply that the molecular pathways inhibiting transcytosis could be targeted to open the BBB and deliver drugs to the CNS. An increasing number of new transcytosis regulators have been discovered, including the aforementioned vitronectin–integrin signaling between PC and neighboring ECs, which suppresses transcytosis for barrier integrity. It is plausible that vitronectin binding to integrin α5 exerts adhesive forces to maintain the plasma membrane tension to ensure low rates of transcytosis in CNS ECs (Ayloo et al. 2022). Together, these findings revealed that unique biophysical properties of the CNS endothelial membrane such as membrane lipid composition and membrane tension are essential for maintaining a low rate of transcytosis, thus BBB integrity.

Leukocyte Adhesion Molecules

In the healthy CNS, there is an extremely low level of immune surveillance, with an almost complete lack of neutrophils and lymphocytes within the parenchyma. Entry of a leukocyte from the blood into a tissue is a multiple step process that includes rolling adhesion, firm adhesion, and extravasation. This requires a series of different LAMs, including selectins (E-selectin, P-selectin) for rolling adhesion and immunoglobulin family members for firm adhesion (Huang et al. 2006; Aird 2007b). The expression of these adhesion molecules is much lower in CNS ECs than in peripheral ECs but is elevated during neuroinflammatory diseases, such as stroke and MS (Henninger et al. 1997; Huang et al. 2006; Engelhardt 2008; Daneman et al. 2010a). Interestingly, different subsets of inflammatory cells are observed infiltrating the CNS in different diseases. For instance, in MS, there is infiltration of T cells, B cells, neutrophils, and macrophages at sites of active lesions, whereas in stroke, neutrophils and macrophages infiltrate but lymphocytes are largely excluded. An important question is whether each cell has a different mechanism for crossing the BBB, and whether the discrimination is done at the level of the activated BBB or the activated immune cell.

Other Components of the BBB

Large-scale genomic and proteomic approaches have identified signaling cascades that are turned on in CNS ECs. In particular, Wnt/β-catenin signaling through Lef1, as well as Sonic hedgehog (SHh) signaling through Gli, have been shown to be important for regulating the formation and function of the BBB (Liebner et al. 2008; Stenman et al. 2008; Daneman et al. 2009; Alvarez et al. 2011a). In addition, vascular metabolism has been implicated in regulating barrier properties of CNS vasculature by metabolizing potential toxins or altering the properties of molecules.

Specific enzymes, including carbonic anhydrase IV and γ-glutamyl transpeptidase, have been identified as enriched in CNS vessels compared with vessels from nonneural tissues (Orlowski et al. 1974; Ghandour et al. 1992).

Omics approaches have provided invaluable resources in understanding the gene expression of the BBB, especially single-cell and cell-type-restricted molecular approaches have been increasingly applied to study the BBB and these techniques have proven crucial for revealing the identity and heterogeneity of vascular cells in the brain. A list of the major databases can be found in a recent review (Iadecola et al. 2023). In the mouse brain, ECs exhibit significant transcriptional heterogeneity, yet form a seamless continuum of shifting transcriptional states along the arteriole–capillary–venous network (Vanlandewijck et al. 2018). Recently, the development of improved vascular-enrichment procedures made it possible to map the vascular and perivascular cells in human postmortem brain tissues using snRNA-seq (Garcia et al. 2022; Yang et al. 2022). These studies revealed that the human vascular cells recapitulate many features of murine arteriole–capillary–venous organization of neurovascular cell types, including heterogeneity of EC types. However, it revealed greater diversity in cell types in humans, such as two PC types dedicated to transport function and matrix regulation. Deeper comparison and validation of some of these data is a critical next step in the field. Emerging studies taking advantage of the growth of scRNA-seq and more recently the spatial transcriptome analysis technologies begin to reveal a wealth of information on vascular changes associated with diseases, such as AD, Huntington's diseases (HDs), and paths forward for the improvement of preclinical models in translational research. Details on brain vascular transcriptome can be found in these recent reviews (Crouch et al. 2023; Wälchli et al. 2023). Future work expanding beyond genomics is aimed at identifying the proteomics, miRNAs, noncoding RNAs, lipids, metabolomics, epigenetics, and other regulatory steps that are important for BBB formation and function.

Finally, with these rich transcriptome databases, cell-type-specific gain- and loss-of-function studies are needed to identify which of these BBB-enriched genes are important for each aspect of the BBB, and whether the expression and function of each protein are dynamically regulated by neuronal function, stress, or diet. The development of several CNS endothelial-specific AAV viruses has made this possible and faster than the traditional mouse genetic approach (Körbelin et al. 2016; Krolak et al. 2022).

REGULATION OF THE BBB FORMATION AND HOMEOSTASIS

Although key properties of the BBB are manifested within the ECs, important transplantation studies have shown that they are regulated by interactions with the microenvironment of the CNS (Stewart and Wiley 1981; Janzer and Raff 1987). The BBB is not one physiology, but a series of physiological properties that either need to be induced (TJs, transporters, metabolic enzymes) or inhibited (transcytosis, LAMs) in CNS ECs. Recent work has dissected the cellular and molecular mechanisms that regulate this process and identified that it is a complex process of induction and maintenance signaling interactions among CNS ECs and PCs, astrocytes, and immune cells.

Regulation of Barrier Properties during Angiogenesis

Recent work in genetic mouse models has shown that there is a unique angiogenic program driving vessel formation in the CNS regulated by Wnt/β-catenin that also induces specific barrier properties in CNS ECs (Liebner et al. 2008; Stenman et al. 2008; Daneman et al. 2009). Comparative microarray analysis has identified that effectors of Wnt/β-catenin signaling, including Lef1, Apcdd1, and tnfrsf19, are enriched CNS ECs compared to ECs in peripheral organs (Daneman et al. 2009, 2010a). Transgenic reporter mice have confirmed that Wnt/β-catenin signaling is activated in CNS ECs during embryonic angiogenesis (Liebner et al. 2008; Stenman et al. 2008; Daneman et al. 2009). Different Wnt ligands are secreted by neural stem cells and neural progenitors in spatially distinct regions, notably

Wnt7a and Wnt7b in ventral regions and Wnt1, Wnt3, Wnt 3a, and Wnt4 in dorsal regions (Stenman et al. 2008; Daneman et al. 2009). Disruption of Wnt signaling in all ECs by conditional depletion of β-catenin leads to widespread CNS angiogenic defects with overtly normal blood vessel formation in peripheral tissues. These defects include a thickening of the vascular plexus, which contains endothelial progenitors, a loss of capillary beds, and the formation of hemorrhagic vascular malformations, which together suggest that Wnt is a migration signal driving vessels into the CNS (Stenman et al. 2008; Daneman et al. 2009). These phenotypes were also observed following the deletion of neural Wnts (Wnt7a/7b), demonstrating that the CNS angiogenic program requires Wnt as well as β-catenin. This CNS-specific angiogenic program was also shown to induce the expression of nutrient transporters, such as glut1, as well as the specific TJ molecules like claudin-3 (Liebner et al. 2008; Stenman et al. 2008; Daneman et al. 2009). Taken together, these data suggest that specific properties of the BBB are induced as vessels invade the CNS by a unique angiogenic program. Different Wnt ligands and Fzd receptors are expressed in spatially distinct regions and appear to be important for the regulation of CNS angiogenesis and BBB formation in those regions. One interesting receptor/ligand pair is Norrin/Fzd4, which is required for the formation of the retinal vasculature. Norrin is a transforming growth factor (TGF)-β family member with no homology with Wnt ligands, which can activate Fzd4 and induce canonical Wnt signaling. Loss of Norrin or Fzd4 produces major retinal vascular defects including a reduction in endothelial proliferation, vascular malformations, crossing of arteries and veins, a loss of venous fate, and leakiness of the blood–retinal barrier (Xu et al. 2004; Ye et al. 2010; Wang et al. 2012). Fzd4 mutants also have region-specific BBB defects in the cerebellum, spinal cord, and olfactory bulb but not the cortex, striatum, or hypothalamus. The more widespread phenotype of Fzd4 mutants suggests that it may also be activated by other ligands. Use of genetic mosaics has shown that Fzd4 is required cell-autonomously for sealing the BBB, and the Fzd4-deficient ECs have a loss of claudin-5 and an increase in PLVAP (Wang et al. 2012). Interestingly, deletion of Fzd4 in adults leads to up-regulation of PLVAP, loss of claudin-5, and leakage of the BBB, whereas reintroduction of Norrin to Norrin-deficient retinas leads to sealing of BBB properties (Wang et al. 2012). These data suggest that canonical Wnt signaling is not only required for BBB induction but also for maintenance of the BBB phenotype in adults, when the ligands are glial derived.

Regulation of the BBB by Pericytes

Analysis of mouse mutants in PDGF-BB/PDGFR-β signaling has identified an important role for PCs in regulating BBB formation and function (Armulik et al. 2010; Daneman et al. 2010b). These mutant mouse models include Pdgfb-null and Pdgfrb-null mice that completely lack CNS PCs and die at birth, as well as ECM-retention motif mutations to Pdgfb or hypomorphic alleles of Pdgfrb in which mice have fewer PCs than their wild-type littermates. Analysis of the BBB in Pdgfrb null mice during embryogenesis revealed a leaky BBB, demonstrating that PCs are required to regulate the formation of the BBB. In particular, a lack of PCs leads to an increase in the rate of transcytosis and an increase in the expression of LAMs resulting in CNS-immune infiltration (Armulik et al. 2010; Daneman et al. 2010b). Further use of mice with Pdgfrb hypomorphic alleles, which have varying numbers of CNS PCs, showed that the total number is important for the relative permeability of the vessels (Armulik et al. 2010; Daneman et al. 2010b). Additionally, work done in adult mice with ECM-retention motif mutations to Pdgfb that contain fewer PCs has identified that PCs are required during adulthood to regulate BBB homeostasis, and particularly do so by inhibiting transcytosis (Armulik et al. 2010). Microarray analysis comparing the transcriptional profile of CNS ECs with pdgfrb-mutant mice and wild-type mice suggests minimal changes in the expression of genes involved in BBB-specific properties, such as TJs, nutrient transport, or efflux transport, but an increase in the expression of genes involved in peripheral EC-specific "leaky" properties, including transcytosis (PLVAP) and leukocyte adhesion (Icam1, Alcam) (Daneman et al. 2010b). Taken together, these data suggest that PCs are not involved in the induction of BBB-specific properties (TJs, transporters), but play an important role in the inhibition of properties normally associated with leaky peripheral vessels (transcytosis, LAMs).

While PCs are embedded within the EC BM, they also secrete BM proteins that are important for BBB function, including laminin (Gautam et al. 2016), as well as the recently characterized vitronectin (Ayloo et al. 2022), which provides the first evidence for a specific mechanism of PC–EC signaling that regulates BBB integrity without ablating PCs or reducing PC coverage.

In addition to their role in BBB formation, PCs are also required for the maintenance of the BBB in adulthood. Specifically, $Pdgfrb^{+/-}$ mice, which do not have reduced PC coverage or increased BBB permeability during development, display an age-dependent gradual loss of PCs beginning at 1 month of age, resulting in increased BBB permeability to exogenous tracers and endogenous serum proteins (Bell et al. 2010). In addition, in a recent study that acutely deleted the PC-specific transcription factor FOXF2 in adult mice, there was an increase in BBB leakage of the tracer Evans Blue (Reyahi et al. 2015).

Regulation of the BBB by Astrocytes

The persistence of a functional BBB throughout adulthood is maintained and regulated by numerous factors unique to the microniche of the NVU (Abbott et al. 2006). Astrocyte–BBB–EC interactions are known to regulate EC morphology, angiogenesis, and to influence the phenotype of the barrier under physiological and pathological conditions (Prat et al. 2001). During the development of FGF-2 deficient mice, astroglial contacts with cerebral blood vessels are delayed compared with wild-type animals (Reuss et al. 2003). However, this delay did not result in changes to the permeability properties of the BBB (Saunders et al. 2016). Conversely, it is possible that developing perivascular astrocytes, in FGF-2 deficient mice, secret factors that play a role in BBB formation (Murakami et al. 2008). Early studies have found that transplanting glial progenitors to peripheral tissues induced BBB properties in vessels and prevented tracer leakage (Janzer and Raff 1987). Moreover, coculture experiments also showed that astrocytes induce enzyme activity at the BBB in ECs (Beck et al. 1986) and increased expression of TJs (Tao-Cheng et al. 1987). During embryonic development, at ~E15, PCs are recruited to the endothelial surface and induce BBB properties such as TJ protein expression and the absence of transcytosis (Cohen-Salmon et al. 2021). During this period, the BBB is established before astrocytes are differentiated and is likely mediated by PCs (Daneman et al. 2010b). However, consistent with in situ experiments (Janzer and Raff 1987), secreted factors from radial glia (astrocyte progenitors) may act on PCs and/or ECs during BBB formation (Mills et al. 2021; Díaz-Castro et al. 2023). For instance, radial glia cells were shown to stabilize developing cortical vessels via inhibition of Wnt signaling and proliferation in EC, a process that is potentially mediated by MMP-2 (Ma et al. 2013). Astrocyte maturation and expansion coincide with vascular growth and maturation (Freitas-Andrade et al. 2023). During mouse cortical development, astrocyte network formation and maturation spread out during prenatal and postnatal development (Clavreul et al. 2019). Mouse cortical astrocytes are from a dual contribution of delaminated embryonic apical progenitors and early postnatal progenitors (radial glia) that both generate pial and protoplasmic astrocytes (Clavreul et al. 2019). During the first postnatal week (P0–P7) both pre- and postnatal progenitors scatter throughout the neocortical wall while proliferating (Clavreul et al. 2019; Rurak et al. 2022; Freitas-Andrade et al. 2023). This is followed by a maturation phase (P7–P21) where expansion and proliferation are significantly reduced while individual astrocytes increase their volume and the complexity of their processes (Clavreul et al. 2019; Freitas-Andrade et al. 2023).

Postnatally, angiogenesis continues to expand the cerebrovascular network, as cortical growth proceeds in the weeks after birth (Coelho-Santos and Shih 2020). At P0, astrocyte cell bodies are closely associated with the microvasculature and they do not exhibit the familiar complex spongiform morphology of the adult (Freitas-Andrade et al. 2023). Within the first week after birth, astrocytes undergo significant spatial and morphological changes; astrocytes migrate away from the microvessels and begin to extend processes, some of which extended back to the vasculature (Freitas-Andrade et al. 2023). During this time, rudimentary endfeet begin to appear around microvessels and, by P14, astrocytic endfeet completely surround the vessel and take on a mature appearance (Gilbert et al. 2019, 2021; Freitas-Andrade et al. 2023). Also, within this period, two transmembrane proteins MLC1 and GlialCAM form a complex at the junction between mature astrocyte endfeet and this is correlated with increased expression of claudin-5 and P-gP (Gilbert et al. 2019). All cortical astrocytes are connected to at least one blood vessel, and astrocytes contact more vessels in deeper cortical layers, where vessel density is higher. Only in the hippocampus, ~2.6% of astrocytes do not contact blood vessels (Hösli et al. 2022). This underscores the close relationship between astrocytes and microvessels of the brain that begins before birth. The signaling mechanisms that induce endfeet process polarity are unknown; however, recent studies implicate HMGB1 and MLC1 as important factors for endfoot placement around blood vessels during postnatal brain development (Gilbert et al. 2021; Freitas-Andrade et al. 2023). Interestingly, PCs may play a role in endfoot properties in astrocyte processes that reach the vasculature (Díaz-Castro et al. 2023). PC-deficient mice exhibit reduced endfoot expression of α-syntrophin, aquaporin-4 (AQP4), and laminin α2 (Armulik et al. 2010). Astrocytic endfeet are a highly complex compartment containing mitochondria, rough endoplasmic reticulum, and vesicles (Díaz-Castro et al. 2023) with protein synthesis and posttranslational modification occurring locally (Boulay et al. 2017). A large portion of the astrocyte proteome is dedicated to endfoot function, including several proteins that regulate and induce the BBB and angiogenesis (Boulay et al. 2017; Yosef et al. 2020; Stokum et al. 2021).

Astrocytes are known to produce factors that modulate endothelial functioning during development and adulthood. Astrocytes play important roles in the homeostatic control of arterial blood pressure and cerebral blood flow (CBF) (Marina et al. 2020). Astrocytes are intimately associated with tens of thousands of synapses through highly ramified branches (Fields et al. 2015) and modulate CBF in response to synaptic activity (Anderson and Nedergaard 2003). They express metabotropic glutamate receptors (mGluRs) and sense glutamate release from synaptic clefts, and activation of mGluRs induces an increase in intracellular Ca^{2+} concentration spreading to the astrocytic endfeet (Zonta et al. 2003). These increases in Ca^{2+} concentration induce the release of vasoactive factors from astrocytic endfeet and are dependent on the metabolic state of the neuronal microenvironment (Mulligan and MacVicar 2006; Gordon et al. 2008). Astrocytes play a significant role in brain angiogenesis both in physiological and pathological conditions (Freitas-Andrade et al. 2020). Astrocytes secrete angiogenic factors that promote vascular growth, such as vascular endothelial growth factor (VEGF). For instance, VEGF from neonatal retinal astrocytes is critical for proper radial migration of ECs (Rattner et al. 2019). During development, VEGF is required for the formation, remodeling, and survival of embryonic blood vessels in the brain. During early embryogenesis, radial glia cells seem to be the source of VEGF needed for vascular development, although ECs have been described to promote cell-autonomous activation of the VEGF signaling (Lee et al. 2007). Although VEGF is a factor mostly known to promote angiogenesis during development, in adulthood, VEGF decreases the stability of the BBB during pathological conditions (Argaw et al. 2009, 2012). Several mechanisms are triggered in astrocytes during late brain development to attenuate angiogenesis and promote vessel stability. Src-suppressed C-kinase substrate (SSeCKS), increases in astrocytes as the brain develops and is linked with decreasing expression of astrocyte-derived VEGF-A and induces expression of angiopoietin-1 (Ang-1) involved in vessel stability and BBB differentiation (Lee et al. 2003). Perivascular astrocytes have been described to express αvβ8 integrin, which activates latent-TGF-β sequestered in the basal lamina, resulting in TGF-β-dependent inhibition of EC migration and reduction

in genes associated with angiogenic pathways (Cambier et al. 2005). Recently, a study showed that α7β1 integrin, expressed by perivascular astroglial cells, adheres to laminins in the vascular BM surrounding cerebral blood vessels and is important for vessel stability and BBB homeostasis (Chen et al. 2023). Genetic ablation of *Itga7* resulted in defective adhesion between astrocytic endfeet and laminins, reduction in TJ expression in ECs, leading to defective BBB maturation and stability as well as faulty cell–cell communication in the NVU (Chen et al. 2023). Taken together, astrocytes, PCs, and ECs delicately regulate a balance between BBB instability and angiogenesis during growth and plasticity versus BBB stability and vessel quiescence.

The role of astrocytes in BBB development and maintenance is not fully elucidated; however, some key pathways have been described. One of these pathways is the SHh signaling cascade known to be involved in embryonic morphogenesis, neuronal guidance, and angiogenesis. Astrocytes secrete SHh (Wang et al. 2008) and BBB ECs express the SHh receptor Patched-1, the signal transducer Smoothened (Smo), as well as transcription factors of the Gli family. Interestingly, transendothelial electrical resistance (TEER) and permeability experiments showed that activation of the Hh pathway induced the expression of junctional proteins and promoted a BBB phenotype. In addition, mice genetically engineered to lose the signal transducer Smo on ECs had a significant increase in BBB permeability that correlated with a decrease in junctional protein expression and disturbed BMs (Alvarez et al. 2011a, 2013), supporting the concept that the Hh pathway has a significant influence on BBB function.

Perivascular cells, including astrocytes, secrete angiopoietins, Ang-1, which participate in the complex process of BBB differentiation by promoting angiogenesis and inducing a time-dependent decrease in endothelial permeability. This occurs through the up-regulation of junctional protein expression (Prat et al. 2001). In contrast, Ang-2 is known to participate in the early phases of BBB breakdown during injury and disease (Nourhaghighi et al. 2003). Interestingly, when factors known to compromise BBB function, such as VEGF, are coexpressed with Ang-1, the barrier integrity is enhanced and neuroprotective properties are induced (Shen et al. 2011). Astrocytes also produce the angiotensin-converting enzyme-1 (ACE-1), which converts angiotensin I into angiotensin II and acts on type 1 angiotensin receptors (AT1) expressed by BBB ECs. Angiotensin II induces tightening of vessels, and, in the CNS, activation of AT1 restricts BBB permeability and stabilizes junctional protein function by promoting their recruitment into lipid rafts. Angiotensinogen (AGT)-deficient mice have an aberrant expression of occludin at the BBB, suggesting that astrocyte-secreted angiotensin II promotes TJ formation (Wosik et al. 2007). Retinoic acid (RA) can be secreted by radial glial cells, and recent findings suggest that RA is also secreted by astrocytes, and its receptor, RA receptor β (RAR-β), is expressed in the developing vasculature. RAR-β activation increases TEER, which correlated with enhanced expression of VE-cadherin, P-gp, and ZO-1. In vivo, pharmacologic modulation of RAR-β resulted in a perturbed BBB (Mizee et al. 2013). Interestingly, RA is known to regulate the Hh, Wnt, and FGF pathways (Halilagic et al. 2007; Paschaki et al. 2012), which implies that RA secretion by radial glial cells could be a master upstream regulator of BBB development. TGF-β is a pleiomorphic cytokine involved in cell growth, differentiation, morphogenesis, apoptosis, and immunomodulation. In the CNS, TGF-β is neuroprotective, and in vitro studies have shown its capacity to induce Mdr1 activity and to reduce BBB permeability (Dohgu et al. 2004). TGF-β is secreted by astrocytes and CNS ECs, and TGF-β is known to down-regulate the extent of leukocyte transmigration across the endothelium. In addition, TGF-β signaling induces the expression of MFSD2A in postnatal brain ECs (Tiwary et al. 2018). Other factors expressed by astrocytic endfeet have been reported to induce BBB maintenance, under physiological conditions. For example, connexin 30 and 43 (Cx30 and Cx43) promote BBB stability and are critical factors in BBB leakiness during inflammation (Ezan et al. 2012; De Bock et al. 2022). Astrocytic neogenin, a member of deleted in colorectal cancer (DCC) family netrin receptors, was shown to play a critical role in vessel structure and function but also BBB integrity (Yao et al. 2020). Recently, Stokum et al. (2021) found that the reversion-inducing cysteine-rich protein with Kazal motifs (RECK), is strongly expressed specifically in astrocytic endfeet; RECK is an important mediator of cerebral vascularization and maintenance of BBB differentiation (Stokum et al. 2021).

Cross talk between ECs and astrocytic endfeet is critical for BBB maintenance. Astroglial endfeet attachment to the vasculature is mediated by reelin-induced deposition of laminin-α4 by ECs to the extracellular compartment, which in turn enables the binding of glial endfeet to microvessels via the activation of integrin-β1 in astrocytes (Segarra et al. 2018). Additionally, laminin produced by astrocytes polarizes astrocytic endfeet, inhibits PC differentiation, and induces/maintains TJ protein expression in ECs (Yao et al. 2014). Interestingly, this phenotype was predominantly found in the deep regions of the brain including, the striatum, basal ganglia, thalamus, and hypothalamus, and was rarely found in the cortex and hippocampus (Yao et al. 2014). This may reflect the heterogeneous nature of astrocytes and/or BM within the vasculature of the brain. It was also shown that conditional laminin γ1 KO mice exhibited a reduction in AQP4 expression (Yao et al. 2014). One of the established functions of astrocyte–EC interactions is the regulation of brain water flux through the astrocytic water channel AQP4, which is expressed in astrocytic endfeet (Nielsen et al. 1997). Notably, AQP4 global or glial-specific KO mice have a decrease in brain edema following experimental conditions that increase brain water uptake (Manley et al. 2000; Haj-Yasein et al. 2011). In dystrophin-deficient *mdx* mice, age-dependent reduction of AQP4 was observed along with ZO-1 and claudin-1 assembly disturbances and BBB breakdown (Nico et al. 2003). A close correlation between AQP4 and BBB has been demonstrated. AQP4 is developmentally expressed during BBB differentiation (Wen et al. 1999; Nico et al. 2001). However, the mechanism between AQP4 and BBB development and maintenance is not fully elucidated.

Currently, several lines of evidence point to the concept that astrocytes may not play a role in BBB formation, but they are important for BBB maintenance. For example, BBB development begins approximately at E15 before astrocyte growth (Ben-Zvi et al. 2014; Biswas et al. 2020). Moreover, in adult mice, live in vivo imaging during and after transient loss of astrocyte endfeet, by laser irradiation, showed striking plasticity where neighboring astrocytes covered exposed microvessels with new endfeet (Kubotera et al. 2019; Mills et al. 2021). However, no leakage of large tracers was observed, indicating no loss of BBB function. In contrast, acute sparse astrocyte ablation in adult mice leads to BBB leakage and loss of ZO-1 expression, demonstrating a requirement of astrocytes for BBB maintenance (Heithoff et al. 2021). However, HMGB1 ablation in astrocytes resulted in aberrant morphological remodeling of both astrocyte endfeet and ECs, without affecting BBB function (Freitas-Andrade et al. 2023).

Under pathological conditions, such as stroke, astrocytes play a detrimental role in BBB function (Freitas-Andrade et al. 2020). However, recent studies are demonstrating more nuanced effects of astrocytes on the BBB such as the role of peripheral inflammation, aging, and depressive disorders (Rajkowska et al. 2013; Mills et al. 2021; Mou et al. 2022). A recent study reported that hyperglycemia resulted in a decrease in Cx43 levels with a concomitant increase in secreted VEGF in astrocytes, resulting in decreased TEER and permeability without affecting TJ protein levels (Garvin et al. 2022). Taken together, it is evident that astrocytes are key players in BBB initiation, maturation, and maintenance; however, the specific pathways linked with these processes remain to be fully resolved.

Convergence of Signaling Events at the BBB

The BBB is regulated by a complex set of cellular signaling mechanisms that regulate both the induction of barrier properties during initial angiogenesis into the CNS, as well as the maintenance of barrier properties in adults. Neural stem cells appear to be the key cell type involved in the early differentiation of the ECs into BBB ECs, and then PCs and astrocytes provide further cues modulating the different barrier properties of these CNS ECs. The number of distinct factors that

are known to impact on BBB permeability highlights the diversity of the CNS inputs needed to generate this physiological barrier. This also emphasizes the redundancy of molecular signals affecting BBB formation and stability, and the need for future work to identify how each of these signals are coordinated to regulate different aspects of the BBB. These signals can, however, be integrated into a general concept (Prat et al. 2001). Key signaling pathways and transcription factors have either barrier-promoting properties (Wnt, Hh, Sox-18, nrf-2, ERG, Nkx2-1, and SP3/YY1) or barrier-disrupting effects (NF-κB, Snail, FoxO1, PKC, and eNOS) (Yuan 2002). Within the signaling pathways promoting BBB functioning, Wnt and Hh seem to be dominant and to cooperate in driving a BBB phenotype. Wnt ligand binding to Frizzled/LRP5/6 activates β-catenin, which leads to the expression and targeting of the junctional proteins claudin-3 and p120 to the cell membrane (Liebner et al. 2008; Hong et al. 2010). β-Catenin also down-regulates the activity of Snail, which has a negative effect on the stability of p120/VE-cadherin complexes and on the expression of TJ molecules occludin and claudin-5. Loss of the Wnt coreceptor Lrp5 causes down-regulation of claudin-5 expression (Chen et al. 2011). The Hh signaling pathway appears to drive the transcription and expression of junctional proteins, but also dampens inflammatory responses on CNS ECs. Activation of Gli-1 by the Hh ligands or Wnt signaling is reported to activate Sox-18 (Alvarez et al. 2011a), which controls claudin-5 expression (Fontijn et al. 2008). Wnt and Hh activation also induces the expression of NR2F2, a transcription factor that promotes Ang-1 expression, inducing junctional protein expression through tie-2. NR2F2 also down-regulates the expression of Ang-2, a factor known to decrease junctional protein expression. In a similar way, activation of the nrf-2 pathway by oxidative stress activates antioxidant response elements (AREs), which are known to stabilize ZO-1, occludin, and claudin-5 expression (Fan et al. 2013). In addition, nrf-2 protects ECs during injury by suppressing the expression of inflammatory genes (Chen et al. 2006). In this sense, signaling pathways and transcription factors supporting barrier function also tend to promote anti-inflammatory responses. Overview of molecular pathways involved in BBB can be found in this review (Langen et al. 2019).

One of the major issues when analyzing previous work is that many different measures have been used to quantify BBB function when analyzing the effect of genetic or environmental perturbations on the barrier, making it difficult to compare and contrast different studies. Furthermore, in many cases, only a small number of measures are used to examine BBB function, whether a single molecular tracer or analysis of a small set of molecular markers. The BBB is not a single entity, but a series of different properties possessed by the CNS ECs and regulated by interactions with different neural, vascular, and immune cells; thus, a more exhaustive approach to understanding how different pathways regulate each aspect of the BBB is required to fully understand this barrier. Thus, future work needs to identify whether each of these signaling pathways regulates all aspects of the BBB, or whether different properties of the BBB are induced and regulated by different pathways, and, if so, how do each of these pathways coordinate to regulate the BBB, allowing proper neuronal function. New genetic tools allow for the manipulation of genes and pathways both in development and in adulthood and, thus, will be able to determine whether the pathways are required for induction during development, maintenance during adulthood, and/or disruption during disease. Furthermore, new intravital imaging techniques in live awake-behaving animals will enable the understanding of how plastic the BBB is and whether different properties of the BBB can be dynamically regulated in response to neuronal activity, diet, infection, or other environmental stimuli.

DYSFUNCTION OF THE BBB IN CNS DISORDERS

Disruption of the BBB is observed in many diseases including neurological disorders like MS, stroke, AD, Parkinson's disease (PD), neurodevelopmental conditions, and TBI (Van Dyken and Lacoste 2018; Freitas-Andrade et al. 2020; Ouellette and Lacoste 2021). Functional imaging of human patients and analysis of postmortem human brain samples has identified pathological breakdown of the BBB. In addition, work with animal models of disease and in vitro BBB models has enabled the identification of several molecular mechanisms that cause changes to the BBB. This dysfunction can include alterations in many different properties of the BBB including TJs, transporters, transcytosis, and LAM expression. This breakdown can lead to edema, disruption of ionic homeostasis, altered signaling, and immune infiltration that can lead to neuronal dysregulation and, ultimately, degeneration. Although BBB dysfunction is often secondary to the primary insult in these diseases, in some cases, it has been a suggested cause, including in MS and AD (Fig. 3).

Modulation of the BBB in Neurodegenerative Diseases

The BBB in Multiple Sclerosis and Related Disorders

In most CNS pathologies, the BBB is affected as a result of the inflammation, injury, or degenerative processes specific to the pathology. However, in a few diseases, the BBB is specifically targeted by the pathogenic process or by the disease determinants. Neuromyelitis optica (NMO) and MS are among these diseases (McQuaid et al. 2009; Cramer et al. 2014). The etiology of MS remains elusive, but it is clear that multiple factors are involved in disease development, including environmental and genetic factors; BBB dysfunction is considered a major hallmark of MS, and is deemed a trigger of disease onset. MS is a T-cell-mediated disease in which CD4 T-helper (Th) cells of the Th17 and Th1 phenotype play a fundamental role in its pathogenesis. B cells are also essential in MS immunopathogenesis, as antibodies produced within the CNS are a fundamental feature of the disease (i.e., oligoclonal bands) and as B-cell-directed therapies provide strong protection against lesion formation. During immune cell infiltration and lesion formation, BBB function becomes compromised (Larochelle et al. 2011), which is characterized by vascular leakage associated with alterations of junctional proteins. Analysis of MS tissue shows

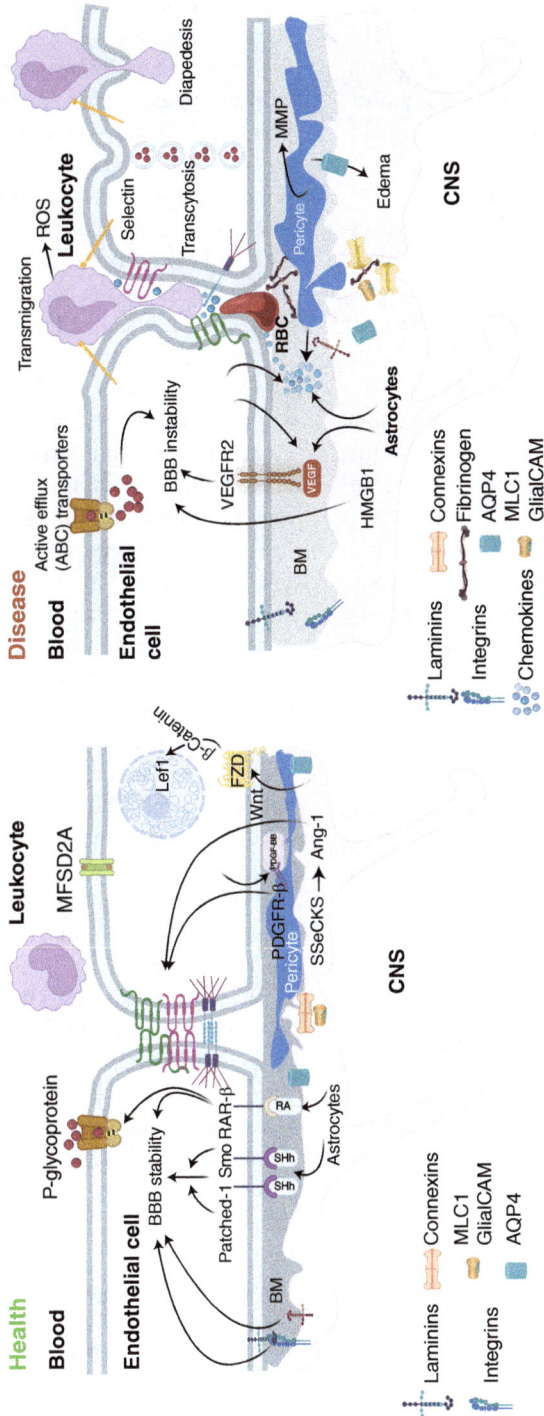

Figure 3. Schematic representation of blood–brain barrier (BBB) regulation in health and disease. (CNS) Central nervous system, (BM) basement membrane, (MMP) matrix metalloproteinase, (ROS) reactive oxygen species.

that abnormalities in the expression of junctional proteins coincide with perivascular astrogliosis, and such changes are detected in the very early stages of lesion formation (Prat et al. 2001). This has been, in part, explored by Luo et al. (2008) when inducing active experimental autoimmune encephalomyelitis (EAE) in mice expressing luciferase under the control of GFAP. Despite showing clinical signs only at day 11, increases in bioluminescence associated with GFAP expression could be detected in the brain of these animals as early as 3 days postinduction, suggesting that astrocytes are activated in the very early stages of EAE and in the absence of clinical signs of the disease.

Besides its primary neuroprotective function, the BBB has also been shown to actively promote neuroinflammation by orchestrating immune responses during CNS-targeted autoimmune aggression. BBB ECs are an important source of proinflammatory chemokines CCL2 (Biernacki et al. 2001; Kebir et al. 2007), CCL5, and CXCL10, which are required for lymphocyte and monocyte recruitment to the CNS (Prat et al. 2001). Immune cell infiltration into the CNS correlates with the production of proinflammatory mediators, such as interleukin (IL)-17, IL-22, granulocyte macrophage colony-stimulating factor (GM-CSF), interferon (IFN)-γ, and tumor necrosis factor (TNF) (Alvarez et al. 2011b). These cytokines have been implicated in the modulation of EC function by up-regulating the expression of proinflammatory mediators and by affecting the expression of junctional proteins and, thus, compromising BBB permeability. Lastly, BBB ECs express intercellular adhesion molecule (ICAM)-1, ICAM-2, vascular CAM (VCAM)-1, activated leukocyte CAM (ALCAM), melanoma CAM (MCAM), and Ninjurin-1, which mediate, at least in part, the adhesion process and transmigration of leukocytes and leukocyte subtypes to the CNS (Cayrol et al. 2008; Dodelet-Devillers et al. 2009; Greenwood et al. 2011). Thus, although the BBB protects against CNS-directed inflammation by restricting immune cell access to the brain, it can also regulate the local inflammatory response by expressing proinflammatory molecules that promote the recruitment of peripheral immune cells into the CNS.

Following migration across ECs, leukocytes cross the endothelial BM and, subsequently, the parenchymal BM to get access into the CNS. The composition of the endothelial BM can regulate the extent of perivascular infiltration as large amounts of leukocytes are detected in vessels expressing laminin 411 and low levels of 511, whereas in the absence of 411, laminin 511 is ubiquitously expressed and associated with low-T-cell infiltration and milder disease. In EAE and MS, immune cell infiltrates are in great part contained to the perivascular space, and the process of leukocyte migration across the parenchymal BM and astrocyte endfeet appears to be more tightly controlled than the diapedesis across ECs (Engelhardt and Sorokin 2009). In EAE, CD45 T-cell infiltration across the parenchymal BM is not laminin dependent, but rather requires focal activation of MMP-2 and MMP-9 to selectively cleave dystroglycan, affecting the BM stability and integrity (Agrawal et al. 2006). Interestingly, parenchymal BM components and other ECM-binding receptors on the astrocyte endfeet remain unaffected, indicating the existence of specific and specialized protective mechanisms under the control of astrocytes and possibly other cells within the NVU (Engelhardt and Sorokin 2009). Thus, further understanding is needed in terms of astrocyte involvement in supporting or inhibiting the activation and migration of immune cells as well as the repair of the affected BBB/NVU during MS/EAE and other CNS disorders.

Reactive astrocytes can also be the source of factors that will negatively affect barrier function at the NVU. In MS and EAE, VEGF-A is expressed by reactive astrocytes, and in vitro/in vivo studies show its capacity to induce BBB breakdown by disrupting claudin-5 and occludin expression and promoting immune cell infiltration to the CNS. Additional studies propose that IL-1 production by microglia induces VEGF-A up-regulation. VEGF-A is released from the astrocytes and binding to its receptor VEGFR2 on BBB-ECs activates eNOS-dependent down-regulation of the junctional proteins claudin-5 and occludin that leads to BBB breakdown. Although reactive astrocytes can produce BBB-promoting (i.e., SHh) or BBB-disrupting (i.e., VEGF) factors, they can also lose or down-regulate factors that have the capacity to promote barrier function. In this regard, astrocytes produce AGT (which is cleaved into angiotensin II), and analysis of MS tissues showed that expression of AGT in astrocytes and occludin in ECs is decreased in MS lesions when compared to normal appearing white matter. This pattern correlates with the down-regulated expression of AGT detected in astrocytes stimulated in vitro with IFN-γ and TNF-α. Interestingly, nonimmunized (non-EAE) AGT-deficient mice have compromised BBB function, which correlates with decreased and disrupted expression of occludin. Therefore, local inflammatory mediators present in perivascular cuffs can also negatively impact on the capacity of reactive astrocytes to promote BBB function by down-regulating their production of BBB-promoting factors (Wosik et al. 2007). For more details on BBB dysfunction in MS, the following reviews can be also consulted (Girolamo et al. 2014; Kamphuis et al. 2015; Xiao et al. 2020).

NMO is also an immune-mediated disease of the CNS affecting predominantly the spinal cord and the optic nerves. In NMO, the production of anti-AQP4 IgG antibodies affects the function of the astrocyte water channel AQP4 directly affecting BBB function. The binding of anti-AQP4 antibodies to their target results in the activation of complement-dependent cytotoxic cell damage that leads to the loss of AQP4, GFAP, and the excitatory amino acid transporter 2 (EAAT2). In addition, the BBB damage is associated with focal areas of perivascular immune cell infiltration and demyelination, particularly granulocytes, and eosinophils that degranulate in the perivascular space causing local damage that includes astrocyte injury. Although oligodendrocytes are affected as a result of the pathophysiological changes, the exact mechanism(s) leading to oligodendrocyte and neuronal damage remains to be determined.

The BBB in Alzheimer's Disease

Early signs of BBB breakdown in AD can be detected before the onset of dementia. Neuroimaging techniques evidenced AD-related BBB breakdown in gray and white matter (Montagne et al. 2016; van de Haar et al. 2016). Aβ and τ pathologies contribute to BBB disruption in AD patients and mouse models (Park et al. 2011; Sagare et al. 2013; Alata et al. 2015). Several players involved in Aβ clearance are reduced in AD patients, including phosphatidylinositol-binding clathrin assembly protein (PICALM), P-glycoprotein, and glucose transporter-1 (GLUT-1) (Mooradian et al. 1997; Chiu et al. 2015; Zhao et al. 2015). Brain microvessels in AD also show reduced expression of LRP1, a major Aβ clearance receptor at the BBB (Deane et al. 2004; Donahue et al. 2006).

Several features lead to increased BBB permeability in AD, including reduced expression of TJ molecules, perivascular accumulation of blood-derived products, degeneration of PCs and ECs, as well as infiltration of circulating leukocytes (Sweeney et al. 2018; Huang et al. 2020). Aβ disrupts TJs and increases vascular permeability by suppressing the expression of ZO-1, claudin-5, and occludin, while increasing the expression of MMP-2 and MMP-9 (Marco and Skaper 2006; Kook et al. 2012; Blair et al. 2015; Wan et al. 2015). Studies have reported leakage of blood-derived proteins (fibrinogen, thrombin, albumin, IgG) around capillaries from postmortem brain tissue in the cerebral cortex and hippocampus of AD patients (Ryu and McLarnon 2009; Hultman et al. 2013; Sengillo et al. 2013). Furthermore, animal studies showed that a lack of PC-derived factors contributes to endothelial degeneration in AD (Bell et al. 2010). PC loss is associated with BBB dysfunction as well as accelerated Aβ buildup and tau pathology (Sagare et al. 2013). There is also evidence of PC loss in the hippocampus and cortex of AD patients (Sagare et al. 2013; Sengillo et al. 2013; Huang et al. 2020).

Cerebral amyloid angiopathy (CAA) is also associated with increased BBB permeability, and Aβ deposition in CAA occurs on the BM of arteries, arterioles, and capillaries (Magaki et al. 2018; Gireud-Goss et al. 2021). Ultrastructural studies measured a thinned endothelium, shrinkage and degeneration of ECs, and vessel occlusions, all of which can lead to microinfarcts (Attems and Jellinger 2004; Thal et al. 2009). TJ proteins in CAA-laden vessels are decreased (Tai et al. 2010), and after exposure to exogenous Aβ human ECs display decreased expression of occludin, while postmortem brain tissue of CAA patients show decreased expression of claudin-5, ZO-1, CD31, and collagen IV (Tai et al. 2010; Carrano et al. 2011; Magaki et al. 2018). In addition, CAA patients displayed increased expression of MMP-2 and MMP-9, which may lead to BM degradation and increased BBB permeability. In the Tg2576 mouse model of CAA, BBB integrity was compromised due to decreased expression of claudin-5 and claudin-1 (Carrano et al. 2011). In another mouse model of CAA (TgSwDI mice), spontaneous hemorrhage and loss of BBB integrity were reported (Davis et al. 2004).

The BBB in Parkinson's Disease

Disruption of the BBB was detected in the substantia nigra pars compacta (SNc) from animal models (Barcia et al. 2005; Rite et al. 2007; Chao et al. 2009). While only a few human studies investigated the BBB in PD patients, there is evidence of BBB dysfunction, for instance in the postcommissural putamen (Kortekaas et al. 2005; Gray and Woulfe 2015). Increased BBB leakage was also revealed using ASL and dynamic contrast enhanced-MRI in brain regions previously associated with PD in patients with known cerebrovascular disease (Wardlaw et al. 2008; Al-Bachari et al. 2020).

Accumulation of α-synuclein in ECs may also contribute to BBB dysfunction and increased cerebrovascular permeability (Elabi et al. 2021). Higher number of EC nuclei was found in the SNc of PD patients (Faucheux et al. 1999). Other EC dysfunctions were reported, such as the down-regulation of TJ proteins (Kuan et al. 2016). In the MPTP mouse model of PD, down-regulation of TJ protein ZO-1 and BBB leakage was measured in the substantia nigra (Patel et al. 2011). An altered BM was also observed in PD mice (Yang et al. 2015). VEGF, a prominent growth factor promoting BBB permeability, was up-regulated in the substantia nigra (but not the striatum) of PD patients, while animal models of PD displayed parkinsonian traits following administration of exogenous VEGF

into the substantia nigra (Barcia et al. 2005; Wada et al. 2006; Rite et al. 2007).

Regulation of the BBB in Traumatic Brain Injury

TBI can induce acute BBB disruption through vascular shear stress, increased transcytosis, hemorrhages, edema, long-term alterations in CBF, and chronic inflammation (Jullienne et al. 2016; Ichkova et al. 2020; Badaut et al. 2015), features that are known to contribute to Aβ deposition and tau pathology (Iadecola 2013; De Silva and Faraci 2016). Autopsies of TBI patients show diffuse Aβ plaques similar to those identified in AD (Perry et al. 2016). The formation of Aβ in perivascular spaces following TBI may lead to an injury cascade consisting of cerebrovascular damage, oxidative stress, and ECs dysfunction (Ramos-Cejudo et al. 2018). Alterations in brain EC survival, BBB integrity, and neuroinflammation are considered early events post-TBI, all of which are characteristic of cerebrovascular damage involved in the progression of AD and impairment of Aβ clearance. These early vascular impairments thus promote the onset of neurodegenerative diseases (Ramos-Cejudo et al. 2018). Considering early vascular injuries in TBI, biomarker studies are including a variety of neuroimaging, molecular techniques, and novel in vitro platforms to better understand the incidence of cerebrovascular dysfunction and the onset of neurodegenerative diseases (Graham and Sharp 2019; Martinez and Stabenfeldt 2019; Bolden et al. 2023).

Modulation of the BBB Following Hypoxia/Ischemia and in Stroke

In vivo and in vitro stroke models have shown that cerebral vascular permeability increases in a time- and hypoxia-dependent manner, affecting ECs, PCs and astrocytes, microglia, and perivascular macrophages at the vessel wall (Freitas-Andrade et al. 2020). This leads to a subsequent increase in cerebral edema; however, the processes involved in the hypoxia-induced BBB permeability are not completely understood. Work in animal models of stroke has identified that there is a biphasic leakage of the BBB, with an early opening within hours following hypoxia/

ischemia, followed by a refractory phase and then a second opening the next day (Kuroiwa et al. 1985; Huang et al. 1999). In addition, analysis in transgenic models has identified that there are stepwise alterations in the BBB, with an increase in transcytosis observed first followed by alterations in the TJs (Knowland et al. 2014). There are also important changes in ion channel and efflux transporter expression and activity.

Focal cerebral ischemia damages elements of the BBB and induces inflammatory processes that alter the relationships of ECs, ECM, and astroglial cells (Freitas-Andrade et al. 2020). This results in profound changes in microvascular permeability. Focal increases in permeability to fibrinogen, IgG, and other large proteins are detected within a few hours following middle cerebral artery occlusion (MCAo) (Knowland et al. 2014). Conversely, and surprisingly, hypoxic conditions induce expression of ZO-1 in vitro and claudin-5 and occludin in vivo. The exact functional consequences of these up-regulations are not clear. Nevertheless, levels of EC-expressed integrins α1β1, α3β1, and α6β1 decrease rapidly after MCAo and MMPs are activated on ischemic insult, which induces basal lamina remodeling, and also chemokine activation (Edwards and Bix 2019). Finally, dystroglycan, expressed by astrocytes, disappears after MCAo, a phenomenon responsible for the detachment of astrocyte endfeet and perivascular edema (Milner et al. 2008; Hawkins et al. 2013); this is further exacerbated by the loss of AQP4 from astrocytic endfeet (Banitalebi et al. 2022). These studies suggest that adhesive interactions between the endothelium and the ECM contribute to the acute cerebrovascular remodeling seen in stroke (Del Zoppo et al. 2006).

Molecular Alterations of the Tight Junctions

TJ disruption is a hallmark of both ischemic and hemorrhagic stroke and is typically associated with increased vascular permeability and homeostatic changes in the neuronal microenvironment. Clinically, strokes are known to cause an increase in vasogenic edema, which can be attributed to an increase in BBB permeability. Recent in vitro studies have begun to elucidate the molecular changes leading to increases in BBB permeability. In studies by Mark and Davis (2002) and Witt et al. (2008), an increase in actin protein levels and actin stress fibers was observed following hypoxic insult, whereas hypoxia alone had no effect on protein expression of the TJs, occludin, claudin-1, or ZO-1/2. Following hypoxia, reoxygenation increases expression of occludin, claudin-1, and ZO-1/2. Changes in the cellular localization of the TJ proteins occludin and ZO-1/2 following hypoxic insult were confirmed with dynamic confocal microscopy recordings. Interestingly, these changes were reversible and returned to control levels on reoxygenation. In an ischemia/reperfusion model, BBB permeability exhibited a biphasic pattern, which was linked to changes in claudin-5, occludin, and ZO-1 protein levels (Jiao et al. 2011).

Changes in junctional structure formation or stability are now known to involve up-regulation in VEGF, and inhibition of VEGF attenuates the hypoxia-induced increase in BBB permeability (Fischer et al. 2002; Schoch et al. 2002; Fischer et al. 2004). In addition, hypoxia increases nitric oxide (NO) release by ECs, and inhibition of NO synthase reduces the effect of hypoxia on cell permeability (Mark et al. 2004). Although the exact mechanisms involved in the VEGF- and NO-mediated changes in EC permeability are still being investigated, some reports have shown that NO may directly modify the TJ proteins by nitrosylation or nitrosation.

MMPs and the BBB

MMPs are zinc-dependent proteases that can degrade fibronectin and laminins. As the basal lamina is composed of collagen, fibronectin, laminin, and heparin sulfate, and serves as an important scaffold for brain ECs, MMPs have been considered as obvious initiators of BBB disruption. Following ischemia/reperfusion, MMPs have been shown to be up-regulated in the brain (Lenglet et al. 2014), either through proinflammatory cytokine pathways (via NF-κB) or through activation of HIF-1α and furin, which convert pro-MT-MMP into activated MT-MMP. More specifically, it has been shown that MMP-9, MMP-3, and MMP-2 levels were increased following ischemia/reperfusion, correlating with the increase in sucrose diffusion across the BBB. Additionally, inhibition of MMP with pharmacological agents or use of MMP KO animals reduced BBB disruption (Rosenberg et al. 1998). It remains unclear whether MMP-mediated BBB disruption occurs at the level of the basal lamina, or at the level of the TJ and AJ, as these junctional proteins were also shown to be substrates of MMPs.

Endothelial Transcytosis following Stroke

Caveolae-mediated transcytosis is normally suppressed in the healthy brain (Drab et al. 2001; Predescu et al. 2001; Schnitzer 2001; Tuma and Hubbard 2003; Ben-Zvi et al. 2014). Under physiological conditions, MFSD2A, expressed in brain ECs, acts as a lipid flippase transporting phospholipids from the outer to inner plasma membrane leaflet and altering the plasma membrane composition in such a way that caveolae vesicles are unable to form (Andreone et al. 2017). However, caveolae-mediated transcytosis is activated following tMCAo, and cav-1 expression increases early following stroke, before TJ disassembly (Knowland et al. 2014). A significant correlation between the extent of BBB disruption following brain ischemia and cav-1 expression was confirmed in mice subjected to focal cortical ischemia induced by photothrombosis (Choi et al. 2016). In summary, ischemia/reperfusion-induced BBB disruption in the peri-infarct region involves (1) up-regulation of caveolae-mediated endothelial transcytosis in the early phase of reperfusion (between 0 h and 12 h) and (2) major TJ remodeling in the late phase (48–60 h). The first phase, which peaks at 6 hours, leads to nonselective vesicular transport of blood-borne molecules across the brain endothelium. The second phase leads to the breakdown of the vessel wall, exacerbating cerebrovascular leakage (Cipolla et al. 2004; Knowland et al. 2014; Nahirney et al. 2016; Haley and Lawrence 2017). BBB disruption via increased caveolae-mediated bulk-flow fluid transcytosis allows free mobility of toxic substances and accumulation into the brain of plasma proteins that notably include immunoglobulins, albumin, laminin, thrombin, and ferritin, collectively leading to neuroinflam-

mation, neuronal death, and functional impairment (Zlokovic 2010).

Modulation of Channels and Transporters

Ion channels and transporters are key components of the BBB, which maintain cerebral physiological and metabolic homeostasis. As one of the major consequences of stroke is the formation of cerebral vasogenic and cytotoxic edema, understanding the effect of stroke on the function of channels and transporters at the BBB could identify important therapeutic targets.

During ischemic stroke, there is an important release of glutamate from neurons that bind to N-methyl-D-aspartate (NMDA) receptors. This excess NMDA receptor activation is largely responsible for cytotoxic edema of neurons (Sharp et al. 2003). Studies have shown that BBB ECs also express both NMDA and metabotropic glutamate receptors (Krizbai et al. 1998; Sharp et al. 2003). Circulating inflammatory mediators have also been shown to stimulate a release of glutamate, which disrupts the BBB via metabotropic receptors (Collard et al. 2002). Interestingly, in vitro studies showed that NMDA receptor activation reduces BBB integrity, whereas activation of metabotropic receptors increased BBB electrical resistance suggesting a tightening of the BBB.

The activity of exchangers and transporters, such as the Na^+/H^+ exchanger (NHE), Na^+/K^+ ATPase, and $Na^+/K^+/Cl^-$ cotransporter, contribute to maintaining ion balance at the BBB and in the brain in general. During stroke, osmotic and ion balance are altered, leading to the activation of ion transporters and exchangers.

Modulation of the BBB in Neurodevelopmental Disorders

Several studies reported a defective BBB in autism spectrum disorders (ASDs), schizophrenia, and HD. A small subset of ASD participants demonstrated higher levels of autoantibodies against brain ECs in the serum compared to typically developing individuals, suggesting an impact on the BBB (Connolly et al. 1999). Another group of children diagnosed with ASD displayed reduced levels of adhesion molecules including soluble platelet endothelial cell adhesion molecule-1 (PECAM-1) and P-selectin. As these molecular factors are essential to modulate BBB permeability through signaling and leukocyte infiltration, this suggests that BBB dysregulation may be at play in ASD (Onore et al. 2012). Furthermore, a postmortem study, with a small sample size, demonstrated altered BBB integrity in ASD with increased gene expression of MMP-9 (Fiorentino et al. 2016). Studies have shown that MMP-9 regulates cell proliferation, adhesion, degradation of laminin and collagen, angiogenesis, and oxidative injury, and is implicated in BBB breakdown (Lepeta and Kaczmarek 2015; Turner and Sharp 2016). Additionally, important components of BBB integrity displayed altered expression in ASD patients, including claudin-5 and claudin-12, as well as tricellulin (MARVD2) a component of TJs involved in decreased permeability to macromolecules in brain ECs (Fiorentino et al. 2016). In a valproic acid rat model of autism, increased BBB permeability to Evans blue was found in the cerebellum, a phenotype attenuated by treatment with memantine, an NMDA receptor modulator. This BBB alteration was also attenuated using minocycline (antibiotic) and agomelatine (melatonin receptor) treatment (Kumar et al. 2015; Kumar and Sharma 2016). Animal studies have also investigated transendothelial transport mechanisms in ASD mouse models. Tărlungeanu et al. (2016) demonstrated that the large neutral amino acid transporter (LAT1, *Slc7a5*) localized at the BBB to maintain normal levels of brain branched-chain amino acid (BCAA) was required for neurotypical development. Mice harboring an endothelial-specific deletion of *Slc7a5* displayed behaviors reminiscent of ASD, including motor dysfunctions consistent with a study in human patients harboring the constitutive mutation (Novarino et al. 2012; Tărlungeanu et al. 2016). Interestingly, the administration of BCAA rescued ASD-like behaviors in $Slc7a5^{\Delta EC}$ mice (Tărlungeanu et al. 2016). A postmortem analysis of brain tissue from individuals diagnosed with ASD revealed significantly higher levels of markers associated with PCs, as well as increased vascular tortuosity, indirectly suggesting impairments in the BBB (Azmitia et al. 2016). A more recent study in $16p11.2^{df/+}$ ASD mice revealed impaired endothelial function in young (P14) $16p11.2^{df/+}$ male mice compared to sex-/age-matched littermates (Ouellette et al. 2020), yet whether the BBB is affected in this ASD mouse model remains to be elucidated.

Dysregulation of the BBB has also been reported in schizophrenia (Müller and Ackenheil 1995; Shcherbakova et al. 1999; Carrier et al. 2021; Crockett et al. 2021; Ouellette and Lacoste 2021). Briefly, evidence of schizophrenia-associated microvascular abnormalities in the neocortex includes thickening of the basal lamina, vacuolation of cytoplasm in ECs, swelling of astrocyte endfeet, as well as atypical vascular arborization (Uranova et al. 2010; Carrier et al. 2021). Specific mutations are associated with schizophrenia, including alterations in the 22q11.2 deletion syndrome (22qDS) and polymorphisms in claudin-5 (Gur et al. 2017; Greene et al. 2018) altogether revealing barrier dysfunction in schizophrenia patients (Greene et al. 2018; Crockett et al. 2021). Postmortem brain sections from 22qDS patients and animal models of 22qDS both demonstrated reduced claudin-5 expression, which compromised BBB function (Nishiura et al. 2017; Guo et al. 2020; Crockett et al. 2021; Usta et al. 2021). Additionally, altered levels of VE-cadherin and occludin in ECs were identified in schizophrenia (Cai et al. 2020). Furthermore, BBB hyperpermeability has been associated with another risk allele for schizophrenia. NDST3, expressed in the brain, encodes an enzyme involved in the metabolism of heparan sulfate, a component of the basal lamina ECM that is required for BBB integrity (Khandaker et al. 2015).

HD is a hereditary, autosomal dominant, and neurodegenerative disorder (Davenport 1915; Bano et al. 2011) leading to altered muscle coordination and declined mental abilities (Ha and Fung 2012). An expansion of trinucleotide CAG repeats in the Huntingtin (HTT) gene results in the production of a mutant Htt (mHtt) protein that aggregates in specific brain regions and leads to neurotoxicity (Zheng and Diamond 2012). Increases in BBB permeability were observed in HD patients and animal models of HD (Steventon et al. 2020). Interestingly, there is evidence that BBB leakage increases alongside HD progression. BBB dysfunction in HD patients has been associated with decreased TJ molecules such as occludin and claudin-5 (Drouin-Ouellet et al. 2015). Other markers associated with BBB permeability, including hepatocyte growth factor, IL-8, and tissue inhibitor of MMP-1, were found elevated in HD patients. An R6/2 transgenic mouse model of HD (Li et al. 2005) confirmed elevated TJ molecules similar to HD patients. R6/2 mice also displayed increased transcytosis and paracellular transport across the brain endothelium (Drouin-Ouellet et al. 2015). In R6/2 mice, TJ imbalance and perturbed BBB homeostasis were perceptible at very early stages of the disease, before symptom onset (Di Pardo et al. 2017).

Summary

BBB disruption has also been observed in a series of other neurological diseases including amyotrophic lateral sclerosis (ALS), epilepsy, edema, brain traumas, PD, as well as systemic diseases, such as liver failure (Daneman 2012). A major question remains whether there is a common mechanism for BBB disruption during all different diseases, or whether the dysfunction in each disease results from different cellular and molecular mechanisms. Several different signaling pathways have been shown to be important for BBB disruption across multiple pathologies including VEGF, reactive oxygen species, TNF-α, and MMP-mediated disruption of TJs and ECM, but it remains unknown how different triggers may engage these pathways, and how different outcomes can be observed in different diseases. As with studying BBB formation, one of the major issues when analyzing studies of BBB dysfunction is that each study uses different measures of the BBB, thus making it difficult to compare the dysfunction during the disease. What is clear is that disruption of the BBB in many of these diseases appears to be multimodal, with increasing vesicle trafficking, disruption of TJ strands, and alterations of endothelial transport and metabolic processes. What remains unclear is how each of these processes interacts with each other. For instance, an increase in transcytosis precedes TJ disruption following stroke, but it is not clear whether this increased vesicle trafficking leads to the removal of junction proteins and transporters from the cell surface. Furthermore, it remains unclear what the cellular signaling events that coordinate these processes are. Is the BBB disruption a result of a loss of PC and astrocyte signals, including Wnt, SHh, and others, or is it because of the reception of disruption signals from neural or immune cells? In the future, a more exhaustive analysis of the BBB at different stages of each disease as well as a large-scale "omics" analysis of changes to the BBB in each disease will clarify this important question.

BBB dysfunction occurs in many different neurological diseases in a wide variety of species, and, thus, this is an evolutionary conserved important feature of these diseases. A critical question moving forward is to understand what aspects of this BBB dysfunction are healing and what aspects are pathological. Like any inflammatory event, a small amount is likely helpful in clearing debris, fighting pathogens, and aiding in wound healing, whereas a large amount can be debilitating causing tissue dysfunction and degeneration. Thus, understanding the molecular mechanisms regulating BBB breakdown and developing methods to appropriately modulate this process will be critical in developing therapeutics to aid in the repair process of these diseases.

CONCLUSIONS

The BBB is an important cellular barrier that tightly controls the microenvironment of the CNS and allows for proper neuronal function. This barrier is an extremely important factor to consider when determining treatments for different neurological diseases, both because disruption of the BBB can lead to severe pathology, but also because crossing the BBB is an essential consideration in the development of CNS therapeutics. Recent work has identified molecules required for BBB function as well as cellular and molecular signaling events that regulate the development of the BBB, its function in adulthood, and its response to injury. Although much progress has been made, many questions remain. Are all of the different BBB properties regulated by the same or different pathways? How are various signaling pathways coordinated to regulate different aspects of the BBB? Which pathways

 Cite this article as *Cold Spring Harb Perspect Biol* doi: 10.1101/cshperspect.a041422

induce BBB properties during development, and which are required throughout life for BBB maintenance? How dynamic is the BBB? Are different BBB properties, including transporters and TJs, dynamically regulated in response to neural activity? How do alterations in the BBB affect neuronal activity, brain function, and behavior? Are there localized specialties of the BBB that regulate regional neuronal development or function? What leads to the loss of BBB properties during neurological disease: a loss of maintenance signals or the presence of disruption signals? Tackling these fundamental questions will allow for the development of therapeutics to modulate the BBB both for restoring its function during neurological disease and for developing methods to bypass the BBB for drug delivery.

ACKNOWLEDGMENTS

We acknowledge Drs. Richard Daneman and Alexandre Prat, authors of "The Blood–Brain Barrier" chapter published in the first edition of *Glia* (doi:10.1101/cshperspect.a020412), upon which this work is based.

REFERENCES

*Reference is also in this subject collection.

Abbott NJ, Khan EU, Rollinson CM, Reichel A, Janigro D, Dombrowski SM, Dobbie MS, Begley DJ. 2002. Drug resistance in epilepsy: the role of the blood–brain barrier. *Novartis Found Symp* **243:** 38–47; discussion 47–53, 180–185. doi:10.1002/0470846356.ch4

Abbott NJ, Rönnbäck L, Hansson E. 2006. Astrocyte–endothelial interactions at the blood–brain barrier. *Nat Rev Neurosci* **7:** 41–53. doi:10.1038/nrn1824

Agrawal S, Anderson P, Durbeej M, van Rooijen N, Ivars F, Opdenakker G, Sorokin LM. 2006. Dystroglycan is selectively cleaved at the parenchymal basement membrane at sites of leukocyte extravasation in experimental autoimmune encephalomyelitis. *J Exp Med* **203:** 1007–1019. doi:10.1084/jem.20051342

Aird WC. 2007a. Phenotypic heterogeneity of the endothelium. II: Representative vascular beds. *Circ Res* **100:** 174–190. doi:10.1161/01.RES.0000255690.03436.ae

Aird WC. 2007b. Phenotypic heterogeneity of the endothelium. I: Structure, function, and mechanisms. *Circ Res* **100:** 158–173. doi:10.1161/01.RES.0000255691.76142.4a

Ajami B, Bennett JL, Krieger C, Tetzlaff W, Rossi FM. 2007. Local self-renewal can sustain CNS microglia maintenance and function throughout adult life. *Nat Neurosci* **10:** 1538–1543. doi:10.1038/nn2014

Alata W, Ye Y, St-Amour I, Vandal M, Calon F. 2015. Human apolipoprotein E ε4 expression impairs cerebral vascularization and blood–brain barrier function in mice. *J Cereb Blood Flow Metab* **35:** 86–94. doi:10.1038/jcbfm.2014.172

Al-Bachari S, Naish JH, Parker GJM, Emsley HCA, Parkes LM. 2020. Blood–brain barrier leakage is increased in Parkinson's disease. *Front Physiol* **11:** 593026. doi:10.3389/fphys.2020.593026

Alvarez JI, Dodelet-Devillers A, Kebir H, Ifergan I, Fabre PJ, Terouz S, Sabbagh M, Wosik K, Bourbonnière L, Bernard M, et al. 2011a. The Hedgehog pathway promotes blood–brain barrier integrity and CNS immune quiescence. *Science* **334:** 1727–1731. doi:10.1126/science.1206936

Alvarez JI, Cayrol R, Prat A. 2011b. Disruption of central nervous system barriers in multiple sclerosis. *Biochim Biophys Acta* **1812:** 252–264. doi:10.1016/j.bbadis.2010.06.017

Alvarez JI, Katayama T, Prat A. 2013. Glial influence on the blood brain barrier. *Glia* **61:** 1939–1958. doi:10.1002/glia.22575

Amasheh S, Schmidt T, Mahn M, Florian P, Mankertz J, Tavalali S, Gitter AH, Schulzke JD, Fromm M. 2005. Contribution of claudin-5 to barrier properties in tight junctions of epithelial cells. *Cell Tissue Res* **321:** 89–96. doi:10.1007/s00441-005-1101-0

Anderson CM, Nedergaard M. 2003. Astrocyte-mediated control of cerebral microcirculation. *Trends Neurosci* **26:** 340–344; author reply 344–345. doi:10.1016/S0166-2236(03)00141-3

Andreone BJ, Chow BW, Tata A, Lacoste B, Ben-Zvi A, Bullock K, Deik AA, Ginty DD, Clish CB, Gu C. 2017. Blood–brain barrier permeability is regulated by lipid transport-dependent suppression of caveolae-mediated transcytosis. *Neuron* **94:** 581–594.e5. doi:10.1016/j.neuron.2017.03.043

Argaw AT, Gurfein BT, Zhang Y, Zameer A, John GR. 2009. VEGF-mediated disruption of endothelial CLN-5 promotes blood–brain barrier breakdown. *Proc Natl Acad Sci* **106:** 1977–1982. doi:10.1073/pnas.0808698106

Argaw AT, Asp L, Zhang J, Navrazhina K, Pham T, Mariani JN, Mahase S, Dutta DJ, Seto J, Kramer EG, et al. 2012. Astrocyte-derived VEGF-A drives blood–brain barrier disruption in CNS inflammatory disease. *J Clin Invest* **122:** 2454–2468. doi:10.1172/JCI60842

Armulik A, Genové G, Mae M, Nisancioglu MH, Wallgard E, Niaudet C, He L, Norlin J, Lindblom P, Strittmatter K, et al. 2010. Pericytes regulate the blood–brain barrier. *Nature* **468:** 557–561. doi:10.1038/nature09522

Armulik A, Genové G, Betsholtz C. 2011. Pericytes: developmental, physiological, and pathological perspectives, problems, and promises. *Dev Cell* **21:** 193–215. doi:10.1016/j.devcel.2011.07.001

Astudillo P, Larraín J. 2014. Wnt signaling and cell-matrix adhesion. *Curr Mol Med* **14:** 209–220. doi:10.2174/1566524014666140128105352

Attems J, Jellinger KA. 2004. Only cerebral capillary amyloid angiopathy correlates with Alzheimer pathology—a pilot study. *Acta Neuropathol* **107:** 83–90. doi:10.1007/s00401-003-0796-9

Attwell D, Buchan AM, Charpak S, Lauritzen M, Macvicar BA, Newman EA. 2010. Glial and neuronal control of brain blood flow. *Nature* **468:** 232–243. doi:10.1038/nature09613

Ayloo S, Lazo CG, Sun S, Zhang W, Cui B, Gu C. 2022. Pericyte-to-endothelial cell signaling via vitronectin-integrin regulates blood-CNS barrier. *Neuron* **110:** 1641–1655.e6. doi:10.1016/j.neuron.2022.02.017

Azmitia EC, Saccomano ZT, Alzoobaee MF, Boldrini M, Whitaker-Azmitia PM. 2016. Persistent angiogenesis in the autism brain: an immunocytochemical study of postmortem cortex, brainstem and cerebellum. *J Autism Dev Disord* **46:** 1307–1318. doi:10.1007/s10803-015-2672-6

Badaut J, Ajao DO, Sorensen DW, Fukuda AM, Pellerin L. 2015. Caveolin expression changes in the neurovascular unit after juvenile traumatic brain injury: signs of blood-brain barrier healing? *Neuroscience* **285:** 215–226. doi:10.1016/j.neuroscience.2014.10.035

Balda MS, Whitney JA, Flores C, González S, Cereijido M, Matter K. 1996. Functional dissociation of paracellular permeability and transepithelial electrical resistance and disruption of the apical-basolateral intramembrane diffusion barrier by expression of a mutant tight junction membrane protein. *J Cell Biol* **134:** 1031–1049. doi:10.1083/jcb.134.4.1031

Banitalebi S, Skauli N, Geiseler S, Ottersen OP, Amiry-Moghaddam M. 2022. Disassembly and mislocalization of AQP4 in incipient scar formation after experimental stroke. *Int J Mol Sci* **23:** 1117. doi:10.3390/ijms23031117

Bano D, Zanetti F, Mende Y, Nicotera P. 2011. Neurodegenerative processes in Huntington's disease. *Cell Death Dis* **2:** e228. doi:10.1038/cddis.2011.112

Barcia C, Bautista V, Sánchez-Bahillo A, Fernández-Villalba E, Faucheux B, Poza y Poza M, Fernandez Barreiro A, Hirsch EC, Herrero MT. 2005. Changes in vascularization in substantia nigra pars compacta of monkeys rendered parkinsonian. *J Neural Transm (Vienna)* **112:** 1237–1248. doi:10.1007/s00702-004-0256-2

Beck DW, Roberts RL, Olson JJ. 1986. Glial cells influence membrane-associated enzyme activity at the blood–brain barrier. *Brain Res* **381:** 131–137. doi:10.1016/0006-8993(86)90700-6

Bell RD, Winkler EA, Sagare AP, Singh I, LaRue B, Deane R, Zlokovic BV. 2010. Pericytes control key neurovascular functions and neuronal phenotype in the adult brain and during brain aging. *Neuron* **68:** 409–427. doi:10.1016/j.neuron.2010.09.043

Benz F, Wichitnaowarat V, Lehmann M, Germano RFV, Mihova D, Macas J, Adams RH, Taketo MM, Guérit S, Plate KH, et al. 2019. Low Wnt/β-catenin signaling determines leaky vessels in the subfornical organ and affects water homeostasis in mice. *eLife* **8:** e43818. doi:10.7554/eLife.43818

Ben-Zvi A, Lacoste B, Kur E, Andreone BJ, Mayshar Y, Yan H, Gu C. 2014. Mfsd2a is critical for the formation and function of the blood-brain barrier. *Nature* **509:** 507–511. doi:10.1038/nature13324

Bergmann S, Lawler SE, Qu Y, Fadzen CM, Wolfe JM, Regan MS, Pentelute BL, Agar NYR, Cho CF. 2018. Blood–brain-barrier organoids for investigating the permeability of CNS therapeutics. *Nat Protoc* **13:** 2827–2843. doi:10.1038/s41596-018-0066-x

Betz AL, Goldstein GW. 1978. Polarity of the blood–brain barrier: neutral amino acid transport into isolated brain capillaries. *Science* **202:** 225–227. doi:10.1126/science.211586

Betz AL, Firth JA, Goldstein GW. 1980. Polarity of the blood-brain barrier: distribution of enzymes between the luminal and antiluminal membranes of brain capillary endothelial cells. *Brain Res* **192:** 17–28. doi:10.1016/0006-8993(80)91004-5

Biernacki K, Prat A, Blain M, Antel JP. 2001. Regulation of Th1 and Th2 lymphocyte migration by human adult brain endothelial cells. *J Neuropathol Exp Neurol* **60:** 1127–1136. doi:10.1093/jnen/60.12.1127

Biswas S, Cottarelli A, Agalliu D. 2020. Neuronal and glial regulation of CNS angiogenesis and barriergenesis. *Development* **147:** dev182279. doi:10.1242/dev.182279

Blair LJ, Frauen HD, Zhang B, Nordhues BA, Bijan S, Lin YC, Zamudio F, Hernandez LD, Sabbagh JJ, Selenica ML, et al. 2015. Tau depletion prevents progressive blood-brain barrier damage in a mouse model of tauopathy. *Acta Neuropathol Commun* **3:** 8. doi:10.1186/s40478-015-0186-2

Bolden CT, Skibber MA, Olson SD, Zamorano Rojas M, Milewicz S, Gill BS, Cox CS Jr. 2023. Validation and characterization of a novel blood–brain barrier platform for investigating traumatic brain injury. *Sci Rep* **13:** 16150. doi:10.1038/s41598-023-43214-7

Boulay AC, Saubaméa B, Adam N, Chasseigneaux S, Mazaré N, Gilbert A, Bahin M, Bastianelli L, Blugeon C, Perrin S, et al. 2017. Translation in astrocyte distal processes sets molecular heterogeneity at the gliovascular interface. *Cell Discov* **3:** 17005. doi:10.1038/celldisc.2017.5

Brightman MW, Reese TS. 1969. Junctions between intimately apposed cell membranes in the vertebrate brain. *J Cell Biol* **40:** 648–677. doi:10.1083/jcb.40.3.648

Cahoy JD, Emery B, Kaushal A, Foo LC, Zamanian JL, Christopherson KS, Xing Y, Lubischer JL, Krieg PA, Krupenko SA, et al. 2008. A transcriptome database for astrocytes, neurons, and oligodendrocytes: a new resource for understanding brain development and function. *J Neurosci* **28:** 264–278. doi:10.1523/JNEUROSCI.4178-07.2008

Cai HQ, Catts VS, Webster MJ, Galletly C, Liu D, O'Donnell M, Weickert TW, Weickert CS. 2020. Increased macrophages and changed brain endothelial cell gene expression in the frontal cortex of people with schizophrenia displaying inflammation. *Mol Psychiatry* **25:** 761–775. doi:10.1038/s41380-018-0235-x

Cambier S, Gline S, Mu D, Collins R, Araya J, Dolganov G, Einheber S, Boudreau N, Nishimura SL. 2005. Integrin αvβ8-mediated activation of transforming growth factor-β by perivascular astrocytes. *Am J Pathol* **166:** 1883–1894. doi:10.1016/S0002-9440(10)62497-2

Carrano A, Hoozemans J, van der Vies S, Rozemuller A, van Horssen J, de Vries HE. 2011. Amyloid β induces oxidative stress-mediated blood–brain barrier changes in capillary amyloid angiopathy. *Antioxid Redox Signal* **15:** 1167–1178. doi:10.1089/ars.2011.3895

Carrier M, Guilbert J, Lévesque JP, Tremblay ME, Desjardins M. 2021. Structural and functional features of developing brain capillaries, and their alteration in schizophrenia. *Front Cell Neurosci* **14:** 595002. doi:10.3389/fncel.2020.595002

Castro Dias M, Coisne C, Baden P, Enzmann G, Garrett L, Becker L, Hölter SM; German Mouse Clinic Consortium; Hrabe de Angelis M, Deutsch U, et al. 2019a. Claudin-12 is not required for blood–brain barrier tight junction function. *Fluids Barriers CNS* **16:** 30. doi:10.1186/s12987-019-0150-9

Castro Dias M, Coisne C, Lazarevic I, Baden P, Hata M, Iwamoto N, Francisco DMF, Vanlandewijck M, He L, Baier FA, et al. 2019b. Claudin-3-deficient C57BL/6J mice display intact brain barriers. *Sci Rep* **9:** 203. doi:10.1038/s41598-018-36731-3

Cater RJ, Chua GL, Erramilli SK, Keener JE, Choy BC, Tokarz P, Chin CF, Quek DQY, Kloss B, Pepe JG, et al. 2021. Structural basis of omega-3 fatty acid transport across the blood–brain barrier. *Nature* **595:** 315–319. doi:10.1038/s41586-021-03650-9

Cayrol R, Wosik K, Berard JL, Dodelet-Devillers A, Ifergan I, Kebir H, Haqqani AS, Kreymborg K, Krug S, Moumdjian R, et al. 2008. Activated leukocyte cell adhesion molecule promotes leukocyte trafficking into the central nervous system. *Nat Immunol* **9:** 137–145. doi:10.1038/ni1551

Chao YX, He BP, Wah Tay SS. 2009. Mesenchymal stem cell transplantation attenuates blood brain barrier damage and neuroinflammation and protects dopaminergic neurons against MPTP toxicity in the substantia nigra in a model of Parkinson's disease. *J Neuroimmunol* **216:** 39–50. doi:10.1016/j.jneuroim.2009.09.003

Chen XL, Dodd G, Thomas S, Zhang X, Wasserman MA, Rovin BH, Kunsch C. 2006. Activation of Nrf2/ARE pathway protects endothelial cells from oxidant injury and inhibits inflammatory gene expression. *Am J Physiol Heart Circ Physiol* **290:** H1862–H1870. doi:10.1152/ajpheart.00651.2005

Chen J, Stahl A, Krah NM, Seaward MR, Dennison RJ, Sapieha P, Hua J, Hatton CJ, Juan AM, Aderman CM, et al. 2011. Wnt signaling mediates pathological vascular growth in proliferative retinopathy. *Circulation* **124:** 1871–1881. doi:10.1161/CIRCULATIONAHA.111.040337

Chen Z, Kelly JR, Morales JE, Sun RC, De A, Burkin DJ, McCarty JH. 2023. The α7 integrin subunit in astrocytes promotes endothelial blood–brain barrier integrity. *Development* **150:** dev201356. doi:10.1242/dev.201356

Chiu C, Miller MC, Monahan R, Osgood DP, Stopa EG, Silverberg GD. 2015. P-glycoprotein expression and amyloid accumulation in human aging and Alzheimer's disease: preliminary observations. *Neurobiol Aging* **36:** 2475–2482. doi:10.1016/j.neurobiolaging.2015.05.020

Choi KH, Kim HS, Park MS, Kim JT, Kim JH, Cho KA, Lee MC, Lee HJ, Cho KH. 2016. Regulation of Caveolin-1 expression determines early brain edema after experimental focal cerebral ischemia. *Stroke* **47:** 1336–1343. doi:10.1161/STROKEAHA.116.013205

Chow BW, Gu C. 2017. Gradual suppression of transcytosis governs functional blood–retinal barrier formation. *Neuron* **93:** 1325–1333.e3. doi:10.1016/j.neuron.2017.02.043

Cipolla MJ, Crete R, Vitullo L, Rix RD. 2004. Transcellular transport as a mechanism of blood–brain barrier disruption during stroke. *Front Biosci* **9:** 777–785. doi:10.2741/1282

Clavreul S, Abdeladim L, Hernández-Garzón E, Niculescu D, Durand J, Ieng SH, Barry R, Bonvento G, Beaurepaire E, Livet J, et al. 2019. Cortical astrocytes develop in a plastic manner at both clonal and cellular levels. *Nat Commun* **10:** 4884. doi:10.1038/s41467-019-12791-5

Coelho-Santos V, Shih AY. 2020. Postnatal development of cerebrovascular structure and the neurogliovascular unit. *Wiley Interdiscip Rev Dev Biol* **9:** e363. doi:10.1002/wdev.363

Cohen-Salmon M, Slaoui L, Mazaré N, Gilbert A, Oudart M, Alvear-Perez R, Elorza-Vidal X, Chever O, Boulay AC. 2021. Astrocytes in the regulation of cerebrovascular functions. *Glia* **69:** 817–841. doi:10.1002/glia.23924

Colegio OR, Van Itallie C, Rahner C, Anderson JM. 2003. Claudin extracellular domains determine paracellular charge selectivity and resistance but not tight junction fibril architecture. *Am J Physiol Cell Physiol* **284:** C1346–C1354. doi:10.1152/ajpcell.00547.2002

Collard CD, Park KA, Montalto MC, Alapati S, Buras JA, Stahl GL, Colgan SP. 2002. Neutrophil-derived glutamate regulates vascular endothelial barrier function. *J Biol Chem* **277:** 14801–14811. doi:10.1074/jbc.M110557200

Connolly AM, Chez MG, Pestronk A, Arnold ST, Mehta S, Deuel RK. 1999. Serum autoantibodies to brain in Landau–Kleffner variant, autism, and other neurologic disorders. *J Pediatr* **134:** 607–613. doi:10.1016/S0022-3476(99)70248-9

Conway EM, Carmeliet P. 2004. The diversity of endothelial cells: a challenge for therapeutic angiogenesis. *Genome Biol* **5:** 207. doi:10.1186/gb-2004-5-2-207

Coomber BL, Stewart PA. 1985. Morphometric analysis of CNS microvascular endothelium. *Microvasc Res* **30:** 99–115. doi:10.1016/0026-2862(85)90042-1

Cordon-Cardo C, O'Brien JP, Casals D, Rittman-Grauer L, Biedler JL, Melamed MR, Bertino JR. 1989. Multidrug-resistance gene (P-glycoprotein) is expressed by endothelial cells at blood–brain barrier sites. *Proc Natl Acad Sci* **86:** 695–698. doi:10.1073/pnas.86.2.695

Cornford EM, Hyman S, Swartz BE. 1994. The human brain GLUT1 glucose transporter: ultrastructural localization to the blood–brain barrier endothelia. *J Cereb Blood Flow Metab* **14:** 106–112. doi:10.1038/jcbfm.1994.15

Courtoy PJ, Boyles J. 1983. Fibronectin in the microvasculature: localization in the pericyte-endothelial interstitium. *J Ultrastruct Res* **83:** 258–273. doi:10.1016/S0022-5320(83)90133-8

Cramer SP, Simonsen H, Frederiksen JL, Rostrup E, Larsson HB. 2014. Abnormal blood-brain barrier permeability in normal appearing white matter in multiple sclerosis investigated by MRI. *Neuroimage Clin* **4:** 182–189. doi:10.1016/j.nicl.2013.12.001

Crockett AM, Ryan SK, Vásquez AH, Canning C, Kanyuch N, Kebir H, Ceja G, Gesualdi J, Zackai E, McDonald-McGinn D, et al. 2021. Disruption of the blood–brain barrier in 22q11.2 deletion syndrome. *Brain* **144:** 1351–1360. doi:10.1093/brain/awab055

Crouch EE, Joseph T, Marsan E, Huang EJ. 2023. Disentangling brain vasculature in neurogenesis and neurodegeneration using single-cell transcriptomics. *Trends Neurosci* **46:** 551–565. doi:10.1016/j.tins.2023.04.007

Cuevas P, Gutierrez-Diaz JA, Reimers D, Dujovny M, Diaz FG, Ausman JI. 1984. Pericyte endothelial gap junctions in human cerebral capillaries. *Anat Embryol (Berl)* **170:** 155–159. doi:10.1007/BF00319000

Daneman R. 2012. The blood–brain barrier in health and disease. *Ann Neurol* **72:** 648–672. doi:10.1002/ana.23648

Daneman R, Agalliu D, Zhou L, Kuhnert F, Kuo CJ, Barres BA. 2009. Wnt/β-catenin signaling is required for CNS, but not non-CNS, angiogenesis. *Proc Natl Acad Sci* **106**: 641–646. doi:10.1073/pnas.0805165106

Daneman R, Zhou L, Agalliu D, Cahoy JD, Kaushal A, Barres BA. 2010a. The mouse blood–brain barrier transcriptome: a new resource for understanding the development and function of brain endothelial cells. *PLoS ONE* **5**: e13741. doi:10.1371/journal.pone.0013741

Daneman R, Zhou L, Kebede AA, Barres BA. 2010b. Pericytes are required for blood–brain barrier integrity during embryogenesis. *Nature* **468**: 562–566. doi:10.1038/nature09513

Davenport CB. 1915. Huntington's chorea in relation to heredity and eugenics. *Proc Natl Acad Sci* **1**: 283–285. doi:10.1073/pnas.1.5.283

Davis J, Xu F, Deane R, Romanov G, Previti ML, Zeigler K, Zlokovic BV, Van Nostrand WE. 2004. Early-onset and robust cerebral microvascular accumulation of amyloid β-protein in transgenic mice expressing low levels of a vasculotropic Dutch/Iowa mutant form of amyloid β-protein precursor. *J Biol Chem* **279**: 20296–20306. doi:10.1074/jbc.M312946200

Deane R, Wu Z, Sagare A, Davis J, Du Yan S, Hamm K, Xu F, Parisi M, LaRue B, Hu HW, et al. 2004. LRP/amyloid β-peptide interaction mediates differential brain efflux of Aβ isoforms. *Neuron* **43**: 333–344. doi:10.1016/j.neuron.2004.07.017

De Bock M, De Smet M, Verwaerde S, Tahiri H, Schumacher S, Van Haver V, Witschas K, Steinhäuser C, Rouach N, Vandenbroucke RE, et al. 2022. Targeting gliovascular connexins prevents inflammatory blood–brain barrier leakage and astrogliosis. *JCI Insight* **7**: e135263. doi:10.1172/jci.insight.135263

Del Zoppo GJ, Milner R, Mabuchi T, Hung S, Wang X, Koziol JA. 2006. Vascular matrix adhesion and the blood–brain barrier. *Biochem Soc Trans* **34**: 1261–1266. doi:10.1042/BST0341261

De Silva TM, Faraci FM. 2016. Microvascular dysfunction and cognitive impairment. *Cell Mol Neurobiol* **36**: 241–258. doi:10.1007/s10571-015-0308-1

De Vivo DC, Trifiletti RR, Jacobson RI, Ronen GM, Behmand RA, Harik SI. 1991. Defective glucose transport across the blood–brain barrier as a cause of persistent hypoglycorrhachia, seizures, and developmental delay. *N Engl J Med* **325**: 703–709. doi:10.1056/NEJM199109053251006

De Vivo DC, Leary L, Wang D. 2002. Glucose transporter 1 deficiency syndrome and other glycolytic defects. *J Child Neurol* **17**: 3S15–3S23; discussion 3S24–3S25. doi:10.1177/088307380201700501

Díaz-Castro B, Robel S, Mishra A. 2023. Astrocyte endfeet in brain function and pathology: open questions. *Annu Rev Neurosci* **46**: 101–121. doi:10.1146/annurev-neuro-091922-031205

Diaz-Flores L, Gutiérrez R, Madrid JF, Varela H, Valladares F, Acosta E, MartınVasallo P, Dıaz-Flores L Jr. 2009. Pericytes. Morphofunction, interactions and pathology in a quiescent and activated mesenchymal cell niche. *Histol Histopathol* **24**: 909–969.

Dion-Albert L, Cadoret A, Doney E, Kaufmann FN, Dudek KA, Daigle B, Parise LF, Cathomas F, Samba N, Hudson N, et al. 2022. Vascular and blood–brain barrier-related changes underlie stress responses and resilience in female mice and depression in human tissue. *Nat Commun* **13**: 164. doi:10.1038/s41467-021-27604-x

Di Pardo A, Amico E, Scalabri F, Pepe G, Castaldo S, Elifani F, Capocci L, De Sanctis C, Comerci L, Pompeo F, et al. 2017. Impairment of blood–brain barrier is an early event in R6/2 mouse model of Huntington disease. *Sci Rep* **7**: 41316. doi:10.1038/srep41316

Dodelet-Devillers A, Cayrol R, van Horssen J, Haqqani AS, de Vries HE, Engelhardt B, Greenwood J, Prat A. 2009. Functions of lipid raft membrane microdomains at the blood–brain barrier. *J Mol Med (Berl)* **87**: 765–774. doi:10.1007/s00109-009-0488-6

Dohgu S, Yamauchi A, Takata F, Naito M, Tsuruo T, Higuchi S, Sawada Y, Kataoka Y. 2004. Transforming growth factor-β1 upregulates the tight junction and P-glycoprotein of brain microvascular endothelial cells. *Cell Mol Neurobiol* **24**: 491–497. doi:10.1023/B:CEMN.0000022776.47302.ce

Donahue JE, Flaherty SL, Johanson CE, Duncan JA, Silverberg GD, Miller MC, Tavares R, Yang W, Wu Q, Sabo E, et al. 2006. RAGE, LRP-1, and amyloid-β protein in Alzheimer's disease. *Acta Neuropathol* **112**: 405–415. doi:10.1007/s00401-006-0115-3

Drab M, Verkade P, Elger M, Kasper M, Lohn M, Lauterbach B, Menne J, Lindschau C, Mende F, Luft FC, et al. 2001. Loss of caveolae, vascular dysfunction, and pulmonary defects in caveolin-1 gene-disrupted mice. *Science* **293**: 2449–2452. doi:10.1126/science.1062688

Drouin-Ouellet J, Sawiak SJ, Cisbani G, Lagacé M, Kuan WL, Saint-Pierre M, Dury RJ, Alata W, St-Amour I, Mason SL, et al. 2015. Cerebrovascular and blood–brain barrier impairments in Huntington's disease: potential implications for its pathophysiology. *Ann Neurol* **78**: 160–177. doi:10.1002/ana.24406

Edwards DN, Bix GJ. 2019. Roles of blood–brain barrier integrins and extracellular matrix in stroke. *Am J Physiol Cell Physiol* **316**: C252–C263. doi:10.1152/ajpcell.00151.2018

Elabi O, Gaceb A, Carlsson R, Padel T, Soylu-Kucharz R, Cortijo I, Li W, Li JY, Paul G. 2021. Human α-synuclein overexpression in a mouse model of Parkinson's disease leads to vascular pathology, blood–brain barrier leakage and pericyte activation. *Sci Rep* **11**: 1120. doi:10.1038/s41598-020-80889-8

Enerson BE, Drewes LR. 2006. The rat blood–brain barrier transcriptome. *J Cereb Blood Flow Metab* **26**: 959–973. doi:10.1038/sj.jcbfm.9600249

Engelhardt B. 2008. Immune cell entry into the central nervous system: involvement of adhesion molecules and chemokines. *J Neurol Sci* **274**: 23–26. doi:10.1016/j.jns.2008.05.019

Engelhardt B, Sorokin L. 2009. The blood–brain and the blood–cerebrospinal fluid barriers: function and dysfunction. *Semin Immunopathol* **31**: 497–511. doi:10.1007/s00281-009-0177-0

Ezan P, André P, Cisternino S, Saubaméa B, Boulay AC, Doutremer S, Thomas MA, Quenech'du N, Giaume C, Cohen-Salmon M. 2012. Deletion of astroglial connexins weakens the blood-brain barrier. *J Cereb Blood Flow Metab* **32**: 1457–1467. doi:10.1038/jcbfm.2012.45

Fan X, Staitieh BS, Jensen JS, Mould KJ, Greenberg JA, Joshi PC, Koval M, Guidot DM. 2013. Activating the Nrf2-mediated antioxidant response element restores barrier function in the alveolar epithelium of HIV-1 transgenic rats. *Am J Physiol Lung Cell Mol Physiol* **305**: L267–L277. doi:10 .1152/ajplung.00288.2012

Faucheux BA, Agid Y, Hirsch EC, Bonnet AM. 1999. Blood vessels change in the mesencephalon of patients with Parkinson's disease. *Lancet* **353**: 981–982. doi:10.1016/ S0140-6736(99)00641-8

Felix L, Stephan J, Rose CR. 2021. Astrocytes of the early postnatal brain. *Eur J Neurosci* **54**: 5649–5672. doi:10 .1111/ejn.14780

Fields RD, Woo DH, Basser PJ. 2015. Glial regulation of the neuronal connectome through local and long-distant communication. *Neuron* **86**: 374–386. doi:10.1016/j .neuron.2015.01.014

Fiorentino M, Sapone A, Senger S, Camhi SS, Kadzielski SM, Buie TM, Kelly DL, Cascella N, Fasano A. 2016. Blood– brain barrier and intestinal epithelial barrier alterations in autism spectrum disorders. *Mol Autism* **7**: 49. doi:10.1186/ s13229-016-0110-z

Fischer S, Wobben M, Marti HH, Renz D, Schaper W. 2002. Hypoxia-induced hyperpermeability in brain microvessel endothelial cells involves VEGF-mediated changes in the expression of zonula occludens-1. *Microvasc Res* **63**: 70– 80. doi:10.1006/mvre.2001.2367

Fischer S, Wiesnet M, Marti HH, Renz D, Schaper W. 2004. Simultaneous activation of several second messengers in hypoxia-induced hyperpermeability of brain derived endothelial cells. *J Cell Physiol* **198**: 359–369. doi:10.1002/jcp .10417

Fontijn RD, Volger OL, Fledderus JO, Reijerkerk A, de Vries HE, Horrevoets AJ. 2008. SOX-18 controls endothelial-specific claudin-5 gene expression and barrier function. *Am J Physiol Heart Circ Physiol* **294**: H891–H900. doi:10 .1152/ajpheart.01248.2007

Freitas-Andrade M, Raman-Nair J, Lacoste B. 2020. Structural and functional remodeling of the brain vasculature following stroke. *Front Physiol* **11**: 948. doi:10.3389/fphys .2020.00948

Freitas-Andrade M, Comin CH, Dyken PV, Ouellette J, Raman-Nair J, Blakeley N, Liu QY, Leclerc S, Pan Y, Liu Z, et al. 2023. Astroglial Hmgb1 regulates postnatal astrocyte morphogenesis and cerebrovascular maturation. *Nat Commun* **14**: 4965 doi:10.1038/s41467-023-40682-3

Furuse M. 2010. Molecular basis of the core structure of tight junctions. *Cold Spring Harb Perspect Biol* **2**: a002907. doi:10.1101/cshperspect.a002907

Furuse M, Sasaki H, Fujimoto K, Tsukita S. 1998. A single gene product, claudin-1 or -2, reconstitutes tight junction strands and recruits occludin in fibroblasts. *J Cell Biol* **143**: 391–401. doi:10.1083/jcb.143.2.391

Furuse M, Furuse K, Sasaki H, Tsukita S. 2001. Conversion of *Zonulae occludentes* from tight to leaky strand type by introducing claudin-2 into Madin–Darby canine kidney I cells. *J Cell Biol* **153**: 263–272. doi:10.1083/jcb.153.2.263

Furuse M, Hata M, Furuse K, Yoshida Y, Haratake A, Sugitani Y, Noda T, Kubo A, Tsukita S. 2002. Claudin-based tight junctions are crucial for the mammalian epidermal barrier: a lesson from claudin-1-deficient mice. *J Cell Biol* **156**: 1099–1111. doi:10.1083/jcb.200110122

Galea I. 2021. The blood–brain barrier in systemic infection and inflammation. *Cell Mol Immunol* **18**: 2489–2501. doi:10.1038/s41423-021-00757-x

Garcia FJ, Sun N, Lee H, Godlewski B, Galani K, Zhou B, Mantero J, Bennett DA, Sahin M, Kellis M, et al. 2022. Single-cell dissection of the human brain vasculature. *Nature* **603**: 893–899. doi:10.1038/s41586-022-04521-7

Garvin J, Semenikhina M, Liu Q, Rarick K, Isaeva E, Levchenko V, Staruschenko A, Palygin O, Harder D, Cohen S. 2022. Astrocytic responses to high glucose impair barrier formation in cerebral microvessel endothelial cells. *Am J Physiol Regul Integr Comp Physiol* **322**: R571–R580. doi:10 .1152/ajpregu.00315.2020

Gautam J, Zhang X, Yao Y. 2016. The role of pericytic laminin in blood brain barrier integrity maintenance. *Sci Rep* **6**: 36450. doi:10.1038/srep36450

Gerhardt H, Wolburg H, Redies C. 2000. N-cadherin mediates pericytic–endothelial interaction during brain angiogenesis in the chicken. *Dev Dyn* **218**: 472–479. doi:10 .1002/1097-0177(200007)218:3<472::AID-DVDY1008> 3.0.CO;2-#

Ghandour MS, Langley OK, Zhu XL, Waheed A, Sly WS. 1992. Carbonic anhydrase IV on brain capillary endothelial cells: a marker associated with the blood–brain barrier. *Proc Natl Acad Sci* **89**: 6823–6827. doi:10.1073/pnas.89.15 .6823

Gilbert A, Vidal XE, Estevez R, Cohen-Salmon M, Boulay AC. 2019. Postnatal development of the astrocyte perivascular MLC1/GlialCAM complex defines a temporal window for the gliovascular unit maturation. *Brain Struct Funct* **224**: 1267–1278. doi:10.1007/s00429-019-01832-w

Gilbert A, Elorza-Vidal X, Rancillac A, Chagnot A, Yetim M, Hingot V, Deffieux T, Boulay AC, Alvear-Perez R, Cisternino S, et al. 2021. Megalencephalic leukoencephalopathy with subcortical cysts is a developmental disorder of the gliovascular unit. *eLife* **10**: e71379. doi:10.7554/eLife .71379

Ginhoux F, Greter M, Leboeuf M, Nandi S, See P, Gokhan S, Mehler MF, Conway SJ, Ng LG, Stanley ER, et al. 2010. Fate mapping analysis reveals that adult microglia derive from primitive macrophages. *Science* **330**: 841–845. doi:10 .1126/science.1194637

Gireud-Goss M, Mack AF, McCullough LD, Urayama A. 2021. Cerebral amyloid angiopathy and blood–brain barrier dysfunction. *Neuroscientist* **27**: 668–684. doi:10.1177/ 1073858420954811

Girolamo F, Coppola C, Ribatti D, Trojano M. 2014. Angiogenesis in multiple sclerosis and experimental autoimmune encephalomyelitis. *Acta Neuropathol Commun* **2**: 84. doi:10.1186/s40478-014-0084-z

Gordon GR, Choi HB, Rungta RL, Ellis-Davies GC, MacVicar BA. 2008. Brain metabolism dictates the polarity of astrocyte control over arterioles. *Nature* **456**: 745–749. doi:10 .1038/nature07525

Gordon GR, Howarth C, MacVicar BA. 2011. Bidirectional control of arteriole diameter by astrocytes. *Exp Physiol* **96**: 393–399. doi:10.1113/expphysiol.2010.053132

Gow A, Southwood CM, Li JS, Pariali M, Riordan GP, Brodie SE, Danias J, Bronstein JM, Kachar B, Lazzarini RA. 1999. CNS myelin and sertoli cell tight junction strands are absent in *Osp/claudin-11* null mice. *Cell* **99**: 649–659. doi:10 .1016/S0092-8674(00)81553-6

Graham NS, Sharp DJ. 2019. Understanding neurodegeneration after traumatic brain injury: from mechanisms to clinical trials in dementia. *J Neurol Neurosurg Psychiatry* **90:** 1221–1233.

Gray MT, Woulfe JM. 2015. Striatal blood–brain barrier permeability in Parkinson's disease. *J Cereb Blood Flow Metab* **35:** 747–750. doi:10.1038/jcbfm.2015.32

Greene C, Kealy J, Humphries MM, Gong Y, Hou J, Hudson N, Cassidy LM, Martiniano R, Shashi V, Hooper SR, et al. 2018. Dose-dependent expression of claudin-5 is a modifying factor in schizophrenia. *Mol Psychiatry* **23:** 2156–2166. doi:10.1038/mp.2017.156

Greenwood J, Heasman SJ, Alvarez JI, Prat A, Lyck R, Engelhardt B. 2011. Review: leucocyte-endothelial cell crosstalk at the blood–brain barrier: a prerequisite for successful immune cell entry to the brain. *Neuropathol Appl Neurobiol* **37:** 24–39. doi:10.1111/j.1365-2990.2010.01140.x

Gu Y, Zheng G, Xu M, Li Y, Chen X, Zhu W, Tong Y, Chung SK, Liu KJ, Shen J. 2012. Caveolin-1 regulates nitric oxide-mediated matrix metalloproteinases activity and blood–brain barrier permeability in focal cerebral ischemia and reperfusion injury. *J Neurochem* **120:** 147–156. doi:10.1111/j.1471-4159.2011.07542.x

Guo Y, Singh LN, Zhu Y, Gur RE, Resnick A, Anderson SA, Alvarez JI. 2020. Association of a functional claudin-5 variant with schizophrenia in female patients with the 22q11.2 deletion syndrome. *Schizophr Res* **215:** 451–452. doi:10.1016/j.schres.2019.09.014

Guo X, Ge T, Xia S, Wu H, Colt M, Xie X, Zhang B, Zeng J, Chen J, Zhu D, et al. 2021. atp13a5 marker reveals pericytes of the central nervous system in mice. bioRxiv doi:10.1101/2021.07.09.451694

Gupta IR, Ryan AK. 2010. Claudins: unlocking the code to tight junction function during embryogenesis and in disease. *Clin Genet* **77:** 314–325. doi:10.1111/j.1399-0004.2010.01397.x

Gur RE, Bassett AS, McDonald-McGinn DM, Bearden CE, Chow E, Emanuel BS, Owen M, Swillen A, Van den Bree M, Vermeesch J, et al. 2017. A neurogenetic model for the study of schizophrenia spectrum disorders: the International 22q11.2 Deletion Syndrome Brain Behavior Consortium. *Mol Psychiatry* **22:** 1664–1672. doi:10.1038/mp.2017.161

Guttenplan KA, Liddelow SA. 2019. Astrocytes and microglia: models and tools. *J Exp Med* **216:** 71–83. doi:10.1084/jem.20180200

Ha AD, Fung VS. 2012. Huntington's disease. *Curr Opin Neurol* **25:** 491–498. doi:10.1097/WCO.0b013e3283550c97

Ha SN, Hochman J, Sheridan RP. 2007. Mini review on molecular modeling of P-glycoprotein (Pgp). *Curr Top Med Chem* **7:** 1525–1529. doi:10.2174/156802607782194806

Hadj-Rabia S, Baala L, Vabres P, Hamel-Teillac D, Jacquemin E, Fabre M, Lyonnet S, De Prost Y, Munnich A, Hadchouel M, et al. 2004. Claudin-1 gene mutations in neonatal sclerosing cholangitis associated with ichthyosis: a tight junction disease. *Gastroenterology* **127:** 1386–1390. doi:10.1053/j.gastro.2004.07.022

Hajal C, Offeddu GS, Shin Y, Zhang S, Morozova O, Hickman D, Knutson CG, Kamm RD. 2022. Engineered human blood-brain barrier microfluidic model for vascular permeability analyses. *Nat Protoc* **17:** 95–128. doi:10.1038/s41596-021-00635-w

Haj-Yasein NN, Vindedal GF, Eilert-Olsen M, Gundersen GA, Skare O, Laake P, Klungland A, Thorén AE, Burkhardt JM, Ottersen OP, et al. 2011. Glial-conditional deletion of aquaporin-4 (*Aqp4*) reduces blood–brain water uptake and confers barrier function on perivascular astrocyte endfeet. *Proc Natl Acad Sci* **108:** 17815–17820. doi:10.1073/pnas.1110655108

Haley MJ, Lawrence CB. 2017. The blood–brain barrier after stroke: structural studies and the role of transcytotic vesicles. *J Cereb Blood Flow Metab* **37:** 456–470. doi:10.1177/0271678X16629976

Halilagic A, Ribes V, Ghyselinck NB, Zile MH, Dollé P, Studer M. 2007. Retinoids control anterior and dorsal properties in the developing forebrain. *Dev Biol* **303:** 362–375. doi:10.1016/j.ydbio.2006.11.021

Hall CN, Reynell C, Gesslein B, Hamilton NB, Mishra A, Sutherland BA, O'Farrell FM, Buchan AM, Lauritzen M, Attwell D. 2014. Capillary pericytes regulate cerebral blood flow in health and disease. *Nature* **508:** 55–60. doi:10.1038/nature13165

Hampson G, Konrad MA, Scoble J. 2008. Familial hypomagnesaemia with hypercalciuria and nephrocalcinosis (FHHNC): compound heterozygous mutation in the claudin 16 (*CLDN16*) gene. *BMC Nephrol* **9:** 12. doi:10.1186/1471-2369-9-12

Hartmann DA, Berthiaume AA, Grant RI, Harrill SA, Koski T, Tieu T, McDowell KP, Faino AV, Kelly AL, Shih AY. 2021. Brain capillary pericytes exert a substantial but slow influence on blood flow. *Nat Neurosci* **24:** 633–645. doi:10.1038/s41593-020-00793-2

Hawkins BT, Gu YH, Izawa Y, Del Zoppo GJ. 2013. Disruption of dystroglycan–laminin interactions modulates water uptake by astrocytes. *Brain Res* **1503:** 89–96. doi:10.1016/j.brainres.2013.01.049

Heithoff BP, George KK, Phares AN, Zuidhoek IA, Munoz-Ballester C, Robel S. 2021. Astrocytes are necessary for blood–brain barrier maintenance in the adult mouse brain. *Glia* **69:** 436–472. doi:10.1002/glia.23908

Henninger DD, Panés J, Eppihimer M, Russell J, Gerritsen M, Anderson DC, Granger DN. 1997. Cytokine-induced VCAM-1 and ICAM-1 expression in different organs of the mouse. *J Immunol* **158:** 1825–1832. doi:10.4049/jimmunol.158.4.1825

Hickey WF, Kimura H. 1988. Perivascular microglial cells of the CNS are bone marrow–derived and present antigen in vivo. *Science* **239:** 290–292. doi:10.1126/science.3276004

Hong JY, Park JI, Cho K, Gu D, Ji H, Artandi SE, McCrea PD. 2010. Shared molecular mechanisms regulate multiple catenin proteins: canonical Wnt signals and components modulate p120-catenin isoform-1 and additional p120 subfamily members. *J Cell Sci* **123:** 4351–4365. doi:10.1242/jcs.067199

Hösli L, Zuend M, Bredell G, Zanker HS, Porto de Oliveira CE, Saab AS, Weber B. 2022. Direct vascular contact is a hallmark of cerebral astrocytes. *Cell Rep* **39:** 110599. doi:10.1016/j.celrep.2022.110599

Hou J, Gomes AS, Paul DL, Goodenough DA. 2006. Study of claudin function by RNA interference. *J Biol Chem* **281:** 36117–36123. doi:10.1074/jbc.M608853200

Huang ZG, Xue D, Preston E, Karbalai H, Buchan AM. 1999. Biphasic opening of the blood–brain barrier following transient focal ischemia: effects of hypothermia. *Can J Neurol Sci* **26:** 298–304. doi:10.1017/S0317167100000421

Huang J, Upadhyay UM, Tamargo RJ. 2006. Inflammation in stroke and focal cerebral ischemia. *Surg Neurol* **66:** 232–245. doi:10.1016/j.surneu.2005.12.028

Huang Z, Wong LW, Su Y, Huang X, Wang N, Chen H, Yi C. 2020. Blood–brain barrier integrity in the pathogenesis of Alzheimer's disease. *Front Neuroendocrinol* **59:** 100857. doi:10.1016/j.yfrne.2020.100857

Hudson LC, Bragg DC, Tompkins MB, Meeker RB. 2005. Astrocytes and microglia differentially regulate trafficking of lymphocyte subsets across brain endothelial cells. *Brain Res* **1058:** 148–160. doi:10.1016/j.brainres.2005.07.071

Hultman K, Strickland S, Norris EH. 2013. The *APOE* ε4/ε4 genotype potentiates vascular fibrin(ogen) deposition in amyloid-laden vessels in the brains of Alzheimer's disease patients. *J Cereb Blood Flow Metab* **33:** 1251–1258. doi:10.1038/jcbfm.2013.76

Iadecola C. 2013. The pathobiology of vascular dementia. *Neuron* **80:** 844–866. doi:10.1016/j.neuron.2013.10.008

Iadecola C, Smith EE, Anrather J, Gu C, Mishra A, Misra S, Perez-Pinzon MA, Shih AY, Sorond FA, van Veluw SJ, et al. 2023. The neurovasculome: key roles in brain health and cognitive impairment: a scientific statement from the American Heart Association/American Stroke Association. *Stroke* **54:** e251–e271. doi:10.1161/STR.0000000000000431

Ichkova A, Rodriguez-Grande B, Zub E, Saudi A, Fournier ML, Aussudre J, Sicard P, Obenaus A, Marchi N, Badaut J. 2020. Early cerebrovascular and long-term neurological modifications ensue following juvenile mild traumatic brain injury in male mice. *Neurobiol Dis* **141:** 104952. doi:10.1016/j.nbd.2020.104952

Janzer RC, Raff MC. 1987. Astrocytes induce blood–brain barrier properties in endothelial cells. *Nature* **325:** 253–257. doi:10.1038/325253a0

Jiao H, Wang Z, Liu Y, Wang P, Xue Y. 2011. Specific role of tight junction proteins claudin-5, occludin, and ZO-1 of the blood–brain barrier in a focal cerebral ischemic insult. *J Mol Neurosci* **44:** 130–139. doi:10.1007/s12031-011-9496-4

Jiang H, Gallet S, Klemm P, Scholl P, Folz-Donahue K, Altmüller J, Alber J, Heilinger C, Kukat C, Loyens A, et al. 2020. MCH neurons regulate permeability of the median eminence barrier. *Neuron* **107:** 306–319.e9. doi:10.1016/j.neuron.2020.04.020

Johnson-Legér CA, Aurrand-Lions M, Beltraminelli N, Fasel N, Imhof BA. 2002. Junctional adhesion molecule-2 (JAM-2) promotes lymphocyte transendothelial migration. *Blood* **100:** 2479–2486. doi:10.1182/blood-2001-11-0098

Jullienne A, Obenaus A, Ichkova A, Savona-Baron C, Pearce WJ, Badaut J. 2016. Chronic cerebrovascular dysfunction after traumatic brain injury. *J Neurosci Res* **94:** 609–622. doi:10.1002/jnr.23732

Kalucka J, de Rooij L, Goveia J, Rohlenova K, Dumas SJ, Meta E, Conchinha NV, Taverna F, Teuwen LA, Veys K, et al. 2020. Single-cell transcriptome atlas of murine endothelial cells. *Cell* **180:** 764–779.e20. doi:10.1016/j.cell.2020.01.015

Kamphuis WW, Derada Troletti C, Reijerkerk A, Romero IA, de Vries HE. 2015. The blood–brain barrier in multiple sclerosis: microRNAs as key regulators. *CNS Neurol Disord Drug Targets* **14:** 157–167. doi:10.2174/1871527314666150116125246

Karow M, Camp JG, Falk S, Gerber T, Pataskar A, Gac-Santel M, Kageyama J, Brazovskaja A, Garding A, Fan W, et al. 2018. Direct pericyte-to-neuron reprogramming via unfolding of a neural stem cell-like program. *Nat Neurosci* **21:** 932–940. doi:10.1038/s41593-018-0168-3

Katsimpardi L, Litterman NK, Schein PA, Miller CM, Loffredo FS, Wojtkiewicz GR, Chen JW, Lee RT, Wagers AJ, Rubin LL. 2014. Vascular and neurogenic rejuvenation of the aging mouse brain by young systemic factors. *Science* **344:** 630–634. doi:10.1126/science.1251141

Katt ME, Mayo LN, Ellis SE, Mahairaki V, Rothstein JD, Cheng L, Searson PC. 2019. The role of mutations associated with familial neurodegenerative disorders on blood–brain barrier function in an iPSC model. *Fluids Barriers CNS* **16:** 20. doi:10.1186/s12987-019-0139-4

Kaur C, Ling EA. 2017. The circumventricular organs. *Histol Histopathol* **32:** 879–892. doi:10.14670/HH-11-881

Kebir H, Kreymborg K, Ifergan I, Dodelet-Devillers A, Cayrol R, Bernard M, Giuliani F, Arbour N, Becher B, Prat A. 2007. Human TH17 lymphocytes promote blood–brain barrier disruption and central nervous system inflammation. *Nat Med* **13:** 1173–1175. doi:10.1038/nm1651

Keuschnigg J, Henttinen T, Auvinen K, Karikoski M, Salmi M, Jalkanen S. 2009. The prototype endothelial marker PALE is a leukocyte trafficking molecule. *Blood* **114:** 478–484. doi:10.1182/blood-2008-11-188763

Khandaker GM, Cousins L, Deakin J, Lennox BR, Yolken R, Jones PB. 2015. Inflammation and immunity in schizophrenia: implications for pathophysiology and treatment. *Lancet Psychiatry* **2:** 258–270. doi:10.1016/S2215-0366(14)00122-9

Knohl SJ, Scheinman SJ. 2004. Inherited hypercalciuric syndromes: Dent's disease (CLC-5) and familial hypomagnesemia with hypercalciuria (paracellin-1). *Semin Nephrol* **24:** 55–60. doi:10.1053/j.semnephrol.2003.08.011

Knowland D, Arac A, Sekiguchi KJ, Hsu M, Lutz SE, Perrino J, Steinberg GK, Barres BA, Nimmerjahn A, Agalliu D. 2014. Stepwise recruitment of transcellular and paracellular pathways underlies blood–brain barrier breakdown in stroke. *Neuron* **82:** 603–617. doi:10.1016/j.neuron.2014.03.003

Koh I, Hagiwara M. 2024. Modular tissue-in-a-CUBE platform to model blood–brain barrier (BBB) and brain interaction. *Commun Biol* **7:** 177. doi:10.1038/s42003-024-05857-8

Kook SY, Hong HS, Moon M, Ha CM, Chang S, Mook-Jung I. 2012. Aβ$_{1–42}$–RAGE interaction disrupts tight junctions of the blood–brain barrier via Ca^{2+}-calcineurin signaling. *J Neurosci* **32:** 8845–8854. doi:10.1523/JNEUROSCI.6102-11.2012

Körbelin J, Dogbevia G, Michelfelder S, Ridder DA, Hunger A, Wenzel J, Seismann H, Lampe M, Bannach J, Pasparakis M, et al. 2016. A brain microvasculature endothelial cell-specific viral vector with the potential to treat neurovascular and neurological diseases. *EMBO Mol Med* **8:** 609–625. doi:10.15252/emmm.201506078

Kortekaas R, Leenders KL, van Oostrom JCH, Vaalburg W, Bart J, Willemsen ATM, Hendrikse NH. 2005. Blood–

brain barrier dysfunction in parkinsonian midbrain in vivo. *Ann Neurol* **57:** 176–179. doi:10.1002/ana.20369

Krizbai IA, Deli MA, Pestenácz A, Siklós L, Szabó CA, András I, Joó F. 1998. Expression of glutamate receptors on cultured cerebral endothelial cells. *J Neurosci Res* **54:** 814–819. doi:10.1002/(SICI)1097-4547(19981215)54:6<814::AID-JNR9>3.0.CO;2-3

Krolak T, Chan KY, Kaplan L, Huang Q, Wu J, Zheng Q, Kozareva V, Beddow T, Tobey IG, Pacouret S, et al. 2022. A high-efficiency AAV for endothelial cell transduction throughout the central nervous system. *Nat Cardiovasc Res* **1:** 389–400. doi:10.1038/s44161-022-00046-4

Kuan WL, Bennett N, He X, Skepper JN, Martynyuk N, Wijeyekoon R, Moghe PV, Willams-Gray CH, Barker RA. 2016. α-Synuclein pre-formed fibrils impair tight junction protein expression without affecting cerebral endothelial cell function. *Exp Neurol* **285:** 72–81. doi:10.1016/j.expneurol.2016.09.003

Kubotera H, Ikeshima-Kataoka H, Hatashita Y, Allegra Mascaro AL, Pavone FS, Inoue T. 2019. Astrocytic endfeet recover blood vessels after removal by laser ablation. *Sci Rep* **9:** 1263. doi:10.1038/s41598-018-37419-4

Kumar H, Sharma B. 2016. Memantine ameliorates autistic behavior, biochemistry and blood–brain barrier impairments in rats. *Brain Res Bull* **124:** 27–39. doi:10.1016/j.brainresbull.2016.03.013

Kumar H, Sharma BM, Sharma B. 2015. Benefits of agomelatine in behavioral, neurochemical and blood–brain barrier alterations in prenatal valproic acid induced autism spectrum disorder. *Neurochem Int* **91:** 34–45. doi:10.1016/j.neuint.2015.10.007

Kuroiwa T, Ting P, Martinez H, Klatzo I. 1985. The biphasic opening of the blood–brain barrier to proteins following temporary middle cerebral artery occlusion. *Acta Neuropathol* **68:** 122–129.

Langen UH, Ayloo S, Gu C. 2019. Development and cell biology of the blood–brain barrier. *Annu Rev Cell Dev Biol* **35:** 591–613. doi:10.1146/annurev-cellbio-100617-062608

Larochelle C, Alvarez JI, Prat A. 2011. How do immune cells overcome the blood–brain barrier in multiple sclerosis? *FEBS Lett* **585:** 3770–3780. doi:10.1016/j.febslet.2011.04.066

Larsen JM, Martin DR, Byrne ME. 2014. Recent advances in delivery through the blood–brain barrier. *Curr Top Med Chem* **14:** 1148–1160. doi:10.2174/1568026614666140329230311

Larson DM, Carson MP, Haudenschild CC. 1987. Junctional transfer of small molecules in cultured bovine brain microvascular endothelial cells and pericytes. *Microvasc Res* **34:** 184–199. doi:10.1016/0026-2862(87)90052-5

Lee SW, Kim WJ, Choi YK, Song HS, Son MJ, Gelman IH, Kim YJ, Kim KW. 2003. SSeCKS regulates angiogenesis and tight junction formation in blood–brain barrier. *Nat Med* **9:** 900–906. doi:10.1038/nm889

Lee S, Chen TT, Barber CL, Jordan MC, Murdock J, Desai S, Ferrara N, Nagy A, Roos KP, Iruela-Arispe ML. 2007. Autocrine VEGF signaling is required for vascular homeostasis. *Cell* **130:** 691–703. doi:10.1016/j.cell.2007.06.054

Lenglet S, Montecucco F, Mach F, Schaller K, Gasche Y, Copin JC. 2014. Analysis of the expression of nine secreted matrix metalloproteinases and their endogenous inhibitors in the brain of mice subjected to ischaemic stroke. *Thromb Haemost* **112:** 363–378. doi:10.1160/TH14-01-0007

Lepeta K, Kaczmarek L. 2015. Matrix metalloproteinase-9 as a novel player in synaptic plasticity and schizophrenia. *Schizophr Bull* **41:** 1003–1009. doi:10.1093/schbul/sbv036

Li JY, Boado RJ, Pardridge WM. 2001. Blood–brain barrier genomics. *J Cereb Blood Flow Metab* **21:** 61–68. doi:10.1097/00004647-200101000-00008

Li JY, Boado RJ, Pardridge WM. 2002. Rat blood–brain barrier genomics. II: *J Cereb Blood Flow Metab* **22:** 1319–1326. doi:10.1097/01.WCB.0000040944.89393.0f

Li JY, Popovic N, Brundin P. 2005. The use of the R6 transgenic mouse models of Huntington's disease in attempts to develop novel therapeutic strategies. *Neurotherapeutics* **2:** 447–464. doi:10.1602/neurorx.2.3.447

Liebner S, Corada M, Bangsow T, Babbage J, Taddei A, Czupalla CJ, Reis M, Felici A, Wolburg H, Fruttiger M, et al. 2008. Wnt/β-catenin signaling controls development of the blood–brain barrier. *J Cell Biol* **183:** 409–417. doi:10.1083/jcb.200806024

Liu LB, Xue YX, Liu YH. 2010. Bradykinin increases the permeability of the blood-tumor barrier by the caveolae-mediated transcellular pathway. *J Neurooncol* **99:** 187–194. doi:10.1007/s11060-010-0124-x

Löscher W, Potschka H. 2005. Blood–brain barrier active efflux transporters: ATP-binding cassette gene family. *NeuroRx* **2:** 86–98. doi:10.1602/neurorx.2.1.86

Lu TM, Houghton S, Magdeldin T, Durán JGB, Minotti AP, Snead A, Sproul A, Nguyen DT, Xiang J, Fine HA, et al. 2021. Pluripotent stem cell-derived epithelium misidentified as brain microvascular endothelium requires ETS factors to acquire vascular fate. *Proc Natl Acad Sci* **118:** e2016950118. doi:10.1073/pnas.2016950118

Ludwig RJ, Zollner TM, Santoso S, Hardt K, Gille J, Baatz H, Johann PS, Pfeffer J, Radeke HH, Schön MP, et al. 2005. Junctional adhesion molecules (JAM)-B and -C contribute to leukocyte extravasation to the skin and mediate cutaneous inflammation. *J Invest Dermatol* **125:** 969–976. doi:10.1111/j.0022-202X.2005.23912.x

Luo J, Ho P, Steinman L, Wyss-Coray T. 2008. Bioluminescence in vivo imaging of autoimmune encephalomyelitis predicts disease. *J Neuroinflammation* **5:** 6. doi:10.1186/1742-2094-5-6

Ma S, Kwon HJ, Johng H, Zang K, Huang Z. 2013. Radial glial neural progenitors regulate nascent brain vascular network stabilization via inhibition of Wnt signaling. *PLoS Biol* **11:** e1001469. doi:10.1371/journal.pbio.1001469

Magaki S, Tang Z, Tung S, Williams CK, Lo D, Yong WH, Khanlou N, Vinters HV. 2018. The effects of cerebral amyloid angiopathy on integrity of the blood–brain barrier. *Neurobiol Aging* **70:** 70–77. doi:10.1016/j.neurobiolaging.2018.06.004

Majesky MW. 2007. Developmental basis of vascular smooth muscle diversity. *Arterioscler Thromb Vasc Biol* **27:** 1248–1258. doi:10.1161/ATVBAHA.107.141069

Manley GT, Fujimura M, Ma T, Noshita N, Filiz F, Bollen AW, Chan P, Verkman AS. 2000. Aquaporin-4 deletion in mice reduces brain edema after acute water intoxication and ischemic stroke. *Nat Med* **6:** 159–163. doi:10.1038/72256

Marco S, Skaper SD. 2006. Amyloid β-peptide$_{1-42}$ alters tight junction protein distribution and expression in brain mi-

Cite this article as *Cold Spring Harb Perspect Biol* doi: 10.1101/cshperspect.a041422

crovessel endothelial cells. *Neurosci Lett* **401**: 219–224. doi:10.1016/j.neulet.2006.03.047

Marina N, Christie IN, Korsak A, Doronin M, Brazhe A, Hosford PS, Wells JA, Sheikhbahaei S, Humoud I, Paton JFR, et al. 2020. Astrocytes monitor cerebral perfusion and control systemic circulation to maintain brain blood flow. *Nat Commun* **11**: 131. doi:10.1038/s41467-019-13956-y

Mark KS, Davis TP. 2002. Cerebral microvascular changes in permeability and tight junctions induced by hypoxia-reoxygenation. *Am J Physiol Heart Circ Physiol* **282**: H1485–H1494. doi:10.1152/ajpheart.00645.2001

Mark KS, Burroughs AR, Brown RC, Huber JD, Davis TP. 2004. Nitric oxide mediates hypoxia-induced changes in paracellular permeability of cerebral microvasculature. *Am J Physiol Heart Circ Physiol* **286**: H174–H180. doi:10.1152/ajpheart.00669.2002

Martinez BI, Stabenfeldt SE. 2019. Current trends in biomarker discovery and analysis tools for traumatic brain injury. *J Biol Eng* **13**: 16. doi:10.1186/s13036-019-0145-8

Martìn-Padura I, Lostaglio S, Schneemann M, Williams L, Romano M, Fruscella P, Panzeri C, Stoppacciaro A, Ruco L, Villa A, et al. 1998. Junctional adhesion molecule, a novel member of the immunoglobulin superfamily that distributes at intercellular junctions and modulates monocyte transmigration. *J Cell Biol* **142**: 117–127. doi:10.1083/jcb.142.1.117

Masuda S, Oda Y, Sasaki H, Ikenouchi J, Higashi T, Akashi M, Nishi E, Furuse M. 2011. LSR defines cell corners for tricellular tight junction formation in epithelial cells. *J Cell Sci* **124**: 548–555. doi:10.1242/jcs.072058

Matsuo K, Engelhardt B, Nishihara H. 2023. Differentiation of human induced pluripotent stem cells to brain microvascular endothelial cell-like cells with a mature immune phenotype. *J Vis Exp*. doi:10.3791/65134

McCarthy KM, Skare IB, Stankewich MC, Furuse M, Tsukita S, Rogers RA, Lynch RD, Schneeberger EE. 1996. Occludin is a functional component of the tight junction. *J Cell Sci* **109**: 2287–2298. doi:10.1242/jcs.109.9.2287

McQuaid S, Cunnea P, McMahon J, Fitzgerald U. 2009. The effects of blood–brain barrier disruption on glial cell function in multiple sclerosis. *Biochem Soc Trans* **37**: 329–331. doi:10.1042/BST0370329

Mills WA 3rd, Woo AM, Jiang S, Martin J, Surendran D, Bergstresser M, Kimbrough IF, Eyo UB, Sofroniew MV, Sontheimer H. 2022. Astrocyte plasticity in mice ensures continued endfoot coverage of cerebral blood vessels following injury and declines with age. *Nat Commun* **13**: 1794. doi:10.1038/s41467-022-29475-2

Milner R, Hung S, Wang X, Spatz M, del Zoppo GJ. 2008. The rapid decrease in astrocyte-associated dystroglycan expression by focal cerebral ischemia is protease-dependent. *J Cereb Blood Flow Metab* **28**: 812–823. doi:10.1038/sj.jcbfm.9600585

Mittapalli RK, Manda VK, Adkins CE, Geldenhuys WJ, Lockman PR. 2010. Exploiting nutrient transporters at the blood–brain barrier to improve brain distribution of small molecules. *Ther Deliv* **1**: 775–784. doi:10.4155/tde.10.76

Miyamoto T, Morita K, Takemoto D, Takeuchi K, Kitano Y, Miyakawa T, Nakayama K, Okamura Y, Sasaki H, Miyachi Y, et al. 2005. Tight junctions in Schwann cells of peripheral myelinated axons: a lesson from claudin-19-deficient mice. *J Cell Biol* **169**: 527–538. doi:10.1083/jcb.200501154

Mizee MR, Wooldrik D, Lakeman KA, van het Hof B, Drexhage JA, Geerts D, Bugiani M, Aronica E, Mebius RE, Prat A, et al. 2013. Retinoic acid induces blood–brain barrier development. *J Neurosci* **33**: 1660–1671. doi:10.1523/JNEUROSCI.1338-12.2013

Montagne A, Nation DA, Pa J, Sweeney MD, Toga AW, Zlokovic BV. 2016. Brain imaging of neurovascular dysfunction in Alzheimer's disease. *Acta Neuropathol* **131**: 687–707. doi:10.1007/s00401-016-1570-0

Mooradian AD, Chung HC, Shah GN. 1997. GLUT-1 expression in the cerebra of patients with Alzheimer's disease. *Neurobiol Aging* **18**: 469–474. doi:10.1016/S0197-4580(97)00111-5

Morita K, Sasaki H, Furuse M, Tsukita S. 1999. Endothelial claudin: claudin-5/TMVCF constitutes tight junction strands in endothelial cells. *J Cell Biol* **147**: 185–194. doi:10.1083/jcb.147.1.185

Mou Y, Du Y, Zhou L, Yue J, Hu X, Liu Y, Chen S, Lin X, Zhang G, Xiao H, et al. 2022. Gut microbiota interact with the brain through systemic chronic inflammation: implications on neuroinflammation, neurodegeneration, and aging. *Front Immunol* **13**: 796288. doi:10.3389/fimmu.2022.796288

Muller N, Ackenheil M. 1995. Immunoglobulin and albumin content of cerebrospinal fluid in schizophrenic patients: relationship to negative symptomatology. *Schizophr Res* **14**: 223–228. doi:10.1016/0920-9964(94)00045-a

Mulligan SJ, MacVicar BA. 2006. VRACs CARVe a path for novel mechanisms of communication in the CNS. *Sci STKE* **2006**: e42. doi:10.1126/stke.3572006pe42

Munji RN, Soung AL, Weiner GA, Sohet F, Semple BD, Trivedi A, Gimlin K, Kotoda M, Korai M, Aydin S, et al. 2019. Profiling the mouse brain endothelial transcriptome in health and disease models reveals a core blood-brain barrier dysfunction module. *Nat Neurosci* **22**: 1892–1902. doi:10.1038/s41593-019-0497-x

Murakami M, Nguyen LT, Zhuang ZW, Moodie KL, Carmeliet P, Stan RV, Simons M. 2008. The FGF system has a key role in regulating vascular integrity. *J Clin Invest* **118**: 3355–3366. doi:10.1172/JCI35298

Nahirney PC, Reeson P, Brown CE. 2016. Ultrastructural analysis of blood–brain barrier breakdown in the peri-infarct zone in young adult and aged mice. *J Cereb Blood Flow Metab* **36**: 413–425. doi:10.1177/0271678X15608396

Nguyen LN, Ma D, Shui G, Wong P, Cazenave-Gassiot A, Zhang X, Wenk MR, Goh EL, Silver DL. 2014. Mfsd2a is a transporter for the essential omega-3 fatty acid docosahexaenoic acid. *Nature* **509**: 503–506. doi:10.1038/nature13241

Nico B, Frigeri A, Nicchia GP, Quondamatteo F, Herken R, Errede M, Ribatti D, Svelto M, Roncali L. 2001. Role of aquaporin-4 water channel in the development and integrity of the blood–brain barrier. *J Cell Sci* **114**: 1297–1307. doi:10.1242/jcs.114.7.1297

Nico B, Frigeri A, Nicchia GP, Corsi P, Ribatti D, Quondamatteo F, Herken R, Girolamo F, Marzullo A, Svelto M, et al. 2003. Severe alterations of endothelial and glial cells in the blood–brain barrier of dystrophic *mdx* mice. *Glia* **42**: 235–251. doi:10.1002/glia.10216

Nielsen S, Nagelhus EA, Amiry-Moghaddam M, Bourque C, Agre P, Ottersen OP. 1997. Specialized membrane domains for water transport in glial cells: high-resolution immunogold cytochemistry of aquaporin-4 in rat brain. *J Neurosci* **17:** 171–180. doi:10.1523/JNEUROSCI.17-01-00171.1997

Nirwane A, Yao Y. 2022. Cell-specific expression and function of laminin at the neurovascular unit. *J Cereb Blood Flow Metab* **42:** 1979–1999. doi:10.1177/0271678X221113027

Nishihara H, Perriot S, Gastfriend BD, Steinfort M, Cibien C, Soldati S, Matsuo K, Guimbal S, Mathias A, Palecek SP, et al. 2022. Intrinsic blood–brain barrier dysfunction contributes to multiple sclerosis pathogenesis. *Brain* **145:** 4334–4348. doi:10.1093/brain/awac019

Nishiura K, Ichikawa-Tomikawa N, Sugimoto K, Kunii Y, Kashiwagi K, Tanaka M, Yokoyama Y, Hino M, Sugino T, Yabe H, et al. 2017. PKA activation and endothelial claudin-5 breakdown in the schizophrenic prefrontal cortex. *Oncotarget* **8:** 93382–93391. doi:10.18632/oncotarget.21850

Nitta T, Hata M, Gotoh S, Seo Y, Sasaki H, Hashimoto N, Furuse M, Tsukita S. 2003. Size-selective loosening of the blood–brain barrier in claudin-5-deficient mice. *J Cell Biol* **161:** 653–660. doi:10.1083/jcb.200302070

Noell S, Wolburg-Buchholz K, Mack AF, Beedle AM, Satz JS, Campbell KP, Wolburg H, Fallier-Becker P. 2011. Evidence for a role of dystroglycan regulating the membrane architecture of astroglial endfeet. *Eur J Neurosci* **33:** 2179–2186. doi:10.1111/j.1460-9568.2011.07688.x

Nourhaghighi N, Teichert-Kuliszewska K, Davis J, Stewart DJ, Nag S. 2003. Altered expression of angiopoietins during blood–brain barrier breakdown and angiogenesis. *Lab Invest* **83:** 1211–1222. doi:10.1097/01.LAB.0000082383.40635.FE

Novarino G, El-Fishawy P, Kayserili H, Meguid NA, Scott EM, Schroth J, Silhavy JL, Kara M, Khalil RO, Ben-Omran T, et al. 2012. Mutations in *BCKD*-kinase lead to a potentially treatable form of autism with epilepsy. *Science* **338:** 394–397. doi:10.1126/science.1224631

O'Brown NM, Megason SG, Gu C. 2019. Suppression of transcytosis regulates zebrafish blood–brain barrier function. *eLife* **8:** e47326. doi:10.7554/eLife.47326

Ohtsuki S, Hirayama M, Ito S, Uchida Y, Tachikawa M, Terasaki T. 2014. Quantitative targeted proteomics for understanding the blood–brain barrier: towards pharmacoproteomics. *Expert Rev Proteomics* **11:** 303–313. doi:10.1586/14789450.2014.893830

Oldendorf WH, Cornford ME, Brown WJ. 1977. The large apparent work capability of the blood–brain barrier: a study of the mitochondrial content of capillary endothelial cells in brain and other tissues of the rat. *Ann Neurol* **1:** 409–417. doi:10.1002/ana.410010502

Onore C, Careaga M, Ashwood P. 2012. The role of immune dysfunction in the pathophysiology of autism. *Brain Behav Immun* **26:** 383–392. doi:10.1016/j.bbi.2011.08.007

Orlowski M, Sessa G, Green JP. 1974. γ-Glutamyl transpeptidase in brain capillaries: possible site of a blood–brain barrier for amino acids. *Science* **184:** 66–68. doi:10.1126/science.184.4132.66

Ouellette J, Lacoste B. 2021. From neurodevelopmental to neurodegenerative disorders: the vascular continuum. *Front Aging Neurosci* **13:** 749026. doi:10.3389/fnagi.2021.749026

Ouellette J, Toussay X, Comin CH, Costa LDF, Ho M, Lacalle-Aurioles M, Freitas-Andrade M, Liu QY, Leclerc S, Pan Y,

et al. 2020. Vascular contributions to 16p11.2 deletion autism syndrome modeled in mice. *Nat Neurosci* **23:** 1090–1101. doi:10.1038/s41593-020-0663-1

Park L, Wang G, Zhou P, Zhou J, Pitstick R, Previti ML, Younkin L, Younkin SG, Van Nostrand WE, Cho S, et al. 2011. Scavenger receptor CD36 is essential for the cerebrovascular oxidative stress and neurovascular dysfunction induced by amyloid-β. *Proc Natl Acad Sci* **108:** 5063–5068. doi:10.1073/pnas.1015413108

Park TE, Mustafaoglu N, Herland A, Hasselkus R, Mannix R, FitzGerald EA, Prantil-Baun R, Watters A, Henry O, Benz M, et al. 2019. Hypoxia-enhanced blood–brain barrier Chip recapitulates human barrier function and shuttling of drugs and antibodies. *Nat Commun* **10:** 2621. doi:10.1038/s41467-019-10588-0

Paschaki M, Lin SC, Wong RL, Finnell RH, Dollé P, Niederreither K. 2012. Retinoic acid-dependent signaling pathways and lineage events in the developing mouse spinal cord. *PLoS ONE* **7:** e32447. doi:10.1371/journal.pone.0032447

Patel A, Toia GV, Colletta K, Bradaric BD, Carvey PM, Hendey B. 2011. An angiogenic inhibitor, cyclic RGDfV, attenuates MPTP-induced dopamine neuron toxicity. *Exp Neurol* **231:** 160–170. doi:10.1016/j.expneurol.2011.06.004

Peppiatt CM, Howarth C, Mobbs P, Attwell D. 2006. Bidirectional control of CNS capillary diameter by pericytes. *Nature* **443:** 700–704. doi:10.1038/nature05193

Perry DC, Sturm VE, Peterson MJ, Pieper CF, Bullock T, Boeve BF, Miller BL, Guskiewicz KM, Berger MS, Kramer JH, et al. 2016. Association of traumatic brain injury with subsequent neurological and psychiatric disease: a meta-analysis. *J Neurosurg* **124:** 511–526. doi:10.3171/2015.2.JNS14503

Persidsky Y, Ghorpade A, Rasmussen J, Limoges J, Liu XJ, Stins M, Fiala M, Way D, Kim KS, Witte MH, et al. 1999. Microglial and astrocyte chemokines regulate monocyte migration through the blood–brain barrier in human immunodeficiency virus-1 encephalitis. *Am J Pathol* **155:** 1599–1611. doi:10.1016/S0002-9440(10)65476-4

Pfau SJ, Langen UH, Fisher TM, Prakash I, Nagpurwala F, Lozoya RA, Lee WCA, Wu Z, Gu C. 2021. Vascular and perivascular cell profiling reveals the molecular and cellular bases of blood–brain barrier heterogeneity. bioRxiv doi:10.1101/2021.04.26.441465

Pfeiffer F, Schäfer J, Lyck R, Makrides V, Brunner S, Schaeren-Wiemers N, Deutsch U, Engelhardt B. 2011. Claudin-1 induced sealing of blood–brain barrier tight junctions ameliorates chronic experimental autoimmune encephalomyelitis. *Acta Neuropathol* **122:** 601–614. doi:10.1007/s00401-011-0883-2

Polfliet MM, Zwijnenburg PJ, van Furth AM, van der Poll T, Döpp EA, Renardel de Lavalette C, van Kesteren-Hendrikx EM, van Rooijen N, Dijkstra CD, van den Berg TK. 2001. Meningeal and perivascular macrophages of the central nervous system play a protective role during bacterial meningitis. *J Immunol* **167:** 4644–4650. doi:10.4049/jimmunol.167.8.4644

Potschka H, Fedrowitz M, Löscher W. 2001. P-glycoprotein and multidrug resistance-associated protein are involved in the regulation of extracellular levels of the major antiepileptic drug carbamazepine in the brain. *Neuroreport* **12:** 3557–3560. doi:10.1097/00001756-200111160-00037

Prat A, Biernacki K, Wosik K, Antel JP. 2001. Glial influence on the human blood–brain barrier. *Glia* **36:** 145–155. doi:10.1002/glia.1104

Predescu SA, Predescu DN, Palade GE. 2001. Endothelial transcytotic machinery involves supramolecular protein-lipid complexes. *Mol Biol Cell* **12:** 1019–1033. doi:10.1091/mbc.12.4.1019

Rajkowska G, Hughes J, Stockmeier CA, Javier Miguel-Hidalgo J, Maciag D. 2013. Coverage of blood vessels by astrocytic endfeet is reduced in major depressive disorder. *Biol Psychiatry* **73:** 613–621. doi:10.1016/j.biopsych.2012.09.024

Ramos-Cejudo J, Wisniewski T, Marmar C, Zetterberg H, Blennow K, de Leon MJ, Fossati S. 2018. Traumatic brain injury and Alzheimer's disease: the cerebrovascular link. *EBioMed* **28:** 21–30. doi:10.1016/j.ebiom.2018.01.021

Ransohoff RM, Engelhardt B. 2012. The anatomical and cellular basis of immune surveillance in the central nervous system. *Nat Rev Immunol* **12:** 623–635. doi:10.1038/nri3265

Rattner A, Williams J, Nathans J. 2019. Roles of HIFs and VEGF in angiogenesis in the retina and brain. *J Clin Invest* **129:** 3807–3820. doi:10.1172/JCI126655

Reese TS, Karnovsky MJ. 1967. Fine structural localization of a blood–brain barrier to exogenous peroxidase. *J Cell Biol* **34:** 207–217. doi:10.1083/jcb.34.1.207

Reuss B, Dono R, Unsicker K. 2003. Functions of fibroblast growth factor (FGF)-2 and FGF-5 in astroglial differentiation and blood–brain barrier permeability: evidence from mouse mutants. *J Neurosci* **23:** 6404–6412. doi:10.1523/JNEUROSCI.23-16-06404.2003

Reyahi A, Nik AM, Ghiami M, Gritli-Linde A, Pontén F, Johansson BR, Carlsson P. 2015. Foxf2 is required for brain pericyte differentiation and development and maintenance of the blood–brain barrier. *Dev Cell* **34:** 19–32. doi:10.1016/j.devcel.2015.05.008

Rite I, Machado A, Cano J, Venero JL. 2007. Blood–brain barrier disruption induces in vivo degeneration of nigral dopaminergic neurons. *J Neurochem* **101:** 1567–1582. doi:10.1111/j.1471-4159.2007.04567.x

Rosenberg GA, Estrada EY, Dencoff JE. 1998. Matrix metalloproteinases and TIMPs are associated with blood–brain barrier opening after reperfusion in rat brain. *Stroke* **29:** 2189–2195. doi:10.1161/01.STR.29.10.2189

Rurak GM, Simard S, Freitas-Andrade M, Lacoste B, Charih F, Van Geel A, Stead J, Woodside B, Green JR, Coppola G, et al. 2022. Sex differences in developmental patterns of neocortical astroglia: a mouse translatome database. *Cell Rep* **38:** 110310. doi:10.1016/j.celrep.2022.110310

Ryu JK, McLarnon JG. 2009. A leaky blood–brain barrier, fibrinogen infiltration and microglial reactivity in inflamed Alzheimer's disease brain. *J Cell Mol Med* **13:** 2911–2925. doi:10.1111/j.1582-4934.2008.00434.x

Sabbagh MF, Nathans J. 2020. A genome-wide view of the dedifferentiation of central nervous system endothelial cells in culture. *eLife* **9:** e51276. doi:10.7554/eLife.51276

Sadeghian H, Lacoste B, Qin T, Toussay X, Rosa R, Oka F, Chung DY, Takizawa T, Gu C, Ayata C. 2018. Spreading depolarizations trigger caveolin-1-dependent endothelial transcytosis. *Ann Neurol* **S84:** 409–423. doi:10.1002/ana.25298

Sagare AP, Bell RD, Zhao Z, Ma Q, Winkler EA, Ramanathan A, Zlokovic BV. 2013. Pericyte loss influences Alzheimer-like neurodegeneration in mice. *Nat Commun* **4:** 2932. doi:10.1038/ncomms3932

Saitou M, Furuse M, Sasaki H, Schulzke JD, Fromm M, Takano H, Noda T, Tsukita S. 2000. Complex phenotype of mice lacking occludin, a component of tight junction strands. *Mol Biol Cell* **11:** 4131–4142. doi:10.1091/mbc.11.12.4131

Saunders NR, Dziegielewska KM, Unsicker K, Ek CJ. 2016. Delayed astrocytic contact with cerebral blood vessels in FGF-2 deficient mice does not compromise permeability properties at the developing blood-brain barrier. *Dev Neurobiol* **76:** 1201–1212. doi:10.1002/dneu.22383

Schinkel AH, Smit JJ, van Tellingen O, Beijnen JH, Wagenaar E, van Deemter L, Mol CA, van der Valk MA, Robanus-Maandag EC, te Riele HP, et al. 1994. Disruption of the mouse *mdr1a* P-glycoprotein gene leads to a deficiency in the blood–brain barrier and to increased sensitivity to drugs. *Cell* **77:** 491–502. doi:10.1016/0092-8674(94)90212-7

Schinkel AH, Wagenaar E, van Deemter L, Mol CA, Borst P. 1995. Absence of the *mdr1a* P-glycoprotein in mice affects tissue distribution and pharmacokinetics of dexamethasone, digoxin, and cyclosporin A. *J Clin Invest* **96:** 1698–1705. doi:10.1172/JCI118214

Schinkel AH, Wagenaar E, Mol CA, van Deemter L. 1996. P-glycoprotein in the blood–brain barrier of mice influences the brain penetration and pharmacological activity of many drugs. *J Clin Invest* **97:** 2517–2524. doi:10.1172/JCI118699

Schnitzer JE. 2001. Caveolae: from basic trafficking mechanisms to targeting transcytosis for tissue-specific drug and gene delivery in vivo. *Adv Drug Deliv Rev* **49:** 265–280. doi:10.1016/S0169-409X(01)00141-7

Schoch HJ, Fischer S, Marti HH. 2002. Hypoxia-induced vascular endothelial growth factor expression causes vascular leakage in the brain. *Brain* **125:** 2549–2557. doi:10.1093/brain/awf257

Segarra M, Aburto MR, Cop F, Llaó-Cid C, Härtl R, Damm M, Bethani I, Parrilla M, Husainie D, Schänzer A, et al. 2018. Endothelial Dab1 signaling orchestrates neuro-glia-vessel communication in the central nervous system. *Science* **361:** eaao2861. doi:10.1126/science.aao2861

Sengillo JD, Winkler EA, Walker CT, Sullivan JS, Johnson M, Zlokovic BV. 2013. Deficiency in mural vascular cells coincides with blood–brain barrier disruption in Alzheimer's disease. *Brain Pathol* **23:** 303–310. doi:10.1111/bpa.12004

Sharp CD, Hines I, Houghton J, Warren A, Jackson TH, Jawahar A, Nanda A, Elrod JW, Long A, Chi A, et al. 2003. Glutamate causes a loss in human cerebral endothelial barrier integrity through activation of NMDA receptor. *Am J Physiol Heart Circ Physiol* **285:** H2592–H2598. doi:10.1152/ajpheart.00520.2003

Shcherbakova I, Neshkova E, Dotsenko V, Platonova T, Shcherbakova E, Yarovaya G. 1999. The possible role of plasma kallikrein-kinin system and leukocyte elastase in pathogenesis of schizophrenia. *Immunopharmacology* **43:** 273–279. doi:10.1016/s0162-3109(99)00099-5

Shen F, Walker EJ, Jiang L, Degos V, Li J, Sun B, Heriyanto F, Young WL, Su H. 2011. Coexpression of angiopoietin-1 with VEGF increases the structural integrity of the blood–brain barrier and reduces atrophy volume. *J Cereb Blood Flow Metab* **31:** 2343–2351. doi:10.1038/jcbfm.2011.97

Shepro D, Morel NM. 1993. Pericyte physiology. *FASEB J* **7:** 1031–1038. doi:10.1096/fasebj.7.11.8370472

Shue EH, Carson-Walter EB, Liu Y, Winans BN, Ali ZS, Chen J, Walter KA. 2008. Plasmalemmal vesicle associated protein-1 (PV-1) is a marker of blood–brain barrier disruption in rodent models. *BMC Neurosci* **9:** 29. doi:10.1186/1471-2202-9-29

* Sierra A, Miron VE, Paolicelli RC, Ransohoff RM. 2024. Microglia in health and diseases: integrative hubs of the central nervous system (CNS). *Cold Spring Harb Perspect Biol* doi:10.1101/cshperspect.a041366

Sims DE. 1986. The pericyte—a review. *Tissue Cell* **18:** 153–174. doi:10.1016/0040-8166(86)90026-1

Song HW, Foreman KL, Gastfriend BD, Kuo JS, Palecek SP, Shusta EV. 2020. Transcriptomic comparison of human and mouse brain microvessels. *Sci Rep* **10:** 12358. doi:10.1038/s41598-020-69096-7

Sorokin L. 2010. The impact of the extracellular matrix on inflammation. *Nat Rev Immunol* **10:** 712–723. doi:10.1038/nri2852

Spitzer D, Khel MI, Pütz T, Zinke J, Jia X, Sommer K, Filipski K, Thorsen F, Freiman TM, Günther S, et al. 2023. A flow cytometry-based protocol for syngenic isolation of neurovascular unit cells from mouse and human tissues. *Nat Protoc* **18:** 1510–1542. doi:10.1038/s41596-023-00805-y

Stenman JM, Rajagopal J, Carroll TJ, Ishibashi M, McMahon J, McMahon AP. 2008. Canonical Wnt signaling regulates organ-specific assembly and differentiation of CNS vasculature. *Science* **322:** 1247–1250. doi:10.1126/science.1164594

Steventon JJ, Furby H, Ralph J, O'Callaghan P, Rosser AE, Wise RG, Busse M, Murphy K. 2020. Altered cerebrovascular response to acute exercise in patients with Huntington's disease. *Brain Commun* **2:** fcaa044. doi:10.1093/braincomms/fcaa044

Stewart PA, Wiley MJ. 1981. Developing nervous tissue induces formation of blood–brain barrier characteristics in invading endothelial cells: a study using quail–chick transplantation chimeras. *Dev Biol* **84:** 183–192. doi:10.1016/0012-1606(81)90382-1

Stokum JA, Shim B, Huang W, Kane M, Smith JA, Gerzanich V, Simard JM. 2021. A large portion of the astrocyte proteome is dedicated to perivascular endfeet, including critical components of the electron transport chain. *J Cereb Blood Flow Metab* **41:** 2546–2560. doi:10.1177/0271678X211004182

Streit WJ, Conde JR, Fendrick SE, Flanary BE, Mariani CL. 2005. Role of microglia in the central nervous system's immune response. *Neurol Res* **27:** 685–691. doi:10.1179/016164105X49463a

Sweeney MD, Kisler K, Montagne A, Toga AW, Zlokovic BV. 2018. The role of brain vasculature in neurodegenerative disorders. *Nat Neurosci* **21:** 1318–1331. doi:10.1038/s41593-018-0234-x

Tai LM, Holloway KA, Male DK, Loughlin AJ, Romero IA. 2010. Amyloid-β-induced occludin down-regulation and increased permeability in human brain endothelial cells is mediated by MAPK activation. *J Cell Mol Med* **14:** 1101–1112.

Tao-Cheng JH, Nagy Z, Brightman MW. 1987. Tight junctions of brain endothelium in vitro are enhanced by astroglia. *J Neurosci* **7:** 3293–3299. doi:10.1523/JNEUROSCI.07-10-03293.1987

Tărlungeanu DC, Deliu E, Dotter CP, Kara M, Janiesch PC, Scalise M, Galluccio M, Tesulov M, Morelli E, Sonmez FM, et al. 2016. Impaired amino acid transport at the blood–brain barrier is a cause of autism spectrum disorder. *Cell* **167:** 1481–1494.e18. doi:10.1016/j.cell.2016.11.013

Thal DR, Capetillo-Zarate E, Larionov S, Staufenbiel M, Zurbruegg S, Beckmann N. 2009. Capillary cerebral amyloid angiopathy is associated with vessel occlusion and cerebral blood flow disturbances. *Neurobiol Aging* **30:** 1936–1948. doi:10.1016/j.neurobiolaging.2008.01.017

Thiebaut F, Tsuruo T, Hamada H, Gottesman MM, Pastan I, Willingham MC. 1989. Immunohistochemical localization in normal tissues of different epitopes in the multidrug transport protein P170: evidence for localization in brain capillaries and crossreactivity of one antibody with a muscle protein. *J Histochem Cytochem* **37:** 159–164. doi:10.1177/37.2.2463300

Tiwary S, Morales JE, Kwiatkowski SC, Lang FF, Rao G, McCarty JH. 2018. Metastatic brain tumors disrupt the blood–brain barrier and alter lipid metabolism by inhibiting expression of the endothelial cell fatty acid transporter Mfsd2a. *Sci Rep* **8:** 8267. doi:10.1038/s41598-018-26636-6

Tuma P, Hubbard AL. 2003. Transcytosis: crossing cellular barriers. *Physiol Rev* **83:** 871–932. doi:10.1152/physrev.00001.2003

Turner RJ, Sharp FR. 2016. Implications of MMP9 for blood–brain barrier disruption and hemorrhagic transformation following ischemic stroke. *Front Cell Neurosci* **10:** 56. doi:10.3389/fncel.2016.00056

Ufnal M, Skrzypecki J. 2014. Blood-borne hormones in a cross-talk between peripheral and brain mechanisms regulating blood pressure, the role of circumventricular organs. *Neuropeptides* **48:** 65–73. doi:10.1016/j.npep.2014.01.003

Unger ER, Sung JH, Manivel JC, Chenggis ML, Blazar BR, Krivit W. 1993. Male donor-derived cells in the brains of female sex-mismatched bone marrow transplant recipients: a Y-chromosome specific in situ hybridization study. *J Neuropathol Exp Neurol* **52:** 460–470. doi:10.1097/00005072-199309000-00004

Uranova NA, Zimina IS, Vikhreva OV, Krukov NO, Rachmanova VI, Orlovskaya DD. 2010. Ultrastructural damage of capillaries in the neocortex in schizophrenia. *World J Biol Psychiatry* **11:** 567–578. doi:10.3109/15622970903414188

Usta A, Kılıç F, Demirdaş A, Işık U, Doğuç DK, Bozkurt M. 2021. Serum zonulin and claudin-5 levels in patients with schizophrenia. *Eur Arch Psychiatry Clin Neurosci* **271:** 767–773. doi:10.1007/s00406-020-01152-9

Van de Haar H, Burgmans S, Jansen J, van Osch M, van Buchem M, Muller M, Hofman P, Verhey F, Backes W. 2016. Blood–brain barrier leakage in patients with early Alzheimer disease. *Radiology* **281:** 527–535. doi:10.1148/radiol.2016152244

Van Dyken P, Lacoste B. 2018. Impact of metabolic syndrome on neuroinflammation and the blood–brain barrier. *Front Neurosci* **12:** 930. doi:10.3389/fnins.2018.00930

Van Itallie CM, Anderson JM. 2006. Claudins and epithelial paracellular transport. *Annu Rev Physiol* **68:** 403–429. doi:10.1146/annurev.physiol.68.040104.131404

Van Itallie CM, Anderson JM. 2013. Claudin interactions in and out of the tight junction. *Tissue Barriers* **1:** e25247. doi:10.4161/tisb.25247

Van Itallie C, Rahner C, Anderson JM. 2001. Regulated expression of claudin-4 decreases paracellular conductance through a selective decrease in sodium permeability. *J Clin Invest* **107:** 1319–1327. doi:10.1172/JCI12464

Van Itallie CM, Holmes J, Bridges A, Gookin JL, Coccaro MR, Proctor W, Colegio OR, Anderson JM. 2008. The density of small tight junction pores varies among cell types and is increased by expression of claudin-2. *J Cell Sci* **121:** 298–305. doi:10.1242/jcs.021485

Vanlandewijck M, He L, Mäe MA, Andrae J, Ando K, Del Gaudio F, Nahar K, Lebouvier T, Laviña B, Gouveia L, et al. 2018. A molecular atlas of cell types and zonation in the brain vasculature. *Nature* **554:** 475–480. doi:10.1038/nature25739

Vass K, Hickey WF, Schmidt RE, Lassmann H. 1993. Bone marrow-derived elements in the peripheral nervous system. An immunohistochemical and ultrastructural investigation in chimeric rats. *Lab Invest* **69:** 275–282.

Villeda SA, Luo J, Mosher KI, Zou B, Britschgi M, Bieri G, Stan TM, Fainberg N, Ding Z, Eggel A, et al. 2011. The ageing systemic milieu negatively regulates neurogenesis and cognitive function. *Nature* **477:** 90–94. doi:10.1038/nature10357

Villeda SA, Plambeck KE, Middeldorp J, Castellano JM, Mosher KI, Luo J, Smith LK, Bieri G, Lin K, Berdnik D, et al. 2014. Young blood reverses age-related impairments in cognitive function and synaptic plasticity in mice. *Nat Med* **20:** 659–663. doi:10.1038/nm.3569

Wada K, Arai H, Takahashi M, Fukae J, Oizumi H, Yasuda T, Mizuno Y, Mochzuki H. 2006. Expression levels of vascular endothelial growth factor and its receptors in Parkinson's disease. *Neuroreport* **17:** 705–709. doi:10.1097/01.wnr.0000215769.71657.65

Wälchli T, Bisschop J, Carmeliet P, Zadeh G, Monnier PP, De Bock K, Radovanovic I. 2023. Shaping the brain vasculature in development and disease in the single-cell era. *Nat Rev Neurosci* **24:** 271–298. doi:10.1038/s41583-023-00684-y

Wan W, Cao L, Liu L, Zhang C, Kalionis B, Tai X, Li Y, Xia S. 2015. Aβ$_{1-42}$ oligomer-induced leakage in an in vitro blood–brain barrier model is associated with up-regulation of RAGE and metalloproteinases, and down-regulation of tight junction scaffold proteins. *J Neurochem* **134:** 382–393. doi:10.1111/jnc.13122

Wang Y, Imitola J, Rasmussen S, O'Connor KC, Khoury SJ. 2008. Paradoxical dysregulation of the neural stem cell pathway Sonic Hedgehog-Gli1 in autoimmune encephalomyelitis and multiple sclerosis. *Ann Neurol* **64:** 417–427. doi:10.1002/ana.21457

Wang Y, Rattner A, Zhou Y, Williams J, Smallwood PM, Nathans J. 2012. Norrin/Frizzled4 signaling in retinal vascular development and blood–brain barrier plasticity. *Cell* **151:** 1332–1344. doi:10.1016/j.cell.2012.10.042

Wang Y, Sabbagh MF, Gu X, Rattner A, Williams J, Nathans J. 2019. β-Catenin signaling regulates barrier-specific gene expression in circumventricular organ and ocular vasculatures. *eLife* **8:** 3221.

Wardlaw JM, Farrall A, Armitage PA, Carpenter T, Chappell F, Doubal A, Chowdhury D, Cvoro V, Dennis MS. 2008. Changes in background blood–brain barrier integrity between lacunar and cortical ischemic stroke subtypes. *Stroke* **39:** 1327–1332. doi:10.1161/STROKEAHA.107.500124

Wen H, Nagelhus EA, Amiry-Moghaddam M, Agre P, Ottersen OP, Nielsen S. 1999. Ontogeny of water transport in rat brain: postnatal expression of the aquaporin-4 water channel. *Eur J Neurosci* **11:** 935–945. doi:10.1046/j.1460-9568.1999.00502.x

Westergaard E, Brightman MW. 1973. Transport of proteins across normal cerebral arterioles. *J Comp Neurol* **152:** 17–44. doi:10.1002/cne.901520103

Witt KA, Mark KS, Sandoval KE, Davis TP. 2008. Reoxygenation stress on blood–brain barrier paracellular permeability and edema in the rat. *Microvasc Res* **75:** 91–96. doi:10.1016/j.mvr.2007.06.004

Wolburg H, Wolburg-Buchholz K, Fallier-Becker P, Noell S, Mack AF. 2011. Structure and functions of aquaporin-4-based orthogonal arrays of particles. *Int Rev Cell Mol Biol* **287:** 1–41. doi:10.1016/B978-0-12-386043-9.00001-3

Wong V, Gumbiner BM. 1997. A synthetic peptide corresponding to the extracellular domain of occludin perturbs the tight junction permeability barrier. *J Cell Biol* **136:** 399–409. doi:10.1083/jcb.136.2.399

Wosik K, Cayrol R, Dodelet-Devillers A, Berthelet F, Bernard M, Moumdjian R, Bouthillier A, Reudelhuber TL, Prat A. 2007. Angiotensin II controls occludin function and is required for blood–brain barrier maintenance: relevance to multiple sclerosis. *J Neurosci* **27:** 9032–9042. doi:10.1523/JNEUROSCI.2088-07.2007

Williams K, Alvarez X, Lackner AA. 2001. Central nervous system perivascular cells are immunoregulatory cells that connect the CNS with the peripheral immune system. *Glia* **36:** 156–164. doi:10.1002/glia.1105

Wood CAP, Zhang J, Aydin D, Xu Y, Andreone BJ, Langen UH, Dror RO, Gu C, Feng L. 2021. Structure and mechanism of blood-brain-barrier lipid transporter MFSD2A. *Nature* **596:** 444–448. doi:10.1038/s41586-021-03782-y

Wu C, Ivars F, Anderson P, Hallmann R, Vestweber D, Nilsson P, Robenek H, Tryggvason K, Song J, Korpos E, et al. 2009. Endothelial basement membrane laminin α5 selectively inhibits T lymphocyte extravasation into the brain. *Nat Med* **15:** 519–527. doi:10.1038/nm.1957

Xiao M, Xiao ZJ, Yang B, Lan Z, Fang F. 2020. Blood-brain barrier: more contributor to disruption of central nervous system homeostasis than victim in neurological disorders. *Front Neurosci* **14:** 764. doi:10.3389/fnins.2020.00764

Xu Q, Wang Y, Dabdoub A, Smallwood PM, Williams J, Woods C, Kelley MW, Jiang L, Tasman W, Zhang K, et al. 2004. Vascular development in the retina and inner ear: control by Norrin and Frizzled-4, a high-affinity ligand-receptor pair. *Cell* **116:** 883–895. doi:10.1016/S0092-8674(04)00216-8

Yamamoto H, Ehling M, Kato K, Kanai K, van Lessen M, Frye M, Zeuschner D, Nakayama M, Vestweber D, Adams RH. 2015. Integrin β1 controls VE-cadherin localization and blood vessel stability. *Nat Commun* **6:** 6429. doi:10.1038/ncomms7429

Yang P, Pavlovic D, Waldvogel H, Dragunow M, Synek B, Turner C, Faull R, Guan J. 2015. String vessel formation is increased in the brain of Parkinson disease. *J Parkinsons Dis* **5**: 821–836. doi:10.3233/JPD-140454

Yang AC, Stevens MY, Chen MB, Lee DP, Stähli D, Gate D, Contrepois K, Chen W, Iram T, Zhang L, et al. 2020. Physiological blood–brain transport is impaired with age by a shift in transcytosis. *Nature* **583**: 425–430. doi:10.1038/s41586-020-2453-z

Yang AC, Vest RT, Kern F, Lee DP, Agam M, Maat CA, Losada PM, Chen MB, Schaum N, Khoury N, et al. 2022. A human brain vascular atlas reveals diverse mediators of Alzheimer's risk. *Nature* **603**: 885–892. doi:10.1038/s41586-021-04369-3

Yao Y, Chen ZL, Norris EH, Strickland S. 2014. Astrocytic laminin regulates pericyte differentiation and maintains blood brain barrier integrity. *Nat Commun* **5**: 3413. doi:10.1038/ncomms4413

Yao LL, Hu JX, Li Q, Lee D, Ren X, Zhang JS, Sun D, Zhang HS, Wang YG, Mei L, et al. 2020. Astrocytic neogenin/netrin-1 pathway promotes blood vessel homeostasis and function in mouse cortex. *J Clin Invest* **130**: 6490–6509. doi:10.1172/JCI132372

Ye X, Wang Y, Nathans J. 2010. The Norrin/Frizzled4 signaling pathway in retinal vascular development and disease. *Trends Mol Med* **16**: 417–425. doi:10.1016/j.molmed.2010.07.003

Yosef N, Xi Y, McCarty JH. 2020. Isolation and transcriptional characterization of mouse perivascular astrocytes.

PLoS ONE **15**: e0240035. doi:10.1371/journal.pone.0240035

Yuan SY. 2002. Protein kinase signaling in the modulation of microvascular permeability. *Vascul Pharmacol* **39**: 213–223. doi:10.1016/S1537-1891(03)00010-7

Zhao LN, Yang ZH, Liu YH, Ying HQ, Zhang H, Xue YX. 2011. Vascular endothelial growth factor increases permeability of the blood–tumor barrier via caveolae-mediated transcellular pathway. *J Mol Neurosci* **44**: 122–129. doi:10.1007/s12031-010-9487-x

Zhao Z, Sagare AP, Ma Q, Halliday MR, Kong P, Kisler K, Winkler EA, Ramanathan A, Kanekiyo T, Bu G, et al. 2015. Central role for PICALM in amyloid-β blood-brain barrier transcytosis and clearance. *Nat Neurosci* **18**: 978–987. doi:10.1038/nn.4025

Zheng Z, Diamond MI. 2012. Huntington disease and the huntingtin protein. *Prog Mol Biol Transl Sci* **107**: 189–214. doi:10.1016/B978-0-12-385883-2.00010-2

Zlokovic BV. 2008. The blood–brain barrier in health and chronic neurodegenerative disorders. *Neuron* **57**: 178–201. doi:10.1016/j.neuron.2008.01.003

Zlokovic BV. 2010. Neurodegeneration and the neurovascular unit. *Nat Med* **16**: 1370–1371. doi:10.1038/nm1210-1370

Zonta M, Angulo MC, Gobbo S, Rosengarten B, Hossmann KA, Pozzan T, Carmignoto G. 2003. Neuron-to-astrocyte signaling is central to the dynamic control of brain microcirculation. *Nat Neurosci* **6**: 43–50. doi:10.1038/nn980

Cite this article as *Cold Spring Harb Perspect Biol* doi: 10.1101/cshperspect.a041422

Satellite Glial Cells: No Longer the Most Overlooked Glia

Susan J. Birren,[1] Lisa V. Goodrich,[2] and Rosalind A. Segal[2,3]

[1]Department of Biology, Brandeis University, Waltham, Massachusetts 02453, USA

[2]Department of Neurobiology, Harvard Medical School, Boston, Massachusetts 02115, USA

[3]Department of Cancer Biology, Dana-Farber Cancer Institute, Boston, Massachusetts 02215, USA

Correspondence: birren@brandeis.edu

Many glial biologists consider glia the neglected cells of the nervous system. Among all the glia of the central and peripheral nervous system, satellite glia may be the most often overlooked. Satellite glial cells (SGCs) are located in ganglia of the cranial nerves and the peripheral nervous system. These small cells surround the cell bodies of neurons in the trigeminal ganglia (TG), spiral ganglia, nodose and petrosal ganglia, sympathetic ganglia, and dorsal root ganglia (DRG). Essential SGC features include their intimate connections with the associated neurons, their small size, and their derivation from neural crest cells. Yet SGCs also exhibit tissue-specific properties and can change rapidly, particularly in response to injury. To illustrate the range of SGC functions, we will focus on three types: those of the spiral, sympathetic, and DRG, and consider both their shared features and those that differ based on location.

Satellite glial cells (SGCs) are best defined by their remarkably close association with neuronal cell bodies. They are small cells that completely ensheath the associated neuronal cell bodies, with the number of SGCs per cell varying from 1 to 20 or more depending on the size of the associated neuron (see Fig. 1; Pannese 1960). SGCs have small nuclei, within a thin cell where the distance between the inner and outer membrane can be as little as 50 nm (Pannese 1960; Peters and Webster 1972). They use cadherins, particularly cadherin 19, and gap junctions, particularly connexin 43 (CX43), to form close contacts with one another and with the nearby neurons. Indeed, the cell membranes of satellite cells can actually interdigitate with the neuronal membranes, and the distance between the neuronal and glial membranes is ~20 nm, rivaling the distance seen at a synaptic cleft (Pannese 1960; Hanani and Spray 2020). Recognized markers for satellite glia include GFAP, S100, fatty acid–binding protein 7/brain lipid–binding protein (FABP7, also known as BFABP), and glutamine synthase (see Tables 1 and 2; Hanani and Spray 2020). Genes that distinguish satellite glia from multiple locations have also been discovered through single-cell RNA sequencing (Avraham et al. 2021a; Tasdemir-Yilmaz et al. 2021; Mapps et al. 2022b).

Given their close association, individual neurons and their associated SGCs can be thought of as a functional unit (Hanani 2010), with SGCs

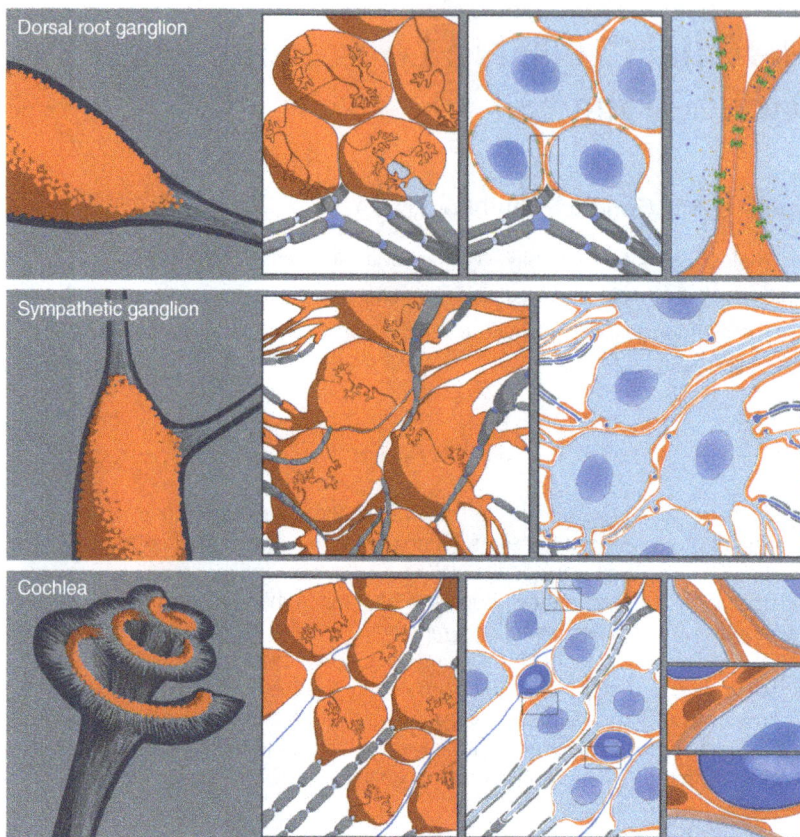

Figure 1. Features of satellite glial cells (SGCs). SGCs are small cells that ensheath cell bodies of neurons in peripheral ganglia, including the sensory (*upper* panel), sympathetic (*middle* panel), and spiral (*bottom* panel) ganglia. Common properties of SCGs include communicating with other SCGs and with neurons via gap junctions (shown in green). Ganglion-specific roles of SCGs include engulfment of synaptic sites formed by spinal cord inputs in the sympathetic ganglia and myelination of spiral ganglion neuronal soma. (*Left*) Whole ganglia, (*middle*), ensheathed neuronal cell bodies, (*right*) highlighted properties of SGC–neuronal interactions.

providing metabolic and trophic support and enabling well-regulated electrical activity of the ensheathed neuron, with little interference from other, nearby neurons. This arrangement allows the SGCs to maintain homeostasis by controlling the microenvironment of the associated neuron and by communicating with that neuron through both gap junctions and the release of extracellular signaling factors (Hanani 2005, 2010; Spray and Hanani 2019; Spray et al. 2019). The neurons, in turn, signal back to the SGCs, thereby coordinating responses within the unit.

Satellite glia exhibit many similarities to astrocytes and are often considered the functional analogs of astrocytes for the peripheral nervous system. Both satellite glia and astrocytes provide trophic support for neurons, buffer the extracellular fluid, and scavenge cellular debris (Hanani and Verkhratsky 2021). In addition, like astrocytes, satellite glia play critical roles in neuronal development, as well as in pathologic conditions, particularly injury or inflammation, and pain. At times the molecular similarities between CNS astrocytes and the peripheral satellite glia may make it difficult to determine whether observed phenotypes are due to changes in the CNS or in the PNS (Agulhon et al. 2013; Clasadonte et al. 2017).

Table 1. Mouselines

Transgenic mouse lines	Name	Promoter	Transgene	Purpose	References
Transgene	Fabp7-CreER2; ROSA26mEGFP	Fatty acid–binding protein-7	eGFP	SCG marker, specific to peripheral glia	Mapps et al. 2022a
Transgene	Fabp7-CreER2; ROSA26eGFP-DTA	Fatty acid–binding protein-7	Diptheria toxin-A	Inducible ablation of SGC, specificity to peripheral glia	Mapps et al. 2022a
Transgene	Gfap-hM3Dq	Glial fibrillary acidic protein	Gq-DREADD	Ca^{2+}-mediated glial activity, all SGC, astrocyte	Xie et al. 2017
Transgene	P0-Cre$^{+/-}$::hM3Dq$^{+/-}$	P0	Gq-DREADD	PNS-specific Ca^{2+}-mediated glial activity, SGC, myelinating Schwann cells	Xie et al. 2017
Transgene	GFAP-Cre::GCaMP6f	Glial fibrillary acidic protein	Gq-DREADD/GCaMP3	Ca^{2+}-mediated glial activity, calcium sensor	Xie et al. 2017
Transgene	GFAP-Cre::GCaMP6f	Glial fibrillary acidic protein	GCaMP6f	Calcium sensor, SGC, astrocytes	Rabah et al. 2020
Transgene	Cx43-CreERT2::GCaMP6f	Connexin 43	GCaMP6f	Calcium sensor, SGC, astrocytes	Rabah et al. 2020
Transgene	CX3CR1-eGFP	C-X3-C motif chemokine receptor 1	eGFP	SGC marker	Rabah et al. 2020
Transgene	Tg(P2rx7-EGFP)FY174Gsat	P2X purinoceptor 7	eGFP	SCG, Schwann cells (cochlear)	Prades et al 2021
Targeted	Npy-tm1(cre)Zman	NPY	Cre	Expression of Cre recombinase in peripheral glia	Milstein et al. 2015
Transgene	Tg(Npy-hrGFP)1Lowl	NPY	(Renilla) GFP	Peripheral glia marker	van den Pol et al. 2009
Targeted	Npytm1.1(flpo)Hze	NPY	FlpO	Expression of FlpO recombinase in peripheral glia	Daigle et al. 2018
Transgene	Tg(Plp1-cre/ERT)3Pop	PLP1	Cre-ERT	Inducible expression of Cre recombinase in myelinating glial cells	Mallon et al. 2002
Transgene	Tg(Plp1-EGFP)10Wmac	PLP1	eGFP	Myelinating glial cell marker	Doerflinger et al. 2003

(SGC) Satellite glial cell, (SCG) superior cervical ganglia, (eGFP) enhanced green fluorescent protein, (PNS) peripheral nervous system.

Table 2. Antibodies

Antigen	Antigen expression: cell types	SGC expression: species	SGC expression: locations	Antibody applications	References (species; location; antibody supplier)
FABP7	Highly expressed in SGCs specifically	Human (high), mouse (high)	DRG, enteric ganglia, nodose ganglion, spiral ganglion, sympathetic ganglia, trigeminal ganglion	Immunofluorescence (IF), immunohistochemistry (IHC), western blot (WB)	Avraham et al. (2022) (human, mouse; DRG; Thermo Fisher, PA5-24949); Callahan et al. (2008) (mouse; sympathetic; proprietary); Hu et al. (2013) (mouse; DRG/trigeminal; Chemicon); Kurtz et al. (1994) (mouse; trigeminal; proprietary); Lowenstein et al. (2023) (mouse; nodose; RNA scope #414651); Mapps et al. (2022a) (mouse; sympathetic; Abcam #ab32423); Mapps et al. (2022b) (mouse; DRG/sympathetic); Schreiner et al. (2007) (mouse; DRG/enteric/sympathetic; proprietary); Shi et al. (2008) (mouse; sympathetic; Chemicon); Tan et al. (2023) (mouse; spiral; Abcam #ab32423)

Continued

Cite this article as *Cold Spring Harb Perspect Biol* doi: 10.1101/cshperspect.a041367

Table 2. *Continued*

Antigen	Antigen expression: cell types	SGC expression: species	SCG expression: locations	Antibody applications	References (species; location; antibody supplier)
GFAP	Peripheral glia, astrocytes, neural crest progenitors	Canine (high), guinea pig (low), human (low), NHP (high), mouse (low without stress/in vivo), rat (low without stress/in vivo)	DRG, sympathetic ganglia, trigeminal ganglion	Immunocytochemistry (ICC), IF, IHC	Avraham et al. (2022) (human, mouse; DRG), Donegan et al. (2013) (rat; trigeminal; Sigma-Aldrich), Elfvin et al. (1987) (rat, guinea pig; sympathetic; proprietary), Feldman-Goriachnik and Hanani (2019) (mouse; sympathetic; Dako), Huang et al. (2021) (dog, mouse; DRG; Sigma-Aldrich #G3893), Koike et al. (2019) (rat; DRG; Santa Cruz), Maniglier et al. (2022) (mouse; DRG; Millipore #MAB3402, Dako #Z0334), Poulsen et al. (2014) (rat; trigeminal; Sigma-Aldrich #G3893), Svenningsen et al. (2004) (rat; DRG; Chemicon), Tongtako et al. (2017) (dog, mouse, NHP; DRG; Dako #Z0334), Wang et al. (2019) (rat; DRG; Abcam), Warwick and Hanani (2013) (mouse; DRG; Dako), Wang et al. (2021) (rat; DRG; Cell Signaling Technology #80788S), Zhang et al. (2009) (rat; DRG; Chemicon)

Continued

Table 2. *Continued*

Antigen	Antigen expression: cell types	SCG expression: species	SCG expression: locations	Antibody applications	References (species; location; antibody supplier)
Glutamine synthetase (GS)	GS	Canine (reported low and high), guinea pig (high), human (low), mouse (high), NHP (high), rat (high)	DRG, spiral ganglion, sympathetic ganglia, trigeminal ganglion	ICC, IF, IHC	Avraham et al. (2022) (human, mouse; DRG), Donegan et al. (2013) (rat; trigeminal; Millipore), Eybalin et al. (1996) (guinea pig; spiral; proprietary), Feldman-Goriachnik and Hanani (2019) (mouse; sympathetic/trigeminal; Santa Cruz), Huang et al. (2021) (dog, mouse; DRG; Thermo Fisher #PA5-28940), Poulsen et al. (2014) (rat; trigeminal; Sigma-Aldrich #G2781), Tang et al. (2010) (mouse; DRG/trigeminal; Millipore), Tasdemir-Yilmaz et al. (2021) (mouse; DRG/spiral), Tongtako et al. (2017) (dog, mouse, NHP; DRG; Santa Cruz #sc-9067); Wang et al. (2019) (rat; DRG; Abcam), Warwick and Hanani (2013) (mouse; DRG; Santa Cruz)

Continued

Cite this article as *Cold Spring Harb Perspect Biol* doi: 10.1101/cshperspect.a041367

Table 2. *Continued*

Antigen	Antigen expression: cell types	SGC expression: species	SCG expression: locations	Antibody applications	References (species; location; antibody supplier)
Kir4.1	Highly expressed in SGCs, and in astrocytes	Canine (high), human (low), mouse (high), rat (high)	DRG, spiral ganglion, sympathetic ganglia, trigeminal ganglion	IF, IHC, Immuno-electron microscopy (EM), WB	Avraham et al. (2022) (human, mouse; DRG), Hibino et al. (1999) (rat; spiral; proprietary), Huang et al. (2021) (dog, mouse; DRG; Alomone #APC-035) Koike et al. (2019) (rat; DRG; Alomone), Mapps et al. (2022a,b) (mouse; DRG), Mapps et al. (2022a,b) (mouse; sympathetic; Alomone # APC-035), Tan et al. (2023) (mouse; spiral; Alomone #AGP-012); Tang et al. (2010) (mouse; DRG/trigeminal; Alomone), Vit et al. (2008) (rat; trigeminal; Alomone)
P75-NGFr	Peripheral glia (SGCs and nmSCs, neurons, neural crest progenitors	Canine (low), human (high), mouse (low), rat (high)	DRG, spiral ganglion, sympathetic ganglia, trigeminal ganglion	ICC, IF, IHC, Immuno-EM	Donegan et al. (2013) (rat; trigeminal; Chemicon), Hu et al. (2013) (mouse; DRG/trigeminal; Promega), Koike et al. (2019) (rat; DRG; Abcam, Santa Cruz, Alomone), Lin et al. (2006) (rat; sympathetic/parasympathetic; proprietary), Liu et al. (2012) (human; spiral; Biosensis #M-011100), Pannese and Procacci (2002) (rat; DRG; Oncogene IgG-192), Tongtako et al. (2017) (dog, mouse; DRG; ATTC Clone HB8737), Wang et al. (2019) (rat; DRG; Abcam), Wang et al. (2021) (rat; DRG; Abcam #ab3125)

Continued

Table 2. *Continued*

Antigen	Antigen expression: cell types	SGC expression: species	SCG expression: locations	Antibody applications	References (species; location; antibody supplier)
Plp1	CNS (oligodendrocytes) and PNS (SGCs, myelinating SCs) glia, Neural crest progenitors	Human, mouse, rat	DRG, spiral ganglion, sympathetic ganglia, trigeminal ganglion	IF	Avraham et al. (2022) (human, mouse; DRG), Chu et al. (2023) (mouse; trigeminal), Mapps et al. (2022a,b) (mouse; DRG/sympathetic), Svenningsen et al. (2004) (rat; DRG; proprietary), Tasdemir-Yilmaz et al. (2021) (mouse; DRG/spiral)
S100	CNS and PNS glia, Neural crest progenitors	Canine (low), human (high), guinea pig (high), mouse (high), rat (high)	DRG, spiral ganglion, sympathetic ganglia, trigeminal ganglion	ICC, IF, IHC, Immuno-EM	Avraham et al. (2022) (human, mouse; DRG), Callahan et al. (2008) (mouse; sympathetic; Dako), Enes et al. (2020) (rat; sympathetic; Dako), Koike et al. (2019) (rat; DRG; Dako; Lav vision), Liu et al. (2014) (human, guinea pig; spiral/ trigeminal; Dako), Svenningsen et al. (2004) (rat; DRG; Chemicon, Dako), Tongtako et al. (2017) (dog; DRG; Sigma-Aldrich #S2644), Wang et al. (2019) (rat; DRG; Abcam), Wang et al. (2021) (rat; DRG; Abcam #ab52642), Whitlon et al. (2009) (mouse; spiral; Sigma-Aldrich), Zhang et al. (2021) (rat; DRG; Abcam #ab52642)

Continued

Cite this article as *Cold Spring Harb Perspect Biol* doi: 10.1101/cshperspect.a041367

Table 2. *Continued*

Antigen	Antigen expression: cell types	SGC expression: species	SCG expression: locations	Antibody applications	References (species; location; antibody supplier)
Sox10	Neural crest progenitors	Chicken (high), mouse (high), rat (high)	DRG, enteric ganglia, embryonic neural crest, nodose ganglion, sympathetic ganglia, spiral ganglion, trigeminal ganglion	ICC, IF	Avraham et al. (2022) (human, mouse; DRG), Callahan et al. (2008) (mouse; sympathetic; Dako), Enes et al. (2020) (rat; sympathetic; Dako), Kim et al. (2013) (mouse; sympathetic; Dako), Koike et al. (2019) (rat; DRG; Dako; Lav vision), Liu et al. (2014) (human, guinea pig; spiral/trigeminal; Dako), Svenningsen et al. (2004) (rat; DRG; Chemicon; Dako), Tongtako et al. (2017) (dog; DRG; Sigma-Aldrich #S2644), Wang et al. (2019) (rat; DRG; Abcam), Wang et al. (2021) (rat; DRG; Abcam #ab52642), Whitlon et al. (2009) (mouse; spiral; Sigma-Aldrich), Zhang et al. (2021) (rat; DRG; Abcam #ab52642)

(SGC) Satellite glial cell, (SCG) superior cervical ganglia, (DRG) dorsal root ganglia, (NHP) nonhuman primate, (CNS) central nervous system, (PNS) peripheral nervous system.

Here we will consider (1) the origins of SGCs during development and how retained plasticity contributes to nerve regeneration and repair, (2) similarities and differences among the distinct types of SGCs, (3) the roles of SGCs in the functioning of the nervous system, and (4) how SGCs influence neuropathologic conditions, particularly inflammation and pain.

SGC DEVELOPMENT AND PLASTICITY

Developing SGCs face the daunting challenges of populating ganglia throughout the body and forming tight neuron–glia units that permit rapid and dynamic bidirectional responses. Unlike the closely related Schwann cells, which arise from the same pool of progenitors, SGCs must also acquire additional tissue-specific properties while retaining plasticity, with the ability to change as needed, such as after injury. These needs are met by the remarkable properties of their progenitors, namely, the neural crest cells (NCCs), which are highly migratory and flexible, so much so that they are sometimes referred to as the fourth germ layer. SGCs seem to retain this plasticity throughout life, adding to their importance not only for supporting neuronal function but also for regeneration and repair.

SGCs and all other peripheral glial cells originate from the neural crest (see Fig. 2). Early in embryogenesis, multipotent NCCs migrate away from the dorsal neural tube and disperse throughout the nervous system, producing neurons, glia, and connective tissue (Bronner and Simões-Costa 2016). NCCs differ based on their location within the rostral–caudal axis (Soldatov et al. 2019) and, as they migrate, NCCs further restrict their fates in response to signals from the environment. NCCs directly give rise to glial precursors (GPs) that have sometimes been designated as Schwann cell precursors (SCPs), although these GP/SCPs produce additional glial cells other than Schwann cells. Other NCCs give rise to a transient pluripotent population of boundary cap cells (BCCs) that can become GPs or differentiate into neuronal cells (Kastriti and Adameyko 2017). The commitment to a glial identity depends on Notch signaling (Taylor et al. 2007) and on the expression of Sox10. De-

letion of *Sox10* leads to a massive loss of SGCs and Schwann cells (Britsch et al. 2001), highlighting the unique importance of this transcription factor in gliogenesis. In vitro analysis of *Sox10* mutant NCCs suggests that Sox10 promotes NCCs to survive and then take on a glial fate (Paratore et al. 2001).

SGCs as well as Schwann cells arise from the NCC-derived GPs (Kastriti and Adameyko 2017; Tasdemir-Yilmaz et al. 2021). The immature Schwann cells subsequently develop either as myelinating or nonmyelinating Schwann cells depending on the diameter of the associated axon as well as on Neuregulin-1 (Nrg1) to ErbB2 and ErbB3 signaling (Michailov et al. 2004; Taveggia et al. 2005). Ultimately, Schwann cells lose the rostral–caudal specializations seen in the NCCs and exhibit a convergent differentiation pathway (Tasdemir-Yilmaz et al. 2021). In contrast, differentiating SGCs begin to express *Fabp7*, followed by GFAP and S100 (Jessen and Mirsky 2005), while still retaining some properties of the NCCs and GPs.

Why some GPs ultimately produce SGCs and others produce Schwann cells is poorly understood (reviewed in Woodhoo and Sommer 2008). This divergence may depend on the microenvironment, as GPs from embryonic dorsal root ganglia (DRG) are able to produce both SGCs and Schwann cells in culture depending on cell–cell contacts (George et al. 2018). One key determinant may be the activation of Neuregulin signaling. Although all of the GPs settling in the ganglia and along the nerves express ErbB3 (Meyer and Birchmeier 1995), SGCs still develop in *ErbB2* and *ErbB3* mutants, in stark contrast to the severe loss of Schwann cells (Riethmacher et al. 1997). Likewise, there are many fewer *ErbB3*-positive Schwann cells in *Nrg1* mutant mice, but no obvious loss of SGCs. In addition to this differential dependence on Neuregulin signaling, in vitro studies suggest that Fabp7-positive SGCs develop first, before immature Schwann cells are detected (Woodhoo et al. 2004). Additionally, SGCs appear to sustain the expression of developmentally regulated genes that are expressed in NCCs and then downregulated in Schwann cells, such as *Cdh19* and *Erm* (Hagedorn et al. 2000; George et al. 2018;

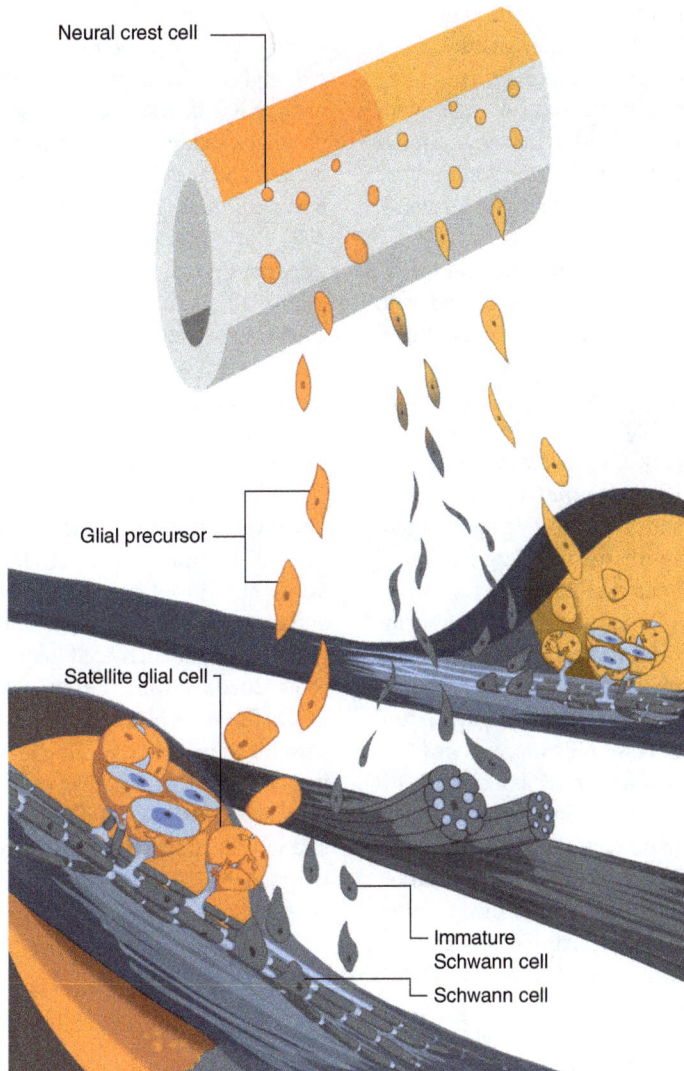

Figure 2. Development of satellite glial cells (SGCs). Neural crest cells (NCCs, orange) migrate and coalesce to form dorsal root ganglia, autonomic ganglia, or join with placode-derived cells to form cranial ganglia. NCCs differentiate to form glial precursors (GPs, orange) that can either become immature Schwann cells, or SGCs. Further maturation of Schwann cells into myelinating or nonmyelinating Schwann cells (gray) depends on interactions with an axon; further differentiation of the GPs into SGCs may similarly depend on the microenvironment and interactions with the neuron.

Tasdemir-Yilmaz et al. 2021). When cultured, SGCs from adult DRG can become more NCC-like, as reflected by the expression of p75 and ErbB receptors, as well as their ability to self-renew (Wang et al. 2021; Lu et al. 2023) and to express many stem cell-related genes (Li et al. 2007). Thus, SGCs are more closely related to

the GPs and NCCs than are the Schwann cells (Tasdemir-Yilmaz et al. 2021).

Molecular analysis of SGCs from different stages and tissues further shows their sustained developmental potential, as well as the influence of the environment on their final maturation. Immature SGCs and GPs express many cell cycle and pluripotency genes (Tasdemir-Yilmaz et al. 2021; van Weperen et al. 2021). As they mature, SGCs express additional genes important for their known interactions with neurons, as well as for their ability to produce lipids (Tasdemir-Yilmaz et al. 2021; van Weperen et al. 2021; Mapps et al. 2022b). The specific cohorts of genes vary depending on where the SGCs reside. For instance, the SGCs that envelop spiral ganglion neuron (SGN) cell bodies express many genes associated with myelination (Tasdemir-Yilmaz et al. 2021), whereas SGCs in the sympathetic ganglia are enriched for expression of genes associated with glutamate turnover (Mapps et al. 2022b). Differences between cochlear and lumbar SGCs become even more obvious in mature SGCs; in contrast, the progenitors that produce Schwann cells go through convergent differentiation and lose tissue-specific gene expression (Tasdemir-Yilmaz et al. 2021).

The apparent plasticity of SGCs likely contributes to their broader effects on nervous system function and repair, while also providing an attractive entry point for regeneration. Mature SGCs retain the expression of many genes that initially define GPs and that are associated with stem cells, including *nestin* (Gallaher et al. 2014), *Sox10*, and *Sox2* (Koike et al. 2014). Likewise, cultured adult SGCs express the BCC marker *Egr2* and can take on Schwann cell–like features or differentiate into nociceptive-like neurons under certain culture conditions (Wang et al. 2021). This apparently latent developmental potential can be uncovered following injury, which stimulates SGCs to reenter the cell cycle. For instance, after sciatic nerve transection, SGCs proliferate around the injured neurons, often in close association with macrophages (Gallaher et al. 2014; Krishnan et al. 2018). Additionally, Nestin⁺ SGCs with stem cell–like properties are present in adult DRGs in vivo, where they can differentiate into SGCs under homeostatic conditions, or into nociceptive sensory neurons or Schwann cells after nerve injury (Maniglier et al. 2022). SGCs have also been proposed to work with immune cells to clear debris and encourage regeneration (van Velzen et al. 2009; Hanani and Spray 2020). Consistent with this role in nerve repair, a subset of adult SGCs express immune response genes (Avraham et al. 2021a; Mapps et al. 2022b).

SGCs from other regions of the PNS show similar flexibility, even after they have taken on strikingly different fates. For instance, early postnatal SGCs in the cochlea also exhibit signs of plasticity, evidenced by maintained expression of Sox2 and the ability to produce Tuj1⁺ neurons when cultured (Kempfle et al. 2020; Chen et al. 2021) or upon overexpression of neurogenic transcription factors Ascl1 and Neurod1 (Noda et al. 2018). As in the DRG, adult cochlear SGCs appear to de-differentiate and take on astrocyte-like properties after injury (Wise et al. 2017). Furthermore, both SGCs and Schwann cells can regenerate in the mature cochlea after diphtheria toxin–mediated ablation (Wan and Corfas 2017), emphasizing the retained ability to divide under certain circumstances and in certain locations (Tasdemir-Yilmaz et al. 2021). Cochlear SGCs also proliferate after ablation of SGNs and can express neuronal markers upon overexpression of the stem cell gene *Lin28* (Kempfle et al. 2020). Given the NCC-like potential of SGCs, these cochlear glia populations may offer a useful source for replacing lost SGNs and thus improving hearing (Meas et al. 2018). The remarkable plasticity of SGCs across systems is likely a reflection of the broader retention of plasticity in NCC-derived adult tissues (Parfejevs et al. 2018).

COMMON FUNCTIONALITIES AND TISSUE-SPECIFIC DIFFERENCES

SGCs are defined by common features even as they diversify to carry out system-specific functions in sensory, sympathetic, and auditory ganglia (Hanani and Spray 2020). Most broadly, satellite glia create a microenvironment within their resident ganglia that allows for coordinated signaling between glial cells and facilitates glial–neuron communication. The outcome of this communication can differ dramatically within and across systems and across development.

Structurally, the association of an individual SGC with one (or occasionally two) neuronal soma (Pannese 2010) is in contrast to the non-overlapping stellate projections of central astrocytes, which enwrap as many as eight neurons and hundreds of dendrites in a nonoverlapping pattern (Bushong et al. 2002; Halassa et al. 2007). In sensory ganglia, SGCs form a continuous sheath around the neuronal soma, with the number of glial cells scaling with the size of the neuron (Pannese 1960). The intimate arrangement of SGCs and neurons within different ganglia allows for a range of context-specific interactions (Fig. 1), including the ensheathment of individual or pairs of cell bodies in rat sensory ganglia (Pannese et al. 1991), enwrapping of cholinergic synaptic sites in sympathetic ganglia (Elfvin 1971), and myelination of neuronal soma in mouse spiral ganglia (Toesca 1996). The tight interaction between neurons and their associated glia is disrupted in disease models such as Fragile X (Avraham et al. 2021a), supporting the functional importance of the neuron–glial structural unit in these ganglia.

Satellite glia communicate with each other via gap junctions to form a glial network within the ganglia. Identification of glial–glial gap junctions was demonstrated in ultrastructural studies in sensory (Sakuma et al. 2001) and sympathetic neurons (Elfvin and Forsman 1978). The connexin proteins that form the gap junction structures between glial cells were identified by immunohistochemistry and RNA analysis in rat and mouse DRG (Garrett and Durham 2008; Procacci et al. 2008) and in human and guinea pig spiral ganglia (Liu et al. 2014). Functional studies showed the flow of injected dyes between SGCs in sympathetic and sensory ganglia (Huang et al. 2005; Hanani et al. 2010). Dye flow tended to be restricted to SGCs forming a structural unit engaging single encapsulated neurons. However, following nerve lesions or the induction of inflammatory processes, the extent of communication expanded to include SGCs associated with additional neuron–glial units (Hanani et al. 2002, 2010). This suggests that pathological processes activate a broader glial network, an idea supported by the finding that connexin proteins, including Cx43, increase following nerve lesions and inflammatory stimuli in cranial, sympathetic, and sensory somatoganglia. Indeed, multiple connexin proteins, including Cx26, 36, 40, and 43, are expressed in sensory ganglia, and up-regulation of different connexins is associated with different models of inflammatory and neuro-

pathic pain (Garrett and Durham 2008), demonstrating both common and pathology-specific changes in the glial network. These changes in gap junctions have been directly implicated in the development of pain as shown by a reduction of chronic constriction-induced pain behaviors following chronic knockdown of the widely studied glial Cx43 in the trigeminal ganglia (TG) (Ohara et al. 2008). Glial communication via Cx43 also contributes to central pain circuits, with loss of Cx43 in GFAP-expressing glia reducing behavioral responses to nerve injury (Luo et al. 2023). These changes in pain responses persist following local depletion of Cx43 in spinal cord astrocytes, suggesting that, in addition the actions of peripheral glia, central glia also contribute to the modulation of pain responses. How these central and peripheral glia coordinate to regulate different types of neuropathic pain is an open area for inquiry.

Gap junction–based communication results in Ca^{2+} signaling across the glial network which, in central astrocytes, is linked to the release of neuroactive molecules (Scemes and Giaume 2006; Bazargani and Attwell 2016). In the periphery, increased glial Ca^{2+} signaling is modulated by interactions with associated neurons (Suadicani et al. 2010), but the question of whether gap junctions between neurons and SGCs provide a direct mode of modulation in the ganglia has been a subject of debate. Some structural (Sakuma et al. 2001) and dye coupling (Zuriel and Devor 2001) experiments showed no evidence for gap junctions between neurons and glia in intact DRG. However, dye coupling between neurons and SGCs has been observed in dissociated TG cultures, in intact TG ganglia isolated from mice following LPS injections, and in DRG from mice treated with CFA (Ledda et al. 2009; Spray et al. 2019). Gap junction–dependent Ca^{2+} signals in TG SGCs can also be evoked by mechanical stimulation of adjacent neurons. This signaling is bidirectional, as mechanical stimulation of SGCs also increased Ca^{2+} signaling in neurons in the same cultures (Suadicani et al. 2010). In addition, Ca^{2+} imaging across a large number of DRG neurons in vivo showed injury-induced gap junction coupling between neurons and between neurons and SGCs in inflammatory and neuropathic pain models (Kim et al. 2016). Blocking gap junctions in this model reduced the coordinate firing of adjacent neurons and reduced pain responses to mechanical stimulation. These studies indicate that, in addition to expanding the gap junction–based glial network, pathological perturbations result in the emergence of direct activation of SGCs by the neurons and contribute to neuropathic and inflammatory pain.

Beyond direct gap junction–mediated communication, neurons release molecules that modulate the activity of SGCs. Activity-dependent release of ATP from sensory neuronal soma, demonstrated using sniffer pipettes, activates glial purinergic P2Y receptors (Zhang et al. 2007; Rozanski et al. 2013). Cultured rat cochlear glia also respond to ATP by activating P2X receptor–mediated currents and metabotropic P2Y receptors and by an increase in Ca^{2+} dynamics (Prades et al. 2021). Along with increased connexin expression and gap junction function, inflammatory changes also increase the sensitivity of P2X receptors in nodose ganglia, enhancing glial Ca^{2+} and dye coupling responses to ATP (Feldman-Goriachnik et al. 2015). Heightened responses to ATP show cell-type-specific properties in different ganglia depending on the specific lesion. Thus, 6-OHDA toxicity selectively targets sympathetic neurons, and leads to increased Ca^{2+} responses to ATP in sympathetic, but not TG SGCs, while LPS damage to sensory ganglia results in altered ATP responses in TG, but not sympathetic SGCs (Feldman-Goriachnik and Hanani 2019). In fact, SGCs from different ganglia can display both common and discrete responses to neuronal factors. Endothelin-1 causes Ca^{2+} increases in sympathetic as well as sensory nodose and TG neurons while decreasing dye coupling in all SGC types (Feldman-Goriachnik and Hanani 2017). In contrast, sympathetic SGCs, which surround cholinergic synapses in the sympathetic superior cervical ganglia (SCG), showed increased Ca^{2+} in response to acetylcholine signaling via muscarinic receptors, whereas no response to acetylcholine was seen in TG and nodose SGCs (Feldman-Goriachnik et al. 2018). In some cases, SGCs respond differently because of differences in the neurons they

associate with. For example, nitric oxide released by sensory neurons (Alm et al. 1995) promotes SGC activation within the DRG and TG (Feldman-Goriachnik and Hanani 2021). These experiments demonstrate that the response profiles of SGCs reflect both common properties and the specific local environments of their ganglia. The common and divergent aspects of SGCs from different locations emerge from the developmental and mature gene expression profiles of SGC from sensory, sympathetic, and spiral ganglia (Tasdemir-Yilmaz et al. 2021; van Weperen et al. 2021; Mapps et al. 2022b).

Once activated by these interactions, how do the SGCs communicate with and regulate the properties of their neighboring neurons? In the CNS, astrocytes release a range of factors, including ATP, cytokines, and neurotrophins, which modify neuronal development and function with different repertoires of glial factors underlying different brain states (Allen and Lyons 2018). Likewise, SGCs also release synaptic and paracrine factors that regulate the activity and structure of nearby neurons. Both DRG and sympathetic SGCs release neurotrophic factors (Zhou et al. 1999; Enes et al. 2020) as well as ATP and cytokines. Several factors, including NGF, IL-1, IL-6, and other inflammatory cytokines, increase following nerve lesions or inflammatory processes (Zhou et al. 1999; Takeda et al. 2007; Dubový et al. 2010; Jager et al. 2020), and are associated with pathological changes that include increased neuronal excitability and sympathetic sprouting within sensory ganglia. Thus, factors released by SGCs result in functional changes in neuronal structure and function both in the intact and damaged nervous system.

SGC EFFECTS ON NERVOUS SYSTEM FUNCTION

Ultimately, glial–neuron interactions modulate the normal development and function of the neurons, with a range of effects reported in different systems. SGCs actively buffer the ionic environment via expression of the Kir4.1 inwardly rectifying K^+ channel in somatosensory (Vit et al. 2008), sympathetic (Mapps et al. 2022a), and spiral ganglia (Hibino et al. 1999; Smith et al.

2021), allowing for tight control of neuronal excitability. In the sympathetic system, where spinal cord inputs form cholinergic synapses that set the sympathetic drive to peripheral organs, SGCs promote synapse formation, influence neurotransmitter synthesis, and support neuronal survival (Enes et al. 2020; Mapps et al. 2022a). Sympathetic SGCs also regulate synaptic activity, increasing transmission in cultured neurons and limiting transmission in vivo (Enes et al. 2020; Mapps et al. 2022a), suggesting a potentially interesting role in setting the balance of sympathetic drive. Further, SGCs can directly modify the impact of sympathetic activity on peripheral physiology, with chemogenetically induced SGC activity acutely increasing heart rate and having long-term effects on blood pressure in vivo (Xie et al. 2017). In the somatosensory system, SGCs provide trophic support (Zhou et al. 1999), prevent excitotoxicity, and maintain neurotransmitter homeostasis by expressing glutamate transporters required for the glutamate–glutamine cycle (Berger and Hediger 2000; Jasmin et al. 2010). While less is known about the roles of SGCs in spiral ganglia, the unusual ability of these satellite glia to myelinate neuronal cell bodies in rodents (Rosenbluth 1962; Toesca 1996), if not in humans (Ota and Kimura 1980; Arnold 1987), supports the need for fast and efficient neurotransmission in the auditory system.

Although SGCs buffer and modulate normal neuronal function in the best of circumstances, glial responses to injury and inflammation can contribute to pathological changes in neuronal functions that underlie pain syndromes (Takeda et al. 2007; Kim et al. 2016; Hossain et al. 2017; Hanani and Spray 2020). The effects of glia in pathological situations reflect their interactions both with damaged axons and with peripheral immune cells associated with inflammation. Macrophages infiltrate and are activated in somatosensory (Hu and McLachlan 2002) and sympathetic ganglia (Schreiber et al. 2002) following peripheral injury and in a rat model of hypertension (Neely et al. 2022), as well as in spiral ganglia following acoustic trauma or treatment with ototoxic drugs (Kaur et al. 2018). Cochlear macrophages also regulate SGC number during postnatal development (Brown et al.

2017). Much work remains to be done to define how SGCs interact with the peripheral immune system to promote the development of pathological states ranging from peripheral pain to heart disease.

SGCs AND NERVOUS SYSTEM PATHOLOGY

Since SGCs are an intrinsic part of the neuronal functional unit, it is perhaps not surprising that they show major changes in response to nerve injury, even though SGCs can be at a great distance from the actual sites of damage. The response to injury has been most often studied in SGCs of the trigeminal and DRG. Different models of injury, including ligation, transection, and crush injury, induce consistent changes in the SGCs, such as decreased expression of Kir4.1, the major K-channel of these glial cells, up-regulation of GFAP and of the gap junction proteins Cx43 and Cx26, as well as increases in the hemichannel component pannexin (Stephenson and Byers 1995; Hanani et al. 2002; Ohara et al. 2008; Vit et al. 2008; Takeda et al. 2011; Zhang et al. 2015). Injury also results in altered expression of multiple genes involved in lipid metabolism, such as fatty acid synthase (Fasn) (Avraham et al. 2020). Similar changes in gene expression to these described for the somatosensory system have been observed in SGCs of the autonomic system in response to injury (Hanani 2010; Feldman-Goriachnik et al. 2015), and may represent common features of these glial cells in multiple locations.

The abundant evidence that nerve injury causes changes in gene expression in SGCs that are far removed from the site of injury (Jager et al. 2020) suggests there must be mechanisms for conveying signals from the injury site to the neuron–glia unit. One pathway enabling this response is likely to be the changes in firing patterns of the injured neurons (Xie et al. 2009). Altered neuronal firing patterns, and the resultant changes in intracellular small second messengers such as Ca^{2+} and cAMP, could directly transmit signals to the SGCs through gap junctions (Suadicani et al. 2010; Spray et al. 2019). Alternatively, the activity-dependent release of paracrine factors may indirectly affect SGC gene expression. Notably, electrically active, damaged sensory neurons release nitric oxide, which can rapidly move across the 20 nm distance to the surrounding SGCs. Nitric oxide can induce many of the changes in SGCs that occur in response to injury, indicating that this is a critical component in SGC activation (Belzer and Hanani 2019). Other signaling molecules, such as chemokines, may also participate in injury-induced activation of SGC (Blum et al. 2014, 2017).

Whether mediated directly or indirectly, injury-induced changes in SGC gene expression have clear repercussions for glial biology. Increased expression of the gap junction protein Cx43 not only causes enhanced coupling of the SGCs surrounding an individual neuron, but also coupling between SGCs and neurons, and even among SGCs that surround distinct neuronal cell bodies (Hanani et al. 2002; Pannese et al. 2003; Kim et al. 2016). These changes in gap junction and cellular coupling can be detected by the passage of dye from one SGC to its neighboring cells, and in Ca^{2+} waves that travel through the SGC network (Kim et al. 2016). Increased expression of pannexin facilitates the release of extracellular modulators such as ATP into the extracellular space, resulting in the activation of purinergic receptors on neurons and on other SGCs (Braun et al. 2004; Hanani 2012; Zhang et al. 2015; Hanstein et al. 2016). Together, all these alterations represent an activated state, in which the SGCs can proliferate and can also release proinflammatory cytokines such as IL-6, and TNF (Ohtori et al. 2004; Dubový et al. 2010; Wu et al. 2015; Afroz et al. 2019). In turn, SGC-derived inflammatory molecules cause increased neuronal firing (Eliav et al. 2009; Black et al. 2018). Thus, injury initiates a feed-forward proinflammatory response involving both SGCs and somatosensory neurons, and this accelerated response is thought to be partially responsible for systemic symptoms and chronic pain after nerve injury.

Systemic inflammation due to the release of proinflammatory chemokines by SGCs and other cells is likely to contribute to both localized and systemic symptoms. Evidence for SGCs contributing to chronic pain includes the finding that inhibition of Kir4.1 in TGs can cause behavior typical of pain responses (Vit et al. 2008), while genetic targeting of the hemichannel component

pannexin in SGCs exerted an analgesic effect in mice following nerve injury (Hanstein et al. 2016). Since SGCs in the TG and DRG are implicated in somatosensory and nociceptive sensations, the contribution of activated SGCs in these locations to chronic pain following nerve injury can be readily understood. While SGCs in autonomic and cranial ganglia also exhibit changes in gene expression after nerve injury, the physiologic and behavioral consequences of glial activation in sympathetic ganglia and in spiral ganglia are less clear (Zhou et al. 1999; Hanani 2010; Li et al. 2018; Hanani and Spray 2020). However, increased sympathetic activity in multiple pain models (Li et al. 2018) and SGC-dependent sprouting of sympathetic fibers in sensory ganglia following nerve injury (Zhou et al. 1999) indicate a role for SGCs in cross-regulation between the sensory and autonomic systems. Injury-induced changes in SGCs have additional critical roles in the response to injury beyond pain and inflammation. Enhanced expression of Fascn and increased lipid synthetic capacity facilitate axon regrowth after injury, while the autophagic capacity of SGCs can remove any debris due to cell death of injured neurons. Conditional knockout of *Fascn* in SGCs diminishes the regenerative capacity of injured axons, demonstrating the importance of the SGC lipid synthetic pathway (Avraham et al. 2021b). Thus, SGCs play critical, multifaceted roles in responses to injury.

CONCLUDING REMARKS

SGCs have critical functions in the development, function, and pathology of the peripheral nervous system. Their intimate association with adjacent neurons, and their plasticity in response to local and distant changes in physiological state, enables these small cells to bring about large variations in the neuronal circuitry of peripheral systems. SGCs share many common properties across the different ganglia within the peripheral sensory, sympathetic, and auditory systems. This allows for the ongoing maintenance of a microenvironment offering trophic support, ionic balance, and regulation of neuronal activity. Yet, remarkably, these same SCG cells contribute to a diversity of neuronal functions and response to damage and inflammation that are specific to distinct classes of gan-

Cite this article as *Cold Spring Harb Perspect Biol* doi: 10.1101/cshperspect.a041367

glia. As the molecular and structural features of these diverse glia are being defined the range of these common and highly specific functions are certain to increase, providing new insights into the glial control of nervous system function in both central circuits and the periphery. The field already benefits from many reliable antibodies and mouse strains that can be used to visualize and manipulate SGCs (see Tables 1 and 2), so more discoveries are on the horizon. As we define the common attributes of vertebrate satellite glia across different regions of the nervous system, we may also uncover similarities between satellite glia and particular types of invertebrate glial cells, such as the cortex glia that surround neuronal cell bodies in *Drosophila* (Corty and Coutinho-Budd 2023). Thus, the biology of satellite glia represents a growing area of scientific interest.

REFERENCES

Afroz S, Arakaki R, Iwasa T, Oshima M, Hosoki M, Inoue M, Baba O, Okayama Y, Matsuka Y. 2019. CGRP induces differential regulation of cytokines from satellite glial cells in trigeminal ganglia and orofacial nociception. *Int J Mol Sci* **20**: 711. doi:10.3390/ijms20030711

Agulhon C, Boyt KM, Xie AX, Friocourt F, Roth BL, McCarthy KD. 2013. Modulation of the autonomic nervous system and behaviour by acute glial cell Gq protein-coupled receptor activation in vivo. *J Physiol* **591**: 5599–5609. doi:10.1113/jphysiol.2013.261289

Allen NJ, Lyons DA. 2018. Glia as architects of central nervous system formation and function. *Science* **362**: 181–185. doi:10.1126/science.aat0473

Alm P, Uvelius B, Ekström J, Holmqvist B, Larsson B, Andersson KE. 1995. Nitric oxide synthase-containing neurons in rat parasympathetic, sympathetic and sensory ganglia: a comparative study. *Histochem J* **27**: 819–831. doi:10.1007/BF02388306

Arnold W. 1987. Myelination of the human spiral ganglion. *Acta Otolaryngol Suppl* **436**: 76–84. doi:10.3109/00016488709124979

Avraham O, Deng PY, Jones S, Kuruvilla R, Semenkovich CF, Klyachko VA, Cavalli V. 2020. Satellite glial cells promote regenerative growth in sensory neurons. *Nat Commun* **11**: 4891. doi:10.1038/s41467-020-18642-y

Avraham O, Deng PY, Maschi D, Klyachko VA, Cavalli V. 2021a. Disrupted association of sensory neurons with enveloping satellite glial cells in Fragile X mouse model. *Front Mol Neurosci* **14**: 796070. doi:10.3389/fnmol.2021.796070

Avraham O, Feng R, Ewan EE, Rustenhoven J, Zhao G, Cavalli V. 2021b. Profiling sensory neuron microenvironment after peripheral and central axon injury reveals key pathways for neural repair. *eLife* **10**: e68457. doi:10.7554/eLife.68457

Avraham O, Chamessian A, Feng R, Yang L, Halevi AE, Moore AM, Gereau RWT, Cavalli V. 2022. Profiling the molecular signature of satellite glial cells at the single cell level reveals high similarities between rodents and humans. *Pain* **163**: 2348–2364. doi:10.1097/j.pain.0000000000002628

Bazargani N, Attwell D. 2016. Astrocyte calcium signaling: the third wave. *Nat Neurosci* **19**: 182–189. doi:10.1038/nn.4201

Belzer V, Hanani M. 2019. Nitric oxide as a messenger between neurons and satellite glial cells in dorsal root ganglia. *Glia* **67**: 1296–1307. doi:10.1002/glia.23603

Berger UV, Hediger MA. 2000. Distribution of the glutamate transporters GLAST and GLT-1 in rat circumventricular organs, meninges, and dorsal root ganglia. *J Comp Neurol* **421**: 385–399. doi:10.1002/(SICI)1096-9861(20000605)421:3<385::AID-CNE7>3.0.CO;2-S

Black BJ, Atmaramani R, Kumaraju R, Plagens S, Romero-Ortega M, Dussor G, Price TJ, Campbell ZT, Pancrazio JJ. 2018. Adult mouse sensory neurons on microelectrode arrays exhibit increased spontaneous and stimulus-evoked activity in the presence of interleukin-6. *J Neurophysiol* **120**: 1374–1385. doi:10.1152/jn.00158.2018

Blum E, Procacci P, Conte V, Hanani M. 2014. Systemic inflammation alters satellite glial cell function and structure. A possible contribution to pain. *Neuroscience* **274**: 209–217. doi:10.1016/j.neuroscience.2014.05.029

Blum E, Procacci P, Conte V, Sartori P, Hanani M. 2017. Long term effects of lipopolysaccharide on satellite glial cells in mouse dorsal root ganglia. *Exp Cell Res* **350**: 236–241. doi:10.1016/j.yexcr.2016.11.026

Braun N, Sévigny J, Robson SC, Hammer K, Hanani M, Zimmermann H. 2004. Association of the ecto-ATPase NTPDase2 with glial cells of the peripheral nervous system. *Glia* **45**: 124–132. doi:10.1002/glia.10309

Britsch S, Goerich DE, Riethmacher D, Peirano RI, Rossner M, Nave KA, Birchmeier C, Wegner M. 2001. The transcription factor Sox10 is a key regulator of peripheral glial development. *Genes Dev* **15**: 66–78. doi:10.1101/gad.186601

Bronner ME, Simões-Costa M. 2016. The neural crest migrating into the twenty-first century. *Curr Top Dev Biol* **116**: 115–134. doi:10.1016/bs.ctdb.2015.12.003

Brown LN, Xing Y, Noble KV, Barth JL, Panganiban CH, Smythe NM, Bridges MC, Zhu J, Lang H. 2017. Macrophage-mediated glial cell elimination in the postnatal mouse cochlea. *Front Mol Neurosci* **10**: 407. doi:10.3389/fnmol.2017.00407

Bushong EA, Martone ME, Jones YZ, Ellisman MH. 2002. Protoplasmic astrocytes in CA1 stratum radiatum occupy separate anatomical domains. *J Neurosci* **22**: 183–192. doi:10.1523/JNEUROSCI.22-01-00183.2002

Callahan T, Young HM, Anderson RB, Enomoto H, Anderson CR. 2008. Development of satellite glia in mouse sympathetic ganglia: GDNF and GFRα1 are not essential. *Glia* **56**: 1428–1437. doi:10.1002/glia.20709

Chen Z, Huang Y, Yu C, Liu Q, Qiu C, Wan G. 2021. Cochlear Sox2+ glial cells are potent progenitors for spiral ganglion neuron reprogramming induced by small molecules. *Front Cell Dev Biol* **9**: 728352. doi:10.3389/fcell.2021.728352

Chu Y, Jia S, Xu K, Liu Q, Mai L, Liu J, Fan W, Huang F. 2023. Single-cell transcriptomic profile of satellite glial cells in trigeminal ganglion. *Front Mol Neurosci* **16:** 1117065. doi:10.3389/fnmol.2023.1117065

Clasadonte J, Scemes E, Wang Z, Boison D, Haydon PG. 2017. Connexin 43-mediated astroglial metabolic networks contribute to the regulation of the sleep-wake cycle. *Neuron* **95:** 1365–1380.e5. doi:10.1016/j.neuron.2017.08.022

Corty MM, Coutinho-Budd J. 2023. *Drosophila* glia take shape to sculpt the nervous system. *Curr Opin Neurobiol* **79:** 102689. doi:10.1016/j.conb.2023.102689

Daigle TL, Madisen L, Hage TA, Valley MT, Knoblich U, Larsen RS, Takeno MM, Huang L, Gu H, Larsen R, et al. 2018. A suite of transgenic driver and reporter mouse lines with enhanced brain-cell-type targeting and functionality. *Cell* **174:** 465–480.e22. doi:10.1016/j.cell.2018.06.035

Doerflinger NH, Macklin WB, Popko B. 2003. Inducible site-specific recombination in myelinating cells. *Genesis* **35:** 63–72. doi:10.1002/gene.10154

Donegan M, Kernisant M, Cua C, Jasmin L, Ohara PT. 2013. Satellite glial cell proliferation in the trigeminal ganglia after chronic constriction injury of the infraorbital nerve. *Glia* **61:** 2000–2008. doi:10.1002/glia.22571

Dubový P, Klusáková I, Svíženská I, Brázda V. 2010. Satellite glial cells express IL-6 and corresponding signal-transducing receptors in the dorsal root ganglia of rat neuropathic pain model. *Neuron Glia Biol* **6:** 73–83. doi:10.1017/S1740925X10000074

Elfvin LG. 1971. Ultrastructural studies on the synaptology of the inferior mesenteric ganglion of the cat. I. Observations on the cell surface of the postganglionic perikarya. *J Ultrastruct Res* **37:** 411–425. doi:10.1016/s0022-5320(71)80135-1

Elfvin LG, Forsman C. 1978. The ultrastructure of junctions between satellite cells in mammalian sympathetic ganglia as revealed by freeze-etching. *J Ultrastruct Res* **63:** 261–274. doi:10.1016/S0022-5320(78)80051-3

Elfvin LG, Björklund H, Dahl D, Seiger A. 1987. Neurofilament-like and glial fibrillary acidic protein-like immunoreactivities in rat and guinea-pig sympathetic ganglia in situ and after perturbation. *Cell Tissue Res* **250:** 79–86. doi:10.1007/BF00214657

Eliav E, Benoliel R, Herzberg U, Kalladka M, Tal M. 2009. The role of IL-6 and IL-1β in painful perineural inflammatory neuritis. *Brain Behav Immun* **23:** 474–484. doi:10.1016/j.bbi.2009.01.012

Enes J, Haburčák M, Sona S, Gerard N, Mitchell AC, Fu W, Birren SJ. 2020. Satellite glial cells modulate cholinergic transmission between sympathetic neurons. *PLoS ONE* **15:** e0218643. doi:10.1371/journal.pone.0218643

Eybalin M, Norenberg MD, Renard N. 1996. Glutamine synthetase and glutamate metabolism in the guinea pig cochlea. *Hear Res* **101:** 93–101. doi:10.1016/S0378-5955(96)00136-0

Feldman-Goriachnik R, Hanani M. 2017. The effects of endothelin-1 on satellite glial cells in peripheral ganglia. *Neuropeptides* **63:** 37–42. doi:10.1016/j.npep.2017.03.002

Feldman-Goriachnik R, Hanani M. 2019. The effects of sympathetic nerve damage on satellite glial cells in the mouse superior cervical ganglion. *Auton Neurosci* **221:** 102584. doi:10.1016/j.autneu.2019.102584

Feldman-Goriachnik R, Hanani M. 2021. How do neurons in sensory ganglia communicate with satellite glial cells? *Brain Res* **1760:** 147384. doi:10.1016/j.brainres.2021.147384

Feldman-Goriachnik R, Belzer V, Hanani M. 2015. Systemic inflammation activates satellite glial cells in the mouse nodose ganglion and alters their functions. *Glia* **63:** 2121–2132. doi:10.1002/glia.22881

Feldman-Goriachnik R, Wu B, Hanani M. 2018. Cholinergic responses of satellite glial cells in the superior cervical ganglia. *Neurosci Lett* **671:** 19–24. doi:10.1016/j.neulet.2018.01.051

Gallaher ZR, Johnston ST, Czaja K. 2014. Neural proliferation in the dorsal root ganglia of the adult rat following capsaicin-induced neuronal death. *J Comp Neurol* **522:** 3295–3307. doi:10.1002/cne.23598

Garrett FG, Durham PL. 2008. Differential expression of connexins in trigeminal ganglion neurons and satellite glial cells in response to chronic or acute joint inflammation. *Neuron Glia Biol* **4:** 295–306. doi:10.1017/S1740925X09990093

George D, Ahrens P, Lambert S. 2018. Satellite glial cells represent a population of developmentally arrested Schwann cells. *Glia* **66:** 1496–1506. doi:10.1002/glia.23320

Hagedorn L, Paratore C, Brugnoli G, Baert JL, Mercader N, Suter U, Sommer L. 2000. The Ets domain transcription factor Erm distinguishes rat satellite glia from Schwann cells and is regulated in satellite cells by neuregulin signaling. *Dev Biol* **219:** 44–58. doi:10.1006/dbio.1999.9595

Halassa MM, Fellin T, Takano H, Dong JH, Haydon PG. 2007. Synaptic islands defined by the territory of a single astrocyte. *J Neurosci* **27:** 6473–6477. doi:10.1523/JNEUROSCI.1419-07.2007

Hanani M. 2005. Satellite glial cells in sensory ganglia: from form to function. *Brain Res Brain Res Rev* **48:** 457–476. doi:10.1016/j.brainresrev.2004.09.001

Hanani M. 2010. Satellite glial cells in sympathetic and parasympathetic ganglia: in search of function. *Brain Res Rev* **64:** 304–327. doi:10.1016/j.brainresrev.2010.04.009

Hanani M. 2012. Intercellular communication in sensory ganglia by purinergic receptors and gap junctions: implications for chronic pain. *Brain Res* **1487:** 183–191. doi:10.1016/j.brainres.2012.03.070

Hanani M, Spray DC. 2020. Emerging importance of satellite glia in nervous system function and dysfunction. *Nat Rev Neurosci* **21:** 485–498. doi:10.1038/s41583-020-0333-z

Hanani M, Verkhratsky A. 2021. Satellite glial cells and astrocytes, a comparative review. *Neurochem Res* **46:** 2525–2537. doi:10.1007/s11064-021-03255-8

Hanani M, Huang TY, Cherkas PS, Ledda M, Pannese E. 2002. Glial cell plasticity in sensory ganglia induced by nerve damage. *Neuroscience* **114:** 279–283. doi:10.1016/S0306-4522(02)00279-8

Hanani M, Caspi A, Belzer V. 2010. Peripheral inflammation augments gap junction-mediated coupling among satellite glial cells in mouse sympathetic ganglia. *Neuron Glia Biol* **6:** 85–89. doi:10.1017/S1740925X10000025

Hanstein R, Hanani M, Scemes E, Spray DC. 2016. Glial pannexin1 contributes to tactile hypersensitivity in a mouse model of orofacial pain. *Sci Rep* **6:** 38266. doi:10.1038/srep38266

Cite this article as *Cold Spring Harb Perspect Biol* doi: 10.1101/cshperspect.a041367

Hibino H, Horio Y, Fujita A, Inanobe A, Doi K, Gotow T, Uchiyama Y, Kubo T, Kurachi Y. 1999. Expression of an inwardly rectifying K$^+$ channel, Kir4.1, in satellite cells of rat cochlear ganglia. *Am J Physiol* **277**: C638–C644. doi:10 .1152/ajpcell.1999.277.4.C638

Hossain MZ, Unno S, Ando H, Masuda Y, Kitagawa J. 2017. Neuron-glia crosstalk and neuropathic pain: involvement in the modulation of motor activity in the orofacial region. *Int J Mol Sci* **18**: 2051. doi:10.3390/ijms18102051

Hu P, McLachlan EM. 2002. Macrophage and lymphocyte invasion of dorsal root ganglia after peripheral nerve lesions in the rat. *Neuroscience* **112**: 23–38. doi:10.1016/ S0306-4522(02)00065-9

Hu ZL, Zhang X, Shi M, Tian ZW, Huang Y, Chen JY, Ding YQ. 2013. Delayed but not loss of gliogenesis in Rbpj-deficient trigeminal ganglion. *Int J Clin Exp Pathol* **6**: 1261–1271.

Huang TY, Cherkas PS, Rosenthal DW, Hanani M. 2005. Dye coupling among satellite glial cells in mammalian dorsal root ganglia. *Brain Res* **1036**: 42–49. doi:10.1016/j.brainres .2004.12.021

Huang B, Zdora I, de Buhr N, Lehmbecker A, Baumgärtner W, Leitzen E. 2021. Phenotypical peculiarities and species-specific differences of canine and murine satellite glial cells of spinal ganglia. *J Cell Mol Med* **25**: 6909–6924. doi:10 .1111/jcmm.16701

Jager SE, Pallesen LT, Richner M, Harley P, Hore Z, McMahon S, Denk F, Vægter CB. 2020. Changes in the transcriptional fingerprint of satellite glial cells following peripheral nerve injury. *Glia* **68**: 1375–1395. doi:10.1002/glia.23785

Jasmin L, Vit JP, Bhargava A, Ohara PT. 2010. Can satellite glial cells be therapeutic targets for pain control? *Neuron Glia Biol* **6**: 63–71. doi:10.1017/S1740925X10000098

Jessen KR, Mirsky R. 2005. The origin and development of glial cells in peripheral nerves. *Nat Rev Neurosci* **6**: 671–682. doi:10.1038/nrn1746

Kastriti ME, Adameyko I. 2017. Specification, plasticity and evolutionary origin of peripheral glial cells. *Curr Opin Neurobiol* **47**: 196–202. doi:10.1016/j.conb.2017.11.004

Kaur T, Ohlemiller KK, Warchol ME. 2018. Genetic disruption of fractalkine signaling leads to enhanced loss of cochlear afferents following ototoxic or acoustic injury. *J Comp Neurol* **526**: 824–835. doi:10.1002/cne.24369

Kempfle JS, Luu NC, Petrillo M, Al-Asad R, Zhang A, Edge ASB. 2020. Lin28 reprograms inner ear glia to a neuronal fate. *Stem Cells* **38**: 890–903. doi:10.1002/stem.3181

Kim YS, Anderson M, Park K, Zheng Q, Agarwal A, Gong C, Saijilafu, Young L, He S, LaVinka PC, et al. 2016. Coupled activation of primary sensory neurons contributes to chronic pain. *Neuron* **91**: 1085–1096. doi:10.1016/j .neuron.2016.07.044

Koike T, Wakabayashi T, Mori T, Takamori Y, Hirahara Y, Yamada H. 2014. Sox2 in the adult rat sensory nervous system. *Histochem Cell Biol* **141**: 301–309. doi:10.1007/ s00418-013-1158-x

Koike T, Tanaka S, Hirahara Y, Oe S, Kurokawa K, Maeda M, Suga M, Kataoka Y, Yamada H. 2019. Morphological characteristics of p75 neurotrophin receptor-positive cells define a new type of glial cell in the rat dorsal root ganglia. *J Comp Neurol* **527**: 2047–2060. doi:10.1002/cne.24667

Krishnan A, Bhavanam S, Zochodne D. 2018. An intimate role for adult dorsal root ganglia resident cycling cells in the generation of local macrophages and satellite glial cells. *J Neuropathol Exp Neurol* **77**: 929–941. doi:10.1093/jnen/ nly072

Kurtz A, Zimmer A, Schnütgen F, Brüning G, Spener F, Müller T. 1994. The expression pattern of a novel gene encoding brain–fatty acid binding protein correlates with neuronal and glial cell development. *Development* **120**: 2637–2649. doi:10.1242/dev.120.9.2637

Ledda M, Blum E, De Palo S, Hanani M. 2009. Augmentation in gap junction-mediated cell coupling in dorsal root ganglia following sciatic nerve neuritis in the mouse. *Neuroscience* **164**: 1538–1545. doi:10.1016/j.neuroscience.2009 .09.038

Li HY, Say EH, Zhou XF. 2007. Isolation and characterization of neural crest progenitors from adult dorsal root ganglia. *Stem Cells* **25**: 2053–2065. doi:10.1634/stemcells.2007-0080

Li AL, Zhang JD, Xie W, Strong JA, Zhang JM. 2018. Inflammatory changes in paravertebral sympathetic ganglia in two rat pain models. *Neurosci Bull* **34**: 85–97. doi:10 .1007/s12264-017-0142-1

Lin G, Bella AJ, Lue TF, Lin CS. 2006. Brain-derived neurotrophic factor (BDNF) acts primarily via the JAK/STAT pathway to promote neurite growth in the major pelvic ganglion of the rat: part 2. *J Sex Med* **3**: 821–829. doi:10 .1111/j.1743-6109.2006.00292.x

Liu W, Glueckert R, Kinnefors A, Schrott-Fischer A, Bitsche M, Rask-Andersen H. 2012. Distribution of P75 neurotrophin receptor in adult human cochlea—an immunohistochemical study. *Cell Tissue Res* **348**: 407–415. doi:10 .1007/s00441-012-1395-7

Liu W, Glueckert R, Linthicum FH, Rieger G, Blumer M, Bitsche M, Pechriggl E, Rask-Andersen H, Schrott-Fischer A. 2014. Possible role of gap junction intercellular channels and connexin 43 in satellite glial cells (SGCs) for preservation of human spiral ganglion neurons: a comparative study with clinical implications. *Cell Tissue Res* **355**: 267–278. doi:10.1007/s00441-013-1735-2

Lowenstein E, Misios A, Buchert S, Ruffault P-L. 2023. Molecular characterization of nodose ganglia development reveals a novel population of Phox2b$^+$ glial progenitors in mice. bioRxiv doi:10.1101/2023.07.25.550402

Lu J, Wang D, Xu J, Zhang H, Yu W. 2023. New insights on the role of satellite glial cells. *Stem Cell Rev Rep* **19**: 358–367. doi:10.1007/s12015-022-10460-7

Luo LL, Wang JW, Yin XL, Chen XY, Zhang XF, Ye ZC. 2023. Astrocytic connexin 43 deletion ameliorates SNI-induced neuropathic pain by reducing microglia activation. *Biochem Biophys Res Commun* **638**: 192–199. doi:10.1016/j .bbrc.2022.11.071

Mallon BS, Shick HE, Kidd GJ, Macklin WB. 2002. Proteolipid promoter activity distinguishes two populations of NG2-positive cells throughout neonatal cortical development. *J Neurosci* **22**: 876–885. doi:10.1523/JNEUROSCI .22-03-00876.2002

Maniglier M, Vidal M, Bachelin C, Deboux C, Chazot J, Garcia-Diaz B, Baron-Van Evercooren A. 2022. Satellite glia of the adult dorsal root ganglia harbor stem cells that yield glia under physiological conditions and neurons in response to injury. *Stem Cell Rep* **17**: 2467–2483. doi:10 .1016/j.stemcr.2022.10.002

Mapps AA, Boehm E, Beier C, Keenan WT, Langel J, Liu M, Thomsen MB, Hattar S, Zhao H, Tampakakis E, et al. 2022a. Satellite glia modulate sympathetic neuron survival, activity, and autonomic function. *eLife* **11**: e74295. doi:10.7554/eLife.74295

Mapps AA, Thomsen MB, Boehm E, Zhao H, Hattar S, Kuruvilla R. 2022b. Diversity of satellite glia in sympathetic and sensory ganglia. *Cell Rep* **38**: 110328. doi:10.1016/j .celrep.2022.110328

Meas SJ, Zhang CL, Dabdoub A. 2018. Reprogramming glia into neurons in the peripheral auditory system as a solution for sensorineural hearing loss: lessons from the central nervous system. *Front Mol Neurosci* **11**: 77. doi:10 .3389/fnmol.2018.00077

Meyer D, Birchmeier C. 1995. Multiple essential functions of neuregulin in development. *Nature* **378**: 386–390. doi:10 .1038/378386a0

Michailov GV, Sereda MW, Brinkmann BG, Fischer TM, Haug B, Birchmeier C, Role L, Lai C, Schwab MH, Nave KA. 2004. Axonal neuregulin-1 regulates myelin sheath thickness. *Science* **304**: 700–703. doi:10.1126/science .1095862

Milstein AD, Bloss EB, Apostolides PF, Vaidya SP, Dilly GA, Zemelman BV, Magee JC. 2015. Inhibitory gating of input comparison in the CA1 microcircuit. *Neuron* **87**: 1274–1289. doi:10.1016/j.neuron.2015.08.025

Neely OC, Domingos AI, Paterson DJ. 2022. Macrophages can drive sympathetic excitability in the early stages of hypertension. *Front Cardiovasc Med* **8**: 807904. doi:10 .3389/fcvm.2021.807904

Noda T, Meas SJ, Nogami J, Amemiya Y, Uchi R, Ohkawa Y, Nishimura K, Dabdoub A. 2018. Direct reprogramming of spiral ganglion non-neuronal cells into neurons: toward ameliorating sensorineural hearing loss by gene therapy. *Front Cell Dev Biol* **6**: 16. doi:10.3389/fcell.2018.00016

Ohara PT, Vit JP, Bhargava A, Jasmin L. 2008. Evidence for a role of connexin 43 in trigeminal pain using RNA interference in vivo. *J Neurophysiol* **100**: 3064–3073. doi:10 .1152/jn.90722.2008

Ohtori S, Takahashi K, Moriya H, Myers RR. 2004. TNF-α and TNF-α receptor type 1 upregulation in glia and neurons after peripheral nerve injury: studies in murine DRG and spinal cord. *Spine (Phila Pa 1976)* **29**: 1082–1088. doi:10.1097/00007632-200405150-00006

Ota CY, Kimura RS. 1980. Ultrastructural study of the human spiral ganglion. *Acta Otolaryngol* **89**: 53–62. doi:10.3109/ 00016488009127108

Pannese E. 1960. Observations on the morphology, submicroscopic structure and biological properties of satellite cells (s.c.) in sensory ganglia of mammals. *Z Zellforsch Mikrosk Anat* **52**: 567–597. doi:10.1007/BF00339847

Pannese E. 2010. The structure of the perineuronal sheath of satellite glial cells (SGCs) in sensory ganglia. *Neuron Glia Biol* **6**: 3–10. doi:10.1017/S1740925X10000037

Pannese E, Procacci P. 2002. Ultrastructural localization of NGF receptors in satellite cells of the rat spinal ganglia. *J Neurocytol* **31**: 755–763. doi:10.1023/A:1025708132119

Pannese E, Ledda M, Arcidiacono G, Rigamonti L. 1991. Clusters of nerve cell bodies enclosed within a common

connective tissue envelope in the spinal ganglia of the lizard and rat. *Cell Tissue Res* **264**: 209–214. doi:10.1007/ BF00313957

Pannese E, Ledda M, Cherkas PS, Huang TY, Hanani M. 2003. Satellite cell reactions to axon injury of sensory ganglion neurons: increase in number of gap junctions and formation of bridges connecting previously separate perineuronal sheaths. *Anat Embryol (Berl)* **206**: 337–347. doi:10.1007/s00429-002-0301-6

Paratore C, Goerich DE, Suter U, Wegner M, Sommer L. 2001. Survival and glial fate acquisition of neural crest cells are regulated by an interplay between the transcription factor Sox10 and extrinsic combinatorial signaling. *Development* **128**: 3949–3961. doi:10.1242/dev.128.20.3949

Parfejevs V, Antunes AT, Sommer L. 2018. Injury and stress responses of adult neural crest-derived cells. *Dev Biol* **444**: S356–365. doi:10.1016/j.ydbio.2018.05.011

Peters APS, Webster HD. 1972. *The fine structure of the nervous system: the neurons and supporting cells*. WB Saunders, Philadelphia.

Poulsen JN, Larsen F, Duroux M, Gazerani P. 2014. Primary culture of trigeminal satellite glial cells: a cell-based platform to study morphology and function of peripheral glia. *Int J Physiol Pathophysiol Pharmacol* **6**: 1–12.

Prades S, Heard G, Gale JE, Engel T, Kopp R, Nicke A, Smith KE, Jagger DJ. 2021. Functional P2X7 receptors in the auditory nerve of hearing rodents localize exclusively to peripheral glia. *J Neurosci* **41**: 2615–2629. doi:10.1523/ JNEUROSCI.2240-20.2021

Procacci P, Magnaghi V, Pannese E. 2008. Perineuronal satellite cells in mouse spinal ganglia express the gap junction protein connexin43 throughout life with decline in old age. *Brain Res Bull* **75**: 562–569. doi:10.1016/j.brainresbull .2007.09.007

Rabah Y, Rubino B, Moukarzel E, Agulhon C. 2020. Characterization of transgenic mouse lines for selectively targeting satellite glial cells and macrophages in dorsal root ganglia. *PLoS ONE* **15**: e0229475. doi:10.1371/journal.pone .0229475

Riethmacher D, Sonnenberg-Riethmacher E, Brinkmann V, Yamaai T, Lewin GR, Birchmeier C. 1997. Severe neuropathies in mice with targeted mutations in the ErbB3 receptor. *Nature* **389**: 725–730. doi:10.1038/39593

Rosenbluth J. 1962. The fine structure of acoustic ganglia in the rat. *J Cell Biol* **12**: 329–359. doi:10.1083/jcb.12.2.329

Rozanski GM, Li Q, Kim H, Stanley EF. 2013. Purinergic transmission and transglial signaling between neuron somata in the dorsal root ganglion. *Eur J Neurosci* **37**: 359–365. doi:10.1111/ejn.12082

Sakuma E, Wang HJ, Asai Y, Tamaki D, Amano K, Mabuchi Y, Herbert DC, Soji T. 2001. Gap junctional communication between the satellite cells of rat dorsal root ganglia. *Kaibogaku Zasshi* **76**: 297–302.

Scemes E, Giaume C. 2006. Astrocyte calcium waves: what they are and what they do. *Glia* **54**: 716–725. doi:10.1002/ glia.20374

Schreiber RC, Vaccariello SA, Boeshore K, Shadiack AM, Zigmond RE. 2002. A comparison of the changes in the non-neuronal cell populations of the superior cervical ganglia following decentralization and axotomy. *J Neurobiol* **53**: 68–79. doi:10.1002/neu.10093

Cite this article as *Cold Spring Harb Perspect Biol* doi: 10.1101/cshperspect.a041367

Schreiner S, Cossais F, Fischer K, Scholz S, Bösl MR, Holtmann B, Sendtner M, Wegner M. 2007. Hypomorphic *Sox10* alleles reveal novel protein functions and unravel developmental differences in glial lineages. *Development* **134:** 3271–3281. doi:10.1242/dev.003350

Shi H, Cui H, Alam G, Gunning WT, Nestor A, Giovannucci D, Zhang M, Ding HF. 2008. Nestin expression defines both glial and neuronal progenitors in postnatal sympathetic ganglia. *J Comp Neurol* **508:** 867–878. doi:10.1002/cne.21719

Smith KE, Murphy P, Jagger DJ. 2021. Divergent membrane properties of mouse cochlear glial cells around hearing onset. *J Neurosci Res* **99:** 679–698. doi:10.1002/jnr.24744

Soldatov R, Kaucka M, Kastriti ME, Petersen J, Chontorotzea T, Englmaier L, Akkuratova N, Yang Y, Häring M, Dyachuk V, et al. 2019. Spatiotemporal structure of cell fate decisions in murine neural crest. *Science* **364:** eaas9536. doi:10.1126/science.aas9536

Spray DC, Hanani M. 2019. Gap junctions, pannexins and pain. *Neurosci Lett* **695:** 46–52. doi:10.1016/j.neulet.2017.06.035

Spray DC, Iglesias R, Shraer N, Suadicani SO, Belzer V, Hanstein R, Hanani M. 2019. Gap junction mediated signaling between satellite glia and neurons in trigeminal ganglia. *Glia* **67:** 791–801. doi:10.1002/glia.23554

Stephenson JL, Byers MR. 1995. GFAP immunoreactivity in trigeminal ganglion satellite cells after tooth injury in rats. *Exp Neurol* **131:** 11–22. doi:10.1016/0014-4886(95)90003-9

Suadicani SO, Cherkas PS, Zuckerman J, Smith DN, Spray DC, Hanani M. 2010. Bidirectional calcium signaling between satellite glial cells and neurons in cultured mouse trigeminal ganglia. *Neuron Glia Biol* **6:** 43–51. doi:10.1017/S1740925X09990408

Svenningsen AF, Colman DR, Pedraza L. 2004. Satellite cells of dorsal root ganglia are multipotential glial precursors. *Neuron Glia Biol* **1:** 85–93. doi:10.1017/S1740925X04000110

Takeda M, Tanimoto T, Kadoi J, Nasu M, Takahashi M, Kitagawa J, Matsumoto S. 2007. Enhanced excitability of nociceptive trigeminal ganglion neurons by satellite glial cytokine following peripheral inflammation. *Pain* **129:** 155–166. doi:10.1016/j.pain.2006.10.007

Takeda M, Takahashi M, Nasu M, Matsumoto S. 2011. Peripheral inflammation suppresses inward rectifying potassium currents of satellite glial cells in the trigeminal ganglia. *Pain* **152:** 2147–2156. doi:10.1016/j.pain.2011.05.023

Tan J, Duron A, Sucov H, Makita T. 2023. Placode and neural crest origins of congenital deafness in mouse models of Waardenburg-Shah syndrome. bioRxiv doi:10.1101/2023.10.27.564370

Tang X, Schmidt TM, Perez-Leighton CE, Kofuji P. 2010. Inwardly rectifying potassium channel Kir4.1 is responsible for the native inward potassium conductance of satellite glial cells in sensory ganglia. *Neuroscience* **166:** 397–407. doi:10.1016/j.neuroscience.2010.01.005

Tasdemir-Yilmaz OE, Druckenbrod NR, Olukoya OO, Dong W, Yung AR, Bastille I, Pazyra-Murphy MF, Sitko AA, Hale EB, Vigneau S, et al. 2021. Diversity of developing peripheral glia revealed by single-cell RNA sequencing. *Dev Cell* **56:** 2516–2535.e8. doi:10.1016/j.devcel.2021.08.005

Taveggia C, Zanazzi G, Petrylak A, Yano H, Rosenbluth J, Einheber S, Xu X, Esper RM, Loeb JA, Shrager P, et al.

2005. Neuregulin-1 type III determines the ensheathment fate of axons. *Neuron* **47:** 681–694. doi:10.1016/j.neuron.2005.08.017

Taylor MK, Yeager K, Morrison SJ. 2007. Physiological Notch signaling promotes gliogenesis in the developing peripheral and central nervous systems. *Development* **134:** 2435–2447. doi:10.1242/dev.005520

Toesca A. 1996. Central and peripheral myelin in the rat cochlear and vestibular nerves. *Neurosci Lett* **221:** 21–24. doi:10.1016/S0304-3940(96)13273-0

Tongtako W, Lehmbecker A, Wang Y, Hahn K, Baumgärtner W, Gerhauser I. 2017. Canine dorsal root ganglia satellite glial cells represent an exceptional cell population with astrocytic and oligodendrocytic properties. *Sci Rep* **7:** 13915. doi:10.1038/s41598-017-14246-7

van den Pol AN, Yao Y, Fu LY, Foo K, Huang H, Coppari R, Lowell BB, Broberger C. 2009. Neuromedin B and gastrin-releasing peptide excite arcuate nucleus neuropeptide Y neurons in a novel transgenic mouse expressing strong *Renilla* green fluorescent protein in NPY neurons. *J Neurosci* **29:** 4622–4639. doi:10.1523/JNEUROSCI.3249-08.2009

van Velzen M, Laman JD, Kleinjan A, Poot A, Osterhaus AD, Verjans GM. 2009. Neuron-interacting satellite glial cells in human trigeminal ganglia have an APC phenotype. *J Immunol* **183:** 2456–2461. doi:10.4049/jimmunol.0900890

van Weperen VYH, Littman RJ, Arneson DV, Contreras J, Yang X, Ajijola OA. 2021. Single-cell transcriptomic profiling of satellite glial cells in stellate ganglia reveals developmental and functional axial dynamics. *Glia* **69:** 1281–1291. doi:10.1002/glia.23965

Vit JP, Ohara PT, Bhargava A, Kelley K, Jasmin L. 2008. Silencing the Kir4.1 potassium channel subunit in satellite glial cells of the rat trigeminal ganglion results in pain-like behavior in the absence of nerve injury. *J Neurosci* **28:** 4161–4171. doi:10.1523/JNEUROSCI.5053-07.2008

Wan G, Corfas G. 2017. Transient auditory nerve demyelination as a new mechanism for hidden hearing loss. *Nat Commun* **8:** 14487. doi:10.1038/ncomms14487

Wang XB, Ma W, Luo T, Yang JW, Wang XP, Dai YF, Guo JH, Li LY. 2019. A novel primary culture method for high-purity satellite glial cells derived from rat dorsal root ganglion. *Neural Regen Res* **14:** 339–345. doi:10.4103/1673-5374.244797

Wang D, Lu J, Xu X, Yuan Y, Zhang Y, Xu J, Chen H, Liu J, Shen Y, Zhang H. 2021. Satellite glial cells give rise to nociceptive sensory neurons. *Stem Cell Rev Rep* **17:** 999–1013. doi:10.1007/s12015-020-10102-w

Warwick RA, Hanani M. 2013. The contribution of satellite glial cells to chemotherapy-induced neuropathic pain. *Eur J Pain* **17:** 571–580. doi:10.1002/j.1532-2149.2012.00219.x

Whitlon DS, Tieu D, Grover M, Reilly B, Coulson MT. 2009. Spontaneous association of glial cells with regrowing neurites in mixed cultures of dissociated spiral ganglia. *Neuroscience* **161:** 227–235. doi:10.1016/j.neuroscience.2009.03.044

Wise AK, Pujol R, Landry TG, Fallon JB, Shepherd RK. 2017. Structural and ultrastructural changes to Type I spiral ganglion neurons and Schwann cells in the deafened guinea pig cochlea. *J Assoc Res Otolaryngol* **18:** 751–769. doi:10.1007/s10162-017-0631-y

Woodhoo A, Sommer L. 2008. Development of the Schwann cell lineage: from the neural crest to the myelinated nerve. *Glia* **56:** 1481–1490. doi:10.1002/glia.20723

Woodhoo A, Dean CH, Droggiti A, Mirsky R, Jessen KR. 2004. The trunk neural crest and its early glial derivatives: a study of survival responses, developmental schedules and autocrine mechanisms. *Mol Cell Neurosci* **25:** 30–41. doi:10.1016/j.mcn.2003.09.006

Wu Z, Wang S, Wu I, Mata M, Fink DJ. 2015. Activation of TLR-4 to produce tumour necrosis factor-α in neuropathic pain caused by paclitaxel. *Eur J Pain* **19:** 889–898. doi:10.1002/ejp.613

Xie W, Strong JA, Zhang JM. 2009. Early blockade of injured primary sensory afferents reduces glial cell activation in two rat neuropathic pain models. *Neuroscience* **160:** 847–857. doi:10.1016/j.neuroscience.2009.03.016

Xie AX, Lee JJ, McCarthy KD. 2017. Ganglionic GFAP[+] glial Gq-GPCR signaling enhances heart functions in vivo. *JCI Insight* **2:** e90565. doi:10.1172/jci.insight.90565

Zhang X, Chen Y, Wang C, Huang LY. 2007. Neuronal somatic ATP release triggers neuron-satellite glial cell communication in dorsal root ganglia. *Proc Natl Acad Sci* **104:** 9864–9869. doi:10.1073/pnas.0611048104

Zhang H, Mei X, Zhang P, Ma C, White FA, Donnelly DF, Lamotte RH. 2009. Altered functional properties of satellite glial cells in compressed spinal ganglia. *Glia* **57:** 1588–1599. doi:10.1002/glia.20872

Zhang Y, Laumet G, Chen SR, Hittelman WN, Pan HL. 2015. Pannexin-1 up-regulation in the dorsal root ganglion contributes to neuropathic pain development. *J Biol Chem* **290:** 14647–14655. doi:10.1074/jbc.M115.650218

Zhang R, Chen S, Wang X, Gu X, Yi S. 2021. Cell populations in neonatal rat peripheral nerves identified by single-cell transcriptomics. *Glia* **69:** 765–778. doi:10.1002/glia.23928

Zhou XF, Deng YS, Chie E, Xue Q, Zhong JH, McLachlan EM, Rush RA, Xian CJ. 1999. Satellite-cell-derived nerve growth factor and neurotrophin-3 are involved in noradrenergic sprouting in the dorsal root ganglia following peripheral nerve injury in the rat. *Eur J Neurosci* **11:** 1711–1722. doi:10.1046/j.1460-9568.1999.00589.x

Zuriel E, Devor M. 2001. Dye coupling does not explain functional crosstalk within dorsal root ganglia. *J Peripher Nerv Syst* **6:** 227–231. doi:10.1046/j.1529-8027.2001.01024.x

Enteric Glia

Meenakshi Rao[1] and Brian D. Gulbransen[2]

[1]Department of Pediatrics, Boston Children's Hospital and Harvard Medical School, Boston, Massachusetts 02115, USA

[2]Department of Physiology, Michigan State University, East Lansing, Michigan 48824, USA

Correspondence: meenakshi.rao@childrens.harvard.edu; gulbrans@msu.edu

Enteric glia are a unique type of peripheral neuroglia that accompany neurons in the enteric nervous system (ENS) of the digestive tract. The ENS displays integrative neural circuits that are capable of governing moment-to-moment gut functions independent of input from the central nervous system. Enteric glia are interspersed with neurons throughout these intrinsic gut neural circuits and are thought to fulfill complex roles directed at maintaining homeostasis in the neuronal microenvironment and at neuroeffector junctions in the gut. Changes to glial functions contribute to a wide range of gastrointestinal diseases, but the precise roles of enteric glia in gut physiology and pathophysiology are still under examination. This review summarizes current concepts regarding enteric glial development, diversity, and functions in health and disease.

The enteric nervous system (ENS) is a major branch of the autonomic nervous system and is comprised of a large number of intramural neurons and glia that form a contiguous network along the entire extent of the digestive tract (Fig. 1). The ENS is implicated in virtually every aspect of digestive tract function, including peristalsis, nutrient absorption, fluid flux, hormone secretion, epithelial turnover, and mucosal immunity. The ENS can mediate many of these behaviors autonomously but often acts in close collaboration with extrinsic projections into the gut from vagal afferent and efferent neurons, visceral afferent neurons, and sympathetic postganglionic neurons.

In mammals, the ENS is organized into two major ganglionated plexuses. The myenteric plexus is larger, extends the full length of the digestive tract from the esophagus to the anus, and is located in between the two orthogonally arranged layers of smooth muscle that surround the bowel. The submucosal plexus contains fewer neurons per ganglion, is absent to sparse in the esophagus and stomach, and is located in the loose connective tissue of the submucosa (Fig. 1). Humans and other large mammals have multiple layers of submucosal ganglia, as well as a third minor plexus in the mucosa. Neuronal soma are restricted to these ganglia, but their projections extend extensively between ganglia, between plexuses, and to myriad targets in the mucosa and the smooth muscle syncytium. These complex microcircuits are still being defined.

Transcriptional profiling and immunohistochemistry studies suggest that there are at least 15

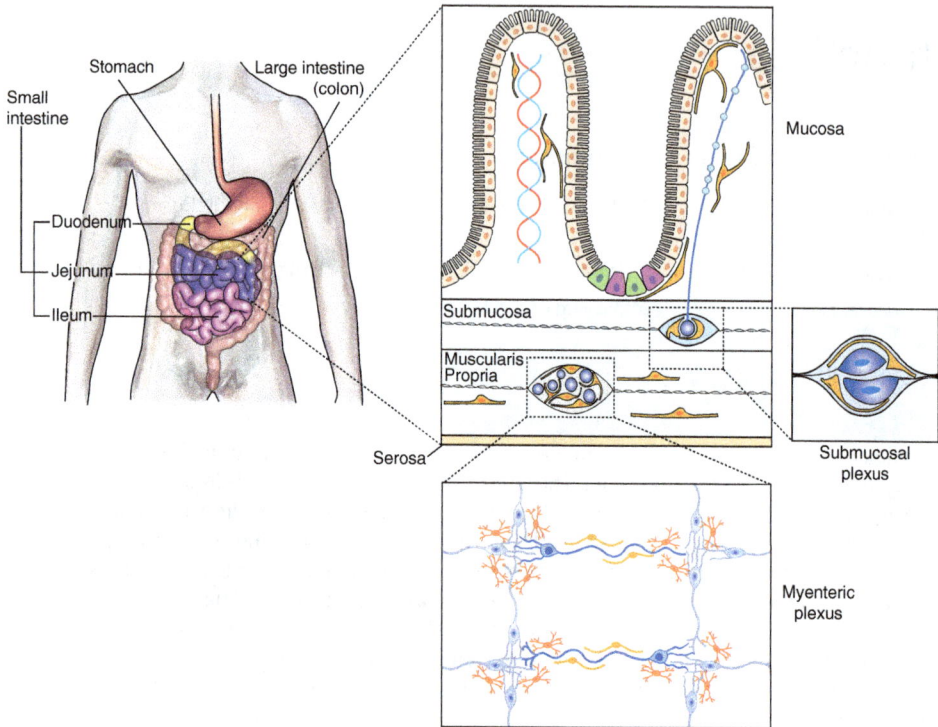

Figure 1. The complex microcircuits of the enteric nervous system (ENS) are integrated into the length and breadth of the digestive tract. The schematic cross section of the small intestine illustrates how the two major networks of enteric ganglia, the myenteric and submucosal plexus, are integrated into the laminar structure of the gut. Glia (orange) closely associate with neuronal soma (blue) in each of these ganglia (*insets*) and nerve fibers in the mucosa.

distinct types of enteric neurons in the myenteric plexus alone that elaborate a variety of neurotransmitters (Sang and Young 1996; Qu et al. 2008; Drokhlyansky et al. 2020; May-Zhang et al. 2021; Morarach et al. 2021). The majority of these neurons are either cholinergic or nitrergic, but often co-release other neurotransmitters or neuropeptides with signaling and/or neuromodulatory capabilities. Glia are abundant within this ENS and found in close association with neurons within the enteric ganglia and with nerve fibers in all of the laminae of the gut, from muscle to mucosa (Fig. 1). These enteric glia are nonmyelinating, and while they share some features with other types of glia elsewhere in the nervous system, they are unique and not directly analogous to any other population. Here, we review the development and diversification of enteric glia, their functional interactions with neurons and nonneuronal cells in the intestine,

and their potential roles in human disease. While space limits comprehensive discussion of all these topics, we invite the interested reader to consult several recent reviews for additional information (Rosenberg and Rao 2021; Seguella and Gulbransen 2021; Boesmans et al. 2022).

ENTERIC GLIAL DEVELOPMENT AND DIVERSIFICATION

In most vertebrates, the ENS is formed by vagal neural crest-derived progenitors that migrate into and populate the developing bowel, with a minor potential contribution from the sacral crest that is restricted to the distal colon. These bipotential ENS progenitors proliferate and give rise to both neurons and glia, building the ENS from the "outside-in," starting with the myenteric plexus followed by the submucosal plexus (Pham et al. 1991). Clonally related cells can thus

Cite this article as *Cold Spring Harb Perspect Biol* doi: 10.1101/cshperspect.a041368

be found organized in columns along the radial axis of the intestine (Lasrado et al. 2017). Cells expressing the calcium-binding protein S100B, a marker of enteric glia, are detectable in the gut as early as embryonic day 14.5 (E14.5) in mice and 7 postconceptual weeks (PCW) in humans (Young et al. 2003; Fu et al. 2004). In both species, glial expansion and differentiation continue well into the postnatal period as the digestive tract functionally matures. In some vertebrates, such as zebrafish, the presence, developmental origin, and molecular identity of glia in the ENS remain controversial. The zebrafish ENS consists of a network of single neurons, rather than dual ganglionated plexuses, and cells expressing markers characteristic of enteric glia in other vertebrates, such as *S100b* and *Gfap* (encoding Glial Fibrillary Acidic Protein), are variably detected at homeostasis (Hagström and Olsson 2010; Baker et al. 2019; El-Nachef and Bronner 2020; McCallum et al. 2020). Enteric neurons are present in invertebrates but generally have not been reported to have associated glia. Thus, this review focuses on the current understanding of enteric glial biology based on observations made primarily in rodents and humans.

Glial Differentiation and Cell Fate Specification in the ENS

Bipotential ENS progenitors in the fetal bowel are characterized by expression of the transcription factor PHOX2B (Paired-like Homeobox 2b), the transcription factor SOX10 (Sex Determining Region Y Box 10), and the receptor tyrosine kinase RET (Rearranged during Transfection). Loss-of-function mutations in the genes encoding any of these proteins lead to congenital aganglionosis, in which the ENS fails to develop in most of the bowel (Schuchardt et al. 1994; Southard-Smith et al. 1998; Pattyn et al. 1999). Genetically defined $Sox10^+$ progenitors in the E12.5 mouse ENS exhibit three transcriptional types that represent putative bipotential, neurogenic, and gliogenic trajectories (Lasrado et al. 2017). Consistent with this possibility, lineage tracing of these $Sox10^+$ cells reveals that some give rise to only neurons, some to only glia, and some to both cell types in the adult bowel (Lasrado et al. 2017).

Disrupting *Ret* in these progenitors diminished neuronal, but not glial, differentiation (Lasrado et al. 2017). Neural crest progenitors elsewhere in the peripheral nervous system (PNS) must maintain *Sox10* expression to generate glia and downregulate it to generate neurons (Kim et al. 2003). These observations converge to suggest that bipotential ENS progenitors down-regulate *Ret* and maintain *Sox10* expression to generate glia. Like SOX10, FOXD3 (Forkhead Box D3) is a transcription factor widely expressed by ENS progenitors, required for their multipotency, and then down-regulated by differentiating neurons while maintained in glia. Whereas *Foxd3* deletion in migrating neural crest cells causes congenital aganglionosis, disruption in postmigratory vagal neural crest derivatives is less severe and associated with defects in both glial differentiation and maintenance (Teng et al. 2008; Mundell et al. 2012). Conversely, disruption of other pathways such as Hedgehog/Notch or the transcription factor NR2F1 (Nuclear Receptor Subfamily 2 Group F Member 1) is sufficient to cause premature gliogenesis at the expense of enteric neurons (Ngan et al. 2011; Bergeron et al. 2016). While these and other factors are individually revealed, an overarching framework for the transcriptional programs that direct enteric glial differentiation remains to be defined.

Glial Heterogeneity in the ENS

The term "enteric glia" encompasses a diverse group of cells distributed across the length and breadth of the gut. These glia exhibit clear morphological, transcriptional, and phenotypic heterogeneity that has led to several proposed classifications (Fig. 2). Early dye-filling studies of individual enteric glia revealed extensively branched "protoplasmic" glia within myenteric ganglia and less branched, more elongated "fibrous" glia in nerve fiber tracts between ganglia, classified as type I and type II, respectively (Hanani and Reichenbach 1994). This initial morphological classification was later broadened to four types including multipolar glia with a limited number of branches, found mostly in the mucosa (type III), and glia with small central soma that extend long, thin, bipolar projections, most-

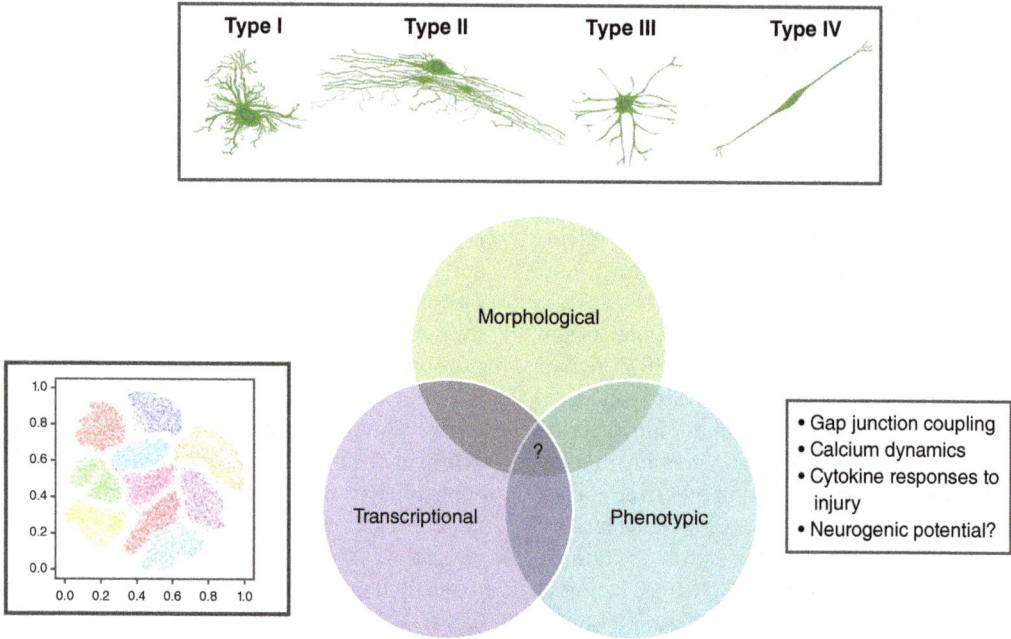

Figure 2. Enteric glia are transcriptionally, morphologically, and phenotypically diverse. The images of four morphological types of glia were created from data in Gulbransen and Sharkey (2012).

ly found along small nerve fibers in the muscle layers (type IV) (Gulbransen and Sharkey 2012). These four morphological types can be found throughout the esophagus and the small and large intestines (Boesmans et al. 2015; Rao et al. 2015; Kapitza et al. 2021).

In addition to dramatically different morphologies, enteric glia also inhabit very distinct microenvironments across the radial axis of the gut. Another potential classification scheme distinguishes enteric glia based on their locations: intraganglionic glia in the myenteric and submucosal ganglia that closely appose neuronal soma, interganglionic glia located along nerve fiber bundles connecting ganglia, intramuscular glia that tightly associate with nerve fibers innervating the smooth muscle syncytium, and mucosal glia located in the immune cell-rich milieu of the lamina propria in association with mucosal nerves. Recent single-cell RNA-sequencing (scRNA-seq) studies of the mouse and human ENS, largely focused on the myenteric plexus, have identified two to seven distinct transcriptional types of enteric glia depending on region,

developmental stage, and species examined (Zeisel et al. 2018; Drokhlyansky et al. 2020; Elmentaite et al. 2021; Fawkner-Corbett et al. 2021; Progatzky et al. 2021). Further work will be necessary to understand how these transcriptional phenotypes map onto the morphological and location-based classification schemes. Taken together, the transcriptional data in combination with immunohistochemical analysis of individual molecular markers suggest that *Sox10*, *S100b*, and *Plp1* (encoding Proteolipid Protein 1) are generally expressed by all classes of enteric glia. In mice, *Lpar1* (encoding Lysophosphatidic Acid Receptor 1) is another strong candidate (Grubišić and Gulbransen 2022). *Gfap*, in contrast, is robustly expressed within the myenteric plexus, but is not consistently detected in mucosal or intramuscular glia (Jessen and Mirsky 1983; Rao et al. 2015). *Gfap* levels seem to fluctuate the most with cell and tissue state (Boesmans et al. 2015) and may be generally unreliable for marking enteric glia in human tissue (Kinchen et al. 2018; Drokhlyansky et al. 2020; Baidoo et al. 2022). This reactivity, however, may render *Gfap* particularly useful in the context of tissue inflammation.

Regardless of how the morphological, spatial, and transcriptional classifications of enteric glia overlap, there is evidence for functional heterogeneity that will need to be integrated into these frameworks (Fig. 2). Dye-filling studies first revealed that even within a single myenteric ganglion, some glia are coupled by gap junctions and some are not (Hanani et al. 1989). Subsequent work has shown that myenteric glia exhibit different types of calcium responses to purinergic stimulation (Boesmans et al. 2015), and some of these responses are functionally coupled to distinct types of circuits (Ahmadzai et al. 2021). Emerging reports indicate that intestinal injury or infection can provoke the emergence of transcriptionally distinct populations of glia in the mucosa and the myenteric plexus that are important for epithelial and immune responses, respectively (Progatzky et al. 2021; Baghdadi et al. 2022). In all of these cases, it is unclear to what extent this functional heterogeneity reflects differing phenotypic states of enteric glia in response to local signals versus different types of enteric glia that have terminally differentiated to perform fundamentally different tasks.

Schwann Cell Precursors, Enteric Glia, and Neurogenesis in the ENS

The vertebrate ENS is populated not just by neural crest–derived progenitors, but also by Schwann cell precursors (SCPs) associated with extrinsic peripheral nerves that innervate the bowel (Uesaka et al. 2015, 2021; El-Nachef and Bronner 2020; Soret et al. 2020). During development, these SCPs give rise to a minor population of submucosal neurons in the small intestine but contribute nearly 20% of all neurons in the mouse colon (Uesaka et al. 2015). Unlike vagal neural crest–derived ENS progenitors, neurogenic SCPs do not seem to pass through a bipotential state (Uesaka et al. 2015). Nevertheless, SCPs can contribute to enteric glial populations in the context of some developmental defects in the ENS (Soret et al. 2020); but it remains unclear to what extent they do so at steady state. Regardless, SCPs express both *Sox10* and *Plp1*, and thus

at least some proportion of cells identified as enteric glia based on these two markers could be neurogenic SCPs.

There is growing evidence that there is neurogenesis in the adult ENS in specific conditions, and that enteric glia and/or SCPs are the cellular sources of new neurons. In adult mice, while the degree of ongoing neurogenesis in the healthy intestine is hotly debated (Kulkarni et al. 2017; Virtanen et al. 2022), available evidence indicates that ~3% of enteric glia are actively cycling (Joseph et al. 2011). While these glia readily generate neurons in vitro, they rarely do so in vivo (Joseph et al. 2011; Laranjeira et al. 2011). In adult zebrafish, nonneuronal cells expressing *Sox10*, *Plp1b*, and *Foxd3* closely associate with enteric neurons and exhibit many ultrastructural features of mammalian enteric glia (McCallum et al. 2020). Unlike their mammalian counterparts, however, more of these cells are cycling and readily give rise to neurons in vivo (McCallum et al. 2020). While neurogenesis is limited in the mouse gut at baseline, it can be provoked by tissue injury or congenital defects and both enteric glia and SCPs can contribute to neurogenesis under these conditions (Belkind-Gerson et al. 2017; Soret et al. 2020; Uesaka et al. 2021). Consistent with the possibility that enteric glia are "on call" to generate neurons in times of need, emerging epigenetic studies suggest that the chromatin state of some mature enteric glia is similar to that of bipotential ENS progenitors and remains poised for neuronal differentiation (Guyer et al. 2022; Laddach et al. 2023).

INTERACTIONS BETWEEN ENTERIC GLIA AND NEURONS

Enteric glia surround neuron cell bodies and processes in the ENS. Glia are always associated with neurons and are never found without being linked to a neural structure. There are numerous points of direct contact between enteric neurons and glia, whereas direct contacts between glia are rare (Gabella 2022). The apparent inseparable anatomical relationship between the two cell types reflects intimate functional ties. Ongoing communication between glia and enteric neurons supports neuronal health, neurotransmission, and modulates neural circuits in the digestive tract (Seguella and Gulbransen 2021). Most of the known mechanisms of interaction between enteric glia and neurons have been identified in mice and are directed at supporting cellular and organ-level homeostasis. These mechanisms can be broadly grouped into two main modes of interaction reflecting supportive roles in which glia supply neurons with metabolic substrates and regulate their extracellular environment and active signaling roles where glia detect transmitters released by neurons and, in turn, release transmitters to modulate neurotransmission (Fig. 3).

Supportive Roles

Neurotransmitter Synthesis

Immunolabeling and genomic data suggest that enteric glia make precursors for the synthesis of neurotransmitters including nitric oxide (NO), glutamate, and γ-aminobutyric acid (GABA) (Jessen and Mirsky 1983; Aoki et al. 1991; Naga-

hama et al. 2001; Delvalle et al. 2018a; Drokhlyansky et al. 2020; Progatzky et al. 2021). NO is one of the main transmitters used by inhibitory motor neurons in the gut and GABA plays a role in modulating enteric neurocircuit activity (Fung and Berghe 2020). The functional significance of glutamate in the normal physiology of the ENS is not clear. Enteric neurons have the capacity to synthesize these and other neurotransmitters themselves. Therefore, it is likely that glia supplement neurotransmitter synthesis but are not required to fully sustain neurotransmission.

Neurotransmitter Degradation

Enteric glia play a central role in regulating the availability of extracellular neuroactive substances such as adenosine triphosphate (ATP), histamine, and GABA. Glia express high levels of the cell surface enzyme nucleoside triphosphate diphosphohydrolase-2 (NTPDase2), which generates nucleosides from the hydrolysis of ATP (Braun

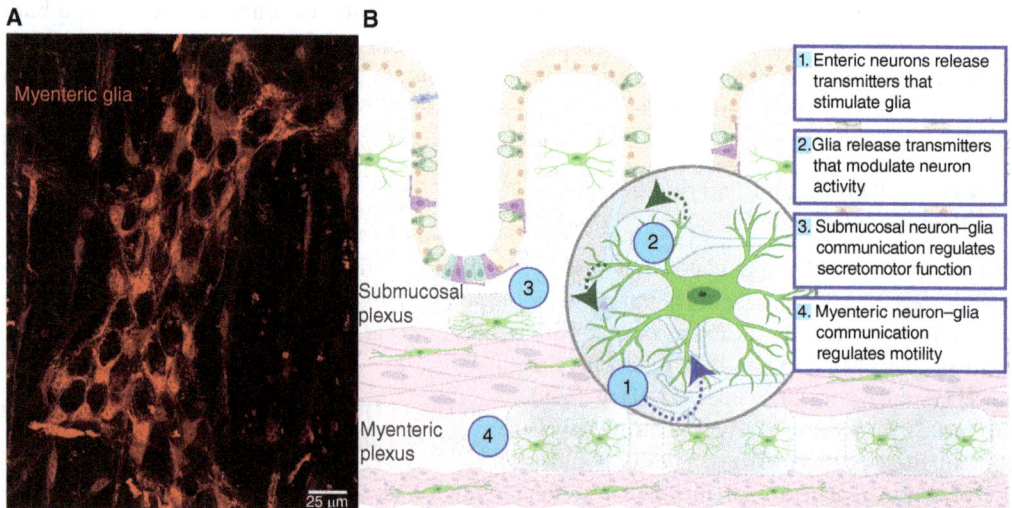

Figure 3. Intercellular communication between enteric glia and neurons modulates gut motility and secretomotor functions. (A) Image of a single myenteric ganglion from the colon of a transgenic mouse expressing the red fluorescent reporter tdTomato (red) in enteric glia. Note that tdTomato+ enteric glia surround enteric neuron cell bodies, which appear as larger unlabeled oval spaces within the ganglion. (B) Schematic diagram depicting neuron-to-glia and subsequent glia-to-neuron communication involved in the control of gut motility and secretions. Enteric glia detect most neuroactive substances in the enteric nervous system (ENS) and specific neurotransmitters that act on glia differ between cells, synaptic circuits, and gut regions. Acetylcholine and purines are prominent neurotransmitters involved in fast synaptic communication in myenteric circuits that stimulate glia. Neuroactive substances produced by enteric glia are also diverse and include purines, γ-aminobutyric acid (GABA), S100B, and endocannabinoids.

et al. 2004; Lavoie et al. 2011; Grubišić et al. 2019). Enteric glia also express histamine N-methyltransferase (HNMT) (McClain et al. 2020), the main enzyme responsible for histamine metabolism in the nervous system, and GAT2, a transporter responsible for removing GABA from the extracellular space (Fletcher et al. 2002). Enteric neurons express NTPDase3 and GAT3 and contribute to the removal of ATP and GABA. The relative contributions of enteric neurons and glia to neurotransmitter removal are still unclear and are likely context dependent, with glial mechanisms playing a prominent role during challenges such as inflammation (Grubišić et al. 2019).

Neuron Survival and Maturation

Enteric glia synthesize several neuroprotective compounds involved in cellular defense such as the prostaglandin derivative 15-deoxy-prostaglandin-J2 (Abdo et al. 2012), multiple neurotrophins (Ibiza et al. 2016; Delvalle et al. 2018a; Kovler et al. 2021), and antioxidants including glutathione (GSH) and its derivative S-nitrosoglutathione (Savidge et al. 2007; Bach-Ngohou et al. 2010; Brown and Gulbransen 2018). Reduced GSH is a prominent endogenous antioxidant that is necessary for neuroprotection in the ENS. Enteric glia are the main site of GSH synthesis in the ENS; although enteric neurons also express the cellular machinery for GSH synthesis and may contribute to GSH production to some extent (Brown and Gulbransen 2018). In addition to promoting survival, glial-derived factors are important cues in enteric neural network maturation (Berre-Scoul et al. 2017). These processes have been studied in cell culture systems, which indicate that enteric glia secrete neurotrophins that increase the complexity and density of neuronal synapses in the ENS. Whether similar processes influence ENS maturation in vivo is unknown.

Active Signaling Roles

Neuron-to-Glia Communication

Enteric glia express receptors for, and are responsive to, most neurotransmitters used in the ENS (Seguella and Gulbransen 2021). These include acetylcholine (ACh), ATP, tachykinins, serotonin (5-HT), histamine, and lipid mediators. Many glial receptors for neuroactive compounds are G-protein-coupled receptors (GPCRs) (Kimball and Mulholland 1996), although there is evidence for some expression of ligand-gated ion channels as well (Schneider et al. 2021). Neurotransmitters acting on glial Gq-coupled GPCRs lead to cellular activity in the form of intracellular calcium (Ca^{2+}) responses (Kimball and Mulholland 1996; Gulbransen and Sharkey 2009; Broadhead et al. 2012). Glial Ca^{2+} responses can be driven through GPCR signaling, influx from the extracellular space through membrane channels, or a combination and summation of both (Kimball and Mulholland 1996; Sarosi et al. 1998; Gulbransen and Sharkey 2009; Schneider et al. 2021). Ca^{2+} responses appear to be the major mode of encoding activity in enteric glia. However, some transmitters such as adenosine act on glial Gs-coupled receptors and elicit intracellular signaling mediated by cAMP (Christofi et al. 1993; Grubišić et al. 2022). Enteric glia are electrically passive cells and do not exhibit action potentials similar to those in neurons that are driven by sodium and potassium gradients (Hanani et al. 2000; Bhave et al. 2017).

Enteric glia respond to synaptic activity within enteric neurocircuits (Gomes et al. 2009; Gulbransen and Sharkey 2009; Broadhead et al. 2012) and to the activity of extrinsic neurons that innervate the gut from dorsal root (Delvalle et al. 2018a) and sympathetic ganglia (Gulbransen et al. 2010). Vagal nerve activity affects enteric glial morphology, but it is unclear whether vagal nerves signal directly to enteric glia or through indirect mechanisms (Costantini et al. 2010). Neuron–glial junctions are abundant in myenteric ganglia when observed by electron microscopy and are more frequently encountered than neuron–neuron synapses in the ENS (Cook and Burnstock 1976; Gabella 1981, 2022). This suggests that enteric glia are specifically targeted by neural pathways as opposed to only passively detecting neurotransmitter overflow from neuronal synapses. Subpopulations of enteric glia are functionally associated with specific pathways in enteric neurocircuits and respond to and modulate neuron activity on a synapse-by-synapse basis (Ahmadzai et al. 2021). Multipolar intrinsic primary afferent neurons activate a large pool of enteric glia, whereas unipolar ascending and descending interneurons activate smaller, discrete pools of enteric glia. It is currently unclear how glial specificity is formed in the ENS and whether functional subtypes display unique genetic profiles or anatomical specializations.

It should be noted that evidence for the above is based on studies in the colon and ileum of rodents such as guinea pigs and mice. The extent to which neuron–glia signaling occurs in other regions of the digestive tract, the types of neurons and transmitters are involved, and the nature of these mechanisms in humans are currently unknown.

Glia-to-Neuron Communication

Enteric glia are active signaling nodes in the ENS and glial activity can result in the release of a number of neuroactive compounds such as ATP, prostaglandin E2 (PGE2), and GABA. These enteric gliotransmitters are released downstream from intracellular Ca^{2+} responses but the mechanisms differ. ATP release is mediated by the opening of membrane channels composed of connexin-43 subunits (McClain et al. 2014; Brown et al. 2016). Connexin-43 hemichannels are a major conduit of transmitter release from enteric glia that are permeable molecules such as ATP, which have direct and indirect effects on neural activity through influences on neurons or neighboring glia. Glial GABA release appears to be an indirect consequence of ATP release through connexin-43 hemichannels. ATP is a negatively charged molecule that increases membrane potential when released, leading to the reversal of the GABA transporter GAT2 and GABA efflux (Fried et al. 2017). PGE2 release can also be evoked downstream from glial Ca^{2+} responses and requires connexin-43 hemichannels, but the specific release mechanisms have not been defined (Murakami et al. 2007; Morales Soto et al. 2023).

Glial activity and gliotransmitter release affect enteric neuron activity and have subsequent effects on gut motor and secretomotor functions. Ca^{2+} imaging experiments in which enteric glia express chemogenetic receptors show that stimulating glial Gq signaling and intracellular Ca^{2+} responses elicits activity in the surrounding enteric neurons (Ahmadzai et al. 2021). These effects are also observed in response to glial activation by native ligands such as the purine adenosine diphosphate (ADP) and the biolipid lysophosphatidic acid (LPA) (Ahmadzai et al. 2021, 2022). The effects of glial activity on neurons differ depending on the glial receptor pathway engaged and the type and synaptic connections of the neuron. Glial activation by purines seems to potentiate the activity of neurons in ascending motor pathways while glial activation by ACh has an inhibitory effect on surrounding neural pathways (Ahmadzai et al. 2021). Glial stimulation through LPAR1 also causes subsequent activity in a subset of enteric neurons, which results in reduced neuromuscular contractions (Ahmadzai et al. 2022).

Glial activity produces excitatory effects on gut motor and secretomotor functions through effects on enteric neurons. Stimulating glial Ca^{2+} responses through Gq receptor pathways evokes contractions in segments of rodent colon and ileum and enhances ongoing neurogenic motor programs referred to as colonic motor complexes (McClain et al. 2015; Delvalle et al. 2018b). These effects are mediated through neural pathways and the effects of glial stimulation on motor function are lost when neuronal action potentials are blocked with tetrodotoxin. Impairing gliotransmission by eliminating glial connexin-43 hemichannels has the converse effect and reduces the efficacy of neuromuscular transmission in the gut (McClain et al. 2014). Impaired neuromuscular transmission is also observed in models in which glial metabolism is disrupted (Nasser et al. 2006) or when glia are ablated (Rao et al. 2017; Kovler et al. 2021). Glial stimulation also has an excitatory effect on secretomotor function that is mediated through effects on enteric neurons (Grubišić and Gulbransen 2017). Vasointestinal polypeptide (VIP), a neurotransmitter involved in secretomotor responses, activates a subset of submucosal neurons that evokes responses in the surrounding enteric glia through purinergic mechanisms (Fung et al. 2017). Glial activation, in turn, promotes activity in secretomotor neurons and drives gut secretions (Grubišić and Gulbransen 2017).

Cite this article as *Cold Spring Harb Perspect Biol* doi: 10.1101/cshperspect.a041368

GLIAL INTERACTIONS WITH NONNEURONAL CELLS IN THE GUT ENVIRONMENT

The cellular composition of the gut microenvironment varies dramatically along the radial axis of the gut, providing a host of nonneuronal interaction partners for enteric glia both within and outside the plexuses. These interactions have major impacts on gut physiology and pathophysiology, which may be independent of glial–neuronal interactions.

Gut Microbes

While enteric glia have not been shown to extend across the epithelium to contact microbes in the gut lumen, their development and activities are influenced by the microbiome. Glia in germ-free mice fail to fully populate the mucosa of the small intestine despite the normal appearance of glia in the myenteric and submucosal plexuses; microbial repletion is sufficient to rescue this deficit (Kabouridis et al. 2015). These observations suggest that mucosal glia are developmentally distinct from other enteric glia and require the presence of microbes to establish a normal population. In adult mice, mucosal glia in the small intestine are continuously repopulated by glia that arise from *Sox10*- and *Gfap*-expressing progenitors located in the myenteric plexus (Kabouridis et al. 2015). It is unclear whether gut microbes are necessary for this homeostatic repopulation. One study suggested that broad-spectrum antibiotic administration depletes mucosa glia (Kabouridis et al. 2015), another suggested that it does so in mice but not humans (Inlender et al. 2021), while a third showed no change in mucosal glia but a small deficit in the myenteric plexus (Vicentini et al. 2021). The microbial burden is orders of magnitude higher in the colon than the small intestine, yet microbial depletion has no effect on glial density in the mucosa or the plexuses despite widespread effects on enteric neurons (Vicentini et al. 2021). Furthermore, unlike in mice, $SOX10^+/S100B^+$ cells are detectable in the mucosa early in fetal development (Fawkner-Corbett et al. 2021), suggesting that there are likely region-, host species-, and microbial species-specific dif-ferences in the impact of microbes on enteric glial development that need further investigation.

Enteric glia are equipped to respond to microbial signals both directly and indirectly. In mice and humans, enteric glia express Toll-like receptors (TLRs), proteins that detect microbe-associated molecular patterns (MAMPs) (Esposito et al. 2014; Turco et al. 2014). Primary human myenteric glia exposed to bacterial lipopolysaccharide, a TLR ligand, up-regulate a variety of transcripts for proinflammatory cytokines (Liñán-Rico et al. 2016). Conversely, interfering with the downstream effector of TLR signaling, immune adaptor protein MYD88 (Myeloid Differentiation Primary Response 88), in *Gfap*-expressing cells impairs cytokine release from some immune cells and alters glial phenotypes in response to a high-fat diet (Ibiza et al. 2016; Liu et al. 2023). As reviewed in detail above and below, enteric glia interact with neurons, immune effector cells, and epithelial cells, all of which are affected by the gut microbiota. Thus, there are many pathways by which microbial signals could indirectly influence glial function.

Epithelial Barrier and Intestinal Stem Cell Regulation

The simple columnar epithelium of the GI tract represents the largest interface between the mammalian host and the outside environment. The close approximation of glial processes to gut epithelial cells, particularly at the level of crypts that house intestinal stem cells, suggests that glia might regulate epithelial barrier integrity and turnover. Studies using the herpes simplex virus thymidine kinase (HSV-TK) system or the adaptive immune system to target *Gfap*-expressing cells for ablation suggested that enteric glial depletion disrupted epithelial barrier integrity and crypt cell proliferation, causing severe inflammation and impaired repair (for review, see Rosenberg and Rao 2021; Seguella and Gulbransen 2021; Prochera and Rao 2023). Recent work using more cell-autonomous approaches to target *Plp1*- or *Sox10*-expressing cells for ablation or functional disruption, however, have not supported these observations. Instead, they have shown that depleting glia is not sufficient to cause inflammation and that enteric glia are dispensable for epithelial barrier regulation and epithelial turnover at homeostasis in vivo (Grubišić and Gulbransen 2017; Rao et al. 2017; Yuan et al. 2020; Baghdadi et al. 2022). One study using a human *GFAP* promoter transgene to target cells for ablation in mice suggested a small *Gfap*[+] subpopulation of mucosal glia is critical for regulating epithelial proliferation but did not examine whether transgene expression was restricted to glia (Baghdadi et al. 2022). Enteric glia secrete a number of soluble factors that can promote epithelial health in vitro including growth factors and small molecules—their role in epithelial regulation in vivo may become most impactful upon specific types of injury (Neunlist et al. 2007; Pochard et al. 2016; Grubišić et al. 2022). Necrotizing enterocolitis, a fulminant condition in neonates characterized by hypoxia and dysbiosis, as well as radiation-induced enteritis are both exacerbated by glial loss (Kovler et al. 2021; Baghdadi et al. 2022). These phenotypes are linked to deficiency of glial-derived signals including brain-derived neurotrophic factor (BDNF) and secreted Wnts, respectively. In contrast, acute or chronic colitis caused by chemical injury is unaffected by glial cell depletion (Rao et al. 2017; Yuan et al. 2020), suggesting that the large intestinal epithelium may be less dependent on glia. Neuronal function was not interrogated in depth, if at all, in these injury models. Given the close partnership between enteric glia and neurons reviewed above and the significance of neuronal inputs for epithelial function (Sharkey et al. 2018), future work probing how glial disruption alters neuro-epithelial communication will be necessary to resolve to what extent observations in these injury models reflect direct glial interactions with the epithelium.

ENTERIC GLIA IN GUT PATHOPHYSIOLOGY

Response to Acute Inflammation

Common gastrointestinal disorders such as inflammatory bowel disease (IBD) and irritable bowel syndrome (IBS) involve neuroplastic changes in the ENS that lead to long-lasting changes in gut function. Enteric neuroplasticity

is driven in many cases by acute or chronic inflammation. Enteric glia play a central role in ENS responses to inflammation and neuroinflammatory processes (Fig. 4). Enteric glia respond to mediators of inflammation such as purines, cytokines, and bacterial components (Gulbransen and Sharkey 2009; Esposito et al. 2014; Stoffels et al. 2014) and, in turn, produce proinflammatory cytokines such as IL-6, IL-1β, and CXCL10 (Rühl et al. 2001; Liñán-Rico et al. 2016; Progatzky et al. 2021) as well as other proinflammatory factors such as S100B, ATP, and PGE2 (Esposito et al. 2007; Murakami et al. 2007; Brown et al. 2016). Enteric glia also exert anti-inflammatory effects by secreting mediators such as *S*-nitrosoglutathione (Savidge et al. 2007), omega-6 fatty acid derivatives (Abdo et al. 2012), polyunsaturated fatty acid metabolites (Pochard et al. 2016), and cytokines/chemokines (Progatzky et al. 2021; Stakenborg et al. 2022). The specific response and phenotype induced in enteric glia largely depends on the nature, severity, and duration of the pathological insult (Seguella and Gulbransen 2021). Glial activation by purines induces a reactive-type phenotype that includes changes to gene expression, morphology, and function. The acute response of glia to neuronal purine release under proinflammatory conditions includes activation of Ca^{2+} responses, NO production from inducible nitric oxide synthase (iNOS), and purine release through connexin-43 hemichannels (Brown et al. 2016; Delvalle et al. 2018a). Under these conditions, glial purine release promotes neuroinflammation and leads to neurodegeneration by activating neuronal P2X7 purine receptors (Gulbransen et al. 2012; Brown et al. 2016).

Enteric glia exert broad effects on inflammatory responses through interactions with immune cells. Immune cell subsets influenced by enteric glia include T lymphocytes, macrophages, innate lymphoid cells, mast cells, and neutrophils. Enteric glia influence T lymphocyte proliferation (Kermarrec et al. 2016) and enteric glia promote the differentiation of T-cell subsets involved in immune tolerance through antigen presentation mechanisms during acute inflammation (Chow et al. 2021). Muscularis macrophage activation involves intercellular signaling with enteric glia mediated by the cytokine macrophage colony-

Cite this article as *Cold Spring Harb Perspect Biol* doi: 10.1101/cshperspect.a041368

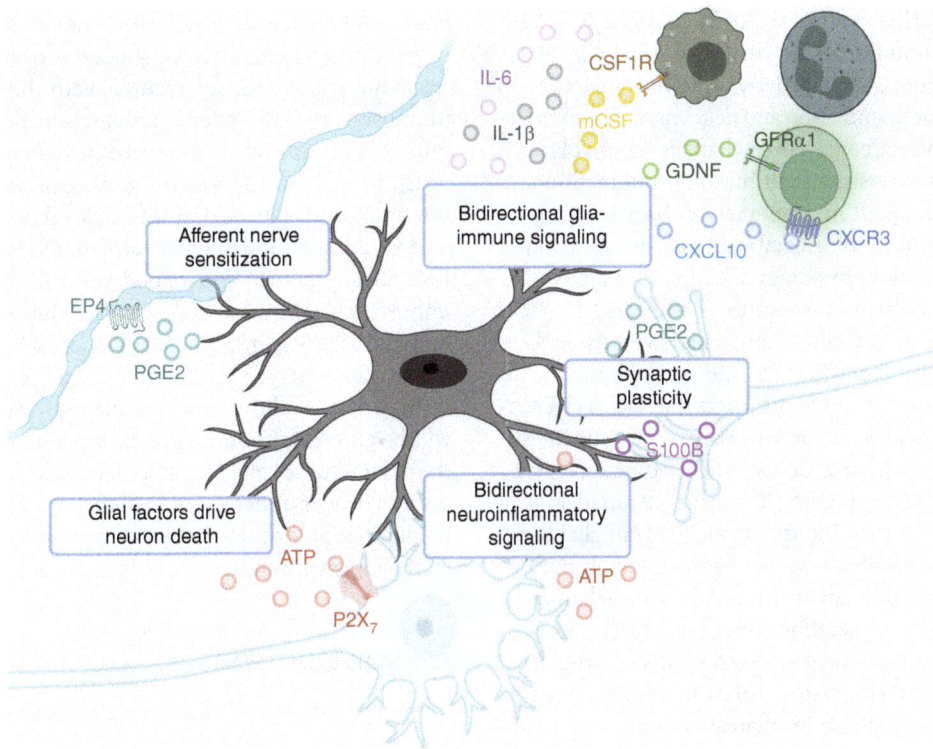

Figure 4. Enteric glia and neuroinflammation. Schematic diagram depicting known roles of enteric glia in neuroinflammatory processes in the enteric nervous system. Enteric glia secrete factors including chemokine (C-X-C motif) ligand 10 (CXCL10), macrophage colony-stimulating factor (mCSF), and glial-derived neurotrophic factor (GDNF) that influence immune responses. Proinflammatory cytokines such as IL-6 and IL-1β are secreted by both glia and immune cells and mediate bidirections signaling. Other glial signals such as prostaglandin E2 (PGE2) are involved in sensitizing intrinsic and extrinsic neurons in the gut. Similarly, S100B is a glial-derived mediator that can modulate neuron activity. Adenosine triphosphate (ATP) is a signal produced by both enteric neurons and glia and contributes to the effects of enteric neuroinflammation.

stimulating factor (mCSF/CSF1) (Grubišić et al. 2020), which is also involved in generating anti-inflammatory macrophages in response to tissue damage (Stakenborg et al. 2022). Immune constraining actions of glia also occur through interactions with type 3 innate lymphoid cells mediated by glial-derived neurotrophic factor (GDNF) (Ibiza et al. 2016) and by secreting the chemokine (C-X-C motif) ligand 10 (CXCL10) (Progatzky et al. 2021). However, glial–immune interactions can also have detrimental actions such as those driven by the glial CXCL1-dependent recruitment of neutrophils that contributes to gut barrier function during the resolution of colitis (Grubišić et al. 2022). Enteric glial–mast cell interactions have also been proposed to have detrimental actions

in disease states (McClain et al. 2020; de-Faria et al. 2021), but the nature of the influence of glia on mast cells remains undefined.

Glia in Disease States

Changes to glial survival, phenotype, and function are associated with nearly every pathological condition in the gut. Glia are activated by bacterial (Turco et al. 2014; Rosenbaum et al. 2016; Chow et al. 2021), viral (Selgrad et al. 2009; Esposito et al. 2017), and helminth infections (Progatzky et al. 2021) and contribute to enacting and regulating defensive immune responses. Likewise, diseases with underlying inflammatory components such as IBD (Liñán-Rico et al.

2016), IBS (Lilli et al. 2018), and obesity (Stenkamp-Strahm et al. 2013; Seguella et al. 2021) promote glial reactivity, which contributes to neuroinflammation and neuroplasticity in enteric neurocircuits. These changes are involved in driving persistent gut motility dysfunction and visceral pain in a number of disorders of gut–brain interaction. Glial reactivity also contributes to the pathophysiology of neurodegenerative diseases such as Parkinson's (Thomasi et al. 2022) but its role in these disorders is currently unclear. Glia also contribute to the pathophysiology of severe gut motility disorders that lack a clear inflammatory component such as chronic intestinal pseudo-obstruction and postoperative ileus. In these conditions, either a loss (Ahmadzai et al. 2022) or gain (Stoffels et al. 2014) of glial functions contributes to changes in the neuromuscular control of gut function. A loss of glial function or glial degeneration contributes to the development of necrotizing enterocolitis (Kovler et al. 2021). In contrast, a glial gain of function seems to contribute to mechanisms underlying carcinogenesis through interactions with stem cells. Enteric glia may influence the activity and/or maturation of intestinal stem cells and epithelial cells under normal conditions (Neunlist et al. 2007; Baghdadi et al. 2022) but are not normally protumorigenic. However, factors produced in the tumor microenvironment transform glia into a protumorigenic phenotype that stimulates further tumorigenesis (Valès et al. 2019; Yuan et al. 2020). Understanding how glia contribute to the pathogenesis of diverse intestinal diseases is an active area of research and holds great potential for developing new therapies.

CONCLUDING REMARKS

A growing body of evidence supports the concept that enteric glia are active signaling nodes that coordinate the actions of multiple cell types in the gut. Processes mediated by enteric glia impact gut functions through effects on neural network activity and immune signaling. Defects in these processes contribute to disease through diverse mechanisms. Despite mounting support for the importance of enteric glia in gut functions and disease, many questions remain regarding the basic mechanisms involved. How glial diversity impacts organ function throughout the digestive tract through unique interactions with the immune system, bidirectional cross talk with the microbiome, and roles in normal neural circuit function and neuroplasticity in disease are all important active areas of research. Mechanisms underlying the fundamental roles of glia in the ENS remain poorly understood and will be an important area for the field to tackle. The development of new genetic tools and approaches to interrogate enteric glial functions without perturbing glia in the rest of the nervous system will be crucial for addressing these gaps. Given the central role of glia in gut homeostasis, developing a deeper understanding of their functions holds a vast potential to harness the power of glia to benefit gastrointestinal health.

ACKNOWLEDGMENTS

Biorender.com was used to generate some of the schematics.

REFERENCES

Abdo HH, Mahé MM, Derkinderen P, Bach-Ngohou K, Neunlist M, Lardeux B. 2012. The omega-6 fatty acid derivative 15-deoxy-$\Delta^{12,14}$-prostaglandin J2 is involved in neuroprotection by enteric glial cells against oxidative stress. *J Physiol* **590:** 2739–2750. doi:10.1113/jphysiol.2011.222935

Ahmadzai MM, Seguella L, Gulbransen BD. 2021. Circuit-specific enteric glia regulate intestinal motor neurocircuits. *Proc Natl Acad Sci* **118:** e2025938118. doi:10.1073/pnas.2025938118

Ahmadzai MM, McClain JL, Dharshika C, Seguella L, Giancola F, De Giorgio R, Gulbransen BD. 2022. LPAR1 regulates enteric nervous system function through glial signaling and contributes to chronic intestinal pseudo-obstruction. *J Clin Invest* **132:** e149464. doi:10.1172/JCI149464

Aoki E, Semba R, Kashiwamata S. 1991. Evidence for the presence of L-arginine in the glial components of the peripheral nervous system. *Brain Res* **559:** 159–162. doi:10.1016/0006-8993(91)90300-K

Bach-Ngohou K, Mahé MM, Aubert P, Abdo HH, Boni S, Bourreille A, Denis MG, Lardeux B, Neunlist M, Masson D. 2010. Enteric glia modulate epithelial cell proliferation and differentiation through 15-deoxy-$\Delta^{12,14}$-prostaglandin J2. *J Physiol* **588:** 2533–2544. doi:10.1113/jphysiol.2010.188409

Baghdadi MB, Ayyaz A, Coquenlorge S, Chu B, Kumar S, Streutker C, Wrana JL, Kim TH. 2022. Enteric glial cell

Cite this article as *Cold Spring Harb Perspect Biol* doi: 10.1101/cshperspect.a041368

heterogeneity regulates intestinal stem cell niches. *Cell Stem Cell* **29:** 86–100.e6. doi:10.1016/j.stem.2021.10.004

Baidoo N, Sanger GJ, Belai A. 2022. Effect of old age on the subpopulations of enteric glial cells in human descending colon. *Glia* **71:** 305–316. doi:10.1002/glia.24272

Baker PA, Meyer MD, Tsang A, Uribe RA. 2019. Immuno-histochemical and ultrastructural analysis of the maturing larval zebrafish enteric nervous system reveals the formation of a neuropil pattern. *Sci Rep* **9:** 6941. doi:10.1038/s41598-019-43497-9

Belkind-Gerson J, Graham HK, Reynolds J, Hotta R, Nagy N, Cheng L, Kamionek M, Shi HN, Aherne CM, Goldstein AM. 2017. Colitis promotes neuronal differentiation of Sox2+ and PLP1+ enteric cells. *Sci Rep* **7:** 2525. doi:10.1038/s41598-017-02890-y

Bergeron KF, Nguyen CM, Cardinal T, Charrier B, Silversides DW, Pilon N. 2016. Upregulation of the Nr2f1-A830082K12Rik gene pair in murine neural crest cells results in a complex phenotype reminiscent of Waardenburg syndrome type 4. *Dis Model Mech* **9:** 1283–1293. doi:10.1242/dmm.026773

Berre-Scoul CL, Chevalier J, Oleynikova E, Cossais F, Talon S, Neunlist M, Boudin H. 2017. A novel enteric neuron-glia coculture system reveals the role of glia in neuronal development. *J Physiol* **595:** 583–598. doi:10.1113/JP271989

Bhave S, Gade A, Kang M, Hauser KF, Dewey WL, Akbarali HI. 2017. Connexin-purinergic signaling in enteric glia mediates the prolonged effect of morphine on constipation. *FASEB J* **31:** 2649–2660. doi:10.1096/fj.201601068R

Boesmans W, Lasrado R, Vanden Berghe P, Pachnis V. 2015. Heterogeneity and phenotypic plasticity of glial cells in the mammalian enteric nervous system. *Glia* **63:** 229–241.

Boesmans W, Nash A, Tasnády KR, Yang W, Stamp LA, Hao MM. 2022. Development, diversity, and neurogenic capacity of enteric glia. *Front Cell Dev Biol* **9:** 775102. doi:10.3389/fcell.2021.775102

Braun N, Sévigny J, Robson SC, Hammer K, Hanani M, Zimmermann H. 2004. Association of the ecto-ATPase NTPDase2 with glial cells of the peripheral nervous system. *Glia* **45:** 124–132. doi:10.1002/glia.10309

Broadhead MJ, Bayguinov PO, Okamoto T, Heredia DJ, Smith TK. 2012. Ca²⁺ transients in myenteric glial cells during the colonic migrating motor complex in the isolated murine large intestine. *J Physiol* **590:** 335–350. doi:10.1113/jphysiol.2011.219519

Brown IAM, Gulbransen BD. 2018. The antioxidant glutathione protects against enteric neuron death in situ, but its depletion is protective during colitis. *Am J Physiol Gastrointest Liver Physiol* **314:** G39–G52. doi:10.1152/ajpgi.00165.2017

Brown IAM, McClain JL, Watson RE, Patel BA, Gulbransen BD. 2016. Enteric glia mediate neuron death in colitis through purinergic pathways that require connexin-43 and nitric oxide. *Cell Mol Gastroenterol Hepatol* **2:** 77–91. doi:10.1016/j.jcmgh.2015.08.007

Chow AK, Grubišić V, Gulbransen BD. 2021. Enteric glia regulate lymphocyte activation via autophagy-mediated MHC-II expression. *Cell Mol Gastroenterol Hepatol* **12:** 1215–1237. doi:10.1016/j.jcmgh.2021.06.008

Christofi FL, Hanani M, Maudlej N, Wood JD. 1993. Enteric glial cells are major contributors to formation of cyclic AMP in myenteric plexus cultures from adult guinea-pig

small intestine. *Neurosci Lett* **159:** 107–110. doi:10.1016/0304-3940(93)90810-8

Cook RD, Burnstock G. 1976. The ultrastructure of Auerbach's plexus in the guinea-pig. II: Non-neuronal elements. *J Neurocytol* **5:** 195–206. doi:10.1007/BF01181656

Costantini TW, Bansal V, Krzyzaniak M, Putnam JG, Peterson CY, Loomis WH, Wolf P, Baird A, Eliceiri BP, Coimbra R. 2010. Vagal nerve stimulation protects against burn-induced intestinal injury through activation of enteric glia cells. *Am J Physiol Gastrointest Liver Physiol* **299:** G1308–G1318. doi:10.1152/ajpgi.00156.2010

de-Faria FM, Casado-Bedmar M, Lindqvist CM, Jones MP, Walter SA, Keita ÅV. 2021. Altered interaction between enteric glial cells and mast cells in the colon of women with irritable bowel syndrome. *Neurogastroenterol Motil* **33:** e14130. doi:10.1111/nmo.14130

Delvalle NM, Dharshika C, Morales-Soto W, Fried DE, Gaudette L, Gulbransen BD. 2018a. Communication between enteric neurons, glia, and nociceptors underlies the effects of tachykinins on neuroinflammation. *Cell Mol Gastroenterol Hepatol* **6:** 321–344. doi:10.1016/j.jcmgh.2018.05.009

Delvalle NM, Fried DE, Rivera-Lopez G, Gaudette L, Gulbransen BD. 2018b. Cholinergic activation of enteric glia is a physiological mechanism that contributes to the regulation of gastrointestinal motility. *Am J Physiol Gastrointest Liver Physiol* **315:** G473–G483. doi:10.1152/ajpgi.00155.2018

Drokhlyansky E, Smillie CS, Van Wittenberghe N, Ericsson M, Griffin GK, Eraslan G, Dionne D, Cuoco MS, Goder-Reiser MN, Sharova T, et al. 2020. The human and mouse enteric nervous system at single-cell resolution. *Cell* **182:** 1606–1622.e23. doi:10.1016/j.cell.2020.08.003

Elmentaite R, Kumasaka N, Roberts K, Fleming A, Dann E, King HW, Kleshchevnikov V, Dabrowska M, Pritchard S, Bolt L, et al. 2021. Cells of the human intestinal tract mapped across space and time. *Nature* **597:** 250–255. doi:10.1038/s41586-021-03852-1

El-Nachef WN, Bronner ME. 2020. De novo enteric neurogenesis in post-embryonic zebrafish from Schwann cell precursors rather than resident cell types. *Development* **147:** dev186619. doi:10.1242/dev.186619

Esposito G, Cirillo C, Sarnelli G, De Filippis D, D'Armiento FP, Rocco A, Nardone G, Petruzzelli R, Grosso M, Izzo P, et al. 2007. Enteric glial-derived S100B protein stimulates nitric oxide production in celiac disease. *Gastroenterology* **133:** 918–925. doi:10.1053/j.gastro.2007.06.009

Esposito G, Capoccia E, Turco F, Palumbo I, Lu J, Steardo A, Cuomo R, Sarnelli G, Steardo L. 2014. Palmitoylethanolamide improves colon inflammation through an enteric glia/Toll like receptor 4-dependent PPAR-α activation. *Gut* **63:** 1300–1312. doi:10.1136/gutjnl-2013-305005

Esposito G, Capoccia E, Gigli S, Pesce M, Bruzzese E, D'Alessandro A, Cirillo C, di Cerbo A, Cuomo R, Seguella L, et al. 2017. HIV-1 Tat-induced diarrhea evokes an enteric glia-dependent neuroinflammatory response in the central nervous system. *Sci Rep* **7:** 7735. doi:10.1038/s41598-017-05245-9

Fawkner-Corbett D, Antanaviciute A, Parikh K, Jagielowicz M, Gerós AS, Gupta T, Ashley N, Khamis D, Fowler D, Morrissey E, et al. 2021. Spatiotemporal analysis of human intestinal development at single-cell resolution. *Cell* **184:** 810–826.e23. doi:10.1016/j.cell.2020.12.016

Fletcher EL, Clark MJ, Furness JB. 2002. Neuronal and glial localization of GABA transporter immunoreactivity in the myenteric plexus. *Cell Tissue Res* **308:** 339–346. doi:10.1007/s00441-002-0566-3

Fried DE, Watson RE, Robson SC, Gulbransen BD. 2017. Ammonia modifies enteric neuromuscular transmission through glial γ-aminobutyric acid signaling. *Am J Physiol Gastrointest Liver Physiol* **313:** G570–G580. doi:10.1152/ajpgi.00154.2017

Fu M, Tam PK, Sham MH, Lui VC. 2004. Embryonic development of the ganglion plexuses and the concentric layer structure of human gut: a topographical study. *Anat Embryol (Berl)* **208:** 33–41. doi:10.1007/s00429-003-0371-0

Fung C, Berghe PV. 2020. Functional circuits and signal processing in the enteric nervous system. *Cell Mol Life Sci* **37:** 487.

Fung C, Boesmans W, Cirillo C, Foong JPP, Bornstein JC, Vanden Berghe P. 2017. VPAC receptor subtypes tune purinergic neuron-to-glia communication in the murine submucosal plexus. *Front Cell Neurosci* **11:** 118. doi:10.3389/fncel.2017.00118

Gabella G. 1981. Ultrastructure of the nerve plexuses of the mammalian intestine: the enteric glial cells. *Neuroscience* **6:** 425–436. doi:10.1016/0306-4522(81)90135-4

Gabella G. 2022. Enteric glia: extent, cohesion, axonal contacts, membrane separations and mitochondria in Auerbach's ganglia of guinea pigs. *Cell Tissue Res* **389:** 409–426. doi:10.1007/s00441-022-03656-3

Gomes P, Chevalier J, Boesmans W, Roosen L, Van den Abbeel V, Neunlist M, Tack J, Vanden Berghe P. 2009. ATP-dependent paracrine communication between enteric neurons and glia in a primary cell culture derived from embryonic mice. *Neurogastroenterol Motil* **21:** 870–e62. doi:10.1111/j.1365-2982.2009.01302.x

Grubišić V, Gulbransen BD. 2017. Enteric glial activity regulates secretomotor function in the mouse colon but does not acutely affect gut permeability. *J Physiol* **595:** 3409–3424. doi:10.1113/JP273492

Grubišić V, Gulbransen BD. 2022. Astrocyte cell surface antigen 2 and other potential cell surface markers of enteric glia in the mouse colon. *ASN Neuro* **14:** 17590914221083203. doi:10.1177/17590914221083203

Grubišić V, Perez-Medina AL, Fried DE, Sévigny J, Robson SC, Galligan JJ, Gulbransen BD. 2019. NTPDase1 and -2 are expressed by distinct cellular compartments in the mouse colon and differentially impact colonic physiology and function after DSS colitis. *Am J Physiol Gastrointest Liver Physiol* **317:** G314–G332. doi:10.1152/ajpgi.00104.2019

Grubišić V, McClain JL, Fried DE, Grants I, Rajasekhar P, Csizmadia E, Ajijola OA, Watson RE, Poole DP, Robson SC, et al. 2020. Enteric glia modulate macrophage phenotype and visceral sensitivity following inflammation. *Cell Rep* **32:** 108100. doi:10.1016/j.celrep.2020.108100

Grubišić V, Bali V, Fried DE, Eltzschig HK, Robson SC, Mazei-Robison MS, Gulbransen BD. 2022. Enteric glial adenosine 2B receptor signaling mediates persistent epithelial barrier dysfunction following acute DSS colitis. *Mucosal Immunol* **15:** 964–976. doi:10.1038/s41385-022-00550-7

Gulbransen BD, Sharkey KA. 2009. Purinergic neuron-to-glia signaling in the enteric nervous system. *Gastroenterology* **136:** 1349–1358. doi:10.1053/j.gastro.2008.12.058

Gulbransen BD, Sharkey KA. 2012. Novel functional roles for enteric glia in the gastrointestinal tract. *Nat Rev Gastroenterol Hepatol* **9:** 625–632.

Gulbransen BD, Bains JS, Sharkey KA. 2010. Enteric glia are targets of the sympathetic innervation of the myenteric plexus in the guinea pig distal colon. *J Neurosci* **30:** 6801–6809. doi:10.1523/JNEUROSCI.0603-10.2010

Gulbransen BD, Bashashati M, Hirota SA, Gui X, Roberts JA, MacDonald JA, Muruve DA, McKay DM, Beck PL, Mawe GM, et al. 2012. Activation of neuronal P2X7 receptor-pannexin-1 mediates death of enteric neurons during colitis. *Nat Med* **18:** 600–604. doi:10.1038/nm.2679

Guyer RA, Stavely R, Robertson K, Bhave S, Mueller JL, Picard NM, Hotta R, Kaltschmidt JA, Goldstein AM. 2023. Single-cell multiome sequencing clarifies enteric glial cell diversity and identifies an intraganglionic population poised for neurogenesis. *Cell Rep* **42:** 112194.

Hagström C, Olsson C. 2010. Glial cells revealed by GFAP immunoreactivity in fish gut. *Cell Tissue Res* **341:** 73–81. doi:10.1007/s00441-010-0979-3

Hanani M, Reichenbach A. 1994. Morphology of horseradish peroxidase (HRP)-injected glial cells in the myenteric plexus of the guinea-pig. *Cell Tissue Res* **278:** 153–160.

Hanani M, Zamir O, Baluk P. 1989. Glial cells in the guinea pig myenteric plexus are dye coupled. *Brain Res* **497:** 245–249. doi:10.1016/0006-8993(89)90269-2

Hanani M, Francke M, Härtig W, Grosche J, Reichenbach A, Pannicke T. 2000. Patch-clamp study of neurons and glial cells in isolated myenteric ganglia. *Am J Physiol Gastrointest Liver Physiol* **278:** G644–G651. doi:10.1152/ajpgi.2000.278.4.G644

Ibiza S, García-Cassani B, Ribeiro H, Carvalho T, Almeida L, Marques R, Misic AM, Bartow-McKenney C, Larson DM, Pavan WJ, et al. 2016. Glial-cell-derived neuroregulators control type 3 innate lymphoid cells and gut defence. *Nature* **535:** 440–443. doi:10.1038/nature18644

Inlender T, Nissim-Eliraz E, Stavely R, Hotta R, Goldstein AM, Yagel S, Gutnick MJ, Shpigel NY. 2021. Homeostasis of mucosal glial cells in human gut is independent of microbiota. *Sci Rep* **11:** 12796. doi:10.1038/s41598-021-92384-9

Jessen KR, Mirsky R. 1983. Astrocyte-like glia in the peripheral nervous system: an immunohistochemical study of enteric glia. *J Neurosci* **3:** 2206–2218. doi:10.1523/JNEUROSCI.03-11-02206.1983

Joseph NM, He S, Quintana E, Kim YG, Núñez G, Morrison SJ. 2011. Enteric glia are multipotent in culture but primarily form glia in the adult rodent gut. *J Clin Invest* **121:** 3398–3411. doi:10.1172/JCI58186

Kabouridis PS, Lasrado R, McCallum S, Chng SH, Snippert HJ, Clevers H, Pettersson S, Pachnis V. 2015. Microbiota controls the homeostasis of glial cells in the gut lamina propria. *Neuron* **85:** 289–295.

Kermarrec L, Durand T, Neunlist M, Naveilhan P, Neveu I. 2016. Enteric glial cells have specific immunosuppressive properties. *J Neuroimmunol* **295–296:** 79–83. doi:10.1016/j.jneuroim.2016.04.011

Cite this article as *Cold Spring Harb Perspect Biol* doi: 10.1101/cshperspect.a041368

Kim J, Lo L, Dormand E, Anderson DJ. 2003. SOX10 maintains multipotency and inhibits neuronal differentiation of neural crest stem cells. *Neuron* **38:** 17–31. doi:10.1016/S0896-6273(03)00163-6

Kimball BC, Mulholland MW. 1996. Enteric glia exhibit P2U receptors that increase cytosolic calcium by a phospholipase C-dependent mechanism. *J Neurochem* **66:** 604–612. doi:10.1046/j.1471-4159.1996.66020604.x

Kinchen J, Chen HH, Parikh K, Antanaviciute A, Jagielowicz M, Fawkner-Corbett D, Ashley N, Cubitt L, Mellado-Gomez E, Attar M, et al. 2018. Structural remodeling of the human colonic mesenchyme in inflammatory bowel disease. *Cell* **175:** 372–386.e17. doi:10.1016/j.cell.2018.08.067

Kovler ML, Gonzalez Salazar AJ, Fulton WB, Lu P, Yamaguchi Y, Zhou Q, Sampah M, Ishiyama A, Prindle T Jr, Wang S, et al. 2021. Toll-like receptor 4-mediated enteric glia loss is critical for the development of necrotizing enterocolitis. *Sci Transl Med* **13:** eabg3459. doi:10.1126/scitranslmed.abg3459

Kulkarni S, Micci MA, Leser J, Shin C, Tang SC, Fu YY, Liu L, Li Q, Saha M, Li C, et al. 2017. Adult enteric nervous system in health is maintained by a dynamic balance between neuronal apoptosis and neurogenesis. *Proc Natl Acad Sci* **114:** E3709–E3718. doi:10.1073/pnas.1619406114

Laddach A, Chng SH, Lasrado R, Progatzky F, Shapiro M, Erickson A, Sampedro Castaneda M, Artemov AV, BonFrauches AC, Amaniti EM, et al. 2023. A branching model of lineage differentiation underpinning the neurogenic potential of enteric glia. *Nat Commun* **14:** 5904. doi:10.1038/s41467-023-41492-3

Laranjeira C, Sandgren K, Kessaris N, Richardson W, Potocnik A, Vanden Berghe P, Pachnis V. 2011. Glial cells in the mouse enteric nervous system can undergo neurogenesis in response to injury. *J Clin Invest* **121:** 3412–3424. doi:10.1172/JCI58200

Lasrado R, Boesmans W, Kleinjung J, Pin C, Bell D, Bhaw L, McCallum S, Zong H, Luo L, Clevers H, et al. 2017. Lineage-dependent spatial and functional organization of the mammalian enteric nervous system. *Science* **356:** 722–726. doi:10.1126/science.aam7511

Lavoie EG, Gulbransen BD, Martín-Satué M, Aliagas E, Sharkey KA, Sévigny J. 2011. Ectonucleotidases in the digestive system: focus on NTPDase3 localization. *Am J Physiol Gastrointest Liver Physiol* **300:** G608–G620. doi:10.1152/ajpgi.00207.2010

Lilli NL, Quénéhervé L, Haddara S, Brochard C, Aubert P, Rolli-Derkinderen M, Durand T, Naveilhan P, Hardouin JB, De Giorgio R, et al. 2018. Glioplasticity in irritable bowel syndrome. *Neurogastroenterol Motil* **30:** e13232. doi:10.1111/nmo.13232

Liñán-Rico A, Turco F, Ochoa-Cortes F, Harzman A, Needleman BJ, Arsenescu R, Abdel-Rasoul M, Fadda P, Grants I, Whitaker E, et al. 2016. Molecular signaling and dysfunction of the human reactive enteric glial cell phenotype: implications for GI infection, IBD, POI, neurological, motility, and GI disorders. *Inflamm Bowel Dis* **22:** 1812–1834. doi:10.1097/MIB.0000000000000854

Liu Z, Sun H, Xu S, Wang H, Zhang Z, Wei Y, Kou Y, Wang Y. 2023. Dietary ingredient change induces a transient MyD88-dependent mucosal enteric glial cell response and promotes obesity. *Nutr Neurosci* **26:** 1183–1193.

May-Zhang AA, Tycksen E, Southard-Smith AN, Deal KK, Benthal JT, Buehler DP, Adam M, Simmons AJ, Monaghan JR, Matlock BK, et al. 2021. Combinatorial transcriptional profiling of mouse and human enteric neurons identifies shared and disparate subtypes in situ. *Gastroenterology* **160:** 755–770.e26. doi:10.1053/j.gastro.2020.09.032

McCallum S, Obata Y, Fourli E, Boeing S, Peddie CJ, Xu Q, Horswell S, Kelsh RN, Collinson L, Wilkinson D, et al. 2020. Enteric glia as a source of neural progenitors in adult zebrafish. *eLife* **9:** e56086. doi:10.7554/eLife.56086

McClain JL, Grubišić V, Fried D, Gomez-Suarez RA, Leinninger GM, Sévigny J, Parpura V, Gulbransen BD. 2014. Ca^{2+} responses in enteric glia are mediated by connexin-43 hemichannels and modulate colonic transit in mice. *Gastroenterology* **146:** 497–507.e1. doi:10.1053/j.gastro.2013.10.061

McClain JL, Fried DE, Gulbransen BD. 2015. Agonist-evoked Ca^{2+} signaling in enteric glia drives neural programs that regulate intestinal motility in mice. *Cell Mol Gastroenterol Hepatol* **1:** 631–645. doi:10.1016/j.jcmgh.2015.08.004

McClain JL, Mazzotta EA, Maradiaga N, Duque-Wilckens N, Grants I, Robison AJ, Christofi FL, Moeser AJ, Gulbransen BD. 2020. Histamine-dependent interactions between mast cells, glia, and neurons are altered following early-life adversity in mice and humans. *Am J Physiol Gastrointest Liver Physiol* **319:** G655–G668. doi:10.1152/ajpgi.00041.2020

Morales-Soto W, Gonzales J, Jackson WF, Gulbransen BD. 2023. Enteric glia promote visceral hypersensitivity during inflammation through intercellular signaling with gut nociceptors. *Sci Signal* **16:** eadg1668. doi:10.1126/scisignal.adg1668

Morarach K, Mikhailova A, Knoflach V, Memic F, Kumar R, Li W, Ernfors P, Marklund U. 2021. Diversification of molecularly defined myenteric neuron classes revealed by single-cell RNA sequencing. *Nat Neurosci* **24:** 34–46. doi:10.1038/s41593-020-00736-x

Mundell NA, Plank JL, LeGrone AW, Frist AY, Zhu L, Shin MK, Southard-Smith EM, Labosky PA. 2012. Enteric nervous system specific deletion of Foxd3 disrupts glial cell differentiation and activates compensatory enteric progenitors. *Dev Biol* **363:** 373–387. doi:10.1016/j.ydbio.2012.01.003

Murakami M, Ohta T, Otsuguro KI, Ito S. 2007. Involvement of prostaglandin E(2) derived from enteric glial cells in the action of bradykinin in cultured rat myenteric neurons. *Neuroscience* **145:** 642–653. doi:10.1016/j.neuroscience.2006.12.052

Nagahama M, Semba R, Tsuzuki M, Aoki E. 2001. L-arginine immunoreactive enteric glial cells in the enteric nervous system of rat ileum. *Biol Signals Recept* **10:** 336–340. doi:10.1159/000046901

Nasser Y, Fernandez E, Keenan CM, Ho W, Oland LD, Tibbles LA, Schemann M, MacNaughton WK, Rühl A, Sharkey KA. 2006. Role of enteric glia in intestinal physiology: effects of the gliotoxin fluorocitrate on motor and secretory function. *Am J Physiol Gastrointest Liver Physiol* **291:** G912–G927. doi:10.1152/ajpgi.00067.2006

Neunlist M, Aubert P, Bonnaud S, Van Landeghem L, Coron E, Wedel T, Naveilhan P, Ruhl A, Lardeux B, Savidge T, et al. 2007. Enteric glia inhibit intestinal epithelial cell proliferation partly through a TGF-β1-dependent pathway. *Am*

J Physiol Gastrointest Liver Physiol **292**: G231–G241. doi:10.1152/ajpgi.00276.2005

Ngan ES, Garcia-Barceló MM, Yip BH, Poon HC, Lau ST, Kwok CK, Sat E, Sham MH, Wong KK, Wainwright BJ, et al. 2011. Hedgehog/Notch-induced premature gliogenesis represents a new disease mechanism for Hirschsprung disease in mice and humans. *J Clin Invest* **121**: 3467–3478. doi:10.1172/JCI43737

Pattyn A, Morin X, Cremer H, Goridis C, Brunet JF. 1999. The homeobox gene Phox2b is essential for the development of autonomic neural crest derivatives. *Nature* **399**: 366–370. doi:10.1038/20700

Pham TD, Gershon MD, Rothman TP. 1991. Time of origin of neurons in the murine enteric nervous system: sequence in relation to phenotype. *J Comp Neurol* **314**: 789–798. doi:10.1002/cne.903140411

Pochard C, Coquenlorge S, Jaulin J, Cenac N, Vergnolle N, Meurette G, Freyssinet M, Neunlist M, Rolli-Derkinderen M. 2016. Defects in 15-HETE production and control of epithelial permeability by human enteric glial cells from patients with Crohn's disease. *Gastroenterology* **150**: 168–180. doi:10.1053/j.gastro.2015.09.038

Prochera A, Rao M. 2023. Mini-review: enteric glial regulation of the gastrointestinal epithelium. *Neurosci Lett* **805**: 137215. doi:10.1016/j.neulet.2023.137215

Progatzky F, Shapiro M, Chng SH, Garcia-Cassani B, Classon CH, Sevgi S, Laddach A, Bon-Frauches AC, Lasrado R, Rahim M, et al. 2021. Regulation of intestinal immunity and tissue repair by enteric glia. *Nature* **599**: 125–130. doi:10.1038/s41586-021-04006-z

Qu ZD, Thacker M, Castelucci P, Bagyánszki M, Epstein ML, Furness JB. 2008. Immunohistochemical analysis of neuron types in the mouse small intestine. *Cell Tissue Res* **334**: 147–161. doi:10.1007/s00441-008-0684-7

Rao M, Nelms BD, Dong L, Salinas-Rios V, Rutlin M, Gershon MD, Corfas G. 2015. Enteric glia express proteolipid protein 1 and are a transcriptionally unique population of glia in the mammalian nervous system. *Glia* **63**: 2040–2057. doi:10.1002/glia.22876

Rao M, Rastelli D, Dong L, Chiu S, Setlik W, Gershon MD, Corfas G. 2017. Enteric glia regulate gastrointestinal motility but are not required for maintenance of the epithelium in mice. *Gastroenterology* **153**: 1068–1081.e7. doi:10.1053/j.gastro.2017.07.002

Rosenbaum C, Schick MA, Wollborn J, Heider A, Scholz CJ, Cecil A, Niesler B, Hirrlinger J, Walles H, Metzger M. 2016. Activation of myenteric glia during acute inflammation in vitro and in vivo. *PLoS ONE* **11**: e0151335. doi:10.1371/journal.pone.0151335

Rosenberg HJ, Rao M. 2021. Enteric glia in homeostasis and disease: from fundamental biology to human pathology. *iScience* **24**: 102863. doi:10.1016/j.isci.2021.102863

Rühl AA, Franzke S, Collins SM, Stremmel W. 2001. Interleukin-6 expression and regulation in rat enteric glial cells. *Am J Physiol Gastrointest Liver Physiol* **280**: G1163–G1171. doi:10.1152/ajpgi.2001.280.6.G1163

Sang Q, Young HM. 1996. Chemical coding of neurons in the myenteric plexus and external muscle of the small and large intestine of the mouse. *Cell Tissue Res* **284**: 39–53. doi:10.1007/s004410050565

Sarosi GA, Barnhart DC, Turner DJ, Mulholland MW. 1998. Capacitative Ca^{2+} entry in enteric glia induced by thapsi-

gargin and extracellular ATP. *Am J Physiol* **275**: G550–G555. doi:10.1152/ajpgi.1998.275.3.G550

Savidge TC, Newman P, Pothoulakis C, Ruhl A, Neunlist M, Bourreille A, Hurst R, Sofroniew MV. 2007. Enteric glia regulate intestinal barrier function and inflammation via release of S-nitrosoglutathione. *Gastroenterology* **132**: 1344–1358. doi:10.1053/j.gastro.2007.01.051

Schneider R, Leven P, Glowka T, Kuzmanov I, Lysson M, Schneiker B, Miesen A, Baqi Y, Spanier C, Grants I, et al. 2021. A novel P2X2-dependent purinergic mechanism of enteric gliosis in intestinal inflammation. *EMBO Mol Med* **13**: e12724. doi:10.15252/emmm.202012724

Schuchardt A, D'Agati V, Larsson-Blomberg L, Costantini F, Pachnis V. 1994. Defects in the kidney and enteric nervous system of mice lacking the tyrosine kinase receptor Ret. *Nature* **367**: 380–383. doi:10.1038/367380a0

Seguella L, Gulbransen BD. 2021. Enteric glial biology, intercellular signalling and roles in gastrointestinal disease. *Nat Rev Gastroenterol Hepatol* **18**: 571–587. doi:10.1038/s41575-021-00423-7

Seguella L, Pesce M, Capuano R, Casano F, Pesce M, Corpetti C, Vincenzi M, Maftei D, Lattanzi R, Del Re A, et al. 2021. High-fat diet impairs duodenal barrier function and elicits glia-dependent changes along the gut-brain axis that are required for anxiogenic and depressive-like behaviors. *J Neuroinflamm* **18**: 115. doi:10.1186/s12974-021-02164-5

Selgrad M, De Giorgio R, Fini L, Cogliandro RF, Williams S, Stanghellini V, Barbara G, Tonini M, Corinaldesi R, Genta RM, et al. 2009. JC virus infects the enteric glia of patients with chronic idiopathic intestinal pseudo-obstruction. *Gut* **58**: 25–32. doi:10.1136/gut.2008.152512

Sharkey KA, Beck PL, McKay DM. 2018. Neuroimmunophysiology of the gut: advances and emerging concepts focusing on the epithelium. *Nat Rev Gastroenterol Hepatol* **15**: 765–784. doi:10.1038/s41575-018-0051-4

Soret R, Schneider S, Bernas G, Christophers B, Souchkova O, Charrier B, Righini-Grunder F, Aspirot A, Landry M, Kembel SW, et al. 2020. Glial cell derived neurotrophic factor induces enteric neurogenesis and improves colon structure and function in mouse models of Hirschsprung disease. *Gastroenterology* **159**: 1824–1838.e17. doi:10.1053/j.gastro.2020.07.018

Southard-Smith EM, Kos L, Pavan WJ. 1998. Sox10 mutation disrupts neural crest development in Dom Hirschsprung mouse model. *Nat Genet* **18**: 60–64. doi:10.1038/ng0198-60

Stakenborg M, Abdurahiman S, De Simone V, Goverse G, Stakenborg N, van Baarle L, Wu Q, Pirottin D, Kim JS, Chappell-Maor L, et al. 2022. Enteric glial cells favor accumulation of anti-inflammatory macrophages during the resolution of muscularis inflammation. *Mucosal Immunol* **15**: 1296–1308. doi:10.1038/s41385-022-00563-2

Stenkamp-Strahm C, Patterson S, Boren J, Gericke M, Balemba O. 2013. High-fat diet and age-dependent effects on enteric glial cell populations of mouse small intestine. *Auton Neurosci* **177**: 199–210. doi:10.1016/j.autneu.2013.04.014

Stoffels B, Hupa KJ, Snoek SA, van Bree S, Stein K, Schwandt T, Vilz TO, Lysson M, Veer CV, Kummer MP, et al. 2014. Postoperative ileus involves interleukin-1 receptor signaling in enteric glia. *Gastroenterology* **146**: 176–187.e1. doi:10.1053/j.gastro.2013.09.030

Teng L, Mundell NA, Frist AY, Wang Q, Labosky PA. 2008. Requirement for Foxd3 in the maintenance of neural crest progenitors. *Development* **135:** 1615–1624. doi:10.1242/dev.012179

Thomasi BBdeM, Valdetaro L, Ricciardi MCG, Hayashide L, Fernandes ACMN, Mussauer A, da Silva ML, da Faria-Melibeu AC, Ribeiro MGL, de Mattos Coelho-Aguiar J, et al. 2022. Enteric glial cell reactivity in colonic layers and mucosal modulation in a mouse model of Parkinson's disease induced by 6-hydroxydopamine. *Brain Res Bull* **187:** 111–121. doi:10.1016/j.brainresbull.2022.06.013

Turco F, Sarnelli G, Cirillo C, Palumbo I, De Giorgi F, D'Alessandro A, Cammarota M, Giuliano M, Cuomo R. 2014. Enteroglial-derived S100B protein integrates bacteria-induced Toll-like receptor signalling in human enteric glial cells. *Gut* **63:** 105–115. doi:10.1136/gutjnl-2012-302090

Uesaka T, Nagashimada M, Enomoto H. 2015. Neuronal differentiation in Schwann cell lineage underlies postnatal neurogenesis in the enteric nervous system. *J Neurosci* **35:** 9879–9888. doi:10.1523/JNEUROSCI.1239-15.2015

Uesaka T, Okamoto M, Nagashimada M, Tsuda Y, Kihara M, Kiyonari H, Enomoto H. 2021. Enhanced enteric neurogenesis by Schwann cell precursors in mouse models of Hirschsprung disease. *Glia* **69:** 2575–2590. doi:10.1002/glia.24059

Valès S, Bacola G, Biraud M, Touvron M, Bessard A, Geraldo F, Dougherty KA, Lashani S, Bossard C, Flamant M, et al. 2019. Tumor cells hijack enteric glia to activate colon cancer stem cells and stimulate tumorigenesis. *EBioMedicine* **49:** 172–188. doi:10.1016/j.ebiom.2019.09.045

Vicentini FA, Keenan CM, Wallace LE, Woods C, Cavin JB, Flockton AR, Macklin WB, Belkind-Gerson J, Hirota SA, Sharkey KA. 2021. Intestinal microbiota shapes gut physiology and regulates enteric neurons and glia. *Microbiome* **9:** 210.

Virtanen H, Garton DR, Andressoo JO. 2022. Myenteric neurons do not replicate in small intestine under normal physiological conditions in adult mouse. *Cell Mol Gastroenterol Hepatol* **14:** 27–34. doi:10.1016/j.jcmgh.2022.04.001

Young HM, Bergner AJ, Müller T. 2003. Acquisition of neuronal and glial markers by neural crest-derived cells in the mouse intestine. *J Comp Neurol* **456:** 1–11. doi:10.1002/cne.10448

Yuan R, Bhattacharya N, Kenkel JA, Shen J, DiMaio MA, Bagchi S, Prestwood TR, Habtezion A, Engleman EG. 2020. Enteric glia play a critical role in promoting the development of colorectal cancer. *Front Oncol* **10:** 595892. doi:10.3389/fonc.2020.595892

Zeisel A, Hochgerner H, Lönnerberg P, Johnsson A, Memic F, van der Zwan J, Häring M, Braun E, Borm LE, La Manno G, et al. 2018. Molecular architecture of the mouse nervous system. *Cell* **174:** 999–1014.e22. doi:10.1016/j.cell.2018.06.021

Glia at Transition Zones

Sarah Kucenas,[1] Pernelle Pulh,[2] Piotr Topilko,[2] and Cody J. Smith[3]

[1]Department of Biology, University of Virginia, Charlottesville, Virginia 22904, USA

[2]Institut Mondor de Recherche Biomédicale, Inserm U955-Team 9, 94010 Créteil, France

[3]Department of Biological Sciences, University of Notre Dame, Notre Dame, Indiana 46556, USA

Correspondence: csmith67@nd.edu

Neural cells are segregated into their distinct central nervous system (CNS) and peripheral nervous system (PNS) domains. However, at specialized regions of the nervous system known as transition zones (TZs), glial cells from both the CNS and PNS are uniquely present with other specialized TZ cells. Herein we review the current understanding of vertebrate TZ cells. The article discusses the distinct cells at vertebrate TZs with a focus on cells that are located on the peripheral side of the spinal cord TZs. In addition to the developmental origin and differentiation of these TZ cells, the functional importance and the role of TZ cells in disease are highlighted. This article also reviews the common and unique features of vertebrate TZs from zebrafish to mice. We propose challenges and open questions in the field that could lead to exciting insights in the field of glial biology.

Transition zones (TZs) are unique and specialized structures within the nervous system where central and peripheral nervous system (CNS and PNS, respectively) cells physically interact (Fig. 1). They play an integral role in connecting the PNS with the CNS, allowing information to travel between the two halves of the nervous system efficiently and uninterrupted. TZs are evolutionarily conserved and are observed from fly to man and can be broadly characterized as motor, sensory, or mixed nerve interfaces (Fraher 1992, 1997; Fontenas and Kucenas 2017; Radomska and Topilko 2017). In vertebrates, both motor and sensory TZs are present with distinct anatomical positions and are present along the body axis, located where cranial nerves cross into and out of the brain and in every somite in the trunk, where sensory and motor nerves enter or exit the spinal cord (Fig. 1; Fraher 1992, 1997; Fontenas and Kucenas 2017).

Although the specific names of the cell types that make up TZs vary by organism, the CNS side of these are usually occupied by astroglia and ensheathing glia (Golding and Cohen 1997; Golding et al. 1997; Suter and Jaworski 2019). On the PNS side, peripheral ensheathing glia form a tight intersection at TZs and often form nodes of Ranvier with CNS ensheathing glia. While it is widely accepted that both CNS and PNS cells interact at these unique locations, most of our understanding of TZs has focused on PNS glia. Therefore, in this article, we review what is known about the origins and roles of PNS cells at TZs from work in vertebrate spinal cord TZs, as well as discuss open questions about the cells

---

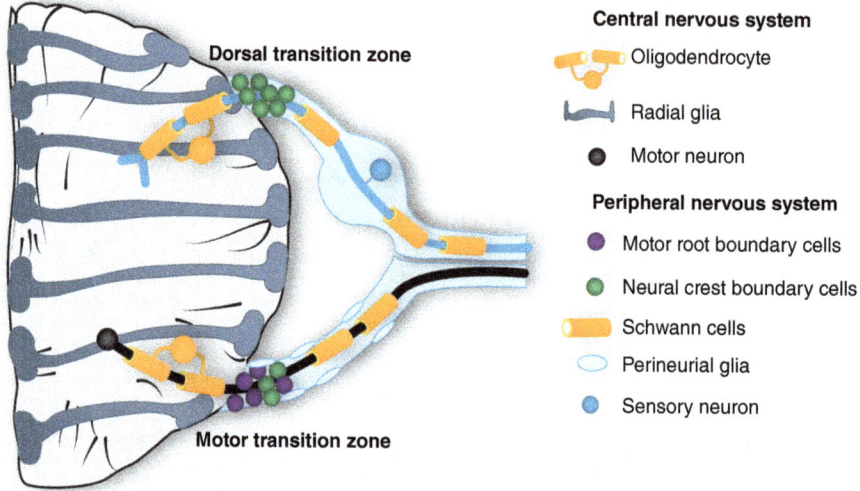

Figure 1. Transition zone (TZ) glial populations. Schematic representation showing the cross section of the embryonic vertebrate spinal cord. Radial glial (blue gray) and oligodendrocytes (orange) are positioned on the central nervous system (CNS) side of the TZ. Boundary cells on the peripheral side of the TZ are diverse on the motor axon but homogenous on the sensory axon. Boundary cells in the peripheral nervous system (PNS) differentiate into a variety of cells including Schwann cells (SCs, orange). Perineurial glial ensheathe the motor axons and the SCs that ensheathe them.

and mechanisms that contribute to the CNS side of vertebrate TZs.

DEVELOPMENTAL ORIGIN AND DIFFERENTIATION

During development, vertebrate spinal cord TZs are assembled and actively occupied by a diverse set of cells in the PNS. The spinal cord motor TZ, referred to as the motor exit point (MEP) TZ, is initially populated by neural crest (NC)-derived cells (Fig. 2; Niederländer and Lumsden 1996; Golding and Cohen 1997; Smith et al. 2014). As development proceeds, the MEP TZ is also occupied by distinct CNS-derived or pia matter–derived cells (Fig. 2; Kucenas et al. 2008; Clark et al. 2014; Smith et al. 2014; Fontenas and Kucenas 2021; Gerschenfeld et al. 2023). Ultimately, these cells produce glia that make up the motor nerve. The sensory neuron TZ, referred to as the dorsal root entry zone (DREZ), however, has significantly less known about it. We know the DREZ is occupied by NC-derived cells in all vertebrates (Niederländer and Lumsden 1996; Golding and

Cohen 1997; Golding et al. 1997; Maro et al. 2004; Smith et al. 2017). But whether there are other populations of cells, including a CNS or pia matter–derived lineage like that observed at the MEP TZ, is not known. Below, we briefly introduce the cell lineages that contribute to TZ assembly and function in the spinal cord of vertebrates.

Neural Crest Cells

All TZs in vertebrates are occupied by NC-derived glia during development (Golding and Cohen 1997; Golding et al. 1997). Across vertebrate phylogeny, these NC cells express markers including *sox10* and *foxd3* (Coulpier et al. 2009; Smith et al. 2014; Furlan and Adameyko 2018; Fontenas and Kucenas 2021). This includes the mixed nerves in the head and the sensory and motor TZs found in the trunk along the spinal cord (Coulpier et al. 2009; Smith et al. 2014; Furlan and Adameyko 2018; Fontenas and Kucenas 2021). Time-lapse imaging in zebrafish has revealed the detailed and dynamic migration path that these cells take during the early stages of development. In the trunk of zebrafish, NC cells

 Cite this article as *Cold Spring Harb Perspect Biol* doi: 10.1101/cshperspect.a041369

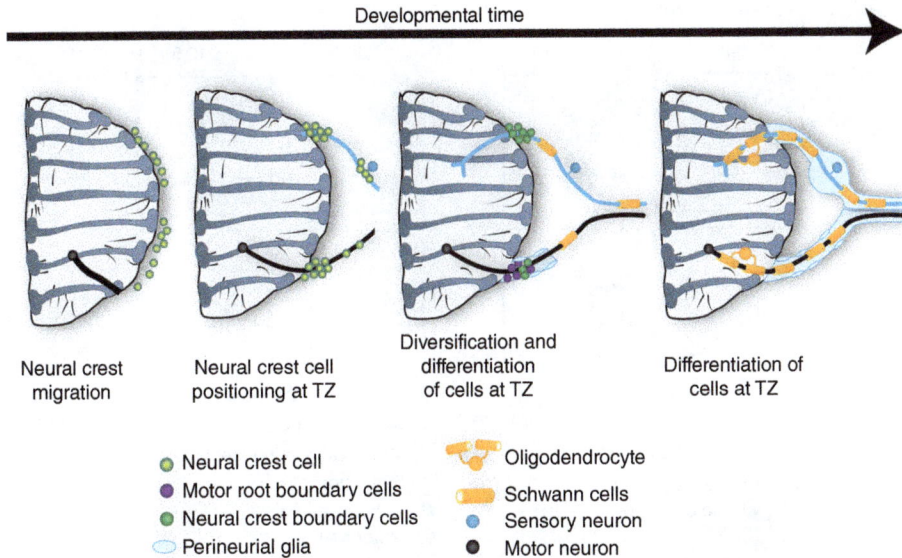

Figure 2. Development of the transition zone (TZ) glia. Schematics depicting the development of the TZs. In the early stages of development, neural crest (NC) cells migrate along the edge of the spinal cord. As development progresses, NC cells locate to the motor and sensory roots. At the motor root, TZ cells diversify with cells from the central nervous system (CNS) (zebrafish) or pia matter (mice). Boundary cap (BC) cells in the dorsal root entry zone (DREZ, green) originate from the NC and differentiate into sensory neurons, glial satellite cells, and Schwann cells (SCs). BC cells in the motor exit point (MEP) have double, NC (green), and pia matter (violet) origin. While the NC-derived BC gives birth to SCs, those from the pia matter are at the origin of perivascular derivatives.

stream from their dorsal origin through each somite until they reach the MEP (Raible et al. 1992). While some NC cells continue to travel along motor axons, in zebrafish, a subpopulation halts migration at the MEP (McGraw et al. 2012; Smith et al. 2016; Zhu et al. 2019). By 48 hours postfertilization (hpf) in zebrafish, motor axons are occupied by NC cells that begin to ensheath them (Kucenas et al. 2008, 2009; Smith et al. 2014; Fontenas and Kucenas 2021). From work in both zebrafish and mouse, we know that the molecular signals that drive Schwann cell (SC) differentiation, like Erbb3/Nrg1 and GPR126, are essential to generate SCs from these NC cells at the motor root (Riethmacher et al. 1997; Britsch et al. 2001; Jessen and Mirsky 2005; Honjo et al. 2008; Monk et al. 2009; Mogha et al. 2013). It is also widely accepted in all vertebrates that these SCs myelinate the motor nerve, expressing molecules like myelin basic protein (MBP) (Fröb et al. 2012; Smith et al. 2014; Fontenas and Kucenas 2021).

The DREZ is also occupied by a population of NC-derived cells in vertebrates (Fig. 2; Golding and Cohen 1997; Golding et al. 1997; Smith et al. 2017). In movies of the migration process in zebrafish, NC cells that migrate to the MEP go on to produce cells of the dorsal root ganglia (DRG) in zebrafish, either through cell proliferation of NC cells at the MEP or direct migration to the DRG (Honjo et al. 2008; McGraw et al. 2012; Smith et al. 2017). In addition to SCs, these NC cells then produce satellite glia and a plethora of sensory neurons in the DRG. NC cells at the DREZ develop with the pioneer axons of the DRG (Smith et al. 2017; Nichols and Smith 2019). As these pioneer axons are navigating to the DREZ in zebrafish, NC cells travel shortly behind the neuronal growth cones (Fig. 3; Smith et al. 2017; Nichols and Smith 2019). Once that DRG pioneer axon enters the spinal cord, a single NC cell transitions from a loose association with the pioneer axon to a tighter ensheathment in zebrafish (Smith et al. 2017). These NC cells then occupy

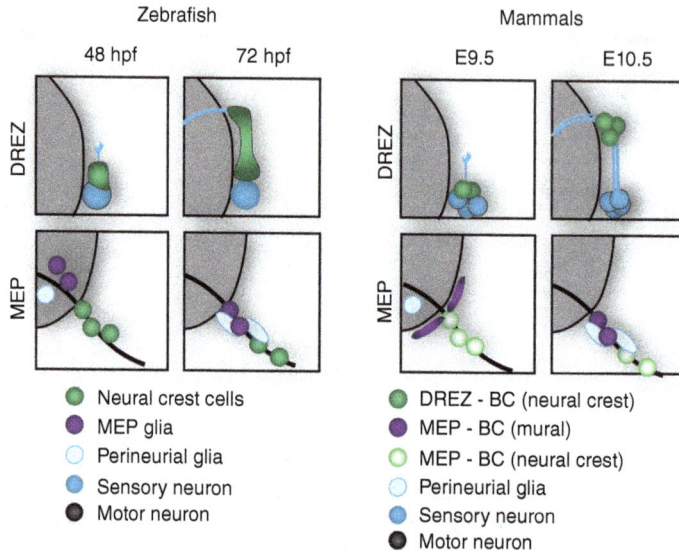

Figure 3. Comparison of vertebrate transition zones (TZs). Depiction of the dorsal root entry zone (DREZ) and motor exit point (MEP) TZs of the vertebrate spinal cord. Schematics show the zebrafish DREZ contains a single neural crest (NC) cell, while the MEP contains the central nervous system (CNS)-derived perineurial glia and MEP glia, and NC cells. In zebrafish, perineurial are derived from the CNS. The mammalian DREZ also contains NC cell–derived populations that denoted boundary cap (BC) cells. At the mammalian MEP, diverse cells are present as in zebrafish but denoted as MEP-BCs that are either pia matter or NC derived. Perineurial glia at the mammalian MEP are at least partially derived from the CNS. (hpf) Hours postfertilization.

the DREZ TZ, functioning as the ensheathing cell of the axon, the transition cell of the DREZ, and the progenitor population for other cells in the ganglia (Fig. 3; Smith et al. 2017; Nichols et al. 2018). Genetic analysis in zebrafish revealed that the entry of the pioneer axon and a *tnfa/tnfr2* signaling cascade are important for the NC TZ cells at the DREZ (Smith et al. 2017). However, most of our knowledge about the signaling pathways required for the differentiation of the DREZ NC cells has been completed in mouse and chick (Radomska and Topilko 2017).

While NC cells in zebrafish are simply characterized as NC, in other animals, a subset of NC cells are described to give birth to discrete clusters of cells defined as boundary cap (BC) cells that occupy the spinal TZs (Fig. 3). BCs were identified during embryogenesis in species starting from amphibians, because of their particular location at the DREZ and MEP TZs of cranial and spinal nerves (except olfactory and optic nerves) (Altman and Bayer 1984; Niederländer

and Lumsden 1996). For nearly three decades, BC characterization was limited to morphological and topological description in different species thanks to the pyknotic aspect of their nuclei (Altman and Bayer 1984). The functional studies of BCs have really begun with the identification BC-specific markers including *Krox20* (*EGR2* in human) and *Prss56* among many others (*Cdh7*, *Sema6A*, *Sema3B*, *Sema3G*, *Cxcr4*, *Netrin5*, *Wif1*, *HeyL*, and *Hey2*) in mouse or chick embryos (Topilko et al. 1994; Bron et al. 2007; Coulpier et al. 2009; Zhu et al. 2015). Expression of many of these genes is initiated in BCs in mice around E10.5 and persists until E15.5, suggesting that from that stage BCs profoundly change their properties by inactivating these markers or are eliminated by natural death. To address this issue, the Charnay laboratory designed two mouse lines (*Prss56^{Cre}* and *Krox20^{Cre}*), carrying the Cre recombinase coding sequence insertion into *Prss56* and *Krox20* loci and used them in combination with a Cre-inducible fluorescent reporter

Cite this article as *Cold Spring Harb Perspect Biol* doi: 10.1101/cshperspect.a041369

line ($Rosa26^{Tom}$) for BC fate mapping. By doing so, they discovered that BCs give rise to derivatives that rapidly migrate along the nerves into the periphery (Maro et al. 2004; Gresset et al. 2015). This rapid migration to the periphery is also observed in time-lapse imaging experiments in zebrafish (Honjo et al. 2008; Kucenas et al. 2009; Zhu et al. 2019).

CNS and Pia Matter–Derived Glia at the Motor Exit Point

It is now appreciated that the motor TZ in vertebrates has a heterogeneous group of cells compared to the DREZ. In addition to the BCs that are NC-derived, it is clear that the organization of the vertebrate MEP is more complex (Figs. 1–3). This concept is supported by data in mice that demonstrate *Krox20* and *Prss56*-expressing cells at the MEP TZ have apparent molecular and functional heterogeneity (Gerschenfeld et al. 2023). The complete origin of the mammalian MEP TZ cells in mice has remained until recently enigmatic. An exciting observation was recently reported by Charnay laboratory using a combination of genetic and single-cell resolution approaches (Gerschenfeld et al. 2023). They observed that once motor axons emigrate from the neural tube, their growth cone crosses the pia matter (component of the future meninges) and activates *Krox20* in pia matter cells. In turn, these *Krox20*-expressing cells (*Sox10⁻*) attach to motor axons and travel to the skin, then detach and differentiate into mural cells (Gerschenfeld et al. 2023).

The increased complexity of the MEP TZ cells is also present in zebrafish but the cells are characterized as a distinct cell type from the NC cells at the MEP (Smith et al. 2014). The zebrafish MEP TZ is known to be populated by a cell type defined as MEP glia (Fig. 3; Smith et al. 2014; Fontenas and Kucenas 2021). From time-lapse imaging experiments in zebrafish, we know the MEP TZ is initially populated by NC-derived cells (Fig. 2; Smith et al. 2014; Fontenas and Kucenas 2018, 2021). But as development proceeds, these cells are displaced by CNS-derived MEP glia (Fig. 3; Smith et al. 2014; Fontenas and Kucenas 2018, 2021). Ultimately, NC-derived SCs reside more distally along the same motor nerve,

separated from the MEP TZ by MEP glia (Smith et al. 2014). MEP glia are first observed at the MEP by the second day of the zebrafish embryo, positioned immediately adjacent to the CNS on the peripheral side of spinal motor nerves (Fig. 3; Smith et al. 2014; Fontenas and Kucenas 2021). Using in vivo, time-lapse imaging, studies revealed that MEP glia originate from precursors within the spinal cord, squeeze through the MEP TZ, and occupy the PNS side of the spinal motor root (Smith et al. 2014; Fontenas and Kucenas 2021). These MEP glia share characteristics of both CNS and PNS glia (Fontenas and Kucenas 2021). Like SCs, MEP glia ensheath and myelinate motor axons in the periphery (Smith et al. 2014; Fontenas and Kucenas 2021). However, as mentioned above, there is a clear segregation of their domains, with MEP glia myelinating the region proximal to the CNS and SCs myelinating axonal segments found more distally along the nerve. MEP glial development is dependent on *erbb3/nrg1* signaling, similar to SCs (Smith et al. 2014). They also express *sox10*, *foxd3*, and *wif1*, similar to NC cells at the TZ (Smith et al. 2014; Fontenas and Kucenas 2021). Interestingly, *Wif1* is also known to mark mammalian BCs (Coulpier et al. 2009). MEP glial development is also dependent on signaling pathways that are characterized in CNS cells. Like oligodendrocytes, MEP glia are derived from the ventral spinal cord, express *olig2* and *nkx2.2*, and require molecules like Shh, which pattern the ventral spinal cord (Smith et al. 2014; Fontenas and Kucenas 2021). While it is not clear whether all MEP glia terminally differentiate or whether a subset remains in a progenitor state, fate mapping identified they persist at the MEP TZ at least until 8 days postfertilization in zebrafish (Smith et al. 2014).

Perineurial Glia

The MEP TZ is also occupied by a second CNS-derived glial population (Fig. 1). At the motor TZ of vertebrates, perineurial glia make up the perineurium, which literally translates to "around the neuron" and was first described by Henle in 1841 and named by Key and Retzius in 1876 (Shanthaveerappa and Bourne 1962, 1966; Shantha et al. 1968; Akert et al. 1976). This structure is composed of many concentric rings of perineurial glia that encase nerve fascicles and eventually individual nerve fibers and terminals (Burkel 1967; Shantha et al. 1968; Akert et al. 1976). Individual perineurial cells are extremely thin, have a double basal lamina, and are fitted together by cell contacts formed by zonulae occludens, or tight junctions, and it is these structures that give the perineurium its blood–nerve-barrier function (Burkel 1967; Shantha et al. 1968; Kristensson and Olsson 1971; Akert et al. 1976). Using in vivo, time-lapse imaging in zebrafish, the origin of these cells for spinal motor nerves was shown to be the neuroectoderm (Fig. 3; Kucenas et al. 2008), as what is observed in invertebrates, including *Drosophila*, where most peripheral glia come from central neuroblasts (Klämbt and Goodman 1991; Schmidt et al. 1997; Sepp et al. 2000, 2001; Freeman and Doherty 2006; Parker and Auld 2006). These motor nerve–associated perineurial glia arise from *nkx2.2a*-positive lateral floor plate precursors and require the presence of motor axons and NC-derived SCs for their migration from the ventral spinal cord via MEP TZs into the periphery (Kucenas et al. 2008; Binari et al. 2013). Similar to the observations in zebrafish, CNS-derived cells also produce mouse perineurial glia (Clark et al. 2014). Interestingly, despite our knowledge that all peripheral nerves are ensheathed by a perineurium, the origin of sensory nerve root perineurial glia still remains a mystery.

Unifying and Contrasting Themes in Vertebrates

In all model systems where vertebrate TZs have been studied, it is clear that they are assembled early in development. In the vertebrate nervous system, it is also widely accepted that NC-derived cells are critical for TZ development (Kucenas et al. 2008, 2009). In all vertebrates studied, DREZ-associated glia are thought to be homogeneous and derived solely from the NC. The MEP TZ, however, has a more heterogeneous population of cells that are derived from both the NC, and pial and spinal neural precursors (Kucenas et al. 2009; Smith et al. 2014; Fontenas and Kucenas 2021; Gerschenfeld et al. 2023).

Despite these unifying points, there are also some contrasting themes between different model systems (Fig. 3). For example, in mice, sensory TZs are occupied by BCs, whereas the cells in zebrafish are simply referred to as NC cells (Niederländer and Lumsden 1996; Kucenas et al. 2009; Smith et al. 2014, 2017). At the motor TZ, CNS-derived cells are present in zebrafish and mice, but the zebrafish cell type is denoted as MEP glia (Smith et al. 2014), whereas the mouse cells are MEP-BCs (Radomska and Topilko 2017). Mice also have a large number of cells at these TZs compared to zebrafish. Despite these defined differences between organisms, it is also possible that the relative size of the animal accounts for these differences. For example, the distance between the DREZ and DRG in zebrafish is only about 18 μm (Nichols and Smith 2019), a distance that can be occupied by a single cell. The zebrafish DRG also has fewer cells compared to the mouse DRG and thus a single cell could meet the proliferative demand required to make the DRG (McGraw et al. 2012; Nichols et al. 2018). It is also possible that these differences between organisms are simply a result of the nomenclature of the systems. As we learn more about TZs, the similarity between TZs in different model systems seems to outweigh the differences between the systems.

FUNCTIONAL IMPORTANCE OF TRANSITION ZONE GLIA

Progenitors

TZ-associated glia function as progenitor cells in multiple species, including the cells that myelinate the PNS nerves. In zebrafish, NC cells at both the DREZ and MEP proliferate to produce SCs that ensheath or myelinate motor and sensory axons (Raible et al. 1992; Raible and Eisen 1994; Honjo et al. 2008; Smith et al. 2017). At the MEP of zebrafish, both NC cells and MEP glia produce myelinating cells (Smith et al. 2014; Fontenas and Kucenas 2021). DREZ cells in zebrafish also produce cells that make the DRG (Smith et al. 2017; Nichols et al. 2018). However, the role of TZ glia as progenitors has been more extensively investigated in mammals and chick

Cite this article as *Cold Spring Harb Perspect Biol* doi: 10.1101/cshperspect.a041369

models. The mammalian and chick DREZ-BCs expressing *Krox20* or *Prss56* migrate into the dorsal root and DRG to give birth to virtually all *Sox10*[+] (marker of NC-derived glia) myelinating and nonmyelinating (Remak) SCs and a subpopulation of satellite glia and sensory neurons (Maro et al. 2004; Gresset et al. 2015). Interestingly, MEP-BCs expressing *Krox20* or *Prss56* give birth to partly overlapping derivatives. While both populations give birth to myelinating SCs engulfing ventral (motor) root axons, only those expressing *Prss56* undergo rapid migration along spinal and cranial nerves to join the skin and give rise to myelinating and Remak SCs, terminal glia, and a small population of melanocytes (Radomska et al. 2019). BC-derived terminal glia include lanceolate glia at the hair follicle innervation and the recently identified subepidermal glia located at the nociceptive nerve endings and directly involved in the initiation of pain sensations (Abdo et al. 2019). Moreover, *Prss56*-expressing BCs are at the origin of the niche of immature stem-like cells, also defined as skin-derived precursors (SKPs) in the adult mouse skin (Gresset et al. 2015). While their precise localization along the nerves remains unknown, they are characterized by their ability to self-renew by forming spheres under floating culture conditions that can be propagated over long periods, thus preserving a broad differentiation potential, including SCs, adipocytes, chondrocytes, SMA[+] cells, among many others (Kang et al. 2011; Gresset et al. 2015).

In contrast, *Krox20*-expressing MEP-BCs are at the origin of derivatives migrating along the same routes over the same period, but once they reach their destination (skin), they detach from the nerves and activate the molecular program of perivascular differentiation to become pericytes and blood vessel smooth muscle cells of the skin vasculature (Gerschenfeld et al. 2023). Such an observation points to a molecular and functional dichotomy of MEP-BCs and addresses questions about the mechanisms governing their potential dual fate. Overall, BCs appear as a second wave of pluripotent stem cells, in addition to the NC, to feed the developing mouse embryo with glial, neuronal, and mural cells.

Establishing a Boundary of Glia

One of the best-characterized functions of these TZ glia is their role in establishing the boundary between the CNS and PNS. In development, these boundaries at TZs are less restrictive before TZ glia are present. For example, time-lapse imaging in zebrafish revealed that oligodendrocyte progenitor cells extend short processes into the PNS before MEP glia differentiate (Smith et al. 2014). Similarly, time-lapse imaging in zebrafish revealed that NC cells can enter the CNS at MEP TZs early during development but are restricted later (Smith et al. 2016; Zhu et al. 2019). However, once TZ glia are present at the boundaries, CNS glia are restricted to the CNS (Golding et al. 1997; Kucenas et al. 2009; Smith et al. 2014). Therefore, under homeostatic conditions in all vertebrates, glia in the CNS and PNS are largely restricted to their respective domains (Golding and Cohen 1997; Kucenas et al. 2009; Smith et al. 2014).

However, experimental results from multiple vertebrate species show that disruption of TZ glial populations can cause ectopic migration of oligodendrocytes and astrocytes into the PNS (Coulpier et al. 2010). These conclusions are supported by experiments in mice where BCs were removed via ablation of BC-derived SCs and through inactivation of Krox20 in BC-derived SCs or by targeted expression of diphtheria toxin (Coulpier et al. 2010). Interestingly, spinal cord injury in adult mice leads to the migration of BC-derived SCs into the CNS, showing that in some pathological conditions, the border can be crossed in both directions (Zujovic et al. 2010).

In zebrafish, genetic perturbation of peripheral glia in mutants of *tfap2a* and *foxd3* demonstrated that PNS TZ cells are essential to restrict oligodendrocytes in the CNS (Kucenas et al. 2009). This phenomenon has also been confirmed in *erbb3b* mutants in zebrafish that lack PNS TZ glia (Smith et al. 2014). Additionally, zebrafish studies that used laser-mediated cell ablation provided strong evidence that MEP glia are specifically essential to keep oligodendrocyte progenitor cells in the CNS (Smith et al. 2014). Our understanding of the molecular mediators of boundary establishment and persistence is less mature. However, recent studies demonstrate that motor neuron (MN) activity, driven by adenosine receptor A2, plays a role in restricting oligodendrocytes to the spinal cord (Fontenas et al. 2019). Collectively, it is clear that glia at TZs function to segregate the migratory glia of the CNS and PNS in the vertebrate spinal cord.

Neuronal Development and Organization

TZ glia are also important contributors to nerve development. While the vast majority of axon guidance literature has neglected the role of glia, it has been proposed that BCs during development function as a positive substrate for the DRG axons to enter the spinal cord (Golding et al. 1997). However, recent evidence in zebrafish shows that the pioneer axons of the DRG do not use TZ cells at the DREZ as a positive substrate (Nichols and Smith 2019). Thus, pioneer axon navigation may use a separate mechanism than BC-dependent navigation of DRG axons at the DREZ.

To assess the role of MEP-BCs in MNs axonal outgrowth, three different experiments were performed in mice and chicks: (1) analysis of *Splotch* mutant mice (mutations in the homeodomain of the *Pax3* gene), which fails to develop BCs due to missing NC cells in the trunk, (2) selective ablation of *Krox20*-expressing BC cells in mouse embryos using the $Krox20^{DT}$ transgene, and (3) surgical ablation of the dorsal neural tube in chick embryos resulting in the absence of NC cells (Vermeren et al. 2003). All three experiments revealed that while the absence of BCs has no effect on MN differentiation nor their axonal growth, progressive escape of MS cells bodies that emigrate to the periphery and death of those cells was observed. These results demonstrate that BCs play a key role as an MEP gatekeeper by preventing MN escapement. The molecular mechanism responsible for this has been partially identified and involves complex interactions between Semaphorin6A and Netrin5, expressed by BCs and Neuropilin-2 and/or Plexin-A2 proteins, as well as DCC protein and/or UNC5 receptor that are expressed by MNs (Bron et al. 2007; Mauti et al. 2007; Garrett et al. 2016).

INJURY AND DISEASE

Nerve Regeneration, Demyelination, and Gatekeeper of CNS/PNS Interface

Given the roles of TZ glia in the development of the nervous system, it is intriguing to hypothesize that such cells could impact the regeneration of the area. The TZ can be injured in brachial plexus injuries, which can occur in development as obstetrical brachial plexus injuries or in adults from severe trauma like car accidents (Tang et al. 2012). These injuries can create both PNS and CNS debris at the roots, resulting in microglia that leave the CNS to clear debris (Tang et al. 2012). In addition, in such injuries that severely alter the DREZ, regeneration of sensory axons into the CNS is highly unsuccessful (Golding et al. 1997; Di Maio et al. 2011; Tang et al. 2012). The rate of this regeneration back into the CNS depends on the age and severity (Golding et al. 1997; Tang et al. 2012), although it is still unsuccessful at most ages. The prevailing model is that regeneration of sensory axons into the CNS fails in part because NC-derived cells like BCs cells are not present at later stages of life and thus cannot instruct regrowing axons into the CNS (Golding et al. 1997; Tang et al. 2012). Additional cells at TZs, including astrocytes and oligodendrocytes are also thought to inhibit regrowth (Golding et al. 1997; Di Maio et al. 2011; Tang et al. 2012). While these components are contributors to the failed regeneration, in zebrafish larvae, regrowth of the sensory axons through TZs can be enhanced by modulating invasive components in the regrowing axons (Nichols and Smith 2020). This new evidence follows the theme of pioneer axon growth in development that is also independent of boundary glia (Nichols and Smith 2019). Regardless, it is clear that glia at TZs can impact the regeneration of some axons. For example, perineurial glia along the motor nerve serve as glial bridges for regenerating axons (Fig. 4; Lewis and Kucenas et al. 2014). Thus, the role of TZ glia populations on peripheral nerve regeneration is context-dependent.

While it is not clear why CNS and PNS cells need to be restricted, the conserved nature of the CNS/PNS boundary strongly suggests it is impor-

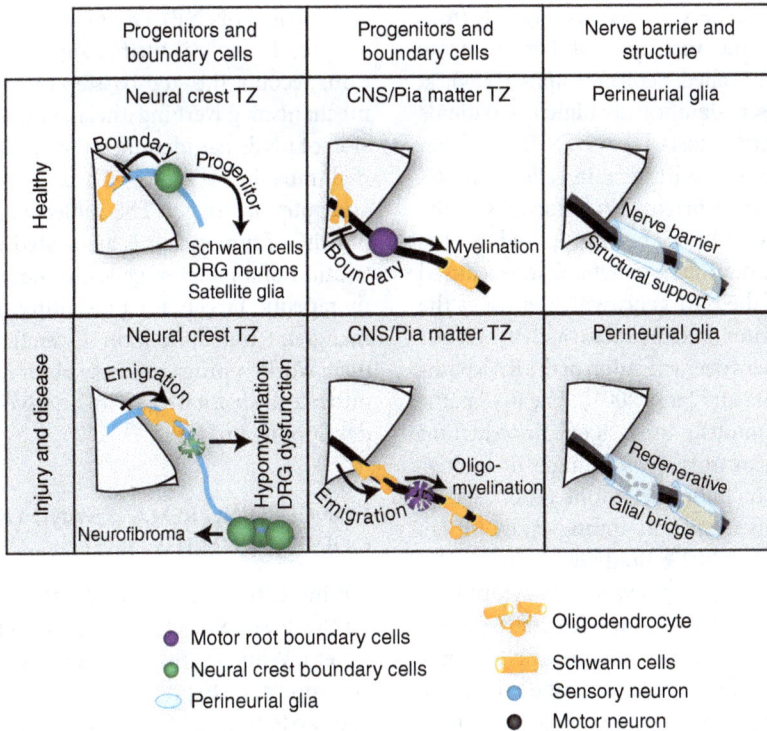

Figure 4. Function and disease implications of transition zone (TZ) glia. Schematic of the TZ cells in healthy, injury, and disease. Schematics show neural crest (NC) TZ cells, central nervous system (CNS)/pia matter TZ cells, and perineurial glia and their corresponding function in the healthy nervous system. NC TZ and CNS/pia matter TZ cells both serve as progenitors and boundary cells; however, their cellular progeny are partially overlapping. Perineurial glia function in structural and nerve barrier roles. In injury and disease contexts, these cells could alter neural function. In particular, defects in NC TZ cells cause hypomyelination and changes in dorsal root ganglia development. Dysregulation of NC TZ cells can cause neurofibromas. Defects in both NC TZ and CNS/pia matter TZ cells cause the emigration of oligodendrocytes, which can then myelinate the peripheral axons. After injury, perineurial glia function as glial bridges for regenerating axons to regrow.

tant. What we do know is that during certain disease contexts, the gatekeeping function of the boundary is disrupted. For example, in peripheral neuropathies, it has been shown that CNS cells can enter into the PNS at the motor root (Coulpier et al. 2010). Whether the ectopic location of these CNS cells is a negative or positive component of peripheral neuropathies is not known. We do know that CNS cells in experimental models can emigrate to the PNS and myelinate peripheral nerves (Fig. 4; Kucenas et al. 2009; Smith et al. 2014). Thus, it is possible that such CNS cells, oligodendrocytes, could fulfill the role of myelination that is normally reserved for SCs. It is also possible that BCs could be beneficial to demyelinating diseases of the CNS. This is because they

serve as progenitors for myelinating cells of the PNS (Fig. 4; Maro et al. 2004). Such an idea is supported by experiments in mice that show BCs that are grafted into a lesioned spinal cord can proliferate and produce myelinating cells in the CNS. In such experimental contexts, the grafted BCl progeny gives rise to mature oligodendrocytes (Zujovic et al. 2010). Nonetheless, future research will need to further dissect the functional consequences or benefits of altering the TZs and their gatekeeping function.

BCs: Cells of Origin of Neurofibromas in NF1

In the adult PNS, virtually all SCs from the nerve roots and a subpopulation of SCs and terminal glia from the skin nerve terminals originate from BCs. Because the incidence of benign nerve sheath tumors, called neurofibromas (NFs), at these two sites, is significantly higher in patients with neurofibromatosis type 1 (NF1) and because mutations in NF1 gene in cells from SCs lineage are at their origin, a role for BCs in the pathogenesis of NFs has been proposed. NF1 is an autosomal-dominant genetic disease caused by the loss of the NF1 gene, which encodes the negative regulator of RAS protein activity, resulting in permanent overactivation of the RAS pathway (Cichowski and Jacks 2001). The RAS pathway is a ubiquitously used signal transduction cascade that controls a wide range of biological activities, including cell proliferation, senescence, and differentiation among many others (Anastasaki et al. 2022; Brady et al. 2022). Patients with NF1 develop a variety of symptoms, including eye and bone lesions, learning disabilities, and NFs. There are two types of NFs: cutaneous NFs (cNFs) developing often in large numbers at the level of nerve endings, and plexiform NFs (pNFs), mainly emerging along the nerve roots. Interestingly, pNFs but not cNFs possess the ability to progress into malignant peripheral nerve sheath tumors (MPNSTs) (Cimino and Gutmann 2018) with no effective treatment to date. Several NF1 mouse models were conceived by targeting *Nf1* into SC lineage at different stages of their maturation. Interestingly, almost all of them develop pNFs but not cNFs nor MPNSTs suggesting their distinct glial origin. An exception is the mouse model (Prss56Cre Nf1-KO) in which simultaneous inactivation of *Nf1* and expression of the fluorescent reporter Tomato were targeted to BCs and their derivatives. These mice develop both types of NFs including the malignant transformation of pNFs pointing to BCs and their derivatives as the population at the origin of nerve sheath tumors in NF1 (Fig. 4; Radomska et al. 2019). Exploration of this model has enabled researchers to make seminal progress in understanding NF pathogenesis (Coulpier et al. 2023) and in designing tools for NF-targeted drug discovery. First, a comparison of different Nf1-KO models identified subepidermal glia, a particular type of BC-derived terminal glia as the potential population

at the origin of cNFs. Ongoing scRNA-seq transcriptomic analysis further supports such function. Second, the step-by-step dissection of the mechanism governing the malignant progression of pNFs has identified the glial–mesenchymal transition of tumor SCs as the earliest event and potential trigger. This molecular "metamorphosis" of tumor cells is associated with the activation of a panel of genes, some of which are therapeutic targets for preventing or reversing malignant transformation. Overall, despite being a work in progress, these observations demonstrate the pivotal role of BC-derived glial in the development of NFs.

CONCLUDING REMARKS AND OPEN QUESTIONS IN THE FIELD

It is increasingly clear that glia at TZs are critical for the development and function of the nervous system. While the field widely agrees that TZ glia are important for the development of the nervous system, their role in later stages of life, requires further investigation, especially in understanding their potential role in disease. Given their progenitor capabilities, it is intriguing to speculate whether TZ glia could be targeted for potential therapies. However, more research is needed to understand the maintenance, differentiation, and function of these cells. There also needs to be a greater understanding of TZs in humans. Even with advances in single-cell sequencing of human tissue, we still do not have an understanding of the boundary cells in humans.

Further, while the majority of this review focused on the PNS side of the TZs, the CNS side of the TZs is likely critical for nervous system function. Beyond the observations that a subpopulation of oligodendrocytes in zebrafish requires the DREZ for its development and that radial glia inhibit migration of PNS cells into the CNS during development (Zhu et al. 2015; Smith et al. 2016; Green et al. 2022), few studies have investigated the CNS cells that occupy TZs. Thus, it is unclear whether specialized subpopulations of CNS cells, like seen in the PNS, populate the TZs and are critical regulators of its development and function. Finally, mammalian BCs also give rise to two stem-like cell populations in the embryonic nerves as well as embryonic and adult skin. Do BC-derived stem cells persist in the adult peripheral nerves and what might be their function in nerve repair remains an important question for future research.

Answering some of these open questions will require the generation of new tools and resources. For example, the identification of distinct molecular markers could advance understanding in model systems and humans. This then would allow BC cells to be distinguished from NC cells and other peripheral neural populations. High-resolution spatial omics, in particular, would provide the important spatial identity of the cells that could be used to identify a comprehensive molecular profile of TZ cells. With new molecular markers, the field could then generate genetic reporters in mice and zebrafish to probe the genetic and cellular mechanisms that are essential for TZ cells. The collective knowledge gained from these model systems can then be used to advance our understanding of human TZs in development and disease.

ACKNOWLEDGMENTS

We thank Lotta Barnes for constructing the first draft of the figures.

REFERENCES

Abdo H, Calvo-Enrique L, Lopez JM, Song J, Zhang MD, Usoskin D, El Manira A, Adameyko I, Hjerling-Leffler J, Ernfors P. 2019. Specialized cutaneous Schwann cells initiate pain sensation. *Science* **365:** 695–699. doi:10.1126/science.aax6452

Akert K, Sandri C, Weibel ER, Peper K, Moor H. 1976. The fine structure of the perineural endothelium. *Cell Tissue Res* **165:** 281–295. doi:10.1007/BF00222433

Altman J, Bayer SA. 1984. The development of the rat spinal cord. *Adv Anat Embryol Cell Biol* **85:** 1–164. doi:10.1007/978-3-642-69537-7

Anastasaki C, Orozco P, Gutmann DH. 2022. RAS and beyond: the many faces of the neurofibromatosis type 1 protein. *Dis Model Mech* **15:** dmm049362. doi:10.1242/dmm.049362

Binari LA, Lewis GM, Kucenas S. 2013. Perineurial glia require notch signaling during motor nerve development but not regeneration. *J Neurosci* **33:** 4241–4252. doi:10.1523/JNEUROSCI.4893-12.2013

Brady DC, Hmeljak J, Dar AC. 2022. Understanding and drugging RAS: 40 years to break the tip of the iceberg.

Dis Model Mech **15:** dmm049519. doi:10.1242/dmm.049519

Britsch S, Goerich DE, Riethmacher D, Peirano RI, Rossner M, Nave KA, Birchmeier C, Wegner M. 2001. The transcription factor Sox10 is a key regulator of peripheral glial development. *Genes Dev* **15:** 66–78. doi:10.1101/gad.186601

Bron R, Vermeren M, Kokot N, Andrews W, Little GE, Mitchell KJ, Cohen J. 2007. Boundary cap cells constrain spinal motor neuron somal migration at motor exit points by a semaphorin-plexin mechanism. *Neural Dev* **2:** 21. doi:10.1186/1749-8104-2-21

Burkel WE. 1967. The histological fine structure of perineurium. *Anat Rec* **158:** 177–189. doi:10.1002/ar.1091580207

Cichowski K, Jacks T. 2001. NF1 tumor suppressor gene function: narrowing the GAP. *Cell* **104:** 593–604. doi:10.1016/S0092-8674(01)00245-8

Cimino PJ, Gutmann DH. 2018. Neurofibromatosis type 1. *Handb Clin Neurol* **148:** 799–811. doi:10.1016/B978-0-444-64076-5.00051-X

Clark JK, O'keefe A, Mastracci TL, Sussel L, Matise MP, Kucenas S. 2014. Mammalian *Nkx2.2*$^+$ perineurial glia are essential for motor nerve development. *Dev Dyn* **243:** 1116–1129. doi:10.1002/dvdy.24158

Coulpier F, Le Crom S, Maro GS, Manent J, Giovannini M, Maciorowski Z, Fischer A, Gessler M, Charnay P, Topilko P. 2009. Novel features of boundary cap cells revealed by the analysis of newly identified molecular markers. *Glia* **57:** 1450–1457. doi:10.1002/glia.20862

Coulpier F, Decker L, Funalot B, Vallat JM, Garcia-Bragado F, Charnay P, Topilko P. 2010. CNS/PNS boundary transgression by central glia in the absence of Schwann cells or Krox20/Egr2 function. *J Neurosci* **30:** 5958–5967. doi:10.1523/JNEUROSCI.0017-10.2010

Coulpier F, Pulh P, Oubrou L, Naudet J, Fertitta L, Gregoire JM, Bocquet A, Schmitt AM, Wolkenstein P, Radomska KJ, Topilko P. 2023. Topical delivery of mitogen-activated protein kinase inhibitor binimetinib prevents the development of cutaneous neurofibromas in neurofibromatosis type 1 mutant mice. *Transl Res* **261:** 16–27. doi:10.1016/j.trsl.2023.06.003

Di Maio A, Skuba A, Himes BT, Bhagat SL, Hyun JK, Tessler A, Bishop D, Son YJ. 2011. In vivo imaging of dorsal root regeneration: rapid immobilization and presynaptic differentiation at the CNS/PNS border. *J Neurosci* **31:** 4569–4582. doi:10.1523/JNEUROSCI.4638-10.2011

Fontenas L, Kucenas S. 2017. Livin' on the edge: glia shape nervous system transition zones. *Curr Opin Neurobiol* **47:** 44–51. doi:10.1016/j.conb.2017.09.008

Fontenas L, Kucenas S. 2018. Motor exit point (MEP) glia: novel myelinating glia that bridge CNS and PNS myelin. *Front Cell Neurosci* **12:** 1–8. doi:10.3389/fncel.2018.00333

Fontenas L, Kucenas S. 2021. Spinal cord precursors utilize neural crest cell mechanisms to generate hybrid peripheral myelinating glia. *eLife* **10:** e64267. doi:10.7554/eLife.64267

Fontenas L, Welsh TG, Piller M, Coughenour P, Gandhi AV, Prober DA, Kucenas S. 2019. The neuromodulator adenosine regulates oligodendrocyte migration at motor exit point transition zones. *Cell Rep* **27:** 115–128.e5. doi:10.1016/j.celrep.2019.03.013

Fraher JP. 1992. The CNS—PNS transitional zone of the rat. morphometric studies at cranial and spinal levels. *Prog Neurobiol* **38**: 261–316. doi:10.1016/0301-0082(92)90022-7

Fraher JP. 1997. Axon-glial relationships in early CNS-PNS transitional zone development: an ultrastructural study. *J Neurocytol* **26**: 41–52. doi:10.1023/A:1018511425126

Freeman MR, Doherty J. 2006. Glial cell biology in *Drosophila* and vertebrates. *Trends Neurosci* **29**: 82–90. doi:10.1016/j.tins.2005.12.002

Fröb F, Bremer M, Finzsch M, Kichko T, Reeh P, Tamm ER, Charnay P, Wegner M. 2012. Establishment of myelinating Schwann cells and barrier integrity between central and peripheral nervous systems depend on *Sox10*. *Glia* **60**: 806–819. doi:10.1002/glia.22310

Furlan A, Adameyko I. 2018. Schwann cell precursor: a neural crest cell in disguise? *Dev Biol* **444**: S25–S35. doi:10.1016/j.ydbio.2018.02.008

Garrett AM, Jucius TJ, Sigaud LPR, Tang FL, Xiong WC, Ackerman SL, Burgess RW. 2016. Analysis of expression pattern and genetic deletion of netrin5 in the developing mouse. *Front Mol Neurosci* **9**: 359. doi:10.3389/fnmol.2016.00003

Gerschenfeld G, Coulpier F, Gresset A, Pulh P, Job B, Topilko T, Siegenthaler J, Kastriti ME, Brunet I, Charnay P, et al. 2023. Neural tube-associated boundary caps are a major source of mural cells in the skin. *eLife* **12**: e69413. doi:10.7554/eLife.69413

Golding JP, Cohen J. 1997. Border controls at the mammalian spinal cord: late-surviving neural crest boundary cap cells at dorsal root entry sites may regulate sensory afferent ingrowth and entry zone morphogenesis. *Mol Cell Neurosci* **9**: 381–396. doi:10.1006/mcne.1997.0647

Golding J, Shewan D, Cohen J. 1997. Maturation of the mammalian dorsal root entry zone—from entry to no entry. *Trends Neurosci* **20**: 303–308. doi:10.1016/S0166-2236(96)01044-2

Green LA, Gallant RM, Brandt JP, Nichols EL, Smith CJ. 2022. A subset of oligodendrocyte lineage cells interact with the developing dorsal root entry zone during its genesis. *Front Cell Neurosci* **16**: 1–20. doi:10.3389/fncel.2022.893629

Gresset A, Coulpier F, Gerschenfeld G, Jourdon A, Matesic G, Richard L, Vallat JM, Charnay P, Topilko P. 2015. Boundary caps give rise to neurogenic stem cells and terminal glia in the skin. *Stem Cell Rep* **5**: 278–290. doi:10.1016/j.stemcr.2015.06.005

Honjo Y, Kniss J, Eisen JS. 2008. Neuregulin-mediated ErbB3 signaling is required for formation of zebrafish dorsal root ganglion neurons. *Development* **135**: 2993. doi:10.1242/dev.027763

Jessen KR, Mirsky R. 2005. The origin and development of glial cells in peripheral nerves. *Nat Rev Neurosci* **6**: 671–682. doi:10.1038/nrn1746

Kang HK, Min SK, Jung SY, Jung K, Jang DH, Kim OB, Chun GS, Lee ZH, Min BM. 2011. The potential of mouse skin-derived precursors to differentiate into mesenchymal and neural lineages and their application to osteogenic induction in vivo. *Int J Mol Med* **28**: 1001–1011. doi:10.3892/ijmm.2011.785

Klämbt C, Goodman CS. 1991. The diversity and pattern of glia during axon pathway formation in the *Drosophila* embryo. *Glia* **4**: 205–213. doi:10.1002/glia.440040212

Kristensson K, Olsson Y. 1971. The perineurium as a diffusion barrier to protein tracers. *Acta Neuropathol* **17**: 127–138. doi:10.1007/BF00687488

Kucenas S, Takada N, Park HC, Woodruff E, Broadie K, Appel B. 2008. CNS-derived glia ensheath peripheral nerves and mediate motor root development. *Nat Neurosci* **11**: 143–151. doi:10.1038/nn2025

Kucenas S, Wang WD, Knapik EW, Appel B. 2009. A selective glial barrier at motor axon exit points prevents oligodendrocyte migration from the spinal cord. *J Neurosci* **29**: 15187–15194. doi:10.1523/JNEUROSCI.4193-09.2009

Lewis GM, Kucenas S. 2014. Perineurial glia are essential for motor axon regrowth following nerve injury. *J Neurosci* **34**: 12762–12777. doi:10.1523/JNEUROSCI.1906-14.2014

Maro GS, Vermeren M, Voiculescu O, Melton L, Cohen J, Charnay P, Topilko P. 2004. Neural crest boundary cap cells constitute a source of neuronal and glial cells of the PNS. *Nat Neurosci* **7**: 930–938. doi:10.1038/nn1299

Mauti O, Domanitskaya E, Andermatt I, Sadhu R, Stoeckli ET. 2007. Semaphorin6A acts as a gate keeper between the central and the peripheral nervous system. *Neural Dev* **2**: 28. doi:10.1186/1749-8104-2-28

McGraw HF, Snelson CD, Prendergast A, Suli A, Raible DW. 2012. Postembryonic neuronal addition in Zebrafish dorsal root ganglia is regulated by Notch signaling. *Neural Dev* **7**: 23. doi:10.1186/1749-8104-7-23

Mogha A, Benesh AE, Patra C, Engel FB, Schöneberg T, Liebscher I, Monk KR. 2013. Gpr126 functions in Schwann cells to control differentiation and myelination via G-protein activation. *J Neurosci* **33**: 17976–17985. doi:10.1523/JNEUROSCI.1809-13.2013

Monk KR, Naylor SG, Glenn TD, Mercurio S, Perlin JR, Dominguez C, Moens CB, Talbot WS. 2009. A G protein-coupled receptor is essential for Schwann cells to initiate myelination. *Science* **325**: 1402–1405. doi:10.1126/science.1173474

Nichols EL, Smith CJ. 2019. Pioneer axons employ Cajal's battering ram to enter the spinal cord. *Nat Commun* **10**: 562. doi:10.1038/s41467-019-08421-9

Nichols EL, Smith CJ. 2020. Functional regeneration of the sensory root via axonal invasion. *Cell Rep* **30**: 9–17.e3. doi:10.1016/j.celrep.2019.12.008

Nichols EL, Green LA, Smith CJ. 2018. Ensheathing cells utilize dynamic tiling of neuronal somas in development and injury as early as neuronal differentiation. *Neural Dev* **13**: 19. doi:10.1186/s13064-018-0115-8

Niederländer C, Lumsden A. 1996. Late emigrating neural crest cells migrate specifically to the exit points of cranial branchiomotor nerves. *Development* **122**: 2367–2374. doi:10.1242/dev.122.8.2367

Parker RJ, Auld VJ. 2006. Roles of glia in the *Drosophila* nervous system. *Semin Cell Dev Biol* **17**: 66–77. doi:10.1016/j.semcdb.2005.11.012

Radomska KJ, Topilko P. 2017. Boundary cap cells in development and disease. *Curr Opin Neurobiol* **47**: 209–215. doi:10.1016/j.conb.2017.11.003

Radomska KJ, Coulpier F, Gresset A, Schmitt A, Debbiche A, Lemoine S, Wolkenstein P, Vallat JM, Charnay P, Topilko

Cite this article as *Cold Spring Harb Perspect Biol* doi: 10.1101/cshperspect.a041369

P. 2019. Cellular origin, tumor progression, and pathogenic mechanisms of cutaneous neurofibromas revealed by mice with *Nf1* knockout in boundary cap cells. *Cancer Discov* 9: 130–147. doi:10.1158/2159-8290.CD-18-0156

Raible DW, Eisen JS. 1994. Restriction of neural crest cell fate in the trunk of the embryonic zebrafish. *Development* 120: 495–503. doi:10.1242/dev.120.3.495

Raible DW, Wood A, Hodsdon W, Henion PD, Weston JA, Eisen JS. 1992. Segregation and early dispersal of neural crest cells in the embryonic zebrafish. *Dev Dyn* 195: 29–42. doi:10.1002/aja.1001950104

Riethmacher D, Sonnenberg-Riethmacher E, Brinkmann V, Yamaai T, Lewin GR, Birchmeier C. 1997. Severe neuropathies in mice with targeted mutations in the ErbB3 receptor. *Nature* 389: 725–730. doi:10.1038/39593

Schmidt H, Rickert C, Bossing T, Vef O, Urban J, Technau GM. 1997. The embryonic central nervous system lineages of *Drosophila melanogaster*. II: Neuroblast lineages derived from the dorsal part of the nueroectoderm. *Dev Biol* 189: 186–204. doi:10.1006/dbio.1997.8660

Sepp KJ, Schulte J, Auld VJ. 2000. Developmental dynamics of peripheral glia in *Drosophila melanogaster*. *Glia* 30: 122–133. doi:10.1002/(sici)1098-1136(200004)30:2<122::aid-glia2>3.0.co;2-b

Sepp KJ, Schulte J, Auld VJ. 2001. Peripheral glia direct axon guidance across the CNS/PNS transition zone. *Dev Biol* 238: 47–63. doi:10.1006/dbio.2001.0411

Shantha TR, Golarz MN, Bourne GH. 1968. Histological and histochemical observations on the capsule of the muscle spindle in normal and denervated muscle. *Acta Anat (Basel)* 69: 632–646. doi:10.1159/000143103

Shanthaveerappa TR, Bourne GH. 1962. A perineural epithelium. *J Cell Biol* 14: 343–346. doi:10.1083/jcb.14.2.343

Shanthaveerappa TR, Bourne GH. 1966. Perineural epithelium: a new concept of its role in the integrity of the peripheral nervous system. *Science* 154: 1464–1467. doi:10.1126/science.154.3755.1464

Smith CJ, Morris AD, Welsh TG, Kucenas S. 2014. Contact-mediated inhibition between oligodendrocyte progenitor cells and motor exit point glia establishes the spinal cord transition zone. *PLoS Biol* 12: e1001961. doi:10.1371/journal.pbio.1001961

Smith CJ, Johnson K, Welsh TG, Barresi MJF, Kucenas S. 2016. Radial glia inhibit peripheral glial infiltration into the spinal cord at motor exit point transition zones. *Glia* 64: 1138–1153. doi:10.1002/glia.22987

Smith CJ, Wheeler MA, Marjoram L, Bagnat M, Deppmann CD, Kucenas S. 2017. TNFa/TNFR2 signaling is required for glial ensheathment at the dorsal root entry zone. *PLoS Genet* 13: e1006712. doi:10.1371/journal.pgen.1006712

Suter TACS, Jaworski A. 2019. Cell migration and axon guidance at the border between central and peripheral nervous system. *Science* 365: eaaw8231. doi:10.1126/science.aaw8231

Tang X, Skuba A, Han S, Kim H, Ferguson T, Son Y. 2012. Sensory nerve regeneration at the CNS-PNS interface. *InTech* doi:10.5772/29384

Topilko P, Schneider-Maunoury S, Levi G, Baron-Van Evercooren A, Chennoufi ABY, Seitanidou T, Babinet C, Charnay P. 1994. Krox-20 controls myelination in the peripheral nervous system. *Nature* 371: 796–799. doi:10.1038/371796a0

Vermeren M, Maro GS, Bron R, McGonnell IM, Charnay P, Topilko P, Cohen J. 2003. Integrity of developing spinal motor columns is regulated by neural crest derivatives at motor exit points. *Neuron* 37: 403–415. doi:10.1016/S0896-6273(02)01188-1

Zhu Y, Matsumoto T, Nagasawa T, Mackay F, Murakami F. 2015. Chemokine signaling controls integrity of radial glial scaffold in developing spinal cord and consequential proper position of boundary cap cells. *J Neurosci* 35: 9211–9224. doi:10.1523/JNEUROSCI.0156-15.2015

Zhu Y, Crowley SC, Latimer AJ, Lewis GM, Nash R, Kucenas S. 2019. Migratory neural crest cells phagocytose dead cells in the developing nervous system. *Cell* 179: 74–89.e10. doi:10.1016/j.cell.2019.08.001

Zujovic V, Thibaud J, Bachelin C, Vidal M, Coulpier F, Charnay P, Topilko P, Baron-Van Evercooren A. 2010. Boundary cap cells are highly competitive for CNS remyelination: fast migration and efficient differentiation in PNS and CNS myelin-forming cells. *Stem Cells* 28: 470–479. doi:10.1002/stem.290

Cellular Contributions to Glymphatic and Lymphatic Waste Clearance in the Brain

Leon C.D. Smyth,[1,2,5] Natalie Beschorner,[3,5] Maiken Nedergaard,[3,4] and Jonathan Kipnis[1,2]

[1]Brain Immunology and Glia (BIG) Center, Washington University in St. Louis, St. Louis, Missouri 63110, USA

[2]Department of Pathology and Immunology, School of Medicine, Washington University in St. Louis, St. Louis, Missouri 63110, USA

[3]Center for Translational Neuromedicine, University of Copenhagen, 2200 Copenhagen, Denmark

[4]Department of Neurosurgery, Center for Translational Neuromedicine, University of Rochester Medical Center, Rochester, New York 14642, USA

Correspondence: maiken.nedergaard@sund.ku.dk; kipnis@wustl.edu

Cerebrospinal fluid (CSF) bathes and cushions the brain; however, it also serves a major role in the clearance of metabolic wastes and in the distribution of glucose, lipids, and amino acids. Unlike every other organ in the body, the brain parenchyma lacks a traditional lymphatic system to drain fluids and central nervous system (CNS) antigens. It was historically assumed that all brain wastes were removed by endogenous processing, such as phagocytosis and autophagy, while excess fluids drained directly into the blood. However, the twin discoveries of the glial-lymphatic (glymphatic) system and meningeal lymphatics have transformed our understanding of brain waste clearance. The glymphatic system describes the movement of fluids through the subarachnoid space (SAS), the influx along periarterial spaces into the brain parenchyma, and the ultimate efflux back into the SAS along perivenous spaces where it comes into direct contact with the meningeal lymphatics. The dura mater of the meninges contains a bona fide lymphatic network that can drain CSF that has entered the dura. Together, these pathways provide insights into the clearance of molecules and fluids from the brain, and show that the CNS is physically connected to the adaptive immune system. Here, we outline the glymphatic and lymphatic systems, and describe the cellular components that are important to their function.

THE ANATOMY OF BRAIN CLEARANCE

The majority of cerebrospinal fluid (CSF) is produced by the choroid plexus, a highly vascularized epithelial tissue that sits within the ventricles of the brain (Lun et al. 2015). Unlike the rest of the brain, the vasculature of the choroid plexus is fenestrated, allowing free leakage of plasma (Wang et al. 2019); however, the abluminal junctions between the epithelial cells of the choroid plexus form the blood–CSF barrier, preventing the free entry of these plasma products into the ventricular circulation (Vong et al. 2021). Choroid plexus epithelial cells then take fluid from the interstitium of the choroid plexus and actively secrete CSF into the ventricles (Steffensen et al. 2018; Shipley et al. 2020). In humans, the volume of CSF produced is estimated to be 430–1000 mL per day (Huang et al. 2004). This is far greater than the total CSF volume of 120–150 mL, meaning that the entire CSF volume is turned over 3–4 times a day (Huang et al. 2004). CSF secretion is dependent on several ion transporters including the sodium/potassium ATPase, and the sodium–potassium–chloride cotransporter (Steffensen et al. 2018; Xu et al. 2021). These are thought to generate an osmotic gradient that drives water into the ventricles, facilitated by aquaporin-1 (Aqp1) water channels (Steffensen et al. 2018; Rasmussen et al. 2022). Recent in vivo imaging of choroid plexus epithelial cells has suggested that they perform apocrine secretion, a process that may be a more significant contributor to the formation of CSF and its composition than has been previously appreciated (Shipley et al. 2020; Courtney et al. 2024). Choroidal CSF secretion can be rapidly modulated through serotonin signaling (Shipley et al. 2020; Courtney et al. 2024), and may play a role in the regulation of CSF formation in response to changes in intracranial pressure or circadian rhythm (Myung et al. 2018). While it is generally thought that the choroid plexus produces the majority of the CSF around the CNS, there are also extrachoroidal sources of CSF, including production through hydrolysis and transport across the blood–brain barrier (BBB) (Roques et al. 2021; Rasmussen et al. 2022). CSF produced in the ventricles has a low protein content, and is comprised mostly of albumin, transthyretin, and other molecules that offer support to brain cells, such as insulin-like growth factor-1 (Salehi et al. 2008; Jang et al. 2022).

After being produced by the choroid plexus, CSF is circulated by ciliated ependymal cells through the lateral, third, and fourth ventricles (Faubel et al. 2016). The cilia on ependymal cells have concerted beating patterns that help propel CSF in a directional manner down the foramen of Magendie into the pontine cistern and foramen of Luschka into the cisterna magna (Faubel et al. 2016; Olstad et al. 2019). Once in the cisterna magna, the CSF has now entered the subarachnoid space (SAS), a potential space that sits between the two innermost meningeal layers (leptomeninges), the pia mater and the arachnoid mater (Mestre et al. 2018b). The pia mater follows the contours of the brain, and is firmly attached to the astrocyte endfeet of the glia limitans (Mestre et al. 2022). On the other hand, the arachnoid mater drapes loosely over the top of the pia mater, to which it is anchored by thin collagenous trabeculae (Mestre et al. 2018b). The arachnoid mater forms a second barrier to free mixture of the CSF with peripheral fluids, with the outermost layer of arachnoid barrier cells (ABCs) forming tight junctions that prevent free exchange of dural interstitial fluid (ISF) and CSF (Derk et al. 2022; Pietilä et al. 2023; Smyth et al. 2024). The SAS is highly vascularized, mainly containing large caliber arteries and veins (Mastorakos and McGavern 2019). Arteries in the SAS branch into penetrating arterioles, which are oriented perpendicular to the brain surface, and dive deep into the parenchyma. These penetrating arterioles bring the pia mater with them as they dive into the brain (Mestre et al. 2022), forming a perivascular space (PVS) between the pia mater and glial limitans at the outer extent, and the blood vessel wall at the inner extent (Wardlaw et al. 2020). Contrary to descriptions in textbooks, the SAS is not a uniformly open fluid-filled space, but forms tent-like structures around large vessels, since these hold the two layers of leptomeninges apart (Mestre et al. 2018b). This leads to the SAS appearing eye-shaped around vessels, while at a distance from vessels, the SAS is relatively collapsed (Hartmann et al. 2019). On the other hand, PVSs in penetrating arteries are circular, with the boundary formed entirely by the pia/glial limitans (Bojarskaite et al. 2023).

In the SAS, CSF flows in a glymphatic manner through PVSs to perfuse the brain. Glymphatic flow of CSF occurs in three phases (Fig. 1A; Iliff et al. 2012):

1. Subarachnoid/periarterial influx of CSF.

2. Parenchymal flow of CSF.

3. Perivenous/subarachnoid efflux of CSF.

Initially, CSF from the cisterna magna follows the large arteries in the circle of Willis,

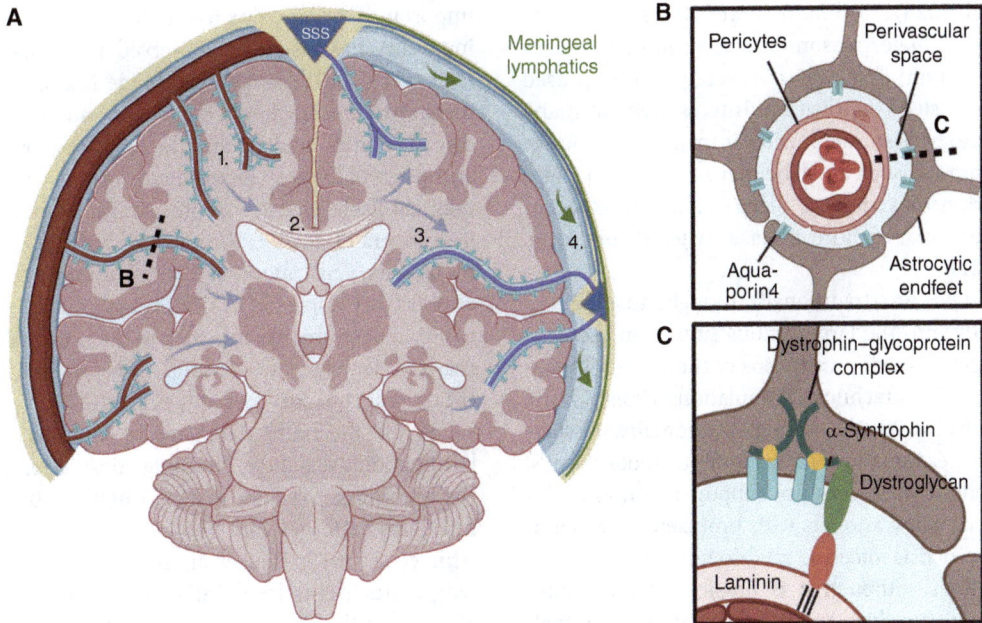

Figure 1. Glymphatic–lymphatic fluid flow within the human brain. (*A*) Glymphatic flow in the human brain (1) cerebrospinal fluid (CSF) within the ventricles and the subarachnoid space (SAS) is in constant motion and partly flows along the perivascular spaces surrounding the major cerebral arteries. CSF can enter the brain parenchyma along the entire length of the periarterial space where CSF influx is facilitated by aquaporin-4 (Aqp4) water channels. (2) In the interstitial space of the brain parenchyma, CSF dilutes the interstitial fluid (ISF) and drives the fluid including potentially toxic metabolic products within the tortuous interstitial space in the parenchyma until it reaches perivenous spaces. (3) Removal of ISFs via venous perivascular spaces, along white matter fiber tracts and beneath the ependymal membrane of the ventricles. From the SAS, a meningeal lymphatic vessel network drains ISF and CSF into the cervical lymph nodes. (*B*) Anatomy of perivascular spaces around penetrating arteries. (*C*) Perivascular localization of Aqp4 in the brain is dependent on the dystrophin–glycoprotein complex. Dystrophin is further connected to the transmembrane proteins β- and α-dystroglycan that are associated with the endothelial basal lamina.

moving along the basal cisterns toward the olfactory bulbs (Iliff et al. 2013a,b; Sweeney et al. 2019). CSF flow then follows the ambient cistern and the olfactory recess, projecting up toward the dorsal portions of the brain (Iliff et al. 2013a). It also projects up along the middle cerebral artery, flowing in the SAS above the cortex (Iliff et al. 2013a). CSF is driven into the PVSs of penetrating arterioles (Iliff et al. 2013b; Mestre et al. 2018b). Along the entire PVS of the penetrating arteriole, CSF can cross the glia limitans to enter the brain parenchyma. This fluid flow across the glia limitans is facilitated by aquaporin-1 (Aqp4) water channels that are clustered on the luminal side of the astrocytic endfeet (Iliff et al. 2012). The influx of CSF from the arterial

PVSs drives the movement of brain ISF (Iliff et al. 2012; Mestre et al. 2020b). CSF influx to the parenchyma is also thought to promote the outflow of a CSF–ISF mixture that includes waste products of metabolic activity. The route for brain ISF to enter the PVS and SAS has been more contentious, with suggestions that it enters at both the penetrating arterioles and ascending venules (Iliff et al. 2012; van veluw et al. 2020). The glymphatic system is most active during sleep, which activates CSF flow to a similar degree as that seen under ketamine anesthesia (Xie et al. 2013). In particular, loss of noradrenergic tone during sleep and the changes to vascular dynamics are thought to drive this phenomenon (Xie et al. 2013; Turner et al. 2020; Bojarskaite et al. 2023). Facilitating brain clearance is likely to be a major reason for sleep and may explain why brain function is severely compromised under sleep deprivation. Interestingly, in many diseases associated with impaired CSF waste clearance, such as Alzheimer's disease (AD) and Parkinson's disease, sleep disturbances occur many years before disease onset (Sabia et al. 2021).

CSF was traditionally thought to drain directly into the large venous sinuses in the dura mater through protrusions of the arachnoid mater called arachnoid granulations (Pollay 2010; Shah et al. 2023). Although this remains the standard explanation in medical textbooks, most modern work does not support this notion. Indeed, many animals with proficient CSF drainage, such as mice and rats, do not have true granulations, although it is possible that primitive granulation-like structures exist in these animals (Radoš et al. 2021). Likewise, human infants also lack granulations but CSF drainage is intact (Radoš et al. 2021). More recently, it has been suggested that arachnoid granulations are conduits for CSF to enter the dura, rather than the blood (Shah et al. 2023). Tracers injected into the CSF accumulate in the lymph nodes before being detected in the blood, suggesting that rapid lymphatic efflux of CSF precedes entry to the blood (Ma et al. 2017). Furthermore, CSF tracers accumulate very rapidly in the dura, in a lymphatic-independent manner (Rustenhoven et al. 2021). Until recently, it was not fully understood how CSF was able to reach the dura, because the SAS and the dura mater are partitioned by the presence of the arachnoid barrier (Nabeshima et al. 1975). However, several reports have shed light on anatomical connections between the dura mater and the SAS that occur through gaps in the arachnoid barrier (Hsu et al. 2022; Spera et al. 2023; Smyth et al. 2024). These are present around bridging veins, which drain blood from the SAS into the dural venous sinuses. Bridging veins create small perivascular discontinuities in the arachnoid barrier, termed arachnoid cuff exit (ACE) points, which funnel CSF toward lymphatics adjacent to the dural venous sinuses (Smyth et al. 2024). This perivenous CSF efflux pathway is linked to the glymphatic system, and

importantly tracer efflux to the dura near bridging veins has also been observed in humans (Meng et al. 2019; Ringstad and Eide 2020). Furthermore, the arachnoid mater has additional discontinuities underneath the olfactory bulb, and potentially near the pituitary gland, which may similarly enable CSF to escape from the SAS (Spera et al. 2023; Yoon et al. 2024). Once in the dura, CSF can signal to dural immune cells and instruct responses to CNS antigens or move to adjacent compartments such as the skull bone marrow to regulate immune cell production (Mazzitelli et al. 2022; Pulous et al. 2022).

While the brain parenchyma contains no lymphatic vasculature, the outermost layer of the meninges, the dura mater, contains a bona fide lymphatic network (Louveau et al. 2015; Ahn et al. 2019; Hsu et al. 2022). Meningeal lymphatics drain local ISF, but because CSF also enters the dura, they drain CSF too (Louveau et al. 2015, 2018; Da Mesquita et al. 2018). The meningeal lymphatics are closely aligned to the large venous sinuses (Rustenhoven et al. 2021). On the dorsal aspect of the dura mater, lymphatic capillaries are clustered around the rostral rhinal vein and transverse sinus, the same regions that receive the largest amount of CSF (Louveau et al. 2018; Rustenhoven et al. 2021). Lymphatic vessels are also present around the petrosal, sigmoid, and cavernous sinuses around the base of the brain, as well as within the dura mater underlying the olfactory bulb, in a continuous network with the rest of the dura (Ahn et al. 2019; Hsu et al. 2022; Spera et al. 2023; Yoon et al. 2024). The spinal cord dura mater also contains lymphatic vessels, mostly clustered around dorsal root ganglia, and although less is known about the biology of spinal cord meningeal lymphatics, it is likely that they play an analogous role to those in the cranial meninges (Louveau et al. 2018; Jacob et al. 2019). Ligation of the afferent lymphatic vessels draining into the deep cervical lymph node (dcLN) and ablation of dorsal lymphatic vessels with visudyne both prevent appropriate drainage of CSF, but surprisingly, also disrupt glymphatic flow (Da Mesquita et al. 2018). This lymphatic drainage of CSF and CSF-surveilling antigen-presenting cells is critical to adaptive

Cite this article as *Cold Spring Harb Perspect Biol* doi: 10.1101/cshperspect.a041370

immune responses to CNS antigens (Louveau et al. 2018; Song et al. 2020; Rustenhoven et al. 2021). Furthermore, drainage of CSF by meningeal lymphatics is also critical to the clearance of wastes from the brain parenchyma, with Aβ deposition exacerbated in mouse models for β-amyloidosis when meningeal lymphatics are disrupted (Da Mesquita et al. 2018, 2021b). Importantly, there is strong evidence from histological and imaging studies that both glymphatic flow (Eide et al. 2018; Ringstad et al. 2018; Meng et al. 2019; Ringstad and Eide 2020) and meningeal lymphatic drainage of CSF (Absinta et al. 2017; Eide et al. 2018; Albayram et al. 2022; Jacob et al. 2022) also occur in humans.

The four major components of CSF flow in the brain are:

1. Aqp4 polarization to vascular endfeet of astrocytes, which increases the water permeability of the glia limitans to facilitate fluid flow through the brain.

2. Vasomotion and vascular pulsatility, which drive perivascular fluid flow in the SAS and PVS.

3. Intracranial pressure, which increases resistance to fluid flow in the SAS, PVS, and the brain.

4. State of brain activity and interstitial volume fraction.

Collectively, cells of the brain and the meninges can influence CSF flow by modifying these factors. We have summarized their contributions below.

ASTROCYTES

The glymphatic system was originally named for the central role of astrocytes in its function and the pseudolymphatic function performed (glial-lymphatic). Much of the work on astrocytes in the glymphatic system has revolved around the role of Aqp4. Astrocytes project endfeet to the vasculature, creating the glia limitans, a boundary between the brain parenchyma and its border tissues including blood vasculature and leptomeninges (Sofroniew 2015). Astrocyte endfeet at both the PVS and projecting to the pia are highly enriched in the water channel Aqp4 (Fig. 1B; Nagelhus et al. 1998). As described above, penetrating arterioles and ascending venules are surrounded by a CSF-filled PVS, which is contacted on its outer limit by astrocyte endfeet. In the original report from 2012, it was found that CSF tracers flowed along the PVSs of penetrating arteries, and arterial pulsations forced these tracers into the parenchyma (Iliff et al. 2012, 2013b). It was hypothesized that Aqp4 enhances the water permeability of the endfeet, enabling the efficient pressure-dependent flow of fluid from the PVS into the brain parenchyma (Iliff et al. 2012). Indeed, Aqp4 knockout animals have similar movement of CSF through the SAS and PVS, but perfusion of these tracers to the parenchyma is abrogated (Gomolka et al. 2022). The critical role for Aqp4 in CSF flow and efflux has been substantiated in multiple different models, including animals where the Aqp4 scaffold, α-syntrophin (*Snta*) is knocked out, as well as with pharmacological inhibitors (Mestre et al. 2018a; Harrison et al. 2020). Many laboratories use the intensity of vascular-associated Aqp4 staining to that not associated with the vasculature, the Aqp4 polarization index, as a proxy for the extent of CSF perfusion of the brain parenchyma (Zeppenfeld et al. 2017; Hablitz et al. 2020). Despite this, it is uncertain how Aqp4 drives the flow of CSF in the parenchyma, since it would only facilitate the movement of water into the astrocyte soma. It is possible that Aqp4 enables the pulsatile entry of water from the PVS into the astrocytic syncytium, enabling it to pulse, and thereby drive exchange between the PVS and the parenchyma. Alternatively, it may alter the biophysical properties of astrocyte endfeet to make them more permissive to the paracellular movement of fluids.

The factors governing the polarization of Aqp4 to the vasculature are poorly understood; however, it is possible that vascular mural cells (pericytes, fibroblasts, and vascular smooth muscle cells [vSMCs]) play an important role in defining the specialization of astrocyte endfeet. Indeed, deletion of the *Pdgfb* retention motif (*Pdgfb*[Ret/Ret] mice) leads to the mislocalization of Aqp4 (Armulik et al. 2010; Munk et al. 2019). In this model, there is a population of non-vascular-associated pericytes, and astrocytic Aqp4 is also localized to these cells, indicating that mural cells guide and stabilize astrocyte endfeet (Armulik et al. 2010). It is unclear what molecular interactions drive the specialization of astrocyte endfeet; however, it is notable that many of the same molecules involved in anchoring muscle fibers to the extracellular matrix are also enriched in astrocyte endfeet (Stokum et al. 2021). This strongly suggests that laminin–dystroglycan interactions are key to astrocyte endfoot specialization (Fig. 1C), and potentially also Aqp4 localization. Interestingly, pial fibroblasts strongly and specifically express many laminin isoforms (Vanlandewijck et al. 2018). Indeed, in $Pdgfb^{Ret/Ret}$ mice, where proper Aqp4 polarization is lost, the glymphatic function is severely impaired (Munk et al. 2019). Recent evidence suggests that glymphatic flow follows a circadian rhythm, independent of sleep, with the highest activity during the rest phase in mice, indicating that sleep is synchronized to circadian rhythms to enhance the efficiency of glymphatic flow (Hablitz et al. 2020). This is thought to be facilitated by the enhanced expression of Aqp4, and the other members of the dystrophin complex that enable its polarization (Hablitz et al. 2020). In the absence of Aqp4, the circadian regulation of glymphatic flow is abolished, further demonstrating that Aqp4 is the main factor driving the circadian regulation of glymphatic flow (Hablitz et al. 2020). It is likely that astrocyte endfeet are specialized at facilitating fluid flow into the brain while maintaining a barrier with the CSF-containing spaces.

Interestingly, efflux of solutes from the brain shows biphasic kinetics, with a critical dependence on size (Plá et al. 2022). This is suggestive of size-gating of molecular transport. This may occur at glia limitans, since the SAS does not appear to impose size limits to flow. Alternatively, efflux along cranial and spinal nerves exhibit a slower time course of efflux (Mestre et al. 2018b; Rasmussen et al. 2022). Influx is also size-restricted with large dextrans entering the brain parenchyma less efficiently, compared to low molecular weight compounds (Iliff et al. 2012). Furthermore, the initial release of molecules from the brain into the periphery is slow,

followed by a more rapid efflux (Plá et al. 2022). The biphasic kinetics of ISF drainage from the brain also suggest a critical role for Aqp4 in brain ISF efflux, with initial slow Aqp4-dependent ISF efflux to the CSF followed by rapid efflux from the SAS to the periphery (Plá et al. 2022). How large molecules and debris enter the CSF is currently unknown, although astrocytes would be well-positioned to transport large cargo within their processes to the PVS or SAS. Although much of the focus of CNS clearance has been on amyloid, because it is deposited extracellularly, there is also increasing evidence that other pathological aggregates, such as tau, can be found in the CSF. Indeed it is possible that CSF is essential to the removal of these from "tombstone" aggregates in tauopathies such as AD and chronic traumatic encephalopathy. Interestingly, tau isolated from the CSF is also capable of seeding further tau aggregation, suggesting that CSF movement may play a role in the spread of tauopathy (Takeda et al. 2016).

Multiple mouse models for β-amyloidosis show redistribution of Aqp4 from astrocyte endfeet to the cell body (Yang et al. 2011; Xu et al. 2015; Smith et al. 2019). There is also evidence from human AD that Aqp4 is also mislocalized during aging and AD (Zeppenfeld et al. 2017). Interestingly, this does not apply only to models of amyloidosis, with mislocalization of Aqp4 in tauopathy, amyotrophic lateral sclerosis, aging, traumatic brain injury, and stroke (Friedman et al. 2009; Iliff et al. 2014; Kress et al. 2014; Early et al. 2020; Harrison et al. 2020). All these conditions are accompanied by reactive gliosis, suggesting that astrocyte quiescence may be important in the normal polarization of Aqp4. It should be noted that human brains show substantially reduced Aqp4 polarization to the vasculature, although this could be confounded by postmortem artifacts (Zeppenfeld et al. 2017).

Even without overexpression of Aβ, $Aqp4^{-/-}$ mice have impaired Aβ clearance (Iliff et al. 2012). Furthermore, when crossed to mouse models of amyloidosis, $Aqp4^{-/-}$ show increased levels of Aβ deposition in the parenchyma, with increased amyloid plaques, particularly diffuse plaques that are closely associated with neuronal dysfunction (Peng et al. 2016). *Aqp4* knockout mice also have increased levels of leptomeningeal amyloid and cerebral amyloid angiopathy, which further suggests that CSF flow is impaired (Xu et al. 2015). This has ramifications for brain function, with worse cognition in the APP/PS1 mouse model (Xu et al. 2015). Likewise, Aqp4 mislocalization in tauopathy models accelerates neurodegeneration (Ishida et al. 2022). On the other hand, in neurological diseases characterized by edema, Aqp4 deficiency improves outcomes by preventing water influx from the SAS into the brain parenchyma (Manley et al. 2000; Mestre et al. 2020a).

NEURONS

Although the primary focus of glymphatic flow has been on astrocytes, neurons are another important contributor to CSF flow. The discovery that sleep is the major physiological driver for glymphatic flow suggested that neurons may be involved in CSF flow (Xie et al. 2013). Indeed, in sleeping mice, CSF flow is similar to ketamine anesthesia, a major CSF flow inducer (Xie et al. 2013). This was attributed to the expansion of the interstitial space in mice, regulated by noradrenergic tone (Fig. 2C; Xie et al. 2013). The glymphatic system provides an important mechanistic link between sleep disturbances

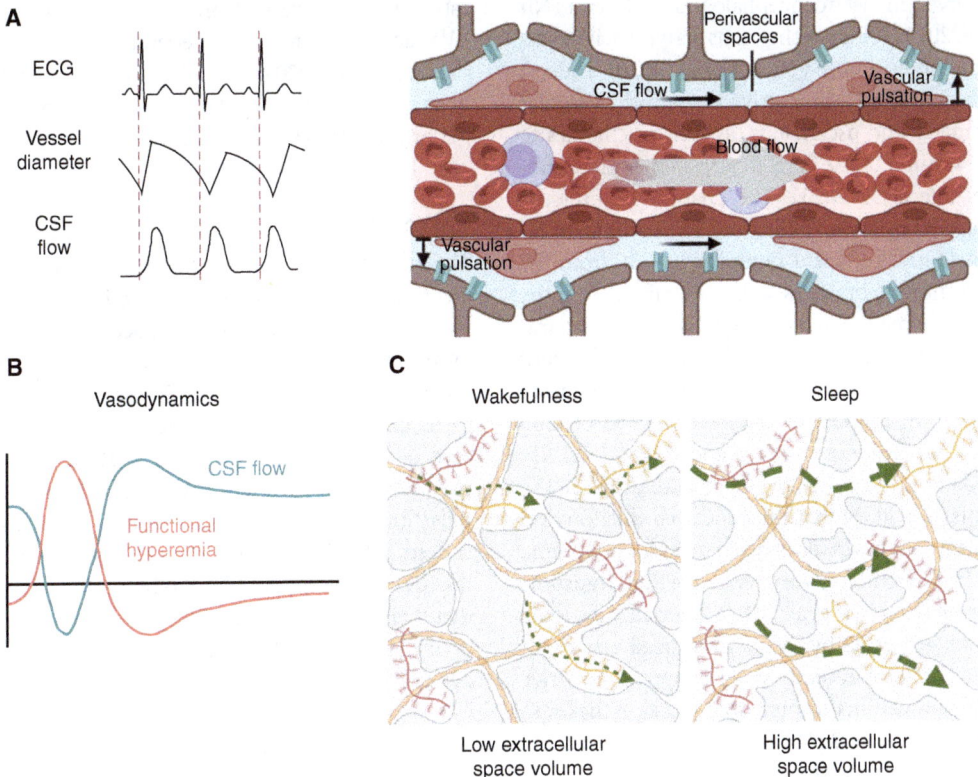

Figure 2. Physiological processes driving fluid flow in the glymphatic system. (*A*) Cerebrospinal fluid (CSF) flow within the perivascular spaces is driven by vascular pumping. During systole (R component in electrocardiograms [ECGs]), the distension of the arterial wall by cardiac pulse wave pumps CSF deep into the brain parenchyma. (*B*) Changes in arterial diameter due to vascular autoregulation correlate with CSF flow. During functional hyperemia, vasoconstriction and loss in cerebral blood volume increase CSF flow. (*C*) During sleep, brain extracellular volume increases causing less hydraulic resistance and improved fluid flow through the brain parenchyma.

and AD through its effect on clearance, and failure of sleep-driven clearance can lead to the accumulation of waste metabolites such as lactate in the brain (Lundgaard et al. 2017).

Neurovascular coupling is the demand-driven change to the blood supply to local brain regions following periods of sustained activity (Kaplan et al. 2020). Unlike neurons, which can respond rapidly to changes, the vasculature responds to signals on the order of seconds, and, as a result, during waking brain activity, most unsynchronized neuronal activity is filtered out as noise by the vasculature. However, large-scale changes to brain activity can drive CSF flow (Holstein-Rønsbo et al. 2023). Furthermore, during sleep, the brain undergoes large-scale synchronized firing, which creates the different rhythms observed in electroencephalograms (EEGs) (Nir et al. 2011). This synchronous activity drives vasomotion en masse (Turner et al. 2020). Indeed, different sleep stages drive changes to vascular diameter and oscillations, as well as changes to the volume of the PVS (Turner et al. 2020; Bojarskaite et al. 2023). The changes in pressure triggered by these oscillations are thought to facilitate more efficient pumping of CSF from the PVS into the brain, as well as its efflux from the parenchyma and into the dura (Fig. 2A; Kedarasetti et al. 2020). It is possible that the different types of vasomotion linked to different sleep stages are coupled to different aspects of CSF flow, for instance, mixing and efflux (Fig. 2B; Kedarasetti et al. 2020; Bojarskaite et al. 2023). Thus, it is likely that the concerted dilation and contraction of vessels drive changes to the volume of the interstitial space to enhance fluid transport during sleep (Xie et al. 2013). Alternatively, ionic waves generated by the synchronized activity of neurons have recently been proposed to drive CSF flow within the interstitial spaces of the brain (Jiang-Xie et al. 2024).

Interestingly, the choice of anesthetic has a profound impact on CSF flow, with ketamine and dexmedetomidine causing the strongest induction of CSF flow (Hablitz et al. 2019; Lilius et al. 2019). Studies of different anesthetics indicate that those that induce the α power band in the EEG produce the strongest induction of CSF flow, and that this correlates with increase α pow-

er in natural sleep (Hablitz et al. 2019). Indeed, in human MRI studies where the blood-oxygen-level-dependent (BOLD) signal and CSF pulsation within the fourth ventricle are examined, EEG signals drive BOLD, which precedes CSF pulsations. Although CSF pulsation within the fourth ventricle is not equivalent to CSF flow through the SAS, it is likely that some of the same processes leading to CSF pulsation within the ventricles are also at play in the SAS, which supports the notion that neuronal activity is coupled to CSF flow through its effects on the vasculature (Fultz et al. 2019). One corollary is that, in pathological conditions where neurons depolarize en masse, they drive an influx of CSF into the brain parenchyma to cause edema (Mestre et al. 2020a). This occurs in stroke, where spreading depolarization waves drive CSF from the SAS into the PVS and brain to trigger edema and swelling (Mestre et al. 2020a). Massive vasoconstriction during spreading depolarization expands the periarterial spaces and accelerates CSF influx.

VASCULAR CELLS

The flow of CSF in both the SAS and penetrating/ascending vessels has been acknowledged since the nineteenth century. These spaces present the lowest resistance path for CSF to travel, avoiding the tortuous and dense parenchyma. The cardiac cycle, but also vasomotion, alters the local pressure of the PVS (Mestre et al. 2018b). This is particularly important for arteries, where both pulsation and vasomotion are more pronounced. Indeed, intravital imaging of particles flowing in the SAS beside the middle cerebral artery shows that their velocity is tightly coupled to the cardiac cycle (Mestre et al. 2018b). Furthermore, optogenetically induced vascular pulsations enhance CSF flow along the middle cerebral artery (Holstein-Rønsbo et al. 2023). Interestingly, a similar phenomenon can also be observed in active brain regions, suggesting that brain activity and neurovascular coupling are critical to regulating CSF flow (Fig. 2B; Holstein-Rønsbo et al. 2023). On the other hand, less is known about longer term and larger scale changes to vessel diameter, such as those that occur in sleep (Turner et al. 2020; Bojarskaite et al. 2023). Modeling suggests that

Cite this article as *Cold Spring Harb Perspect Biol* doi: 10.1101/cshperspect.a041370

vasomotion triggered by neurovascular coupling is indeed linked to CSF flow in PVSs (Kedarasetti et al. 2020). Both PVSs around penetrating vessels and SAS are large enough for the flow of solutes, but also large particles. However, in principle, the flow of solutes and molecules through these compartments follows the same principles, being driven by the cardiac cycle and vasomotion.

Intriguingly, many vascular pathologies are risk factors for AD, including hypertension, diabetes, and obesity (Gottesman et al. 2017). Furthermore, the largest genetic risk factor for AD, APOE4, has a major impact on blood vessels, even in the absence of AD pathology (Montagne et al. 2021; Yamazaki et al. 2021). It is likely that many of these AD risk factors act through their effects on glymphatic clearance. Hypertension causes vascular remodeling, with increased fibrosis and vSMC coverage of vessels, which increases vascular stiffness (Baumbach and Heistad 1989; van Hespen et al. 2021). Likewise, the increased pressure within the blood vessel reduces its compliance. Collectively these decrease the amplitude of vascular pulsations during the cardiac cycle. Indeed, in a spontaneous rat model of hypertension, CSF flow is decreased (Mortensen et al. 2019). Furthermore, acute induction of hypertension through the administration of angiotensin II reduces particle velocity and increases the backflow of particles in the SAS (Mestre et al. 2018b). On the other hand, processes that improve vascular function and increase pulsation, such as exercise, improve glymphatic flow (von Holstein-Rathlou et al. 2018). While the cardiac cycle is critical to the movement of particles along arteries, less attention has been paid to what features may drive flow along bridging veins, which regulate CSF efflux. Indeed, although many veins are relatively static in mice, bridging veins have a layer of venous smooth muscle cells capable of contracting and regulating diameter (Smyth et al. 2024). The tone of the vessel will strongly impact the size of the space at the ACE point, with vasodilation decreasing its size and increasing resistance to outflow and vice versa. Hence, it is likely that factors regulating the oscillations of bridging veins are important to regulating CSF efflux.

It is possible that failure of glymphatic flow can also trigger a vicious cycle that further impairs CSF efflux. In AD, failure of glymphatic flow may contribute to amyloidosis in the brain. Interestingly, in AD, the vasculature accumulates amyloid-β (Aβ) in a process termed cerebral amyloid angiopathy (Greenberg et al. 2020). These Aβ deposits mainly occur around arteries, but the accumulation of Aβ serves to further stiffen vessels, and in severe cases can destroy their structure completely, leading to hemorrhage (Vonsattel et al. 1991). In a rat model of CAA, glymphatic transport and efflux are impaired, with a striking redistribution away from the middle cerebral artery to the ventral arteries around the circle of Willis (Chen et al. 2022). Furthermore, vessel segments covered by amyloid show reduced rhythmicity in their pulsation (Kim et al. 2020). This amyloid-induced dysfunction, much like hypertension, prevents proper CSF flow (Kim et al. 2020). Interestingly, in mouse models of amyloidosis, amyloid accumulates around bridging veins, and may be "clogging" ACE points to disrupt CSF clearance (Antila et al. 2024). Furthermore, it is possible that similar "clogging" processes may occur in other neurological diseases, such as subarachnoid hemorrhage. Such "clogging" would increase resistance to outflow, and increase intracranial pressure (ICP).

MACROPHAGES

The leptomeninges and PVSs of the brain are highly enriched in scavenger macrophages that constantly surveil the CSF (Van Hove et al. 2019; Drieu et al. 2022; Siret et al. 2022). These macrophages are closely aligned with blood vessels, but their role in CSF flow was previously unknown. Recent work indicates that the removal of CNS macrophages, either pharmacologically or genetically, impairs CSF flow within the SAS (Drieu et al. 2022). Perivascular macrophages have previously been reported to regulate arterial stiffness and pulsatility through degradation of ECM (Lim et al. 2018), and indeed macrophage depletion in the brain led to increased levels of ECM deposition that impaired vascular responses to neuronal activity. Furthermore, macrophage phenotype is also critical to the maintenance of vascular dynamics, with Lyve1⁺ scavenger macrophages particularly important to the removal of both brain wastes and ECM (Drieu et al. 2022). Many conditions, particularly infections, but also autoimmunity, neurodegeneration, and aging lead to the activation of macrophages and it would be interesting to determine whether therapeutic modification of macrophage phenotype may improve brain clearance in these conditions (Van Hove et al. 2019; De Vlaminck et al. 2022; Drieu et al. 2022; Rebejac et al. 2022).

MENINGEAL LYMPHATICS

Although it sits outside the brain, the dura mater contains a bona fide lymphatic vasculature (Aspelund et al. 2015; Louveau et al. 2015). After flowing around the brain, CSF is flushed into the dura where it mixes with dural ISF and is removed by meningeal lymphatics (Louveau et al. 2015, 2018; Rustenhoven et al. 2021; Smyth et al. 2024). Meningeal lymphatics are, therefore, able to drain CSF. This is important immunologically, allowing lymph nodes to collect brain antigens and surveil the immunologic state of the brain. In mice where meningeal lymphatic vessels were either surgically ligated, or when ablated with photoactivated drug (visudyne), glymphatic flow is reduced (Da Mesquita et al. 2018). At the time, the connection between the glymphatic and meningeal lymphatic systems was not appreciated, but this provided the first evidence that these two systems cooperate in the removal of wastes from the brain. While initial studies indicated that there is no effect of visudyne ablation on intracranial pressure, loss of drainage routes may lead to transient increases in ICP, which are then compensated by reduced CSF production, leading to an overall failure of CSF flow. Indeed, elevated ICP would also increase resistance to flow in the SAS, meaning fluid flow is less efficient (Eide et al. 2021; Xiang et al. 2022). Interestingly, meningeal lymphatic function is highly plastic, with decreased drainage in multiple models of neurological disorders including infection, brain tumors, injury, stroke, and neurodegenerative disease (Bolte et al. 2020; Hu et al. 2020; Yanev et al. 2020; Da Mesquita et al. 2021b;

Li et al. 2022). Improving meningeal lymphatic function in mouse models for β-amyloidosis improves cognitive function, and reduces amyloid load through improved glymphatic function (Da Mesquita et al. 2021b). Unlike other lymphatic beds, meningeal lymphatics are highly dependent on VEGF-C for their maintenance, and modulating the expression of this can modulate their coverage and function (Antila et al. 2017; Da Mesquita et al. 2018; Song et al. 2020). Interestingly, it has recently been shown that the accumulation of T cells in the meninges during aging leads to increased levels of the cytokine interferon γ, which in turn triggers lymphatic dysfunction (Fig. 2C; Rustenhoven et al. 2023). Meningeal lymphatic dysfunction impairs glymphatic flow but it does not affect Aqp4 polarization or expression, further suggesting that the flow through the SAS is Aqp4 independent and while CSF influx into the parenchyma is Aqp4-dependent. Interestingly, under experimental conditions when meningeal immune cells could not leave via lymphatics, for instance, CCR7⁻/⁻ mice, creating a mild local inflammatory response, the glymphatic flow was impaired and Aqp4 polarization and expression were altered (Da Mesquita et al. 2021a).

LEPTOMENINGEAL STROMA

Less work has been done on the relationship between the stromal cells of the leptomeninges and the glymphatic system. Fibroblasts sit within the SAS and PVS and are frequently associated with the vasculature (Vanlandewijck et al. 2018). In the developing brain, fibroblasts colonize the perivascular niche, arising from the meninges (Jones et al. 2023). This corresponds with the opening of the PVS, and it is possible that the arrival of fibroblasts and the resulting shift in ECM deposition around vessels enables the formation of a PVS. Although fibroblasts are likely to play multiple roles, in many disease models, their activation results in fibrosis (Kelly et al. 2016; Dorrier et al. 2021). If deposited around the vasculature, changes to ECM can increase the stiffness and reduce the compliance of vessels (REFs) (Lim et al. 2018; Drieu et al. 2022). On the other hand, deposition within the SAS and accumulation of ECM around trabeculae may increase the tortuosity of the SAS, making it more difficult for large molecules or particles to move. It is also possible that macrophages suppress activated fibroblast phenotypes, as there appear to be strong signaling connections between fibroblasts and macrophages, and macrophage removal by clodronate injection leads to fibrosis-related changes in fibroblasts (Drieu et al. 2022).

CONCLUSIONS

Glymphatic flow of CSF is a critical mechanism through which the brain removes wastes, allowing them to be drained by meningeal lymphatic vessels. In the past decade, the major processes underpinning CSF flow have been unearthed, and it is now thought that the intracranial pressure, vasomotion, and Aqp4 polarization are major players in CSF flow. CSF flow is disrupted in many neurological diseases, and it is known that inflammation interacts with many of the processes described above, providing a rational link between disease processes and failure of brain clearance. Here, we have focused studies on cells and their effects on glymphatic flow. While initially focused on the role of astroglia and the vasculature, several new cellular players are emerging as important to regulating different aspects of CSF flow including neurons, macrophages, and fibroblasts. It will be interesting to see the contribution of further cell types in the coming years including what, if any, role other brain cell types such as microglia and oligodendrocytes may play in CSF flow.

REFERENCES

Absinta M, Ha SK, Nair G, Sati P, Luciano NJ, Palisoc M, Louveau A, Zaghloul KA, Pittaluga S, Kipnis J, et al. 2017. Human and nonhuman primate meninges harbor lymphatic vessels that can be visualized noninvasively by MRI. *eLife* 6: e29738. doi:10.7554/eLife.29738

Ahn JH, Cho H, Kim JH, Kim SH, Ham JS, Park I, Suh SH, Hong SP, Song JH, Hong YK, et al. 2019. Meningeal lymphatic vessels at the skull base drain cerebrospinal fluid. *Nature* 572: 62–66. doi:10.1038/s41586-019-1419-5

Albayram MS, Smith G, Tufan F, Tuna IS, Bostancıklıoğlu M, Zile M, Albayram O. 2022. Non-invasive MR imaging of human brain lymphatic networks with connections to cervical lymph nodes. *Nat Commun* 13: 203. doi:10.1038/s41467-021-27887-0

Antila S, Karaman S, Nurmi H, Airavaara M, Voutilainen MH, Mathivet T, Chilov D, Li Z, Koppinen T, Park JH, et al. 2017. Development and plasticity of meningeal lymphatic vessels. *J Exp Med* 214: 3645–3667. doi:10.1084/jem.20170391

Antila S, Chilov D, Nurmi H, Li Z, Näsi A, Gotkiewicz M, Sitnikova V, Jäntti H, Acosta N, Koivisto H, et al. 2024. Sustained meningeal lymphatic vessel atrophy or expansion does not alter Alzheimer's disease-related amyloid pathology. *Nat Cardiovasc Res* 3: 474–491. doi:10.1038/s44161-024-00445-9

Armulik A, Genové G, Mäe M, Nisancioglu MH, Wallgard E, Niaudet C, He L, Norlin J, Lindblom P, Strittmatter K, et al. 2010. Pericytes regulate the blood-brain barrier. *Nature* 468: 557–561. doi:10.1038/nature09522

Aspelund A, Antila S, Proulx ST, Karlsen TV, Karaman S, Detmar M, Wiig H, Alitalo K. 2015. A dural lymphatic vascular system that drains brain interstitial fluid and macromolecules. *J Exp Med* 212: 991–999. doi:10.1084/jem.20142290

Baumbach GL, Heistad DD. 1989. Remodeling of cerebral arterioles in chronic hypertension. *Hypertension* 13: 968–972. doi:10.1161/01.HYP.13.6.968

Bojarskaite L, Vallet A, Bjørnstad DM, Gullestad Binder KM, Cunen C, Heuser K, Kuchta M, Mardal KA, Enger R. 2023. Sleep cycle-dependent vascular dynamics enhance perivascular cerebrospinal fluid flow and solute transport. *Nat Commun* 14: 953. doi:10.1038/s41467-023-36643-5

Bolte AC, Dutta AB, Hurt ME, Smirnov I, Kovacs MA, McKee CA, Ennerfelt HE, Shapiro D, Nguyen BH, Frost EL, et al. 2020. Meningeal lymphatic dysfunction exacerbates traumatic brain injury pathogenesis. *Nat Commun* 11: 4524. doi:10.1038/s41467-020-18113-4

Chen X, Liu X, Koundal S, Elkin R, Zhu X, Monte B, Xu F, Dai F, Pedram M, Lee H, et al. 2022. Cerebral amyloid angiopathy is associated with glymphatic transport reduction and time-delayed solute drainage along the neck arteries. *Nat Aging* 2: 214–223. doi:10.1038/s43587-022-00181-4

Courtney Y, Head JP, Yimer ED, Dani N, Shipley FB, Libermann TA, Lehtinen MK. 2024. A choroid plexus apocrine secretion mechanism shapes CSF proteome and embryonic brain development. bioRxiv doi:10.1101/2024.01.08.574486

Da Mesquita S, Louveau A, Vaccari A, Smirnov I, Cornelison RC, Kingsmore KM, Contarino C, Onengut-Gumuscu S, Farber E, Raper D, et al. 2018. Functional aspects of meningeal lymphatics in ageing and Alzheimer's disease. *Nature* 560: 185–191. doi:10.1038/s41586-018-0368-8

Da Mesquita S, Herz J, Wall M, Dykstra T, de Lima KA, Norris GT, Dabhi N, Kennedy T, Baker W, Kipnis J. 2021a. Aging-associated deficit in CCR7 is linked to worsened glymphatic function, cognition, neuroinflammation, and β-amyloid pathology. *Sci Adv* 7: eabe4601. doi:10.1126/sciadv.abe4601

Da Mesquita S, Papadopoulos Z, Dykstra T, Brase L, Farias FG, Wall M, Jiang H, Kodira CD, de Lima KA, Herz J, et al. 2021b. Meningeal lymphatics affect microglia responses and anti-Aβ immunotherapy. *Nature* **593**: 255–260. doi:10.1038/s41586-021-03489-0

Derk J, Como CN, Jones HE, Joyce LR, Kim S, Spencer BL, Bonney S, O'Rourke R, Pawlikowski B, Doran KS, Siegenthaler JA. 2022. Formation and function of the meningeal arachnoid barrier around the developing mouse brain. *Dev Cell* **58**: 635–644.e4. doi:10.1016/j.devcel.2023.03.005

De Vlaminck K, Van Hove H, Kancheva D, Scheyltjens I, Pombo Antunes AR, Bastos J, Vara-Perez M, Ali L, Mampay M, Deneyer L, et al. 2022. Differential plasticity and fate of brain-resident and recruited macrophages during the onset and resolution of neuroinflammation. *Immunity* **55**: 2085–2102.e9. doi:10.1016/j.immuni.2022.09.005

Dorrier CE, Aran D, Haenelt EA, Sheehy RN, Hoi KK, Pintarić L, Chen Y, Lizama CO, Cautivo KM, Weiner GA, et al. 2021. CNS fibroblasts form a fibrotic scar in response to immune cell infiltration. *Nat Neurosci* **24**: 234–244. doi:10.1038/s41593-020-00770-9

Drieu A, Du S, Storck SE, Rustenhoven J, Papadopoulos Z, Dykstra T, Zhong F, Kim K, Blackburn S, Mamuladze T, et al. 2022. Parenchymal border macrophages regulate the flow dynamics of the cerebrospinal fluid. *Nature* **611**: 585–593. doi:10.1038/s41586-022-05397-3

Early AN, Gorman AA, Van Eldik LJ, Bachstetter AD, Morganti JM. 2020. Effects of advanced age upon astrocyte-specific responses to acute traumatic brain injury in mice. *J Neuroinflammation* **17**: 115. doi:10.1186/s12974-020-01800-w

Eide PK, Vatnehol SAS, Emblem KE, Ringstad G. 2018. Magnetic resonance imaging provides evidence of glymphatic drainage from human brain to cervical lymph nodes. *Sci Rep* **8**: 7194. doi:10.1038/s41598-018-25666-4

Eide PK, Pripp AH, Ringstad G, Valnes LM. 2021. Impaired glymphatic function in idiopathic intracranial hypertension. *Brain Commun* **3**: fcab043. doi:10.1093/braincomms/fcab043

Faubel R, Westendorf C, Bodenschatz E, Eichele G. 2016. Cilia-based flow network in the brain ventricles. *Science* **353**: 176–178. doi:10.1126/science.aae0450

Friedman B, Schachtrup C, Tsai PS, Shih AY, Akassoglou K, Kleinfeld D, Lyden PD. 2009. Acute vascular disruption and aquaporin 4 loss after stroke. *Stroke* **40**: 2182–2190. doi:10.1161/STROKEAHA.108.523720

Fultz NE, Bonmassar G, Setsompop K, Stickgold RA, Rosen BR, Polimeni JR, Lewis LD. 2019. Coupled electrophysiological, hemodynamic, and cerebrospinal fluid oscillations in human sleep. *Science* **366**: 628–631. doi:10.1126/science.aax5440

Gomolka RS, Hablitz L, Mestre H, Giannetto M, Du T, Hauglund N, Xie L, Peng W, Martinez PM, Nedergaard M, et al. 2022. Loss of aquaporin-4 results in glymphatic system dysfunction via brain-wide interstitial fluid stagnation. *eLife* **12**: e82232. doi:10.7554/eLife.82232.sa0

Gottesman RF, Albert MS, Alonso A, Coker LH, Coresh J, Davis SM, Deal JA, McKhann GM, Mosley TH, Sharrett AR, et al. 2017. Associations between midlife vascular risk factors and 25-year incident dementia in the atherosclerosis risk in communities (ARIC) cohort. *JAMA Neurol* **74**: 1246–1254. doi:10.1001/jamaneurol.2017.1658

Greenberg SM, Bacskai BJ, Hernandez-Guillamon M, Pruzin J, Sperling R, van Veluw SJ. 2020. Cerebral amyloid angiopathy and Alzheimer disease—one peptide, two pathways. *Nat Rev Neurol* **16**: 30–42. doi:10.1038/s41582-019-0281-2

Hablitz LM, Vinitsky HS, Sun Q, Stæger FF, Sigurdsson B, Mortensen KN, Lilius TO, Nedergaard M. 2019. Increased glymphatic influx is correlated with high EEG δ power and low heart rate in mice under anesthesia. *Sci Adv* **5**: eaav5447. doi:10.1126/sciadv.aav5447

Hablitz LM, Plá V, Giannetto M, Vinitsky HS, Stæger FF, Metcalfe T, Nguyen R, Benrais A, Nedergaard M. 2020. Circadian control of brain glymphatic and lymphatic fluid flow. *Nat Commun* **11**: 4411. doi:10.1038/s41467-020-18115-2

Harrison IF, Ismail O, Machhada A, Colgan N, Ohene Y, Nahavandi P, Ahmed Z, Fisher A, Meftah S, Murray TK, et al. 2020. Impaired glymphatic function and clearance of tau in an Alzheimer's disease model. *Brain* **143**: 2576–2593. doi:10.1093/brain/awaa179

Hartmann K, Stein KP, Neyazi B, Sandalcioglu IE. 2019. First in vivo visualization of the human subarachnoid space and brain cortex via optical coherence tomography. *Ther Adv Neurol Disord* **12**: 1756286419843040. doi:10.1177/1756286419843040

Holstein-Rønsbo S, Gan Y, Giannetto MJ, Rasmussen MK, Sigurdsson B, Beinlich FRM, Rose L, Untiet V, Hablitz LM, Kelley DH, et al. 2023. Glymphatic influx and clearance are accelerated by neurovascular coupling. *Nat Neurosci* **26**: 1042–1053. doi:10.1038/s41593-023-01327-2

Hsu M, Laaker C, Madrid A, Herbath M, Choi YH, Sandor M, Fabry Z. 2022. Neuroinflammation creates an immune regulatory niche at the meningeal lymphatic vasculature near the cribriform plate. *Nat Immunol* **23**: 581–593. doi:10.1038/s41590-022-01158-6

Hu X, Deng Q, Ma L, Li Q, Chen Y, Liao Y, Zhou F, Zhang C, Shao L, Feng J, et al. 2020. Meningeal lymphatic vessels regulate brain tumor drainage and immunity. *Cell Res* **30**: 229–243. doi:10.1038/s41422-020-0287-8

Huang TY, Chung HW, Chen MY, Giiang LH, Chin S-C, Lee CS, Chen CY, Liu YJ. 2004. Supratentorial cerebrospinal fluid production rate in healthy adults: quantification with two-dimensional cine phase-contrast MR imaging with high temporal and spatial resolution. *Radiology* **233**: 603–608. doi:10.1148/radiol.2332030884

Iliff JJ, Wang M, Liao Y, Plogg BA, Peng W, Gundersen GA, Benveniste H, Vates GE, Deane R, Goldman SA, et al. 2012. A paravascular pathway facilitates CSF flow through the brain parenchyma and the clearance of interstitial solutes, including amyloid β. *Sci Transl Med* **4**: 147ra111. doi:10.1126/scitranslmed.3003748

Iliff JJ, Lee H, Yu M, Feng T, Logan J, Nedergaard M, Benveniste H. 2013a. Brain-wide pathway for waste clearance captured by contrast-enhanced MRI. *J Clin Invest* **123**: 1299–1309. doi:10.1172/JCI67677

Iliff JJ, Wang M, Zeppenfeld DM, Venkataraman A, Plog BA, Liao Y, Deane R, Nedergaard M. 2013b. Cerebral arterial pulsation drives paravascular CSF–interstitial fluid exchange in the murine brain. *J Neurosci* **33**: 18190–18199. doi:10.1523/JNEUROSCI.1592-13.2013

Iliff JJ, Chen MJ, Plog BA, Zeppenfeld DM, Soltero M, Yang L, Singh I, Deane R, Nedergaard M. 2014. Impairment of glymphatic pathway function promotes tau pathology after traumatic brain injury. *J Neurosci* **34:** 16180–16193. doi:10.1523/JNEUROSCI.3020-14.2014

Ishida K, Yamada K, Nishiyama R, Hashimoto T, Nishida I, Abe Y, Yasui M, Iwatsubo T. 2022. Glymphatic system clears extracellular tau and protects from tau aggregation and neurodegeneration. *J Exp Med* **219:** e20211275. doi:10.1084/jem.20211275

Jacob L, Boisserand LSB, Geraldo LHM, de Brito Neto J, Mathivet T, Antila S, Barka B, Xu Y, Thomas J-M, Pestel J, et al. 2019. Anatomy and function of the vertebral column lymphatic network in mice. *Nat Commun* **10:** 4594. doi:10.1038/s41467-019-12568-w

Jacob L, de Brito Neto J, Lenck S, Corcy C, Benbelkacem F, Geraldo LH, Xu Y, Thomas JM, El Kamouh MR, Spajer M, et al. 2022. Conserved meningeal lymphatic drainage circuits in mice and humans. *J Exp Med* **219:** e20220035. doi:10.1084/jem.20220035

Jang A, Petrova B, Cheong TC, Zawadzki ME, Jones JK, Culhane AJ, Shipley FB, Chiarle R, Wong ET, Kanarek N, et al. 2022. Choroid plexus-CSF-targeted antioxidant therapy protects the brain from toxicity of cancer chemotherapy. *Neuron* **110:** 3288–3301.e8. doi:10.1016/j.neuron.2022.08.009

Jiang-Xie LF, Drieu A, Bhasiin K, Quintero D, Smirnov I, Kipnis J. 2024. Neuronal dynamics direct cerebrospinal fluid perfusion and brain clearance. *Nature* **627:** 157–164. doi:10.1038/s41586-024-07108-6

Jones HE, Coelho-Santos V, Bonney SK, Abrams KA, Shih AY, Siegenthaler JA. 2023. Meningeal origins and dynamics of perivascular fibroblast development on the mouse cerebral vasculature. *Development* **150:** dev.201805. doi:10.1242/dev.201805

Kaplan L, Chow BW, Gu C. 2020. Neuronal regulation of the blood–brain barrier and neurovascular coupling. *Nat Rev Neurosci* **21:** 416–432. doi:10.1038/s41583-020-0322-2

Kedarasetti RT, Turner KL, Echagarruga C, Gluckman BJ, Drew PJ, Costanzo F. 2020. Functional hyperemia drives fluid exchange in the paravascular space. *Fluids Barriers CNS* **17:** 52. doi:10.1186/s12987-020-00214-3

Kelly KK, MacPherson AM, Grewal H, Strnad F, Jones JW, Yu J, Pierzchalski K, Kane MA, Herson PS, Siegenthaler JA. 2016. Col1a1$^+$ perivascular cells in the brain are a source of retinoic acid following stroke. *BMC Neurosci* **17:** 49. doi:10.1186/s12868-016-0284-5

Kim SH, Ahn JH, Yang H, Lee P, Koh GY, Jeong Y. 2020. Cerebral amyloid angiopathy aggravates perivascular clearance impairment in an Alzheimer's disease mouse model. *Acta Neuropathol Commun* **8:** 181. doi:10.1186/s40478-020-01042-0

Kress BT, Iliff JJ, Xia M, Wang M, Wei HS, Zeppenfeld D, Xie L, Kang H, Xu Q, Liew JA, et al. 2014. Impairment of paravascular clearance pathways in the aging brain. *Ann Neurol* **76:** 845–861. doi:10.1002/ana.24271

Li X, Qi L, Yang D, Hao S, Zhang F, Zhu X, Sun Y, Chen C, Ye J, Yang J, et al. 2022. Meningeal lymphatic vessels mediate neurotropic viral drainage from the central nervous system. *Nat Neurosci* **25:** 577–587. doi:10.1038/s41593-022-01063-z

Lilius TO, Blomqvist K, Hauglund NL, Liu G, Stæger FF, Bærentzen S, Du T, Ahlström F, Backman JT, Kalso EA, et al. 2019. Dexmedetomidine enhances glymphatic brain

delivery of intrathecally administered drugs. *J Controlled Release* **304:** 29–38. doi:10.1016/j.jconrel.2019.05.005

Lim HY, Lim SY, Tan CK, Thiam CH, Goh CC, Carbajo D, Chew SHS, See P, Chakarov S, Wang XN, et al. 2018. Hyaluronan receptor LYVE-1-expressing macrophages maintain arterial tone through hyaluronan-mediated regulation of smooth muscle cell collagen. *Immunity* **49:** 326–341.e7. doi:10.1016/j.immuni.2018.06.008

Louveau A, Smirnov I, Keyes TJ, Eccles JD, Rouhani SJ, Peske JD, Derecki NC, Castle D, Mandell JW, Lee KS, et al. 2015. Structural and functional features of central nervous system lymphatic vessels. *Nature* **523:** 337–341. doi:10.1038/nature14432

Louveau A, Herz J, Alme MN, Salvador AF, Dong MQ, Viar KE, Herod SG, Knopp J, Setliff JC, Lupi AL, et al. 2018. CNS lymphatic drainage and neuroinflammation are regulated by meningeal lymphatic vasculature. *Nat Neurosci* **21:** 1380–1391. doi:10.1038/s41593-018-0227-9

Lun MP, Monuki ES, Lehtinen MK. 2015. Development and functions of the choroid plexus–cerebrospinal fluid system. *Nat Rev Neurosci* **16:** 445–457. doi:10.1038/nrn3921

Lundgaard I, Lu ML, Yang E, Peng W, Mestre H, Hitomi E, Deane R, Nedergaard M. 2017. Glymphatic clearance controls state-dependent changes in brain lactate concentration. *J Cereb Blood Flow Metab* **37:** 2112–2124. doi:10.1177/0271678X16661202

Ma Q, Ineichen BV, Detmar M, Proulx ST. 2017. Outflow of cerebrospinal fluid is predominantly through lymphatic vessels and is reduced in aged mice. *Nat Commun* **8:** 1434. doi:10.1038/s41467-017-01484-6

Manley GT, Fujimura M, Ma T, Noshita N, Filiz F, Bollen AW, Chan P, Verkman AS. 2000. Aquaporin-4 deletion in mice reduces brain edema after acute water intoxication and ischemic stroke. *Nat Med* **6:** 159–163. doi:10.1038/72256

Mastorakos P, McGavern D. 2019. The anatomy and immunology of vasculature in the central nervous system. *Sci Immunol* **4:** eaav0492. doi:10.1126/sciimmunol.aav0492

Mazzitelli JA, Smyth LCD, Cross KA, Dykstra T, Sun J, Du S, Mamuladze T, Smirnov I, Rustenhoven J, Kipnis J. 2022. Cerebrospinal fluid regulates skull bone marrow niches via direct access through dural channels. *Nat Neurosci* **25:** 555–560. doi:10.1038/s41593-022-01029-1

Meng Y, Abrahao A, Heyn CC, Bethune AJ, Huang Y, Pople CB, Aubert I, Hamani C, Zinman L, Hynynen K, et al. 2019. Glymphatics visualization after focused ultrasound-induced blood–brain barrier opening in humans. *Ann Neurol* **86:** 975–980. doi:10.1002/ana.25604

Mestre H, Hablitz LM, Xavier AL, Feng W, Zou W, Pu T, Monai H, Murlidharan G, Castellanos Rivera RM, Simon MJ, et al. 2018a. Aquaporin-4-dependent glymphatic solute transport in the rodent brain. *eLife* **7:** e40070. doi:10.7554/eLife.40070

Mestre H, Tithof J, Du T, Song W, Peng W, Sweeney AM, Olveda G, Thomas JH, Nedergaard M, Kelley DH. 2018b. Flow of cerebrospinal fluid is driven by arterial pulsations and is reduced in hypertension. *Nat Commun* **9:** 4878. doi:10.1038/s41467-018-07318-3

Mestre H, Du T, Sweeney AM, Liu G, Samson AJ, Peng W, Mortensen KN, Stæger FF, Bork PAR, Bashford L, et al. 2020a. Cerebrospinal fluid influx drives acute ischemic tissue swelling. *Science* **367**: eaax7171. doi:10.1126/sci ence.aax7171

Mestre H, Mori Y, Nedergaard M. 2020b. The brain's glymphatic system: current controversies. *Trends Neurosci* **43**: 458–466. doi:10.1016/j.tins.2020.04.003

Mestre H, Verma N, Greene TD, Lin LA, Ladron-de-Guevara A, Sweeney AM, Liu G, Thomas VK, Galloway CA, de Mesy Bentley KL, et al. 2022. Periarteriolar spaces modulate cerebrospinal fluid transport into brain and demonstrate altered morphology in aging and Alzheimer's disease. *Nat Commun* **13**: 3897. doi:10.1038/s41467-022-31257-9

Montagne A, Nikolakopoulou AM, Huuskonen MT, Sagare AP, Lawson EJ, Lazic D, Rege SV, Grond A, Zuniga E, Barnes SR, et al. 2021. APOE4 accelerates advanced-stage vascular and neurodegenerative disorder in old Alzheimer's mice via cyclophilin A independently of amyloid-β. *Nat Aging* **1**: 506–520. doi:10.1038/s43587-021-00073-z

Mortensen KN, Sanggaard S, Mestre H, Lee H, Kostrikov S, Xavier ALR, Gjedde A, Benveniste H, Nedergaard M. 2019. Impaired glymphatic transport in spontaneously hypertensive rats. *J Neurosci* **39**: 6365–6377. doi:10.1523/JNEUROSCI.1974-18.2019

Munk AS, Wang W, Bèchet NB, Eltanahy AM, Cheng AX, Sigurdsson B, Benraiss A, Mäe MA, Kress BT, Kelley DH, et al. 2019. PDGF-B is required for development of the glymphatic system. *Cell Rep* **26**: 2955–2969.e3. doi:10.1016/j.celrep.2019.02.050

Myung J, Schmal C, Hong S, Tsukizawa Y, Rose P, Zhang Y, Holtzman MJ, De Schutter E, Herzel H, Bordyugov G, et al. 2018. The choroid plexus is an important circadian clock component. *Nat Commun* **9**: 1062. doi:10.1038/s41467-018-03507-2

Nabeshima S, Reese TS, Landis DMD, Brightman MW. 1975. Junctions in the meninges and marginal glia. *J Comp Neurol* **164**: 127–169. doi:10.1002/cne.901640202

Nagelhus EA, Veruki ML, Torp R, Haug FM, Laake JH, Nielsen S, Agre P, Ottersen OP. 1998. Aquaporin-4 water channel protein in the rat retina and optic nerve: polarized expression in Müller cells and fibrous astrocytes. *J Neurosci* **18**: 2506–2519. doi:10.1523/JNEUROSCI.18-07-02506.1998

Nir Y, Staba RJ, Andrillon T, Vyazovskiy VV, Cirelli C, Fried I, Tononi G. 2011. Regional slow waves and spindles in human sleep. *Neuron* **70**: 153–169. doi:10.1016/j.neuron.2011.02.043

Olstad EW, Ringers C, Hansen JN, Wens A, Brandt C, Wachten D, Yaksi E, Jurisch-Yaksi N. 2019. Ciliary beating compartmentalizes cerebrospinal fluid flow in the brain and regulates ventricular development. *Curr Biol* **29**: 229–241.e6. doi:10.1016/j.cub.2018.11.059

Peng W, Achariyar TM, Li B, Liao Y, Mestre H, Hitomi E, Regan S, Kasper T, Peng S, Ding F, et al. 2016. Suppression of glymphatic fluid transport in a mouse model of Alzheimer's disease. *Neurobiol Dis* **93**: 215–225. doi:10.1016/j.nbd.2016.05.015

Pietilä R, Del Gaudio F, He L, Vázquez-Liébanas E, Vanlandewijck M, Muhl L, Mocci G, Björnholm KD, Lindblad C, Fletcher-Sandersjöö A, et al. 2023. Molecular anatomy of adult mouse leptomeninges. *Neuron* **111**: 3745–3764.e7. doi:10.1016/j.neuron.2023.09.002

Plá V, Bork P, Harnpramukkul A, Olveda G, Ladrón-de-Guevara A, Giannetto MJ, Hussain R, Wang W, Kelley DH, Hablitz LM, et al. 2022. A real-time in vivo clearance assay for quantification of glymphatic efflux. *Cell Rep* **40**: 111320. doi:10.1016/j.celrep.2022.111320

Pollay M. 2010. The function and structure of the cerebrospinal fluid outflow system. *Cerebrospinal Fluid Res* **7**: 9. doi:10.1186/1743-8454-7-9

Pulous FE, Cruz-Hernández JC, Yang C, Kaya Z, Paccalet A, Wojtkiewicz G, Capen D, Brown D, Wu JW, Schloss MJ, et al. 2022. Cerebrospinal fluid can exit into the skull bone marrow and instruct cranial hematopoiesis in mice with bacterial meningitis. *Nat Neurosci* **25**: 567–576. doi:10.1038/s41593-022-01060-2

Radoš M, Živko M, Periša A, Orešković D, Klarica M. 2021. No arachnoid granulations—no problems: number, size, and distribution of arachnoid granulations from birth to 80 years of age. *Front Aging Neurosci* **13**: 698865. doi:10.3389/fnagi.2021.698865

Rasmussen MK, Mestre H, Nedergaard M. 2022. Fluid transport in the brain. *Physiol Rev* **102**: 1025–1151. doi:10.1152/physrev.00031.2020

Rebejac J, Eme-Scolan E, Arnaud Paroutaud L, Kharbouche S, Teleman M, Spinelli L, Gallo E, Roussel-Queval A, Zarubica A, Sansoni A, et al. 2022. Meningeal macrophages protect against viral neuroinfection. *Immunity* **55**: 2103–2117.e10. doi:10.1016/j.immuni.2022.10.005

Ringstad G, Eide PK. 2020. Cerebrospinal fluid tracer efflux to parasagittal dura in humans. *Nat Commun* **11**: 354. doi:10.1038/s41467-019-14195-x

Ringstad G, Valnes LM, Dale AM, Pripp AH, Vatnehol SAS, Emblem KE, Mardal KA, Eide PK. 2018. Brain-wide glymphatic enhancement and clearance in humans assessed with MRI. *JCI Insight* **3**: e121537. doi:10.1172/jci.insight.121537

Roques M, De Barros A, Bonneville F. 2021. Rethink the classical view of cerebrospinal fluid production. *Nat Rev Neurol* **17**: 590. doi:10.1038/s41582-021-00538-0

Rustenhoven J, Drieu A, Mamuladze T, de Lima KA, Dykstra T, Wall M, Papadopoulos Z, Kanamori M, Salvador AF, Baker W, et al. 2021. Functional characterization of the dural sinuses as a neuroimmune interface. *Cell* **184**: 1000–1016.e27. doi:10.1016/j.cell.2020.12.040

Rustenhoven J, Pavlou G, Storck SE, Dykstra T, Du S, Wan Z, Quintero D, Scallan JP, Smirnov I, Kamm RD, et al. 2023. Age-related alterations in meningeal immunity drive impaired CNS lymphatic drainage. *J Exp Med* **220**: e20221929. doi:10.1084/jem.20221929

Sabia S, Fayosse A, Dumurgier J, van Hees VT, Paquet C, Sommerlad A, Kivimäki M, Dugravot A, Singh-Manoux A. 2021. Association of sleep duration in middle and old age with incidence of dementia. *Nat Commun* **12**: 2289. doi:10.1038/s41467-021-22354-2

Salehi Z, Mashayekhi F, Naji M. 2008. Insulin like growth factor-1 and insulin like growth factor binding proteins in the cerebrospinal fluid and serum from patients with Alzheimer's disease. *Biofactors* **33**: 99–106. doi:10.1002/biof.5520330202

Shah T, Leurgans SE, Mehta RI, Yang J, Galloway CA, de Mesy Bentley KL, Schneider JA, Mehta RI. 2023. Arachnoid granulations are lymphatic conduits that communi-

Cite this article as *Cold Spring Harb Perspect Biol* doi: 10.1101/cshperspect.a041370

cate with bone marrow and dura-arachnoid stroma. *J Exp Med* **220**: e20220618. doi:10.1084/jem.20220618

Shipley FB, Dani N, Xu H, Deister C, Cui J, Head JP, Sadegh C, Fame RM, Shannon ML, Flores VI, et al. 2020. Tracking calcium dynamics and immune surveillance at the choroid plexus blood-cerebrospinal fluid interface. *Neuron* **108**: 623–639.e10. doi:10.1016/j.neuron.2020.08.024

Siret C, van Lessen M, Bavais J, Jeong HW, Reddy Samawar SK, Kapupara K, Wang S, Simic M, de Fabritus L, Tchoghandjian A, et al. 2022. Deciphering the heterogeneity of the Lyve1+ perivascular macrophages in the mouse brain. *Nat Commun* **13**: 7366. doi:10.1038/s41467-022-35166-9

Smith AJ, Duan T, Verkman AS. 2019. Aquaporin-4 reduces neuropathology in a mouse model of Alzheimer's disease by remodeling peri-plaque astrocyte structure. *Acta Neuropathol Commun* **7**: 74. doi:10.1186/s40478-019-0728-0

Smyth LCD, Xu D, Okar SV, Dykstra T, Rustenhoven J, Papadopoulos Z, Bhasiin K, Kim MW, Drieu A, Mamuladze T, et al. 2024. Identification of direct connections between the dura and the brain. *Nature* **627**: 165–173. doi:10.1038/s41586-023-06993-7

Sofroniew MV. 2015. Astrocyte barriers to neurotoxic inflammation. *Nat Rev Neurosci* **16**: 249–263. doi:10.1038/nrn3898

Song E, Mao T, Dong H, Boisserand LSB, Antila S, Bosenberg M, Alitalo K, Thomas JL, Iwasaki A. 2020. VEGF-C-driven lymphatic drainage enables immunosurveillance of brain tumours. *Nature* **577**: 689–694. doi:10.1038/s41586-019-1912-x

Spera I, Cousin N, Ries M, Kedracka A, Castillo A, Aleandri S, Vladymyrov M, Mapunda JA, Engelhardt B, Luciani P, et al. 2023. Open pathways for cerebrospinal fluid outflow at the cribriform plate along the olfactory nerves. *EBioMedicine* **91**: 104558. doi:10.1016/j.ebiom.2023.104558

Steffensen AB, Oernbo EK, Stoica A, Gerkau NJ, Barbuskaite D, Tritsaris K, Rose CR, MacAulay N. 2018. Cotransporter-mediated water transport underlying cerebrospinal fluid formation. *Nat Commun* **9**: 2167. doi:10.1038/s41467-018-04677-9

Stokum JA, Shim B, Huang W, Kane M, Smith JA, Gerzanich V, Simard JM. 2021. A large portion of the astrocyte proteome is dedicated to perivascular endfeet, including critical components of the electron transport chain. *J Cereb Blood Flow Metab* **41**: 2546–2560. doi:10.1177/0271678X211004182

Sweeney AM, Pla V, Du T, Liu G, Sun Q, Peng S, Plog BA, Kress BT, Wang X, Mestre H, et al. 2019. In vivo imaging of cerebrospinal fluid transport through the intact mouse skull using fluorescence macroscopy. *J Vis Exp* **29**: 10.3791/59774. doi:10.3791/59774

Takeda S, Commins C, DeVos SL, Nobuhara CK, Wegmann S, Roe AD, Costantino I, Fan Z, Nicholls SB, Sherman AE, et al. 2016. Seed-competent high-molecular-weight tau species accumulates in the cerebrospinal fluid of Alzheimer's disease mouse model and human patients. *Ann Neurol* **80**: 355–367. doi:10.1002/ana.24716

Turner KL, Gheres KW, Proctor EA, Drew PJ. 2020. Neurovascular coupling and bilateral connectivity during NREM and REM sleep eds. *eLife* **9**: e62071. doi:10.7554/eLife.62071

van Hespen KM, Mackaaij C, Waas ISE, de Bree MP, Zwanenburg JJM, Kuijf HJ, Daemen MJAP, Hendrikse J, Hermkens DMA. 2021. Arterial remodeling of the intracranial arteries in patients with hypertension and controls. *Hypertension* **77**: 135–146. doi:10.1161/HYPERTENSIONAHA.120.16029

Van Hove H, Martens L, Scheyltjens I, De Vlaminck K, Pombo Antunes AR, De Prijck S, Vandamme N, De Schepper S, Van Isterdael G, Scott CL, et al. 2019. A single-cell atlas of mouse brain macrophages reveals unique transcriptional identities shaped by ontogeny and tissue environment. *Nat Neurosci* **22**: 1021–1035. doi:10.1038/s41593-019-0393-4

Vanlandewijck M, He L, Mäe MA, Andrae J, Ando K, Del Gaudio F, Nahar K, Lebouvier T, Laviña B, Gouveia L, et al. 2018. A molecular atlas of cell types and zonation in the brain vasculature. *Nature* **554**: 475–480. doi:10.1038/nature25739

van Veluw SJ, Hou SS, Calvo-Rodriguez M, Arbel-Ornath M, Snyder AC, Frosch MP, Greenberg SM, Bacskai BJ. 2020. Vasomotion as a driving force for paravascular clearance in the awake mouse brain. *Neuron* **105**: 549–561.e5. doi:10.1016/j.neuron.2019.10.033

von Holstein-Rathlou S, Petersen NC, Nedergaard M. 2018. Voluntary running enhances glymphatic influx in awake behaving, young mice. *Neurosci Lett* **662**: 253–258. doi:10.1016/j.neulet.2017.10.035

Vong KI, Ma TC, Li B, Leung TCN, Nong W, Ngai SM, Hui JHL, Jiang L, Kwan KM. 2021. SOX9-COL9A3-dependent regulation of choroid plexus epithelial polarity governs blood–cerebrospinal fluid barrier integrity. *Proc Natl Acad Sci* **118**: e2009568118. doi:10.1073/pnas.2009568118

Vonsattel JP, Myers RH, Tessa Hedley-Whyte E, Ropper AH, Bird ED, Richardson EP. 1991. Cerebral amyloid angiopathy without and with cerebral hemorrhages: a comparative histological study. *Ann Neurol* **30**: 637–649. doi:10.1002/ana.410300503

Wang Y, Sabbagh MF, Gu X, Rattner A, Williams J, Nathans J. 2019. β-Catenin signaling regulates barrier-specific gene expression in circumventricular organ and ocular vasculatures. *eLife* **8**: e43257. doi:10.7554/eLife.43257

Wardlaw JM, Benveniste H, Nedergaard M, Zlokovic BV, Mestre H, Lee H, Doubal FN, Brown R, Ramirez J, MacIntosh BJ, et al. 2020. Perivascular spaces in the brain: anatomy, physiology and pathology. *Nat Rev Neurol* **16**: 137–153. doi:10.1038/s41582-020-0312-z

Xiang T, Feng D, Zhang X, Chen Y, Wang H, Liu X, Gong Z, Yuan J, Liu M, Sha Z, et al. 2022. Effects of increased intracranial pressure on cerebrospinal fluid influx, cerebral vascular hemodynamic indexes, and cerebrospinal fluid lymphatic efflux. *J Cereb Blood Flow Metab* **42**: 2287–2302. doi:10.1177/0271678X221119855

Xie L, Kang H, Xu Q, Chen MJ, Liao Y, Thiyagarajan M, O'Donnell J, Christensen DJ, Nicholson C, Iliff JJ, et al. 2013. Sleep drives metabolite clearance from the adult brain. *Science* **342**: 373–377. doi:10.1126/science.1241224

Xu Z, Xiao N, Chen Y, Huang H, Marshall C, Gao J, Cai Z, Wu T, Hu G, Xiao M. 2015. Deletion of aquaporin-4 in APP/PS1 mice exacerbates brain Aβ accumulation and memory deficits. *Mol Neurodegener* **10:** 58. doi:10.1186/s13024-015-0056-1

Xu H, Fame RM, Sadegh C, Sutin J, Naranjo C, Syau D, Cui J, Shipley FB, Vernon A, Gao F, et al. 2021. Choroid plexus NKCC1 mediates cerebrospinal fluid clearance during mouse early postnatal development. *Nat Commun* **12:** 447. doi:10.1038/s41467-020-20666-3

Yamazaki Y, Liu CC, Yamazaki A, Shue F, Martens YA, Chen Y, Qiao W, Kurti A, Oue H, Ren Y, et al. 2021. Vascular ApoE4 impairs behavior by modulating gliovascular function. *Neuron* **109:** 438–447.e6. doi:10.1016/j.neuron.2020.11.019

Yanev P, Poinsatte K, Hominick D, Khurana N, Zuurbier KR, Berndt M, Plautz EJ, Dellinger MT, Stowe AM. 2020. Impaired meningeal lymphatic vessel development worsens stroke outcome. *J Cereb Blood Flow Metab* **40:** 263–275. doi:10.1177/0271678X18822921

Yang J, Lunde LK, Nuntagij P, Oguchi T, Camassa LMA, Nilsson LNG, Lannfelt L, Xu Y, Amiry-Moghaddam M, Ottersen OP, et al. 2011. Loss of astrocyte polarization in the Tg-ArcSwe mouse model of Alzheimer's disease. *J Alzheimers Dis* **27:** 711–722. doi:10.3233/JAD-2011-110725

Yoon J-H, Jin H, Kim HJ, Hong SP, Yang MJ, Ahn JH, Kim YC, Seo J, Lee Y, McDonald DM, et al. 2024. Nasopharyngeal lymphatic plexus is a hub for cerebrospinal fluid drainage. *Nature* **625:** 768–777. doi:10.1038/s41586-023-06899-4

Zeppenfeld DM, Simon M, Haswell JD, D'Abreo D, Murchison C, Quinn JF, Grafe MR, Woltjer RL, Kaye J, Iliff JJ. 2017. Association of perivascular localization of aquaporin-4 with cognition and Alzheimer disease in aging brains. *JAMA Neurol* **74:** 91–99. doi:10.1001/jamaneurol.2016.4370

Cite this article as *Cold Spring Harb Perspect Biol* doi: 10.1101/cshperspect.a041370

Remyelination in the Central Nervous System

Robin J.M. Franklin,[1] Benedetta Bodini,[2,3] and Steven A. Goldman[4,5]

[1]Altos Labs Cambridge Institute of Science, Cambridge CB21 6GH, United Kingdom

[2]Sorbonne Université, Paris Brain Institute, CNRS, INSERM, Paris 75013, France

[3]Saint-Antoine Hospital, APHP, Paris 75012, France

[4]Center for Translational Neuromedicine, University of Rochester Medical Center, Rochester, New York 14642, USA

[5]University of Copenhagen Faculty of Medicine, Copenhagen 2200, Denmark

Correspondence: rfranklin@altoslabs.com

The inability of the mammalian central nervous system (CNS) to undergo spontaneous regeneration has long been regarded as a central tenet of neurobiology. However, while this is largely true of the neuronal elements of the adult mammalian CNS, save for discrete populations of granule neurons, the same is not true of its glial elements. In particular, the loss of oligodendrocytes, which results in demyelination, triggers a spontaneous and often highly efficient regenerative response, remyelination, in which new oligodendrocytes are generated and myelin sheaths are restored to denuded axons. Yet remyelination in humans is not without limitation, and a variety of demyelinating conditions are associated with sustained and disabling myelin loss. In this work, we will (1) review the biology of remyelination, including the cells and signals involved; (2) describe when remyelination occurs and when and why it fails, including the consequences of its failure; and (3) discuss approaches for therapeutically enhancing remyelination in demyelinating diseases of both children and adults, both by stimulating endogenous oligodendrocyte progenitor cells and by transplanting these cells into demyelinated brain.

IDENTIFYING REMYELINATION IN ANIMAL MODELS

Remyelination is the process in which new myelin sheaths are restored to axons that have lost their myelin sheaths as a result of primary demyelination (Franklin and ffrench-Constant 2017).[6] Primary demyelination is the term used to describe the loss of myelin from an otherwise intact axon and should be distinguished from myelin loss secondary to axonal loss—a process called Wallerian degeneration or, misleadingly, secondary demyelination. Remyelination is sometimes referred to as myelin repair. However, this term suggests a damaged but otherwise intact myelin internode being "patched up," a process for which there is no evidence, and which does not emphasize the

---

[6]This is an update to a previous article published in Cold Spring Harbor Perspectives in Biology [Franklin and Goldman (2015). Cold Spring Harb Perspect Biol 7: a020594. doi:10.1101/cshperspect.a020594].

truly regenerative nature of remyelination, in which the pre-lesion cytoarchitecture is all but fully restored. Remyelinated tissue very closely resembles normally myelinated tissue but differs in one important aspect—the newly generated myelin sheaths are typically shorter and thinner than the original myelin sheaths. When myelin is initially formed in the peri- and postnatal periods, there is a striking correlation between axon diameter and myelin sheath thickness and length established during myelination, which is less apparent in remyelination. Instead, myelin sheath thickness and length show little change with increasing axonal diameter with the result that the myelin is generally thinner and shorter than would be expected for a given diameter of axon (Fig. 1). Although some remodeling of the new myelin internode occurs, the original dimensions are rarely regained (Powers et al. 2013). The relationship between axon diameter and myelin sheath is expressed as the G ratio, which is the fraction of the axonal circumference to the axon plus myelin sheath circumference. The identification of abnormally thin myelin sheaths (> than normal G ratio) remains the "gold standard" for unequivocally identifying remyelination and is most reliably identified in resin-embedded tissue, viewed by light microscopy following toluidine blue staining, or by electron microscopy. This effect is obvious when large diameter axons are remyelinated but is less clear with smaller diameter axons such as those of the corpus callosum, where G ratios of remyelinated axons can be difficult to distinguish from those of normally myelinated axons (Stidworthy et al. 2003). How is the relationship between myelin parameters and axon size established in myelination, and why is it disengaged in remyelination? In the peripheral nervous system (PNS), axonally expressed neuregulin (NRG)1-type III plays a key role; reduced expression results in thinner myelin sheath (increased G ratio), whereas overexpression leads to a thicker than expected myelin sheath (decreased G ratio) (Michailov et al. 2004). In the central nervous system (CNS), however, the role of neuregulins in controlling myelin sheath length and thickness is less clear (Brinkmann et al. 2008), although they play a

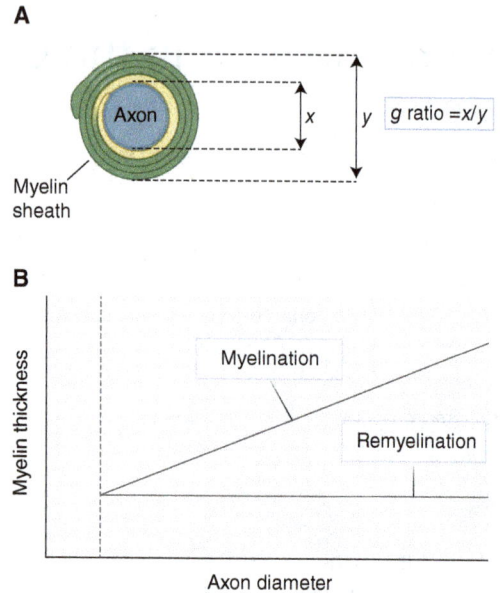

Figure 1. The relationship between myelin sheath and the axon diameter in myelination and remyelination. (A) The relationship between the thickness of the myelin sheath (determined by the number of wraps or lamellae) and the axon diameter is expressed as the G ratio, calculated by dividing the diameter of axon by the diameter of the axon plus the myelin sheath. The higher the G ratio, the thinner the myelin sheath, where a G ratio of 1 corresponds to an un- or demyelinated axon. (B) In developmental myelination, there is an increase in myelin sheath thickness with increasing axonal diameter. In remyelination, however, the myelin sheath thickness remains the same regardless of the diameter. (This figure is based on data in Stidworthy et al. 2003 and Franklin et al. 2012; adapted and reprinted with permission from Franklin and ffrench-Constant 2017.)

role of rendering axons dependent on electrical activity for myelination. The factors that govern the G ratio in remyelination would seem to be distinct from those operating in developmental myelination, and an explanation for the increased G ratio in remyelination remain elusive. For example, overexpression of Nrg leads to CNS hypermyelination in development but not during remyelination (Brinkmann et al. 2008). Similarly, activation of the Akt pathway in the CNS, which results in thicker than expected myelin sheaths in development (Flores et al.

2008), does not result in thicker remyelinated sheaths following demyelination in the adult (Harrington et al. 2010). One hypothesis is that whereas the myelinating oligodendrocyte associates with a dynamically changing axon yet to achieve its full length and diameter, the remyelinating oligodendrocyte engages an axon that is comparatively static having already reached it mature size (Franklin and Hinks 1999). As a result, the remyelinating oligodendrocyte is not subjected to the same dynamic stresses encountered by the myelinating oligodendrocyte.

IDENTIFYING REMYELINATION IN HUMANS

Identifying remyelination in humans has proven to be particularly challenging, as several of the imaging techniques proposed to measure myelin content changes in the CNS do not fully meet the required criteria of being sensitive and specific to myelin while also being reproducible and clinically meaningful. The key objective of myelin imaging applied to humans with demyelinating diseases is to quantitatively measure myelin loss and myelin regeneration, as well as to distinguish between repaired and partially demyelinated tissues.

Measuring Remyelination in Humans with Magnetic Resonance Imaging

Several advanced magnetic resonance imaging (MRI) techniques have been introduced to explore myelin content changes in humans, including magnetization transfer imaging (MTI), T2 relaxometry, and diffusion tensor imaging (DTI). Compared with histology, either in animal or in human studies, these techniques have been shown to present a reasonably accurate and comparable correlation with myelin content (Mancini et al. 2020). A few have been already employed to specifically measure remyelination in patients with multiple sclerosis (MS) (Bodini et al. 2021).

Changes in magnetization transfer ratio (MTR), an MTI-based metric reflecting the exchange rate between the bound protons to mac-romolecules and the unbound protons in free water, are thought to be highly sensitive to myelin content variations due to the significant contribution of myelin to the macromolecules involved in the MT phenomenon (Schmierer et al. 2004; Moll et al. 2011; Moccia et al. 2020). Significantly reduced MTR values in newly forming lesions, possibly reflecting acute demyelination, have been found in the white matter of patients with MS, followed by a progressive, although partial recovery over the following months, likely to indicate a subsequent phase of remyelination (Chen et al. 2008). More recently, MTR has been used to generate MS-patient-specific maps of demyelination and remyelination at the cerebral cortex (Derakhshan et al. 2014; Lazzarotto et al. 2022). A promising development of MTI is inhomogeneous MTI, a refined technique that selectively images tissues with long dipolar relaxation time components, such as myelin-rich structures, allowing a more specific measure of myelin content in healthy individuals and of myelin content changes in patients with demyelinating disorders (Duhamel et al. 2019; Lee et al. 2022). Another established technique with good sensitivity and specificity for myelin is myelin water imaging, which measures the myelin water fraction (MWF), defined as the quickly decaying signal arising from water trapped between myelin sheaths (Mackay et al. 1994; MacKay and Laule 2016). MWF has been employed to investigate myelin damage and repair in MS, as well as other diseases in which myelin pathology is suspected (MacKay and Laule 2016).

Besides MTI and myelin water imaging, myelin loss and regeneration have also been investigated in MS using changes in radial diffusivity (RD), a DTI-based metric describing microscopic water movements perpendicular to axonal tracts that had previously been shown to mainly reflect myelin content in experimental models of white matter damage (Song et al. 2002, 2005). In particular, changes in RD values in demyelinating lesions, possibly reflecting myelin loss and repair, have been found to be able to discriminate the functional outcome following optic neuritis and spinal cord relapses in patients with MS (Freund et al. 2010; Naismith et al. 2010).

Measuring Remyelination in Humans with Positron Emission Tomography

While being very sensitive to microstructural injury, all these MRI-derived metrics only allow an indirect and insufficiently specific measure of myelin content changes in humans, since they are significantly affected by other pathological processes such as inflammation, edema, and axonal loss (Petiet et al. 2019; Moccia et al. 2020). An alternative quantitative technique offering a more direct and specific access to the myelin compartment in humans is positron-emission tomography (PET) with myelin-binding radiotracers (Bodini et al. 2021). While several PET tracers, mainly belonging to the stilbene- and benzothiazole-derivative classes, have been shown to bind selectively to the CNS myelin structure in experimental models, so far only ^{11}C-PiB, ^{18}F-florbetaben, ^{18}F-florbetapir, and ^{11}C-MEDAS have been employed in clinical studies to explore myelin content changes in patients with MS (Bodini et al. 2021). Although a group-level analysis of all myelin PET studies in patients with MS indicated a decrease in tracer binding in white matter lesions compared to healthy controls, indicating demyelination, the use of individual lesion- or voxel-based analysis strategies revealed significant heterogeneity in the extent of myelin loss within the lesions (Bodini et al. 2016; Carotenuto et al. 2020; Pytel et al. 2020; Zhang et al. 2021; van der Weijden et al. 2022). The measure of dynamic myelin content changes over time with myelin PET revealed highly heterogeneous remyelination profiles across patients and lesions, which have been found to significantly correlate with neurological disability and brain atrophy (Bodini et al. 2016; Tonietto et al. 2022).

Taken together, MRI and PET studies exploring changes in myelin content in MS unequivocally indicate that myelin loss and repair are heterogeneous processes across patients. Moreover, demonstrating in vivo that individual profiles of remyelination are key determinants of atrophy and clinical evolution, these studies support the notion that an effective reparative response to demyelination can protect neurons from degeneration and, ultimately, patients with MS from experiencing an irreversible accumulation of clinical disability.

REMYELINATION IS THE NORMAL RESPONSE TO DEMYELINATION

Remyelination as a regenerative process shares many common features with regenerative processes occurring in other tissues of the body, and is the expected or default response to demyelination. The evidence for this comes from both experimentally induced and clinical demyelination. When demyelination is induced by toxins injurious to oligodendrocytes and myelin (for example, by dietary cuprizone or direct delivery of lysolecithin or ethidium bromide), then remyelination in white matter usually proceeds to completion, albeit in an age-dependent manner (Sim et al. 2002b). An interesting situation arises in the cerebral cortex where lengths of axon can remain unmyelinated during developmental myelination (Gibson et al. 2014). Using time-lapse, two-photon microscopy it has been shown that following cuprizone-induced demyelination of the adult mouse somatosensory cortex, the pattern of remyelination that occurs is distinct from the pre-demyelination pattern (Orthmann-Murphy et al. 2020). There is evidence that axons undergoing primary demyelination in experimental or clinical traumatic injury undergo complete remyelination, and that the persistence of chronically demyelinated axons is unusual (Lasiene et al. 2008). An exception is when demyelination is induced by or associated with the adaptive immune response, such as occurs in the autoimmune-mediated condition MS and in a laboratory animal model, experimental autoimmune encephalomyelitis (EAE). In this context, remyelination occurs in an environment intrinsically hostile to the oligodendrocyte lineage, although recent evidence has challenged the dogma that such an environment will be inimical to remyelination (Yilmaz et al. 2023). Thus, remyelination failure in MS (and EAE) is not inevitable, but rather a feature of the specific disease environment of MS (Mezydlo et al. 2023). Nevertheless, even in MS—a disease prototypically associated with failed or inadequate remyelination—remyelination can proceed to

Cite this article as *Cold Spring Harb Perspect Biol* doi: 10.1101/cshperspect.a041371

completion, and can be anatomically extensive (Patrikios et al. 2006; Patani et al. 2007; Goldschmidt et al. 2009; Piaton et al. 2009). Remyelination can also be extensive in EAE, and models with significant persistent demyelination are unusual (Linington et al. 1992; Hampton et al. 2008). Remyelination is especially efficient following demyelination of cerebral cortical gray matter, in both experimental models (Merkler et al. 2006) and clinical disease (Albert et al. 2007; Strijbis et al. 2017), although the reason for this is unclear.

REMYELINATION RESTORES FUNCTION AND PROTECTS AXONS

Remyelination restores saltatory conduction and reverses functional deficits (Smith et al. 1979; Jeffery et al. 1999; Liebetanz and Merkler 2006; Traka et al. 2010). Compelling evidence in support of functional restoration by remyelination is provided by an unusual demyelinating condition in cats, in which the reversal of clinical signs is associated with spontaneous remyelination (Duncan et al. 2009).

An additional and key function of remyelination is the protective effect it has on the underlying axon (Irvine and Blakemore 2008). Axonal and eventually even neuronal loss is a major cause of the progressive nature of chronic demyelinating disease, such as occurs in MS (Trapp and Nave 2008), and is likely primarily due to the absence of the myelin sheath, rather than the direct damage by inflammation that accounts for axonal loss in acute lesions. Thus, patients on appropriate immunosuppressive therapy and with apparently quiescent disease still show monotonically increasing disability and clinical progression, as these patients manifest persistent demyelination regardless of their lack of active disease. Indeed, remyelination is not the principal reason for the resolution of clinical signs following an acute relapse, which rather likely results from the resolution of inflammation, paired with adaptive responses by affected axons that serve to restore conduction.

Evidence that myelin is required for axon survival was first obtained using genetic mouse models, and is supported by subsequent studies of human pathology (Nave and Trapp 2008). Transgenic mice lacking the enzyme 2′,3′-cyclic nucleotide 3′ phosphodiesterase (CNP) or the myelin transmembrane protein proteolipid protein (PLP) show long-term axonal degeneration, even in the presence of myelin sheaths that are either ultrastructurally normal or show only minor abnormalities (Griffiths et al. 1998; Lappe-Siefke et al. 2003). Further analysis of the *Plp* mutant mice has revealed a disturbance in axoplasmic transport in the absence of PLP (Edgar et al. 2004) and has led to the identification of myelin-associated Sirtuin 2 as a potential mediator of long-term axonal stability (Werner et al. 2007). Myelin is also important for axon survival in humans, as patients with Pelizaeus–Merzbacher disease (PMD) caused by mutations in *PLP* show axon loss (Garbern et al. 2002), and studies of MS autopsy tissue show that axon preservation is seen in those areas where remyelination has occurred (although whether this is because remyelination has occurred where axons persist or that axons persist because of remyelination is unclear) (Kornek et al. 2000). Axon degeneration also occurs as a consequence of genetically induced, oligodendrocyte-specific ablation, even in *Rag 1*-deficient mice that have no functional lymphocytes (Pohl et al. 2011). These observations offer compelling evidence that axonal survival is dependent on intact oligodendrocytes, and that axonal degeneration in chronically demyelinated lesions can occur independently of inflammation. The nature of the "trophic" exchange from oligodendrocyte to axon remains to be fully elucidated but at least in part rests on transfer of energy metabolites from oligodendroglia to axons through monocarboxylate transporter 1 (MCT1) (Lee et al. 2012; Morrison et al. 2013; Schäffner et al. 2023), a dependency that increases with adult aging (Philips et al. 2021).

MECHANISMS OF REMYELINATION

OPCs Are the Principal Source of New Myelin-Forming Oligodendrocytes

In most instances following experimental demyelination in animal models, remyelination involves the generation of new mature oligodendrocytes. In the vast majority of cases, the new oligodendrocytes that mediate remyelination are derived from a population of adult CNS progenitor cells, most often referred to as adult oligodendrocyte progenitor cells (OPCs), that in humans arise largely from fetal forebears in the outer subventricular zone (SVZ) (Huang et al. 2020). More broadly, these cells are also referred to as glial progenitor cells (GPCs), since in humans they are bipotential for oligodendrocytes and astrocytes alike and retain neurogenic competence as well under appropriate conditions (Nunes et al. 2003). These multiprocessed, proliferating OPCs are widespread throughout the adult CNS, occurring in both the white matter and gray matter (3%–5% of cells in the adult human forebrain, and 5%–8% in rodents) (Horner et al. 2000; Dawson et al. 2003; Richardson et al. 2011). Adult and fetal OPCs share many similarities, although the adult cells have a longer basal cell cycle time and migrate less rapidly than do their fetal counterparts (Wolswijk and Noble 1989) in rodents as well as humans (Windrem et al. 2004). However, following inflammatory demyelination, OPCs become activated by the inflamed environment and assume a transcriptomic profile more akin to that of the replicating OPCs of the neonate, rather than that of the resting OPCs of the intact adult CNS (Moyon et al. 2015).

Evidence obtained using Cre-lox fate mapping in transgenic mice following experimental demyelination has shown that OPCs produce the vast majority of remyelinating oligodendrocytes (Tripathi et al. 2010; Zawadzka et al. 2010). Remyelinating oligodendrocytes may also come from the stem and progenitor cells of the adult SVZ, either from the progenitor cells contributing to the rostral migratory stream (RMS) (Nait-Oumesmar et al. 1999) or from the GFAP-expressing neural stem cells (NSCs) of the SVZ per se (Menn et al. 2006). However, the contribution that SVZ-derived cells make relative to that from local OPCs may be small, and their contribution to repair beyond the periventricular white matter is likely negligible (Kazanis et al. 2017).

Although experimental evidence indicates that oligodendrocytes themselves do not give rise to new remyelinating oligodendrocytes (Crawford et al. 2016), evidence from experimental models indicates that in some instances new myelin sheaths can be generated from oligodendrocytes that survive within areas of demyelination, albeit shorn of their original myelin sheath–associated processes (Duncan et al. 2018). This type of remyelination has been shown in cat, mouse, and zebrafish models of demyelination, and in the latter is rare compared to OPC-mediated remyelination and is often aberrant, involving, for example, myelin of neuronal cell bodies (Duncan et al. 2018; Bacmeister et al. 2020; Neely et al. 2022). Carbon dating studies have been used to suggest that this form of remyelination occurs in humans in the MS brain (Yeung et al. 2019). However, difficulties in unequivocally establishing whether demyelination and subsequent remyelination has occurred in the human brain make it difficult to be certain how extensive this form of remyelination really is in MS (Neumann et al. 2020). The foci within MS tissue with above background levels of de novo oligodendrogenesis suggest that OPC-driven remyelination is also occurring (Yeung et al. 2019). Remyelination by surviving oligodendrocytes nevertheless represents an interesting new form of remyelination deserving of further investigation (Franklin et al. 2021).

Remyelination Requires the Activation, Recruitment, and Differentiation of Adult OPCs

In response to injury, local OPCs undergo a switch from an essentially quiescent state to a regenerative phenotype. This activation is the first step in the remyelination process and involves not only changes in morphology but also up-regulation of several genes, many associated with the generation of oligodendrocytes during development such as the transcription factors *Olig2*, *Nkx2.2*, *MyT1*, and *Sox2* (Fancy et al. 2004; Watanabe et al. 2004; Shen et al. 2008). The activation of OPCs is likely to be in response to acute injury-induced changes in microglia and astrocytes, two cell types exquisitely sensitive to disturbance in tissue homeostasis (Glezer et al. 2006; Rhodes et al. 2006). These

Cite this article as *Cold Spring Harb Perspect Biol* doi: 10.1101/cshperspect.a041371

two cell types, themselves activated by injury, are the major source of factors that induce the rapid proliferative response of OPCs to demyelinating injury (Fig. 2). This response is modulated by levels of the cell cycle regulatory proteins p27Kip-1 and Cdk2 (Crockett et al. 2005; Caillava et al. 2011) and is promoted by the growth factors PDGF and FGF (Woodruff et al. 2004; Murtie et al. 2005), Endothelin-1 (Gadea et al. 2009), and many other factors associated with acute inflammatory lesions and demonstrated to have OPC mitogenic activity in tissue culture (Vela et al. 2002). Semaphorins are important regulators of OPC migration following demyelination: semaphorin 3A impairs OPC recruitment to the demyelinated area, while semaphorin 3F overexpression accelerates not only OPC recruitment, but also remyelination rate (Piaton et al. 2011). The population of areas of demyelination by OPCs is referred to as the "recruitment phase" of remyelination and involves OPC migration in addition to the ongoing proliferation.

For remyelination to be complete, the recruited OPC must next differentiate into remye-linating oligodendrocytes—the *differentiation phase* (Fig. 2). This phase encompasses three distinct steps—establishing contact with the axon to be remyelinated, expression of myelin genes and generation of myelin membrane, and, finally, wrapping and compaction to form the sheath. Despite these being fundamental properties of oligodendrocytes, we still have an incomplete understanding about how axo-glial contact is established and how this interaction then regulates, within each individual cell process, the morphological changes that constitute myelination. Nevertheless, some molecules have been shown to contribute to the regulation of differentiation, and it is clear that the differentiation of OPCs into myelinating oligodendrocytes in development and during the regenerative process share many similarities (Fancy et al. 2011). FGF plays a key role in inhibiting differentiation as well as promoting recruitment and thereby regulates the correct transition from the recruitment to the differentiation phases (Armstrong et al. 2002), and IGF-I is another factor that plays major roles in both processes (Mason et al. 2003). Semaphorin 3A, in addition to its role in

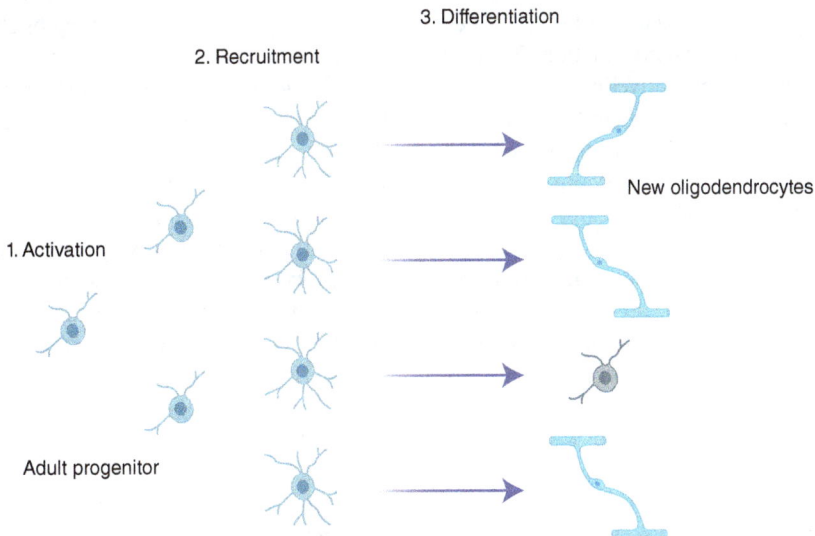

Figure 2. The stage of oligodendrocyte progenitor cell (OPC)-mediated remyelination. Endogenous OPCs (adult progenitors) become activated in response to the change in environment triggered by demyelination. In the activated state, OPCs divide and migrate to populate the area of demyelination at a density much greater than the surrounding tissue during the recruitment phase. In a timely manner, recruited OPCs exit the cell cycle and differentiate into myelin-forming oligodendrocytes—the differentiation phase. Note, a number of OPCs remain undifferentiated to restore the pre-lesion density of OPCs.

OPC recruitment (Piaton et al. 2011), is also an inhibitor of OPC differentiation (Syed et al. 2011). LINGO-1, a component of the trimolecular Nogo receptor, has been found to be a negative regulator of oligodendrocyte differentiation in development (Mi et al. 2005), while mice deficient in LINGO-1 or treated with an antibody antagonist against LINGO-1 exhibited increased remyelination and functional recovery from EAE (Mi et al. 2007). The canonical Wnt pathway is as a very powerful negative regulator of oligodendrocyte differentiation in both development and remyelination (Fancy et al. 2009; Ye et al. 2009). The nuclear receptor retinoid X receptor-γ (RXRγ) is a key positive regulator of oligodendrocyte differentiation directly from the analysis of remyelinating tissue (Huang et al. 2011). Electrical activity in demyelinated axons and synaptic signaling to OPCs via AMPA receptors also play an important role in their differentiation (Gautier et al. 2015), while direct electrical stimulation of demyelinated axons can enhance remyelination (Ortiz et al. 2019).

However, differences in the regulation of development and regeneration of myelin do occur; the transcription factor Olig1, although nonessential for developmental myelination (Xin et al. 2005), is required for remyelination where it plays a pivotal permissive role in OPC differentiation (Arnett et al. 2004). In contrast, the Notch signaling pathway, a negative (Wang et al. 1998) or positive (Hu et al. 2003) regulator of differentiation in development (depending on the ligand), is redundant during remyelination, since conditional knockout of the *Notch1* gene in OPCs has no or a limited effect on remyelination (Stidworthy et al. 2004; Zhang et al. 2009). The differentiation inhibitory function of endothelin-1 has recently been shown to operate via activation of the Notch pathway, supporting a view that on balance this pathway is inhibitory (Hammond et al. 2014).

Inflammation and Remyelination

The innate immune response to demyelination is important for creating an environment conducive to remyelination (Franklin and Simons 2022). The relationship between inflammation and regeneration is well recognized in many other tissues. However, its involvement in myelin regeneration has been obscured in a field dominated by the immune-mediated pathology of MS and its various animal models such as EAE, where it is unquestionably true that the adaptive immune response mediates tissue damage.

The innate immune component of remyelination involves both CNS resident microglia and monocytes recruited from the circulation, both of which can differentiate into phagocytic macrophages. That the innate immune system plays an important role in remyelination is now firmly established (Kotter et al. 2001; Lloyd and Miron 2019). These roles are many and complex, needing fine tuning so as not to be insufficient to activate a regenerative response nor overly extensive to be the cause of additional damage. The innate immune response during remyelination includes (1) recruitment of inflammatory cells; (2) astrocyte activation; (3) activating and recruiting OPCs and their subsequent differentiation through the expression of a range of growth factors and cytokines; (4) removal of myelin debris that contains inhibitors of OPC differentiation and is required for recycling of cholesterol; (5) extracellular matrix (ECM) production and modification; and, ultimately, (6) resolving inflammation (Franklin and Simons 2022). An activated macrophage phenotype is associated with the recruitment phase of remyelination and the switch to an inflammatory environment dominated by alternatively activated macrophages classically associated with tissue regeneration is causally related to the initiation of differentiation, in part via the production of activin-A (Miron et al. 2013). An area of emerging interest is how microglia register that demyelination has occurred and assume the activation states necessary to contribute to remyelination, with particular focus on the role of pattern-recognition factors such as Toll-like receptors and C-type lectin receptors, and triggering receptors expressed on myeloid cells (TREM2) (Cantoni et al. 2015; Poliani et al. 2015; Cignarella et al. 2020; Cunha et al. 2020; Bosch-Queralt et al. 2021; Gouna et al. 2021). Signaling via these receptors activates microglia via downstream mediators such

Cite this article as *Cold Spring Harb Perspect Biol* doi: 10.1101/cshperspect.a041371

as nuclear factor-κB (NF-κB) and myeloid differentiation primary response 88 (MyD88) (Cunha et al. 2020). Evidence is now emerging that the adaptive immune response may also contribute to successful remyelination, and in particular regulatory T cells through the production of the ECM-related protein CCN3 (Dombrowski et al. 2017).

DEMYELINATED CNS AXONS CAN ALSO BE REMYELINATED BY SCHWANN CELLS

CNS remyelination can also be mediated by Schwann cells, the myelin-forming cells of the PNS; this occurs in several experimental animal models of demyelination as well as in human demyelinating disease (Snyder et al. 1975; Itoyama et al. 1983, 1985; Dusart et al. 1992; Felts et al. 2005). Schwann cell remyelination occurs preferentially where astrocytes are absent—for example, where they have been killed along with oligodendrocytes by the demyelinating agent (Blakemore 1975; Itoyama et al. 1985). Remyelinating Schwann cells within the CNS were generally thought to migrate into the CNS from PNS sources such as spinal and cranial roots, meningeal fibers, or autonomic nerves following a breach in the *glia limitans* (Franklin and Blakemore 1993). In support of this idea, CNS Schwann cell remyelination typically occurs in proximity to spinal/cranial nerves or around blood vessels (Snyder et al. 1975; Duncan and Hoffman 1997; Sim et al. 2002a). However, recent genetic fate mapping studies have revealed that very few CNS remyelinating Schwann cells are derived from PNS Schwann cells but instead the majority derive from OPCs (Zawadzka et al. 2010), revealing a remarkable capacity of these cells to differentiate into cells of neural crest lineage as well as other neuroepithelial lineages (astrocytes and oligodendrocytes). This process occurs preferentially in the perivascular niche and is aided by high levels of BMP4 and an absence of astrocyte-derived BMP-antagonist Soctdc1 (Ulanska-Poutanen et al. 2018). Thus, the OPC may be more appropriately considered a broadly multilineage-competent neuroectodermal progenitor capable of producing not only astrocytes and oligodendro-

cytes, but also Schwann cells, in a context-dependent fashion (Crawford et al. 2014).

The implications of Schwann cell remyelination of CNS axons are unclear (Chen et al. 2021). While both Schwann cell and oligodendrocyte remyelination are associated with a return of saltatory conduction (Smith et al. 1979), their relative abilities to promote axon survival, a major function of myelin (Nave and Trapp 2008), have yet to be established. Thus, from a clinical perspective we do not yet know whether OPC differentiation into Schwann cells has a beneficial or deleterious effect compared to oligodendrocyte remyelination.

CAUSES OF REMYELINATION FAILURE

The efficiency of remyelination is affected by the non-disease-related factors of age and sex (Sim et al. 2002b; Li et al. 2006). These generic factors will have a bearing on the efficiency of remyelination regardless of the disease process involved and will be discussed first.

Like all other regenerative processes, the efficiency of remyelination decreases with age. This manifests as a decrease in the rate at which it occurs and is likely to have a profound bearing on the outcome of a disease process that in the case of MS can occur over many decades. The age-associated effects on remyelination are due to a decrease in the efficiency of both OPC recruitment and differentiation (Sim et al. 2002b). Of these two events, the impairment of differentiation is rate-determining since increasing the provision of OPCs by the overexpression of the OPC mitogen and recruitment factor PDGF following demyelination in old mice does not accelerate remyelination (Woodruff et al. 2004). The impairment of OPC differentiation in aging mirrors the failure of oligodendrocyte differentiation associated with many chronically demyelinated MS plaques (Wolswijk 1998a; Kuhlmann et al. 2008).

The development of protocols for isolating OPCs from aged rodents has allowed a detailed study of the changes that occur in OPCs with aging. In short, OPCs acquire all the hallmarks associated with adult stem cell aging, including mitochondrial dysfunction, loss of DNA repair mechanisms, changes in the epigenome, and dysregulated nutrient signaling (Neumann et al. 2019a). Interestingly, the drivers of OPC aging are not cell intrinsic but rather are determined by features of the niche and especially its biomechanical properties—stiff substrates induce an aged phenotype, while soft substrates induce a youthful phenotype (Segel et al. 2019). The mechanisms governing the OPC age-state involve regulation of the levels of the transcription factor c-Myc, declining levels of which are associated with aging (Neumann et al. 2021). The effects of aging on remyelination are not all cell autonomous as aging will have deleterious consequences for the function of all cell types contributing to the remyelination environment. Best studied are the age-related changes in the innate immune component of remyelination, where aging results in a diminished ability to clear and process myelin debris (Kotter et al. 2006; Natrajan et al. 2015; Cantuti-Castelvetri et al. 2018; Rawji et al. 2018, 2020a).

A key question relating to the development of remyelination therapies is the extent to which age-associated changes can be reversed. This has been demonstrated using the heterochronic parabiosis model—by parabiotic union of a young adult animal to an old adult animal, the old adult animal can be made to remyelinate with the efficiency of a young adult (Ruckh et al. 2012). This is achieved, in part, by the recruitment of circulating young monocytes to bolster the myelin debris clearance function of the old macrophages. Reversing the age-related decline in remyelination efficiency has been achieved by a number of interventions including (1) rejuvenating aged OPCs by disabling their ability to sense a stiff aged niche through the deletion of the mechanoreceptor Piezo-1 or by increasing levels of c-Myc in aged OPCs; (2) by calorie restriction; and (3) by administration of the calorie-restriction mimetic, metformin (Fig. 3; Neumann et al. 2019a,b; Segel et al. 2019).

In addition to these generic factors, remyelination could also be incomplete or fail for disease-specific reasons. The strongest evidence for remyelination failure is provided by MS, and the subsequent discussion will specifically relate to this disease, although the issues discussed will be relevant to other diseases with a demyelinating component. Theoretically, remyelination via generation of new oligodendrocytes from OPCs could fail because of (1) a primary deficiency in progenitor cells; (2) a failure of progenitor cell recruitment; or (3) a failure of progenitor cell differentiation and maturation.

Early speculation on remyelination failure focused on the first of these mechanisms, that the process of remyelination itself would deplete an area of CNS of its progenitor cells so that subsequent episodes of demyelination occurring at or around the same site would fail to remyelinate due to a lack of OPCs. However, data from experimental studies indicate that OPCs are remarkably efficient at repopulating regions from which they have been depleted (Chari and Blakemore 2002), and that repeat episodes of focal demyelination in the same area neither depletes OPCs nor prevents subsequent remyelination (Penderis et al. 2003). The situation may be different, however, when the same area of tissue is exposed to a sustained demyelinating insult, where remyelination impairment seems to be due, at least in part, to a deficiency in OPC availability (Mason et al. 2004; Armstrong et al. 2006).

In the second mechanism, MS lesions fail to remyelinate not because of a shortage of available progenitor cells but rather because of a failure of OPC recruitment: proliferation, migration, and repopulation of areas of demyelination. Here, descriptions of demyelinated areas from which oligodendrocyte lineage (OL) cells are absent do indicate that this may account for failure of remyelination in at least a proportion of lesions. Why lesions should become deficient in OPCs is not clear but one possibility is that they are direct targets of the disease process within the lesion. The identification of patients with antibodies recognizing OPC-expressed antigens (NG2) support this possibility (Niehaus et al. 2000). Failure of OPC recruitment into areas of demyelination may arise due to disturbances in the local levels of the OPC migration guidance cues semaphorin 3A and 3F (Williams et al. 2007a).

The best evidence at present supports the third mechanism, a failure of differentiation and maturation, as several sets of observations

Cite this article as *Cold Spring Harb Perspect Biol* doi: 10.1101/cshperspect.a041371

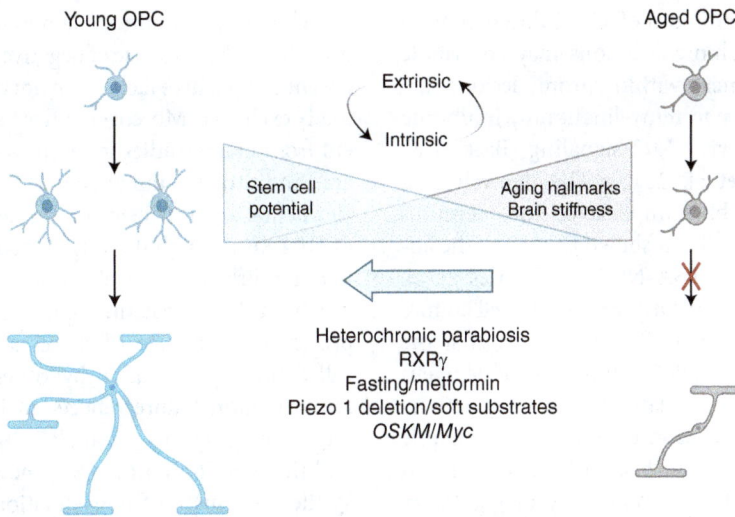

Figure 3. Oligodendrocyte progenitor cell (OPC), aging, and rejuvenation. OPCs, like other adult stem cells, undergo functional decline with aging: they have a diminished ability to self-renew and to differentiate. The dysfunction of aged OPCs is underlined by the acquisition of hallmarks of ageing, like DNA damage and mitochondrial dysfunction. The functional capacity of an OPC is determined by its environment (niche) where the physical properties (stiffness) of the brain play a key role in the aging process. Interventions, such as heterochronic parabiosis, activation of the retinoid X receptor-γ (RXRγ), calorie restriction or treatment with metformin, manipulation of substrate (or capacity to sense substrate properties), or partial reprogramming, can reinstate the stem cell potential of OPCs and thereby their capacity for remyelination. Alternatively, aged OPCs can be reprogrammed to a more youthful state. The deletion of Piezo 1 prevents OPCs from sensing the stiffness of the niche. Thus, aged OPCs behave as young OPCs that are normally exposed to a soft environment. Therefore, strategies that restore a more youthful environment or that make OPCs impervious to extracellular changes that occur with ageing lead to a functional rejuvenation of OPCs and thereby store the capacity of aged animals for remyelination. (Figure adapted from Neumann et al. 2019b, and reprinted under the Creative Commons Attribution 4.0 International license [CC BY 4.0 DEED].)

based on the detection of oligodendrocyte lineage cells within areas of demyelination indicate that this stage of remyelination is most vulnerable to failure in MS. The presence of OPCs apparently unable to differentiate within MS lesions was initially shown with the OL marker O4 (Wolswijk 1998b) and subsequently with NG2 (Chang et al. 2000), PLP (to reveal premyelinating oligodendrocytes) (Chang et al. 2002), and with Olig2 and Nkx2.2 (Kuhlmann et al. 2008). Even though the density of OPCs within chronic lesions is on average lower than in normal white matter, the density can be as high as that in normal white matter or remyelinated lesions, showing that OPC availability is not a limiting factor for remyelination.

One possible explanation for this failure of differentiation is that chronically demyelinated lesions contain factors that inhibit progenitor differentiation. First implicated was the Notch–Jagged pathway, a negative regulator of OPC differentiation. Notch and its downstream activator Hes5 were detected in OPCs and Jagged in astrocytes within chronic demyelinated MS lesions (John et al. 2002). However, the expression of Notch by OPCs and Jagged by other cells within lesions undergoing remyelination and, more informatively, the limited remyelination phenotype in experimental models following conditional deletion of *Notch* in OL cells, suggest that Notch–Jagged signaling is a redundant, nonessential negative regulator of remyelination (Stidworthy et al. 2004).

Other potential inhibitory factors have been identified in different experimental and pathological studies. The accumulation of the glycosaminoglycan inhibitor of OPC differentiation hyaluronan within MS lesions may contribute to an environment within chronic lesions that is not conducive to remyelination by inhibiting OPC function via TLR2 signaling (Back et al. 2005; Sloane et al. 2010). The demyelinated axon itself has been implicated, since demyelinated axons have been shown to express the adhesion molecule PSA-NCAM (Charles et al. 2002), which inhibits myelination in cell culture (Charles et al. 2000). The possibility that the properties of the OPC within areas of demyelination might be regulated by electrical activity (or lack of) in (Gibson et al. 2014) or synaptic input from (Etxeberria et al. 2010) demyelinated axons highlights the complexity of regulatory factors that govern remyelination and by extension account for remyelination failure.

While many studies have concentrated on putative inhibitory signals to account for the failure of OPCs to undergo complete differentiation within demyelinated MS plaques, an alternative explanation is that these lesions fail to remyelinate because of a deficiency of signals that induce differentiation. This hypothesis, based on the absence of factors, is difficult to prove but is consistent with a model of remyelination in which the acute inflammatory events play a key role in progenitor activation and creating an environment conducive to remyelination (see above). While MS lesions are rarely devoid of any inflammatory activity, chronic lesions are relatively noninflammatory compared to the acute lesions and constitute a less active environment in which OPC differentiation might become quiescent. Acute inflammatory lesions are characterized by reactive astrocytes that are the source of many pro-remyelination signaling factors (Williams et al. 2007b; Moore et al. 2011; Molina-Gonzalez and Miron 2019; Rawji et al. 2020b). In contrast, chronic quiescent lesions are characterized by scarring astrocytes that are transcriptionally quiet compared to reactive astrocytes. The scarring astrocyte is better viewed as a consequence of remyelination failure and not its cause. Thus, neither the reactive nor the scarring astrocytes—both contributing to astrogliosis—are likely drivers of remyelination failure.

The two possibilities that remyelination failure reflects the presence of negative factors or the absence of positive factors are not of course mutually exclusive. Moreover, it has become apparent from many studies in recent years that there are a multitude of interacting factors, both environmental and intrinsic, that guide the behavior of OL cells through the various stages of remyelination. Efficient remyelination may depend as much on the precise timing of action as on the presence or absence of these factors. This is called the "dysregulation hypothesis" in which remyelination failure reflects an inappropriate sequence of events (Franklin 2002; Franklin and ffrench-Constant 2008; Fancy et al. 2011). While the causes of remyelination failure in a disease as complex and variable as MS are likely to be many, we still regard this hypothesis as useful for understanding remyelination failure in the majority of cases.

ENHANCING ENDOGENOUS REMYELINATION

Since remyelination can occur completely and the cells responsible are abundant throughout the adult CNS, even within demyelinated lesions, a conceptually attractive approach to enhancing remyelination is to target the endogenous regenerative process (Franklin and Gallo 2014). This approach is predicated on the view that if the mechanisms of remyelination can be understood and non-redundant pathways described, then the causes of remyelination failure and hence plausible therapeutic targets will be identified. From the preceding sections it will be clear that remyelination failure is associated with either insufficient OPC recruitment or, more commonly, failed OPC/oligodendrocyte differentiation. However, the underlying biology of these two phases of remyelination is different and sometimes mutually exclusive. The implication is therefore that pro-recruitment therapies may not promote remyelination where the primary problem is OPC differentiation and vice versa.

A further consideration in the development of remyelination enhancement therapies is the use of appropriate animal models. In the chronic demyelinated plaques of MS, remyelination is

Cite this article as *Cold Spring Harb Perspect Biol* doi: 10.1101/cshperspect.a041371

assumed to have failed; hence, the requirement is for an intervention that will reactivate a dormant process. In contrast, in many of the demyelination models used to test enhancement of remyelination, such as the toxin-based models, remyelination does not fail. In those models, one can only achieve the acceleration of an already effective ongoing process. That said, assessing the temporal dynamics of remyelination in MS tissue, and whether it has stopped or merely slowed, is difficult to assess from the "snapshots" provided by biopsy or postmortem tissue. Nevertheless, this problem can in part be overcome in two ways. First, by using aged animals in which the slow rate of remyelination is suboptimal presents an opportunity for assessing its enhancement. Second, by modifying standard lesion models in which the endogenous repair process is compromised, such as in the chronic cuprizone model (Armstrong et al. 2006). Theiler's virus-induced demyelination model has also proven useful for demonstrating enhanced remyelination (Njenga et al. 1999). However, assessment of remyelination has proven especially complex in EAE, in which the processes of demyelination and remyelination can occur concurrently. This can make it very difficult to distinguish an effect that renders the environment less hostile to remyelination, allowing it to proceed at its natural rate, from one in which the rate of remyelination is actually accelerated. For example, systemic delivery of putative remyelination-enhancing factors can affect the balance of myelin damage and regeneration via effects on cells other than oligodendroglial cells, such as those of the immune system.

Despite caveats regarding models and methods of analysis, over the last ~15 years several proof-of-principle studies have allowed the identification of promising pharmacological compounds for enhancing remyelination, both repurposed and newly developed, that modulate key pathways involved in endogenous myelin repair.

Several studies investigating humanized monoclonal antibodies against LINGO-1 (opicinumab) have demonstrated the potential of these compounds to increase axonal myelination both in vitro and in animal models (Mi et al. 2007;

Rudick et al. 2008). A promising remyelination potential has also been demonstrated for the antagonists of muscarininc M1 receptor (clemastine, benztropine, quetiapine). In primary OPC cultures, these molecules have been found to be effective in orienting OPCs toward a differentiated phenotype, promoting the formation of mature, myelinating oligodendrocytes (Angelis et al. 2012; Mei et al. 2014; Abiraman et al. 2015; Li et al. 2015). Moreover, in EAE models, a significant improvement in the efficacy of remyelination has been obtained following the removal of the muscarinic acetylcholine receptor M1 in oligodendrocytes (Mei et al. 2016). Several studies have explored the promyelinating potential of modulators of RXRγ (bexarotene, IRX4204), a nuclear receptor with a key role in immune regulation and OPC differentiation and maturation (Huang et al. 2011; Natrajan et al. 2015). The treatment of rodents with a pan-RXR agonist, 9-cis-retinoic acid, following experimental demyelination, resulted in enhanced remyelination, while RXRγ knockout mice showed a significantly reduced number of oligodendrocytes in the lesion area (Huang et al. 2011). Besides RXRγ, other nuclear hormone receptors can be targeted to enhance remyelination, including the vitamin D receptor, the peroxisome proliferator-activated receptor (PPAR), and the liver X receptor (de la Fuente et al. 2015; Veloz et al. 2022). Another class of compounds showing promise as promyelinating treatments are the histamine H3 receptor antagonists (GSK239512, GSK247246), which have been shown to enhance remyelination through increased mature oligodendrocyte differentiation in the cuprizone mouse model (Chen et al. 2017). Several lines of evidence also pointed toward the promyelinating potential of steroid hormone-based therapies (androgens, estrogen, thyroid hormones analogs), which have been shown to enhance myelin repair, possibly via the induction of OPC differentiation (Sutiwisesak et al. 2021). Furthermore, steroids (clobetasol) and antifungals (miconazole) have also demonstrated notable remyelinating properties in experimental models, although results have not been completely consistent across models. Both clobetasol and miconazole have been found to promote oligodendrocyte differentiation and increase remyelination in a rodent model of demyelination (Najm et al. 2015). Interestingly, many of these compounds (clemastine, benzatropine, tamoxifen, bazedoxifene, and miconazole), while targeting completely different pathways, all share the property of promoting oligodendrocyte maturation through the same signaling mechanism, which is the increased production of 8,9-unsaturated sterols (Hubler et al. 2018).

Of note are compounds that combine an anti-inflammatory with a remyelinating mechanism of action, such as siponimod, a modulator of the sphingosine-1 receptor (S1P-R) that crosses the blood–brain barrier and selectively targets S1P1-R and S1P5-R. While significantly reducing lymphocyte migration in the CNS through S1PR1, siponimod may also promote remyelination via S1PR5 expressed on oligodendrocytes (Behrangi et al. 2022).

Despite the significant number of molecules known to have promyelinating properties, so far, only a limited subgroup has advanced to being tested in phase 2 clinical trials in patients with MS. While most of the already completed and ongoing clinical trials testing promyelinating drugs are conducted in patients with acute or chronic optic neuritis, and employ the change in the latency of visual evoked potentials (VEPs) as an outcome measure (with a reduced VEP latency interpreted as reflecting remyelination), a few of them investigate patients with relapsing MS, with an either relapsing-remitting or progressive disease course (Lubetzki et al. 2020). In this case, changes in lesion MTR, with increasing MTR values over time being interpreted as possibly reflecting remyelination, or changes in multicomponent scores of clinical function, are employed as outcome measures.

While preclinical studies identified opicinumab, a monoclonal antibody directed against LINGO-1, as a promising molecule, the results of clinical trials testing this molecule in patients with MS have been disappointing (Cadavid et al. 2017, 2019). The safety and efficacy of the retinoid receptor agonist bexarotene have been tested in the clinical trial CCMR One, in which 52 patients with RRMS received either 300 mg/m^2 of BSA per day of oral bexarotene or oral placebo for 6 months. In this study, the change in mean lesional MTR over the 6-month follow-up was employed as the primary efficacy outcome (Brown et al. 2021). While bexarotene was found to be associated with an increased number of adverse events, the change in mean lesional MTR did not significantly differ between bexarotene and placebo. Nevertheless, this clinical trial demonstrated that MS lesions are heterogeneous in their capacity to remyelinate following treatment with bexarotene. In particular, lesions that were more demyelinated at baseline as well as lesions that were localized in the gray matter showed greater remyelination following this treatment. Although the results of this study do not support the clinical use of bexarotene due to its poor tolerability, future clinical trials may investigate the remyelinating properties of other RXR agonists with better tolerability profiles.

Among the few clinical trials testing promyelinating treatments that have been already completed, there is also ReBUILD, a double-blind, crossover trial, which compared a group of patients with chronic optic neuritis on the antagonist of the muscarinic M1 receptor clemastine (5.36 mg orally twice daily) for 90 days followed by placebo for 60 days, with a group of patients on placebo for 90 days followed by clemastine (5.36 mg orally twice daily) for 60 days (Green et al. 2017). In this study, clemastine was shown to reduce the VEP latency delay without being associated with serious adverse events. Based on the promising results of ReBUILD, clemastine is currently being tested in the context of acute optic neuritis (NCT02521311) and as combinatory remyelinating therapy (NCT03109288).

The histamine H3 receptor blocker GSK239512 has also been tested in a phase II clinical trial in patients with relapsing-remitting MS, showing a limited yet significant effect in increasing the MTR value of white matter lesions in treated patients compared to patients on placebo (Schwartzbach et al. 2017). The results of two clinical trials exploring the remyelinating properties of nanocrystalline gold (NCT03536559) and of the estrogenic compound bazedoxifene (NCT04002934), which have been tested in patients with MS chronic optic neuritis, are expected very soon. Finally, the remyelinating effects of testosterone in relapsing-remitting MS are being tested in a phase II clinical trial that is currently recruiting (NCT03910738).

In addition to the opportunities provided by the pharmacological manipulation of different pathways involved in remyelination, there is another intriguing therapeutic strategy to enhance remyelination, based on the growing evidence suggesting that axonal electrical activity has a strong influence on oligodendrogliogenesis and (re)myelination. In particular, silencing electrical activity using tetrodotoxin has been demonstrated to reduce the percentage of myelinated fibers, in vitro in myelinating cocultures and in vivo in the optic nerve after intravitreal injection (Demerens et al. 1996). Conversely, through the optogenetic stimulation of neurons of the murine premotor cortex, it has been demonstrated that neuronal activity promotes the proliferation of OPCs, as well as oligodendrocyte maturation and myelination, with functional improvement (Gibson et al. 2014). In demyelinating conditions, axonal electrical activity remains essential to promote remyelination. Indeed, it has been demonstrated that, after demyelination, axons still have the potential to conduct electrical impulses. This neuronal activity regulates OPC differentiation by synaptic release of glutamate. Blocking action potentials with tetrodotoxin leads to increased OPC apoptosis and failure of OPC differentiation into mature oligodendrocytes (Gautier et al. 2015). Based on this evidence, electrical stimulation is currently being considered as a promising therapeutic approach to enhance myelin repair in demyelinating conditions. In this perspective, transorbital electrical stimulation is currently being tested as a means to shorten VEP latency in MS patients with acute optic neuritis (NCT04042363).

REMYELINATION BY GLIAL PROGENITOR CELL TRANSPLANTATION

The mobilization of endogenous oligodendrocyte progenitor cells may be appropriate for a variety of disorders of progressive dys- or demyelination. However, a broad swath of neurological disease involves the structural loss of white matter, reflecting frank loss of both oligodendrocytes and their progenitors, and often of their associated fibrous astrocytes. These disorders include ischemic and traumatic demyelination, as well as demyelination associated with sustained autoimmune inflammation, and insults as diverse as chemotherapy and radiotherapy. In addition, many of the neurodegenerative disorders —including Huntington's disease, Alzheimer's disease, and schizophrenia, among others—are attended by progressive demyelination, as is nominally healthy aging itself. Moreover, the childhood hereditary and metabolic disorders of white matter, the hypomyelinating leukodystrophies and lysosomal storage disorders in particular, are intrinsically refractory to any approach aimed at activating resident progenitors —which themselves carry the culpable mutation. Together, these diverse disorders of myelin require extensive tissue repair, and in many cases even whole neuraxis myelination. Thus, any practical cell therapeutics for the myelin disorders must provide large numbers of progenitors biased to oligodendrocyte differentiation and myelinogenesis.

This requirement prompted the development of human OPCs from a variety of sources, including human tissue (Roy et al. 1999; Dietrich et al. 2002; Windrem et al. 2002, 2004) and later from embryonic and induced pluripotent stem cells (Izrael et al. 2007; Hu et al. 2009; Stacpoole et al. 2013; Wang et al. 2013; Douvaras et al. 2014; Piao et al. 2015). Each of these sources has now been shown able to generate potentially myelinogenic oligodendrocytes (Fig. 4), and each has its own strengths and weaknesses as a cellular therapeutic (Goldman et al. 2012, 2021). Of note, OPCs may also be generated directly from mouse and human fibroblasts as well (Najm et al. 2013; Yang et al. 2013; Ehrlich et al. 2017; Tanabe et al. 2022), suggesting the potential development of yet another source of clinically appropriate OPCs (Goldman 2013).

To serve as practical, safe, and effective therapeutic vectors, OPCs must be deliverable in both reliable purity and significant quantity. To

Cell sources

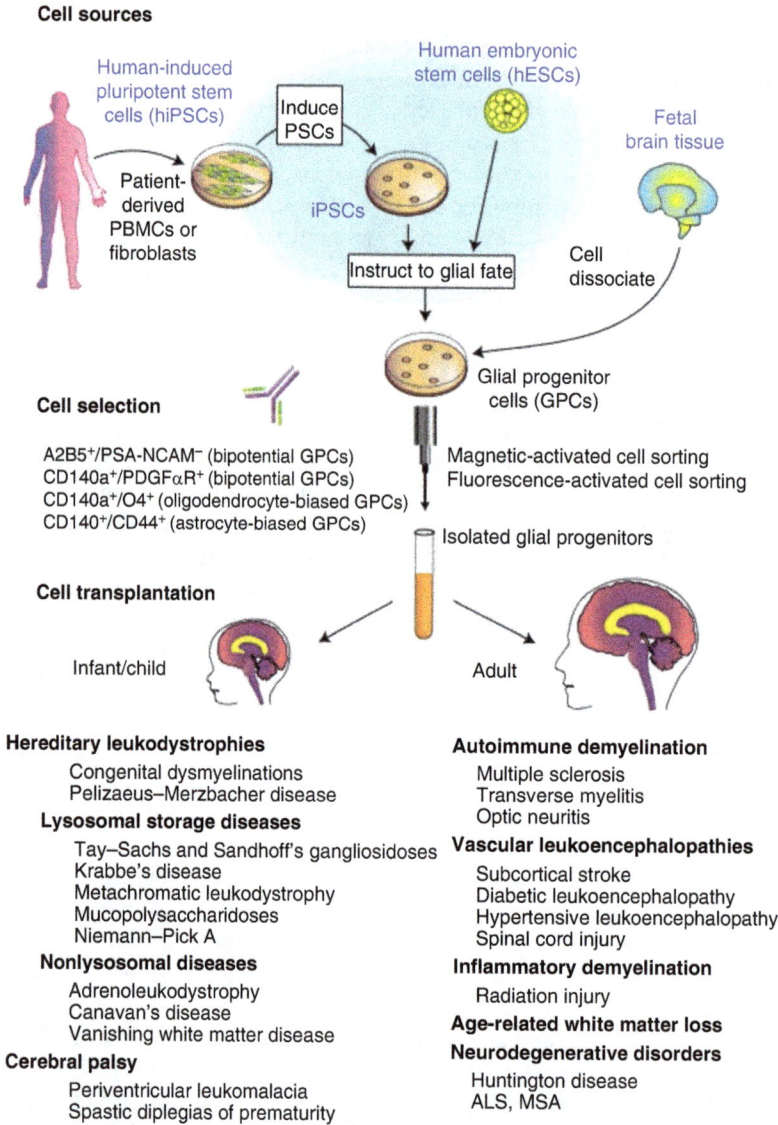

Figure 4. Glial progenitor cell (GPC) sources, phenotypes, and clinical targets. GPCs may be directly sorted from tissue or produced from either human embryonic stem cells (hESCs) or human-induced pluripotential cells (hiPSCs) and then immunoselected based on their expression of either gangliosides recognized by monoclonal antibody (mAb) A2B5 or of CD140a/PDGFαR. The CD140a phenotype includes all potential oligodendrocytes, while the tetraspanin CD9 and sulfatide-directed mAb O4 identify progressively more oligodendrocyte-biased fractions (Sim et al. 2011; Douvaras et al. 2014). In contrast, CD44 recognizes a more astrocyte-biased fraction (Liu et al. 2004). The choice of tissue-, hESC-, or iPSC-derived GPCs depends upon whether allogeneic or autologous grafts are desired. Whereas autologous grafts of iPSC-derived GPCs might obviate the need for immunosuppression, their generation may take months, and their use in the hereditary leukodystrophies would first require correction of the underlying genetic disorder in the donor cell pool; at present, such genetic disorders of myelin would be better approached with allografted tissue- or hESC-derived GPCs. (This figure is adapted and reprinted, with permission, from Goldman et al. 2012.)

address this need, several methods for isolating human OPCs (hOPCs) from mixed cell populations have been developed, which use fluorescence activated or magnetic bead-based cell sorting targeting OPC surface antigens (Roy et al. 1999; Nunes et al. 2003; Sim et al. 2011). These hOPC isolates were first obtained from human brain tissue, and their ability to robustly remyelinate demyelinated tissue was initially established in animal models of both congenital and adult dysmyelination (Windrem et al. 2002, 2004). In *Shiverer* mice, a dysmyelinated mouse deficient in Myelin basic protein, fetal and adult tissue-derived hOPCs were noted to behave differently after neonatal xenograft. In particular, isolates derived from adult white matter myelinated recipient brain much more rapidly than fetal hOPCs; adult-derived progenitors achieved widespread myelination just 4 weeks after graft, while cells derived from second trimester fetuses took over 3 months to do so. On the other hand, these fetal hOPCs expanded in vivo, emigrated more widely, engrafted more efficiently, and differentiated as both astrocytes and oligodendrocytes, all suggesting their greater potential as therapeutic vectors (Fig. 5; Windrem et al. 2004; Sim et al. 2011). Nonetheless, despite the mitotic competence of these cells, they remain finite in both initial number and in expansion competence, necessitating their periodic reacquisition from new donors, and thus rendering their routine use impracticable. Recent efforts have thus focused on the generation of myelinogenic OPCs from pluripotent human stem cells as preferred sources for the scalable development of therapeutic hOPCs.

PLURIPOTENT STEM CELLS AS A SOURCE OF MYELINOGENIC PROGENITORS

Following the first report of myelination in the injured spinal cord by implanted murine embryonic stem cells (mESCs) (Brüstle et al. 1999), oligodendrocytes derived from human ESCs (hESCs) were reported to similarly myelinate demyelinated foci in spinal cord contusions (Nistor et al. 2005). hESC-derived hOPCs were then shown to generate myelin in the rodent brain (Zhang et al. 2001), and subsequent studies

with improved differentiation protocols reported more efficient glial and oligodendrocytic differentiation (Izrael et al. 2007; Hu et al. 2009). Nonetheless, none of these studies isolated hOPCs prior to transplant, nor did any follow animals for the periods of time required to ensure the stability and safety of the engrafted cells. This was problematic, since persistent undifferentiated hES cells in the donor pool may retain the potential to generate teratomas or neuroepithelial tumors (Roy et al. 2006; Pruszak et al. 2009). To address these considerations, Wang et al. (2013) developed a more stringent protocol for differentiating hOPCs from pluripotent human stem cells, which yielded highly enriched and myelinogenic populations of hOPCs, with no detectable residual stem cells nor evidence of tumorigenicity after transplant. In particular, Wang and colleagues noted that the risk of tumorigenesis and unintended expansion from hESCs could be effectively mitigated by extended differentiation protocols and rigorous quality assessments during cell production. Nonetheless, hESC lines are derived from human embryos, which has hindered their use in a number of both political jurisdictions and cultures. Enthusiasm has thus developed for the potential use of induced pluripotent stem cells (iPSCs) as a more universally acceptable and clinically scalable source of OPCs.

iPSCs are pluripotent cells that have been generated by the reprogramming of somatic cells to a less phenotypically committed stem cell ground state, by the forced expression of transcription factors that permit the self-renewing stem cell phenotype (Stadtfeld and Hochedlinger 2010). Most typically, iPSCs are generated from somatic cells cotransduced with a set of stem cell–associated transcription factors, including POU5F1/OCT4, SOX2, MYC, KLF4, and NANOG (Yamanaka 2007, 2008), or alternatively, with miRNAs that modulate the expression of their encoded proteins, or small molecules that mimic their actions (Lin et al. 2009; Anokye-Danso et al. 2011). iPSCs have the advantage over hESCs of yielding lineage-restricted cells that can be autologously transplanted back to the subjects from which they were generated, potentially obviating the need for posttransplant

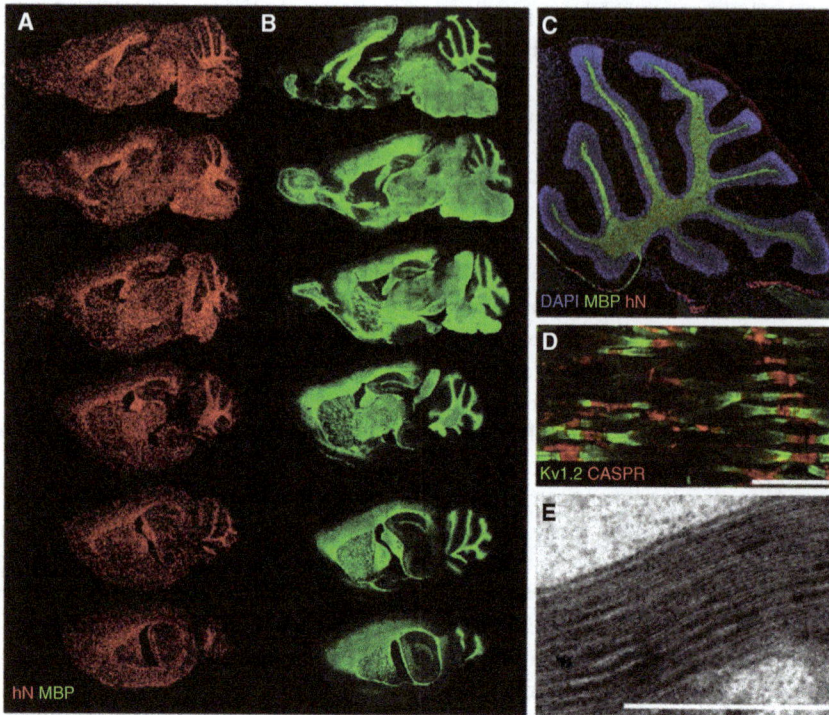

Figure 5. Transplanted fetal human glial progenitor cells (hGPCs) myelinate the congenitally unmyelinated shiverer brain. (*A–E*) Myelination of congenitally hypomyelinated *Shiverer* (*Shi*) mice by human fetal tissue-derived glial progenitor cells. (*A,B*) Representative sagittal images of an engrafted Shi/Shi × *Rag2*$^{-/-}$ brain, sacrificed at 1 yr of age. (*A*) Human donor cells identified by an antihuman nuclear antibody (hN; red). (*B*) Donor-derived myelin basic protein (MBP; green) in sections adjacent or nearly so to matched sections in *A*. All major white matter tracts heavily express MBP (which is all donor-derived in *Mbp*-null *Shiverer* mice). (*C*) Sagittal view through cerebellum of a year-old engrafted Shi/Shi × *Rag2*$^{-/-}$ brain. All cells were stained with DAPI (blue); donor cells were identified by human nuclear antigen (hN, red) and donor-derived myelin by MBP (green). (*D*) Reconstituted nodes of Ranvier in the cervical spinal cord of a transplanted and rescued 1-yr-old shi/shi × rag2$^{-/-}$ mouse, showing paranodal Caspr protein and juxtaparanodal potassium channel Kv1.2, symmetrically flanking each node. Untransplanted *Shiverer* brains do not have organized nodes of Ranvier and, hence, cannot support saltatory conduction (Caspr, red; Kv1.2, green). (*E*) Electron micrograph of a 16-wk-old *Shiverer* mouse implanted perinatally with hGPCs shows a *Shiverer* axon with a densely compacted myelin sheath. Scale bars, 5 µm (*D*); 1 µm (*E*). (Images *A* and *B* from Windrem et al. 2008, *C* and *D* from Goldman et al. 2012, and *E* from Windrem et al. 2004, reprinted with permission. This figure is adapted, with permission, from Osorio and Goldman 2016 and Mariani et al. 2019.)

immune suppression. The differentiation strategies for generating hOPCs from hESCs have proven adaptable to iPSCs as well, and yield both glial progenitors and terminally differentiated oligodendrocytes that are highly myelinogenic in vivo (Czepiel et al. 2011). More recent advances in optimizing the methods for generating hOPCs from human iPSCs and ESCs have led to the production of highly enriched populations of hOPCs that efficiently myelinate in vivo,

while manifesting no evidence of tumorigenic potential (Fig. 6; Wang et al. 2013; Douvaras et al. 2014; Piao et al. 2015).

The capability to now produce scalable quantities of highly myelinogenic iPSC-derived hOPCs allows us to realistically consider their use for treatment of myelin disorders. These iPSC hOPCs have the potential advantage of autologous use, so as to avoid the immunogenicity of OPCs derived from an otherwise unmatched allogeneic line. Such an autologous strategy, while not readily scalable to larger patient populations, may nonetheless be appropriate for individual cases of vascular, traumatic, and inflammatory demyelination. In this latter set of disorders, no underlying genetic lesion may exist, and sufficient axonal numbers may remain for remyelination to be beneficial. Indeed, after genetic correction using contemporary genome editing (Gaj et al. 2013), hOPCs may be generated from edited iPSCs derived from patients harboring pathological mutations, and then transplanted back into those same patients. Such genetically corrected autologous grafts may permit the treatment of genetic disorders of myelin with curative intent, yet without the need for immune suppression.

CHALLENGES IN THE USE OF PLURIPOTENT STEM CELL–DERIVED hOPCS

The autologous transplantation of iPSC-derived hOPCs generated from individual patients is an attractive proposition, but in practice one that would be difficult to scale up to large patient populations. As such, whether OPCs are derived from hESCs or iPSCs, it is likely that they will need to be delivered as allogeneic grafts to genetically unmatched recipients. Because of the likelihood of rejection of incompatible immunophenotypes, these strategies may thus be limited by a need for long-term immunosuppression in graft recipients. Yet chronic immunosuppression carries its own significant risks and morbidities, suggesting the need for developing donor cells able to avoid postgraft immunodetection and rejection. To this end, a number of laboratories have developed strategies for producing hypoimmune pluripotent stem cell lines whose derivatives might avoid immune recognition, whether by HLA class 1 and 2 antigen deletion, substitution, or knockdown (Gornalusse et al. 2017; Deuse et al. 2019; Xu et al. 2019), or the expression of checkpoint inhibitors such as PDL1 or CD47 (Gornalusse et al. 2017; Deuse et al. 2019), or combinations thereof (Lanza et al. 2019; Malik et al. 2019). Most recently, the concurrent use of CD47 overexpression as a means of suppressing dendritic cell activity, while knocking out HLA

class 1 and 2 antigens by the concurrent knockout of β2 microglobulin and CIITA, has emerged as a compelling strategy for establishing immunoavoidance. Somatic cells derived from iPSCs edited by this approach have durably engrafted in both blood and solid organs, after both allogeneic and xenogeneic transplantation (Hu et al. 2023a,b,c).

These "off-the-shelf" universal hypoimmune pluripotent stem cells may avoid or minimize the need for host immunosuppression, and have the added advantage of reducing treatment cost, as single master cell banks might be employed to produce a range of desired target cell phenotypes. That said, none of these strategies has yet been studied well in the CNS, and so none have yet been shown to prevent the immune rejection of cells transplanted into the adult brain or spinal cord. Furthermore, the attractiveness and potential utility of these second-generation approaches to immunoavoidance are not without risk; all carry the risk of enabling expansion or unintended differentiation that cannot be checked by the immune system, should tumorigenic or otherwise undesired cell types arise after transplant. As a result, some means of conditionally ablating undesired donor cells might be prudent, as might engineering the cells to undergo induced death upon pharmacological direction (Liang et al. 2018). A number of such strategies are now under development and appear promising (Jones et al. 2014; Sheikh et al. 2021), although the long-term safety and efficacy of each still needs to be demonstrated in the CNS. In that regard, while initial studies using an inducible Caspase9 system proved successful in eliminating undifferentiated cells from NSC grafts (Itakura et al. 2017), the brain presents a number of unique issues to the use of such a "suicide switch" strategy. These concerns include both the blood–brain barrier permeability of prodrugs needed to trigger donor cell death and the potential for cerebritis and cerebral edema in the setting of targeted cell clearance; the latter presents the possibility of an immune effector cell-associated neurotoxicity syndrome (ICANS)-like morbidity (Siegler and Kenderian 2020; Sheth and Gauthier 2021). Thus, while the development of immunoavoidant lines and tar-

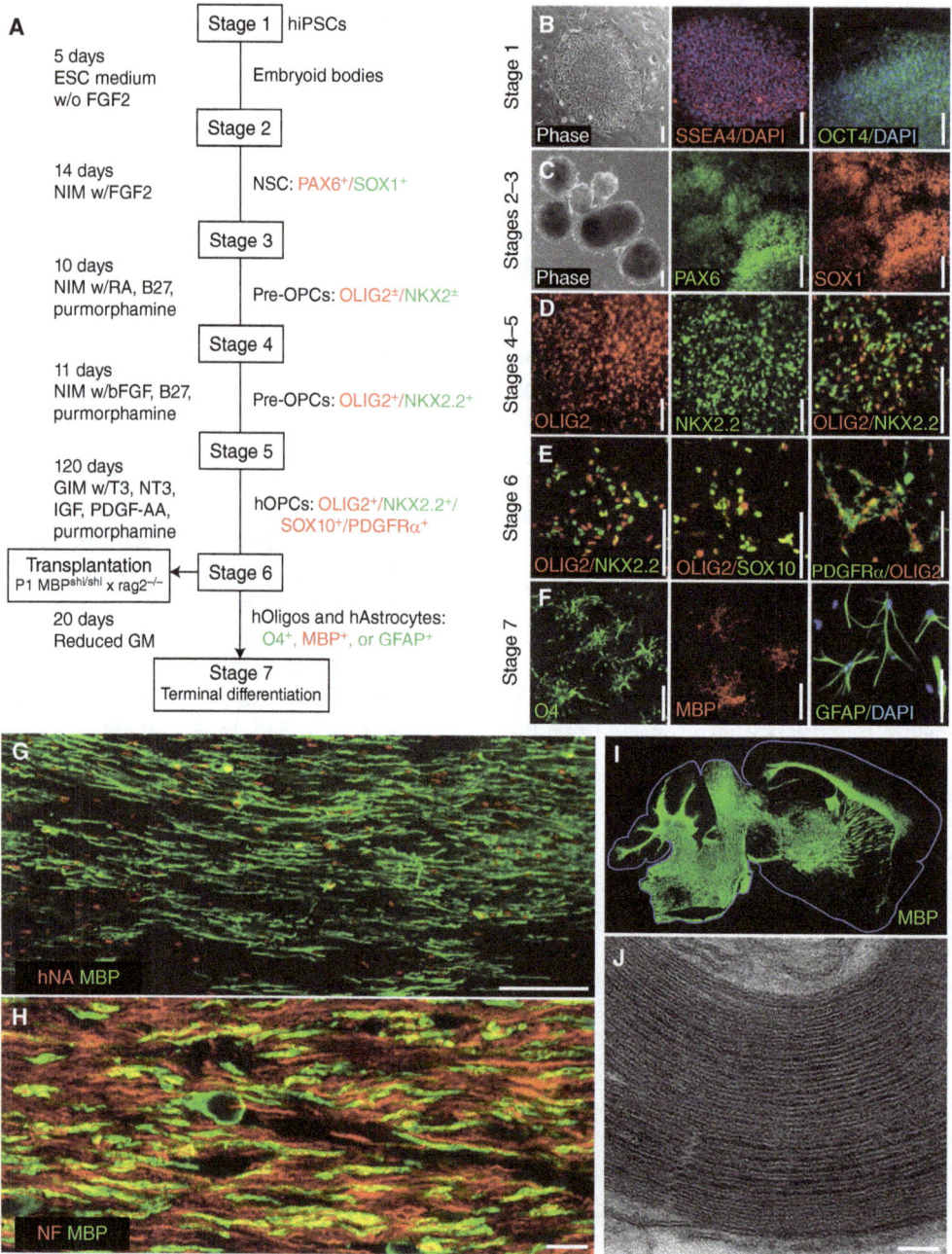

Figure 6. (*See following page for legend.*)

geted donor cell elimination strategies may be crucial to the development of safe and effective clinical vectors, neither is without risks that will need to be considered carefully in patient selection and treatment.

Aside from their pluripotency and the risks as well as opportunities that this may entail, iPSCs may retain some of the epigenetic marks —the DNA methylation and histone acetylation patterns of chromosomal architecture—of the donor cells from which they are derived (Stadtfeld et al. 2010). As a result, their age and cell type of origin may influence the differentiation competencies of iPS cells derived from different tissue sources and from subjects of different ages (Polo et al. 2010). This is of potential concern in that if iPSCs do not undergo complete reprogramming, then their derived hOPCs might be expected to differ from their tissue or hESC-derived homologs (Kim et al. 2010; Polo et al. 2010). Whether any such differences will prove meaningful in clinical practice remains to be seen, yet it would seem most wise to pursue the clinical use of hOPCs produced from iPSCs derived from the youngest subjects possible, with RNA expression patterns, DNA and histone methylation marks, and chromatin occupancies all as close to those of naive hES cells as possible.

It is worth noting here that hOPCs may be generated not only from pluripotent cells, but directly from somatic cells, using phenotype-defining transcription factors. As noted, several recent studies have reported the direct induction of both OPCs and oligodendrocytes from fibroblasts, using targeted overexpression of defined proglial transcripts (Najm et al. 2013; Yang et al. 2013; Ehrlich et al. 2017; Tanabe et al. 2022). Such avoidance of pluripotent intermediates in the generation of glial progenitors may mitigate the risk of tumorigenesis; yet the mitotic potential, in vivo differentiation efficiencies, and durabilities of such directly induced OPCs and their derived oligodendrocytes still need to be established. Indeed, the phenotypic stability of

such cells remains unclear, since at least some lineages generated via direct conversion may retain the ability, and perhaps a proclivity, to revert back to their parental phenotype. For instance, Scholer and colleagues (Kim et al. 2021) demonstrated the metastability of OPCs directly induced from pericytes, which reverted back to pericyte lineage after in vivo transplantation and in vivo residence. The extent to which such phenotypic instability is a function of time, source phenotype, reprogramming strategy employed, and host environment all remain to be established, and will need to be done before OPCs generated via direct induction might be considered as potential clinical vectors. That said, the already available panoply of human pluripotent stem cells, both embryonic and induced, and the potential use of directly reprogrammed cells as well, augur well for the availability of scalable sources of human OPCs going forward.

PEDIATRIC TARGETS OF PROGENITOR CELL–BASED THERAPIES FOR MYELINOGENESIS

Tens of thousands of children in the United States suffer from diseases of myelin failure or loss. These include the metabolic demyelinations such as adrenoleukodystrophy; the lysosomal storage disorders, such as metachromatic leukodystrophy, neuronal ceroid lipofuscinoses, mucopolysaccharidoses and gangliosidoses, Niemann–Pick disease, and Krabbe's disease; the hypomyelinating diseases, such as

Figure 6. Human pluripotent stem cell–derived glial progenitor cells (GPCs) can remyelinate dysmyelinated hosts. (A) This schematic outlines the multistage protocol by which GPCs, oligodendrocytes, and astrocytes may be generated from human pluripotent stem cells, whether human embryonic stem cells (hESCs) or induced pluripotent stem cells (iPSCs). (B–F) Representative images taken at serial stages of glial differentiation, with the serial expression of selected marker proteins noted at each stage. (G) 3 mo after neonatal transplant into hypomyelinated Shiverer mice, human-induced PSC (hiPSC)-derived GPCs have matured as myelinating, myelin basic protein (MBP)-expressing oligodendrocytes (MBP, green; human nuclear antigen, red). (H) The hiPSC–derived oligodendrocytes ensheath mouse axons (neurofilament, red; MBP, green). (I) hiPSC-derived oligodendrocytes can myelinate the entire brain of Shiverer mice, which do not otherwise express MBP (green). (J) The myelin generated by hiPSC oligodendrocytes is ultrastructurally normal, exhibiting major dense lines and thick myelin sheaths. The use of such serial and distinct stages of growth factor exposure, paired with more extended periods of differentiation, have led to the production of highly enriched populations of human GPCs (hGPCs), that are highly efficient at myelinogenesis in vivo while manifesting no evident tumorigenesis. Scale bars, 100 μm (B–E); 25 μm (F); 100 μm (G); 10 μm (H); 100 nm (J). (This figure is adapted, with permission, from Wang et al. 2013 and Mariani et al. 2019.)

Pelizaeus–Merzbacher disease and hereditary spastic paraplegia; the myelinoclastic disorders, vanishing white matter disease and Canavan's disease; dysmyelinating disorders such as Alexander's disease (Brenner et al. 2001; Dietrich et al. 2005; Li et al. 2005; Bugiani et al. 2011); and most commonly of all, periventricular leukomalacia, the most common form of cerebral palsy (Back and Rivkees 2004; Follett et al. 2004; Robinson et al. 2005). Their mechanistic heterogeneity notwithstanding, all of these conditions include the prominent loss of oligodendrocytes and central myelin, highlighting their potential attractiveness as targets for cell replacement (Helman et al. 2015; Goldman et al. 2021).

Neonatal Delivery of OPCs for Enzymatic Reconstitution and Myelin Preservation

Since OPC engraftment is both widespread and associated with astrocytic as well as oligodendrocytic production, glial progenitors would seem an especially promising vehicle for dispersing functionally competent glia throughout otherwise diseased and/or enzyme-deficient brain parenchyma. The lysosomal storage disorders present especially attractive targets in this regard (Snyder et al. 1995), since wild-type lysosomal enzymes may be released by integrated donor cells, and taken up by deficient host cells through the mannose-6-phosphate receptor pathway (Urayama et al. 2004). As a result, a relatively small number of donor glia may provide sufficient enzymatic activity to correct the underlying catalytic deficit and storage disorder of a much larger number of host cells (Lacorazza et al. 1996; Jeyakumar et al. 2005). The intracerebral delivery of OPCs would thus seem an especially attractive approach for treating those demyelinating diseases associated with enzyme deficiencies specific to brain. By way of example, metachromatic leukodystrophy (MLD) is characterized by deficient expression of *Arylsulfatase A* (*ARSA*), which results in sulfatide misaccumulation and oligodendrocyte loss. Mesenchymal and hematopoietic stem cell grafts have proven unable to correct the CNS manifestations of this disorder (Koç et al. 2002). In contrast, experimental models of MLD have responded well to OPC grafts

(Givogri et al. 2006), suggesting that the broader dispersal competence and greater histiotypic appropriateness of orthotopically engrafted OPCs might provide significant therapeutic advantages relative to nonneural phenotypes. Similarly, while asymptomatic Krabbe patients transplanted with umbilical cord stem cells manifested slower disease progression than untreated controls, the benefits of transplantation to children engrafted after symptom onset have been limited (Escolar et al. 2005). Yet the intracerebral parenchymal infiltration of stromal derivatives is minimal, suggesting that treatment of these children with native, brain-penetrant NSCs or glial progenitors might comprise a more promising treatment strategy (Pellegatta et al. 2006). More broadly, these early efforts speak to the potential of engrafted NSCs and OPCs as vehicles for intracerebral enzyme replacement, in the lysosomal storage disorders as well as in the broader category of disorders of brain metabolism with associated dysmyelination.

Neonatal Delivery of OPCs for Structural Myelin Replacement

To assess the potential of OPC-based treatment for congenital dysmyelination, Windrem and colleagues transplanted sorted hOPCs of both fetal and adult origin into newborn hypomyelinated *Shiverer* mice (Windrem et al. 2002, 2004). This work followed similar studies in which both native and immortalized murine NSCs had been transplanted into *Shiverers*; each yielded some degree of context-dependent differentiation and myelination (Yandava et al. 1999; Mitome et al. 2001). On that basis, Windrem and colleagues extracted fetal hOPCs from second-trimester forebrain as well as adult OPCs from surgically resected subcortical white matter; both were isolated by either fluorescence-activated or immunomagnetic sorting, based on the phenotype A2B5$^+$/PSA-NCAM$^-$. When transplanted into recipient *Shiverer* mice, both fetal and adult hOPCs spread widely throughout the brain and developed as both astrocytes and oligodendrocytes (Fig. 5). The cells did so in a highly context-dependent fashion, such that those donor cells that engrafted presumptive white matter developed as myelinogenic oligodendrocytes and fibrous astrocytes, while those invading cortical and subcortical gray either remained as progenitors or differentiated largely as protoplasmic astrocytes (Windrem et al. 2004).

Following a single neonatal intracallosal injection, the majority of donor cells engrafted the presumptive white matter, such that the corpus callosum and capsules densely expressed myelin basic protein throughout the forebrain white matter tracts (Windrem et al. 2004). Donor-derived myelin effectively ensheathed host *Shiverer* axons, as validated by both confocal imaging and by the ultrastructural observation of donor-derived compact myelin—of which native *Shiverer* oligodendrocytes are incapable. Confocal analysis also revealed nodes of Ranvier between donor-derived myelinated segments, while transcallosal conduction velocities were normalized in the OPC-transplanted mice. Using a multisite delivery approach intended to achieve broader cell dispersal in the recipient CNS, Windrem et al. (2008) then observed cell colonization throughout the entire neuraxis, with effective whole-neuraxis myelination that included the entire brain, brainstem, cerebellum, and cranial nerve roots, along with much of the spinal cord and roots. This was associated with significantly and substantially prolonged survivals in transplanted *Shiverer* mice, with frank rescue and phenotypic recovery of a large minority (Windrem et al. 2008); whereas untreated *Shiverers* invariably die by 20 weeks of age, some engrafted animals achieved normal murine life spans. These data strongly suggested the feasibility of neonatal OPC implantation as a strategy for treating the congenital disorders of myelin formation and maintenance. Later studies refined the criteria for selecting myelinogenic progenitors, by identifying the PDGFa receptor epitope CD140a as recognizing the population of potentially oligodendrocytic cells (Sim et al. 2011). OPCs selected on the basis of CD140a expression proved superior to those selected on the basis of A2B5 in their efficiency, rapidity, and fidelity of differentiation and ultimate extent of myelination; as such, they have supplanted the latter as a preferred cellular vector for therapeutic remyelination.

Challenges in the Use of Glial Progenitor Grafts for Childhood Myelin Disorders

Cell-based therapies comprise a broad platform for the potential amelioration of enzymatic and storage disorders, in that a common strategy of intracerebral delivery of OPCs may prove broadly applicable across a variety of specific enzymatic disorders. In practice though, given treatment regimens will need to be tightly calibrated to specific disease phenotypes and stages. Little data are available as to the number or proportion of wild-type cells required to achieve local correction of enzymatic activity and substrate clearance in any storage disorder, and these values will likely need to be empirically derived for different disease targets. Similarly, the efficiency and extent of myelination required to achieve significant benefit in hypomyelinating disorders remains unclear, and will depend upon disease extent and duration as much as donor cell dispersal and myelinogenic competence in the disease environment. These caveats notwithstanding, there is considerable reason for optimism that cell-based therapy of the pediatric myelin disorders—including not only the storage disorders, but also the primary dysmyelinations such as Pelizaeus–Merzbacher disease (Osorio et al. 2017; Osório and Goldman 2018) vanishing white matter disease (Bugiani et al. 2011; Dooves et al. 2019), and other hypomyelinating leukodystrophies (Helman et al. 2015; Wolf et al. 2021) may soon prove feasible (Goldman 2011). In that regard, a phase 1 trial of implanted NSCs reported the long-term safety of NSC implants into children with Pelizaeus–Merzbacher disease, with some evidence on imaging of new myelin formation at the sites of implantation (Gupta et al. 2012, 2019). One may expect that with the development of improved and scalable preparations of lineage-restricted hGPCs, that this treatment strategy may offer tangible benefit to these especially needy pediatric patient populations.

Adult Disease Targets of Glial Progenitor Cell-Based Treatment

In adults, oligodendrocytic loss is causally involved in diseases as diverse as traumatic brain

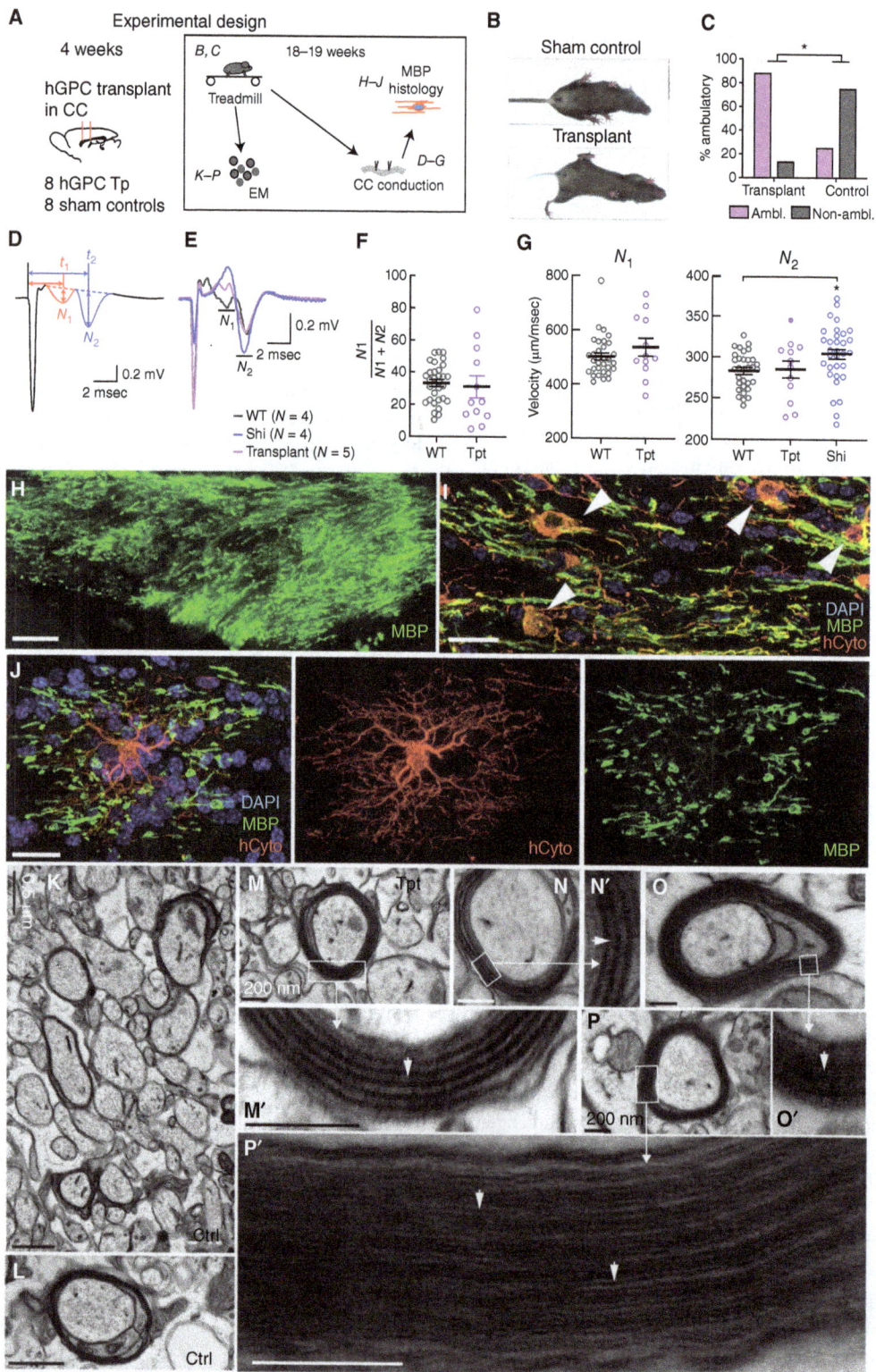

Figure 7. (*See following page for legend.*)

injury, MS and its variants, and hypertensive and diabetic white matter loss. In addition, the prominent role of oligodendrocytic loss in both vascular dementia and age-related white matter loss is becoming increasingly recognized (Kalaria 2018). All of these disorders are potential targets of glial progenitor cell replacement therapy, although the adult disease environment may limit the feasibility of this approach in ways not encountered in pediatric targets (Franklin and ffrench-Constant 2008). For instance, the chronically ischemic brain tissue of diabetic and hypertensive patients with small vessel disease may present inhospitable environments for graft acceptance, and require aggressive treatment of the underlying vascular disorders before any cell replacement strategy may be considered. Similarly, the inflammatory disease environment of patients with autoimmune demyelination presents its own challenges, which need to be overcome before cell-based remyelination can succeed. Nonetheless, current disease-modifying strategies for treating both vascular and autoimmune

diseases have advanced to the point where transplant-based remyelination of adult targets may now be feasible (Goldman et al. 2012, 2021).

Multiple Sclerosis and Autoimmune Demyelination

Most experimental models of cell-based remyelination have focused on MS, a debilitating disease characterized by both inflammatory myelinolysis and degenerative axonal loss. The attraction of MS as a therapeutic target derives from its high incidence and extraordinary prevalence, given its typical onset in youth and long disease course. In the United States alone, over 200 young adults are diagnosed weekly, with more than 300,000 affected nationally. Within 10 years, roughly one-half of MS patients develop the progressive neurodegeneration of secondary progressive MS. In the past, MS had not been an actively researched target for cell therapy, since there was limited enthusiasm for introducing new cells into an active disease environ-

---

Figure 7. Myelination by adult-transplanted human glial progenitor cells (hGPCs) restores function in *Shiverer* mice. (*A*) Mice were injected bilaterally in the corpus callosum and striatum at 4 wk of age with either cells or vehicle. When 18 wk old, the mice were examined sequentially in a series of tests spanning behavior to anatomy. The arrows indicate the sequence of tests and italics indicate the relevant figures. (*B*) Sham-injected (*top*) and hGPC-transplanted (*bottom*) *Shiverers* were videoed from below on a treadmill, using DigiGait. (*C*) Of the eight *Shiverer* controls and eight transplanted *Shiverers* tested on the treadmill, only two of eight untreated mice were able to walk on the treadmill, while seven of eight engrafted mice did so on their first try (*P* = 0.012, chi-squared), and all eight did so on a second attempt. (*D*) Schematic of the measurements taken to assess the *trans*-callosal response to electrical stimulation of sampled slices. (*E*) Representative traces of normally myelinated shiverer heterozygous mice (wild-type [WT]), sham-injected *Shiverers* (sham), and transplanted *Shiverers* (Tpt) demonstrating restoration of the fast conduction N1 component in the transplanted mice. (*F*) N1 amplitude shows no significant difference between slice preparations derived from normal WT and transplanted (Tpt) *Shiverer* brains, suggesting transplant-mediated normalization of the ratio of myelinated to unmyelinated axons in the transplanted *Shiverer* brain. Untreated *Shiverers* have effective N1 velocities of zero, and cannot be assessed as such. (*G*) The slow component, N2 conduction velocity of transplanted callosal slices was significantly more rapid than that of untreated *Shiverers*, and no different from that of WT. (*H–P′*) All images are taken from 18-wk-old *Shiverers* injected intracallosally at 4 wk with hGPCs. (*H*) Low-magnification image of the corpus callosum of one hemisphere of an adult-engrafted *Shiverer*. hN (red), (myelin basic protein [MBP]; green). (*I*) Human oligodendrocytes, coimmunostained for human cytoplasmic antigen (hCyto) and MBP; the latter is necessarily human, as *Shiverer* mice do not make MBP. (*J*) A single mature donor-derived oligodendrocyte. *Right-hand* images show respective color splits for human cytoplasm (red) and MBP (green), showing the many myelinating sheaths produced by a single hGPC-derived oligodendrocyte. (*K–L*). Electron micrographs (EMs) of sham control corpus callosum demonstrating the cytoplasmic inclusions in the myelin sheath characteristic of *Shiverers*. (*M–P′*) EMs of transplanted shiverers. The *insets* show the major dense lines (arrowheads), characteristic of compact myelin. Scale bars, 100 μm (*H*); 20 μm (*I,J*); 500 nm (*K,L*); 200 nm (*M–P*); 100 nm (*M′–P′*). (This figure is adapted from Windrem et al. 2020 and reprinted under the terms of the Creative Commons CC-BY license.)

---

ment—one essentially primed for allograft rejection. Yet the continuous improvement of approaches toward CNS immune modulation have so substantially diminished disease recurrence, as to make the use of cell replacement strategies for restoring myelin to demyelinated lesions more tenable.

In that regard both NSCs and OPCs have been assessed as potential cell therapeutics for myelin repair, in a variety of models of acquired demyelination (Pluchino et al. 2004; Franklin and Kotter 2008). As cellular vectors for remyelination, NSCs have thus far disappointed in preclinical studies, as their efficiency of oligodendrocytic differentiation and myelination appears low. In contrast, the intracerebral delivery of OPCs into demyelinated brain has been associated with substantial oligodendrocytic differentiation and myelination in a variety of models of adult demyelination (Windrem et al. 2002). Most notably (Duncan et al. 2009), fetal tissue–derived hOPCs transplanted into cuprizone-demyelinated adult mouse brain have been shown to effectively remyelinate host axons, with both neurological improvement and rescue of callosal conduction velocity (Fig. 7; Windrem et al. 2020), while PSC-derived OPCs have been shown to effectively remyelinate demyelinated tissue in rodent models of both spinal cord injury (Kawabata et al. 2016) and stroke (Martinez-Curiel et al. 2023). Thus, when provided a permissive axonal substrate, donor hOPCs serve as effective cellular vectors for remyelination.

Challenges in the Use of OPC Transplantation for Adult Disorders

Regardless of the cell-autonomous capabilities of donor OPCs, the complexity of the adult disease environment, which may include latent inflammatory activity and underlying axonal loss, vascular insufficiency, local gliosis, and chronic inflammatory cells, may all conspire to make adult targets less approachable than their pediatric counterparts (Franklin 2002). Indeed, as noted previously, in some disease settings endogenous progenitors may be intrinsically competent to remyelinate adult demyelinated brain, but may be impeded from doing so in aged an-

imals by a deficient innate immune system, which might in turn be rescued by immune reconstitution from a younger parabiotic partner (Ruckh et al. 2012). In that regard, human microglia may now be produced from PSCs as well, and can engraft host brains efficiently (Abud et al. 2017; Shibuya et al. 2022), raising the possibility of co-engrafting young human OPCs and microglia together, so as to provide a more permissive immune environment for OPC engraftment and remyelination.

Clearly then, any cell-based therapeutic strategies for adult demyelination, especially those intended to remyelinate both acute and chronic lesions of MS, will require aggressive disease modification, as well as rigorous stratification to define those patients with sufficient axonal integrity as to potentially benefit from this approach. That said, these data offer hope that OPC delivery may soon serve as a viable treatment option for the restoration of both lost myelin and compromised function to afflicted patients.

SYNOPSIS

Our understanding of the biology of oligodendrocytes and their progenitors, as well as of myelin and its disorders, has increased tremendously over the past several years, aided by new technologies in genomics and stem cell biology, transgenic and chimeric animal models of disease, and imaging and cell signaling analysis. These new insights provide real hope that triggering remyelination from pharmacologically mobilized endogenous progenitors as well as using tissue- and stem cell–derived OPCs as transplantable agents for myelin replacement are each viable strategies for myelin repair, which may each soon be poised for clinical translation. We must remain cautious though, as the associated risks of these new approaches are also only now becoming appreciated. That said, given the rapidly accelerating growth in remyelination biology over the past decade, progress over the next decade seems assured to be as scientifically exciting as it promises to be therapeutically meaningful.

Cite this article as *Cold Spring Harb Perspect Biol* doi: 10.1101/cshperspect.a041371

COMPETING INTEREST STATEMENT

R.J.M.F is an employee of Altos Labs. S.A.G. is a part-time employee and stockholder of Sana Biotechnology, from which his laboratory receives sponsored research support. S.A.G. is also a cofounder of CNS2, Inc., in which he holds equity and from which his laboratory receives research support.

ACKNOWLEDGMENTS

Work discussed from the Franklin laboratory was mainly supported by The UK Multiple Sclerosis Society, the National Multiple Sclerosis Society, and the Adelson Medical Research Foundation. Work discussed from the Goldman laboratory was supported by the National Institutes of Health (NINDS, NIMH, and NIA), as well as by the Adelson Medical Research Foundation, Lundbeck Foundation, CNS2, Inc., and Sana Biotechnology, Inc., with past relevant funding from the Novo Nordisk Foundation, National Multiple Sclerosis Society, and the New York State Stem Cell Research Program (NYSTEM). We are especially grateful to Dr. Khalil Rawji for his help in putting this work together.

REFERENCES

Abiraman K, Pol SU, O'Bara MA, Chen GD, Khaku ZM, Wang J, Thorn D, Vedia BH, Ekwegbalu EC, Li JX, et al. 2015. Anti-muscarinic adjunct therapy accelerates functional human oligodendrocyte repair. *J Neurosci* 35: 3676–3688. doi:10.1523/JNEUROSCI.3510-14.2015

Abud EM, Ramirez RN, Martinez ES, Healy LM, Nguyen CHH, Newman SA, Yeromin AV, Scarfone VM, Marsh SE, Fimbres C, et al. 2017. iPSC-derived human microglia-like cells to study neurological diseases. *Neuron* 94: 278–293.e9. doi:10.1016/j.neuron.2017.03.042

Albert M, Antel J, Brück W, Stadelmann C. 2007. Extensive cortical remyelination in patients with chronic multiple sclerosis. *Brain Pathol* 17: 129–138. doi:10.1111/j.1750-3639.2006.00043.x

Angelis FD, Bernardo A, Magnaghi V, Minghetti L, Tata AM. 2012. Muscarinic receptor subtypes as potential targets to modulate oligodendrocyte progenitor survival, proliferation, and differentiation. *Dev Neurobiol* 72: 713–728. doi:10.1002/dneu.20976

Anokye-Danso F, Trivedi CM, Juhr D, Gupta M, Cui Z, Tian Y, Zhang Y, Yang W, Gruber PJ, Epstein JA, et al. 2011. Highly efficient miRNA-mediated reprogramming of mouse and human somatic cells to pluripotency. *Cell Stem Cell* 8: 376–388. doi:10.1016/j.stem.2011.03.001

Armstrong RC, Le TQ, Frost EE, Borke RC, Vana AC. 2002. Absence of fibroblast growth factor 2 promotes oligodendroglial repopulation of demyelinated white matter. *J Neurosci* 22: 8574–8585. doi:10.1523/JNEUROSCI.22-19-08574.2002

Armstrong RC, Le TQ, Flint NC, Vana AC, Zhou YX. 2006. Endogenous cell repair of chronic demyelination. *J Neuropathol Exp Neurol* 65: 245–256. doi:10.1097/01.jnen.0000205142.08716.7e

Arnett HA, Fancy SPJ, Alberta JA, Zhao C, Plant SR, Kaing S, Raine CS, Rowitch DH, Franklin RJM, Stiles CD. 2004. bHLH transcription factor Olig1 is required to repair demyelinated lesions in the CNS. *Science* 306: 2111–2115. doi:10.1126/science.1103709

Back SA, Rivkees SA. 2004. Emerging concepts in periventricular white matter injury. *Semin Perinatol* 28: 405–414. doi:10.1053/j.semperi.2004.10.010

Back SA, Tuohy TMF, Chen H, Wallingford N, Craig A, Struve J, Luo NL, Banine F, Liu Y, Chang A, et al. 2005. Hyaluronan accumulates in demyelinated lesions and inhibits oligodendrocyte progenitor maturation. *Nat Med* 11: 966–972. doi:10.1038/nm1279

Bacmeister CM, Barr HJ, McClain CR, Thornton MA, Nettles D, Welle CG, Hughes EG. 2020. Motor learning promotes remyelination via new and surviving oligodendrocytes. *Nat Neurosci* 23: 819–831. doi:10.1038/s41593-020-0637-3

Behrangi N, Heinig L, Frintrop L, Santrau E, Kurth J, Krause B, Atanasova D, Clarner T, Fragoulis A, Joksch M, et al. 2022. Siponimod ameliorates metabolic oligodendrocyte injury via the sphingosine-1 phosphate receptor 5. *Proc Natl Acad Sci* 119: e2204509119. doi:10.1073/pnas.2204509119

Blakemore WF. 1975. Remyelination by Schwann cells of axons demyelinated by intraspinal injection of 6-aminonicotinamide in the rat. *J Neurocytol* 4: 745–757. doi:10.1007/BF01181634

Bodini B, Veronese M, García-Lorenzo D, Battaglini M, Poirion E, Chardain A, Freeman L, Louapre C, Tchikviladze M, Papeix C, et al. 2016. Dynamic imaging of individual remyelination profiles in multiple sclerosis. *Ann Neurol* 79: 726–738. doi:10.1002/ana.24620

Bodini B, Tonietto M, Airas L, Stankoff B. 2021. Positron emission tomography in multiple sclerosis—straight to the target. *Nat Rev Neurol* 17: 663–675. doi:10.1038/s41582-021-00537-1

Bosch-Queralt M, Cantuti-Castelvetri L, Damkou A, Schiferer M, Schlepckow K, Alexopoulos I, Lütjohann D, Klose C, Vaculčiaková L, Masuda T, et al. 2021. Diet-dependent regulation of TGFβ impairs reparative innate immune responses after demyelination. *Nat Metab* 3: 211–227. doi:10.1038/s42255-021-00341-7

Brenner M, Johnson AB, Boespflug-Tanguy O, Rodriguez D, Goldman JE, Messing A. 2001. Mutations in GFAP, encoding glial fibrillary acidic protein, are associated with Alexander disease. *Nat Genet* 27: 117–120. doi:10.1038/83679

Brinkmann BG, Agarwal A, Sereda MW, Garratt AN, Müller T, Wende H, Stassart RM, Nawaz S, Humml C, Velanac V, et al. 2008. Neuregulin-1/ErbB signaling serves distinct functions in myelination of the peripheral and central nervous system. *Neuron* 59: 581–595. doi:10.1016/j.neuron.2008.06.028

Brown JWL, Cunniffe NG, Prados F, Kanber B, Jones JL, Needham E, Georgieva Z, Rog D, Pearson OR, Overell J, et al. 2021. Safety and efficacy of bexarotene in patients with relapsing-remitting multiple sclerosis (CCMR one): a randomised, double-blind, placebo-controlled, parallel-group, phase 2a study. *Lancet Neurol* 20: 709–720. doi:10.1016/S1474-4422(21)00179-4

Brüstle O, Jones KN, Learish RD, Karram K, Choudhary K, Wiestler OD, Duncan ID, McKay RDG. 1999. Embryonic stem cell-derived glial precursors: a source of myelinating transplants. *Science* 285: 754–756. doi:10.1126/science.285.5428.754

Bugiani M, Boor I, van Kollenburg B, Postma N, Polder E, van Berkel C, van Kesteren RE, Windrem MS, Hol EM, Scheper GC, et al. 2011. Defective glial maturation in vanishing white matter disease. *J Neuropathol Exp Neurol* 70: 69–82. doi:10.1097/NEN.0b013e318203ae74

Cadavid D, Balcer L, Galetta S, Aktas O, Ziemssen T, Vanopdenbosch L, Frederiksen J, Skeen M, Jaffe GJ, Butzkueven H, et al. 2017. Safety and efficacy of opicinumab in acute optic neuritis (RENEW): a randomised, placebo-controlled, phase 2 trial. *Lancet Neurol* 16: 189–199. doi:10.1016/S1474-4422(16)30377-5

Cadavid D, Mellion M, Hupperts R, Edwards KR, Calabresi PA, Drulović J, Giovannoni G, Hartung HP, Arnold DL, Fisher E, et al. 2019. Safety and efficacy of opicinumab in patients with relapsing multiple sclerosis (SYNERGY): a randomised, placebo-controlled, phase 2 trial. *Lancet Neurol* 18: 845–856. doi:10.1016/S1474-4422(19)30137-1

Caillava C, Vandenbosch R, Jablonska B, Deboux C, Spigoni G, Gallo V, Malgrange B, Evercooren ABV. 2011. Cdk2 loss accelerates precursor differentiation and remyelination in the adult central nervous system. *J Cell Biol* 193: 397–407. doi:10.1083/jcb.201004146

Cantoni C, Bollman B, Licastro D, Xie M, Mikesell R, Schmidt R, Yuede CM, Galimberti D, Olivecrona G, Klein RS, et al. 2015. TREM2 regulates microglial cell activation in response to demyelination in vivo. *Acta Neuropathol* 129: 429–447. doi:10.1007/s00401-015-1388-1

Cantuti-Castelvetri L, Fitzner D, Bosch-Queralt M, Weil MT, Su M, Sen P, Ruhwedel T, Mitkovski M, Trendelenburg G, Lütjohann D, et al. 2018. Defective cholesterol clearance limits remyelination in the aged central nervous system. *Science* 359: 684–688. doi:10.1126/science.aan4183

Carotenuto A, Giordano B, Dervenoulas G, Wilson H, Veronese M, Chappell Z, Polychronis S, Pagano G, Mackewn J, Turkheimer FE, et al. 2020. [^{18}F]Florbetapir PET/MR imaging to assess demyelination in multiple sclerosis. *Eur J Nucl Med Mol Imaging* 47: 366–378. doi:10.1007/s00259-019-04533-y

Chang A, Nishiyama A, Peterson J, Prineas J, Trapp BD. 2000. NG2-Positive oligodendrocyte progenitor cells in adult human brain and multiple sclerosis lesions. *J Neurosci* 20: 6404–6412. doi:10.1523/JNEUROSCI.20-17-06404.2000

Chang A, Tourtellotte WW, Rudick R, Trapp BD. 2002. Premyelinating oligodendrocytes in chronic lesions of multiple sclerosis. *N Engl J Med* 346: 165–173. doi:10.1056/NEJMoa010994

Chari DM, Blakemore WF. 2002. Efficient recolonisation of progenitor-depleted areas of the CNS by adult oligodendrocyte progenitor cells. *Glia* 37: 307–313. doi:10.1002/glia.10038

Charles P, Hernandez MP, Stankoff B, Aigrot MS, Colin C, Rougon G, Zalc B, Lubetzki C. 2000. Negative regulation of central nervous system myelination by polysialylated-neural cell adhesion molecule. *Proc Natl Acad Sci* 97: 7585–7590. doi:10.1073/pnas.100076197

Charles P, Reynolds R, Seilhean D, Rougon G, Aigrot MS, Niezgoda A, Zalc B, Lubetzki C. 2002. Re-expression of PSA-NCAM by demyelinated axons: an inhibitor of remyelination in multiple sclerosis? *Brain* 125: 1972–1979. doi:10.1093/brain/awf216

Chen JT, Collins DL, Atkins HL, Freedman MS, Arnold DL, Group the CMS. 2008. Magnetization transfer ratio evolution with demyelination and remyelination in multiple sclerosis lesions. *Ann Neurol* 63: 254–262. doi:10.1002/ana.21302

Chen Y, Zhen W, Guo T, Zhao Y, Liu A, Rubio JP, Krull D, Richardson JC, Lu H, Wang R. 2017. Histamine receptor 3 negatively regulates oligodendrocyte differentiation and remyelination. *PLoS ONE* 12: e0189380. doi:10.1371/journal.pone.0189380

Chen CZ, Neumann B, Förster S, Franklin RJM. 2021. Schwann cell remyelination of the central nervous system: why does it happen and what are the benefits? *Open Biol* 11: 200352.

Cignarella F, Filipello F, Bollman B, Cantoni C, Locca A, Mikesell R, Manis M, Ibrahim A, Deng L, Benitez BA, et al. 2020. TREM2 activation on microglia promotes myelin debris clearance and remyelination in a model of multiple sclerosis. *Acta Neuropathol* 140: 513–534. doi:10.1007/s00401-020-02193-z

Crawford AH, Stockley JH, Tripathi RB, Richardson WD, Franklin RJM. 2014. Oligodendrocyte progenitors: adult stem cells of the central nervous system? *Exp Neurol* 260: 50–55. doi:10.1016/j.expneurol.2014.04.027

Crawford AH, Tripathi RB, Foerster S, McKenzie I, Kougioumtzidou E, Grist M, Richardson WD, Franklin RJM. 2016. Pre-existing mature oligodendrocytes do not contribute to remyelination following toxin-induced spinal cord demyelination. *Am J Pathol* 186: 511–516. doi:10.1016/j.ajpath.2015.11.005

Crockett DP, Burshteyn M, Garcia C, Muggironi M, Casaccia-Bonnefil P. 2005. Number of oligodendrocyte progenitors recruited to the lesioned spinal cord is modulated by the levels of the cell cycle regulatory protein p27Kip-1. *Glia* 49: 301–308. doi:10.1002/glia.20111

Cunha MI, Su M, Cantuti-Castelvetri L, Müller SA, Schifferer M, Djannatian M, Alexopoulos I, van der Meer F, Winkler A, van Ham TJ, et al. 2020. Pro-inflammatory activation following demyelination is required for myelin clearance and oligodendrogenesis. *J Exp Med* 217: e20191390. doi:10.1084/jem.20191390

Czepiel M, Balasubramaniyan V, Schaafsma W, Stancic M, Mikkers H, Huisman C, Boddeke E, Copray S. 2011. Differentiation of induced pluripotent stem cells into functional oligodendrocytes. *Glia* 59: 882–892. doi:10.1002/glia.21159

Dawson MRL, Polito A, Levine JM, Reynolds R. 2003. NG2-expressing glial progenitor cells: an abundant and widespread population of cycling cells in the adult rat CNS. *Mol Cell Neurosci* **24:** 476–488. doi:10.1016/S1044-7431(03)00210-0

de la Fuente AG, Errea O, van Wijngaarden P, Gonzalez GA, Kerninon C, Jarjour AA, Lewis HJ, Jones CA, Nait-Oumesmar B, Zhao C, et al. 2015. Vitamin D receptor–retinoid X receptor heterodimer signaling regulates oligodendrocyte progenitor cell differentiation. *J Cell Biol* **211:** 975–985. doi:10.1083/jcb.201505119

Demerens C, Stankoff B, Logak M, Anglade P, Allinquant B, Couraud F, Zalc B, Lubetzki C. 1996. Induction of myelination in the central nervous system by electrical activity. *Proc Natl Acad Sci* **93:** 9887–9892. doi:10.1073/pnas.93.18.9887

Derakhshan M, Caramanos Z, Narayanan S, Arnold DL, Collins DL. 2014. Surface-based analysis reveals regions of reduced cortical magnetization transfer ratio in patients with multiple sclerosis: a proposed method for imaging subpial demyelination. *Hum Brain Mapp* **35:** 3402–3413. doi:10.1002/hbm.22410

Deuse T, Hu X, Gravina A, Wang D, Tediashvili G, De C, Thayer WO, Wahl A, Garcia JV, Reichenspurner H, et al. 2019. Hypoimmunogenic derivatives of induced pluripotent stem cells evade immune rejection in fully immunocompetent allogeneic recipients. *Nat Biotechnol* **37:** 252–258. doi:10.1038/s41587-019-0016-3

Dietrich J, Noble M, Mayer-Proschel M. 2002. Characterization of A2B5+ glial precursor cells from cryopreserved human fetal brain progenitor cells. *Glia* **40:** 65–77. doi:10.1002/glia.10116

Dietrich J, Lacagnina M, Gass D, Richfield E, Mayer-Pröschel M, Noble M, Torres C, Pröschel C. 2005. EIF2B5 mutations compromise GFAP+ astrocyte generation in vanishing white matter leukodystrophy. *Nat Med* **11:** 277–283. doi:10.1038/nm1195

Dombrowski Y, O'Hagan T, Dittmer M, Penalva R, Mayoral SR, Bankhead P, Fleville S, Eleftheriadis G, Zhao C, Naughton M, et al. 2017. Regulatory T cells promote myelin regeneration in the central nervous system. *Nat Neurosci* **20:** 674–680. doi:10.1038/nn.4528

Dooves S, Leferink PS, Krabbenborg S, Breeuwsma N, Bots S, Hillen AEJ, Jacobs G, van der Knaap MS, Heine VM. 2019. Cell replacement therapy improves pathological hallmarks in a mouse model of leukodystrophy vanishing white matter. *Stem Cell Rep* **12:** 441–450.

Douvaras P, Wang J, Zimmer M, Hanchuk S, O'Bara MA, Sadiq S, Sim FJ, Goldman J, Fossati V. 2014. Efficient generation of myelinating oligodendrocytes from primary progressive multiple sclerosis patients by induced pluripotent stem cells. *Stem Cell Rep* **3:** 250–259.

Duhamel G, Prevost VH, Cayre M, Hertanu A, Mchinda S, Carvalho VN, Varma G, Durbec P, Alsop DC, Girard OM. 2019. Validating the sensitivity of inhomogeneous magnetization transfer (ihMT) MRI to myelin with fluorescence microscopy. *Neuroimage* **199:** 289–303. doi:10.1016/j.neuroimage.2019.05.061

Duncan ID, Hoffman RL. 1997. Schwann cell invasion of the central nervous system of the myelin mutants. *J Anat* **190:** 35–49. doi:10.1046/j.1469-7580.1997.19010035.x

Duncan ID, Brower A, Kondo Y, Curlee JF Jr, Schultz RD. 2009. Extensive remyelination of the CNS leads to functional recovery. *Proc Natl Acad Sci* **106:** 6832–6836. doi:10.1073/pnas.0812500106

Duncan ID, Radcliff AB, Heidari M, Kidd G, August BK, Wierenga LA. 2018. The adult oligodendrocyte can participate in remyelination. *Proc Natl Acad Sci* **115:** E11807–E11816.

Dusart I, Marty S, Peschanski M. 1992. Demyelination, and remyelination by Schwann cells and oligodendrocytes after kainate-induced neuronal depletion in the central nervous system. *Neuroscience* **51:** 137–148. doi:10.1016/0306-4522(92)90478-K

Edgar JM, McLaughlin M, Yool D, Zhang SC, Fowler JH, Montague P, Barrie JA, McCulloch MC, Duncan ID, Garbern J, et al. 2004. Oligodendroglial modulation of fast axonal transport in a mouse model of hereditary spastic paraplegia. *J Cell Biol* **166:** 121–131. doi:10.1083/jcb.200312012

Ehrlich M, Mozafari S, Glatza M, Starost L, Velychko S, Hallmann A-L, Cui Q-L, Schambach A, Kim K-P, Bachelin C, et al. 2017. Rapid and efficient generation of oligodendrocytes from human induced pluripotent stem cells using transcription factors. *Proc Natl Acad Sci* **114:** E2243–E2252.

Escolar ML, Poe MD, Provenzale JM, Richards KC, Allison J, Wood S, Wenger DA, Pietryga D, Wall D, Champagne M, et al. 2005. Transplantation of umbilical-cord blood in babies with infantile Krabbe's disease. *N Engl J Med* **352:** 2069–2081. doi:10.1056/NEJMoa042604

Etxeberria A, Mangin JM, Aguirre A, Gallo V. 2010. Adult-born SVZ progenitors receive transient synapses during remyelination in corpus callosum. *Nat Neurosci* **13:** 287–289. doi:10.1038/nn.2500

Fancy SPJ, Zhao C, Franklin RJM. 2004. Increased expression of Nkx2.2 and Olig2 identifies reactive oligodendrocyte progenitor cells responding to demyelination in the adult CNS. *Mol Cell Neurosci* **27:** 247–254. doi:10.1016/j.mcn.2004.06.015

Fancy SPJ, Baranzini SE, Zhao C, Yuk DI, Irvine KA, Kaing S, Sanai N, Franklin RJM, Rowitch DH. 2009. Dysregulation of the Wnt pathway inhibits timely myelination and remyelination in the mammalian CNS. *Genes Dev* **23:** 1571–1585. doi:10.1101/gad.1806309

Fancy SPJ, Chan JR, Baranzini SE, Franklin RJM, Rowitch DH. 2011. Myelin regeneration: a recapitulation of development? *Annu Rev Neurosci* **34:** 21–43. doi:10.1146/annurev-neuro-061010-113629

Felts PA, Woolston AM, Fernando HB, Asquith S, Gregson NA, Mizzi OJ, Smith KJ. 2005. Inflammation and primary demyelination induced by the intraspinal injection of lipopolysaccharide. *Brain* **128:** 1649–1666. doi:10.1093/brain/awh516

Flores AI, Narayanan SP, Morse EN, Shick HE, Yin X, Kidd G, Avila RL, Kirschner DA, Macklin WB. 2008. Constitutively active akt induces enhanced myelination in the CNS. *J Neurosci* **28:** 7174–7183. doi:10.1523/JNEUROSCI.0150-08.2008

Follett PL, Deng W, Dai W, Talos DM, Massillon LJ, Rosenberg PA, Volpe JJ, Jensen FE. 2004. Glutamate receptor-mediated oligodendrocyte toxicity in periventricular leukomalacia: a protective role for topiramate. *J Neurosci* **24**: 4412–4420. doi:10.1523/JNEUROSCI.0477-04.2004

Franklin RJM. 2002. Why does remyelination fail in multiple sclerosis? *Nat Rev Neurosci* **3**: 705–714. doi:10.1038/nrn917

Franklin RJM, Blakemore WF. 1993. Requirements for schwann cell migration within cns environments: a viewpoint. *Int J Dev Neurosci* **11**: 641–649. doi:10.1016/0736-5748(93)90052-F

Franklin RJM, ffrench-Constant C. 2008. Remyelination in the CNS: from biology to therapy. *Nat Rev Neurosci* **9**: 839–855. doi:10.1038/nrn2480

Franklin RJM, ffrench-Constant C. 2017. Regenerating CNS myelin—from mechanisms to experimental medicines. *Nat Rev Neurosci* **18**: 753–769. doi:10.1038/nrn.2017.136

Franklin RJM, Gallo V. 2014. The translational biology of remyelination: past, present, and future. *Glia* **62**: 1905–1915. doi:10.1002/glia.22622

Franklin RJM, Hinks GL. 1999. Understanding CNS remyelination: clues from developmental and regeneration biology. *J Neurosci Res* **58**: 207–213. doi:10.1002/(SICI)1097-4547(19991015)58:2<207::AID-JNR1>3.0.CO;2-1

Franklin RJM, Kotter MR. 2008. The biology of CNS remyelination: the key to therapeutic advances. *J Neurol* **255**: 19–25. doi:10.1007/s00415-008-1004-6

Franklin RJM, Simons M. 2022. CNS remyelination and inflammation: from basic mechanisms to therapeutic opportunities. *Neuron* **110**: 3549–3565. doi:10.1016/j.neuron.2022.09.023

Franklin RJM, Zhao C, Lubetzki C, ffrench-Constant C. 2012. Endogenous remyelination in the CNS. In *Myelin repair and neuroprotection in multiple sclerosis* (ed. Duncan ID, Franklin RJM), pp. 71–92. Springer, New York.

Franklin RJM, Frisén J, Lyons DA. 2021. Revisiting remyelination: towards a consensus on the regeneration of CNS myelin. *Semin Cell Dev Biol* **116**: 3–9. doi:10.1016/j.semcdb.2020.09.009

Freund P, Wheeler-Kingshott C, Jackson J, Miller D, Thompson A, Ciccarelli O. 2010. Recovery after spinal cord relapse in multiple sclerosis is predicted by radial diffusivity. *Mult Scler J* **16**: 1193–1202. doi:10.1177/1352458510376180

Gadea A, Aguirre A, Haydar TF, Gallo V. 2009. Endothelin-1 regulates oligodendrocyte development. *J Neurosci* **29**: 10047–10062. doi:10.1523/JNEUROSCI.0822-09.2009

Gaj T, Gersbach CA, Barbas CF III. 2013. ZFN, TALEN, and CRISPR/Cas-based methods for genome engineering. *Trends Biotechnol* **31**: 397–405. doi:10.1016/j.tibtech.2013.04.004

Garbern JY, Yool DA, Moore GJ, Wilds IB, Faulk MW, Klugmann M, Nave K, Sistermans EA, van der Knaap MS, Bird TD, et al. 2002. Patients lacking the major CNS myelin protein, proteolipid protein 1, develop length-dependent axonal degeneration in the absence of demyelination and inflammation. *Brain* **125**: 551–561. doi:10.1093/brain/awf043

Gautier HOB, Evans KA, Volbracht K, James R, Sitnikov S, Lundgaard I, James F, Lao-Peregrin C, Reynolds R, Franklin RJM, et al. 2015. Neuronal activity regulates remyelination via glutamate signalling to oligodendro-cyte progenitors. *Nat Commun* **6**: 8518. doi:10.1038/ncomms9518

Gibson EM, Purger D, Mount CW, Goldstein AK, Lin GL, Wood LS, Inema I, Miller SE, Bieri G, Zuchero JB, et al. 2014. Neuronal activity promotes oligodendrogenesis and adaptive myelination in the mammalian brain. *Science* **344**: 1252304. doi:10.1126/science.1252304

Givogri MI, Galbiati F, Fasano S, Amadio S, Perani L, Superchi D, Morana P, Carro UD, Marchesini S, Brambilla R, et al. 2006. Oligodendroglial progenitor cell therapy limits central neurological deficits in mice with metachromatic leukodystrophy. *J Neurosci* **26**: 3109–3119. doi:10.1523/JNEUROSCI.4366-05.2006

Glezer I, Lapointe A, Rivest S. 2006. Innate immunity triggers oligodendrocyte progenitor reactivity and confines damages to brain injuries. *FASEB J* **20**: 750–752. doi:10.1096/fj.05-5234fje

Goldman SA. 2011. Progenitor cell-based treatment of the pediatric myelin disorders. *Arch Neurol* **68**: 848–856. doi:10.1001/archneurol.2011.46

Goldman SA. 2013. White matter from fibroblasts. *Nat Biotechnol* **31**: 412–413. doi:10.1038/nbt.2570

Goldman SA, Nedergaard M, Windrem MS. 2012. Glial progenitor cell-based treatment and modeling of neurological disease. *Science* **338**: 491–495. doi:10.1126/science.1218071

Goldman SA, Mariani JN, Madsen PM. 2021. Glial progenitor cell-based repair of the dysmyelinated brain: progression to the clinic. *Semin Cell Dev Biol* **116**: 62–70.

Goldschmidt T, Antel J, König FB, Brück W, Kuhlmann T. 2009. Remyelination capacity of the MS brain decreases with disease chronicity. *Neurology* **72**: 1914–1921. doi:10.1212/WNL.0b013e3181a8260a

Gornalusse GG, Hirata RK, Funk SE, Riolobos L, Lopes VS, Manske G, Prunkard D, Colunga AG, Hanafi LA, Clegg DO, et al. 2017. HLA-E-expressing stem cells escape allogeneic responses and lysis by NK cells. *Nat Biotechnol* **35**: 765–772. doi:10.1038/nbt.3860

Gouna G, Klose C, Bosch-Queralt M, Liu L, Gokce O, Schiferer M, Cantuti-Castelvetri L, Simons M. 2021. TREM2-dependent lipid droplet biogenesis in phagocytes is required for remyelination. *J Exp Med* **218**: e20210227. doi:10.1084/jem.20210227

Green AJ, Gelfand JM, Cree BA, Bevan C, Boscardin WJ, Mei F, Inman J, Arnow S, Devereux M, Abounasr A, et al. 2017. Clemastine fumarate as a remyelinating therapy for multiple sclerosis (ReBUILD): a randomised, controlled, double-blind, crossover trial. *Lancet* **390**: 2481–2489. doi:10.1016/S0140-6736(17)32346-2

Griffiths I, Klugmann M, Anderson T, Yool D, Thomson C, Schwab MH, Schneider A, Zimmermann F, McCulloch M, Nadon N, et al. 1998. Axonal swellings and degeneration in mice lacking the major proteolipid of myelin. *Science* **280**: 1610–1613. doi:10.1126/science.280.5369.1610

Gupta N, Henry RG, Strober J, Kang SM, Lim DA, Bucci M, Caverzasi E, Gaetano L, Mandelli ML, Ryan T, et al. 2012. Neural stem cell engraftment and myelination in the human brain. *Sci Transl Med* **4**: 155ra137. doi:10.1126/scitranslmed.3004373

Gupta N, Henry RG, Kang SM, Strober J, Lim DA, Ryan T, Perry R, Farrell J, Ulman M, Rajalingam R, et al. 2019.

Cite this article as *Cold Spring Harb Perspect Biol* doi: 10.1101/cshperspect.a041371

Long-term safety, immunologic response, and imaging outcomes following neural stem cell transplantation for Pelizaeus–Merzbacher disease. *Stem Cell Rep* **13:** 254–261. doi:10.1016/j.stemcr.2019.07.002

Hammond TR, Gadea A, Dupree J, Kerninon C, Nait-Oumesmar B, Aguirre A, Gallo V. 2014. Astrocyte-derived endothelin-1 inhibits remyelination through notch activation. *Neuron* **81:** 1442. doi:10.1016/j.neuron.2014.03.007

Hampton DW, Anderson J, Pryce G, Irvine KA, Giovannoni G, Fawcett JW, Compston A, Franklin RJM, Baker D, Chandran S. 2008. An experimental model of secondary progressive multiple sclerosis that shows regional variation in gliosis, remyelination, axonal and neuronal loss. *J Neuroimmunol* **201–202:** 200–211. doi:10.1016/j.jneuroim.2008.05.034

Harrington EP, Zhao C, Fancy SPJ, Kaing S, Franklin RJM, Rowitch DH. 2010. Oligodendrocyte PTEN is required for myelin and axonal integrity, not remyelination. *Ann Neurol* **68:** 703–716. doi:10.1002/ana.22090

Helman G, et al. 2015. Disease specific therapies in leukodystrophies and leukoencephalopathies. *Mol Genet Metab* **114:** 527–536.

Horner PJ, Power AE, Kempermann G, Kuhn HG, Palmer TD, Winkler J, Thal LJ, Gage FH. 2000. Proliferation and differentiation of progenitor cells throughout the intact adult rat spinal cord. *J Neurosci* **20:** 2218–2228. doi:10.1523/JNEUROSCI.20-06-02218.2000

Hu QD, Ang BT, Karsak M, Hu WP, Cui XY, Duka T, Takeda Y, Chia W, Sankar N, Ng YK, et al. 2003. F3/Contactin acts as a functional ligand for notch during oligodendrocyte maturation. *Cell* **115:** 163–175. doi:10.1016/s0092-8674(03)00810-9

Hu BY, Du ZW, Zhang SC. 2009. Differentiation of human oligodendrocytes from pluripotent stem cells. *Nat Protoc* **4:** 1614–1622. doi:10.1038/nprot.2009.186

Hu X, Manner K, DeJesus R, White K, Gattis C, Ngo P, Bandoro C, Tham E, Chu EY, Young C, et al. 2023a. Hypoimmune anti-CD19 chimeric antigen receptor T cells provide lasting tumor control in fully immunocompetent allogeneic humanized mice. *Nat Commun* **14:** 2020. doi:10.1038/s41467-023-37785-2

Hu X, White K, Olroyd AG, DeJesus R, Dominguez AA, Dowdle WE, Friera AM, Young C, Wells F, Chu EY, et al. 2023b. Hypoimmune induced pluripotent stem cells survive long term in fully immunocompetent, allogeneic rhesus macaques. *Nat Biotechnol* 1–11. doi:10.1038/s41587-023-01784-x

Hu X, Gattis C, Olroyd AG, Friera AM, White K, Young C, Basco R, Lamba M, Wells F, Ankala R, et al. 2023c. Human hypoimmune primary pancreatic islets avoid rejection and autoimmunity and alleviate diabetes in allogeneic humanized mice. *Sci Transl Med* **15:** eadg5794. doi:10.1126/scitranslmed.adg5794

Huang JK, Jarjour AA, Oumesmar BN, Kerninon C, Williams A, Krezel W, Kagechika H, Bauer J, Zhao C, Evercooren ABV, et al. 2011. Retinoid X receptor γ signaling accelerates CNS remyelination. *Nat Neurosci* **14:** 45–53. doi:10.1038/nn.2702

Huang W, Bhaduri A, Velmeshev D, Wang S, Wang L, Rottkamp CA, Alvarez-Buylla A, Rowitch DH, Kriegstein AR. 2020. Origins and proliferative states of human oligoden-drocyte precursor cells. *Cell* **182:** 594–608.e11. doi:10.1016/j.cell.2020.06.027

Hubler Z, Allimuthu D, Bederman I, Elitt MS, Madhavan M, Allan KC, Shick HE, Garrison E, Karl MT, Factor DC, et al. 2018. Accumulation of 8,9-unsaturated sterols drives oligodendrocyte formation and remyelination. *Nature* **560:** 372–376. doi:10.1038/s41586-018-0360-3

Irvine KA, Blakemore WF. 2008. Remyelination protects axons from demyelination-associated axon degeneration. *Brain* **131:** 1464–1477. doi:10.1093/brain/awn080

Itakura G, Kawabata S, Ando M, Nishiyama Y, Sugai K, Ozaki M, Iida T, Ookubo T, Kojima K, Kashiwagi R, et al. 2017. Fail-safe system against potential tumorigenicity after transplantation of iPSC derivatives. *Stem Cell Rep* **8:** 673–684. doi:10.1016/j.stemcr.2017.02.003

Itoyama Y, Webster HDF, Richardson EP, Trapp BD. 1983. Schwann cell remyelination of demyelinated axons in spinal cord multiple sclerosis lesions. *Ann Neurol* **14:** 339–346. doi:10.1002/ana.410140313

Itoyama Y, Ohnishi A, Tateishi J, Kuroiwa Y, de Webster HF. 1985. Spinal cord multiple sclerosis lesions in Japanese patients: Schwann cell remyelination occurs in areas that lack glial fibrillary acidic protein (GFAP). *Acta Neuropathol* **65:** 217–223. doi:10.1007/BF00687001

Izrael M, Zhang P, Kaufman R, Shinder V, Ella R, Amit M, Itskovitz-Eldor J, Chebath J, Revel M. 2007. Human oligodendrocytes derived from embryonic stem cells: effect of noggin on phenotypic differentiation in vitro and on myelination in vivo. *Mol Cell Neurosci* **34:** 310–323. doi:10.1016/j.mcn.2006.11.008

Jeffery ND, Crang AJ, O'Leary MT, Hodge SJ, Blakemore WF. 1999. Behavioural consequences of oligodendrocyte progenitor cell transplantation into experimental demyelinating lesions in the rat spinal cord. *Eur J Neurosci* **11:** 1508–1514. doi:10.1046/j.1460-9568.1999.00564.x

Jeyakumar M, Dwek RA, Butters TD, Platt FM. 2005. Storage solutions: treating lysosomal disorders of the brain. *Nat Rev Neurosci* **6:** 713–725. doi:10.1038/nrn1725

John GR, Shankar SL, Shafit-Zagardo B, Massimi A, Lee SC, Raine CS, Brosnan CF. 2002. Multiple sclerosis: re-expression of a developmental pathway that restricts oligoden-drocyte maturation. *Nat Med* **8:** 1115–1121. doi:10.1038/nm781

Jones BS, Lamb LS, Goldman F, Stasi AD. 2014. Improving the safety of cell therapy products by suicide gene transfer. *Front Pharmacol* **5:** 254. doi:10.3389/fphar.2014.00254

Kalaria RN. 2018. The pathology and pathophysiology of vascular dementia. *Neuropharmacology* **134:** 226–239. doi:10.1016/j.neuropharm.2017.12.030

Kawabata S, Takano M, Numasawa-Kuroiwa Y, Itakura G, Kobayashi Y, Nishiyama Y, Sugai K, Nishimura S, Iwai H, Isoda M, et al. 2016. Grafted human iPS cell-derived oligodendrocyte precursor cells contribute to robust remyelination of demyelinated axons after spinal cord injury. *Stem Cell Rep* **6:** 1–8.

Kazanis I, Evans KA, Andreopoulou E, Dimitriou C, Koutsakis C, Karadottir RT, Franklin RJM. 2017. Subependymal zone-derived oligodendroblasts respond to focal demyelination but fail to generate myelin in young and aged mice. *Stem Cell Rep* **8:** 685–700. doi:10.1016/j.stemcr.2017.01.007

Kim K, Doi A, Wen B, Ng K, Zhao R, Cahan P, Kim J, Aryee MJ, Ji H, Ehrlich LI, et al. 2010. Epigenetic memory in induced pluripotent stem cells. *Nature* **467:** 285–290. doi:10.1038/nature09342

Kim KP, Li C, Bunina D, Jeong HW, Ghelman J, Yoon J, Shin B, Park H, Han DW, Zaugg JB, et al. 2021. Donor cell memory confers a metastable state of directly converted cells. *Cell Stem Cell* **28:** 1291–1306.e10. doi:10.1016/j.stem .2021.02.023

Koç ON, Day J, Nieder M, Gerson SL, Lazarus HM, Krivit W. 2002. Allogeneic mesenchymal stem cell infusion for treatment of metachromatic leukodystrophy (MLD) and Hurler syndrome (MPS-IH). *Bone Marrow Transplant* **30:** 215–222. doi:10.1038/sj.bmt.1703650

Kornek B, Storch MK, Weissert R, Wallstroem E, Stefferl A, Olsson T, Linington C, Schmidbauer M, Lassmann H. 2000. Multiple sclerosis and chronic autoimmune encephalomyelitis: a comparative quantitative study of axonal injury in active, inactive, and remyelinated lesions. *Am J Pathol* **157:** 267–276. doi:10.1016/S0002-9440(10)64537-3

Kotter MR, Setzu A, Sim FJ, Rooijen NV, Franklin RJM. 2001. Macrophage depletion impairs oligodendrocyte remyelination following lysolecithin-induced demyelination. *Glia* **35:** 204–212. doi:10.1002/glia.1085

Kotter MR, Li WW, Zhao C, Franklin RJM. 2006. Myelin impairs CNS remyelination by inhibiting oligodendrocyte precursor cell differentiation. *J Neurosci* **26:** 328–332. doi:10.1523/JNEUROSCI.2615-05.2006

Kuhlmann T, Miron V, Cui Q, Cuo Q, Wegner C, Antel J, Brück W. 2008. Differentiation block of oligodendroglial progenitor cells as a cause for remyelination failure in chronic multiple sclerosis. *Brain* **131:** 1749–1758. doi:10 .1093/brain/awn096

Lacorazza HD, Flax JD, Snyder EY, Jendoubi M. 1996. Expression of human β-hexosaminidase α-subunit gene (the gene defect of Tay–Sachs disease) in mouse brains upon engraftment of transduced progenitor cells. *Nat Med* **2:** 424–429. doi:10.1038/nm0496-424

Lanza R, Russell DW, Nagy A. 2019. Engineering universal cells that evade immune detection. *Nat Rev Immunol* **19:** 723–733. doi:10.1038/s41577-019-0200-1

Lappe-Siefke C, Goebbels S, Gravel M, Nicksch E, Lee J, Braun PE, Griffiths IR, Nave KA. 2003. Disruption of Cnp1 uncouples oligodendroglial functions in axonal support and myelination. *Nat Genet* **33:** 366–374. doi:10.1038/ng1095

Lasiene J, Shupe L, Perlmutter S, Horner P. 2008. No evidence for chronic demyelination in spared axons after spinal cord injury in a mouse. *J Neurosci* **28:** 3887–3896. doi:10.1523/JNEUROSCI.4756-07.2008

Lazzarotto A, Tonietto M, Poirion E, Battaglini M, Palladino R, Benoit C, Ricigliano VA, Maillart E, Stefano ND, Stankoff B, et al. 2022. Clinically relevant profiles of myelin content changes in patients with multiple sclerosis: a multimodal and multicompartment imaging study. *Mult Scler J* **28:** 1881–1890. doi:10.1177/13524585221096975

Lee Y, Morrison BM, Li Y, Lengacher S, Farah MH, Hoffman PN, Liu Y, Tsingalia A, Jin L, Zhang PW, et al. 2012. Oligodendroglia metabolically support axons and contribute to neurodegeneration. *Nature* **487:** 443–448. doi:10.1038/nature11314

Lee CH, Walczak P, Zhang J. 2022. Inhomogeneous magnetization transfer MRI of white matter structures in the hypomyelinated shiverer mouse brain. *Magn Reson Med* **88:** 332–340. doi:10.1002/mrm.29207

Li R, Johnson AB, Salomons G, Goldman JE, Naidu S, Quinlan R, Cree B, Ruyle SZ, Banwell B, D'Hooghe M, et al. 2005. Glial fibrillary acidic protein mutations in infantile, juvenile, and adult forms of Alexander disease. *Ann Neurol* **57:** 310–326. doi:10.1002/ana.20406

Li WW, Penderis J, Zhao C, Schumacher M, Franklin RJM. 2006. Females remyelinate more efficiently than males following demyelination in the aged but not young adult CNS. *Exp Neurol* **202:** 250–254. doi:10.1016/j.expneurol .2006.05.012

Li Z, He Y, Fan S, Sun B. 2015. Clemastine rescues behavioral changes and enhances remyelination in the cuprizone mouse model of demyelination. *Neurosci Bull* **31:** 617–625. doi:10.1007/s12264-015-1555-3

Liang Q, Monetti C, Shutova MV, Neely EJ, Hacibekiroglu S, Yang H, Kim C, Zhang P, Li C, Nagy K, et al. 2018. Linking a cell-division gene and a suicide gene to define and improve cell therapy safety. *Nature* **563:** 701–704. doi:10.1038/s41586-018-0733-7

Liebetanz D, Merkler D. 2006. Effects of commissural de- and remyelination on motor skill behaviour in the cuprizone mouse model of multiple sclerosis. *Exp Neurol* **202:** 217–224. doi:10.1016/j.expneurol.2006.05.032

Lin T, Ambasudhan R, Yuan X, Li W, Hilcove S, Abujarour R, Lin X, Hahm HS, Hao E, Hayek A, et al. 2009. A chemical platform for improved induction of human iPSCs. *Nat Methods* **6:** 805–808. doi:10.1038/nmeth.1393

Linington C, Engelhardt B, Kapocs G, Lassman H. 1992. Induction of persistently demyelinated lesions in the rat following the repeated adoptive transfer of encephalitogenic T cells and demyelinating antibody. *J Neuroimmunol* **40:** 219–224. doi:10.1016/0165-5728(92)90136-9

Liu Y, Han SSW, Wu Y, Tuohy TMF, Xue H, Cai J, Back SA, Sherman LS, Fischer I, Rao MS. 2004. CD44 expression identifies astrocyte-restricted precursor cells. *Dev Biol* **276:** 31–46.

Lloyd AF, Miron VE. 2019. The pro-remyelination properties of microglia in the central nervous system. *Nat Rev Neurol* **15:** 447–458. doi:10.1038/s41582-019-0184-2

Lubetzki C, Zalc B, Williams A, Stadelmann C, Stankoff B. 2020. Remyelination in multiple sclerosis: from basic science to clinical translation. *Lancet Neurol* **19:** 678–688. doi:10.1016/S1474-4422(20)30140-X

MacKay AL, Laule C. 2016. Magnetic resonance of myelin water: an in vivo marker for myelin. *Brain Plast* **2:** 71–91. doi:10.3233/BPL-160033

MacKay A, Whittall K, Adler J, Li D, Paty D, Graeb D. 1994. In vivo visualization of myelin water in brain by magnetic resonance. *Magn Reson Med* **31:** 673–677. doi:10.1002/ mrm.1910310614

Malik NN, Jenkins AM, Mellon J, Bailey G. 2019. Engineering strategies for generating hypoimmunogenic cells with high clinical and commercial value. *Regen Med* **14:** 983–989. doi:10.2217/rme-2019-0117

Mancini M, Karakuzu A, Cohen-Adad J, Cercignani M, Nichols TE, Stikov N. 2020. An interactive meta-analysis of MRI biomarkers of myelin. *eLife* **9:** e61523. doi:10 .7554/eLife.61523

 Cite this article as *Cold Spring Harb Perspect Biol* doi: 10.1101/cshperspect.a041371

Mariani JN, Zou L, Goldman SA. 2019. Human glial chimeric mice to define the role of glial pathology in human disease. *Methods Mol Biol* **1936**: 311–331. doi:10.1007/978-1-4939-9072-6_18

Martinez-Curiel R, Jansson L, Tsupykov O, Avaliani N, Aretio-Medina C, Hidalgo I, Monni E, Bengzon J, Skibo G, Lindvall O, et al. 2023. Oligodendrocytes in human induced pluripotent stem cell-derived cortical grafts remyelinate adult rat and human cortical neurons. *Stem Cell Rep* **18**: 1643–1656.

Mason JL, Xuan S, Dragatsis I, Efstratiadis A, Goldman JE. 2003. Insulin-like growth factor (IGF) signaling through type 1 IGF receptor plays an important role in remyelination. *J Neurosci* **23**: 7710–7718. doi:10.1523/JNEUROSCI.23-20-07710.2003

Mason JL, Toews A, Hostettler JD, Morell P, Suzuki K, Goldman JE, Matsushima GK. 2004. Oligodendrocytes and progenitors become progressively depleted within chronically demyelinated lesions. *Am J Pathol* **164**: 1673–1682. doi:10.1016/S0002-9440(10)63726-1

Mei F, Fancy SPJ, Shen YAA, Niu J, Zhao C, Presley B, Miao E, Lee S, Mayoral SR, Redmond SA, et al. 2014. Micropillar arrays as a high-throughput screening platform for therapeutics in multiple sclerosis. *Nat Med* **20**: 954–960. doi:10.1038/nm.3618

Mei F, Lehmann-Horn K, Shen YAA, Rankin KA, Stebbins KJ, Lorrain DS, Pekarek K, Sagan SA, Xiao L, Teuscher C, et al. 2016. Accelerated remyelination during inflammatory demyelination prevents axonal loss and improves functional recovery. *eLife* **5**: e18246. doi:10.7554/eLife.18246

Menn B, Garcia-Verdugo JM, Yaschine C, Gonzalez-Perez O, Rowitch D, Alvarez-Buylla A. 2006. Origin of oligodendrocytes in the subventricular zone of the adult brain. *J Neurosci* **26**: 7907–7918. doi:10.1523/JNEUROSCI.1299-06.2006

Merkler D, Ernsting T, Kerschensteiner M, Brück W, Stadelmann C. 2006. A new focal EAE model of cortical demyelination: multiple sclerosis-like lesions with rapid resolution of inflammation and extensive remyelination. *Brain* **129**: 1972–1983. doi:10.1093/brain/awl135

Mezydlo A, Treiber N, Gavilanes EMU, Eichenseer K, Ancău M, Wens A, Carral CA, Schifferer M, Snaidero N, Misgeld T, Kerschensteiner M. 2023. Remyelination by surviving oligodendrocytes is inefficient in the inflamed mammalian cortex. *Neuron* **111**: 1748–1759.e8. doi:10.1016/j.neuron.2023.03.031

Mi S, Miller RH, Lee X, Scott ML, Shulag-Morskaya S, Shao Z, Chang J, Thill G, Levesque M, Zhang M, et al. 2005. LINGO-1 negatively regulates myelination by oligodendrocytes. *Nat Neurosci* **8**: 745–751. doi:10.1038/nn1460

Mi S, Hu B, Hahm K, Luo Y, Kam Hui ES, Yuan Q, Wong WM, Wang L, Su H, Chu TH, et al. 2007. LINGO-1 antagonist promotes spinal cord remyelination and axonal integrity in MOG-induced experimental autoimmune encephalomyelitis. *Nat Med* **13**: 1228–1233. doi:10.1038/nm1664

Michailov GV, Sereda MW, Brinkmann BG, Fischer TM, Haug B, Birchmeier C, Role L, Lai C, Schwab MH, Nave KA. 2004. Axonal neuregulin-1 regulates myelin sheath thickness. *Science* **304**: 700–703. doi:10.1126/science.1095862

Miron VE, Boyd A, Zhao JW, Yuen TJ, Ruckh JM, Shadrach JL, van Wijngaarden P, Wagers AJ, Williams A, Franklin RJM, et al. 2013. M2 microglia and macrophages drive oligodendrocyte differentiation during CNS remyelination. *Nat Neurosci* **16**: 1211–1218. doi:10.1038/nn.3469

Mitome M, Low HP, van deb Pol A, Nunnari JJ, Wolf MK, Billings-Gagliardi S, Schwartz WJ. 2001. Towards the reconstruction of central nervous system white matter using neural precursor cells. *Brain* **124**: 2147–2161. doi:10.1093/brain/124.11.2147

Moccia M, van de Pavert S, Eshaghi A, Haider L, Pichat J, Yiannakas M, Ourselin S, Wang Y, Wheeler-Kingshott C, Thompson A, et al. 2020. Pathologic correlates of the magnetization transfer ratio in multiple sclerosis. *Neurology* **95**: e2965–e2976. doi:10.1212/WNL.0000000000010909

Molina-Gonzalez I, Miron VE. 2019. Astrocytes in myelination and remyelination. *Neurosci Lett* **713**: 134532. doi:10.1016/j.neulet.2019.134532

Moll NM, Rietsch AM, Thomas S, Ransohoff AJ, Lee J, Fox R, Chang A, Ransohoff RM, Fisher E. 2011. Multiple sclerosis normal-appearing white matter: pathology–imaging correlations. *Ann Neurol* **70**: 764–773. doi:10.1002/ana.22521

Moore CS, Abdullah SL, Brown A, Arulpragasam A, Crocker SJ. 2011. How factors secreted from astrocytes impact myelin repair. *J Neurosci Res* **89**: 13–21. doi:10.1002/jnr.22482

Morrison BM, Lee Y, Rothstein JD. 2013. Oligodendroglia: metabolic supporters of axons. *Trends Cell Biol* **23**: 644–651. doi:10.1016/j.tcb.2013.07.007

Moyon S, Dubessy AL, Aigrot MS, Trotter M, Huang JK, Dauphinot L, Potier MC, Kerninon C, Parsadaniantz SM, Franklin RJM, et al. 2015. Demyelination causes adult CNS progenitors to revert to an immature state and express immune cues that support their migration. *J Neurosci* **35**: 4–20. doi:10.1523/JNEUROSCI.0849-14.2015

Murtie JC, Zhou YX, Le TQ, Vana AC, Armstrong RC. 2005. PDGF and FGF2 pathways regulate distinct oligodendrocyte lineage responses in experimental demyelination with spontaneous remyelination. *Neurobiol Dis* **19**: 171–182. doi:10.1016/j.nbd.2004.12.006

Naismith RT, Xu J, Tutlam NT, Scully PT, Trinkaus K, Snyder AZ, Song SK, Cross AH. 2010. Increased diffusivity in acute multiple sclerosis lesions predicts risk of black hole. *Neurology* **74**: 1694–1701. doi:10.1212/WNL.0b013e3181e042c4

Nait-Oumesmar B, Decker L, Lachapelle F, Avellana-Adalid V, Bachelin C, Evercooren ABV. 1999. Progenitor cells of the adult mouse subventricular zone proliferate, migrate and differentiate into oligodendrocytes after demyelination. *Eur J Neurosci* **11**: 4357–4366. doi:10.1046/j.1460-9568.1999.00873.x

Najm FJ, Lager AM, Zaremba A, Wyatt K, Caprariello AV, Factor DC, Karl RT, Maeda T, Miller RH, Tesar PJ. 2013. Transcription factor-mediated reprogramming of fibroblasts to expandable, myelinogenic oligodendrocyte progenitor cells. *Nat Biotechnol* **31**: 426–433. doi:10.1038/nbt.2561

Najm FJ, Madhavan M, Zaremba A, Shick E, Karl RT, Factor DC, Miller TE, Nevin ZS, Kantor C, Sargent A, et al. 2015.

Drug-based modulation of endogenous stem cells promotes functional remyelination in vivo. *Nature* **522:** 216–220. doi:10.1038/nature14335

Natrajan MS, de la Fuente AG, Crawford AH, Linehan E, Nuñez V, Johnson KR, Wu T, Fitzgerald DC, Ricote M, Bielekova B, et al. 2015. Retinoid X receptor activation reverses age-related deficiencies in myelin debris phagocytosis and remyelination. *Brain* **138:** 3581–3597. doi:10.1093/brain/awv289

Nave KA, Trapp BD. 2008. Axon-glial signaling and the glial support of axon function. *Neuroscience* **31:** 535–561.

Neely SA, Williamson JM, Klingseisen A, Zoupi L, Early JJ, Williams A, Lyons DA. 2022. New oligodendrocytes exhibit more abundant and accurate myelin regeneration than those that survive demyelination. *Nat Neurosci* **25:** 415–420. doi:10.1038/s41593-021-01009-x

Neumann B, Baror R, Zhao C, Segel M, Dietmann S, Rawji KS, Foerster S, McClain CR, Chalut K, van Wijngaarden P, et al. 2019a. Metformin restores CNS remyelination capacity by rejuvenating aged stem cells. *Cell Stem Cell* **25:** 473–485.e8. doi:10.1016/j.stem.2019.08.015

Neumann B, Segel M, Chalut KJ, Franklin RJ. 2019b. Remyelination and ageing: reversing the ravages of time. *Mult Scler J* **25:** 1835–1841. doi:10.1177/1352458519884006

Neumann B, Foerster S, Zhao C, Bodini B, Reich DS, Bergles DE, Káradóttir RT, Lubetzki C, Lairson LL, Zalc B, et al. 2020. Problems and pitfalls of identifying remyelination in multiple sclerosis. *Cell Stem Cell* **26:** 617–619. doi:10.1016/j.stem.2020.03.017

Neumann B, Segel M, Ghosh T, Zhao C, Tourlomousis P, Young A, Förster S, Sharma A, Chen CZY, Cubillos JF, et al. 2021. Myc determines the functional age state of oligodendrocyte progenitor cells. *Nat Aging* **1:** 826–837. doi:10.1038/s43587-021-00109-4

Niehaus A, Shi J, Grzenkowski M, Diers-Fenger M, Archelos J, Hartung HP, Toyka K, Brück W, Trotter J. 2000. Patients with active relapsing-remitting multiple sclerosis synthesize antibodies recognizing oligodendrocyte progenitor cell surface protein: implications for remyelination. *Ann Neurol* **48:** 362–371. doi:10.1002/1531-8249(200009)48:3<362::AID-ANA11>3.0.CO;2-6

Nistor GI, Totoiu MO, Haque N, Carpenter MK, Keirstead HS. 2005. Human embryonic stem cells differentiate into oligodendrocytes in high purity and myelinate after spinal cord transplantation. *Glia* **49:** 385–396. doi:10.1002/glia.20127

Njenga MK, Murray PD, McGavern D, Lin X, Drescher KM, Rodriguez M. 1999. Absence of spontaneous central nervous system remyelination in class II-deficient mice infected with Theiler's virus. *J Neuropathol Exp Neurol* **58:** 78–91. doi:10.1097/00005072-199901000-00009

Nunes MC, Roy NS, Keyoung HM, Goodman RR, McKhann G, Jiang L, Kang J, Nedergaard M, Goldman SA. 2003. Identification and isolation of multipotential neural progenitor cells from the subcortical white matter of the adult human brain. *Nat Med* **9:** 439–447. doi:10.1038/nm837

Orthmann-Murphy J, Call CL, Molina-Castro GC, Hsieh YC, Rasband MN, Calabresi PA, Bergles DE. 2020. Remyelination alters the pattern of myelin in the cerebral cortex. *eLife* **9:** e56621. doi:10.7554/eLife.56621

Ortiz FC, Habermacher C, Graciarena M, Houry P-Y, Nishiyama A, Nait-Oumesmar B, Angulo MC. 2019. Neuronal activity in vivo enhances functional myelin repair. *JCI Insight* **4:** e123434.

Osorio MJ, Goldman SA. 2016. Cell therapy for pediatric disorders of glia. In *Translational neuroscience* (ed. Tuszynski MH), pp. 275–296. Springer, Boston.

Osório MJ, Goldman SA. 2018. Chapter 45: Neurogenetics of Pelizaeus–Merzbacher disease. *Handb Clin Neurol* **148:** 701–722.

Patani R, Balaratnam M, Vora A, Reynolds R. 2007. Remyelination can be extensive in multiple sclerosis despite a long disease course. *Neuropathol Appl Neurobiol* **33:** 277–287. doi:10.1111/j.1365-2990.2007.00805.x

Patrikios P, Stadelmann C, Kutzelnigg A, Rauschka H, Schmidbauer M, Laursen H, Sorensen PS, Brück W, Lucchinetti C, Lassmann H. 2006. Remyelination is extensive in a subset of multiple sclerosis patients. *Brain* **129:** 3165–3172. doi:10.1093/brain/awl217

Pellegatta S, Tunici P, Poliani PL, Dolcetta D, Cajola L, Colombelli C, Ciusani E, Donato SD, Finocchiaro G. 2006. The therapeutic potential of neural stem/progenitor cells in murine globoid cell leukodystrophy is conditioned by macrophage/microglia activation. *Neurobiol Dis* **21:** 314–323. doi:10.1016/j.nbd.2005.07.016

Penderis J, Shields SA, Franklin RJM. 2003. Impaired remyelination and depletion of oligodendrocyte progenitors does not occur following repeated episodes of focal demyelination in the rat central nervous system. *Brain* **126:** 1382–1391. doi:10.1093/brain/awg126

Petiet A, Adanyeguh I, Aigrot M, Poirion E, Nait-Oumesmar B, Santin M, Stankoff B. 2019. Ultrahigh field imaging of myelin disease models: toward specific markers of myelin integrity? *J Comp Neurol* **527:** 2179–2189. doi:10.1002/cne.24598

Philips T, Mironova YA, Jouroukhin Y, Chew J, Vidensky S, Farah MH, Pletnikov MV, Bergles DE, Morrison BM, Rothstein JD. 2021. MCT1 deletion in oligodendrocyte lineage cells causes late-onset hypomyelination and axonal degeneration. *Cell Rep* **34:** 108610.

Piao J, Major T, Auyeung G, Policarpio E, Menon J, Droms L, Gutin P, Uryu K, Tchieu J, Soulet D, Tabar V. 2015. Human embryonic stem cell-derived oligodendrocyte progenitors remyelinate the brain and rescue behavioral deficits following radiation. *Cell Stem Cell* **16:** 198–210.

Piaton G, Williams A, Seilhean D, Lubetzki C. 2009. Remyelination in multiple sclerosis. *Prog Brain Res* **175:** 453–464. doi:10.1016/S0079-6123(09)17530-1

Piaton G, Aigrot MS, Williams A, Moyon S, Tepavcevic V, Moutkine I, Gras J, Matho KS, Schmitt A, Soellner H, et al. 2011. Class 3 semaphorins influence oligodendrocyte precursor recruitment and remyelination in adult central nervous system. *Brain* **134:** 1156–1167. doi:10.1093/brain/awr022

Pluchino S, Furlan R, Martino G. 2004. Cell-based remyelinating therapies in multiple sclerosis: evidence from experimental studies. *Curr Opin Neurol* **17:** 247–255. doi:10.1097/00019052-200406000-00003

Pohl HBF, Porcheri C, Mueggler T, Bachmann LC, Martino G, Riethmacher D, Franklin RJM, Rudin M, Suter U. 2011. Genetically induced adult oligodendrocyte cell death is associated with poor myelin clearance, reduced remyelination, and axonal damage. *J Neurosci* **31**: 1069–1080. doi:10.1523/JNEUROSCI.5035-10.2011

Poliani PL, Wang Y, Fontana E, Robinette ML, Yamanishi Y, Gilfillan S, Colonna M. 2015. TREM2 sustains microglial expansion during aging and response to demyelination. *J Clin Invest* **125**: 2161–2170. doi:10.1172/JCI77983

Polo JM, Liu S, Figueroa ME, Kulalert W, Eminli S, Tan KY, Apostolou E, Stadtfeld M, Li Y, Shioda T, et al. 2010. Cell type of origin influences the molecular and functional properties of mouse induced pluripotent stem cells. *Nat Biotechnol* **28**: 848–855. doi:10.1038/nbt.1667

Powers BE, Sellers DL, Lovelett EA, Cheung W, Aalami SP, Zapertov N, Maris DO, Horner PJ. 2013. Remyelination reporter reveals prolonged refinement of spontaneously regenerated myelin. *Proc Natl Acad Sci* **110**: 4075–4080. doi:10.1073/pnas.1210293110

Pruszak J, Ludwig W, Blak A, Alavian K, Isacson O. 2009. CD15, CD24, and CD29 define a surface biomarker code for neural lineage differentiation of stem cells. *Stem Cells* **27**: 2928–2940. doi:10.1002/stem.211

Pytel V, Matias-Guiu JA, Matías-Guiu J, Cortés-Martínez A, Montero P, Moreno-Ramos T, Arrazola J, Carreras JL, Cabrera-Martín MN. 2020. Amyloid PET findings in multiple sclerosis are associated with cognitive decline at 18 months. *Mult Scler Relat Disord* **39**: 101926. doi:10.1016/j.msard.2020.101926

Rawji KS, Kappen J, Tang W, Teo W, Plemel JR, Stys PK, Yong VW. 2018. Deficient surveillance and phagocytic activity of myeloid cells within demyelinated lesions in aging mice visualized by ex vivo live multiphoton imaging. *J Neurosci* **38**: 1973–1988. doi:10.1523/JNEUROSCI.2341-17.2018

Rawji KS, Young AMH, Ghosh T, Michaels NJ, Mirzaei R, Kappen J, Kolehmainen KL, Alaeiilkhchi N, Lozinski B, Mishra MK, et al. 2020a. Niacin-mediated rejuvenation of macrophage/microglia enhances remyelination of the aging central nervous system. *Acta Neuropathol* **139**: 893–909. doi:10.1007/s00401-020-02129-7

Rawji KS, Martinez GAG, Sharma A, Franklin RJM. 2020b. The role of astrocytes in remyelination. *Trends Neurosci* **43**: 596–607. doi:10.1016/j.tins.2020.05.006

Rhodes KE, Raivich G, Fawcett JW. 2006. The injury response of oligodendrocyte precursor cells is induced by platelets, macrophages and inflammation-associated cytokines. *Neuroscience* **140**: 87–100. doi:10.1016/j.neuroscience.2006.01.055

Richardson WD, Young KM, Tripathi RB, McKenzie I. 2011. NG2-glia as multipotent neural stem cells: fact or fantasy? *Neuron* **70**: 661–673. doi:10.1016/j.neuron.2011.05.013

Robinson S, Petelenz K, Li Q, Cohen ML, Dechant A, Tabrizi N, Bucek M, Lust D, Miller RH. 2005. Developmental changes induced by graded prenatal systemic hypoxic-ischemic insults in rats. *Neurobiol Dis* **18**: 568–581. doi:10.1016/j.nbd.2004.10.024

Roy NS, Wang S, Harrison-Restelli C, Benraiss A, Fraser RAR, Gravel M, Braun PE, Goldman SA. 1999. Identification, isolation, and promoter-defined separation of mitotic oligodendrocyte progenitor cells from the adult human subcortical white matter. *J Neurosci* **19**: 9986–9995. doi:10.1523/JNEUROSCI.19-22-09986.1999

Roy NS, Cleren C, Singh SK, Yang L, Beal MF, Goldman SA. 2006. Functional engraftment of human ES cell–derived dopaminergic neurons enriched by coculture with telomerase-immortalized midbrain astrocytes. *Nat Med* **12**: 1259–1268. doi:10.1038/nm1495

Ruckh JM, Zhao JW, Shadrach JL, van Wijngaarden P, Rao TN, Wagers AJ, Franklin RJM. 2012. Rejuvenation of regeneration in the aging central nervous system. *Cell Stem Cell* **10**: 96–103. doi:10.1016/j.stem.2011.11.019

Rudick RA, Mi S, Sandrock AW. 2008. LINGO-1 antagonists as therapy for multiple sclerosis: in vitro and in vivo evidence. *Expert Opin Biol Ther* **8**: 1561–1570. doi:10.1517/14712598.8.10.1561

Schäffner E, Bosch-Queralt M, Edgar JM, Lehning M, Strauß J, Fleischer N, Kungl T, Wieghofer P, Berghoff Stefan A, Tilo Reinert, et al. 2023. Myelin insulation as a risk factor for axonal degeneration in autoimmune demyelinating disease. *Nat Neurosci* **26**: 1218–1228.

Schmierer K, Scaravilli F, Altmann DR, Barker GJ, Miller DH. 2004. Magnetization transfer ratio and myelin in postmortem multiple sclerosis brain. *Ann Neurol* **56**: 407–415. doi:10.1002/ana.20202

Schwartzbach CJ, Grove RA, Brown R, Tompson D, Bergh FT, Arnold DL. 2017. Lesion remyelinating activity of GSK239512 versus placebo in patients with relapsing-remitting multiple sclerosis: a randomised, single-blind, phase II study. *J Neurol* **264**: 304–315. doi:10.1007/s00415-016-8341-7

Segel M, Neumann B, Hill MFE, Weber IP, Viscomi C, Zhao C, Young A, Agley CC, Thompson AJ, Gonzalez GA, et al. 2019. Niche stiffness underlies the ageing of central nervous system progenitor cells. *Nature* **573**: 130–134. doi:10.1038/s41586-019-1484-9

Sheikh S, Ernst D, Keating A. 2021. Prodrugs and prodrug-activated systems in gene therapy. *Mol Ther* **29**: 1716–1728. doi:10.1016/j.ymthe.2021.04.006

Shen S, Sandoval J, Swiss VA, Li J, Dupree J, Franklin RJM, Casaccia-Bonnefil P. 2008. Age-dependent epigenetic control of differentiation inhibitors is critical for remyelination efficiency. *Nat Neurosci* **11**: 1024–1034. doi:10.1038/nn.2172

Sheth VS, Gauthier J. 2021. Taming the beast: CRS and ICANS after CAR T-cell therapy for ALL. *Bone Marrow Transplant* **56**: 552–566. doi:10.1038/s41409-020-01134-4

Shibuya Y, Kumar KK, Mader MM, Yoo Y, Ayala LA, Zhou M, Mohr MA, Neumayer G, Kumar I, Yamamoto R, et al. 2022. Treatment of a genetic brain disease by CNS-wide microglia replacement. *Sci Transl Med* **14**: eabl9945. doi:10.1126/scitranslmed.abl9945

Siegler EL, Kenderian SS. 2020. Neurotoxicity and cytokine release syndrome after chimeric antigen receptor T cell therapy: insights into mechanisms and novel therapies. *Front Immunol* **11**: 1973. doi:10.3389/fimmu.2020.01973

Sim FJ, Zhao C, Li WW, Lakatos A, Franklin RJM. 2002a. Expression of the POU-domain transcription factors SCIP/Oct-6 and Brn-2 is associated with Schwann cell but not oligodendrocyte remyelination of the CNS. *Mol Cell Neurosci* **20**: 669–682. doi:10.1006/mcne.2002.1145

Sim FJ, Zhao C, Penderis J, Franklin RJM. 2002b. The age-related decrease in CNS remyelination efficiency is attributable to an impairment of both oligodendrocyte progenitor recruitment and differentiation. *J Neurosci* **22**: 2451–2459. doi:10.1523/JNEUROSCI.22-07-02451.2002

Sim FJ, McClain CR, Schanz SJ, Protack TL, Windrem MS, Goldman SA. 2011. CD140a identifies a population of highly myelinogenic, migration-competent and efficiently engrafting human oligodendrocyte progenitor cells. *Nat Biotechnol* **29:** 934–941. doi:10.1038/nbt.1972

Sloane JA, Batt C, Ma Y, Harris ZM, Trapp B, Vartanian T. 2010. Hyaluronan blocks oligodendrocyte progenitor maturation and remyelination through TLR2. *Proc Natl Acad Sci* **107:** 11555–11560. doi:10.1073/pnas.1006496107

Smith KJ, Blakemore WF, McDonald WI. 1979. Central remyelination restores secure conduction. *Nature* **280:** 395–396. doi:10.1038/280395a0

Snyder DH, Valsamis MP, Stone SH, Raine CS. 1975. Progressive demyelination and reparative phenomena in chronic experimental allergic encephalomyelitis. *J Neuropathol Exp Neurol* **34:** 209–221. doi:10.1097/00005072-197505000-00001

Snyder EY, Taylor RM, Wolfe JH. 1995. Neural progenitor cell engraftment corrects lysosomal storage throughout the MRS VII mouse brain. *Nature* **374:** 367–370. doi:10.1038/374367a0

Song SK, Sun SW, Ramsbottom MJ, Chang C, Russell J, Cross AH. 2002. Dysmyelination revealed through MRI as increased radial (but unchanged axial) diffusion of water. *Neuroimage* **17:** 1429–1436. doi:10.1006/nimg.2002.1267

Song SK, Yoshino J, Le TQ, Lin SJ, Sun SW, Cross AH, Armstrong RC. 2005. Demyelination increases radial diffusivity in corpus callosum of mouse brain. *Neuroimage* **26:** 132–140. doi:10.1016/j.neuroimage.2005.01.028

Stacpoole SRL, Spitzer S, Bilican B, Compston A, Karadottir R, Chandran S, Franklin RJM. 2013. High yields of oligodendrocyte lineage cells from human embryonic stem cells at physiological oxygen tensions for evaluation of translational biology. *Stem Cell Rep* **1:** 437–450.

Stadtfeld M, Hochedlinger K. 2010. Induced pluripotency: history, mechanisms, and applications. *Genes Dev* **24:** 2239–2263. doi:10.1101/gad.1963910

Stadtfeld M, Apostolou E, Akutsu H, Fukuda A, Follett P, Natesan S, Kono T, Shioda T, Hochedlinger K. 2010. Aberrant silencing of imprinted genes on chromosome 12qF1 in mouse induced pluripotent stem cells. *Nature* **465:** 175–181. doi:10.1038/nature09017

Stidworthy MF, Genoud S, Suter U, Mantei N, Franklin RJM. 2003. Quantifying the early stages of remyelination following cuprizone-induced demyelination. *Brain Pathol* **13:** 329–339. doi:10.1111/j.1750-3639.2003.tb00032.x

Stidworthy MF, Genoud S, Li WW, Leone DP, Mantei N, Suter U, Franklin RJM. 2004. Notch1 and Jagged1 are expressed after CNS demyelination, but are not a major rate-determining factor during remyelination. *Brain* **127:** 1928–1941. doi:10.1093/brain/awh217

Strijbis EMM, Kooi E-J, van der Valk P, Geurts JJG. 2017. Cortical remyelination is heterogeneous in multiple sclerosis. *J Neuropathol Exp Neurol* **76:** 390–401.

Sutiwisesak R, Burns TC, Rodriguez M, Warrington AE. 2021. Remyelination therapies for multiple sclerosis: optimizing translation from animal models into clinical trials. *Expert Opin Investig Drugs* **30:** 857–876. doi:10.1080/13543784.2021.1942840

Syed YA, Hand E, Möbius W, Zhao C, Hofer M, Nave KA, Kotter MR. 2011. Inhibition of CNS remyelination by the presence of semaphorin 3A. *J Neurosci* **31:** 3719–3728. doi:10.1523/JNEUROSCI.4930-10.2011

Tanabe K, Nobuta H, Yang N, Ang CE, Huie P, Jordan S, Oldham MC, Rowitch DH, Wernig M. 2022. Generation of functional human oligodendrocytes from dermal fibroblasts by direct lineage conversion. *Development* **149:** dev199723.

Tonietto M, Poirion E, Lazzarotto A, Ricigliano V, Papeix C, Bottlaender M, Bodini B, Stankoff B. 2022. Periventricular remyelination failure in multiple sclerosis: a substrate for neurodegeneration. *Brain* **146:** 182–194. doi:10.1093/brain/awac334

Traka M, Arasi K, Avila RL, Podojil JR, Christakos A, Miller SD, Soliven B, Popko B. 2010. A genetic mouse model of adult-onset, pervasive central nervous system demyelination with robust remyelination. *Brain* **133:** 3017–3029.

Trapp BD, Nave KA. 2008. Multiple sclerosis: an immune or neurodegenerative disorder? *Annu Rev Neurosci* **31:** 247–269. doi:10.1146/annurev.neuro.30.051606.094313

Tripathi RB, Rivers LE, Young KM, Jamen F, Richardson WD. 2010. NG2 glia generate new oligodendrocytes but few astrocytes in a murine experimental autoimmune encephalomyelitis model of demyelinating disease. *J Neurosci* **30:** 16383–16390. doi:10.1523/JNEUROSCI.3411-10.2010

Ulanska-Poutanen J, Mieczkowski J, Zhao C, Konarzewska K, Kaza B, Pohl HB, Bugajski L, Kaminska B, Franklin RJ, Zawadzka M. 2018. Injury-induced perivascular niche supports alternative differentiation of adult rodent CNS progenitor cells. *eLife* **7:** e30325. doi:10.7554/eLife.30325

Urayama A, Grubb JH, Sly WS, Banks WA. 2004. Developmentally regulated mannose 6-phosphate receptor-mediated transport of a lysosomal enzyme across the blood-brain barrier. *Proc Natl Acad Sci* **101:** 12658–12663. doi:10.1073/pnas.0405042101

van der Weijden CWJ, Meilof JF, van der Hoorn A, Zhu J, Wu C, Wang Y, Willemsen ATM, Dierckx RAJO, Lammertsma AA, de Vries EFJ. 2022. Quantitative assessment of myelin density using [^{11}C]MeDAS PET in patients with multiple sclerosis: a first-in-human study. *Eur J Nucl Med Mol Imaging* **49:** 3492–3507. doi:10.1007/s00259-022-05770-4

Vela JM, Molina-Holgado E, Arévalo-Martín Á, Almazán G, Guaza C. 2002. Interleukin-1 regulates proliferation and differentiation of oligodendrocyte progenitor cells. *Mol Cell Neurosci* **20:** 489–502. doi:10.1006/mcne.2002.1127

Veloz RIZ, McKenzie T, Palacios BE, Hu J. 2022. Nuclear hormone receptors in demyelinating diseases. *J Neuroendocrinol* **34:** e13171. doi:10.1111/jne.13171

Wang S, Sdrulla AD, diSibio G, Bush G, Nofziger D, Hicks C, Weinmaster G, Barres BA. 1998. Notch receptor activation inhibits oligodendrocyte differentiation. *Neuron* **21:** 63–75. doi:10.1016/S0896-6273(00)80515-2

Wang S, Bates J, Li X, Schanz S, Chandler-Militello D, Levine C, Maherali N, Studer L, Hochedlinger K, Windrem M, et al. 2013. Human iPSC-derived oligodendrocyte progenitor cells can myelinate and rescue a mouse model of congenital hypomyelination. *Cell Stem Cell* **12:** 252–264. doi:10.1016/j.stem.2012.12.002

Cite this article as *Cold Spring Harb Perspect Biol* doi: 10.1101/cshperspect.a041371

Watanabe M, Hadzic T, Nishiyama A. 2004. Transient up-regulation of Nkx2.2 expression in oligodendrocyte lineage cells during remyelination. *Glia* **46**: 311–322. doi:10.1002/glia.20006

Werner HB, Kuhlmann K, Shen S, Uecker M, Schardt A, Dimova K, Orfaniotou F, Dhaunchak A, Brinkmann BG, Möbius W, et al. 2007. Proteolipid protein is required for transport of sirtuin 2 into CNS myelin. *J Neurosci* **27**: 7717–7730. doi:10.1523/JNEUROSCI.1254-07.2007

Williams A, Piaton G, Aigrot MS, Belhadi A, Théaudin M, Petermann F, Thomas JL, Zalc B, Lubetzki C. 2007a. Semaphorin 3A and 3F: key players in myelin repair in multiple sclerosis? *Brain* **130**: 2554–2565. doi:10.1093/brain/awm202

Williams A, Piaton G, Lubetzki C. 2007b. Astrocytes—friends or foes in multiple sclerosis? *Glia* **55**: 1300–1312. doi:10.1002/glia.20546

Windrem MS, Roy NS, Wang J, Nunes M, Benraiss A, Goodman R, McKhann GM II, Goldman SA. 2002. Progenitor cells derived from the adult human subcortical white matter disperse and differentiate as oligodendrocytes within demyelinated lesions of the rat brain. *J Neurosci Res* **69**: 966–975. doi:10.1002/jnr.10397

Windrem MS, Nunes MC, Rashbaum WK, Schwartz TH, Goodman RA, McKhann G, Roy NS, Goldman SA. 2004. Fetal and adult human oligodendrocyte progenitor cell isolates myelinate the congenitally dysmyelinated brain. *Nat Med* **10**: 93–97. doi:10.1038/nm974

Windrem MS, Schanz SJ, Guo M, Tian GF, Washco V, Stanwood N, Rasband M, Roy NS, Nedergaard M, Havton LA, et al. 2008. Neonatal chimerization with human glial progenitor cells can both remyelinate and rescue the otherwise lethally hypomyelinated shiverer mouse. *Cell Stem Cell* **2**: 553–565. doi:10.1016/j.stem.2008.03.020

Windrem MS, Schanz SJ, Zou L, Chandler-Militello D, Kuypers NJ, Nedergaard M, Lu Y, Mariani JN, Goldman SA. 2020. Human glial progenitor cells effectively remyelinate the demyelinated adult brain. *Cell Rep* **31**: 107658. doi:10.1016/j.celrep.2020.107658

Wolswijk G. 1998a. Oligodendrocyte regeneration in the adult rodent CNS and the failure of this process in multiple sclerosis. *Prog Brain Res* **117**: 233–247. doi:10.1016/S0079-6123(08)64019-4

Wolswijk G. 1998b. Chronic stage multiple sclerosis lesions contain a relatively quiescent population of oligodendrocyte precursor cells. *J Neurosci* **18**: 601–609. doi:10.1523/JNEUROSCI.18-02-00601.1998

Wolswijk G, Noble M. 1989. Identification of an adult-specific glial progenitor cell. *Development* **105**: 387–400. doi:10.1242/dev.105.2.387

Woodruff RH, Fruttiger M, Richardson WD, Franklin RJM. 2004. Platelet-derived growth factor regulates oligodendrocyte progenitor numbers in adult CNS and their response following CNS demyelination. *Mol Cell Neurosci* **25**: 252–262. doi:10.1016/j.mcn.2003.10.014

Xin M, Yue T, Ma Z, Wu F, Gow A, Lu QR. 2005. Myelinogenesis and axonal recognition by oligodendrocytes in brain are uncoupled in olig1-null mice. *J Neurosci* **25**: 1354–1365. doi:10.1523/JNEUROSCI.3034-04.2005

Xu H, Wang B, Ono M, Kagita A, Fujii K, Sasakawa N, Ueda T, Gee P, Nishikawa M, Nomura M, et al. 2019. Targeted disruption of HLA genes via CRISPR-Cas9 generates iPSCs with enhanced immune compatibility. *Cell Stem Cell* **24**: 566–578.e7. doi:10.1016/j.stem.2019.02.005

Yamanaka S. 2007. Strategies and new developments in the generation of patient-specific pluripotent stem cells. *Cell Stem Cell* **1**: 39–49. doi:10.1016/j.stem.2007.05.012

Yamanaka S. 2008. Pluripotency and nuclear reprogramming. *Philos Trans R Soc B* **363**: 2079–2087. doi:10.1098/rstb.2008.2261

Yandava BD, Billinghurst LL, Snyder EY. 1999. "Global" cell replacement is feasible via neural stem cell transplantation: evidence from the dysmyelinated shiverer mouse brain. *Proc Natl Acad Sci* **96**: 7029–7034. doi:10.1073/pnas.96.12.7029

Yang N, Zuchero JB, Ahlenius H, Marro S, Ng YH, Vierbuchen T, Hawkins JS, Geissler R, Barres BA, Wernig M. 2013. Generation of oligodendroglial cells by direct lineage conversion. *Nat Biotechnol* **31**: 434–439. doi:10.1038/nbt.2564

Ye F, Chen Y, Hoang T, Montgomery RL, Zhao X, Bu H, Hu T, Taketo MM, van Es JH, Clevers H, et al. 2009. HDAC1 and HDAC2 regulate oligodendrocyte differentiation by disrupting the β-catenin–TCF interaction. *Nat Neurosci* **12**: 829–838. doi:10.1038/nn.2333

Yeung MSY, Djelloul M, Steiner E, Bernard S, Salehpour M, Possnert G, Brundin L, Frisén J. 2019. Dynamics of oligodendrocyte generation in multiple sclerosis. *Nature* **566**: 538–542. doi:10.1038/s41586-018-0842-3

Yilmaz EN, Albrecht S, Groll K, Thomas C, Wallhorn L, Herold M, Hucke S, Klotz L, Kuhlmann T. 2023. Influx of T cells into corpus callosum increases axonal injury, but does not change the course of remyelination in toxic demyelination. *Glia* **71**: 991–1001.

Zawadzka M, Rivers LE, Fancy SPJ, Zhao C, Tripathi R, Jamen F, Young K, Goncharevich A, Pohl H, Rizzi M, et al. 2010. CNS-resident glial progenitor/stem cells produce Schwann cells as well as oligodendrocytes during repair of CNS demyelination. *Cell Stem Cell* **6**: 578–590. doi:10.1016/j.stem.2010.04.002

Zhang SC, Wernig M, Duncan ID, Brüstle O, Thomson JA. 2001. In vitro differentiation of transplantable neural precursors from human embryonic stem cells. *Nat Biotechnol* **19**: 1129–1133. doi:10.1038/nbt1201-1129

Zhang Y, Argaw AT, Gurfein BT, Zameer A, Snyder BJ, Ge C, Lu QR, Rowitch DH, Raine CS, Brosnan CF, et al. 2009. Notch1 signaling plays a role in regulating precursor differentiation during CNS remyelination. *Proc Natl Acad Sci* **106**: 19162–19167. doi:10.1073/pnas.0902834106

Zhang M, Ni Y, Zhou Q, He L, Meng H, Gao Y, Huang X, Meng H, Li P, Chen M, et al. 2021. 18F-florbetapir PET/MRI for quantitatively monitoring myelin loss and recovery in patients with multiple sclerosis: a longitudinal study. *EClinicalMedicine* **37**: 100982. doi:10.1016/j.eclinm.2021.100982

Glial Malignancies

Suzanne J. Baker,[1] Hui Zong,[2] and Michelle Monje[3]

[1]Center of Excellence in Neuro-Oncology Sciences, Department of Developmental Neurobiology, St. Jude Children's Research Hospital, Memphis, Tennessee 38105-3678, USA

[2]Department of Microbiology, Immunology, and Cancer Biology, University of Virginia, Charlottesville, Virginia 22908, USA

[3]Department of Neurology and Neurological Sciences and Howard Hughes Medical Institute, Stanford University, Palo Alto, California 94305, USA

Correspondence: mmonje@stanford.edu

Gliomas comprise a diverse spectrum of related tumor subtypes with varying biological and molecular features and clinical outcomes. Advances in detailed genetic and epigenetic characterizations along with an appreciation that subtypes associated with developmental origins, including brain location and patient age, have shifted glioma classification from the historical reliance on histopathological features to updated categories incorporating molecular signatures and spatiotemporal incidence. Within a subtype, individual gliomas show cellular heterogeneity, generally containing subpopulations resembling different types of normal glial and progenitor cells. In addition to tumor-autonomous mechanisms of aberrant growth regulation driven by genetic mutations and signaling between tumor cells, interactions with the tumor microenvironment, including neurons, astrocytes, oligodendrocyte precursor cells, and the immune microenvironment play important roles in driving glioma growth and influencing response to treatment. The emerging understanding of the complex contributions of normal brain to glioma growth represents new opportunities for therapeutic advances.

GLIOMAS

Gliomas are a heterogeneous family of primary brain cancers so named because of a resemblance to normal glial cells. Gliomas are classified by grade as low-grade (World Health Organization [WHO], grades 1 and 2) or high-grade (WHO, grades 3 and 4) (Louis et al. 2021) by the typical age at which a glioma subtype tends to occur (pediatric or adult), anatomical location (hemispheric or midline), and by characteristic oncogenic mutations (e.g., oncogenic

mutations in the genes encoding histone H3, H3K27M mutation) (Khuong-Quang et al. 2012; Schwartzentruber et al. 2012; Wu et al. 2012). Low-grade gliomas (LGGs) may be debilitating and in adults may progress to higher grade over time. High-grade gliomas (HGGs) are lethal in both children and adults, and effective therapy for HGGs has been elusive to date.

Glioma classification has evolved dramatically to leverage advances in molecular characterization of brain tumors. The WHO publishes criteria for classification and grading of tumors to provide a shared framework for research and clinical trials worldwide. In 2007, WHO classification of glioma tumor grade was based on tumor cell cytological features and mitotic index, along with microvascular proliferation and necrosis (Louis et al. 2007). In 2016, these guidelines incorporated molecular features in addition to histopathological features to distinguish between glioblastoma with or without a key mutation in genes encoding isocitrate dehydrogenase (IDH) 1 and 2, and to recognize H3 K27M mutant glioma as a distinct entity (Louis et al. 2016). The current guidelines in the 2021 WHO classification include major changes to classifications that incorporate molecular diagnostics along with histology (Louis et al. 2021). Additionally, DNA methylome profiling has emerged as a powerful and robust tool for brain tumor classification that groups tumors together based on their epigenetic profiles that likely represent signatures of cellular and developmental origins further modified by tumor-specific aberrations. This method has proven especially useful for brain tumors with unclear diagnoses based on histopathologic features (Capper et al. 2018). The resulting revised glioma classification guidelines include an increased number of glioma families, each with multiple subtypes to represent the heterogeneity of this group of brain tumors. The updated classification should provide more accurate diagnosis, facilitate greater insights into clinical trial development and interpretation, and improve the relevance of comparisons of incidence and clinical trial results between different studies worldwide.

The selective association of specific mutations with spatiotemporal glioma development highlights distinctions in glioma pathogenesis, with the most striking examples including "oncohistone" mutations. Histone H3 K27M mutations are found in ~80% of diffuse midline glioma, which arise predominantly in children, and H3.3 G34R/V mutations occur in more than 30% of adolescent hemispheric gliomas and in younger adult glioblastoma patients (Khuong-Quang et al. 2012; Schwartzentruber et al. 2012; Sturm et al. 2012; Wu et al. 2012). Diffuse hemispheric gliomas arising in infants also have distinct mutation signatures, with 80% of tumors containing a fusion gene that generates a ligand-independent activated version of the RTKs (receptor tyrosine kinases), NTRK, MET, ROS, or ALK in the context of an otherwise very low mutation burden (Wu et al. 2014; Guerreiro Stucklin et al. 2019; Clarke et al. 2020). These subtypes are now clearly subclassified as diffuse midline glioma, H3 K27-altered, diffuse hemispheric glioma, H3 G34-mutant, or infant-type hemispheric glioma. Such classifications are meaningful to understanding tumor biology and therapeutic vulnerability. Infant-type hemispheric glioma patients are also distinguished by significantly greater survival compared to other pediatric or adult diffuse HGG patients. Extremely encouraging responses of glioma patients with NTRK fusion genes suggest that this outcome can be further improved by NTRK inhibitor treatment and holds promise for RTK-selective inhibitors in infant gliomas with other RTK fusion genes (Cocco et al. 2018; Desai et al. 2022). Many glioma-associated mutations also impact tumor cell metabolism, which can be further modulated by signals within the tumor microenvironment. Metabolic aberrations can enhance tumor survival, and also serve as therapeutic vulnerabilities (Venneti and Thompson 2017; Bi et al. 2020; Chung et al. 2020; Golbourn et al. 2022).

While many gliomas were formerly referred to as "astrocytomas" due to subpopulations of tumor cells that express astrocytic makers such as GFAP, most evidence points to neural stem and precursor cells in the oligodendrocyte lineage (Liu et al. 2011, 2022; Monje et al. 2011; Funato et al. 2014; Galvao et al. 2014; Alcantara Llaguno et al. 2015; Nagaraja et al. 2017, 2019; Filbin et al. 2018; Wang et al. 2020; Jessa et al. 2022) as the cellular origins of most gliomas. An intriguing exception is the diffuse hemispheric glioma, H3 G34-mutant, that likely arises from interneuron precursor cells rather than glial precursor cells (Chen et al. 2020; Bressan et al. 2021; Funato et al. 2021). Oligodendrocyte precursor cells (OPCs) maintain sibling contacts to achieve homeostatic constraint: they form a tiled pattern in the brain, remain quiescent, and only enter cell cycle when neighboring OPCs either die or differentiate into oligodendrocytes (Hughes et al. 2013; also see Emery and Wood 2024; Hill et al.

Cite this article as *Cold Spring Harb Perspect Biol* doi: 10.1101/cshperspect.a041373

2024; Simons et al. 2024). It remains puzzling how gliomagenic mutant OPCs break this constraint to progress toward malignancy.

GLIOMA CELLULAR SUBPOPULATIONS MIMICKING GLIA

Gliomas are composed of cellular subpopulations that mimic normal glial cells and their precursors. Single-cell transcriptomics in HGGs reveal subpopulations of oligodendroglial-like cells and astrocyte-like cells (Tirosh et al. 2016; Filbin et al. 2018; Neftel et al. 2019). In adult hemispheric HGGs (wild-type for isocitrate dehydrogenase [IDH WT]), tumors are composed of oligodendroglial-like cells, astrocyte-like cells, neural stem–like cells, and cells in a mesenchymal state, with neural stem–like cells as the tumor-initiating, "cancer stem cell" population. The proportion of these cell states in a given tumor shift over time and in response to therapies (Neftel et al. 2019). In pediatric diffuse midline gliomas characterized by H3K27M mutation, OPC-like cells are the tumor-initiating, "cancer stem cell" population, and oligodendrocyte-like and astrocyte-like cells comprise the remaining cellular states in the tumor (Filbin et al. 2018; Liu et al. 2022). This cellular hierarchy varies in H3-K27M diffuse midline glioma from older patients, which contain subpopulations of mesenchymal-like cells not observed in pediatric patients (Liu et al. 2022). One of the major effects of H3 K27M as an epigenetic oncogene is aberrant regulation of cell state to stall or reduce glial differentiation (Funato et al. 2014; Jessa et al. 2019; Larson et al. 2019). Removing H3 K27M from patient-derived DMG tumor cells delayed or completely abrogated tumor growth, and was associated with increased glial differentiation of tumor cells enriching for signatures of astrocytes or oligodendrocytes, with different propensities for glial subtypes in different DMG tumors (Harutyunyan et al. 2019; Silveira et al. 2019).

GLIOMA ARCHITECTURE AND PHYSIOLOGY

Malignant glioma cells engage with and modulate neural circuitry, much as their normal glial counterparts do. Just as synapses form between neurons and healthy OPCs (Bergles et al. 2000), bona fide synapses also form between neurons and malignant glioma cells (Venkataramani et al. 2019, 2022; Venkatesh et al. 2019). Such neuron-to-glioma synapses are enriched in glioma cell substrates most closely resembling OPCs (Venkatesh et al. 2019). Astrocyte-like glioma cells form a gap junction–coupled network (Osswald et al. 2015), much like healthy astrocytes do. This glioma-to-glioma cell gap junction–coupled network also couples to nonmalignant astrocytes through gap junction–mediated connections (Venkataramani et al. 2022). In these ways, glioma cell networks integrate into neuronal circuitry synaptically and into astrocytic networks through gap junctions (Venkataramani et al. 2019, 2022; Venkatesh et al. 2019). Reciprocally, astrocyte-like glioma cells secrete synaptogenic factors such as glypican-3 and thrombospondin-1, contributing to glioma-induced neuronal hyperexcitability that further augments the growth-promoting effects of neuron–glioma interactions (Fig. 1; John Lin et al. 2017; Yu et al. 2020; Krishna et al. 2023).

This connectivity of the tumor is crucially important for glioma pathophysiology. Synaptic communication from neurons to glioma cells promotes glioma cell proliferation and invasion (Venkataramani et al. 2019, 2022; Venkatesh et al. 2019). Gap junction–coupling between glioma cells facilitates synchronous calcium transients within the tumor (Venkataramani et al. 2019, 2022; Venkatesh et al. 2019), which appear to be very functionally important to tumor growth (Venkataramani et al. 2019, 2022; Venkatesh et al. 2019; Hausmann et al. 2023). Such connectivity also promotes therapeutic resistance (Osswald et al. 2015), and coupling to normal astrocytes appears to provide metabolic support through mitochondrial transfer as noted above (Watson et al. 2023). Taken together, such malignant glioma hijacking of normal glial mechanisms enables integration of the tumor into the brain structures it invades. Intraoperative electrophysiological studies have demonstrated that the degree to which the tumor is functionally connected to the rest of the brain is strongly inversely correlated with overall patient survival (Krishna

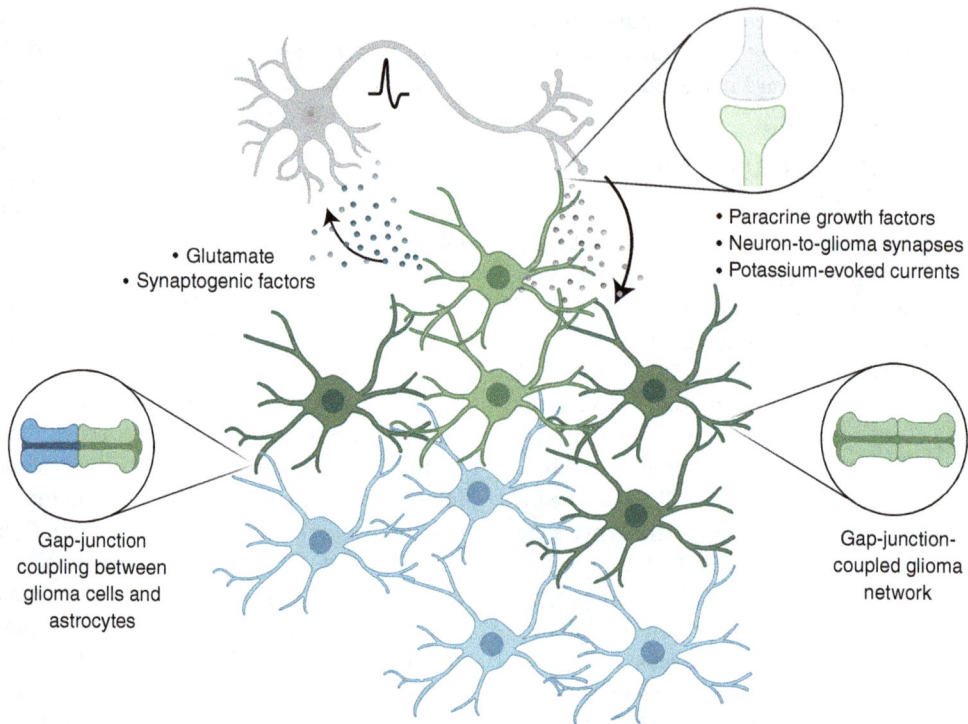

Figure 1. Neuron–glioma interactions. Neuronal (gray) activity drives glioma growth and invasion through paracrine factors, neuron-to-glioma synapses, and activity-regulated, potassium-evoked currents (Venkatesh et al. 2015, 2019; Pan et al. 2021; Venkataramani et al. 2022). This results in membrane depolarization and consequent calcium transients in glioma cells (green) that are amplified and propagated in a glioma-to-glioma cell gap junction–coupled network (Venkataramani et al. 2019; Venkatesh et al. 2019). This network on glioma cells also connects via gap junctions to nonmalignant astrocytes (blue) (Venkataramani et al. 2022). (Figure created with BioRender.)

et al. 2023). Therapeutic targeting of neuron-to-glioma synapses with medications that inhibit AMPARs, synaptogenic factors, or gap junctions shows promise in preclinical models (Venkataramani et al. 2019; Venkatesh et al. 2019; Krishna et al. 2023).

GLIOMA INTERACTIONS WITH THE BRAIN MICROENVIRONMENT

Clinical features of gliomas—including the spatiotemporal and spatiomolecular patterns of pediatric gliomagenesis, the strict CNS specificity of glioma metastases, and the diffusely infiltrating nature of pediatric gliomas—have long suggested a central role for the tumor microenvironment in disease pathogenesis. Recent studies have begun

to demonstrate the importance of interactions between glioma cells and other cell types of the brain that often mirror normal neuron–glial and glial–glial interactions (Fig. 2). Understanding these microenvironmental interactions elucidates key signaling mechanisms by which neural precursor cells (NPCs), neurons, and glia each contribute to glioma progression.

Neurons

Neuronal activity powerfully modulates neurodevelopment and ongoing brain plasticity. While the influence of activity on neuronal development and connectivity has been recognized for some time, a recently recognized dimension of activity-dependent oligodendroglial development and plasticity also contributes to the dynamic adapt-

Cite this article as *Cold Spring Harb Perspect Biol* doi: 10.1101/cshperspect.a041373

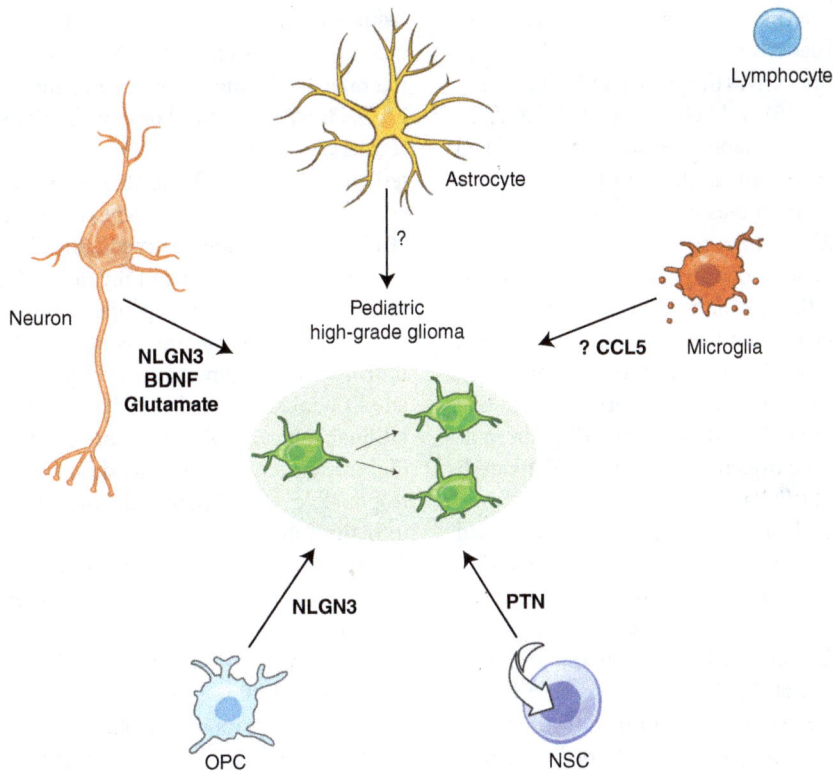

Figure 2. The high-grade glioma (HGG) microenvironment. HGGs are chiefly diffusely infiltrating malignancies that interact with numerous neural cell types present in the microenvironment. Neuronal activity robustly regulates the proliferation of glioma cells (green), through activity-regulated secretion of brain-derived neurotrophic factor (BDNF) and of neuroligin-3 (NLGN3) from both neurons (orange) and oligodendrocyte precursor cells (OPCs, light blue) (Venkatesh et al. 2015, 2017). The role that astrocytes (yellow) play in pediatric HGG (pHGG) pathophysiology remains to be determined. Neural stem cells (NSCs, purple) of the subventricular zone (SVZ) stem cell niche secrete pleiotrophin, which forms a complex with NSCs-derived HSP90, SPARC/SPARCL1, and functions as a chemoattractant factor, promoting pHGG invasion of the SVZ niche (Qin et al. 2017). Glioma-associated microglia (red), while present in substantial numbers in the pHGG microenvironment, express an intermediate, noninflammatory state of reactivity distinct from glioma-associated microglia in adult HGG (aHGG) (Lieberman et al. 2019). The degree to which glioma-associated microglia promote tumor progression is yet to be determined for pHGG, although the important roles glioma-associated microglia play in aHGG (Pyonteck et al. 2013) and pediatric low-grade glioma (pLGG) (Solga et al. 2015), where microglia-derived CCL5 regulates pLGG progression, highlight this as an important unanswered question. Consistent with the "immune cold" tumor microenvironment of pHGG (Lieberman et al. 2019), lymphocytes (royal blue) are largely absent from the glioma microenvironment with the exception of "hypermutator" tumors (Mackay et al. 2018). (Figure illustration created with SciStories.)

ability of the vertebrate brain (discussed in Simons et al. 2024; for additional reviews, see Mount and Monje 2017; Knowles et al. 2022). Neuronal activity robustly stimulates the proliferation of normal OPCs and earlier neural precursors in the healthy brain (Gibson et al. 2014). This activity-regulated response of NPCs in the oligo-

dendroglial lineage is part of "adaptive myelination," a process through which experience modulates myelin structure and, thus, neural circuit function (for review, see Mount and Monje 2017). Malignant gliomas, including H3K27M-altered diffuse midline gliomas and hemispheric pediatric glioblastoma, respond similarly to their normal counterparts and proliferate in response to neuronal activity (Venkatesh et al. 2015). Neural activity thus drives the growth of HGGs (Venkatesh et al. 2015) and LGGs (Pan et al. 2021). In addition to the synaptic mechanisms described above, an important mechanism by which neuronal activity promotes pediatric high-grade glioma (pHGG) proliferation and overall growth is through the release of activity-regulated paracrine factors into the glioma microenvironment. Recent studies have identified two important glioma mitogens secreted in response to neuronal activity—the neurotrophin brain–derived neurotrophic factor (BDNF) and a synaptic adhesion molecule called neuroligin-3 (NLGN3) (Venkatesh et al. 2015, 2017; Pan et al. 2021).

While pediatric HGGs are molecularly and clinically distinct from adult HGGs (aHGGs), this response to neuronal activity-regulated secreted factors in general and to NLGN3 in particular is conserved across glioma subtypes (Venkatesh et al. 2015). Unexpectedly, NLGN3 appears to be required for glioma growth, as patient-derived xenografts of both pediatric and adult glioma fail to progress in the microenvironment of the NLGN3 knockout mouse brain (Venkatesh et al. 2017). While this surprising dependency is conserved across glioma subtypes, including pediatric glioblastoma (histone wild-type), H3K27M$^+$ diffuse midline glioma, adult glioblastoma (Venkatesh et al. 2017), and NF1-associated optic glioma (Pan et al. 2021), it does not extend to a patient-derived model of breast cancer brain metastasis (Venkatesh et al. 2017). Whether NLGN3 is an important factor in the microenvironment of ependymoma or other nonglial pediatric brain malignancies has not yet been studied.

NLGN3 exerts numerous effects on glioma cells. After binding to an as-of-yet unidentified glioma-binding partner, NLGN3 activates numerous oncogenic signaling pathways including focal adhesion kinase (FAK) and downstream PI3K-mTOR, SRC, and RAS pathways (Venkatesh et al. 2015, 2017). While stimulation of oncogenic signaling helps to explain the growth-promoting effect of this important microenvironmental molecule, it does not explain the striking dependency, which suggests that NLGN3

regulates a process fundamental to glioma progression. The effects of NLGN3 exposure in glioma offer intriguing clues. A prominent effect of NLGN3 exposure on glioma is feedforward upregulation of glioma NLGN3 expression (Venkatesh et al. 2015, 2017), together with a number of additional synapse-related genes (Venkatesh et al. 2017). The mechanisms mediating this striking dependency remain to be fully defined, but may elucidate important therapeutic opportunities.

NLGN3 cleavage and secretion into the tumor microenvironment is strictly activity regulated (Venkatesh et al. 2017). Careful computational, pharmacological, and genetic studies identified the ADAM10 sheddase as the enzyme responsible for NLGN3 cleavage and release into the tumor microenvironment. Excitingly, ADAM10 inhibitor therapy substantially reduces growth of pHGGs in patient-derived orthotopic xenograft models (Venkatesh et al. 2017) and of NF1-associated LGGs in genetically engineered mouse models (Pan et al. 2021).

In a mouse model of olfactory bulb HGG, IGF1 from olfactory neurons is critical for the progression of gliomas originating from OPCs (Chen et al. 2022). Like NLGN3, IGF1 stimulates Ras/MAPK and PI3K-mTOR pathways, suggesting convergent evolution of glioma cells that gain similar support from neurons through distinct factors. This also underscores the principle that different neuronal subtypes can provide different paracrine factor support for gliomas, indicating that therapeutic strategies to block such interactions may need to be somewhat region or circuit specific.

Oligodendroglial Lineage Cells

The activity-regulated nature of NLGN3 and known role at synapses suggests that neurons are an important source of secreted NLGN3. While this is the case (Venkatesh et al. 2017), conditional genetic mouse modeling experiments demonstrate that OPCs are also a major source of secreted NLGN3 (Venkatesh et al. 2017). OPCs are the only normal glial cell type to form bona fide synapses with neurons (Bergles et al. 2000; Káradóttir et al. 2005), and OPCs express very high levels of NLGN3 (Zhang et al.

Cite this article as *Cold Spring Harb Perspect Biol* doi: 10.1101/cshperspect.a041373

2014). While the role of neuron-OPC synapses in the normal brain remains to be fully defined, the activity-regulated release of this postsynaptic adhesion molecule from OPCs as a result of neuronal activity positions OPCs as important players in the glioma microenvironment.

Neural Stem Cells

The lateral ventricle subventricular zone (SVZ) neural stem cell niche is a frequent and consequential site of glioma spread; glioma colonization of this niche is strongly linked to tumor recurrence in adult glioblastoma. The same propensity for lateral ventricle SVZ spread holds true for pediatric gliomas, even those arising at anatomically distant sites such as the brainstem (Caretti et al. 2014). NPC:glioma cell communication underpins this propensity of both adult and pediatric gliomas to colonize the SVZ through secretion of chemoattractant signals toward which glioma cells home. Biochemical, proteomic, and functional analyses of SVZ NPC-secreted factors revealed the neurite outgrowth-promoting factor pleiotrophin, along with required binding partners SPARC/SPARCL1 and HSP90B, as key mediators of this chemoattractant effect (Qin et al. 2017).

Microglia, Macrophages, and Lymphocytes

At the same time, recent studies have underscored the glioma subtype–specific composition of the immune microenvironment and have begun to untangle the nature of immune-glioma interactions and the key differences between pediatric and adult glioma immune microenvironments. pHGGs are largely "immune cold," with minimal lymphocytic infiltrates and microglia in an intermediate, noninflammatory state of reactivity (Lin et al. 2018; Lieberman et al. 2019), except for hypermutator tumors and pleiomorphic xanthroastrocytomas (Mackay et al. 2018). Contributing to the stark differences observed in the immune microenvironment between adult and pediatric HGGs, patient-derived glioma cells from aHGGs secrete vastly higher levels of proinflammatory cytokines and chemokines than do pediatric HGGs (Lin et al. 2018). These observa-

tions have important implications for therapy. Response to the anti-VEGF agent Avastin, for example, correlates with the extent of CD8[+] T-cell infiltration (Mackay et al. 2018), a finding that should be viewed in light of the influence of Avastin not only on blood vessel angiogenesis but also meningeal lymphangiogenesis (Song et al. 2020). The paucity of PD-L1 expression in H3K27M[+] diffuse midline gliomas such as DIPG argues against use of checkpoint inhibitors (Lin et al. 2018; Lieberman et al. 2019) and rather for adoptive T-cell strategies such as CAR T-cell therapy (Mount et al. 2018; Majzner et al. 2022).

The extent to which glioma-associated microglia promote HGG progression and limit antitumor immunity remains to be fully understood, but microglia and other myeloid cells appear to play important roles in limiting antitumor immunity in adult glioblastoma models (Pyonteck et al. 2013). In models of pediatric LGGs, microglia are required to establish a microenvironment that supports glioma initiation. Models of pediatric optic pathway gliomagenesis occurring in association with the neurofibromatosis type I tumor predisposition syndrome requires Nf1 heterozygous microglia (Daginakatte and Gutmann 2007; Daginakatte et al. 2008; Pong et al. 2013), which promote gliomagenesis through Ccl5 secretion (Solga et al. 2015). In BRAF fusion[+] pilocytic astrocytoma, gliomagenesis depends upon recruitment of glioma-associated microglia through a Ccl2-Ccr2 glioma-to-microglia signaling axis (Chen et al. 2019).

Astrocytes

Astrocytes represent a large fraction of brain cells (von Bartheld et al. 2016), and yet the role of astrocytes in the pediatric glioma microenvironment is largely unexplored. Astrocytes are an important source of numerous growth factors and also express the neuroligins (Stogsdill et al. 2017), but whether astrocytes secrete NLGN3 or contribute other important growth factors to pediatric glioma cells remains to be explored. Astrocyte reactivity is regulated by the state of reactive microglia (Liddelow et al. 2017), so differential glioma-associated microglial cell states (Lin et al. 2018) may strongly influence the state and function of glioma-associated astrocytes. Astrocytic influences in pediatric glioma may also be region dependent, as astrocytes display prominent regional heterogeneity (Molofsky et al. 2014; John Lin et al. 2017). In adult gliomas, the role of glioma-associated astrocytes is an emerging area of interest (Brandao et al. 2019). Recent evidence indicates that astrocyte-like glioma cells form gap junction–coupled connections with nonmalignant astrocytes (Venkataramani et al. 2022), and these glioma–glial connections may allow transfer of mitochondria from astrocytes to glioma cells, supporting tumor cell self-renewal and proliferation (Watson et al. 2023).

CONCLUDING REMARKS

Recent advances in glioma research have deepened our understanding of the molecular mechanisms driving cell-autonomous features of glioma formation, including genetic, epigenetic, and metabolic alterations that are associated with specific glioma subtypes arising from distinct cellular and developmental origins. The glioma cell heterogeneity reflecting hierarchies of abnormal tumor cells recapitulating aspects of normal glial and progenitor cell types contributes to disease pathogenesis. Exciting new findings show that tumor cells integrate with the tumor microenvironment through signaling mechanisms underlying normal communication between healthy glia and the diverse cell types and signals in the brain microenvironment. An appreciation of this greater network highlights the need to approach glioma as an integrated holistic brain disease rather than an isolated entity. To translate these new findings into clinical benefit, focused efforts are needed to precisely target molecular aberrations, and to leverage knowledge of glial cell biology to target cellular vulnerabilities. Disruption of the glioma–brain microenvironment network represents an innovative new therapeutic opportunity. While exciting, the intricate and sophisticated complexity of the brain poses many challenges and will likely require a careful balance to achieve a therapeutic window to improve patient survival and quality of life. From the perspective of glial biology, these various and complex intercellular interactions elucidate opportunities for

targeting in new microenvironmental and immunotherapeutic therapeutic strategies that may prove central to improved outcomes for these intractable brain cancers.

ACKNOWLEDGMENTS

The authors acknowledge support from the National Cancer Institute (P01CA096832 to S.J.B.).

REFERENCES

*Reference is also in this subject collection.

Alcantara Llaguno SR, Wang Z, Sun D, Chen J, Xu J, Kim E, Hatanpaa KJ, Raisanen JM, Burns DK, Johnson JE, et al. 2015. Adult lineage-restricted CNS progenitors specify distinct glioblastoma subtypes. *Cancer Cell* **28:** 429–440. doi:10.1016/j.ccell.2015.09.007

Bergles DE, Roberts JD, Somogyi P, Jahr CE. 2000. Glutamatergic synapses on oligodendrocyte precursor cells in the hippocampus. *Nature* **405:** 187–191. doi:10.1038/35012083

Bi J, Chowdhry S, Wu S, Zhang W, Masui K, Mischel PS. 2020. Altered cellular metabolism in gliomas—an emerging landscape of actionable co-dependency targets. *Nat Rev Cancer* **20:** 57–70. doi:10.1038/s41568-019-0226-5

Brandao M, Simon T, Critchley G, Giamas G. 2019. Astrocytes, the rising stars of the glioblastoma microenvironment. *Glia* **67:** 779–790. doi:10.1002/glia.23520

Bressan RB, Southgate B, Ferguson KM, Blin C, Grant V, Alfazema N, Wills JC, Marques-Torrejon MA, Morrison GM, Ashmore J, et al. 2021. Regional identity of human neural stem cells determines oncogenic responses to histone H3.3 mutants. *Cell Stem Cell* **28:** 877–893.e9. doi:10.1016/j.stem.2021.01.016

Capper D, Jones DTW, Sill M, Hovestadt V, Schrimpf D, Sturm D, Koelsche C, Sahm F, Chavez L, Reuss DE, et al. 2018. DNA methylation-based classification of central nervous system tumours. *Nature* **555:** 469–474. doi:10.1038/nature26000

Caretti V, Bugiani M, Freret M, Schellen P, Jansen M, van Vuurden D, Kaspers G, Fisher PG, Hulleman E, Wesseling P, et al. 2014. Subventricular spread of diffuse intrinsic pontine glioma. *Acta Neuropathol* **128:** 605–607. doi:10.1007/s00401-014-1307-x

Chen R, Keoni C, Waker CA, Lober RM, Chen YH, Gutmann DH. 2019. KIAA1549-BRAF expression establishes a permissive tumor microenvironment through NFκB-mediated CCL2 production. *Neoplasia* **21:** 52–60. doi:10.1016/j.neo.2018.11.007

Chen CCL, Deshmukh S, Jessa S, Hadjadj D, Lisi V, Andrade AF, Faury D, Jawhar W, Dali R, Suzuki H, et al. 2020. Histone H3.3G34-mutant interneuron progenitors co-opt PDGFRA for gliomagenesis. *Cell* **183:** 1617–1633. e22. doi:10.1016/j.cell.2020.11.012

Chen P, Wang W, Liu R, Lyu J, Zhang L, Li B, Qiu B, Tian A, Jiang W, Ying H, et al. 2022. Olfactory sensory experience regulates gliomagenesis via neuronal IGF1. *Nature* **606:** 550–556. doi:10.1038/s41586-022-04719-9

Chung C, Sweha SR, Pratt D, Tamrazi B, Panwalkar P, Banda A, Bayliss J, Hawes D, Yang F, Lee HJ, et al. 2020. Integrated metabolic and epigenomic reprograming by H3K27M mutations in diffuse intrinsic pontine gliomas. *Cancer Cell* **38:** 334–349.e9. doi:10.1016/j.ccell.2020.07.008

Clarke M, Mackay A, Ismer B, Pickles JC, Tatevossian RG, Newman S, Bale TA, Stoler I, Izquierdo E, Temelso S, et al. 2020. Infant high-grade gliomas comprise multiple subgroups characterized by novel targetable gene fusions and favorable outcomes. *Cancer Discov* **10:** 942–963. doi:10.1158/2159-8290.CD-19-1030

Cocco E, Scaltriti M, Drilon A. 2018. NTRK fusion-positive cancers and TRK inhibitor therapy. *Nat Rev Clin Oncol* **15:** 731–747. doi:10.1038/s41571-018-0113-0

Daginakatte GC, Gutmann DH. 2007. Neurofibromatosis-1 (Nf1) heterozygous brain microglia elaborate paracrine factors that promote Nf1-deficient astrocyte and glioma growth. *Hum Mol Genet* **16:** 1098–1112. doi:10.1093/hmg/ddm059

Daginakatte GC, Gianino SM, Zhao NW, Parsadanian AS, Gutmann DH. 2008. Increased c-Jun-NH2-kinase signaling in neurofibromatosis-1 heterozygous microglia drives microglia activation and promotes optic glioma proliferation. *Cancer Res* **68:** 10358–10366. doi:10.1158/0008-5472.CAN-08-2506

Desai AV, Robinson GW, Gauvain K, Basu EM, Macy ME, Maese L, Whipple NS, Sabnis AJ, Foster JH, Shusterman S, et al. 2022. Entrectinib in children and young adults with solid or primary CNS tumors harboring *NTRK*, *ROS1*, or *ALK* aberrations (STARTRK-NG). *Neuro Oncol* **24:** 1776–1789. doi:10.1093/neuonc/noac087

* Emery B, Wood TL. 2024. Regulators of oligodendrocyte differentiation. *Cold Spring Harb Perspect Biol* **16:** a041358. doi:10.1101/cshperspect.a041358

Filbin MG, Tirosh I, Hovestadt V, Shaw ML, Escalante LE, Mathewson ND, Neftel C, Frank N, Pelton K, Hebert CM, et al. 2018. Developmental and oncogenic programs in H3K27M gliomas dissected by single-cell RNA-seq. *Science* **360:** 331–335. doi:10.1126/science.aao4750

Funato K, Major T, Lewis PW, Allis CD, Tabar V. 2014. Use of human embryonic stem cells to model pediatric gliomas with H3.3K27M histone mutation. *Science* **346:** 1529–1533. doi:10.1126/science.1253799

Funato K, Smith RC, Saito Y, Tabar V. 2021. Dissecting the impact of regional identity and the oncogenic role of human-specific NOTCH2NL in an hESC model of H3.3G34R-mutant glioma. *Cell Stem Cell* **28:** 894–905.e7. doi:10.1016/j.stem.2021.02.003

Galvao RP, Kasina A, McNeill RS, Harbin JE, Foreman O, Verhaak RG, Nishiyama A, Miller CR, Zong H. 2014. Transformation of quiescent adult oligodendrocyte precursor cells into malignant glioma through a multistep reactivation process. *Proc Natl Acad Sci* **111:** E4214–E4223. doi:10.1073/pnas.1414389111

Gibson EM, Purger D, Mount CW, Goldstein AK, Lin GL, Wood LS, Inema I, Miller SE, Bieri G, Zuchero JB, et al. 2014. Neuronal activity promotes oligodendrogenesis and adaptive myelination in the mammalian brain. *Science* **344:** 1252304. doi:10.1126/science.1252304

Golbourn BJ, Halbert ME, Halligan K, Varadharajan S, Krug B, Mbah NE, Kabir N, Stanton AJ, Locke AL, Casillo SM, et al. 2022. Loss of MAT2A compromises methionine metabolism and represents a vulnerability in H3K27M mutant glioma by modulating the epigenome. *Nat Cancer* **3:** 629–648. doi:10.1038/s43018-022-00348-3

Guerreiro Stucklin AS, Ryall S, Fukuoka K, Zapotocky M, Lassaletta A, Li C, Bridge T, Kim B, Arnoldo A, Kowalski PE, et al. 2019. Alterations in ALK/ROS1/NTRK/MET drive a group of infantile hemispheric gliomas. *Nat Commun* **10:** 4343. doi:10.1038/s41467-019-12187-5

Harutyunyan AS, Krug B, Chen H, Papillon-Cavanagh S, Zeinieh M, De Jay N, Deshmukh S, Chen CCL, Belle J, Mikael LG, et al. 2019. H3k27m induces defective chromatin spread of PRC2-mediated repressive H3K27me2/me3 and is essential for glioma tumorigenesis. *Nat Commun* **10:** 1262. doi:10.1038/s41467-019-09140-x

Hausmann D, Hoffmann DC, Venkataramani V, Jung E, Horschitz S, Tetzlaff SK, Jabali A, Hai L, Kessler T, Azorín DD, et al. 2023. Autonomous rhythmic activity in glioma networks drives brain tumour growth. *Nature* **613:** 179–186. doi:10.1038/s41586-022-05520-4

* Hill RA, Nishiyama A, Hughes EG. 2024. Features, fates, and functions of oligodendrocyte precursor cells. *Cold Spring Harb Perspect Biol* **16:** a041425. doi:10.1101/cshperspect.a041425

Hughes EG, Kang SH, Fukaya M, Bergles DE. 2013. Oligodendrocyte progenitors balance growth with self-repulsion to achieve homeostasis in the adult brain. *Nat Neurosci* **16:** 668–676. doi:10.1038/nn.3390

Jessa S, Blanchet-Cohen A, Krug B, Vladoiu M, Coutelier M, Faury D, Poreau B, De Jay N, Hébert S, Monlong J, et al. 2019. Stalled developmental programs at the root of pediatric brain tumors. *Nat Genet* **51:** 1702–1713. doi:10.1038/s41588-019-0531-7

Jessa S, Mohammadnia A, Harutyunyan AS, Hulswit M, Varadharajan S, Lakkis H, Kabir N, Bashardanesh Z, Hébert S, Faury D, et al. 2022. K27m in canonical and noncanonical H3 variants occurs in distinct oligodendroglial cell lineages in brain midline gliomas. *Nat Genet* **54:** 1865–1880. doi:10.1038/s41588-022-01205-w

John Lin CC, Yu K, Hatcher A, Huang TW, Lee HK, Carlson J, Weston MC, Chen F, Zhang Y, Zhu W, et al. 2017. Identification of diverse astrocyte populations and their malignant analogs. *Nat Neurosci* **20:** 396–405. doi:10.1038/nn.4493

Káradóttir R, Cavelier P, Bergersen LH, Attwell D. 2005. NMDA receptors are expressed in oligodendrocytes and activated in ischaemia. *Nature* **438:** 1162–1166. doi:10.1038/nature04302

Khuong-Quang DA, Buczkowicz P, Rakopoulos P, Liu XY, Fontebasso AM, Bouffet E, Bartels U, Albrecht S, Schwartzentruber J, Letourneau L, et al. 2012. K27m mutation in histone H3.3 defines clinically and biologically distinct subgroups of pediatric diffuse intrinsic pontine gliomas. *Acta Neuropathol* **124:** 439–447. doi:10.1007/s00401-012-0998-0

Knowles JK, Batra A, Xu H, Monje M. 2022. Adaptive and maladaptive myelination in health and disease. *Nat Rev Neurol* **18:** 735–746. doi:10.1038/s41582-022-00737-3

Krishna S, Choudhury A, Keough MB, Seo K, Ni L, Kakaizada S, Lee A, Aabedi A, Popova G, Lipkin B, et al. 2023.

Glioblastoma remodelling of human neural circuits decreases survival. *Nature* **617**: 599–607. doi:10.1038/s41586-023-06036-1

Larson JD, Kasper LH, Paugh BS, Jin H, Wu G, Kwon CH, Fan Y, Shaw TI, Silveira AB, Qu C, et al. 2019. Histone H3.3 K27M accelerates spontaneous brainstem glioma and drives restricted changes in bivalent gene expression. *Cancer Cell* **35**: 140–155.e7. doi:10.1016/j.ccell.2018.11.015

Liddelow SA, Guttenplan KA, Clarke LE, Bennett FC, Bohlen CJ, Schirmer L, Bennett ML, Münch AE, Chung WS, Peterson TC, et al. 2017. Neurotoxic reactive astrocytes are induced by activated microglia. *Nature* **541**: 481–487. doi:10.1038/nature21029

Lieberman NAP, DeGolier K, Kovar HM, Davis A, Hoglund V, Stevens J, Winter C, Deutsch G, Furlan SN, Vitanza NA, et al. 2019. Characterization of the immune microenvironment of diffuse intrinsic pontine glioma: implications for development of immunotherapy. *Neuro Oncol* **21**: 83–94. doi:10.1093/neuonc/noy145

Lin GL, Nagaraja S, Filbin MG, Suvà ML, Vogel H, Monje M. 2018. Non-inflammatory tumor microenvironment of diffuse intrinsic pontine glioma. *Acta Neuropathol Commun* **6**: 51. doi:10.1186/s40478-018-0553-x

Liu C, Sage JC, Miller MR, Verhaak RG, Hippenmeyer S, Vogel H, Foreman O, Bronson RT, Nishiyama A, Luo L, et al. 2011. Mosaic analysis with double markers reveals tumor cell of origin in glioma. *Cell* **146**: 209–221. doi:10.1016/j.cell.2011.06.014

Liu I, Jiang L, Samuelsson ER, Marco Salas S, Beck A, Hack OA, Jeong D, Shaw ML, Englinger B, LaBelle J, et al. 2022. The landscape of tumor cell states and spatial organization in H3-K27M mutant diffuse midline glioma across age and location. *Nat Genet* **54**: 1881–1894. doi:10.1038/s41588-022-01236-3

Louis DN, Ohgaki H, Wiestler OD, Cavenee WK, Burger PC, Jouvet A, Scheithauer BW, Kleihues P. 2007. The 2007 WHO classification of tumours of the central nervous system. *Acta Neuropathol* **114**: 97–109. doi:10.1007/s00401-007-0243-4

Louis DN, Perry A, Reifenberger G, von Deimling A, Figarella-Branger D, Cavenee WK, Ohgaki H, Wiestler OD, Kleihues P, Ellison DW. 2016. The 2016 World Health Organization classification of tumors of the central nervous system: a summary. *Acta Neuropathol* **131**: 803–820. doi:10.1007/s00401-016-1545-1

Louis DN, Perry A, Wesseling P, Brat DJ, Cree IA, Figarella-Branger D, Hawkins C, Ng HK, Pfister SM, Reifenberger G, et al. 2021. The 2021 WHO classification of tumors of the central nervous system: a summary. *Neuro Oncol* **23**: 1231–1251. doi:10.1093/neuonc/noab106

Mackay A, Burford A, Molinari V, Jones DTW, Izquierdo E, Brouwer-Visser J, Giangaspero F, Haberler C, Pietsch T, Jacques TS, et al. 2018. Molecular, pathological, radiological, and immune profiling of non-brainstem pediatric high-grade glioma from the HERBY phase II randomized trial. *Cancer Cell* **33**: 829–842.e5. doi:10.1016/j.ccell.2018.04.004

Majzner RG, Ramakrishna S, Yeom KW, Patel S, Chinnasamy H, Schultz LM, Richards RM, Jiang L, Barsan V, Mancusi R, et al. 2022. GD2-CAR T cell therapy for H3K27M-mutated diffuse midline gliomas. *Nature* **603**: 934–941. doi:10.1038/s41586-022-04489-4

Molofsky AV, Kelley KW, Tsai HH, Redmond SA, Chang SM, Madireddy L, Chan JR, Baranzini SE, Ullian EM, Rowitch DH. 2014. Astrocyte-encoded positional cues maintain sensorimotor circuit integrity. *Nature* **509**: 189–194. doi:10.1038/nature13161

Monje M, Mitra SS, Freret ME, Raveh TB, Kim J, Masek M, Attema JL, Li G, Haddix T, Edwards MS, et al. 2011. Hedgehog-responsive candidate cell of origin for diffuse intrinsic pontine glioma. *Proc Natl Acad Sci* **108**: 4453–4458. doi:10.1073/pnas.1101657108

Mount CW, Monje M. 2017. Wrapped to adapt: experience-dependent myelination. *Neuron* **95**: 743–756. doi:10.1016/j.neuron.2017.07.009

Mount CW, Majzner RG, Sundaresh S, Arnold EP, Kadapakkam M, Haile S, Labanieh L, Hulleman E, Woo PJ, Rietberg SP, et al. 2018. Potent antitumor efficacy of anti-GD2 CAR T cells in H3-K27M$^+$ diffuse midline gliomas. *Nat Med* **24**: 572–579. doi:10.1038/s41591-018-0006-x

Nagaraja S, Vitanza NA, Woo PJ, Taylor KR, Liu F, Zhang L, Li M, Meng W, Ponnuswami A, Sun W, et al. 2017. Transcriptional dependencies in diffuse intrinsic pontine glioma. *Cancer Cell* **31**: 635–652.e6. doi:10.1016/j.ccell.2017.03.011

Nagaraja S, Quezada MA, Gillespie SM, Arzt M, Lennon JJ, Woo PJ, Hovestadt V, Kambhampati M, Filbin MG, Suva ML, et al. 2019. Histone variant and cell context determine H3K27M reprogramming of the enhancer landscape and oncogenic state. *Mol Cell* **76**: 965–980.e12. doi:10.1016/j.molcel.2019.08.030

Neftel C, Laffy J, Filbin MG, Hara T, Shore ME, Rahme GJ, Richman AR, Silverbush D, Shaw ML, Hebert CM, et al. 2019. An integrative model of cellular states, plasticity, and genetics for glioblastoma. *Cell* **178**: 835–849.e21. doi:10.1016/j.cell.2019.06.024

Osswald M, Jung E, Sahm F, Solecki G, Venkataramani V, Blaes J, Weil S, Horstmann H, Wiestler B, Syed M, et al. 2015. Brain tumour cells interconnect to a functional and resistant network. *Nature* **528**: 93–98. doi:10.1038/nature16071

Pan Y, Hysinger JD, Barron T, Schindler NF, Cobb O, Guo X, Yalçın B, Anastasaki C, Mulinyawe SB, Ponnuswami A, et al. 2021. NF1 mutation drives neuronal activity-dependent initiation of optic glioma. *Nature* **594**: 277–282. doi:10.1038/s41586-021-03580-6

Pong WW, Higer SB, Gianino SM, Emnett RJ, Gutmann DH. 2013. Reduced microglial CX3CR1 expression delays neurofibromatosis-1 glioma formation. *Ann Neurol* **73**: 303–308. doi:10.1002/ana.23813

Pyonteck SM, Akkari L, Schuhmacher AJ, Bowman RL, Sevenich L, Quail DF, Olson OC, Quick ML, Huse JT, Teijeiro V, et al. 2013. CSF-1R inhibition alters macrophage polarization and blocks glioma progression. *Nat Med* **19**: 1264–1272. doi:10.1038/nm.3337

Qin EY, Cooper DD, Abbott KL, Lennon J, Nagaraja S, Mackay A, Jones C, Vogel H, Jackson PK, Monje M. 2017. Neural precursor-derived pleiotrophin mediates subventricular zone invasion by glioma. *Cell* **170**: 845–859.e19. doi:10.1016/j.cell.2017.07.016

Schwartzentruber J, Korshunov A, Liu XY, Jones DT, Pfaff E, Jacob K, Sturm D, Fontebasso AM, Quang DA, Tönjes M, et al. 2012. Driver mutations in histone H3.3 and chromatin remodelling genes in paediatric glioblastoma. *Nature* 482: 226–231. doi:10.1038/nature10833

Silveira AB, Kasper LH, Fan Y, Jin H, Wu G, Shaw TI, Zhu X, Larson JD, Easton J, Shao Y, et al. 2019. H3.3 K27M depletion increases differentiation and extends latency of diffuse intrinsic pontine glioma growth in vivo. *Acta Neuropathol* 137: 637–655. doi:10.1007/s00401-019-01975-4

* Simons M, Gibson EM, Nave K-A. 2024. Oligodendrocytes: myelination, plasticity, and axonal support. *Cold Spring Harb Perspect Biol* doi:10.1101/cshperspect.a041359

Solga AC, Pong WW, Kim KY, Cimino PJ, Toonen JA, Walker J, Wylie T, Magrini V, Griffith M, Griffith OL, et al. 2015. RNA sequencing of tumor-associated microglia reveals Ccl5 as a stromal chemokine critical for neurofibromatosis-1 glioma growth. *Neoplasia* 17: 776–788. doi:10.1016/j.neo.2015.10.002

Song E, Mao T, Dong H, Boisserand LSB, Antila S, Bosenberg M, Alitalo K, Thomas JL, Iwasaki A. 2020. VEGF-C-driven lymphatic drainage enables immunosurveillance of brain tumours. *Nature* 577: 689–694. doi:10.1038/s41586-019-1912-x

Stogsdill JA, Ramirez J, Liu D, Kim YH, Baldwin KT, Enustun E, Ejikeme T, Ji RR, Eroglu C. 2017. Astrocytic neuroligins control astrocyte morphogenesis and synaptogenesis. *Nature* 551: 192–197. doi:10.1038/nature24638

Sturm D, Witt H, Hovestadt V, Khuong-Quang DA, Jones DT, Konermann C, Pfaff E, Tönjes M, Sill M, Bender S, et al. 2012. Hotspot mutations in H3F3A and IDH1 define distinct epigenetic and biological subgroups of glioblastoma. *Cancer Cell* 22: 425–437. doi:10.1016/j.ccr.2012.08.024

Tirosh I, Venteicher AS, Hebert C, Escalante LE, Patel AP, Yizhak K, Fisher JM, Rodman C, Mount C, Filbin MG, et al. 2016. Single-cell RNA-seq supports a developmental hierarchy in human oligodendroglioma. *Nature* 539: 309–313. doi:10.1038/nature20123

Venkataramani V, Tanev DI, Strahle C, Studier-Fischer A, Fankhauser L, Kessler T, Körber C, Kardorff M, Ratliff M, Xie R, et al. 2019. Glutamatergic synaptic input to glioma cells drives brain tumour progression. *Nature* 573: 532–538. doi:10.1038/s41586-019-1564-x

Venkataramani V, Yang Y, Schubert MC, Reyhan E, Tetzlaff SK, Wißmann N, Botz M, Soyka SJ, Beretta CA, Pramatarov RL, et al. 2022. Glioblastoma hijacks neuronal mechanisms for brain invasion. *Cell* 185: 2899–2917.e31. doi:10.1016/j.cell.2022.06.054

Venkatesh HS, Johung TB, Caretti V, Noll A, Tang Y, Nagaraja S, Gibson EM, Mount CW, Polepalli J, Mitra SS, et al. 2015. Neuronal activity promotes glioma growth through neuroligin-3 secretion. *Cell* 161: 803–816. doi:10.1016/j.cell.2015.04.012

Venkatesh HS, Tam LT, Woo PJ, Lennon J, Nagaraja S, Gillespie SM, Ni J, Duveau DY, Morris PJ, Zhao JJ, et al. 2017. Targeting neuronal activity-regulated neuroligin-3 dependency in high-grade glioma. *Nature* 549: 533–537. doi:10.1038/nature24014

Venkatesh HS, Morishita W, Geraghty AC, Silverbush D, Gillespie SM, Arzt M, Tam LT, Espenel C, Ponnuswami A, Ni L, et al. 2019. Electrical and synaptic integration of glioma into neural circuits. *Nature* 573: 539–545. doi:10.1038/s41586-019-1563-y

Venneti S, Thompson CB. 2017. Metabolic reprogramming in brain tumors. *Annu Rev Pathol* 12: 515–545. doi:10.1146/annurev-pathol-012615-044329

von Bartheld CS, Bahney J, Herculano-Houzel S. 2016. The search for true numbers of neurons and glial cells in the human brain: a review of 150 years of cell counting. *J Comp Neurol* 524: 3865–3895. doi:10.1002/cne.24040

Wang Z, Sun D, Chen YJ, Xie X, Shi Y, Tabar V, Brennan CW, Bale TA, Jayewickreme CD, Laks DR, et al. 2020. Cell lineage-based stratification for glioblastoma. *Cancer Cell* 38: 366–379.e8. doi:10.1016/j.ccell.2020.06.003

Watson DC, Bayik D, Storevik S, Moreino SS, Sprowls SA, Han J, Augustsson MT, Lauko A, Sravya P, Røsland GV, et al. 2023. GAP43-dependent mitochondria transfer from astrocytes enhances glioblastoma tumorigenicity. *Nat Cancer* 4: 648–664. doi:10.1038/s43018-023-00556-5

Wu G, Broniscer A, McEachron TA, Lu C, Paugh BS, Becksfort J, Qu C, Ding L, Huether R, Parker M, et al. 2012. Somatic histone H3 alterations in pediatric diffuse intrinsic pontine gliomas and non-brainstem glioblastomas. *Nat Genet* 44: 251–253. doi:10.1038/ng.1102

Wu G, Diaz AK, Paugh BS, Rankin SL, Ju B, Li Y, Zhu X, Qu C, Chen X, Zhang J, et al. 2014. The genomic landscape of diffuse intrinsic pontine glioma and pediatric non-brainstem high-grade glioma. *Nat Genet* 46: 444–450. doi:10.1038/ng.2938

Yu K, Lin CJ, Hatcher A, Lozzi B, Kong K, Huang-Hobbs E, Cheng YT, Beechar VB, Zhu W, Zhang Y, et al. 2020. PIK3CA variants selectively initiate brain hyperactivity during gliomagenesis. *Nature* 578: 166–171. doi:10.1038/s41586-020-1952-2

Zhang Y, Chen K, Sloan SA, Bennett ML, Scholze AR, O'Keeffe S, Phatnani HP, Guarnieri P, Caneda C, Ruderisch N, et al. 2014. An RNA-sequencing transcriptome and splicing database of glia, neurons, and vascular cells of the cerebral cortex. *J Neurosci* 34: 11929–11947. doi:10.1523/JNEUROSCI.1860-14.2014

Glia in Neurodegenerative Disease

Gerard Crowley,[1] David Attwell,[2] Hemali Phatnani,[3,4] Harald Sontheimer,[5] and Soyon Hong[1]

[1]UK Dementia Research Institute, Institute of Neurology, University College London, London W1T 7NF, United Kingdom

[2]Department of Neuroscience, Physiology & Pharmacology, University College London, London WC1N 3AZ, United Kingdom

[3]Department of Neurology, Columbia University Medical Center, New York, New York 10032, USA

[4]Center for Genomics of Neurodegenerative Disease, New York Genome Center, New York, New York 10013, USA

[5]Department of Neuroscience, School of Medicine, University of Virginia, Charlottesville, Virginia 22903, USA

Correspondence: soyon.hong@ucl.ac.uk

It is becoming increasingly clear that the dominant, century-old neurocentric view of neurodegeneration is insufficient to explain why certain neurons degenerate, in particular with aging. Genetic studies in patient populations as well as mechanistic and functional studies in animal models altogether implicate nonneuronal cells, especially glia, to play more than bystander roles in neurodegeneration. Throughout the life span, neuronal function and homeostasis are modulated by glia, the functions of which become even more critical with aging. This review highlights key emerging concepts of the role of glia in neurodegeneration.

Neurodegenerative diseases are characterized by region-specific and selective dysfunction and degeneration of neurons, which are associated with pathological buildup of protein aggregates. The process of neurodegeneration had been assumed to be largely a neuron-intrinsic process. With further research, however, came the understanding that such a "cell-autonomous" explanation is often insufficient. Rather, other brain cells, including astrocytes and microglia, can contribute substantially to disease etiology, as can decreases of cerebral blood flow. Almost all neurodegenerative diseases present with changes in glial morphology (reactive gliosis/microgliosis), function, and gene expression. In many examples, such glial changes contribute to disease presentation and severity, and, in a few instances, glial cells have even been shown to cause the disease. This distinction is important as future glia-targeted therapies would pursue different objectives based on their relative contribution to the disease. Given that our knowledge of non-cell-autonomous mechanisms of neurodegenerative disease involving glia is still evolving, we have elected to show a few select examples where a glial contribution to disease is well substantiated, and the underlying mechanism(s) reasonably well understood. We begin with the critical concept of glial heterogeneity, which is crucial when considering which mechanisms and pathways go awry in disease. Then, we delve into various glial and immune cell functions and contribution to disease susceptibility, in particular their influence on neuronal function and homeostasis.

GLIAL HETEROGENEITY

Based on their morphology and function, glial cells in the central nervous system (CNS) can be broadly divided into four types: astrocytes, oligodendrocyte lineage cells, microglia, and ependymal cells (see Herculano-Houzel 2009 for a review on the human brain in numbers). Recent technological advances have made it possible to further differentiate between subpopulations of glial cells depending on regional localization and gene expression, enabling more detailed investigations into finer function and correlation with disease susceptibility. Here, we focus on how these methods have facilitated the characterization of astrocytes, oligodendrocyte lineage cells, and microglia, how subpopulations of these cell types have been described, and how these subpopulations relate to disease susceptibility. We propose that a common understanding of glial subtypes and substates will be required to appropriately interrogate the role of glia in neurodegenerative disease.

How Do We Find and Describe Heterogeneity?

A range of novel technologies has emerged in the last few years that allows us to catalog cellular states with spatial and single-cell (sc) resolution, accelerating the discovery of neural cell heterogeneity (for reviews, see Ochocka and Kaminska 2021; Sun et al. 2022). Spatial resolution is enabled by spatial transcriptomics methods that are based either on sequencing or imaging; each method has its advantages and disadvantages (reviewed in Maniatis et al. 2021). Cellular resolution is obtained with sc-RNAseq (Tang et al. 2009), single-nucleus (sn)-RNAseq (Lake et al. 2016), and sc-proteomics (Wilson and Nairn 2018). Together, such technologies provide insights into cell heterogeneity based on differences in the transcriptome; further data analysis then divides the classes of glial cells into distinct subclusters characterized by up-regulation of molecular marker genes (for review, see Sun et al. 2022).

For example, based on extensive sc-RNAseq data from mice, astrocytes have been divided into seven subclusters (Endo et al. 2022), microglia into nine subclusters (Hammond et al. 2019), and oligodendrocyte lineage cells have been divided into 13 subclusters (including six mature oligodendrocyte subtypes as well as progenitor, precursor, myelin forming, and newly formed oligodendrocytes) (Marques et al. 2016). This has been further validated by sn-RNAseq analyses on human dorsolateral prefrontal cortex samples that similarly identified seven astrocytes, 13 oligodendrocytes, and nine microglia subtypes (Fujita et al. 2024). Such subclusters—defined on the basis of sc or sn transcriptional profiles—can be further divided according to spatial localization and age with the application of spatially resolved methods (e.g., Maniatis et al. 2019). One caveat associated with using sc transcriptomic workflows to assign cell-type-specific subclusters is that this process is necessarily subjective, which may lead to "over-/underclustering." Nonetheless, an appreciation of such heterogeneity highlights the need for in-depth functional analyses and suggests new directions for further exploration (for review, see Paolicelli et al. 2022), but it is important to note that changes of protein expression may not always be detectable at the mRNA level (Caldwell et al. 2022; Johnson et al. 2022).

Intrinsic and Extrinsic Factors Determining Heterogeneity

Both internal (developmental in origin) as well as external (derived from the environment) cues combine to determine the transcriptional heterogeneity of glia; depending on the type of glial cell, these cues may carry different weights. For example, microglia heterogeneity does not appear until after distribution to specific brain regions early in development, implicating extrinsic factors (for review, see Stratoulias et al. 2019); intriguingly, the nine microglia subpopulations identified have been shown to decrease in heterogeneity over the life span (Hammond et al. 2019). However, certain subclusters of microglia were found to be significantly enriched during discrete developmental stages, including *Ms4a7*-expressing microglia detected in the embryonic mouse brain and *Spp1*-expressing axon tract-associated microglia densely occupying the early postnatal brain (Hammond et al. 2019). Masuda et al. (2019) corroborated the finding that microglial heterogeneity is greatest during development and becomes more limited upon progression into adulthood. This study identified distinct microglial signatures associated with demyelination and remyelination, respectively, emphasizing the strong influence of local molecular cues on microglial substates (Masuda et al. 2019). These distinct signatures were also detected in humans with multiple sclerosis (MS), underlining the important point that subclusters should be validated across species to maximize physiological relevance and therapeutic potential in disease contexts.

Unlike microglia, astrocytes and oligodendrocytes stem from multiple developmental sources (for reviews, see Clavreul et al. 2022; Seeker and Williams 2022), but regional differences in astrocyte subpopulations are also likely to be dependent on the local environment—upstream regulators associated with astrocyte subclusters are related to cues derived from the local environment, such as neurotransmitters, growth factors, and cytokines (Endo et al. 2022). For oligodendrocytes, both intrinsic and extrinsic cues appear to be crucial—white matter oligodendrocyte progenitor cells (OPCs) produced more mature oligodendrocytes when transplanted to gray matter compared to gray matter OPCs; however, this was still at a slower rate compared to OPCs in white matter (Viganò et al. 2013).

What Are the Implications for Disease?

It is becoming increasingly clear that glia are heterogeneous and play diverse roles in health and disease. Insight into which functions of microglia change during aging and contribute to Alzheimer's disease (AD) risk will yield insight for therapeutic design (i.e., how and when to target microglia to change disease prognosis) (Lewcock et al. 2020). Besides the normal functions of glial cell heterogeneity, changes in homeostasis have been implicated in neurodegenerative diseases. It is important to note that neurodegenerative diseases are often associated with formation of protein aggregates, which also increase with aging. Glial cell subclusters that aid in aggregate clearance therefore tend to increase both in aging and neurodegenerative disease populations.

Alzheimer's Disease

A recent review has summarized the transcriptomics data generated from 13 studies comparing brain tissue from AD and control individuals (Luquez et al. 2022) showing significant changes in heterogeneity in all glial cell subtypes. Notably, in microglia, a switch from a homeostatic to a more activated profile is seen (Keren-Shaul et al. 2017; Grubman et al. 2019; Mathys et al. 2019), suggesting microglial dysregulation in AD. Throughout the life span, microglia display remarkable plasticity to adapt to various functional demands of their microenvironment. These heterogeneous functional states of microglia, which can be captured at the RNA level, vary across development (Keren-Shaul et al. 2017; Hammond et al. 2019; Li et al. 2019) and in distinct brain environments (Grabert et al. 2016; de Biase et al. 2017; Ayata et al. 2018; Masuda et al. 2019; Stogsdill et al. 2022). In neurodegeneration, microglial cell states are fundamentally altered, especially when surrounding pathological aggregates such as amyloid plaques (Chen et al. 2020). A microglial subtype associated with neurodegenerative disease (disease-associated microglia [DAM]) was found using sc-RNAseq on whole mouse brain in an AD model (Keren-Shaul et al. 2017), and this subtype is partially preserved in humans (Mathys et al. 2019; Zhou et al. 2020). Apart from DAM, there are other disease-associated microglial cell states, including type I interferon-I microglia (summarized in Chen and Colonna 2021). How these cell states correlate with function, and the factors mediating these cell states, remains unclear.

Astrocytes adopt so-called "reactive" states during pathology, tilting the balance from their homeostatic neurotrophic function to a more neurotoxic influence (Liddelow and Barres 2017; Linnerbauer et al. 2020; Han et al. 2021). Such reactive astrocytes induced by microglial factors have been identified in multiple neurodegenerative diseases, including AD, Parkinson's disease (PD), amyotrophic lateral sclerosis (ALS), MS, and Huntington's disease (HD) (Liddelow et al. 2017). Astrocytic signatures that alter in human and mouse models of neurodegeneration include genes associated with autophagy and synaptic homeostasis (Diaz-Castro et al. 2019; Habib et al. 2020; Lee et al. 2022). Interestingly, astrocytes in microglia-dense plaque niches appear to acquire a more neurotoxic phenotype with decreased synaptic signaling (Mallach et al. 2024). The overarching importance of astrocytes in neuroinflammatory and neurodegenerative processes is reviewed in Patani et al. (2023).

During AD, there was a general reduction in myelinating subtypes of oligodendrocytes as well as up-regulation of stress-response signatures (Grubman et al. 2019; Lau et al. 2020). Notably, a recent study suggests that increased levels of dysfunctional myelin with age can exacerbate amyloid-β deposition in AD mouse models (Depp et al. 2023). Enhancing myelin regeneration can even ameliorate cognitive decline in an AD mouse model (Chen et al. 2021). Indeed, emerging data indicate a requirement of microglia to regulate myelin dynamics throughout the life span in both mice and humans (McNamara et al. 2023). Interestingly, a picture is emerging of a dysregulation of microglia–oligodendrocyte cross talk in AD, where TREM2 variants associated with decreased microglial activity showed an up-regulation of a subcluster of oligodendrocytes with elevated expression of the myelin biosynthesis repressor TFEB (for review, see Luquez et al. 2022). Nasu–Hakola disease, punctuated by bone cysts and early-onset dementia, serves as a prime example of the deleterious effect of white matter neurodegeneration in the context of disturbed microglial signaling via abrogated TREM2 (Humphrey et al. 2015). Oligodendrocyte lineage cells have been relatively understudied in AD; therefore, these recent findings highlight the need to closely examine how the functions of these cell populations are disrupted during neurodegeneration.

Amyotrophic Lateral Sclerosis

Spatial transcriptomic analyses of the spinal cord from an ALS mouse model and human postmortem tissue identified distinct astrocyte and microglial subpopulations that exhibited region-specific changes in gene expression over time (Maniatis et al. 2019). An analysis of bulk RNAseq data from postmortem ALS-FTD (frontotemporal dementia) samples showed an increase in microglia and astrocyte gene expression and decreased expression of oligodendrocyte genes, while several activated microglia modules correlated negatively with retrospective disease duration (Humphrey et al. 2023).

Multiple Sclerosis

Sn-RNAseq studies on white matter from healthy individuals and MS patients identified OPCs and six mature oligodendrocyte subclusters that differed significantly in MS patients. Analyses of the composition of these clusters suggested an increase in transcriptional programs associated with myelination (Jäkel et al. 2019).

In additional to individual studies, the accumulation of transcriptional data from the brain and open data-sharing platforms such as the Allen Brain Bank and Human Cell Atlas also afford the opportunity to combine data sets. For example, sn-RNAseq data from human brain samples of AD patients and age-matched controls, combined with data from three published AD studies (using RNAseq and spatial transcriptomics), revealed previously unidentified astrocyte subpopulations distributed across cortical layers that up-regulate genes involved in processes such as cell death, oxidative stress, lipid storage, and fatty acid oxidation in AD (Sadick et al. 2022). However, to compile data sets, it is important that the way subclusters of cells are defined is consistent across studies. This emphasizes the importance of open data sharing and at the same time, presages exciting times for data compiling and mining that can further our understanding of the intricate networks in the human body at unprecedented cellular resolution. In conclusion, glial heterogeneity has emerged as a critical determinant of the cellular response to neurodegeneration with the potential to modulate both disease susceptibility and outcome.

GLIAL PHAGOCYTOSIS, LIPID METABOLISM, AND ENDOLYSOSOMAL PATHWAYS

Genetic modifiers of risk in late-onset AD converge on phagocytosis, lipid metabolism, and endolysosomal pathways, including *APOE*, *PLCG2*, *CD33*, *CR1*, *GRN*, and *CLU* (Andrews et al. 2023). Many of these genes are enriched or uniquely expressed in glia and immune cells. Relevance of the immune and endolysosomal pathways was further corroborated when cognitively healthy centenarians were found to be enriched with the protective alleles of many of these genetic variants (Tesi et al. 2024). Astrocytes are the highest expressors of apolipoprotein E (ApoE), which is the strongest genetic link to sporadic late-onset AD, under homeostatic conditions. The epsilon 4 (ε4) allele confers increased susceptibility to AD, while the ε2 allele is protective (Corder et al. 1993). The most common allelic variant, ε3, is defined as neutral in its contribution to AD risk and age of onset (for a detailed review on ApoE in AD, see Jackson et al. 2024). ApoE has both cell-autonomous and nonautonomous functions in glial phagocytosis and lipid metabolism. For instance, ApoE functions as a transporter of high-density-like lipoproteins and has been shown to control cholesterol distribution and homeostasis in astrocytes, microglia, and oligodendrocytes—a role that is adversely impacted in ApoE4 carriers (Blanchard et al. 2022; Tcw et al. 2022; Wang et al. 2022; Andrews et al. 2023). Interestingly, ApoE expression is increased strongly by microglia in the vicinity of β-amyloid (Aβ) plaques (Sala Frigerio et al. 2019); similarly, in models of acute demyelination and other CNS pathologies, microglia strongly up-regulate ApoE as an integral component of the DAM signature (Keren-Shaul et al. 2017). ApoE can also modulate phagocytosis of synapses and plaques by microglia and astrocytes. For example, there are isoform-dependent differences in astrocyte-synapse engulfment, where ApoE4 knockin mice display reduced synaptic turnover and increased numbers of C1q-bound "senescent" synapses when compared to their ApoE2- and ApoE3-expressing counterparts (Chung et al. 2016).

Similar emphases on glial phagocytosis, metabolism, and endolysosomal pathways have been reported for other genetic variants associated with modulating AD risk, including TREM2, PLCG2, and ABI3. With aging, microglia become metabolically more inactive (von Bernhardi et al. 2015), leading to a rise of TREM2-dependent DAM subsets (for review, see Deczkowska et al. 2018) and white matter–associated microglia (WAM) (Safaiyan et al. 2021). With aging, the presence of lipid droplet-associated microglia (LDAM) also increases (Marschallinger et al. 2020). LDAM are defective in phagocytosis, produce high levels of reactive oxygen species (ROS), and secrete proinflammatory cytokines. TREM2, a major risk factor for AD (Guerreiro et al. 2013; Jonsson et al. 2013), in particular, has been shown to regulate ApoE and microglial lipid metabolism (Nugent et al. 2020; Gouna et al. 2021). In line with this, an antibody that activates TREM2 increased microglial lipid metabolism in mouse models of AD (van Lengerich et al. 2023). Moreover, in mouse models of AD that carry loss-of-function mutations in TREM2, microglia fail to recognize and phagocytose Aβ plaques (Lewcock et al. 2020), although there are conflicting claims on this point related to the timing of TREM2 modulation (Jay et al. 2015, 2017; Wang et al. 2015; Parhizkar et al. 2019; Meilandt et al. 2020; for a detailed review on TREM2, please see Schlepckow et al. 2023). Furthermore, PLCG2, where a mild gain-of-function genetic variant (P522R) is associated with lowering the risk of AD, dementia with Lewy bodies, and FTD (Sims et al. 2017), has been shown to act downstream from TREM2 to boost microglial metabolism and lipid processing (Andreone et al. 2020; Takalo et al. 2020). Conversely, a loss-of-function PLCG2 variant, potentially associated with elevated AD risk (M28L) (Kunkle et al. 2019), leads to a distinct microglial response and an exacerbation of amyloid plaque phenotypes (Tsai et al. 2023). These studies altogether highlight glial phagocytosis and metabolism as key functions that determine AD risk. Additional research is needed to clarify the specific effects on different cell types (De Strooper and Karran 2016) and understand how various glial and immune cells work together to restore neuronal homeostasis. Achieving precise spatiotemporal resolution in these investigations is essential.

Role of Glia in Plaque Formation/Clearance

Studies in mouse models of AD where microglial function is manipulated altogether suggest that microglia contribute to Aβ clearance. Microglia express various receptors to promote clearance and phagocytosis of Aβ, including CD33 and TAM (Simard et al. 2006; Huang et al. 2021). Apart from direct phagocytosis, microglia also help clear aggregates by secreting degrading enzymes, including neprilysin and insulin-degrading enzyme (for review, see Hansen et al. 2018). Microglia have also been demonstrated to form intercellular connections called tunneling nanotubes to degrade pathogenic aggregates (Scheiblich et al. 2021). Interestingly, this ability to form nanotubes was impaired in microglia carrying a PD risk mutation in LRRK2, suggesting a potential role of nanotubes in neurodegeneration. These data altogether suggest that microglia, as professional phagocytes of the brain, use various mechanisms to clear insoluble aggregates from the brain parenchyma. Microglia also surround plaques, likely constituting a protective barrier around toxic Aβ fibrils (Condello et al. 2015).

However, emerging data also suggest an apparent contradictory role for microglia in amyloidosis in that microglia in fact may be responsible for contributing to plaque growth and/or propagation (Heneka et al. 2013; Baik et al. 2016; d'Errico et al. 2022). NLRP3, a major immune sensor expressed on microglia in the brain and an initiator of the inflammasome (Heneka et al. 2013), exacerbated plaque deposition via release of ASC specks that cross-seed Aβ deposition (Venegas et al. 2017). Several other studies depleting microglia (Sosna et al. 2018; Spangenberg et al. 2019; Kiani Shabestari et al. 2022) or targeting phagocytic receptors such as TAM (Huang et al. 2021) also report decreased, not increased, formation of Aβ plaques. For example, specific depletion of microglia by deletion of the fms-intronic regulatory element (FIRE) in the *Csf1r* locus resulted in a striking transition from Aβ plaque to vascular Aβ pathology resembling cerebral amyloid angiopathy, suggesting a requirement of microglia for the seeding and growth of amyloid plaques in the brain paren-

chyma (Kiani Shabestari et al. 2022). These studies thus suggest that microglia promote plaque development. The crucial aspects may not lie in the overall quantity of plaques, however, but in the specific types of plaques that form in these manipulation studies. Elegant biochemical studies using patient brain-derived aggregates had indeed suggested that plaque cores themselves are largely inactive and may even play protective roles in sequestering the diffusible Aβ fibrils (Shankar et al. 2008; Spires-Jones and Hyman 2014; Stern et al. 2023). Hence, microglia may be responsible for promoting plaque cores themselves, perhaps to sequester and contain the pathogenic Aβ species (Schlepckow et al. 2020; Hou et al. 2022; Lemke and Huang 2022). In support of this, in the TAM-deficient mouse model of AD, there were fewer dense-core plaques but higher levels of diffuse plaques and a corresponding exacerbation of dystrophic neurites (Huang et al. 2021). Altogether, these data suggest microglia playing an integral but complex role in amyloidosis. Detailed mechanistic insight into this however is still lacking. How aging or other known genetic AD risk variants alter such functions in microglia are also not known. Further studies are warranted to delineate the role of microglia in amyloidosis especially in context of aging.

Glia-Synapse Phagocytosis in Neurodegeneration

Perhaps unique to microglia as resident macrophages in the brain parenchyma, as compared to other tissue-resident macrophages, is their phagocytosis of neuronal synapses. In AD, one role of microglia that is becoming increasingly clear is their involvement in synapse loss (Bartels et al. 2020). This is of critical relevance in AD, where region-specific synapse loss and dysfunction is an early pathological hallmark that precedes amyloid plaque formation (Selkoe 2002). Microglia can act as cellular mediators of synapse loss throughout the life span (Paolicelli et al. 2011; Tremblay et al. 2011; Schafer et al. 2012; Squarzoni et al. 2014; Bartels et al. 2020). One mechanism that mediates synaptic engulfment by microglia is the classical complement cascade

Cite this article as *Cold Spring Harb Perspect Biol* doi: 10.1101/cshperspect.a041375

involving C1q and C3 (Stevens et al. 2007); blocking microglia-synapse engulfment leads to supernumerary synapses in the developing thalamus (Schafer et al. 2012). Interestingly, this synaptic engulfment pathway in microglia is aberrantly reactivated in various models of AD in a region-specific manner to mediate synapse loss and dysfunction (Hong et al. 2016; Lui et al. 2016; Shi et al. 2017; Dejanovic et al. 2018; Wu et al. 2019). Similar complement-mediated mechanisms have been reported in other models of neurodegenerative diseases and a virus-induced memory impairment model (Vasek et al. 2016; Werneburg et al. 2020; Wilton et al. 2023), suggesting microglia-complement signaling as a potential broad mechanism underlying synaptopathy. It is important to note that, whereas complement has received much of the spotlight as a microglia-mediated synapse elimination pathway, various other molecular pathways have also been proposed to modulate this process, including fractalkine signaling (Paolicelli et al. 2011; Gunner et al. 2019). These data altogether raise a crucial question of whether microglia-mediated synapse engulfment itself is a pathological process that needs to be therapeutically targeted in disease.

However, it is still unclear which synapses are targeted for elimination in neurodegeneration, and the identity of triggers that reactivate this process in microglia is also elusive. In the case of C1q and C3, multiple studies in animal models suggest that there is an overactivation of the microglia-synapse pruning pathway: blocking the classical complement cascade ameliorates synapse loss and cognitive impairment in these mouse models of neurodegeneration (Hong et al. 2016; Shi et al. 2017; Wu et al. 2019). On the other hand, microglia can target damaged or dysfunctional synapses for engulfment. An example of such an upstream determinant for synapse specificity is phosphatidylserine (Scott-Hewitt et al. 2020; Li et al. 2021a). Diffusible Aβ oligomers can lead to focal synaptic apoptosis in the absence of cell death (reviewed in Sokolova et al. 2021). Aβ oligomers bind to synapses, leading to dysfunctional neuronal activity and a local externalization of phosphatidylserine, a classic "eat me" signal on dendritic spines (Rueda-Car-

rasco et al. 2023). In this context, the specific targeting and engulfment of those spines by TREM2$^+$ microglia result in amelioration of neuronal hyperactivity, which may suggest a beneficial role for microglia-synapse engulfment in earliest stages of amyloidosis.

Aside from microglia, astrocytes and OPCs can also partake in synapse engulfment (Chung et al. 2013; Auguste et al. 2022). While astrocytes are not considered bona fide immune cells, they express immune phagocytic genes, including MERTK, MEFG10, and IL-33, which have been shown to have a direct or indirect role in synaptic pruning, depending on the context (Chung et al. 2013; Vainchtein et al. 2018). Similar to microglia, astrocytes engulf synapses in an activity-dependent manner in the developing visual thalamus via MERTK and MEFG10 (Chung et al. 2013). This pathway has further been shown to play a role in astrocytic synapse elimination in adulthood during normal plasticity (Lee et al. 2021). Astrocytes have also been shown to indirectly regulate synaptic pruning by modulating microglial engulfment via the secretion of IL-33 in the developing spinal cord (Vainchtein et al. 2018). The relevance of these processes in neurodegeneration still needs to be evaluated. However, these data also raise the question of whether these multicellular processes are redundant or synergistic and whether there is cross talk and cooperation between astrocytes and microglia (and other cell types) in regulating synapse density. Interestingly, astrocytes and microglia have also been shown to remove distinct parts of dying neurons in an orchestrated and synchronized fashion (Damisah et al. 2020). In mouse models of tauopathy, inhibitory synaptic markers were found more often inside microglial lysosomes than in astrocytes, while more excitatory synaptic markers localized to astrocytic lysosomes than microglial ones, suggesting a potential cell-type-specific delineation of synaptic engulfment (Dejanovic et al. 2022). However, when microglial sensing and engulfment of synapses were impaired upon TREM2 deletion in another mouse model of AD, astrocytes were found to compensate for the lack of microglia-synapse engulfment (Dejanovic et al. 2022). Finally, perivascular macrophages (PVMs), which share their primitive yolk-sac origin with microglia and occupy the vascular niche as early as E10.5 (van Hove et al. 2019), have been shown to modulate synapse engulfment of neighboring microglial cells via SPP1 (De Schepper et al. 2023), suggesting an intriguing cross talk between two distinct brain macrophages impacting synapse fate. Collectively, these studies suggest a dynamic cell–cell cross talk at the synapse where activation of a pathway in one cell type influences the function of the other cells. These data raise the intriguing prospect of whether this interaction and collaboration among specialized glial cells becomes dysfunctional with aging and neurodegeneration and acts to heighten the vulnerability of synapses to loss and dysfunction.

Glia and Neuroinflammation

Microglia, as the primary brain-resident macrophages, sense acute changes in their local microenvironment (including alterations in neuronal activity), respond to "eat-me" and "don't eat-me" signals to perform selective phagocytosis and degradation, and release cytokines, chemokines, and various other factors to resolve injury and help restore tissue homeostasis. Microglia also use a complex sensome that encodes more than 100 genes to act as primary damage sensors of the brain (Hickman et al. 2013). Interestingly, with aging, transcripts for pathways involved in neuroprotection are increased in the microglial sensome, while those involved in endogenous ligand recognition appear to be decreased, suggesting a potential vulnerability could result from microglia themselves becoming senescent in the aging brain. Another key feature of microglia is that they are constantly motile, in part to survey and sample their microenvironment for danger and injury (Davalos et al. 2005; Nimmerjahn et al. 2005; Meyer-Luehmann et al. 2008; Condello et al. 2015). This movement of microglia appears targeted (e.g., toward synaptic junctures [Wake et al. 2009], neuronal somata [Cserép et al. 2020], and amyloid plaques [Condello et al. 2015]). During acute injury, microglial processes rapidly converge to the target site of injury (Davalos et al. 2005), responding to purinergic agonists such as ATP and ADP in a manner depending on the K^+

channel THIK-1 (Madry et al. 2018). Such dynamic plasticity and microenvironment sensing require significant energy expenditure and metabolic flexibility (Bernier et al. 2020). Interestingly, this immunometabolic switching by microglia requires TREM2 (Ulland et al. 2017; Piers et al. 2020); further, microglia carrying TREM2 loss-of-function variants display a compromised ability for rapid directional chemotaxis (Mazaheri et al. 2017; Jairaman et al. 2022) as well as reduced clustering of microglial cells around plaques (Wang et al. 2015; Yuan et al. 2016). These studies suggest that overall, microglial responses in disease risk-carrying brains may be inadequate for sensing and responding to tissue injury, thus potentially contributing to chronic dyshomeostasis in the aging and diseased brain.

Cross Talk between Microglia and Other Immune Cells Modulating Neurodegeneration

Emerging data suggest intriguing immune-immune cross talk on and across brain borders that influences microglial cell states and function. $CD8^+$ T cells infiltrate the CNS in AD (Itagaki et al. 1988; Togo et al. 2002; Gate et al. 2020) and exacerbate various aspects of AD neuropathology including neurodegeneration and white matter damage by directly engaging microglia (Chen et al. 2023; Jorfi et al. 2023; Kedia et al. 2024). Furthermore, depletion of regulatory T cells in the circulation of a mouse model of AD reduced plaque burden and ameliorated spatial memory defects by facilitating peripheral immune cell entry via the choroid plexus, an understudied site in the context of neurodegeneration (Baruch et al. 2015). These studies altogether suggest a bidirectional microglia T-cell cross talk in modulating development and progression of neurodegeneration. Furthermore, in aging and in AD, there appears a rise of a monocyte-derived macrophage subpopulation called disease inflammatory macrophages (DIMs) (Silvin et al. 2022). What DIMs do in the aging and diseased brain is still unclear; however, it suggests yet another potential contribution of peripherally sourced immune cells in the aging and diseased brain. With emerging insight into the brain's own meningeal vasculature and innate-adaptive cross talk in AD models (Da Mesquita et al. 2018), there is now a crucial need to elucidate how peripheral central cross talk is facilitated and impaired in the aging brain. Insight into this interaction will likely yield new avenues for therapeutic intervention.

These studies altogether highlight important intercellular interactions that influence glial function in neurodegeneration, raising a further need to explore cell–cell cross talk in the local milieu of diseased tissue. Utilization of spatial "omics" tools as mentioned previously (Chen et al. 2020; Stickels et al. 2021; Bugeon et al. 2022; Stogsdill et al. 2022) will likely yield the much-needed spatiotemporal resolution to describe cell-type-specific pathways as well as cell–cell cross talk in the immediate microenvironment of microglia (Kedia et al. 2024). Which of these cell states reflect the failure to resolve tissue inflammation, and which reflect a continuous effort by our brain's immune system to compensate for the senescent microglia, will be critical to understanding how we can target glia in neurodegeneration (Fig. 1).

Studies on FIRE mice also depict significant AD- and age-related vascular alterations in response to the absence of microglia throughout the life span (Kiani Shabestari et al. 2022; Chadarevian et al. 2024; Munro et al. 2024). The next section will delineate other key functions of glia,

in particular, specialized roles that glia adopt to regulate CNS blood flow and thus, the energy supply of the brain.

GLIAL REGULATION OF CNS ENERGY SUPPLY AND ITS CONTRIBUTION TO NEURODEGENERATION

Many neurodegenerative diseases involve a loss of energy supply to the brain (Sweeney et al. 2018), including stroke, AD, Down syndrome, PD, MS, HD, and ALS. Since glia play a key role in regulating cerebral blood flow (Attwell et al. 2010) and may contribute to passing metabolic substrates to synapses and axons (Pellerin and Magistretti 2012; Nave and Werner 2021), it is almost inevitable that glia play an important role in the defects of energy supply that trigger neurodegeneration (Fig. 2).

Glial Regulation of Cerebral Blood Flow

Neurovascular coupling is the process by which increased neuronal activity dilates local blood vessels to increase the energy supply to active neurons. In physiological conditions, neuronally evoked intracellular calcium ($[Ca^{2+}]_i$) elevations in astrocytes, generated by release of ATP or glutamate (Glu) from neurons, lead to the formation in astrocytes (by phospholipase

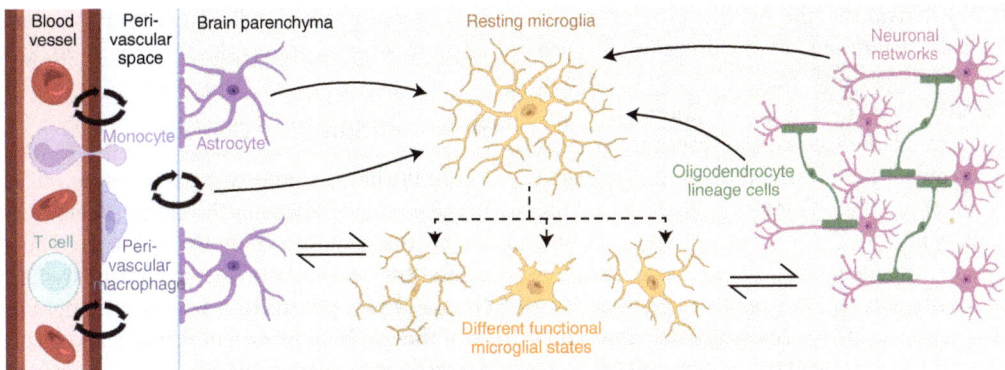

Figure 1. Microglia integrate signals from multiple cell types to determine functional response. Resting microglia receive local signals from neurons, astrocytes, and oligodendrocyte lineage cells as well as long-range signals from the vasculature, including perivascular macrophages and T cells. Microglia integrate these signals and alter their transcriptional signature to appropriately respond. Each functional cluster then feeds back to the relevant effector cells, modulating their behavior. (Image generated using Biorender.)

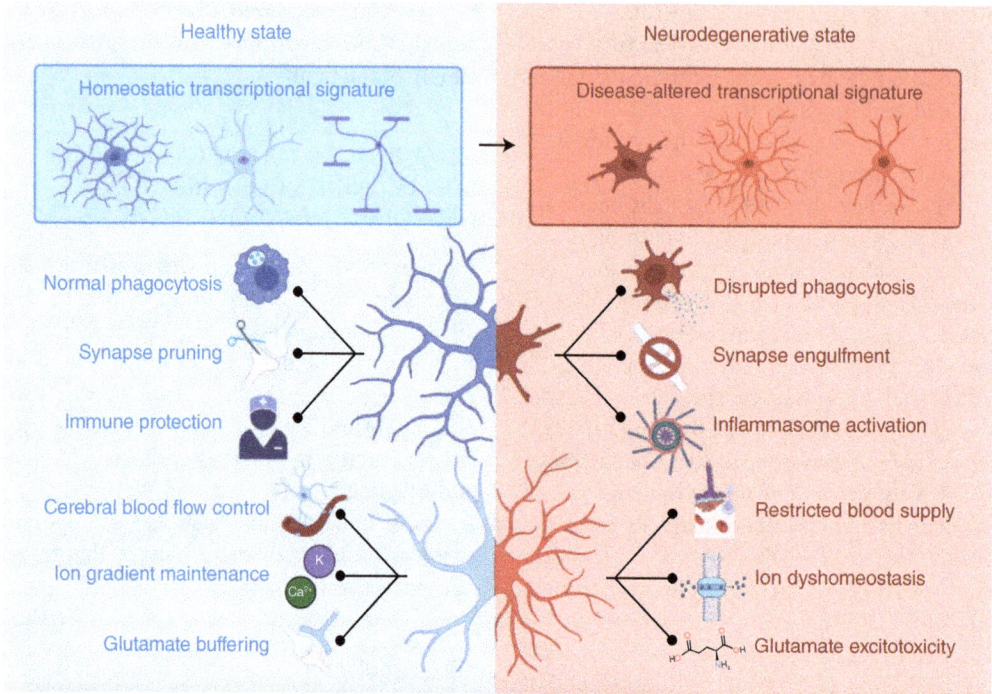

Figure 2. During homeostasis, glia retain transcriptional signatures that allow execution of duties indicative of their normal, healthy function. During neurodegeneration, however, glial transcriptional signatures are altered, leading to induction of dysfunctional processes that both initiate and further exacerbate disease. (Image generated using Biorender.)

D_2) of arachidonic acid (Bezzi et al. 1998; Zonta et al. 2003; Mishra et al. 2016). This is then converted into vasodilating (prostaglandin E_2, epoxyeicosatrenoic acids) or constricting (20-HETE) derivatives, the relative proportions of which are determined by the prevailing oxygen level (Mulligan and MacVicar 2004; Gordon et al. 2008; Mishra et al. 2016). Control of vessel diameter, and hence vascular resistance and cerebral blood flow, can occur at the arteriole (Zonta et al. 2003; Gordon et al. 2008) or the capillary pericyte (Peppiatt et al. 2006) level, and it has been suggested that the astrocyte signaling pathway targets capillary pericytes preferentially (Mishra et al. 2016). Microglia can also contribute to neurovascular coupling. Dilation is evoked via a mechanism involving purinergic signaling from neurons to microglial $P2Y_{12}$ receptors, and subsequent generation of nitric oxide and/or adenosine (Császár et al. 2022), while fractalkine-induced microglia-evoked angioten-

sin II signaling evokes constriction, at least in the retina (Mills et al. 2021). Finally, a preprint has reported that white matter oligodendrocytes sense axonal activity using NMDA receptors and release prostaglandin E_2 to evoke a slow dilation of blood vessels (Restrepo et al. 2022).

Stroke and Spreading Depression

When the blood supply to the brain is cut off, as in ischemic stroke when a thrombus occludes an artery, the resulting rundown of ion gradients caused by inhibition of the sodium-potassium (Na^+/K^+) pump leads to deleterious changes in both the vasculature, and in neurons and glia.

In the vasculature, the loss of ion pumping will lead to a contraction-inducing rise of $[Ca^{2+}]_i$ in both arteriolar smooth muscle cells (SMCs) and capillary pericytes. Surprisingly, after arterial blood flow is restored, pericyte contraction is maintained for hours (Hall et al. 2014), thus

Cite this article as *Cold Spring Harb Perspect Biol* doi: 10.1101/cshperspect.a041375

restricting blood flow and contributing to the postischemic "no-reflow" phenomenon in the cerebral microvasculature (Ames et al. 1968). The long duration of this contraction may reflect pericytes dying in rigor, in which condition they are unable to relax (Hall et al. 2014), thus presumably decreasing blood flow until they are removed by microglia. The rise of $[Ca^{2+}]_i$ in pericytes could partly reflect the failure to pump out Ca^{2+} ions that enter the cell passively; however, it is also possible that glial-evoked signaling contributes. As noted above, astrocyte-derived arachidonic acid (which will also be produced when cell $[Ca^{2+}]$ rises in ischemia) can be converted into 20-HETE, which evokes constriction by closing SMC and pericyte K^+ channels, evoking depolarization and activation of voltage-gated Ca^{2+} channels (Gonzalez-Fernandez et al. 2021). Indeed, 20-HETE production has been shown to contribute to the decrease of cerebral blood flow occurring both in the penumbra region of a simulated ischemic stroke (Li et al. 2021b) and in cortical spreading depression (when there is a similar rundown of ion gradients; Fordsmann et al. 2013).

The profound rundown of ion gradients triggered by ischemia (with $[K^+]_o$ and $[Na^+]_i$ both rising by about 60 mM) has acute effects on the energetics of Glu transport. Normally, neurotransmitter Glu is cleared from the extracellular space by cotransport into cells with 3 Na^+ ions and 1 H^+, and the countertransport of a K^+ ion (Zerangue and Kavanaugh 1996; Levy et al. 1998), with the great majority of this transport occurring into astrocytes (Danbolt 2001). When the ion gradients and transmembrane voltage gradient run down in ischemia, Glu uptake is predicted from thermodynamics and observed to reverse and release Glu into the extracellular space (Szatkowski et al. 1990; Rossi et al. 2000). Selective knockout of different transporters has shown that, although the cessation of uptake by astrocyte transporters (GLT-1 and GLAST) must play an important role in allowing the extracellular Glu concentration to rise, at early times the majority of the released Glu appears to be exported by the neuronal transporter EAAC1, possibly because the intracellular concentration of Glu is higher in neurons than in astrocytes (Gebhardt et al. 2002; Hamann et al.

2002). The effects on neurons and glia of the Glu released by failure of this uptake process are considered later in this review.

Alzheimer's Disease

Imaging studies have shown that, in affected areas, cerebral blood flow is decreased by ~45% early in AD (Asllani et al. 2008). By examining biopsy tissue from cognitively impaired humans and imaging AD model mice in vivo, it has been shown that this reduction of blood flow is generated exclusively by pericyte-mediated constriction of capillaries (Nortley et al. 2019), while arterioles and venules do not change their diameter. This constriction was shown to be mimicked by application of soluble oligomers of Aβ, which act by evoking ROS release from microglia and pericytes (Nortley et al. 2019). The ROS, in turn, release endothelin-1 from a cell type that is as yet unknown in the context of AD, but which may be astrocytes as has been shown for a demyelination model (Hammond et al. 2014). Thus, glial cells may play an unexpected role early in the time course of AD, by reducing cerebral blood flow and thus contributing to the deleterious changes discussed elsewhere in this review.

Parkinson's Disease

Like AD, PD is associated with reduced cerebral blood flow (Fernández-Seara et al. 2012), and with increased microglial ROS production (Belarbi et al. 2017). This raises the possibility that the glial-based mechanism established for the effects of Aβ in reducing cerebral blood flow, involving microglial ROS production and endothelin-1 release, might also be active in Parkinson's. Indeed, an increase in endothelin-1 level in patients with PD has been reported (Makarov et al. 2013).

Disorders Caused by a Decrease in Glial Energetic Support of Neurons

Both in the gray matter (Pellerin and Magistretti 2012) and the white matter (Nave and Werner 2021), it has been suggested that glial cells support neuronal function by supplying neurons with substrates for ATP generation. In the gray matter, glucose provided to astrocytes from the blood is proposed to be converted to lactate and passed to neuronal synapses and somata to support information processing (Pellerin and Magistretti 2012). In the white matter, the insulation of axons from the extracellular space by myelinated internodes hinders the provision of glucose to axons, and it is suggested that, in oligodendrocytes, glucose is converted into lactate or a related compound and passed to the ensheathed axon by monocarboxylate transporters (MCTs) (Fünfschilling et al. 2012; Lee et al. 2012; Saab et al. 2016; Meyer et al. 2018). A failure of this transfer mechanism by virtue of reduced MCT expression may contribute to causing ALS (Lee et al. 2012). Similarly, a decrease of energy supply from oligodendrocytes to axons, especially in aging, is proposed to induce axonal damage and demyelination in AD and in multiple system atrophy (Mot et al. 2018; Philips et al. 2021). It is unclear whether similar deficits of astrocyte-neuron metabolic cooperation in the gray matter contribute to neurodegeneration.

Overall, it is clear that glia perform indispensable support roles in neurovascular coupling and CNS energy metabolism. When this trophic presence becomes dysfunctional, pathology and neurodegeneration can take hold. Not only do astrocytes facilitate the supply of blood to neurons, they are also vital in maintaining ion gradients and a low extracellular Glu concentration, thus allowing neurons to fire action potentials on demand. This supportive influence will be further explored in the next section.

GLIAL REGULATION OF NEURONAL EXCITABILITY

In 1964, Steven Kuffler impaled astrocytes in the optic nerve of the salamander to record their resting potential (Kuffler and Potter 1964). Much to his surprise, with each flash of light to the retina, the astrocytes depolarized by a few millivolts. As subsequent studies with K^+ selective microelectrodes would show, these depolarizations were caused by the temporary increase in extracellular K^+ concentration resulting from the repolariza-

tion of the neuronal action potential. The interpretation was that astrocytes had a highly K^+ selective membrane, in large part due to expression of K^+ channels, which were eventually molecularly identified as Kir4.1 (encoded by the *KCNJ10* gene) (Kofuji et al. 2000). Years later, Eric Newman recorded from astrocytes in the retina and showed that these K^+ channels were unevenly expressed along the astrocytes with the highest density in their endfoot processes (Newman 1986). This allows astrocytes to take up potassium in regions of high neuronal activity, and therefore elevated K^+, and release it onto distant blood vessels in a process called K^+ syphoning. Alternatively, as K^+ can also travel through gap junctions to adjacent cells, K^+ can be moved across multiple cells in a process called spatial buffering (Orkand 1980). The ability of astrocytes to effectively take up and sequester potassium requires an inward directed K^+ gradient, which is established by an active Na^+/K^+ ATPase and the syncytial coupling between adjacent cells, which will make them almost isopotential even when an astrocytic process experiences elevated extracellular K^+. This rather elegant design allows for the rapid clearance of extracellular K^+ to ensure continued neuronal firing. Maintenance of K^+ homeostasis is one of the cardinal functions of astrocytes in the healthy brain and, as knockout of Kir4.1 demonstrates, relies critically on the expression of Kir4.1 channels. Consequently, any change in Kir4.1 expression that occurs in the context of disease can impair K^+ homeostasis and profoundly affect neuronal signaling. This has now been demonstrated to occur in a number of injury and disease conditions as further discussed below. Note, however, that conclusive evidence for Kir4.1 dysfunction in disease only exists for animal models of human disease.

Seizures and Epilepsy

Any acute injury and many neurodegenerative diseases present with structural and functional changes collectively called reactive gliosis (Escartin et al. 2021). Many of these conditions also present with seizures. In almost all instances, gliotic astrocytes show a loss of Kir4.1 expression and function (Nwaobi et al. 2016) and a causal link to seizure is quite likely. Experimental deletion of *KCNJ10*, the gene encoding Kir4.1, causes mice to have tremors and seizures (Neusch et al. 2006). Moreover, the sclerotic part of the hippocampus of mesial temporal lobe epilepsy (MTLE) patients that is surgically removed to cure seizures harbors astrocytes with reduced Kir4.1 expression. In a genetic mouse model of epilepsy where the β1 integrin receptor was deleted in astrocytes, animals develop spontaneous seizures at 2–3 months of age, and this is associated with a deficiency in Kir4.1-mediated K⁺ uptake (Robel et al. 2015). The most convincing disease showcasing that astrocytic loss of Kir4.1 is sufficient to cause seizures and other morbidities in humans is a rare recessive developmental disorder called SeSAME syndrome, so named because it is characterized by early-onset seizures, sensorineuronal deafness, ataxia, mental disability, and electrolyte imbalance. This severe developmental disorder is caused by a number of mutations in the *KCNJ10* gene (Scholl et al. 2009). Dependent on the particular mutation, Kir4.1 channels are either absent or altered in their physiological properties.

Alzheimer's Disease and Huntington's Disease

Postmortem tissue from AD patients shows a loss of Kir4.1 as do many of the mouse models of AD generated through overexpression of genes implicated in AD pathology (Wilcock et al. 2009). This particularly affects perivascular astrocytic endfeet, which become largely devoid of Kir4.1, suggesting a possible link to abnormal glia–vascular interactions in AD. In contrast to AD, where known familial genes that cause disease are rare, HD is always caused by the expression of a single gene, mutant huntingtin (mHTT), making it easier to study in mouse models. HD is characterized by the progressive loss of neurons, primarily in the striatum causing characteristic involuntary movements. Interestingly, those striatal astrocytes show greatly reduced Kir4.1 expression, and expression of the mHTT gene in astrocytes alone is sufficient to induce neuronal death in the striatum (Tong et al. 2014). Hence, astrocytic loss of Kir4.1 in HD is pathogenic. Interestingly, viral overexpression of the *KCNJ10* gene encoding Kir4.1 in striatal astrocytes of HD mice causes a partial rescue of the disease symptoms, suggesting a strong glial participation in the disease etiology.

Depression

In a recent study, a change in astrocytic Kir4.1 expression has even been causally implicated in major depressive disorders (Cui et al. 2018). Here, neurons in the lateral habenula nucleus, often called the antireward center, show a peculiar change in burst activity in animals that display depressive symptoms. This change in electrical activity is secondary to an increase in astrocytic Kir4.1 expression in this region resulting in a hyperpolarized neuronal membrane potential (Cui et al. 2018). Hence, depression may be the first disorder where an increase, rather than decrease, in Kir4.1 expression contributes to disease. Each of these examples shows a noncell-autonomous contribution of astrocytes to disease, in each instance with the consequences being secondary to altered extracellular K⁺ homeostasis.

Astrocytes and Glutamate Homeostasis

A second, equally important homeostatic function of astrocytes that is profoundly impaired in neurodegenerative diseases pertains to Glu. First introduced by Olney in the 1960s (Blood et al. 1969; Olney 1969), the idea that Glu excitotoxicity appears to be the final arbiter of neuronal death in neurological illnesses is now widely accepted (Rothstein 1996). Any neuron that expresses NMDA-type Glu receptors is vulnerable to Glu toxicity if these channels are activated in a sustained, nonphysiological way. Normally, the pore of the NMDA receptor, which is highly Ca^{2+} permeable, is blocked by an intracellular Mg^{2+} ion. If the cell depolarizes, this Mg^{2+} is dislodged, permitting Ca^{2+} entry. In stroke, this occurs as a result of energy failure, depolarizing the cell. In other diseases, it could be the sustained presence of excessive extracellular Glu, which would chronically depolarize neurons by activating AMPA receptors. To prevent Glu excitotoxicity, glutamatergic synapses throughout the brain are covered by astrocytic leaflets, which highly express the Na^+-dependent Glu transporter EAAT2. As already mentioned, this transporter is electrogenic and couples the influx of three Na^+ and one H^+ to the uptake of one Glu^- and efflux of one K^+ (Danbolt 2001). Therefore, the astrocyte must have a negative resting membrane potential to fully harness the Na^+ gradient for Glu transport. This explains why depolarized astrocytes (e.g., those lacking Kir4.1) are less efficient in Glu uptake. EAAT2 accounts for 1% of total brain protein to assure that Glu clearance by astrocytes will not fail (Danbolt 2001). However, it appears to fail quite frequently during neurological disease as shown by the examples below.

Stroke

Stroke is the disease most unequivocally associated with Glu excitotoxicity (as outlined in the earlier section on Stroke and Spreading Depression). Energy loss due to focal or global ischemia causes the catastrophic release of Glu from chronically depolarized glutamatergic terminals. It also raises extracellular K^+, thereby depolarizing neurons and astrocytes alike. As both neurons and astrocytes lose all energy substrates, and since Glu uptake relies on the transmembrane gradient for Na^+ established in an energy consuming way via the Na^+/K^+ ATPase, astrocytic depolarization completely impairs astrocytic Glu reuptake, thus establishing a sustained and catastrophic increase of Glu in the affected brain region (Katayama et al. 1990).

Amyotrophic Lateral Sclerosis

Autopsy tissue from ALS patients shows a dramatic 70%–90% decrease in EAAT2 expression in the motor cortex and spinal cord (Rothstein et al. 1995). Unsurprisingly, the only available treatment that shows some clinical benefit, Riluzole, limits the amount of Glu released into the spinal cord. One of the first gene mutations associated with familial forms of ALS is the gene encoding superoxide dismutase-1 (SOD-1), an enzyme that detoxifies cellular free radicals (Beckman et al. 2001). Transgenic mice with

this mutation replicate a majority of the salient clinical features of the disease, and also show a reduction in astrocytic EAAT2 expression. Importantly, restoring just the astrocytic EAAT2 expression is sufficient to delay disease onset, although it does not prevent disease (Guo et al. 2003). This suggests that astrocytic impairment of EAAT2 contributes to disease onset. Enhancement of EAAT2 expression by ceftriaxone has been explored in clinical trials to treat ALS (Clinicaltrials.gov identifier: NCT00349622) and PD (ClinicalTrials.gov identifier: NCT03413384).

Alzheimer's Disease

Amyloid plaques are frequently surrounded by astrocytes, which change their morphology suggestive of a reactive phenotype. These reactive astrocytes show decreased expression of EAAT2 along with impaired Glu uptake in mouse models of AD (Hefendehl et al. 2016). Interestingly, loss of just one allele of the EAAT2 gene causes memory decline in wild-type mice and in transgenic AD mice, while overexpressing the EAAT2 gene is sufficient to attenuate their decline in memory function (Mookherjee et al. 2011). These data suggest that astrocytic EAAT2 is neuroprotective and, unsurprisingly, EAAT2 has become a potential therapeutic target to slow AD progression (Wood et al. 2022). Loss of EAAT2 also contributes to neuronal hyperexcitability that may well explain seizures as a frequent comorbidity observed in AD patients (Zott et al. 2019).

Huntington's Disease

As already mentioned, in HD, the disease-causing mHTT is found in neurons and astrocytes alike. In astrocytes, mHTT prevents binding of the transcription factor SP1 to the gene encoding EAAT2, resulting in down-regulation of the transporter, which is sufficient to impair motor function (Bradford et al. 2009). Importantly, EAAT2 expression and neurodegeneration (Wood et al. 2019) are ameliorated by expression of wild-type HTT in astrocytes even in the continued presence of mHTT in neurons.

Seizures and Epilepsy

Seizures and epilepsy are typically associated with an impaired excitation-inhibition balance. Given the importance of Glu homeostasis by astrocytes, it is not surprising that EAAT2 expression is decreased in reactive astrocytes associated with seizure foci in mouse and human (Coulter and Eid 2012). Interestingly, EAAT2 transport not only serves to clear excessive Glu from the extracellular space but is also essential to supply glutamine to neurons for the synthesis of GABA. Following reuptake by astrocytes, Glu is deaminated to glutamine, which is then shuttled to inhibitory neurons for de novo synthesis of GABA (Ortinski et al. 2010). AAV-induced gliosis, resulting in down-regulation of EAAT2 alone, can disrupt neuronal GABA synthesis causing seizures.

Multiple Sclerosis

MS is characterized by the progressive loss of myelin around axons followed by axonal degeneration and, hence, can be considered a secondary neurodegenerative disease. While initially the axonal injury was thought to be resulting from a loss of oligodendrocytes, it is clear that Glu excitotoxicity plays a major role in the final demise of the axon. Here, Glu is released onto axons where it activates axonal NMDA receptors. The influx of Na^+ in turn activates a Na^+/Ca^{2+} exchanger that causes a pathological increase of Ca^{2+} in the axon, causing white matter ischemia (Trapp and Stys 2009).

In light of the above-illustrated importance of astrocytic Glu homeostasis, great efforts have been made in identifying drugs that could enhance EAAT2 expression. Notable examples include 17β-estradiol, tamoxifen, and the antibiotics minocycline and ceftriaxone.

CONCLUDING REMARKS

Based on the wealth of evidence presented herein, the importance of glia in neurodegenerative processes cannot be understated. We have highlighted the vital roles that glia undertake to maintain neuronal health at homeostasis, which is achieved by region-specific heterogeneity of cellular states and tightly regulated molecular pathways. When this support becomes compromised during aging or disease, glia can be the drivers of neuronal dysfunction and death. This means, however, that targeting glia with therapeutics can prove fruitful in staving off neurodegeneration, further emphasizing the urgent need to understand in detail not only the glial mechanisms responsible for disease initiation and progression, but also those mechanisms performed in the healthy steady state to protect neuronal function.

ACKNOWLEDGMENTS

G.C. and S.H. are supported by the UK Dementia Research Institute (UKDRI1011), which receives its funding from UK DRI Ltd, funded by the UK Medical Research Council, Alzheimer's Society, and Alzheimer's Research UK. S.H. is also supported by the Chan Zuckerberg Initiative Neurodegeneration Challenge Network, Alzheimer's Association, and the Bright Focus Foundation. D.A. is supported by a Wellcome Trust Senior Investigator Award (219366) and funding from the British Heart Foundation/UK DRI Centre for Vascular Dementia Research. H.P. is supported by the National Institutes of Health (NS116350, NS118183, NS118570, HG011014, and AG066831), the ALS Association, the Tow Foundation, and Target ALS. H.S. is supported by the National Institutes of Health (NS036692, AG065836, and CA227149).

REFERENCES

Ames A, Wright RL, Kowada M, Thurston JM, Majno G. 1968. Cerebral ischemia. II: The no-reflow phenomenon. *Am J Pathol* **52:** 437–453.

Andreone BJ, Przybyla L, Llapashtica C, Rana A, Davis SS, van Lengerich B, Lin K, Shi J, Mei Y, Astarita G, et al. 2020. Alzheimer's-associated PLCγ2 is a signaling node required for both TREM2 function and the inflammatory response in human microglia. *Nat Neurosci* **23:** 927–938. doi:10.1038/s41593-020-0650-6

Andrews SJ, Renton AE, Fulton-Howard B, Podlesny-Drabiniok A, Marcora E, Goate AM. 2023. The complex genetic architecture of Alzheimer's disease: novel insights and future directions. *EBioMedicine* **90:** 104511. doi:10.1016/j.ebiom.2023.104511

Asllani I, Habeck C, Scarmeas N, Borogovac A, Brown TR, Stern Y. 2008. Multivariate and univariate analysis of continuous arterial spin labeling perfusion MRI in Alzheimer's disease. *J Cereb Blood Flow Metab* **28:** 725–736. doi:10.1038/sj.jcbfm.9600570

Attwell D, Buchan AM, Charpak S, Lauritzen M, MacVicar BA, Newman EA. 2010. Glial and neuronal control of brain blood flow. *Nature* **468:** 232–243. doi:10.1038/nature09613

Auguste YSS, Ferro A, Kahng JA, Xavier AM, Dixon JR, Vrudhula U, Nichitiu AS, Rosado D, Wee TL, Pedmale UV, et al. 2022. Oligodendrocyte precursor cells engulf synapses during circuit remodeling in mice. *Nat Neurosci* **25:** 1273–1278. doi:10.1038/s41593-022-01170-x

Ayata P, Badimon A, Strasburger HJ, Duff MK, Montgomery SE, Loh Y-HE, Ebert A, Pimenova AA, Ramirez BR, Chan AT, et al. 2018. Epigenetic regulation of brain region-specific microglia clearance activity. *Nat Neurosci* **21:** 1049–1060. doi:10.1038/s41593-018-0192-3

Baik SH, Kang S, Son SM, Mook-Jung I. 2016. Microglia contributes to plaque growth by cell death due to uptake of amyloid β in the brain of Alzheimer's disease mouse model. *Glia* **64:** 2274–2290. doi:10.1002/glia.23074

Bartels T, De Schepper S, Hong S. 2020. Microglia modulate neurodegeneration in Alzheimer's and Parkinson's diseases. *Science* **370:** 66–69. doi:10.1126/science.abb8587

Baruch K, Rosenzweig N, Kertser A, Deczkowska A, Sharif AM, Spinrad A, Tsitsou-Kampeli A, Sarel A, Cahalon L, Schwartz M. 2015. Breaking immune tolerance by targeting Foxp3$^+$ regulatory T cells mitigates Alzheimer's disease pathology. *Nat Commun* **6:** 7967. doi:10.1038/ncomms8967

Beckman JS, Estévez AG, Crow JP, Barbeito L. 2001. Superoxide dismutase and the death of motoneurons in ALS. *Trends Neurosci* **24:** S15–S20. doi:10.1016/S0166-2236(00)01981-0

Belarbi K, Cuvelier E, Destée A, Gressier B, Chartier-Harlin M-C. 2017. NADPH oxidases in Parkinson's disease: a systematic review. *Mol Neurodegener* **12:** 84. doi:10.1186/s13024-017-0225-5

Bernier LP, York EM, MacVicar BA. 2020. Immunometabolism in the brain: how metabolism shapes microglial function. *Trends Neurosci* **43:** 854–869. doi:10.1016/j.tins.2020.08.008

Bezzi P, Carmignoto G, Pasti L, Vesce S, Rossi D, Rizzini BL, Pozzan T, Volterra A. 1998. Prostaglandins stimulate calcium-dependent glutamate release in astrocytes. *Nature* **391:** 281–285. doi:10.1038/34651

Blanchard JW, Akay LA, Davila-Velderrain J, von Maydell D, Mathys H, Davidson SM, Effenberger A, Chen CY, Maner-Smith K, Hajjar I, et al. 2022. APOE4 impairs myelination via cholesterol dysregulation in oligodendrocytes. *Nature* **611:** 769–779. doi:10.1038/s41586-022-05439-w

Blood FR, Oser BL, White PL, Olney JW. 1969. Monosodium glutamate. *Science* **165:** 1028–1029. doi:10.1126/science.165.3897.1028

Bradford J, Shin J-Y, Roberts M, Wang C-E, Li X-J, Li S. 2009. Expression of mutant huntingtin in mouse brain astrocytes causes age-dependent neurological symptoms. *Proc Natl Acad Sci* **106:** 22480–22485. doi:10.1073/pnas.0911503106

Bugeon S, Duffield J, Dipoppa M, Ritoux A, Prankerd I, Nicoloutsopoulos D, Orme D, Shinn M, Peng H, Forrest H, et al. 2022. A transcriptomic axis predicts state modulation of cortical interneurons. *Nature* **607:** 330–338. doi:10.1038/s41586-022-04915-7

Caldwell ALM, Sancho L, Deng J, Bosworth A, Miglietta A, Diedrich JK, Shokhirev MN, Allen NJ. 2022. Aberrant astrocyte protein secretion contributes to altered neuronal development in multiple models of neurodevelopmental disorders. *Nat Neurosci* **25:** 1163–1178. doi:10.1038/s41593-022-01150-1

Chadarevian JP, Hasselmann J, Lahian A, Spitale RC, Davtyan H, Blurton-jones M, Chadarevian JP, Hasselmann J, Lahian A, Capocchi JK, et al. 2024. Therapeutic potential of human microglia transplantation in a chimeric model of CSF1R- related leukoencephalopathy. *Neuron* doi:10.1016/j.neuron.2024.05.023

Chen Y, Colonna M. 2021. Microglia in Alzheimer's disease at single-cell level. Are there common patterns in humans and mice? *J Exp Med* **218:** e20202717. doi:10.1084/jem.20202717

Chen WT, Lu A, Craessaerts K, Pavie B, Sala Frigerio C, Corthout N, Qian X, Laláková J, Kühnemund M, Voytyuk I, et al. 2020. Spatial transcriptomics and in situ sequencing to study Alzheimer's disease. *Cell* **182:** 976–991.e19. doi:10.1016/j.cell.2020.06.038

Chen JF, Liu K, Hu B, Li RR, Xin W, Chen H, Wang F, Chen L, Li RX, Ren SY, et al. 2021. Enhancing myelin renewal reverses cognitive dysfunction in a murine model of Alzheimer's disease. *Neuron* **109:** 2292–2307.e5. doi:10.1016/j.neuron.2021.05.012

Chen X, Firulyova M, Manis M, Herz J, Smirnov I, Aladyeva E, Wang C, Bao X, Finn MB, Hu H, et al. 2023. Microglia-mediated T cell infiltration drives neurodegeneration in tauopathy. *Nature* **615:** 668–677. doi:10.1038/s41586-023-05788-0

Chung W, Clarke LE, Wang GX, Stafford BK, Sher A, Chakraborty C, Joung J, Foo LC, Thompson A, Chen C, et al. 2013. Astrocytes mediate synapse elimination through MEGF10 and MERTK pathways. *Nature* **504:** 394–400. doi:10.1038/nature12776

Chung WS, Verghese PB, Chakraborty C, Joung J, Hyman BT, Ulrich JD, Holtzman DM, Barres BA. 2016. Novel allele-dependent role for APOE in controlling the rate of synapse pruning by astrocytes. *Proc Natl Acad Sci* **113:** 10186–10191. doi:10.1073/pnas.1609896113

Clavreul S, Dumas L, Loulier K. 2022. Astrocyte development in the cerebral cortex: complexity of their origin, genesis, and maturation. *Front Neurosci* **16:** 916055. doi:10.3389/fnins.2022.916055

Condello C, Yuan P, Schain A, Grutzendler J. 2015. Microglia constitute a barrier that prevents neurotoxic protofibrillar Aβ42 hotspots around plaques. *Nat Commun* **6:** 6176. doi:10.1038/ncomms7176

Corder EH, Saunders AM, Strittmatter WJ, Schmechel DE, Gaskell PC, Small GW, Roses AD, Haines JL, Pericak-Vance MA. 1993. Gene dose of apolipoprotein E type 4 allele and the risk of Alzheimer's disease in late onset families. *Science* **261:** 921–923. doi:10.1126/science.8346443

Coulter DA, Eid T. 2012. Astrocytic regulation of glutamate homeostasis in epilepsy. *Glia* **60:** 1215–1226. doi:10.1002/glia.22341

Császár E, Lénárt N, Cserép C, Környei Z, Fekete R, Pósfai B, Balázsfi D, Hangya B, Schwarcz AD, Szabadits E, et al. 2022. Microglia modulate blood flow, neurovascular coupling, and hypoperfusion via purinergic actions. *J Exp Med* **219:** e20211071. doi:10.1084/jem.20211071

Cserép C, Pósfai B, Lénárt N, Fekete R, László ZI, Lele Z, Orsolits B, Molnár G, Heindl S, Schwarcz AD, et al. 2020. Microglia monitor and protect neuronal function through specialized somatic purinergic junctions. *Science* **367:** 528–537. doi:10.1126/science.aax6752

Cui Y, Yang Y, Ni Z, Dong Y, Cai G, Foncelle A, Ma S, Sang K, Tang S, Li Y, et al. 2018. Astroglial Kir4.1 in the lateral habenula drives neuronal bursts in depression. *Nature* **554:** 323–327. doi:10.1038/nature25752

Da Mesquita S, Louveau A, Vaccari A, Smirnov I, Cornelison RC, Kingsmore KM, Contarino C, Onengut-Gumuscu S, Farber E, Raper D, et al. 2018. Functional aspects of meningeal lymphatics in ageing and Alzheimer's disease. *Nature* **560:** 185–191. doi:10.1038/s41586-018-0368-8

Damisah EC, Hill RA, Rai A, Chen F, Rothlin CV, Ghosh S, Grutzendler J. 2020. Astrocytes and microglia play orchestrated roles and respect phagocytic territories during neuronal corpse removal in vivo. *Sci Adv* **6:** eaba3239. doi:10.1126/sciadv.aba3239

Danbolt NC. 2001. Glutamate uptake. *Prog Neurobiol* **65:** 1–105. doi:10.1016/S0301-0082(00)00067-8

Davalos D, Grutzendler J, Yang G, Kim JV, Zuo Y, Jung S, Littman DR, Dustin ML, Gan WB. 2005. ATP mediates rapid microglial response to local brain injury in vivo. *Nat Neurosci* **8:** 752–758. doi:10.1038/nn1472

De Biase LM, Schuebel KE, Fusfeld ZH, Jair K, Hawes IA, Cimbro R, Zhang H-Y, Liu Q-R, Shen H, Xi Z-X, et al. 2017. Local cues establish and maintain region-specific phenotypes of basal ganglia microglia. *Neuron* **95:** 341–356.e6. doi:10.1016/j.neuron.2017.06.020

Deczkowska A, Keren-Shaul H, Weiner A, Colonna M, Schwartz M, Amit I. 2018. Disease-associated microglia: a universal immune sensor of neurodegeneration. *Cell* **173:** 1073–1081. doi:10.1016/j.cell.2018.05.003

Dejanovic B, Huntley MA, De Mazière A, Meilandt WJ, Wu T, Srinivasan K, Jiang Z, Gandham V, Friedman BA, Ngu H, et al. 2018. Changes in the synaptic proteome in tauopathy and rescue of tau-induced synapse loss by C1q antibodies. *Neuron* **100:** 1322–1336.e7. doi:10.1016/j.neuron.2018.10.014

Dejanovic B, Wu T, Tsai MC, Graykowski D, Gandham VD, Rose CM, Bakalarski CE, Ngu H, Wang Y, Pandey S, et al. 2022. Complement C1q-dependent excitatory and inhibitory synapse elimination by astrocytes and microglia in Alzheimer's disease mouse models. *Nat Aging* **2:** 837–850. doi:10.1038/s43587-022-00281-1

Depp C, Sun T, Sasmita AO, Spieth L, Berghoff SA, Nazarenko T, Overhoff K, Steixner-Kumar AA, Subramanian S, Arinrad S, et al. 2023. Myelin dysfunction drives amyloid-β deposition in models of Alzheimer's disease. *Nature* **618:** 349–357. doi:10.1038/s41586-023-06120-6

d'Errico P, Ziegler-Waldkirch S, Aires V, Hoffmann P, Mezö C, Erny D, Monasor LS, Liebscher S, Ravi VM, Joseph K, et al. 2022. Microglia contribute to the propagation of Aβ into unaffected brain tissue. *Nat Neurosci* **25:** 20–25. doi:10.1038/s41593-021-00951-0

De Schepper S, Ge JZ, Crowley G, Ferreira LSS, Garceau D, Toomey CE, Sokolova D, Rueda-Carrasco J, Shin SH, Kim JS, et al. 2023. Perivascular cells induce microglial phagocytic states and synaptic engulfment via SPP1 in mouse models of Alzheimer's disease. *Nat Neurosci* **26:** 406–415. doi:10.1038/s41593-023-01257-z

De Strooper B, Karran E. 2016. The cellular phase of Alzheimer's disease. *Cell* **164:** 603–615. doi:10.1016/j.cell.2015.12.056

Diaz-Castro B, Gangwani MR, Yu X, Coppola G, Khakh BS. 2019. Astrocyte molecular signatures in Huntington's disease. *Sci Transl Med* **11:** 1–12. doi:10.1126/scitranslmed.aaw8546

Endo F, Kasai A, Soto JS, Yu X, Qu Z, Hashimoto H, Gradinaru V, Kawaguchi R, Khakh BS. 2022. Molecular basis of astrocyte diversity and morphology across the CNS in health and disease. *Science* **378:** eadc9020. doi:10.1126/science.adc9020

Escartin C, Galea E, Lakatos A, O'Callaghan JP, Petzold GC, Serrano-Pozo A, Steinhäuser C, Volterra A, Carmignoto G, Agarwal A, et al. 2021. Reactive astrocyte nomenclature, definitions, and future directions. *Nat Neurosci* **24:** 312–325. doi:10.1038/s41593-020-00783-4

Fernández-Seara MA, Mengual E, Vidorreta M, Aznárez-Sanado M, Loayza FR, Villagra F, Irigoyen J, Pastor MA. 2012. Cortical hypoperfusion in Parkinson's disease assessed using arterial spin labeled perfusion MRI. *Neuroimage* **59:** 2743–2750. doi:10.1016/j.neuroimage.2011.10.033

Fordsmann JC, Ko RWY, Choi HB, Thomsen K, Witgen BM, Mathiesen C, Lønstrup M, Piilgaard H, MacVicar BA, Lauritzen M. 2013. Increased 20-HETE synthesis explains reduced cerebral blood flow but not impaired neurovascular coupling after cortical spreading depression in rat cerebral cortex. *J Neurosci* **33:** 2562–2570. doi:10.1523/JNEUROSCI.2308-12.2013

Fujita M, Gao Z, Zeng L, McCabe C, White CC, Ng B, Green GS, Rozenblatt-Rosen O, Phillips D, Amir-Zilberstein L, et al. 2024. Cell subtype-specific effects of genetic variation in the Alzheimer's disease brain. *Nat Genet* **56:** 605–614. doi:10.1038/s41588-024-01685-y

Fünfschilling U, Supplie LM, Mahad D, Boretius S, Saab AS, Edgar J, Brinkmann BG, Kassmann CM, Tzvetanova ID, Möbius W, et al. 2012. Glycolytic oligodendrocytes maintain myelin and long-term axonal integrity. *Nature* **485:** 517–521. doi:10.1038/nature11007

Gate D, Saligrama N, Leventhal O, Yang AC, Unger MS, Middeldorp J, Chen K, Lehallier B, Channappa D, De Los Santos MB, et al. 2020. Clonally expanded CD8T cells patrol the cerebrospinal fluid in Alzheimer's disease. *Nature* **577:** 399–404. doi:10.1038/s41586-019-1895-7

Gebhardt C, Körner R, Heinemann U. 2002. Delayed anoxic depolarizations in hippocampal neurons of mice lacking the excitatory amino acid carrier 1. *J Cereb Blood Flow Metab* **22:** 569–575. doi:10.1097/00004647-200205000-00008

Gonzalez-Fernandez E, Liu Y, Auchus AP, Fan F, Roman RJ. 2021. Vascular contributions to cognitive impairment and dementia: the emerging role of 20-HETE. *Clin Sci (London)* **135:** 1929–1944. doi:10.1042/CS20201033

Gordon GRJ, Choi HB, Rungta RL, Ellis-Davies GCR, Mac-Vicar BA. 2008. Brain metabolism dictates the polarity of astrocyte control over arterioles. *Nature* **456:** 745–749. doi:10.1038/nature07525

Gouna G, Klose C, Bosch-Queralt M, Liu L, Gokce O, Schifferer M, Cantuti-Castelvetri L, Simons M. 2021. TREM2-dependent lipid droplet biogenesis in phagocytes is required for remyelination. *J Exp Med* **218:** e20210227. doi:10.1084/jem.20210227

Grabert K, Michoel T, Karavolos MH, Clohisey S, Baillie JK, Stevens MP, Freeman TC, Summers KM, Mccoll BW. 2016. Microglial brain region–dependent diversity and selective regional sensitivities to ageing. *Nat Neurosci* **19:** 504–516. doi:10.1038/nn.4222

Grubman A, Chew G, Ouyang JF, Sun G, Choo XY, McLean C, Simmons RK, Buckberry S, Vargas-Landin DB, Poppe D, et al. 2019. A single-cell atlas of entorhinal cortex from individuals with Alzheimer's disease reveals cell-type-specific gene expression regulation. *Nat Neurosci* **22:** 2087–2097. doi:10.1038/s41593-019-0539-4

Guerreiro R, Wojtas A, Bras J, Carrasquillo M, Rogaeva E, Majounie E, Cruchaga C, Sassi C, Kauwe JSK, Younkin S, et al. 2013. *TREM2* variants in Alzheimer's disease. *N Eng J Med* **368:** 117–127. doi:10.1056/NEJMoa1211851

Gunner G, Cheadle L, Johnson KM, Ayata P, Badimon A, Mondo E, Nagy MA, Liu L, Bemiller SM, Kim KW, et al. 2019. Sensory lesioning induces microglial synapse elimination via ADAM10 and fractalkine signaling. *Nat Neurosci* **22:** 1075–1088. doi:10.1038/s41593-019-0419-y

Guo H, Lai L, Butchbach MER, Stockinger MP, Shan X, Bishop GA, Lin CG. 2003. Increased expression of the glial glutamate transporter EAAT2 modulates excitotoxicity and delays the onset but not the outcome of ALS in mice. *Hum Mol Genet* **12:** 2519–2532. doi:10.1093/hmg/ddg267

Habib N, McCabe C, Medina S, Varshavsky M, Kitsberg D, Dvir-Szternfeld R, Green G, Dionne D, Nguyen L, Marshall JL, et al. 2020. Disease-associated astrocytes in Alzheimer's disease and aging. *Nat Neurosci* **23:** 701–706. doi:10.1038/s41593-020-0624-8

Hall CN, Reynell C, Gesslein B, Hamilton NB, Mishra A, Sutherland BA, O'Farrell FM, Buchan AM, Lauritzen M, Attwell D. 2014. Capillary pericytes regulate cerebral blood flow in health and disease. *Nature* **508:** 55–60. doi:10.1038/nature13165

Hamann M, Rossi DJ, Marie H, Attwell D. 2002. Knocking out the glial glutamate transporter GLT-1 reduces glutamate uptake but does not affect hippocampal glutamate dynamics in early simulated ischaemia. *Eur J Neurosci* **15:** 308–314. doi:10.1046/j.0953-816x.2001.01861.x

Hammond TR, Gadea A, Dupree J, Kerninon C, Nait-Oumesmar B, Aguirre A, Gallo V. 2014. Astrocyte-derived endothelin-1 inhibits remyelination through Notch activation. *Neuron* **81:** 588–602. doi:10.1016/j.neuron.2013.11.015

Hammond TR, Dufort C, Dissing-Olesen L, Giera S, Young A, Wysoker A, Walker AJ, Gergits F, Segel M, Nemesh J, et al. 2019. Single-cell RNA sequencing of microglia throughout the mouse lifespan and in the injured brain reveals complex cell-state changes. *Immunity* **50:** 253–271.e6. doi:10.1016/j.immuni.2018.11.004

Han RT, Kim RD, Molofsky AV, Liddelow SA. 2021. Astrocyte-immune cell interactions in physiology and pathology. *Immunity* **54:** 211–224. doi:10.1016/j.immuni.2021.01.013

Hansen DV, Hanson JE, Sheng M. 2018. Microglia in Alzheimer's disease. *J Cell Biol* **217:** 459–472. doi:10.1083/jcb.201709069

Hefendehl JK, LeDue J, Ko RWY, Mahler J, Murphy TH, MacVicar BA. 2016. Mapping synaptic glutamate transporter dysfunction in vivo to regions surrounding Aβ plaques by iGluSnFR two-photon imaging. *Nat Commun* **7:** 13441. doi:10.1038/ncomms13441

Heneka MT, Kummer MP, Stutz A, Delekate A, Schwartz S, Vieira-Saecker A, Griep A, Axt D, Remus A, Tzeng T-C, et al. 2013. NLRP3 is activated in Alzheimer's disease and contributes to pathology in APP/PS1 mice. *Nature* **493:** 674–678. doi:10.1038/nature11729

Herculano-Houzel S. 2009. The human brain in numbers: a linearly scaled-up primate brain. *Front Hum Neurosci* **3:** 31. doi:10.3389/neuro.09.031.2009

Hickman SE, Kingery ND, Ohsumi TK, Borowsky ML, Wang L, Means TK, El Khoury J. 2013. The microglial sensome revealed by direct RNA sequencing. *Nat Neurosci* **16:** 1896–1905. doi:10.1038/nn.3554

Hong S, Beja-Glasser VF, Nfonoyim BM, Frouin A, Li S, Ramakrishnan S, Merry KM, Shi Q, Rosenthal A, Barres BA, et al. 2016. Complement and microglia mediate early synapse loss in Alzheimer mouse models. *Science* **352:** 712–716. doi:10.1126/science.aad8373

Hou J, Chen Y, Grajales-Reyes G, Colonna M. 2022. TREM2 dependent and independent functions of microglia in Alzheimer's disease. *Mol Neurodegener* **17:** 1–19. doi:10.1186/s13024-021-00511-x

Huang Y, Happonen KE, Burrola PG, O'Connor C, Hah N, Huang L, Nimmerjahn A, Lemke G. 2021. Microglia use TAM receptors to detect and engulf amyloid β plaques. *Nat Immunol* **22:** 586–594. doi:10.1038/s41590-021-00913-5

Humphrey MB, Xing J, Titus A. 2015. The TREM2-DAP12 signaling pathway in Nasu–Hakola disease: a molecular genetics perspective. *Res Rep Biochem* **5:** 89–100. doi:10.2147/RRBC.S58057

Humphrey J, Venkatesh S, Hasan R, Herb JT, de Paiva Lopes K, Küçükali F, Byrska-Bishop M, Evani US, Narzisi G, Fagegaltier D, et al. 2023. Integrative transcriptomic analysis of the amyotrophic lateral sclerosis spinal cord implicates glial activation and suggests new risk genes. *Nat Neurosci* **26:** 150–162. doi:10.1038/s41593-022-01205-3

Itagaki S, McGeer PL, Akiyama H. 1988. Presence of T-cytotoxic suppressor and leucocyte common antigen positive cells in Alzheimer's disease brain tissue. *Neurosci Lett* **91:** 259–264. doi:10.1016/0304-3940(88)90690-8

Jackson RJ, Hyman BT, Serrano-Pozo A. 2024. Multifaceted roles of APOE in Alzheimer disease. *Nat Rev Neurol* doi:10.1038/s41582-024-00988-2

Jairaman A, McQuade A, Granzotto A, Kang YJ, Chadarevian JP, Gandhi S, Parker I, Smith I, Cho H, Sensi SL, et al. 2022. TREM2 regulates purinergic receptor-mediated calcium signaling and motility in human iPSC-derived microglia. *eLife* **11:** e73021. doi:10.7554/eLife.73021

Jäkel S, Agirre E, Mendanha Falcão A, van Bruggen D, Lee KW, Knuesel I, Malhotra D, Ffrench-Constant C,

Williams A, Castelo-Branco G. 2019. Altered human oligodendrocyte heterogeneity in multiple sclerosis. *Nature* **566:** 543–547. doi:10.1038/s41586-019-0903-2

Jay TR, Miller CM, Cheng PJ, Graham LC, Bemiller S, Broihier ML, Xu G, Margevicius D, Karlo JC, Sousa GL, et al. 2015. TREM2 deficiency eliminates TREM2+ inflammatory macrophages and ameliorates pathology in Alzheimer's disease mouse models. *J Exp Med* **212:** 287–295. doi:10.1084/jem.20142322

Jay TR, Hirsch AM, Broihier ML, Miller CM, Neilson LE, Ransohoff RM, Lamb BT, Landreth GE. 2017. Disease progression-dependent effects of TREM2 deficiency in a mouse model of Alzheimer's disease. *J Neurosci* **37:** 637–647. doi:10.1523/JNEUROSCI.2110-16.2016

Johnson ECB, Carter EK, Dammer EB, Duong DM, Gerasimov ES, Liu Y, Liu J, Betarbet R, Ping L, Yin L, et al. 2022. Large-scale deep multi-layer analysis of Alzheimer's disease brain reveals strong proteomic disease-related changes not observed at the RNA level. *Nat Neurosci* **25:** 213–225. doi:10.1038/s41593-021-00999-y

Jonsson T, Stefansson H, Steinberg S, Jonsdottir I, Jonsson PV, Snaedal J, Bjornsson S, Huttenlocher J, Levey AI, Lah JJ, et al. 2013. Variant of TREM2 associated with the risk of Alzheimer's disease. *N Eng J Med* **368:** 107–116. doi:10.1056/NEJMoa1211103

Jorfi M, Park J, Hall CK, Lin CCJ, Chen M, von Maydell D, Kruskop JM, Kang B, Choi Y, Prokopenko D, et al. 2023. Infiltrating CD8+ T cells exacerbate Alzheimer's disease pathology in a 3D human neuroimmune axis model. *Nat Neurosci* **26:** 1489–1504. doi:10.1038/s41593-023-01415-3

Katayama Y, Becker DP, Tamura T, Hovda DA. 1990. Massive increases in extracellular potassium and the indiscriminate release of glutamate following concussive brain injury. *J Neurosurg* **73:** 889–900. doi:10.3171/jns.1990.73.6.0889

Kedia S, Ji H, Feng R, Androvic P, Spieth L, Liu L, Franz J, Zdiarstek H, Anderson KP, Kaboglu C, et al. 2024. T cell-mediated microglial activation triggers myelin pathology in a mouse model of amyloidosis. *Nat Neurosci* doi:10.1038/s41593-024-01682-8

Keren-Shaul H, Spinrad A, Weiner A, Matcovitch-Natan O, Dvir-Sziternfeld R, Ulland TK, David E, Baruch K, Lara-Astaiso D, Toth B, et al. 2017. A unique microglia type associated with restricting development of Alzheimer's disease. *Cell* **169:** 1276–1290.e17. doi:10.1016/j.cell.2017.05.018

Kiani Shabestari S, Morabito S, Danhash EP, McQuade A, Sanchez JR, Miyoshi E, Chadarevian JP, Claes C, Coburn MA, Hasselmann J, et al. 2022. Absence of microglia promotes diverse pathologies and early lethality in Alzheimer's disease mice. *Cell Rep* **39:** 110961. doi:10.1016/j.celrep.2022.110961

Kofuji P, Ceelen P, Zahs KR, Surbeck LW, Lester HA, Newman EA. 2000. Genetic inactivation of an inwardly rectifying potassium channel (Kir4.1 subunit) in mice: phenotypic impact in retina. *J Neurosci* **20:** 5733–5740. doi:10.1523/JNEUROSCI.20-15-05733.2000

Kuffler SW, Potter DD. 1964. Glia in the leech central nervous system: physiological properties and neuron-glia relationship. *J Neurophysiol* **27:** 290–320. doi:10.1152/jn.1964.27.2.290

Kunkle BW, Grenier-Boley B, Sims R, Bis JC, Damotte V, Naj AC, Boland A, Vronskaya M, van der Lee SJ, Amlie-Wolf A, et al. 2019. Genetic meta-analysis of diagnosed Alzheimer's disease identifies new risk loci and implicates Aβ, tau, immunity and lipid processing. *Nat Genet* **51:** 414–430. doi:10.1038/s41588-019-0358-2

Lake BB, Ai R, Kaeser GE, Salathia NS, Yung YC, Liu R, Wildberg A, Gao D, Fung H-L, Chen S, et al. 2016. Neuronal subtypes and diversity revealed by single-nucleus RNA sequencing of the human brain. *Science* **352:** 1586–1590. doi:10.1126/science.aaf1204

Lau S-F, Cao H, Fu AKY, Ip NY. 2020. Single-nucleus transcriptome analysis reveals dysregulation of angiogenic endothelial cells and neuroprotective glia in Alzheimer's disease. *Proc Natl Acad Sci* **117:** 25800–25809. doi:10.1073/pnas.2008762117

Lee Y, Morrison BM, Li Y, Lengacher S, Farah MH, Hoffman PN, Liu Y, Tsingalia A, Jin L, Zhang P-W, et al. 2012. Oligodendroglia metabolically support axons and contribute to neurodegeneration. *Nature* **487:** 443–448. doi:10.1038/nature11314

Lee JH, Kim JY, Noh S, Lee H, Lee SY, Mun JY, Park H, Chung WS. 2021. Astrocytes phagocytose adult hippocampal synapses for circuit homeostasis. *Nature* **590:** 612–617. doi:10.1038/s41586-020-03060-3

Lee E, Jung YJ, Park YR, Lim S, Choi YJ, Lee SY, Kim CH, Mun JY, Chung WS. 2022. A distinct astrocyte subtype in the aging mouse brain characterized by impaired protein homeostasis. *Nat Aging* **2:** 726–741. doi:10.1038/s43587-022-00257-1

Lemke G, Huang Y. 2022. The dense-core plaques of Alzheimer's disease are granulomas. *J Exp Med* **219:** 1–11. doi:10.1084/jem.20212477

Levy LM, Warr O, Attwell D. 1998. Stoichiometry of the glial glutamate transporter GLT-1 expressed inducibly in a Chinese hamster ovary cell line selected for low endogenous Na+-dependent glutamate uptake. *J Neurosci* **18:** 9620–9628. doi:10.1523/JNEUROSCI.18-23-09620.1998

Lewcock JW, Schlepckow K, Di Paolo G, Tahirovic S, Monroe KM, Haass C. 2020. Emerging microglia biology defines novel therapeutic approaches for Alzheimer's disease. *Neuron* **108:** 801–821. doi:10.1016/j.neuron.2020.09.029

Li Q, Cheng Z, Zhou L, Darmanis S, Neff NF, Okamoto J, Gulati G, Bennett ML, Sun LO, Clarke LE, et al. 2019. Developmental heterogeneity of microglia and brain myeloid cells revealed by deep single-cell RNA sequencing. *Neuron* **101:** 207–223.e10. doi:10.1016/j.neuron.2018.12.006

Li T, Yu D, Oak HC, Zhu B, Wang L, Jiang X, Molday RS, Kriegstein A, Piao X. 2021a. Phospholipid-flippase chaperone CDC50A is required for synapse maintenance by regulating phosphatidylserine exposure. *EMBO J* **40:** e107915. doi:10.15252/embj.2021107915

Li Z, McConnell HL, Stackhouse TL, Pike MM, Zhang W, Mishra A. 2021b. Increased 20-HETE signaling suppresses capillary neurovascular coupling after ischemic stroke in regions beyond the infarct. *Front Cell Neurosci* **15:** 762843. doi:10.3389/fncel.2021.762843

Liddelow SA, Barres BA. 2017. Reactive astrocytes: production, function, and therapeutic potential. *Immunity* **46:** 957–967. doi:10.1016/j.immuni.2017.06.006

Liddelow SA, Guttenplan KA, Clarke LE, Bennett FC, Bohlen CJ, Schirmer L, Bennett ML, Münch AE, Chung WS, Peterson TC, et al. 2017. Neurotoxic reactive astrocytes are induced by activated microglia. *Nature* **541:** 481–487. doi:10.1038/nature21029

Linnerbauer M, Wheeler MA, Quintana FJ. 2020. Astrocyte crosstalk in CNS inflammation. *Neuron* **108:** 608–622. doi:10.1016/j.neuron.2020.08.012

Lui H, Zhang J, Makinson SR, Cahill MK, Kelley KW, Huang HY, Shang Y, Oldham MC, Martens LH, Gao F, et al. 2016. Progranulin deficiency promotes circuit-specific synaptic pruning by microglia via complement activation. *Cell* **165:** 921–935. doi:10.1016/j.cell.2016.04.001

Luquez T, Gaur P, Kosater IM, Lam M, Lee DI, Mares J, Paryani F, Yadav A, Menon V. 2022. Cell type-specific changes identified by single-cell transcriptomics in Alzheimer's disease. *Genome Med* **14:** 136. doi:10.1186/s13073-022-01136-5

Madry C, Kyrargyri V, Arancibia-Cárcamo IL, Jolivet R, Kohsaka S, Bryan RM, Attwell D. 2018. Microglial ramification, surveillance, and interleukin-1β release are regulated by the two-pore domain K$^+$ channel THIK-1. *Neuron* **97:** 299–312.e6. doi:10.1016/j.neuron.2017.12.002

Makarov NS, Spiridonova SV, Nikitina VV, Voskresenskaia ON, Zakharova NB. 2013. Molecular markers of endothelial damage in patients with Parkinson's disease. *Zh Nevrol Psikhiatr Im S S Korsakova* **113:** 61–64.

Mallach A, Zielonka M, van Lieshout V, An Y, Khoo JH, Vanheusden M, Chen WT, Moechars D, Arancibia-Carcamo IL, Fiers M, et al. 2024. Microglia-astrocyte crosstalk in the amyloid plaque niche of an Alzheimer's disease mouse model, as revealed by spatial transcriptomics. *Cell Rep* **43:** 114216. doi:10.1016/j.celrep.2024.114216

Maniatis S, Äijö T, Vickovic S, Braine C, Kang K, Mollbrink A, Fagegaltier D, Andrusivová Ž, Saarenpää S, Saiz-Castro G, et al. 2019. Spatiotemporal dynamics of molecular pathology in amyotrophic lateral sclerosis. *Science* **364:** 89–93. doi:10.1126/science.aav9776

Maniatis S, Petrescu J, Phatnani H. 2021. Spatially resolved transcriptomics and its applications in cancer. *Curr Opin Genet Dev* **66:** 70–77. doi:10.1016/j.gde.2020.12.002

Marques S, Zeisel A, Codeluppi S, van Bruggen D, Mendanha Falcão A, Xiao L, Li H, Häring M, Hochgerner H, Romanov RA, et al. 2016. Oligodendrocyte heterogeneity in the mouse juvenile and adult central nervous system. *Science* **352:** 1326–1329. doi:10.1126/science.aaf6463

Marschallinger J, Iram T, Zardeneta M, Lee SE, Lehallier B, Haney MS, Pluvinage JV, Mathur V, Hahn O, Morgens DW, et al. 2020. Lipid-droplet-accumulating microglia represent a dysfunctional and proinflammatory state in the aging brain. *Nat Neurosci* **23:** 194–208. doi:10.1038/s41593-019-0566-1

Masuda T, Sankowski R, Staszewski O, Böttcher C, Amann L, Sagar, Scheiwe C, Nessler S, Kunz P, van Loo G, et al. 2019. Spatial and temporal heterogeneity of mouse and human microglia at single-cell resolution. *Nature* **566:** 388–392. doi:10.1038/s41586-019-0924-x

Mathys H, Davila-Velderrain J, Peng Z, Gao F, Mohammadi S, Young JZ, Menon M, He L, Abdurrob F, Jiang X, et al. 2019. Single-cell transcriptomic analysis of Alzheimer's disease. *Nature* **570:** 332–337. doi:10.1038/s41586-019-1195-2

Mazaheri F, Snaidero N, Kleinberger G, Madore C, Daria A, Werner G, Krasemann S, Capell A, Trümbach D, Wurst W, et al. 2017. TREM2 deficiency impairs chemotaxis and microglial responses to neuronal injury. *EMBO Rep* **18:** 1186–1198. doi:10.15252/embr.201743922

McNamara NB, Munro DAD, Bestard-Cuche N, Uyeda A, Bogie JFJ, Hoffmann A, Holloway RK, Molina-Gonzalez I, Askew KE, Mitchell S, et al. 2023. Microglia regulate central nervous system myelin growth and integrity. *Nature* **613:** 120–129. doi:10.1038/s41586-022-05534-y

Meilandt WJ, Ngu H, Gogineni A, Lalehzadeh G, Lee S-H, Srinivasan K, Imperio J, Wu T, Weber M, Kruse AJ, et al. 2020. Trem2 deletion reduces late-stage amyloid plaque accumulation, elevates the Aβ42:Aβ40 Ratio, and exacerbates axonal dystrophy and dendritic spine loss in the PS2APP Alzheimer's mouse model. *J Neurosci* **40:** 1956–1974. doi:10.1523/JNEUROSCI.1871-19.2019

Meyer N, Richter N, Fan Z, Siemonsmeier G, Pivneva T, Jordan P, Steinhäuser C, Semtner M, Nolte C, Kettenmann H. 2018. Oligodendrocytes in the mouse corpus callosum maintain axonal function by delivery of glucose. *Cell Rep* **22:** 2383–2394. doi:10.1016/j.celrep.2018.02.022

Meyer-Luehmann M, Spires-Jones TL, Prada C, Garcia-Alloza M, de Calignon A, Rozkalne A, Koenigsknecht-Talboo J, Holtzman DM, Bacskai BJ, Hyman BT. 2008. Rapid appearance and local toxicity of amyloid-beta plaques in a mouse model of Alzheimer's disease. *Nature* **451:** 720–724. doi:10.1038/nature06616

Mills SA, Jobling AI, Dixon MA, Bui BV, Vessey KA, Phipps JA, Greferath U, Venables G, Wong VHY, Wong CHY, et al. 2021. Fractalkine-induced microglial vasoregulation occurs within the retina and is altered early in diabetic retinopathy. *Proc Natl Acad Sci* **118:** e2112561118. doi:10.1073/pnas.2112561118

Mishra A, Reynolds JP, Chen Y, Gourine AV, Rusakov DA, Attwell D. 2016. Astrocytes mediate neurovascular signaling to capillary pericytes but not to arterioles. *Nat Neurosci* **19:** 1619–1627. doi:10.1038/nn.4428

Mookherjee P, Green PS, Watson GS, Marques MA, Tanaka K, Meeker KD, Meabon JS, Li N, Zhu P, Olson VG, et al. 2011. GLT-1 loss accelerates cognitive deficit onset in an Alzheimer's disease animal model. *J Alzheimers Dis* **26:** 447–455. doi:10.3233/JAD-2011-110503

Mot AI, Depp C, Nave K-A. 2018. An emerging role of dysfunctional axon-oligodendrocyte coupling in neurodegenerative diseases. *Dialogues Clin Neurosci* **20:** 283–292. doi:10.31887/dcns.2018.20.4/amot

Mulligan SJ, MacVicar BA. 2004. Calcium transients in astrocyte endfeet cause cerebrovascular constrictions. *Nature* **431:** 195–199. doi:10.1038/nature02827

Munro DAD, Bestard-Cuche N, McQuaid C, Chagnot A, Shabestari SK, Chadarevian JP, Maheshwari U, Szymkowiak S, Morris K, Mohammad M, et al. 2024. Microglia protect against age-associated brain pathologies. *Neuron* doi:10.1016/j.neuron.2024.05.018

Nave K-A, Werner HB. 2021. Ensheathment and myelination of axons: evolution of glial functions. *Annu Rev Neurosci* **44:** 197–219. doi:10.1146/annurev-neuro-100120-122621

Neusch C, Papadopoulos N, Müller M, Maletzki I, Winter SM, Hirrlinger J, Handschuh M, Bähr M, Richter DW, Kirchhoff F, et al. 2006. Lack of the Kir4.1 channel subunit abolishes K⁺ buffering properties of astrocytes in the ventral respiratory group: impact on extracellular K⁺ regulation. *J Neurophysiol* **95**: 1843–1852. doi:10.1152/jn.00996.2005

Newman EA. 1986. High potassium conductance in astrocyte endfeet. *Science* **233**: 453–454. doi:10.1126/science.3726539

Nimmerjahn A, Kirchhoff F, Helmchen F. 2005. Resting microglial cells are highly dynamic surveillants of brain parenchyma in vivo. *Science* **308**: 1314–1318. doi:10.1126/science.1110647

Nortley R, Korte N, Izquierdo P, Hirunpattarasilp C, Mishra A, Jaunmuktane Z, Kyrargyri V, Pfeiffer T, Khennouf L, Madry C, et al. 2019. Amyloid β oligomers constrict human capillaries in Alzheimer's disease via signaling to pericytes. *Science* **365**: eaav9518. doi:10.1126/science.aav9518

Nugent AA, Lin K, van Lengerich B, Lianoglou S, Przybyla L, Davis SS, Llapashtica C, Wang J, Kim DJ, Xia D, et al. 2020. TREM2 regulates microglial cholesterol metabolism upon chronic phagocytic challenge. *Neuron* **105**: 837–854.e9. doi:10.1016/j.neuron.2019.12.007

Nwaobi SE, Cuddapah VA, Patterson KC, Randolph AC, Olsen ML. 2016. The role of glial-specific Kir4.1 in normal and pathological states of the CNS. *Acta Neuropathol* **132**: 1–21. doi:10.1007/s00401-016-1553-1

Ochocka N, Kaminska B. 2021. Microglia diversity in healthy and diseased brain: insights from single-cell omics. *Int J Mol Sci* **22**: 3027. doi:10.3390/ijms22063027

Olney JW. 1969. Brain lesions, obesity, and other disturbances in mice treated with monosodium glutamate. *Science* **164**: 719–721. doi:10.1126/science.164.3880.719

Orkand RK. 1980. Extracellular potassium accumulation in the nervous system. *Fed Proc* **39**: 1515–1518.

Ortinski PI, Dong J, Mungenast A, Yue C, Takano H, Watson DJ, Haydon PG, Coulter DA. 2010. Selective induction of astrocytic gliosis generates deficits in neuronal inhibition. *Nat Neurosci* **13**: 584–591. doi:10.1038/nn.2535

Paolicelli RC, Bolasco G, Pagani F, Maggi L, Scianni M, Panzanelli P, Giustetto M, Ferreira TA, Guiducci E, Dumas L, et al. 2011. Synaptic pruning by microglia is necessary for normal brain development. *Science* **333**: 1456–1458. doi:10.1126/science.1202529

Paolicelli RC, Sierra A, Stevens B, Tremblay M-E, Aguzzi A, Ajami B, Amit I, Audinat E, Bechmann I, Bennett M, et al. 2022. Microglia states and nomenclature: a field at its crossroads. *Neuron* **110**: 3458–3483. doi:10.1016/j.neuron.2022.10.020

Parhizkar S, Arzberger T, Brendel M, Kleinberger G, Deussing M, Focke C, Nuscher B, Xiong M, Ghasemigharagoz A, Katzmarski N, et al. 2019. Loss of TREM2 function increases amyloid seeding but reduces plaque-associated ApoE. *Nat Neurosci* **22**: 191–204. doi:10.1038/s41593-018-0296-9

Pasciuto E, Burton OT, Roca CP, Lagou V, Rajan WD, Theys T, Mancuso R, Tito RY, Kouser L, Callaerts-Vegh Z, et al. 2020. Microglia require CD4T cells to complete the fetal-to-adult transition. *Cell* **182**: 625–640.e24. doi:10.1016/j.cell.2020.06.026

Patani R, Hardingham GE, Liddelow SA. 2023. Functional roles of reactive astrocytes in neuroinflammation and neurodegeneration. *Nat Rev Neurol* **19**: 395–409. doi:10.1038/s41582-023-00822-1

Pellerin L, Magistretti PJ. 2012. Sweet sixteen for ANLS. *J Cereb Blood Flow Metab* **32**: 1152–1166. doi:10.1038/jcbfm.2011.149

Peppiatt CM, Howarth C, Mobbs P, Attwell D. 2006. Bidirectional control of CNS capillary diameter by pericytes. *Nature* **443**: 700–704. doi:10.1038/nature05193

Philips T, Mironova YA, Jouroukhin Y, Chew J, Vidensky S, Farah MH, Pletnikov MV, Bergles DE, Morrison BM, Rothstein JD. 2021. MCT1 deletion in oligodendrocyte lineage cells causes late-onset hypomyelination and axonal degeneration. *Cell Rep* **34**: 108610. doi:10.1016/j.celrep.2020.108610

Piers TM, Cosker K, Mallach A, Johnson GT, Guerreiro R, Hardy J, Pocock JM. 2020. A locked immunometabolic switch underlies TREM2 R47H loss of function in human iPSC-derived microglia. *FASEB J* **34**: 2436–2450. doi:10.1096/fj.201902447R

Restrepo A, Trevisiol A, Restrepo-Arango C, Depp C, Sasmita AO, Keller A, Tzvetanova ID, Hirrlinger J, Nave K-A. 2022. Axo-vascular coupling mediated by oligodendrocytes. bioRxiv doi:10.1101/2022.06.16.495900

Robel S, Buckingham SC, Boni JL, Campbell SL, Danbolt NC, Riedemann T, Sutor B, Sontheimer H. 2015. Reactive astrogliosis causes the development of spontaneous seizures. *J Neurosci* **35**: 3330–3345. doi:10.1523/jneurosci.1574-14.2015

Rossi DJ, Oshima T, Attwell D. 2000. Glutamate release in severe brain ischaemia is mainly by reversed uptake. *Nature* **403**: 316–321. doi:10.1038/35002090

Rothstein JD. 1996. Excitotoxicity hypothesis. *Neurology* **47**: S19–S25, discussion S26. doi:10.1212/WNL.47.4_Suppl_2.19S

Rothstein JD, van Kammen M, Levey AI, Martin LJ, Kuncl RW. 1995. Selective loss of glial glutamate transporter GLT-1 in amyotrophic lateral sclerosis. *Ann Neurol* **38**: 73–84. doi:10.1002/ana.410380114

Rueda-Carrasco J, Sokolova D, Lee SE, Childs T, Jurčáková N, Crowley G, De Schepper S, Ge JZ, Lachica JI, Toomey CE, et al. 2023. Microglia-synapse engulfment via PtdSer-TREM2 ameliorates neuronal hyperactivity in Alzheimer's disease models. *EMBO J* **42**: e113246. doi:10.15252/embj.2022113246. Erratum for: *EMBO J*, 2024, doi:10.1038/s44318-024-00159-5

Saab AS, Tzvetavona ID, Trevisiol A, Baltan S, Dibaj P, Kusch K, Möbius W, Goetze B, Jahn HM, Huang W, et al. 2016. Oligodendroglial NMDA receptors regulate glucose import and axonal energy metabolism. *Neuron* **91**: 119–132. doi:10.1016/j.neuron.2016.05.016

Sadick JS, O'Dea MR, Hasel P, Dykstra T, Faustin A, Liddelow SA. 2022. Astrocytes and oligodendrocytes undergo subtype-specific transcriptional changes in Alzheimer's disease. *Neuron* **110**: 1788–1805.e10. doi:10.1016/j.neuron.2022.03.008

Safaiyan S, Besson-Girard S, Kaya T, Cantuti-Castelvetri L, Liu L, Ji H, Schifferer M, Gouna G, Usifo F, Kannaiyan N, et al. 2021. White matter aging drives microglial diversity.

Neuron **109:** 1100–1117.e10. doi:10.1016/j.neuron.2021 .01.027

Sala Frigerio C, Wolfs L, Fattorelli N, Thrupp N, Voytyuk I, Schmidt I, Mancuso R, Chen WT, Woodbury ME, Srivastava G, et al. 2019. The major risk factors for Alzheimer's disease: age, sex, and genes modulate the microglia response to Aβ plaques. *Cell Rep* **27:** 1293–1306.e6. doi:10 .1016/j.celrep.2019.03.099

Schafer DP, Lehrman EK, Kautzman AG, Koyama R, Mardinly AR, Yamasaki R, Ransohoff RM, Greenberg ME, Barres BA, Stevens B. 2012. Microglia sculpt postnatal neural circuits in an activity and complement-dependent manner. *Neuron* **74:** 691–705. doi:10.1016/j.neuron.2012 .03.026

Scheiblich H, Dansokho C, Mercan D, Schmidt SV, Bousset L, Wischhof L, Eikens F, Odainic A, Spitzer J, Griep A, et al. 2021. Microglia jointly degrade fibrillar α-synuclein cargo by distribution through tunneling nanotubes. *Cell* **184:** 5089–5106.e21. doi:10.1016/j.cell.2021.09.007

Schlepckow K, Monroe KM, Kleinberger G, Cantuti-Castelvetri L, Parhizkar S, Xia D, Willem M, Werner G, Pettkus N, Brunner B, et al. 2020. Enhancing protective microglial activities with a dual function TREM2 antibody to the stalk region. *EMBO Mol Med* **12:** 1–22. doi:10.15252/ emmm.201911227

Schlepckow K, Morenas-Rodríguez E, Hong S, Haass C. 2023. Stimulation of TREM2 with agonistic antibodies —an emerging therapeutic option for Alzheimer's disease. *Lancet Neurol* **22:** 1048–1060. doi:10.1016/S1474-4422(23)00247-8

Scholl UI, Choi M, Liu T, Ramaekers VT, Häusler MG, Grimmer J, Tobe SW, Farhi A, Nelson-Williams C, Lifton RP. 2009. Seizures, sensorineural deafness, ataxia, mental retardation, and electrolyte imbalance (SeSAME syndrome) caused by mutations in *KCNJ10*. *Proc Natl Acad Sci* **106:** 5842–5847. doi:10.1073/pnas.0901749106

Scott-Hewitt N, Perrucci F, Morini R, Erreni M, Mahoney M, Witkowska A, Carey A, Faggiani E, Schuetz LT, Mason S, et al. 2020. Local externalization of phosphatidylserine mediates developmental synaptic pruning by microglia. *EMBO J* **39:** 1–20. doi:10.15252/embj.2020105380

Seeker LA, Williams A. 2022. Oligodendroglia heterogeneity in the human central nervous system. *Acta Neuropathol* **143:** 143–157. doi:10.1007/s00401-021-02390-4

Selkoe DJ. 2002. Alzheimer's disease is a synaptic failure. *Science* **298:** 789–791. doi:10.1126/science.1074069

Shankar GM, Li S, Mehta TH, Garcia-Munoz A, Shepardson NE, Smith I, Brett FM, Farrell MA, Rowan MJ, Lemere CA, et al. 2008. Amyloid-beta protein dimers isolated directly from Alzheimer's brains impair synaptic plasticity and memory. *Nat Med* **14:** 837–842. doi:10.1038/ nm1782

Shi Q, Chowdhury S, Ma R, Le KX, Hong S, Caldarone BJ, Stevens B, Lemere CA. 2017. Complement C3 deficiency protects against neurodegeneration in aged plaque-rich APP/PS1 mice. *Sci Transl Med* **9:** eaaf6295. doi:10.1126/ scitranslmed.aaf6295

Silvin A, Uderhardt S, Piot C, Da Mesquita S, Yang K, Geirsdottir L, Mulder K, Eyal D, Liu Z, Bridlance C, et al. 2022. Dual ontogeny of disease-associated microglia and disease inflammatory macrophages in aging and neurode-

generation. *Immunity* **55:** 1448–1465.e6. doi:10.1016/j .immuni.2022.07.004

Simard AR, Soulet D, Gowing G, Julien JP, Rivest S. 2006. Bone marrow-derived microglia play a critical role in restricting senile plaque formation in Alzheimer's disease. *Neuron* **49:** 489–502. doi:10.1016/j.neuron.2006.01.022

Sims R, Van Der Lee SJ, Naj AC, Bellenguez C, Badarinarayan N, Jakobsdottir J, Kunkle BW, Boland A, Raybould R, Bis JC, et al. 2017. Rare coding variants in PLCG2, ABI3, and TREM2 implicate microglial-mediated innate immunity in Alzheimer's disease. *Nat Genet* **49:** 1373–1384. doi:10.1038/ng.3916

Sokolova D, Childs T, Hong S. 2021. Insight into the role of phosphatidylserine in complement-mediated synapse loss in Alzheimer's disease. *Fac Rev* **10:** 19. doi:10 .12703/r/10-19

Sosna J, Philipp S, Albay R, Reyes-Ruiz JM, Baglietto-Vargas D, LaFerla FM, Glabe CG. 2018. Early long-term administration of the CSF1R inhibitor PLX3397 ablates microglia and reduces accumulation of intraneuronal amyloid, neuritic plaque deposition and pre-fibrillar oligomers in 5XFAD mouse model of Alzheimer's disease. *Mol Neurodegener* **13:** 11. doi:10.1186/s13024-018-0244-x

Spangenberg E, Severson PL, Hohsfield LA, Crapser J, Zhang J, Burton EA, Zhang Y, Spevak W, Lin J, Phan NY, et al. 2019. Sustained microglial depletion with CSF1R inhibitor impairs parenchymal plaque development in an Alzheimer's disease model. *Nat Commun* **10:** 3758. doi:10 .1038/s41467-019-11674-z

Spires-Jones TL, Hyman BT. 2014. The intersection of amyloid β and tau at synapses in Alzheimer's disease. *Neuron* **82:** 756–771. doi:10.1016/j.neuron.2014.05.004

Squarzoni P, Oller G, Hoeffel G, Pont-Lezica L, Rostaing P, Low D, Bessis A, Ginhoux F, Garel S. 2014. Microglia modulate wiring of the embryonic forebrain. *Cell Rep* **8:** 1271–1279. doi:10.1016/j.celrep.2014.07.042

Stephan AH, Madison DV, Mateos JM, Fraser DA, Lovelett EA, Coutellier L, Kim L, Tsai HH, Huang EJ, Rowitch DH, et al. 2013. A dramatic increase of C1q protein in the CNS during normal aging. *J Neurosci* **33:** 13460–13474. doi:10 .1523/JNEUROSCI.1333-13.2013

Stern AM, Yang Y, Jin S, Yamashita K, Meunier AL, Liu W, Cai Y, Ericsson M, Liu L, Goedert M, et al. 2023. Abundant Aβ fibrils in ultracentrifugal supernatants of aqueous extracts from Alzheimer's disease brains. *Neuron* **111:** 2012–2020.e4. doi:10.1016/j.neuron.2023.04.007

Stevens B, Allen NJ, Vazquez LE, Howell GR, Christopherson KS, Nouri N, Micheva KD, Mehalow AK, Huberman AD, Stafford B, et al. 2007. The classical complement cascade mediates CNS synapse elimination. *Cell* **131:** 1164–1178. doi:10.1016/j.cell.2007.10.036

Stickels RR, Murray E, Kumar P, Li J, Marshall JL, di Bella DJ, Arlotta P, Macosko EZ, Chen F. 2021. Highly sensitive spatial transcriptomics at near-cellular resolution with Slide-seqV2. *Nat Biotechnol* **39:** 313–319. doi:10.1038/ s41587-020-0739-1

Stogsdill JA, Kim K, Binan L, Farhi SL, Levin JZ, Arlotta P. 2022. Pyramidal neuron subtype diversity governs microglia states in the neocortex. *Nature* **608:** 750–756. doi:10.1038/s41586-022-05056-7

Stratoulias V, Venero JL, Tremblay M-È, Joseph B. 2019. Microglial subtypes: diversity within the microglial community. *EMBO J* **38**: e101997. doi:10.15252/embj.2019101997

Sun J, Song Y, Chen Z, Qiu J, Zhu S, Wu L, Xing L. 2022. Heterogeneity and molecular markers for CNS glial cells revealed by single-cell transcriptomics. *Cell Mol Neurobiol* **42**: 2629–2642. doi:10.1007/s10571-021-01159-3

Sweeney MD, Kisler K, Montagne A, Toga AW, Zlokovic BV. 2018. The role of brain vasculature in neurodegenerative disorders. *Nat Neurosci* **21**: 1318–1331. doi:10.1038/s41593-018-0234-x

Szatkowski M, Barbour B, Attwell D. 1990. Non-vesicular release of glutamate from glial cells by reversed electrogenic glutamate uptake. *Nature* **348**: 443–446. doi:10.1038/348443a0

Takalo M, Wittrahm R, Wefers B, Parhizkar S, Jokivarsi K, Kuulasmaa T, Mäkinen P, Martiskainen H, Wurst W, Xiang X, et al. 2020. The Alzheimer's disease-associated protective Plcγ2-P522R variant promotes immune functions. *Mol Neurodegener* **15**: 1–14. doi:10.1186/s13024-020-00402-7

Tang F, Barbacioru C, Wang Y, Nordman E, Lee C, Xu N, Wang X, Bodeau J, Tuch BB, Siddiqui A, et al. 2009. mRNA-seq whole-transcriptome analysis of a single cell. *Nat Methods* **6**: 377–382. doi:10.1038/nmeth.1315

Tcw J, Qian L, Pipalia NH, Chao MJ, Liang SA, Shi Y, Jain BR, Bertelsen SE, Kapoor M, Marcora E, et al. 2022. Cholesterol and matrisome pathways dysregulated in astrocytes and microglia. *Cell* **185**: 2213–2233.e25. doi:10.1016/j.cell.2022.05.017

Tesi N, van der Lee S, Hulsman M, van Schoor NM, Huisman M, Pijnenburg Y, van der Flier WM, Reinders M, Holstege H. 2024. Cognitively healthy centenarians are genetically protected against Alzheimer's disease. *Alzheimers Dement* **20**: 3864–3875. doi:10.1002/alz.13810

Togo T, Akiyama H, Iseki E, Kondo H, Ikeda K, Kato M, Oda T, Tsuchiya K, Kosaka K. 2002. Occurrence of T cells in the brain of Alzheimer's disease and other neurological diseases. *J Neuroimmunol* **124**: 83–92. doi:10.1016/S0165-5728(01)00496-9

Tong X, Ao Y, Faas GC, Nwaobi SE, Xu J, Haustein MD, Anderson MA, Mody I, Olsen ML, Sofroniew MV, et al. 2014. Astrocyte Kir4.1 ion channel deficits contribute to neuronal dysfunction in Huntington's disease model mice. *Nat Neurosci* **17**: 694–703. doi:10.1038/nn.3691

Trapp BD, Stys PK. 2009. Virtual hypoxia and chronic necrosis of demyelinated axons in multiple sclerosis. *Lancet Neurol* **8**: 280–291. doi:10.1016/S1474-4422(09)70043-2

Tremblay M-È, Stevens B, Sierra A, Wake H, Bessis A, Nimmerjahn A. 2011. The role of microglia in the healthy brain. *J Neurosci* **31**: 16064–16069. doi:10.1523/JNEUROSCI.4158-11.2011

Tsai AP, Dong C, Lin PBC, Oblak AL, Viana Di Prisco G, Wang N, Hajicek N, Carr AJ, Lendy EK, Hahn O, et al. 2023. Genetic variants of phospholipase C-γ2 alter the phenotype and function of microglia and confer differential risk for Alzheimer's disease. *Immunity* **56**: 2121–2136.e6. doi:10.1016/j.immuni.2023.08.008

Ulland TK, Song WM, Huang SC-C, Ulrich JD, Sergushichev A, Beatty WL, Loboda AA, Zhou Y, Cairns NJ, Kambal A, et al. 2017. TREM2 maintains microglial metabolic fitness in Alzheimer's disease. *Cell* **170**: 649–663.e13. doi:10.1016/j.cell.2017.07.023

Vainchtein ID, Chin G, Cho FS, Kelley KW, Miller JG, Chien EC, Liddelow SA, Nguyen PT, Nakao-Inoue H, Dorman LC, et al. 2018. Astrocyte-derived interleukin-33 promotes microglial synapse engulfment and neural circuit development. *Science* **359**: 1269–1273. doi:10.1126/science.aal3589

van Hove H, Martens L, Scheyltjens I, de Vlaminck K, Pombo Antunes AR, de Prijck S, Vandamme N, de Schepper S, van Isterdael G, Scott CL, et al. 2019. A single-cell atlas of mouse brain macrophages reveals unique transcriptional identities shaped by ontogeny and tissue environment. *Nat Neurosci* **22**: 1021–1035. doi:10.1038/s41593-019-0393-4

van Lengerich B, Zhan L, Xia D, Chan D, Joy D, Park JI, Tatarakis D, Calvert M, Hummel S, Lianoglou S, et al. 2023. A TREM2-activating antibody with a blood–brain barrier transport vehicle enhances microglial metabolism in Alzheimer's disease models. *Nat Neurosci* **26**: 416–429. doi:10.1038/s41593-022-01240-0

Vasek MJ, Garber C, Dorsey D, Durrant DM, Bollman B, Soung A, Yu J, Perez-Torres C, Frouin A, Wilton DK, et al. 2016. A complement-microglial axis drives synapse loss during virus-induced memory impairment. *Nature* **534**: 538–543. doi:10.1038/nature18283

Venegas C, Kumar S, Franklin BS, Dierkes T, Brinkschulte R, Tejera D, Vieira-Saecker A, Schwartz S, Santarelli F, Kummer MP, et al. 2017. Microglia-derived ASC specks cross-seed amyloid-β in Alzheimer's disease. *Nature* **552**: 355–361. doi:10.1038/nature25158

Viganò F, Möbius W, Götz M, Dimou L. 2013. Transplantation reveals regional differences in oligodendrocyte differentiation in the adult brain. *Nat Neurosci* **16**: 1370–1372. doi:10.1038/nn.3503

von Bernhardi R, Eugenín-von Bernhardi L, Eugenín J. 2015. Microglial cell dysregulation in brain aging and neurodegeneration. *Front Aging Neurosci* **7**: 124. doi:10.3389/fnagi.2015.00124

Wake H, Moorhouse AJ, Jinno S, Kohsaka S, Nabekura J. 2009. Resting microglia directly monitor the functional state of synapses in vivo and determine the fate of ischemic terminals. *J Neurosci* **29**: 3974–3980. doi:10.1523/JNEUROSCI.4363-08.2009

Wang Y, Cella M, Mallinson K, Ulrich JD, Young KL, Robinette ML, Gilfillan S, Krishnan GM, Sudhakar S, Zinselmeyer BH, et al. 2015. TREM2 lipid sensing sustains the microglial response in an Alzheimer's disease model. *Cell* **160**: 1061–1071. doi:10.1016/j.cell.2015.01.049

Wang N, Wang M, Jeevaratnam S, Rosenberg C, Ikezu TC, Shue F, Doss SV, Alnobani A, Martens YA, Wren M, et al. 2022. Opposing effects of apoE2 and apoE4 on microglial activation and lipid metabolism in response to demyelination. *Mol Neurodegener* **17**: 1–20. doi:10.1186/s13024-021-00511-x

Werneburg S, Jung J, Kunjamma RB, Ha SK, Luciano NJ, Willis CM, Gao G, Biscola NP, Havton LA, Crocker SJ, et al. 2020. Targeted complement inhibition at synapses prevents microglial synaptic engulfment and synapse loss in demyelinating disease. *Immunity* **52**: 167–182.e7. doi:10.1016/j.immuni.2019.12.004

Wilcock DM, Vitek MP, Colton CA. 2009. Vascular amyloid alters astrocytic water and potassium channels in mouse models and humans with Alzheimer's disease. *Neurosci-*

ence **159**: 1055–1069. doi:10.1016/j.neuroscience.2009.01.023

Wilson RS, Nairn AC. 2018. Cell-type-specific proteomics: a neuroscience perspective. *Proteomes* **6**: 51. doi:10.3390/proteomes6040051

Wilton DK, Mastro K, Heller MD, Gergits FW, Willing CR, Fahey JB, Frouin A, Daggett A, Gu X, Kim YA, et al. 2023. Microglia and complement mediate early corticostriatal synapse loss and cognitive dysfunction in Huntington's disease. *Nat Med* **29**: 2866–2884. doi:10.1038/s41591-023-02566-3

Wood TE, Barry J, Yang Z, Cepeda C, Levine MS, Gray M. 2019. Mutant huntingtin reduction in astrocytes slows disease progression in the BACHD conditional Huntington's disease mouse model. *Hum Mol Genet* **28**: 487–500. doi:10.1093/hmg/ddy363

Wood OWG, Yeung JHY, Faull RLM, Kwakowsky A. 2022. EAAT2 as a therapeutic research target in Alzheimer's disease: a systematic review. *Front Neurosci* **16**: 952096. doi:10.3389/fnins.2022.952096

Wu T, Dejanovic B, Gandham VD, Gogineni A, Edmonds R, Schauer S, Srinivasan K, Huntley MA, Wang Y, Wang TM, et al. 2019. Complement C3 is activated in human AD brain and is required for neurodegeneration in mouse models of amyloidosis and tauopathy. *Cell Rep* **28**: 2111–2123.e6. doi:10.1016/j.celrep.2019.07.060

Yuan P, Condello C, Keene CD, Wang Y, Bird TD, Paul SM, Luo W, Colonna M, Baddeley D, Grutzendler J. 2016. TREM2 haplodeficiency in mice and humans impairs the microglia barrier function leading to decreased amyloid compaction and severe axonal dystrophy. *Neuron* **90**: 724–739. doi:10.1016/j.neuron.2016.05.003

Zerangue N, Kavanaugh MP. 1996. Flux coupling in a neuronal glutamate transporter. *Nature* **383**: 634–637. doi:10.1038/383634a0

Zhou Y, Song WM, Andhey PS, Swain A, Levy T, Miller KR, Poliani PL, Cominelli M, Grover S, Gilfillan S, et al. 2020. Human and mouse single-nucleus transcriptomics reveal TREM2-dependent and TREM2-independent cellular responses in Alzheimer's disease. *Nat Med* **26**: 131–142. doi:10.1038/s41591-019-0695-9. Erratum for: *Nat Med*, 2020, **26**: 981. doi:10.1038/s41591-020-0922-4

Zonta M, Angulo MC, Gobbo S, Rosengarten B, Hossmann KA, Pozzan T, Carmignoto G. 2003. Neuron-to-astrocyte signaling is central to the dynamic control of brain microcirculation. *Nat Neurosci* **6**: 43–50. doi:10.1038/nn980

Zott B, Simon MM, Hong W, Unger F, Chen-Engerer HJ, Frosch MP, Sakmann B, Walsh DM, Konnerth A. 2019. A vicious cycle of β amyloid–dependent neuronal hyperactivation. *Science* **365**: 559–565. doi:10.1126/science.aay0198

Multiple Sclerosis and Other Acquired Demyelinating Diseases of the Central Nervous System

Michael D. Kornberg[1] and Peter A. Calabresi[1,2]

[1]Department of Neurology, Johns Hopkins University, Baltimore, Maryland 21287, USA

[2]Solomon H. Snyder Department of Neuroscience, Johns Hopkins University, Baltimore, Maryland 21205, USA

Correspondence: michael.kornberg@jhmi.edu; pcalabr1@jhmi.edu

Acquired demyelinating diseases of the central nervous system (CNS) comprise inflammatory conditions, including multiple sclerosis (MS) and related diseases, as well as noninflammatory conditions caused by toxic, metabolic, infectious, traumatic, and neurodegenerative insults. Here, we review the spectrum of diseases producing acquired CNS demyelination before focusing on the prototypical example of MS, exploring the pathologic mechanisms leading to myelin injury in relapsing and progressive MS and summarizing the mechanisms and modulators of remyelination. We highlight the complex interplay between the immune system, oligodendrocytes and oligodendrocyte progenitor cells (OPCs), and other CNS glia cells such as microglia and astrocytes in the pathogenesis and clinical course of MS. Finally, we review emerging therapeutic strategies that exploit our growing understanding of disease mechanisms to limit progression and promote remyelination.

Acquired demyelinating diseases are nonhereditary conditions that impact myelin health and produce neurologic disability as a result of pathologic insults to the myelin sheath and/or the oligodendrocytes (OLs) from which they arise. Within the central nervous system (CNS), the prototypical acquired demyelinating disease is multiple sclerosis (MS), in which myelin injury occurs alongside inflammation, namely, a presumed immune-mediated attack on OLs. However, the pathological spectrum of acquired CNS demyelination is broad and includes more rare inflammatory conditions as well as toxic, metabolic, infectious, and traumatic causes. Evidence now suggests that some neurodegenerative diseases, which were not traditionally thought to involve a role for myelin, may be associated with OL dysfunction as either a cause or consequence of the underlying pathology.

In this review, we first provide an overview of acquired demyelinating diseases impacting the CNS, dividing our discussion into inflammatory and noninflammatory causes. Although we define acquired demyelinating diseases as nonhereditary, it is important to note that some of these conditions are, in fact, associated with genetic risk factors. However, the contribution from individual gene variants is small, and none of these conditions is inherited in a simple Mendelian fashion. Following an initial overview of the spectrum of acquired CNS demyelinating disease, we focus primarily on MS, discussing the pathologic mechanisms producing myelin injury and limiting remyelination. We highlight the complex interplay between OLs, the immune system, and other CNS glia cells (namely, microglia and astrocytes) and the neurodegenerative consequences of chronic demyelination and reactive gliosis. Finally, we review how advances in our collective understanding of the pathobiology of demyelination and remyelination have contributed to current therapeutic approaches to limit myelin injury and promote myelin repair in MS.

OVERVIEW OF ACQUIRED CNS DEMYELINATING DISEASES

Inflammatory Demyelinating Diseases

Several autoimmune or inflammatory conditions produce demyelination as a primary pathological consequence. These diseases span both pediatric and adult populations and produce varying degrees of concomitant neuroaxonal injury. In some cases, a specific self-antigen has been identified as the target of immune reactivity, whereas in others the existence or identity of self-antigens remains elusive. With improvements in diagnostics and biomarker development, our pathologic classification of these disorders is becoming more refined, with consequences for prognosis and treatment. Below, we will review the major inflammatory demyelinating diseases, beginning with the prototypical example of MS.

Multiple Sclerosis

MS is by far the most common of the inflammatory demyelinating diseases, with an estimated prevalence in the United States of nearly one million individuals (Wallin et al. 2019a). Worldwide estimates range between two and three million people, although these are likely vast underestimates based on underdiagnosis in many parts of the world (Thompson et al. 2018; Wallin et al. 2019b; Hwang et al. 2022). The median age of diagnosis is 30 years, although a small proportion of new cases occurs in pediatric populations or in those over the age of 50

(Brenton 2022; Ward and Goldman 2022). As such, MS is the most common cause of nontraumatic neurologic disability in young adults.

Epidemiologic studies have demonstrated a female-to-male predominance between 2:1 and 3:1, with the ratio of females to males possibly increasing over time (Koch-Henriksen et al. 2018; Magyari and Sorensen 2019; Wallin et al. 2019a). The reasons underlying the female predominance in MS susceptibility and its increase over time have not been definitively established, although several factors have been suggested (Voskuhl 2020). Most autoimmune diseases have a higher incidence in females, which has been linked both to stronger baseline immune responses and greater susceptibility to the breakdown of immune tolerance (Rubtsova et al. 2015). In MS as well as other autoimmune diseases, sex hormones and differences in X-linked gene dosage (due to the presence of two X chromosomes in females compared to only one in males) appear to play primary roles in increased female susceptibility (Smith-Bouvier et al. 2008; Rubtsova et al. 2015; Itoh et al. 2019; Voskuhl 2020). Interestingly, the long noncoding RNA Xist, which is expressed only in females and forms ribonucleoprotein complexes (Xist-RNP) to silence one X chromosome for gene dosage compensation, has itself been implicated as a target of autoantibodies that drives sex differences in autoimmunity (Dou et al. 2024). The increasing female predominance over time appears to be driven specifically by the increasing incidence in females (Voskuhl 2020). Although the reasons for this epidemiologic shift remain uncertain, one suggestion has been a societal trend toward later and fewer pregnancies in Western countries; pregnancy plays a protective role in MS, and fewer pregnancies have been linked with a higher risk of MS attack (Ponsonby et al. 2012).

The risk of developing MS depends on a combination of genetic and environmental factors (Waubant et al. 2019). More than 200 risk alleles have been identified, most of which reside in or near genes associated with immune functions (Sawcer et al. 2014; Mitrović et al. 2018; Patsopoulos et al. 2019). However, genetic predisposition accounts for only a small proportion of risk, and a number of environmental risk factors have been identified. These environmental factors include viral infection (particularly with Epstein–Barr virus [EBV]), vitamin D level, smoking, and obesity, among others.

The pathologic hallmark of MS is the plaque, a focal area of demyelination associated with varying degrees of inflammation, neuroaxonal loss, remyelination, and gliosis (Reich et al. 2018). Although classically considered a white matter disease, demyelination also occurs in gray matter. For most individuals with MS, the disease begins with a "relapsing–remitting" phase characterized by discrete episodes of focal inflammatory demyelination in the CNS. These focal attacks of inflammation, which produce the characteristic plaques (also called lesions) visible on magnetic resonance imaging (MRI), can be asymptomatic or produce clinical symptoms depending on their location within the CNS. When symptomatic, these discrete attacks are termed "relapses." As acute inflammation resolves over days to weeks, a variable degree of clinical recovery occurs (remission). After a period of time, many individuals with relapsing–remitting MS will convert to a progressive phase (termed "secondary progressive MS," [SPMS]) characterized by insidious progression of disability despite the absence of discrete relapses. In the current era of early diagnosis and treatment, fewer individuals are developing SPMS compared to historical cohorts (Cree et al. 2016; Tintore et al. 2021). In 10%–15% of individuals, MS presents with insidious progression from onset without relapses, which is termed "primary progressive MS" [PPMS]. Disability, both physical and cognitive, accrues from a combination of incomplete recovery from relapse as well as insidious progression. Many therapies exist for MS, which primarily target the peripheral immune system to prevent inflammatory demyelinating attacks (Cross and Riley 2022). Although these therapies effectively prevent relapses and new plaque formation, which limits disability in individuals with relapsing–remitting MS, none has been shown to directly augment remyelination or slow noninflammatory progression.

MS is considered a demyelinating disease because plaques are characterized pathologically by the destruction of OLs and myelin with relative sparing of axons. However, relative sparing should be emphasized, and it is important to note that neuroaxonal loss does occur in MS

and contributes to permanent disability. Variable axonal loss occurs during acute plaque formation, and progressive MS is often characterized by diffuse atrophy and neuroaxonal loss in both white and gray matter (Trapp et al. 1998; Mahad et al. 2015; Reich et al. 2018; Thompson et al. 2018). The pathologic characteristics and distinctions between relapsing and progressive MS are discussed in detail in later sections.

Neuromyelitis Optica Spectrum Disorder

Neuromyelitis optica spectrum disorder (NMOSD) is a rare inflammatory disease with an estimated worldwide prevalence of 0.5–4 per 100,000 people—although prevalence varies based on geographic region and ethnic background (Costello 2022). In comparison to MS, the mean age of onset is greater (40 years), as is the female predominance (~9:1) (Hor et al. 2020; Papp et al. 2021). Also, in contrast to MS, a pathogenic self-antigen has been identified in the majority (~70%) of cases, in which antibodies against the water channel aquaporin-4 (AQP4) can be detected in patient sera.

Although brain (and particularly brainstem) involvement occurs in NMOSD, the disease has a predisposition for affecting the optic nerves and spinal cord in a longitudinally extensive manner. Although classically considered a demyelinating disease, the primary target of the immune attack appears to be astrocytes, which express AQP4 on end feet. Injury to OLs and myelin, in addition to neuroaxonal destruction, occurs secondary to the autoimmune astrocytopathy. Similar to MS, NMOSD is characterized by recurrent relapses of focal inflammatory attacks. However, NMOSD relapses are associated with greater disability and less recovery, often producing blindness and paralysis, and relapse-independent progression has not been described. Fortunately, a number of therapies have been shown to prevent relapses in AQP4-positive NMOSD (Costello 2022).

Myelin Oligodendrocyte Glycoprotein Antibody Disease

Myelin OL glycoprotein (MOG) antibody disease (MOGAD) refers to a spectrum of inflammatory demyelinating disorders associated with autoantibodies against MOG, a protein expressed on the outer myelin sheath (Longbrake 2022). First recognized as a distinct pathologic entity in 2015, the epidemiology, pathology, and clinical course of MOGAD are not fully understood. Unlike MS and NMOSD, MOGAD has a bimodal age distribution, with children representing approximately one-third of new cases, without a clear predilection based on biological sex (Jurynczyk et al. 2017; Brill et al. 2021).

MOGAD is associated with discrete episodes of inflammatory demyelination, although the clinical manifestations and disease course vary. Involvement of optic nerves and spinal cord is common, and individuals previously diagnosed with AQP4 autoantibody-negative NMOSD or other inflammatory demyelinating disorders (such as acute disseminated encephalomyelitis [ADEM], see below) are now classified as MOGAD based on the presence of MOG autoantibodies. Unlike MS and NMOSD, MOGAD can be a monophasic condition for at least a sizable minority of individuals, although relapse is common. Immunomodulatory therapies appear to decrease the risk of relapse, but the efficacy of specific therapies differs for MOGAD compared to MS or NMOSD (Chen et al. 2020, 2022; Barreras et al. 2022; Longbrake 2022).

Acute Disseminated Encephalomyelitis

In contrast to the inflammatory disorders described above, ADEM is primarily a disease of children, with the typical age of onset <8 years (Brenton 2022). As the name implies, it is defined by encephalopathy (altered level of arousal with cognitive dysfunction) from brain involvement but can also affect the optic nerves and spinal cord. ADEM commonly occurs following a preceding infection and can be associated with systemic symptoms such as fever. The majority of cases are monophasic and associated with favorable recovery, with symptoms and MRI findings (large lesions of both white and gray matter) resolving over a period of days to weeks. MOG autoantibodies are detected in nearly half of pediatric ADEM patients at presentation, including in most of the rare individuals who experience a

relapsing course (Bruijstens et al. 2020; Brenton 2022).

Noninflammatory Demyelinating Diseases

A wide spectrum of noninflammatory insults can produce demyelination as a primary pathologic consequence. This includes toxic and metabolic stressors to which OLs are particularly vulnerable, as well as OL-tropic viral infections and traumatic injury. Furthermore, animal models suggest that alterations in OL neurotrophic functions might directly contribute to neurodegeneration observed in aging and degenerative diseases such as amyotrophic lateral sclerosis (ALS) (Lee et al. 2012; Kang et al. 2013; Philips and Rothstein 2017; Henn et al. 2022). Mature OLs support axon health by providing lactate, which is shuttled through the monocarboxylate transporter 1 (MCT1) (Fünfschilling et al. 2012; Lee et al. 2012). OL expression of MCT1 decreases with age, and OL-specific knockout allows for normal developmental myelination but produces axonal degeneration in later life (Philips et al. 2021). In a superoxide dismutase 1 (SOD1) gain-of-function mouse model of ALS, OL-specific SOD1 deletion substantially delays disease onset and prolongs survival (Kang et al. 2013), supporting a potential primary role for OLs in the disease. In addition, aging and microvascular injury are associated with white matter pathology characterized by demyelination. In such cases, the senescence of OLs and increased susceptibility to metabolic stressors have been implicated. However, such mechanisms are difficult to completely deconvolute from inflammation, and recent studies have shown that interferon γ–secreting cytotoxic CD8 T cells may be involved in white matter aging (Kaya et al. 2022a).

A list of the most common and/or well-described noninflammatory causes of CNS demyelination is included in Table 1, and several salient examples are discussed in greater detail below.

Vitamin B12 (Cobalamin) Deficiency

Vitamin B12, or cobalamin, is a water-soluble vitamin obtained through dietary sources. It is present exclusively in animal-derived foods and is absent from plant-based nutritional sources

Table 1. Common causes of noninflammatory acquired demyelinating disease

Category	Etiology	Features/mechanism/pathology	References
Metabolic	Vitamin B12 (Cobalamin) deficiency	Vitamin B12 is an essential cofactor in methionine and folate metabolism. Deficiency causes swelling/disruption of the myelin sheath, followed by macrophage infiltration and axonal degeneration. Classic central nervous system manifestations include subacute combined degeneration, optic neuropathy, and cognitive dysfunction.	Ammouri et al. 2019; Nawaz et al. 2020; Parks 2021; Bhattacharyya et al. 2024
	Folate deficiency	Similar to vitamin B12 deficiency.	Parks 2021; Bhattacharyya et al. 2024
	Nitrous oxide exposure	Causes inactivation of vitamin B12, leading to functional B12 deficiency and related demyelinating manifestations.	Parks 2021; Bhattacharyya et al. 2024
	Copper deficiency	Mimics vitamin B12 deficiency.	Parks 2021; Bhattacharyya et al. 2024
	Osmotic myelinolysis (central pontine or extrapontine)	Demyelination due to osmotic stress, most commonly rapid correction of hyponatremia. Most often affects central pons but can involve cerebral hemispheres.	Brown 2000; Singh et al. 2014; Sarbu et al. 2016; Jacoby 2020
Toxic	Heroin inhalation (Chasing the Dragon)	Associated with inhaling the vapor of heated heroin. Sometimes fatal. Precise mechanism remains uncertain. Characterized by vacuolization of myelin with less frequent axonal degeneration.	Tormoehlen 2011
	High-dose methotrexate	Demyelination with variable degrees of axonal injury, associated with intrathecal or high-dose intravenous administration of methotrexate for cancer treatment.	Filley and Kleinschmidt-DeMasters 2001; Sarbu et al. 2016; Parks 2021
	MDMA (Ecstasy)	Axonal injury with secondary demyelination.	Bertram et al. 1999; Filley and Kleinschmidt-DeMasters 2001
	Marchiafava–Bignami disease	Associated with chronic, heavy alcohol consumption. Demyelination primarily impacting the corpus callosum.	Sarbu et al. 2016
Infectious	Progressive multifocal leukoencephalopathy (PML)	Due to infection and lysis of oligodendrocytes by John Cunningham (JC) virus. Seen in immunosuppressed individuals, such as those with acquired immunodeficiency syndrome (AIDS), lymphoma, or treated with immunosuppressive medications.	Aksamit 2012; Bernard-Valnet et al. 2021; Cortese et al. 2021b; Schweitzer et al. 2023

Continued

Table 1. *Continued*

Category	Etiology	Features/mechanism/pathology	References
	Human immunodeficiency virus (HIV) encephalopathy	HIV infection of microglia/macrophages causes demyelination and variable neuronal loss, astrogliosis, and vacuolar changes.	Sarbu et al. 2016
Traumatic/other	Delayed posthypoxic leukoencephalopathy (hypoxic/ischemic or carbon monoxide induced)	Rare demyelinating disease following hypoxia or carbon monoxide exposure. Characterized by recovery from initial insult followed by neurological deterioration over days to weeks, associated with diffuse demyelination and reactive microglia/macrophages and astrocytes.	Tormoehlen 2011
	Traumatic brain injury (TBI)	TBI causes oligodendrocyte apoptosis and myelin loss independent of axonal shearing. Mechanisms remain incompletely understood.	Marion et al. 2018; Mira et al. 2021

unless they have been artificially fortified (Stabler 2013). Vitamin B12 is protein-bound when ingested, and proper absorption requires gastric acids and pancreatic enzymes to dissociate the vitamin, which then binds a gastric-derived protein (called intrinsic factor [IF]) before the vitamin–IF complex is absorbed in the small intestines. As such, vitamin B12 deficiency can occur through either lack of nutritional access or as a result of conditions that prevent proper absorption. In developed countries, inadequate intake is most commonly associated with vegan diet. Malabsorption can occur due to gastric bypass, intestinal resection, inflammatory bowel disease, or chronic pancreatitis. Several medications can also decrease vitamin B12 absorption (Allen 2009; Stabler 2013). Vitamin B12 deficiency is more common among older individuals, as a result of gastric atrophy or pernicious anemia, which is an autoimmune disease targeting IF. In the United States and the United Kingdom, vitamin B12 deficiency was found in ∼6% of adults aged 60 or greater, whereas deficiency rates in children and adults in underresourced areas have been reported at 40% or higher (Allen 2009).

The classic neurologic manifestation of vitamin B12 deficiency is subacute combined degeneration, a condition that affects the dorsal col-

umns and lateral corticospinal tracts of the spinal cord (Parks 2021; Bhattacharyya et al. 2024). Symptoms generally progress over months and are characterized by sensory loss in the limbs followed by spastic weakness. Pathologically, subacute combined degeneration is characterized by myelin sheath swelling and vacuolization followed at later stages by secondary axonal degeneration (Russell et al. 1900; Pant et al. 1968; Dinn et al. 1978). Peripheral nerve involvement also commonly occurs, characterized by a combination of demyelination and axonal loss (Healton et al. 1991; Hemmer et al. 1998). Brain involvement in severe or long-standing cases of deficiency can cause cognitive dysfunction, and optic nerve involvement has also been described (Healton et al. 1991; Ammouri et al. 2019). Infants who suffer from vitamin B12 deficiency have been reported to experience cognitive delay or regression due to hypomyelination (Graham et al. 1992; Venkatramanan et al. 2016; Feraco et al. 2021). The rate of neurologic complications among individuals with vitamin B12 deficiency has not been precisely defined but has been estimated to occur in ∼15%–30% of these individuals, with recovery possible upon vitamin B12 supplementation (Ammouri et al. 2019; Bhattacharyya et al. 2024).

Cite this article as *Cold Spring Harb Perspect Biol* doi: 10.1101/cshperspect.a041374

Vitamin B12 serves as an essential cofactor for two human enzymes: methionine synthase and methylmalonyl-CoA mutase (Stabler 2013). Methionine synthase converts homocysteine to methionine, which contributes to one-carbon metabolism and is critical for purine/pyrimidine synthesis, methylation reactions, and other cellular functions. Methylmalonyl-CoA mutase converts methylmalonyl-CoA to succinyl-CoA, and its inhibition leads to the accumulation of odd-chain and branched-chain fatty acids. The precise mechanisms linking vitamin B12 deficiency to OL and myelin injury remain uncertain, and animal models are limited (Metz 1992). Decreased methylation of lipids and myelin basic protein (MBP) from methionine synthase dysfunction has been implicated in myelin injury (Metz 1992; Ammouri et al. 2019), and impairment of methylmalonyl-CoA mutase leads to the incorporation of excess odd-chain and branched-chain fatty acids into membrane lipids, potentially compromising myelin integrity (Frenkel et al. 1973; Metz 1992).

Osmotic Myelinolysis

Osmotic myelinolysis, which is also called osmotic demyelination syndrome, occurs as a result of rapid fluid shifts in the brain that are thought to cause shear stress on vulnerable OLs (Brown 2000; Jacoby 2020). The majority of cases occur when a chronically low serum sodium concentration (hyponatremia) is rapidly corrected (Singh et al. 2014). Hyponatremia places osmotic stress on OLs and other CNS glia, such as astrocytes. In response, these glia cells adapt over several days by exuding organic osmolytes to restore the osmotic gradient. When hyponatremia is rapidly corrected, these organic osmolytes cannot be repleted quickly enough, creating a reverse osmotic gradient that causes rapid fluid shifts from the intracellular to the extracellular compartment (Jacoby 2020). OLs appear to be particularly vulnerable to these fluid shifts, producing a primarily demyelinating pathology in both humans and animals (Laureno 1983). The precise incidence of osmotic myelinolysis is unknown. The condition is often iatrogenic, when chronic hyponatremia is corrected too quickly in a hospital setting. Most patients have an underlying condition that predisposes to hyponatremia and/or causes a baseline depletion of organic osmolytes, such as alcoholism, malnutrition, or liver cirrhosis (Singh et al. 2014; Jacoby 2020). Renal failure is another frequently comorbid condition, likely because of its effects on fluid and electrolyte management. In some cases in which these comorbidities exist, osmotic myelinolysis can occur even in the absence of hyponatremia.

OLs within the central pons are particularly susceptible to osmotic demyelination, such that classically the condition was called central pontine myelinolysis. The symptoms of central pontine myelinolysis can be particularly devastating, causing weakness or paralysis of all four limbs as well as severe deficits with speaking, eating, and eye movements. However, extrapontine myelinolysis is now widely appreciated, with pathology affecting the cerebellum and cerebral hemispheres either in isolation or together with central pontine myelinolysis and causing ataxia, encephalopathy, and seizures (Singh et al. 2014). The diagnosis is made based on an appropriate clinical setting along with typical findings on MRI. Symptoms typically develop within 1 week of rapid hyponatremia correction but can have delayed onset up to 2 weeks (Omari et al. 2002). There are no specific treatments. Functional recovery is possible and has been reported to occur in ~50% of cases, although the condition is fatal in ~25% of cases with fatality even greater in the setting of major comorbidities such as liver transplant (Singh et al. 2014).

Progressive Multifocal Leukoencephalopathy

Progressive multifocal leukoencephalopathy (PML), is an opportunistic infection affecting brain OLs that occurs most commonly in individuals with severe and persistent immunosuppression (Cortese et al. 2021b). The infection is caused by the John Cunningham virus (JC virus), a polyomavirus named for the individual from whom it was first isolated and identified (Aksamit 2012). The JC virus is ubiquitous in nature, producing a persistent but asymptomatic infection in immunocompetent hosts. The precise means of transmission is unclear but thought to occur through oropharynx entry following either human-to-human transmission or contact with contaminated food, water, and surfaces (Cortese et al. 2021b). Serological studies have found evidence of asymptomatic infection in 50%–80% of the adult population, with rates increasing with age (Schweitzer et al. 2023). The virus creates latent reservoirs in the kidneys and possibly other organs, including the brain. In the setting of immunosuppression, the JC virus can reactivate, and it is believed that viral evolution following reactivation produces pathogenic mutations allowing infection of OLs. Worldwide, acquired immunodeficiency syndrome (AIDS) is the most common cause of immunosuppression leading to PML, which occurs in an estimated 1%–4% of AIDS cases (Aksamit 2012). In geographic areas with access to antiretroviral therapy, PML incidence has decreased accordingly. Non-AIDS-related risk factors for PML include hematologic malignancies and chronic treatment with immunosuppressive therapies in the setting of autoimmune disease, cancer, or organ transplant (Joly et al. 2023). A number of MS therapies have been linked with cases of PML, most notably natalizumab, which prevents lymphocyte entry into the CNS and, therefore, impairs immune surveillance (Rindi et al. 2024).

In PML, the JC virus causes a lytic infection of OLs, producing progressive areas of demyelination (Aksamit 2012; Cortese et al. 2021b). PML typically affects the cerebral hemispheres and cerebellum, sparing the spinal cord and optic nerves, and produces severe, progressive neurologic disability. More rare, variant clinical presentations have also been described, such as a granule cell neuronopathy (Du Pasquier et al. 2003). PML causes typical white matter lesions on MRI, but diagnosis is usually confirmed by the detection of the JC virus in cerebrospinal fluid (CSF). When immunosuppression cannot be readily reversed, PML is often fatal; before the advent of antiretroviral therapy, the 1-year mortality rate in AIDS-associated PML was over 90% (Berger et al. 1998). When occurring in the context of immunosuppressive therapies that can be discontinued, mortality has been closer to 20% (Joly et al. 2023).

Currently, there are no effective treatments for PML, so the mainstay is reconstitution of the immune system when possible (such as by stopping immunosuppressive medications) (Bernard-Valnet et al. 2021). Immune reconstitution can halt the progression of disability but does not reverse it; recovery is typically minimal and disability is permanent, making PML a devastating condition. Immune reconstitution can itself be dangerous, as an exuberant inflammatory response can cause an initial clinical worsening associated with brain swelling and sometimes requiring treatment with corticosteroids—a condition called immune reconstitution inflammatory syndrome (IRIS) (Tan et al. 2009; Bowen et al. 2018). A number of therapeutic approaches to rapidly control the JC virus have been investigated (Bernard-Valnet et al. 2021), including immune checkpoint inhibitors (Cortese et al. 2019; Boumaza et al. 2023), adoptive transfer of T cells (Cortese et al. 2021a), and treatment with interleukin 7 (Lajaunie et al. 2022), although none has shown clear efficacy.

MECHANISMS OF MYELIN INJURY IN INFLAMMATORY DEMYELINATING DISEASE

Whereas the inciting factor driving myelin injury is straightforward in noninflammatory demyelinating conditions (e.g., known toxic exposure or nutritional deficiency), the pathologic mechanisms causing demyelination in inflammatory conditions are more complex. Using MS as the prototypical example, mechanisms of injury change based on age and disease stage (i.e., relapsing vs. progressive MS), and the initial inciting event remains uncertain. Below, we review current knowledge of the causes of demyelination in MS.

"Inside-Out" versus "Outside-In" Hypotheses

Although actively demyelinating MS lesions are invariably associated with inflammation, there has historically been debate regarding whether inflammation is a primary inciting factor or, alternatively, a secondary response to myelin injury. The so-called "inside-out" hypothesis of MS posits that pathology begins within the CNS, with intrinsic disruption of the normal myelin sheath (i.e., a primary oligodendrogliopathy) triggering a secondary immune response and infiltration of inflammatory cells across the blood–brain barrier (BBB). Conversely, the "outside-in" hypothesis suggests that the breakdown of tolerance mechanisms and aberrant activation of myelin-reactive peripheral immune cells represents the primary pathophysiologic driver of MS, causing inflammatory injury to otherwise normal CNS myelin.

There is abundant evidence that, whether primary or secondary, peripheral immune infiltration into the CNS plays a pathogenic role in MS contributing to demyelination, plaque formation, and neurologic disability. Two lines of evidence are particularly convincing in this regard. First, among the >200 allelic variants associated with the risk of MS, almost all occur within immune-related genes (Sawcer et al. 2014; Mitrović et al. 2018). Second, currently approved disease-modifying therapies for MS act by targeting the peripheral immune system and preventing immune activation and/or trafficking into the CNS. These therapies are well established to reduce or prevent new plaque formation, relapses causing neurologic symptoms, and both short- and long-term disability when initiated during the relapsing–remitting stage of MS (Cree et al. 2016; Tintore et al. 2021; Cross and Riley 2022).

However, the pathogenic role of the immune system does not preclude the possibility that a primary oligodendrogliopathy represents the proximate pathologic event in at least some patients with MS, triggering the immune response in susceptible individuals through the release of inflammatory damage signals or unmasking of neoantigens ('t Hart et al. 2021). Although the majority of active MS lesions demonstrate evidence of inflammatory cells closely associating with myelin (patterns I and II, described further below), a minority of patients exhibit an early pattern of pathology characterized by preferential loss of myelin-associated glycoprotein (MAG) and OL injury possibly independent of inflammation (Lucchinetti et al. 2000; Henderson et al. 2009; Metz et al. 2014). This pathological pattern (termed pattern III) demonstrates OL process retraction and apoptosis suggestive of a "dying-back" oligodendrogliopathy. A more rarely found pattern involves OL death in otherwise normal-appearing white matter (NAWM) in MS brains, often in regions adjacent to expanding plaques (so-called pattern IV pathology).

Myelin Injury in Relapsing Multiple Sclerosis

Regardless of the inciting events, the relapsing phase of MS is characterized by peripheral immune cells crossing the BBB and forming dense, inflammatory plaques associated with OL loss and demyelination (Fig. 1). As detailed above, there is abundant evidence that inflammation plays a direct role in myelin loss, supporting a consensus that OLs and/or mature myelin sheaths are the target of an autoimmune response. However, the precise cellular and molecular mechanisms that mediate myelin injury remain incompletely understood. Moreover, as discussed earlier, histopathologic studies suggest that the mechanisms of myelin loss may differ between people with MS, although each individual exhibits a consistent histopathologic pattern between active lesions (Lucchinetti et al. 2000; Metz et al. 2014). Despite this heterogeneity, some common pathologic principles apply to all or most individuals with relapsing MS. Namely, demyelination during relapsing MS results from interactions between infiltrating adaptive immune cells (e.g., CD4 and CD8 T lymphocytes, B lymphocytes, antibody-producing plasma cells, monocytes/monocyte-derived macrophages) and local glial cells (microglia, astrocytes, and inflammatory OL progenitor cells (OPCs)/OLs) (Dendrou et al. 2015; Grigoriadis and van Pesch 2015; Reich et al. 2018).

Adaptive Immunity

Opening of the BBB with extravasation of peripheral leukocytes into the CNS parenchyma is an early, defining characteristic of actively forming demyelinating lesions. Pathological studies demonstrate dense CD8 T-cell infiltrates and abundant macrophages/microglia, with variable numbers of CD4 cell, B-cell, and plasma cell infiltration, along with reactive astrocytes (Dendrou et al. 2015). Infiltrating autoreactive lymphocytes are reactivated by antigen-presenting cells within the CNS. Animal models suggest that CNS-resident dendritic cells are primarily

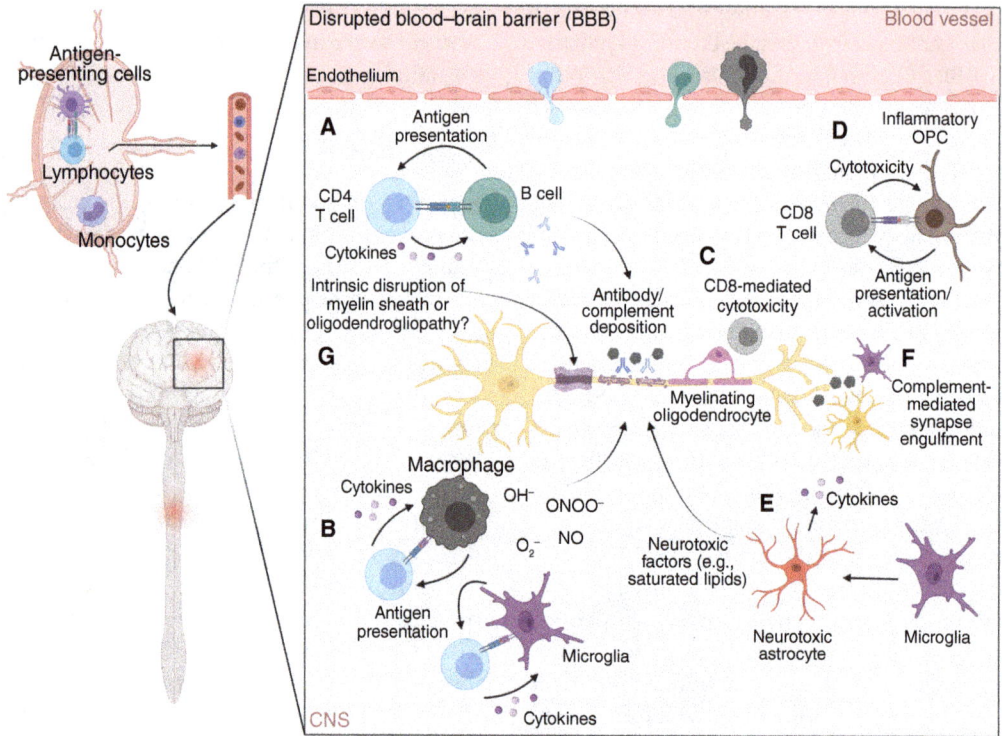

Figure 1. Mechanisms of myelin injury in relapsing multiple sclerosis (MS). Discrete episodes of acute, inflammatory demyelination within the central nervous system (CNS) are the hallmark of relapsing MS, characterized by peripheral immune cells invading the CNS parenchyma across a disrupted blood–brain barrier. Mechanisms of demyelination and neuronal injury in relapsing MS are summarized here. (*A*) Interactions between CD4 T lymphocytes and B lymphocytes. The most common histologic lesion pattern (pattern II) involves immunoglobulin and complement deposition at sites of myelin injury, suggesting a pathologic role for antibodies produced by mature B cells and plasma cells. However, B cells appear to have pathologic functions independent of antibody production, perhaps as antigen-presenting cells. (*B*) Inflammatory microglia and macrophages produce myelin injury through the secretion of reactive oxygen/nitrogen species and propagate inflammation via antigen presentation and cytokine release. (*C*) CD8 T lymphocytes are abundant within active lesions and may injure oligodendrocytes (OLs) through direct cytotoxic interactions. (*D*) OL progenitor cells (OPCs) adopt an inflammatory phenotype and cross-present antigen via major histocompatibility complex (MHC) expression. (*E*) Inflammatory microglia induce a neurotoxic astrocyte subset that produces factors injurious to OLs and neurons, as well as proinflammatory cytokines. (*F*) Microglia contribute to gray matter injury through pathologic complement-mediated synapse engulfment. (*G*) Although abundant evidence demonstrates that peripherally derived inflammation is pathogenic in MS, some evidence suggests that structural abnormalities of myelin may serve as the primary trigger of inflammation in at least a subset of individuals. Some people with MS display a histologic pattern of demyelination associated with a "dying-back" oligodendrogliopathy (pattern III). (Figure prepared with BioRender.com.)

responsible for antigen presentation (Giles et al. 2018; Manouchehri et al. 2021), although microglia and peripherally derived macrophages may also play this role.

The abundance of CD8 cells within actively demyelinating lesions suggests they play a major role in OL and myelin injury. CD8 lymphocytes are cytotoxic and, therefore, presumably contribute to demyelination through direct cytotoxic interactions with OLs, although the relative contribution of this mechanism has not been definitively established (Saxena et al. 2011; Denic et al. 2013;

Cite this article as *Cold Spring Harb Perspect Biol* doi: 10.1101/cshperspect.a041374

Dendrou et al. 2015). A pathogenic role for CD4 cells is best supported by genetic evidence, as the susceptibility allele carrying the greatest risk of MS lies within a human leukocyte antigen (HLA) class II gene responsible for antigen presentation to CD4 lymphocytes (Waubant et al. 2019). CD4 cells induce proinflammatory and toxic functions in macrophages and microglia and may further contribute to demyelination through interactions with CD8 cells and B lymphocytes. B lymphocytes clearly play a role in mediating demyelination, as evidenced by the great success of B-cell-depleting therapies in preventing demyelinating events (Comi et al. 2021; Cross and Riley 2022). However, it remains uncertain whether B cells induce myelin injury directly or via indirect interactions with other inflammatory cells. The most commonly observed histologic pattern in active MS lesions (so-called pattern II) involves immunoglobulin and complement deposition at sites of active demyelination, suggesting a direct role for antibody-mediated myelin injury (Lucchinetti et al. 2000). However, B-cell-depleting therapies do not target long-lived plasma cells and produce benefit in MS without effects on immunoglobulin production.

Several potential myelin autoantigens have been suggested, including MBP epitopes that share high sequence homology with EBV (Lang et al. 2002; Holmøy et al. 2004). These findings support the notion of molecular mimicry and are particularly intriguing given that EBV infection is a major risk factor for MS (Bjornevik et al. 2022). Peripheral blood myelin-specific CD4 T cells are frequently HLA-DRB1*1501 (class II) restricted, supporting the notion that the strongest gene variant associated with MS risk confers susceptibility to CD4 T-cell activation by antigen-presenting cells expressing HLA (MHC) class II molecules that present peptides to these helper T cells (Wucherpfennig et al. 1995, 1997; Chitnis 2007). There is an extensive literature on myelin-specific CD4$^+$ T cells having an effector memory phenotype including being CCR7 negative, costimulation independent, having somatic mutations, and being clonally expanded in CNS tissues of people with MS (Kivisäkk et al. 2004; Chitnis 2007). Peptides from several different myelin proteins including MBP, MOG, and proteolipid protein (PLP) have been shown to activate these T cells in MS patients

(Ota et al. 1990; Pette et al. 1990; Martin et al. 1991; Valli et al. 1993; Pelfrey et al. 1994; Zhang et al. 1994; Chitnis 2007). Because several groups have shown impaired T-cell tolerance with a deficiency of regulatory T cells in people with MS, attempts to exploit the known encephalitogenic region of MBP were undertaken using altered peptide ligands (APLs) (Karin et al. 1994; Nicholson et al. 1995; Brocke et al. 1996). In this approach, the critical T-cell receptor contact residues of MBP were modified with neutral amino acids in a manner that was predicted to be tolerizing in vitro. Unfortunately, in a subset of patients, the APL resulted in paradoxical T-cell activation and dramatic clinical worsening with cross-reactive APL-native peptide clones being isolated from the CSF (Bielekova et al. 2000). This study, nonetheless, provides support that myelin reactive T cells are likely pathogenic in MS.

The CSF of people with MS contains oligoclonal IgG bands in 85%–90% of cases, supporting a role for clonally expanded antibody-secreting cells such as class-switched B cells or differentiated plasma cells (Reich et al. 2018). A specific antigen to which these bands are directed has not been identified. However, reports of pathogenic antibodies to proteins expressed at the paranodes such as contactin-2 could play a role in some cases (Derfuss et al. 2009). Intriguingly, antibodies against the EBV protein EBV nuclear antigen 1 (EBNA1) were shown to cross-react with the OL protein glial cell adhesion molecule (GlialCAM) (Lanz et al. 2021). However, no myelin antigens have been consistently identified as inflammatory triggers across individuals with MS.

Microglia

Microglia, the resident mononuclear phagocytes of the CNS, are believed to play a role in both the relapsing and progressive phases of MS (Dendrou et al. 2015; Voet et al. 2019; Absinta et al. 2020; Yong and Yong 2021; Healy et al. 2022; Yong 2022b). In the context of MS and experimental models, microglia can be difficult to distinguish from peripheral monocyte-derived macrophages and CNS border-associated macrophages, such that the relative contributions of these cell populations remain uncertain. Microglia/macrophages appear to serve both adaptive and maladaptive (or beneficial and deleterious) functions in MS. Whereas an initial inflammatory response might serve to constrain injury, for instance through synaptic pruning of excitatory synapses on metabolically challenged neurons (Hong et al. 2016b), overexuberant or chronic activation propagates inflammation and exacerbates injury through multiple mechanisms. Conversely, as discussed in further detail below, microglia/macrophage functions are critical to the resolution of inflammation and regenerative processes through phagocytosis of inhibitory debris and the release of trophic factors (Franklin and Simons 2022). The advent of single-cell RNA sequencing has demonstrated vast phenotypic heterogeneity among this cell population in MS and other neurologic diseases (Keren-Shaul et al. 2017; Krasemann et al. 2017; Hammond et al. 2019; Absinta et al. 2021; Young et al. 2021; Paolicelli et al. 2022; Zia et al. 2022; Hou et al. 2023), such that simple characterization is difficult. Furthermore, the consequences of microglia/macrophage activation in human MS have been largely inferred based on clinical-pathologic correlation, although direct evidence of the contribution of these cells (both helpful and harmful) has been demonstrated in animal models.

In the active demyelinating plaques characteristic of relapsing MS, microglia/macrophages comprise the most abundant cell type regardless of the histologic pattern (Lucchinetti et al. 2000). Further, several lines of evidence suggest that microglia/macrophage activation might be an early step in lesion formation. In pathologic samples from the brains of individuals with MS, aggregates of activated microglia (microglia nodules) are observed in "prelesion" white matter (Van Der Valk and Amor 2009; van Horssen et al. 2012; Singh et al. 2013). In a series of fulminant demyelinating cases in which death occurred rapidly after onset, the pathological evaluation revealed reactive microglia and caspase-independent OL apoptosis without T-cell involvement (Barnett and Prineas 2004). Positron emission tomography (PET) imaging has shown increased expression of the microglia/macrophage marker 18 kDa translocator protein (TSPO) in developing lesions before evidence of

BBB breakdown on MRI (Oh et al. 2011). In the experimental autoimmune encephalomyelitis (EAE) mouse model of MS, microglia/macrophages associate with dysfunctional myelinated axons within the spinal cord as an early event, and without T-cell involvement (Nikić et al. 2011).

Although depletion of microglia/macrophages can worsen demyelination in some preclinical settings (Kotter et al. 2001; Rubino et al. 2018; Tanabe et al. 2019; Plemel et al. 2020), which highlights the potential beneficial roles of these cells, microglia/macrophages have also been shown to directly contribute to OL and neuronal injury in EAE and other models through multiple mechanisms. These include antigen presentation to infiltrating lymphocytes, inflammatory cytokine secretion, production of reactive oxygen and nitrogen species, release of cytotoxic molecules, pathologic synapse engulfment, and induction of neurotoxic astrocytes (Hong et al. 2016a; Voet et al. 2019; Werneburg et al. 2020; Yong 2022a). Further, several of the gene variants associated with MS risk are expressed in microglia, implicating a direct role for these cells in the disease process (Patsopoulos et al. 2019).

Astrocytes

Although the presence of reactive astrocytes within MS lesions has been long known, the extent of active astrocyte participation in myelin injury and repair has been recognized only recently (Lee et al. 2022). Early formulations of astrocyte involvement in MS limited their role to the formation of a "glial scar" serving to wall off lesions and limit regeneration. However, we now know that these cells play a much richer role. Similar to microglia/macrophages, in the context of demyelinating lesions astrocytes can exhibit diverse phenotypes and functions, with both adaptive and maladaptive activities. Further, astrocytes participate in extensive cross talk with other cell types, with interactions between astrocytes, microglia, and infiltrating peripheral immune cells determining the fate of newly forming lesions.

Analogous to microglia/macrophages, astrocyte depletion can either attenuate or exacerbate

Cite this article as *Cold Spring Harb Perspect Biol* doi: 10.1101/cshperspect.a041374

inflammatory demyelination in animal models, depending on the specific model and timing of depletion (Voskuhl et al. 2009; Toft-Hansen et al. 2011; Mayo et al. 2014). As briefly alluded to above, inflammatory microglia have been shown to secrete factors (namely, C1q, IL-1α, and TNF-α) that induce a neurotoxic phenotype in astrocytes, which can be found in a number of CNS disorders including MS (Liddelow et al. 2017) and mediate toxicity to neurons in part through the release of saturated lipids (Guttenplan et al. 2021). Other microglia–astrocyte signaling pathways, such as semaphorin4D–plexinB1/2 and ephrinB3–EPHB3, have similarly been identified in EAE and MS (Clark et al. 2021). Reactive astrocytes are toxic to mature OLs and inhibit OPC differentiation suggesting they play a role in propagating demyelination and impairing remyelination. Astrocytes themselves both respond to cytokines and damage-associated molecular pattern molecules (DAMPs) and secrete inflammatory cytokines and chemokines, playing a role in recruitment of peripheral immune cells into the CNS (Kim et al. 2014; Mills Ko et al. 2014; Moreno et al. 2014; Wheeler et al. 2019; Linnerbauer et al. 2020). Conversely, a population of "anti-inflammatory" astrocytes has been described in EAE, which express lysosome-associated membrane glycoprotein 1 (LAMP1) and tumor necrosis factor–related apoptosis-inducing ligand (TRAIL) and mediate T-cell apoptosis (Sanmarco et al. 2021). Astrocyte-derived heparin-binding EGF-like growth factor (HB-EGF) has similarly been shown to have tissue-protective and anti-inflammatory effects in the context of neuroinflammation (Linnerbauer et al. 2024). The complexity of astrocyte functions and an incomplete understanding of their impact at various stages of lesion formation represent challenges to therapeutically targeting astrocytes in MS.

Inflammatory Oligodendrocyte Progenitor Cells and Oligodendrocytes

Following myelin injury, OLs are replaced by the proliferation and differentiation of OPCs, as discussed in detail below in the context of remyelination. Beyond these roles, both OPCs and OLs have been shown to directly modulate CNS inflammation and myelin injury. In response to inflammatory cytokines, OPCs and OLs express immune-related genes, including genes associated with antigen processing and presentation. These inflammatory OPCs (iOPCs) and inflammatory OLs (iOLs) can be found in both human MS lesions and the mouse model EAE (Falcão et al. 2018a; Jäkel et al. 2019; Kirby et al. 2019; Schirmer et al. 2019; Absinta et al. 2021). Moreover, iOPCs phagocytose debris and cross-present antigen to cytotoxic CD8 (Kirby et al. 2019) and CD4 (Falcão et al. 2018a) T lymphocytes. Oligodendroglia-specific knockout of the phagocytic receptor low-density lipoprotein receptor-related protein 1 (LRP1) limits CD8 cell activation, dampens inflammation, and attenuates myelin injury in both EAE and the cuprizone-induced model of demyelination (Fernández-Castañeda et al. 2020). The discovery of iOPCs and iOLs shows the mechanistic complexity of myelin injury in MS and, as discussed further below, may have implications for myelin repair. Further, the recent discovery of interferon-responsive OLs with abundant CD8 T-cell infiltrates in aging (Kaya et al. 2022b) and in inherited leukodystrophies highlights that immune pathways may serve as a secondary mechanism of injury in classical noninflammatory diseases. Such a phenomenon has been described in adults with adrenoleukodystrophy who have inflammatory lesions in the course of this genetic disease (Schlüter et al. 2018).

Myelin Injury in Progressive Forms of Multiple Sclerosis

Although demyelination and associated neuro-axonal injury occur most robustly during acute lesion formation in relapsing–remitting MS, these pathological processes continue in progressive forms of MS. However, despite some overlap, the pathological mechanisms of injury differ between relapsing and progressive MS (Fig. 2; Mahad et al. 2015; Faissner et al. 2019; Absinta et al. 2020; Yong and Yong 2021; Kuhlmann et al. 2022). Whereas the peripheral, adaptive immune system plays a primary role in demyelination in relapsing MS by crossing a leaky BBB and forming dense inflam-

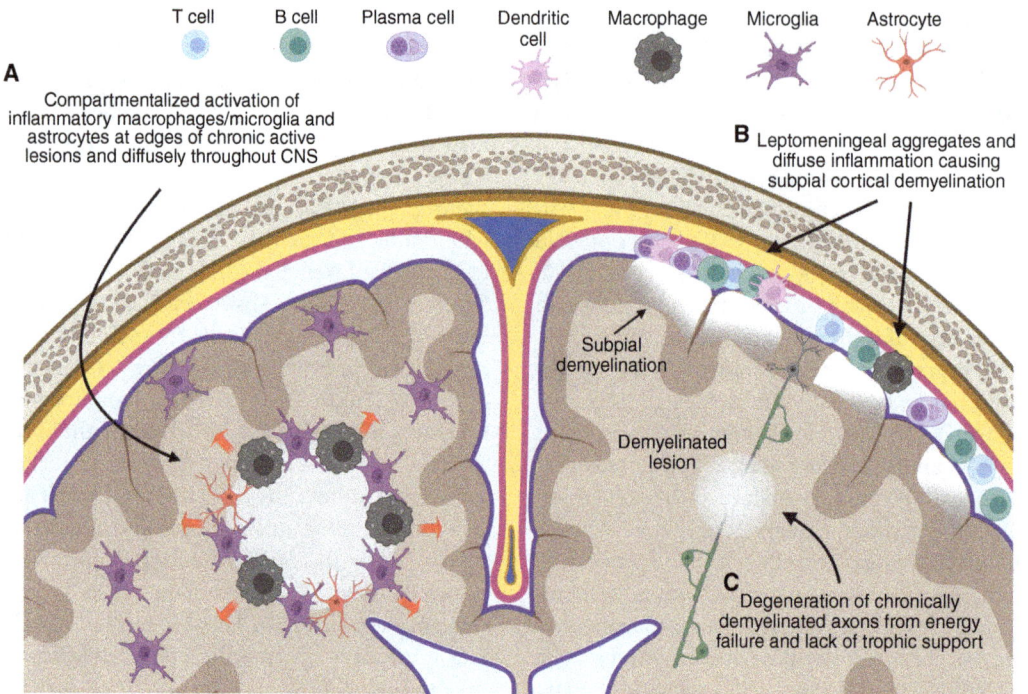

Figure 2. Mechanisms of myelin injury in progressive multiple sclerosis (MS). Demyelination and neuroaxonal injury continue in progressive MS, but the mechanisms of injury differ from relapsing MS and occur behind an intact blood–brain barrier, as summarized here. (*A*) Chronic, maladaptive activation of macrophages, microglia, and astrocytes continues at the borders of slowly expanding chronic lesions, as well as diffusely throughout the normal-appearing gray and white matter. (*B*) Leptomeningeal inflammation, both diffuse and as organized lymphoid aggregates, becomes more pronounced in progressive MS and associates with subpial demyelination of the adjacent cortex. (*C*) Independent of inflammation, chronic demyelination produces progressive axonal degeneration due to axonal energy failure and lack of trophic support. (Figure prepared with BioRender.com.)

matory plaques, BBB permeability and large collections of peripheral immune cells are largely absent from the CNS parenchyma in progressive MS. Instead, inflammation in progressive MS persists behind an apparently intact BBB, characterized by chronic activation of microglia/macrophages and astrocytes at the edges of chronic active (or "smoldering") lesions and diffusely throughout the CNS. Inflammation involving the leptomeninges, including organization into lymphoid follicle-like structures, also becomes more pronounced in progressive MS and can be associated with accumulating areas of cortical demyelination. Finally, chronic demyelination itself promotes neurodegeneration and progressive disability by decreasing trophic support and raising the metabolic demands on demyelinated axons. These pathological mechanisms will be reviewed here.

Compartmentalized Inflammation and the Role of Microglia/Macrophages and Astrocytes

After resolution of the acute inflammation associated with relapse and new plaque formation, some MS lesions continue to demonstrate a rim of iron-laden, activated microglia/macrophages at the lesion edge that can persist for years. Multiple longitudinal studies have found that such lesions, which are termed "chronic active" or "smoldering," slowly expand over time and can be identified on specific MRI pulse sequences by virtue of a paramagnetic rim reflective of microglia/macrophage iron content (Absinta et al. 2016, 2019; Dal-Bianco et al. 2017, 2021; Elliott et al. 2019b). The rim of activated microglia/macrophages is associated with persistent, active demyelination at the lesion edge, providing the basis for slow expansion. Chronic active lesions are found in greater numbers in progressive MS, although importantly they can also exist in people with relapsing–remitting MS at the earliest stages of the disease (Frischer et al. 2015; Luchetti et al. 2018). Critically, chronic active lesions are a biomarker of poor outcome; individuals with a larger number of slowly expanding or paramagnetic rim lesions on MRI experience greater disability over time (Absinta et al. 2016, 2019; Elliott et al. 2019a; Preziosa et al. 2022).

In addition to persistence at the chronic active lesion edge, microglia/macrophage activation can also be observed diffusely throughout the white and gray matter in MS, becoming more pronounced in progressive forms of MS (Politis et al. 2012; Herranz et al. 2016; Healy et al. 2022). The degree of diffuse microglia/macrophage activation in progressive MS, as measured by TSPO PET imaging, similarly shows a positive correlation with disability.

It is important to note that the presumed role of microglia/macrophages in driving ongoing demyelination, neurodegeneration, and disability accrual in progressive MS remains correlative. Microglia/macrophage activation could theoretically represent a secondary response to injury rather than a cause. Definitive evidence of a primary role for these cells would require therapies that robustly modify their chronic activation, which do not exist. Nonetheless, the dual roles of microglia/macrophages as either drivers of maladaptive inflammation or agents of repair likely remain relevant in progressive MS. The chronic activation of microglia/macrophages observed at the edges of chronic active lesions and throughout the CNS likely reflects a maladaptive phenotype that promotes injury and precludes repair. Transcriptional profiling of cells at the chronic active lesion edge revealed populations of microglia and astrocytes with characteristics of proinflammatory, degeneration-associated subsets—termed "microglia inflamed in MS" (MIMS) and "astrocytes inflamed in MS" (AIMS) (Absinta et al. 2021). MIMS demonstrated increased expression of complement component 1q (C1q), which could be therapeutically targeted in EAE to modulate microglia–astrocyte cross talk.

Leptomeningeal Inflammation

Although classically considered a disease of white matter, gray matter pathology (both demyelination and neuronal loss) occurs in MS, becoming more pronounced in progressive forms of MS and correlating with disability (Kutzelnigg et al. 2005; Lucchinetti et al. 2011; Lassmann et al. 2012; Eshaghi et al. 2018a,b; Colato et al. 2021). Relatedly, inflammatory infiltrates of the leptomeninges (the fibrous connective tissue abutting the pial surface) can be found in all stages of MS but are observed more frequently in progressive MS (Magliozzi et al. 2007, 2010; Howell et al. 2011; Choi et al. 2012; Popescu and Lucchinetti 2012). These infiltrates are comprised of B cells, T cells, plasma cells, and antigen-presenting cells such as dendritic cells and macrophages, sometimes forming organized structures reminiscent of tertiary lymphoid structures. Critically, areas of leptomeningeal inflammation are often associated with demyelination of adjacent gray matter cortex in a pattern described as "subpial," with a gradient of myelin loss and neuroaxonal injury extending from the pial surface. This pattern of gray matter pathology suggests that soluble factors produced by leptomeningeal infiltrates might diffuse through the upper layers of the cortex to produce the observed injury. As such, there is substantial interest in targeting leptomeningeal inflammation as a potential therapy for progressive MS. To date, no therapies have been shown to modify leptomeningeal infiltrates in humans, including B-cell-depleting antibody therapies that may be limited by drug access to the meninges (Bhargava et al. 2019; Bonnan et al. 2021). However, leptomeningeal inflammation can be modeled in rodents, and a class of drugs known as Bruton's tyrosine kinase (BTK) inhibitors effectively attenuated leptomeningeal infiltrates in these mice (Bhargava et al. 2021) (discussed below in the Therapeutic Strategies section), setting the stage for human studies.

Consequences of Chronic Demyelination

As described earlier, OLs provide critical metabolic support and trophic factors to axons (Philips and Rothstein 2017; Henn et al. 2022), such that progressive mitochondrial dysfunction and energy failure lead to slowly progressive injury and degeneration of chronically demyelinated axons (Mahad et al. 2015). Metabolic failure of chronically denuded axons thus likely contributes to insidious neurodegeneration and disability in progressive MS, identifying an additional critical role for preserving or restoring myelin to slow progression in individuals with MS.

REMYELINATION IN HEALTH AND DISEASE

Whereas demyelination represents the pathologic hallmark of MS, repair, and replacement of damaged myelin (i.e., remyelination) occurs to varying degrees between individuals with MS and through the course of the disease, becoming less efficient with age. Remyelination is critical to restoring neurologic function and preventing the ongoing neurodegeneration associated with progressive disability. While substantial progress has been made in preventing relapses and new lesion formation, no approved therapies promote remyelination. Remyelinating therapies, therefore, represent a critical unmet need.

Before approaching the reasons why remyelination often fails in MS, it is first necessary to understand the biological processes underlying developmental and adaptive myelination and, most importantly, successful remyelination in the adult CNS. Below, we will review the current understanding of the dynamics of remyelination followed by a discussion of how these processes might fail in MS.

Normal Myelinogenesis and Successful Remyelination

During normal development, myelinating OLs are derived from OPCs, which migrate throughout the CNS before differentiating into terminal OLs (Cristobal and Lee 2022). Developmental myelination associated with new OL generation continues into early adulthood in both animals and humans, with OL numbers peaking at ~10 months of age in rodents (Hughes et al. 2018). Once formed, mature OLs are extremely stable and long-lived, with minimal turnover (Hughes et al. 2018; Yeung et al. 2019). After the period of developmental myelination, new OL generation

appears to be rare under homeostatic conditions. Nonetheless, OPCs comprise 5%–8% of CNS cells in the adult brain (Hughes et al. 2013; Lubetzki et al. 2020), and OL generation from OPCs increases during experience-driven cortical plasticity in adult mice (Hughes et al. 2018).

The dynamics of successful remyelination after injury have largely been deduced from animal models. Following demyelination in mice and lower vertebrates, remyelination is mediated by newly formed OLs generated from OPCs, which proliferate and migrate to areas of injury before differentiating into myelinating OLs (Tripathi et al. 2010; Zawadzka et al. 2010; Neely et al. 2022). However, evidence also exists that surviving, mature OLs can contribute to remyelination by extending new processes to wrap denuded axons (Duncan et al. 2018). At least in zebrafish, remyelination mediated by surviving OLs is not as efficient or robust as that generated by OLs newly formed following OPC migration and differentiation (Neely et al. 2022).

Due to obvious limitations with regard to tissue accessibility and tools for in vivo investigation, the biological processes associated with successful remyelination in humans are more difficult to ascertain. In individuals with MS, the histological correlate of remyelination is thought to be the so-called "shadow plaque," defined by thinly myelinated axons that produce positive but pale myelin staining compared to surrounding white matter. The frequency of shadow plaques varies widely between individuals and across the stages of the disease (Patrikios et al. 2006; Frischer et al. 2015), possibly reflecting the observed heterogeneity in relapse recovery and disability accrual among MS patients. Nonetheless, the conclusion that shadow plaques represent successfully remyelinated lesions remains somewhat speculative, and it is possible that completely remyelinated lesions are difficult to distinguish from NAWM (Neumann et al. 2020). Definitively identifying the source of new myelin is similarly difficult in humans. Using ^{14}C dating, OL turnover was found to be low in many individuals with MS within both shadow plaques and NAWM, suggesting that surviving OLs might be the source of remyelination in these patients. However, a subset (~25%) of the cohort

Cite this article as *Cold Spring Harb Perspect Biol* doi: 10.1101/cshperspect.a041374

with aggressive MS demonstrated more than a threefold higher turnover of OLs compared to age-matched controls, demonstrating that new OL generation occurs in at least some individuals with the disease (Yeung et al. 2019).

Disease-Associated Modulators of Remyelination

The challenge of studying remyelination in humans makes it difficult to pinpoint the cellular processes that fail in MS and the stages of lesion formation and disease chronicity at which these processes fail. Further complicating the understanding of remyelination failure is the likelihood that the causes of this failure differ between individuals and through different phases of the disease. While proliferation and migration of OPCs might be deficient in some cases, studies suggest that OPCs are present in MS lesions but are blocked from differentiating into mature OLs (Chang et al. 2000; Wolswijk 2000; Kuhlmann et al. 2008). Alternatively, one pivotal immunohistological study of MS lesions identified premyelinating OLs with extended processes that associated with but failed to myelinate dystrophic axons, suggesting that unreceptive axons rather than OL differentiation blockade may be a major driver of remyelination failure (Chang et al. 2002). In peripheral nerves, neuregulin provides a critical signal to Schwann cells to myelinate axons (Birchmeier and Bennett 2016), but its role or counterpart in the CNS is less clear. Nonetheless, keeping in mind these uncertainties and the caveats inherent in extrapolating from experimental models to human disease, much has been learned about factors that modulate remyelination and may contribute to the failure of myelin repair in disease. Such factors include those intrinsic to OPCs, which appear to be the likely source of remyelinating OLs, as well as extrinsic factors within the lesion environment, including chemoattractants, inhibitory substances, input from denuded axons, and the functional state of microglia/macrophages (Fig. 3).

Inhibitory Signals within Demyelinated Lesions

Many extrinsic factors have been identified that either positively or negatively regulate the vari-

ous stages of myelin repair, and many of these are altered within MS lesions (Plemel et al. 2017). Recruitment of OPCs is augmented by the chemoattractant semaphorin 3F, whereas semaphorin 3A and netrin 1 act as repellants to OPC migration (Piaton et al. 2011; Tepavčević et al. 2014). Notch and WNT signaling within OPCs and engagement of the OPC membrane proteins LINGO1 and the M1 muscarinic receptor inhibit differentiation (Wang et al. 1998; Mi et al. 2005; Fancy et al. 2009; Zhang et al. 2009; Dai et al. 2014; Hammond et al. 2014; Mei et al. 2014; Mathieu et al. 2019), as do several extracellular matrix components such as hyaluronan (Back et al. 2005; Sloane et al. 2010) and chondroitin sulfate proteoglycans (Keough et al. 2016; Pu et al. 2018). Myelin debris generated following demyelination is a potent inhibitor of OPC differentiation (Kotter et al. 2006; Baer et al. 2009; Plemel et al. 2013), suggesting the critical importance of effective debris clearance by phagocytic cells. The mechanical properties of the OPC microenvironment change with age, which inhibits OPC functions (Segel et al. 2019). Axonal activity can promote myelination, highlighting the need to maintain receptive axons (Gibson et al. 2014; Hines et al. 2015; Mitew et al. 2018).

The coagulation factor fibrinogen has been shown to activate the bone morphogenetic protein (BMP) signaling pathway in OPCs and suppress remyelination. Fibrinogen inhibits OPC differentiation into myelinating OLs through phosphorylation of Smad 1/5/8, while promoting an astrocytic fate in vitro. Therapeutic depletion of fibrinogen can rescue remyelination in in vivo animal models (Petersen et al. 2017).

Dysfunctional Immune Environment

Remyelination in MS must be accomplished in the context of an inflammatory environment, which contributes to the success or failure of myelin repair (Franklin and Simons 2022). Inflammation is, of course, a broad term, and the consequences of immune activation for tissue regeneration and myelin repair are complex, incompletely understood, and likely distinct depending on the lesion stage. On the one hand, coordinated immune responses are critical to

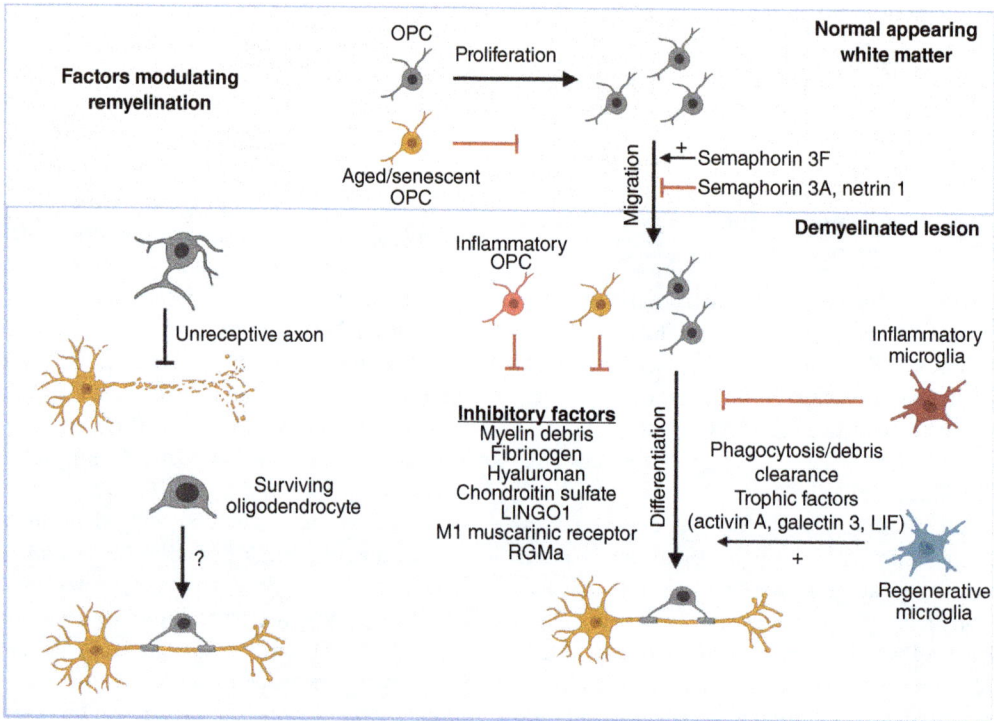

Figure 3. Mechanisms and modulators of myelin repair following pathologic demyelination. Factors impacting successful remyelination are summarized in schematic form. Experimental models suggest that remyelination depends on the proliferation and recruitment of oligodendrocyte (OL) progenitor cells (OPCs) to areas of injury, followed by differentiation into myelinating OLs. An array of factors modulates this process, including some intrinsic to OPCs (such as aging/senescence) and some associated with the local microenvironment (such as inhibitory factors and inflammation). In some cases, remyelination may be limited by damaged, unreceptive axons. Although OPCs are believed to be the primary source of new myelin, some evidence suggests that mature surviving OLs maintain the ability to extend processes and remyelinate axons. (Figure prepared with BioRender.com.)

wound repair in general (Karin and Clevers 2016; Wynn and Vannella 2016). Inflammatory cytokines and chemokines can stimulate OPC proliferation and recruitment (Arnett et al. 2001; Mason et al. 2001; Bieber et al. 2003; Patel et al. 2010; El Behi et al. 2017; Cunha et al. 2020), and microglia/macrophages play an indispensable role in debris clearance (particularly scavenging of myelin debris) (Neumann et al. 2009; Lloyd and Miron 2019; Franklin and Simons 2022) and the secretion of trophic factors such as activin A (Miron et al. 2013), galectin 3 (Pasquini et al. 2011), and leukemia inhibitory factor (Marriott et al. 2008; Deverman and Patterson 2012; Fischer et al. 2014) that stimulate OL maturation. On the other hand, inflammatory cytokines and

maladaptive immune responses directly contribute to OPC and OL injury (Ruijs et al. 1990, 1993; Andrews et al. 1998; Moore et al. 2015; Jamann et al. 2022), produce iOPCs and iOLs that cannot participate in repair (Falcão et al. 2018b; Jäkel et al. 2019; Kirby et al. 2019), and directly inhibit remyelination in experimental models (Baxi et al. 2015).

Following inflammatory demyelination, it is likely that a well-coordinated series of immune events is necessary for successful myelin repair, which fails and becomes maladaptive in many individuals with MS. In support of this notion, waves of distinct microglia/macrophage phenotypes have been shown to be necessary for successful remyelination in animal models (Lloyd et al.

Cite this article as *Cold Spring Harb Perspect Biol* doi: 10.1101/cshperspect.a041374

2019). Similar to festering wounds, some MS lesions may fail to transition to regenerative immune environments, possibly represented by the slowly expanding, poorly remyelinated lesions described earlier, which are rimmed with iron-laden, inflammatory microglia/macrophages (Frischer et al. 2015; Absinta et al. 2016). In this regard, it is notable that microglia/macrophage phagocytic functions decrease with age (Cantuti-Castelvetri et al. 2018; Pluvinage et al. 2019), and remyelination is enhanced in aged rodents by parabiotic monocytes derived from younger animals (Ruckh et al. 2012)—consistent with the observation that remyelination is less efficient in aged animals and older individuals with MS (Sim et al. 2002; Neumann et al. 2019b).

OPC-Intrinsic Factors Contributing to Remyelination Failure

In addition to external factors including inflammation, factors intrinsic to OPCs may contribute to remyelination failure in MS. One such factor may be the propensity to form iOPCs that propagate injury rather than serving as a source for new OLs. Another important OPC-intrinsic factor may be a decreased capacity for proliferation and differentiation associated with aging and senescence. OPCs derived from aged animals display decreased remyelinating capacity, which may further contribute to the failure of remyelination with advancing age. Intriguingly, the age-related regenerative capacity of OPCs was shown to be dependent on Myc signaling (Neumann et al. 2021) and reversible pharmacologically with metformin (Neumann et al. 2019a).

THERAPEUTIC STRATEGIES IN INFLAMMATORY DEMYELINATING DISEASE

Many therapies are approved or used off-label to treat MS and other inflammatory demyelinating diseases, including NMOSD and MOGAD (for reviews, see Costello 2022; Cross and Riley 2022; Longbrake 2022). Although a full accounting of these therapies is beyond the scope of this discussion, to date they include immunomodulatory and immunosuppressive medications designed to prevent inflammatory attacks within the CNS. Current therapies principally target the peripheral immune system, and several are highly effective at preventing clinical relapse and new lesion formation. As such, these disease-modifying treatments prevent step-wise disability accrual in individuals with relapsing disease. Moreover, there is growing evidence that early prevention of relapses lowers long-term disability as well as the risk of conversion to progressive MS (Cree et al. 2016; Tintore et al. 2021). As a result, early diagnosis and treatment are a mainstay of current clinical practice and key to preventing myelin injury in the first place.

Despite the many available therapies, major therapeutic gaps exist. Most notably, no current treatments promote remyelination or slow the insidious disability associated with progressive forms of MS. As described in earlier sections, the processes of remyelination and progression are intertwined—chronic demyelination contributes to progressive axonal degeneration, while many of the maladaptive microglia/macrophage functions associated with progression drive ongoing demyelination and prevent myelin repair. Building on the understanding of these pathological processes, we will review emerging therapeutic strategies to slow progression and promote repair.

Immune-Independent Approaches to Promote Remyelination

One major strategy for promoting remyelination in MS is to augment the intrinsic capacity of OPCs to differentiate into myelinating OLs, given that differentiation blockade may contribute to poor repair. Several targets and early candidate drugs have been identified in this regard, some of which have shown tempered promise in early clinical studies. One promising approach has been to screen existing drugs for the ability to enhance OPC differentiation in high-throughput assays and accessible experimental models such as zebrafish, which has identified several candidate drugs including the antifungals miconazole and clobetasol and the antimuscarinic agents clemastine and benztropine (Deshmukh et al. 2013; Mei et al. 2014; Ackerman et al. 2015; Najm et al. 2015; Ackerman and Monk 2016;

Häberlein et al. 2022). Building on these findings, the M1 muscarinic acetylcholine receptor has been identified as a key target, the antagonism of which promotes differentiation in vitro and promotes remyelination in animal models (Deshmukh et al. 2013; Mei et al. 2014). Clemastine was evaluated in a phase 2 clinical study of patients with chronic optic nerve demyelination from MS and was found to produce a modest but measurable increase in transmission of visual inputs through the optic nerve (Green et al. 2017). The retinoid X receptor (RXR) γ has been identified as a regulator of OPC differentiation (Huang et al. 2011), and the nonselective RXR agonist bexarotene has been studied in both animal models and humans (Natrajan et al. 2015; Chiang et al. 2020; He et al. 2020; Brown et al. 2021; Santos-Gil et al. 2021). In a phase 2 study of relapsing–remitting MS patients, bexarotene produced benefits in some exploratory measures of remyelination but was too poorly tolerated for further development (Brown et al. 2021). Another promising finding has been the critical role of cholesterol metabolism in OPC differentiation, with an accumulation of 8,9-unsaturated sterols as a result of cholesterol biosynthesis inhibition representing a potentially shared mechanism of many differentiation-promoting drugs (Hubler et al. 2018). The cholesterol biosynthetic enzyme Δ8,7-sterol isomerase (or emopamil-binding protein, EBP) has been identified as a particularly attractive drug target within this pathway (Allimuthu et al. 2019). As described earlier, metformin was found to reverse OPC senescence in vitro as well as in rodents (Neumann et al. 2019a), leading to ongoing studies in MS patients (NCT05349474, NCT05131828).

In addition to enhancing the intrinsic differentiation capacity of OPCs, another approach is to neutralize or overcome inhibitory signals within demyelinated lesions. Although candidate drugs have been identified to neutralize hyaluronan and chondroitin sulfate proteoglycans, these have not progressed to clinical trials (Plemel et al. 2017; Stephenson et al. 2019; Oh and Bar-Or 2022). A monoclonal antibody against LINGO1 was evaluated in multiple early clinical studies with mixed results (Cadavid et al. 2017,

2019; NCT03222973) but ultimately not enough promise to support further development. Repulsive guidance molecule A (RGMa) is another inhibitor of remyelination and repair found within lesions, and the anti-RGMa monoclonal antibody elezanumab showed promise in animal models (Demicheva et al. 2015) but failed to show an effect in early clinical trials (conferences .medicom-publishers.com/specialisation/neur ology/ectrims-2021/elezanumab-did-not-outper form-placebo-in-progressive-and-relapsing-ms).

Finally, given the necessity of maintaining intact axons to serve as the substrate for remyelination, neuroprotective therapies may prove equally important to the goal of myelin repair.

Creating a Favorable Immune Environment for Repair

Although early clinical studies have shown modest promise, it remains uncertain whether treatments targeting the differentiation capacity of OPCs will be successful without simultaneously addressing the maladaptive inflammatory processes that restrain myelin repair. In the presence of inflammatory cytokines and other factors, many of the potent OPC differentiating agents are no longer effective (Liddelow et al. 2017; Petersen et al. 2021; Smith et al. 2022). MS disease-modifying therapies do not directly target reactive gliosis and, therefore, ongoing production of CNS cytokines and consequential reactive oxygen products have not been addressed (Yong and Yong 2021). In this regard, much focus has been placed on modulating microglia/macrophages, which as described earlier can be either beneficial or detrimental in myelin repair and remain pathologically relevant in all stages of MS. No currently approved therapies are known to robustly target microglia/macrophages within the CNS, let alone tip the scales toward regenerative functions that promote repair. One promising class of drugs in clinical development is the BTK inhibitors discussed earlier (Schneider and Oh 2022). In addition to B cells, BTK inhibitors are well established to modulate myeloid cell function (which includes macrophages and microglia). Given that several of these drugs appear to be CNS penetrant, there is hope that

BTK inhibitors might alter the compartmentalized inflammation associated with ongoing demyelination and failure of remyelination in progressive MS. Prior work demonstrated a positive impact of these drugs on leptomeningeal inflammation in an animal model (Bhargava et al. 2021). Two other drugs with putative effects on CNS microglia/macrophages, ibudilast and masitinib, have shown promise in early clinical studies in progressive MS (Fox et al. 2018; Vermersch et al. 2022).

Given the critical importance of myelin debris clearance in remyelination, there is substantial interest in augmenting the phagocytic and scavenging functions of microglia/macrophages (Franklin and Simons 2022). Several potential protein targets and signaling pathways have been identified in this regard, including those involving TREM2 (Cignarella et al. 2020; Gouna et al. 2021), MerTK (Healy et al. 2016; Shen et al. 2021), and CD36 (Rawji et al. 2020).

Cell-Based Therapies

Although an extensive discussion is beyond the scope of this review, emerging cell therapies for MS and other demyelinating diseases merit discussion. Stem cell–based therapies can broadly be divided into hematopoietic and nonhematopoietic varieties, with a primary goal either of immune reconstitution or promotion of endogenous repair mechanisms.

Hematopoietic stem cells have been investigated in MS in the context of autologous hematopoietic stem cell transplant (aHSCT), which involves an immunoablative regimen of chemotherapy followed by injection of autologous, bone marrow–derived stem cells to reconstitute the immune system (Bose et al. 2021). The goal of this approach is to ablate the self-reactive peripheral immune cells that drive the disease, in essence "rebooting" the immune system. As such, aHSCT is primarily an anti-inflammatory treatment approach that has shown the most promise in young patients with highly active relapsing MS, with limited potential as a treatment for individuals with noninflammatory progressive MS. The risks of aHSCT remain substantial, and it is currently being investigated in several randomized trials of patients with active and/or relapsing MS (Peterson et al. 2022).

A variety of nonhematopoietic stem cells have been investigated in preclinical models of MS, with some advancing to clinical trials in humans (Smith et al. 2021). In contrast to aHSCT, the goal of these therapies is to promote regeneration and repair within the CNS either indirectly (through the generation of a permissive lesion microenvironment) or directly (through differentiation into myelinating OLs and direct integration). Many studies have focused on mesenchymal stem cells, which are pluripotent cells derived most commonly from bone marrow or adipose tissue. Preclinically, mesenchymal stem cells have been shown to track to sites of CNS injury and act primarily through modulation of the local immune environment. A phase 2 study of mesenchymal stem cell transplant produced negative results in relapsing MS (Uccelli et al. 2021), but further studies utilizing both intravenous and intrathecal administration are ongoing. Strategies under investigation for direct tissue repair include injection of neural stem cells and OPCs derived from autologous induced pluripotent stem cells (iPSCs). These latter therapies remain far from clinical practice, and it remains uncertain whether regenerative cell therapies will be able to overcome the inhibitory, nonpermissive environment of MS lesions discussed in detail above.

CONCLUDING REMARKS

Acquired demyelinating disease of the CNS comprises a diverse set of inflammatory and noninflammatory pathologies. Much focus has been placed on understanding the pathophysiology of inflammatory demyelinating conditions, in particular, MS. Accordingly, substantial progress has been made toward unraveling the complex interactions between peripheral immunity, neurons, OPCs/OLs, and other glial cells that form the pathological basis of MS. Although this progress has translated to effective treatments to prevent clinical relapse and new plaque formation, the focus now must turn to the compartmentalized processes within the CNS that dictate progression and regeneration.

ACKNOWLEDGMENTS

We thank our colleagues within the Johns Hopkins Division of Neuroimmunology and Neurological Infections, as well as our collaborators both within and outside Johns Hopkins. M.D.K. acknowledges funding from the National Institute of Neurological Disorders and Stroke, Department of Defense Congressionally Directed Medical Research Programs, Doris Duke Charitable Foundation, Race to Erase MS, and Herbert R. Mayer and Jeanne C. Mayer Foundation. P.A.C. acknowledges funding from the National Institute of Neurological Disorders and Stroke, the Department of Defense Congressionally Directed Medical Research Programs, and the National MS Society. Figures 1–3 were created using BioRender.com.

REFERENCES

Absinta M, Sati P, Schindler M, Leibovitch EC, Ohayon J, Wu T, Meani A, Filippi M, Jacobson S, Cortese ICM, et al. 2016. Persistent 7-tesla phase rim predicts poor outcome in new multiple sclerosis patient lesions. *J Clin Invest* **126**: 2597–2609. doi:10.1172/JCI86198

Absinta M, Sati P, Masuzzo F, Nair G, Sethi V, Kolb H, Ohayon J, Wu T, Cortese ICM, Reich DS. 2019. Association of chronic active multiple sclerosis lesions with disability in vivo. *JAMA Neurol* **76**: 1474–1483. doi:10.1001/jamaneurol.2019.2399

Absinta M, Lassmann H, Trapp B. 2020. Mechanisms underlying progression in multiple sclerosis. *Curr Opin Neurol* **33**: 277–285. doi:10.1097/WCO.0000000000000818

Absinta M, Maric D, Gharagozloo M, Garton T, Smith MD, Jin J, Fitzgerald KC, Song A, Liu P, Lin JP, et al. 2021. A lymphocyte–microglia–astrocyte axis in chronic active multiple sclerosis. *Nature* **597**: 709–714. doi:10.1038/s41586-021-03892-7

Ackerman SD, Monk KR. 2016. The scales and tales of myelination: using zebrafish and mouse to study myelinating glia. *Brain Res* **1641**: 79–91. doi:10.1016/j.brainres.2015.10.011

Ackerman SD, Garcia C, Piao X, Gutmann DH, Monk KR. 2015. The adhesion GPCR Gpr56 regulates oligodendrocyte development via interactions with Gα12/13 and RhoA. *Nat Commun* **6**: 6122. doi:10.1038/ncomms7122

Aksamit AJ. 2012. Progressive multifocal leukoencephalopathy. *Continuum (Minneap Minn)* **18**: 1374–1391. doi:10.1212/01.CON.0000423852.70641.de

Allen LH. 2009. How common is vitamin B-12 deficiency? *Am J Clin Nutr* **89**: 693S–696S. doi:10.3945/ajcn.2008.26947A

Allimuthu D, Hubler Z, Najm FJ, Tang H, Bederman I, Seibel W, Tesar PJ, Adams DJ. 2019. Diverse chemical scaffolds enhance oligodendrocyte formation by inhibiting CYP51,

TM7SF2, or EBP. *Cell Chem Biol* **26**: 593–599.e4. doi:10.1016/j.chembiol.2019.01.004

Ammouri W, Harmouche H, Khibri H, Benkirane S, Azlarab M, Mouatassim N, Maamar M, Mezalek Tazi Z, Adnaoui M. 2019. Neurological manifestations of cobalamin deficiency. *Open Access J Neurol Neurosurg* **12**: 555834. doi:10.19080/OAJNN.2019.12.555834

Andrews T, Zhang P, Bhat NR. 1998. TNFα potentiates IFNγ-induced cell death in oligodendrocyte progenitors. *J Neurosci Res* **54**: 574–583. doi:10.1002/(SICI)1097-4547(19981201)54:5<574::AID-JNR2>3.0.CO;2-0

Arnett HA, Mason J, Marino M, Suzuki K, Matsushima GK, Ting JPY. 2001. TNFα promotes proliferation of oligodendrocyte progenitors and remyelination. *Nat Neurosci* **4**: 1116–1122. doi:10.1038/nn738

Back SA, Tuohy TMF, Chen H, Wallingford N, Craig A, Struve J, Ning LL, Banine F, Liu Y, Chang A, et al. 2005. Hyaluronan accumulates in demyelinated lesions and inhibits oligodendrocyte progenitor maturation. *Nat Med* **11**: 966–972. doi:10.1038/nm1279

Baer AS, Syed YA, Kang SU, Mitteregger D, Vig R, Ffrench-Constant C, Franklin RJM, Altmann F, Lubec G, Kotter MR. 2009. Myelin-mediated inhibition of oligodendrocyte precursor differentiation can be overcome by pharmacological modulation of Fyn-RhoA and protein kinase C signalling. *Brain* **132**: 465–481. doi:10.1093/brain/awn334

Barnett MH, Prineas JW. 2004. Relapsing and remitting multiple sclerosis: pathology of the newly forming lesion. *Ann Neurol* **55**: 458–468. doi:10.1002/ana.20016

Barreras P, Vasileiou ES, Filippatou AG, Fitzgerald KC, Levy M, Pardo CA, Newsome SD, Mowry EM, Calabresi PA, Sotirchos ES. 2022. Long-term effectiveness and safety of rituximab in neuromyelitis optica spectrum disorder and MOG antibody disease. *Neurology* **99**: e2504–e2516. doi:10.1212/WNL.0000000000201260

Baxi EG, DeBruin J, Tosi DM, Grishkan IV, Smith MD, Kirby LA, Strasburger HJ, Fairchild AN, Calabresi PA, Gocke AR. 2015. Transfer of myelin-reactive Th17 cells impairs endogenous remyelination in the central nervous system of cuprizone-fed mice. *J Neurosci* **35**: 8626–8639. doi:10.1523/JNEUROSCI.3817-14.2015

Berger JR, Pall L, Lanska D, Whiteman M. 1998. Progressive multifocal leukoencephalopathy in patients with HIV infection. *J Neurovirol* **4**: 59–68. doi:10.3109/13550289809113482

Bernard-Valnet R, Koralnik IJ, Du Pasquier R. 2021. Advances in treatment of progressive multifocal leukoencephalopathy. *Ann Neurol* **90**: 865–873. doi:10.1002/ana.26198

Bertram M, Egelhoff T, Schwarz S, Schwab S. 1999. Toxic leukencephalopathy following "ecstasy" ingestion. *J Neurol* **246**: 617–618. doi:10.1007/s004150050416

Bhargava P, Wicken C, Smith MD, Strowd RE, Cortese I, Reich DS, Calabresi PA, Mowry EM. 2019. Trial of intrathecal rituximab in progressive multiple sclerosis patients with evidence of leptomeningeal contrast enhancement. *Mult Scler Relat Disord* **30**: 136–140. doi:10.1016/j.msard.2019.02.013

Bhargava P, Kim S, Reyes AA, Grenningloh R, Boschert U, Absinta M, Pardo C, Van Zijl P, Zhang J, Calabresi PA. 2021. Imaging meningeal inflammation in CNS autoimmunity identifies a therapeutic role for BTK inhibition. *Brain* **144:** 1396–1408. doi:10.1093/brain/awab045

Bhattacharyya S, Holroyd KB, Berkowitz AL. 2024. Metabolic and toxic myelopathies. *Continuum (Minneap Minn)* **30:** 199–223. doi:10.1212/CON.0000000000001376

Bieber AJ, Kerr S, Rodriguez M. 2003. Efficient central nervous system remyelination requires T cells. *Ann Neurol* **53:** 680–684. doi:10.1002/ana.10578

Bielekova B, Goodwin B, Richert N, Cortese I, Kondo T, Afshar G, Gran B, Eaton J, Antel J, Frank JA, et al. 2000. Encephalitogenic potential of the myelin basic protein peptide (amino acids 83-99) in multiple sclerosis: results of a phase II clinical trial with an altered peptide ligand. *Nat Med* **6:** 1167–1175. doi:10.1038/80516

Birchmeier C, Bennett DLH. 2016. Neuregulin/ErbB signaling in developmental myelin formation and nerve repair. *Curr Top Dev Biol* **116:** 45–64. doi:10.1016/bs.ctdb.2015.11.009

Bjornevik K, Cortese M, Healy BC, Kuhle J, Mina MJ, Leng Y, Elledge SJ, Niebuhr DW, Scher AI, Munger KL, et al. 2022. Longitudinal analysis reveals high prevalence of Epstein–Barr virus associated with multiple sclerosis. *Science* **375:** 296–301. doi:10.1126/science.abj8222

Bonnan M, Ferrari S, Courtade H, Money P, Desblache P, Barroso B, Debeugny S. 2021. No early effect of intrathecal rituximab in progressive multiple sclerosis (EFFRITE clinical trial). *Mult Scler Int* **2021:** 1–7. doi:10.1155/2021/8813498

Bose G, Thebault S, Rush CA, Atkins HL, Freedman MS. 2021. Autologous hematopoietic stem cell transplantation for multiple sclerosis: a current perspective. *Mult Scler* **27:** 167–173. doi:10.1177/1352458520917936

Boumaza X, Bonneau B, Roos-Weil D, Pinnetti C, Rauer S, Nitsch L, Del Bello A, Jelcic I, Sühs KW, Gasnault J, et al. 2023. Progressive multifocal leukoencephalopathy treated by immune checkpoint inhibitors. *Ann Neurol* **93:** 257–270. doi:10.1002/ana.26512

Bowen L, Nath A, Smith B. 2018. CNS immune reconstitution inflammatory syndrome. *Handb Clin Neurol* **152:** 167–176. doi:10.1016/B978-0-444-63849-6.00013-X

Brenton JN. 2022. Pediatric acquired demyelinating disorders. *Continuum (Minneap Minn)* **28:** 1104–1130. doi:10.1212/CON.0000000000001128

Brill L, Ganelin-Cohen E, Dabby R, Rabinowicz S, Zohar-Dayan E, Rein N, Aloni E, Karmon Y, Vaknin-Dembinsky A. 2021. Age-related clinical presentation of MOG-IgG seropositivity in Israel. *Front Neurol* **11:** 1739. doi:10.3389/fneur.2020.612304

Brocke S, Gijbels K, Allegretta M, Ferber I, Piercy C, Blankenstein T, Martin R, Utz U, Karin N, Mitchell D, et al. 1996. Treatment of experimental encephalomyelitis with a peptide analogue of myelin basic protein. *Nature* **379:** 343–346. doi:10.1038/379343a0

Brown WD. 2000. Osmotic demyelination disorders: central pontine and extrapontine myelinolysis. *Curr Opin Neurol* **13:** 691–697. doi:10.1097/00019052-200012000-00014

Brown JWL, Cunniffe NG, Prados F, Kanber B, Jones JL, Needham E, Georgieva Z, Rog D, Pearson OR, Overell J, et al. 2021. Safety and efficacy of bexarotene in patients with relapsing-remitting multiple sclerosis (CCMR one): a randomised, double-blind, placebo-controlled, parallel-

group, phase 2a study. *Lancet Neurol* **20:** 709–720. doi:10.1016/S1474-4422(21)00179-4

Bruijstens AL, Lechner C, Flet-Berliac L, Deiva K, Neuteboom RF, Hemingway C, Wassmer E; E.U. paediatric MOG consortium; Baumann M, Bartels F, et al. 2020. E.U. paediatric MOG consortium consensus: part 1 - classification of clinical phenotypes of paediatric myelin oligodendrocyte glycoprotein antibody-associated disorders. *Eur J Paediatr Neurol* **29:** 2–13. doi:10.1016/j.ejpn.2020.10.006

Cadavid D, Balcer L, Galetta S, Aktas O, Ziemssen T, Vanopdenbosch L, Frederiksen J, Skeen M, Jaffe GJ, Butzkueven H, et al. 2017. Safety and efficacy of opicinumab in acute optic neuritis (RENEW): a randomised, placebo-controlled, phase 2 trial. *Lancet Neurol* **16:** 189–199. doi:10.1016/S1474-4422(16)30377-5

Cadavid D, Mellion M, Hupperts R, Edwards KR, Calabresi PA, Drulović J, Giovannoni G, Hartung HP, Arnold DL, Fisher E, et al. 2019. Safety and efficacy of opicinumab in patients with relapsing multiple sclerosis (SYNERGY): a randomised, placebo-controlled, phase 2 trial. *Lancet Neurol* **18:** 845–856. doi:10.1016/S1474-4422(19)30137-1

Cantuti-Castelvetri L, Fitzner D, Bosch-Queralt M, Weil MT, Su M, Sen P, Ruhwedel T, Mitkovski M, Trendelenburg G, Lütjohann D, et al. 2018. Defective cholesterol clearance limits remyelination in the aged central nervous system. *Science* **359:** 684–688. doi:10.1126/science.aan4183

Chang A, Nishiyama A, Peterson J, Prineas J, Trapp BD. 2000. NG2-positive oligodendrocyte progenitor cells in adult human brain and multiple sclerosis lesions. *J Neurosci* **20:** 6404–6412. doi:10.1523/JNEUROSCI.20-17-06404.2000

Chang A, Tourtellotte WW, Rudick R, Trapp BD. 2002. Premyelinating oligodendrocytes in chronic lesions of multiple sclerosis. *N Engl J Med* **346:** 165–173. doi:10.1056/NEJMoa010994

Chen JJ, Flanagan EP, Bhatti MT, Jitprapaikulsan J, Dubey D, Lopez Chiriboga ASS, Fryer JP, Weinshenker BG, McKeon A, Tillema JM, et al. 2020. Steroid-sparing maintenance immunotherapy for MOG-IgG associated disorder. *Neurology* **95:** E111–E120.

Chen JJ, Huda S, Hacohen Y, Levy M, Lotan I, Wilf-Yarkoni A, Stiebel-Kalish H, Hellmann MA, Sotirchos ES, Henderson AD, et al. 2022. Association of maintenance intravenous immunoglobulin with prevention of relapse in adult myelin oligodendrocyte glycoprotein antibody-associated disease. *JAMA Neurol* **79:** 518–525. doi:10.1001/jamaneurol.2022.0489

Chiang ACA, Seua AV, Singhmar P, Arroyo LD, Mahalingam R, Hu J, Kavelaars A, Heijnen CJ. 2020. Bexarotene normalizes chemotherapy-induced myelin decompaction and reverses cognitive and sensorimotor deficits in mice. *Acta Neuropathol Commun* **8:** 1–15. doi:10.1186/s40478-020-01061-x

Chitnis T. 2007. The role of CD4 T cells in the pathogenesis of multiple sclerosis. *Int Rev Neurobiol* **79:** 43–72. doi:10.1016/S0074-7742(07)79003-7

Choi SR, Howell OW, Carassiti D, Magliozzi R, Gveric D, Muraro PA, Nicholas R, Roncaroli F, Reynolds R. 2012. Meningeal inflammation plays a role in the pathology of primary progressive multiple sclerosis. *Brain* **135:** 2925–2937. doi:10.1093/brain/aws189

Cignarella F, Filipello F, Bollman B, Cantoni C, Locca A, Mikesell R, Manis M, Ibrahim A, Deng L, Benitez BA, et al. 2020. TREM2 activation on microglia promotes myelin debris clearance and remyelination in a model of multiple sclerosis. *Acta Neuropathol* 1: 3.

Clark IC, Gutiérrez-Vázquez C, Wheeler MA, Li Z, Rothhammer V, Linnerbauer M, Sanmarco LM, Guo L, Blain M, Zandee SEJ, et al. 2021. Barcoded viral tracing of single-cell interactions in central nervous system inflammation. *Science* 372: eabf1230. doi:10.1126/science.abf1230

Colato E, Stutters J, Tur C, Narayanan S, Arnold DL, Gandini Wheeler-Kingshott CAM, Barkhof F, Ciccarelli O, Chard DT, Eshaghi A. 2021. Predicting disability progression and cognitive worsening in multiple sclerosis using patterns of grey matter volumes. *J Neurol Neurosurg Psychiatry* 92: 995–1006. doi:10.1136/jnnp-2020-325610

Comi G, Bar-Or A, Lassmann H, Uccelli A, Hartung HP, Montalban X, Sørensen PS, Hohlfeld R, Hauser SL. 2021. Role of B cells in multiple sclerosis and related disorders. *Ann Neurol* 89: 13–23. doi:10.1002/ana.25927

Cortese I, Muranski P, Enose-Akahata Y, Ha SK, Smith B, Monaco M, Ryschkewitsch C, Major EO, Ohayon J, Schindler MK, et al. 2019. Pembrolizumab treatment for progressive multifocal leukoencephalopathy. *N Engl J Med* 380: 1597–1605. doi:10.1056/NEJMoa1815039

Cortese I, Beck ES, Al-Louzi O, Ohayon J, Andrada F, Osuorah I, Dwyer J, Billioux BJ, Dargah-zada N, Schindler MK, et al. 2021a. BK virus-specific T cells for immunotherapy of progressive multifocal leukoencephalopathy: an open-label, single-cohort pilot study. *Lancet Neurol* 20: 639–652. doi:10.1016/S1474-4422(21)00174-5

Cortese I, Reich DS, Nath A. 2021b. Progressive multifocal leukoencephalopathy and the spectrum of JC virus-related disease. *Nat Rev Neurol* 17: 37–51. doi:10.1038/s41582-020-00427-y

Costello F. 2022. Neuromyelitis optica spectrum disorders. *Continuum (Minneap Minn)* 28: 1131–1170. doi:10.1212/CON.0000000000001168

Cree BAC, Gourraud PA, Oksenberg JR, Bevan C, Crabtree-Hartman E, Gelfand JM, Goodin DS, Graves J, Green AJ, Mowry E, et al. 2016. Long-term evolution of multiple sclerosis disability in the treatment era. *Ann Neurol* 80: 499–510. doi:10.1002/ana.24747

Cristobal CD, Lee HK. 2022. Development of myelinating glia: an overview. *Glia* 70: 2237–2259. doi:10.1002/glia.24238

Cross A, Riley C. 2022. Treatment of multiple sclerosis. *Continuum (Minneap Minn)* 28: 1025–1051. doi:10.1212/CON.0000000000001170

Cunha MI, Su M, Cantuti-Castelvetri L, Müller SA, Schifferer M, Djannatian M, Alexopoulos I, van der Meer F, Winkler A, van Ham TJ, et al. 2020. Pro-inflammatory activation following demyelination is required for myelin clearance and oligodendrogenesis. *J Exp Med* 217: e20191390. doi:10.1084/jem.20191390

Dai ZM, Sun S, Wang C, Huang H, Hu X, Zhang Z, Lu QR, Qiu M. 2014. Stage-specific regulation of oligodendrocyte development by Wnt/β-catenin signaling. *J Neurosci* 34: 8467–8473. doi:10.1523/JNEUROSCI.0311-14.2014

Dal-Bianco A, Grabner G, Kronnerwetter C, Weber M, Höftberger R, Berger T, Auff E, Leutmezer F, Trattnig S, Lassmann H, et al. 2017. Slow expansion of multiple sclerosis

iron rim lesions: pathology and 7 T magnetic resonance imaging. *Acta Neuropathol* 133: 25–42. doi:10.1007/s00401-016-1636-z

Dal-Bianco A, Grabner G, Kronnerwetter C, Weber M, Kornek B, Kasprian G, Berger T, Leutmezer F, Rommer PS, Trattnig S, et al. 2021. Long-term evolution of multiple sclerosis iron rim lesions in 7 T MRI. *Brain* 144: 833–847. doi:10.1093/brain/awaa436

Demicheva E, Cui YF, Bardwell P, Barghorn S, Kron M, Meyer AH, Schmidt M, Gerlach B, Leddy M, Barlow E, et al. 2015. Targeting repulsive guidance molecule A to promote regeneration and neuroprotection in multiple sclerosis. *Cell Rep* 10: 1887–1898. doi:10.1016/j.celrep.2015.02.048

Dendrou CA, Fugger L, Friese MA. 2015. Immunopathology of multiple sclerosis. *Nat Rev Immunol* 15: 545–558. doi:10.1038/nri3871

Denic A, Wootla B, Rodriguez M. 2013. CD8$^+$ T cells in multiple sclerosis. *Expert Opin Ther Targets* 17: 1053–1066. doi:10.1517/14728222.2013.815726

Derfuss T, Parikh K, Velhin S, Braun M, Mathey E, Krumbholz M, Kümpfel T, Moldenhauer A, Rader C, Sonderegger P, et al. 2009. Contactin-2/TAG-1-directed autoimmunity is identified in multiple sclerosis patients and mediates gray matter pathology in animals. *Proc Natl Acad Sci* 106: 8302–8307. doi:10.1073/pnas.0901496106

Deshmukh VA, Tardif V, Lyssiotis CA, Green CC, Kerman B, Kim HJ, Padmanabhan K, Swoboda JG, Ahmad I, Kondo T, et al. 2013. A regenerative approach to the treatment of multiple sclerosis. *Nature* 502: 327–332. doi:10.1038/nature12647

Deverman BE, Patterson PH. 2012. Exogenous leukemia inhibitory factor stimulates oligodendrocyte progenitor cell proliferation and enhances hippocampal remyelination. *J Neurosci* 32: 2100–2109. doi:10.1523/JNEUROSCI.3803-11.2012

Dinn JJ, Mccann S, Wilson P, Reed B, Weir D, Scott J. 1978. Animal model for subacute combined degeneration. *Lancet* 312: 1154. doi:10.1016/S0140-6736(78)92314-0

Dou DR, Zhao Y, Belk JA, Zhao Y, Casey KM, Chen DC, Li R, Yu B, Srinivasan S, Abe BT, et al. 2024. Xist ribonucleoproteins promote female sex-biased autoimmunity. *Cell* 187: 733–749.e16. doi:10.1016/j.cell.2023.12.037

Du Pasquier RA, Corey S, Margolin DH, Williams K, Pfister LA, De Girolami U, Mac Key JJ, Wüthrich C, Joseph JT, Koralnik IJ. 2003. Productive infection of cerebellar granule cell neurons by JC virus in an HIV$^+$ individual. *Neurology* 61: 775–782. doi:10.1212/01.WNL.0000081306.86961.33

Duncan ID, Radcliff AB, Heidari M, Kidd G, August BK, Wierenga LA. 2018. The adult oligodendrocyte can participate in remyelination. *Proc Natl Acad Sci* 115: E11807–E11816.

El Behi M, Sanson C, Bachelin C, Guillot-Noël L, Fransson J, Stankoff B, Maillart E, Sarrazin N, Guillemot V, Abdi H, et al. 2017. Adaptive human immunity drives remyelination in a mouse model of demyelination. *Brain* 140: 967–980. doi:10.1093/brain/awx008

Elliott C, Belachew S, Wolinsky JS, Hauser SL, Kappos L, Barkhof F, Bernasconi C, Fecker J, Model F, Wei W, et al. 2019a. Chronic white matter lesion activity predicts clinical progression in primary progressive multiple sclerosis. *Brain* 142: 2787–2799. doi:10.1093/brain/awz212

Elliott C, Wolinsky JS, Hauser SL, Kappos L, Barkhof F, Bernasconi C, Wei W, Belachew S, Arnold DL. 2019b. Slowly expanding/evolving lesions as a magnetic resonance imaging marker of chronic active multiple sclerosis lesions. *Mult Scler* **25:** 1915–1925. doi:10.1177/13524585188 14117

Eshaghi A, Marinescu RV, Young AL, Firth NC, Prados F, Jorge Cardoso M, Tur C, De Angelis F, Cawley N, Brownlee WJ, et al. 2018a. Progression of regional grey matter atrophy in multiple sclerosis. *Brain* **141:** 1665–1677. doi:10.1093/brain/awy088

Eshaghi A, Prados F, Brownlee WJ, Altmann DR, Tur C, Cardoso MJ, De Angelis F, van de Pavert SH, Cawley N, De Stefano N, et al. 2018b. Deep gray matter volume loss drives disability worsening in multiple sclerosis. *Ann Neurol* **83:** 210–222. doi:10.1002/ana.25145

Faissner S, Plemel JR, Gold R, Yong VW. 2019. Progressive multiple sclerosis: from pathophysiology to therapeutic strategies. *Nat Rev Drug Discov* **18:** 905–922. doi:10.1038/s41573-019-0035-2

Falcão AM, van Bruggen D, Marques S, Meijer M, Jäkel S, Agirre E, Samudyata, Floriddia EM, Vanichkina DP, ffrench-Constant C, et al. 2018a. Disease-specific oligodendrocyte lineage cells arise in multiple sclerosis. *Nat Med* **24:** 1837–1844. doi:10.1038/s41591-018-0236-y

Falcão AM, van Bruggen D, Marques S, Meijer M, Jäkel S, Agirre E, Samudyata, Floriddia EM, Vanichkina DP, Ffrench-Constant C, et al. 2018b. Disease-specific oligodendrocyte lineage cells arise in multiple sclerosis. *Nat Med* **24:** 1837–1844. doi:10.1038/s41591-018-0236-y

Fancy SPJ, Baranzini SE, Zhao C, Yuk DI, Irvine KA, Kaing S, Sanai N, Franklin RJM, Rowitch DH. 2009. Dysregulation of the Wnt pathway inhibits timely myelination and remyelination in the mammalian CNS. *Genes Dev* **23:** 1571–1585. doi:10.1101/gad.1806309

Feraco P, Incandela F, Franceschi R, Gagliardo C, Bellizzi M. 2021. Clinical and brain imaging findings in a child with vitamin B12 deficiency. *Pediatr Rep* **13:** 583–588. doi:10.3390/pediatric13040069

Fernández-Castañeda A, Chappell MS, Rosen DA, Seki SM, Beiter RM, Johanson DM, Liskey D, Farber E, Onengut-Gumuscu S, Overall CC, et al. 2020. The active contribution of OPCs to neuroinflammation is mediated by LRP1. *Acta Neuropathol* **139:** 365–382. doi:10.1007/s00401-019-02073-1

Filley CM, Kleinschmidt-DeMasters BK. 2001. Toxic leukoencephalopathy. *N Engl J Med* **345:** 425–432. doi:10.1056/NEJM200108093450606

Fischer R, Wajant H, Kontermann R, Pfizenmaier K, Maier O. 2014. Astrocyte-specific activation of TNFR2 promotes oligodendrocyte maturation by secretion of leukemia inhibitory factor. *Glia* **62:** 272–283. doi:10.1002/glia.22605

Fox RJ, Coffey CS, Conwit R, Cudkowicz ME, Gleason T, Goodman A, Klawiter EC, Matsuda K, McGovern M, Naismith RT, et al. 2018. Phase 2 trial of ibudilast in progressive multiple sclerosis. *N Engl J Med* **379:** 846–855. doi:10.1056/NEJMoa1803583

Franklin RJM, Simons M. 2022. CNS remyelination and inflammation: from basic mechanisms to therapeutic opportunities. *Neuron* **110:** 3549–3565. doi:10.1016/j.neuron.2022.09.023

Frenkel EP, Kitchens RL, Johnston JM. 1973. The effect of vitamin B12 deprivation on the enzymes of fatty acid synthesis. *J Biol Chem* **248:** 7540–7546. doi:10.1016/S0021-9258(19)43324-3

Frischer JM, Weigand SD, Guo Y, Kale N, Parisi JE, Pirko I, Mandrekar J, Bramow S, Metz I, Brück W, et al. 2015. Clinical and pathological insights into the dynamic nature of the white matter multiple sclerosis plaque. *Ann Neurol* **78:** 710–721. doi:10.1002/ana.24497

Fünfschilling U, Supplie LM, Mahad D, Boretius S, Saab AS, Edgar J, Brinkmann BG, Kassmann CM, Tzvetanova ID, Möbius W, et al. 2012. Glycolytic oligodendrocytes maintain myelin and long-term axonal integrity. *Nature* **485:** 517–521. doi:10.1038/nature11007

Gibson EM, Purger D, Mount CW, Goldstein AK, Lin GL, Wood LS, Inema I, Miller SE, Bieri G, Zuchero JB, et al. 2014. Neuronal activity promotes oligodendrogenesis and adaptive myelination in the mammalian brain. *Science* **344:** 1252304. doi:10.1126/science.1252304

Giles DA, Duncker PC, Wilkinson NM, Washnock-Schmid JM, Segal BM. 2018. CNS-resident classical DCs play a critical role in CNS autoimmune disease. *J Clin Invest* **128:** 5322–5334. doi:10.1172/JCI123708

Gouna G, Klose C, Bosch-Queralt M, Liu L, Gokce O, Schifferer M, Cantuti-Castelvetri L, Simons M. 2021. TREM2-dependent lipid droplet biogenesis in phagocytes is required for remyelination. *J Exp Med* **218:** e20210227. doi:10.1084/jem.20210227

Graham SM, Arvela OM, Wise GA. 1992. Long-term neurologic consequences of nutritional vitamin B12 deficiency in infants. *J Pediatr* **121:** 710–714. doi:10.1016/S0022-3476(05)81897-9

Green AJ, Gelfand JM, Cree BA, Bevan C, Boscardin WJ, Mei F, Inman J, Arnow S, Devereux M, Abounasr A, et al. 2017. Clemastine fumarate as a remyelinating therapy for multiple sclerosis (ReBUILD): a randomised, controlled, double-blind, crossover trial. *Lancet* **390:** 2481–2489. doi:10.1016/S0140-6736(17)32346-2

Grigoriadis N, van Pesch V. 2015. A basic overview of multiple sclerosis immunopathology. *Eur J Neurol* **22** (Suppl 2): 3–13. doi:10.1111/ene.12798

Guttenplan KA, Weigel MK, Prakash P, Wijewardhane PR, Hasel P, Rufen-Blanchette U, Münch AE, Blum JA, Fine J, Neal MC, et al. 2021. Neurotoxic reactive astrocytes induce cell death via saturated lipids. *Nature* **599:** 102–107. doi:10.1038/s41586-021-03960-y

Häberlein F, Mingardo E, Merten N, Schulze Köhling NK, Reinoß P, Simon K, Japp A, Nagarajan B, Schrage R, Pegurier C, et al. 2022. Humanized zebrafish as a tractable tool for in vivo evaluation of pro-myelinating drugs. *Cell Chem Biol* **29:** 1541–1555.e7. doi:10.1016/j.chembiol.2022.08.007

Hammond TR, Gadea A, Dupree J, Kerninon C, Nait-Oumesmar B, Aguirre A, Gallo V. 2014. Astrocyte-derived endothelin-1 inhibits remyelination through notch activation. *Neuron* **81:** 588–602. doi:10.1016/j.neuron.2013.11.015

Hammond TR, Dufort C, Dissing-Olesen L, Giera S, Young A, Wysoker A, Walker AJ, Gergits F, Segel M, Nemesh J, et al. 2019. Single-cell RNA sequencing of microglia throughout the mouse lifespan and in the injured brain reveals complex cell-state changes. *Immunity* 50: 253–271.e6. doi:10.1016/j.immuni.2018.11.004

He J, Huang Y, Liu H, Sun X, Wu J, Zhang Z, Liu L, Zhou C, Jiang S, Huang Z, et al. 2020. Bexarotene promotes microglia/macrophages—specific brain-derived neurotrophic factor expression and axon sprouting after traumatic brain injury. *Exp Neurol* 334: 113462. doi:10.1016/j.expneurol.2020.113462

Healton E, Savage DG, Brust JC, Garrett TJ, Lindenbaum J. 1991. Neurologic aspects of cobalamin deficiency. *Medicine (Baltimore)* 70: 229–245. doi:10.1097/00005792-199107000-00001

Healy LM, Perron G, Won SY, Michell-Robinson MA, Rezk A, Ludwin SK, Moore CS, Hall JA, Bar-Or A, Antel JP. 2016. MerTK is a functional regulator of myelin phagocytosis by human myeloid cells. *J Immunol* 196: 3375–3384. doi:10.4049/jimmunol.1502562

Healy LM, Stratton JA, Kuhlmann T, Antel J. 2022. The role of glial cells in multiple sclerosis disease progression. *Nat Rev Neurol* 18: 237–248. doi:10.1038/s41582-022-00624-x

Hemmer B, Glocker FX, Schumacher M, Deuschl G, Lücking CH. 1998. Subacute combined degeneration: clinical, electrophysiological, and magnetic resonance imaging findings. *J Neurol Neurosurg Psychiatry* 65: 822–827. doi:10.1136/jnnp.65.6.822

Henderson APD, Barnett MH, Parratt JDE, Prineas JW. 2009. Multiple sclerosis: distribution of inflammatory cells in newly forming lesions. *Ann Neurol* 66: 739–753. doi:10.1002/ana.21800

Henn RE, Noureldein MH, Elzinga SE, Kim B, Savelieff MG, Feldman EL. 2022. Glial-neuron crosstalk in health and disease: a focus on metabolism, obesity, and cognitive impairment. *Neurobiol Dis* 170: 105766. doi:10.1016/j.nbd.2022.105766

Herranz E, Giannì C, Louapre C, Treaba CA, Govindarajan ST, Ouellette R, Loggia ML, Sloane JA, Madigan N, Izquierdo-Garcia D, et al. 2016. Neuroinflammatory component of gray matter pathology in multiple sclerosis. *Ann Neurol* 80: 776–790. doi:10.1002/ana.24791

Hines JH, Ravanelli AM, Schwindt R, Scott EK, Appel B. 2015. Neuronal activity biases axon selection for myelination in vivo. *Nat Neurosci* 18: 683–689. doi:10.1038/nn.3992

Holmøy T, Kvale EØ, Vartdal F. 2004. Cerebrospinal fluid CD4+ T cells from a multiple sclerosis patient cross-recognize Epstein-Barr virus and myelin basic protein. *J Neurovirol* 10: 278–283. doi:10.1080/13550280490499524

Hong S, Beja-Glasser VF, Nfonoyim BM, Frouin A, Li S, Ramakrishnan S, Merry KM, Shi Q, Rosenthal A, Barres BA, et al. 2016a. Complement and microglia mediate early synapse loss in Alzheimer mouse models. *Science* 352: 712–716. doi:10.1126/science.aad8373

Hong S, Dissing-Olesen L, Stevens B. 2016b. New insights on the role of microglia in synaptic pruning in health and disease. *Curr Opin Neurobiol* 36: 128–134. doi:10.1016/j.conb.2015.12.004

Hor JY, Asgari N, Nakashima I, Broadley SA, Leite MI, Kissani N, Jacob A, Marignier R, Weinshenker BG, Paul F, et al. 2020. Epidemiology of neuromyelitis optica spectrum disorder and its prevalence and incidence worldwide. *Front Neurol* 11: 1–13. doi:10.3389/fneur.2020.00501

Hou J, Zhou Y, Cai Z, Terekhova M, Swain A, Andhey PS, Guimaraes RM, Ulezko Antonova A, Qiu T, Sviben S, et al.

2023. Transcriptomic atlas and interaction networks of brain cells in mouse CNS demyelination and remyelination. *Cell Rep* 42: 112293. doi:10.1016/j.celrep.2023.112293

Howell OW, Reeves CA, Nicholas R, Carassiti D, Radotra B, Gentleman SM, Serafini B, Aloisi F, Roncaroli F, Magliozzi R, et al. 2011. Meningeal inflammation is widespread and linked to cortical pathology in multiple sclerosis. *Brain* 134: 2755–2771. doi:10.1093/brain/awr182

Huang JK, Jarjour AA, Oumesmar BN, Kerninon C, Williams A, Krezel W, Kagechika H, Bauer J, Zhao C, Van Evercooren AB, et al. 2011. Retinoid X receptor γ signaling accelerates CNS remyelination. *Nat Neurosci* 14: 45–53. doi:10.1038/nn.2702

Hubler Z, Allimuthu D, Bederman I, Elitt MS, Madhavan M, Allan KC, Shick HE, Garrison E, T. Karl M, Factor DC, et al. 2018. Accumulation of 8,9-unsaturated sterols drives oligodendrocyte formation and remyelination. *Nature* 560: 372–376. doi:10.1038/s41586-018-0360-3

Hughes EG, Kang SH, Fukaya M, Bergles DE. 2013. Oligodendrocyte progenitors balance growth with self-repulsion to achieve homeostasis in the adult brain. *Nat Neurosci* 16: 668–676. doi:10.1038/nn.3390

Hughes EG, Orthmann-Murphy JL, Langseth AJ, Bergles DE. 2018. Myelin remodeling through experience-dependent oligodendrogenesis in the adult somatosensory cortex. *Nat Neurosci* 21: 696–706. doi:10.1038/s41593-018-0121-5

Hwang S, Garcia-Dominguez MA, Kathryn, Fitzgerald C, Saylor DR. 2022. Association of multiple sclerosis prevalence with sociodemographic, health systems, and lifestyle factors on a national and regional level. *Neurology* 99: e1813–e1823. doi:10.1212/WNL.0000000000200962

Itoh Y, Golden LC, Itoh N, Matsukawa MA, Ren E, Tse V, Arnold AP, Voskuhl RR. 2019. The X-linked histone demethylase Kdm6a in CD4+ T lymphocytes modulates autoimmunity. *J Clin Invest* 129: 3852–3863. doi:10.1172/JCI126250

Jacoby N. 2020. Electrolyte disorders and the nervous system. *Continuum (Minneap Minn)* 26: 632–658. doi:10.1212/CON.0000000000000872

Jäkel S, Agirre E, Falcão AM, Van Bruggen D, Lee KW, Knuesel I, Malhotra D. 2019. Altered human oligodendrocyte heterogeneity in multiple sclerosis. *Nature* 566: 543–547. doi:10.1038/s41586-019-0903-2

Jamann H, Cui QL, Desu HL, Pernin F, Tastet O, Halaweh A, Farzam-kia N, Mamane VH, Ouédraogo O, Cleret-Buhot A, et al. 2022. Contact-dependent granzyme B-mediated cytotoxicity of Th17-polarized cells toward human oligodendrocytes. *Front Immunol* 13: 1–17. doi:10.3389/fimmu.2022.850616

Joly M, Conte C, Cazanave C, Le Moing V, Tattevin P, Delobel P, Sommet A, Martin-Blondel G. 2023. Progressive multifocal leukoencephalopathy: epidemiology and spectrum of predisposing conditions. *Brain* 146: 349–358. doi:10.1093/brain/awac237

Jurynczyk M, Messina S, Woodhall MR, Raza N, Everett R, Roca-Fernandez A, Tackley G, Hamid S, Sheard A, Reynolds G, et al. 2017. Clinical presentation and prognosis in MOG-antibody disease: a UK study. *Brain* 140: 3128–3138. doi:10.1093/brain/awx276

Kang SH, Li Y, Fukaya M, Lorenzini I, Cleveland DW, Ostrow LW, Rothstein JD, Bergles DE. 2013. Degeneration and impaired regeneration of gray matter oligodendrocytes in amyotrophic lateral sclerosis. *Nat Neurosci* **16:** 571–579. doi:10.1038/nn.3357

Karin M, Clevers H. 2016. Reparative inflammation takes charge of tissue regeneration. *Nature* **529:** 307–315. doi:10.1038/nature17039

Karin N, Mitchell DJ, Brocke S, Ling N, Steinman L. 1994. Reversal of experimental autoimmune encephalomyelitis by a soluble peptide variant of a myelin basic protein epitope: T cell receptor antagonism and reduction of interferon γ and tumor necrosis factor α production. *J Exp Med* **180:** 2227–2237. doi:10.1084/jem.180.6.2227

Kaya T, Mattugini N, Liu L, Ji H, Cantuti-Castelvetri L, Wu J, Schifferer M, Groh J, Martini R, Besson-Girard S, et al. 2022a. CD8⁺ T cells induce interferon-responsive oligodendrocytes and microglia in white matter aging. *Nat Neurosci* **25:** 1446–1457. doi:10.1038/s41593-022-01183-6

Kaya T, Mattugini N, Liu L, Ji H, Cantuti-Castelvetri L, Wu J, Schifferer M, Groh J, Martini R, Besson-Girard S, et al. 2022b. CD8⁺ T cells induce interferon-responsive oligodendrocytes and microglia in white matter aging. *Nat Neurosci* **25:** 1446–1457. doi:10.1038/s41593-022-01183-6

Keough MB, Rogers JA, Zhang P, Jensen SK, Stephenson EL, Chen T, Hurlbert MG, Lau LW, Rawji KS, Plemel JR, et al. 2016. An inhibitor of chondroitin sulfate proteoglycan synthesis promotes central nervous system remyelination. *Nat Commun* **7:** 11312. doi:10.1038/ncomms11312

Keren-Shaul H, Spinrad A, Weiner A, Matcovitch-Natan O, Dvir-Szternfeld R, Ulland TK, David E, Baruch K, Lara-Astaiso D, Toth B, et al. 2017. A unique microglia type associated with restricting development of Alzheimer's disease. *Cell* **169:** 1276–1290.e17. doi:10.1016/j.cell.2017.05.018

Kim RY, Hoffman AS, Itoh N, Ao Y, Spence R, Sofroniew MV, Voskuhl RR. 2014. Astrocyte CCL2 sustains immune cell infiltration in chronic experimental autoimmune encephalomyelitis. *J Neuroimmunol* **274:** 53–61. doi:10.1016/j.jneuroim.2014.06.009

Kirby L, Jin J, Cardona JG, Smith MD, Martin KA, Wang J, Strasburger H, Herbst L, Alexis M, Karnell J, et al. 2019. Oligodendrocyte precursor cells present antigen and are cytotoxic targets in inflammatory demyelination. *Nat Commun* **10:** 1–20. doi:10.1038/s41467-019-11638-3

Kivisäkk P, Mahad DJ, Callahan MK, Sikora K, Trebst C, Tucky B, Wujek J, Ravid R, Staugaitis SM, Lassmann H, et al. 2004. Expression of CCR7 in multiple sclerosis: implications for CNS immunity. *Ann Neurol* **55:** 627–638. doi:10.1002/ana.20049

Koch-Henriksen N, Thygesen LC, Stenager E, Laursen B, Magyari M. 2018. Incidence of MS has increased markedly over six decades in Denmark particularly with late onset and in women. *Neurology* **90:** e1954–e1963. doi:10.1212/WNL.0000000000005612

Kotter MR, Setzu A, Sim FJ, Van Rooijen N, Franklin RJM. 2001. Macrophage depletion impairs oligodendrocyte remyelination following lysolecithin-induced demyelination. *Glia* **35:** 204–212. doi:10.1002/glia.1085

Kotter MR, Li WW, Zhao C, Franklin RJM. 2006. Myelin impairs CNS remyelination by inhibiting oligodendrocyte precursor cell differentiation. *J Neurosci* **26:** 328–332. doi:10.1523/JNEUROSCI.2615-05.2006

Krasemann S, Madore C, Cialic R, Baufeld C, Calcagno N, El Fatimy R, Beckers L, O'Loughlin E, Xu Y, Fanek Z, et al. 2017. The TREM2-APOE pathway drives the transcriptional phenotype of dysfunctional microglia in neurodegenerative diseases. *Immunity* **47:** 566–581.e9. doi:10.1016/j.immuni.2017.08.008

Kuhlmann T, Miron V, Cuo Q, Wegner C, Antel J, Brück W. 2008. Differentiation block of oligodendroglial progenitor cells as a cause for remyelination failure in chronic multiple sclerosis. *Brain* **131:** 1749–1758. doi:10.1093/brain/awn096

Kuhlmann T, Moccia M, Coetzee T, Cohen JA, Correale J, Graves J, Marrie RA, Montalban X, Yong VW, Thompson AJ, et al. 2022. Multiple sclerosis progression: time for a new mechanism-driven framework. *Lancet Neurol* **22:** 78–88. doi:10.1016/S1474-4422(22)00289-7

Kutzelnigg A, Lucchinetti CF, Stadelmann C, Brück W, Rauschka H, Bergmann M, Schmidbauer M, Parisi JE, Lassmann H. 2005. Cortical demyelination and diffuse white matter injury in multiple sclerosis. *Brain* **128:** 2705–2712. doi:10.1093/brain/awh641

Lajaunie R, Mainardi I, Gasnault J, Rousseau V, Tarantino AG, Sommet A, Cinque P, Martin-Blondel G, Debard A, Delobel P, et al. 2022. Outcome of progressive multifocal leukoencephalopathy treated by interleukin-7. *Ann Neurol* **91:** 496–505. doi:10.1002/ana.26307

Lang HLE, Jacobsen H, Ikemizu S, Andersson C, Harlos K, Madsen L, Hjorth P, Sondergaard L, Svejgaard A, Wucherpfennig K, et al. 2002. A functional and structural basis for TCR cross-reactivity in multiple sclerosis. *Nat Immunol* **3:** 940–943. doi:10.1038/ni835

Lanz TV, Camille Brewer R, Ho PP, Moon JS, Jude KM, Bartley M, Schubert RD, Hawes IA, Vazquez SE, Iyer M, et al. 2021. Clonally expanded B cells in multiple sclerosis bind EBV EBNA1 and GlialCAM. *Nature* **603:** 321–327. doi:10.1038/s41586-022-04432-7

Lassmann H, Van Horssen J, Mahad D. 2012. Progressive multiple sclerosis: pathology and pathogenesis. *Nat Rev Neurol* **8:** 647–656. doi:10.1038/nrneurol.2012.168

Laureno R. 1983. Central pontine myelinolysis following rapid correction of hyponatremia. *Ann Neurol* **13:** 232–242. doi:10.1002/ana.410130303

Lee Y, Morrison BM, Li Y, Lengacher S, Farah MH, Hoffman PN, Liu Y, Tsingalia A, Jin L, Zhang PW, et al. 2012. Oligodendroglia metabolically support axons and contribute to neurodegeneration. *Nature* **487:** 443–448. doi:10.1038/nature11314

Lee H-G, Wheeler MA, Quintana FJ. 2022. Function and therapeutic value of astrocytes in neurological diseases. *Nat Rev Drug Discov* **21:** 339–358. doi:10.1038/s41573-022-00390-x

Liddelow SA, Guttenplan KA, Clarke LE, Bennett FC, Bohlen CJ, Schirmer L, Bennett ML, Münch AE, Chung WS, Peterson TC, et al. 2017. Neurotoxic reactive astrocytes are induced by activated microglia. *Nature* **541:** 481–487. doi:10.1038/nature21029

Linnerbauer M, Wheeler MA, Quintana FJ. 2020. Astrocyte crosstalk in CNS inflammation. *Neuron* **108:** 608–622. doi:10.1016/j.neuron.2020.08.012

Linnerbauer M, Lößlein L, Vandrey O, Peter A, Han Y, Tsaktanis T, Wogram E, Needhamsen M, Kular L, Nagel L, et al. 2024. The astrocyte-produced growth factor HB-EGF limits autoimmune CNS pathology. *Nat Immunol* **25:** 432–447. doi:10.1038/s41590-024-01756-6

Lloyd AF, Miron VE. 2019. The pro-remyelination properties of microglia in the central nervous system. *Nat Rev Neurol* **15:** 447–458. doi:10.1038/s41582-019-0184-2

Lloyd AF, Davies CL, Holloway RK, Labrak Y, Ireland G, Carradori D, Dillenburg A, Borger E, Soong D, Richardson JC, et al. 2019. Central nervous system regeneration is driven by microglia necroptosis and repopulation. *Nat Neurosci* **22:** 1046–1052. doi:10.1038/s41593-019-0418-z

Longbrake E. 2022. Myelin oligodendrocyte glycoprotein-associated disorders. *Continuum (Minneap Minn)* **28:** 1171–1193. doi:10.1212/CON.0000000000001127

Lubetzki C, Zalc B, Williams A, Stadelmann C, Stankoff B. 2020. Remyelination in multiple sclerosis: from basic science to clinical translation. *Lancet Neurol* **19:** 678–688. doi:10.1016/S1474-4422(20)30140-X

Lucchinetti C, Brück W, Parisi J, Scheithauer B, Rodriguez M, Lassmann H. 2000. Heterogeneity of multiple sclerosis lesions: implications for the pathogenesis of demyelination. *Ann Neurol* **47:** 707–717. doi:10.1002/1531-8249 (200006)47:6<707::aid-ana3>3.0.co;2-q

Lucchinetti CF, Popescu BFG, Bunyan RF, Moll NM, Roemer SF, Lassmann H, Brück W, Parisi JE, Scheithauer BW, Giannini C, et al. 2011. Inflammatory cortical demyelination in early multiple sclerosis. *N Engl J Med* **365:** 2188–2197. doi:10.1056/NEJMoa1100648

Luchetti S, Fransen NL, van Eden CG, Ramaglia V, Mason M, Huitinga I. 2018. Progressive multiple sclerosis patients show substantial lesion activity that correlates with clinical disease severity and sex: a retrospective autopsy cohort analysis. *Acta Neuropathol* **135:** 511–528. doi:10.1007/s00401-018-1818-y

Magliozzi R, Howell O, Vora A, Serafini B, Nicholas R, Puopolo M, Reynolds R, Aloisi F. 2007. Meningeal B-cell follicles in secondary progressive multiple sclerosis associate with early onset of disease and severe cortical pathology. *Brain* **130:** 1089–1104. doi:10.1093/brain/awm038

Magliozzi R, Howell OW, Reeves C, Roncaroli F, Nicholas R, Serafini B, Aloisi F, Reynolds R. 2010. A gradient of neuronal loss and meningeal inflammation in multiple sclerosis. *Ann Neurol* **68:** 477–493. doi:10.1002/ana.22230

Magyari M, Sorensen PS. 2019. The changing course of multiple sclerosis: rising incidence, change in geographic distribution, disease course, and prognosis. *Curr Opin Neurol* **32:** 320–326. doi:10.1097/WCO.0000000000000695

Mahad DH, Trapp BD, Lassmann H. 2015. Pathological mechanisms in progressive multiple sclerosis. *Lancet Neurol* **14:** 183–193. doi:10.1016/S1474-4422(14)70256-X

Manouchehri N, Hussain RZ, Cravens PD, Esaulova E, Artyomov MN, Edelson BT, Wu GF, Cross AH, Doelger R, Loof N, et al. 2021. CD11c⁺CD88⁺CD317⁺ myeloid cells are critical mediators of persistent CNS autoimmunity. *Proc Natl Acad Sci* **118:** e2014492118. doi:10.1073/pnas.2014492118

Marion CM, Radomski KL, Cramer NP, Galdzicki Z, Armstrong RC. 2018. Experimental traumatic brain injury identifies distinct early and late phase axonal conduction deficits of white matter pathophysiology, and reveals intervening recovery. *J Neurosci* **38:** 8723–8736. doi:10.1523/JNEUROSCI.0819-18.2018

Marriott MP, Emery B, Cate HS, Binder MD, Kemper D, Wu Q, Kolbe S, Gordon IR, Wang H, Egan G, et al. 2008. Leukemia inhibitory factor signaling modulates both central nervous system demyelination and myelin repair. *Glia* **56:** 686–698. doi:10.1002/glia.20646

Martin R, Howell MD, Jaraquemada D, Flerlage M, Richert J, Brostoff S, Long EO, McFarlin DE, McFarland HF. 1991. A myelin basic protein peptide is recognized by cytotoxic T cells in the context of four HLA-DR types associated with multiple sclerosis. *J Exp Med* **173:** 19–24. doi:10.1084/jem.173.1.19

Mason JL, Suzuki K, Chaplin DD, Matsushima GK. 2001. Interleukin-1β promotes repair of the CNS. *J Neurosci* **21:** 7046–7052. doi:10.1523/JNEUROSCI.21-18-07046.2001

Mathieu PA, Almeira Gubiani MF, Rodríguez D, Gómez Pinto LI, Calcagno MDL, Adamo AM. 2019. Demyelination-remyelination in the central nervous system: ligand-dependent participation of the Notch signaling pathway. *Toxicol Sci* **171:** 172–192. doi:10.1093/toxsci/kfz130

Mayo L, Trauger SA, Blain M, Nadeau M, Patel B, Alvarez JI, Mascanfroni ID, Yeste A, Kivisäkk P, Kallas K, et al. 2014. Regulation of astrocyte activation by glycolipids drives chronic CNS inflammation. *Nat Med* **20:** 1147–1156. doi:10.1038/nm.3681

Mei F, Fancy SPJ, Shen YAA, Niu J, Zhao C, Presley B, Miao E, Lee S, Mayoral SR, Redmond SA, et al. 2014. Micropillar arrays as a high-throughput screening platform for therapeutics in multiple sclerosis. *Nat Med* **20:** 954–960. doi:10.1038/nm.3618

Metz J. 1992. Cobalamin deficiency and the pathogenesis of nervous system disease. *Annu Rev Nutr* **12:** 59–79. doi:10.1146/annurev.nu.12.070192.000423

Metz I, Weigand SD, Popescu BFG, Frischer JM, Parisi JE, Guo Y, Lassmann H, Brück W, Lucchinetti CF. 2014. Pathologic heterogeneity persists in early active multiple sclerosis lesions. *Ann Neurol* **75:** 728–738. doi:10.1002/ana.24163

Mi S, Miller RH, Lee X, Scott ML, Shulag-Morskaya S, Shao Z, Chang J, Thill G, Levesque M, Zhang M, et al. 2005. LINGO-1 negatively regulates myelination by oligodendrocytes. *Nat Neurosci* **8:** 745–751. doi:10.1038/nn1460

Mills Ko E, Ma JH, Guo F, Miers L, Lee E, Bannerman P, Burns T, Ko D, Sohn J, Soulika AM, et al. 2014. Deletion of astroglial CXCL10 delays clinical onset but does not affect progressive axon loss in a murine autoimmune multiple sclerosis model. *J Neuroinflammation* **11:** 1–11. doi:10.1186/1742-2094-11-105

Mira RG, Lira M, Cerpa W. 2021. Traumatic brain injury: mechanisms of glial response. *Front Physiol* **12:** 740939. doi:10.3389/fphys.2021.740939

Miron VE, Boyd A, Zhao JW, Yuen TJ, Ruckh JM, Shadrach JL, Van Wijngaarden P, Wagers AJ, Williams A, Franklin RJM, et al. 2013. M2 microglia and macrophages drive oligodendrocyte differentiation during CNS remyelination. *Nat Neurosci* **16:** 1211–1218. doi:10.1038/nn.3469

Mitew S, Gobius I, Fenlon LR, McDougall SJ, Hawkes D, Xing YL, Bujalka H, Gundlach AL, Richards LJ, Kilpatrick TJ, et al. 2018. Pharmacogenetic stimulation of neuronal activity increases myelination in an axon-specific manner. *Nat Commun* **9:** 306. doi:10.1038/s41467-017-02719-2

Cite this article as *Cold Spring Harb Perspect Biol* doi: 10.1101/cshperspect.a041374

Mitrovič M, Patsopoulos NA, Beecham AH, Dankowski T, Goris A, Dubois B, D'hooghe MB, Lemmens R, Van Damme P, Søndergaard HB, et al. 2018. Low-frequency and rare-coding variation contributes to multiple sclerosis risk. *Cell* **175:** 1679–1687.e7. doi:10.1016/j.cell.2018.09.049

Moore CS, Cui Q-L, Warsi NM, Durafourt BA, Zorko N, Owen DR, Antel JP, Bar-Or A. 2015. Direct and indirect effects of immune and central nervous system-resident cells on human oligodendrocyte progenitor cell differentiation. *J Immunol* **194:** 761–772. doi:10.4049/jimmunol.1401156

Moreno M, Bannerman P, Ma J, Guo F, Miers L, Soulika AM, Pleasure D. 2014. Conditional ablation of astroglial CCL2 suppresses CNS accumulation of M1 macrophages and preserves axons in mice with MOG peptide EAE. *J Neurosci* **34:** 8175–8185. doi:10.1523/JNEUROSCI.1137-14.2014

Najm FJ, Madhavan M, Zaremba A, Shick E, Karl RT, Factor DC, Miller TE, Nevin ZS, Kantor C, Sargent A, et al. 2015. Drug-based modulation of endogenous stem cells promotes functional remyelination in vivo. *Nature* **522:** 216–220. doi:10.1038/nature14335

Natrajan MS, De La Fuente AG, Crawford AH, Linehan E, Nuñez V, Johnson KR, Wu T, Fitzgerald DC, Ricote M, Bielekova B, et al. 2015. Retinoid X receptor activation reverses age-related deficiencies in myelin debris phagocytosis and remyelination. *Brain* **138:** 3581–3597. doi:10.1093/brain/awv289

Nawaz A, Khattak NN, Khan MS, Nangyal H, Sabri S, Shakir M. 2020. Deficiency of vitamin B12 and its relation with neurological disorders: a critical review. *J Basic Appl Zool* **81:** 1–9. doi:10.1186/s41936-020-00148-0

Neely SA, Williamson JM, Klingseisen A, Zoupi L, Early JJ, Williams A, Lyons DA. 2022. New oligodendrocytes exhibit more abundant and accurate myelin regeneration than those that survive demyelination. *Nat Neurosci* **25:** 415–420. doi:10.1038/s41593-021-01009-x

Neumann H, Kotter MR, Franklin RJM. 2009. Debris clearance by microglia: an essential link between degeneration and regeneration. *Brain* **132:** 288–295. doi:10.1093/brain/awn109

Neumann B, Baror R, Zhao C, Segel M, Dietmann S, Rawji KS, Foerster S, McClain CR, Chalut K, van Wijngaarden P, et al. 2019a. Metformin restores CNS remyelination capacity by rejuvenating aged stem cells. *Cell Stem Cell* **25:** 473–485.e8. doi:10.1016/j.stem.2019.08.015

Neumann B, Segel M, Chalut KJ, Franklin RJM. 2019b. Remyelination and ageing: reversing the ravages of time. *Mult Scler J* **25:** 1835–1841. doi:10.1177/1352458519884006

Neumann B, Foerster S, Zhao C, Bodini B, Reich DS, Bergles DE, Káradóttir RT, Lubetzki C, Lairson LL, Zalc B, et al. 2020. Problems and pitfalls of identifying remyelination in multiple sclerosis. *Cell Stem Cell* **26:** 617–619. doi:10.1016/j.stem.2020.03.017

Neumann B, Segel M, Ghosh T, Zhao C, Tourlomousis P, Young A, Förster S, Sharma A, Chen CZ-Y, Cubillos JF, et al. 2021. Myc determines the functional age state of oligodendrocyte progenitor cells. *Nat Aging* **1:** 826–837. doi:10.1038/s43587-021-00109-4

Nicholson LB, Greer JM, Sobel RA, Lees MB, Kuchroo VK. 1995. An altered peptide ligand mediates immune deviation and prevents autoimmune encephalomyelitis. *Immunity* **3:** 397–405. doi:10.1016/1074-7613(95)90169-8

Nikić I, Merkler D, Sorbara C, Brinkoetter M, Kreutzfeldt M, Bareyre FM, Brück W, Bishop D, Misgeld T, Kerschensteiner M. 2011. A reversible form of axon damage in experimental autoimmune encephalomyelitis and multiple sclerosis. *Nat Med* **17:** 495–499. doi:10.1038/nm.2324

Oh J, Bar-Or A. 2022. Emerging therapies to target CNS pathophysiology in multiple sclerosis. *Nat Rev Neurol* **18:** 1–10.

Oh U, Fujita M, Ikonomidou VN, Evangelou IE, Matsuura E, Harberts E, Ohayon J, Pike VW, Zhang Y, Zoghbi SS, et al. 2011. Translocator protein PET imaging for glial activation in multiple sclerosis. *J Neuroimmune Pharmacol* **6:** 354–361. doi:10.1007/s11481-010-9243-6

Omari A, Kormas N, Field M. 2002. Delayed onset of central pontine myelinolysis despite appropriate correction of hyponatraemia. *Intern Med J* **32:** 273–274. doi:10.1046/j.1445-5994.2002.00220.x

Ota K, Matsui M, Milford EL, Mackin GA, Weiner HL, Hafler DA. 1990. T-cell recognition of an immuno-dominant myelin basic protein epitope in multiple sclerosis. *Nature* **346:** 183–187. doi:10.1038/346183a0

Pant SS, Asbury AK, Richardson EP Jr. 1968. The myelopathy of pernicious anemia. A neuropathological reappraisal. *Acta Neurol Scand* **44:** 7–36.

Paolicelli RC, Sierra A, Stevens B, Tremblay ME, Aguzzi A, Ajami B, Amit I, Audinat E, Bechmann I, Bennett M, et al. 2022. Microglia states and nomenclature: a field at its crossroads. *Neuron* **110:** 3458–3483. doi:10.1016/j.neuron.2022.10.020

Papp V, Magyari M, Aktas O, Berger T, Broadley SA, Cabre P, Jacob A, Kira JI, Leite MI, Marignier R, et al. 2021. Worldwide incidence and prevalence of neuromyelitis optica: a systematic review. *Neurology* **96:** 59–77. doi:10.1212/WNL.0000000000011153

Parks NE. 2021. Metabolic and toxic myelopathies. *Continuum (Minneap Minn)* **27:** 143–162. doi:10.1212/CON.0000000000000963

Pasquini LA, Millet V, Hoyos HC, Giannoni JP, Croci DO, Marder M, Liu FT, Rabinovich GA, Pasquini JM. 2011. Galectin-3 drives oligodendrocyte differentiation to control myelin integrity and function. *Cell Death Differ* **18:** 1746–1756. doi:10.1038/cdd.2011.40

Patel JR, McCandless EE, Dorsey D, Klein RS. 2010. CXCR4 promotes differentiation of oligodendrocyte progenitors and remyelination. *Proc Natl Acad Sci* **107:** 11062–11067. doi:10.1073/pnas.1006301107

Patrikios P, Stadelmann C, Kutzelnigg A, Rauschka H, Schmidbauer M, Laursen H, Sorensen PS, Brück W, Lucchinetti C, Lassmann H. 2006. Remyelination is extensive in a subset of multiple sclerosis patients. *Brain* **129:** 3165–3172. doi:10.1093/brain/awl217

Patsopoulos NA, Baranzini SE, Santaniello A, Shoostari P, Cotsapas C, Wong G, Beecham AH, James T, Replogle J, Vlachos IS, et al. 2019. Multiple sclerosis genomic map implicates peripheral immune cells and microglia in susceptibility. *Science* **365:** eaav7188. doi:10.1126/science.aav7188

Pelfrey CM, Trotter JL, Tranquill LR, McFarland HF. 1994. Identification of a second T cell epitope of human proteolipid protein (residues 89–106) recognized by proliferative and cytolytic CD4$^+$ T cells from multiple sclerosis patients.

J Neuroimmunol **53**: 153–161. doi:10.1016/0165-5728(94)90025-6

Petersen MA, Ryu JK, Chang KJ, Etxeberria A, Bardehle S, Mendiola AS, Kamau-Devers W, Fancy SPJ, Thor A, Bushong EA, et al. 2017. Fibrinogen activates BMP signaling in oligodendrocyte progenitor cells and inhibits remyelination after vascular damage. *Neuron* **96**: 1003–1012.e7. doi:10.1016/j.neuron.2017.10.008

Petersen MA, Tognatta R, Meyer-Franke A, Bushong EA, Mendiola AS, Yan Z, Muthusamy A, Merlini M, Meza-Acevedo R, Cabriga B, et al. 2021. BMP receptor blockade overcomes extrinsic inhibition of remyelination and restores neurovascular homeostasis. *Brain* **144**: 2291–2301. doi:10.1093/brain/awab106

Peterson S, Jalil A, Beard K, Kakara M, Sriwastava S. 2022. Updates on efficacy and safety outcomes of new and emerging disease modifying therapies and stem cell therapy for multiple sclerosis: a review. *Mult Scler Relat Disord* **68**: 104125. doi:10.1016/j.msard.2022.104125

Pette M, Fujita K, Wilkinson D, Altmann DM, Trowsdale J, Giegerich G, Hinkkanen A, Epplen JT, Kappos L, Wekerle H. 1990. Myelin autoreactivity in multiple sclerosis: recognition of myelin basic protein in the context of HLA-DR2 products by T lymphocytes of multiple-sclerosis patients and healthy donors. *Proc Natl Acad Sci* **87**: 7968–7972. doi:10.1073/pnas.87.20.7968

Philips T, Rothstein JD. 2017. Oligodendroglia: metabolic supporters of neurons. *J Clin Invest* **127**: 3271–3280. doi:10.1172/JCI90610

Philips T, Mironova YA, Jouroukhin Y, Chew J, Vidensky S, Farah MH, Pletnikov MV, Bergles DE, Morrison BM, Rothstein JD. 2021. MCT1 deletion in oligodendrocyte lineage cells causes late-onset hypomyelination and axonal degeneration. *Cell Rep* **34**: 108610. doi:10.1016/j.celrep.2020.108610

Piaton G, Aigrot MS, Williams A, Moyon S, Tepavcevic V, Moutkine I, Gras J, Matho KS, Schmitt A, Soellner H, et al. 2011. Class 3 semaphorins influence oligodendrocyte precursor recruitment and remyelination in adult central nervous system. *Brain* **134**: 1156–1167. doi:10.1093/brain/awr022

Plemel JR, Manesh SB, Sparling JS, Tetzlaff W. 2013. Myelin inhibits oligodendroglial maturation and regulates oligodendrocytic transcription factor expression. *Glia* **61**: 1471–1487. doi:10.1002/glia.22535

Plemel JR, Liu WQ, Yong VW. 2017. Remyelination therapies: a new direction and challenge in multiple sclerosis. *Nat Rev Drug Discov* **16**: 617–634. doi:10.1038/nrd.2017.115

Plemel JR, Stratton JA, Michaels NJ, Rawji KS, Zhang E, Sinha S, Baaklini CS, Dong Y, Ho M, Thorburn K, et al. 2020. Microglia response following acute demyelination is heterogeneous and limits infiltrating macrophage dispersion. *Sci Adv* **6**: 1–16. doi:10.1126/sciadv.aay6324

Pluvinage JV, Haney MS, Smith BAH, Sun J, Iram T, Bonanno L, Li L, Lee DP, Morgens DW, Yang AC, et al. 2019. CD22 blockade restores homeostatic microglial phagocytosis in ageing brains. *Nature* **568**: 187–192. doi:10.1038/s41586-019-1088-4

Politis M, Giannetti P, Su P, Turkheimer F, Keihaninejad S, Wu K, Waldman A, Malik O, Matthews PM, Reynolds R, et al. 2012. Increased PK11195 PET binding in the cortex of patients with MS correlates with disability. *Neurology* **79**: 523–530. doi:10.1212/WNL.0b013e3182635645

Ponsonby AL, Lucas RM, Van Der Mei IA, Dear K, Valery PC, Pender MP, Taylor BV, Kilpatrick TJ, Coulthard A, Chapman C, et al. 2012. Offspring number, pregnancy, and risk of a first clinical demyelinating event: the AusImmune Study. *Neurology* **78**: 867–874. doi:10.1212/WNL.0b013e31824c4648

Popescu BF, Lucchinetti CF. 2012. Meningeal and cortical grey matter pathology in multiple sclerosis. *BMC Neurol* **12**: 11. doi:10.1186/1471-2377-12-11

Preziosa P, Pagani E, Meani A, Moiola L, Rodegher M, Filippi M, Rocca MA. 2022. Slowly expanding lesions predict 9-year multiple sclerosis disease progression. *Neurol Neuroimmunol Neuroinflamm* **9**: e1139. doi:10.1212/NXI.0000000000001139

Pu A, Stephenson EL, Yong VW. 2018. The extracellular matrix: focus on oligodendrocyte biology and targeting CSPGs for remyelination therapies. *Glia* **66**: 1809–1825. doi:10.1002/glia.23333

Rawji KS, Young AMH, Ghosh T, Michaels NJ, Mirzaei R, Kappen J, Kolehmainen KL, Alaeiilkhchi N, Lozinski B, Mishra MK, et al. 2020. Niacin-mediated rejuvenation of macrophage/microglia enhances remyelination of the aging central nervous system. *Acta Neuropathol* **139**: 893–909. doi:10.1007/s00401-020-02129-7

Reich DS, Lucchinetti CF, Calabresi PA. 2018. Multiple sclerosis. *N Engl J Med* **378**: 169–180. doi:10.1056/NEJMra1401483

Rindi LV, Zaçe D, Braccialarghe N, Massa B, Barchi V, Iannazzo R, Fato I, De Maria F, Kontogiannis D, Malagnino V, et al. 2024. Drug-induced progressive multifocal leukoencephalopathy (PML): a systematic review and meta-analysis. *Drug Saf* **47**: 333–354. doi:10.1007/s40264-023-01383-4

Rubino SJ, Mayo L, Wimmer I, Siedler V, Brunner F, Hametner S, Madi A, Lanser A, Moreira T, Donnelly D, et al. 2018. Acute microglia ablation induces neurodegeneration in the somatosensory system. *Nat Commun* **9**: 1–13. doi:10.1038/s41467-018-05929-4

Rubtsova K, Marrack P, Rubtsov AV. 2015. Sexual dimorphism in autoimmunity. *J Clin Invest* **125**: 2187–2193. doi:10.1172/JCI78082

Ruckh JM, Zhao JW, Shadrach JL, Van Wijngaarden P, Rao TN, Wagers AJ, Franklin RJM. 2012. Rejuvenation of regeneration in the aging central nervous system. *Cell Stem Cell* **10**: 96–103. doi:10.1016/j.stem.2011.11.019

Ruijs TCG, Freedman MS, Grenier YG, Olivier A, Antel JP. 1990. Human oligodendrocytes are susceptible to cytolysis by major histocompatibility complex class I-restricted lymphocytes. *J Neuroimmunol* **27**: 89–97. doi:10.1016/0165-5728(90)90058-U

Ruijs TCG, Louste K, Brown EA, Antel JP. 1993. Lysis of human glial cells by major histocompatibility complex-unrestricted CD4⁺ cytotoxic lymphocytes. *J Neuroimmunol* **42**: 105–111. doi:10.1016/0165-5728(93)90217-M

Russell JSR, Batten FE, Collier J, Risien Eussell JS. 1900. Subacute combined degeneration of the spinal cord. *Brain* **23**: 39–110. doi:10.1093/brain/23.1.39

Sanmarco LM, Wheeler MA, Gutiérrez-Vázquez C, Polonio CM, Linnerbauer M, Pinho-Ribeiro FA, Li Z, Giovannoni F, Batterman KV, Scalisi G, et al. 2021. Gut-licensed IFNγ⁺

NK cells drive LAMP1⁺TRAIL⁺ anti-inflammatory astrocytes. *Nature* **590**: 473–479. doi:10.1038/s41586-020-03116-4

Santos-Gil DF, Arboleda G, Sandoval-Hernández AG. 2021. Retinoid X receptor activation promotes re-myelination in a very old triple transgenic mouse model of Alzheimer's disease. *Neurosci Lett* **750**: 135764. doi:10.1016/j.neulet.2021.135764

Sarbu N, Shih RY, Jones RV, Horkayne-Szakaly I, Oleaga L, Smirniotopoulos JG. 2016. White matter diseases with radiologic-pathologic correlation. *Radiographics* **36**: 1426–1447. doi:10.1148/rg.2016160031

Sawcer S, Franklin RJM, Ban M. 2014. Multiple sclerosis genetics. *Lancet Neurol* **13**: 700–709. doi:10.1016/S1474-4422(14)70041-9

Saxena A, Martin-Blondel G, Mars LT, Liblau RS. 2011. Role of CD8⁺ T cell subsets in the pathogenesis of multiple sclerosis. *FEBS Lett* **585**: 3758–3763. doi:10.1016/j.febslet.2011.08.047

Schirmer L, Velmeshev D, Holmqvist S, Kaufmann M, Werneburg S, Jung D, Vistnes S, Stockley JH, Young A, Steindel M, et al. 2019. Neuronal vulnerability and multilineage diversity in multiple sclerosis. *Nature* **573**: 75–82. doi:10.1038/s41586-019-1404-z

Schlüter A, Sandoval J, Fourcade S, Díaz-Lagares A, Ruiz M, Casaccia P, Esteller M, Pujol A. 2018. Epigenomic signature of adrenoleukodystrophy predicts compromised oligodendrocyte differentiation. *Brain Pathol* **28**: 902–919. doi:10.1111/bpa.12595

Schneider R, Oh J. 2022. Bruton's tyrosine kinase inhibition in multiple sclerosis. *Curr Neurol Neurosci Rep* **22**: 721–734. doi:10.1007/s11910-022-01229-z

Schweitzer F, Laurent S, Cortese I, Fink GR, Silling S, Skripuletz T, Metz I, Wattjes MP, Warnke C. 2023. Progressive multifocal leukoencephalopathy: pathogenesis, diagnostic tools, and potential biomarkers of response to therapy. *Neurology* **101**: 700–713. doi:10.1212/WNL.0000000000207622

Segel M, Neumann B, Hill MFE, Weber IP, Viscomi C, Zhao C, Young A, Agley CC, Thompson AJ, Gonzalez GA, et al. 2019. Niche stiffness underlies the ageing of central nervous system progenitor cells. *Nature* **573**: 130–134. doi:10.1038/s41586-019-1484-9

Shen K, Reichelt M, Kyauk RV, Ngu H, Shen YAA, Foreman O, Modrusan Z, Friedman BA, Sheng M, Yuen TJ. 2021. Multiple sclerosis risk gene Mertk is required for microglial activation and subsequent remyelination. *Cell Rep* **34**: 108835. doi:10.1016/j.celrep.2021.108835

Sim FJ, Zhao C, Penderis J, Franklin RJM. 2002. The age-related decrease in CNS remyelination efficiency is attributable to an impairment of both oligodendrocyte progenitor recruitment and differentiation. *J Neurosci* **22**: 2451–2459. doi:10.1523/JNEUROSCI.22-07-02451.2002

Singh S, Metz I, Amor S, Van Der Valk P, Stadelmann C, Brück W. 2013. Microglial nodules in early multiple sclerosis white matter are associated with degenerating axons. *Acta Neuropathol* **125**: 595–608. doi:10.1007/s00401-013-1082-0

Singh TD, Fugate JE, Rabinstein AA. 2014. Central pontine and extrapontine myelinolysis: a systematic review. *Eur J Neurol* **21**: 1443–1450. doi:10.1111/ene.12571

Sloane JA, Batt C, Ma Y, Harris ZM, Trapp B, Vartanian T. 2010. Hyaluronan blocks oligodendrocyte progenitor maturation and remyelination through TLR2. *Proc Natl Acad Sci* **107**: 11555–11560. doi:10.1073/pnas.1006496107

Smith JA, Nicaise AM, Ionescu RB, Hamel R, Peruzzotti-Jametti L, Pluchino S. 2021. Stem cell therapies for progressive multiple sclerosis. *Front Cell Dev Biol* **9**: 10–15.

Smith MD, Chamling X, Gill AJ, Martinez H, Li W, Fitzgerald KC, Sotirchos ES, Moroziewicz D, Bauer L, Paull D, et al. 2022. Reactive astrocytes derived from human induced pluripotent stem cells suppress oligodendrocyte precursor cell differentiation. *Front Mol Neurosci* **15**: 1–14. doi:10.3389/fnmol.2022.874299

Smith-Bouvier DL, Divekar AA, Sasidhar M, Du S, Tiwari-Woodruff SK, King JK, Arnold AP, Singh RR, Voskuhl RR. 2008. A role for sex chromosome complement in the female bias in autoimmune disease. *J Exp Med* **205**: 1099–1108. doi:10.1084/jem.20070850

Stabler SP. 2013. Vitamin B₁₂ deficiency. *N Engl J Med* **368**: 149–160. doi:10.1056/NEJMcp1113996

Stephenson EL, Zhang P, Ghorbani S, Wang A, Gu J, Keough MB, Rawji KS, Silva C, Yong VW, Ling CC. 2019. Targeting the chondroitin sulfate proteoglycans: evaluating fluorinated glucosamines and xylosides in screens pertinent to multiple sclerosis. *ACS Cent Sci* **5**: 1223–1234. doi:10.1021/acscentsci.9b00327

Tan K, Roda R, Ostrow L, McArthur J, Nath A. 2009. PML-IRIS in patients with HIV infection: clinical manifestations and treatment with steroids. *Neurology* **72**: 1458–1464. doi:10.1212/01.wnl.0000343510.08643.74

Tanabe S, Saitoh S, Miyajima H, Itokazu T, Yamashita T. 2019. Microglia suppress the secondary progression of autoimmune encephalomyelitis. *Glia* **67**: 1694–1704. doi:10.1002/glia.23640

Tepavčević V, Kerninon C, Aigrot MS, Meppiel E, Mozafari S, Arnould-Laurent R, Ravassard P, Kennedy TE, Nait-Oumesmar B, Lubetzki C. 2014. Early netrin-1 expression impairs central nervous system remyelination. *Ann Neurol* **76**: 252–268. doi:10.1002/ana.24201

't Hart BA, Luchicchi A, Schenk GJ, Stys PK, Geurts JJG. 2021. Mechanistic underpinning of an inside-out concept for autoimmunity in multiple sclerosis. *Ann Clin Transl Neurol* **8**: 1709–1719. doi:10.1002/acn3.51401

Thompson AJ, Baranzini SE, Geurts J, Hemmer B, Ciccarelli O. 2018. Multiple sclerosis. *Lancet* **391**: 1622–1636. doi:10.1016/S0140-6736(18)30481-1

Tintore M, Cobo-Calvo A, Carbonell P, Arrambide G, Otero-Romero S, Río J, Tur C, Comabella M, Nos C, Arévalo MJ, et al. 2021. Effect of changes in MS diagnostic criteria over 25 years on time to treatment and prognosis in patients with clinically isolated syndrome. *Neurology* **97**: e1641–e1652. doi:10.1212/WNL.0000000000012726

Toft-Hansen H, Füchtbauer L, Owens T. 2011. Inhibition of reactive astrocytosis in established experimental autoimmune encephalomyelitis favors infiltration by myeloid cells over T cells and enhances severity of disease. *Glia* **59**: 166–176. doi:10.1002/glia.21088

Tormoehlen LM. 2011. Toxic leukoencephalopathies. *Neurol Clin* **29**: 591–605. doi:10.1016/j.ncl.2011.05.005

Trapp BD, Peterson J, Ransohoff RM, Rudick R, Mörk S, Bö L. 1998. Axonal transection in the lesions of multiple sclerosis. *N Engl J Med* **338**: 278–285. doi:10.1056/NEJM199801293380502

Tripathi RB, Rivers LE, Young KM, Jamen F, Richardson WD. 2010. NG2 glia generate new oligodendrocytes but few astrocytes in a murine experimental autoimmune encephalomyelitis model of demyelinating disease. *J Neurosci* **30**: 16383–16390. doi:10.1523/JNEUROSCI.3411-10.2010

Uccelli A, Laroni A, Ali R, Battaglia MA, Blinkenberg M, Brundin L, Clanet M, Fernandez O, Marriot J, Muraro P, et al. 2021. Safety, tolerability, and activity of mesenchymal stem cells versus placebo in multiple sclerosis (MESEMS): a phase 2, randomised, double-blind crossover trial. *Lancet Neurol* **20**: 917–929. doi:10.1016/S1474-4422(21)00301-X

Valli A, Sette A, Kappos L, Oseroff C, Sidney J, Miescher G, Hochberger M, Albert ED, Adorini L. 1993. Binding of myelin basic protein peptides to human histocompatibility leukocyte antigen class II molecules and their recognition by T cells from multiple sclerosis patients. *J Clin Invest* **91**: 616–628. doi:10.1172/JCI116242

Van Der Valk P, Amor S. 2009. Preactive lesions in multiple sclerosis. *Curr Opin Neurol* **22**: 207–213. doi:10.1097/WCO.0b013e32832b4c76

van Horssen J, Singh S, van der Pol S, Kipp M, Lim JL, Peferoen L, Gerritsen W, Kooi EJ, Witte ME, Geurts JJG, et al. 2012. Clusters of activated microglia in normal-appearing white matter show signs of innate immune activation. *J Neuroinflammation* **9**: 1–9. doi:10.1186/1742-2094-9-156

Venkatramanan S, Armata IE, Strupp BJ, Finkelstein JL. 2016. Vitamin B-12 and cognition in children. *Adv Nutr* **7**: 879–888. doi:10.3945/an.115.012021

Vermersch P, Brieva-Ruiz L, Fox RJ, Paul F, Ramio-Torrenta L, Schwab M, Moussy A, Mansfield C, Hermine O, Maciejowski M, et al. 2022. Efficacy and safety of masitinib in progressive forms of multiple sclerosis: a randomized, phase 3, clinical trial. *Neurol Neuroimmunol Neuroinflamm* **9**: e1148. doi:10.1212/NXI.0000000000001148

Voet S, Prinz M, van Loo G. 2019. Microglia in central nervous system inflammation and multiple sclerosis pathology. *Trends Mol Med* **25**: 112–123. doi:10.1016/j.molmed.2018.11.005

Voskuhl RR. 2020. The effect of sex on multiple sclerosis risk and disease progression. *Mult Scler J* **26**: 554–560. doi:10.1177/1352458519892491

Voskuhl RR, Peterson RS, Song B, Ao Y, Morales LBJ, Tiwari-Woodruff S, Sofroniew MV. 2009. Reactive astrocytes form scar-like perivascular barriers to leukocytes during adaptive immune inflammation of the CNS. *J Neurosci* **29**: 11511–11522. doi:10.1523/JNEUROSCI.1514-09.2009

Wallin MT, Culpepper WJ, Campbell JD, Nelson LM, Langer-Gould A, Ann Marrie R, Cutter GR, Kaye WE, Wagner L, Tremlett H, et al. 2019a. The prevalence of MS in the United States: a population-based estimate using health claims data. *Neurology* **92**: e1029–e1040. doi:10.1212/WNL.0000000000007035

Wallin MT, Culpepper WJ, Nichols E, Bhutta ZA, Gebrehiwot TT, Hay SI, Khalil IA, Krohn KJ, Liang X, Naghavi M, et al. 2019b. Global, regional, and national burden of multiple sclerosis 1990–2016: a systematic analysis for the Global Burden of Disease Study 2016. *Lancet Neurol* **18**: 269–285. doi:10.1016/S1474-4422(18)30443-5

Wang S, Sdrulla AD, DiSibio G, Bush G, Nofziger D, Hicks C, Weinmaster G, Barres BA. 1998. Notch receptor activation inhibits oligodendrocyte differentiation. *Neuron* **21**: 63–75. doi:10.1016/S0896-6273(00)80515-2

Ward M, Goldman MD. 2022. Epidemiology and pathophysiology of multiple sclerosis. *Continuum (Minneap Minn)* **28**: 988–1005. doi:10.1212/CON.0000000000001136

Waubant E, Lucas R, Mowry E, Graves J, Olsson T, Alfredsson L, Langer-Gould A. 2019. Environmental and genetic risk factors for MS: an integrated review. *Ann Clin Transl Neurol* **6**: 1905–1922. doi:10.1002/acn3.50862

Werneburg S, Jung J, Kunjamma RB, Popko B, Reich DS, Schafer Correspondence DP. 2020. Targeted complement inhibition at synapses prevents microglial synaptic engulfment and synapse loss in demyelinating disease. *Immunity* **52**: 167–182.e7. doi:10.1016/j.immuni.2019.12.004

Wheeler MA, Jaronen M, Covacu R, Zandee SEJ, Scalisi G, Rothhammer V, Tjon EC, Chao CC, Kenison JE, Blain M, et al. 2019. Environmental control of astrocyte pathogenic activities in CNS inflammation. *Cell* **176**: 581–596.e18. doi:10.1016/j.cell.2018.12.012

Wolswijk G. 2000. Oligodendrocyte survival, loss and birth in lesions of chronic-stage multiple sclerosis. *Brain* **123**: 105–115. doi:10.1093/brain/123.1.105

Wucherpfennig KW, Hafler DA, Strominger JL. 1995. Structure of human T-cell receptors specific for an immunodominant myelin basic protein peptide: positioning of T-cell receptors on HLA-DR2/peptide complexes. *Proc Natl Acad Sci* **92**: 8896–8900. doi:10.1073/pnas.92.19.8896

Wucherpfennig KW, Catz I, Hausmann S, Strominger JL, Steinman L, Warren KG. 1997. Recognition of the immunodominant myelin basic protein peptide by autoantibodies and HLA-DR2-restricted T cell clones from multiple sclerosis patients. Identity of key contact residues in the B-cell and T-cell epitopes. *J Clin Invest* **100**: 1114–1122. doi:10.1172/JCI119622

Wynn TA, Vannella KM. 2016. Macrophages in tissue repair, regeneration, and fibrosis. *Immunity* **44**: 450–462. doi:10.1016/j.immuni.2016.02.015

Yeung MSY, Djelloul M, Steiner E, Bernard S, Salehpour M, Possnert G, Brundin L, Frisén J. 2019. Dynamics of oligodendrocyte generation in multiple sclerosis. *Nature* **566**: 538–542. doi:10.1038/s41586-018-0842-3

Yong VW. 2022a. Microglia in multiple sclerosis: protectors turn destroyers. *Neuron* **110**: 3534–3548. doi:10.1016/j.neuron.2022.06.023

Yong VW. 2022b. Microglia in multiple sclerosis: protectors turn destroyers. *Neuron* **110**: 3534–3548. doi:10.1016/j.neuron.2022.06.023

Yong HYF, Yong VW. 2022. Mechanism-based criteria to improve therapeutic outcomes in progressive multiple sclerosis. *Nat Rev Neurol* **18**: 40–55. doi:10.1038/s41582-021-00581-x

Young AMH, Kumasaka N, Calvert F, Hammond TR, Kundu K, Segel M, Murphy NA, Mcmurran CE, Trivedi R, Kirollos R, et al. 2021. A map of transcriptional heterogeneity and regulatory variation in human microglia. *Nat Genet* **53**: 861–868. doi:10.1038/s41588-021-00875-2

Zawadzka M, Rivers LE, Fancy SPJ, Zhao C, Tripathi R, Jamen F, Young K, Goncharevich A, Pohl H, Rizzi M, et al. 2010. CNS-resident glial progenitor/stem cells produce Schwann cells as well as oligodendrocytes during repair of CNS demyelination. *Cell Stem Cell* **6:** 578–590. doi:10 .1016/j.stem.2010.04.002

Zhang J, Markovic-Plese S, Lacet B, Raus J, Weiner HL, Hafler DA. 1994. Increased frequency of interleukin 2-responsive T cells specific for myelin basic protein and proteolipid protein in peripheral blood and cerebrospinal fluid of patients with multiple sclerosis. *J Exp Med* **179:** 973–984. doi:10.1084/jem.179.3.973

Zhang Y, Argaw AT, Gurfein BT, Zameer A, Snyder BJ, Ge C, Lu QR, Rowitch DH, Raine CS, Brosnan CF, et al. 2009. Notch1 signaling plays a role in regulating precursor differentiation during CNS remyelination. *Proc Natl Acad Sci* **106:** 19162–19167. doi:10.1073/pnas.0902 834106

Zia S, Hammond BP, Zirngibl M, Sizov A, Baaklini CS, Panda SP, Ho MFS, Lee KV, Mainali A, Burr MK, et al. 2022. Single-cell microglial transcriptomics during demyelination defines a microglial state required for lytic carcass clearance. *Mol Neurodegener* **17:** 1–24. doi:10.1186/s13 024-022-00584-2

Glial Origins of Inherited White Matter Disorders

Anjana Sevagamoorthy,[1] Adeline Vanderver,[1,2] Jamie L. Fraser,[3]
and Jennifer Orthmann-Murphy[2]

[1]Division of Neurology, Department of Pediatrics, Children's Hospital of Philadelphia, Philadelphia, Pennsylvania 19104, USA

[2]Department of Neurology, Perelman School of Medicine, University of Pennsylvania, Philadelphia, Pennsylvania 19104, USA

[3]Division of Genetics and Metabolism, Rare Disease Institute, Center for Genetic Medicine Research, Children's National Hospital, Washington, D.C. 20012, USA

Correspondence: Jennifer.Orthmann-Murphy@pennmedicine.upenn.edu

Inherited white matter disorders (IWMDs) are a phenotypically and genotypically heterogeneous group of disorders affecting the central nervous system (CNS) with or without peripheral neuropathy. They are classified either as leukodystrophies (LDs), with primary glial abnormalities, or genetic leukoencephalopathies (gLEs), where other CNS cells are involved. As a group, these disorders are common, with an incidence of 1 in 7500 births. However, IWMDs often go undiagnosed or suffer delayed or misdiagnosis due to their heterogeneous presentation. Many of these disorders present with lethal secondary manifestations that can be prevented through early disease recognition, periodic surveillance, and preventative management. Emerging therapeutics, including gene therapy trials for metachromatic leukodystrophy (MLD) and adrenoleukodystrophy (ALD), suggest disease progression may be slowed or even prevented if treated early. Therapies for IWMDs that target glial cells or the peripheral immune system may provide novel insights for treating acquired disorders of white matter.

Inherited white matter disorders (IWMDs) are a group of diseases with abnormalities in the white matter of the central nervous system (CNS) with or without the involvement of the peripheral nerves (Vedolin 2011; Van Haren et al. 2015; Helman et al. 2018). White matter is comprised of bundles of axons that are myelinated by the myelin-forming cells of the CNS, oligodendrocytes. The high density of myelin, and dense lipid content of myelin itself, leads to the "white" appearance of white matter on gross pathology (Stadelman et al. 2019). In addition to myelinating oligodendrocytes, white matter also contains other glial cells, including oligodendrocyte precursor cells, astrocytes, microglia, as well as microvasculature. Inherited abnormalities in white matter arise from defects in glial cells, leading to altered myelin formation during development (dys- and/or hypomyelination) or active loss of acquired myelin (demyelination) (Fig. 1). IWMDs are genetically and phenotypically highly heterogeneous, including a wide range of ages of onset, from birth to late adulthood (Vedolin 2011; Van Haren et al.

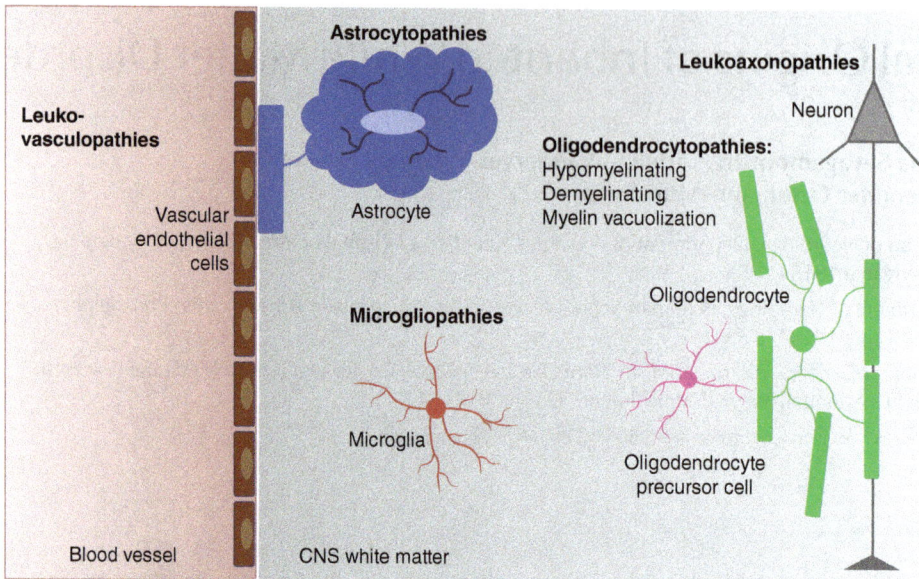

Figure 1. Inherited white matter disorders (IWMDs) grouped by cellular pathogenesis. Schematic depicting organization of IWMDs by cellular pathogenesis in central nervous system (CNS) white matter and the vasculature.

2015; Helman et al. 2018). This heterogeneity poses diagnostic challenges and results in misdiagnosis or missed diagnosis of affected individuals. While emerging evidence suggests that IWMDs are relatively common, with an incidence of 1 in 7500 live births (Bonkowsky et al. 2010), this could be a gross underestimation due to a lack of accurate and timely diagnosis in many individuals. Advancements in molecular diagnostic techniques, particularly next-generation sequencing, have greatly improved diagnostic yield by up to 85% (Helman et al. 2020; Vanderver et al. 2020b) and are an increasingly affordable and accessible approach for the timely diagnosis in affected individuals. Further, these advanced molecular genetic techniques have increased our understanding of disease mechanisms and informed clinical management and development of therapeutics. Some leukodystrophies (LDs) and genetic leukoencephalopathies (gLEs) have associated clinical manifestations that are treatable, but necessitate periodic surveillance and comprehensive preventative care to decrease morbidity and mortality (Table 1; Adang et al. 2017). In addi-

tion, a growing number of these conditions have disease-modifying therapies (Table 2; Sessa et al. 2016; Eichler et al. 2017; Vanderver et al. 2020a). Therefore, advanced molecular diagnostic technology and increased diagnostic yields improve patient care, advance our understanding of underlying pathogenic mechanisms, and may lead to a better understanding of glial cell contributions to healthy brain function (Nowacki et al. 2022).

TYPES OF INHERITED WHITE MATTER DISORDERS

IWMDs are classified into two broad heterogeneous groups—LDs and gLEs (Di Rocco et al. 2004; Vanderver et al. 2015b). LDs are a group of disorders with primary glial and myelin abnormalities resulting in prominent defects in the white matter of the CNS, with or without peripheral nervous system (PNS) involvement (Schiffmann and van der Knaap 2009; Steenweg et al. 2010; Vanderver et al. 2015b; van der Knaap and Bugiani 2017; van der Knaap et al. 2019). A more inclusive group of IWMDs are

Table 1. Examples of clinically relevant, treatable nonneurologic manifestations of leukodystrophies and leukoencephalopathies

Disease	Gene	Clinical manifestations of disease requiring surveillance and preventative management
Hypomyelination		
4H leukodystrophy	POLR1C; POLR3K; POLR3A; POLR3B	Hypothyroidism, growth failure, osteopenia, obstructive sleep apnea, severe myopia, and retinal detachment
HIKESHI-associated hypomyelination (HLD13)	HIKESHI	Sepsis-shock-like presentation in the context of mild viral infection, requiring acute supportive care
Salla disease; infantile sialic acid storage disease	SLC17A5	Hepatosplenomegaly, cardiomegaly, failure to thrive, recurrent respiratory infections
RNF220-associated hypomyelinating leukodystrophy (HLD23)	RNF220	DCM, heart failure, death in second decade
Epilepsy, hearing loss, and mental retardation syndrome	SPATA5	Thrombocytopenia and immunodeficiency
Neurodevelopmental disorder with spasticity, hypomyelinating leukodystrophy, and brain abnormalities	PI4KA	Immunodeficiency and anemia
Demyelination		
Adrenoleukodystrophy	ABCD1	Adrenal insufficiency
Metachromatic leukodystrophy	ARSA	High risk of gall bladder carcinoma
Cerebrotendinous xanthomatosis (CTX)	CYP27A1	Myocardial infarction
D-bifunctional protein deficiency	HSD17B4	Hemosiderosis, adrenal cortical atrophy
Classic Refsum disease	PHYH	Cardiomyopathy, cardiomegaly, and heart failure (sudden death)
Ataxia–pancytopenia syndrome	SAMD9L	Bone marrow failure, myelodysplastic syndrome, immunodeficiency, hypogammaglobulinemia, increased risk of AML
Peroxisome biogenesis disorders	PEX1, PEX10, PEX12, PEX13, PEX14, PEX16, PEX19, PEX2, PEX3, PEX5, PEX6, PEX7	May present with cardiac defects, liver, and kidney abnormalities; PEX19-associated subtype can cause DIC
Photosensitive trichothiodystrophy (TTD) and xeroderma pigmentosum	ERCC2, ERCC3	Increase the risk of skin cancer: basal cell, squamous, and malignant melanoma
Cockayne syndrome subtypes	ERCC6, ERCC8	High risk of pneumonia, cardiac arrhythmias, renal failure and cryptorchidism (increased risk of testicular cancer)

Continued

Table 1. *Continued*

Disease	Gene	Clinical manifestations of disease requiring surveillance and preventative management
Myelin vacuolation		
Mitochondrial DNA depletion syndrome subtypes 4A and 4B; mitochondrial recessive ataxia syndrome; progressive external ophthalmoplegia with mitochondrial deletions—AD and AR subtypes	POLG	DCM and cardiac defects
Multiple mitochondrial dysfunctions syndrome 2	BOLA3	DCM, HCM
Mitochondrial complex 4 deficiency nuclear subtypes 3 and 5	COX10, LRPPRC	HCM, subtype 5 additionally presents with glycemia dysregulation during crisis
D-2-hydroxyglutaric aciduria 1	D2HGDH	Cardiomyopathy
Infantile-onset progressive leukoencephalopathy	KARS1	HCM, renal tubular acidosis and microlytic anemia
Charcot–Marie–Tooth disease, axonal type 2EE; mitochondrial DNA depletion syndrome 6	MPV17	Hepatic cirrhosis and failure; CMT type 2EE additionally presents with restrictive lung disease.
Combined oxidative phosphorylation deficiency 5	MRPS22	HCM
Combined oxidative phosphorylation deficiency 15	MTFMT	Wolf–Parkinson–White syndrome
Mitochondrial complex 1 or complex 2 deficiency nuclear subtypes	Complex 1—NDUFA10, NDUFA11, NDUFA2, NDUFAF1, NDUFAF4, NDUFAF3, NDUFS2 Complex 2—SDHA, SDHB, SDHD	Present with HCM and cardiac insufficiency; an exception is NDUFAF3, which presents with hydronephrosis and hydroureter
Mitochondrial DNA depletion syndrome 1	TYMP	Increased risk of intestinal perforation
Astrocytopathies		
Oculodentodigital dysplasia	GJA1	Cardiac defects
Aicardi–Goutieres syndrome (AGS)	ADAR, RNASEH2A, SAMHD1, TREX1, IFIH1, RNU7-1	One or more hematological abnormalities such as hemolytic anemia, thrombocytopenia, hypergammaglobulinemia, hypercomplementemia, or pancytopenia hepatitis; pulmonary hypertension in particular in IFIH1 and TREX1, cardiomyopathy, and other cardiovascular complications; additionally, renal abnormalities have been identified in AGS with RNU7-1, IFIH1, and TREX1 genotypes

Continued

Table 1. *Continued*

Disease	Gene	Clinical manifestations of disease requiring surveillance and preventative management
Leukoaxonopathies		
Congenital disorder of glycosylation type IIm	SLC35A2	Coagulation defects, recurrent infections
Mowat–Wilson syndrome	ZEB2	Cardiac defects, cryptorchidism with increased risk of testicular cancer
Leukovasculopathies		
Cerebral arteriopathy with subcortical infarcts and leukoencephalopathy 1	NOTCH3	Cardiac defects, cryptorchidism
Hereditary angiopathy with nephropathy, aneurysms, and muscle cramps; brain small vessel disease 1; pontine autosomal-dominant microangiopathy and leukoencephalopathy (PADMAL)	COL4A1	Supraventricular arrhythmias, internal carotid and middle cerebral artery aneurysms, renal failure, increased risk of cerebral hemorrhage, and hemolytic anemia in PADMAL
Brain small vessel disease 2	COL4A2	Intracranial hemorrhage
Cerebroretinal microangiopathy with calcifications and cysts 1 (COATS plus syndrome)	CTC1	Increased risk of intestinal bleeding and less commonly bone marrow failure
Cerebroretinal microangiopathy with calcifications and cysts 2	STN1	Gastrointestinal (GI) bleeding and esophageal varices; hypocellular bone marrow and pancytopenia

(DCM) Dilated cardiomyopathy, (HCM) hypertrophic cardiomyopathy, (AML) acute myeloid leukemia, (DIC) disseminated intra-vascular coagulation.

Table 2. Summary of approved or investigational disease course-altering therapies for leukodystrophies and leukoencephalopathies (including disorders not discussed in the text)

Disease	Gene	Approved or emerging disease-modifying therapies
Hypomyelination		
Pelizaeus–Merzbacher disease	PLP1	Antisense oligonucleotides (ASOs) NCT06150716 Stem cell transplantation: human neural stem cell (HuCNS-SCs; NCT01005004) and human umbilical cord blood–derived oligodendrocyte-like cells (NCT02254863) (Gupta et al. 2012, 2019)
Allan–Herndon–Dudley syndrome	SLC16A2	Investigation thyroid analogs: tiratricol (TRIAC) and diiodothyropropionic acid (DIPTA)
Demyelination		
Metachromatic leukodystrophy	ARSA	Approved treatment with ex vivo gene therapy with HSC transplantation; trials for in vivo gene therapies and enzyme replacement therapies
X-linked adrenoleukodystrophy (X-ALD)	ABCD1	Approved treatment with ex vivo gene therapy with HSC transplantation in CC-ALD; ongoing trials in AMN
Tay Sach's disease	HEXA	Investigational drugs: venglustat (NCT04221451), N-acetyl-L-leucine (NCT03759665); gene therapy trials: TSHA-101 (infantile-onset form) and AXO-GM2-001; HSC or UCB-derived oligodendrocyte-like cells augmented UCB transplantation
GM1 gangliosidosis	GLB1	Gene therapy trials: LYS-GM101, PBGM01, AAV9-GLB1 HSC transplantation
Sjogren–Larsson syndrome (SLS)	ALDH3A2	Investigational drug—ADX-629
Cerebrotendinous xanthomatosis (CTX)	CYP27A1	Investigational drug—chenodeoxycholic acid (NCT04270682)
D-bifunctional protein deficiency	HSD17B4	Allogenic HSC transplantation
Peroxisome biogenesis disorders	PEX1, PEX10, PEX12, PEX13, PEX14, PEX16, PEX19, PEX2, PEX3, PEX5, PEX6, PEX7	Cholic acid (compassionate treatment); trial for allogenic HSC transplantation
Myelin vacuolation		
Phenylketonuria	PAH	Approved therapies: kuvan (sapropterin) and pegvaliase enzyme therapy Trials for liver transplantation, gene therapies, and investigational drugs to reduce phenylalanine accumulation such as PTC923 (NCT05166161), CDX6114 (NCT04085666), SYNB1618 (NCT04534842), SYNB1934 (NCTO4534842)

Continued

Cite this article as *Cold Spring Harb Perspect Biol* doi: 10.1101/cshperspect.a041457

Table 2. *Continued*

Disease	Gene	Approved or emerging disease-modifying therapies
Canavan disease	ASPA	Gene therapy trials—AAV9 BBP-812 and rAAV-Olig001-ASPA (NCT04998396 and NCT04833907); ASO approaches
Astrocytopathies		
Alexander disease	GFAP	Investigational drug with ASO approaches: zilganersen (ION373) (NCT04849741)
Vanishing white matter diseases	EIF2B1-5	Ongoing trial for guanabenz in early childhood-onset form (UDRACT #2017-001438-25)
Giant axonal neuropathy 1	GAN	Gene therapy trial (scAAV9/JeT-GAN)
Aicardi–Goutieres syndrome (AGS)	ADAR, RNASEH2A, RNASEH2C, SAMHD1, TREX1, IFIH1, LSM11, RNU7-1	Investigational drugs: reverse transcriptase inhibitors (RTI) such as tenofovir, emtricitabine, lamivudine, zidovudine, and abacavir (NCT02363452, NCT03304717, NCT04731103, and NCT05613868; trial for new RTI—censavudine (TPN-101) inhibitors such as baricitinib; trial for AGS from biallelic mutations in TREX1, RNASEH2A, RNASEH2B, RNASEH2C, or dominant mutations in TREX1
Microgliopathies		
Krabbe disease (globoid cell leukodystrophies)	GALC	Trials for stem cell transplant (hematopoietic stem cells and human placental-derived stem cells) and gene therapies FBX-101 (NCT04693598) and PBKR03 (NCT04771416)
CSF1R-related leukoencephalopathy (adult-onset leukoencephalopathy with axonal spheroids and pigmented glia; ALSP)	CSF1R	Phase 2 trial with TREM2 agonist (NCT05677659); hematopoietic stem cell transplant (NCT04503213)
Leukoaxonopathies		
Adult polyglucosan body disease	GBE1	Urgent Compassionate Use Program (Israel) for GHF-201, to polyglucosan bodies based on preclinical testing
Congenital disorder of glycosylation type IIm	SLC35A2	Ongoing trials on the use of fenfluramine to reduce seizures frequency
Developmental and epileptic encephalopathy 4	STXBP1	Ongoing trial use of fenfluramine to reduce seizures frequency
Leukovasculopathies		
Alzheimer disease 3	PSEN1	Repurposing of bromocriptine, a dopamine D1 receptor agonist A new phase 2 trial for APH-1105, an amyloid precursor protein secretase modulator

(HSC) Hematopoietic stem cells, (UCB) umbilical cord blood.

gLEs, which consist of conditions that are not restricted to primary glial cell or white matter abnormalities (Vanderver et al. 2015b; van der Knaap and Bugiani 2017; van der Knaap et al. 2019) but are typically associated with CNS white matter imaging abnormalities. While multiple differentiating factors have been defined in the literature, the primary underlying criteria that distinguish gLEs from LDs are their lack of specificity for the CNS white matter; gLEs often also have systemic manifestations if caused by mitochondrial disorders or inherited metabolic disorders (Vanderver et al. 2015b; van der Knaap and Bugiani 2017; van der Knaap et al. 2019). Other gLEs primarily affect neuronal function and have secondary myelin abnormalities (e.g., myelination delay seen in some early epileptic encephalopathies). Some types of complex hereditary spastic paraparesis (HSP) mimic IWMD in clinical presentation, but are a distinct group of genetically diverse disorders, characterized by length-dependent degeneration of axons of the corticospinal tract and the dorsal column (Hedera 1993; McDermott et al. 2000; Lo Giudice et al. 2014); affected individuals typically present with progressive lower limb spasticity and abnormal gait (Hedera 1993; McDermott et al. 2000). A subset of these HSP-related genes cause more complex neurological phenotypes involving additional CNS regions beyond those anticipated for HSP. In some of these cases, different variants in the same genes may cause HSP, IWMD, or complex HSPs with white matter involvement (i.e., HSP due to *PLP1* or *GJC2* pathogenic variants, which are primarily expressed by oligodendrocytes). Notably, complex HSPs with white matter involvement have a relatively lower impact on survival and may present later in life compared to many IWMDs.

DIAGNOSIS OF LEUKODYSTROPHIES AND GENETIC LEUKOENCEPHALOPATHIES IN CHILDREN AND ADULTS

LDs and gLEs are a diagnostically challenging group of disorders due to phenotypic and genotypic heterogeneity. LDs are often life span disorders, and individual conditions can present at variable ages: early infantile (typically character-

ized as onset within 12 months of life), late-infantile (1 to typically 5 years), juvenile (5–12 years), and adult-onset (12 years and above) (Parikh et al. 2015). More severe and lethal presentations are typically associated with early-onset (prenatal–infantile) presentations, while later-onset (adult) forms may have lower mortality (Parikh et al. 2015; Lynch et al. 2019; Resende et al. 2019). Depending on the age of onset, patients present with a range of neurologic symptoms such as motor, neurocognitive, and behavioral abnormalities. Motor symptoms, such as delayed acquisition (or regression) of milestones, abnormal gait, ataxia, tone and movement abnormalities, fine motor impairments, and coordination difficulties, are prominent features in early-onset forms (Parikh et al. 2015). In juvenile and adult types, cognitive and behavioral abnormalities may be more pronounced, and patients may present with varying systemic manifestations (Federico and Gallus 1993; Brienza et al. 2015). Affected individuals with LDs and gLEs all have abnormal white matter signals on brain magnetic resonance imaging (MRI) sequences (Parikh et al. 2015; Vanderver et al. 2015b). The most common MRI finding is the presence of T2 hyperintense lesions associated with variable-intensity T1 signal within the white matter (Vanderver et al. 2015b). Based on the pattern of white matter MRI changes, one may be able to narrow down to certain broad groups of LDs or gLEs, or in some cases the exact form of LD (Schiffmann and van der Knaap 2009; Steenweg et al. 2010; Vedolin 2011; Finsterer and Zarrouk Mahjoub 2012; Yang and Prabhu 2014; Vanderver et al. 2015a; Barkovich and Deon 2016; Resende et al. 2019; Roosendaal et al. 2021; Huisman et al. 2022). However, clinical phenotype and imaging pattern are shared among some LDs, and so these features alone are not always sufficient to obtain a specific diagnosis. It is important to exclude treatable acquired disorders. Genetic testing (ranging from single gene or panel testing to broad exome or genome sequencing) is definitive to diagnose LDs and gLEs (van der Knaap et al. 2019; Vanderver et al. 2020b). In cases of diagnostic uncertainty, after molecular testing is complete, additional biochemical, electrophysiologic, and nerve con-

Cite this article as *Cold Spring Harb Perspect Biol* doi: 10.1101/cshperspect.a041457

duction tests may be performed to determine the pathogenicity of variants of uncertain significance (Parikh et al. 2015). Given that novel noncoding variants are increasingly being identified to cause LDs and gLEs (Jenkinson et al. 2016; Helman et al. 2020), whole genome sequencing is recommended if other molecular testing fails to identify a molecular diagnosis.

CLASSIFICATION OF LEUKODYSTROPHIES AND LEUKOENCEPHALOPATHIES BY CELLULAR PATHOLOGY

LDs and gLEs can be classified into five broad groups based on cellular pathology (Bugiani and van der Knaap 2017; van der Knaap and Bugiani 2017). These include oligodendrocytopathies, astrocytopathies, microgliocytopathies, leukoaxonopathies, and leukovasculopathies (Fig. 1). This review reports on genes curated, in association with IWMDs, by gene curation experts as part of an ongoing project in ClinGen (Rehm et al. 2015). LDs or gLEs with clinically relevant secondary manifestations or with an approved or investigational therapy are summarized in Tables 1 and 2, respectively.

OLIGODENDROCYTOPATHIES

Oligodendrocytopathies are primary myelin disorders in which oligodendrocytes and/or myelin are intrinsically affected and are categorized as (1) hypomyelination (insufficient myelin formation), (2) demyelination (loss of myelin), and (3) myelin vacuolation (loss of myelin integrity) (van der Knaap and Bugiani 2017).

Hypomyelination

Hypomyelinating disorders are the largest group of established LDs; the common pathology in these disorders is disrupted oligodendrocyte formation and function, leading to reduced developmental deposition of CNS myelin.

Pelizaeus–Merzbacher disease (PMD) is one of the best-known hypomyelinating LDs and is caused by missense pathogenic variants or copy number variations of the X-linked proteolipin protein 1 (*PLP1*) gene (Raskind et al. 1991;

Hodes et al. 1993; Wolf et al. 1993, 2005; Cailloux et al. 2000). PLP, and its smaller splice isoform DM20 (or myelin protein PLP), is a hydrophobic tetraspan transmembrane protein expressed by oligodendrocytes in the CNS and Schwann cells in the PNS. PLP is the most abundant protein in CNS myelin (Jahn et al. 2020; Gargareta et al. 2022) and is thought to adhere and stabilize myelin membranes via its extracellular domains, but the exact mechanisms underlying these interactions are not fully understood (Griffiths et al. 1998; Bakhti et al. 2014). Although preclinical studies suggest that cholesterol mediates PLP trafficking and myelin membrane formation (Simons et al. 2002; Saher et al. 2012), dietary interventions have not yet shown clinical evidence of success in patients with PMD. Other approaches tested in preclinical models included iron chelation (Nobuta et al. 2019) and cell-based therapy (Gruenenfelder et al. 2020). To date, genetic strategies addressing PLP1 duplications have been prioritized in clinical trials (see Table 2).

Affected patients present with variable severity, ranging from a severe connatal onset to mild spastic paraplegia, with some genotype/phenotype correlation (Boulloche and Aicardi 1986; Hodes et al. 1993; Cailloux et al. 2000). Both missense and gene duplication variants result in intracellular accumulation of PLP1, causing oligodendrocyte cell death by a variety of mechanisms (Boulloche and Aicardi 1986; Hodes et al. 1993; Cailloux et al. 2000). There are multiple therapies under investigation for PMD, which include transplantation of stem cells that could be induced to differentiate into oligodendrocyte lineage cells and preclinical studies exploring whether down-regulation of *PLP1* expression using antisense oligonucleotide therapies may provide rescue from severe phenotypes given that *PLP1*-null variants lead to milder disease (Table 2; Elitt et al. 2020).

POLR3-related disorder (also known as 4H LD for its characteristic findings of hypomyelination, hypodontia, and hypogonadotropic hypogonadism) is a relatively common hypomyelinating disorder, comprising 7% of all LDs diagnosed by exome sequencing (Schmidt et al. 2020). POLR3-related disorder is caused by autosomal recessive pathogenic variants in subunits of the polymerase III (Pol III) complex, which are encoded by *POLR3A, POLR3B, POLR1C,* and *POLR3K* (Bernard et al. 2011; Tétreault et al. 2011; Thiffault et al. 2015; Dorboz et al. 2018). POLR3 is responsible for the transcription of noncoding RNAs (ncRNAs), and is essential for global protein synthesis (White 2011) and myelin development (Feinstein et al. 2010; Taft et al. 2013; Wolf et al. 2014; Mendes et al. 2018; Ognjenović and Simonović 2018). Oligodendrocytes are proposed to be particularly vulnerable to a deficiency in Pol III function due to the high metabolic requirement of myelin formation throughout white matter during early postnatal development, leading to hypomyelination (Bernard et al. 2011; Thiffault et al. 2015; Choquet et al. 2019). Whether there are additional defects specific to oligodendrocyte lineage cells and their ability to differentiate is still under investigation (Lata et al. 2021).

POLR3-related disorders have no treatment and are life-threatening. Individuals often present in early childhood with gross motor delay or regression. Rare variants include severe infantile and milder late-onset forms. Neurologic manifestations include pyramidal and extrapyramidal tone abnormalities and motor impairment, while systemic issues include hypogonadotropic hypogonadism (delayed/absent/arrested puberty) and other endocrine disturbances, hypodontia (missing teeth and other dental abnormalities), and early severe myopia. Life expectancy correlates with the severity of the disease and typically progresses in the teenage years to impair independent mobility, speech, and feeding. Although the underlying pathogenesis is currently poorly understood, potential avenues for disease-modifying therapies include cell-based transplantation of stem cells expressing functional POL3R or gene therapy to express functional POL3R (Perrier et al. 2021).

Monocarboxylate transporter 8 (MCT8) deficiency syndrome, previously known as Allan–Herndon–Dudley syndrome (AHDS), is an X-linked gLE due to pathogenic variants in *SLC16A2* leading to thyroid hormone transport dysfunction. Males with this disorder exhibit hypotonia, dystonic motor abnormalities, devel-

opmental delay, and a hypomyelination pattern on MRI (Gika et al. 2010). In affected individuals, thyroid hormone dysregulation is exacerbated by thyroxine supplementation (Sarret et al. 1993). Because thyroid hormone (T3) induces oligodendrocyte precursor cell differentiation into oligodendrocytes (Billon et al. 2002), dysregulated thyroid signaling in MCT8 deficiency likely disrupts oligodendrocyte differentiation and myelination, but may also alter MCT8-expressing neuronal populations that are present during development (Heuer et al. 2005). Investigational thyroid hormone analogs that do not require MCT8 transport, such as tiratricol (TRIAC) (Bauer 2019; Emerson 2019; Groeneweg et al. 2019) and diiodothyropropionic acid (DIPTA) (Verge et al. 2012; Leung et al. 2016), have been shown to regulate thyroid hormone levels and significantly decrease morbidity and mortality in MCT8 deficiency. It is possible that early diagnosis and treatment could prevent hypomyelination and developmental delay in this disorder.

In some hypomyelinating disorders, the causative gene is expressed by both oligodendrocytes and Schwann cells. For example, pathogenic variants in genes (*CLDN11, NKX6-2, SLC16A2, SLC17A5,* and *CNTNAP1*) lead to hypomyelinating disorders with secondary systemic manifestations including neuropathy. In addition, there are pathogenic variants (in genes like *SPATA5* or *PI4KA*) that present with systemic findings (secondary immunodeficiency, thrombocytopenia, and/or anemia) due to expression in cells outside of the CNS (Tanaka et al. 2015; Salter et al. 2021; Verdura et al. 2021; Baple et al. 2022). Future therapeutic approaches developed for the disorders with the involvement of cells outside the CNS may need to incorporate strategies that target both oligodendrocytes and Schwann cells and/or relevant systemic organs.

Leukoaxonopathies manifest with both hypomyelination as well as neuronal dysfunction. Leukoaxonopathies are a group of genetic conditions caused by pathogenic variants in genes altering both neuronal and oligodendrocyte lineage function or have secondary myelin defects due to neuronal/axonal dysfunction (van der

Cite this article as *Cold Spring Harb Perspect Biol* doi: 10.1101/cshperspect.a041457

Knaap and Bugiani 2017). Several epileptic encephalopathies associated with white matter abnormalities are included in this group of conditions.

The most common hypomyelinating LD, TUBB4A-related LD, is an example of an LD with pathogenic variants in the gene that alters both neuronal and oligodendrocyte function. It is caused by autosomal-dominant pathogenic variants in *TUBB4A* (OMIM #612438) (Schmidt et al. 2020), which encodes β-tubulin 4. Abnormalities in β-tubulin 4 likely alter TUBB4A incorporation into microtubules (Savage et al. 1994) and cause abnormalities of tubulin assembly and microtubule polymerization (Krajka et al. 2022). *TUBB4A* pathogenic variants impact oligodendrocytes and, variably, specific neuronal populations in the cerebellar granule layer and striatum (Curiel et al. 2017; Sase et al. 2020). TUBB4A-related LD represents a variable clinical spectrum: an early-onset encephalopathy with severe motor impairment; an intermediate form typically with late-infantile onset known as hypomyelination with atrophy of basal ganglia and cerebellum (H-ABC); and a milder juvenile or adult-onset phenotype. H-ABC typically presents with delayed motor milestones, cognitive dysfunction, hypertonia, choreoathetosis, gait dysfunction including ataxia, nystagmus, dysarthria, and dysphagia (van der Knaap et al. 2002; Kumar et al. 2015). While there are no currently approved disease-modifying approaches for this disorder, rodent models with aberrant *TUBB4A* expression have been developed (Sase et al. 2020), and studies are ongoing to better understand pathogenic mechanisms and develop potential therapeutics.

Adult polyglucosan body disease (APBD) is an adult-onset glycogen type IV storage disorder (OMIM #607839) and another example of a leukoaxonopathy because the phenotype and pathology are characterized by widespread and progressive neuronal dysfunction in both the CNS and PNS, along with extensive leukoencephalopathy on MRI and myelin loss on pathology. APBD is caused by autosomal-recessive pathogenic variants in *GBE1*, which encode the glycogen branching enzyme, the enzyme that catalyzes the last step in glycogen biosynthesis

(Koch et al. 2023). Defects in GBE1 lead to the accumulation of poorly branched glycogen and the formation of polyglucosan bodies in neurons and astrocytes in the CNS (Robitaille et al. 1980). Phenotypic onset is typically in those between their 50s and 60s, and is variable; it can include cognitive impairment, parkinsonism, spastic paraplegia, and peripheral neuropathy, representing the wide range of neuronal dysfunction that can occur due to impaired glycogen formation. It is not currently known why impaired glycogen formation leads to demyelination in the CNS, nor whether polyglucosan bodies are directly toxic to oligodendrocyte lineage cells. However, we speculate polyglucosan body formation leads to the loss of dysfunctional myelinated axons and then secondary loss of oligodendrocytes. There are no approved disease-modifying approaches for this disorder, but there are ongoing studies to target biochemical pathways that lead to the removal of polyglucosan bodies (Kakhlon et al. 2021) as well as gene therapy strategies.

Demyelination

The demyelinating disorders demonstrate loss of oligodendrocytes and myelin in the CNS; the pathology of many of these disorders often involve other CNS cells, as well as PNS myelin and Schwann cells (van der Knaap and Bugiani 2017). Most of the genes affected are not exclusively expressed by oligodendrocyte lineage cells —however, oligodendrocyte pathology is always involved.

X-linked adrenoleukodystrophy (X-ALD) is the most common LD, and the first LD with an Food and Drug Administration (FDA)-approved treatment, ex vivo gene therapy (Eichler et al. 2017). X-ALD is caused by pathogenic variants in *ABCD1*, which encodes the peroxisomal very-long-chain fatty acid (VLCFA) transporter ABCD1; expression of ABCD1 pathogenic variants causes accumulation of VLCFAs and peroxisomal dysfunction. Studies in animal models of ALD indicate that healthy peroxisomal function is important for the maintenance of both myelin and the underlying axon (Kassmann 2014), although it is not yet fully understood why cerebral white matter and spinal cord long tracts are particularly vulnerable to VLCFA accumulation. X-ALD variably presents with wide-ranging phenotypes, that can start in early childhood through adulthood—myeloneuropathy, adrenal insufficiency, and cerebral adrenoleukodystrophy (cALD, the most severe form). Individuals with a single X chromosome are typically more severely clinically impacted by X-ALD, although individuals with two X chromosomes may develop milder disease phenotypes. cALD is characterized by white matter lesions on MRIs that appear before clinical or neurological symptoms, and are typically associated with a leading edge of gadolinium-enhancement (Engelen et al. 2012). The white matter lesions visualized on MRIs represent areas of inflammatory demyelination, with infiltration of lymphocytes and activated macrophages (Ferrer et al. 2010). In childhood-onset cALD, symptoms typically manifest between 4 and 8 years of age as attention-deficit disorder or hyperactivity; this is followed by the onset of cognitive, visual, and motor abnormalities (Moser et al. 2019). Treatment is effective in presymptomatic or early symptomatic children, leading to the widespread implementation of newborn screening using 26-lysophosphatidyl choline (26-LysoPC) (Sandlers et al. 2012). Lifelong and age-specific surveillance for cALD is essential (Engelen et al. 2022), as the lifetime prevalence for developing cALD with pathogenic ABCD1 variants may be as high as 30%. The goal of both ex vivo gene therapy and hematopoietic stem cell transplant (HSCT) is to halt disease progression and prevent the accumulation of additional brain lesions in individuals with cALD. There are no current treatment options for AMN or adult-onset cALD, although clinical trials of similar therapeutic approaches are ongoing.

Metachromatic leukodystrophy (MLD) and its biochemical phenocopies arise from defects in ARSA (OMIM #607574), PSAP (OMIM #176801), or SUMF1 (OMIM #607939, multiple sulfatase deficiency [MSD]). These genes encode different proteins and have varying functions; they all exhibit excess lysosomal accumulation of sulfatides from primary or secondary sulfatase deficiency (Stein et al. 1989) or posttranslational

defect of cellular sulfatases in MSD (Dierks et al. 2003, 2005; Cosma et al. 2004; Sardiello et al. 2005). Lysosomal accumulation of sulfatides is hypothesized to lead to oligodendrocyte injury and demyelination. Patients may present across the life span, but most present in the late-infantile period with a decline in previously acquired motor milestones over a period of weeks to months (Mahmood et al. 2010; Kehrer et al. 2011). Early presentation with loss of reflexes may sometimes be confused with acquired peripheral acute demyelinating inflammatory polyneuropathy, or Guillain–Barre syndrome (Modesti et al. 2022). Some individuals with MLD may present with isolated symptoms such as strabismus (Beerepoot et al. 2022) or gall bladder disease (Garavelli et al. 2009; van Rappard et al. 2016) before the onset of subacute neurologic disease. Without treatment, this disorder invariably leads to rapid decline after the onset of motor symptoms (Kehrer et al. 2011). Ex vivo gene therapy (Sessa et al. 2016) with replacement of ARSA in cells of myeloid lineage was recently approved in both the European Union for early-onset MLD and the United States; enzyme replacement therapies are also under investigation (Cable et al. 2011; Solders et al. 2014; Boucher et al. 2015; Hironaka et al. 2015; Boelens and van Hasselt 2016; Groeschel et al. 2016; Sessa et al. 2016). Therapies are effective only in presymptomatic individuals, and newborn screening approaches will be important to the management of children with MLD (Table 2). No therapies are currently approved for juvenile or adult-onset MLD; however, bone marrow transplant is considered in early symptomatic individuals (Adang et al. 2024).

Cerebrotendinous xanthomatosis (CTX) is an autosomal-recessive disorder caused by pathogenic variants in CYP27A1, which encodes the mitochondrial enzyme sterol 27-hydroxylase. Dysfunctional enzyme leads to impaired ability to convert cholesterol to the bile acid chenodeoxycholic acid, causing cholesterol and cholestanol to progressively accumulate in many organs, including the CNS. CTX has a wide phenotypic spectrum over the life span. Neurologic deterioration predominantly occurs in adults. Early in the disease, children manifest with diarrhea, cholestasis (in infantile form), and cataracts after age 1 year (in juvenile forms) (Cruysberg et al. 1995; Cruysberg 2002; Mignarri et al. 2014; Gong et al. 2017; Freedman et al. 2019; Atilla et al. 2021; Zhang et al. 2021). Cardiovascular injury and abnormal cholesterol accumulation are typical in CTX and are associated with a high risk of myocardial infarction, requiring periodic screening for blood cholesterol and EKG (Valdivielso et al. 2004; Androdias et al. 2012). Supplementation with primary bile acids, such as chenodeoxycholic acid, is an available and effective therapy that reduces levels of cholestanol, bile alcohols, and cholesterol. Chenodeoxycholic acid administration and reduction of cholestanol and cholesterol minimizes or prevents additional damage in both the CNS and other involved organs, especially if given early in the disease course (Yahalom et al. 2013; Degrassi et al. 2020). Available treatments have made early identification through newborn screening an important goal for the CTX community (Vaz et al. 2023). Although there is some evidence for partial improvement of cognitive impairment after starting chenodeoxycholic acid therapy (Bonnot et al. 2010), it is not known whether this effect is due to improved neuronal or oligodendrocyte lineage cell function (De Stefano et al. 2001).

Myelin Vacuolation

Myelin vacuolation is the disruption of myelin integrity due to the formation of spaces (vacuoles) between myelin sheath layers that likely represent fluid accumulation (van der Knaap and Bugiani 2017), and is a key pathologic finding in a subset of LDs caused by a variety of gene defects.

Canavan disease is a myelin vacuolating LD caused by pathogenic variants in *ASPA* (OMIM #271900). *ASPA* encodes aspartoacylase, which catalyzes the deacetylation of *N*-acetylaspartate (NAA) to generate free acetate in oligodendrocytes. This biochemical pathway is required for myelin lipid synthesis (Madhavarao et al. 2005). Pathogenic variants in *ASPA* cause reduced NAA-derived acetate and accumulation of NAA in oligodendrocytes (Madhavarao et al.

2005), leading to abnormal lipid and osmotic gradient formation (Grønbæck-Thygesen and Hartmann-Petersen 2024), which could be the basis of formation of myelin vacuoles and formation of spongiform white matter found in pathology. The infantile form of Canavan disease is the most severe subtype and presents with developmental delay, macrocephaly, and visual abnormalities, progressing to severe feeding difficulties and death in the first decade of life (Hoshino and Kubota 2014). The juvenile form has a milder presentation in infancy and early childhood (Matalon et al. 1993; Hoshino and Kubota 2014). No FDA-approved treatment for Canavan disease exists, but in vivo gene therapy using the delivery of an ASPA transgene is under investigation (Table 2).

Phenylketonuria (PKU) is a treatable form of myelin vacuolating leukoencephalopathy associated with cellular accumulation of phenylalanine. PKU is due to pathogenic variants in the *PAH* gene, which encodes phenylalanine hydroxylase (PAH, OMIM #261600) and other nonclassical disease forms caused by other protein deficiencies, historically classified as nonclassical PKU. Individuals with classical PKU typically present with intellectual disability (Moyle et al. 2007; Waisbren et al. 2007; Antshel 2010; Burton et al. 2013), reduced attention span, slower information-processing time, slower motor reaction time (Channon et al. 2007; Moyle et al. 2007), and behavioral concerns such as anxiety, depression, or phobias (Koch et al. 2002). Timely diagnosis of PKU is key due to available therapies. Dietary restriction via a structured low-protein diet remains the mainstay of therapy (Burgard et al. 1999; Singh et al. 2014). For individuals with specific *PAH* genetic variants, PAH activation is an option with FDA-approved therapy KUVAN (sapropterin) if there is an improvement in phenylalanine levels during a clinical response trial period. Notably, sapropterin does not improve phenylalanine levels in all individuals with PKU. Alternately, enzyme replacement with pegvaliase has recently been approved in adults (see Table 2 for additional trials; Bernegger and Blau 2002; Pérez-Dueñas et al. 2004; Zurflüh et al. 2006; Ho and Christodoulou 2014; Vockley et al. 2014).

Effective adoption of newborn screening and postnatal dietary management in classical and nonclassical PKU has prevented the formation of characteristic myelin vacuolization in those individuals identified at birth.

ASTROCYTOPATHIES

Astrocytopathies are IWMDs that result from defects in genes associated with astrocyte function (van der Knaap and Bugiani 2017). Alexander disease (AxD), the most well-known astrocytopathy, is due to autosomal-dominant pathogenic gain-of-function *GFAP* genetic variants (OMIM #203450). Glial fibrillary acidic protein (GFAP) is an intermediate filament protein primarily expressed by astrocytes (Reeves et al. 1989), and is typically up-regulated when astrocytes respond to a variety of types of damage in the CNS (Sofroniew 2009). Autosomal-dominant pathogenic variants in *GFAP* lead to overexpression and accumulation of GFAP, along with sequestration of protein chaperones, forming aggregates called Rosenthal fibers on neuropathology (Messing et al. 1998). It is not yet clear why this primary astrocytopathy leads to an LD; however, recent in vitro studies of patient IPSCs suggest that the dysfunctional astrocyte could directly impair OPC proliferation and differentiation or modify the extracellular environment to indirectly impair myelination in white matter tracts (Li et al. 2018).

Affected individuals may present in the infantile period with macrocephaly, seizures, recurrent vomiting, and developmental delay or developmental regression. Affected individuals presenting in older childhood or adulthood may have swallowing difficulties, behavioral difficulties, and gait abnormalities (Stumpf et al. 2003; Prust et al. 2011; Yoshida et al. 2011). Reduction of GFAP expression in AxD-affected astrocytes remains the target of current investigational therapeutic strategies under development (Table 2).

Vanishing white matter disease (VWMD, OMIM #603896) is caused by autosomal-recessive pathogenic variants in *EIF2B1-5*, which encode the subunits of eukaryotic translation initiation factor 2B. VWMD is characterized by episodic motor or developmental regression in the context of physiologic stressors attributed to an exaggerated integrated stress response (ISR). VWMD can present across the life span but is more severe in individuals with infantile and early childhood onset (Hamilton et al. 2018). As suggested by the name, white matter is diffusely abnormal, with characteristic hypomyelination and cystic degeneration on brain MRI over time (van der Knaap et al. 2006). Preclinical studies of VWMD rodent models suggest VWMD astrocytes drive disease, because they impair oligodendrocyte differentiation (Dooves et al. 2016) and damage motor neurons (Klok et al. 2018). In human pathology, the number of GFAP+ astrocytes is reduced, and those that are present have abnormal morphology and gene expression signatures (Dietrich et al. 2005; Leferink et al. 2019; Man et al. 2022). Guanabenz is being investigated as a repurposing strategy to limit the ISR following functional and neuropathological improvement in VWMD mouse models (Witkamp et al. 2022).

Aicardi–Goutieres syndrome (AGS) is one of the most common LDs and has been shown to be associated with pathogenic variants in several genes that contribute to sensing or metabolism of nucleic acids, including *ADAR*, *RNASEH2A*, *RNASEH2C*, *SAMHD1*, *TREX1*, *IFIH1*, *LSM11*, and *RNU7-1* (Rice et al. 2007, 2009, 2012, 2014; Haaxma et al. 2010; Livingston et al. 2014a; Crow et al. 2015; Uggenti et al. 2020). Rodent models and human AGS pathology suggest that chronic overproduction of interferon-α by astrocytes leads to chronic neuroinflammation (Akwa et al. 1998; van Heteren et al. 2008), although more recent models suggest a more complex picture with potential involvement of multiple cell types (Guo et al. 2022). Early reports characterized AGS as a neonatal or early infantile-onset disorder in which individuals typically present with irritability, intellectual disability, microcephaly, fevers of nonspecific origin, loss of acquired skills, and systemic issues, including hepatomegaly, thrombocytopenia, and chilblains (Goutières et al. 1998; Rice et al. 2007). More recent clinical reports suggest that AGS may have milder disease presentations with isolated spastic paraplegia or later-onset presentations after the first year of life (Piccoli et al. 2021). While no regulatory body-approved treatments for AGS are available, multiple clinical trials for reverse transcriptase inhibitors (RTIs) (tenofovir, emtricitabine, lamivudine, zidovudine, abacavir, and censavudine) in AGS have been initiated, although none have definitively shown therapeutic benefit (Table 2). Janus kinase (JAK) inhibitors such as baricitinib have shown improvement in skin and neurologic symptoms of AGS (Meesilpavik-kai et al. 2019; Vanderver et al. 2020a; Casas-Alba et al. 2022; Cetin Gedik et al. 2022). JAK inhibitors were recently recommended as standard-of-care disease-modifying therapies in AGS (Cetin Gedik et al. 2022). AGS is associated with increased morbidity due to its varying systemic manifestations including hematologic abnormalities (hemolytic anemia, thrombocytopenia, and pancytopenia) and renal abnormalities including renal failure among some genotypes (He et al. 2021; Adang et al. 2022). These factors necessitate early identification of AGS to reduce morbidity and increase benefit from enrollment in ongoing trials for therapeutics. Additional studies of these potential therapies in animal models may demonstrate how they alter the role of astrocytes in mediating the pathogenesis of AGS.

MICROGLIOPATHIES

Abnormal microglial function is the hallmark feature of microgliopathies. Microglia are the resident macrophages in the brain and are derived from yolk-sac macrophages that migrate through the embryonic circulation during the first state of hematopoiesis (Bennett and Bennett 2020). Dysregulated or dysfunctional microglia injure white matter cells, leading to neurodegeneration (Konno et al. 2018).

Adult-onset leukoencephalopathy with axonal spheroids and pigmented glia (ALSP; OMIM #221820) is caused by autosomal-dominant pathogenic variants in *CSF1R*, and is a primary microgliopathy (Rademakers et al. 2012) because microglia are the main brain cells to express CSF1R (along with other cells of monocyte lineage). ALSP is a devastating disorder with variable presentation (including behavioral, cognitive, pyramidal, and extrapyramidal motor symptoms) with often rapid progression to death following diagnosis (Konno et al. 2018). Pathology in affected white matter is notable for microglial and myelin loss, reactive astrocytes, and axonal spheroids. Autosomal-recessive pathogenic variants in *CSF1R* cause a condition called brain abnormalities, neurodegeneration, and dysosteosclerosis (BANDDOS) (Guo et al. 2019; Oosterhof et al. 2019), an ultra-rare disorder that presents in childhood and has overlapping MRI features with ALSP, including patchy periventricular white matter lesions, calcifications, and areas of diffusion restriction (Konno et al. 2018). Many of the pathogenic variants in *CSF1R* that cause ALSP alter tyrosine kinase domain phosphorylation, and therefore the downstream canonical signaling of CSF1R that is critical for the survival and proliferation of monocytes (Rademakers et al. 2012). Current therapeutic strategies under investigation boosting TREM2 signaling to augment downstream CSF1R signaling (Table 2). Other potential treatments in preclinical testing include cell-based microglial replacement strategies (Chadarevian et al. 2023). Although some evidence exists that HSCT may slow disease progression in ALSP (Eichler et al. 2016; Tipton et al. 2021; Dulski et al. 2022), the disease stage at which this approach would be most beneficial for ALSP and the effects on brain pathology/microglial function are not known. Nasu–Hakola disease (i.e., polycystic lipomembranous osteodysplasia with sclerosis leukoencephalopathy) is a microgliopathy that is phenotypically and pathologically similar to CSF1R-related disorders (particularly the BANDDOS phenotype), but is caused by autosomal-recessive pathogenic variants in *TYROBP* (OMIM #221770) or *TREM2* (OMIM #605086) genes. As TREM2 and CSF1R have overlapping downstream signaling pathways in monocytes (Cheng et al. 2021), there may be similar pathogenic mechanisms leading to dysfunctional microglia and disrupted white matter in both ALSP and Nasu–Hakola disease, and potentially similar treatment approaches.

Krabbe (or globoid cell LD) disease is a demyelinating LD caused by pathogenic variants in *GALC* (OMIM #245200), which encodes the enzyme glucocerebrosidase. GALC enzyme deficiency and toxic accumulation of the galactosylceramide psychosine cause globoid cell formation, microglial dysfunction, and ultimately oligodendrocyte death (Ida et al. 1994; Potter et al. 2013; Scott-Hewitt et al. 2018; Sirkis et al. 2021). The infantile form presents with irritability, feeding difficulties with weight loss, neurologic deterioration, visual abnormalities, and death by 24 months of age on average (Lieberman et al. 1980; Orsini et al. 1993; Korn-Lubetzki et al. 2003). Presymptomatic early infantile Krabbe disease may be treated with HSCT, with the goal to prevent neurologic damage (Yoon et al. 2021). Therapeutic options are more limited in symptomatic infants and affected individuals with later onset, as HSCT does not reverse pathology. The remaining uncertainties around risk stratification based on biochemical and genotypic data continue to limit the implementation of newborn screening. Investigational in vivo gene therapies to transfer functional *GALC* genes either as monotherapy or in combination with HSCT are underway based on preclinical efficacy studies (see Table 2; Bradbury et al. 2018, 2020).

LEUKOVASCULOPATHIES

Leukovasculopathies are a group of genetic disorders presenting with white matter abnormalities secondary to the pathology of the small blood vessels in the brain, including small arteries, arterioles, venules, and capillaries. Leukovasculopathies are typically present in adulthood and are progressive. None of these have a disease-modifying therapy.

Cerebral arteriopathy, autosomal-dominant, with subcortical infarcts and leukoencephalopathy (CADASIL) type 1 (OMIM #125310) is caused by autosomal-dominant pathogenic variants in *NOTCH3*, which encodes the transmembrane receptor NOTCH3. NOTCH3 is expressed by vascular smooth muscle cells and pericytes. Abnormal NOTCH3 protein accumulates intracellularly, but pathogenic mechanisms leading to the clinical phenotype are not well understood (Locatelli et al. 2020). CADASIL typically presents in middle-aged adults with transient ischemic strokes, cognitive deficits, migraine, depression, and acute encephalopathy (Dichgans et al. 1998; Dichgans 2009; Reyes et al. 2009; Adib-Samii et al. 2010; Valenti et al. 2011; Ragno et al. 2013; Guey et al. 2016; Tan and Markus 2016). Recently, biallelic pathogenic variants in *NOTCH3* were reported to cause early-onset infantile leukoencephalopathy, calcifications, and lacunae on imaging, and developmental delay with stroke-like events (Stellingwerff et al. 2022). Currently, the management for CADASIL is limited to supportive care, and may include clinical management of complications such as stroke and hemorrhage. The efficacy of standard-of-care secondary stroke prevention treatments is unknown in CADASIL. CADASIL2 (OMIM #616779) is an autosomal-dominant disorder caused by pathogenic variants in *HTRA1*. The presentation of CADASIL2 mirrors CADASIL type 1, although CADASIL may present at a later age (Verdura et al. 2015). Biallelic *HTRA1* pathogenic variants cause autosomal-recessive cerebral arteriopathy with subcortical infarcts and leukoencephalopathy (CARASIL), which presents in the second to third decade of life (OMIM #600142). Much less is known about the pathophysiology of CADASIL2, however. *HTRA1* is expressed by astrocytes (Chen et al. 2018), so possibly influences vascular pathology via astrocyte interaction with the blood–brain barrier.

Pathogenic variants in α collagen chains also cause leukovasculopathies. *COL4A1*-related pathogenic variants cause either brain small vessel disease (BSVD) type 1 (OMIM #175780) or pontine autosomal-dominant microangiopathy and leukoencephalopathy (PADMAL) (OMIM #618564); *COL4A2*-related pathogenic variants cause BSVD2 (OMIM #614483). *COL4A1* and *COL4A2* encode the α-1 and α-2 subunits of the type 4 collagen in the vascular basement membrane, respectively, and are ubiquitously expressed in the early stages of life (Yoneda et al. 2012). Pathogenic variants in COL4A1 and COL4A2 lead to structural deficiencies in the basement membrane, impairing the integrity of blood vessels (Meuwissen et al. 2015). Individuals affected by these disorders typically present with increased susceptibility to gastrointestinal and intracranial hemorrhage and stroke (Meuwissen et al. 2015). Age at presentation is highly variable, from prenatal intracranial hemorrhage and evolution to porencephaly to later-onset presentations, and there is significant intrafamilial variability. Thus, appropriate genetic counseling is an essential component of disease management.

Several novel leukovasculopathies have been recently described. In one, autosomal-dominant pathogenic variants in cathepsin A (*CTSA*) cause cathepsin A–related arteriopathy with strokes and leukoencephalopathy (CARASAL) in adults with ischemic and hemorrhagic strokes, cognitive impairment, hypertension, and severe leukoencephalopathy on MRI. Cathepsin A contributes to endothelin-1 degradation. In white matter histopathology from patients with pathogenic *CTSA* pathogenic variants, demyelination was observed, astrocytes showed increased endothelin-1 expression, and an increased number of oligodendrocyte precursor cells were identified (Bugiani et al. 2016). It is not currently known, however, whether the endothelin-1 accumulation in astrocytes drives disease or whether a more direct cathepsin A-mediated effect on the vasculature exists in this disorder.

In a second disorder, pathogenic variants in the *CTC1* (CST telomere replication complex component 1) gene were recently in found to cause cerebroretinal microangiopathy with calcifications and cysts 1 (CRMCC1, OMIM #612199). CRMCC1 is an autosomal-recessive disorder that presents with tone abnormalities, seizure, ataxia, and cognitive decline (Briggs et al. 2008). The pathogenesis is not well understood but some people with the disorder have abnormally short telomeres. Affected individuals may also develop pancytopenia due to bone marrow failure (Briggs et al. 2008) and require regular clinical surveillance. A partial phenocopy disorder, leukoencephalopathy with calcifications and cysts (LCC, OMIM #614561) presents with CNS features that closely resemble CRMCC1 without apparent systemic involvement. LCC is caused by biallelic pathogenic variants in *SNORD118*, which encodes the U8 ribosomal Box C/D small nucleolar RNA, and is necessary for 60S ribosomal biogenesis (Livingston et al. 2014b; Jenkinson et al. 2016; Badrock et al. 2020; Crow et al. 2021). As a non-protein-coding gene, *SNORD118* is often excluded from exome sequencing reporting, and genome sequencing should be considered if LCC is suspected. LCC is also proposed to predominantly cause cerebrovascular microangiopathy with subsequent white matter injury. While no approved therapies or clinical trials for these leukovasculopathies currently exist, the morbidity and mortality associated with the secondary manifestations detailed above necessitate an accurate diagnosis of these disorders.

EMERGING THERAPEUTIC APPROACHES

Our discussion of the current landscape of active therapeutic development for LDs and gLEs reveals multiple strategic approaches. In current practice, most IWMDs are managed symptomatically due to the lack of targeted disease-modifying therapies. With the advent of advanced neuroimaging, neuropathological analyses, and molecular genetic sequencing techniques, our understanding of the disease mechanisms underlying these disorders informs therapeutic development toward novel targets. The emerging approaches include altering glial cell gene expression with gene therapy or antisense oligonucleotides (ASOs) administration, replacing dysfunctional or lost glial cells (particularly for oligodendrocytopathies or microgliopathies), augmenting disrupted signaling pathways in glial cells (CSF1R-related disorders), targeting the peripheral immune system (hematopoietic bone marrow transplant), or drug repurposing based on increased understanding of biochemical and metabolic mechanisms.

ASOs are being explored to address LDs with gain-of-function disease mechanisms. ASOs have been successfully used in preclinical mouse models for PMD (Elitt et al. 2020), Canavan disease (Hull et al. 2020), and H-ABC, and in a phase 3 clinical trial for AxD, all of which show promising results as potential therapeutics in future clinical trials.

Gene therapies, which involve the correction of genetic variants through ex vivo or in vivo gene transfer to glial or hematopoietic stem cells, are a therapeutic breakthrough for LD and gLE. Ex vivo gene therapy has been approved by the FDA for children with MLD and X-ALD and appears to arrest of disease progression in presymptomatic individuals (Sessa et al. 2016; Eichler et al. 2017). In both of these cases, the addition of gene therapy builds upon prior experience with targeting the peripheral immune system with hematopoietic bone marrow transplant, which alone is insufficient to improve symptoms (but may delay or halt progression for cerebral ALD, MLD, and even ALSP). A growing number of ongoing clinical trials are underway for LDs (Helman et al. 2015; Jensen et al. 2021; von Jonquieres et al. 2021; Kurtzberg 2022; Nowacki et al. 2022) promising significant advances in disease modulation going forward.

A growing understanding of disease mechanisms has also enabled the repurposing of drugs for disease-modifying therapy in LDs and leukoencephalopathies. Current examples include JAK inhibitors (Vanderver et al. 2020a; Cetin Gedik et al. 2022) and RTI (Rice et al. 2018) for AGS, venglustat, and N-acetyl-L-leucine for Tay-Sachs disease, cholic acid for CTX and peroxisomal biogenesis disorders, and Guanabenz for VWMD. Additional approaches approved and under investigation are outlined further in Table 2.

CONCLUDING REMARKS

LDs and leukoencephalopathies are a group of phenotypically and genotypically heterogeneous disorders that are challenging to diagnose and manage. These IWMDs are associated with a high degree of morbidity and mortality across the life span. A better understanding of clinical courses and pathophysiologies has contributed to a better understanding of the role of glial cells in the pathogenesis of these disorders. While most of these disorders do not have approved therapies, breakthroughs in therapeutic development and clinical trials are ongoing. Our improved understanding of the underlying pathology and therapeutic interventions will ultimately translate, either directly or indirectly, to acquired disorders that affect glial cells.

ACKNOWLEDGMENTS

J.O.-M. is a site principal investigator (PI) for Vigil Neurosciences sponsored studies. The authors would like to acknowledge the patients and families affected by these leukodystrophies, the patient advocacy organizations who inspire and encourage our work, and the Global Leukodystrophy Initiative Clinical Trials Network, which has funded research to A.V., J.L.F., and J.O.-M.

REFERENCES

Adang LA, Sherbini O, Ball L, Bloom M, Darbari A, Amartino H, DiVito D, Eichler F, Escolar M, Evans SH, et al. 2017. Revised consensus statement on the preventive and symptomatic care of patients with leukodystrophies. *Mol Genet Metab* **122:** 18–32. doi:10.1016/j.ymgme.2017.08 .006

Adang LA, Gavazzi F, D'Aiello R, Isaacs D, Bronner N, Arici ZS, Flores Z, Jan A, Scher C, Sherbini O, et al. 2022. Hematologic abnormalities in Aicardi Goutières syndrome. *Mol Genet Metab* **136:** 324–329. doi:10.1016/j .ymgme.2022.06.003

Adang LA, Bonkowsky JL, Boelens JJ, Mallack E, Ahrens-Nicklas R, Bernat JA, Bley A, Burton B, Darling A, Eichler F, et al. 2024. Consensus guidelines for the monitoring and management of metachromatic leukodystrophy in the United States. *Cytotherapy.* doi:10.1016/j.jcyt.2024 .03.487

Adib-Samii P, Brice G, Martin RJ, Markus HS. 2010. Clinical spectrum of CADASIL and the effect of cardiovascular risk factors on phenotype: study in 200 consecutively recruited individuals. *Stroke* **41:** 630–634. doi:10.1161/ STROKEAHA.109.568402

Akwa Y, Hassett DE, Eloranta ML, Sandberg K, Masliah E, Powell H, Whitton JL, Bloom FE, Campbell IL. 1998. Transgenic expression of IFN-α in the central nervous system of mice protects against lethal neurotropic viral infection but induces inflammation and neurodegeneration. *J Immunol* **161:** 5016–5026. doi:10.4049/jimmunol .161.9.5016

Androdias G, Vukusic S, Gignoux L, Boespflug-Tanguy O, Acquaviva C, Zabot MT, Couvert P, Carrie A, Confavreux C, Labauge P. 2012. Leukodystrophy with a cerebellar cystic aspect and intracranial atherosclerosis: an atypical presentation of cerebrotendinous xanthomatosis. *J Neurol* **259:** 364–366. doi:10.1007/s00415-011-6167-x

Antshel KM. 2010. ADHD, learning, and academic performance in phenylketonuria. *Mol Genet Metab* **99** (Suppl. 1): S52–S58. doi:10.1016/j.ymgme.2009.09.013

Atilla H, Coskun T, Elibol B, Kadayifcilar S, Altinel S; GEN-EYE-I Working Group. 2021. Prevalence of cerebrotendinous xanthomatosis in cases with idiopathic bilateral juvenile cataract in ophthalmology clinics in Turkey. *J AAPOS* **25:** 269.e1–269.e6. doi:10.1016/j.jaapos.2021.04 .015

Badrock AP, Uggenti C, Wacheul L, Crilly S, Jenkinson EM, Rice GI, Kasher PR, Lafontaine DLJ, Crow YJ, O'Keefe

RT. 2020. Analysis of U8 snoRNA variants in zebrafish reveals how bi-allelic variants cause leukoencephalopathy with calcifications and cysts. *Am J Hum Genet* **106:** 694–706. doi:10.1016/j.ajhg.2020.04.003

Bakhti M, Aggarwal S, Simons M. 2014. Myelin architecture: zippering membranes tightly together. *Cell Mol Life Sci* **71:** 1265–1277. doi:10.1007/s00018-013-1492-0

Baple EL, Salter C, Uhlig H, Wolf NI, Crosby AH. 2022. PI4KA-related disorder. In *GeneReviews* (ed. Adam MP, et al.). University of Washington, Seattle

Barkovich AJ, Deon S. 2016. Hypomyelinating disorders: an MRI approach. *Neurobiol Dis* **87:** 50–58. doi:10.1016/j.nbd.2015.10.015

Bauer AJ. 2019. Triac in the treatment of Allan–Herndon–Dudley syndrome. *Lancet Diabetes Endocrinol* **7:** 661–663. doi:10.1016/S2213-8587(19)30217-7

Beerepoot S, Wolf NI, Wehner K, Bender B, van der Knaap MS, Krägeloh-Mann I, Groeschel S. 2022. Acute-onset paralytic strabismus in toddlers is important to consider as a potential early sign of late-infantile metachromatic leukodystrophy. *Eur J Paediatr Neurol* **37:** 87–93. doi:10.1016/j.ejpn.2022.01.020

Bennett ML, Bennett FC. 2020. The influence of environment and origin on brain resident macrophages and implications for therapy. *Nat Neurosci* **23:** 157–166. doi:10.1038/s41593-019-0545-6

Bernard G, Chouery E, Putorti ML, Tétreault M, Takanohashi A, Carosso G, Clément I, Boespflug-Tanguy O, Rodriguez D, Delague V, et al. 2011. Mutations of POLR3A encoding a catalytic subunit of RNA polymerase Pol III cause a recessive hypomyelinating leukodystrophy. *Am J Hum Genet* **89:** 415–423. doi:10.1016/j.ajhg.2011.07.014

Bernegger C, Blau N. 2002. High frequency of tetrahydrobiopterin-responsiveness among hyperphenylalaninemias: a study of 1,919 patients observed from 1988 to 2002. *Mol Genet Metab* **77:** 304–313. doi:10.1016/S1096-7192(02)00171-3

Billon N, Jolicoeur C, Tokumoto Y, Vennstrom B, Raff M. 2002. Normal timing of oligodendrocyte development depends on thyroid hormone receptor α1 (TRα1). *EMBO J* **21:** 6452–6460. doi:10.1093/emboj/cdf662

Boelens JJ, van Hasselt PM. 2016. Neurodevelopmental outcome after hematopoietic cell transplantation in inborn errors of metabolism: current considerations and future perspectives. *Neuropediatrics* **47:** 285–292. doi:10.1055/s-0036-1584602

Bonkowsky JL, Nelson C, Kingston JL, Filloux FM, Mundorff MB, Srivastava R. 2010. The burden of inherited leukodystrophies in children. *Neurology* **75:** 718–725. doi:10.1212/WNL.0b013e3181eee46b

Bonnot O, Fraidakis MJ, Lucanto R, Chauvin D, Kelley N, Plaza M, Dubourg O, Lyon-Caen O, Sedel F, Cohen D. 2010. Cerebrotendinous xanthomatosis presenting with severe externalized disorder: improvement after one year of treatment with chenodeoxycholic acid. *CNS Spectr* **15:** 231–237. doi:10.1017/S1092852900000067

Boucher AA, Miller W, Shanley R, Ziegler R, Lund T, Raymond G, Orchard PJ. 2015. Long-term outcomes after allogeneic hematopoietic stem cell transplantation for metachromatic leukodystrophy: the largest single-institution cohort report. *Orphanet J Rare Dis* **10:** 94. doi:10.1186/s13023-015-0313-y

Boulloche J, Aicardi J. 1986. Pelizaeus-Merzbacher disease: clinical and nosological study. *J Child Neurol* **1:** 233–239. doi:10.1177/088307388600100310

Bradbury AM, Rafi MA, Bagel JH, Brisson BK, Marshall MS, Pesayco Salvador J, Jiang X, Swain GP, Prociuk ML, ODonnell PA, et al. 2018. AAVrh10 gene therapy ameliorates central and peripheral nervous system disease in canine globoid cell leukodystrophy (Krabbe disease). *Hum Gene Ther* **29:** 785–801. doi:10.1089/hum.2017.151

Bradbury AM, Bagel JH, Nguyen D, Lykken EA, Pesayco Salvador J, Jiang X, Swain GP, Assenmacher CA, Hendricks IJ, Miyadera K, et al. 2020. Krabbe disease successfully treated via monotherapy of intrathecal gene therapy. *J Clin Invest* **130:** 4906–4920. doi:10.1172/JCI133953

Brienza M, Fiermonte G, Cambieri C, Mignarri A, Dotti MT, Fiorelli M. 2015. Enlarging brain xanthomas in a patient with cerebrotendinous xanthomatosis. *J Inherit Metab Dis* **38:** 981–982. doi:10.1007/s10545-014-9805-5

Briggs TA, Abdel-Salam GM, Balicki M, Baxter P, Bertini E, Bishop N, Browne BH, Chitayat D, Chong WK, Eid MM, et al. 2008. Cerebroretinal microangiopathy with calcifications and cysts (CRMCC). *Am J Med Genet A* **146A:** 182–190. doi:10.1002/ajmg.a.32080

Bugiani M, van der Knaap MS. 2017. Childhood white matter disorders: much more than just diseases of myelin. *Acta Neuropathol* **134:** 329–330. doi:10.1007/s00401-017-1750-6

Bugiani M, Kevelam SH, Bakels HS, Waisfisz Q, Ceuterick-de Groote C, Niessen HW, Abbink TE, Lesnik Oberstein SA, van der Knaap MS. 2016. Cathepsin A-related arteriopathy with strokes and leukoencephalopathy (CARASAL). *Neurology* **87:** 1777–1786. doi:10.1212/WNL.0000000000003251

Burgard P, Bremer HJ, Bührdel P, Clemens PC, Mönch E, Przyrembel H, Trefz FK, Ullrich K. 1999. Rationale for the German recommendations for phenylalanine level control in phenylketonuria 1997. *Eur J Pediatr* **158:** 46–54. doi:10.1007/s004310051008

Burton BK, Leviton L, Vespa H, Coon H, Longo N, Lundy BD, Johnson M, Angelino A, Hamosh A, Bilder D. 2013. A diversified approach for PKU treatment: routine screening yields high incidence of psychiatric distress in phenylketonuria clinics. *Mol Genet Metab* **108:** 8–12. doi:10.1016/j.ymgme.2012.11.003

Cable C, Finkel RS, Lehky TJ, Biassou NM, Wiggs EA, Bunin N, Pierson TM. 2011. Unrelated umbilical cord blood transplant for juvenile metachromatic leukodystrophy: a 5-year follow-up in three affected siblings. *Mol Genet Metab* **102:** 207–209. doi:10.1016/j.ymgme.2010.10.002

Cailloux F, Gauthier-Barichard F, Mimault C, Isabelle V, Courtois V, Giraud G, Dastugue B, Boespflug-Tanguy O. 2000. Genotype–phenotype correlation in inherited brain myelination defects due to proteolipid protein gene mutations. *Eur J Hum Genet* **8:** 837–845. doi:10.1038/sj.ejhg.5200537

Casas-Alba D, Darling A, Caballero E, Mensa-Vilaró A, Bartrons J, Antón J, García-Cazorla A, Vanderver A, Armangué T. 2022. Efficacy of baricitinib on chronic pericardial effusion in a patient with Aicardi–Goutieres syndrome. *Rheumatology (Oxford)* **61:** e87–e89. doi:10.1093/rheumatology/keab860

Cetin Gedik K, Lamot L, Romano M, Demirkaya E, Piskin D, Torreggiani S, Adang LA, Armangue T, Barchus K, Cordova DR, et al. 2022. The 2021 European Alliance of Associations for Rheumatology/American College of Rheumatology points to consider for diagnosis and management of autoinflammatory type I interferonopathies: CANDLE/PRAAS, SAVI, and AGS. *Arthritis Rheumatol* **74:** 735–751. doi:10.1002/art.42087

Chadarevian JP, Lombroso SI, Peet GC, Hasselmann J, Tu C, Marzan DE, Capocchi J, Purnell FS, Nemec KM, Lahian A, et al. 2023. Engineering an inhibitor-resistant human CSF1R variant for microglia replacement. *J Exp Med* **220:** e20220857. doi:10.1084/jem.20220857

Channon S, Goodman G, Zlotowitz S, Mockler C, Lee PJ. 2007. Effects of dietary management of phenylketonuria on long-term cognitive outcome. *Arch Dis Child* **92:** 213–218. doi:10.1136/adc.2006.104786

Chen J, Van Gulden S, McGuire TL, Fleming AC, Oka C, Kessler JA, Peng CY. 2018. BMP-responsive protease HtrA1 is differentially expressed in astrocytes and regulates astrocytic development and injury response. *J Neurosci* **38:** 3840–3857. doi:10.1523/JNEUROSCI.2031-17.2018

Cheng B, Xin L, Kai D, Shengshun D, Zhouyi R, Yingmin C, Liangcheng L, Zhaoji L, Huang X, Huaxi X, et al. 2021. Triggering receptor expressed on myeloid cells-2 (TREM2) interacts with colony-stimulating factor 1 receptor (CSF1R) but is not necessary for CSF1/CSF1R-mediated microglial survival. *Front Immunol* **12:** 633796. doi:10.3389/fimmu.2021.633796

Choquet K, Forget D, Meloche E, Dicaire MJ, Bernard G, Vanderver A, Schiffmann R, Fabian MR, Teichmann M, Coulombe B, et al. 2019. Leukodystrophy-associated POLR3A mutations down-regulate the RNA polymerase III transcript and important regulatory RNA BC200. *J Biol Chem* **294:** 7445–7459. doi:10.1074/jbc.RA118.006271

Cosma MP, Pepe S, Parenti G, Settembre C, Annunziata I, Wade-Martins R, Di Domenico C, Di Natale P, Mankad A, Cox B, et al. 2004. Molecular and functional analysis of SUMF1 mutations in multiple sulfatase deficiency. *Hum Mutat* **23:** 576–581. doi:10.1002/humu.20040

Crow YJ, Chase DS, Lowenstein Schmidt J, Szynkiewicz M, Forte GM, Gornall HL, Oojageer A, Anderson B, Pizzino A, Helman G, et al. 2015. Characterization of human disease phenotypes associated with mutations in *TREX1, RNASEH2A, RNASEH2B, RNASEH2C, SAMHD1,* and *IFIH1. Am J Med Genet A* **167A:** 296–312. doi:10.1002/ajmg.a.36887

Crow YJ, Marshall H, Rice GI, Seabra L, Jenkinson EM, Baranano K, Battini R, Berger A, Blair E, Blauwblomme T, et al. 2021. Leukoencephalopathy with calcifications and cysts: genetic and phenotypic spectrum. *Am J Med Genet A* **185:** 15–25. doi:10.1002/ajmg.a.61907

Cruysberg JR. 2002. Cerebrotendinous xanthomatosis: juvenile cataract and chronic diarrhea before the onset of neurologic disease. *Arch Neurol* **59:** 1975. doi:10.1001/archneur.59.12.1975-a

Cruysberg JR, Wevers RA, van Engelen BG, Pinckers A, van Spreeken A, Tolboom JJ. 1995. Ocular and systemic manifestations of cerebrotendinous xanthomatosis. *Am J Ophthalmol* **120:** 597–604. doi:10.1016/S0002-9394(14)72206-8

Curiel J, Rodríguez Bey G, Takanohashi A, Bugiani M, Fu X, Wolf NI, Nmezi B, Schiffmann R, Bugaighis M, Pierson T, et al. 2017. TUBB4A mutations result in specific neuronal and oligodendrocytic defects that closely match clinically distinct phenotypes. *Hum Mol Genet* **26:** 4506–4518. doi:10.1093/hmg/ddx338

Degrassi I, Amoruso C, Giordano G, Del Puppo M, Mignarri A, Dotti MT, Naturale M, Nebbia G. 2020. Case report: early treatment with chenodeoxycholic acid in cerebrotendinous xanthomatosis presenting as neonatal cholestasis. *Front Pediatr* **8:** 382. doi:10.3389/fped.2020.00382

De Stefano N, Dotti MT, Mortilla M, Federico A. 2001. Magnetic resonance imaging and spectroscopic changes in brains of patients with cerebrotendinous xanthomatosis. *Brain* **124:** 121–131. doi:10.1093/brain/124.1.121

Dichgans M. 2009. Cognition in CADASIL. *Stroke* **40:** S45–S47.

Dichgans M, Mayer M, Uttner I, Brüning R, Müller-Höcker J, Rungger G, Ebke M, Klockgether T, Gasser T. 1998. The phenotypic spectrum of CADASIL: clinical findings in 102 cases. *Ann Neurol* **44:** 731–739. doi:10.1002/ana.410440506

Dierks T, Schmidt B, Borissenko LV, Peng J, Preusser A, Mariappan M, von Figura K. 2003. Multiple sulfatase deficiency is caused by mutations in the gene encoding the human C_α-formylglycine generating enzyme. *Cell* **113:** 435–444. doi:10.1016/S0092-8674(03)00347-7

Dierks T, Dickmanns A, Preusser-Kunze A, Schmidt B, Mariappan M, von Figura K, Ficner R, Rudolph MG. 2005. Molecular basis for multiple sulfatase deficiency and mechanism for formylglycine generation of the human formylglycine-generating enzyme. *Cell* **121:** 541–552. doi:10.1016/j.cell.2005.03.001

Dietrich J, Lacagnina M, Gass D, Richfield E, Mayer-Pröschel M, Noble M, Torres C, Pröschel C. 2005. EIF2B5 mutations compromise GFAP⁺ astrocyte generation in vanishing white matter leukodystrophy. *Nat Med* **11:** 277–283. doi:10.1038/nm1195

Di Rocco M, Biancheri R, Rossi A, Filocamo M, Tortori-Donati P. 2004. Genetic disorders affecting white matter in the pediatric age. *Am J Med Genet B Neuropsychiatr Genet* **129B:** 85–93. doi:10.1002/ajmg.b.30029

Dooves S, Bugiani M, Postma NL, Polder E, Land N, Horan ST, van Deijk A-LF, van de Kreeke A, Jacobs G, Vuong C, et al. 2016. Astrocytes are central in the pathomechanisms of vanishing white matter. *J Clin Invest* **126:** 1512–1524. doi:10.1172/JCI83908

Dorboz I, Dumay-Odelot H, Boussaid K, Bouyacoub Y, Barreau P, Samaan S, Jmel H, Eymard-Pierre E, Cances C, Bar C, et al. 2018. Mutation in POLR3K causes hypomyelinating leukodystrophy and abnormal ribosomal RNA regulation. *Neurol Genet* **4:** e289. doi:10.1212/NXG.0000000000000289

Dulski J, Heckman MG, White LJ, Żur-Wyrozumska K, Lund TC, Wszolek ZK. 2022. Hematopoietic stem cell transplantation in CSF1R-related leukoencephalopathy: retrospective study on predictors of outcomes. *Pharmaceutics* **14:** 2778. doi:10.3390/pharmaceutics14122778

Eichler FS, Li J, Guo Y, Caruso PA, Bjonnes AC, Pan J, Booker JK, Lane JM, Tare A, Vlasac I, et al. 2016.

CSF1R mosaicism in a family with hereditary diffuse leukoencephalopathy with spheroids. *Brain* 139: 1666–1672. doi:10.1093/brain/aww066

Eichler F, Duncan C, Musolino PL, Orchard PJ, De Oliveira S, Thrasher AJ, Armant M, Dansereau C, Lund TC, Miller WP, et al. 2017. Hematopoietic stem-cell gene therapy for cerebral adrenoleukodystrophy. *N Engl J Med* 377: 1630–1638. doi:10.1056/NEJMoa1700554

Elitt MS, Barbar L, Shick HE, Powers BE, Maeno-Hikichi Y, Madhavan M, Allan KC, Nawash BS, Gevorgyan AS, Hung S, et al. 2020. Suppression of proteolipid protein rescues Pelizaeus–Merzbacher disease. *Nature* 585: 397–403. doi:10.1038/s41586-020-2494-3

Emerson CH. 2019. TRIAC may ameliorate T$_3$ thyrotoxicosis in the Allan–Herndon–Dudley syndrome with MCT8 deficiency. *Clinl Thyroidol* 31: 458–462. doi:10.1089/ct.2019;31.458-462

Engelen M, Kemp S, de Visser M, van Geel BM, Wanders RJ, Aubourg P, Poll-The BT. 2012. X-linked adrenoleukodystrophy (X-ALD): clinical presentation and guidelines for diagnosis, follow-up and management. *Orphanet J Rare Dis* 7: 51. doi:10.1186/1750-1172-7-51

Engelen M, van Ballegoij WJC, Mallack EJ, Van Haren KP, Köhler W, Salsano E, van Trotsenburg ASP, Mochel F, Sevin C, Regelmann MO, et al. 2022. International recommendations for the diagnosis and management of patients with adrenoleukodystrophy: a consensus-based approach. *Neurology* 99: 940–951. doi:10.1212/WNL.0000000000201374

Federico A, Gallus GN. 1993. Cerebrotendinous xanthomatosis. In *GeneReviews* (ed. Adam MP, et al.). University of Washington, Seattle.

Feinstein M, Markus B, Noyman I, Shalev H, Flusser H, Shelef I, Liani-Leibson K, Shorer Z, Cohen I, Khateeb S, et al. 2010. Pelizaeus–Merzbacher–like disease caused by AIMP1/p43 homozygous mutation. *Am J Hum Genet* 87: 820–828. doi:10.1016/j.ajhg.2010.10.016

Ferrer I, Aubourg P, Pujol A. 2010. General aspects and neuropathology of X-linked adrenoleukodystrophy. *Brain Pathol* 20: 817–830. doi:10.1111/j.1750-3639.2010.00390.x

Finsterer J, Zarrouk Mahjoub S. 2012. Leukoencephalopathies in mitochondrial disorders: clinical and MRI findings. *J Neuroimaging* 22: e1–11. doi:10.1111/j.1552-6569.2011.00693.x

Freedman SF, Brennand C, Chiang J, DeBarber A, Del Monte MA, Duell PB, Fiorito J, Marshall R. 2019. Prevalence of cerebrotendinous xanthomatosis among patients diagnosed with acquired juvenile-onset idiopathic bilateral cataracts. *JAMA Ophthalmol* 137: 1312–1316. doi:10.1001/jamaophthalmol.2019.3639

Garavelli L, Rosato S, Mele A, Wischmeijer A, Rivieri F, Gelmini C, Sandonà F, Sassatelli R, Carlinfante G, Giovanardi F, et al. 2009. Massive hemobilia and papillomatosis of the gallbladder in metachromatic leukodystrophy: a life-threatening condition. *Neuropediatrics* 40: 284–286. doi:10.1055/s-0030-1248246

Gargareta VI, Reuschenbach J, Siems SB, Sun T, Piepkorn L, Mangana C, Späte E, Goebbels S, Huitinga I, Möbius W, et al. 2022. Conservation and divergence of myelin proteome and oligodendrocyte transcriptome profiles between humans and mice. *eLife* 11: e77019. doi:10.7554/eLife.77019

Gika AD, Siddiqui A, Hulse AJ, Edward S, Fallon P, McEntagart ME, Jan W, Josifova D, Lerman-Sagie T, Drummond J, et al. 2010. White matter abnormalities and dystonic motor disorder associated with mutations in the *SLC16A2* gene. *Dev Med Child Neurol* 52: 475–482. doi:10.1111/j.1469-8749.2009.03471.x

Gong JY, Setchell KDR, Zhao J, Zhang W, Wolfe B, Lu Y, Lackner K, Knisely AS, Wang NL, Hao CZ, et al. 2017. Severe neonatal cholestasis in cerebrotendinous xanthomatosis: genetics, immunostaining, mass spectrometry. *J Pediatr Gastroenterol Nutr* 65: 561–568. doi:10.1097/MPG.0000000000001730

Goutières F, Aicardi J, Barth PG, Lebon P. 1998. Aicardi-Goutières syndrome: an update and results of interferon-α studies. *Ann Neurol* 44: 900–907. doi:10.1002/ana.410440608

Griffiths I, Klugmann M, Anderson T, Thomson C, Vouyiouklis D, Nave KA. 1998. Current concepts of PLP and its role in the nervous system. *Microsc Res Tech* 41: 344–358. doi:10.1002/(SICI)1097-0029(19980601)41:5<344::AID-JEMT2>3.0.CO;2-Q

Groeneweg S, Peeters RP, Moran C, Stoupa A, Auriol F, Tonduti D, Dica A, Paone L, Rozenkova K, Malikova J, et al. 2019. Effectiveness and safety of the tri-iodothyronine analogue Triac in children and adults with MCT8 deficiency: an international, single-arm, open-label, phase 2 trial. *Lancet Diabetes Endocrinol* 7: 695–706. doi:10.1016/S2213-8587(19)30155-X

Groeschel S, Kühl JS, Bley AE, Kehrer C, Weschke B, Döring M, Böhringer J, Schrum J, Santer R, Kohlschütter A, et al. 2016. Long-term outcome of allogeneic hematopoietic stem cell transplantation in patients with juvenile metachromatic leukodystrophy compared with nontransplanted control patients. *JAMA Neurol* 73: 1133–1140. doi:10.1001/jamaneurol.2016.2067

Grønbæk-Thygesen M, Hartmann-Petersen R. 2024. Cellular and molecular mechanisms of aspartoacylase and its role in Canavan disease. *Cell Biosci* 14: 45. doi:10.1186/s13578-024-01224-6

Gruenenfelder FI, McLaughlin M, Griffiths IR, Garbern J, Thomson G, Kuzman P, Barrie JA, McCulloch ML, Penderis J, Stassart R, et al. 2020. Neural stem cells restore myelin in a demyelinating model of Pelizaeus–Merzbacher disease. *Brain* 143: 1383–1399. doi:10.1093/brain/awaa080

Guey S, Mawet J, Hervé D, Duering M, Godin O, Jouvent E, Opherk C, Alili N, Dichgans M, Chabriat H. 2016. Prevalence and characteristics of migraine in CADASIL. *Cephalalgia* 36: 1038–1047. doi:10.1177/0333102415620909

Guo L, Bertola DR, Takanohashi A, Saito A, Segawa Y, Yokota T, Ishibashi S, Nishida Y, Yamamoto GL, Franco J, et al. 2019. Bi-allelic CSF1R mutations cause skeletal dysplasia of dysosteosclerosis–Pyle disease spectrum and degenerative encephalopathy with brain malformation. *Am J Hum Genet* 104: 925–935. doi:10.1016/j.ajhg.2019.03.004

Guo X, Steinman RA, Sheng Y, Cao G, Wiley CA, Wang Q. 2022. An AGS-associated mutation in ADAR1 catalytic domain results in early-onset and MDA5-dependent encephalopathy with IFN pathway activation in the brain. *J*

Neuroinflammation **19**: 285. doi:10.1186/s12974-022-02646-0

Gupta N, Henry RG, Strober J, Kang SM, Lim DA, Bucci M, Caverzasi E, Gaetano L, Mandelli ML, Ryan T, et al. 2012. Neural stem cell engraftment and myelination in the human brain. *Sci Transl Med* **4**: 155ra137. doi:10.1126/scitranslmed.3004373

Gupta N, Henry RG, Kang SM, Strober J, Lim DA, Ryan T, Perry R, Farrell J, Ulman M, Rajalingam R, et al. 2019. Long-term safety, immunologic response, and imaging outcomes following neural stem cell transplantation for Pelizaeus–Merzbacher disease. *Stem Cell Rep* **13**: 254–261. doi:10.1016/j.stemcr.2019.07.002

Haaxma CA, Crow YJ, van Steensel MA, Lammens MM, Rice GI, Verbeek MM, Willemsen MA. 2010. A de novo p.Asp18Asn mutation in *TREX1* in a patient with Aicardi-Goutières syndrome. *Am J Med Genet A* **152a**: 2612–2617. doi:10.1002/ajmg.a.33620

Hamilton EMC, van der Lei HDW, Vermeulen G, Gerver JAM, Lourenço CM, Naidu S, Mierzewska H, Gemke R, de Vet HCW, Uitdehaag BMJ, et al. 2018. Natural history of vanishing white matter. *Ann Neurol* **84**: 274–288. doi:10.1002/ana.25287

He T, Xia Y, Yang J. 2021. Systemic inflammation and chronic kidney disease in a patient due to the RNASEH2B defect. *Ped Rheumatol* **19**: 9. doi:10.1186/s12969-021-00497-2

Hedera P. 1993. Hereditary spastic paraplegia overview. In *GeneReviews* (ed. Adam MP, et al.). University of Washington, Seattle.

Helman G, Van Haren K, Escolar ML, Vanderver A. 2015. Emerging treatments for pediatric leukodystrophies. *Pediatr Clin North Am* **62**: 649–666. doi:10.1016/j.pcl.2015.03.006

Helman G, Venkateswaran S, Vanderver A. 2018. The spectrum of adult-onset heritable white-matter disorders. *Handb Clin Neurol* **148**: 669–692. doi:10.1016/B978-0-444-64076-5.00043-0

Helman G, Lajoie BR, Crawford J, Takanohashi A, Walkiewicz M, Dolzhenko E, Gross AM, Gainullin VG, Bent SJ, Jenkinson EM, et al. 2020. Genome sequencing in persistently unsolved white matter disorders. *Ann Clin Transl Neurol* **7**: 144–152. doi:10.1002/acn3.50957

Heuer H, Maier MK, Iden S, Mittag J, Friesema EC, Visser TJ, Bauer K. 2005. The monocarboxylate transporter 8 linked to human psychomotor retardation is highly expressed in thyroid hormone-sensitive neuron populations. *Endocrinology* **146**: 1701–1706. doi:10.1210/en.2004-1179

Hironaka K, Yamazaki Y, Hirai Y, Yamamoto M, Miyake N, Miyake K, Okada T, Morita A, Shimada T. 2015. Enzyme replacement in the CSF to treat metachromatic leukodystrophy in mouse model using single intracerebroventricular injection of self-complementary AAV1 vector. *Sci Rep* **5**: 13104. doi:10.1038/srep13104

Ho G, Christodoulou J. 2014. Phenylketonuria: translating research into novel therapies. *Transl Pediatr* **3**: 49–62.

Hodes ME, Pratt VM, Dlouhy SR. 1993. Genetics of Pelizaeus–Merzbacher disease. *Dev Neurosci* **15**: 383–394. doi:10.1159/000111361

Hoshino H, Kubota M. 2014. Canavan disease: clinical features and recent advances in research. *Pediatr Int* **56**: 477–483. doi:10.1111/ped.12422

Huisman TAGM, Kralik SF, Desai NK, Serrallach BL, Orman G. 2022. Neuroimaging of primary mitochondrial disorders in children: a review. *J Neuroimaging* **32**: 191–200. doi:10.1111/jon.12976

Hull V, Wang Y, Burns T, Zhang S, Sternbach S, McDonough J, Guo F, Pleasure D. 2020. Antisense oligonucleotide reverses leukodystrophy in Canavan disease mice. *Ann Neurol* **87**: 480–485. doi:10.1002/ana.25674

Ida H, Rennert OM, Watabe K, Eto Y, Maekawa K. 1994. Pathological and biochemical studies of fetal Krabbe disease. *Brain Dev* **16**: 480–484. doi:10.1016/0387-7604(94)90013-2

Jahn O, Siems SB, Kusch K, Hesse D, Jung RB, Liepold T, Uecker M, Sun T, Werner HB. 2020. The CNS myelin proteome: deep profile and persistence after post-mortem delay. *Front Cell Neurosci* **14**: 239. doi:10.3389/fncel.2020.00239

Jenkinson EM, Rodero MP, Kasher PR, Uggenti C, Oojageer A, Goosey LC, Rose Y, Kershaw CJ, Urquhart JE, Williams SG, et al. 2016. Mutations in SNORD118 cause the cerebral microangiopathy leukoencephalopathy with calcifications and cysts. *Nat Genet* **48**: 1185–1192. doi:10.1038/ng.3661

Jensen TL, Gøtzsche CR, Woldbye DPD. 2021. Current and future prospects for gene therapy for rare genetic diseases affecting the brain and spinal cord. *Front Mol Neurosci* **14**: 695937. doi:10.3389/fnmol.2021.695937

Kakhlon O, Vaknin H, Mishra K, D'Souza J, Marisat M, Sprecher U, Wald-Altman S, Dukhovny A, Raviv Y, Daadoosh B, et al. 2021. Alleviation of a polyglucosan storage disorder by enhancement of autophagic glycogen catabolism. *EMBO Mol Med* **13**: e14554. doi:10.15252/emmm.202114554

Kassmann C. 2014. Myelin peroxisomes—essential organelles for the maintenance of white matter in the nervous system. *Biochimie* **98**: 111–118. doi:10.1016/j.biochi.2013.09.020

Kehrer C, Blumenstock G, Gieselmann V, Krägeloh-Mann I, German L. 2011. The natural course of gross motor deterioration in metachromatic leukodystrophy. *Dev Med Child Neurol* **53**: 850–855. doi:10.1111/j.1469-8749.2011.04028.x

Klok MD, Bugiani M, de Vries SI, Gerritsen W, Breur M, van der Sluis S, Heine VM, Kole MHP, Baron W, van der Knaap MS. 2018. Axonal abnormalities in vanishing white matter. *Ann Clin Transl Neurol* **5**: 429–444. doi:10.1002/acn3.540

Koch R, Burton B, Hoganson G, Peterson R, Rhead W, Rouse B, Scott R, Wolff J, Stern AM, Guttler F, et al. 2002. Phenylketonuria in adulthood: a collaborative study. *J Inherit Metab Dis* **25**: 333–346. doi:10.1023/A:1020158631102

Koch R, Soler-Alfonso C, Kiely BT, Asai A, Smith AL, Bali DS, Kang PB, Landstrom AP, Akman HO, Burrow TA, et al. 2023. Diagnosis and management of glycogen storage disease type IV, including adult polyglucosan body disease: a clinical practice resource. *Mol Genet Metab* **138**: 107525. doi:10.1016/j.ymgme.2023.107525

Konno T, Kasanuki K, Ikeuchi T, Dickson DW, Wszolek ZK. 2018. *CSF1R*-related leukoencephalopathy: a major player in primary microgliopathies. *Neurology* **91:** 1092–1104. doi:10.1212/WNL.0000000000006642

Korn-Lubetzki I, Dor-Wollman T, Soffer D, Raas-Rothschild A, Hurvitz H, Nevo Y. 2003. Early peripheral nervous system manifestations of infantile Krabbe disease. *Pediatr Neurol* **28:** 115–118. doi:10.1016/S0887-8994(02)00489-7

Krajka V, Vulinovic F, Genova M, Tanzer K, Jijumon AS, Bodakuntla S, Tennstedt S, Mueller-Fielitz H, Meier B, Janke C, et al. 2022. H-ABC- and dystonia-causing *TUBB4A* mutations show distinct pathogenic effects. *Sci Adv* **8:** eabj9229. doi:10.1126/sciadv.abj9229

Kumar KR, Vulinovic F, Lohmann K, Park JS, Schaake S, Sue CM, Klein C. 2015. Mutations in *TUBB4A* and spastic paraplegia. *Mov Disord* **30:** 1857–1858. doi:10.1002/mds.26444

Kurtzberg J. 2022. Gene therapy offers new hope for children with metachromatic leukodystrophy. *Lancet* **399:** 338–339. doi:10.1016/S0140-6736(22)00057-5

Lata E, Choquet K, Sagliocco F, Brais B, Bernard G, Teichmann M. 2021. RNA polymerase III subunit mutations in genetic diseases. *Front Mol Biosci* **8:** 696438. doi:10.3389/fmolb.2021.696438

Leferink PS, Dooves S, Hillen AEJ, Watanabe K, Jacobs G, Gasparotto L, Cornelissen-Steijger P, van der Knaap MS, Heine VM. 2019. Astrocyte subtype vulnerability in stem cell models of vanishing white matter. *Ann Neurol* **86:** 780–792. doi:10.1002/ana.25585

Leung EK, Yi X, Refetoff S, Yeo KT. 2016. Diiodothyropropionic acid (DITPA) cross-reacts with thyroid function assays on different immunoassay platforms. *Clin Chim Acta* **453:** 203–204. doi:10.1016/j.cca.2015.12.008

Li L, Tian E, Chen X, Chao J, Klein J, Qu Q, Sun G, Sun G, Huang Y, Warden CD, et al. 2018. GFAP mutations in astrocytes impair oligodendrocyte progenitor proliferation and myelination in an hiPSC model of Alexander disease. *Cell Stem Cell* **23:** 239–251.e6. doi:10.1016/j.stem.2018.07.009

Lieberman JS, Oshtory M, Taylor RG, Dreyfus PM. 1980. Perinatal neuropathy as an early manifestation of Krabbe's disease. *Arch Neurol* **37:** 446–447. doi:10.1001/archneur.1980.00500560076012

Livingston JH, Lin JP, Dale RC, Gill D, Brogan P, Munnich A, Kurian MA, Gonzalez-Martinez V, De Goede CG, Falconer A, et al. 2014a. A type I interferon signature identifies bilateral striatal necrosis due to mutations in *ADAR1*. *J Med Genet* **51:** 76–82. doi:10.1136/jmedgenet-2013-102038

Livingston JH, Mayer J, Jenkinson E, Kasher P, Stivaros S, Berger A, Cordelli DM, Ferreira P, Jefferson R, Kutschke G, et al. 2014b. Leukoencephalopathy with calcifications and cysts: a purely neurological disorder distinct from coats plus. *Neuropediatrics* **45:** 175–182. doi:10.1055/s-0033-1364180

Locatelli M, Padovani A, Pezzini A. 2020. Pathophysiological mechanisms and potential therapeutic targets in cerebral autosomal dominant arteriopathy with subcortical infarcts and leukoencephalopathy (CADASIL). *Front Pharmacol* **11:** 321. doi:10.3389/fphar.2020.00321

Lo Giudice T, Lombardi F, Santorelli FM, Kawarai T, Orlacchio A. 2014. Hereditary spastic paraplegia: clinical-genetic characteristics and evolving molecular mechanisms.

Exp Neurol **261:** 518–539. doi:10.1016/j.expneurol.2014.06.011

Lynch DS, Wade C, Paiva ARB, John N, Kinsella JA, Merwick A, Ahmed RM, Warren JD, Mummery CJ, Schott JM, et al. 2019. Practical approach to the diagnosis of adult-onset leukodystrophies: an updated guide in the genomic era. *J Neurol Neurosurg Psychiatry* **90:** 543–555. doi:10.1136/jnnp-2018-319481

Madhavarao CN, Arun P, Moffett JR, Szucs S, Surendran S, Matalon R, Garbern J, Hristova D, Johnson A, Jiang W, et al. 2005. Defective N-acetylaspartate catabolism reduces brain acetate levels and myelin lipid synthesis in Canavan's disease. *Proc Natl Acad Sci* **102:** 5221–5226. doi:10.1073/pnas.0409184102

Mahmood A, Berry J, Wenger DA, Escolar M, Sobeih M, Raymond G, Eichler FS. 2010. Metachromatic leukodystrophy: a case of triplets with the late infantile variant and a systematic review of the literature. *J Child Neurol* **25:** 572–580. doi:10.1177/0883073809341669

Man JHK, van Gelder CAGH, Breur M, Okkes D, Molenaar D, van der Sluis S, Abbink T, Altelaar M, van der Knaap MS, Bugiani M. 2022. Cortical pathology in vanishing white matter. *Cells* **11:** 3581. doi:10.3390/cells11223581

Matalon R, Delgado L, Michals-Matalon K. 1993. Canavan disease. In *GeneReviews* (ed. Adam MP, et al.). University of Washington, Seattle.

McDermott C, White K, Bushby K, Shaw P. 2000. Hereditary spastic paraparesis: a review of new developments. *J Neurol Neurosurg Psychiatry* **69:** 150–160. doi:10.1136/jnnp.69.2.150

Meesilpavikkai K, Dik WA, Schrijver B, van Helden-Meeuwsen CG, Versnel MA, van Hagen PM, Bijlsma EK, Ruivenkamp CAL, Oele MJ, Dalm V. 2019. Efficacy of baricitinib in the treatment of chilblains associated with Aicardi–Goutières syndrome, a type I interferonopathy. *Arthritis Rheumatol* **71:** 829–831. doi:10.1002/art.40805

Mendes MI, Gutierrez Salazar M, Guerrero K, Thiffault I, Salomons GS, Gauquelin L, Tran LT, Forget D, Gauthier MS, Waisfisz Q, et al. 2018. Bi-allelic mutations in EPRS, encoding the glutamyl-prolyl-aminoacyl-tRNA synthetase, cause a hypomyelinating leukodystrophy. *Am J Hum Genet* **102:** 676–684. doi:10.1016/j.ajhg.2018.02.011

Messing A, Head MW, Galles K, Galbreath EJ, Goldman JE, Brenner M. 1998. Fatal encephalopathy with astrocyte inclusions in GFAP transgenic mice. *Am J Pathol* **152:** 391–398.

Meuwissen ME, Halley DJ, Smit LS, Lequin MH, Cobben JM, de Coo R, van Harssel J, Sallevelt S, Woldringh G, van der Knaap MS, et al. 2015. The expanding phenotype of COL4A1 and COL4A2 mutations: clinical data on 13 newly identified families and a review of the literature. *Genet Med* **17:** 843–853. doi:10.1038/gim.2014.210

Mignarri A, Gallus GN, Dotti MT, Federico A. 2014. A suspicion index for early diagnosis and treatment of cerebrotendinous xanthomatosis. *J Inherit Metab Dis* **37:** 421–429. doi:10.1007/s10545-013-9674-3

Modesti NB, Evans SH, Jaffe N, Vanderver A, Gavazzi F. 2022. Early recognition of patients with leukodystrophies. *Curr Probl Pediatr Adolesc Health Care* **52:** 101311. doi:10.1016/j.cppeds.2022.101311

Moser HW, Smith KD, Watkins PA, Powers J, Moser AB. 2019. X-linked adrenoleukodystrophy. In *The online met-*

abolic and molecular bases of inherited disease (ed. Valle DL, et al.). McGraw-Hill, New York.

Moyle JJ, Fox AM, Arthur M, Bynevelt M, Burnett JR. 2007. Meta-analysis of neuropsychological symptoms of adolescents and adults with PKU. *Neuropsychol Rev* **17:** 91–101. doi:10.1007/s11065-007-9021-2

Nobuta H, Yang N, Ng YH, Marro SG, Sabeur K, Chavali M, Stockley JH, Killilea DW, Walter PB, Zhao C, et al. 2019. Oligodendrocyte death in Pelizaeus–Merzbacher disease is rescued by iron chelation. *Cell Stem Cell* **25:** 531–541.e6. doi:10.1016/j.stem.2019.09.003

Nowacki JC, Fields AM, Fu MM. 2022. Emerging cellular themes in leukodystrophies. *Front Cell Dev Biol* **10:** 902261. doi:10.3389/fcell.2022.902261

Ognjenović J, Simonović M. 2018. Human aminoacyl-tRNA synthetases in diseases of the nervous system. *RNA Biol* **15:** 623–634. doi:10.1080/15476286.2017.1330245

Oosterhof N, Chang IJ, Karimiani EG, Kuil LE, Jensen DM, Daza R, Young E, Astle L, van der Linde HC, Shivaram GM, et al. 2019. Homozygous mutations in CSF1R cause a pediatric-onset leukoencephalopathy and can result in congenital absence of microglia. *Am J Hum Genet* **104:** 936–947. doi:10.1016/j.ajhg.2019.03.010

Orsini JJ, Escolar ML, Wasserstein MP, Caggana M. 1993. Krabbe disease. In *GeneReviews* (ed. Adam MP, et al.). University of Washington, Seattle.

Parikh S, Bernard G, Leventer RJ, van der Knaap MS, van Hove J, Pizzino A, McNeill NH, Helman G, Simons C, Schmidt JL, et al. 2015. A clinical approach to the diagnosis of patients with leukodystrophies and genetic leukoencephalopathies. *Mol Genet Metab* **114:** 501–515. doi:10.1016/j.ymgme.2014.12.434

Pérez-Dueñas B, Vilaseca MA, Mas A, Lambruschini N, Artuch R, Gómez L, Pineda J, Gutiérrez A, Mila M, Campistol J. 2004. Tetrahydrobiopterin responsiveness in patients with phenylketonuria. *Clin Biochem* **37:** 1083–1090. doi:10.1016/j.clinbiochem.2004.09.005

Perrier S, Michell-Robinson MA, Bernard G. 2021. POLR3-related leukodystrophy: exploring potential therapeutic approaches. *Front Cell Neurosci* **14:** 631802. doi:10.3389/fncel.2020.631802

Piccoli C, Bronner N, Gavazzi F, Dubbs H, De Simone M, De Giorgis V, Orcesi S, Fazzi E, Galli J, Masnada S, et al. 2021. Late-onset Aicardi–Goutières syndrome: a characterization of presenting clinical features. *Pediatr Neurol* **115:** 1–6. doi:10.1016/j.pediatrneurol.2020.10.012

Potter GB, Santos M, Davisson MT, Rowitch DH, Marks DL, Bongarzone ER, Petryniak MA. 2013. Missense mutation in mouse GALC mimics human gene defect and offers new insights into Krabbe disease. *Hum Mol Genet* **22:** 3397–3414. doi:10.1093/hmg/ddt190

Prust M, Wang J, Morizono H, Messing A, Brenner M, Gordon E, Hartka T, Sokohl A, Schiffmann R, Gordish-Dressman H, et al. 2011. GFAP mutations, age at onset, and clinical subtypes in Alexander disease. *Neurology* **77:** 1287–1294. doi:10.1212/WNL.0b013e3182309f72

Rademakers R, Baker M, Nicholson AM, Rutherford NJ, Finch N, Soto-Ortolaza A, Lash J, Wider C, Wojtas A, DeJesus-Hernandez M, et al. 2012. Mutations in the colony stimulating factor 1 receptor (CSF1R) gene cause hereditary diffuse leukoencephalopathy with spheroids. *Nat Genet* **44:** 200–205. doi:10.1038/ng.1027

Ragno M, Berbellini A, Cacchiò G, Manca A, Di Marzio F, Pianese L, De Rosa A, Silvestri S, Scarcella M, De Michele G. 2013. Parkinsonism is a late, not rare, feature of CADASIL: a study on Italian patients carrying the R1006C mutation. *Stroke* **44:** 1147–1149. doi:10.1161/STROKEAHA.111.000458

Raskind WH, Williams CA, Hudson LD, Bird TD. 1991. Complete deletion of the proteolipid protein gene (PLP) in a family with X-linked Pelizaeus–Merzbacher disease. *Am J Hum Genet* **49:** 1355–1360.

Reeves SA, Helman LJ, Allison A, Israel MA. 1989. Molecular cloning and primary structure of human glial fibrillary acidic protein. *Proc Natl Acad Sci* **86:** 5178–5182. doi:10.1073/pnas.86.13.5178

Rehm HL, Berg JS, Brooks LD, Bustamante CD, Evans JP, Landrum MJ, Ledbetter DH, Maglott DR, Martin CL, Nussbaum RL, et al. 2015. Clingen—the clinical genome resource. *N Engl J Med* **372:** 2235–2242. doi:10.1056/NEJMsr1406261

Resende LL, de Paiva ARB, Kok F, da Costa Leite C, Lucato LT. 2019. Adult leukodystrophies: a step-by-step diagnostic approach. *Radiographics* **39:** 153–168. doi:10.1148/rg.2019180081

Reyes S, Viswanathan A, Godin O, Dufouil C, Benisty S, Hernandez K, Kurtz A, Jouvent E, O'Sullivan M, Czernecki V, et al. 2009. Apathy: a major symptom in CADASIL. *Neurology* **72:** 905–910. doi:10.1212/01.wnl.0000344166.03470.f8

Rice G, Patrick T, Parmar R, Taylor CF, Aeby A, Aicardi J, Artuch R, Montalto SA, Bacino CA, Barroso B, et al. 2007. Clinical and molecular phenotype of Aicardi–Goutières syndrome. *Am J Hum Genet* **81:** 713–725. doi:10.1086/521373

Rice GI, Bond J, Asipu A, Brunette RL, Manfield IW, Carr IM, Fuller JC, Jackson RM, Lamb T, Briggs TA, et al. 2009. Mutations involved in Aicardi–Goutières syndrome implicate SAMHD1 as regulator of the innate immune response. *Nat Genet* **41:** 829–832. doi:10.1038/ng.373

Rice GI, Kasher PR, Forte GM, Mannion NM, Greenwood SM, Szynkiewicz M, Dickerson JE, Bhaskar SS, Zampini M, Briggs TA, et al. 2012. Mutations in ADAR1 cause Aicardi–Goutières syndrome associated with a type I interferon signature. *Nat Genet* **44:** 1243–1248. doi:10.1038/ng.2414

Rice GI, Del Toro Duany Y, Jenkinson EM, Forte GM, Anderson BH, Ariaudo G, Bader-Meunier B, Baildam EM, Battini R, Beresford MW, et al. 2014. Gain-of-function mutations in IFIH1 cause a spectrum of human disease phenotypes associated with upregulated type I interferon signaling. *Nat Genet* **46:** 503–509. doi:10.1038/ng.2933

Rice GI, Meyzer C, Bouazza N, Hully M, Boddaert N, Semeraro M, Zeef LAH, Rozenberg F, Bondet V, Duffy D, et al. 2018. Reverse-transcriptase inhibitors in the Aicardi–Goutières syndrome. *N Engl J Med* **379:** 2275–2277. doi:10.1056/NEJMc1810983

Robitaille Y, Carpenter S, Karpati G, DiMauro S. 1980. A distinct form of adult polyglucosan body disease with massive involvement of central and peripheral neuronal processes and astrocytes: a report of four cases and a review of the occurrence of polyglucosan bodies in other conditions such as Lafora's disease and normal ageing. *Brain* **103:** 315–336. doi:10.1093/brain/103.2.315

Cite this article as *Cold Spring Harb Perspect Biol* doi: 10.1101/cshperspect.a041457

Roosendaal SD, van de Brug T, Alves C, Blaser S, Vanderver A, Wolf NI, van der Knaap MS. 2021. Imaging patterns characterizing mitochondrial leukodystrophies. *AJNR Am J Neuroradiol* **42:** 1334–1340. doi:10.3174/ajnr.A7097

Saher G, Rudolphi F, Corthals K, Ruhwedel T, Schmidt KF, Löwel S, Dibaj P, Barrette B, Möbius W, Nave KL. 2012. Therapy of Pelizaeus–Merzbacher disease in mice by feeding a cholesterol-enriched diet. *Nat Med* **18:** 1130–1135. doi:10.1038/nm.2833

Salter CG, Cai Y, Lo B, Helman G, Taylor H, McCartney A, Leslie JS, Accogli A, Zara F, Traverso M, et al. 2021. Biallelic *PI4KA* variants cause neurological, intestinal and immunological disease. *Brain* **144:** 3597–3610. doi:10.1093/brain/awab313

Sandlers Y, Moser AB, Hubbard WC, Kratz LE, Jones RO, Raymond GV. 2012. Combined extraction of acyl carnitines and 26:0 lysophosphatidylcholine from dried blood spots: prospective newborn screening for X-linked adrenoleukodystrophy. *Mol Genet Metab* **105:** 416–420. doi:10.1016/j.ymgme.2011.11.195

Sardiello M, Annunziata I, Roma G, Ballabio A. 2005. Sulfatases and sulfatase modifying factors: an exclusive and promiscuous relationship. *Hum Mol Genet* **14:** 3203–3217. doi:10.1093/hmg/ddi351

Sarret C, Oliver Petit I, Tonduti D. 1993. Allan–Herndon–Dudley syndrome. In *GeneReviews* (ed. Adam MP, et al.). University of Washington, Seattle.

Sase S, Almad AA, Boecker CA, Guedes-Dias P, Li JJ, Takanohashi A, Patel A, McCaffrey T, Patel H, Sirdeshpande D, et al. 2020. TUBB4A mutations result in both glial and neuronal degeneration in an H-ABC leukodystrophy mouse model. *eLife* **9:** e52986. doi:10.7554/eLife.52986

Savage C, Xue Y, Mitani S, Hall D, Zakhary R, Chalfie M. 1994. Mutations in the *Caenorhabditis elegans* β-tubulin gene *mec-7*: effects on microtubule assembly and stability and on tubulin autoregulation. *J Cell Sci* **107:** 2165–2175. doi:10.1242/jcs.107.8.2165

Schiffmann R, van der Knaap MS. 2009. Invited article: an MRI-based approach to the diagnosis of white matter disorders. *Neurology* **72:** 750–759. doi:10.1212/01.wnl.0000343049.00540.c8

Schmidt JL, Pizzino A, Nicholl J, Foley A, Wang Y, Rosenfeld JA, Mighion L, Bean L, da Silva C, Cho MT, et al. 2020. Estimating the relative frequency of leukodystrophies and recommendations for carrier screening in the era of next-generation sequencing. *Am J Med Genet A* **182:** 1906–1912. doi:10.1002/ajmg.a.61641

Scott-Hewitt NJ, Folts CJ, Noble MD. 2018. Heterozygous carriers of galactocerebrosidase mutations that cause Krabbe disease have impaired microglial function and defective repair of myelin damage. *Neural Regen Res* **13:** 393–401. doi:10.4103/1673-5374.228712

Sessa M, Lorioli L, Fumagalli F, Acquati S, Redaelli D, Baldoli C, Canale S, Lopez ID, Morena F, Calabria A, et al. 2016. Lentiviral haemopoietic stem-cell gene therapy in early-onset metachromatic leukodystrophy: an ad-hoc analysis of a non-randomised, open-label, phase 1/2 trial. *Lancet* **388:** 476–487. doi:10.1016/S0140-6736(16)30374-9

Simons M, Krämer E-M, Macchi P, Rathke-Hartlieb S, Trotter J, Nave KA, Schulz JB. 2002. Overexpression of the myelin proteolipid protein leads to accumulation of cholesterol and proteolipid protein in endosome/lysosomes: implications for Pelizaeus–Merzbacher disease. *J Cell Biol* **157:** 327–336. doi:10.1083/jcb.200110138

Singh RH, Rohr F, Frazier D, Cunningham A, Mofidi S, Ogata B, Splett PL, Moseley K, Huntington K, Acosta PB, et al. 2014. Recommendations for the nutrition management of phenylalanine hydroxylase deficiency. *Genet Med* **16:** 121–131. doi:10.1038/gim.2013.179

Sirkis DW, Bonham LW, Yokoyama JS. 2021. The role of microglia in inherited white-matter disorders and connections to frontotemporal dementia. *Appl Clin Genet* **14:** 195–207. doi:10.2147/TACG.S245029

Sofroniew MV. 2009. Molecular dissection of reactive astrogliosis and glial scar formation. *Trends Neurosci* **32:** 638–647. doi:10.1016/j.tins.2009.08.002

Solders M, Martin DA, Andersson C, Remberger M, Andersson T, Ringdén O, Solders G. 2014. Hematopoietic SCT: a useful treatment for late metachromatic leukodystrophy. *Bone Marrow Transplant* **49:** 1046–1051. doi:10.1038/bmt.2014.93

Stadelman C, Timmler S, Barrantes-Freer A, Simons M. 2019. Myelin in the central nervous system: structure, function, and pathology. *Physiological Rev* **99:** 1381–1431. doi:10.1152/physrev.00031.2018

Steenweg ME, Vanderver A, Blaser S, Bizzi A, de Koning TJ, Mancini GM, van Wieringen WN, Barkhof F, Wolf NI, van der Knaap MS. 2010. Magnetic resonance imaging pattern recognition in hypomyelinating disorders. *Brain* **133:** 2971–2982. doi:10.1093/brain/awq257

Stein C, Gieselmann V, Kreysing J, Schmidt B, Pohlmann R, Waheed A, Meyer HE, O'Brien JS, von Figura K. 1989. Cloning and expression of human arylsulfatase A. *J Biol Chem* **264:** 1252–1259. doi:10.1016/S0021-9258(19)85079-2

Stellingwerff MD, Nulton C, Helman G, Roosendaal SD, Benko WS, Pizzino A, Bugiani M, Vanderver A, Simons C, van der Knaap MS. 2022. Early-onset vascular leukoencephalopathy caused by bi-allelic NOTCH3 variants. *Neuropediatrics* **53:** 115–121. doi:10.1055/a-1739-2722

Stumpf E, Masson H, Duquette A, Berthelet F, McNabb J, Lortie A, Lesage J, Montplaisir J, Brais B, Cossette P. 2003. Adult Alexander disease with autosomal dominant transmission: a distinct entity caused by mutation in the glial fibrillary acid protein gene. *Arch Neurol* **60:** 1307–1312. doi:10.1001/archneur.60.9.1307

Taft RJ, Vanderver A, Leventer RJ, Damiani SA, Simons C, Grimmond SM, Miller D, Schmidt J, Lockhart PJ, Pope K, et al. 2013. Mutations in DARS cause hypomyelination with brain stem and spinal cord involvement and leg spasticity. *Am J Hum Genet* **92:** 774–780. doi:10.1016/j.ajhg.2013.04.006

Tan RY, Markus HS. 2016. CADASIL: migraine, encephalopathy, stroke and their inter-relationships. *PLoS ONE* **11:** e0157613. doi:10.1371/journal.pone.0157613

Tanaka AJ, Cho MT, Millan F, Juusola J, Retterer K, Joshi C, Niyazov D, Garnica A, Gratz E, Deardorff M, et al. 2015. Mutations in SPATA5 are associated with microcephaly, intellectual disability, seizures, and hearing loss. *Am J Hum Genet* **97:** 457–464. doi:10.1016/j.ajhg.2015.07.014

Tétreault M, Choquet K, Orcesi S, Tonduti D, Balottin U, Teichmann M, Fribourg S, Schiffmann R, Brais B, Vanderver A, et al. 2011. Recessive mutations in POLR3B, encoding the second largest subunit of Pol III, cause a rare hypomyelinating leukodystrophy. *Am J Hum Genet* **89:** 652–655. doi:10.1016/j.ajhg.2011.10.006

Thiffault I, Wolf NI, Forget D, Guerrero K, Tran LT, Choquet K, Lavallée-Adam M, Poitras C, Brais B, Yoon G, et al. 2015. Recessive mutations in POLR1C cause a leukodystrophy by impairing biogenesis of RNA polymerase III. *Nat Commun* **6:** 7623. doi:10.1038/ncomms8623

Tipton PW, Kenney-Jung D, Rush BK, Middlebrooks EH, Nascene D, Singh B, Holtan S, Ayala E, Broderick DF, Lund T, Wszolek ZK. 2021. Treatment of CSF1R-related leukoencephalopathy: breaking new ground. *Mov Disord* **36:** 2901–2909. doi:10.1002/mds.28734

Uggenti C, Lepelley A, Depp M, Badrock AP, Rodero MP, El-Daher MT, Rice GI, Dhir S, Wheeler AP, Dhir A, et al. 2020. cGAS-mediated induction of type I interferon due to inborn errors of histone pre-mRNA processing. *Nat Genet* **52:** 1364–1372. doi:10.1038/s41588-020-00737-3

Valdivielso P, Calandra S, Durán JC, Garuti R, Herrera E, González P. 2004. Coronary heart disease in a patient with cerebrotendinous xanthomatosis. *J Intern Med* **255:** 680–683. doi:10.1111/j.1365-2796.2004.01316.x

Valenti R, Pescini F, Antonini S, Castellini G, Poggesi A, Bianchi S, Inzitari D, Pallanti S, Pantoni L. 2011. Major depression and bipolar disorders in CADASIL: a study using the DSM-IV semi-structured interview. *Acta Neurol Scand* **124:** 390–395. doi:10.1111/j.1600-0404.2011 .01512.x

van der Knaap MS, Bugiani M. 2017. Leukodystrophies: a proposed classification system based on pathological changes and pathogenetic mechanisms. *Acta Neuropathol* **134:** 351–382. doi:10.1007/s00401-017-1739-1

van der Knaap MS, Naidu S, Pouwels PJ, Bonavita S, van Coster R, Lagae L, Sperner J, Surtees R, Schiffmann R, Valk J. 2002. New syndrome characterized by hypomyelination with atrophy of the basal ganglia and cerebellum. *AJNR Am J Neuroradiol* **23:** 1466–1474.

van der Knaap MS, Pronk JC, Scheper GC. 2006. Vanishing white matter disease. *Lancet Neurol* **5:** 413–423. doi:10 .1016/S1474-4422(06)70440-9

van der Knaap MS, Schiffmann R, Mochel F, Wolf NI. 2019. Diagnosis, prognosis, and treatment of leukodystrophies. *Lancet Neurol* **18:** 962–972. doi:10.1016/S1474-4422(19) 30143-7

Vanderver A, Prust M, Kadom N, Demarest S, Crow YJ, Helman G, Orcesi S, La Piana R, Uggetti C, Wang J, et al. 2015a. Early-onset Aicardi–Goutières syndrome: magnetic resonance imaging (MRI) pattern recognition. *J Child Neurol* **30:** 1343–1348. doi:10.1177/0883073814 562252

Vanderver A, Prust M, Tonduti D, Mochel F, Hussey HM, Helman G, Garbern J, Eichler F, Labauge P, Aubourg P, et al. 2015b. Case definition and classification of leukodystrophies and leukoencephalopathies. *Mol Genet Metab* **114:** 494–500. doi:10.1016/j.ymgme.2015.01.006

Vanderver A, Adang L, Gavazzi F, McDonald K, Helman G, Frank DB, Jaffe N, Yum SW, Collins A, Keller SR, et al. 2020a. Janus kinase inhibition in the Aicardi–Goutières syndrome. *N Engl J Med* **383:** 986–989. doi:10.1056/ NEJMc2001362

Vanderver A, Bernard G, Helman G, Sherbini O, Boeck R, Cohn J, Collins A, Demarest S, Dobbins K, Emrick L, et al.

2020b. Randomized clinical trial of first-line genome sequencing in pediatric white matter disorders. *Ann Neurol* **88:** 264–273. doi:10.1002/ana.25757

Van Haren K, Bonkowsky JL, Bernard G, Murphy JL, Pizzino A, Helman G, Suhr D, Waggoner J, Hobson D, Vanderver A, et al. 2015. Consensus statement on preventive and symptomatic care of leukodystrophy patients. *Mol Genet Metab* **114:** 516–526. doi:10.1016/j.ymgme.2014 .12.433

van Heteren JT, Rozenberg F, Aronica E, Troost D, Lebon P, Kuijpers TW. 2008. Astrocytes produce interferon-α and CXCL10, but not IL-6 or CXCL8, in Aicardi–Goutières syndrome. *Glia* **56:** 568–578. doi:10.1002/glia.20639

van Rappard DF, Bugiani M, Boelens JJ, van der Steeg AF, Daams F, de Meij TG, van Doorn MM, van Hasselt PM, Gouma DJ, Verbeke JI, et al. 2016. Gallbladder and the risk of polyps and carcinoma in metachromatic leukodystrophy. *Neurology* **87:** 103–111. doi:10.1212/WNL .0000000000002811

Vaz FM, Jamal Y, Barto R, Gelb MH, DeBarber AE, Wevers RA, Nelen MR, Verrips A, Bootsma AH, Bouva MJ, et al. 2023. Newborn screening for cerebrotendinous xanthomatosis: a retrospective biomarker study using both flow-injection and UPLC-MS/MS analysis in 20,000 newborns. *Clin Chim Acta* **539:** 170–174. doi:10.1016/j.cca.2022.12 .011

Vedolin L. 2011. Inherited white matter disorders of childhood: a magnetic resonance imaging- based pattern recognition approach. *Top Magn Reson Imaging* **22:** 215–222. doi:10.1097/RMR.0b013e318295b416

Verdura E, Hervé D, Scharrer E, Amador Mdel M, Guyant-Maréchal L, Philippi A, Corlobé A, Bergametti F, Gazal S, Prieto-Morin C, et al. 2015. Heterozygous *HTRA1* mutations are associated with autosomal dominant cerebral small vessel disease. *Brain* **138:** 2347–2358. doi:10.1093/ brain/awv155

Verdura E, Rodríguez-Palmero A, Vélez-Santamaria V, Planas-Serra L, de la Calle I, Raspall- Chaure M, Roubertie A, Benkirane M, Saettini F, Pavinato L, et al. 2021. Biallelic *PI4KA* variants cause a novel neurodevelopmental syndrome with hypomyelinating leukodystrophy. *Brain* **144:** 2659–2669. doi:10.1093/brain/awab124

Verge CF, Konrad D, Cohen M, Di Cosmo C, Dumitrescu AM, Marcinkowski T, Hameed S, Hamilton J, Weiss RE, Refetoff S. 2012. Diiodothyropropionic acid (DITPA) in the treatment of MCT8 deficiency. *J Clin Endocrinol Metab* **97:** 4515–4523. doi:10.1210/jc.2012-2556

Vockley J, Andersson HC, Antshel KM, Braverman NE, Burton BK, Frazier DM, Mitchell J, Smith WE, Thompson BH, Berry SA, et al. 2014. Phenylalanine hydroxylase deficiency: diagnosis and management guideline. *Genet Med* **16:** 188–200. doi:10.1038/gim.2013.157

von Jonquieres G, Rae CD, Housley GD. 2021. Emerging concepts in vector development for glial gene therapy: implications for leukodystrophies. *Front Cell Neurosci* **15:** 661857. doi:10.3389/fncel.2021.661857

Waisbren SE, Noel K, Fahrbach K, Cella C, Frame D, Dorenbaum A, Levy H. 2007. Phenylalanine blood levels and clinical outcomes in phenylketonuria: a systematic literature review and meta-analysis. *Mol Genet Metab* **92:** 63–70. doi:10.1016/j.ymgme.2007.05.006

White RJ. 2011. Transcription by RNA polymerase III: more complex than we thought. *Nat Rev Genet* **12**: 459–463. doi:10.1038/nrg3001

Witkamp D, Oudejans E, Hu-A-Ng GV, Hoogterp L, Krzywańska AM, Žnidaršič M, Marinus K, de Veij Mestdagh CF, Bartelink I, Bugiani M, et al. 2022. Guanabenz ameliorates disease in vanishing white matter mice in contrast to sephin1. *Ann Clin Transl Neurol* **9**: 1147–1162. doi:10.1002/acn3.51611

Wolf NI, van Spaendonk RML, Hobson GM, Kamholz J. 1993. PLP1 disorders. In *GeneReviews* (ed. Adam MP, et al.). University of Washington, Seattle.

Wolf NI, Sistermans EA, Cundall M, Hobson GM, Davis-Williams AP, Palmer R, Stubbs P, Davies S, Endziniene M, Wu Y, et al. 2005. Three or more copies of the proteolipid protein gene PLP1 cause severe Pelizaeus-Merzbacher disease. *Brain* **128**: 743–751. doi:10.1093/brain/awh409

Wolf NI, Salomons GS, Rodenburg RJ, Pouwels PJ, Schieving JH, Derks TG, Fock JM, Rump P, van Beek DM, van der Knaap MS, et al. 2014. Mutations in *RARS* cause hypomyelination. *Ann Neurol* **76**: 134–139. doi:10.1002/ana.24167

Yahalom G, Tsabari R, Molshatzki N, Ephraty L, Cohen H, Hassin-Baer S. 2013. Neurological outcome in cerebrotendinous xanthomatosis treated with chenodeoxycholic acid: early versus late diagnosis. *Clin Neuropharmacol* **36**: 78–83. doi:10.1097/WNF.0b013e318288076a

Yang E, Prabhu SP. 2014. Imaging manifestations of the leukodystrophies, inherited disorders of white matter. *Radiol Clin North Am* **52**: 279–319. doi:10.1016/j.rcl.2013.11.008

Yoneda Y, Haginoya K, Arai H, Yamaoka S, Tsurusaki Y, Doi H, Miyake N, Yokochi K, Osaka H, Kato M, et al. 2012. De novo and inherited mutations in COL4A2, encoding the type IV collagen α2 chain cause porencephaly. *Am J Hum Genet* **90**: 86–90. doi:10.1016/j.ajhg.2011.11.016

Yoon IC, Bascou NA, Poe MD, Szabolcs P, Escolar ML. 2021. Long-term neurodevelopmental outcomes of hematopoietic stem cell transplantation for late-infantile Krabbe disease. *Blood* **137**: 1719–1730. doi:10.1182/blood.202000 05477

Yoshida T, Sasaki M, Yoshida M, Namekawa M, Okamoto Y, Tsujino S, Sasayama H, Mizuta I, Nakagawa M; Alexander Disease Study Group in Japan. 2011. Nationwide survey of Alexander disease in Japan and proposed new guidelines for diagnosis. *J Neurol* **258**: 1998–2008. doi:10.1007/s00415-011-6056-3

Zhang P, Zhao J, Peng XM, Qian YY, Zhao XM, Zhou WH, Wang JS, Wu BB, Wang HJ. 2021. Cholestasis as a dominating symptom of patients with CYP27A1 mutations: an analysis of 17 Chinese infants. *J Clin Lipidol* **15**: 116–123. doi:10.1016/j.jacl.2020.12.004

Zurflüh MR, Fiori L, Fiege B, Ozen I, Demirkol M, Gärtner KH, Thöny B, Giovannini M, Blau N. 2006. Pharmacokinetics of orally administered tetrahydrobiopterin in patients with phenylalanine hydroxylase deficiency. *J Inherit Metab Dis* **29**: 725–731. doi:10.1007/s10545-006-0425-6

Peripheral Nervous System (PNS) Myelin Diseases

Steven S. Scherer[1] and John Svaren[2]

[1]Department of Neurology, The Perelman School of Medicine at the University of Pennsylvania, Philadelphia, Pennsylvania 19104, USA

[2]Department of Comparative Biosciences, Waisman Center, University of Wisconsin-Madison, Madison, Wisconsin 53705, USA

Correspondence: sscherer@pennmedicine.upenn.edu

This is a review of inherited and acquired causes of human demyelinating neuropathies and a subset of disorders that affect axon–Schwann cell interactions. Nearly all inherited demyelinating neuropathies are caused by mutations in genes that are expressed by myelinating Schwann cells, affecting diverse functions in a cell-autonomous manner. The most common acquired demyelinating neuropathies are Guillain–Barré syndrome and chronic, inflammatory demyelinating polyneuropathy, both of which are immune-mediated. An additional group of inherited and acquired disorders affect axon–Schwann cell interactions in the nodal region. Overall, these disorders affect the formation of myelin and its maintenance, with superimposed axonal loss that is clinically important.

Peripheral neuropathy, or just neuropathy, are terms for diseases that damage axons in the peripheral nervous system (PNS). The clinical manifestations of a given kind of neuropathy depend on the kinds of axons that are affected —sensory, motor, and/or autonomic. Demyelinating neuropathies are the focus of this review; these are disorders in which myelinating Schwann cells are the chief target of the disease process. This feature usually distinguishes them from "axonal" neuropathies, in which axons are primarily affected. Axonal loss is a key feature of demyelinating neuropathies, and accounts for much of the clinical disability.

THE (CONFUSING) NOMENCLATURE OF INHERITED NEUROPATHIES

Charcot–Marie–Tooth disease (CMT) is the most frequently used name for inherited neuropathies that are not part of a larger syndrome (Shy et al. 2005). With an estimated prevalence of 1 in 2500 persons, CMT is one of the commonest neurogenetic disorders. "Classical" CMT has an insidious onset in the first or second decade and progresses slowly. Motor and sensory nerve function are affected in a length-dependent manner, with distal muscle weakness and atrophy, impaired sensation (particularly "large fiber" modalities), and diminished or absent deep tendon reflexes. Most dominantly inherited, demyelinating forms are called CMT1 (Table 1) and are characterized by nerve conduction velocities (NCVs) <38 m/sec (50 m/sec is the lower limit of normal), which are typically assessed in upper limb nerves and segmental demyelination and remyelination with onion bulb formations (supernumerary Schwann cells surrounding dysmyelinated axons) in nerve biopsies. The term CMT4 was originally restricted to recessive, demyelinating forms of CMT, but in recent years, some authors used this term for both demyelinating and axonal forms of recessive CMT.

The name Déjérine–Sottas syndrome (DSS) was historically applied to children who have a severe demyelinating neuropathy (Baets et al. 2011). They have delayed motor development before 3 yr of age, sometimes extending to infancy, with more severe clinical manifestations than seen in CMT1. The motor responses have reduced amplitudes and very slow (<10 m/sec) NCVs. Nerves are often enlarged, and biopsies reveal a complete absence of axons with normal (thick) myelin sheaths, and sometimes prominent onion bulbs.

Congenital hypomyelinating neuropathy (CHN) is a term applied to infants who have hypotonic weakness at birth caused by a severe neuropathy (Baets et al. 2011). NCVs are severely reduced (<5 m/sec), and biopsies show similar features as described for DSS. The most severe forms of CHN are associated with arthrogryposis at birth and include lethal congenital contracture syndrome (LCCS) and arthrogryposis multiplex congenita, neurogenic, with myelin defects (AMCNMY), both of which are characterized by multiple joint contractures resulting from reduced fetal movements.

DEMYELINATING NEUROPATHIES ARE SCHWANN CELL–AUTONOMOUS

The genetic cause can now be found for most cases of inherited demyelinating neuropathy (Fridman et al. 2015). Nearly all mutations that cause inherited demyelinating neuropathies affect genes that are expressed by myelinating Schwann cells. Transplanting nerve segments of *Trembler* mice (which have a dominantly inherited demyelinating neuropathy caused by a missense *Pmp22* mutation) into normal mice provided the initial experimental proof of this concept (Aguayo et al. 1977), which has been subsequently shown by conditionally expressing mutant alleles in myelinating Schwann cells in rodent models. Thus, most therapies for these disorders will need to target myelinating Schwann cells, which are separated from the circulation by the perineurium and the blood–nerve barrier.

GENETIC DISRUPTIONS OF THE MYELIN SHEATH

Myelin is a multilamellar spiral of specialized cell membrane that ensheaths axons larger than 1 micron in diameter in the PNS (see Salzer et al. 2023). By reducing the capacitance, myelin reduces current flow across the internodal axonal membrane, thereby facilitating saltatory conduction at nodes (see Rasband and Peles 2023). Figure 1A shows two internodes; one has been unrolled to reveal its trapezoidal shape. The myelin sheath itself can be divided into two domains —compact and noncompact myelin—each containing a nonoverlapping set of proteins (Fig. 1B). Compact myelin forms the bulk of the myelin sheath. It is largely composed of lipids, mainly cholesterol and sphingolipids (including galactocerebroside and sulfatide), the intrinsic membrane proteins, Myelin protein zero (MPZ or P_0) and peripheral myelin protein 22 kDa (PMP22), and the cytosolic proteins myelin basic protein (MBP) and myelin protein 2 (PMP2). P_0 is the most abundant myelin protein component and is a homophilic cell adhesion molecule that forms tetramers that interact in *cis* and in *trans* to stabilize the apposed cell membranes (Shapiro et al. 1996). The function of PMP22 has been partially elucidated—it is a member of a large family of tetraspan proteins and has been associated with specific, ordered phase membrane domains/lipid rafts (Lee et al. 2014; Marinko et al. 2020b). Both too much and too little PMP22 affects the stability of compact myelin (Vallat et al. 1996). Noncompact myelin is found

Table 1. Inherited demyelinating neuropathies

Gene (OMIM)	Disease (OMIM)	Clinical
Components of Schwann cell myelin sheath		
PMP22 (601097) dominant, deletion	**HNPP** (162500)	Focal neuropathies, sometimes severe, transient palsies
Dominant, duplication	**CMT1A** (118220)	CMT1 phenotypes, with range of onset and severity
Dominant, usually missense mutations	**CMT1E** (118300)	Phenotypes range from CHN and DSS to typical CMT1, hearing loss
Dominant or recessive mutations	**DSS** (145900)	Severe, demyelinating neuropathy with an onset in early childhood
MPZ (159440) dominant mutations	**CMT1B** (118200)	Phenotypes range from severe CMT to typical CMT1
Dominant or recessive mutations	**CHN2** (618184) **DSS** (145900)	Severe, demyelinating neuropathy with an onset in infancy (CHN2) or early childhood (DSS)
Dominant, usually missense mutations	**CMT2I** (607677) **CMT2J** (607736)	Painful, adult-onset axonal neuropathy with (CMT2J) or without (CMT2I) unreactive pupils and hearing loss
PMP2 (170715) dominant mutations	**CMT1G** (618279)	CMT1 phenotype
GJB1 (304040) X-linked mutations	CMTX1 (302800)	Males more affected than females, intermediate conduction slowing; sometimes CNS manifestations
CADM3 (609743) dominant mutations	**CMT2FF** (619519)	A recurrent p.Y172C mutation, upper limb predominance, intermediate conduction slowing
MAG (159460) recessive mutations	SPG75 (616680)	Infantile-onset Pelizaeus–Merzbacher disease-like phenotype that slowly evolves into complicated HSP/encephalopathy, dysarthria, optic atrophy
Nodal region		
DCNTNAP1 (602346) recessive mutations	CHN3 (618186) LCCS7 (616286)	Polyhydramnios, distal arthrogryposis, hypotonia, respiratory distress, facial diplegia, abnormally myelinated axons
GLDN (608603) recessive mutations	LCSS11 (617194)	Arthrogryposis, widened nodes of Ranvier
LGI4 (608303) recessive mutations	AMCNMY (617468)	Arthrogryposis, severe lack of myelinated axons
NFASC (609145) recessive mutations	NEDCPMD (618356)	Variable severity: from hypotonia with developmental delay to infantile-onset progressive ataxia and neuropathy
NRCAM (601581)	NEDNMS (619833)	Severe neurodevelopmental disorder, neuropathy, skeletal abnormalities
Receptors, signal transduction, and endosomal trafficking		
ERBB3 (190151) recessive mutations	LCCS2 (607598)	Hydramnios, fetal akinesia, limb contractures, distended urinary bladder, myopia, and vitreoretinopathy
Recessive mutations	VSCN1 (243180)	Intestinal ganglionosis or hypoganglionosis, intestinal pseudoobstruction, neuropathy, and arthrogryposis
ERBB2 (164870) recessive mutations	VSCN2 (619465)	Intestinal dysmotility, hypotonia, mild developmental delay, sensorineural hearing loss, and clubfeet
ADGRG6 (612243) recessive mutations	LCSS9 (616503)	Arthrogryposis, atrophy, lack of myelinated axons
ADCY6 (600294) recessive mutations	LCCS8 (616287)	Distal arthrogryposis, hypotonia, respiratory distress, facial diplegia, no motor responses, no myelinated axons

Continued

Table 1. *Continued*

Gene (OMIM)	Disease (OMIM)	Clinical
PRX (605725) recessive mutations	CMT4F (614895)	DSS (145900) or severe CMT1 phenotypes
MTMR2 (603557) recessive mutations	CMT4B1 (601382)	Severe CMT1 or DSS; focally folded myelin sheaths
SBF2 (607697) recessive mutations	CMT4B2 (604563)	Severe CMT1 or DSS, glaucoma; focally folded myelin sheaths
FIG4 (609390) recessive mutations	CMT4J (611228)	Most cases caused by Ile41Thr allele in *trans* with other alleles; variable onset and severity, with non-length-dependent weakness, can rapidly progress
SH3TC2 (608206) recessive mutations	CMT4C (601596)	CMT phenotype, scoliosis, hearing loss
NDRG1 (605262) recessive mutations	CMT4D (601455)	Severe CMT, hearing loss; founder mutation in Roma
LITAF (603795) dominant mutations	**CMT1C** (601098)	CMT1 phenotype
Transcription factors		
EGR2 (129010) dominant mutations	**CMT1D** (607678)	Phenotypes range from DSS (145900) to CMT1
Dominant and recessive mutations	**CHN1** (605253)	Severe, demyelinating neuropathy, infantile onset
SOX10 (602229) dominant mutations	**Waardenburg syndrome 2E** (611584)	Hypopigmentation, deafness, hypogonadotropic hypogonadism, anosmia, agenesis of the olfactory bulbs
Dominant mutations	**PCWH** (609136)	CNS and PNS dysmyelination, Hirschsprung disease
Metabolic or mitochondrial		
ARSA (607574) recessive mutations	Metachromatic leukodystrophy (250100)	Optic atrophy, mental retardation, hypotonia; possibly treatable with bone marrow transplant
GALC (606890) recessive mutations	Globoid cell leukodystrophy, Krabbe disease (245200)	Spasticity, optic atrophy, mental retardation; possibly treatable with bone marrow transplant
ABHD12 (613599) recessive mutations	PHARC (612674)	Polyneuropathy, hearing loss, ataxia, retinitis pigmentosa, and cataract
SURF1 (185620) recessive mutations	Leigh syndrome variant (256000)	Leigh syndrome variant presenting with a demyelinating neuropathy
TYMP (131222) recessive mutations	MTDPS1 (603041)	MNGIE phenotype
PHYH (602026) recessive mutations	Refsum disease (266500)	Deafness, ataxia, retinitis pigmentosa, ichthyosis, heart failure; treatable with dietary restriction
Other		
FBLN5 (604580) dominant mutations	**CMT1H** (619764)	The p.R373C mutation has variable severity and age of onset, intermediate slowing
POLR3B (614366) dominant mutations	**CMT1I** (619742)	Intellectual disabilities, spasticity, ataxia; demyelinating neuropathy not well documented
ITPR3 (147267) dominant mutations	**CMT1J** (620111)	Highly variable severity neuropathy
FGD4 (11104) recessive mutations	CMT4H (609311)	CMT1 phenotype, often early onset, focally folded myelin sheaths
SLC12A6 (604878) recessive mutations	ACCPN (218000)	Seizures, malformed corpus callosum, mental retardation

Continued

Cite this article as *Cold Spring Harb Perspect Biol* doi: 10.1101/cshperspect.a041376

Table 1. *Continued*

Gene (OMIM)	Disease (OMIM)	Clinical
FAM126A (610531) recessive mutations	HLD5 (610532)	Congenital cataracts, abnormal MRI signal in CNS white matter
CTDP1 (604927) recessive mutations	CCFDN (604168)	Vlax Roma, congenital cataracts, microcornea, delayed psychomotor development, skeletal anomalies, hypogonadism, facial dysmorphism

These are grouped by their cell biology, and largely follow the classification in OMIM (www.ncbi.nlm.nih.gov/Omim). The genes are named by the HUGO gene nomenclature (www.genenames.org). Dominant mutations are in bold (see also Rossor et al. 2017).

(ACCPN) Agenesis of the corpus callosum with peripheral neuropathy, (AMCNMY) arthrogryposis multiplex congenita, neurogenic, with myelin defect, (CCFDN) congenital cataracts, facial dysmorphism, and neuropathy, (CHN) congenital hypomyelinating neuropathy, (DSS) Déjérine–Sottas syndrome, (HD) hypomyelinating leukodystrophy, (HNPP) hereditary neuropathy with liability to pressure palsies, (LCCS) lethal congenital contracture syndrome, (MNGIE) mitochondrial neurogastrointestinal encephalopathy syndrome, (MTDPS1) mitochondrial DNA depletion syndrome 1, (NEDCPMD) neurodevelopmental disorder with hyptonia, neuropathy, and deafness, (NEDNMS) neurodevelopmental disorder with neuromuscular and skeletal abnormalities, (PCWH) peripheral demyelinating syndrome, central dysmyelinating leukodystrophy, Waardenburg syndrome, and Hirschprung disease, (PHARC) polyneuropathy, hearing loss, ataxia, retinitis pigmentosa, and cataract, (SPG) spastic paraplegia, (VSCN) visceral neuropathy, familial.

in the paranodes and incisures and contains the gap junction protein connexin32 (Cx32/GJB1), myelin-associated glycoprotein (MAG), two claudins (CLDN2, CLDN5, tight junction–forming molecules), and E-cadherin (CDH1, part of adherens junctions).

PMP22 DUPLICATIONS CAUSE CMT1A

Deletion or duplication of a 1.5 megabase segment of chromosome 17, containing the *PMP22* gene, causes hereditary liability to pressure palsies (HNPP) or CMT1A, respectively, establishing a singular genetic mechanism as the cause (Lupski and Garcia 2001). Two homologous DNA sequences that flank the *PMP22* gene are the molecular basis for its frequent deletion/duplication; their high degree of homology promotes unequal crossing over during meiosis. Most CMT1A patients have three copies of *PMP22*. Rare patients have four copies of PMP22, resulting in a more severe neuropathy. Conversely, patients harboring a CMT1A duplication on one chromosome and an HNPP-associated deletion is present on the other may be unaffected (Hirt et al. 2015), indicating that gene dosage rather than the duplication per se is the cause of the disease.

CMT1A is the most common cause of CMT (Fridman et al. 2015). Most patients are affected by the end of the second decade (Baets et al. 2011). There is considerable variability in the degree of neurological deficits within families, and in a genome-wide association study, polymorphisms of the *SIPA1L2* locus were found to modify the severity of distal leg weakness (Tao et al. 2019). Motor NCVs are slowed, averaging around 20 m/sec, even before the clinical onset of disease, and do not change significantly even over two decades, whereas the motor amplitudes and the number of motor units decrease slowly, and this correlates with clinical disability. Thus, axonal loss and not reduced conduction velocity per se causes weakness, a general principle for all inherited demyelinating neuropathies.

The amount of PMP22 protein in compact myelin is increased in CMT1A (Vallat et al. 1996; Katona et al. 2009), although overexpression also affects trafficking dynamics (Marinko et al. 2020a). Overexpressing PMP22 in myelinating Schwann cells of rodents leads to dose-dependent alterations in myelination, with nearly complete failure to myelinate in the highest expressing lines (Fledrich et al. 2012). Decreasing PMP22 levels in myelinating Schwann cells is an obvious target for treating CMT1A (Pantera et al. 2020), and antisense oligonucleotides or RNA interference against *Pmp22* mRNA showed efficacy in rodent models of CMT1A (Zhao et al. 2018; Boutary et al. 2021; Stavrou et al. 2022). Other pharmacological approaches to reduce

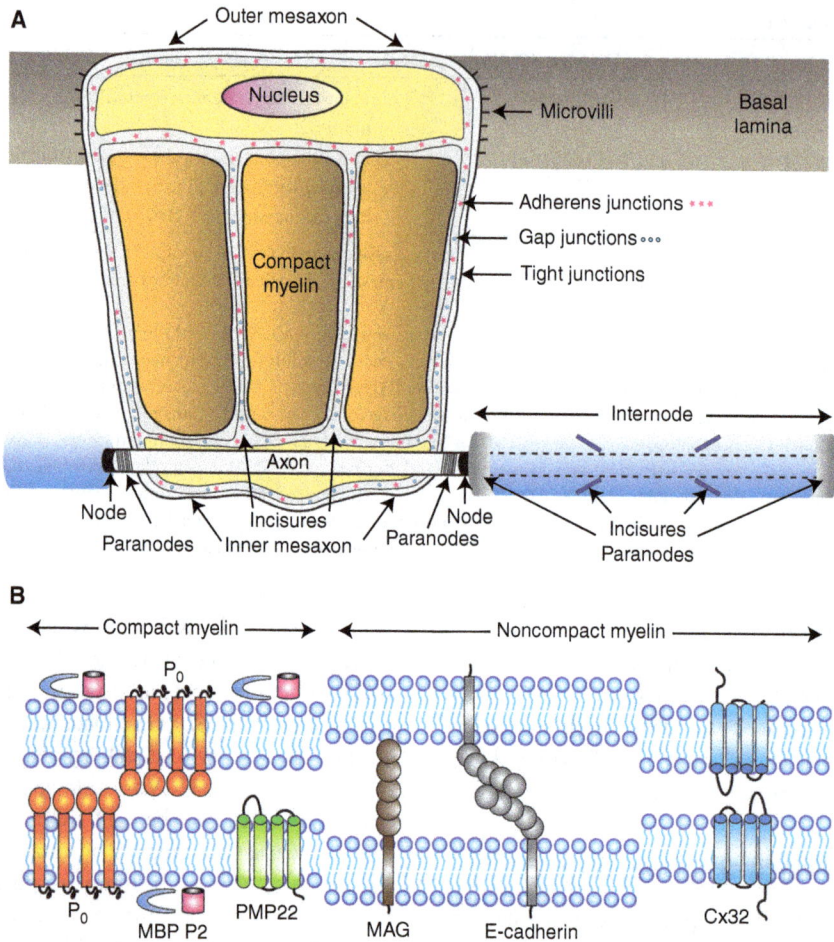

Figure 1. The architecture of the myelinated axon in the peripheral nervous system (PNS). (*A*) One myelinating Schwann cell has been "unrolled" to reveal the regions apposing the basal lamina (abaxonal surface) and the axon (adaxonal surface). Note that regions of noncompact myelin (paranodes and incisures) contain tight junctions, gap junctions, and adherens junctions. In *B*, note that P_0, PMP22, myelin basic protein (MBP), and P2 are found in compact myelin, whereas Cx32, myelin-associated glycoprotein (MAG), and E-cadherin are localized in noncompact myelin.

PMP22 have also improved the neuropathy in rodent models of CMT1A (Sereda et al. 2003; Chumakov et al. 2014).

PMP22 DELETIONS CAUSE HNPP

Given the shared mechanism of duplication and deletion, HNPP should be at least as common as CMT1A, but is much less commonly reported (Fridman et al. 2015), presumably because most patients do not seek medical attention.

Electrophysiological slowing and/or conduction block and at common sites of nerve compression are the hallmarks of HNPP (Li et al. 2002). Biopsies of sensory nerves show focal thickenings (tomaculae) caused by folding of the myelin sheath, as well as segmental demyelination and remyelination. PMP22 protein is reduced in the compact myelin from individuals with HNPP (Vallat et al. 1996). $Pmp22^{+/-}$ mice, an animal model of HNPP, also develop a demyelinating neuropathy with many features of HNPP, including compression-induced neuropathy (Bai et al. 2010), and there is a correlation of neuropathy hallmarks with dose-dependent reduction of PMP22 levels (Pantera et al. 2020). Homozygous deletions of *PMP22* cause DSS in humans, and *Pmp22*-null mice also have a severe demyelinating neuropathy, demonstrating that myelinated axons require PMP22 (Adlkofer et al. 1995).

OTHER *PMP22* MUTATIONS

In addition to *PMP22* duplications and deletions, more than 100 different heterozygous *PMP22* mutations have been described. The ones that cause HNPP presumably result in loss of function, some mutations are benign or hypomorphic (e.g., p.T118M), and others cause phenotypes that are more severe than HNPP (often described as DSS or CMT1), and thus have a toxic gain of function. Most dominant PMP22 mutant proteins do not reach the compact myelin; they appear to be retained in the endoplasmic reticulum (ER) and/or an intermediate compartment between the ER and the Golgi (Naef and Suter 1999). The PMP22 mutants that are retained in the ER have a prolonged half-life and sequester chaperones that are required for the proper processing of glycosylated proteins (Dickson et al. 2002). In addition, some PMP22 mutants can trigger a maladaptive ER unfolded protein response. *Trembler* and *Trembler^J* mice are the classic animal models harboring missense *Pmp22* mutations, pG150D and p.L16P, respectively.

MPZ MUTATIONS CAUSE CHN, DSS, CMT1B, OR A CMT2-LIKE PHENOTYPE

Mutations in *MPZ* account for ~5% of all CMT (Fridman et al. 2015). More than 200 *MPZ* mutations have been reported, most of which cause either a demyelinating neuropathy with a variable age of onset and severity (CHN, DSS, or CMT1) and/or an adult-onset, axonal neuropathy (Callegari et al. 2019; Fridman et al. 2023). The reasons for these highly divergent phenotypes are unknown, although there seem to be distinct structural effects (Ptak et al. 2023). Individuals who are heterozygous for loss-of-func-

tion mutations (e.g., p.D104Tframeshift) have a milder demyelinating neuropathy, demonstrating that haploinsufficiency of *MPZ* causes neuropathy. In contrast, people who are heterozygous for most *MPZ* mutations have a more severe phenotype, demonstrating that these mutations cause a toxic gain of function. Some of these *MPZ* mutations cause an unfolded protein response, which has adaptive and maladaptive effects in myelinating Schwann cells (Callegari et al. 2019), and manipulating these effects is a strategy for treating some CMT1B patients (Das et al. 2015).

Some dominant mutations in *MPZ* cause CMT2 phenotype, an "axonal" neuropathy. How mutations in a gene that is expressed by myelinating Schwann cells and not by neurons causes an axonal neuropathy is an important, unsolved problem (Callegari et al. 2019). In these patients, motor NCVs are mildly slowed, and nerve biopsies show axonal loss and clusters of regenerated axons; these are strikingly different from what one sees in demyelinating neuropathies associated with the "demyelinating" *MPZ* mutations. In spite of their late onset, these patients may progress to the point of using a wheelchair.

There are mouse models of loss-of-function *MPZ* mutations ($Mpz^{+/-}$ and $Mpz^{-/-}$) as well as mutations that are associated with demyelinating (p.S63del, p.R98C, p.Q215x) and axonal (p.T124M) phenotypes (Pennuto et al. 2008; Saporta et al. 2012; Shackleford et al. 2022).

DOMINANT *PMP2* MUTATIONS CAUSE CMT1G

CMT1G is a rare form of CMT1, caused by specific, dominant mutations in *PMP2* (Motley et al. 2016). The pathogenic *PMP2* mutations are predicted to affect the pocket the binds fatty acids, and how dominant mutations cause demyelination remains to be determined.

RECESSIVE *MAG* MUTATIONS CAUSE HSP75

Recessive mutations in *MAG* cause HSP75; this is one of the few forms of hereditary spastic paraplegias (HSPs) caused by mutations in genes that are expressed by oligodendrocytes. In both the central nervous system (CNS) and PNS, MAG is localized to the adaxonal (facing the axon) membrane of the myelin sheath (Fig. 2) and interacts with gangliosides on the axonal membrane (Schnaar 2010). *Mag*-null mice have a mild phenotype (Yin et al. 1998).

DOMINANT *CADM3* MUTATIONS CAUSE CMT2FF

A recurrent, de novo p.Y172C mutation in *CADM3* causes a dominant axonal neuropathy (CMT2FF), characterized by childhood onset and upper limb predominance (Rebelo et al. 2021). The p.Y172C mutation results in aberrant disulfide bonds and retention in the ER. CADM3 is a cell-adhesion molecule, and is localized to the axonal membrane, where it may interact in *trans* with CADM4 on the adaxonal Schwann cell membrane (Fig. 2). The p.Y172C mutant has diminished binding to CADM4, but how this leads to neuropathy is unknown. Deletion of *Cadm4* in mice leads to a neuropathy (Golan et al. 2013).

GJB1 MUTATIONS CAUSE CMTX1

CMTX1 is caused by mutations in *GJB1*, the gene that encodes the gap junction protein connexin 32 (Cx32) (Scherer and Kleopa 2012). CMTX1 is the

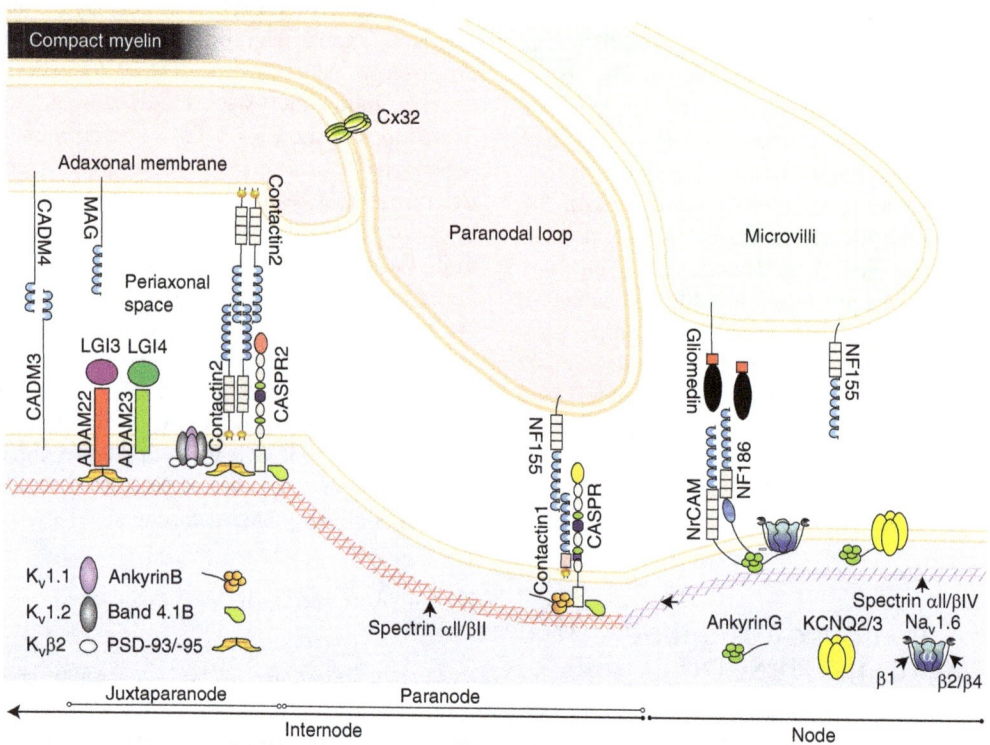

Figure 2. Molecular organization of the nodal region. This schematic drawing depicts nodes, paranodes, and juxtaparanodes, emphasizing molecules that have been implicated in human diseases. At nodes, gliomedin (from Schwann cells) binds to Nr-CAM, and NF186 binds to ankyrinG, which binds to the spectrin cytoskeleton (comprised of αII/βIV spectrin). The α-subunits of Na$_v$ channels, as well as KCNQ2 and KCNQ3 channels also bind to ankyrinG. At paranodes, contactin-associated protein (CASPR) and contactin1 heterodimers interact in *trans* with NF155. At juxtaparanodes, contactin2 dimers interact homophilically in *trans*, and form a complex with CASPR2, LGI3, LGI4, ADAM22, ADAM23, tetramers of K$_v$1.1/K$_v$1.2 channels, and PSD-93/-95. Protein 4.1B links the cytoplasmic tail of CASPR and CASPR2 to the spectrin cytoskeleton. In internodes, CADM3 is depicted to interact with CADM4. Homotypic gap junctions comprised Cx32 link the paranodal membranes of the myelin sheath.

Cite this article as *Cold Spring Harb Perspect Biol* doi: 10.1101/cshperspect.a041376

second-most common form of CMT, accounting for ~10% of patients. Because *GJB1* is an X-linked gene, there is a sexual dimorphism in the clinical presentation. In males, the clinical onset is between 5 and 20 yr, whereas women are variably affected presumably because of the inactivation of either allele of the affected X chromosome. Motor NCVs in the arms show "intermediate" slowing (25–40 m/sec); motor amplitudes become reduced as the disease progresses. Nerve biopsies show less demyelination and remyelination and more axonal degeneration and regeneration than in other demyelinating neuropathies.

More than 700 different *GJB1* variants have been identified, most of which probably cause CMTX1, but the lack of clinical information makes this uncertain for many mutations (Record et al. 2023). About 10% of *GJB1* variants are likely benign polymorphisms. Mutations that cause neuropathy, including deletions of the entire *GJB1* gene, appear to cause a similar phenotype, indicating that loss of function is the common denominator. Expressing mutant Cx32 proteins in heterologous cells reveals that most CMTX1-associated mutants do not form functional gap junctions, indicating the importance of each amino acid in the function of this highly conserved protein. It is suspected that the loss of gap junctions formed by Cx32 in the myelin sheath is the cause of the neuropathy. There are several, genetically authentic animal models of CMTX1, including p.R15W, p.T55I, p.R75W, p.R220x, and deletion of the *Gjb1* gene, all of which have similar phenotypes (Abrams et al. 2023).

DISRUPTION OF COMPONENTS OF THE NODAL REGION

Nodes of Ranvier have molecular specializations that enable saltatory conduction—the key function of myelinated axons (see Rasband and Peles 2023). Mutations in many genes that encode components of nodes result in neuropathy or related issues that are not necessarily caused by altered myelination (Table 1). Some nodal components are unique to PNS nodes; others are also found in the CNS, and this could account for at least some of the CNS manifestations of their

mutant alleles. Loss-of-function mutations to nearly all of the genes discussed in this section have been generated in mice.

The molecular architecture of nodes is shown in Figure 2. Gliomedin, a cell-adhesion molecule that is secreted by Schwann cells into the extracellular matrix, binds to two cell-adhesion molecules on the axolemma, neurofascin-186 (NF186), and NrCAM. NF186, in turn, recruits the nodal isoform of ankyrinG, which is linked to the βIV-spectrin cytoskeleton (Rasband and Peles 2021). Recessive *GLDN* mutations cause a severe, dysmyelinating neuropathy that may lead to fetal death. Recessive *NRCAM* mutations cause a syndrome of developmental delay, hypotonia, neuropathy, and spasticity. Recessive mutations in *ANK3* (encodes ankyrinG) cause a severe CNS phenotype but neuropathy was not reported, perhaps because ankyrinR can substitute for ankyrinG (Ho et al. 2014). Recessive mutations in *SPNB4* (encodes spectrin β4) cause a severe neurodevelopmental disorder and neuropathy. Conditional deletion of *Sptan1* (encodes spectrin α2) in mouse sensory neurons results in axonal degeneration (Huang et al. 2017); it is possible that neuropathy has been overlooked in children with de novo dominant mutations in *SPTAN1* owing to their severe neurodevelopmental disorder.

Most if not all nodes are highly enriched in $Na_v1.6$ (Krzemien et al. 2000), a voltage-gated Na^+ channel, as well as KCNQ2 and KNCQ3, voltage-gated K^+ channels (Pan et al. 2006), all of which bind to ankyrinG (Fig. 2). Recessive mutations in the gene encoding $Na_v1.6$ (*SCN8A*) have not been reported in humans, but recessive *Scn8a* mutations in mice lead to paralysis and death, indicating that (myelinated) motor axons require this Na_v channel for conduction (Burgess et al. 1995). KCNQ2 and KCNQ3 channels likely contribute to repolarization, and one dominant *KCNQ2* mutation causes excessive repetitive firing of myelinated motor axons but does not cause neuropathy (Dedek et al. 2001). Dominant mutations in *KCNA1* (encodes $K_v1.1$) and recessive mutations in *LGI4* and *LGI3* also cause excessive repetitive firing of myelinated motor axons, but not neuropathy (Browne et al. 1994; Marafi et al. 2022). LGI3 and

LGI4 bind to ADAM22 and ADAM23, and form a complex in the juxtaparanodal region along with $K_v1.1$ and $K_v1.2$ (Fig. 2). Mutations in the genes encoding $Na_v1.7$ (*SCN9A*), $Na_v1.8$ (*SCN10A*), and $Na_v1.9$ (*SCN11A*) cause sensory neuropathies and/or altered pain sensation, but these Na^+ channels are mostly expressed by nociceptive sensory neurons, the majority of which have unmyelinated axons (Chiu et al. 2014; Usoskin et al. 2015).

At paranodes, a heterodimer of contactin1 and contactin-associated protein (CASPR) on the axolemma interacts in *trans* with NF155 on myelinating Schwann cells, forming the septate-like junctions (see Rasband and Peles 2023). Recessive mutations in *CNTNAP1* (the gene that encodes CASPR) cause a severe, dysmyelinating neuropathy associated with CNS findings. Recessive mutations of *NFASC* (the gene that encodes NF155 and NF186) cause a severe, dysmyelinating neuropathy and accompanying CNS issues. Recessive mutations in *CNTN1* have been reported in one family with congenital myopathy but not neuropathy. The mutations in *CASPR2*, *NFASC*, and *CNTN1* result in the loss of septate-like junctions; this permits the $K_v1.1$ and $K_v1.2$ channels to occupy the paranodal region and results in the shunting of current during the propagation of action potentials.

RECEPTORS, SIGNALING, AND ENDOSOMAL TRAFFICKING

Receptors mediate the axon–Schwann cell interactions that shape the development of myelinating Schwann cells (see Rasband and Peles 2023). The most well characterized of these is a neuregulin receptor formed by a heterodimer of ERBB2 and ERBB3, which governs many aspects of Schwann cell development. Recessive *ERBB3* mutations have been reported to cause a severe dysmyelinating neuropathy in one family (Narkis et al. 2007), but subsequent reports describe a complex phenotype that includes a progressive axonal neuropathy (Le et al. 2021). In mice, deletion of *Nrg1*, *Erbb2*, or *Erbb3* results in the arrested development of Schwann cells, whereas deleting *Nrg1* in adult PNS neurons or *Erbb2* in Schwann cells had no discernable effects on the

maintenance of myelinated axons (Atanasoski et al. 2006; Fricker et al. 2013).

Recessive *ADGRG6* mutations cause LCCS9 (Ravenscroft et al. 2015); *ADGRG6* encodes GPR126/ADGRG6, which is a member of the adhesion G protein–coupled receptor family. GPR126 binds components of the extracellular matrix and signals though cAMP. Recessive *Adgrg6* mutations cause arrested development of Schwann cells in a cell-autonomous manner, indicating that this is a highly conserved pathway (Monk et al. 2009; Mogha et al. 2013). *ADCY6* encodes adenylyl cyclase 6, and recessive *ADCY6* mutations cause LCCS8 (Agolini et al. 2020). Thus, activation of adenylyl cyclase 6 by GPR126 may be essential for the normal development of myelinating Schwann cells.

Myelinating Schwann cells express two laminin-2 receptors—dystroglycan and α6β4 integrin (Fig. 3). Dystroglycan binds to dystroglycan-related protein 2 (DRP2) and periaxin (PRX) in regions that directly appose the myelin sheath. Mutations that may cause neuropathy include (1) recessive mutations in *LAMA2*, which encodes a subunit of laminin-2 and cause a mild (and probably not demyelinating) neuropathy and a severe, congenital myopathy, (2) recessive mutations in *PRX* cause CMT4F (Guilbot et al. 2001; Sherman et al. 2001), (3) mutations in *DRP2* may cause neuropathy (Brennan et al. 2015), and (4) recessive mutations in *NDRG1*, which encode a phosphoprotein that is downstream of integrin signaling, cause CMT4D (Heller et al. 2014). In mice, Schwann cell–specific deletion of β4 integrin, dystroglycan, or both, cause a mild neuropathy (Nodari et al. 2008).

Recessive mutations in *MTMR2* and *SBF2* (encodes MTMR13) cause CMT4B1 and CMT4B2, respectively (Previtali et al. 2007). CMT4B3 is caused by recessive mutations in *MTMR5*, which have a complex CNS phenotype and an axonal neuropathy; recessive *MTMR5* mutations likely cause neuropathy through cell-autonomous effects in neurons but not Schwann cells. These genes belong to the family of myotubularin-related dual-specific phosphatases whose substrates are phosphoinositides; MTMR13 lacks phosphatase activity but interacts directly with MTMR2 to form a complex that

Cite this article as *Cold Spring Harb Perspect Biol* doi: 10.1101/cshperspect.a041376

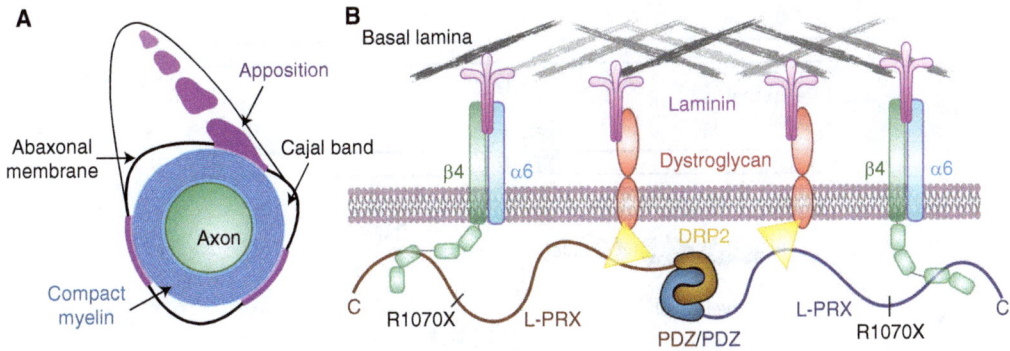

Figure 3. Extracellular matrix receptors of myelinating Schwann cells. (*A*) Cross-section of a myelinated peripheral axon, showing compact myelin, the abaxonal membrane, Cajal bands, and appositions. Although dystrophin-related protein 2 (DRP2) is restricted to the appositions, and PRX is concentrated in them, α6β4 integrin and dystroglycan are localized around the entire circumference. (*B*) A possible architecture for L-PRX-based protein scaffolds at the Schwann cell abaxonal membrane. The position of the truncating R1070X mutation in L-PRX is indicated, and the β4 integrin binding site lies carboxy-terminal to it, although an exact binding site remains to be identified. (Figure reprinted from Raasakka et al. 2019 under the terms of the Creative Commons Attribution License (CC BY).)

dephosphorylates the 3′ phosphate of PI(3)P or PI(3,5)P2. Genetic knockouts in mice demonstrate that myelinating Schwann cells require MTMR2 and MTMR13 to maintain normal myelin sheaths and have an unusual pathological finding—focally folded myelin sheaths—that is also found in nerve biopsies from patients with CMT4B1 and CMT4B2. Reduced dephosphorylation of PI(3)P or PI(3,5)P2 may cause abnormal vesicular trafficking as well as altered AKT, mTORC1, and RHOA signaling in myelinating Schwann cells (Bolino et al. 2016; Sawade et al. 2020). Further, recessive mutations in *FIG4*, ostensibly a PI(3,5)P2 5′ phosphatase, also cause a demyelinating neuropathy (CMT4J), further supporting a special role for PI(3,5)P2 in myelinating Schwann cells. FIG4 does have phosphatase activity like MTMR2/13, but when complexed with VAC14 and PIKFYVE kinase, FIG4 activates PI(3,5)P2 production. Thus, loss of FIG4 or VAC14 function produces less, not more, PI(3,5)P2 in cells—the opposite of what is expected in CMT4B1/B2. Consistent with its opposing effects on PI(3,5)P2 levels, *Fig4* deficiency rescues the loss of MTMR2 function in Schwann cells (Vaccari et al. 2011). In addition to its effects on endosomal trafficking, PI(3,5)P2 activates the release of lysosomal Ca^{2+} through the cation channel TRPML1, which is required to maintain the homeostasis of endosomes and lysosomes in mammalian cells. FIG4-deficient cells have enlarged lysosomes, and a small molecule activator of TRPML1, ML-SA1, rescues the vacuolation phenotype of FIG4-deficient cells (Zou et al. 2015; Edgar et al. 2020).

A key function of phosphoinositides is regulating endosomal trafficking (Posor et al. 2022), and this is affected in several neuropathies (Markworth et al. 2021). In addition to the effects of recessive *MTMR2*, *SBF2*, and *FIG4* mutations (above), this process is also affected by mutations in *SH3TC2* (CMT4C) and *LITAF* (CMT1C), as shown in Figure 4. SH3TC2 is an RAB11 effector, interacting directly with RAB11 and ERBB3, and may regulate Neuregulin-1 (and potentially other receptors) signaling in Schwann cells. *SH3TC2* mutations that cause neuropathy disrupt interactions with RAB11, thereby reducing endosomal recycling and likely ERBB2/ERBB3 signaling (Roberts et al. 2010; Stendel et al. 2010; Gouttenoire et al. 2013). LITAF/SIMPLE interacts with phosphoethanolamine head groups of the surface of intracellular membranes (Ho et al. 2016) and is required for efficient recruitment of ESCRT components to endosomal membranes and for regulating endosomal trafficking and signaling attenuation of ERBB receptors. CMT-linked SIMPLE muta-

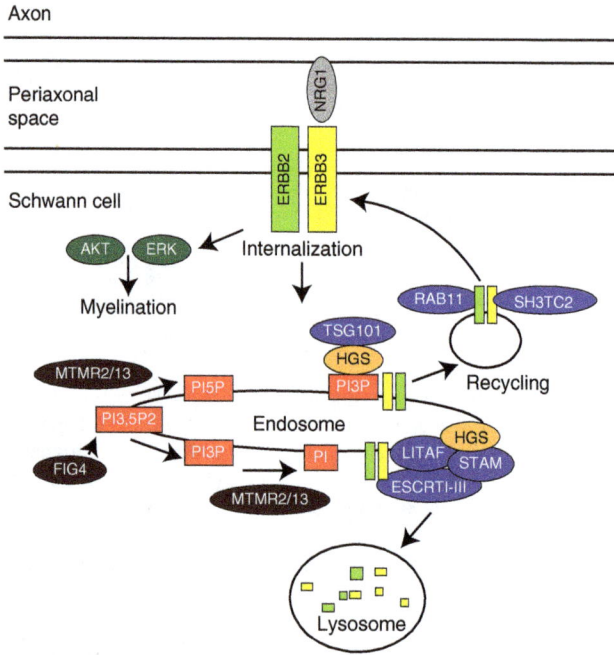

Figure 4. Model of how Charcot–Marie–Tooth (CMT) mutations effect endosomal signaling in Schwann cells. The Schwann cell ERBB2/ERBB3 neuregulin receptor promotes myelination via AKT and ERK signaling. The magnitude of this signaling is modulated by the extent to which internalized receptors are recycled to the plasma membrane or degraded. HGS functions in a complex with TSG101 to recycle ERBB2/ERBB3 receptors to the cell surface via recycling endosomes, or complexes with STAM and ESCRTI-III to promote their degradation in lysosomes. SH3TC2 is an adaptor protein for RAB9, which facilitates endosome recycling; recessive *SH3TC2* mutations reduce ERBB2/ERBB3 recycling to the plasma membrane. Dominant *LITAF* mutations reduce the recruitment of ESCRT components to endosomal membranes and result in more ERBB2/ERBB3 receptor signaling. FIG4 enhances PI3,5P2 formation; MTMR2/13 removes the 3' phosphate from PI3,5P2 and PI3P.

tions impair ERBB trafficking via a loss-of-function, dominant-negative mechanism, resulting in prolonged activation of ERK1/2 signaling (Lee et al. 2012). Similar to its effects on FIG4-deficient cells, ML-SA1 (the small molecule activator of TRPML1) was able to rescue the vacuolated lysosomes of LITAF knockout and CMT1C patient fibroblasts, showing a convergent phenotype (Edgar et al. 2020).

METABOLISM

Recessive mutations in *ARSA* and *GALC* cause metachromatic leukodystrophy and Krabbe disease, with accumulation of sulfatide and galactoceramide, respectively, in myelinating Schwann cells and oligodendrocytes. A demyelinating neuropathy is part of MNGIE (mitochondri-al neurogastrointestinal encephalopathy) syndrome caused by recessive *TYMP* mutations and some *SURF1* mutations (usually associated with Leigh syndrome), indicating that abnormal mitochondrial function can affect myelination, as has been directly demonstrated in mice by deleting the TFAM regulator of genes encoding mitochondrial proteins in Schwann cells (Viader et al. 2011).

TRANSCRIPTION FACTORS AND PROMOTER MUTATIONS

EGR2 (KROX-20) and SOX10 are two transcription factors (see Salzer et al. 2023) that bind to the regulatory elements on many myelin-related genes and thereby coordinately regulate the development of myelinating Schwann cells (Meijer

Cite this article as *Cold Spring Harb Perspect Biol* doi: 10.1101/cshperspect.a041376

and Svaren 2013). Dominant and recessive *EGR2* mutations cause demyelinating forms of CMT (CMT1D and DSS/CHN in some cases), but most of the mutations are dominant mutations within the zinc finger DNA-binding domain (Lupo et al. 2020). Several rodent models of EGR2 loss of function have severe early-onset neuropathies, and EGR2 is required for the high-level activation of many of the Schwann cell proteins mentioned above (Topilko et al. 1994; Le et al. 2005).

SOX10 is an HMG box transcription factor that is required for Schwann cell specification, but also continues to be expressed in adult Schwann cells. SOX10 is also required for development of oligodendrocytes (Weider et al. 2013), and dominant *SOX10* mutations cause a syndromic condition—PCWH (peripheral demyelinating neuropathy, central demyelinating leukodystrophy, Waardenburg–Shah syndrome, and Hirschsprung disease) or partial variants. EGR2 and SOX10 coordinately activate expression of myelin genes such as *MPZ* and *PMP22* (Pantera et al. 2020), and some CMTX1 patients have mutations within the EGR2 and SOX10 binding sites in the *GJB1* promoter (Tomaselli et al. 2017).

IMMUNE-MEDIATED DEMYELINATING NEUROPATHIES

Guillain–Barré syndrome (GBS) is a group of acute, inflammatory neuropathies, and different subtypes are associated with IgG antibodies against gangliosides that are localized in the nodal region (Yuki and Hartung 2012; Uncini 2023). Elevated titers of IgG against GM1 are found in patients with acute motor axonal neuropathy (AMAN) and acute motor and sensory neuropathy (AMSAN). Elevated titers of IgG against GD1b are found in patients with acute sensory neuropathy (ASAN). Elevated IgG titers against GQ1b are strikingly associated with acute, ataxic neuropathies. These gangliosides are localized to the nodal region (McGonigal and Willison 2022), and antibodies against them can disrupt the integrity of the glial and/or axonal membrane, leading to an influx of Ca^{2+} (Cunningham et al. 2023; McGonigal et al. 2023).

Acute, inflammatory demyelinating polyneuropathy (AIDP) is a common form of GBS. Similar to other forms of GBS, there may be an antecedent infection, but unlike other forms of GBS, antibodies against specific gangliosides have not been detected (Yuki and Hartung 2012; Uncini 2023). Autopsies from severely affected patients reveal an activated complement on myelinating Schwann cells that likely results in demyelination (Hafer-Macko et al. 1996). Like other form of GBS, plasma exchange and intravenous gammaglobulin are effective therapies for AIDP.

Chronic inflammatory demyelinating polyneuropathy (CIDP) is a spectrum of autoimmune diseases that result in a demyelinating neuropathy (Eftimov et al. 2020). The pathogenesis of typical cases of CIDP remains obscure, but a minority of cases have been distinguished from CIDP by their association with an IgM monoclonal gammopathy that binds to a carbohydrate epitope (Latov 2014), a lymphoproliferative disorder with a monoclonal lambda light chain and high VEGF levels (Mauermann 2018), or antibodies (usually IgG4) against paranodal proteins contactin1, CASPR, and especially NF155, as well as the nodal protein NF186 and the juxtaparanodal protein LGI4 (Pascual-Goñi et al. 2021; Querol et al. 2023; Uncini 2023). The antibodies are thought to be pathogenic, best demonstrated by the finding that transferring IgG from patients with NF155 antibodies results in disrupted paranodes (Manso et al. 2019). Antibodies against CASPR2 cause acquired neuromyotonia, in which spontaneous activity of distal motor axons (likely from the diminished function of $K_v1.1$ and $K_v1.2$ channels) results in muscle activation (Irani et al. 2010; Lancaster et al. 2011). Elevated titers of IgM against GQ1b and GD1b are found in chronic ataxic neuropathies, and elevated IgM titers against GM1 are found in patients with chronic multifocal motor neuropathy (MMN). Various immune-modulating therapies improve these disorders.

SUMMARY

It is easier to recognize neuropathy in people than in other animals, but it is often challenging to determine the cause. Most of the monogenic causes of inherited, demyelinating neuropathies have been discovered, and the pathogenic mutations typically disrupt key functions of myelinating Schwann cells, including their interactions with axons. These genes have conserved functions in myelinated axons, and can be manipulated in cells and in organisms, leading to the illumination of the molecular mechanisms of demyelination and even therapeutic targets. Clinical trials for some forms of inherited demyelinating neuropathies are being planned, which is a daunting task (Reilly et al. 2023). In spite of the gaps in understanding their pathogenesis, acquired demyelinating neuropathies are largely treatable. With the important exception of AIDP, various gangliosides in the nodal region appear to be the molecular targets of IgG autoantibodies in GBS. Pathogenic antibodies account for a minority of chronic, acquired demyelinating neuropathies, and these target proteins (IgG4) and gangliosides (and IgM) in the nodal region. Understanding the molecular mechanisms of demyelination, and finding better and safer treatments for CIDP remain important, unsolved issues.

ACKNOWLEDGMENTS

We thank Mark Lancaster and Dr. Scott Wilson for help with the figures, and Dr. Alessandra Bolino for her comments on the manuscript. S.S.S. and J.S. have received support from the Inherited Neuropathy Consortium, Rare Disease Clinical Research Consortium funded by the National Institutes of Health, the National Multiple Sclerosis Society, The Muscular Dystrophy Association, and the Charcot–Marie–Tooth Association; S.S.S. is also supported by the Judy Seltzer Levenson Memorial Fund for CMT Research.

REFERENCES

*Reference is also in this subject collection.

Abrams CK, Lancaster E, Li JJ, Dungan G, Gong D, Scherer SS, Freidin M. 2023. Knock-in mouse models for CMTX1 show a loss of function phenotype in the peripheral ner-

vous system. *Exp Neurol* 360: 112477. doi:10.1016/j.expneurol.2022.114277

Adlkofer K, Martini R, Aguzzi A, Zielasek J, Toyka KV, Suter U. 1995. Hypermyelination and demyelinating peripheral neuropathy in Pmp22-deficient mice. *Nat Genet* 11: 274–280. doi:10.1038/ng1195-274

Agolini E, Cherchi C, Bellacchio E, Martinelli D, Cocciadiferro D, Cutrera R, Chiarini Testa MB, Barone C, Bianca S, Novelli A. 2020. Expanding the clinical and molecular spectrum of lethal congenital contracture syndrome 8 associated with biallelic variants of *ADCY6*. *Clin Genet* 97: 649–654. doi:10.1111/cge.13691

Aguayo AJ, Attiwell M, Trecarten J, Perkins CS, Bray CM. 1977. Abnormal myelination in transplanted Trembler mouse Schwann cells. *Nature* 265: 73–75. doi:10.1038/265073a0

Atanasoski S, Scherer SS, Sirkowski EE, Leone D, Garratt A, Birchmeier C, Suter U. 2006. Erbb2 signaling in Schwann cells is largely dispensable for maintenance of myelinated peripheral nerves and proliferation of adult Schwann cells following injury. *J Neurosci* 26: 2124–2131. doi:10.1523/JNEUROSCI.4594-05.2006

Baets J, Deconinck T, De Vriendt E, Zimon M, Yperzeele L, Van Hoorenbeeck K, Peeters K, Spiegel R, Parman Y, Ceulemans B, et al. 2011. Genetic spectrum of hereditary neuropathies with onset in the first year of life. *Brain* 134: 2664–2676. doi:10.1093/brain/awr184

Bai Y, Zhang X, Katona I, Saporta MA, Shy ME, O'Malley HA, Isom LL, Suter U, Li J. 2010. Conduction block in PMP22 deficiency. *J Neurosci* 30: 600–608. doi:10.1523/JNEUROSCI.4264-09.2010

Bolino A, Piguet F, Alberizzi V, Pellegatta M, Rivellini C, Guerrero-Valero M, Noseda R, Brombin C, Nonis A, D'Adamo P, et al. 2016. Niacin-mediated Tace activation ameliorates CMT neuropathies with focal hypermyelination. *EMBO Mol Med* 8: 1438–1454. doi:10.15252/emmm.201606349

Boutary S, Caillaud M, El Madani M, Vallat JM, Loisel-Duwattez J, Rouyer A, Richard L, Gracia C, Urbinati G, Desmaële D, et al. 2021. Squalenoyl siRNA PMP22 nanoparticles are effective in treating mouse models of Charcot–Marie–Tooth disease type 1A. *Commun Biol* 4: 317. doi:10.1038/s42003-021-01839-2

Brennan KM, Bai Y, Pisciotta C, Wang S, Feely SM, Hoegger M, Gutmann L, Moore SA, Gonzalez M, Sherman DL, et al. 2015. Absence of dystrophin related protein-2 disrupts Cajal bands in a patient with Charcot–Marie–Tooth disease. *Neuromuscul Disord* 25: 786–793. doi:10.1016/j.nmd.2015.07.001

Browne DL, Gancher ST, Nutt JG, Brunt ERP, Smith EA, Kramer P, Litt M. 1994. Episodic ataxia/myokymia syndrome is associated with point mutations in the human potassium channel gene, *KCNA1*. *Nat Genet* 8: 136–140. doi:10.1038/ng1094-136

Burgess DL, Kohrman DC, Galt J, Plummer NW, Jones JM, Spear B, Meisler MH. 1995. Mutation of a new sodium channel gene, *Scn8a*, in the mouse mutant "motor endplate disease." *Nat Genet* 10: 461–465. doi:10.1038/ng0895-461

Callegari I, Gemelli C, Geroldi A, Veneri F, Mandich P, D'Antonio M, Pareyson D, Shy ME, Schenone A, Prada V, et al. 2019. Mutation update for myelin protein zero-related neuropathies and the increasing role of variants causing a late-onset phenotype. *J Neurol* **266**: 2629–2645. doi:10.1007/s00415-019-09453-3

Chiu IM, Barrett LB, Williams EK, Strochlic DE, Lee S, Weyer AD, Lou S, Bryman GS, Roberson DP, Ghasemlou N, et al. 2014. Transcriptional profiling at whole population and single cell levels reveals somatosensory neuron molecular diversity. *eLife* **3**: e04660. doi:10.7554/eLife.04660

Chumakov I, Milet A, Cholet N, Primas G, Boucard A, Pereira Y, Graudens E, Mandel J, Laffaire J, Foucquier J, et al. 2014. Polytherapy with a combination of three repurposed drugs (PXT3003) down-regulates Pmp22 over-expression and improves myelination, axonal and functional parameters in models of CMT1A neuropathy. *Orphanet J Rare Dis* **9**: 201. doi:10.1186/s13023-014-0201-x

Cunningham ME, McGonigal R, Barrie JA, Campbell CI, Yao D, Willison HJ. 2023. Axolemmal nanoruptures arising from paranodal membrane injury induce secondary axon degeneration in murine Guillain-Barré syndrome. *J Peripher Nerv Syst* **28**: 17–31. doi:10.1111/jns.12532

Das I, Krzyzosiak A, Schneider K, Wrabetz L, D'Antonio M, Barry N, Sigurdardottir A, Bertolotti A. 2015. Preventing proteostasis diseases by selective inhibition of a phosphatase regulatory subunit. *Science* **348**: 239–242. doi:10.1126/science.aaa4484

Dedek K, Kunath B, Kananura C, Reuner U, Jentsch T, Steinlein OK. 2001. Myokymia and neonatal epilepsy caused by a mutation in the voltage sensor of the KCNQ2 K$^+$ channel. *Proc Natl Acad Sci* **98**: 12272–12277. doi:10.1073/pnas.211431298

Dickson KM, Bergeron JJM, Shames I, Colby J, Nguyen DT, Chevet E, Thomas DY, Snipes GJ. 2002. Association of calnexin with mutant peripheral myelin protein-22 ex vivo: a basis for "gain-of-function" ER diseases. *Proc Natl Acad Sci* **99**: 9852–9857. doi:10.1073/pnas.152621799

Edgar JR, Ho AK, Laurá M, Horvath R, Reilly MM, Luzio JP, Roberts RC. 2020. A dysfunctional endolysosomal pathway common to two sub-types of demyelinating Charcot–Marie–Tooth disease. *Acta Neuropathol Commun* **8**: 165. doi:10.1186/s40478-020-01043-z

Eftimov F, Lucke IM, Querol LA, Rajabally YA, Verhamme C. 2020. Diagnostic challenges in chronic inflammatory demyelinating polyradiculoneuropathy. *Brain* **143**: 3214–3224. doi:10.1093/brain/awaa265

Fledrich R, Stassart RM, Sereda MW. 2012. Murine therapeutic models for Charcot–Marie–Tooth (CMT) disease. *Br Med Bull* **102**: 89–113. doi:10.1093/bmb/lds010

Fricker FR, Antunes-Martins A, Galino J, Paramsothy R, La Russa F, Perkins J, Goldberg R, Brelstaff J, Zhu N, McMahon SB, et al. 2013. Axonal neuregulin 1 is a rate limiting but not essential factor for nerve remyelination. *Brain* **136**: 2279–2297. doi:10.1093/brain/awt148

Fridman V, Bundy B, Reilly MM, Pareyson D, Bacon C, Burns J, Day JW, Feely S, Finkel RS, Grider T, et al. 2015. CMT subtypes and disease burden in patients enrolled in the inherited neuropathies consortium natural history study: a cross-sectional analysis. *J Neurol Neurosurg Psychiat* **86**: 873–878. doi:10.1136/jnnp-2014-308826

Fridman V, Sillau S, Bockhorst J, Smith K, Moroni I, Pagliano E, Pisciotta C, Piscosquito G, Laurá M, Muntoni F, et al. 2023. Disease progression in Charcot–Marie–Tooth disease related to MPZ mutations: a longitudinal study. *Ann Neurol* **93**: 563–576. doi:10.1002/ana.26518

Golan N, Kartvelishvily E, Spiegel I, Salomon D, Sabanay H, Rechav K, Vainshtein A, Frechter S, Maik-Rachline G, Eshed-Eisenbach Y, et al. 2013. Genetic deletion of Cadm4 results in myelin abnormalities resembling Charcot–Marie–Tooth neuropathy. *J Neurosci* **33**: 10950–10961. doi:10.1523/JNEUROSCI.0571-13.2013

Gouttenoire EA, Lupo V, Calpena E, Bartesaghi L, Schüpfer F, Médard J-J, Maurer F, Beckmann JS, Senderek J, Palau F, et al. 2013. Sh3tc2 deficiency affects neuregulin-1/ErbB signaling. *Glia* **61**: 1041–1051. doi:10.1002/glia.22493

Guilbot A, Williams A, Ravisé N, Verny C, Brice A, Sherman DL, Brophy PJ, LeGuern E, Delague V, Bareil C, et al. 2001. A mutation in periaxin is responsible for CMT4F, an autosomal recessive form of Charcot–Marie–Tooth disease. *Hum Mol Genet* **10**: 415–421. doi:10.1093/hmg/10.4.415

Hafer-Macko CE, Sheikh KA, Li CY, Ho TW, Cornblath DR, McKhann GM, Asbury AK, Griffin JW. 1996. Immune attack on the Schwann cell surface in acute inflammatory demyelinating polyneuropathy. *Ann Neurol* **39**: 627–637. doi:10.1002/ana.410390512

Heller BA, Ghidinelli M, Voelkl J, Einheber S, Smith R, Grund E, Morahan G, Chandler D, Kalaydjieva L, Giancotti F, et al. 2014. Functionally distinct PI 3-kinase pathways regulate myelination in the peripheral nervous system. *J Cell Biol* **204**: 1219–1236. doi:10.1083/jcb.201307057

Hirt N, Eggermann K, Hyrenbach S, Lambeck J, Busche A, Fischer J, Rudnik-Schöneborn S, Gaspar H. 2015. Genetic dosage compensation via co-occurrence of PMP22 duplication and PMP22 deletion. *Neurology* **84**: 1605–1606. doi:10.1212/WNL.0000000000001470

Ho TS, Zollinger DR, Chang KJ, Xu M, Cooper EC, Stankewich MC, Bennett V, Rasband MN. 2014. A hierarchy of ankyrin-spectrin complexes clusters sodium channels at nodes of Ranvier. *Nat Neurosci* **17**: 1664–1672. doi:10.1038/nn.3859

Ho AK, Wagstaff JL, Manna PT, Wartosch L, Qamar S, Garman EF, Freund SM, Roberts RC. 2016. The topology, structure and PE interaction of LITAF underpin a Charcot–Marie–Tooth disease type 1C. *BMC Biol* **14**: 109. doi:10.1186/s12915-016-0332-8

Huang CY-M, Zhang C, Zollinger DR, Leterrier C, Rasband MN. 2017. An αII spectrin-based cytoskeleton protects large-diameter myelinated axons from degeneration. *J Neurosci* **37**: 11323–11334. doi:10.1523/JNEUROSCI.2113-17.2017

Irani SR, Alexander S, Waters P, Kleopa KA, Pettingill P, Zuliani L, Peles E, Buckley C, Lang B, Vincent A. 2010. Antibodies to Kv1 potassium channel-complex proteins leucine-rich, glioma inactivated 1 protein and contactin-associated protein-2 in limbic encephalitis, Morvan's syndrome and acquired neuromyotonia. *Brain* **133**: 2734–2748. doi:10.1093/brain/awq213

Katona I, Wu X, Feely SM, Sottile S, Siskind CE, Miller LJ, Shy ME, Li J. 2009. PMP22 expression in dermal nerve myelin from patients with CMT1A. *Brain* **132**: 1734–1740. doi:10.1093/brain/awp113

Krzemien DM, Schaller KL, Levinson SR, Caldwell JH. 2000. Immunolocalization of sodium channel isoform NaCh6 in the nervous system. *J Comp Neurol* **420**: 70–83. doi:10.1002/(SICI)1096-9861(20000424)420:1<70::AID-CNE5>3.0.CO;2-P

Lancaster E, Huijbers MGM, Bar V, Boronat A, Wong A, Martinez-Hernandez E, Wilson C, Jacobs D, Lai M, Walker RW, et al. 2011. Investigations of Caspr2, an autoantigen of encephalitis and neuromyotonia. *Ann Neurol* **69**: 303–311. doi:10.1002/ana.22297

Latov N. 2014. Diagnosis and treatment of chronic acquired demyelinating polyneuropathies. *Nat Rev Neurol* **10**: 435–446. doi:10.1038/nrneurol.2014.117

Le N, Nagarajan R, Wang JY, Araki T, Schmidt RE, Milbrandt J. 2005. Analysis of congenital hypomyelinating Egr2Lo/Lo nerves identifies Sox2 as an inhibitor of Schwann cell differentiation and myelination. *Proc Natl Acad Sci* **102**: 2596–2601. doi:10.1073/pnas.0407836102

Le T-L, Galmiche L, Levy J, Suwannarat P, Hellebrekers DMEI, Morarach K, Boismoreau F, Theunissen TEJ, Lefebvre M, Martinovic AP, et al. 2021. Dysregulation of the NRG1/ERBB pathway causes a developmental disorder with gastrointestinal dysmotility in humans. *J Clin Invest* **131**: e145837. doi:10.1172/JCI145837

Lee SM, Chin LS, Li L. 2012. Charcot–Marie–Tooth disease-linked protein SIMPLE functions with the ESCRT machinery in endosomal trafficking. *J Cell Biol* **199**: 799–816. doi:10.1083/jcb.201204137

Lee S, Amici S, Tavori H, Zeng WM, Freeland S, Fazio S, Notterpek L. 2014. PMP22 is critical for actin-mediated cellular functions and for establishing lipid rafts. *J Neurosci* **34**: 16140–16152. doi:10.1523/JNEUROSCI.1908-14.2014

Li J, Krajewski K, Shy ME, Lewis RA. 2002. Hereditary neuropathy with liability to pressure palsy: the electrophysiology fits the name. *Neurology* **58**: 1769–1773. doi:10.1212/WNL.58.12.1769

Lupo V, Won S, Frasquet M, Schnitzler MS, Komath SS, Pascual-Pascual SI, Espinós C, Svaren J, Sevilla T. 2020. Bi-allelic mutations in *EGR2* cause autosomal recessive demyelinating neuropathy by disrupting the EGR2-NAB complex. *Eur J Neurol* **27**: 2662–2667. doi:10.1111/ene.14512

Lupski JR, Garcia CA. 2001. Charcot–Marie–Tooth peripheral neuropathies and related disorders. In *The metabolic & molecular basis of inherited disease* (ed. Scriver CR, Beaudet AL, Sly WS, Valle D, Childs B, Kinzler KW), pp. 5759–5788. McGraw-Hill, New York.

Manso C, Querol L, Lleixà C, Poncelet M, Mekaouche M, Vallat JM, Illa I, Devaux JJ. 2019. Anti-neurofascin-155 IgG4 antibodies prevent paranodal complex formation in vivo. *J Clin Invest* **129**: 2222–2236. doi:10.1172/JCI124694

Marafi D, Kozar N, Duan R, Bradley S, Yokochi K, Al Mutairi F, Saadi NW, Whalen S, Brunet T, Kotzaeridou U, et al. 2022. A reverse genetics and genomics approach to gene paralog function and disease: myokymia and the juxtaparanode. *Am J Hum Genet* **109**: 1713–1723. doi:10.1016/j.ajhg.2022.07.006

Marinko JT, Carter BD, Sanders CR. 2020a. Direct relationship between increased expression and mistrafficking of the Charcot–Marie–Tooth–associated protein PMP22. *J Biol Chem* **295**: 11963–11970. doi:10.1074/jbc.AC120.014940

Marinko JT, Kenworthy AK, Sanders CR. 2020b. Peripheral myelin protein 22 preferentially partitions into ordered phase membrane domains. *Proc Natl Acad Sci* **117**: 14168–14177. doi:10.1073/pnas.2000508117

Markworth R, Bähr M, Burk K. 2021. Held up in traffic—defects in the trafficking machinery in Charcot–Marie–Tooth disease. *Front Mol Neurosci* **14**: 695294. doi:10.3389/fnmol.2021.695294

Mauermann ML. 2018. The peripheral neuropathies of POEMS syndrome and Castleman disease. *Hematol Oncol Clin North Am* **32**: 153–163. doi:10.1016/j.hoc.2017.09.012

McGonigal R, Willison HJ. 2022. The role of gangliosides in the organisation of the node of Ranvier examined in glycosyltransferase transgenic mice. *J Anat* **241**: 1259–1271. doi:10.1111/joa.13562

McGonigal R, Cunningham ME, Smyth D, Chou M, Barrie JA, Wilkie A, Campbell C, Saatman KE, Lunn M, Willison HJ. 2023. The endogenous calpain inhibitor calpastatin attenuates axon degeneration in murine Guillain–Barré syndrome. *J Peripher Nerv Syst* **28**: 4–16. doi:10.1111/jns.12520

Meijer D, Svaren J. 2013. Specification of macroglia by transcription factors: Schwann cells in patterning and cell type specification in the developing CNS and PNS. In *Comprehensive developmental neuroscience* (ed. Alvarez-Buylla A, Rowitch D), pp. 759–770. Elsevier, Amsterdam.

Mogha A, Benesh AE, Patra C, Engel FB, Schöneberg T, Liebscher I, Monk KR. 2013. Gpr126 functions in Schwann cells to control differentiation and myelination via G-protein activation. *J Neurosci* **33**: 17976–17985. doi:10.1523/JNEUROSCI.1809-13.2013

Monk KR, Naylor SG, Glenn TD, Mercurio S, Perlin JR, Dominguez C, Moens CB, Talbot WS. 2009. A G protein-coupled receptor is essential for Schwann cells to initiate myelination. *Science* **325**: 1402–1405. doi:10.1126/science.1173474

Motley WW, Palaima P, Yum SW, Gonzalez MA, Tao F, Wanschitz JV, Strickland AV, Löscher WN, Vriendt ED, Koppi S, et al. 2016. De novo *PMP2* mutations in families with type 1 Charcot–Marie–Tooth disease. *Brain* **139**: 1649–1656. doi:10.1093/brain/aww055

Naef R, Suter U. 1999. Impaired intracellular trafficking is a common disease mechanism of PMP22 point mutations in peripheral neuropathies. *Neurobiol Dis* **6**: 1–14. doi:10.1006/nbdi.1998.0227

Narkis G, Ofir R, Manor E, Landau D, Elbedour K, Birk OS. 2007. Lethal congenital contractural syndrome type 2 (LCCS2) is caused by a mutation in ERBB3 (Her3), a modulator of the phosphatidylinositol-3-kinase/Akt pathway. *Am J Hum Genet* **81**: 589–595. doi:10.1086/520770

Nodari A, Previtali SC, Dati G, Occhi S, Court FA, Colombelli C, Zambroni D, Dina G, Del Carro U, Campbell KP, et al. 2008. α6β4 integrin and dystroglycan cooperate to stabilize the myelin sheath. *J Neurosci* **28**: 6714–6719. doi:10.1523/JNEUROSCI.0326-08.2008

Cite this article as *Cold Spring Harb Perspect Biol* doi: 10.1101/cshperspect.a041376

Pan Z, Kao T, Horvath Z, Lemos J, Sul JY, Cranstoun SD, Bennett V, Scherer SS, Cooper EC. 2006. A common ankyrin-G-based mechanism retains KCNQ and Na$_V$ channels at electrically active domains of the axon. *J Neurosci* **26**: 2599–2613. doi:10.1523/JNEUROSCI.4314-05.2006

Pantera H, Shy ME, Svaren J. 2020. Regulating PMP22 expression as a dosage sensitive neuropathy gene. *Brain Res* **1726**: 146491. doi:10.1016/j.brainres.2019.146491

Pascual-Goñi E, Fehmi J, Lleixà C, Martín-Aguilar L, Devaux J, Höftberger R, Delmont E, Doppler K, Sommer C, Radunovic A, et al. 2021. Antibodies to the Caspr1/contactin-1 complex in chronic inflammatory demyelinating polyradiculoneuropathy. *Brain* **144**: 1183–1196. doi:10.1093/brain/awab014

Pennuto M, Tinelli E, Malaguti M, Del Carro U, D'Antonio M, Ron D, Quattrini A, Feltri ML, Wrabetz L. 2008. Ablation of the UPR-mediator CHOP restores motor function and reduces demyelination in Charcot–Marie–Tooth 1B mice. *Neuron* **57**: 393–405. doi:10.1016/j.neuron.2007.12.021

Posor Y, Jang W, Haucke V. 2022. Phosphoinositides as membrane organizers. *Nat Rev Mol Cell Biol* **23**: 797–816. doi:10.1038/s41580-022-00490-x

Previtali SC, Quattrini A, Bolino A. 2007. Charcot–Marie–Tooth type 4B demyelinating neuropathy: deciphering the role of MTMR phosphatases. *Expert Reviews Mol Med* **9**: 1–16. doi:10.1017/S1462399407000439

Ptak CP, Peterson TA, Hopkins JB, Ahern CA, Shy ME, Piper RC. 2023. Homomeric interactions of the MPZ Ig domain and their relation to Charcot–Marie–Tooth disease. *Brain* doi:10.1093/brain/awad258

Querol L, Delmont E, Lleixa C. 2023. The autoimmune vulnerability of the node of Ranvier. *J Peripher Nerv Syst* **28**: S12–S22.

Raasakka A, Linxweiler H, Brophy PJ, Sherman DL, Kursula P. 2019. Direct binding of the flexible C-terminal segment of periaxin to β4 integrin suggests a molecular basis for CMT4F. *Front Mol Neurosci* **12**: 84. doi:10.3389/fnmol.2019.00084

Rasband MN, Peles E. 2021. Mechanisms of node of Ranvier assembly. *Nat Rev Neurosci* **22**: 7–20. doi:10.1038/s41583-020-00406-8

* Rasband MN, Peles E. 2023. The nodes of Ranvier: molecular assembly and maintenance. *Cold Spring Harb Perspect Biol* doi:10.1101/cshperspect.a041361

Ravenscroft G, Nolent F, Rajagopalan S, Meireles AM, Paavola KJ, Gaillard D, Alanio E, Buckland M, Arbuckle S, Krivanek M, et al. 2015. Mutations of GPR126 are responsible for severe arthrogryposis multiplex congenita. *Am J Hum Genet* **96**: 955–961. doi:10.1016/j.ajhg.2015.04.014

Rebelo AP, Cortese A, Abraham A, Eshed-Eisenbach Y, Shner G, Vainshtein A, Buglo E, Camarena V, Gaidosh G, Shiekhattar R, et al. 2021. A *CADM3* variant causes Charcot–Marie–Tooth disease with marked upper limb involvement. *Brain* **144**: 1197–1213. doi:10.1093/brain/awab019

Record CJ, Skorupinska M, Laura M, Rossor AM, Pareyson D, Pisciotta C, Feely SME, Lloyd TE, Horvath R, Sadjadi R, et al. 2023. Genetic analysis and natural history of Charcot–Marie–Tooth disease CMTX1 due to *GJB1* variants. *Brain* doi:10.1093/brain/awad187

Reilly MM, Herrmann DN, Pareyson D, Scherer SS, Finkel RS, Zuchner S, Burns J, Shy ME. 2023. Trials for slowly progressive neurogenetic diseases need surrogate endpoints. *Ann Neurol* **93**: 906–910. doi:10.1002/ana.26633

Roberts RC, Peden AA, Buss F, Bright NA, Latouche M, Reilly MM, Kendrick-Jones J, Luzio JP. 2010. Mistargeting of SH3TC2 away from the recycling endosome causes Charcot–Marie–Tooth disease type 4C. *Hum Mol Genet* **19**: 1009–1018. doi:10.1093/hmg/ddp565

Rossor AM, Carr AS, Devine H, Chandrashekar H, Pelayo-Negro AL, Pareyson D, Shy ME, Scherer SS, Reilly MM. 2017. Peripheral neuropathy in complex inherited diseases: an approach to diagnosis. *J Neurol Neurosurg Psychiatry* **88**: 846–863. doi:10.1136/jnnp-2016-313960

* Salzer J, Feltri ML, Jacob C. 2023. Schwann cell development and myelination. *Cold Spring Harb Perspect Biol* doi:10.1101/cshperspect.a041360

Saporta MA, Shy BR, Patzko A, Bai Y, Pennuto M, Ferri C, Tinelli E, Saveri P, Kirschner D, Crowther M, et al. 2012. Mpzr98c arrests Schwann cell development in a mouse model of early-onset Charcot–Marie–Tooth disease type 1B. *Brain* **135**: 2032–2047. doi:10.1093/brain/aws140

Sawade L, Grandi F, Mignanelli M, Patiño-López G, Klinkert K, Langa-Vives F, Di Guardo R, Echard A, Bolino A, Haucke V. 2020. Rab35-regulated lipid turnover by myotubularins represses mTORC1 activity and controls myelin growth. *Nat Commun* **11**: 2835. doi:10.1038/s41467-020-16696-6

Scherer SS, Kleopa KA. 2012. X-linked Charcot–Marie–Tooth disease. *J Periph Nerv Syst* **17**: 9–13. doi:10.1111/j.1529-8027.2012.00424.x

Schnaar RL. 2010. Brain gangliosides in axon-myelin stability and axon regeneration. *Febs Lett* **584**: 1741–1747. doi:10.1016/j.febslet.2009.10.011

Sereda MW, Meyer zu Hörste G, Suter U, Uzma N, Nave KA. 2003. Therapeutic administration of progesterone antagonist in a model of Charcot–Marie–Tooth disease (CMT-1A). *Nat Med* **9**: 1533–1537. doi:10.1038/nm957

Shackleford GG, Marziali LN, Sasaki Y, Claessens A, Ferri C, Weinstock NI, Rossor AM, Silvestri NJ, Wilson ER, Hurley E, et al. 2022. A new mouse model of Charcot–Marie–Tooth 2J neuropathy replicates human axonopathy and suggests alteration in axo-glia communication. *PLoS Genet* **18**: e1010477. doi:10.1371/journal.pgen.1010477

Shapiro L, Doyle JP, Hensley P, Colman DR, Hendrickson WA. 1996. Crystal structure of the extracellular domain from P$_0$, the major structural protein of peripheral nerve myelin. *Neuron* **17**: 435–449. doi:10.1016/S0896-6273(00)80176-2

Sherman DL, Fabrizi C, Gillespie CS, Brophy PJ. 2001. Specific disruption of a Schwann cell dystrophin-related protein complex in a demyelinating neuropathy. *Neuron* **30**: 677–687. doi:10.1016/S0896-6273(01)00327-0

Shy ME, Lupski JR, Chance PF, Klein CJ, Dyck PJ. 2005. Hereditary motor and sensory neuropathies: an overview of clinical, genetic, electrophysiologic, and pathologic features. In *Peripheral neuropathy* (ed. Dyck PJ, Thomas PK), pp. 1623–1658. Saunders, London.

Stavrou M, Kagiava A, Choudury SG, Jennings MJ, Wallace LM, Fowler AM, Heslegrave A, Richter J, Tryfonos C, Christodoulou C, et al. 2022. A translatable RNAi-driven gene therapy silences PMP22/Pmp22 genes and improves neuropathy in CMT1A mice. *J Clin Invest* **132**: e159814. doi:10.1172/JCI159814

Stendel C, Roos A, Kleine H, Arnaud E, Özçelik M, Sidiropoulos PNM, Zenker J, Schüpfer F, Lehmann U, Sobota RM, et al. 2010. SH3TC2, a protein mutant in Charcot–Marie–Tooth neuropathy, links peripheral nerve myelination to endosomal recycling. *Brain* **133**: 2462–2474. doi:10.1093/brain/awq168

Tao F, Beecham GW, Rebelo AP, Svaren J, Blanton SH, Moran JJ, Lopez-Anido C, Morrow JM, Abreu L, Rizzo D, et al. 2019. Variation in *SIPA1L2* is correlated with phenotype modification in Charcot–Marie–Tooth disease type 1A. *Ann Neurol* **85**: 316–330. doi:10.1002/ana .25426

Tomaselli PJ, Rossor AM, Horga A, Jaunmuktane Z, Carr A, Saveri P, Piscosquito G, Pareyson D, Laura M, Blake JC, et al. 2017. Mutations in noncoding regions of *GJB1* are a major cause of X-linked CMT. *Neurology* **88**: 1445–1453. doi:10.1212/WNL.0000000000003819

Topilko P, Schneider-Maunoury S, Levi G, Baron-Van Evercooren A, Chennoufi AB, Seitanidou T, Babinet C, Charnay P. 1994. Krox-20 controls myelination in the peripheral nervous system. *Nature* **371**: 796–799. doi:10.1038/ 371796a0

Uncini A. 2023. Autoimmune nodo-paranodopathies 10 years later: clinical features, pathophysiology and treatment. *J Peripher Nerv Syst* **28**: S23–S35. doi:10.1111/jns .12569

Usoskin D, Furlan A, Islam S, Abdo H, Lönnerberg P, Lou D, Hjerling-Leffler J, Haeggström J, Kharchenko O, Kharchenko PV, et al. 2015. Unbiased classification of sensory neuron types by large-scale single-cell RNA sequencing. *Nat Neurosci* **18**: 145–153. doi:10.1038/nn.3881

Vaccari I, Dina G, Tronchère H, Kaufman E, Chicanne G, Cerri F, Wrabetz L, Payrastre B, Quattrini A, Weisman LS, et al. 2011. Genetic interaction between MTMR2 and

FIG4 phospholipid phosphatases involved in Charcot–Marie–Tooth neuropathies. *PLoS Genet* **7**: e1002319. doi:10.1371/journal.pgen.1002319

Vallat JM, Sindou P, Preux PM, Tabaraud F, Milor AM, Couratier P, Le Guern E, Brice A. 1996. Ultrastructural PMP22 expression in inherited demyelinating neuropathies. *Ann Neurol* **39**: 813–817. doi:10.1002/ana .410390621

Viader A, Golden JP, Baloh RH, Schmidt RE, Hunter DA, Milbrandt J. 2011. Schwann cell mitochondrial metabolism supports long-term axonal survival and peripheral nerve function. *J Neurosci* **31**: 10128–10140. doi:10.1523/ JNEUROSCI.0884-11.2011

Weider M, Reiprich S, Wegner M. 2013. Sox appeal—Sox10 attracts epigenetic and transcriptional regulators in myelinating glia. *Biol Chem* **394**: 1583–1593. doi:10.1515/hsz-2013-0146

Yin XH, Crawford TO, Griffin JW, Tu PH, Lee VMY, Li CM, Roder J, Trapp BD. 1998. Myelin-associated glycoprotein is a myelin signal that modulates the caliber of myelinated axons. *J Neurosci* **18**: 1953–1962. doi:10.1523/JNEURO SCI.18-06-01953.1998

Yuki N, Hartung HP. 2012. Guillain–Barré syndrome. *N Engl J Med* **366**: 2294–2304. doi:10.1056/NEJMra 1114525

Zhao HT, Damle S, Ikeda-Lee K, Kuntz S, Li J, Mohan A, Kim A, Hung G, Scheideler MA, Scherer SS, et al. 2018. *PMP22* antisense oligonucleotides reverse Charcot–Marie–Tooth disease type 1A features in rodent models. *J Clin Invest* **128**: 359–368. doi:10.1172/JCI96499

Zou J, Hu B, Arpag S, Yan Q, Hamilton A, Zeng YS, Vanoye CG, Li J. 2015. Reactivation of lysosomal Ca^{2+} efflux rescues abnormal lysosomal storage in FIG4-deficient cells. *J Neurosci* **35**: 6801–6812. doi:10.1523/JNEUROSCI.4442-14.2015

Cite this article as *Cold Spring Harb Perspect Biol* doi: 10.1101/cshperspect.a041376

Index

www.ingramcontent.com/pod-product-compliance
Lightning Source LLC
Chambersburg PA
CBHW061928190326
41458CB00009B/2693